The Oxford Handbook of Random Matrix Theory

ÆDES CHRISTI
in Academia Oxoniensi

The Oxford Handbook of Random Matrix Theory

Editors

Gernot Akemann, Jinho Baik and Philippe Di Francesco

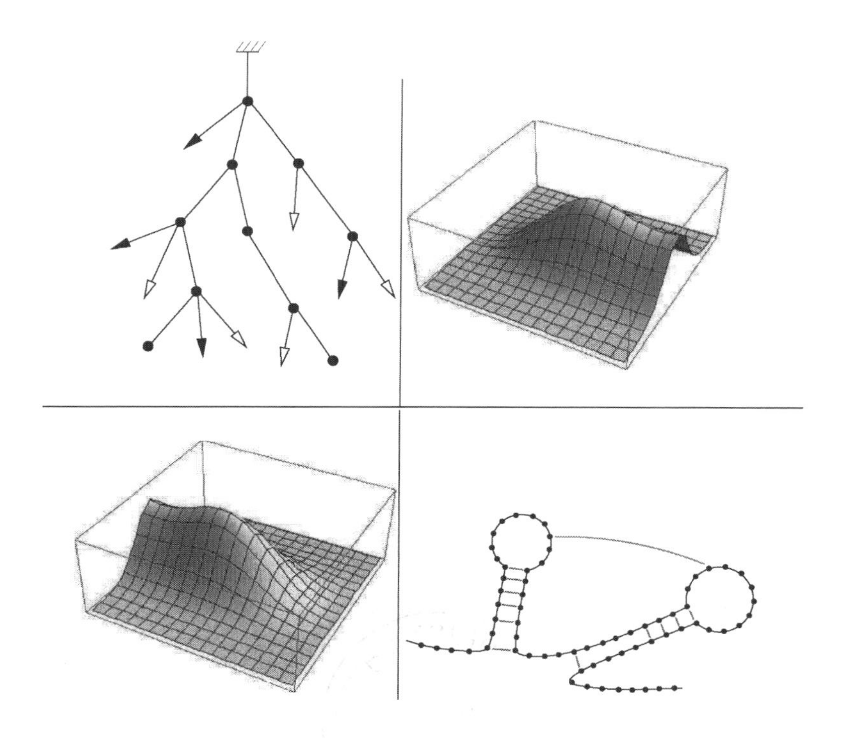

OXFORD
UNIVERSITY PRESS

Great Clarendon Street, Oxford, OX2 6DP,
United Kingdom

Oxford University Press is a department of the University of Oxford.
It furthers the University's objective of excellence in research, scholarship,
and education by publishing worldwide. Oxford is a registered trade mark of
Oxford University Press in the UK and in certain other countries

First published 2011
First published in paperback 2015

Published in the United States of America by Oxford University Press
198 Madison Avenue, New York, NY 10016, United States of America

British Library Cataloguing in Publication Data
Data available

Library of Congress Cataloging in Publication Data
Data available

ISBN 978–0–19–874419–1

Dedicated to the memory of Madan Lal Mehta

Foreword

Fifty years ago, the world-wide community of experts in the theory of random matrices consisted of about ten people. There was our leader Eugene Wigner who first invented the subject. Wigner observed that a heavy nucleus is a liquid drop composed of many particles with unknown strong interactions, and so a random matrix would be a possible model for the Hamiltonian of a heavy nucleus. At the beginning, random matrix theory was a branch of nuclear physics. The purpose of the exercise was to use random matrices to learn about nuclei. Other members of the community in those days were Charles Porter at Brookhaven, Norbert Rosenzweig at Argonne, Robert Thomas at Los Alamos, Madan Lal Mehta and Michel Gaudin at Saclay in France, and the experimenters Bill Havens and James Rainwater at Columbia. Charles Porter edited the predecessor of this book in 1965.

Havens and Rainwater made heroic efforts to measure the energies of states of neutron-capture nuclei, for example states of uranium 239 obtained by capturing resonance neutrons in uranium 238. They could fix the energies accurately by using a sharp pulse of neutrons and measuring the time of flight of the neutrons between the source and the absorber. After years of effort, they had some beautiful data, lists of accurate energies of about twenty different nuclei. But the data had two fatal flaws. First, the levels had widely varying strengths and there was always a chance that the weakest levels would be missed. Second, there were never more than a hundred levels for any one nucleus. The best of all the data was for Erbium 166 which had 82 measured levels. While Rainwater struggled to make the experimental data as clean as possible, Mehta and I struggled to make the theoretical analysis as sharp as possible. We worked out statistical tests to find out whether the observed level series agreed or disagreed with the random matrix model, including the chance that one or two weak levels might be missed. All of our struggles were in vain. 82 levels were too few to give a statistically significant test of the model. As a contribution to the understanding of nuclear physics, random matrix theory was a dismal failure. By 1970 we had decided that random matrix theory was a beautiful piece of pure mathematics having nothing to do with physics. Random matrix theory went temporarily to sleep.

The reawakening of random matrix theory began in 1973, when the mathematician Hugh Montgomery made his brilliant conjecture, that the pair-correlation function of zeros of the Riemann zeta function is identical with the pair-correlation function of eigenvalues of a random Hermitian matrix. His idea embedded random matrices deeply in the purest of pure mathematics. There is overwhelming numerical as well as analytical evidence supporting the conjecture. Instead of having 82 laboriously detected levels of Erbium, we have 70 million zeros of the zeta function accurately computed by Andrew Odlyzko to verify the conjecture. The fact that the conjecture is still unproved after 37 years makes it even more attractive to mathematicians. After Montgomery, a steady stream of mathematicians of many different kinds has flowed into random matrix theory, with the results that are summarized in this book. The rapid march of new ideas and new discoveries has long ago left me behind.

I have not attempted to read all the chapters in this book or to assess their relative merits. I mention only one that gave me particular pleasure, Chapter 24 by Keating and Snaith, describing some further connections between random matrix theory and number theory that grew out of Montgomery's conjecture. This is the kind of mathematics that I enjoy, with many conjectures and few proofs. Deep mysteries remain, and the best is still to come.

One recent application of random matrix theory to real life seems to be missing from this book. Either it is here and I missed it, or the editors of the book missed it. This is the work of two physicists from the Czech Republic, Milan Krbalek and Petr Seba, who studied the bus system of the city of Cuernavaca in Mexico. Unlike other metropolitan bus systems, this system is totally decentralized, with no central authority and no time tables. Each bus is the property of the driver. At each bus stop, the drivers obtain information from bystanders about the time when the previous bus left. Using this information, the drivers independently adjust their speeds so as to maximize their incomes. Adjustments are made using mental arithmetic without any advanced algorithms. Krbalek and Seba recorded the actual departure times of buses at various stops for a month while the system was in normal operation. They found that the spacings between buses agree accurately with the Gaussian unitary ensemble of random matrix theory. The Gaussian unitary ensemble gives the best approximation to uniform spacing that the bus drivers can achieve, based on the limited information available to them.

The benefit of the bus drivers' self-regulation to the public is measured by R, the ratio between the average waiting-time of a passenger and the average spacing-time between buses. The best possible R is 0.5, which would occur if the bus spacings were exactly equal. If the buses are uncorrelated so that their spacings have a Poisson distribution, then R=1. In Cuernavaca, with the buses correlated according to the Gaussian unitary ensemble, the value of R is $[(3\pi)/16] = 0.589$, much closer to the equal-spacing value than to the Poisson distribution value. It is remarkable that so large a public benefit results from so simple an optimization process. I am not able to determine whether the application of random matrix theory to financial markets, as described in Chapter 40 of this book by Bouchaud and Potters, yields any comparable benefits. When an expert on markets tells me that some piece of financial wizardry is sure to benefit mankind, I am inclined to believe that a Cuernavaca bus driver might do the job better.

Freeman Dyson

Institute for Advanced Study
Princeton, New Jersey, USA

Contents

Detailed Contents

Part II Properties of random matrix theory

List of Contributors

Mark Adler, Department of Mathematics, Brandeis University, 415 South Street, MS 050, Waltham, MA 02454, USA, adler@brandeis.edu

Gernot Akemann, Fakultät für Physik, Universität Bielefeld, Postfach 100131, D-33501 Bielefeld, Germany, akemann@physik.uni-bielefeld.de

Greg W. Anderson, School of Mathematics, University of Minnesota, Minneapolis, MN 55455, USA, gwanders@math.umn.edu

Jinho Baik, 530 Church Street, Department of Mathematics, University of Michigan, Ann Arbor, MI 48109-1043, USA, baik@umich.edu

Carlo W. J. Beenakker, Instituut-Lorentz for Theoretical Physics, P.O. Box 9506, NL-2300 RA Leiden, The Netherlands, beenakker@lorentz.leidenuniv.nl

Gérard Ben Arous, Department of Mathematics, Courant Institute of Mathematical Sciences, New York University, Room 702, 251 Mercer St., New York, NY 10012, USA, benarous@cims.nyu.edu

Marco Bertola, Concordia University, Dept. of Math. & Stat., SGW Campus, LB-901-29, 1455 de Maisonneuve W., H3G 1M8, Montreal (QC), Canada, bertola@mathstat.concordia.ca

Oriol Bohigas, Laboratoire de Physique Théorique et Modeles Statistiques, bâtiment 100, Université Paris-Sud, Centre scientifique d'Orsay, 15 rue Georges Clemenceau, 91405 Orsay cedex, France, bohigas@lptms.u-psud.fr

Alexei Borodin, Department of Mathematics, California Institute of Technology, Mathematics 253–37, Pasadena, CA 91125, USA, borodin@caltech.edu

Jean-Philippe Bouchaud, Science & Finance, Capital Fund Management, 6, boulevard Haussmann, F- 75009 Paris, France, jean-philippe.bouchaud@cfm.fr

Jérémie Bouttier, Institut de Physique Théorique, CEA/Saclay, Orme des Merisiers, F-91191 Gif-sur-Yvette Cedex, France, jeremie.bouttier@cea.fr

Edouard Brézin, Laboratoire de Physique Théorique de l'Ecole Normale Supérieure 24, rue Lhomond, F-75231 Paris Cedex 05, France, edouard.brezin@ens.fr

Zdzisław Burda, Department of Theory of Complex Systems, M. Smoluchowski Institute of Physics, Jagiellonian University, ul. Reymonta 4, 30-059 Krakow, Poland, zdzislaw.burda@uj.edu.pl

Leonid O. Chekhov, Steklov Mathematical Institute, Department of Theoretical Physics, Gubkina str. 8, 119991, Moscow, Russia, chekhov@mi.ras.ru

Giovanni M. Cicuta, Dipartimento di Fisica, Università degli studi di Parma, Viale G.P. Usberti n.7/A, Parco Area delle Scienze, I-43100 Parma, Italy, giovanni.cicuta@fis.unipr.it

Philippe Di Francesco, Institut de Physique Théorique, CEA/Saclay, Orme des Merisiers, F-91191 Gif-sur-Yvette Cedex, France, philippe.di-francesco@cea.fr

Freeman Dyson, Institute for Advanced Study, School of Natural Sciences, Einstein Drive, Princeton, NJ 08540, USA, dyson@ias.edu

Noureddine El Karoui, University of California, Berkeley, Department of Statistics, 367 Evans Hall, Berkeley, CA 94720-3860, USA, nkaroui@stat.berkeley.edu

Patrik L. Ferrari, Institut für Angewandte Mathematik Abteilung Stochastische Analysis, Endenicher Allee 60, D-53115 Bonn, Germany, ferrari@unibonn.de

Peter J. Forrester, Department of Mathematics and Statistics, The University of Melbourne, Parkville, Vic 3010, Australia, p.forrester@ms.unimelb.edu.au

Yan V. Fyodorov, School of Mathematical Sciences, University of Nottingham, University Park, Nottingham NG7 2RD, United Kingdom, yan.fyodorov@nottingham.ac.uk

Thomas Guhr, Fachbereich Physik, Universität Duisburg-Essen, Lotharstrasse 1, D-47048 Duisburg, Germany, thomas.guhr@uni-duisburg-essen.de

Alice Guionnet, Unité de Mathématiques Pures et Appliquées, École Normale Supérieure, Lyon 69007, France, Alice.Guionnet@umpa.ens-lyon.fr

Shinobu Hikami, Okinawa Institute of Science and Technology 1919–1 Tancha, Onna-son, Kunigami-gun, Okinawa, 904-0412 Japan, tsurumaki3032003@yahoo.co.jp

Alexander R. Its, Department of Mathematical Sciences, 402 N. Blackford Street, LD270, Indianapolis, IN 46202-3216, USA, itsa@math.iupui.edu

Jerzy Jurkiewicz, Department of Theory of Complex Systems, M. Smoluchowski Institute of Physics, Jagiellonian University, ul. Reymonta 4, 30-059 Krakow, Poland, jurkiewicz@th.if.uj.edu.pl

Eugene Kanzieper, Department of Applied Mathematics, School of Sciences, H.I.T.—Holon Institute of Technology, 52 Golomb Street, POB 305, Holon 58102, Israel, eugene.kanzieper@hit.ac.il

Jon P. Keating, School of Mathematics, University of Bristol, University Walk, Clifton, Bristol BS8 1TW, UK, j.p.keating@bristol.ac.uk

Boris A. Khoruzhenko, School of Mathematical Sciences, Queen Mary University of London, Mile End Road, London E1 4NS, UK, b.khoruzhenko@qmw.ac.uk

Ivan Kostov, Institut de Physique Théorique, CEA/Saclay, Orme des Merisiers, F-91191 Gif-sur-Yvette Cedex, France, ivan.kostov@cea.fr

Vladimir E. Kravtsov, The Abdus Salam International Centre for Theoretical Physics (ICTP), Condensed Matter Group, P.O. Box 586, Strada Costiera 11, I-34100 Trieste, Italy, kravtsov@ictp.it

Arno B. J. Kuijlaars, Department of Mathematics, Katholieke Universiteit Leuven, Celestijnenlaan 200 B, B-3001 Leuven (Heverlee), Belgium, arno.kuijlaars@wis.kuleuven.be

Satya N. Majumdar, Laboratoire de Physique Théorique et Modèles Statistiques, bâtiment 100, Université Paris-Sud, Centre scientifique d'Orsay, 15 rue Georges Clemenceau, 91405 Orsay cedex, France, satya.majumdar@u-psud.fr

Marcos Mariño, Département de Physique Théorique, Université de Genève, 24 quai E. Ansermet, CH-1211 Genève 4, Switzerland, marcos.marino@unige.ch

Luca G. Molinari, Dipartimento di Fisica, Università di Milano, Via Celoria 16, I-20133 Milano, Italy, luca.molinari@mi.infn.it

Alexei Morozov, Institute of Theoretical and Experimental Physics (ITEP), B. Cheremushkinskaya, 25, Moscow, 117259, Russia, morozov@itep.ru

Sebastian Müller, School of Mathematics, University of Bristol, University Walk, Clifton, Bristol BS8 1TW, UK, sebastian.muller@bristol.ac.uk

Taro Nagao, Graduate School of Mathematics, Nagoya University, Chikusa-ku, Nagoya 464–8602, Japan, nagao@math.nagoya-u.ac.jp

Grigori Olshanski, Institute for Information Transmission Problems, Bolshoy Karetny 19, Moscow 127994, GSP-4, Russia, olsh2007@gmail.com

Nicolas Orantin, CAMGSD, Departamento de Matemática, Instituto Superior Técnico, Av. Rovisco Pais, 1049-001 Lisboa, Portugal, norantin@math.ist.utl.pt

Henri Orland, Institut de Physique Théorique, CEA/Saclay, Orme des Merisiers, F-91191 Gif-sur-Yvette Cedex, France, Henri.ORLAND@cea.fr

Marc Potters, Science & Finance, Capital Fund Management, 6, boulevard Haussmann, F- 75009 Paris, France, marc.potters@cfm.fr

Geoff J. Rodgers, Department of Mathematical Sciences, John Crank Building, Brunel University, West London, Uxbridge UB8 3PH, United Kingdom, g.j.rodgers@brunel.ac.uk

Dmitry V. Savin, Department of Mathematical Sciences, John Crank Building, Brunel University, West London, Uxbridge UB8 3PH, United Kingdom, dmitry.savin@brunel.ac.uk

Martin Sieber, School of Mathematics, University of Bristol, University Walk, Clifton, Bristol BS8 1TW, UK, m.sieber@bristol.ac.uk

Nina C. Snaith, School of Mathematics, University of Bristol, University Walk, Clifton, Bristol BS8 1TW, UK, n.c.snaith@bristol.ac.uk

Hans-Jürgen Sommers, Fachbereich Physik, Universität Duisburg-Essen, D-47048 Duisburg, Germany, h.j.sommers@uni-due.de

Roland Speicher, Universität des Saarlandes, Fachrichtung Mathematik, Postfach 151150, 66041 Saarbrücken, Germany, speicher@math.uni-sb.de

Thomas Spencer, Institute for Advanced Study, Einstein Drive, Princeton, NJ 08540, USA, spencer@ias.edu

Herbert Spohn, Zentrum Mathematik, Bereich M5, Technische Universität München, D-85747 Garching, Germany, spohn@ma.tum.de

Antonia M. Tulino, Department of Electrical Engineering, Universiá egli Sudi di Napoli "Federico II", I-80125 Napoli, Italy, atulino@princeton.edu

Pierre van Moerbeke, Département de Mathématique, Université Catholique de Louvain, Bât. M. de Hemptinne, Chemin du Cyclotron, 2, 1348 Louvain-la-Neuve, Belgium, Pierre.Vanmoerbeke@uclouvain.be

Jacobus J. M. Verbaarschot, Department of Physics and Astronomy, State University of New York, Stony Brook, NY 11794, USA, verbaarschot@cs.physics.sunysb.edu

Sergio Verdú, B-308 Engineering Quadrangle, Department of Electrical Engineering, Princeton University, Princeton, NJ 08544, USA, verdu@princeton.edu

Graziano Vernizzi, Physics and Astronomy, Siena College, 515 Loudon Rd., Loudonville, NY 12211-1462, USA, gvernizzi@siena.edu

Hans A. Weidenmüller, Max-Planck-Institut für Kernphysik, Saupfercheckweg 1, D-69117 Heidelberg, Germany, Hans.Weidenmueller@mpi-hd.mpg.de

Anton Zabrodin, Institute of Biochemical Physics, 4 Kosygina st., 119991 Moscow, Russia, and Institute of Theoretical and Experimental Physics (ITEP), B. Cheremushkinskaya, 25, Moscow 117259, Russia, zabrodin@itep.ru

Paul Zinn-Justin, LPTHE, tour 13/14, Université Pierre et Marie Curie - Paris 6, F-75252 Paris cedex, France, pzinn@lpthe.jussieu.fr

Martin R. Zirnbauer, Department of Physics, Cologne University, Zülpicher Strasse 77, D-50937 Cologne, Germany, zirn@thp.uni-koeln.de

Jean-Bernard Zuber, Laboratoire de Physique Théorique et Hautes Energies, Tour 13–14, 5ème étage, Université Pierre et Marie Curie, 4 place Jussieu, F-75005 Paris, France, jean-bernard.zuber@upmc.fr

PART I
Introduction

·1·

Introduction and guide to the handbook

G. Akemann, J. Baik and P. Di Francesco

Abstract

In this chapter we provide a first idea of what Random Matrix Theory is about, what its main features are, and why it is so successful in applications. This handbook is an attempt to cover all its major aspects and modern applications. We will explain how the other 42 chapters on mathematical properties and applications of random matrices are related and built one upon the other. A short list of topics that are not covered or only marginally covered in this handbook is given. We very briefly mention recent new developments since the first edition of this handbook and we conclude with a brief survey over the existing introductory literature.

1.1 Random matrix theory in a nutshell

Before we describe some general features of Random Matrix Theory (RMT) we would like to first introduce the simplest and maybe most frequently used standard example, the Gaussian Unitary Ensembles (GUE) of random matrices. Many of the general concepts and features of the subject will become clearer when having this example in mind. We do not attempt to give a derivation here of the theory, but merely state a minimal set of facts (and examples) as mini summary of what is to come.

Consider a Hermitian matrix H of size $N \times N$, $H = H^{\dagger} = U \Lambda U^{\dagger}$, where U is the diagonalizing unitary matrix $U \in U(N)$, and $\Lambda = \text{diag}(\lambda_1, \ldots, \lambda_N)$ is the diagonal matrix containing the (real) eigenvalues λ_j of H. The space of all N^2 independent matrix elements of such a matrix, that is, all upper triangular complex elements $H_{k<l} = H^R_{k<l} + iH^I_{k<l} \in \mathbb{C}$ and all diagonal real elements H_{ii}, can be endowed with a probability measure, which we choose to be Gaussian:

$$dP(H) = \pi^{-N^2/2} \prod_{1 \le i<j \le N} dH^R_{ij} dH^I_{ij} \prod_{k=1}^{N} dH_{kk} \exp\left[-\sum_{k,l=1}^{N} |H_{kl}|^2 \right] \equiv dH\, \mathrm{e}^{-\mathrm{Tr} H^2}. \tag{1.1.1}$$

The distribution is invariant under unitary conjugation, $H \to H' = WHW^{\dagger}$, with W unitary. The Gaussian distribution and the unitary invariance give this ensemble its name. Note that $dP(H)$ completely factorises into a product of independent Gaussian random variables[1]. As in statistical physics, integration over $dP(H)$ gives the partition function,

$$Z \equiv \int_{\mathbb{R}^{N2}} dH \exp[-\mathrm{Tr}H^2] \equiv \exp[N^2 F], \tag{1.1.2}$$

as well as the free energy F. The N^2-scaling in the definition of F is made so that, as we will see, at large N the leading contribution to F is of order 1.

The spectral representation $H = U\Lambda U^\dagger$ gives rise to a change of variables $H \to (\Lambda, U)$ in Eq. (1.1.2). It turns out that under $dP(H)$, Λ and U are statistically independent, so that the integral separates:

$$Z = c_N \prod_{j=1}^{N} \int_{-\infty}^{\infty} d\lambda_j \, e^{-\lambda_j^2} \prod_{1\leq k<l\leq N} |\lambda_l - \lambda_k|^\beta. \tag{1.1.3}$$

Here $c_N = \int_{U(N)} dU$ is the volume of the unitary group $U(N)$ and $\beta = 2$. Using simple linear algebra the integrand (also called joint probability distribution function) can be written as single determinant. Alternatively, rasing everything into the exponent the GUE has the following action S

$$S(\{\lambda_j\}) \equiv \sum_{j=1}^{N} \lambda_j^2 - \frac{\beta}{2} \sum_{1\leq k\neq l\leq N} \ln|\lambda_l - \lambda_l|,$$

$$Z = c_N \prod_{j=1}^{N} \int_{-\infty}^{\infty} d\lambda_j \exp[-S(\{\lambda_j\})]. \tag{1.1.4}$$

The eigenvalues are thus no longer independent variables but are coupled and repel each other. They behave like charged Coulomb particles in two dimensions at inverse temperature $\beta = 2$ confined to the real line and subject to a confining quadratic potential.

Other values of β which is the so-called Dyson index are possible. Had we taken real symmetric matrices $H = H^T$, or quaternion real self-dual matrices instead, and constructed our RMT as above, the index would have been $\beta = 1$ or $\beta = 4$ respectively. Due to the corresponding invariance of such matrices these ensembles are named the Gaussian Orthogonal (GOE) and Gaussian Symplectic Ensembles (GSE) respectively.

To complete our RMT mini introduction we give two results that follow from the above. First, the probability (density) $\rho(\lambda)$ to find an eigenvalue at value λ, which is obtained as in eq. (1.1.4) by integrating over all but one eigenvalue, is given for $\beta = 2$ and for any N by

$$\rho(\lambda) = \frac{1}{2^N N! \sqrt{\pi}} \exp[-\lambda^2] \Big(2NH_{N-1}(\lambda)^2 - 2(N-1)H_{N-2}(\lambda)H_N(\lambda) \Big), \tag{1.1.5}$$

[1] Note that for this ensemble the variance of the diagonal elements is different from the non-diagonal ones.

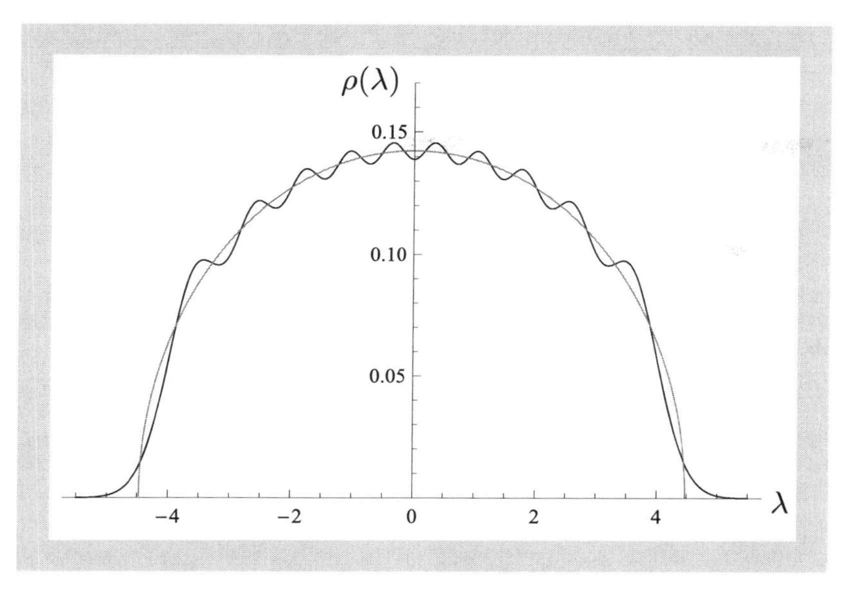

Fig. 1.1 The spectral density $\rho(\lambda)$ eq. (1.1.5) of the GUE for $N = 10$ vs. the Wigner semi-circle.

in terms of the standard Hermite polynomials $H_j(\lambda)$. The density $\rho(\lambda)$ is plotted in Figure 1.1 for $N = 10$. The local maxima indicate the average position of the 10 eigenvalues in this case. When taking $N \gg 1$ large such that the oscillations are smoothed out, the density approaches the famous Wigner-semi circle $\rho(\lambda) \sim \frac{1}{\pi N}\sqrt{2N - \lambda^2}$, as can be already seen from Figure 1.1.

The second example is the rescaled distribution of the largest eigenvalue λ_*, which is given by the Tracy–Widom distribution $F_2(t)$

$$F_2(t) \equiv \text{Prob}\Big(N^{1/6}(\lambda_* - 2\sqrt{N}) \leq t\Big) = \exp\left[-\int_t^{\infty} ds(s-t)q(s)^2\right]. \qquad (1.1.6)$$

Here $q(s)$ is the Hastings-McLeod solution to the Painlevé II equation

$$q(s)'' = sq(s) + 2q(s)^3. \qquad (1.1.7)$$

These two distributions play a crucial role in the theory and applications of RMT.

1.2 What is random matrix theory about?

Several types of applications of RMT are commonly investigated. A few examples are as follows:

Random operators In many cases one is interested in the spectral properties of a certain operator, e.g. a Hamilton- or Dirac-operator, or more generally in a matrix describing correlations of some kind. Typically such operators are too complicated to analyse explicitly. In the RMT approach the above deterministic operators are replaced by finite-dimensional matrices whose elements are random variables. This renders the problem solvable in an apropriate sense while keeping some of its important characteristics. The relevance of RMT as an alternative to a deterministic operator, rests crucially on the universality of the spectral properties of random matrices: establishing universality for matrix ensembles is a key technical element in RMT. Here the relevant matrix ensemble is sometimes taken to be as broad as possible (such as a class of unitary invariant ensembles or an ensemble of banded matrices with independent and identically distributed (iid) entries), only reflecting certain global symmetries of the system. Sometimes the ensemble is taken to be of more specific form, such as GUE, to make the calculation as simple as possible.

On the other hand, in other cases, the entries of a random matrix itself have a meaning as data. For example, in multivariate statistics, a random matrix represents a sample covariant matrix. The large dimensional limit of the eigenvalues and the eigenvectors of such matrices, then casts a light on high dimensional data analysis.

Counting devices RMT is a very efficient tool regarding problems in enumerative combinatorics. This applies both to matrices of finite and infinite size. For example, GUE generates Catalan numbers by simply taking moments of the celebrated semi-circular density. For large size N, RMT provides genus expansions for graphs: such interpretations were key to many developments in 2D quantum gravity, by providing discrete solvable statistical models of matter on random surfaces (Chapter 30). Other interpretations include: string theory (Chapter 31), folding problems (Chapter 42), semi-classical orbits (Chapter 33), knots and maps (Chapters 27, 26), and topological invariants in algebraic geometry (Chapter 29).

The origin of the ease of counting in RMT is that viewed as a zero-dimensional field theory its Feynman rules are very simple and many diagrams can be computed explicitly.

RMT without matrices Random matrix theory nowadays goes beyond the analysis of matrices. Many probabilistic ensembles which did not arise from any matrix models have been analysed by random matrix techniques. For example, in analysing the models in random permutations (Chapter 25) and random growth models (Chapter 38), one does not have any random matrix ensemble. But instead the Coulomb gas ensemble arises and hence RMT techniques can be successfully applied. The key role, then, is played by an associated determinantal point process (Chapter 11). Another celebrated example concerns the zeros of Riemann zeta functions (Chapter 24). These zeros are certainly deterministic, and moreover, do not (yet) have any known spectral interpretation of an operator (see however [Ber99]). It has been observed, however, and verified under certain restrictions, that their statistical behaviour matches the eigenvalues of a random matrix.

1.3 Why is random matrix theory so successful in applications?

RMT has been very successful and continues to enjoy great interest among physicists, mathematicians and other scientists. Here are some reasons:

Flexibility It allows one to build in extra global (anti-unitary) symmetries, such as time reversal, spin, chiral or particle-hole symmetry, as well as some extra structures such as coupling to an external field, treating multi-matrix models ("colourings") or adding extra Fermions – while maintaining exact solvability for all the correlation functions of the eigenvalues.

Universality This has three broad meanings:

(i) Generically the precise choice of probability distribution for the matrix elements or the couplings is not important: in the large N limit, the correlation functions depend on this choice only through a few parameters, typically the mean energy scale or the separation from the spectral edge. Once these parameters are taken into account, the behaviour of the correlation functions does not depend on the choice of the distribution. Of course fine tuning to generate extra zeros of the spectral density $\rho(\lambda)$ allows to generate different universality classes of various degrees of criticality, which themselves are again universal.

(ii) In a more physical sense RMT in the large-N limit becomes equivalent to other more physically motivated effective field theories or sigma models, that depend on the same global symmetries (broken or unbroken): thus RMT can often be used as the simplest, solvable model that captures the essential degrees of freedom of the theory.

(iii) The limiting distributions that arise in the large-N limit in RMT also appear as the limits of other statistical ensembles which are not necessarily related to random matrices. In a certain sense, the role of the normal distribution in the classical central limit theorem is played by the distributions that arise in RMT (Tracy–Widom distribution, Sine distribution,...) in non-commutative settings which may or may not involve random matrices.

Predictive capacity The correlations functions typically depend on few (sometimes zero) parameters. In applications, RMT of course does not know of the physical scale of the problem at hand. It has to be identified by comparing to the theory to be described. Turning the problem around, the scale or physical coupling constants can be extracted very efficiently by fitting data to RMT predictions in their regime of applicability.

Rich mathematical structure This comes from the many facets of the large-N limit, depending on how the limit is taken. Already the simplest ensemble, the GUE, exhibits a variety of different special distributions such as the Sine- and Airy-distributions depending on which location or detail of the spectral correlations is the focus of investigation. These distributions, in turn, are related to the Painlevé transcendents and integrable systems. The multiple connections of RMT to various areas of mathematics make it the ideal bridge between otherwise unrelated fields (probability and analysis, algebra, algebraic geometry, differential systems, combinatorics). More generally, the techniques developed to solve problems in RMT are flexible enough to be applied to other branches of sciences as well.

1.4 Guide to this handbook

This section is meant to be a mini-guide to the handbook. We start by explaining how the main parts of this book are structured, giving some details of which topics are covered, why they are structured in the given way, and how they are related. These relations – there are possibly more – are summarised in Table 1.1 as well as in a graphical way for chapters dealing with applications in Figure 1.2 below. In Table 1.1 we also provide a list of all chapter numbers with their abbreviated titles. Necessarily this mini guide contains some jargon that is explained in the respective chapters that follow. In the past 2 decades, in particular, we have seen a sharp rise of interest in the mathematics community in RMT. This is why contributions to the book have been written both by mathematicians and physicists, which we believe are our main audience. The book is divided into three parts.

Part I (including this chapter) contains an overview of the early years of RMT in Chapter 2, covering approximately the first three decades of RMT. Here the motivations are given why RMT was developed in the context of Nuclear Physics, and early results and applications in various fields of Physics and Mathematics are described. This chapter can be read independently as an introduction, with many details provided in later chapters. Part II is mainly devoted to fundamental properties of RMT as such, whereas Part III covers most modern applications of RMT. This division of II and III is not strict.

Part II can be divided into three sub-parts: A) methods of solving RMT, B) basic properties and fundamental objects in RMT, and C) different models and symmetry classes in RMT.

Part II A is constructed as follows: In Chapters 4 and 5 classical orthogonal polynomials (OP) and skew-OP are used to solve exactly RMT ensembles with

Table 1.1 Chapter numbers with short titles and links "$\rightarrow x$" to chapter number x. We only display links within Part II, links from Part III to Part II, as well as links from Part I to all other chapters.

	Part I		
2	History $\rightarrow$ 3,7,8,23,24,32,33,34,35		
	Part II		Part III
3	Symmetry classes	24	Number theory $\rightarrow$ 4,5,17,19
4	Unitary ensembles $\rightarrow$ 6,11	25	Random permutations $\rightarrow$ 6,10,11,19
5	Orthogonal & symplectic $\rightarrow$ 6,11	26	Enumeration of maps $\rightarrow$ 4,5,15,16
6	Universality $\rightarrow$ 4,5,9,14,15,16,21	27	Knot theory $\rightarrow$ 4,5,15,16
7	Supersymmetry	28	Multivariate statistics $\rightarrow$ 11,22
8	Replicas $\rightarrow$ 9	29	Algebraic geometry $\rightarrow$ 15,16,17,19
9	Painlevé transcendents $\rightarrow$ 6,8,10	30	2D Quantum gravity $\rightarrow$ 6,10,15,16
10	Integrable systems $\rightarrow$ 9,17	31	String theory $\rightarrow$ 4,5,12,15,16
11	Determinantal processes $\rightarrow$ 4,5	32	QCD $\rightarrow$ 4,5,6,17,18
12	Critical statistics $\rightarrow$ 14,23	33	Quantum chaos & graphs $\rightarrow$ 6,7
13	Heavy tails $\rightarrow$ 22	34	Resonance scattering $\rightarrow$ 7,18
14	Phase transitions $\rightarrow$ 6,12,20	35	Condensed matter $\rightarrow$ 12,23
15	Two-matrix model $\rightarrow$ 6	36	Optics $\rightarrow$ 12
16	Loop equations $\rightarrow$ 6	37	Entanglement $\rightarrow$ 4,5,14,20
17	Unitary integrals $\rightarrow$ 10	38	Random growth models $\rightarrow$ 5,6,10,16
18	Non-Hermitian ensembles	39	Laplacian growth $\rightarrow$ 10,15,16,18
19	Characteristic polynomials	40	Finance $\rightarrow$ 13,18
20	Beta ensembles $\rightarrow$ 14	41	Information theory $\rightarrow$ 21,22
21	Wigner ensembles $\rightarrow$ 6	42	RNA folding $\rightarrow$ 4,5,14,15,16
22	Free probability $\rightarrow$ 13	43	Complex networks $\rightarrow$ 8
23	Banded & sparse matrices $\rightarrow$ 12		

unitary, and orthogonal or symplectic invariance respectively, all at finite matrix size. In Chapter 10 these are extended to multiple OP's and a link to the theory of classical integrable systems is established. In Chapter 7 the supersymmetric method and Chapter 8 the replica method are reviewed. Both methods show how to deal with determinants that are needed to generate resolvents and correlations functions in an alternative way to Chapters 4 and 5. In Chapters 15 and 16 two- (and multi-)matrix models are solved using bi-OP and loop equations, respectively. The loop equation method is a systematic expansion of the saddle-point approximation that is first introduced in Chapter 14. In Chapter 17 the character expansion method is introduced for the ensemble of unitary matrices.

Part II B on fundamental properties and basic objects of the theory contains a chapter (Chapter 11) on determinantal point processes. For such processes the joint probability distribution function (jpdf) of the positions of the particles in

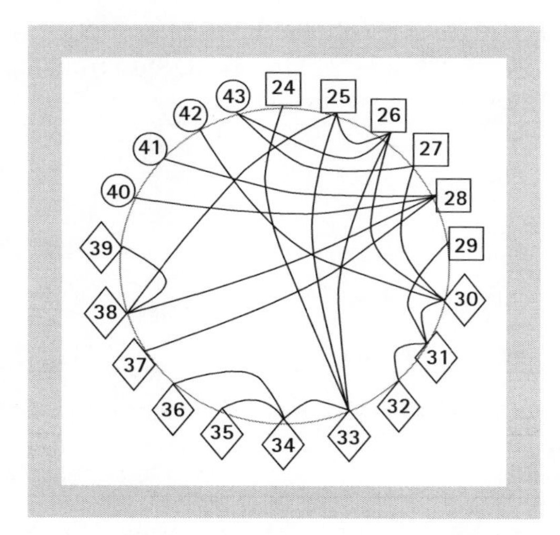

Fig. 1.2 Links among the applications of RMT in Part III. Chapters in squares denote applications to mathematics, chapters in rhombi to physics and chapters in circles to other sciences.

the process, can be computed explicitly. The correlation kernel for (skew) OP's and the generating functions for resolvents can be expressed in terms of characteristic polynomials which serve as building blocks for the theory and which are reviewed in Chapter 19. The limit of large matrix size $N \to \infty$ is one of the main interests in RMT. However, such a limit is not necessarily unique, and a heuristic picture of different phases is given in Chapter 14. One of the most powerful machines to solve invariant RMT in the large-N limit is the Riemann–Hilbert method reviewed in Chapter 6. Here the special functions appearing in various limits are related to Painlevé transcendents, to which the Chapter 9 is devoted. Chapter 6 (and 14) particularly illustrates the fundamental property of RMT called universality mentioned above. This expresses the fact that many results obtained in the large-N limit do not depend on the special choice of a Gaussian distribution of the matrix elements described earlier. Universality is also covered in Chapter 21 from a different mathematical point of view. It deals with generically non-invariant distributions of matrix elements, the so-called Wigner ensembles. The mathematical concept of free random variables is introduced in Chapter 22.

Part II C provides a list of the main matrix models and their classification. In Chapter 3 an ordering principle of these models into symmetry classes is offered. The three classical Wigner-Dyson classes (see Chapters 4 and 5) and their extensions are mapped to ten possible Riemannian symmetric spaces. Loosely speaking these classify the possible spaces on which the corresponding models can live. The two- and multi-matrix models are discussed in Chapters 15 and 16 and the related external field model in Chapter 19. Non-Hermitian invariant RMT's are reviewed in Chapter 18, with some additional information in Chapter 17. Other extensions of the classical one-matrix model are considered in Chapter 12 when we analyze a weakly confining non-Gaussian potential leading to critical statistics, and in Chapter 13 we allow the distribution of the matrix elements to have heavy tails. The understanding of critical statistics is the principal goal of the study of banded and sparse matrices in Chapter 23. Several of these models and concepts can also be studied in the context of the Wigner ensembles, see Chapter 21. An extension of the Dyson index $\beta = 1, 2, 4$ to to the general case $\beta > 0$ is described in Chapter 20, both in terms of eigenvalue and tridiagonal matrix realizations.

Part III on the applications of RMT is the most diverse part of this book. Here we shall be brief and only outline the main content. The first chapters 24 to 29 cover applications to mathematics from number theory to statistics. Chapters 30 to 39 cover applications to physics, from high energies to lower energies in quantum mechanical systems and growth processes. The last chapters cover even broader applications to other sciences, including economics in Chapter 40, engineering in Chapter 41, biology in Chapter 42, and complex networks in Chapter 43. In Figure 1.2 we give a graphical depiction of how these chapters in Part III are related, while Table 1.1 indicates to which chapters from part II they relate.

1.5 What is not covered in detail

Here we list some topics which are related to RMT, but not discussed in chapters of their own right. New developments are mentioned at the end of this section.

- Yang-Mills integrals: The zero-dimensional reduction of (Super) Yang Mills theory leads to a complicated RMT with a number of different matrices (and spinors). Due to its Yang-Mills structure, the model is different from simpler multi-matrix model chains, and standard analytical RMT methods typically do not apply. The model is treated briefly in Section 31.5, see also Section 32.4 for the Eguchi-Kawai reduction. A similar remark applies to RMT in non-commutative geometry.
- Statistical Mechanics: RMT can be viewed as a new statistical mechanics in its own right [Guh98]. Some examples of RMT applied to entanglement and extreme eigenvalues are covered in Chapter 37, but there are other examples, e.g. spin glasses in so-called random energy models, or random Euclidian matrices by Parisi and other authors (see his review in [Bre04]).
- Hankel and Toeplitz determinants: Powerful theorems exist for such determinants depending only on the difference or sum $i \pm j$ of the indices of the matrix elements. They play an important role in RMT, appearing e.g. in the jpdf of the eigenvalues (see circular ensembles below). For a sample of these results, see Sections 41.3.7 and 24.4.
- Parametric correlations: Because of the presence or absence of time reversal in Hamiltonians described by the GOE/GSE or GUE respectively,

transitions between these symmetry classes were studied early on by Mehta and Pandey. The eigenvalues moving from one to another (or the same) symmetry class can be described as parametric Brownian motions of particles. For more details on this approach and its relation to integrability, see Chapter 10.

- Finite rank perturbations: The addition of an external field to a random matrix can also be interpreted in terms of (usually finite-rank) perturbation theory. In the context of Wigner matrices this aspect is briefly covered in Section 21.4.6, as well as in the context of sample covariant matrices it is presented in Section 28.3.2 under the name 'spiked models'. We note also the related topic of contractions or truncations of RMT treated in Section 18.3.
- Circular ensembles: These are the classical counterparts of the 3 Wigner-Dyson ensembles, with eigenvalues living on the unit circle instead of the infinite real line. While they are extensively covered in [Meh67], they also appear briefly in Chapter 24.
- Condition numbers and signal processing: the ratio of the largest and the smallest singular values of a matrix represents the condition number of the associated linear system. The condition numbers and the invertibility of a high-dimensional random matrix have been studied in both numerical analysis and signal processing. See, for example, [Ver10] for an introduction to the application of RMT in compressed sensing.
- Bus system in Cuernavaca: As mentioned in the foreword written by Dyson, the curious empirical fact that the spacings between buses at a fixed stop in the city of Cuernavaca in Mexico follow GUE, is an illustration of the universality of RMT in complex systems. We refer the readers to [Krb00, Krb09] as well as [Bai06] and the references within. This may be viewed as part of a class of more general transport or parking problems, or some class of random walkers, all of which display RMT statistics.

Since the appearance of the first edition of this handbook several new developments have taken place. Because of space constraints we only mention some recent review papers on these topics, rather than updating or adding new chapters.

The understanding of beta ensembles (Chapter 20), notably of its universality and representation, has seen substantial progress. Universality has been shown in different parts of the spectrum; see [Bou14] and references therein. Ideas from stochastic differential operators and Brownian motion have lead to more explicit realisations of the correlation functions at the edge of the spectrum [Kri13], and to a different representation of these ensembles [All12].

In the application of RMT to Quantum Gravity (Chapters 30 and 31) random tensor models have been constructed, aimed at a description in three or higher dimensions. For recent progress in the $1/N$ expansion we refer to [Gur12, Riv14].

The realm of RMT applications in condensed matter physics (Chapter 35) has been broadened to include Majorana Fermions and topological superconductors, see [Bee14] for a review.

In the context of growth models (Chapter 38) new integrable MacDonald processes have been analyzed, see [Bor14, Cor14] for reviews. This and other developments have lead to a deeper understanding of the universality class of the Kadar-Parisi-Zhang (KPZ) equation as reviewed in [Qua15].

Products of random matrices appear naturally e.g. in applications to telecommunications (Chapter 41). Recently they have been shown to lead to integrable determinantal point processes. This gives rise to new universality classes at the origin in the large N limit, which relate to Cauchy multi-matrix models (see also Chapter 15). For a review of these developments, see [Ake15].

1.6 Some existing introductory literature

It is not our intention to give even a short overview over the vast literature on RMT to which we refer in individual chapters. Rather we would like to point to a few classical and influential works, reviews that cover much of the subject, as well as to some recent pedagogical introductions.

The standard reference is the textbook by Mehta on random matrices [Meh67] in its third edition from 2004 – the first edition appeared already in the early days of RMT in 1967. The book focuses on Gaussian ensembles, in particular on the nowadays called classical Gaussian Orthogonal, Unitary and Symplectic Ensemble, as well as their circular counterparts. It is probably still the most complete reference on results for these three Gaussian ensembles at finite matrix size, including all correlation functions, spacing distributions and gap probabilities – which we have not attempted to fully reproduce.

As regards applications, a few of which are sketched in the introduction of [Meh67], several long review articles have been published. The first of which, [Bro81] has as its explicit general aim "a study of the spectrum and the transition strengths in complicated systems" – with its motivation still coming mainly from nuclear physics. The Physics Reports [Guh98] published in 1998 is probably the first attempt to cover all aspects of RMT applications in quantum physics, ranging from chaotic and disordered systems over many-body quantum systems to quantum field theory (QFT).

The application to QFT had grown independently out of ideas from 't Hooft about the dominance of planar Feynman diagrams for matrix valued fields.

The problem of counting such diagrams was then studied in zero- and one-dimensional matrix valued QFT with quartic or cubic interaction in an influential paper [Bre78]. Subsequent developments, including two-dimensional quantum gravity, were covered in the collection of papers [Bre93] and the Physics Reports [DiF95].

Out of the vast body of conferences, schools or special issues on RMT such as [For03], we would like to single out the Les Houches summer school proceedings, in particular the sessions in 1994 [Dav94] and 2004 [Bre04] where an attempt was made to bring together different communities using the same tools.

The most recent publication that aims at a complete overview over RMT is the large volume of Forrester [For10] which even includes (research) exercises.

The lecture notes [Dei00, Dei09] for graduate students in mathematics focus specifically on the issue of universality of invariant ensembles using the orthogonal polynomial method. Another book at the graduate-student level is [And09] which gives an excellent introduction to RMT from the view point of probabilists. Readers with interest in statistics will benefit from [Bai10].

Our short survey survey of the literature is certainly not complete and reflects our own taste and experience.

Acknowledgements

We are indebted to all our colleagues who have helped us with comments and criticism of this project, in particular Percy Deift for the improvement of the introduction. We would also like to thank Sunčana Dulić for help with the graphics.

References

[Ake15] G.Akemann and J.R. Ipsen, *Recent exact and asymptotic results for products of independent random matrices*, arXiv:1502.01667 [math-ph].

[All12] R. Allez, J.-P. Bouchaud, and A. Guionnet, *Invariant β-ensembles and the Gauss-Wigner crossover*, Phys. Rev. Lett. **109** (2012) 094102 [arXiv:1205.3598v2 [math.PR]].

[And09] G.W. Anderson, A. Guionnet, and O. Zeitouni, *An Introduction to Random Matrices*, Cambridge Studies in Advanced Mathematics no. 118, 506 pages, Cambridge University Press (2009).

[Bai10] Z. Bai and J. Silverstein, *Spectral analysis of large dimensional random matrices*, Springer Series in Statistics, Springer, New York, (2010).

[Bai06] J. Baik, A. Borodin, P. Deift and T. Suidan, *A model for the bus system in Cuernavaca (Mexico)*, J. Phys. A: Math. Gen. **39** (2006) 8965 [arXiv:math/0510414 [math.PR]].

[Bee14] C.W.J. Beenakker, *Random-matrix theory of Majorana fermions and topological superconductors*, arXiv:1407.2131v2 [cond-mat.mes-hall].

[Ber99] M.V. Berry and J.P. Keating, *The Riemann zeros and Eigenvalue Asymptotics, SIAM Rev.*, **41** (1999) 236–266.

[Bor14] A. Borodin and L. Petrov, *Integrable probability: From representation theory to Macdonald processes*, Prob. Surv. **11** (2014) 1–58 [arXiv:1310.8007 [math.PR]].

[Bou14] P. Bourgade, L. Erdös, and H.-T. Yau, *Edge Universality of Beta Ensembles*, Comm. Math. Phys. **332** (2014) 261–353 [arXiv:1306.5728v2 [math.PR]].

[Bre78] E. Brézin, C. Itzykson, G. Parisi, and J. B. Zuber, *Planar diagrams*, Comm. Math. Phys. **59**, (1978) 35–51.

[Bre93] E. Brézin and S. Wadia, *The large N expansion in quantum field theory and statistical physics*, World Scientific, Singapore (1993).

[Bre04] E. Brézin, V. Kazakov, D. Serban, P. Wiegmann, and A. Zabrodin (Eds.), *Applications of Random Matrices in Physics*, Proceedings of the NATO Advanced Study Institute on Applications of Random Matrices in Physics, Les Houches, France, 6–25 June 2004, Series: NATO Science Series II: Mathematics, Physics and Chemistry, Vol. 221 (2004).

[Bro81] T.A. Brody, J. Flores, J.B. French, P.A. Mello, A. Pandey, and S.S.M. Wong, *Random-matrix physics: spectrum and strength fluctuations*, Rev. Mod. Phys. **53** (1981) 385–479.

[Cor14] I. Corwin, *Macdonald processes, quantum integrable systems and the Kardar-Parisi-Zhang universality class*, Proceedings of the ICM, arXiv:1403.6877 [math-ph].

[Dav94] F. David, P. Ginsparg, and J. Zinn-Justin (Eds.), *Fluctuating Geometries in Statistical Mechanics and Field Theory*, Proceedings of the NATO Advanced Study Institute, Les Houches, France, 2 August - 9 September 1994, Session LXII, 1092 pages, Elsevier, Amsterdam 1996.

[Dei00] P. Deift, *Orthogonal Polynomials and Random Matrices: A Riemann–Hilbert Approach*, Courant Lecture Notes, 261 pages, American Mathematical Society (7 Dec 2000).

[Dei09] P. Deift and D. Gioev, *Random Matrix Theory: Invariant Ensembles and Universality*, Courant Lecture Notes, 217 pages, American Mathematical Society (15 Aug 2009).

[DiF95] P. Di Francesco, P. Ginsparg and J. Zinn-Justin, *2D Gravity and Random Matrices*, Physics Reports **254** (1995) 1–133.

[For03] P.J. Forrester, N.C. Snaith, and J.J.M. Verbaarschot (Eds.), J. Phys. **A36** Number 12, (2003) R1-R10, 2859–3646 Special Issue: Random Matrix Theory.

[For10] P.J. Forrester, *Log-Gases and Random Matrices*, London Mathematical Society Monographs no. 34, 808 pages, Princeton University Press (26 July 2010).

[Guh98] T. Guhr, A. Müller-Groeling, and H.A. Weidenmüller, *Random Matrix Theories in Quantum Physics: Common Concepts*, Phys. Rept. **299** (1998) 189–425 [arXiv:cond-mat/9707301v1].

[Gur12] R. Gurau, *A review of the 1/N expansion in random tensor models*, Proceedings of the Int. Congress on Math. Phys. 2012, arXiv:1209.3252 [math-ph]

[Krb00] M. Křbalek and P. Šeba, *Statistical properties of the city transport in Cuernevaca (Mexico) and random matrix theory*, J. Phys. A **33** (2000) 229234.

[Krb09] M. Křbalek and P. Šeba, *Spectral rigidity of vehicular streams (random matrix theory approach)*, J. Phys. A **42** (2009) no. 34, 345001.

[Kri13] M. Krishnapur, B. Rider, and B. Virag, *Universality of the Stochastic Airy Operator*, arXiv:1306.4832 [math.PR].

[Meh67] M.L. Mehta, *Random Matrices*, Academic Press, New York, 1967, 259 pages; 2nd Edition 1991, 562 pages; 3rd Edition, Elsevier, Amsterdam 2004, 688 pages.

[Qua15] J. Quastel and H. Spohn, *The one-dimensional KPZ equation and its universality class*, arXiv:1503.06185v1 [math-ph].

[Riv14] V. Rivasseau, *The tensor track, III*, Fortschr. d. Physik **62** (2014) 81–107 [arXiv:1311.1461 [hep-th]].

[Ver10] R. Vershynin, *Introduction to the non-asymptotic analysis of random matrices*, Chapter 5 in *Compressed Sensing: Theory and Applications*, Y.C. Eldar and G. Kutyniok (Eds.), Cambridge University Press (2012) [arXiv:1011.3027v4 [math.PR]].

·2·

History – an overview

O. Bohigas and H. A. Weidenmüller

Abstract

Starting from its inception, we sketch the history of random matrix theory until about 1990, using only published material. Later developments are only partially covered.

2.1 Preface

When asked by the editors of this handbook to write an overview over the history of random matrix theory (RMT), we faced the difficulty that in the last 20 years RMT has experienced very rapid development. As witnessed by the table of contents of this handbook, RMT has expanded into a number of areas of physics and mathematics. It has turned out to be impossible to account for this development in the space of the 20 or so pages allocated to us. We decided to focus attention on the first four decades of the history of RMT, and to follow only some lines of development until recent times. Our choice was determined by personal preference and subject knowledge. We have omitted, for instance, the connections between RMT and low-energy field theory (see Chapter 32), those between RMT and integrability (see Chapter 18 but also Chapters 8, 32, 17, etc.), works on extreme-value statistics (see Chapter 37), and applications of RMT to wireless communication (see Chapter 41) and to stock-market data (see Chapter 40). For our reconstruction of the historical line of development, we have only used published literature on the subject. In addition to the original references given in this paper, there are several reviews and/or reprint collections [Por65, Meh67, Bro81, Boh84b, Gia91, Guh98, Wei09] that are helpful in studying the history of the field.

2.2 Bohr's concept of the compound nucleus

The discovery of narrow resonances in the scattering of slow neutrons by E. Fermi *et al.* [Fer34, Fer35] and others in the 1930s came as a big surprise to nuclear physicists. It contradicted earlier ideas of independent-particle motion in nuclei and led Bohr to formulate the idea of the "compound nucleus" [Boh37]. Bohr wrote

'In the atom and in the nucleus we have indeed to do with two extreme cases of mechanical many-body problems for which a procedure of approximation resting on a combination of one-body problems, so effective in the former case, loses any validity in the latter where we, from the very beginning, have to do with essential collective aspects of the interplay between the constituent particles'.

And:

'The phenomena of neutron capture thus force us to assume that a collision between a... neutron and a heavy nucleus will in the first place result in the formation of a compound system of remarkable stability. The possible later breaking up of this intermediate system by the ejection of a material particle, or its passing with the emission of radiation to a final stable state, must in fact be considered as separate competing processes which have no immediate connection with the first stage of the encounter'.

Bohr's view was generally adopted. Attempts at developing a theory of compound-nuclear reactions could, thus, not be based upon a single-particle approach but had instead to assume that in the compound nucleus, the constituents interact strongly. That fact and the almost complete lack of knowledge of the nucleon-nucleon interaction led to the development of formal theories of nuclear resonance reactions. The most influential of these was the R–matrix theory due to Wigner and Eisenbud [Wig47, Lan58]. The cross section was parameterized in terms of an R–matrix or, in the single-channel case, an R–function. That function contains the eigenvalues E_μ of the nuclear Hamiltonian as unknown parameters and is singular whenever the collision energy E equals one of the E_μ, giving rise to a resonance in the cross section. At the time (and even now) it seemed hopeless to determine the E_μ from a dynamical calculation. Instead of such a calculation, in a series of papers published around 1951 (see the reprint collection [Por65]) Wigner sought to determine characteristic features of the R–function such as its average and its fluctuation about the mean in terms of the distribution of the E_μ. That search and, as he says (see page 225 of Ref. [Por65]), the 'accidental' discovery of the Wishart ensemble of random matrices in an early version of the book by Wilks [Wil62], motivated Wigner to use random matrices [Wig55, Wig57a]. Thus Bohr's idea of the compound nucleus is at the root of the use of random matrices in physics.

2.3 Spectral properties

The years following Wigner's introduction of random matrices saw a rapid development of the theory of spectral fluctuations. The Wishart ensemble [Wis28] consists of matrices H that can be written as $H = AA^T$ where T denotes the transpose, and A is real and Gaussian distributed. That ensemble has only positive eigenvalues. In addition to the Wishart ensemble, Wigner considered also an ensemble of real and symmetric matrices H with elements that have a Gaussian zero-centered distribution. Gaussian ensembles with a probability density proportional to $\exp[-(N/\lambda^2)\mathrm{Tr}H^2]$ have since played a dominant role in physics applications of random matrices. Here $N \gg 1$ is the matrix dimension and λ a parameter that scales the average level density $\rho(E)$. As a function of energy E and for $N \to \infty$, $\rho(E)$ has the form [Wig55, Wig58]

$$\rho(E) = \frac{\pi\lambda}{N}\sqrt{1 - (E/2\lambda)^2}. \tag{2.3.1}$$

This is Wigner's 'semicircle law'. That law can be derived in a number of ways. The method used by Pastur [Pas72] is particularly illuminating and can be generalized. Calculating the distribution $\mathcal{P}(s)$ of spacings of neighbouring levels (the 'nearest-neighbour spacing (NNS) distribution') turned out to be much more difficult. Using the result for random matrices of dimension $N = 2$, Wigner [Wig57b] guessed that $\mathcal{P}(s)$ has the form

$$\mathcal{P}(s) = (\pi/2)s\exp[-(\pi/4)s^2]. \tag{2.3.2}$$

Here s is the actual spacing in units of the mean spacing $d = 1/\rho$. The linear rise of $\mathcal{P}(s)$ for small s is due to quantum-mechanical level repulsion first considered in 1929 by von Neumann and Wigner [Von29] and, in the present context, by Landau and Smorodinsky, see [Por65]. The 'Wigner surmise' (2.3.2) was in agreement with results of computer simulations [Ros58] but a definitive experimental test was not possible at the time. There were not enough data. In heavy atoms there was evidence [Ros60] in favour of the distribution (2.3.2).

For a full theoretical analysis of the eigenvalue distribution, one rewrites the integration measure $\prod_{\mu\leq\nu} \mathrm{d}H_{\mu\nu}$ of the Gaussian real ensemble in terms of the N eigenvalues λ_μ and of the elements of the diagonalizing orthogonal matrix. Wigner [Wig55, Wig57a] obtained

$$\prod_{\mu\leq\nu} \mathrm{d}H_{\mu\nu} \propto \mathrm{d}\mathcal{O} \prod_{\mu<\nu} |\lambda_\mu - \lambda_\nu| \prod_\sigma \mathrm{d}\lambda_\sigma. \tag{2.3.3}$$

Here $\mathrm{d}\mathcal{O}$ is the Haar measure of the orthogonal group in N dimensions. This transition to 'polar coordinates' was first considered by Hua [Hua53] but his book [Hua63] only became available in English in 1963. Equation (2.3.3) shows that the eigenvalues and eigenfunctions of the Gaussian ensemble are statistically uncorrelated. The Haar measure implies [Por60] that for $N \gg 1$ the projections of the eigenvectors onto an arbitrary direction in Hilbert space have a Gaussian distribution centred at zero, and that the partial widths of the neutron resonances have a χ-squared distribution with one degree of freedom (the 'Porter-Thomas distribution'). This fact had been inferred earlier both from the analysis of neutron resonance data [Por56] and from numerical simulations [Blu58].

Calculation of the NNS distribution in the limit $N \to \infty$ was made possible by introduction of the method of orthogonal polynomials by Mehta and Gaudin [Meh60a, Meh60b, Meh60c] (see Chapters 4, 5, and also 6). The key was the recognition that the factor $\prod_{\mu<\nu} |\lambda_\mu - \lambda_\nu|$ in expression (2.3.3) can be expressed in terms of the Vandermonde determinant of the eigenvalues, and that that determinant can be rewritten in terms of Hermite polynomials $H_m(x)$. The latter are mutually orthogonal, $\int dx H_m(x) H_n(x) \exp[-x^2] \propto \delta_{mn}$, with respect to the Gaussian weight. That weight factor arises when the probability density $\exp[-(N/\lambda^2)\mathrm{Tr}H^2]$ for the Gaussian ensemble is expressed in terms of the eigenvalues. The determinantal structure so essential for the method is most simply displayed for the ensemble of Hermitian (non-real) random matrices H of dimension N, defined by a Gaussian weight factor and the measure $\prod_{\mu \le \nu} \mathrm{dRe}H_{\mu\nu} \prod_{\mu<\nu} \mathrm{dIm}H_{\mu\nu}$. After transforming to polar coordinates, that measure is $\mathrm{d}U \prod_{\mu<\nu}(\lambda_\mu - \lambda_\nu)^2 \prod_\sigma \mathrm{d}\lambda_\sigma$, where $\mathrm{d}U$ is the Haar measure of the unitary group in N dimensions. The product of the Gaussian weight factor and the factor $\prod_{\mu<\nu}(\lambda_\mu - \lambda_\nu)^2$ can be written in the form $\det K_N(\lambda_\mu, \lambda_\nu)_{\mu,\nu=1,\ldots,N}$, and the function $K_N(x, y) = \sum_{k=0}^{N-1} \phi_k(x)\phi_k(y)$ is given in terms of the harmonic oscillator functions $\phi_k(x)$ (products of Hermite polynomials and a Gaussian). Results for $N \gg 1$ are obtained by using the asymptotic form of ϕ_k for large k. In this way, Gaudin [Gau61] obtained the NNS distribution for the Gaussian ensemble of real symmetric matrices in the form of an infinite product. He found that the Wigner surmise (2.3.2) is a very good approximation to the exact answer. The method of orthogonal polynomials [Meh67] is not restricted to a Gaussian probability density and can be used for other cases provided the orthogonal polynomials for those cases are known.

In a series of papers, Dyson [Dys62a, Dys62b, Dys62c, Dys62d, Dys62e] contributed to the rapid development of random-matrix theory. Dyson introduced 'circular ensembles' of unitary matrices of dimension N. The eigenvalues are located on the unit circle in the complex plane and in the limit of large matrix dimension have the same spectral fluctuation properties as the eigenvalues of the corresponding Gaussian ensemble that are located on the real axis. Using group theory he showed that there can be only three types of such ensembles (the 'threefold way'). The orthogonal ensemble ($\beta = 1$) applies when time-reversal invariance holds, the unitary ensemble ($\beta = 2$) when time-reversal invariance is violated, and the symplectic ensemble ($\beta = 4$) when the system is time-reversal invariant, has half-odd integral spin and is not rotationally invariant. These statements are not restricted to the circular ensembles but likewise apply to Gaussian ensembles (the GOE with $\beta = 1$, the GUE with $\beta = 2$, and the GSE with $\beta = 4$) and to ensembles with non-Gaussian probability densities. Level repulsion in the three ensembles is governed by the factor $\prod_{\mu<\nu} |\lambda_\mu - \lambda_\nu|^\beta$ which arises from the polar-coordinate representation of the invariant measure. Using orthogonal polynomials, Dyson found the n-level correlation functions for the eigenvalues of the three circular ensembles. He paid special attention to the two-level correlation function. That function relates to the rigidity of the spectrum. In the form of the Δ_3-statistic introduced by Mehta and Dyson [Meh63], that function has, in addition to the NNS distribution, become an important measure for the statistical analysis of empirical or numerically generated eigenvalue distributions. Dyson rederived Gaudin's result for the NNS distribution and calculated the distribution of spacings of next-nearest neighbours. Dyson connected the distribution of the eigenvalues in the Gaussian ensembles with the properties of a classical 'Coulomb gas'. The product $\exp[-(N/\lambda^2)\sum_\mu \lambda_\mu^2]\prod_{\mu<\nu}|\lambda_\mu - \lambda_\nu|^\beta$ is written as $\exp[-\beta W]$ with $W = -\sum_{\mu<\nu}\ln|\lambda_\mu - \lambda_\nu| + \sum_\mu (N/\beta\lambda^2)\lambda_\mu^2$. That distribution is identical to the thermodynamic equilibrium distribution at temperature $kT = 1/\beta$ of the positions of N point charges moving in one dimension under the influence of mutual two-dimensional Coulomb repulsion and an attractive harmonic oscillator potential. Thus the term 'level repulsion' gains direct physical meaning. Dyson also generalized this static analogy to a dynamic one: the eigenvalues undergo Brownian motion.

For a long time Dyson's classification leading to the three canonical ensembles GOE, GUE, GSE was considered complete. Since the beginning of the 1990s, however, new ensembles (the 'chiral' ensembles) of random matrices have been studied both in disordered systems [Gad91, Gad93, Sle93] and in elementary-particle physics [Shu93, Ver94]. The Hamiltonians have the form

$$H = \begin{pmatrix} 0 & h \\ h^\dagger & 0 \end{pmatrix}. \tag{2.3.4}$$

There are three symmetry classes as the ensemble may be orthogonally, unitarily, or symplectically invariant. Depending on the topology of the problem the matrix h may be rectangular. Another four classes (the 'Bogoliubov-de Gennes' ensembles) were discovered [Opp90, Alt97] in the context of superconducting systems. The Hamiltonians have the form

$$H = \begin{pmatrix} h & \Delta \\ -\Delta^* & -h^T \end{pmatrix}. \tag{2.3.5}$$

Here $h = h^\dagger$ and $\Delta = -\Delta^T$. There are four such ensembles because spin rotations have an impact even when time-reversal invariance is violated. Dyson's canonical ensembles are the only ones that occur if the ensembles are postulated to be stationary in energy. The new ensembles arise in systems with a special and distinct energy value like the origin or the Fermi energy. They play a role when the energy of the system is near that energy. These altogether ten matrix ensembles are complete. That was shown [Zir96, Hei05] with the help of Cartan's classification of symmetric spaces first used in that context by Dyson [Dys70]. (When h in Eq. (2.3.4) is rectangular rather than square, that classification scheme will not suffice). A thorough discussion of symmetry classes is given in Chapter 3.

For a meaningful comparison with experimental or numerically generated data, the spectral fluctuation measures of random-matrix theory (RMT) must be stationary, ergodic, and universal. Both stationarity and ergodicity for RMT were first addressed in [Fre78, Pan79], see also [Bro81]. Stationarity holds if the fluctuation measures (typically local entities defined over an energy interval small in comparison with the total range of the spectrum) do not depend on the centroid energy for which they are calculated. Without this property one would face the (arbitrary) choice of the centroid energy. Stationarity holds [Pan79] for both the Gaussian and the circular ensembles because all n-level correlation functions depend on s, the ratio of the actual level spacing and the mean spacing, and have the same analytical form throughout the spectrum. This statement likewise applies to the S-matrix correlation functions considered in Section 2.8. For a given observable $\mathcal{O}(E)$, ergodicity assures the equality of the ensemble average $\langle \mathcal{O}(E) \rangle$ (a theoretically accessible entity) and the running average $\overline{\mathcal{O}(E)}$ of $\mathcal{O}$ over the spectrum of a single realization of the ensemble. Only the latter is accessible experimentally as one always deals with a specific system. Within RMT, the equality $\overline{\mathcal{O}(E)} = \langle \mathcal{O}(E) \rangle$ cannot be proved as there is no way to evaluate the left-hand side theoretically. But the weaker condition [Fre78, Pan79]

$$\overline{\left(\overline{\mathcal{O}(E)} - \langle \mathcal{O}(E) \rangle \right)^2} = 0 \tag{2.3.6}$$

involves ensemble averages throughout, can be worked out theoretically, and implies that ergodicity holds for almost all members of the ensemble (i.e., for all except for a set of measure zero, the measure being the integration measure defining the ensemble). Equation (2.3.6) is met if the autocorrelation function of $\mathcal{O}(E)$ falls off sufficiently rapidly with increasing energy difference. That condition applies [Pan79] for all n-level correlation functions, and the spectral fluctuations of RMT are, therefore, ergodic. For the S-matrix, ergodicity was first proved [Ric77] under the assumption that the underlying random-matrix ensemble has that property. From Eq. (2.3.6), ergodicity for the S-matrix has been demonstrated only in the Ericson regime of strongly overlapping resonances [Fre78, Bro81]. As regards universality, the weight factor defining the Gaussian ensembles can be justified by a maximum-entropy argument [Bal68] but implies the physically implausible form (2.3.1) of the average spectrum. More realistic forms of the spectrum can be obtained [Bal68] by replacing the Gaussian weight factor by $\exp[-V(H)]$ where $V(H)$ is some invariant and positive-definite function of H. It is, therefore, important to ascertain that the local spectral fluctuation measures derived for the Gaussian ensembles hold universally, i.e., also for the non-Gaussian ensembles, except for the replacement of the local average level density (2.3.1) by its counterpart defined by the weight function $\exp[-V(H)]$. In [Fox64] that fact was demonstrated for those unitary ensembles for which orthogonal polynomials were available. It was conjectured repeatedly (see [Dys72, Boh92, Meh67]) that universality holds quite generally. In [Bre93, Bee93], universality was proved for the local two-level correlation function and for $\beta = 1, 2, 4$ provided that function $V(H)$ confines the spectrum to a finite interval of the real energy axis. For a very general class of observables, a proof valid for all universality classes and for weight functions $V(H)$ obeying that same constraint was given in [Hac95]. That proof clearly displays the root of universality: for $N \to \infty$, global and local spectral properties become independent. The technique of proof used in that paper contained in embryonic form what is now referred to as superbosonization [Bun07, Lit08]. For a mathematically rigorous proof of universality see Chapter 6.

The breaking of a symmetry or an invariance has been an important issue for RMT from the early years. In the simplest case the generic Hamiltonian matrix model for symmetry violation [Ros60] has two block-diagonal entries. Both belong to the GOE and are uncorrelated. They model states with two different quantum numbers. The entries in the non-diagonal blocks have smaller variances and model symmetry breaking. The model has been widely employed in nuclear physics to describe isospin violation by the Coulomb interaction. The violation of time-reversal invariance is simulated by writing the Hamiltonian as $H_{\mu\nu} + iA_{\mu\nu}$. Here $H_{\mu\nu}$ is a member of the GOE and A is a real and antisymmetric Gaussian random matrix. Significant violation of symmetry or invariance (perceptible in the spectral fluctuation properties) occurs when the root-mean-square (rms) values of the perturbing matrix elements are of the order of the mean level spacing, i.e. of relative order $1/\sqrt{N}$ (with respect to the unperturbed matrix elements). Following [Dys62d] this extreme sensitivity of the spectral fluctuation measures was first investigated in [Pan81]. It was used [Fre85] to deduce from the NNS distribution of energy levels an upper bound on the strength of time-reversal-invariance breaking in nuclei.

Ginibre [Gin65] extended random matrix theory to the case of non-Hermitian matrices (see the overview in Chapter 18). If the elements are Gaussian-distributed zero-centred uncorrelated random variables, the eigenvalues uniformly fill a circle centred at the origin of the complex plane.

2.4 Data

Precise data on nuclear resonances started to become available in the 1960s from Rainwater's group at Columbia working at the Nevis synchrocyclotron. The authors wrote [Rai60]

> *"Although the Nevis synchrocyclotron is primarily a tool for high-energy physics, ..., it has also been organized from its inception as an exceptionally strong source of pulsed neutrons for slow neutron spectroscopy".*

With the help of time-of-flight spectroscopy, the authors measured the total neutron cross section versus energy for a number of medium-weight and heavy nuclei and identified up to 200 resonances per nucleus. For a spin zero target nucleus, all resonances are expected to have identical quantum numbers because a slow neutron carrying zero angular momentum can excite only resonances with spin 1/2. For target nuclei with spins different from zero, the resonances can have two different total spins, and in some cases these can be separated. The number 200 was too small for a statistically meaningful comparison with RMT predictions. In addition, spurious levels (strong *p*-wave resonances mistaken for weak *s*-wave resonances) and missing levels (*s*-wave resonances with very small widths) could cause errors in the analysis. Thus, in 1963 Dyson and Mehta [Dys63] wrote: 'Unfortunately, our model is as yet neither proved nor disproved'. In the 1970s more data [Lio72a, Lio72b], and also some from proton scattering on light nuclei below the Coulomb barrier [Wil75, Wat81], became available. In 1982 Bohigas, Haq, and Pandey [Haq82, Boh83] combined all data then available into what they called the "nuclear data ensemble" comprising 1726 levels. The analysis of that data set showed very good agreement with the RMT predictions for the nearest-neighbour spacing distribution, for spectral rigidity, and for other statistical measures. This work established the view that complex nuclear spectra obey Wigner-Dyson statistics. Since 1982, more sophisticated experimental techniques have become available. This opens up the possibility for renewed critical analysis of a new data set.

2.5 Many-body theory

In the early 1970s, a link between many-body theory in the form of the nuclear shell model and random matrix theory was established. The shell model is based upon a mean-field description (a set of single-particle states and single-particle energies) supplemented by a 'residual interaction'. The latter is usually assumed to comprise two-body interactions only. French and collaborators had successfully promoted statistical nuclear spectroscopy as a tool to calculate average spectral properties of nuclei (an early summary of this work is given in [Fre66]). These are essentially determined by the normalized traces of powers of the shell-model Hamiltonian. It was found that the spectral density of the shell model is nearly Gaussian in shape (while it is semicircular for RMT). This fact motivated French and Wong [Fre70] and Bohigas and Flores [Boh71] to introduce and investigate the two-body random ensemble (TBRE), a random matrix version of the nuclear shell model: the matrix elements of the residual interaction are not determined from the nuclear dynamics but taken to be uncorrelated Gaussian random variables. Implementation of that model into existing computer codes for shell-model calculations was straightforward. The calculations were done in the *sd*–shell. Here the maximum number of particles (holes) is six. The spectral density of the TBRE was found to be Gaussian. Increasing the rank k of the k-body interaction resulted in a slow transition to semicircular spectral shape. In all cases the spectral fluctuation properties agreed with those of RMT. This suggested agreement of spectral fluctuation properties with RMT predictions both in the dilute and in the dense limits (number of particles or holes small compared to or of the same order as the number of shell-model single-particle states, respectively).

In [Mon73a] a simplified version of the TBRE was proposed that avoids the complexities due to angular momentum of this ensemble. In the embedded k-body random ensemble EGE(k) one considers m spinless fermions in Ω degenerate single-particle states that carry no further quantum numbers. The particles interact via a random k-body interaction that is orthogonally, unitarily, or symplectically invariant, as the case may be. Averaging traces of powers of EGE(k) one finds that with increasing k the form of the average spectrum changes from Gaussian to semicircle. This is because both cases are dominated by different Wick contraction patterns. For $k = 2$, results from the orthogonal EGE(2) can be used to reliably predict average results of shell-model calculations [Bro81]. It is not known whether EGE(k) is stationary, ergodic, or universal. Numerical simulations suggest that the spectral fluctuation properties coincide with those of the canonical ensembles but an analytical proof of that assertion does not exist.

2.6 Chaos

In 1890, Poincaré became aware of the existence of chaotic motion in the astronomical three-body problem. Around the middle of the twentieth century, Kolmogorov and his school and other mathematicians and astronomers investigated classical chaos in considerable depth. But it was not until the computer became a universal research tool that physicists at large became familiar with chaotic motion in classical systems. It came as a surprise that chaos occurs in systems with few degrees of freedom, so that complexity is not synonymous with many degrees of freedom. The insight that chaos is a generic feature of classical systems spread in the 1960s and 1970s and naturally led to the question whether there is a difference between quantum systems that are integrable and those that are fully chaotic in the classical limit. Einstein [Ein17] had anticipated that question when he realized that semiclassical quantization can be applied only to integrable systems. The question was mainly studied on systems with few degrees of freedom. While Chirikov and collaborators [Cas87] studied the time evolution of wave packets and discovered the quantum suppression of chaos, others focused attention on spectral properties of quantum systems. Using the correspondence principle and semiclassical arguments, Percival [Per73] put forward the notion that each discrete level of a quantum system can be associated with regular or chaotic classical motion. Berry and Tabor [Ber77] showed that the eigenvalues of a regular quantum system are uncorrelated and, thus, generically possess a Poissonian spectrum. Chaotic quantum systems also received much attention [McD79, Cas80, Ber81]. Using numerical results on about 1000 eigenvalues of the Sinai billiard (classically, a fully chaotic system) and the refined statistical analysis first developed for neutron resonances, Bohigas, Giannoni, and Schmit [Boh84a] demonstrated agreement of some fluctuation measures with RMT predictions and formulated the following conjecture: the spectral fluctuation measures of a generic classically chaotic system coincide with those of the canonical random matrix ensemble that has the same symmetry (unitary, orthogonal, or symplectic). That conjecture was soon supported by numerical studies of other chaotic systems.

Insight into the validity of the conjecture and the difference between integrable and chaotic systems came mainly from Gutzwiller's trace formula [Gut70, Gut90], see also [Bal70]. The study of manifolds with negative curvature by Schmit (see [Gia91]) and by Balazs and Voros [Bal86] played a particular role. Spectral fluctuation measures can be written in terms of the level density and the latter, in turn, as a Feynman path integral. Using the stationary-phase approximation in the path integral, one expresses the level density in terms of a sum over periodic orbits. In its essentials, Gutzwiller's trace formula has the form

$$\sum_i \delta(E - E_i) \approx \sum_{\text{p. o.}} A \exp[iS]. \tag{2.6.1}$$

The left-hand side is the level density (a quantum object), written as the sum over the eigenvalues E_i of the system, with E the energy. Except for the appearance of Planck's constant $\hbar$, the right-hand side contains only classical information: the sum is over all periodic orbits of the system, S is the action of the trajectory (in units of $\hbar$), and A is an amplitude that contains information on the period and the stability of the orbit. Regular systems possess families of periodic orbits while in fully chaotic systems the periodic orbits are isolated. As shown in [Han84] with the help of a sum rule, in chaotic systems only the short periodic orbits are system-specific while the long periodic orbits have universal properties. It is these features that led to an understanding [Ber85] and, eventually, to a demonstration [Sie01, Heu07] of the Bohigas-Giannoni-Schmit conjecture, see Chapter 33. On the scale of the mean level spacing, the long periodic orbits give rise to universal spectral fluctuation properties. By the uncertainty relation, the shortest periodic orbit determines the scale of the energy interval within which these fluctuations coincide with those of RMT.

2.7 Number theory

In the 1970s a connection of RMT with number theory was discovered. The Riemann zeta function, an analytic function in the entire complex s-plane (except for the point $s = 1$) and defined in terms of prime numbers, plays a fundamental role in number theory. With trivial exceptions, the zeros of this function are known to lie in the strip limited by the two lines $0 + i\alpha$ and $1 + i\alpha$ with α real and $-\infty < \alpha < +\infty$. The Riemann hypothesis says that in fact all these zeros are located on the line $(1/2) + iE$ with $-\infty < E < \infty$. Using that hypothesis, Montgomery [Mon73b] calculated the asymptotic form of the two-point correlation function Y_2 of these zeros. Dyson realized that the expression for Y_2 is the same as for the two–level correlation function of the unitary ensemble in the limit of large matrix dimension. Starting in the 1980s, Odlyzko [Odl87] determined numerically large numbers of zeros of the Riemann zeta function with large imaginary parts. All of these obey the Riemann conjecture. As a function of E, the fluctuating part of the density of the zeros on the line $(1/2) + iE$ is given by

$$-\frac{1}{\pi}\sum_{p}\sum_{r=1}^{\infty}\frac{\ln p}{p^{r/2}}\cos(Er\ln p). \tag{2.7.1}$$

Here p denotes the prime numbers and r counts the repetitions. Odlyzko found that the local fluctuation properties of the zeros agree with those of the unitary ensemble of RMT (Montgomery-Odlyzko conjecture) except for modifications due to small prime numbers. Using the analogy to Gutzwiller's trace formula (2.6.1), Berry and Keating [Ber99] proposed a dynamical interpretation of expression (2.7.1). The sum over p is like the sum over all periodic orbits, $\ln p$ stands for the period of the periodic orbit, $p^{r/2}$ for the stability, and $E\ln p$ for the action. The deviations found by Odlyzko are then interpreted as system-specific departures from universal behaviour. The connection between RMT and the Riemann zeta function is strengthened by the fact that Bogomolny and Keating [Bog95] have shown that the Hardy-Littlewood conjecture (existence of weak correlations between prime numbers) combined with an expansion of the type (2.7.1) implies the Montgomery-Odlyzko conjecture. More recently Keating and Snaith [Kea00] have shown that RMT is able to suggest exact formulas for moments of the Riemann zeta function that were not known and that still have to be proved, see Chapter 24.

2.8 Scattering theory

Parallel to the developments in statistical spectroscopy triggered by Bohr's paper, there was also important work in nuclear reaction theory. Bohr had postulated the independence of formation and decay of the compound nucleus (CN), see Section 2.2. Hauser and Feshbach [Hau52] used that postulate to write the compound-nucleus cross section in factorized form,

$$\sigma_{ab} \propto T_a \frac{T_b}{\sum_c T_c}. \tag{2.8.1}$$

One factor (the "transmission coefficient" T_a) gives the probability of formation of the CN from the entrance channel a, the other factor $T_b / \sum_c T_c$ gives the normalized probability of decay of the CN into one of the available exit channels b. Both factors are linked by microscopic reversibility, $\sigma_{ab} = \sigma_{ba}$. Originally the CN was considered a black box and the formation probability of the CN was put equal to unity, $T_a = 1$ for all channels. That view changed with the advent of the nuclear shell model. The concept of a black box was replaced by that of a partially transparent nucleus, formulated in terms of the optical model of elastic scattering [Fes54]. The imaginary part of the optical-model potential (an extension of the concept of the shell-model potential to scattering processes) describes absorption and, thus, formation of the CN. The transmission coefficients can be calculated from the optical model and may be smaller than unity. The average total width Γ of CN resonances was estimated in terms of these formation probabilities as $\Gamma = (d/(2\pi)) \sum_c T_c$ ("Weisskopf estimate" [Bla52]). Here d is the average spacing of the CN resonances

Isolated CN resonances as studied in the time-of-flight experiments by Rainwater and collaborators occur in the scattering of very slow neutrons: only elastic scattering is possible, and Γ is small compared to d. As the bombarding energy is increased, ever more channels open up (inelastic neutron scattering leaving the residual nucleus in an excited state becomes possible, other breakup channels open up). As a result, Γ grows strongly with energy. At the same time, the average spacing d of resonances shrinks because (like in any many-body fermionic system) the average density $\rho(E) = 1/d$ of states in the CN grows with energy E like $\exp\{\sqrt{aE}\}$ where a is a mass-dependent constant. Thus, with increasing energy the resonances begin to overlap, and for neutrons with bombarding energies of several MeV the CN is in the regime of strongly overlapping resonances ($\Gamma \gg d$). It had been held for a long time [Bla52] that in this regime, the numerous resonances contributing randomly at each energy would yield a scattering amplitude that is smooth in energy. Ericson [Eri60, Eri63] and Brink and Stephen [Bri63] realized that this is not the case. They argued that for $\Gamma \gg d$ the CN cross section should display strong fluctuations with energy (fluctuations that are as big as the average cross section, with a correlation width given by Γ), that the Bohr assumption and the Hauser-Feshbach formula Eq. (2.8.1) hold only for the average cross section (as opposed to the cross section at fixed bombarding energy), and that the elements of the scattering matrix are complex random processes with a Gaussian probability distribution. With the advent of electrostatic accelerators of sufficient energy resolution, such fluctuations of the CN cross section could actually be measured, and the theoretical predictions were verified [Eri66]. Ericson fluctuations (as the phenomenon has come to be called) have since surfaced in many areas of physics. These fluctuations are now understood to describe the generic features of wave scattering by chaotic systems in the regime $\Gamma \gg d$.

These theoretical developments, although inspired by Bohr's idea and informed by statistical arguments, were not linked directly to RMT. The obvious question was: is it possible to derive the Hauser-Feshbach formula (2.8.1), the Weisskopf estimate for Γ, and Ericson fluctuations from a random-matrix model that is consistent with known or anticipated properties of isolated resonances, i.e., from the GOE? The answer could obviously be given only on the basis of a theory of nuclear resonance reactions. Because of its dependence on a large number of arbitrary parameters (channel radii and boundary condition parameters), the formal *R*-matrix theory of Wigner and Eisenbud [Wig47] did not qualify for that purpose. The development of nuclear-structure theory in terms of the shell model led to a parameter-free dynamical theory of nuclear resonance reactions [Mah69]. In that theory, the scattering matrix $S_{ab}(E)$ (a unitary symmetric matrix in the space of Λ open channels) is given explicitly in terms of the nuclear Hamiltonian $H_{\mu\nu}$, a matrix in the N-dimensional space of quasibound states. In the simplest case S and H are related by

$$S_{ab}(E) = \delta_{ab} - 2i\pi \sum_{\mu\nu} W_{a\mu}(D^{-1})_{\mu\nu} W_{\nu b} \tag{2.8.2}$$

where

$$D_{\mu\nu}(E) = \delta_{\mu\nu} E - H_{\mu\nu} + i\pi \sum_{c} W_{\mu c} W_{c\nu}. \tag{2.8.3}$$

Here $W_{\mu a} = W_{a\mu}$ are real matrix elements that couple the quasibound states $\mu = 1, \ldots, N$ to the open channels $a = 1, \ldots, \Lambda$. The dynamical S-matrix of Eqs. (2.8.2) and (2.8.3) becomes a stochastic matrix when we replace the real and symmetric Hamiltonian matrix $H_{\mu\nu}$ in Eq. (2.8.3) by the N-dimensional GOE matrix $H^{\mathrm{GOE}}_{\mu\nu}$. The cross section σ_{ab} is proportional to $|S_{ab}(E)|^2$, and it is the aim of the theory to calculate average values, fluctuations, and correlation functions of cross sections. This is done by averaging over the ensemble defined in Eqs. (2.8.2) and (2.8.3) by the replacement $H_{\mu\nu} \to H^{\mathrm{GOE}}_{\mu\nu}$. The limit $N \to \infty$ is always taken. Ensemble averages are related to experimentally accessible energy averages by ergodicity.

The S-matrix of Eq. (2.8.2) is written as $S_{ab} = \langle S_{ab} \rangle + S^{\mathrm{fl}}_{ab}$. The average part $\langle S_{ab} \rangle$ and the fluctuating part S^{fl}_{ab} relate to very different time scales of the reaction. By ergodicity $\langle S_{ab} \rangle$ corresponds to an average over energy encompassing very many resonances and, thus, to the fast part of the reaction. Therefore, simple models involving only a few degrees of freedom like the optical model of elastic scattering can be used to calculate $\langle S_{ab} \rangle$ which serves as input for the statistical model of Eqs. (2.8.2) and (2.8.3). The fluctuating part $S^{\rm fl}_{ab}$ describes the slow processes (formation and decay of the N resonances) and is the object of interest of the statistical model. The characteristic time scale is the average lifetime of the resonances.

Equations (2.8.2) and (2.8.3) contain the $W_{\mu a}$ as parameters. Because of the orthogonal invariance of the GOE in the space of quasibound states, the distribution of S-matrix elements depends only on the orthogonal invariants $(1/N) \sum_\mu W_{a\mu} W_{\mu b}$. For simplicity we choose a basis in channel space for which $\langle S_{ab} \rangle$ is diagonal. Then $(1/N) \sum_\mu W_{a\mu} W_{\mu b} = \delta_{ab} v_a^2$ where v_a^2 has dimension energy and measures the strength of the average coupling of the quasibound states to channel a. The elements of the S-matrix are dimensionless, and the dimensionless invariants have the form $x_a = \pi N v_a^2 / \lambda = v_a^2 / d$. Here 2λ is the radius of the GOE semicircle and d is the average level spacing (taken in the centre of that semicircle). The average S-matrix can be worked out and has the form $\langle S_{ab} \rangle = \delta_{ab}(1 - x_a)/(1 + x_a)$, and the transmission coefficients are given by $T_a = 1 - |\langle S_{aa} \rangle|^2 = 4x_a/(1 + x_a)^2$. As mentioned above, the transmission coefficients are determined phenomenologically in terms of the optical model and serve as input parameters. The distribution of S-matrix elements defined by Eqs. (2.8.2) and (2.8.3) is then completely determined. The task of theory consists in working out that distribution explicitly.

Finding the complete distribution of S-matrix elements involves an integration over the $N(N+1)/2$ random variables of the GOE Hamiltonian $H^{\rm GOE}$. That is a difficult task which has not been fully accomplished. The method of orthogonal polynomials that is so successful for the calculation of spectral fluctuation properties, fails for the scattering problem. A reduced aim is to determine low moments and correlation functions of the S-matrix as these can be compared directly to experimental data. Analyticity and causality imply that for an arbitrary set of real energies E_i and pairs of channels $\{a_i b_i\}$ with $i = 1, \ldots, n$ we have $\langle \prod_i S_{a_i b_i}(E_i) \rangle = \prod_i \langle S_{a_i b_i}(E_i) \rangle$. Therefore, only correlation functions involving at least one factor S and one factor S^*, need be calculated. For isolated ($\Gamma \ll d$) and strongly overlapping ($\Gamma \gg d$) resonances first results were obtained by rewriting Eqs. (2.8.2) and (2.8.3) in the diagonal representation of $H^{\rm GOE}$. Let $\mathcal{O}$ be the orthogonal transformation that diagonalizes $H^{\rm GOE}$, let E_μ with $\mu = 1, \ldots, N$ be the eigenvalues, and let $\tilde{W}_{a\mu} = \sum_\rho \mathcal{O}_{\mu\rho} W_{a\rho}$ be the transformed coupling matrix elements. The latter are Gaussian-distributed random variables. The S-matrix is expanded in a Born series with respect to the non-diagonal elements of the transformed width matrix $2\pi \sum_a \tilde{W}_{\mu a} \tilde{W}_{a\nu}$, see Eq. (2.8.2). For $\Gamma \ll d$ these are small in comparison with d, and correlations among the E_μ are irrelevant. Thus, the average cross section is calculated easily [Lan57, Mol64]. For $\Gamma \gg d$, it is assumed [Aga75] that the eigenvalues

E_μ have constant spacings d (this 'picket fence model' neglects GOE correlations among the E_μ). The average over the Gaussian-distributed $\tilde{W}_{\mu a}$ is calculated with the help of Wick contraction. The contraction patterns can be written diagrammatically and are ordered according to the power in d/Γ to which they contribute. The terms of low order can be resummed and given explicitly. (That same scheme was later used by Brézin and Zee [Bre94] for the calculation of level correlations in disordered systems.) As a result, one obtains [Aga75] the Hauser-Feshbach formula (2.8.1) as the leading term in the asymptotic expansion of the average cross section. To leading order in d/Γ, the S-matrix elements are found to be Gaussian-distributed random processes with a Lorentzian correlation function. The correlation width Γ is given by the Weisskopf estimate. Thus, for $\Gamma \gg d$ the distribution of S-matrix elements is fully known and consistent with the predictions of [Eri60, Eri63, Bri63]. The neglect of GOE level correlations in these calculations is physically plausible and was later justified when the replica trick was used [Wei84] to calculate the S-matrix correlation function $\langle S_{ab}(E) S^*_{cd}(E+\varepsilon)\rangle$ as a function of the energy difference ε. Without further approximation, that calculation generates an asymptotic expansion in powers of d/Γ the first few terms of which agree with the results of [Aga75]. Exact calculation of the full correlation function $\langle S_{ab}(E) S^*_{cd}(E+\varepsilon)\rangle$ for all values of Γ/d was possible only with the help of the supersymmetry technique [Ver85a]. The result is an integral representation of the correlation function in terms of a threefold integral over real integration variables. For $\Gamma \ll d$ and $\Gamma \gg d$ the result agrees [Ver86] with the perturbative [Lan57, Mol64] and asymptotic results [Aga75, Wei84] obtained earlier. The result [Ver85a] has been widely used in the analysis of chaotic scattering, i.e. for compound-nucleus reactions, for electron transport through disordered mesoscopic samples [Alh00, Bee97], and for the passage of electromagnetic waves through microwave cavities [Fyo05]. (Some of these applications are discussed in Chapters 34 and 35.) Higher moments and correlation functions of S are not known in general. The supersymmetry method becomes too complex for such a calculation.

Dyson's circular ensembles of unitary S-matrices introduced in Section 2.3 correspond to the case of a very large number of channels all with $T_c = 1$. They have not been used in nuclear reaction theory.

2.9 Replica trick and supersymmetry

The method of orthogonal polynomials, so successful for the calculation of spectral fluctuation measures, faced severe difficulties in the case of scattering problems, and for a long time the calculation of the correlation function of the S-matrix given by Eqs. (2.8.2) and (2.8.3) remained an open problem. The situation changed when it was recognized [Ver84] that random-matrix theory corresponds to an Anderson model of dimensionality zero. This established a link between random-matrix theory and the theory of disordered solids that has become ever more important since. In particular, the link suggested that methods developed for calculating ensemble averages in the theory of disordered solids could be used to advantage also in random-matrix problems. The two important methods taken from the theory of disordered solids were the replica trick due to Edwards and Anderson [Edw75] and the supersymmetry method due to Efetov [Efe83b] (for reviews, see Chapters 8 and 7, respectively). Both methods show that in the limit of large matrix dimension, local fluctuation measures (taken on the scale of the mean level spacing) and global (mean) properties of the spectrum separate. This is the basic reason why random-matrix results for the spectral fluctuation measures are universal while that is not generally true for the global properties (see Chapters 30 and 16).

In both approaches it is assumed that the observable $\mathcal{O}$ that is to be averaged, can be written as a suitable derivative of the logarithm of a determinant,

$$\mathcal{O} = \frac{\partial}{\partial j} \ln \det(E\delta_{\mu\nu} - H_{\mu\nu} + M_{\mu\nu}(j)) \bigg|_{j=0}, \tag{2.9.1}$$

or as the derivative of the logarithm of a product of such determinants. The observable usually involves the resolvent of the Hamiltonian, and the logarithm is needed to remove the determinant from the final expression. We have used the same notation as in Eq. (2.8.3). The matrix $M(j)$ specifies the observable. In addition, $M(j)$ may describe the coupling to open channels, etc. The inverse determinant can be written as a Gaussian integral over N complex integration variables ψ_μ. Averaging $\mathcal{O}$ over the ensemble is then tantamount to averaging the logarithm of the generating functional

$$Z(j) = \int_{-\infty}^{+\infty} \prod_{\rho=1}^{N} \mathrm{d}\psi_\rho \, \exp\{(i/2) \sum_{\mu\nu} \psi_\mu^* [E\delta_{\mu\nu} - H_{\mu\nu} + M_{\mu\nu}(j)] \psi_\nu\}, \tag{2.9.2}$$

or of a product of such functionals. Formally speaking, Eq. (2.9.2) establishes the connection to a field theory with commuting (bosonic) fields. Instead of the commuting integration variables ψ_μ appearing in Eq. (2.9.2), one may use anticommuting (fermionic) integration variables χ_μ, and the average of $\mathcal{O}$ is obtained by averaging a generating functional of the form (2.9.2) but with anticommuting integration variables. That yields a relation to a field theory with fermionic fields. The use of anticommuting integration variables goes back to Berezin [Ber61].

Averaging the logarithm of $Z(j)$ is very complicated for both, commuting and anticommuting integration variables. The difficulty is circumvented with the help of the replica trick, originally introduced in [Edw75] for spin-glass problems. Here one determines $\langle \ln Z \rangle$ from the identity

$$\langle \ln Z \rangle = \lim_{n \to 0} \left(\frac{\langle Z^n \rangle - 1}{n} \right). \tag{2.9.3}$$

Equation (2.9.3) holds if $\langle Z^n \rangle$ is known analytically as a function of complex n in the vicinity of $n = 0$. Actually one calculates $\langle Z^n \rangle$ for positive or negative integer n (using fermionic or bosonic integration variables, respectively) and then uses Eq. (2.9.3). Thus, the approach is not exact and is called the replica trick (rather than the replica method). The replica trick with commuting variables was first used in random-matrix theory to calculate the shape of the average spectrum of the GOE (the 'Wigner semicircle law') [Edw76]. In two seminal papers, Wegner [Weg79] and Schäfer and Wegner [Sch80] analysed the structure of the underlying theory for bosonic fields. Evaluating the integrals with the help of the saddle-point approximation, they showed that for the two-point function (an average involving both the resolvent and its complex conjugate), the symmetry of the theory is broken, leading to a Goldstone boson. The single saddle point encountered for the one-point function becomes a saddle-point manifold with hyperbolic symmetry. After integration over the massive modes (the modes orthogonal to the saddle-point manifold), the theory is equivalent to a non-linear sigma model. When applied to the two-point function of the stochastic scattering matrix, again with commuting integration variables, the replica trick yields an asymptotic expansion in powers of d/Γ [Wei84] but, unfortunately, not the full answer.

A new and different approach to ensemble averaging in the theory of disordered metals was initiated by Efetov [Efe82], first applied to disordered systems by Efetov and Larkin [Efe83a]) and then extended by Efetov [Efe83b]. In that approach the use of the replica trick is avoided by writing (similarly to Eq. (2.9.1)) the observable as the derivative of the ratio of two determinants or, equivalently, of the product of a generating functional $Z^{(1)}$ with fermionic integration variables and of a generating functional $Z^{(-1)}$ with bosonic integration variables. (It seems that fermionic integration variables were first used for condensed-matter problems in [Efe80]). At $j = 0$ that product equals unity, the observable is simply the ordinary derivative of the product, and averaging the latter is simple. The price one has to pay consists in working with both commuting and anticommuting integration variables. A similar combination of bosonic and fermionic fields occurs in the theory of elementary particles [Wes74]. Here, a special relativistic fermion-boson symmetry is denoted as "supersymmetry". That same term is used for Efetov's method although the method lacks that special relativistic symmetry. The method can be formulated for the cases of unitary, orthogonal, and symplectic invariance. (For supersymmetry in RMT, see Chapter 7. For symmetry classes, see Chapter 3.)

Efetov's approach leads generally to a supersymmetric non-linear sigma model. With the help of Efetov's method, the two-point function of the scattering matrix defined in Eqs. (2.8.2) and (2.8.3) could be worked out exactly [Ver85a]. In [Efe83b, Ver85a], the results of [Sch80] on the structure of the saddle-point manifold were extended to the supersymmetry approach. The saddle-point manifold was shown to involve both compact and non-compact integration manifolds, stemming from the fermionic and bosonic integration variables, respectively. The supersymmetry method has since seen an ever growing number of applications to problems in random-matrix theory [Efe97]. It applies to all ten random-matrix ensembles [Zir96, Eve08].

The supersymmetry method gave rise to a critique of the replica trick. In [Ver85b] it was argued that the asymptotic result found in [Wei84] is generic in the sense that the replica trick always yields an asymptotic or perturbative (and not the exact) result. The claim was substantiated by analysis of the replica-trick calculation of the spectral two-point function for the GUE. It was shown that the use of only bosonic or only fermionic degrees of freedom fails to account for the full complexity of the saddle-point manifold which is obtained in the framework of the supersymmetry method and is needed to obtain the exact result. The paper was followed by an animated debate in the community on the validity and limitations of the replica trick. In [Kam99, Yur99] a cure for its failure was seen in the breaking of the replica symmetry and hitherto neglected saddle-point contributions, see [Zir99]. Reference [Kan02] cast a new light on the problem. In the framework of matrix ensembles with unitary symmetry it was argued that the deficiencies of the replica trick were caused by an approximate evaluation of $\langle Z^{(n)} \rangle$ for integer n. The problem and the explicit calculation of these functions were circumvented by showing that in the fermionic case (n positive integer) the $\langle Z^{(n)} \rangle$ are closely related to functions $\tau_n(E)$ that are known from the theory of integrable systems. These functions obey a set of coupled non-linear differential equations in the parameter E (the energy), the Toda lattice equations. The same functions also appear in the Hamiltonian formulation of the six Painlevé equations. There, however, the replica index n plays the role of a parameter. That connection shows that it is possible and meaningful to continue the Toda lattice equations analytically in n. Extrapolating the result to $n = 0$ yields the exact result for the one-point and two-point functions. In addition to showing that the replica trick can be made exact (at least for the case of unitary symmetry), the work of [Kan02] established a connection between random-matrix theory and integrable non-linear systems. The arguments in [Kan02] were much simplified in [Spl04]. The authors showed that $\langle Z^{(n)} \rangle$ obeys the Toda lattice equations separately for positive and for negative integer values of n and assumed that these equations also hold at $n = 0$. Then $\langle Z^{(0)} \rangle$ (the quantity of interest) is given by the product $\langle Z^{(1)} \rangle \langle Z^{(-1)} \rangle$.

This shows that '*the factorization of the two-point function into a bosonic and a fermionic partition function is not an accident but rather the consequence of the relation between random-matrix theories and integrable hierarchies*' [Spl04].

The assumption made in [Spl04] – validity of the Toda lattice equations across the point $n = 0$ – was later justified with the help of orthogonal polynomials [Ake05]. Thus, it is now established (at least for the unitary case) that when handled properly the replica trick yields exact results, too, see Chapter 8.

2.10 Disordered solids

In order to describe the response of small disordered metallic particles to external fields, Gor'kov and Eliashberg [Gor65] needed the two-level eigenvalue correlation function. They argued that the Wigner-Dyson statistic should apply in this case as well as in atomic nuclei. That was the first application of RMT in condensed-matter physics. Almost 20 years later Efetov [Efe83b] derived their statistical hypothesis from a generic model for a quasi one-dimensional disordered solid in the limit of small system length using the supersymmetry technique. Independently, in [Ver84] it was recognized that random-matrix theory as used in nuclear physics corresponds to an Anderson model of dimensionality zero. The resulting link between random-matrix theory and the theory of disordered solids has become ever more important since. As described in the following paragraphs, that link has played a role in designing random-matrix models for Anderson localization and for electron transport through disordered mesoscopic samples. It has likewise been important in formulating a theoretical approach to Andreev scattering in disordered conductors where it gave rise to the discovery of the four Bogoliubov-de Gennes ensembles. But the first and obvious question was: How is the RMT behaviour valid for small disordered samples modified when we consider larger samples? Al'tshuler and Shklovskii [Alt86] considered disordered systems in the diffusive regime and showed that the range of validity of RMT in energy is limited by the Thouless energy $E_c = \hbar\mathcal{D}/L^2$. Here $\mathcal{D}$ is the diffusion constant and L is the length of the sample. They also calculated corrections to RMT predictions.

In 1959, Anderson showed that the eigenfunctions of non-interacting electrons in disordered solids may be localized (i.e. do not extend uniformly over the entire system). An important link to chaotic quantum systems was established when Fishman *et al.* [Fis82], see also [Chi81], recognized that such systems (like, for instance, the kicked rotor) display localization in momentum space in a manner closely related to the localization that occurs in ordinary space for disordered solids. In [Sel86] it was proposed to use random band matrices (for which the variances of the non-diagonal elements that have a distance $\geq b$ from the diagonal, are suppressed) as vehicles to study localization phenomena in quasi one-dimensional geometries. In his original papers on RMT, Wigner had named such matrices 'bordered matrices'. As models for random linear chains (which play a role in a variety of physical situations) they were studied early on by Dyson [Dys53], Wigner [Wig55, Wig57a] and Engleman [Eng58]. After being used numerically both for the study of the kicked rotor and in the form of transfer matrices for the Anderson localization problem, random band matrices were first investigated analytically in [Fyo91]. With the help of supersymmetry it was shown that random band matrices are fully equivalent to random quasi one-dimensional systems. The parameter relevant for localization is the square of the band width b divided by the matrix dimension N, as had been expected on numerical grounds. Since then power-law random band matrices (where the variances of the off-diagonal matrix elements $\langle H_{ij}^2 \rangle$ decrease with an inverse power of $|i-j|$) have been used to model the metal-insulator transition in disordered solids [Mir96], see also [Eve08]. (Banded and sparse random matrices are treated in Chapter 23.)

Interest in the energy eigenvalue statistics of disordered mesoscopic samples (metallic or semiconducting devices of sufficiently small size so that at low temperature the electron's inelastic mean free path is larger than the sample size) was stimulated by the experimental discovery of universal conductance fluctuations (UCF). Following the seminal work of Al'tshuler and Shklovskii [Alt86] and Imry's application [Imr86] of RMT to UCF, a random-matrix description of quantum transport was developed in [Mut87]. The energy levels are linked to the transmission eigenvalues. Alternatively, electron transport was described in terms of the scattering approach of Eqs. (2.8.2) and (2.8.3). The linear extension of the sample was accounted for by replacing the GOE Hamiltonian by a random band matrix [Iid90]. In either of these forms, RMT has been widely applied to electron transport through mesoscopic systems [Bee97], see also Chapter 35.

2.11 Interacting fermions and field theory

A dynamical extension of Dyson's Coulomb gas (see Section 2.3) was considered by Calogero [Cal69a, Cal69b]. The Hamiltonian

$$H_C = -\frac{\hbar^2}{2m}\left[\sum_{i=1}^{N}\frac{\partial^2}{\partial\lambda_i^2} - \frac{1}{2}\sum_{i<j}\frac{\beta(\beta-2)}{(\lambda_i-\lambda_j)^2}\right] + \sum_i V(\lambda_i) \tag{2.11.1}$$

describes the dynamics of N particles in one dimension with position variables λ_i interacting via Coulomb forces and under the influence of a common potential V. For a harmonic-oscillator potential V, that Hamiltonian is integrable. The ground-state wave function defines a probability distribution of the particle positions which for $\beta = 1, 2, 4$ coincides with the joint probability densities for the eigenvalues (in units of the mean level spacing) of Dyson's three canonical Gaussian ensembles. An analogous construction for the circular ensembles is due to Sutherland [Sut71]. A link between the parametric level correlation functions of RMT and the time-dependent particle density correlation functions of the Sutherland Hamiltonian was conjectured and later proved in [Sim93a, Sim93b, Sim93c, Sim94]. In recent years, the connection between interacting fermions in one dimension and RMT has given rise to substantial research activity, see [Kor93].

Random-matrix theory has been useful in model studies of power-series expansions of field theories with internal SU(N) symmetry. The terms arising in these expansions can be expressed as Feynman diagrams, and counting the leading terms in an expansion in inverse powers of N (the 'planar diagrams') can be accomplished with the help of RMT [Bre78]. An analogous statement applies to quantum gravity in two dimensions. See Chapters 30 and 16.

Starting from the low-energy effective Lagrangean of quantum chromodynamis (QCD), Leutwyler and Smilga [Leu92] derived sum rules for the inverse eigenvalues of the Dirac operator. These same sum rules can also be derived from chiral random matrix theories [Shu93, Ver93, Hal95]. This fact gave rise to the insight that in the low-energy limit, QCD is equivalent to a chiral random matrix theory that has the same symmetries, see Chapter 32. Similar equivalence relations exist for other field theories, see Chapter 31.

Acknowledgements

The authors are grateful to J. J. M. Verbaarschot for a critical reading of parts of the manuscript and for helpful suggestions.

References

[Aga75] D. Agassi, H. A. Weidenmüller, and G. Mantzouranis, Phys. Rep. **22** (1975) 145.
[Ake05] G. Akemann, J. C. Osborn, K. Splittorf, J. J. M. Verbaarschot, Nucl. Phys. B **712** (2005) 287.
[Alh00] Y. Alhassid, Rev. Mod. Phys. **72** (2000) 895.
[Alt97] A. Altland and M. R. Zirnbauer, Phys. Rev. **B 55** (1997) 1142.
[Alt86] B. L. Altshuler and B. I. Shklovskii, Zh. Eksp. Teor. Fiz. **91** (1986) 220 [Sov. Phys. JETP **64** (1986) 127].
[Bal86] N. L. Balazs and A. Voros, Phys. Rep. **143** (1986) 109.
[Bal68] R. Balian, Nuov. Cim. **LVII B** (1968) 183.
[Bal70] R. Balian and C. Bloch, Ann. Phys. (N. Y.) **60** (1970) 401.
[Bee93] C. W. Beenakker, Phys. Rev. Lett. **70** (1993) 1155.
[Bee97] C. W. Beenakker, Rev. Mod. Phys. **69** (1997) 731.
[Ber61] F. A. Berezin, Dokl. Akad. Nauk SSR **137** (1961) 31.
[Ber77] M. V. Berry and M. Tabor, Proc. R. Soc. London, Ser. A **356** (1977) 375.
[Ber81] M. V. Berry, Ann. Phys. (N.Y.) **131** (1981) 163.
[Ber85] M. V. Berry, Proc. R. Soc. Ser. A **400** (1985) 229.
[Ber99] M. V. Berry and J. P. Keating, SIAM Rev. **41** (1999) 236.
[Bla52] J. M. Blatt and V. F. Weisskopf, *Theoretical Nuclear Physics*, Wiley and Sons, New York (1952).

[Blu58] S. Blumberg and C. E. Porter, Phys. Rev. **110** (1958) 786.

[Bog95] E. B. Bogomolny and J. P. Keating, Nonlinearity **8** (1995) 1115; ibid. **9** (1996) 911.

[Boh71] O. Bohigas and J. Flores, Phys. Lett. **34 B** (1971) 261.

[Boh83] O. Bohigas, R. U. Haq, and A. Pandey, in *Nuclear Data for Science and Technology*, K. H. Böckhoff, editor, Reidel, Dordrecht, 1983, p. 809.

[Boh84a] O. Bohigas, M.-J. Giannoni, and C. Schmit, Phys. Rev. Lett. **52** (1984) 1.

[Boh84b] O. Bohigas and M.-J. Giannoni, in *Lecture Notes in Physics*, Springer–Verlag vol. **209** (1984) p. 1.

[Boh92] O. Bohigas, in: *Chaos and Quantum Physics*, M.-J. Giannoni, A. Voros, and J. Zinn-Justin, editors, Elsevier, Amsterdam (1992).

[Boh37] N. Bohr, Nature **137** (1936) 344.

[Bre78] E. Brézin, C. Itzykson, G. Parisi, J. Zuber, Comm. Math. Phys. **59** (1978) 35.

[Bre93] E. Brézin and A. Zee, Nucl. Phys. **B 402** (1993) 613.

[Bre94] E. Brézin and A. Zee, Phys. Rev. **E 49** (1994) 2588.

[Bri63] D. M. Brink and R. O. Stephen, Phys. Lett. **5** (1963) 77.

[Bro81] T. A. Brody, J. Flores, J. B. French, P. A. Mello, A. Pandey, and S.S. M. Wong, Rev. Mod. Phys. **53** (1981) 385.

[Bun07] J. E. Bunder, K. B. Efetov, V. E. Kravtsov, O. M. Yevtushenko, and M. R. Zirnbauer, J. Stat. Phys. **129** (2007) 809.

[Cas87] G. Casati, B. V. Chirikov, I Guarneri, and D. L. Shepelyansky, Phys. Rep. **154** (1987) 77.

[Cas80] G. Casati, F. Valz-Gris, and I. Guarneri, Lett. Nuovo Cimento Soc. Ital. Fis. **28** (1980) 279.

[Chi81] B. V. Chirikov, F. M. Izrailev, and D. L. Shepelyansky, Sov. Sci. Rev. **C2** (1981) 209.

[Cal69a] F. Calogero, J. Math. Phys. **10** (1969) 2191.

[Cal69b] F. Calogero, J. Math. Phys. **10** (1969) 2197.

[Dys53] F. J. Dyson, Phys. Rev. **92** (1953) 1331.

[Dys62a] F. J. Dyson, J. Math. Phys. **3** (1962) 140.

[Dys62b] F. J. Dyson, J. Math. Phys. **3** (1962) 157.

[Dys62c] F. J. Dyson, J. Math. Phys. **3** (1962) 166.

[Dys62d] F. J. Dyson, J. Math. Phys. **3** (1962) 1191.

[Dys62e] F. J. Dyson, J. Math. Phys. **3** (1962) 1199.

[Dys63] F. J. Dyson and M. L. Mehta, J. Math. Phys. **4** (1963) 701.

[Dys70] F. J. Dyson, Commun. Math. Phys. **19** (1970) 235.

[Dys72] F. J. Dyson, J. Math. Phys. **13** (1972) 90.

[Edw75] S. F. Edwards and P. W. Anderson, J. Phys. F: Metal Physics **5** (1975) 965.

[Edw76] S. F. Edwards and R. C. Jones, J. Phys. A: Math. Gen. **9** (1976) 1595.

[Efe80] K. B. Efetov, A. I. Larkin, and D. E. Khmelnitzky, Zh. Exp. Teor. Fiz. **79** (1980) 1120 (Sov. Phys. JETP **52** (1980) 568).

[Efe82] K. B. Efetov, Zh. Exp. Teor. Fiz. **82** (1982) 872 (Sov. Phys. JETP **55** (1982) 514).

[Efe83a] K. B. Efetov and A. I. Larkin, Zh. Exp. Teor. Fiz. **85** (1983) 764 (Sov. Phys. JETP **58** (1980) 444).

[Efe83b] K. B. Efetov, Adv. Phys. **32** (1983) 53.

[Efe97] K. B. Efetov, *Supersymmetry in Disorder and Chaos*, Cambridge University Press, Cambridge (1997).

[Ein17] A. Einstein, Verh. Deutsch. Phys. Ges. **19** (1917) 82.

[Eng58] R. Engleman, Nuovo Cimento **10** (1958) 615.

[Eri60] T. Ericson, Phys. Rev. Lett. **5** (1960) 430.

[Eri63] T. Ericson, Ann. Phys. (N.Y.) **23** (1963) 390.

[Eri66] T. Ericson and T. Mayer–Kuckuk, Ann. Rev. Nucl. Sci. **16** (1966) 183.
[Eve08] F. Evers and A. D. Mirlin, Rev. Mod. Phys. **80** (2008) 1355.
[Fer34] E. Fermi *et al.*, Proc. Roy. Soc. **A 148** (1934) 483.
[Fer35] E. Fermi *et al.*, Proc. Roy. Soc. **A 149** (1935) 522.
[Fes54] H. Feshbach, C. E. Porter, and V. F. Weisskopf, Phys. Rev. **96** (1954) 448.
[Fis82] S. Fishman, D. R. Grempel, and R. E. Prange, Phys. Rev. Lett. **49** (1982) 509.
[Fox64] D. Fox and P. B. Kahn, Phys. Rev. **134** (1964) B1151.
[Fre66] J. B. French, in *Many–body Description of Nuclei and Reactions*, International School of Physics Enrico Fermi, Course **36**, edited by C. Bloch, Academic Press, New York, 1966, p. 278.
[Fre70] J. B. French and S. S. M. Wong, Phys. Lett. **B 33** (1970) 449.
[Fre78] J. B. French, P. A. Mello, and A. Pandey, Phys. Lett. **B 80** (1978) 17.
[Fre85] J. B. French, V. K. B. Kota, A. Pandey, and S. Tomsovic, Phys. Rev. Lett. **54** (1985) 2313.
[Fyo91] Y. V. Fyodorov and A. D. Mirlin, Phys. Rev. Lett. **67** (1991) 2405.
[Fyo05] Y. V. Fyodorov, D. V. Savin, and H. J. Sommers, J. Phys. A: Math. Gen. **38** (2005) 10731.
[Gad91] R. Gade and F. Wegner, Nucl. Phys. **B 360** (1991) 213.
[Gad93] R. Gade, Nucl. Phys. **B 398** (1993) 499.
[Gau61] M. Gaudin, Nucl. Phys. **25** (1961) 447.
[Gia91] *Chaos and Quantum Physics*, M.-J. Giannoni, A. Voros, and J. Zinn–Justin, editors, North Holland, Amsterdam, 1991.
[Gin65] J. Ginibre, J. Math. Phys. **6** (1965) 440.
[Gor65] L. P. Gor'kov and G. M. Eliashberg, Zh. Exp. Teor. Fiz. **48** (1965) 1407.
[Guh98] T. Guhr, A. Müller–Groeling, and H. A. Weidenmüller, Phys. Rep. **299** (1998) 189.
[Gut70] M. Gutzwiller, J. Math. Phys. **11** (1970) 1791.
[Gut90] M. Gutzwiller, *Chaos in Classical and Quantum Mechanics*, Springer–Verlag, Berlin, 1990.
[Hac95] G. Hackenbroich and H. A. Weidenmüller, Phys. Rev. Lett. **74** (1995) 4118.
[Hal95] M. A. Halasz and J. J. M. Verbaarschot, Phys. Rev. **D 3** (1995) 2563.
[Han84] J. H. Hannay and A. M. Ozorio de Almeida, J. Phys. A **17** (1984) 3429.
[Haq82] R. U. Haq, A. Pandey, and O. Bohigas, Phys. Rev. Lett. **48** (1982) 1086.
[Hau52] W. Hauser and H. Feshbach, Phys. Rev. **87** (1952) 366.
[Hei05] P. Heinzner, A. Huckleberry, and M. R. Zirnbauer, Comm. Math. Phys. **257** (2005) 725.
[Heu07] S. Heusler, S. Müller, A. Altland, P. Braun, and F. Haake, Phys. Rev. Lett. **98** (2007) 044103.
[Hua53] L. K. Hua, J. Chinese Math. Soc. **2** (1953) 288.
[Hua63] L. K. Hua, Harmonic Analysis, American Mathematical Society, Rhode Island (1963) (originally published in Chinese in 1958, translated from the Russian translation that appeared in 1959).
[Iid90] S. Iida, H. A. Weidenmüller, and J. A. Zuk, Ann. Phys. (N.Y.) **200** (1990) 219.
[Imr86] Y. Imry, Europhys. Lett. **1** (1986) 249.
[Kam99] A. Kamenev and M. Mezard, Phys. Rev. **B 60** (1999) 3944.
[Kan02] E. Kanzieper, Phys. Rev. Lett. **89** (2002) 250201.
[Kea00] J. P. Keating and N. C. Snaith, Comm. Math. Phys. **214** (2000) 57.
[Kor93] V. E. Korepin, N. M. Bogoljubov, and A. G. Izergin, *Quantum Inverse Scattering Method and Correlation Functions*, Cambridge University Press, Cambridge (1993).
[Lan57] A. M. Lane and J. E. Lynn, Proc. Roy. Soc. **A 70** (1957) 557.

[Lan58] A. M. Lane and R. G. Thomas, Rev. Mod. Phys. **30** (1958) 257.
[Leu92] H. Leutwyler and A. Smilga, Phys. Rev. **D 46** (1992) 5607.
[Lio72a] H. L. Liou, H. S. Camarda, S. Wynchank, M. Slagowitz, G. Hacken, F. Rahn, and J. Rainwater, Phys. Rev, **C 5** (1972) 974.
[Lio72b] H. L. Liou, H. S. Camarda, and F. Rahn, Phys. Rev. **C 5** (1972) 1002.
[Lit08] M. Littelmann, H.–J. Sommers, and M. R. Zirnbauer, Comm. Math. Phys. **283** (2008) 343.
[Mah69] C. Mahaux and H. A. Weidenmüller, *Shell-Model Approach to Nuclear Reactions*, North–Holland Publishing Co., Amsterdam (1969).
[McD79] S. W. McDonald and A. N. Kaufman, Phys. Rev. Lett. **42** (1979) 1189.
[Meh60a] M. L. Mehta, Nucl. Phys. **18** (1960) 395.
[Meh60b] M. L. Mehta and M. Gaudin, Nucl. Phys. **18** (1960) 420.
[Meh60c] M. L. Mehta, Rapport S. P. H. (Saclay) No. 658 (1960) (reprinted in Ref. [Por65]).
[Meh63] M. L. Mehta and F. J. Dyson, J. Math. Phys. **4** (1963) 713.
[Meh67] M. L. Mehta, *Random Matrices*, Academic Press, New York, 1967; 2nd edition 1991; 3rd edition Elsevier, Amsterdam, 2004.
[Mir96] A. D. Mirlin, Y. V. Fyodorov, F.-M. Dittes, J. Quesada, and T. H. Seligman, Phys. Rev. **E 54** (1996) 3221.
[Mol64] P. A. Moldauer, Phys. Rev. **135 B** (1964) 642.
[Mon73a] K. F. Mon and J. B. French, Ann. Phys. (N.Y.) **95** (1975) 90.
[Mon73b] H. Montgomery, Proc. Sympos. Pure Math. **24** (1973) 181.
[Mut87] K. A. Muttalib, J.-L. Pichard, and A. D. Stone, Phys. Rev. Lett. **59** (1987) 2475.
[Odl87] A. M. Odlyzko, Math. Comp. **48** (1987) 273.
[Opp90] R. Oppermann, Physica **A 167** (1990) 301.
[Pan79] A. Pandey, Ann. Phys. (N.Y.) **119** (1979) 170.
[Pan81] A. Pandey, Ann. Phys. (N.Y.) **134** (1981) 110.
[Pas72] L. A. Pastur, Theor. Math. Phys. **10** (1972) 67.
[Per73] I. C. Percival, J. Phys. **B6** (1973) L229.
[Por56] C. E. Porter and R. G. Thomas, Phys. Rev. **104** (1956) 483.
[Por60] C. E. Porter and N. Rosenzweig, Suomalaisen Tiedeakatemian Toimituksia (Ann. Akad. Sci. Fennicae) **AVI** (1960) No. 44, reprinted in Ref. [Por65].
[Por65] C. E. Porter, Statistical Theories of Spectra: Fluctuations, Academic Press, New York (1965).
[Rai60] J. Rainwater, W. W. Havens, Jr., D. S. Desjardins, and J. L. Rosen, Rev. Sci. Instr. **31** (1960) 481.
[Ric77] J. Richert and H. A. Weidenmüller, Phys. Rev. **C 16** (1977) 1309.
[Ros58] N. Rosenzweig, Phys. Rev. Lett. **1** (1958) 24.
[Ros60] N. Rosenzweig and C. E. Porter, Phys. Rev. **120** (1960) 1698.
[Sch80] L. Schäfer and F. Wegner, Z. Phys. **B 38** (1980) 113.
[Sel86] T. H. Seligman and J. J. M. Verbaarschot, in: Proceedings of the Fourth International Conference on Quantum Chaos and the 2nd Colloquium on Statistical Nuclear Physics, edited by T. H. Seligman and H. Nishioka, Springer, Berlin (1986) p. 131.
[Shu93] E. V. Shuryak and J. J. M. Verbaarschot, Nucl. Phys. **A 560** (1993) 306.
[Sie01] M. Sieber and K. Richter, Physica Scripta **T 90** (2001) 128.
[Sim93a] B. D. Simons, P. A. Lee, B. L. Altshuler, Phys. Rev. **B 48** (1993) 11450.
[Sim93b] B. D. Simons, P. A. Lee, B. L. Altshuler, Phys. Rev. Lett. **70** (1993) 4122.
[Sim93c] B. D. Simons, P. A. Lee, B. L. Altshuler, Nucl. Phys. **B 409** (1993) 487.
[Sim94] B. D. Simons, P. A. Lee, B. L. Altshuler, Phys. Rev. Lett. **72** (1993) 64.
[Sle93] K. Slevin and T. Nagao, Phys. Rev. Lett. **70** (1993) 635.

[Spl04] K. Splittorf and J. J. M. Verbaarschot, Nucl. Phys. **B 683** (2004) 467.
[Sut71] B. Sutherland, J. Math. Phys. **12** (1971) 246.
[Ver86] J. J. M. Verbaarschot, Ann. Phys, (N.Y.) **168** (1986) 368.
[Ver84] J. J. M. Verbaarschot, H. A. Weidenmüller, and M. R. Zirnbauer, Phys. Rev. Lett. **52** (1984) 1597.
[Ver85a] J. J. M. Verbaarschot, H. A. Weidenmüller, and M. R. Zirnbauer, Phys. Rep. **129** (1985) 367.
[Ver85b] J. J. M. Verbaarschot and M. R. Zirnbauer, J. Phys. A: Math. Gen. **17** (1985) 1093.
[Ver93] J. J. M. Verbaarschot and I. Zahed, Phys. Rev. Lett. **70** (1993) 3852.
[Ver94] J. J. M. Verbaarschot, Phys. Rev. Lett. **72** (1994) 2531.
[Von29] J. von Neumann and E. P. Wigner, Phys. Zschr. **30** (1929) 465.
[Wat81] W. A. Watson III, E. G. Bilpuch, and G. E. Mitchell, Z. Phys. **A 300** (1981) 89.
[Weg79] F. Wegner, Phys. Rev. **B 19** (1979) 783.
[Wei84] H. A. Weidenmüller, Ann. Phys. (N.Y.) **158** (1984) 120.
[Wei09] H. A. Weidenmüller and G. E. Mitchell, Rev. Mod. Phys. **81** (2009) 539.
[Wes74] G. Wess and B. Zumino, Nucl. Phys. **B 70** (1974) 39.
[Wig55] E. P. Wigner, Ann. Math. **62** (1955) 548.
[Wig57a] E. P. Wigner, Ann. Math. **65** (1957) 203.
[Wig57b] E. P. Wigner, in: *Conference on Neutron Physics by Time–of–Flight*, Oak Ridge National Laboratory Report No. 2309 (1957) p. 59.
[Wig58] E. P. Wigner, Ann. Math. **67** (1958) 325.
[Wig47] E. P. Wigner and L. Eisenbud, Phys. Rev. **72** (1947) 29.
[Wil62] S. S. Wilks, *Mathematical Statistics*, John Wiley and Sons, New York (1962).
[Wil75] W. M. Wilson, E. G. Bilpuch, and G. E. Mitchell, Nucl. Phys. **A 245** (1975) 285.
[Wis28] J. Wishart, Biometrika **20** A (1928) 32.
[Yur99] I. V. Yurkevich and I. V. Lerner, Phys. Rev. **B 60** (1999) 3955.
[Zir96] M. R. Zirnbauer, J. Math. Phys. **37** (1996) 4986.
[Zir99] M. R. Zirnbauer, preprint cond-mat/9903338 (1999).

PART II
Properties of random matrix theory

·3·

Symmetry classes

Martin R. Zirnbauer

Abstract

Physical systems exhibiting stochastic or chaotic behaviour are often amenable to treatment by random matrix models. In deciding on a good choice of model, random matrix physics is constrained and guided by symmetry considerations. The notion of 'symmetry class' (not to be confused with 'universality class') expresses the relevance of symmetries as an organizational principle. Dyson, in his 1962 paper referred to as the threefold way, gave the prime classification of random matrix ensembles based on a quantum mechanical setting with symmetries. In this article we review Dyson's threefold way from a modern perspective. We then describe a minimal extension of Dyson's setting to incorporate the physics of chiral Dirac fermions and disordered superconductors. In this minimally extended setting, where Hilbert space is replaced by Fock space equipped with the anti-unitary operation of particle-hole conjugation, symmetry classes are in one-to-one correspondence with the large families of Riemannian symmetric spaces.

3.1 Introduction

In Chapter 2 of this handbook, the historical narrative by Bohigas and Weidenmüller describes how random matrix models emerged from quantum physics, more precisely from a statistical approach to the strongly interacting many-body system of the atomic nucleus. Although random matrix theory is nowadays understood to be of relevance to numerous areas of physics, mathematics, and beyond, quantum mechanics is still where many of its applications lie. Quantum mechanics also provides a natural framework in which to classify random matrix ensembles.

In this thrust of development, a symmetry classification of random matrix ensembles was put forth by Dyson in his 1962 paper *The threefold way: algebraic structure of symmetry groups and ensembles in quantum mechanics*, where he proved (quote from the abstract of [Dys62])

"that the most general matrix ensemble, defined with a symmetry group which may be completely arbitrary, reduces to a direct product of independent irreducible ensembles each of which belongs to one of the three known types".

The three types known to Dyson were ensembles of matrices which are either complex Hermitian, or real symmetric, or quaternion self-dual. It is widely acknowledged that Dyson's threefold way has become fundamental to various areas of theoretical physics, including the statistical theory of complex many-body systems, mesoscopic physics, disordered electron systems, and the field of quantum chaos.

Over the last decade, a number of random matrix ensembles beyond Dyson's classification have come to the fore in physics and mathematics. On the physics side these emerged from work [Ver94] on the low-energy Dirac spectrum of quantum chromodynamics, and also from the mesoscopic physics of low-energy quasi-particles in disordered superconductors [AZ97]. In the mathematical research area of number theory, the study of statistical correlations of the values of Riemann zeta and related L-functions has prompted some of the same generalizations [KS99]. It was observed early on [AZ97] that these post-Dyson ensembles, or rather the underlying symmetry classes, are in one-to-one correspondence with the large families of symmetric spaces.

The prime emphasis of the present handbook article will be on describing Dyson's fundamental result from a modern perspective. A second goal will be to introduce the post-Dyson ensembles. While there seems to exist no unanimous view on how these fit into a systematic picture, here we will follow [HHZ05] to demonstrate that they emerge from Dyson's setting upon replacing the plain structure of Hilbert space by the more refined structure of Fock space.[1] The reader is advised that some aspects of this story are treated in a more leisurely manner in the author's encyclopedia article [Zir04].

To preclude any misunderstanding, let us issue a clarification of language right here: 'symmetry class' must not be confused with 'universality class'! Indeed, inside a symmetry class as understood in this article various types of physical behaviour are possible. (For example, random matrix models for weakly disordered time-reversal invariant metals belong to the so-called Wigner-Dyson symmetry class of real symmetric matrices, and so do Anderson tight-binding models with real hopping and strong disorder. The former are believed to exhibit the universal energy level statistics given by the Gaussian orthogonal ensemble, whereas the latter have localized eigenfunctions and hence level statistics which is expected to approach the Poisson limit when the system size goes to infinity.) For this reason the present article must refrain from writing down explicit formulas for joint eigenvalue distributions, which are available only in certain universal limits.

[1] We mention in passing that a classification of Dirac Hamiltonians in two dimensions has been proposed in [BL02]. Unlike ours, this is not a symmetry classification in Dyson's sense.

3.2 Dyson's threefold way

Dyson's classification is formulated in a general and simple mathematical setting which we now describe. First of all, the framework of quantum theory calls for the basic structure of a complex vector space V carrying a Hermitian scalar product $\langle \cdot, \cdot \rangle : V \times V \to \mathbb{C}$. (Dyson actually argues [Dys62] in favour of working over the real numbers, but we will not follow suit in this respect.) For technical simplicity, we do join Dyson in taking V to be finite-dimensional. In applications, $V \simeq \mathbb{C}^n$ will usually be the truncated Hilbert space of a family of disordered or quantum chaotic Hamiltonian systems.

The Hermitian structure of the vector space V determines a group $\mathrm{U}(V)$ of unitary transformations of V. Let us recall that the elements $g \in \mathrm{U}(V)$ are $\mathbb{C}$-linear operators satisfying the condition $\langle gv, gv' \rangle = \langle v, v' \rangle$ for all $v, v' \in V$.

Building on the Hermitian vector space V, Dyson's setting stipulates that V should be equipped with a unitary group action

$$G_0 \times V \to V, \quad (g, v) \mapsto \rho_V(g)v, \quad \rho_V(g) \in \mathrm{U}(V). \tag{3.2.1}$$

In other words, there is some group G_0 whose elements g are represented on V by unitary operators $\rho_V(g)$. This group G_0 is meant to be the group of joint (unitary) symmetries of a family of quantum mechanical Hamiltonian systems with Hilbert space V. We will write $\rho_V(g) \equiv g$ for short.

Now, not every symmetry of a quantum system is of the canonical unitary kind. The prime counterexample is the operation, T, of inverting the time direction, called time reversal for short. It is represented on Hilbert space V by an *anti*-unitary operator $T \equiv \rho_V(T)$, which is to say that T is complex antilinear and preserves the Hermitian scalar product up to complex conjugation:

$$T(zv) = \bar{z}\, Tv, \qquad \langle Tv, Tv' \rangle = \overline{\langle v, v' \rangle} \qquad (z \in \mathbb{C};\ v, v' \in V). \tag{3.2.2}$$

Another operation of this kind is charge conjugation in relativistic theories such as the Dirac equation for the electron and its anti-particle, the positron.

Thus in Dyson's general setting one has a so-called symmetry group $G = G_0 \cup G_1$ where the subgroup G_0 is represented on V by unitaries, while G_1 (not a group) is represented by anti-unitaries. By the definition of what is meant by a 'symmetry', the generator of time evolution, the Hamiltonian H, of the quantum system is fixed by conjugation $gHg^{-1} = H$ with any $g \in G$.

The set G_1 may be empty. When it is not, the composition of any two elements of G_1 is unitary, so every $g \in G_1$ can be obtained from a fixed element of G_1, say T, by right multiplication with some $U \in G_0$: $g = TU$. The same goes for left multiplication, i.e. for every $g \in G_1$ there also exists $U' \in G_0$ so that $g = U'T$. In other words, when G_1 is non-empty, $G_0 \subset G$ is a proper normal subgroup and the factor group $G/G_0 \simeq \mathbb{Z}_2$ consists of exactly two elements, G_0 and $TG_0 = G_1$. For future use we record that conjugation $U \mapsto TUT^{-1} =: a(U)$ by time reversal is an automorphism of G_0.

Following Dyson [Dys62] we assume that the special element T represents an *inversion* symmetry such as time reversal or charge conjugation. T must then be a (projective) involution, i.e., $T^2 = z \times \mathrm{Id}_V$ with $0 \neq z \in \mathbb{C}$, so that conjugation by T^2 is the identity operation. Since T is anti-unitary, z must have modulus $|z| = 1$, and by the $\mathbb{C}$-antilinearity of T the associative law

$$zT = T^2 \cdot T = T \cdot T^2 = Tz = \bar{z}T \tag{3.2.3}$$

forces z to be real, which leaves only two possibilities: $T^2 = \pm \mathrm{Id}_V$.

Let us record here a concrete example of some historical importance: the Hilbert space V might be the space of totally anti-symmetric wave functions of n particles distributed over the shell-model space of an atom or an atomic nucleus, and the symmetry group G might be $G = \mathrm{O}_3 \cup T\mathrm{O}_3$, the full rotation group O_3 (including parity) together with its translate by time reversal T.

In summary, Dyson's setting assumes two pieces of data:

- a finite-dimensional complex vector space V with Hermitian structure,
- a group $G = G_0 \cup G_1$ acting on V by unitary and anti-unitary operators.

It should be stressed that, in principle, the primary object is the Hamiltonian, and the symmetries G are secondary objects derived from it. However, adopting Dyson's standpoint we now turn tables to view the symmetries as fundamental and given and the Hamiltonians as derived objects. Thus, fixing any pair (V, G) our goal is to elucidate the structure of the space of all *compatible* Hamiltonians, i.e. the self-adjoint operators H on V which commute with the G-action. Such a space is reducible in general: the G-compatible Hamiltonian matrices decompose as a direct sum of blocks. The goal of classification is to enumerate the irreducible blocks that occur in this setting.

While the main objects to classify are the spaces of compatible Hamiltonians H, we find it technically convenient to perform some of the discussion at the integrated level of *time evolutions* $U_t = \mathrm{e}^{-\mathrm{i}tH/\hbar}$ instead. This change of focus results in no loss, as the Hamiltonians can always be retrieved by linearization in t at $t = 0$. The compatibility conditions for $U \equiv U_t$ read

$$U = g_0 U g_0^{-1} = g_1 U^{-1} g_1^{-1} \quad \text{(for all } g_\sigma \in G_\sigma\text{)}. \tag{3.2.4}$$

The strategy will be to make a reduction to the case of the trivial group $G_0 = \{\mathrm{Id}\}$. The situation with trivial G_0 can then be handled by enumeration of a finite number of possibilities.

3.2.1 Reduction to the case of $G_0 = \{\mathrm{Id}\}$

To motivate the technical reduction procedure below, we begin by elaborating the example of the rotation group O_3 acting on a Hilbert space of shell-model states. Any Hamiltonian which commutes with $G_0 = O_3$ conserves total angular momentum, L, and parity, π, which means that all Hamiltonian matrix elements connecting states in sectors of different quantum numbers (L, π) vanish identically. Thus, the matrix of the Hamiltonian with respect to a basis of states with definite values of (L, π) has diagonal block structure. O_3-symmetry further implies that the Hamiltonian matrix is diagonal with respect to the orthogonal projection, M, of total angular momentum on some axis in position space. Moreover, for a suitable choice of basis the matrix will be the same for each M-value of a given sector (L, π).

To put these words into formulas, we employ the mathematical notions of orthogonal sum and tensor product to decompose the shell-model space as

$$V \simeq \bigoplus_{L\geq 0;\ \pi=\pm 1} V_{(L,\pi)}\,, \qquad V_{(L,\pi)} = \mathbb{C}^{m(L,\pi)} \otimes \mathbb{C}^{2L+1}\,, \tag{3.2.5}$$

where $m(L,\pi)$ is the multiplicity in V of the O_3-representation with quantum numbers (L, π). The statement above is that all symmetry operators and compatible Hamiltonians are diagonal with respect to this direct sum, and within a fixed block $V_{(L,\pi)}$ the Hamiltonians act on the first factor $\mathbb{C}^{m(L,\pi)}$ and are trivial on the second factor $\mathbb{C}^{2L+1}$ of the tensor product, while the symmetry operators act on the second factor and are trivial on the first factor. Thus we may picture each sector $V_{(L,\pi)}$ as a rectangular array of states where the Hamiltonians act, say, horizontally and are the same in each row of the array, while the symmetries act vertically and are the same in each column.

This concludes our example, and we now move on to the general case of any group G_0 acting reductively on V. To handle this, we need some language and notation as follows. A G_0-representation X is a $\mathbb{C}$-vector space carrying a G_0-action $G_0 \times X \to X$ by $(g, x) \mapsto \rho_X(g)x$. If X and Y are G_0-representations, then by the space $\mathrm{Hom}_{G_0}(X, Y)$ of G_0-equivariant homomorphisms from X to Y one means the complex vector space of $\mathbb{C}$-linear maps $\psi : X \to Y$ with the intertwining property $\rho_Y(g)\psi = \psi\rho_X(g)$ for all $g \in G_0$. If X is an irreducible G_0-representation, then Schur's lemma says that $\mathrm{Hom}_{G_0}(X, X)$ is one-dimensional, being spanned by the identity, Id_X. For two irreducible G_0-representations X and Y, the dimension of $\mathrm{Hom}_{G_0}(X, Y)$ is either zero or one, by an easy corollary of Schur's lemma. In the latter case X and Y are said to belong to the same isomorphism class.

Using the symbol λ to denote the isomorphism classes of irreducible G_0-representations, we fix for each λ a standard representation space R_λ. Note that $\dim \mathrm{Hom}_{G_0}(R_\lambda, V)$ counts the multiplicity in V of the irreducible representation of isomorphism class λ. In our shell-model example with $G_0 = O_3$ we have $\lambda = (L, \pi)$, $R_\lambda = \mathbb{C}^{2L+1}$, and $\dim \mathrm{Hom}_{G_0}(R_\lambda, V) = m(L, \pi)$.

The following statement can be interpreted as saying that the example adequately reflects the general situation.

Lemma 3.2.1 *Let G_0 act reductively on V. Then*

$$\bigoplus_\lambda \mathrm{Hom}_{G_0}(R_\lambda, V) \otimes R_\lambda \to V, \qquad \bigoplus_\lambda (\psi_\lambda \otimes r_\lambda) \mapsto \sum_\lambda \psi_\lambda(r_\lambda)$$

is a G_0-equivariant isomorphism.

Remark: *The decomposition offered by this lemma separates perfectly the unitary symmetry multiplets from the dynamical degrees of freedom and thus gives an immediate view of the structure of the space of G_0-compatible Hamiltonians. Indeed, the direct sum over isomorphism classes (or 'sectors') λ is preserved by the symmetries G_0 as well as the compatible Hamiltonians H; and G_0 is trivial on* $\mathrm{Hom}_{G_0}(R_\lambda, V)$ *while the Hamiltonians are trivial on R_λ.*

Next, we remove the time-evolution trivial factors R_λ from the picture. To do so, we need to go through the step of transferring all given structure to the spaces $E_\lambda := \mathrm{Hom}_{G_0}(R_\lambda, V)$.

Transfer of structure

We first transfer the Hermitian structure of V. In the present setting of a unitary G_0-action, the Hermitian scalar product of V reduces to a Hermitian scalar product on each sector of the direct-sum decomposition of Lemma 3.2.1, by orthogonality of the sum. Hence, we may focus attention on a definite sector $E \otimes R \equiv E_\lambda \otimes R_\lambda$. Fixing a G_0-invariant Hermitian scalar product $\langle \cdot, \cdot \rangle_R$ on $R = R_\lambda$ we define such a product $\langle \cdot, \cdot \rangle_E : E \times E \to \mathbb{C}$ by

$$\langle \psi, \psi' \rangle_E := \langle \psi(r), \psi'(r) \rangle_V / \langle r, r \rangle_R, \tag{3.2.6}$$

which is easily checked to be independent of the choice of $r \in R$, $r \neq 0$.

Before carrying on, we note that for any Hermitian vector space V there exists a canonically defined $\mathbb{C}$-antilinear bijection $C_V : V \to V^*$ to the dual vector space V^* by $C_V(v) := \langle v, \cdot \rangle_V$. (In Dirac's language this is the conversion from 'ket' vector to 'bra' vector.) By naturalness of the transfer of Hermitian structure we have the relation $C_{E \otimes R} = C_E \otimes C_R$.

Turning to the more involved step of transferring time reversal T, we begin with a preparation. If $L : V \to W$ is a linear mapping between vector spaces, we denote by $L^t : W^* \to V^*$ the canonical transpose defined by $(L^t f)(v) = f(Lv)$. Let now V be our Hilbert space with ket-bra bijection $C \equiv C_V$. Then for any $g \in \mathrm{U}(V)$ we have the relation $CgC^{-1} = (g^{-1})^t$ because

$$C(gv) = \langle gv, \cdot \rangle = \langle v, g^{-1} \cdot \rangle = (g^{-1})^t C(v) \qquad (v \in V) . \tag{3.2.7}$$

Moreover, recalling the automorphism $G_0 \ni g \mapsto a(g) = TgT^{-1}$ of G_0 we obtain

$$CTg = a(g^{-1})^t\, CT \qquad (g \in G_0)\,. \tag{3.2.8}$$

Thus, since C and T are bijective, the $\mathbb{C}$-linear mapping $CT : V \to V^*$ is a G_0-equivariant isomorphism interchanging the given G_0-representation on V with the representation on V^* by $g \mapsto a(g^{-1})^t$. In particular, it follows that T stabilizes the decomposition $V = \oplus_\lambda V_\lambda \simeq \oplus_\lambda (E_\lambda \otimes R_\lambda)$ of Lemma 3.2.1.

If T exchanges different sectors V_λ, the situation is very easy to handle (see below). The more challenging case is $TV_\lambda = V_\lambda$, which we now assume.

Lemma 3.2.2 *Let $TV_\lambda = V_\lambda$. Under the isomorphism $V_\lambda \simeq E_\lambda \otimes R_\lambda$ the time-reversal operator transfers to a pure tensor*

$$T = \alpha \otimes \beta\,, \quad \alpha :\; E_\lambda \to E_\lambda\,, \quad \beta :\; R_\lambda \to R_\lambda\,,$$

with anti-unitary α and β.

Proof Writing $E_\lambda \equiv E$ and $R_\lambda \equiv R$ for short, we consider the transferred mapping $CT :\; E \otimes R \to E^* \otimes R^*$, which expands as $CT = \sum \phi_i \otimes \psi_i$ with $\mathbb{C}$-linear mappings $\phi_i\,, \psi_i$. Since CT is known to be a G_0-equivariant isomorphism, so is every map $\psi_i :\; R \to R^*$. By the irreducibility of R and Schur's lemma, there exists only one such map (up to scalar multiples). Hence CT is a pure tensor: $CT = \phi \otimes \psi$. Using $C = C_{E\otimes R} = C_E \otimes C_R$ we obtain $T = \alpha \otimes \beta$ with $\mathbb{C}$-antilinear $\alpha = C_E^{-1}\phi$ and $\beta = C_R^{-1}\psi$. Since the tensor product lets you move scalars between factors, the maps α and β are not uniquely defined. We may use this freedom to make β anti-unitary. Because T is anti-unitary, it then follows from the definition (3.2.6) of the Hermitian structure of E that α is anti-unitary. □

Remark: *By an elementary argument, which was spelled out for the anti-unitary operator T in Eq. (3.2.3), it follows that $\alpha^2 = \epsilon_\alpha$, Id_E and $\beta^2 = \epsilon_\beta$, Id_R with $\epsilon_\alpha, \epsilon_\beta \in \{\pm 1\}$. Writing $T^2 = \epsilon_T \mathrm{Id}_V$ we have the relation $\epsilon_\alpha \epsilon_\beta = \epsilon_T$. Thus when $\epsilon_\beta = -1$ the parity $\epsilon_\alpha = -\epsilon_T$ of the transferred time-reversal operator α is opposite to that of the original time reversal T.*

This change of parity occurs, e.g. in the case of $G_0 = \mathrm{SU}_2$. Indeed, let $R \equiv R_n$ be the irreducible SU_2-representation of dimension $n+1$. It is a standard fact of representation theory that R_n is SU_2-equivariantly isomorphic to R_n^* by a symmetric isomorphism $\psi = \psi^t$ for even n and skew-symmetric isomorphism $\psi = -\psi^t$ for odd n. From $(-1)^n \psi^t = \psi = C_R\,\beta$ and

$$\psi(v)(v') = \langle \beta v, v' \rangle_R = \overline{\langle \beta^2 v, \beta v' \rangle_R} = \langle \beta v', \beta^2 v \rangle_R = \psi(v')(\beta^2 v)\,, \tag{3.2.9}$$

we conclude that $\beta^2 = (-1)^n \mathrm{Id}_{R_n}$.

3.2.2 Classification

By the decomposition of Lemma 3.2.1 the space $Z_{\mathrm{U}(V)}(G_0)$ of G_0-compatible time evolutions in $\mathrm{U}(V)$ is a direct product of unitary groups,

$$Z_{\mathrm{U}(V)}(G_0) \simeq \prod_\lambda \mathrm{U}(E_\lambda) \,. \tag{3.2.10}$$

We now fix a sector $V_\lambda \simeq E_\lambda \otimes R_\lambda$ and run through the different situations (of which there exist three, essentially) due to the absence or presence of a transferred time-reversal symmetry $\alpha : E_\lambda \to E_\lambda$.

Class A

The first type of situation occurs when the set G_1 of anti-unitary symmetries is either empty or else maps $V_\lambda \simeq E_\lambda \otimes R_\lambda$ to a different sector $V_{\lambda'}$, $\lambda \neq \lambda'$. In both cases, the G-compatible time-evolution operators restricted to V_λ constitute a unitary group $\mathrm{U}(E_\lambda) \simeq \mathrm{U}_N$ with $N = \dim E_\lambda$ being the multiplicity of the irreducible representation R_λ in V. The unitary groups U_N or to be precise, their simple parts SU_N, are symmetric spaces (compare Section 3.3.4) of the A family or A series in Cartan's notation – hence the name Class A. In random matrix theory, the Lie group U_N equipped with Haar measure is commonly referred to as the circular unitary ensemble, CUE_N [Dys62a].

The Hamiltonians H in Class A are represented by complex Hermitian $N \times N$ matrices. By putting a U_N-invariant Gaussian probability measure

$$d\mu(H) = c_0\, \mathrm{e}^{-\mathrm{Tr}\, H^2/2\sigma^2} dH, \quad dH = \prod_{i=1}^{N} dH_{ii} \prod_{j<k} dH_{jk}\, dH_{kj}, \tag{3.2.11}$$

on that space, one gets what is called the GUE – the Gaussian unitary ensemble – defining the Wigner-Dyson *universality class* of unitary symmetry. An important physical realization of that class is by electrons in a disordered metal with time-reversal symmetry broken by the presence of a magnetic field.

Classes AI and AII

We now turn to the cases where T is present and $TV_\lambda = V_\lambda \simeq E_\lambda \otimes R_\lambda$. We abbreviate $E_\lambda \equiv E$. From Lemma 3.2.1 we know that $T = \alpha \otimes \beta$ is a pure tensor with anti-unitary α, and we have $\alpha^2 = \epsilon_\alpha\, \mathrm{Id}_E$ with parity $\epsilon_\alpha = \epsilon_T\, \epsilon_\beta$.

Using conjugation by α to define an automorphism

$$\tau : \mathrm{U}(E) \to \mathrm{U}(E), \quad u \mapsto \alpha u \alpha^{-1}, \tag{3.2.12}$$

we transfer the conditions (3.2.4) to V_λ and describe the set $Z_{\mathrm{U}(E)}(G)$ of G-compatible time evolutions in $\mathrm{U}(E)$ as

$$Z_{\mathrm{U}(E)}(G) = \{x \in \mathrm{U}(E) \mid \tau(x) = x^{-1}\}. \tag{3.2.13}$$

Now let $U \equiv \mathrm{U}(E)$ for short and denote by $K \subset U$ the subgroup of τ-fixed elements $k = \tau(k) \in U$. The set $Z_U(G)$ is analytically diffeomorphic to the coset space U/K by the mapping

$$U/K \to Z_U(G) \subset U, \quad uK \mapsto u\tau(u^{-1}), \tag{3.2.14}$$

which is called the Cartan embedding of U/K into U. The remaining task is to determine K. This is done as follows.

Recalling the definition $C_E\,\alpha = \phi$ and using $C_E\,k = (k^{-1})^t C_E$ we express the fixed-point condition $k = \tau(k) = \alpha k \alpha^{-1}$ as $(k^{-1})^t = \phi k \phi^{-1}$ or, equivalently, $\phi = k^t \phi\, k$, which means that the bilinear form associated with $\phi : E \to E^*$,

$$Q_\phi : E \times E \to \mathbb{C}, \quad (e, e') \mapsto \phi(e)(e'), \tag{3.2.15}$$

is preserved by $k \in K$. By running the argument around Eq. (3.2.9) in reverse order (with the obvious substitutions $\psi \to \phi$ and $\beta \to \alpha$), we see that the non-degenerate form Q_ϕ is symmetric if $\epsilon_\alpha = +1$ and skew if $\epsilon_\alpha = -1$. In the former case it follows that $K = \mathrm{O}(E) \simeq \mathrm{O}_N$ is an orthogonal group, while in the latter case, which occurs only if $N \in 2\mathbb{N}$, $K = \mathrm{USp}(E) \simeq \mathrm{USp}_N$ is unitary symplectic. In both cases the coset space U/K is a symmetric space (compare Section 3.3.4) – a fact first noticed by Dyson in [Dys70].

Thus in the present setting of $T V_\lambda = V_\lambda$ we have the following dichotomy for the sets of G-compatible time evolutions $Z_{\mathrm{U}(E)}(G) \simeq U/K$:

Class *AI*: $U/K \simeq \mathrm{U}_N/\mathrm{O}_N \qquad (\epsilon_\alpha = +1)$,
Class *AII*: $U/K \simeq \mathrm{U}_N/\mathrm{USp}_N \quad (\epsilon_\alpha = -1, \; N \in 2\mathbb{N})$.

Again we are referring to symmetric spaces by the names they – or rather their simple parts $\mathrm{SU}_N/\mathrm{SO}_N$ and $\mathrm{SU}_N/\mathrm{USp}_N$ – have in the Cartan classification. In random matrix theory, the symmetric space $\mathrm{U}_N/\mathrm{O}_N$ (or its Cartan embedding into U_N as the symmetric unitary matrices) equipped with U_N-invariant probability measure is called the circular orthogonal ensemble, COE_N, while the Cartan embedding of $\mathrm{U}_N/\mathrm{USp}_N$ equipped with U_N-invariant probability measure is known as the circular symplectic ensemble, CSE_N [Dys62a]. (Note the confusing fact that the naming goes by the subgroup which is divided out.)

Examples for Class *AI* are provided by time-reversal invariant systems with symmetry $G_0 = (\mathrm{SU}_2)_{\mathrm{spin}}$. Indeed, by the fundamental laws of quantum physics time reversal T squares to $(-1)^{2S}$ times the identity on states with spin S. Such states transform according to the irreducible SU_2-representation of dimension $2S + 1$, and from $\beta^2 = (-1)^{2S}$ (see the Remark after Lemma 3.2.1) it follows that $\epsilon_\alpha = \epsilon_T\,\epsilon_\beta = (-1)^{2S}(-1)^{2S} = +1$ in all cases. A historically important realization of Class *AI* is furnished by the highly excited states of atomic nuclei as observed by neutron scattering just above the neutron threshold.

By breaking SU_2-symmetry (i.e. by taking $G_0 = \{\mathrm{Id}\}$) while maintaining T-symmetry for states with half-integer spin (say single electrons, which carry spin $S = 1/2$), one gets $\epsilon_a = \epsilon_T = (-1)^{2S} = -1$, thereby realizing Class AII. An experimental observation of this class and its characteristic wave interference phenomena was first reported in the early 1980s [Ber84] for disordered metallic magnesium films with strong spin-orbit scattering caused by gold impurities.

The Hamiltonians H, obtained by passing to the tangent space of U/K at unity, are represented by Hermitian matrices with entries that are real numbers (Class AI) or real quaternions (Class AII). The simplest random matrix models result from putting K-invariant Gaussian probability measures on these spaces; they are called the Gaussian orthogonal ensemble and Gaussian symplectic ensemble, respectively. Their properties delineate the Wigner-Dyson universality classes of orthogonal and symplectic symmetry.

3.3 Symmetry classes of disordered fermions

While Dyson's threefold way is fundamental and complete in its general Hilbert space setting, the early 1990s witnessed the discovery of various new types of strong universality, which were begging for an extended scheme:

- The introduction of QCD-motivated chiral random matrix ensembles (reviewed by Verbaarschot in Chapter 32 of this handbook) mimicked Dyson's scheme but also transcended it.
- Number theorists had introduced and studied ensembles of L-functions akin to the Riemann zeta function (see the review by Keating and Snaith in Chapter 24 of this handbook). These display random matrix phenomena which are absent in classes A, AI, or AII.
- The proximity effect due to Andreev reflection, a particle-hole conversion process in mesoscopic hybrid systems involving metallic as well as superconducting components, was found [AZ97] to give rise to post-Dyson mechanisms of quantum interference (compare Chapter 35 by Beenakker).

By the middle of the 1990s, it had become clear that there exists a unifying mathematical principle governing these post-Dyson random matrix phenomena. This principle will be explained in the present section. We mention in passing that a fascinating recent development [Kit08, SRF09] uses the same principle in the context of a homotopy classification of topological insulators and superconductors.

3.3.1 Fock space setting

We now describe an extended setting, which replaces the Hermitian vector space V by its exterior algebra $\wedge(V)$ but otherwise retains Dyson's setting to the fullest extent possible. In physics language we say that we pass from the (single-particle) Hilbert space V to the fermionic Fock space $\wedge(V)$ generated by V.

The $\mathbb{Z}$-grading $\wedge(V) = \oplus_n \wedge^n (V)$ by the degree n has the physical meaning of particle number. Thus $\wedge^0(V) \equiv \mathbb{C}$ is the vacuum, $\wedge^1(V) \equiv V$ is the one-particle space, $\wedge^2(V)$ is the two-particle space, and so on. We adhere to the assumption of finite-dimensional V. Particle number n then is in the range $0 \le n \le N := \dim V$. Note that $\dim \wedge^n (V) = \binom{N}{n}$.

The n-particle subspace $\wedge^n(V)$ of the Fock space of a Hermitian vector space V carries an induced Hermitian scalar product defined by

$$\langle u_1 \wedge \cdots \wedge u_n \,,\, v_1 \wedge \cdots \wedge v_n \rangle_{\wedge^n(V)} = \mathrm{Det} \begin{pmatrix} \langle u_1, v_1\rangle_V & \ldots & \langle u_1, v_n\rangle_V \\ \vdots & \ddots & \vdots \\ \langle u_n, v_1\rangle_V & \ldots & \langle u_n, v_n\rangle_V \end{pmatrix}. \quad (3.3.1)$$

Relevant operations on $\wedge(V)$ are exterior multiplication (or particle creation) $\varepsilon(v) : \wedge^n(V) \to \wedge^{n+1}(V)$ by $v \in V$ and contraction (or particle annihilation) $\iota(f) : \wedge^n(V) \to \wedge^{n-1}(V)$ by $f \in V^*$. The standard physics convention is to fix some orthonormal basis $\{e_k\}_{k=1,\ldots,N}$ of V and write $a_k^\dagger := \varepsilon(e_k)$ for the particle creation operators and $a_k := \iota(\langle e_k, \cdot\rangle_V)$ for the particle annihilation operators. This notation reflects the fact that Hermitian conjugation $\dagger$ in Fock space relates $\varepsilon(v)$ and $\iota(f)$ by $\varepsilon(v)^\dagger = \iota(f)$ where $f = \langle v, \cdot\rangle_V$. The operators $a_k^\dagger$ and a_k satisfy the so-called canonical anti-commutation relations

$$a_k a_l + a_l a_k = 0 = a_k^\dagger a_l^\dagger + a_l^\dagger a_k^\dagger, \qquad a_k^\dagger a_l + a_l a_k^\dagger = \delta_{kl}. \quad (3.3.2)$$

These represent the defining relations of the Clifford algebra $\mathrm{Cl}(W)$ of the vector space $W = V \oplus V^*$ with quadratic form $(v \oplus f, v' \oplus f') \mapsto f(v') + f'(v)$.

Having introduced the basic Fock space structure, we now turn to what is going to be our definition of a symmetry group G acting on Fock space $\wedge(V)$. As before, we assume that we are given a normal subgroup $G_0 \subset G$. The action of the elements $g \in G_0$ is defined by unitary operators on V which are extended to $\wedge(V)$ by

$$g(v_1 \wedge \cdots \wedge v_n) := (g v_1) \wedge \cdots \wedge (g v_n). \quad (3.3.3)$$

Similarly, the anti-unitary operator of time reversal T is defined on V and is extended to $\wedge(V)$ by $T(v_1 \wedge \cdots \wedge v_n) := (T v_1) \wedge \cdots \wedge (T v_n)$.

Now the $\mathbb{Z}$-grading of Fock space offers the natural option of introducing another kind of anti-unitary operator, which is a close cousin of the Hodge star operator for the deRham complex: particle-hole conjugation, C, transforms an n-particle state into an $(N - n)$-particle state or a state with n holes. (Note the change of meaning of the symbol C as compared to Section 3.2.1.)

Definition 3.3.1 *Fix a generator* $\Omega \in \wedge^N(V)$, $N = \dim V$, *with normalization* $\langle \Omega, \Omega\rangle_{\wedge^N(V)} = 1$. *Then particle-hole conjugation* $C : \wedge^n(V) \to \wedge^{N-n}(V)$ *is the anti-unitary operator defined by*

$$(C\psi) \wedge \psi' = \langle \psi, \psi' \rangle_{\wedge^n(V)} \, \Omega \, .$$

Thus the definition of the operator C uses the Hermitian scalar product of Fock space and a choice of fully occupied state Ω. An elementary calculation shows that $C^2|_{\wedge^n(V)} = (-1)^{n(N-n)}$.

What are the commutation relations of C with T and $g \in G_0$? To answer this question, we observe that by $\dim \wedge^N(V) = 1$ we may always choose Ω to be T-invariant (i.e., $T\Omega = \Omega$). Then from the following computation,

$$\begin{aligned}(TC\psi) \wedge T\psi' &= T\left((C\psi) \wedge \psi'\right) = T\left(\langle \psi, \psi' \rangle \, \Omega\right) \\ &= \overline{\langle \psi, \psi' \rangle} \, T\Omega = \langle T\psi, T\psi' \rangle \, \Omega = (CT\psi) \wedge T\psi',\end{aligned}$$

we have $CT = TC$. Also, making the natural assumption that both the vacuum space $\wedge^0(V)$ and the fully occupied space $\wedge^N(V)$ transform as the trivial G_0-representation (i.e., $g\Omega = \Omega$ for $g \in G_0$), a similar calculation gives $Cg = gC$.

In order to enlarge the set of possible symmetries and hence the scope of the theory, we now introduce a 'twisted' variant of particle-hole conjugation. Let $S \in \mathrm{U}(V)$ be an involution ($S^2 = \mathrm{Id}$) and extend S to $\wedge(V)$ by Eq. (3.3.3). To obtain an extension of the group G_0, we require that S commutes with T, satisfies $S\Omega = \Omega$, and normalizes G_0, i.e., $SG_0S^{-1} = G_0$. (Here we identify G_0 with its action on Fock space.) By twisted particle-hole conjugation we then mean the operator $\tilde{C} = CS = SC$. Note that $\tilde{C}G_0\tilde{C}^{-1} = G_0$ and $\tilde{C}T = T\tilde{C}$.

Definition 3.3.2 *On the Fock space $\wedge(V)$ over a Hermitian vector space V, let there be the action of a group $G = G_0 \cup TG_0 \cup \tilde{C}G_0 \cup \tilde{C}TG_0$ with G_0 a normal subgroup and $\tilde{C}T = T\tilde{C}$. We call this a* minimal extension *of Dyson's setting if G_0 acts by unitary operators defined on V, time reversal T acts as an anti-unitary operator also defined on V, and twisted particle-hole conjugation $\tilde{C}$ is an anti-unitary bijection $\wedge^n(V) \to \wedge^{N-n}(V)$.*

3.3.2 Classification goal

The simplest question to ask now is this: what is the structure of the set of Hamiltonians that operate on Fock space $\wedge(V)$ and commute with the given G-action on $\wedge(V)$? Since this question ignores the grading of Fock space by particle number, it takes us back to Dyson's setting and the answer is, in fact, provided by Dyson's threefold way. (Note that in the absence of restrictions, the most general Hamiltonian in Fock space involves n-body interactions of arbitrary rank $n = 1, 2, 3, \ldots, N$.) So there is nothing new to discover here.

We shall, however, be guided to a new and interesting answer by asking a somewhat different question: what is the structure of the set of *one-body* time evolutions of $\wedge(V)$ which commute with the given G-action? Here by a one-body time evolution we mean any unitary operator obtained by exponentiating a self-adjoint Hamiltonian H which is *quadratic* in the particle creation and annihilation operators:

$$H = \sum_{kl} W_{kl}\, a_k^\dagger a_l + \frac{1}{2}\sum_{kl} \left(Z_{kl}\, a_k^\dagger a_l^\dagger + \overline{Z}_{kl}\, a_l a_k \right) . \tag{3.3.4}$$

These operators $U = \mathrm{e}^{-\mathrm{i}tH/\hbar} \in \mathrm{U}(\wedge V)$ form what is called the spin group, $\mathrm{Spin}(W_\mathbb{R})$, of the $2N$-dimensional Euclidean $\mathbb{R}$-vector space $W_\mathbb{R}$ spanned by the Majorana operators $a_k + a_k^\dagger$, $\mathrm{i}a_k - \mathrm{i}a_k^\dagger$ $(k = 1, \ldots, N)$. $\mathrm{Spin}(W_\mathbb{R}) \simeq \mathrm{Spin}_{2N}$ is a double covering of the real orthogonal group $\mathrm{SO}(W_\mathbb{R}) \simeq \mathrm{SO}_{2N}$. The spin group of most prominence in physics is $\mathrm{Spin}_3 \equiv \mathrm{SU}_2$, a double covering of SO_3. (This double covering is known to physicists by the statement that a rotation by 2π, which acts as the neutral element of SO_3, changes the sign of a spinor.)

Thus, our interest is now in the set

$$Z_{\mathrm{Spin}}(G) := Z_{\mathrm{U}(\wedge V)}(G) \cap \mathrm{Spin}(W_\mathbb{R}) \tag{3.3.5}$$

of G-compatible time evolutions in $\mathrm{Spin}(W_\mathbb{R}) \subset \mathrm{U}(\wedge V)$. By adaptation of the earlier definition (3.2.4), the G-compatibility conditions are

$$U = g_0 U g_0^{-1} = g_1 U^{-1} g_1^{-1} \tag{3.3.6}$$

for all $g_0 \in G_0 \cup \tilde{C}TG_0$ and $g_1 \in TG_0 \cup \tilde{C}G_0$.

3.3.3 Reduction to Nambu space

To investigate the set $Z_{\mathrm{Spin}}(G)$ we use the following fact. Any invertible element $A \in \mathrm{Cl}(W)$ determines an automorphism $\gamma \mapsto A\gamma A^{-1}$ of the Clifford algebra $\mathrm{Cl}(W)$ by conjugation. This conjugation action restricts to a representation

$$\tau(g)w := gwg^{-1} \tag{3.3.7}$$

of $\mathrm{Spin}(W_\mathbb{R}) \subset \mathrm{Cl}(W)$ on $W_\mathbb{R} \subset \mathrm{Cl}(W)$. Phrased in physics language, the set of Majorana field operators $w = \sum_k (z_k a_k^\dagger + \bar{z}_k a_k) \in W_\mathbb{R}$ is closed under conjugation $w \mapsto gwg^{-1}$ by one-body time evolution operators $g \in \mathrm{Spin}(W_\mathbb{R})$. In fact, by elementary considerations one finds that $\tau(g) : w \mapsto gwg^{-1}$ for $g \in \mathrm{Spin}(W_\mathbb{R})$ is an orthogonal transformation $\tau(g) \in \mathrm{SO}(W_\mathbb{R})$ of the Euclidean vector space $W_\mathbb{R}$. The mapping $\tau : \mathrm{Spin}(W_\mathbb{R}) \to \mathrm{SO}(W_\mathbb{R})$, $g \mapsto \tau(g) = \tau(-g)$ is two-to-one. It is a covering map, which amounts to saying that any path in $\mathrm{SO}(W_\mathbb{R})$ lifts uniquely to a path in $\mathrm{Spin}(W_\mathbb{R})$. Note also that the linear mapping $\tau(g) : W_\mathbb{R} \to W_\mathbb{R}$ extends to a linear mapping $\tau(g) : W \to W$ by $\mathbb{C}$-linearity.

Thus, instead of studying $\mathrm{Spin}(W_\mathbb{R})$ as a group of operators on the full Fock space $\wedge(V)$, we may simplify our work by studying its representation $\tau : \mathrm{Spin} \to \mathrm{SO}$ on the smaller space $W = V \oplus V^*$, here referred to as Nambu space. Of course the object of interest is not $\mathrm{Spin}(W_\mathbb{R})$ but its intersection with the G-compatibility conditions. To keep track of the latter conditions, we now transfer the G-action from $\wedge(V)$ to $W = V \oplus V^*$.

It is immediately clear how to transfer the actions of G_0 and T, as these are defined on V. In the case of the twisted particle-hole conjugation operator $\tilde{C}$, we do the following computation. Let $\psi \in \wedge^n(V)$ and $\psi' \in \wedge^{n+1}(V)$. Then

$$(\tilde{C}a_k^\dagger \psi) \wedge \psi' = \langle S a_k^\dagger \psi, \psi' \rangle \, \Omega = \langle S\psi, S a_k S^{-1} \psi' \rangle \, \Omega$$
$$= (\tilde{C}\psi) \wedge (S a_k S^{-1})\psi' = (-1)^{N-n+1} (S a_k S^{-1} \tilde{C}\psi) \wedge \psi'.$$

Thus the twisted particle-hole conjugate of $a_k^\dagger \in V \subset \mathrm{Cl}(W)$ is $\tilde{C} a_k^\dagger \tilde{C}^{-1} = \pm S a_k S^{-1} \in V^* \subset \mathrm{Cl}(W)$ where the sign alternates with particle number. Note that the operation $a_k^\dagger \mapsto \pm S a_k S^{-1}$ is anti-unitary. Note also that the untwisted particle-hole conjugation $a_k^\dagger \mapsto a_k$ is none other than the $\mathbb{C}$-antilinear bijection $C_V : V \to V^*, v \mapsto \langle v, \cdot \rangle_V$.

To sum up, we have the following induced structures on Nambu space:

- One-body time evolutions $g \in \mathrm{Spin}(W_\mathbb{R})$ act on $W = V \oplus V^*$ by orthogonal transformations $\tau(g) \in \mathrm{SO}(W_\mathbb{R})$.
- G_0 is defined on V and acts on $W = V \oplus V^*$ by $g(v \oplus f) = gv \oplus (g^{-1})^t f$. The same goes for time reversal T.
- The operator $\tilde{C}$ of twisted particle-hole conjugation induces on $W = V \oplus V^*$ an anti-unitary involution $V \leftrightarrow V^*$.

The goal of symmetry classification now is to characterize the set $Z_{\mathrm{SO}}(G)$ of elements in $\mathrm{SO}(W_\mathbb{R})$ which are compatible with the induced G-action on W. This problem was posed and solved in [HHZ05], by using an elaboration of the algebraic tools of Section 3.2 to make a reduction to the case of the trivial group $G_0 = \{\mathrm{Id}\}$. (The involution $V \leftrightarrow V^*$ given by twisted particle-hole conjugation is called a *mixing* symmetry in [HHZ05].) The outcome is as follows.

Theorem 3.3.1 *The space $Z_{\mathrm{SO}}(G)$ is a direct product of factors each of which is a classical irreducible compact symmetric space. Conversely, every classical irreducible compact symmetric space occurs in this setting.*

There is no space to reproduce the proof here, but in order to turn the theorem into an intelligible statement we now record a few basic facts from the theory of symmetric spaces [Hel78, CM04].

3.3.4 Symmetric spaces

Let M be a connected m-dimensional Riemannian manifold and p a point of M. In some open subset N_p of a neighbourhood of p there exists a map $s_p : N_p \to N_p$, the geodesic inversion with respect to p, which sends a point $x \in N_p$ with normal coordinates $(x_1, \ldots, x_m)$ to the point with normal coordinates $(-x_1, \ldots, -x_m)$. The Riemannian manifold M is called locally symmetric if the geodesic inversion is an isometry (i.e. is distance preserving), and is called globally symmetric if s_p extends to an isometry $s_p : M \to M$, for all $p \in M$. A globally symmetric Riemannian manifold is called a symmetric space for short.

The Riemann curvature tensor of a symmetric space is covariantly constant, which leads to three distinct cases: the scalar curvature can be positive, zero, or negative, and the symmetric space is said to be of compact type, Euclidean type, or non-compact type, respectively. In random matrix theory each type plays a role: the first one provides us with the scattering matrices and time evolutions, the second one with the Hamiltonians, and the third one with the transfer matrices. Our focus here will be on compact type, as it is this type that houses the unitary time evolution operators of quantum mechanics.

Symmetric spaces of compact type arise in the following way. Let U be a connected compact Lie group equipped with a Cartan involution, i.e. an automorphism $\tau : U \to U$ with the involutive property $\tau^2 = \mathrm{Id}$. Let $K \subset U$ be the subgroup of τ-fixed points $u = \tau(u)$. Then the coset space U/K is a compact symmetric space in a geometry defined as follows. Writing $\mathfrak{u} := \mathrm{Lie}(U)$ and $\mathfrak{k} := \mathrm{Lie}(K)$ for the Lie algebras of the Lie groups involved, let $\mathfrak{u} = \mathfrak{k} \oplus \mathfrak{p}$ be the decomposition into positive and negative eigenspaces of the involution $d\tau : \mathfrak{u} \to \mathfrak{u}$ induced by linearization of τ at unity. Fix on $\mathfrak{p}$ a Euclidean scalar product $\langle \cdot, \cdot \rangle_{\mathfrak{p}}$ which is invariant under the adjoint K-action $\mathrm{Ad}(k) : \mathfrak{p} \to \mathfrak{p}$ by $X \mapsto kXk^{-1}$. Then the Riemannian metric g_{uK} evaluated on vectors v, v' tangent to the coset uK is $g_{uK}(v, v') := \langle dL_u^{-1}(v), dL_u^{-1}(v') \rangle_{\mathfrak{p}}$ where dL_u denotes the differential of the operation of left translation on U/K by $u \in U$.

It is important for us that the coset space U/K can be realized as a subset

$$M := \{x \in U \mid \tau(x) = x^{-1}\} \tag{3.3.8}$$

by the Cartan embedding $U/K \to M \subset U$, $uK \mapsto u\,\tau(u^{-1})$. The metric tensor in this realization is given (in a self-explanatory notation) by $g = \mathrm{Tr}\, dx\, dx^{-1}$. It is invariant under the K-action $M \to M$ by twisted conjugation $x \mapsto ux\tau(u^{-1})$. The geodesic inversion with respect to $y \in M$ is $s_y : M \to M$, $x \mapsto yx^{-1}y$.

We note that special examples of compact symmetric spaces are afforded by compact Lie groups K. For these examples, one takes $U = K \times K$ with flip involution $\tau(k, k') = (k', k)$ leading to $U/K = (K \times K)/K \simeq K$. Cartan's list of classical (or large families of) compact symmetric spaces is presented in Table 3.1. The form of the generator H of time evolutions $u = \mathrm{e}^{-\mathrm{i}tH/\hbar}$ is indicated in the third column, $H = \begin{pmatrix} W & Z \\ Z^\dagger & -W^t \end{pmatrix}$ and $W = W^\dagger$. In the case of class D this notation is consistent with the Fock space expression (3.3.4) by the covering map $\tau : \mathrm{Spin} \to SO$.

Table 3.1 The Cartan table of classical symmetric spaces

Family	Compact type	Euclidean type
A	U_N	H complex Hermitian
AI	$\mathrm{U}_N/\mathrm{O}_N$	H real symmetric
AII	$\mathrm{U}_{2N}/\mathrm{USp}_{2N}$	H quaternion self-dual
C	USp_{2N}	Z complex symmetric
CI	$\mathrm{USp}_{2N}/\mathrm{U}_N$	Z complex sym., $W=0$
$B,\ D$	SO_N	Z complex skew
$DIII$	$\mathrm{SO}_{2N}/\mathrm{U}_N$	Z complex skew, $W=0$
$AIII$	$\mathrm{U}_{p+q}/(\mathrm{U}_p \times \mathrm{U}_q)$	Z complex $p \times q,\ W=0$
BDI	$\mathrm{O}_{p+q}/(\mathrm{O}_p \times \mathrm{O}_q)$	Z real $p \times q,\ W=0$
CII	$\mathrm{USp}_{2p+2q}/(\mathrm{USp}_{2p} \times \mathrm{USp}_{2q})$	Z quaternion $2p \times 2q,\ W=0$

3.3.5 Post-Dyson classes

We now run through the symmetry classes beyond those of Wigner-Dyson. As was mentioned before, these appear in various areas of physics and in the random matrix theory of L-functions. For brevity we concentrate on their physical realization by quasi-particles in disordered metals and superconductors.

Class D

Consider a superconductor with no symmetries in its quasi-particle dynamics, so $G = \{\mathrm{Id}\}$. (Some concrete physical examples follow below.) The time evolutions $u = \mathrm{e}^{-\mathrm{i}tH/\hbar}$ in this case are constrained only by the condition $u \in \mathrm{Spin}(W_{\mathbb{R}})$ in Fock space and $\tau(u) \in \mathrm{SO}(W_{\mathbb{R}})$ in Nambu space. The orthogonal group $\mathrm{SO}(W_{\mathbb{R}}) \simeq \mathrm{SO}_{2N}$ is a symmetric space of the D family – hence the name class D. In a basis of Majorana fermions $a_k + a_k^\dagger$, $\mathrm{i}a_k - \mathrm{i}a_k^\dagger$, the matrix of $\mathrm{i}H \in \mathfrak{so}_{2N}$ is real skew, and that of H is imaginary skew.

Concrete realizations are found in superconductors where the order parameter transforms under rotations as a spin triplet in spin space and as a p-wave in real space. A recent candidate for a quasi-2d (or layered) spin-triplet p-wave superconductor is the compound $\mathrm{Sr_2Ru\,O_4}$. (A non-charged analogue is the A-phase of superfluid ^{3}He.) Time-reversal symmetry in such a system may be broken spontaneously, or else can be broken by an external magnetic field creating vortices in the superconductor.

The simplest random matrix model for class D, the SO-invariant Gaussian ensemble of imaginary skew matrices, is analysed in Mehta's book [Meh04]. From the expressions given there it is seen that the level correlation functions at high energy coincide with those of the Wigner-Dyson universality class of unitary symmetry (Class A). The level correlations at low energy, however, show different behaviour defining a separate universality class. This universal behaviour at low energies has immediate physical relevance, as it is precisely the low-energy quasi-particles that determine the thermal transport properties of the superconductor at low temperatures.

*Class **DIII***

Now, let time reversal T be a symmetry: $G = \{\mathrm{Id}, T\}$. Physically speaking this implies the absence of magnetic fields, magnetic impurities and other agents which distinguish between the forward and backward directions of time. Our physical degrees of freedom are spin-1/2 particles, so $T^2 = -\mathrm{Id}_V$.

According to (3.3.6) we are looking for the intersection $Z_{\mathrm{SO}}(G)$ of the condition $u^{-1} = TuT^{-1}$ with $\mathrm{Spin}(W_{\mathbb{R}})$, or after transfer to Nambu space, $\mathrm{SO}(W_{\mathbb{R}})$. By introducing the involution $\tau(u) := TuT^{-1}$ we express the wanted set as

$$Z_{\mathrm{SO}}(G) = \{u \in \mathrm{SO}(W_{\mathbb{R}}) \mid u^{-1} = \tau(u)\}. \tag{3.3.9}$$

Following the discussion around Eq. (3.3.8) we have $Z_{\mathrm{SO}}(G) \simeq U/K$ where $U = \mathrm{SO}(W_{\mathbb{R}})$ and $K \subset U$ is the subgroup of τ-fixed points in U.

In order to identify K we note that time reversal $T : W \to W$ preserves the real subspace $W_{\mathbb{R}}$ of Majorana operators $a_k + a_k^\dagger$, $\mathrm{i}a_k - \mathrm{i}a_k^\dagger$. Because $T^2 = -\mathrm{Id}$, the $\mathbb{R}$-linear operator $T : W_{\mathbb{R}} \to W_{\mathbb{R}}$ is a complex structure of the real vector space $W_{\mathbb{R}} \simeq \mathbb{R}^{2N} \simeq \mathbb{C}^N$. In other words, there exists a basis $\{e_{1,j}, e_{2,j}\}_{j=1,\ldots,N}$ of $W_{\mathbb{R}}$ such that $Te_{1,j} = e_{2,j}$ and $Te_{2,j} = -e_{1,j}$. Now the τ-fixed point condition $k = \tau(k)$ says that $k \in K$ commutes with the complex linear extension $J : W \to W$ of $T : W_{\mathbb{R}} \to W_{\mathbb{R}}$ by $Je_{\pm,j} = \pm\mathrm{i}e_{\pm,j}$ where $e_{\pm,j} = e_{1,j} \pm \mathrm{i}e_{2,j}$. The general element k with this property is a U_N-transformation which acts on $\mathrm{span}_{\mathbb{C}}\{e_{+,1}, \ldots, e_{+,N}\}$ as k and on $\mathrm{span}_{\mathbb{C}}\{e_{-,1}, \ldots, e_{-,N}\}$ as $\overline{k} = (k^{-1})^t$. Hence $K = \mathrm{U}_N$ and

$$Z_{\mathrm{SO}}(G) \simeq U/K \simeq \mathrm{SO}_{2N}/\mathrm{U}_N, \tag{3.3.10}$$

which is a symmetric space in the DIII family.

Known realizations of this symmetry class exist in gapless superconductors, say with spin-singlet pairing, but with a sufficient concentration of spin-orbit impurities to break spin-rotation symmetry. In order for quasi-particle excitations to exist at low energy, the spatial symmetry of the order parameter should be different from s-wave. A non-charged realization occurs in the B-phase of ^{3}He, where the order parameter is spin-triplet without breaking time-reversal symmetry. Other candidates are heavy-fermion superconductors, where spin-orbit scattering often happens to be strong owing to the presence of elements with large atomic weights such as uranium and cerium.

Class C

Next let the spin of the quasi-particles be conserved, but let time-reversal symmetry be broken instead. Thus magnetic fields (or some equivalent T-breaking agent) are now present, while the effect of spin-orbit scattering is absent. The symmetry group of the physical system then is $G = G_0 = \mathrm{Spin}_3 = \mathrm{SU}_2$. Such a situation is realized in spin-singlet superconductors in the vortex phase. Prominent examples are the cuprate superconductors, which are layered and exhibit an order parameter with d-wave symmetry in their copper-oxide planes.

The symmetry-compatible time evolutions are identified by going through the process of eliminating the unitary symmetries $G_0 = G$. For that, we decompose the Hilbert space as $V = E \otimes R$, $E = \mathrm{Hom}_G(R, V)$, where $R := \mathbb{C}^2$ is the fundamental representation of $G = \mathrm{SU}_2$. Now there exists a skew-symmetric SU_2-equivariant isomorphism (known in physics by the name of spin-singlet pairing) between the vector space R and its dual R^*. Therefore we have $W = V \oplus V^* \simeq (E \oplus E^*) \otimes R$, and elimination of the conserved factor R transfers the canonical symmetric form of $W = V \oplus V^*$ to the canonical alternating form $(e \oplus f, e' \oplus f') \mapsto f(e') - f'(e)$ of $E \oplus E^*$. On transferring also the Hermitian scalar product from $V \oplus V^*$ to $E \oplus E^*$, one sees that the G-compatible time evolutions form a unitary symplectic group,

$$Z_{\mathrm{SO}}(G) = \mathrm{SO}(W_{\mathbb{R}})^G \simeq \mathrm{USp}(E \oplus E^*), \tag{3.3.11}$$

which is a compact symmetric space of the C family.

Class **C***I*

The next class is obtained by taking spin rotations as well as the time reversal T to be symmetries of the quasi-particle system. Thus the symmetry group now is $G = G_0 \cup TG_0$ with $G_0 = \mathrm{Spin}_3 = \mathrm{SU}_2$. As in the previous symmetry class, physical realizations are provided by the low-energy quasi-particles of unconventional spin-singlet superconductors. The superconductor must now be in the Meissner phase where magnetic fields are expelled by screening currents.

To identify the relevant symmetric space, we again transfer from $V \oplus V^* = (E \oplus E^*) \otimes R$ to the reduced space $E \oplus E^*$. By this reduction, the canonical form undergoes a change of type from symmetric to alternating as before. We must also transfer time reversal; because the fundamental representation $R = \mathbb{C}^2$ of SU_2 is self-dual by a skew-symmetric isomorphism, the parity of the time-reversal operator changes from $T^2 = -\mathrm{Id}_{V \oplus V^*}$ to $T^2 = +\mathrm{Id}_{E \oplus E^*}$ by the mechanism explained at the end of Section 3.2.2.

We have $Z_{\mathrm{SO}}(G) \simeq U/K$ where $U = \mathrm{USp}(E \oplus E^*)$ and K is the subgroup of elements fixed by conjugation with T. Because the reduced T squares to $+1$, we may realize it on matrices as the complex conjugation operator by working in a basis of T-fixed vectors of $E \oplus E^*$. The Lie algebra elements $X \in \mathfrak{usp}(E \oplus E^*)$ have the form $X = \begin{pmatrix} A & B \\ -\overline{B} & \overline{A} \end{pmatrix}$ with anti-Hermitian A and complex symmetric B. They commute with the operation of complex conjugation if A is real skew and B real symmetric. Matrices X with such A and B span the Lie algebra $\mathfrak{u}_N$, $N = \dim(E)$. At the Lie group level it follows that $K = \mathrm{U}_N$. Hence $Z_{SO}(G) \simeq \mathrm{USp}_{2N}/\mathrm{U}_N$ – a symmetric space in the CI family.

Class AIII

So far, we have made no use of twisted particle-hole conjugation $\tilde{C}$ as a symmetry, but now let the symmetry group be $G = G_0 \cup \tilde{C} G_0$ where $G_0 = \mathrm{U}_1$ acts on $W = V \oplus V^*$ by $v \oplus f \mapsto zv \oplus z^{-1} f$ (for $z \in \mathbb{C}$, $|z| = 1$).

In order for the elements $u \in \mathrm{SO}(W_\mathbb{R})$ to commute with the G_0-action, they must be of the block-diagonal form $u = k \oplus (k^{-1})^t$, $k \in \mathrm{U}(V)$. Therefore $Z_{\mathrm{SO}}(G_0) \simeq \mathrm{U}(V)$. The wanted set then is $Z_{\mathrm{SO}}(G) \simeq U/K$ with $U \equiv \mathrm{U}(V)$ and K the subgroup of elements which are fixed by conjugation with $\tilde{C}$.

Recall from Section 3.3.1 that $\tilde{C}|_V = CS$ where $S \in U$, $S^2 = \mathrm{Id}$, and untwisted particle-hole conjugation C coincides (up to an irrelevant sign) with the canonical bijection $C_V : V \to V^*$, $C_V(v) = \langle v, \cdot \rangle_V$. The condition for $u = k \oplus (k^{-1})^t$ to belong to K reads

$$(k^{-1})^t = \tilde{C} k \, \tilde{C}^{-1}. \tag{3.3.12}$$

Since $k^{-1} = k^\dagger$ and $C^{-1} k^t C = k^\dagger$, this condition is equivalent to $k = SkS$.

Now let $V = V_+ \oplus V_-$ where $V_\pm$ are orthogonal subspaces with projection operators $\Pi_\pm$. Then if $S = \Pi_+ - \Pi_-$ we have $K = \mathrm{U}(V_+) \times \mathrm{U}(V_-)$ and hence

$$Z_{\mathrm{SO}}(G) \simeq \mathrm{U}(V)/(\mathrm{U}(V_+) \times \mathrm{U}(V_-)) \tag{3.3.13}$$

or $Z_{\mathrm{SO}}(G) \simeq \mathrm{U}_N/(\mathrm{U}_p \times \mathrm{U}_{N-p})$ with $p = \dim V_+$.

The space (3.3.13) is a symmetric space of the AIII family. Its symmetry class is commonly associated with random-matrix models for the low-energy Dirac spectrum of quantum chromodynamics with massless quarks [Ver94]. An alternative realization exists [ASZ02] in T-invariant spin-singlet superconductors with d-wave pairing and soft impurity scattering.

*Classes **BDI** and **CII***

Finally, let the symmetry group G have the full form of Definition 3.3.2, with $G_0 = \mathrm{U}_1$ and $\tilde{C}$ as before (Class AIII) and a time-reversal symmetry T, $T^2 = \pm\mathrm{Id}$. We recall that the elements of $Z_{\mathrm{SO}}(G_0)$ are $u = k \oplus (k^{-1})^t$, $k \in \mathrm{U}(V)$. The requirement of commutation with the product $\phi := \tilde{C}T : V \to V^*$ of anti-unitary symmetries is equivalent to the condition $\phi = k^t \phi \, k$.

Let $U := Z_{\mathrm{SO}}(G_0 \cup \tilde{C} T G_0)$. To identify U, we use that $\phi(v)(v') = \langle STv, v' \rangle$ and $ST = TS$. By the computation of (3.2.9), it follows that the parity of T equals the parity of the isomorphism $\phi : V \to V^*$. In other words, if $T^2 = \epsilon\,\mathrm{Id}$ then $\phi^t = \epsilon \phi$. Thus the condition $\phi = k^t \phi k$ singles out an orthogonal group $U = \mathrm{O}(V) \simeq \mathrm{O}_N$ in the symmetric case ($\epsilon = +1$) and a unitary symplectic group $U = \mathrm{USp}(V) \simeq \mathrm{USp}_N$ in the alternating case ($\epsilon = -1$).

In both cases, the wanted set is $Z_{\mathrm{SO}}(G) = U/K$ with K the subgroup of fixed points $k = \tilde{C}^{-1}(k^{-1})^t \tilde{C} = SkS$. In the former case we have $K \simeq \mathrm{O}_p \times \mathrm{O}_{N-p}$, and in the latter case $K \simeq \mathrm{USp}_p \times \mathrm{USp}_{N-p}$ (with even N, p). Thus we arrive at the final two entries of Cartan's list:

$$\text{Class } BDI: U/K \simeq \mathrm{O}_N/(\mathrm{O}_p \times \mathrm{O}_{N-p}) \qquad (T^2 = +1),$$
$$\text{Class } CII: U/K \simeq \mathrm{USp}_N/(\mathrm{USp}_p \times \mathrm{USp}_{N-p}) \quad (T^2 = -1).$$

These occur as symmetry classes in the context of the massless Dirac operator [Ver94]. Class BDI is realized by taking the gauge group to be SU_2 or USp_{2n}, Class CII by taking fermions in the adjoint representation or gauge group SO_n.

3.4 Discussion

Given the classification scheme for disordered fermions, it is natural to ask whether an analogous scheme can be developed for the case of bosons. Although there exists no published account of this (see, however, [LSZ06]), we now briefly outline the answer to this question.

The mathematical model for the bosonic Fock space is a symmetric algebra $\mathrm{S}(V)$. It is still equipped with a canonical Hermitian structure induced by that of V. The real form $W_{\mathbb{R}}$ of Nambu space $W = V \oplus V^*$ for bosons has an interpretation as a classical phase space spanned by positions $q_j = (a_j + a_j^\dagger)/\sqrt{2}$ and momenta $p_j = (a_j - a_j^\dagger)/\sqrt{2}\,\mathrm{i}$. At the level of one-body unitary time evolutions in Fock space, the role of the spin group $\mathrm{Spin}(W_{\mathbb{R}})$ for fermions is handed over to the metaplectic group $\mathrm{Mp}(W_{\mathbb{R}})$ for bosons.

By the quantum-classical correspondence, a one-parameter group of time evolutions $u_t = \mathrm{e}^{-\mathrm{i}tH/\hbar} \in \mathrm{Mp}(W_{\mathbb{R}})$ in Fock space gets assigned to a linear symplectic flow $\tau(u_t) \in \mathrm{Sp}(W_{\mathbb{R}})$ in classical phase space. This correspondence $\tau : \mathrm{Mp}(W_{\mathbb{R}}) \to \mathrm{Sp}(W_{\mathbb{R}})$ is still two-to-one (reflecting, e.g. the well-known fact that the sign of the harmonic oscillator wave function is reversed by time evolution over one period). An important difference as compared to fermions is that the classical flow $\tau(u_t) \in \mathrm{Sp}(W_{\mathbb{R}})$ is not unitary in any natural sense.

In nuclear physics, the differential equation of the flow $\tau(u_t)$ is called the RPA equation. For example, in the case without symmetries this equation reads

$$\frac{d}{dt} a_k^\dagger = \sum_j \left(a_j^\dagger A_{jk} + a_j B_{jk}\right), \quad \frac{d}{dt} a_k = \sum_j \left(a_j^\dagger C_{jk} + a_j D_{jk}\right), \tag{3.4.1}$$

where one requires $B = B^t$, $C = C^t$, and $D = -A^t$ in order for the canonical commutation relations of the boson operators $a^\dagger$, a to be conserved. Unitarity of the flow (as a time evolution in Fock space) requires $A = -A^\dagger$ and $C = B^\dagger$. This should be compared with the fermion problem in Class C, where one has exactly the same set of equations but for a single sign change: $C = -B^\dagger$. Thus the corresponding generator of time evolution is $X = \begin{pmatrix} A & B \\ \pm\overline{B} & \overline{A} \end{pmatrix}$ where the plus sign applies to bosons and the minus sign to fermions. In either case X belongs to the *same* complex Lie algebra, $\mathfrak{sp}(W)$. The difference is that the generator for fermions lies in a *compact* real form $\mathfrak{usp}(W) \subset \mathfrak{sp}(W)$, whereas the generator for bosons lies in a *non-compact* real form $\mathfrak{sp}(W_\mathbb{R}) \subset \mathfrak{sp}(W)$.

This remains true in the general case with symmetries. Thus if the word 'symmetry class' is understood in the complex sense, then the bosonic setting does not lead to any new symmetry classes; it just leads to different real forms of the known symmetry classes viewed as complex spaces. The same statement applies to the non-Hermitian situation. Indeed, all of the spaces of [Mag08, BL02] are complex or non-compact real forms of the symmetric spaces of Cartan's table. Here we must reiterate that the notion of symmetry class is an algebraic one whose prime purpose is to inject an organizational principle into the multitude of possibilities. It must not be misunderstood as a cheap vehicle to produce immediate predictions of eigenvalue distributions and universal behaviour!

Let us end with a few historical remarks. The disordered harmonic chain, a model in the post-Dyson Class BDI, was first studied by Dyson [Dys53]. The systematic field-theoretic study of models with sublattice symmetry (later recognized as members of the chiral classes $AIII$, BDI) was initiated by Oppermann, Wegner, and Gade [OW79, Gad93, GW91]. Gapless superconductors were the subject of numerous papers by Oppermann; e.g. [Opp90] computes the one-loop beta function of the non-linear sigma model for Class CI.

The ten-way classification of Section 3.3 was originally discovered by a very different reasoning: the mapping of random matrix problems to effective field-theory models [Zir96] combined with the fact that closure of the renormalization group flow takes place for non-linear sigma models where the target is a symmetric space. A less technical early confirmation of the ten-way classification came from Wegner's flow equations [Weg94]. These take the form of a double-commutator flow for Hamiltonians H belonging to a matrix space $\mathfrak{p}$; if the double commutator $[\mathfrak{p}, [\mathfrak{p}, \mathfrak{p}]]$ closes in $\mathfrak{p}$, so does Wegner's flow. The closure condition is satisfied precisely if $\mathfrak{p}$ is the odd part of a Lie algebra $\mathfrak{u} = \mathfrak{k} \oplus \mathfrak{p}$ with involution, i.e. the infinitesimal model of a symmetric space.

Last but not least, let us mention the viewpoint of Volovik (see, e.g. [Vol03]) who advocates classifying single-particle Green's functions rather than Hamiltonians. That viewpoint in fact has the advantage that it is not tied to non-interacting systems but offers a natural framework in which to include (weak) interactions.

References

[ASZ02] A. Altland, B.D. Simons, and M.R. Zirnbauer, "Theories of low-energy quasi-particle states in disordered d-wave superconductors" Phys. Rep. **359** (2002) 283–354.

[AZ97] A. Altland and M.R. Zirnbauer, "Non-standard symmetry classes in mesoscopic normal-/superconducting hybrid systems", Phys. Rev. B **55** (1997) 1142–1161.

[Ber84] G. Bergmann, "Weak localization in thin films, a time-of-flight experiment with conduction electrons", Phys. Rep. **107** (1984) 1–58.

[BL02] D. Bernard and A. LeClair, "A classification of random Dirac fermions", J. Phys. A **35** (2002) 2555–2567.

[CM04] M. Caselle and U. Magnea, "Random matrix theory and symmetric spaces", Phys. Rep. **394** (2004) 41–156.

[Dys53] F.J. Dyson, "The dynamics of a disordered linear chain", Phys. Rev. **92** (1953) 1331–1338.

[Dys62] F.J. Dyson, "The threefold way: algebraic structure of symmetry groups and ensembles in quantum mechanics", J. Math. Phys. **3** (1962) 1199–1215.

[Dys62a] F.J. Dyson, "Statistical theory of energy levels of complex systems", J. Math. Phys. **3** (1962) 140–156.

[Dys70] F.J. Dyson, "Correlations between eigenvalues of a random matrix", Commun. Math. Phys. **19** (1970) 235–250.

[Gad93] R. Gade, "Anderson localization for sublattic models", Nucl. Phys. B **398** (1993) 499–515.

[GW91] R. Gade and F. Wegner, "The $n = 0$ replica limit of $\mathrm{U}(n)$ and $\mathrm{U}(n)/\mathrm{SO}(n)$ models", Nucl. Phys. B **360** (1991) 213–218.

[HHZ05] P. Heinzner, A.H. Huckleberry, and M.R. Zirnbauer, "Symmetry classes of disordered fermions", Commun. Math. Phys. **257** (2005) 725–771.

[Hel78] S. Helgason, *Differential geometry, Lie groups and symmetric spaces*, Academic Press, New York (1978).

[Kit08] A. Kitaev, "Periodic table for topological insulators and superconductors", AIP Conf. Proc. 1134 (2009) 22–30 [arXiv:0901.2686].

[KS99] N. Katz and P. Sarnak, *Random matrices, Frobenius eigenvalues, and monodromy*, American Mathematical Society, Providence, R.I., (1999).

[LSZ06] T. Lueck, H.-J. Sommers, and M.R. Zirnbauer, "Energy correlations for a random matrix model of disordered bosons", J. Math. Phys. **47** (2006) 103304.

[Mag08] U. Magnea, "Random matrices beyond the Cartan classification", J. Phys. A **41** (2008) 045203.

[Meh04] M.L. Mehta, *Random Matrices*, 3rd Edition, Academic Press, London (2004).

[Opp90] R. Oppermann, "Anderson localization problems in gapless superconducting phases", Physica A **167** (1990) 301–312.

[OW79] R. Oppermann and F. Wegner, "Disordered system with n orbitals per site – $1/n$ expansion", Z. Phys. B **34** (1979) 327–348.

[SRF09] A.P. Schnyder, S. Ryu, A. Furusaki, A.W.W. Ludwig, "Classification of topological insulators and superconductors", AIP Conf. Proc. 1134 (2009) 10–21 [arXiv:0905.2029].

[Ver94] J. Verbaarschot, "Spectrum of the QCD Dirac operator and chiral random-matrix theory", Phys. Rev. Lett. **72** (1994) 2531–2533.

[Vol03] G.E. Volovik, *The Universe in a Helium Droplet*, Clarendon Press, Oxford (2003).

[Weg94] F. Wegner, "Flow equations for Hamiltonians", Annalen d. Physik **3** (1994) 77–91.

[Zir96] M.R. Zirnbauer, "Riemannian symmetric superspaces and their origin in random matrix theory", J. Math. Phys. **37** (1996) 4986–5018.

[Zir04] M.R. Zirnbauer, "Symmetry classes in random matrix theory", *Encyclopedia of Mathematical Physics*, vol. 5, pp. 204–212, Academic Press, Oxford (2006) [arXiv:math-ph/0404058].

·4·

Spectral statistics of unitary ensembles

Greg W. Anderson

Abstract

We review the orthogonal polynomial method for computing correlation functions, cluster functions, gap probabilities, Janossy densities and spacing distributions for the eigenvalues of matrix ensembles with unitary-invariant probability law. We briefly review the classical families of orthogonal polynomials (Hermite, Laguerre, and Jacobi) and some corresponding matrix ensembles.

4.1 Introduction

Given a potential $V : \mathbb{R} \to [-\infty, \infty]$ growing rapidly enough, consider the unitary-invariant probability measure

$$\mathbb{P}_{N,V}(dX) = \frac{1}{\widetilde{Z}_{N,V}} \exp(-\operatorname{tr} V(X)) \prod_{i=1}^{n} dX_{ii} \prod_{1 \le i < j \le n} d\mathrm{Re}X_{ij} d\mathrm{Im}X_{ij} \tag{4.1.1}$$

on the space of N-by-N Hermitian matrices, where $\widetilde{Z}_{N,V}$ is a normalizing constant. It is well known that the randomly enumerated eigenvalues $\lambda_1, \cdots, \lambda_N$ of a random matrix with law $\mathbb{P}_{N,V}$ satisfy

$$\mathbb{P}((\lambda_1, \ldots, \lambda_N) \in S) = \frac{1}{Z_{N,V}} \int_S \Delta_N(x)^2 \prod_{i=1}^{N} \exp(-V(x_i)) dx_i \tag{4.1.2}$$

for sets $S \subset \mathbb{R}^N$, where $Z_{N,V}$ is a normalizing constant and

$$\Delta_N = \Delta_N(x) = \prod_{1 \le i < j \le N} (x_j - x_i) = \begin{vmatrix} x_1^0 & \ldots & x_N^0 \\ \vdots & & \vdots \\ x_1^{N-1} & \ldots & x_N^{N-1} \end{vmatrix} = \det_{i,j=1}^{N} x_j^{i-1} \tag{4.1.3}$$

is the *Vandermonde determinant.*

For another example involving $\Delta(x)^2$, let $V : [0, \infty) \to [-\infty, \infty]$ grow rapidly enough, let $0 < p \le q$ be integers, and consider the probability measure

$$\mathbb{P}_{p,q,V}(dX) = \frac{1}{\widetilde{Z}_{p,q,V}} \exp(-\operatorname{tr} V(XX^*)) \prod_{i=1}^{p} \prod_{j=1}^{q} d\mathrm{Re}X_{ij} d\mathrm{Im}X_{ij} \tag{4.1.4}$$

on the space of p-by-q matrices with complex entries, where $\widetilde{Z}_{p,q,V}$ is a normalizing constant. Then, if X is a random matrix with law $\mathbb{P}_{p,q,V}$, the sample covariance matrix XX^* has unitary-invariant law and the randomly enumerated eigenvalues $\lambda_1, \ldots, \lambda_p$ of XX^* satisfy

$$\mathbb{P}((\lambda_1, \ldots, \lambda_p) \in S) = \frac{1}{Z_{p,q,V}} \int_S \Delta_p(x)^2 \prod_{i=1}^{p} \exp(-V(x_i)) x_i^{q-p} dx_i \qquad (4.1.5)$$

for sets $S \subset (0, \infty)^p$, where $Z_{p,q,V}$ is a further normalizing constant.

Our last example involving $\Delta(x)^2$, rather than involving a potential, is based on a Haar measure on the unitary group. Let $n = 2p + r + s$ for integers $p > 0$ and $r, s \geq 0$. Let U be a Haar-distributed random n-by-n unitary matrix, and put $X = U \begin{bmatrix} -I_{p+r} & 0 \\ 0 & I_{p+s} \end{bmatrix} U^*$, where I_k is the k-by-k identity matrix. Let $\lambda_1, \ldots, \lambda_p$ be the randomly enumerated eigenvalues of the upper left p-by-p block of X. It is known that

$$\mathbb{P}((\lambda_1, \ldots, \lambda_p) \in S) = \frac{1}{Z_{p,r,s}} \int_S |\Delta_p(x)|^2 \prod_{i=1}^{p} (1 - x_i)^r (1 + x_i)^s dx_i \qquad (4.1.6)$$

for sets $S \subset (-1, 1)^p$, where $Z_{p,r,s}$ is a normalizing constant.

Here are some references for the joint distribution results mentioned above. The results (4.1.2) and (4.1.5) are treated in many books, e.g., [And09], [For10], and [Meh04]. Furthermore, Eqn. (4.1.5) for $V(x) = x$ (albeit in its real rather than complex version) is well known in multivariate statistics, see e.g. [And03]. Proofs of Eqn. (4.1.6) can be found, e.g. in [And09] and [For10].

In this chapter we will study the statistical properties of N-tuples of real numbers under a probability measure of a form general enough to encompass all the laws of form (4.1.2), (4.1.5), and (4.1.6). We will review the definitions of basic statistical quantities and explain how their distributions can be made explicit in terms of orthogonal polynomials. Fredholm determinants and resolvent kernels (but only for finite-rank kernels) will have an important role to play. In our discussion, which is essentially algebraic and combinatorial, N is fixed; see many other chapters in this volume for discussion of limiting behaviour as $N \to \infty$. We also briefly review the Hermite (resp. Laguerre, resp. Jacobi) polynomials arising in connection with the law (4.1.2) in the case $V(x) = x^2$ (resp. (4.1.5) in the case $V(x) = x$, resp. (4.1.6)).

Let us be clear that these notes are purely expository, and are merely a translation into probability-orientated language of basic material covered in a more physics-orientated way in [Meh04], [Tra98], and [For10]. For help in translation, and especially for the framework provided by the theory of (finite and determinantal) point processes, we have relied heavily on [And09, Chap. 4],

[Dal02, Chap. 5], and [Hou09]. The paper [Bor03] was also a big influence. Our brief discussion of spacings follows [Dei99] albeit with simplifications.

Finally, we wish to emphasize that many if not most of the methods of calculation reviewed here carry over to the setting of general determinantal point processes. For discussion of the latter, and a listing of essentially all important classes of examples, see Chapter 11 of this volume.

4.2 The orthogonal polynomial method: the setup

At the outset, we fix a measure μ on $\mathbb{R}$ with absolute moments of all orders, and a (large) positive integer N. We allow the possibility that μ is not absolutely continuous with respect to Lebesgue measure, but we assume that the support of μ is an infinite set, so that the powers of x are linearly independent in $L^2(\mu)$. Although N and μ are mostly fixed in the following discussion, we occasionally need to make comparisons between different choices for the pair (N, μ), notably in Proposition 4.8.1 below, in connection with spacings. We therefore tend to keep track of N in the notation and, to a lesser degree, μ.

Let $\Omega_N \subset \mathbb{R}^N$ be the subset consisting of N-tuples of distinct real numbers. Let $\mathbb{P}_{N,\mu}$ be the probability measure on Ω_N which assigns probabilities according to the rule

$$\mathbb{P}_{N,\mu}(S) = \frac{1}{Z_{N,\mu}} \int_S \Delta_N^2 d\mu^{\otimes N}, \text{ for } S \subset \Omega_N, \text{ where } Z_{N,\mu} = \int \Delta_N^2 d\mu^{\otimes N}, \tag{4.2.1}$$

with Δ_N as defined in (4.1.3). Let $\mathbf{E}_{N,\mu}$ and $\mathbf{Var}_{N,\mu}$ denote expectation and variance, respectively, under $\mathbb{P}_{N,\mu}$. Let Λ_N be the function on Ω_N sending each N-tuple $(x_1, \ldots, x_N)$ to the N-element set $\{x_1, \ldots, x_N\}$. Viewed as a random variable under $\mathbb{P}_{N,\mu}$, the map Λ_N becomes a random N-element set of real numbers. Even though random matrices are absent from our setup, we still refer to elements of Λ_N as *eigenvalues*.

Let the orthogonal polynomials $\{\phi_n\}_{n=0}^\infty$ be obtained by applying the Gram-Schmidt process to the sequence $\{x^n\}_{n=0}^\infty$ in $L^2(\mu)$, with signs adjusted so that

$$\phi_n(x) = \phi_{n,\mu}(x) = \gamma_n x^n + \text{lower degree terms} \quad (\gamma_n = \gamma_{n,\mu} > 0). \tag{4.2.2}$$

It will later be proved that

$$N! = Z_{N,\mu} \prod_{n=0}^{N-1} \gamma_{n,\mu}^2. \tag{4.2.3}$$

(See the proof of Lemma 4.4.1 below.)

The remarkable fact we wish to explain is that all probabilities of interest involving Λ_N, computed under $\mathbb{P}_{N,\mu}$, can be expressed cleanly and efficiently in terms of the kernel

$$K_N(x, y) = K_{N,\mu}(x, y) = \sum_{n=0}^{N-1} \phi_n(x)\phi_n(y) \tag{4.2.4}$$

which represents orthogonal projection in $L^2(\mu)$ to the span of $\{x^n\}_{n=0}^{N-1}$. (**Warning**: we write our kernels just so, without normalizing factors, in order not to exclude from consideration measures μ failing to be absolutely continuous with respect to Lebesgue measure, and also to facilitate a later treatment of spacing by conditioning.) To write down such formulas is to prepare oneself very well for studying probabilistic limits as $N \to \infty$, because much is known about the limiting behaviour of the kernel $K_{N,\mu}$. Knowledge has increased substartially much since the introduction of the Riemann–Hilbert method. For more information about the latter, see many chapters in this volume.

Let us now briefly review the most basic properties of the kernel $K_{N,\mu}$. To begin with, $K_{N,\mu}$ is real-valued, symmetric and *reproducing*, i.e.

$$\int K_N(x, t)t^i \mu(dt) = x^i \quad \text{for } i = 0, \ldots, N-1, \text{ hence} \tag{4.2.5}$$

$$\int K_N(x, t)\, K_N(t, y)\mu(dt) = K_N(x, y). \tag{4.2.6}$$

Furthermore, the kernel K_N has a simple expression solely in terms of ϕ_N and ϕ_{N-1}, which we now quickly derive. Toward that end, note first that $\phi_n(x)$ is orthogonal in $L^2(\mu)$ to $x\phi_i(x)$ for $i < n-1$, and hence we have the *three-term recurrence*

$$x\phi_n(x) = \frac{\gamma_n}{\gamma_{n+1}}\phi_{n+1}(x) + \left(\int t\phi_n(t)^2\mu(dt)\right)\phi_n(x) + \frac{\gamma_{n-1}}{\gamma_n}\phi_{n-1}(x). \tag{4.2.7}$$

(By convention $\gamma_i = 0$ and $\phi_i \equiv 0$ for $i < 0$.) The *Christoffel–Darboux identity*

$$(x - y)\, K_N(x, y) = \frac{\gamma_{N-1}}{\gamma_N}\left(\phi_N(x)\phi_{N-1}(y) - \phi_{N-1}(x)\phi_N(y)\right) \tag{4.2.8}$$

follows immediately.

4.3 Examples: classical orthogonal polynomials

Before proceeding with our exposition of the orthogonal polynomial method, we review key properties of the classical families of orthogonal polynomials bearing the names of Hermite, Laguerre and Jacobi, respectively. All the formulas below are taken from the Bateman manuscript [Erd53, Chap. X].

4.3.1 Hermite polynomials $\leftrightarrow$ weight: $\mu(dx) = e^{-x^2}dx$

The *Hermite polynomials* are defined by the formula

$$(-1)^n H_n(x) = e^{x^2}\frac{d^n}{dx^n}e^{-x^2} \tag{4.3.1}$$

of Rodrigues type and satisfy

$$\int_{-\infty}^{\infty} H_m(x)H_n(x)e^{-x^2}dx = \pi^{1/2}2^n n!\delta_{mn}, \tag{4.3.2}$$

$$H_n(x) = (2x)^n + \text{lower degree terms.} \tag{4.3.3}$$

Using (4.2.8), (4.3.1), (4.3.2), and (4.3.3), one can make $\phi_{n,\mu}(x)$, $\gamma_{n,\mu}$ and $K_{n,\mu}(x, y)$ in our setup explicit for $\mu(dx) = e^{-x^2}dx$ in terms of the Hermite polynomials. Note that with μ so chosen, the general machinery developed below will compute probabilities of the form (4.1.2) for the potential $V(x) = x^2$.

For the asymptotic study of $K_{n,\mu}$ by classical methods one gets a foothold by using the generating function identity

$$\sum_{n=0}^{\infty} H_n(x)\frac{z^n}{n!} = e^{2xz-z^2} \tag{4.3.4}$$

and the differential equation

$$y'' - 2xy' + 2ny = 0 \tag{4.3.5}$$

satisfied by $y = H_n(x)$. See [And09, Chap. 3] for a "bare-handed" and detailed working out of the implications of (4.3.4) and (4.3.5) for asymptotics.

We will see presently that the other types of classical orthogonal polynomials have analogous generating function identities and differential equations.

4.3.2 Laguerre polynomials $\leftrightarrow$ weight: $\mu(dx) = \mathbf{1}_{x>0}e^{-x}x^a dx$

Let $a > -1$ be a parameter. The *Laguerre polynomials* with parameter a are defined by the formula

$$n!L_n^a(x) = e^x x^{-a}\frac{d^n}{dx^n}\left(e^{-x}x^{n+a}\right) \tag{4.3.6}$$

of Rodrigues type and satisfy

$$\int_0^{\infty} L_m^a(x)L_n^a(x)x^a e^{-x}dx = \frac{(n+a)!}{n!}\delta_{mn}, \tag{4.3.7}$$

$$L_n^a(x) = \frac{(-x)^n}{n!} + \text{lower degree terms.} \tag{4.3.8}$$

Using (4.2.8), (4.3.6), (4.3.7), and (4.3.8), one can make $\phi_{n,\mu}(x)$, $\gamma_{n,\mu}$ and $K_{n,\mu}(x, y)$ in our setup explicit for the weight $\mu(dx) = \mathbf{1}_{x>0}e^{-x}x^a dx$. Then for μ so chosen, the general machinery developed below will compute probabilities of the form (4.1.5) for the potential $V(x) = x$ provided that $a = q - p$. The generating function identity

$$\sum_{n=0}^{\infty} L_n^a(x) z^n = (1-z)^{-a-1} \exp\left(\frac{xz}{z-1}\right) \tag{4.3.9}$$

and differential equation

$$xy'' + (a+1-x)y' + ny = 0 \tag{4.3.10}$$

satisfied by $y = L_n^a(x)$ can be used for the study of asymptotics.

4.3.3 Jacobi polynomials $\leftrightarrow$ weight: $\mu(dx) = \mathbf{1}_{|x|<1}(1-x)^a(1+x)^b dx$

Let $a, b > -1$ be parameters. The *Jacobi polynomials* with parameters a and b are defined by the formula

$$(-1)^n 2^n n!\, P_n^{(a,b)}(x) = \frac{\frac{d^n}{dx^n}\left((1-x)^{a+n}(1+x)^{b+n}\right)}{(1-x)^a(1+x)^b} \tag{4.3.11}$$

of Rodrigues type. They satisfy

$$\int_{-1}^{1} P_m^{(a,b)}(x)\, P_n^{(a,b)}(x)(1-x)^a(1+x)^b dx = \frac{2^{a+b+1}(n+a)!(n+b)!}{(2n+a+b+1)n!(n+a+b)!}\delta_{mn}, \tag{4.3.12}$$

$$P_n^{(a,b)}(x) = \frac{(2n+a+b)!}{n!(n+a+b)!}\left(\frac{x}{2}\right)^n + \text{lower degree terms.} \tag{4.3.13}$$

Using (4.2.8), (4.3.11), (4.3.12), and (4.3.13), one can make $\phi_{n,\mu}(x)$, $\gamma_{n,\mu}$ and $K_{n,\mu}(x, y)$ in our setup explicit for the weight $\mu(dx) = \mathbf{1}_{|x|<1}(1-x)^a(1+x)^b dx$. Then with μ so chosen, the general machinery developed below will compute probabilities of the form (4.1.6) provided that $r = a$ and $s = b$. The generating function identity

$$\sum_{n=0}^{\infty} P_n^{(a,b)}(x) z^n = 2^{a+b} R^{-1}(1-z+R)^{-a}(1+z+R)^{-b} \tag{4.3.14}$$

where $R = (1-2xz+z^2)^{1/2}$, and the differential equation

$$(1-x^2)y'' + (b-a-(a+b+2)x)y' + n(n+a+b+1)y = 0 \tag{4.3.15}$$

satisfied by $y = P_n^{(a,b)}(x)$ can be used for the study of asymptotics.

4.4 The *k*-point correlation function

We now begin general study of the orthogonal polynomial method. The function

$$\det_{i,j=1}^{k} K_N(x_i, x_j) \tag{4.4.1}$$

turns out to be the *k-point correlation function* (or *joint intensity*) for the process Λ_N (with respect to the reference measure μ). The following result explains why the determinant (4.4.1) is in fact the k-point correlation function. To state the result, we first introduce some terminology. We define a *k-element sample* of a set A to be a sequence of k distinct elements of A. (For our purposes samples are always taken without replacement.)

Proposition 4.4.1 (Joint distribution of k eigenvalues) *The law under $\mathbb{P}_{N,\mu}$ of a random k-element sample of Λ_N has density*

$$\frac{(N-k)!}{N!} \det_{i,j=1}^{k} K_{N,\mu}(x_i, x_j) \tag{4.4.2}$$

with respect to $\mu^{\otimes k}$.

Before giving the proof we present two corollaries. To that end we introduce more terminology. Given a set $S \subset \mathbb{R}^k$, let Occ_S^k be the number of k-element samples of Λ_N belonging to S. For $A \subset \mathbb{R}$, we write $\mathrm{Occ}_A = \mathrm{Occ}_A^1$. We have simply $\mathrm{Occ}_A = \#(A \cap \Lambda_N)$, where $\#S$ denotes the cardinality of a set S. We call Occ_A the *occupancy* of A.

Corollary 4.4.1 (Counting principle)

$$\mathbf{E}_{N,\mu} \mathrm{Occ}_S^k = \int_S \det_{i,j=1}^{k} K_{N,\mu}(x_i, x_j) \prod_{i=1}^{k} \mu(dx_i), \tag{4.4.3}$$

for $S \subset \mathbb{R}^k$.

The corollary is merely a reformulation of the proposition. No proof is needed. The advantage of the corollary over the proposition is that the combinatorial factor $\frac{(N-k)!}{N!}$ has been made to disappear. Formula (4.4.3) is superior to formula (4.4.2) when it comes time to study limits as $N \to \infty$.

Corollary 4.4.2 (Mixed factorial moments of occupancies)

$$\mathbf{E}_{N,\mu} \left(\prod_{i=1}^{r} k_i! \binom{\mathrm{Occ}_{A_i}}{k_i} \right) = \int_{\prod_{i=1}^{r} A_i^{k_i}} \det_{i,j=1}^{k} K_{N,\mu}(x_i, x_j) \prod_{i=1}^{k} \mu(dx_i), \tag{4.4.4}$$

for disjoint sets $A_1, \ldots, A_r \subset \mathbb{R}$ and integers $k_1, \ldots, k_r > 0$ summing to k.

In particular,

$$\mathbf{Var}_{N,\mu}(\mathrm{Occ}_A) = -\int_{A^2} K_N(x, y)^2 \mu(dx)\mu(dy) + \int_A K_N(t, t)\mu(dt) \tag{4.4.5}$$

for $A \subset \mathbb{R}$.

Proof By Corollary 4.4.1, the right side of (4.4.4) equals the expected number of k-element samples from Λ_N belonging to the set $\prod_{i=1}^r A_i^{k_i}$, calculated under the probability measure $\mathbb{P}_{N,\mu}$. It is clear that the left side has the same interpretation. □

We turn now to the proof of Proposition 4.4.1. Toward that end we introduce the notation

$$x[a..b] = (x_a, \ldots, x_b), \quad x(a..b] = (x_{a+1}, \ldots, x_b), \quad x[a..b) = (x_a, \ldots, x_{b-1}), \tag{4.4.6}$$

$$K_N \begin{pmatrix} x[1..k] \\ y[1..k] \end{pmatrix} = K_N \begin{pmatrix} x_1, \ldots, x_k \\ y_1, \ldots, y_k \end{pmatrix} = \det_{i,j=1}^k K_N(x_i, y_j) \tag{4.4.7}$$

which will be useful throughout this chapter. Note that

$$K_{N,\mu} \begin{pmatrix} x[1..k] \\ y[1..k] \end{pmatrix} \equiv 0 \quad \text{for } k > N. \tag{4.4.8}$$

We will consider many seemingly infinite sums below which in fact break off after finitely many terms on account of (4.4.8).

To prove Proposition 4.4.1 we need the following lemma.

Lemma 4.4.1 ("Integrating out")

$$\int K_N(t, t)\mu(dt) = N, \tag{4.4.9}$$

and for $k > 1$,

$$\int K_N \begin{pmatrix} x[1..k) & t \\ y[1..k) & t \end{pmatrix} \mu(dt) = (N - k + 1) K_N \begin{pmatrix} x[1..k) \\ y[1..k) \end{pmatrix}. \tag{4.4.10}$$

As emphasized in [Meh04], this is the fundamental insight for the statistical study of eigenvalues of unitary ensembles. But (I am indebted to A. Borodin for pointing this out) Mehta came to this insight only later – in the first edition of his book (see Section 6.1.1) Mehta derives what we call Proposition 4.4.1 by a direct method which one may prefer to the method based on (4.4.10).

Proof Formula (4.4.9) follows immediately from the the orthogonality relations $\int \phi_i \phi_j d\mu = \delta_{ij}$. Assume that $k > 1$ now. With a square, b a column vector, c a row vector, and d a scalar, one has a determinant identity

$$\det\begin{bmatrix} a & b \\ c & d \end{bmatrix} = d\det a - ca^{\star}b = d\det a - \mathrm{tr}\,(bca^{\star}), \tag{4.4.11}$$

where $a^{\star}$ denotes the transpose of the matrix of cofactors of a, which is obtained by iterating the Laplace expansion. Let us now take

$$\begin{bmatrix} a & b \\ c & d \end{bmatrix} = \begin{bmatrix} a(t) & b(t) \\ c(t) & d(t) \end{bmatrix} = \begin{bmatrix} \left[K_N(x_i, y_j)\right]_{i,j=1}^{k-1} & \left[K_N(x_i, t)\right]_{i=1}^{k-1} \\ \left[K_N(t, y_j)\right]_{j=1}^{k-1} & K_N(t, t) \end{bmatrix}.$$

Then we have $\int d(t)\mu(dt) = N$ by (4.4.9), and $\int b(t)c(t)\mu(dt) = a$ by the reproducing property (4.2.6) of K_N. Finally, by (4.4.11), we get the desired result (4.4.10). □

Proof of Proposition 4.4.1 Lemma 4.4.1 links up with the definition $\mathbb{P}_{N,\mu}$ through the formula

$$\begin{aligned} \det_{i,j=1}^{k} K_N(x_i, x_j) &= \det_{i,j=1}^{N} \sum_{m=0}^{N-1} \phi_m(x_i)\phi_m(x_j) = \left(\det_{i,j=1}^{N} \phi_{i-1}(x_j)\right)^2 \\ &= \Delta_N(x_1, \dots, x_N)^2 \prod_{n=0}^{N-1} \gamma_n^2. \end{aligned} \tag{4.4.12}$$

By integrating the extreme terms in (4.4.12) using Lemma 4.4.1, we obtain the formula (4.2.3). It follows from (4.2.3) and (4.4.12) that the density of $\mathbb{P}_{N,\mu}$ with respect to $\mu^{\otimes N}$ is $\frac{1}{N!}\det_{i,j=1}^{N} K_N(x_i, x_j)$. For $i = 1, \dots, N$, let $\lambda_i : \Omega_N \to \mathbb{R}$ be the projection to the ith coordinate. Then the random vector $(\lambda_1, \dots, \lambda_k)$ under the probability measure $\mathbb{P}_{N,\mu}$ is a random k-element sample of Λ_N, and the density with respect to $\mu^{\otimes k}$ of its law is a marginal density we can calculate using Lemma 4.4.1. Doing so, we get the claimed result.

Remark 4.4.1 *Having proved formula* (4.2.3), *we will not have any further need to mention the constants $Z_{N,\mu}$ and $\gamma_{N,\mu}$. Strikingly enough, all bookkeeping below can be handled in terms of $K_{N,\mu}(x, y)$ alone.*

4.5 Cluster functions

This section can be safely skipped by the reader. The results discussed here will not be used in later sections.

The *k-point cluster function* is defined by the formula

$$C_N(x_1, \dots, x_k) = C_{N,\mu}(x_1, \dots, x_k) = \sum_{\substack{\sigma \in S_k \\ \sigma \text{ is a } k\text{-cycle}}} (-1)^{k-1} \prod_{i=1}^{k} K_N(x_{\sigma^i(1)}, x_{\sigma^{i+1}(1)}). \tag{4.5.1}$$

Notice that the sum on the right is a subsum of that defining $\det_{i,j=1}^{k} K_N(x_i, x_j)$. Recall that the *joint cumulant* of real random variables $X_1, \ldots, X_k$ is defined by the formula

$$\mathbf{C}(X_1, \ldots, X_k) = \mathrm{i}^{-k} \frac{\partial^k}{\partial t_1 \cdots \partial t_k} \log \mathbf{E}(e^{\mathrm{i}(t_1 X_1 + \cdots + t_k X_k)})|_{t_1 = \cdots = t_k = 0}. \tag{4.5.2}$$

The joint cumulant functional $\mathbf{C}$ detects independence in the sense that if $\{1, \ldots, k\}$ is the disjoint union of nonempty sets I and J, and if the σ-fields $\sigma(\{X_i\}_{i \in I})$ and $\sigma(\{X_j\}_{j \in J})$ are independent, then $\mathbf{C}(X_1, \ldots, X_k) = 0$. Recall also that $\mathbf{C}(X_1, X_2) = \mathbf{E}(X_1 X_2) - (\mathbf{E}X_1)(\mathbf{E}X_2) = \mathrm{Cov}(X_1, X_2)$.

The next result at least partly explains the significance of the k-point cluster function.

Proposition 4.5.1 (Joint cumulants of occupancies)

$$\mathbf{C}(\mathrm{Occ}_{A_1}, \ldots, \mathrm{Occ}_{A_k}) = \int_{A_1 \times \cdots \times A_k} C_N(x_1, \ldots, x_k) \prod_{i=1}^{k} \mu(dx_i) \tag{4.5.3}$$

for any disjoint sets $A_1, \ldots, A_k \subset \mathbb{R}$.

In particular, for disjoint sets $A, B \subset \mathbb{R}$, the random variables Occ_A and Occ_B can never be positively correlated, because $C_N(x_1, x_2) = -K_N(x_1, x_2)^2 \leq 0$.

Proof We first need to recall the combinatorial description of joint cumulants. A *set partition* of $\{1, \ldots, k\}$ is a disjoint family Σ of nonempty subsets of $\{1, \ldots, k\}$ whose union is $\{1, \ldots, k\}$. Let $\mathrm{Part}(k)$ denote the family of set partitions of $\{1, \ldots, k\}$. Given

$$A = \{i_1 < \cdots < i_r\} \subset \{1, \ldots, k\}, \tag{4.5.4}$$

and real random variables $X_1, \ldots, X_k$, we define

$$\mathbf{C}(\{X_i\}_{i \in A}) = \mathbf{C}(X_{i_1}, \ldots, X_{i_r}).$$

In turn, given $\Sigma \in \mathrm{Part}(k)$, we define

$$\mathbf{C}_\Sigma(X_1, \ldots, X_k) = \prod_{A \in \Sigma} \mathbf{C}(\{X_i\}_{i \in A}).$$

Given $\Sigma, \Pi \in \mathrm{Part}(k)$ we say that Σ *refines* Π if for every $A \in \Sigma$ there exists $B \in \Pi$ such that $A \subset B$, and in this situation for short we write $\Sigma \leq \Pi$. It is well known that one can find the joint cumulant of $X_1, \ldots, X_k$ by solving the unitriangular system of linear equations

$$\prod_{A \in \Pi} \mathbf{E} \prod_{i \in A} X_i = \sum_{\substack{\Sigma \in \mathrm{Part}(k) \\ \Sigma \leq \Pi}} \mathbf{C}_\Sigma(X_1, \ldots, X_k) \quad (\Pi \in \mathrm{Part}(k)). \tag{4.5.5}$$

To bring these general definitions into contact with the k-point cluster functions, given A as in (4.5.4), we define

$$C_{N,A}(x_1,\ldots,x_k) = C_N(x_{i_1},\ldots,x_{i_r}),$$
$$\det_{i,j\in A} K_N(x_i,x_j) = \det_{\alpha,\beta=1}^{r} K_N(x_{i_\alpha},x_{i_\beta}),$$

and then given $\Pi \in \mathrm{Part}(k)$, we define

$$C_{N,\Pi}(x_1,\ldots,x_k) = \prod_{A\in\Pi} C_{N,A}(x_1,\ldots,x_k),$$
$$K_{N,\Pi}(x_1,\ldots,x_k) = \prod_{A\in\Pi} \det_{i,j\in A} K_N(x_i,x_j).$$

We claim that

$$K_{N,\Pi}(x_1,\ldots,x_k) = \sum_{\substack{\Sigma\in\mathrm{Part}(k)\\ \Sigma\leq\Pi}} C_{N,\Sigma}(x_1,\ldots,x_k). \tag{4.5.6}$$

To prove that the claim it is enough to treat the extreme case $\Pi = \{\{1,\ldots,k\}\}$. But then the left side is a sum of terms indexed by permutations σ of $\{1,\ldots,k\}$, and the right side is merely the grouping of those terms according to the set partition of $\{1,\ldots,k\}$ determined by the cycle structure of σ. The claim is proved. Formula (4.5.6) then follows from Corollary 4.4.2, Fubini's theorem, and the fact that the system of equations (4.5.5) characterizes joint cumulants.

□

4.6 Gap probabilities and Fredholm determinants

We next explain how to compute *gap probabilities*, i.e. probabilities of the form $\mathbb{P}_{N,\mu}(\mathrm{Occ}_A = 0)$ for sets $A \subset \mathbb{R}$, and more generally we compute the joint distribution of $\mathrm{Occ}_{A_1},\ldots,\mathrm{Occ}_{A_r}$ for disjoint $A_1,\ldots,A_r \subset \mathbb{R}$.

To express such probabilities, we first introduce Fredholm determinants of finite-rank kernels. Given $f,g:\mathbb{R}\to\mathbb{C}$, we define $f\otimes g:\mathbb{R}\times\mathbb{R}\to\mathbb{C}$ by $(f\otimes g)(x,y) = f(x)g(y)$. We define a *kernel* $K:\mathbb{R}\times\mathbb{R}\to\mathbb{C}$ to be a function of the form $K=\sum_{i=1}^{r} f_i\otimes g_i$ for some finitely many functions $f_i, g_i:\mathbb{R}\to\mathbb{C}$ belonging to $L^2(\mu)$. Given a kernel K we denote again by K the operator on $L^2(\mu)$ defined by the formula $(Kf)(x) = \int K(x,y)f(y)\mu(dy)$. Since kernels under our narrow working definition are always of finite rank, for any kernel K, there exists some finite-dimensional subspace $V\subset L^2(\mu)$ containing the image of the operator K, and for any such subspace V we define the *Fredholm determinant* by the formula

$$\det(1-K|L^2(\mu)) = \det((f\mapsto f-Kf): V\to V), \tag{4.6.1}$$

which is independent of the choice of V. To calculate such determinants we have the following result which in the finite-rank situation has an easy proof.

Proposition 4.6.1 (Fredholm formula)

$$\det(1 - K|L^2(\mu)) = 1 + \sum_{k=1}^{\infty} \frac{(-1)^k}{k!} \int \cdots \int \det_{i,j=1}^{k} K(x_i, x_j) \prod_{i=1}^{k} \mu(dx_i). \quad (4.6.2)$$

There are no convergence issues here – on account of our finite-rank assumption, the sum breaks off after finitely many terms.

Proof Write $K = \sum_{i=1}^{r} f_i \otimes g_i$, noting that the sum on the right side of (4.6.2) breaks off after the rth term. Let $V \subset L^2(\mu)$ be a finite-dimensional subspace to which all the functions f_i and g_i belong. Let $\{v_i\}_{i=1}^{n}$ be an orthonormal basis for V, noting that $n \geq r$. Let $\langle f, g\rangle = \int \bar{f} g d\mu$ be the inner product in $L^2(\mu)$. Recall that for rectangular matrices A and B which are $p \times q$ and $q \times p$, respectively, $\det(I_p - AB) = \det(I_q - BA)$ holds. Now using the definition of the Fredholm determinant and the preceding observation we can take the first two steps of the following calculation:

$$\begin{aligned}
\text{LHS of (4.6.2)} &= \det_{i,j=1}^{n} \left(\delta_{ij} - \langle v_i, K v_j\rangle\right) = \det_{i,j=1}^{r} \left(\delta_{ij} - \langle \bar{g}_i, f_j\rangle\right) \\
&= 1 + \sum_{k=1}^{r} (-1)^k \sum_{1 \leq j_1 < \cdots < j_k \leq r} \det_{\alpha,\beta=1}^{k} \langle \bar{g}_{j_\alpha}, f_{j_\beta}\rangle \\
&= 1 + \sum_{k=1}^{r} \frac{(-1)^k}{k!} \sum_{j_1=1}^{r} \cdots \sum_{j_k=1}^{r} \det_{\alpha,\beta=1}^{k} \langle \bar{g}_{j_\alpha}, f_{j_\beta}\rangle \\
&= \text{RHS of (4.6.2)}.
\end{aligned}$$

At the third step we use the procedure of expansion by principal minors. The remaining two steps of the calculation are trivial. □

Given $A \subset \mathbb{R}$, let $\chi_A(x) = \begin{cases} 1 & \text{if } x \in A, \\ 0 & \text{if } x \notin A, \end{cases}$ for $x \in \mathbb{R}$. Given kernels K and L, let $K \cdot L$ denote the result of multiplying K and L as functions (not operators), i.e. by definition $(K \cdot L)(x, y) = K(x, y)L(x, y)$. For any K and bounded L, again $K \cdot L$ is a kernel. The basic result is the following.

Proposition 4.6.2 (Fredholm formula for gap probabilities)

$$\mathbb{P}_{N,\mu}(\mathrm{Occ}_A = 0) = \det(1 - K_{N,\mu} \cdot \chi_A \otimes 1 | L^2(\mu)) \quad (4.6.3)$$

for any set $A \subset \mathbb{R}$.

Proof We obtain this result by combining Proposition 4.4.1, (4.6.2) and the case $k = 0$ of Lemma 4.6.1 below. □

For sets $A \subset \mathbb{R}$ we write $A^c = \mathbb{R} \setminus A$.

Lemma 4.6.1 (Refinement of "integrating out")

$$\frac{1}{(N-k)!}\int_{(A^c)^{N-k}} K_{N,\mu}\begin{pmatrix} x[1..k] & t[1..N-k] \\ y[1..k] & t[1..N-k]\end{pmatrix}\prod_{i=1}^{N-k}\mu(dt_i)$$
$$=\sum_{n=0}^{\infty}\frac{(-1)^n}{n!}\int_{A^n} K_{N,\mu}\begin{pmatrix} x[1..k] & t[1..n] \\ y[1..k] & t[1..n]\end{pmatrix}\prod_{i=1}^{n}\mu(dt_i), \tag{4.6.4}$$

for integers $0 \le k \le N$ *and sets* $A \subset \mathbb{R}$.

The sum on the right side, evaluated at $x_1 = y_1, \ldots, x_k = y_k$, is known in the theory of point processes as the *(k-point) Janossy density* associated with the process Λ_N (with respect to the set A and the reference measure μ). Concerning this topic, see [Dal02, Chap. 5] for details; alternatively, for quick overviews, see [And09, Chap. 4], [Bor03], or Chapter 11 of this volume.

Proof We proceed by descending induction on k. Note that the sum on the right breaks off at $n = N - k$. Thus, in particular, the case $k = N$ holds. Suppose now that the proposition holds for a given $0 < k \le N$. To abbreviate further, we write $\mathbf{x}_k = x[1..k]$ and $\mathbf{y}_k = y[1..k]$. Let $L_k\binom{\mathbf{x}_k}{\mathbf{y}_k}$ and $R_k\binom{\mathbf{x}_k}{\mathbf{y}_k} = \sum_{n=0}^{\infty}\frac{(-1)^n}{n!}R_{k,n}\binom{\mathbf{x}_k}{\mathbf{y}_k}$ denote the left and right sides of (4.6.4), respectively. Then, on the one hand,

$$\int_{A^c} L_k\begin{pmatrix}\mathbf{x}_{k-1}, t\\ \mathbf{y}_{k-1}, t\end{pmatrix}\mu(dt) = (N-k+1)L_{k-1}\begin{pmatrix}\mathbf{x}_{k-1}\\ \mathbf{y}_{k-1}\end{pmatrix}$$

holds trivially. But, on the other hand,

$$\int_{A^c} R_{k,n}\begin{pmatrix}\mathbf{x}_{k-1}, t\\ \mathbf{y}_{k-1}, t\end{pmatrix}\mu(dt) = \left(\int - \int_A\right) R_{k,n}\begin{pmatrix}\mathbf{x}_{k-1}, t\\ \mathbf{y}_{k-1}, t\end{pmatrix}\mu(dt)$$
$$= (N-k-n+1)R_{k-1,n}\begin{pmatrix}\mathbf{x}_{k-1}\\ \mathbf{y}_{k-1}\end{pmatrix} - R_{k-1,n+1}\begin{pmatrix}\mathbf{x}_{k-1}\\ \mathbf{y}_{k-1}\end{pmatrix}$$

holds by (4.4.10), whence the result for $k-1$, completing the induction. □

Next, we compute a generating function for the joint distribution of occupancies of disjoint sets. For variety's sake, we prove it using Corollary 4.4.2.

Proposition 4.6.3 (Joint distribution of occupancies)

$$\mathbf{E}_{N,\mu}\left(\prod_{i=1}^{r}(1-s_i)^{\mathrm{Occ}_{A_i}}\right) = \det\left(1 - K_{N,\mu}\cdot\sum_{i=1}^{r}s_i\chi_{A_i}\otimes 1\,\Big|\, L^2(\mu)\right) \tag{4.6.5}$$

for disjoint sets $A_1, \ldots, A_r \subset \mathbb{R}$ *and* $s_1, \ldots, s_r \in \mathbb{C}$.

Proof Plugging into (4.6.2) and rearranging terms yields the following evaluation of the right side of (4.6.5):

$$1+\sum_{k=1}^{\infty}\frac{(-1)^k}{k!}\sum_{\substack{k_1,\dots,k_r\geq 0\\ k=k_1+\cdots+k_r}}\frac{k!\prod_{i=1}^r s_i^{k_i}}{k_1!\cdots k_r!}\int_{\prod_{i=1}^r A_i^{k_i}}\det_{i,j=1}^k K_N(x_i,x_j)\prod_{i=1}^k \mu(dx_i).$$

The latter expression can by Corollary 4.4.2 be rewritten as

$$\sum_{k_1=0}^{\infty}\cdots\sum_{k_r=0}^{\infty}\mathbf{E}_{N,\mu}\prod_{i=1}^r(-s_i)^{k_i}\binom{\mathrm{Occ}_{A_i}}{k_i},$$

which then sums to the left side of (4.6.5). □

Let $\lambda^1>\cdots>\lambda^N$ be the eigenvalues (elements of Λ_N) arranged in descending order. We now apply (4.6.5) to calculate the distribution of the random variable λ^k under $\mathbb{P}_{N,\mu}$.

Corollary 4.6.1 (Distribution of the kth largest eigenvalue)
Put

$$D(s,x)=\det\left(1-K_{N,\mu}\cdot s\chi_{(x,\infty)}\otimes 1\,\middle|\,L^2(\mu)\right) \tag{4.6.6}$$

for $x\in\mathbb{R}$ and $s\in\mathbb{C}$. Then

$$\mathbb{P}_{N,\mu}(\lambda^k\leq x)\;=\;\sum_{i=0}^{k-1}\frac{(-1)^i}{i!}\frac{\partial^i D}{\partial s^i}(1,x) \tag{4.6.7}$$

for $x\in\mathbb{R}$.

Proof The probability on the left equals $\mathbb{P}_{N,\mu}(\mathrm{Occ}_{(x,\infty)}<k)$. The result follows after a trivial manipulation of formula (4.6.5). □

4.7 Resolvent kernels and Janossy densities

Fix a set $A\subset\mathbb{R}$ for which

$$\mathbb{P}_{N,\mu}(\mathrm{Occ}_A=0)>0. \tag{4.7.1}$$

We wish to compute

$$\frac{\mathbf{E}_{N,\mu}\mathrm{Occ}^N_{S\times(A^c)^{N-k}}}{k!(N-k)!}=\mathbb{P}\begin{pmatrix}\#\Lambda_N\cap A=k\text{ and a random}\\ \text{enumeration of }\Lambda_N\cap A\text{ belongs to }S\end{pmatrix} \tag{4.7.2}$$

for sets $S\subset A^k$. Toward the goal of computing (4.7.2), we introduce the *resolvent kernel*, which is defined as

$$\mathbb{P}_{N,\mu}(\mathrm{Occ}_A = 0)\, R_N(x, y) \tag{4.7.3}$$
$$= K_N(x, y) + \sum_{k=1}^{\infty} \frac{(-1)^k}{k!} \int_{A^k} K_N \begin{pmatrix} x & t[1..k] \\ y & t[1..k] \end{pmatrix} \prod_{i=1}^{k} \mu(dt_i).$$

As in previous cases, the sum breaks off after finitely many terms by (4.4.8). The next two lemmas characterize $R_N(x, y) = R^A_{N,\mu}(x, y)$ in other useful ways.

Lemma 4.7.1 (The resolvent equation)

$$\int_A R_N(x, t) K_N(t, y)\mu(dt) = \int_A K_N(x, t) R_N(t, y)\mu(dt) = R_N(x, y) - K_N(x, y), \tag{4.7.4}$$

provided that (4.7.1) *holds, so that* $R_N(x, y)$ *is well defined.*

We will prove Lemma 4.7.1 a little later. The lemma characterizes $R_N(x, y)$ uniquely within the class of polynomial kernels. Indeed, given another polynomial solution $\widehat{R}_N(x, y)$ of (4.7.4),

$$K_N(x, y) + \int_A R_N(x, t)\widehat{R}_N(t, y)\mu(dt)$$
$$- \int_A \int_A R_N(x, t) K_N(t, t')\widehat{R}_N(t', y)\mu(dt)\mu(dt')$$

equals both $R_N(x, y)$ and $\widehat{R}_N(x, y)$. Lemma 4.7.1 justifies our appropriation of the terminology of resolvent kernels from the theory of integral equations. For background on the latter see [Tri57].

The next result will not be used at any later point in the chapter, but is so striking and simple that it deserves mention. For $A \subset \mathbb{R}$, let $\mu_A(dx) = \chi_A(x)\mu(dx)$, where χ_A is as defined before Proposition 4.6.2.

Lemma 4.7.2 (Borodin-Soshnikov [Bor03])
One has

$$R^A_{N,\mu}(x, y) = K_{N,\mu_{A^c}}(x, y), \tag{4.7.5}$$

provided that μ_{A^c} *has infinite support so that the kernel on the right is defined.*

Proof (Lemma 4.7.1 granted) Write $\int_A = \int - \int_{A^c}$. Then use the characterization of the resolvent kernel provided by Lemma 4.7.1, along with the reproducing properties (4.2.5) and (4.2.6). □

We will prove the following result which determines the probability (4.7.2).

Proposition 4.7.1 (Joint distribution conditioned on occupancy)

$$\frac{\mathbf{E}_{N,\mu}\mathrm{Occ}^N_{S\times(A^c)^{N-k}}}{(N-k)!\mathbb{P}_{N,\mu}(\mathrm{Occ}_A = 0)} = \int_S \det_{i,j=1}^{k} R_N(x_i, x_j) \prod_{i=1}^{k} \mu(dx_i) \tag{4.7.6}$$

for sets $S \subset A^k \subset \mathbb{R}^k$, *provided that* (4.7.1) *holds.*

The proof relies on Lemma 4.7.1 stated above, and a second lemma which we next present. For the statement of the latter, we introduce the "two line notation"

$$R_N \begin{pmatrix} x[1..k] \\ y[1..k] \end{pmatrix} = R_N \begin{pmatrix} x_1 & \dots & x_k \\ y_1 & \dots & y_k \end{pmatrix} = \det_{i,j=1}^{k} R_N(x_i, y_j) \tag{4.7.7}$$

parallel to the notation (4.4.7).

Lemma 4.7.3

$$\begin{aligned} &\mathbb{P}_{N,\mu}(\mathrm{Occ}_A = 0)\, R_N \begin{pmatrix} x[1..k] \\ y[1..k] \end{pmatrix} \\ &= K_N \begin{pmatrix} x[1..k] \\ y[1..k] \end{pmatrix} + \sum_{n=1}^{\infty} \frac{(-1)^n}{n!} \int_{A^n} K_N \begin{pmatrix} x[1..k] & t[1..n] \\ y[1..k] & t[1..n] \end{pmatrix} \prod_{i=1}^{n} \mu(dt_i), \end{aligned} \tag{4.7.8}$$

for $A \subset \mathbb{R}$.

Note that the right side above is the k-point Janossy density associated with the process Λ_N under $\mathbb{P}_{N,\mu}$, the set A, and the reference measure μ.

Proof of Proposition 4.7.1 (Lemma 4.7.3 granted.) The right sides of (4.6.4) and (4.7.8) are the same. We therefore get the result by Corollary 4.4.1.

Proofs of Lemmas 4.7.1 and 4.7.3 We prove the two lemmas together because their proofs have a certain calculation in common, and in any case, Lemma 4.7.1 is needed for the proof of Lemma 4.7.3. The somewhat complicated manipulations of integral kernels below are standard in the theory of integral equations, and have been adapted from [Tri57].

To abbreviate slightly, put $K = K_N$ and $R = R_N$. Then put

$$H_n \begin{pmatrix} x[1..k] \\ y[1..k] \end{pmatrix} = \int_{A^n} K \begin{pmatrix} x[1..k] & t[1..n] \\ y[1..k] & t[1..n] \end{pmatrix} \prod_{i=1}^{n} \mu(dt_i). \tag{4.7.9}$$

The definition is meaningful for $k = 0$, in which case we write $\Delta_n = H_n()$. Also we write $H_n(x, y) = H_n \binom{x}{y}$.

Now expand the integrand on the right above by minors of the first row, to obtain the relation

$$\begin{aligned} K \begin{pmatrix} x[1..k] & t[1..n] \\ y[1..k] & t[1..n] \end{pmatrix} &= \sum_{i=1}^{k} (-1)^{i+1} K(x_1, y_i) K \begin{pmatrix} & x(1..k] & t[1..n] \\ y[1..i) & y(i..k] & t[1..n] \end{pmatrix} \\ &- \sum_{j=1}^{n} K(x_1, t_j) K \begin{pmatrix} t_j & x(1..k] & t[1..j) & t(j..n] \\ & y[1..k] & t[1..j) & t(j..n] \end{pmatrix}. \end{aligned}$$

Then, integrate $t[1..n]$ over A^n, to obtain the relation

$$H_n\begin{pmatrix} x[1..k] \\ y[1..k] \end{pmatrix} = \sum_{i=1}^{k}(-1)^{i+1}K(x_1, y_i)\, H_n\begin{pmatrix} x(1..k] \\ y[1..i)\; y(i..k] \end{pmatrix} - n\int_A K(x_1, t)\, H_{n-1}\begin{pmatrix} t & x(1..k] \\ & y[1..k] \end{pmatrix}\mu(dt). \tag{4.7.10}$$

The latter is the key to the proofs of both lemmas.

In the case where $k = 1$ the preceding specializes to

$$H_n(x, y) - \Delta_n \cdot K(x, y) = -n\int_A K(x, t)\, H_{n-1}(t, y)\mu(dt). \tag{4.7.11}$$

Then (4.7.11) proves (4.7.4), since

$$\mathbb{P}_{N,\mu}(\mathrm{Occ}_A = 0) = \sum_{n=0}^{\infty}\frac{(-1)^n}{n!}\Delta_n, \quad \mathbb{P}_{N,\mu}(\mathrm{Occ}_A = 0)\, R(x, y) = \sum_{n=0}^{\infty}\frac{(-1)^n}{n!}H_n(x, y).$$

(The first equality of (4.7.4) is trivial by symmetry of K and R.)

We turn to the proof of (4.7.8).

Proof We proceed by induction on k. The induction base $k = 1$ holds by definition of R, so assume that $k > 1$. Express the right side of (4.7.8) as $\mathbb{P}_{N,\mu}(\mathrm{Occ}_A = 0)\,\widetilde{R}\begin{pmatrix} x[1..k] \\ y[1..k] \end{pmatrix}$. Now multiply both sides of the identity (4.7.10) by $\frac{(-1)^n}{n!}$, sum on n from 0 to ∞, and divide through by $\mathbb{P}_{N,\mu}(\mathrm{Occ}_A = 0)$, thus obtaining the relation

$$\widetilde{R}\begin{pmatrix} x[1..k] \\ x[1..k] \end{pmatrix} = \sum_{i=1}^{k}(-1)^{i+1}K(x_1, y_i)\,\widetilde{R}\begin{pmatrix} x(1..k] \\ y[1..i)\; y(i..k] \end{pmatrix} + \int_A K(x_1, u)\,\widetilde{R}\begin{pmatrix} u & x(1..k] \\ & y[1..k] \end{pmatrix}\mu(du).$$

Next, replace x_1 in the identity above by t, then multiply both sides of the relation above by $R(x_1, t)$, integrate t over A, and apply (4.7.4) on the right side. We get

$$\int_A R(x_1, t)\,\widetilde{R}\begin{pmatrix} t & x(1..k] \\ & y[1..k] \end{pmatrix}\mu(dt) = \sum_{i=1}^{k}(-1)^{i+1}(R-K)(x_1, y_i)\,\widetilde{R}\begin{pmatrix} x(1..k] \\ y[1..i)\; y(i..k] \end{pmatrix} + \int_A (R-K)(x_1, t)\,\widetilde{R}\begin{pmatrix} t & x(1..k] \\ & y[1..k] \end{pmatrix}\mu(dt).$$

Finally, add the last two identities and make the evident cancellations, thus obtaining the relation

$$\tilde{R}\begin{pmatrix} x[1..k] \\ y[1..k] \end{pmatrix} = \sum_{i=1}^{k} (-1)^{i+1} R(x_1, y_i)\, \tilde{R}\begin{pmatrix} x(1..k] \\ y[1..i) \; y(i..k] \end{pmatrix}.$$

By induction on k, the right side is the expansion of $R\begin{pmatrix} x[1..k] \\ y[1..k] \end{pmatrix}$ by minors of its first row, which proves (4.7.8). □

4.8 Spacings

We follow, albeit with somewhat different language and with significant simplifications, the treatment of spacings in [Dei99]. For an approach to spacings far more sophisticated than the one here, see [Kat99].

For $s > 0$ and $A \subset \mathbb{R}$, let

$$S_N(s, A) = \{(x_1, \ldots, x_N) \in \mathbb{R}^N \mid A \cap (x_N, x_N + s] \cap \{x_1, \ldots, x_{N-1}\} = \emptyset, \quad x_N \in A\}.$$

We wish to compute

$$\begin{aligned} &\frac{1}{N!} \mathbf{E}_{N,\mu} \mathrm{Occ}^N_{S_N(s,A)} \\ &= \mathbb{P}\begin{pmatrix} \text{a randomly sampled element of } \Lambda_N \\ \text{belongs to } A \text{ and has no neighbours} \\ \text{on the right in } \Lambda_N \cap A \text{ at distance} \le s \end{pmatrix}. \end{aligned} \tag{4.8.1}$$

Toward the goal of computing (4.8.1), we introduce the technique of conditioning on an eigenvalue, which is of independent interest. For any $t \in \mathbb{R}$ put

$$\mu^t(dx) = (x - t)^2 \mu(dx),$$

and given a set $S \subset \mathbb{R}^N$, let

$$S^t = \{(x_1, \ldots, x_{N-1}) \in \mathbb{R}^{N-1} : (x_1, \ldots, x_{N-1}, t) \in S\}.$$

Proposition 4.8.1 (Conditioning on an eigenvalue)

$$\mathbf{E}_{N,\mu} \mathrm{Occ}^N_S = \int \mathbf{E}_{N-1,\mu^t} \mathrm{Occ}^{N-1}_{S^t} \, K_{N,\mu}(t,t) \mu(dt) \tag{4.8.2}$$

for sets $S \subset \mathbb{R}^N$.

Proof Use Corollary 4.4.1, Fubini's theorem, and Lemma 4.8.1 immediately below. □

Lemma 4.8.1

$$K_{N,\mu}\begin{pmatrix} x[1..N) & t \\ x[1..N) & t \end{pmatrix} = K_{N-1,\mu^t}\begin{pmatrix} x[1..N) \\ x[1..N) \end{pmatrix} K_{N,\mu}(t,t) \prod_{i=1}^{N-1} (x_i - t)^2. \tag{4.8.3}$$

Proof The functions on both sides of (4.8.3) are polynomial, and $K_{N,\mu}(t,t)$ does not vanish identically. So there is no loss of generality in assuming that $K_{N,\mu}(t,t) \neq 0$. Put $\psi_n(x) = (x-t)\phi_{n-1,\mu^t}(x)$ for $n > 0$, where $\{\phi_{n,\mu^t}\}_{n=0}^{\infty}$ is the family of orthogonal polynomials associated with μ^t. Let $\psi_0(x) = \frac{K_{N,\mu}(x,t)}{\sqrt{K_{N,\mu}(t,t)}}$. Then, using the reproducing properties (4.2.5) and (4.2.6) of $K_N(x,y)$, and the orthonormality of $\{\phi_{n,\mu^t}\}$ in $L^2(\mu^t)$, one can verify that $\{\psi_i\}_{i=0}^{N-1}$ is an orthonormal basis for the span of $\{x^i\}_{i=0}^{N-1}$ in $L^2(\mu)$. It follows that $K_{N,\mu}(x,y) = \sum_{i=0}^{N-1} \psi_i(x)\psi_i(y)$, whence (4.8.3), after a calculation similar to (4.4.12). □

The next and final result of these notes makes the probability (4.8.1) explicit.

Proposition 4.8.2

$$\frac{1}{(N-1)!}\mathbf{E}_{N,\mu}\mathrm{Occ}^{N}_{S_N(s,A)} \tag{4.8.4}$$
$$= \int_A \left(\sum_{n=0}^{\infty} \frac{(-1)^n}{n!} \int_{(A\cap(t,t+s])^n} K_{N,\mu}\begin{pmatrix} x[1..n] & t \\ x[1..n] & t \end{pmatrix} \prod_{i=1}^{n} \mu(dx_i) \right) \mu(dt)$$

for $s > 0$.

Proof Consider (4.8.2) in the special case $S = S_N(s,A)$. We have

$$\mathbf{E}_{N-1,\mu^t}\mathrm{Occ}^{N-1}_{S_N(s,A)^t} = \begin{cases} (N-1)!\mathbb{P}_{N-1,\mu^t}(\mathrm{Occ}_{A\cap(t,s+t]} = 0) & \text{if } t \in A, \\ 0 & \text{if } t \notin A. \end{cases} \tag{4.8.5}$$

Use (4.6.7) for $k = 1$ to rewrite the right side of (4.8.5) in terms of K_{N-1,μ^t}, μ^t, $K_{N,\mu}$ and μ. Finally, use (4.8.3) to rewrite the integral on the right side of (4.8.2) solely in terms of $K_{N,\mu}$ and μ, thus obtaining (4.8.4). □

References

[And09] G. W. Anderson, A. Guionnet, and O. Zeitouni, *An Introduction to Random Matrices*, Cambridge studies in advanced mathematics **118**, Cambridge University Press (2010).

[And03] T. W. Anderson, *An Introduction to Multivariate Statistical Analysis*, Wiley-Interscience, 3rd Edition, Hoboken, NJ, 2003.

[Bor03] A. Borodin, A. Soshnikov, J. Statist. Phys. **113** (2003), 595 [arXiv:math-ph/0212063].

[Dei99] P. Deift, *Orthogonal Polynomials and Random Matrices: a Riemann–Hilbert Approach*, Courant Lecture Notes in Mathematics, **3**, New York University, Courant Institute of Mathematical Sciences, New York; American Mathematical Society, Providence, RI (1999).

[Hou09] J. B. Hough, M. Krishnapur, Y. Peres, and B. Virag, *Zeros of Gaussian Analytic Functions and Determinantal Point Processes*, University Lecture Series, **51**, American Mathematical Society, Providence, RI (2009).

[Dal02] D. J. Daley, D. Vere-Jones, *An Introduction to the Theory of Point Processes. Vol. I: Elementary Theory and Methods*, Springer, 2nd Edition, New York (2002).

[Erd53] A. Erdelyi, W. Magnus, F. Oberhettinger, F. Tricomi, *Higher Transcendental Functions, Vol. II,* reprint of the 1953 original, Robert E. Krieger Publishing Co., Inc., Melbourne, Fla. (1981).

[For10] P. Forrester, *Log-gases and Random Matrices (LMS-34)*, Princeton University Press, Princeton (2010).

[Kat99] N. M. Katz, P. Sarnak, *Random Matrices, Frobenius Eigenvalues, and Monodromy,* American Mathematical Society Colloquium Publications, **45**, American Mathematical Society, Providence, RI (1999).

[Meh04] M. L. Mehta, *Random Matrices,* Academic Press, 3rd Edition, London (2004).

[Tra98] C. Tracy and H. Widom, J. Stat. Phys. **92** (1998) 809 [arXiv:solv-int/9804004].

[Tri57] F. Tricomi, *Integral Equations,* reprint of the 1957 original, Dover Publications, New York, NY (1985).

·5·

Spectral statistics of orthogonal and symplectic ensembles

Mark Adler

Abstract

We focus on providing a direct approach to the computing of scalar and matrix kernels, respectively for the unitary ensembles on the one hand and the orthogonal and symplectic ensembles on the other hand, rather than surveying many different techniques and historical developments. This leads to correlation functions and gap probabilities. In the classical cases (Hermite, Laguerre, and Jacobi) we express the matrix kernels for the orthogonal and symplectic ensemble in terms of the scalar kernel for the unitary case, using the relation between the classical orthogonal polynomials going with the unitary ensembles and the skew-orthogonal polynomials going with the orthogonal and symplectic ensembles.

5.1 Introduction

In Chapter 4, it was shown that given a probability distribution function (pdf) of the form

$$P_N^{(\beta)}(x_1, \ldots, x_N) = Z_N^{-1}|\Delta_N(x_1, \ldots, x_N)|^\beta \prod_{j=1}^{N} w(x_j), \quad \beta = 2, \tag{5.1.1}$$

for $(x_1, \ldots, x_N) \in \mathbb{R}^N$, Δ_N the Vandermonde determinant, $w(x)$ a non-negative weight function on $\mathbb{R}$, then it can be rewritten

$$\begin{aligned} P_N^{(2)}(x_1, \ldots, x_N) &= \frac{1}{N!}\det[K_N^{(2)}(x_i, x_j)]_{1\le i,j\le N} \\ K_N^{(2)}(x, y) &= \sum_{i=0}^{N-1} \varphi_i(x)\varphi_i(y), \quad \varphi_i(x) = (w(x))^{\frac{1}{2}} p_i(x), \end{aligned} \tag{5.1.2}$$

with $p_i(x)$ being an i-th degree polynomial, orthonormal with respect to the $\mathbb{R}$-weight $w(x)$, i.e.

$$\int_{\mathbb{R}} p_i(x) p_j(x) w(x)\, dx = \int_{\mathbb{R}} \varphi_i(x)\varphi_j(x)\, dx = \delta_{ij}. \tag{5.1.3}$$

It was also mentioned in Chapter 4 that this example included many interesting examples in random matrix theory and in particular N-by-N random Hermitian matrices, and that the set of random matrices handled by a pdf of the form (5.1.1) are known as the unitary ensembles. It was also shown that as a consequence of the reproducing property of the kernels, the n-point correlation function

$$R_n(x_1,\ldots,x_n) := \frac{N!}{(N-n)!}\int\cdots\int P_N^{(2)}(x_1,\ldots,x_N)\,dx_{n+1}\ldots dx_N$$
$$= \det[K_N^{(2)}(x_i,x_j)]_{1\le i,j\le n}, \tag{5.1.4}$$

which roughly speaking is the probability density that n of the eigenvalues, irrespective of order, lie in infinitesimal neighbourhoods of $x_1,\ldots,x_n$. (Since it integrates out to $N!/(N-n)!$, it is not a probability density.)

Now because of the Weyl integration formula [Hel62], [Hel84], when considering the case of conjugation invariant pdf on the ensembles of real symmetric or self-dual Hermitian quaternionic matrices,[1] viewed as the tangent space at the identity of the associated symmetric spaces, one finds formula (5.1.1) with $\beta = 1$ or 4, respectively, for the pdf.

In particular they also come up in the so-called chiral models in the physics literature, in which case the weight $w(x)$ contains the factor x^a, see for example [Sen98], [Ake05].

In the next section we deal with all cases $\beta = 1, 2, 4$, and although we essentially follow the elegant articles of Tracy–Widom [Tra98] and [Wid99], the reader should bear in mind that major difficulties in the theory have been resolved gradually in developments since the 1960s. I am indebted to the referee for the following brief summary of these developments (as well as for finding numerous misprints). The Fredholm determinant method for treating the gap probability was invented in [Gau61]. The Pfaffian forms of the correlation functions for the orthogonal and symplectic ensembles were discovered in [Dys70] for the circular ensembles, and then applied to the Gaussian ensembles in [Meh71] and extended to the cases with an arbitrary weight $w(x)$ in [Dys72]. The studies on the classical cases (besides the Gaussian ensembles) were initiated in [Meh76] on the Legendre ensembles. Then the general Jacobi and Laguerre ensembles were studied (and Equation (5.3.5) was found) in [Nag91] and [Nag95] by T. Nagao and M. Wadati and T. Nagao and P.J. Forrester, respectively.

[1] The N-by-N matrices H with quaternionic elements q_{ij} are realized: Eq. $H = [q_{ij}]_{1\le i,j\le N}$, $q_{ij} \mapsto \begin{bmatrix} z_{ij} & w_{ij} \\ -\bar{w}_{ij} & \bar{z}_{ij}\end{bmatrix}$, $q_{ji} \mapsto \begin{bmatrix} \bar{z}_{ij} & -w_{ij} \\ \bar{w}_{ij} & z_{ij}\end{bmatrix}$, $i \le j$ in which case the eigenvalues are doubly degenerate [Meh04].

5.2 Direct approach to the kernel

In this section, we give a general method which generalizes the results of Chapter 4 and works for all three cases $\beta = 1, 2, 4$; following the approach of Tracy–Widom [Tra98], [Wid99]. We now state:

Theorem 5.2.1 *Consider the pdf of (5.1.1) for the cases $\beta = 1, 2, 4$. Then we have for the expectation*

$$\mathbf{E}\left(\prod_{j=1}^{N}(1+f(x_j))\right) = \int \dots \int P_N^{(\beta)}(x_1, \dots, x_N) \prod_{j=1}^{N}(1+f(x_j))\, dx_j$$
$$= \begin{cases} \det(I + K_N^{(\beta)} f) & \beta = 2, \\ (\det(I + K_N^{(\beta)} f))^{\frac{1}{2}} & \beta = 1, 4, \end{cases} \tag{5.2.1}$$

where $K_N^{(\beta)}$ is for $\beta = 2$ an operator on $L^2(\mathbb{R})$ with kernel $K_N^{(2)}(x, y)$ and f is the operator, multiplication by f, while for $\beta = 1, 4$, $K_N^{(\beta)}$ is a matrix kernel on $L^2(\mathbb{R}) \oplus L^2(\mathbb{R})$. The kernels are specified below:

$$K_N^{(2)}(x, y) = S_N^{(2)}(x, y), \qquad \underline{\beta = 2} \tag{5.2.2}$$

$$K_N^{(\beta)}(x, y) = \begin{pmatrix} S_N^{(\beta)}(x, y) & S_N^{(\beta)} D(x, y) \\ I S_N^{(\beta)}(x, y) - \delta_{\beta,1}\epsilon(x-y) & S_N^{(\beta)}(y, x) \end{pmatrix}, \qquad \underline{\beta = 1, 4} \tag{5.2.3}$$

with

$$S_N^{(2)}(x, y) = \sum_{i,j=0}^{N-1} \varphi_i(x) \mu_{ij}^{(2)} \varphi_j(y), \tag{5.2.4}$$

$\varphi_i(x) = (w(x))^{1/2} p_i(x)$, the $p_i(x)$ any polynomials of degree i, with the symmetric matrix $\mu^{(2)}$ given by

$$[(\mu^{(2)})^{-1}]_{ij} = \int_{\mathbb{R}} \varphi_i(x)\varphi_j(x)\, dx =: \langle \varphi_i, \varphi_j \rangle_{(2)}, \tag{5.2.5}$$

and (for N even)

$$S_N^{(1)}(x, y) = -\sum_{i,j=0}^{N-1} \varphi_i(x) \mu_{ij}^{(1)} \epsilon\varphi_j(y), \tag{5.2.6}$$

$$I S_N^{(1)}(x, y) = -\sum_{i,j=0}^{N-1} \epsilon\varphi_i(x) \mu_{ij}^{(1)} \epsilon\varphi_j(y) = \epsilon S_N^{(1)}(x, y),$$

$$S_N^{(1)} D(x, y) = \sum_{i,j=0}^{N-1} \varphi_i(x)\mu_{ij}^{(1)}\varphi_j(y) = -\frac{\partial}{\partial y} S_N^{(1)}(x, y),$$

with $\varphi_i(x) = w(x)p_i(x)$*, the* $p_i(x)$ *being any polynomials of degree* i, $\epsilon(x) = \frac{1}{2}sgn(x)$, ϵ = *the integral operator with kernel* $\epsilon(x-y)$ *and the skew-symmetric matrix* $\mu^{(1)}$ *is given by*

$$[(\mu^{(1)})^{-1}]_{ij} = \iint \epsilon(x-y)\varphi_i(x)\varphi_j(y)\,dxdy =: \langle \varphi_i, \varphi_j \rangle_{(1)}, \tag{5.2.7}$$

and finally ($f'(x) = \frac{d}{dx}f(x)$)

$$2S_N^{(4)}(x, y) = \sum_{i,j=0}^{2N-1} \varphi_i'(x)\mu_{i,j}^{(4)}\varphi_j(y), \tag{5.2.8}$$

$$2IS_N^{(4)}(x, y) = \sum_{i,j=0}^{2N-1} \varphi_i(x)\mu_{i,j}^{(4)}\varphi_j(y) = 2\int_y^x S_N^{(4)}(v, y)\,dv,$$

$$2S_N^{(4)} D(x, y) = -\sum_{i,j=0}^{2N-1} \varphi_i'(x)\mu_{i,j}^{(4)}\varphi_j'(y) = -2\frac{\partial}{\partial y} S_N^{(4)}(x, y),$$

with $\varphi_i(x) = (w(x))^{1/2}p_i(x)$*, where the* $p_i(x)$ *are arbitrary polynomials of degree* i *and* $\mu^{(4)}$ *is the skew-symmetric matrix given by*

$$[(\mu^{(4)})^{-1}]_{ij} = \frac{1}{2}\int (\varphi_i(x)\varphi_j'(x) - \varphi_i'(x)\varphi_j(x))\,dx =: \langle \varphi_i, \varphi_j \rangle_{(4)}. \tag{5.2.9}$$

Remark 5.2.1 *Note by the definition of* $\mu^{(2)}$ *(5.2.5), we have the following reproducing property:*

$$\langle S_N^{(2)}(x, \cdot), \varphi_k \rangle_{(2)} \equiv \varphi_k(x), \qquad 0 \le k \le N-1,$$

which, given its degree modulo $(w(x)w(y))^{1/2}$*, uniquely characterizes* $S_N^{(2)}(x, y)$ *as the Christoffel–Darboux kernel of* $\langle\ ,\ \rangle_2$*, i.e.,*

$$S_N^{(2)}(x, y) = \sum_{i=0}^{N-1} \varphi_i(x)\varphi_j(y), \qquad \langle \varphi_i, \varphi_j \rangle_{(2)} = \delta_{ij}.$$

In other words, the kernel is insensitive to the choice of $\mu^{(2)}$, so we may as well take $\mu^{(2)} = I_N$.

Similarly, we have for $\beta = 1$, the reproducing property

$$\langle S_N^{(1)} D(x, \cdot), \varphi_k \rangle_{(1)} = \varphi_k(x), \qquad 0 \le k \le N-1,$$

which now forces $S_N^{(1)} D(x, y)$ to be the Christoffel–Darboux kernel for the skew-symmetric inner product $\langle\ ,\ \rangle_{(1)}$ namely[2]

$$S_N^{(1)} D(x, y) = \sum_{i=0}^{N-1} (\varphi_{2i}(y)\varphi_{2i+1}(x) - \varphi_{2i}(x)\varphi_{2i+1}(y)), \quad \langle \varphi_i, \varphi_j \rangle_{(1)} = J_{ij},$$

and so again the kernel is insensitive to the choice of $\mu^{(1)}$, so we may as well set $\mu^{(1)} = J_N$.

Finally, we also have for $\beta = 4$ the reproducing property

$$\langle I S_N^{(4)}(x, \cdot), \varphi_k \rangle_{(4)} = \varphi_k(x), \quad 0 \le k \le 2N - 1,$$

which as before forces $I S_N^{(4)}(x, y)$ to be the Christoffel–Darboux kernel for the skew-symmetric inner product $\langle\ ,\ \rangle_{(4)}$, i.e.

$$I S_N^{(4)}(x, y) = \sum_{i=0}^{2N-1} (\varphi_{2i}(y)\varphi_{2i+1}(x) - \varphi_{2i}(x)\varphi_{2i+1}(y)), \quad \langle \varphi_i, \varphi_j \rangle_{(4)} = J_{ij}.$$

Thus in all three cases, $\beta = 1, 2, 4$, the Christoffel–Darboux kernel of the inner product matrix $\mu^{(\beta)}$, completely determines the kernel $K_N^{(\beta)}(x, y)$, and in fact for the "classical" cases, it is easily determined. In particular the $K_N^{(\beta)}(x, y)$ are insensitive to the choice of the polynomials $p_i(x)$ and for the classical cases we shall see they are all closely related to the $\beta = 2$ kernel.

We shall now sketch a proof of this theorem, following [Tra98], first providing a necessary lemma, of de Bruijn, found in [deB55].

Lemma 5.2.1 *We have the following three identities involving N-fold integrals with determinantal entries*

$$\int \cdots \int \det[\varphi_i(x_j)]_{1\le i,j\le N} \det[\psi_i(x_j)]_{1\le i,j\le N}\, dx_1 \ldots dx_N = N! \det\left[\int \varphi_i(x)\psi_j(x)\, dx\right]_{1\le i,j\le N}$$

$$\int \cdots \int_{x_1 \le x_2 \le \ldots \le x_N} \det[\varphi_i(x_j)]_{1\le i,j\le N}\, dx_1 \ldots dx_N = \mathrm{Pf}\left[\int\int sgn(y - x)\varphi_i(x)\varphi_j(y)\, dy dx\right]_{1\le i,j\le N}$$

$$\int \cdots \int \det[(\varphi_i(x_j), \psi_i(x_j))]_{\substack{1\le i\le 2N \\ 1\le j\le N}}\, dx_1 \ldots dx_N = N! \mathrm{Pf}\left[\int (\varphi_i(x)\psi_j(x) - \varphi_j(x)\psi_i(x))\, dx\right]_{1\le i,j\le 2N}$$

[2] $J = J_N = I_N \otimes \begin{bmatrix} 0 & 1 \\ -1 & 0 \end{bmatrix}$ = the $2N$-by-$2N$ symplectic matrix, while the φ_{2i+1} is a well-defined module $\varphi_{2i+1} \mapsto \frac{\varphi_{2i+1} + \gamma_i \varphi_{2i}}{\delta_i}$, $\varphi_{2i} \mapsto \delta_i \varphi_{2i}$.

where $PfA = (\det A)^{1/2}$*, for* A *a skew-symmetric matrix, and the second identity requires* N *to be even, although a version exists for* N *odd (see [Meh04], chapter 5.5).*

Sketch of Proof of Theorem 5.2.1 We first do the $\beta = 2$ case, the other cases being technically more complicated but conceptually no different. From (5.1.1) and Lemma 5.2.1,

$$\mathbf{E}\left(\prod_{j=1}^{N}(1+f(x_j))\right)$$
$$= \frac{1}{Z_N}\int\cdots\int \det[x_j^{i-1}]_{1\le i,j\le N}\det[x_j^{i-1}w(x_j)(1+f(x_j))]_{1\le i,j\le N}\, dx_1\ldots dx_N$$
$$= \frac{1}{Z'_N}\det\left[\int x^{i+j}w(x)(1+f(x))\, dx\right]_{0\le i,j\le N-1}$$

(after replacing $x^i(w(x))^{1/2} \mapsto \varphi_i(x) = p_i(x)(w(x))^{1/2}$, the $p_i(x)$ being arbitrary polynomials of degree i)

$$= C_N\det[\textstyle\int \varphi_i(x)\varphi_j(x)(1+f(x))\, dx]_{0\le i,j\le N-1}$$
$$= C'_N\det\left[\delta_{ij} + \int\sum_{k=0}^{N-1}\mu_{ik}^{(2)}\varphi_k(x)\varphi_j(x)f(x)\, dx\right]_{0\le i,j\le N-1}$$
$$(C'_N = 1, \text{ since the L.H.S.} = 1 \text{ when } f = 0)$$
$$= \det(I + K_N^{(2)}f).$$

In the last step, we have applied the fundamental identity $\det(I + AB) = \det(I + BA)$ for arbitrary Hilbert-Schmidt operators, true as long as the products make sense. Indeed, set $A : L^2(\mathbb{R}) \mapsto \mathbb{R}^N$, $B : \mathbb{R}^N \mapsto L^2(\mathbb{R})$, with

$$A(i,x) = \sum_{k=0}^{N-1}\mu_{ik}^{(2)}\varphi_k(x), \qquad B(x,j) = \varphi_j(x)f(x)$$

i.e.

$$Ah(x) = \left(\int\sum_{k=0}^{N-1}\mu_{ik}^{(2)}\varphi_k(x)h(x)\, dx\right)_{i=0}^{N-1}, \quad B(v) = \sum_{j=0}^{N-1}v_j\varphi_j(x)f(x)$$

so

$$AB(i,j) = \int\sum_{k=0}^{N-1}\mu_{ik}^{(2)}\varphi_k(x)\varphi_j(x)\, dx$$

$$BA(x,y) = \sum_{i,j=0}^{N-1}\varphi_i(x)\mu_{ij}^{(2)}\varphi_j(y) = K_N^{(2)}(x,y),$$

yielding the $\beta = 2$ case.

Now consider the $\beta = 4$ case, and observe the crucial identity

$$\Delta_N^4(x) = \det[(x_j^i, (x_j^i)')]_{\substack{0\le i\le 2N-1\\ 1\le j\le N}}$$

(a consequence of L'Hôpital's rule), but replacing in the above $x^i \mapsto p_i(x)$, the $p_i(x)$ being arbitrary polynomials of degree i, in which case $\Delta_N(x) \mapsto$ constant $\Delta_N(x)$. Then find, using (5.1.1) for $\beta = 4$ and Lemma 5.2.1, upon setting $\varphi_i(x) = (w(x))^{1/2} p_i(x)$, that

$$\begin{aligned}
&\left(E\left(\prod_{i=1}^N (1 + f(x_i))\right)\right)^2 \\
&\quad = C_N \det\left[\int \frac{1}{2}(\varphi_i(x)\varphi_j'(x) - \varphi_i'(x)\varphi_j(x))(1 + f(x))\, dx\right]_{0\le i,j\le 2N-1} \\
&\quad = C_N' \det\left[\delta_{ij} + \int (\tilde{\varphi}_i(x)\varphi_j'(x) - \tilde{\varphi}_i'(x)\varphi_j(x))\frac{f(x)}{2}\, dx\right]_{0\le i,j\le 2N-1} \\
&\quad \left(\tilde{\varphi}_i(x) = \sum_{k=0}^{2N-1} \mu_{ik}^{(4)}\varphi_k(x) \quad C_N' = 1 \text{ by setting } f = 0\right) \\
&\quad = \det(I + K_N^{(4)} f),
\end{aligned}$$

once again using $\det(I + AB) = \det(I + BA)$. Indeed, set

$$A : L^2(\mathbb{R}) \oplus L^2(\mathbb{R}) \mapsto \mathbb{R}^{2N}, \qquad B : \mathbb{R}^{2N} \mapsto L^2(\mathbb{R}) \oplus L^2(\mathbb{R})$$

with

$$A(i, x) = \frac{f(x)}{2}(\tilde{\varphi}_i(x), -\tilde{\varphi}_i'(x)), \qquad B(x, i) = \begin{pmatrix}\varphi_i'(x)\\ \varphi_i(x)\end{pmatrix}, \qquad 0 \le i \le 2N - 1,$$

and so

$$A\begin{pmatrix}h_1\\ h_2\end{pmatrix} = \frac{1}{2}\left(\int f(x)\tilde{\varphi}_i(x)h_1(x)\, dx \quad - \int f(x)\tilde{\varphi}_i'(x)h_2(x)\, dx\right)_{i=0}^{2N-1},$$

$$B(v_0, \ldots, v_{2N-1})^T = \sum_{i=0}^{2N-1} v_i \begin{pmatrix}\varphi_i'(x)\\ \varphi_i(x)\end{pmatrix},$$

hence

$$\begin{aligned}
AB(i, j) &= \int \frac{f(x)}{2}(\tilde{\varphi}_i(x), -\tilde{\varphi}_i'(x))\begin{pmatrix}\varphi_j'(x)\\ \varphi_j(x)\end{pmatrix} dx \\
&= \int \frac{f(x)}{2}(\tilde{\varphi}_i(x)\varphi_j'(x) - \tilde{\varphi}_i'(x)\varphi_j(x))\, dx,
\end{aligned}$$

while

$$B A(x, y) = \sum_{i=0}^{2N-1} B(x, i) A(i, y) = \frac{1}{2} \begin{pmatrix} \sum_{i=0}^{2N-1} \varphi_i'(x)\tilde{\varphi}_i(y), & -\sum_{i=0}^{2N-1} \varphi_i'(x)\tilde{\varphi}_i'(y) \\ \sum_{i=0}^{2N-1} \varphi_i(x)\tilde{\varphi}_i(y), & -\sum_{i=0}^{2N-1} \varphi_i(x)\tilde{\varphi}_i'(y) \end{pmatrix} f(y),$$

yielding the $\beta = 4$ case.

Finally consider the $\beta = 1$ case, with N even, and so from (5.1.1) and Lemma 5.2.1, find

$$\left(\mathbf{E} \left(\prod_{j=1}^{N} (1 + f(x_j)) \right) \right)^2$$
$$= \left(\frac{1}{Z_N} \int \cdots \int_{x_1 \le \ldots \le x_N} \prod_{i<j} (x_j - x_i) \prod_{j=1}^{N} (w(x_j)(1 + f(x_j)) \, dx_1 \ldots dx_N \right)^2$$
$$= C_N \left(\int \cdots \int \det[p_{i-1}(x_j) w(x_j)(1 + f(x_j))]_{1 \le i, j \le N} \, dx_1 \ldots dx_N \right)^2$$
$$= C_N' \det \left[\iint \epsilon(x - y) \varphi_i(x) \varphi_j(y)(1 + f(x))(1 + f(y)) \, dx dy \right]_{0 \le i, j \le N-1}$$

(setting $\varphi_i(x) = w(x) p_i(x)$, $p_i(x)$ an arbitrary polynomial of degree i, $\tilde{\varphi}_i(x) = \sum_{j=0}^{N-1} \mu_{ij}^{(1)} \varphi_j(x)$)

$$= \det \left[\delta_{ij} + \int (f \tilde{\varphi}_i \epsilon \varphi_j - f \varphi_j \epsilon \tilde{\varphi}_i - f \varphi_j \epsilon (f \tilde{\varphi}_i)) \, dx \right]_{0 \le i, j \le N-1}$$

(remember $\epsilon f = \int \epsilon(x - y) f(y) \, dy = \frac{1}{2} \int \operatorname{sgn}(x - y) f(y) \, dy$)

$$= \det(I + K_N^{(1)} f),$$

once again using $\det(I + AB) = \det(I + BA)$. Indeed set

$$A : L^2(\mathbb{R}) \oplus L^2(\mathbb{R}) \mapsto \mathbb{R}^N, \quad B : \mathbb{R}^N \mapsto L^2(\mathbb{R}) \oplus L^2(\mathbb{R})$$

with

$$A(i, x) = f(-\epsilon \tilde{\varphi}_i - \epsilon (f \tilde{\varphi}_i), \tilde{\varphi}_i), \quad B(x, i) = \begin{pmatrix} \varphi_i \\ \epsilon \varphi_i \end{pmatrix}$$

and so

$$\begin{aligned}
\det(I+BA) &= \det\left(I+\sum_{i=0}^{N-1} B(x,i)A(i,y)\right)\\
&= \det\begin{pmatrix} I-\sum \varphi_i\otimes(f\epsilon\tilde{\varphi}_i+f\epsilon(f\tilde{\varphi}_i)), & \sum\varphi_i\otimes f\tilde{\varphi}_i\\ -\sum\epsilon\varphi_i\otimes(f\epsilon\tilde{\varphi}_i+f\epsilon(f\tilde{\varphi}_i)), & I+\sum\epsilon\varphi_i\otimes f\tilde{\varphi}_i\end{pmatrix}\\
&= \det\left(\begin{pmatrix} I-\sum\varphi_i\otimes f\epsilon\tilde{\varphi}_i, & \sum\varphi_i\otimes f\tilde{\varphi}_i\\ -\sum\epsilon\varphi_i\otimes f\epsilon\tilde{\varphi}_i-\epsilon f, & I+\sum\epsilon\varphi_i\otimes f\tilde{\varphi}_i\end{pmatrix}\begin{pmatrix} I & 0\\ \epsilon f & I\end{pmatrix}\right)\\
&= \det\begin{pmatrix} I-\sum\varphi_i\otimes f\epsilon\tilde{\varphi}_i, & \sum\varphi_i\otimes f\tilde{\varphi}_i\\ -\sum\epsilon\varphi_i\otimes f\epsilon\tilde{\varphi}_i-\epsilon f, & I+\sum\epsilon\varphi_i\otimes f\tilde{\varphi}_i\end{pmatrix}
\end{aligned}$$

(using $\epsilon^T=-\epsilon$, $\det XY=\det X\cdot\det Y$, $\det\left(\begin{smallmatrix} I & 0\\ \epsilon f & I\end{smallmatrix}\right)=1$)

$$\begin{aligned}
&= \det\left(I+\begin{pmatrix} -\sum\varphi_i\otimes\epsilon\tilde{\varphi}_i, & \sum\varphi_i\otimes\tilde{\varphi}_i\\ -\sum\epsilon\varphi_i\otimes\epsilon\tilde{\varphi}_i-\epsilon, & \sum\epsilon\varphi_i\otimes\tilde{\varphi}_i\end{pmatrix} f\right)\\
&= \det(I+K_N^{(1)}f),
\end{aligned}$$

concluding the case $\beta=1$ and the proof of Theorem 5.2.1.

Remark 5.2.2 *The above methods also work for the circular ensembles, see [Tra98].*

Remark 5.2.3 *For $\beta=2$, we have shown*

$$\mathbf{E}\left(\prod_{j=1}^{N}(1+f(x_j))\right)=\det(I+K_N^{(2)}f). \tag{5.2.10}$$

Setting $f(x)=\sum_{r=1}^{n}z_r\delta(x-y_r)$, we find

$$\det(I+K_N^{(2)}f)=\det[\delta_{ij}+K_N^{(2)}(y_i,y_j)z_j]_{1\le i,j\le n}$$

and so it is easy to see from the definition (5.1.4) that

$$\begin{aligned}
R_n(y_1,\dots,y_n) &= \mathrm{coeff}_{z_1\cdots z_n}\mathbf{E}\left(\prod_{j=1}^{N}\left(1+\sum_{r=1}^{n}z_r\delta(x_j-y_r)\right)\right)\\
&= \mathrm{coeff}_{z_1\cdots z_n}\det[\delta_{ij}+K_N^{(2)}(y_i,y_j)z_j]_{1\le i,j\le n}\\
&= \det[K_N^{(2)}(y_i,y_j)]_{1\le i,j\le n},
\end{aligned} \tag{5.2.11}$$

which we saw in Chapter 4. The probability that no eigenvalues lie in $J \in \mathbb{R}$, $\mathbf{E}(0; J)$ is clearly:[3]

$$\mathbf{E}(0; J) = \mathbf{E}\left(\prod_{j=1}^{N}(1 - \chi_J(x_j))\right) = \det(I - K_N^{(2)}\chi_J), \tag{5.2.12}$$

and more generally the probability of n_i eigenvalues in J_i, $1 \le i \le m$ is given by

$$\begin{aligned} E(n_1, \ldots, n_m; J_1, \ldots, J_m) &= \underset{\begin{pmatrix} n_i \text{ of } x_j \in J_i,\ 1\le i\le m \\ \text{all other } x_j \in \left(\cup_{i=1}^m J_i\right)^c \end{pmatrix}}{\int \cdots \int} P_N^{(2)}(x_1, \ldots, x_N)\, dx_1 \ldots dx_N \\ &= \underset{\prod_{i=1}^m (z_i+1)^{n_i}}{\text{coeff}} \mathbf{E}\left(\prod_{j=1}^{N}\left(\left(1 - \sum_{i=1}^{m}\chi_{J_i}(x_j)\right) + \sum_{i=1}^{m}(z_i + 1)\chi_{J_i}(x_j)\right)\right) \\ &= \frac{1}{n_1! \ldots n_m!}\frac{\partial^{\sum n_i}}{\partial z_1^{n_1} \ldots \partial z_m^{n_m}} \det\left(I + K_N^{(2)}\sum_{i=1}^{m} z_i\chi_{J_i}\right)\Bigg|_{z_1=\ldots=z_m=-1}. \end{aligned} \tag{5.2.13}$$

For $\beta = 1$ and 4, we have shown

$$\mathbf{E}\left(\prod_{j=1}^{N}(1 + f(x_j))\right) = (\det(I + K_N^{(\beta)} f))^{1/2}, \tag{5.2.14}$$

and so as before we have

$$\begin{aligned} R_n(y_1, \ldots, y_n) &= \text{coeff}_{z_1\cdots z_n}\mathbf{E}\left(\prod_{j=1}^{N}\left(1 + \sum_{r=1}^{n} z_r\delta(x_j - y_r)\right)\right) \\ &= \text{coeff}_{z_1\cdots z_n}(\det[\delta_{ij} + K_N^{(\beta)}(y_i, y_j)z_j]_{1\le i,j\le n})^{\frac{1}{2}} \end{aligned} \tag{5.2.15}$$

and that

$$E(0, J) = \mathbf{E}\left(\prod_{j=1}^{N}(1 - \chi_J(x_j))\right) = (\det(I - K_N^{(\beta)}\chi_J))^{\frac{1}{2}},$$

while

$$E(n_1, \ldots, n_m; J_1, \ldots, J_m) = \frac{1}{n_1! \ldots n_m!}\frac{\partial^{\sum n_i}}{\partial z_1^{n_1} \ldots \partial z_m^{n_m}} \det\left(I + K_N^{(\beta)}\sum_{i=1}^{m} z_i\chi_{J_i}\right)^{\frac{1}{2}}\Bigg|_{z_1=\ldots z_n=-1}.$$

[3] $\chi_J(x)$ is the indicator function for the set J.

While $R_n(y_1, \ldots, y_n)$ is much more complicated in the $\beta = 1, 4$ case, than the $\beta = 2$ case, that is not true for the so-called cluster functions, see [Tra98].

5.3 Relations between $K_N^{(2)}$ and $K_N^{(1)}$, $K_N^{(4)}$ via skew-orthogonal polynomials

Theorem 5.2.1 describes the kernels in terms of

$$\hat{S}_N^{(\beta)}(x, y) := \sum_{i,j=0}^{\sigma N-1} \varphi_i(x)\mu_{ij}^{(\beta)}\varphi_j(y),$$

with $\varphi_i(x) = w(x)p_i(x)$, $\beta = 1$, $\varphi_i(x) = (w(x))^{\frac{1}{2}} p_i(x)$, $\beta = 2, 4$ while $\sigma = 1$, $\beta = 1, 2$, and $\sigma = 2$, $\beta = 4$, with $p_i(x)$ arbitrary polynomials of degree i and

$$\mu_{ij}^{(\beta)} \text{ given by } [(\mu^{(\beta)})^{-1}]_{ij} = \langle \varphi_i, \varphi_j \rangle_\beta,$$

with

$$\langle f, g\rangle_1 = \iint \epsilon(x-y)f(x)g(y)\, dxdy, \quad \langle f, g\rangle_4 = \int (f(x)g'(x) - f'(x)g(x))\, dx,$$
$$\langle f, g\rangle_2 = \int f(x)g(x)\, dx.$$

Note $\langle\ ,\ \rangle_1$ and $\langle\ ,\ \rangle_4$ are skew-symmetric inner products, while $\langle\ ,\ \rangle_2$ is a symmetric inner product. In Remark 5.2.1, it was mentioned that the $\hat{S}_N^{(\beta)}(x, y)$ were insensitive to the choice of polynomials $p_i(x)$ and in all three cases the $\hat{S}_N^{\beta}(x, y)$ is the Christoffel–Darboux kernel corresponding to the inner product $\langle\ ,\ \rangle_\beta$. Thus the case $\beta = 2$ seems related to orthonormal polynomials, while the cases $\beta = 1$ and 4, seem related to skew-orthonormal polynomials, due to the canonical form of the Christoffel–Darboux kernels in these two cases, i.e. upon picking $\mu^{(2)} = I_N$, $\mu^{(1)} = \mu^{(4)} = J_N$. If we do this, it turns out, at least for the classical weights, the $S_N^{(\beta)}(x, y)$ for $\beta = 1, 4$ can be described using the $S_{N'}^{(2)}(x, y)$ for appropriate N', plus a rank 1 perturbation.

Before stating the fundamental theorem relating the orthonormal and skew-orthonormal polynomials that enter into the Christoffel–Darboux kernels $\hat{S}_N^{(\beta)}(x, y)$, we need some preliminary observations. Indeed, given a weight $w_2(x)$, perhaps with support on an interval I, it can be represented as $\widetilde{w}_2(x)\chi_I(x)$; however, we shall suppress the $\chi_I(x)$ and the $\sim$, while still integrating over $\mathbb{R}$ and making the assumption that $w_2'/w_2 = -g/f$, with g and f polynomials with no common factor such that $f(x)w_2(x)$ vanishes at the endpoints of the support interval I (in the limiting sense for endpoints at $\pm\infty$) and $f > 0$ in the interior of I.

Then given the inner product:

$$(\varphi, \psi)_2 = \int_{\mathbb{R}} \varphi(x)\psi(x)w_2(x)\, dx = \int_I \varphi(x)\psi(x)w_2(x)\, dx,$$

we have two natural operators (on the space of polynomials in x) going with $w_2(x)$. The first being the operator multiplication by x, and the second the first-order operator (see [Adl02])

$$n := f\frac{d}{dx} + \frac{f' - g}{2} = \left(\frac{f}{w_2}\right)^{\frac{1}{2}} \frac{d}{dx}(f w_2)^{\frac{1}{2}} \tag{5.3.1}$$

and we have

$$(x\varphi, \psi)_2 = (\varphi, x\psi)_2, \quad (n\varphi, \psi)_2 = (-\varphi, n\psi)_2,$$

i.e. x is a symmetric operator and n a skew-symmetric operator with respect to $(\, ,\,)_2$. The operator n is unique up to a constant, but we can require $\pm f$ to be monic, making n unique. The existence of x forces a 3-term recursion relation involving x on the orthonormal polynomials with respect to $(\, ,\,)_2$, and in the case of the classical weights, it forces a 3-term recursion relation involving n. This follows from the fact that x and n, in the basis of orthonormal polynomials, are represented respectively by a 3-band symmetric matrix L and a 3-band skew-symmetric matrix N. In general N has $2d + 1$ bands, with $d = \max(\text{degree f-1}, \text{degree}(f' - g))$, giving rise to a $2d + 1$ skew-symmetric recursion relation involving n.

Let us now define ($\chi_I(x)$ suppressed as usual):

$$w_1(x) := \left(\frac{w_2(x)}{f(x)}\right)^{\frac{1}{2}}, \qquad w_4(x) := w_2(x) f(x), \tag{5.3.2}$$

and the associated inner products:

$$\begin{aligned}
(\varphi, \psi)_1 &:= \iint_{\mathbb{R}^2} \varphi(x)\psi(y)\epsilon(x - y) w_1(x) w_1(y)\, dx dy, \qquad (5.3.3)\\
(\varphi, \psi)_4 &:= \int_{\mathbb{R}} \frac{1}{2}(\varphi(x)\psi'(x) - \varphi'(x)\psi(x)) w_4(x)\, dx,\\
(\varphi, \psi)_2 &= \int_{\mathbb{R}} \varphi(x)\psi(x) w_2(x)\, dx.
\end{aligned}$$

This brings us to the following theorem, whose proof is found in [Adl02]; [4]relating the above inner products, and in the classical cases, relating the skew-symmetric orthonormal polynomials going with $(\, ,\,)_1$ and $(\, ,\,)_4$ with the orthonormal polynomials going with $(\, ,\,)_2$. This theorem generalizes work of [Bre91].

Theorem 5.3.1 *Given the three weights $w_\beta(x)$ of (5.3.2) and inner products $(\, ,\,)_\beta$ of (5.3.3), and the operator n of (5.3.1), then all the inner products are determined by $w_2(x)$ and n as follows:*

[4] In fact in [Adl02], in Sections 6 and 7, (5.3.5) is proven along the way in getting a different choice of skew-orthonormal polynomials. There is a small error in the $\beta = 4$ case.

$$(\varphi, n^{-1}\psi)_2 = (\varphi, \psi)_1, \quad (\varphi, n\psi)_2 = (\varphi, \psi)_4. \tag{5.3.4}$$

The mapping of orthonormal polynomials $p_i(x)$ with respect to $(\ ,\ )_2$ into a specific set of skew-orthonormal polynomials $q_i(x)$ with respect to $(\ ,\ )_\beta$, $\beta = 1, 4$ is given by, in the three classical cases:

$$q_{2n} = p_{2n}, \qquad q_{2n+1} = c_{2n}p_{2n+1} - c_{2n-1}p_{2n-1}, \qquad \beta = 1, \tag{5.3.5}$$

$$q_{2n} = p_{2n} + \sum_{\ell=0}^{n-1}\prod_{k=\ell+1}^{n}\left(\frac{c_{2k-1}}{c_{2k-2}}\right)p_{2\ell}, \qquad q_{2n+1} = \frac{p_{2n+1}}{c_{2n}}, \quad \beta = 4,$$

with the c_ks defined by the operator n as follows:

$$np_k = c_{k-1}p_{k-1} - c_k p_{k+1}, \quad n = f\frac{d}{dx} + \frac{f' - g}{2}, \quad \frac{w_2'}{w_2} = \frac{-g}{f}, \tag{5.3.6}$$

with f and g polynomials having no common root.

Remark 5.3.1 *In the three classical cases of Hermite, Laguerre, and Jacobi one finds for the orthonormal polynomials $p_k(x)$ that*[5]

$$xp_k = a_{k-1}p_{k-1} + b_k p_k + a_k p_{k+1}, \quad np_k = c_{k-1}p_{k-1} - c_k p_{k+1}, \tag{5.3.7}$$

with

$$Hermite: w_2(x) = e^{-x^2}, \quad f = 1, \quad a_{n-1} = \sqrt{n/2}, \quad b_n = 0, \quad c_n = a_n, \tag{5.3.8}$$

Laguerre: $w_2(x) = e^{-x}x^{\alpha}\chi_{[0,\infty)}(x)$, $f = x$,

$$a_{n+1} = \sqrt{n(n+\alpha)}, \quad b_n = 2n + \alpha + 1, \quad c_n = \frac{a_n}{2},$$

Jacobi: $w_2(x) = (1-x)^{\alpha}(1+x)^{\beta}\chi_{[-1,1]}(x)$, $f = 1 - x^2$,

$$a_{n-1} = \left(\frac{4n(n+\alpha+\beta)(n+\alpha)(n+\beta)}{(2n+\alpha+\beta)^2(2n+\alpha+\beta+1)(2n+\alpha+\beta-1)}\right)^{\frac{1}{2}},$$

$$b_n = \frac{\alpha^2 - \beta^2}{(2n+\alpha+\beta)(2n+\alpha+\beta+2)}, \quad c_n = a_n\left(\frac{\alpha+\beta}{2} + n + 1\right).$$

Sketch of Proof of Theorem 5.3.1 Formula (5.3.4) is a consequence of

$$\left(\frac{d}{dx}\right)^{-1}\varphi(x) = \int_{\mathbb{R}} \epsilon(x-y)\varphi(y)\,dy, \quad \frac{d}{dx}\varphi(x) = \int_{\mathbb{R}} \delta'(x-y)\varphi(y)\,dy.$$

[5] Here we include χ rather than suppressing it.

The map O which takes $p := (p_i(x))_{i\geq 0}$ into $q := (q_i(x))_{i\geq 0}$, i.e. $q = Op$, with O a lower triangular semi-infinite matrix is given respectively for the cases $\beta = 1, 4$ by performing the skew-Borel decomposition

$$-N^{-1} = O^{-1}J_\infty(O^{-1})^T, \beta = 1, \quad -N = O^{-1}J_\infty(O^{-1})^T, \beta = 4, \qquad (5.3.9)$$

with N the skew-symmetric 3-band semi-infinite matrix, which expresses the operator n in the orthonormal basis $\{p_k\}$ (given in (5.3.6) and (5.3.7) in terms of the c_k, $k \geq 0$) and J_∞ is the semi-infinite symplectic matrix, $I_\infty \otimes \left[\begin{smallmatrix} 0 & 1 \\ -1 & 0 \end{smallmatrix}\right]$. One makes use of the non-uniqueness of the skew-orthonormal polynomials

$$(q_{2n}, q_{2n+1}) \mapsto \left(\delta_n q_{2n}, \frac{1}{\delta_n}(q_{2n+1} + \gamma_n q_{2n})\right),$$

to maximize the simplicity of the transformation $p \mapsto q$.

For general m semi-infinite and skew-symmetric, one performs the "skew-Borel decomposition", $m = O^{-1}J_\infty(O^{-1})^T$, for O lower triangular, by forming the skew-orthogonormal polynomials $(h_i(z))_{i\geq 0}$ going with the skew-symmetric inner product defined by $\langle z^i, z^j\rangle = m_{ij}$, $i, j \geq 0$ and setting

$$O:\; O(1, z, z^2 \ldots, \;)^T = (h_0(z), h_1(z), \ldots, \;)^T.$$

This is fully explained in [Adl99] and is an immediate generalization of the case where m is symmetric (see [Adl97]). In [Adl99] the recipe for the $h_i(z)$ is given as follows:

$$h_{2n}(z) = \frac{1}{(\mathrm{Pf}(m_{2n})\mathrm{Pf}(m_{2n+2}))^{\frac{1}{2}}}\mathrm{Pf}\left(\begin{array}{c|c} & 1 \\ & z \\ m_{2n+1} & \vdots \\ & z^{2n} \\ \hline -1, -z, \ldots, \;, -z^{2n} & 0 \end{array}\right)$$

$$h_{2n+1}(z) = \frac{1}{(\mathrm{Pf}(m_{2n})\mathrm{Pf}(m_{2n+2}))^{\frac{1}{2}}}$$

$$\mathrm{Pf}\left(\begin{array}{c|cc} & 1 & m_{0,2n+1} \\ & z & m_{1,2n+1} \\ m_{2n} & \vdots & \vdots \\ & z^{2n-1} & m_{2n-1,2n+1} \\ \hline -1, -z, \ldots, \;\; -z^{2n-1} & 0 & -z^{2n+1} \\ -m_{0,2n+1}, \ldots, -m_{2n-1,2n+1} & z^{2n+1} & 0 \end{array}\right)$$

with $m_k := (m_{ij})_{0\leq i,j\leq k-1}$ and $\mathrm{Pf}(A) = (\det A)^{\frac{1}{2}}$ for A a skew-symmetric matrix.

Remark 5.3.2 *In the nonclassical case, one still has the same recipe for O, (5.3.9), but in general N will have $2d + 1$ bands, $d > 1$, and so O will increase in complexity with increasing $d > 1$.*

We can apply Theorem 5.3.1 to compute $\hat{S}_N^{(\beta)}(x, y)$ and hence $S_N^{(\beta)}(x, y)$ for $\beta = 1, 4$ by setting $\mu_{ij}^{(\beta)} = J_{ij}$, so that (up to the weight factor) the $\varphi_i(x)$ are skew-orthogonal polynomials. This leads to the following theorem, found in [Adl00].

Theorem 5.3.2 *In the case of the three classical weights of Remark 5.3.1, the $\beta = 1, 4$ kernel is given in terms of the $\beta = 2$ kernel as follows:*

$$S_N^{(1)}(x, y) = \left(\frac{f(y)}{f(x)}\right)^{\frac{1}{2}} S_{N-1}^{(2)}(x, y) + c_{N-2} \frac{\varphi_{N-1}^{(2)}(x)}{(f(x))^{\frac{1}{2}}} \epsilon \left(\frac{\varphi_{N-2}^{(2)}}{(f)^{\frac{1}{2}}}\right)(y), \quad N \text{ even} \tag{5.3.10}$$

$$S_N^{(4)}(x, y) = \left(\frac{f(y)}{f(x)}\right)^{\frac{1}{2}} S_{2N}^{(2)}(x, y) - c_{2N-1} \frac{\varphi_{2N}^{(2)}(x)}{(f(x))^{\frac{1}{2}}} \int_y^\infty \frac{\varphi_{2N-1}^{(2)}(t)}{(f(t))^{\frac{1}{2}}} dt, \tag{5.3.11}$$

where $\varphi_k^{(2)} = (w_2)^{\frac{1}{2}} p_k$, with p_k and $S_k^{(2)}(x, y)$ being the usual orthonormal polynomials and Christoffel–Darboux kernel with respect to the weight w_2. Given the weight $w(x)$ appearing in $P_N^{(\beta)}(x_1, \dots, x_n)$, $\beta = 1, 4$, (5.1.1), pick $w_2(x)$ such that

$$w(x) = \left(\frac{w_2(x)}{f(x)}\right)^{\frac{1}{2}}, \quad \beta = 1 \quad \textit{and} \quad w(x) = w_2(x) f(x), \quad \beta = 4, \tag{5.3.12}$$

with

$$\frac{w_2'(x)}{w_2(x)} = -\frac{g(x)}{f(x)}, \quad f(x) = 1, x, 1 - x^2,$$

respectively for the Hermite, Laguerre, and Jacobi cases of Remark 5.3.1. The c_k are defined by

$$n p_k = c_{k-1} p_{k-1} - c_k p_{k+1}, \quad n = f \frac{d}{dx} + \frac{f' - g}{2},$$

and are given explicitly in Remark 5.3.1.

Remark 5.3.3 *In the special case of the Gaussian potential, $w_2(x) = e^{-x^2}$, [Tra98] showed that $S_N^{(\beta)}(x, y)$ also have the above representation:*

$$S_N^{(1)}(x, y) = S_N^{(2)}(x, y) + \left(\frac{N}{2}\right)^{\frac{1}{2}} \varphi_{N-1}(x) \epsilon \varphi_N(y),$$

$$2 S_N^{(4)}(x, y) = S_{2N+1}^{(2)}(x, y) + \left(N + \frac{1}{2}\right)^{\frac{1}{2}} \varphi_{2N}(x) \epsilon \varphi_{2N+1}(y),$$

but in order to obtain one formula for all three classical cases we need the above theorem. Indeed, the above formula and different formulas for the Laguerre case due to [For99] are found in [Adl00], section 4, and shown to agree with the theorem at the end of the paper.

Proof of Theorem 5.3.2 The proof uses (5.3.5), which is a consequence of (5.3.12). Set $\mu_{ij}^{(\beta)} = J_{ij}$, $\beta = 1, 4$ in (5.2.6) and (5.2.8), and then substitute in $S_N^{(\beta)}(x, y)$, $\beta = 1, 4$ respectively that

$$\begin{aligned}
\epsilon\varphi_k(v) &= \frac{1}{w(v)}\langle\delta(x-v), q_k\rangle_1, \qquad \varphi_k = wq_k,\\
\epsilon\varphi_k'(v) &= \frac{1}{(w(v))^{\frac{1}{2}}}\langle\delta(x-v), q_k\rangle_4, \qquad \varphi_k = w^{\frac{1}{2}}q_k,\\
\delta(x-y) &= \sum_{n=0}^{\infty}\varphi_n^{(2)}(x)\varphi_n^{(2)}(y).
\end{aligned}$$

Then using both (5.3.5) and the inverse map, finally yields the theorem.

References

[Adl97] M. Adler and P. van Moerbeke, String-orthogonal polynomials, string equations, and 2-Toda symmetries, Commun. on Pure and Appl. Math. **L** (1997) 241–290.

[Adl99] M. Adler, E. Horozov, and P. van Moerbeke, The Pfaff lattice and skew-orthogonal polynomials, Internat. Math. Res. Notices (1999) 569–588.

[Adl00] M. Adler, P. J. Forrester, T. Nagao, and P. van Moerbeke, Classical skew orthogonal polynomials and random matrices, J. Stat. Phys. **99**, 1/2 (2000) 141–170.

[Adl02] M. Adler and P. van Moerbeke, Toda versus Pfaff lattice and related polynomials, Duke Math. J. **112**, 1 (2002) 1–58.

[Ake05] G. Akemann, The Complex Laguerre Symplectic Ensemble of Non-Hermitian Matrices Nucl. Phys. **B730** (2005) 253–299.

[Bre91] E. Brézin and H. Neuberger, Multicritical points of unoriented random surfaces, Nucl. Phys. **B350** (1991) 513–553.

[deB55] N. G. de Bruijn, On some multiple integrals involving determinants, J. Indian Math. Soc. **19** (1955) 133–151.

[Dys70] F. J. Dyson, Correlations between eigenvalues of a random matrix, Commun. Math. Phys. **19** (1970) 235–250.

[Dys72] F. J. Dyson, A class of matrix ensembles, J. Math. Phys. **13** (1972) 90–97.

[For99] P. J. Forrester, T. Nagao, and G. Honner, Correlations for the orthogonal-unitary and symplectic-unitary transitions at the hard and soft edges, Nucl. Phys. **B553** (1999) 601–643.

[Gau61] M. Gaudin, Sur la loi limite de l'espacement des valeurs propres d'une matrices aléatoire, Nucl. Phys. **25** (1961) 447–458.

[Hel62] S. Helgason, *Differential Geometry and Symmetric Spaces*, Academic Press, New York 1962.

[Hel84] S. Helgason, Groups and geometric analysis, in: *Integral Geometry, Invariant Differential Operators, and Spherical Functions*, Academic Press, New York 1984.

[Meh71] M. L. Mehta, A note on correlations between eigenvalues of a random matrix, Commun. Math. Phys. **20** (1971) 245–250.

[Meh76] M. L. Mehta, A note on certain multiple integrals, J. Math. Phys. **17** (1976) 2198–2202.

[Meh04] M. L. Mehta, *Random Matrices*, Academic Press, 3rd Edition, London 2004.

[Nag91] T. Nagao and M. Wadati, Correlation functions of random matrix ensembles related to classical orthogonal polynomials, J. Phys. Soc. Jpn. **60** (1991) 3298–3322; **61** (1992) 78–88; **61** (1992) 1910–1918.

[Nag95] T. Nagao and P. J. Forrester, Asymptotic correlations at the spectrum edge of random matrices, Nucl. Phys. B**435** (1995) 401–420.

[Sen98] M. K. Sener and J. J. M. Verbaarschot, Universality in Chiral Random Matrix Theory at $\beta = 1$ and $\beta = 4$, Phys. Rev. Lett. **81** (2) (1998) 248–251.

[Tra98] C. A. Tracy and H. Widom, Correlation functions, cluster functions, and spacing distributions for random matrices, J. Stat. Phys. **92** (1998) 809–835.

[Wid99] H. Widom, On the relation between orthogonal, symplectic and unitary random matrices, J. Stat. Phys. **94** (1999) 601–643.

·6·

Universality

A. B. J. Kuijlaars

Abstract

Universality of eigenvalue spacings is one of the basic characteristics of random matrices. We give the precise meaning of universality and discuss the standard universality classes (sine, Airy, Bessel) and their appearance in unitary, orthogonal, and symplectic ensembles. The Riemann–Hilbert problem for orthogonal polynomials is one possible tool to derive universality in unitary random matrix ensembles. An overview is presented of the Deift/Zhou steepest descent analysis of the Riemann–Hilbert problem in the one-cut regular case. Non-standard universality classes that arise at singular points in the spectrum are discussed at the end.

6.1 Heuristic meaning of universality

Universality of eigenvalue spacings is one of the basic characteristic features of random matrix theory. It was the main motivation for Wigner to introduce random matrix ensembles to model energy levels in quantum systems.

On a local scale the eigenvalues of random matrices show a repulsion which implies that it is highly unlikely that eigenvalues are very close to each other. Also very large gaps between neighbouring eigenvalues are unlikely. This is in sharp contrast to points that are sampled independently from a given distribution. Such points exhibit Poisson statistics in contrast to what is now called GUE/GOE/GSE statistics. These new random matrix theory type statistics have been observed (and sometimes proved) in many other mathematical and physical systems as well, see [Dei07d], and are therefore universal in a sense that goes beyond random matrix theory. Examples outside of random matrix theory are discussed in various chapters of this handbook. In this chapter the discussion of universality is however restricted to random matrix theory.

The universality conjecture in random matrix theory says that the local eigenvalue statistics of many random matrix ensembles are the same, that is, they do not depend on the exact probability distribution that is put on a set of matrices, but only on some general characteristics of the ensemble. The universality of local eigenvalue statistics is a phenomenon that takes place for large random matrices. In a proper mathematical formulation it is a statement about a certain limiting behaviour as the size of the matrices tends to infinity.

The characteristics that play a role in the determination of the universality classes are the following.

- Invariance properties of the ensemble, of which the prototype is invariance with respect to orthogonal, unitary, or unitary-symplectic conjugation. As discussed in Chapters 4 and 5 of this handbook, these lead to random matrix ensembles with an explicit joint eigenvalue density of the form

$$p(x_1, \ldots, x_n) = \frac{1}{Z_{n,\beta}} \prod_{1 \le i < j \le n} |x_j - x_i|^\beta \prod_{j=1}^{n} e^{-V(x_j)} \tag{6.1.1}$$

  where $\beta = 1, 2, 4$ corresponds to orthogonal, unitary, and symplectic ensembles, respectively. The case $V(x) = \frac{1}{2}x^2$ gives the Gaussian ensembles GOE, GUE, and GSE. The local eigenvalue repulsion increases with β and different values of β give rise to different universality results.
- An even more basic characteristic is the distinction between random matrix ensembles of real symmetric or complex Hermitian matrices. Even without the invariance assumption, the universality of local eigenvalue statistics is conjectured to hold. This has been proved recently for large classes of Wigner ensembles, i.e. random matrix ensembles with independent, identically distributed entries, see [Erd10a, Erd10b, Tao11] and the survey [Erd11]. See Chapter 21 of this handbook for more details. Universality is also expected to exist in classes of non-Hermitian matrices, see Chapter 18.
- Another main characteristic is the nature of the point around which the local eigenvalue statistics are considered. A typical point in the spectrum is such that, possibly after appropriate scaling, the limiting mean eigenvalue density is positive. Such a point is in the bulk of the spectrum, and universality around such a point is referred to as bulk universality.

  At edge points of the limiting spectrum the limiting mean eigenvalue density typically vanishes like a square root, and at such points a different type of universality is expected, which is known as (soft) edge universality. At natural edges of the spectrum, such as the point zero for ensembles of positive definite matrices, the limiting mean eigenvalue density typically blows up like the inverse of a square root, and universality at such a point is known as hard edge universality.
- At very special points in the limiting spectrum the limiting mean eigenvalue density may exhibit singular behaviour. For example, the density may vanish at an isolated point in the interior of the limiting spectrum, or it may vanish to higher order than square root at soft edge points. This

non-generic behaviour may take place in ensembles of the form (6.1.1) with a varying potential NV

$$p(x_1, \ldots, x_n) = \frac{1}{Z_{n,\beta}} \prod_{1 \le i < j \le n} |x_j - x_i|^\beta \prod_{j=1}^{n} e^{-NV(x_j)} \tag{6.1.2}$$

as n, $N \to \infty$ with $n/N \to 1$.
At such special points one expects other universality classes determined by the nature of the vanishing of the limiting mean eigenvalue density at that point; see also a discussion in Chapter 14.

In the rest of this chapter we first give the precise meaning of the notion of universality and we discuss the limiting kernels (sine, Airy, and Bessel) associated with the bulk universality and the edge universality for the unitary, orthogonal, and symplectic universality classes. In Section 6.3 we discuss unitary matrix ensembles in more detail and we show that universality in these ensembles comes down to the convergence of the properly scaled eigenvalue correlation kernels. In Section 6.4 we discuss the Riemann–Hilbert method in some detail. The Riemann–Hilbert method is one of the main methods to prove universality in unitary ensembles. In the final Section 6.5 we discuss certain non-standard universality classes that arise at singular points in the limiting spectrum. We describe the limiting kernels for each of the three types of singular points, namely interior singular points, singular edge points, and exterior singular points.

Other approaches to universality are detailed in Chapter 21 for Wigner ensembles and in Chapter 16 using loop equation techniques.

6.2 Precise statement of universality

For a probability density $p_n(x_1, \ldots, x_n)$ on $\mathbb{R}^n$ that is symmetric (i.e. invariant under permutations of coordinates), let us define the k-point correlation function by

$$R_{n,k}(x_1, \ldots, x_k) = \frac{n!}{(n-k)!} \int \cdots \int p_n(x_1, \ldots, x_n)\, dx_{k+1} \cdots dx_n. \tag{6.2.1}$$

Up to the factor $n!/(n-k)!$ this is the kth marginal distribution of p_n.

Fix a reference point x^* and a constant $c_n > 0$. We centre the points around x^* and scale by a factor c_n, so that $(x_1, \ldots, x_n)$ is mapped to

$$(c_n(x_1 - x^*), \ldots, c_n(x_n - x^*)).$$

These centred and scaled points have the following rescaled k-point correlation functions

$$\frac{1}{c_n^k} R_{n,k}\left(x^* + \frac{x_1}{c_n}, x^* + \frac{x_2}{c_n}, \ldots, x^* + \frac{x_k}{c_n}\right). \tag{6.2.2}$$

The universality is a property of a sequence (p_n) of symmetric probability density functions. Universality at x^* means that for a suitably chosen sequence (c_n) the rescaled k-point correlation functions (6.2.2) have a specific limit as $n \to \infty$. The precise limit determines the universality class.

As discussed in Chapter 4 for determinantal point processes, the main spectral statistics such as gap probabilities and eigenvalue spacings can be expressed in terms of the k-point correlation functions. If the limits (6.2.2) exist and belong to a certain universality class, then this also leads to the universal behaviour of the local eigenvalue statistics.

6.2.1 Unitary universality classes

The unitary universality classes are characterized by the fact that the limit of (6.2.2) can be expressed as a $k \times k$ determinant

$$\det\left[K(x_i, x_j)\right]_{1 \le i,j \le k}$$

involving a kernel $K(x, y)$, which is called the eigenvalue correlation kernel.

Bulk universality A sequence of symmetric probability density functions (p_n) then has bulk universality at a point x^* if there exists a sequence (c_n), so that for every k, the limit of (6.2.2) is given by

$$\det\left[K^{\sin}(x_i, x_j)\right]_{1 \le i,j \le k}$$

where

$$K^{\sin}(x, y) = \frac{\sin \pi(x-y)}{\pi(x-y)} \tag{6.2.3}$$

is the sine kernel. Bulk universality is also known as GUE statistics.

Edge universality The (soft) edge universality holds at x^* if the limit of (6.2.2) is equal to

$$\det\left[K^{\mathrm{Ai}}(x_i, x_j)\right]_{1 \le i,j \le k}$$

where

$$K^{\mathrm{Ai}}(x, y) = \frac{\mathrm{Ai}(x)\,\mathrm{Ai}'(y) - \mathrm{Ai}'(x)\,\mathrm{Ai}(y)}{x-y} \tag{6.2.4}$$

is the Airy kernel. Here Ai denotes the Airy function, which is the solution of the Airy differential equation $y''(x) = xy(x)$ with asymptotics

$$\mathrm{Ai}(x) = \frac{1}{2\sqrt{\pi}x^{1/4}} e^{-\frac{2}{3}x^{3/2}} \left(1 + O(x^{-3/2})\right), \qquad \text{as } x \to +\infty.$$

The Airy kernel is intimately related to the Tracy–Widom distribution for the largest eigenvalue in random matrix theory [Tra94].

Hard edge universality A hard edge is a boundary for the eigenvalues that is part of the model. For example, if the random matrices are real symmetric (or complex Hermitian) positive definite then all eigenvalues are non-negative and zero is a hard edge.

The hard edge universality holds at x^* if the limit of (6.2.2) is equal to

$$\det\left[K^{\mathrm{Bes},\alpha}(x_i, x_j)\right]_{1\le i,j\le k}$$

where

$$K^{\mathrm{Bes},\alpha}(x,y) = \frac{J_\alpha(\sqrt{x})\sqrt{y}J'_\alpha(\sqrt{y}) - \sqrt{x}J'_\alpha(\sqrt{x})J_\alpha(\sqrt{y})}{2(x-y)}, \qquad x, y > 0, \qquad (6.2.5)$$

and J_α is the usual Bessel function of order α.

The Bessel kernels depend on the parameter $\alpha > -1$ which may be interpreted as a measure of the interaction with the hard edge. The bigger α, the more repulsion from the hard edge.

6.2.2 Orthogonal and symplectic universality classes

The orthogonal and symplectic universality classes are characterized by the fact that the limit of (6.2.2) is expressed as a Pfaffian

$$\mathrm{Pf}\left[K(x_i, x_j)\right]_{1\le i,j\le k} \qquad (6.2.6)$$

of a 2×2 matrix kernel

$$K(x,y) = \begin{pmatrix} K_{11}(x,y) & K_{12}(x,y) \\ K_{21}(x,y) & K_{22}(x,y) \end{pmatrix}. \qquad (6.2.7)$$

Recall that the Pfaffian $\mathrm{Pf}(A)$ of a $2k\times 2k$ skew symmetric matrix A is such that

$$\mathrm{Pf}(A)^2 = \det(A).$$

The matrix in (6.2.6) is a $2k\times 2k$ matrix written as a $k\times k$ block matrix with 2×2 blocks. It is skew symmetric provided that $K(y,x) = -K(x,y)^T$ where ^{T} denotes the matrix transpose.

Bulk universality A sequence of symmetric probability density functions (p_n) has orthogonal/symplectic bulk universality at a point x^* if there exists a sequence (c_n), so that for every k, the limit of (6.2.2) is given by

$$\mathrm{Pf}\left[K^{\sin,\beta}(x_i, x_j)\right]_{1\le i,j\le k}$$

with

$$K^{\sin,\beta=1}(x,y) = \begin{pmatrix} -\dfrac{\partial}{\partial x}\dfrac{\sin\pi(x-y)}{\pi(x-y)} & \dfrac{\sin\pi(x-y)}{\pi(x-y)} \\ -\dfrac{\sin\pi(x-y)}{\pi(x-y)} & \displaystyle\int_0^{x-y}\frac{\sin\pi t}{\pi t}dt - \frac{1}{2}\operatorname{sgn}(x-y) \end{pmatrix}$$

where $\operatorname{sgn}(t) = 1, 0, -1$ depending on whether $t > 0, t = 0$, or $t < 0$, in the case of orthogonal (i.e. $\beta = 1$) bulk universality, and

$$K^{\sin,\beta=4}(x,y) = \begin{pmatrix} -\dfrac{\partial}{\partial x}\dfrac{\sin 2\pi(x-y)}{2\pi(x-y)} & \dfrac{\sin 2\pi(x-y)}{2\pi(x-y)} \\ -\dfrac{\sin 2\pi(x-y)}{2\pi(x-y)} & \displaystyle\int_0^{x-y}\frac{\sin 2\pi t}{2\pi t}dt \end{pmatrix}$$

in the case of symplectic (i.e., $\beta = 4$) bulk universality.

Edge universality The orthogonal/symplectic (soft) edge universality holds at x^* if the limit of (6.2.2) is equal to

$$\operatorname{Pf}\left[K^{\mathrm{Ai},\beta}(x_i,x_j)\right]_{1\le i,j\le k}$$

with

$$\begin{aligned} K_{11}^{\mathrm{Ai},\beta=1}(x,y) &= \frac{\partial}{\partial y}K^{\mathrm{Ai}}(x,y) + \frac{1}{2}\operatorname{Ai}(x)\operatorname{Ai}(y) \\ K_{12}^{\mathrm{Ai},\beta=1}(x,y) &= K^{\mathrm{Ai}}(x,y) + \frac{1}{2}\operatorname{Ai}(x)\cdot\int_{-\infty}^{y}\operatorname{Ai}(t)\,dt \\ K_{21}^{\mathrm{Ai},\beta=1}(x,y) &= -K_{12}^{\mathrm{Ai},\beta=1}(x,y) \\ K_{22}^{\mathrm{Ai},\beta=1}(x,y) &= -\int_x^{\infty}K^{\mathrm{Ai}}(t,y)\,dt - \frac{1}{2}\int_x^{y}\operatorname{Ai}(t)\,dt \\ &\quad + \frac{1}{2}\int_x^{\infty}\operatorname{Ai}(t)\,dt\cdot\int_y^{\infty}\operatorname{Ai}(t)\,dt - \frac{1}{2}\operatorname{sgn}(x-y) \end{aligned} \tag{6.2.8}$$

in the case of orthogonal edge universality, and

$$\begin{aligned} K_{11}^{\mathrm{Ai},\beta=4}(x,y) &= \frac{1}{2}\frac{\partial}{\partial y}K^{\mathrm{Ai}}(x,y) + \frac{1}{4}\operatorname{Ai}(x)\operatorname{Ai}(y) \\ K_{12}^{\mathrm{Ai},\beta=4}(x,y) &= \frac{1}{2}K^{\mathrm{Ai}}(x,y) - \frac{1}{4}\operatorname{Ai}(x)\cdot\int_y^{\infty}\operatorname{Ai}(t)\,dt \\ K_{21}^{\mathrm{Ai},\beta=4}(x,y) &= -K_{12}^{\mathrm{Ai},\beta=4}(x,y) \\ K_{22}^{\mathrm{Ai},\beta=4}(x,y) &= -\frac{1}{2}\int_x^{\infty}K^{\mathrm{Ai}}(t,y)\,dt + \frac{1}{4}\int_x^{\infty}\operatorname{Ai}(t)\,dt\cdot\int_y^{\infty}\operatorname{Ai}(t)\,dt \end{aligned} \tag{6.2.9}$$

in the case of symplectic edge universality. The kernel K^{Ai} that appears in (6.2.8) and (6.2.9) is the Airy kernel from (6.2.4).

Hard edge universality The orthogonal/symplectic hard edge universality is expressed in terms of 2×2 kernels with Bessel functions. See Forrester [For10] for the precise statement.

6.2.3 Determinantal and Pfaffian point processes

The universality limits have the characteristic properties of determinantal or Pfaffian point processes, see [For10, Sos00] and Chapter 11 of this handbook.

If the probability densities p_n on $\mathbb{R}^n$ also arise from determinantal point processes, then the statement of universality can be expressed as the convergence of the corresponding correlation kernels after appropriate scaling.

The probability density p_n arises from a determinantal point process, if there exist (scalar) kernels K_n so that for every n and k,

$$R_{n,k}(x_1, \ldots, x_k) = \det\left[K_n(x_i, x_j)\right]_{1 \le i,j \le k}.$$

In particular, one then has for the 1-point function (particle density),

$$R_{n,1}(x) = K_n(x, x)$$

and for the probability density p_n itself

$$p_n(x_1, \ldots, x_n) = \frac{1}{n!}\det\left[K_n(x_i, x_j)\right]_{1 \le i,j \le n}.$$

In this setting the bulk universality comes down to the statement that the centered and rescaled kernels

$$\frac{1}{c_n} K_n\left(x^* + \frac{x}{c_n}, x^* + \frac{y}{c_n}\right) \tag{6.2.10}$$

tend to the sine kernel (6.2.3) as $n \to \infty$, and likewise for the edge universality.

Similarly, the probability densities p_n come from Pfaffian point processes, if there exist 2×2 matrix kernels K_n so that for every n and k,

$$R_{n,k}(x_1, \ldots, x_k) = \mathrm{Pf}\left[K_n(x_i, x_j)\right]_{1 \le i,j \le k}.$$

A proof of universality in a sequence of Pfaffian point processes then comes down to the proof of a scaling limit for the matrix kernels as $n \to \infty$.

Unitary random matrix ensembles are basic examples of determinantal point processes, while orthogonal and symplectic matrix ensembles are examples of Pfaffian point processes.

For the classical ensembles that are associated with Hermite, Laguerre, and Jacobi polynomials the existence and the form of the limiting kernels has been proven using the explicit formulas that are available for these classical orthogonal polynomials, see e.g. [For10, Meh04].

For non-classical ensembles, the results about universality are fairly complete for unitary ensembles, due to their connection with orthogonal polynomials.

This will be discussed in more detail in the next section, and we give references there.

The first rigorous results on bulk universality for orthogonal and symplectic ensembles are due to Stojanovic [Sto00] who discussed ensembles with a quartic potential. This was extended by Deift and Gioev [Dei07a] for ensembles (6.1.1) with polynomial V. Their techniques are extended to treat edge universality in [Dei07b, Dei07c], see also the recent monograph [Dei09]. Varying weights are treated by Shcherbina in [Shc08, Shc09a, Shc09b].

6.3 Unitary random matrix ensembles

We explain in more detail how the universality classes arise for the eigenvalues of a unitary invariant ensemble

$$\frac{1}{\tilde{Z}_{n,N}} e^{-N\mathrm{Tr}\, V(M)} dM \tag{6.3.1}$$

defined on the space of $n \times n$ Hermitian matrices, where

$$dM = \prod_{i=1}^{n} dM_{ii} \prod_{1\leq i<j\leq n} d\,\mathrm{Re}\, M_{ij}\, d\,\mathrm{Im}\, M_{ij},$$

and $\tilde{Z}_{n,N}$ is a normalization constant, see also Chapter 4. The potential function V in (6.3.1) is typically a polynomial, but could be more general as well. To ensure that (6.3.1) is well defined as a probability measure, we assume that

$$\lim_{x\to\pm\infty} \frac{V(x)}{\log(1+x^2)} = +\infty. \tag{6.3.2}$$

The factor N in (6.3.1) will typically be proportional to n. This is needed in the large n limit in order to balance the repulsion among eigenvalues due to the Vandermonde factor in (6.3.3) and the confinement of eigenvalues due to the potential V.

6.3.1 Orthogonal polynomial kernel

The joint probability density for the eigenvalues of a matrix M of the ensemble (6.3.1) has the form

$$p_{n,N}(x_1,\ldots,x_n) = \frac{1}{Z_{n,N}} \prod_{1\leq i<j\leq n} (x_i - x_j)^2 \prod_{j=1}^{n} e^{-NV(x_j)}. \tag{6.3.3}$$

Introduce the monic polynomials $P_{k,N}$, $P_{k,N}(x) = x^k + \cdots$, that are orthogonal with respect to the weight $e^{-NV(x)}$ on $\mathbb{R}$:

$$\int_{-\infty}^{\infty} P_{k,N}(x)x^j e^{-NV(x)}dx = \begin{cases} 0, & \text{for } j = 0, \ldots, k-1, \\ \gamma_{k,N}^2 > 0, & \text{for } j = k, \end{cases} \tag{6.3.4}$$

and the orthogonal polynomial kernel

$$K_{n,N}(x, y) = \sqrt{e^{-NV(x)}}\sqrt{e^{-NV(y)}} \sum_{k=0}^{n-1} \frac{P_{k,N}(x)\, P_{k,N}(y)}{\gamma_{k,N}^2}. \tag{6.3.5}$$

Using the formula for a Vandermonde determinant and performing elementary row operations we write (6.3.3) as

$$\begin{aligned} p_{n,N}(x_1, \ldots, x_n) &= \frac{1}{Z_{n,N}} \left(\det\left[P_{k-1,N}(x_j)\right]_{1\le j,k\le n}\right)^2 \prod_{j=1}^{n} e^{-NV(x_j)} \\ &= \frac{\prod_{k=0}^{n-1} \gamma_{k,N}^2}{Z_{n,N}} \left(\det\left[\sqrt{e^{-NV(x_j)}}\, \frac{P_{k-1,N}(x_j)}{\gamma_{k-1,N}}\right]_{1\le j,k\le n}\right)^2. \end{aligned}$$

Evaluating the square of the determinant using the rule $(\det A)^2 = \det(AA^T)$ we obtain that (6.3.3) can be written as

$$p_{n,N}(x_1, \ldots, x_n) = \frac{1}{n!}\det\left[K_{n,N}(x_i, x_j)\right]_{1\le i,j\le n}, \tag{6.3.6}$$

since it can be shown that $Z_{n,N} = n! \prod_{k=0}^{n-1} \gamma_{k,N}^2$. The orthogonality condition (6.3.4) is then used to prove that the k-point correlation functions (6.2.1) are also determinants

$$R_{n,k}(x_1, \ldots, x_k) = \det\left[K_{n,N}(x_i, x_j)\right]_{1\le i,j\le k},$$

which shows that the eigenvalues are a determinantal point process with correlation kernel $K_{n,N}$. For the above calculation, see also Chapter 4.

The eigenvalues of orthogonal and symplectic ensembles of random matrices have joint probability density

$$\frac{1}{Z_{n,N}} \prod_{1\le i<j\le n} |x_i - x_j|^\beta \prod_{j=1}^{n} e^{-NV(x_j)}$$

with $\beta = 1$ for orthogonal ensembles and $\beta = 4$ for symplectic ensembles. These probability densities are basic examples of Pfaffian ensembles, as follows from the calculations in e.g. [Dei09, Tra98]; see also Chapter 5 of this handbook.

6.3.2 Global eigenvalue regime

As $n,\ N \to \infty$ such that $n/N \to 1$ the eigenvalues have a limiting distribution. A weak form of this statement is expressed by the fact that

$$\lim_{\substack{n,N\to\infty\\ n/N\to 1}} \frac{1}{n} K_{n,N}(x,x) = \rho_V(x), \qquad x \in \mathbb{R} \tag{6.3.7}$$

exists. The density ρ_V describes the global or macroscopic eigenvalue regime.

The probability measure $d\mu_V(x) = \rho_V(x)dx$ minimizes the weighted energy

$$\iint \log \frac{1}{|x-y|} d\mu(x) d\mu(y) + \int V(x) d\mu(x) \tag{6.3.8}$$

among all Borel probability measures μ on $\mathbb{R}$. Heuristically, it is easy to understand why (6.3.8) is relevant. Indeed, from (6.3.3) we find after taking logarithms and dividing by $-n^2$, that the most likely distribution of eigenvalues $x_1, \ldots, x_n$ for (6.3.3) is the one that minimizes

$$\frac{1}{n^2} \sum_{\substack{i,j=1\\ i\neq j}}^{n} \log \frac{1}{|x_i - x_j|} + \frac{N}{n^2} \sum_{j=1}^{n} V(x_j).$$

This discrete minimization problem leads to the minimization of (6.3.8) in the continuum limit as $n,\ N \to \infty$ with $n/N \to 1$.

The minimizer of (6.3.8) is unique and has compact support. It is called the equilibrium measure in the presence of the external field V, because of its connections with logarithmic potential theory [Saf97]. The proof of (6.3.7) with $\rho_V(x)dx$ minimizing (6.3.8) is in [Dei99a, Joh98]. See also the remark at the end of Section 6.4.8 below. It follows as well from the more general large deviation principle that is associated with the weighted energy (6.3.8), see [Ben97] and also [And10, §2.6]. See also Chapter 14 of this handbook.

If V is real analytic on $\mathbb{R}$ then μ_V is supported on a finite union of intervals, say

$$\mathrm{supp}(\mu_V) = \bigcup_{j=1}^{m} [a_j, b_j],$$

and the density ρ_V takes the form

$$\rho_V(x) = \frac{1}{\pi} h_j(x) \sqrt{(b_j - x)(x - a_j)}, \qquad x \in [a_j, b_j], \quad j = 1, \ldots, m \tag{6.3.9}$$

where h_j is real analytic and non-negative on $[a_j, b_j]$, see [Dei98]. Another useful representation is that

$$\rho_V(x) = \frac{1}{\pi} \sqrt{q_V^-(x)}, \qquad x \in \mathbb{R}, \tag{6.3.10}$$

where q_V^- is the negative part of

$$q_V(x) = \left(\frac{V'(x)}{2}\right)^2 - \int \frac{V'(x) - V'(s)}{x - s} d\mu_V(s). \tag{6.3.11}$$

If V is a polynomial then q_V is a polynomial of degree $2(\deg V - 1)$. In that case the number of intervals in the support is bounded by $\frac{1}{2} \deg V$, see again [Dei98]. See also a discussion in Chapters 14 and 16.

6.3.3 Local eigenvalue regime

In the context of Hermitian matrix models, the universality may be stated as the fact that the global eigenvalue regime determines the local eigenvalue regime. The universality results take on a different form depending on the nature of the reference point x^*.

Bulk universality A regular point in the bulk is an interior point x^* of $\text{supp}(\mu_V)$ such that $\rho_V(x^*) > 0$. At a regular point in the bulk one has

$$\lim_{n\to\infty} \frac{1}{cn} K_{n,N}\left(x^* + \frac{x}{cn}, x^* + \frac{y}{cn}\right) = K^{\sin}(x, y) \tag{6.3.12}$$

where $c = \rho_V(x^*)$ and $K^{\sin}$ is the sine kernel (6.2.3).

Convincing heuristic arguments for the universality of the sine kernel were given by Brézin and Zee [Bre93] by the method of orthogonal polynomials. Supersymmetry arguments were used in [Hac95]. Rigorous results for bulk universality (6.3.12) beyond the classical ensembles were first given by Pastur and Shcherbina [Pas97], and later by Bleher and Its [Ble99] and by Deift *et al.* [Dei99b, Dei99c] for real analytic V. In these papers the Riemann–Hilbert techniques were introduced in the study of orthogonal polynomials, which we will review in Section 6.4 below. Conditions on V in (6.1.2) were further weakened in [McL08, Pas08].

In more recent work of Lubinsky [Lub08a, Lub09a, Lub09b] and Levin and Lubinsky [Lev08, Lev09] it was shown that bulk universality holds under extremely weak conditions. One of the results is that in an ensemble (6.1.1) restricted to a compact interval, bulk universality holds at each interior point where V is continuous.

Soft edge universality An edge point $x^* \in \{a_1, b_1, \ldots, a_m, b_m\}$ is a regular edge point of $\text{supp}(\mu_V)$ if $h_j(x^*) > 0$ in (6.3.9). In that case the density ρ_V vanishes as a square root at x^*. The scaling limit is then the Airy kernel (6.2.4). If $x^* = b_j$ is a right-edge point then for a certain constant $c > 0$,

$$\lim_{n\to\infty} \frac{1}{(cn)^{2/3}} K_{n,N}\left(x^* + \frac{x}{(cn)^{2/3}}, x^* + \frac{y}{(cn)^{2/3}}\right) = K^{\text{Ai}}(x, y) \tag{6.3.13}$$

while if $x^* = a_j$ is a left-edge point, we find the same limit after a change of sign

$$\lim_{n\to\infty} \frac{1}{(cn)^{2/3}} K_{n,N}\left(x^* - \frac{x}{(cn)^{2/3}}, x^* - \frac{y}{(cn)^{2/3}}\right) = K^{\mathrm{Ai}}(x, y). \tag{6.3.14}$$

Thus the regular edge points belong to the Airy universality class.

The Airy kernel was implicitly derived in [Bow91]. For classical ensembles the soft edge universality (6.3.13) is derived in [For93, Nag93], see also [For10]. For quartic and sextic potentials V, it is derived in [Kan97] by the so-called Shohat method. For real analytic potentials it is implicit in [Dei99c] and made explicit in [Dei07b].

Hard edge universality The Bessel universality classes arise for eigenvalues of positive definite matrices. Let $a > -1$ be a parameter and consider

$$\frac{1}{\widetilde{Z}_{n,N}}(\det M)^a e^{-N\mathrm{Tr}\,V(M)} dM$$

as a probability measure on the space of $n \times n$ Hermitian positive definite matrices. For the case $V(x) = x$ this is the Laguerre unitary ensemble, which is also known as a complex Wishart ensemble, see Chapter 4.

Assuming that a remains fixed, the global eigenvalue regime does not depend on a. The eigenvalue density ρ_V lives on $[0, \infty)$, and for $x \in [0, \infty)$ it continues to have the representation (6.3.10) but now q_V is modified to

$$q_V(x) = \left(\frac{V'(x)}{2}\right)^2 - \int \frac{V'(x) - V'(s)}{x - s} d\mu_V(s) - \frac{1}{x}\int V'(s) d\mu_V(s). \tag{6.3.15}$$

If $\int V'(s)d\mu_V(s) > 0$ then 0 is in the support of μ_V and ρ_V has a square-root singularity at $x = 0$.

The effect of the parameter a is noticeable in the local eigenvalue regime near $x^* = 0$. If $\int V'(s)d\mu_V(s) > 0$, then the limiting kernel is the Bessel kernel (6.2.5) of order a. For $x, y > 0$ and for an appropriate constant $c > 0$, we have

$$\lim_{n\to\infty} \frac{1}{(cn)^2} K_{n,N}\left(\frac{x}{(cn)^2}, \frac{y}{(cn)^2}\right) = K^{\mathrm{Bes},a}(x, y). \tag{6.3.16}$$

The hard edge universality (6.3.16) was proven in [Kui02, Lub08b].

Spectral singularity A related class of Bessel kernels

$$\widehat{K}^{\mathrm{Bes},a}(x, y) = \pi\sqrt{x}\sqrt{y}\frac{J_{a+\frac{1}{2}}(\pi x)J_{a-\frac{1}{2}}(\pi y) - J_{a-\frac{1}{2}}(\pi x)J_{a+\frac{1}{2}}(\pi y)}{2(x - y)} \tag{6.3.17}$$

appears as scaling limits in Hermitian matrix models of the form

$$\frac{1}{\tilde{Z}_{n,N}} \left|\det M\right|^{2a} e^{-N\mathrm{Tr}\,V(M)} dM \tag{6.3.18}$$

with $a > -1/2$.

The extra factor $|\det M|^{2a}$ is referred to as a spectral singularity and the matrix model (6.3.18) is relevant in quantum chromodynamics, where it is referred to as a chiral ensemble; see Chapter 32 of this handbook. The spectral singularity does not change the global density ρ_V of eigenvalues, but it does have an influence on the local eigenvalue statistics at the origin. Assuming that $c = \rho_V(0) > 0$, one now finds

$$\lim_{n\to\infty} \frac{1}{cn} K_{n,N}\left(\frac{x}{cn}, \frac{y}{cn}\right) = \widehat{K}^{\mathrm{Bes},a}(x, y), \qquad x, y > 0, \tag{6.3.19}$$

instead of (6.3.12).

The universality of the Bessel kernels (6.2.5) and (6.3.17) was discussed by Akemann *et al.* [Ake97b] for integer values of a. It was extended to non-integer a in [Kan98]. The model (6.3.18) was analysed in [Kui03] with the Riemann–Hilbert method.

See [Kle00] for the spectral universality in orthogonal ensembles.

Weight with jump discontinuity More recently [Fou11], a new class of limiting kernels was identified for unitary ensembles of the form

$$\frac{1}{\tilde{Z}_n} e^{-\mathrm{Tr}\,V(M)} dM$$

defined on Hermitian matrices with eigenvalues in $[-1, 1]$ in cases where the weight function $w(x) = e^{-V(x)}$, $x \in [-1, 1]$, has a jump discontinuity at $x = 0$. The limiting kernels in [Fou11] are constructed out of confluent hypergeometric functions, see also [Its08].

6.4 Riemann–Hilbert method

A variety of methods have been developed to prove the above universality results in varying degrees of generality. One of these methods will be described here, namely the steepest descent analysis for the Riemann–Hilbert problem (RH problem). This is a powerful method for obtaining strong and uniform asymptotics for orthogonal polynomials. One of the outcomes is the limiting behaviour of the eigenvalue correlation kernels. However, the method gives much more. It is also able to give asymptotics of the recurrence coefficients, Hankel determinants and other notions associated with orthogonal polynomials. In this section we closely follow the paper [Dei99c] of Deift *et al*, see also [Dei99a].

6.4.1 Statement of the Riemann–Hilbert problem

A Riemann–Hilbert problem is a jump problem for a piecewise analytic function in the complex plane. The RH problem for orthogonal polynomials asks for a 2×2 matrix valued function Y satisfying

(1) $Y : \mathbb{C} \setminus \mathbb{R} \to \mathbb{C}^{2\times 2}$ is analytic,

(2) Y has boundary values on the real line, denoted by $Y_\pm(x)$, where $Y_+(x)$ ($Y_-(x)$) denotes the limit of $Y(z)$ as $z \to x \in \mathbb{R}$ with $\operatorname{Im} z > 0$ ($\operatorname{Im} z < 0$), and

$$Y_+(x) = Y_-(x) \begin{pmatrix} 1 & e^{-NV(x)} \\ 0 & 1 \end{pmatrix},$$

(3) $Y(z) = (I + O(1/z)) \begin{pmatrix} z^n & 0 \\ 0 & z^{-n} \end{pmatrix}$ as $z \to \infty$.

The unique solution is given in terms of the orthogonal polynomials (6.3.4) by

$$Y(z) = \begin{pmatrix} P_{n,N}(z) & \dfrac{1}{2\pi i} \displaystyle\int_{-\infty}^{\infty} \frac{P_{n,N}(x) e^{-NV(x)}}{x - z} dx \\ -2\pi i \gamma_{n-1,N}^2 P_{n-1,N}(z) & -\gamma_{n-1,N}^2 \displaystyle\int_{-\infty}^{\infty} \frac{P_{n-1,N}(x) e^{-NV(x)}}{x - z} dx \end{pmatrix}. \tag{6.4.1}$$

The RH problem for orthogonal polynomials and its solution (6.4.1) are due to Fokas, Its, and Kitaev [Fok92].

By the Christoffel–Darboux formula for orthogonal polynomials, we have that the orthogonal polynomial kernel (6.3.5) is equal to

$$K_{n,N}(x, y) = \sqrt{e^{-NV(x)}} \sqrt{e^{-NV(y)}} \gamma_{n-1,N}^2 \frac{P_{n,N}(x) P_{n-1,N}(y) - P_{n-1,N}(x) P_{n,N}(y)}{x - y} \tag{6.4.2}$$

which in view of (6.4.1) and the fact that $\det Y(z) \equiv 1$ can be rewritten as

$$K_{n,N}(x, y) = \frac{1}{2\pi i (x - y)} \sqrt{e^{-NV(x)}} \sqrt{e^{-NV(y)}} \begin{pmatrix} 0 & 1 \end{pmatrix} Y_+(y)^{-1} Y_+(x) \begin{pmatrix} 1 \\ 0 \end{pmatrix}, \tag{6.4.3}$$

for $x, y \in \mathbb{R}$. Other notions related to the orthogonal polynomials are also contained in the solution of the Riemann–Hilbert problem. The monic polynomials satisfy a three-term recurrence relation

$$x P_{n,N}(x) = P_{n+1,N}(x) + b_{n,N} P_{n,N}(x) + a_{n,N} P_{n-1,N}(x)$$

with recurrence coefficients $a_{n,N} > 0$ and $b_{n,N} \in \mathbb{R}$. The recurrence coefficients can be found from the solution of the RH problem by expanding Y around ∞:

$$Y(z) = \left(I + \frac{1}{z} Y_1 + \frac{1}{z^2} Y_2 + \cdots \right) \begin{pmatrix} z^n & 0 \\ 0 & z^{-n} \end{pmatrix}$$

as $z \to \infty$. The 2×2 matrices Y_1 and Y_2 do not depend on z, but do depend on n and N. Then

$$a_{n,N} = (Y_1)_{12} (Y_1)_{21}, \qquad b_{n,N} = \frac{(Y_2)_{12}}{(Y_1)_{12}} - (Y_1)_{22}. \tag{6.4.4}$$

6.4.2 Outline of the steepest descent analysis

The steepest descent analysis of RH problems is due to Deift and Zhou [Dei93]. It produces a number of explicit and invertible transformations leading to a RH problem for a new matrix-valued function R with identity asymptotics at infinity. Also R depends on n and N, and as $n, N \to \infty$ with $n/N \to 1$, the jump matrices tend to the identity matrix, in regular cases typically at a rate of $O(1/n)$. Then it can be shown that $R(z)$ tends to the identity matrix as n, $N \to \infty$ with $n/N \to 1$ and also at a rate of $O(1/n)$ in regular cases.

Following the transformations in the steepest descent analysis we can then find asymptotic formulas for Y and in particular for the orthogonal polynomial $P_{n,N}$, the recurrence coefficients, and the correlation kernel.

With more work one may be able to obtain more precise asymptotic information on R. For example, if $n = N$ and if we are in the one-cut regular case, then there is an asymptotic expansion for $R(z)$:

$$R(z) \sim I + \frac{R^{(1)}(z)}{n} + \frac{R^{(2)}(z)}{n^2} + \cdots$$

with explicitly computable matrices $R^{(j)}(z)$. This in turn leads to asymptotic expansions for the orthogonal polynomials as well. We will not go into this aspect here.

Here we present the typical steps in the Deift-Zhou steepest descent analysis. We focus on the one-cut case, that is on the situation where V is real analytic and the equilibrium measure μ_V is supported on one interval $[a, b]$. We also assume that we are in a regular case, which means that

$$\rho_V(x) = \frac{d\mu_V(x)}{dx} = \frac{1}{\pi} h(x)\sqrt{(b-x)(x-a)}, \qquad x \in [a, b],$$

with $h(x) > 0$ for $x \in [a, b]$, and strict inequality holds in the variational condition (6.4.6) below. Generically these regularity conditions are satisfied, see [Kui00]. The singular cases lead to different universality classes, and this will be commented on in Section 6.5.

For convenience we also take $N = n$.

6.4.3 First transformation: normalization at infinity

The equilibrium measure μ_V is used in the first transformation of the RH problem. The equilibrium measure satisfies for some constant ℓ

$$2 \int \log \frac{1}{|x-y|} d\mu_V(y) + V(x) = \ell, \qquad x \in [a, b], \tag{6.4.5}$$

$$2 \int \log \frac{1}{|x-y|} d\mu_V(y) + V(x) \geq \ell, \qquad x \in \mathbb{R} \setminus [a, b]. \tag{6.4.6}$$

These are the Euler-Lagrange variational conditions associated with the minimization of the weighted energy (6.3.8). Since (6.3.8) is a strictly convex functional on finite energy probability measures, the conditions (6.4.5)–(6.4.6) characterize the minimizer μ_V.

The equilibrium measure μ_V leads to the g-function

$$g(z) = \int \log(z - x)\, d\mu_V(x), \qquad z \in \mathbb{C} \setminus (-\infty, b], \tag{6.4.7}$$

which is used in the first transformation $Y \mapsto T$. We put

$$T(z) = \begin{pmatrix} e^{-n\ell/2} & 0 \\ 0 & e^{n\ell/2} \end{pmatrix} Y(z) \begin{pmatrix} e^{-n(g(z)-\ell/2)} & 0 \\ 0 & e^{n(g(z)-\ell/2)} \end{pmatrix}. \tag{6.4.8}$$

The jumps in the RH problem for T are conveniently stated in terms of the functions

$$\phi(z) = \int_b^z h(s)((s-b)(s-a))^{1/2}\, ds \tag{6.4.9}$$

$$\widetilde{\phi}(z) = \int_a^z h(s)((s-b)(s-a))^{1/2}\, ds, \tag{6.4.10}$$

where we use the fact that V is real analytic and therefore h has an analytic continuation to a neighbourhood of the real line. Then T satisfies the RH problem

(1) T is analytic in $\mathbb{C} \setminus \mathbb{R}$.
(2) On $\mathbb{R}$ we have the jump $T_+ = T_- J_T$ where

$$J_T(x) = \begin{cases} \begin{pmatrix} e^{2n\phi_+(x)} & 1 \\ 0 & e^{2n\phi_-(x)} \end{pmatrix} & x \in (a, b), \\ \begin{pmatrix} 1 & e^{-2n\phi(x)} \\ 0 & 1 \end{pmatrix} & x > b, \\ \begin{pmatrix} 1 & e^{-2n\widetilde{\phi}(x)} \\ 0 & 1 \end{pmatrix} & x < a. \end{cases} \tag{6.4.11}$$

(3) $T(z) = I + O(1/z)$ as $z \to \infty$.

The RH problem is now normalized at infinity. The jump matrices on $(-\infty, a)$ and (b, ∞) tend to the identity matrix as $n \to \infty$, since $\phi(x) > 0$ for $x > b$ and $\widetilde{\phi}(x) > 0$ for $x < a$, which is due to the assumption that strict inequality holds in (6.4.6). We have to deal with the jump on (a, b). The diagonal entries in the jump matrix on (a, b) are highly oscillatory. The goal of the next transformation is to turn these highly oscillatory entries into exponentially decaying ones.

6.4.4 Second transformation: opening of lenses

We open up a lens around the interval $[a, b]$ as in Figure 6.1 and define

$$S = \begin{cases} T \begin{pmatrix} 1 & 0 \\ -e^{2n\phi} & 0 \end{pmatrix} & \text{in the upper part of the lens,} \\ T \begin{pmatrix} 1 & 0 \\ e^{2n\phi} & 0 \end{pmatrix} & \text{in the lower part of the lens.} \\ T & \text{elsewhere.} \end{cases} \tag{6.4.12}$$

Then S is defined and analytic in $\mathbb{C} \setminus \Sigma_S$ where Σ_S is the contour consisting of the real line, and the upper and lower lips of the lens, with orientation as indicated by the arrows in Figure 6.1. The orientation induces a $+$ side and a $-$ side on each part of Σ_S, where the $+$ side is on the left as one traverses the contour according to its orientation and the $-$ side is on the right. We use $S_\pm$ to denote the limiting values of S on Σ_S when approaching Σ_S from the $\pm$ side. This convention about $\pm$ limits, depending on the orientation of the contour is usual in Riemann–Hilbert problems and will also be used later on.

Then S satisfies the following RH problem:

(1) S is analytic in $\mathbb{C} \setminus \Sigma_S$.
(2) On Σ_S we have the jump $S_+ = S_- J_S$ where

$$J_S(x) = \begin{cases} \begin{pmatrix} 0 & 1 \\ -1 & 0 \end{pmatrix} & \text{for } x \in (a, b), \\ \begin{pmatrix} 1 & 0 \\ e^{2n\phi(x)} & 1 \end{pmatrix} & \text{on the lips of the lens,} \\ J_T(x) & \text{for } x < a \text{ or } x > b. \end{cases} \tag{6.4.13}$$

(3) $S(z) = I + O(1/z)$ as $z \to \infty$.

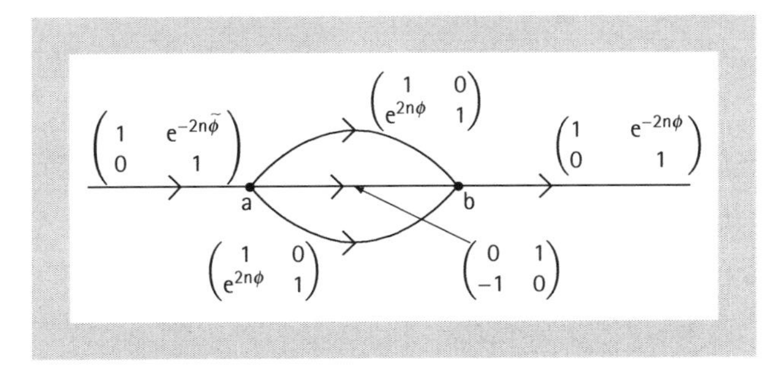

Fig. 6.1 Contour Σ_S and jump matrices for the RH problem for S. This figure is reproduced from [Kui09].

If the lens is taken sufficiently small, we can guarantee that $\mathrm{Re}\,\phi < 0$ on the lips of the lens. Then the jump matrices for S tend to the identity matrix as $n \to \infty$ on the lips of the lens and on the unbounded intervals $(-\infty, a)$ and (b, ∞).

6.4.5 Outside parametrix

The next step is to build an approximations to S, valid for large n, the so-called parametrix. The parametrix consists of two parts, an outside or global parametrix that will model S away from the endpoints and local parametrices that are good approximations to S in a neighbourhood of the endpoints.

The outside parametrix M satisfies

(1) M is analytic in $\mathbb{C} \setminus [a, b]$,

(2) M has the jump

$$M_+(x) = M_-(x) \begin{pmatrix} 0 & 1 \\ -1 & 0 \end{pmatrix} \quad \text{for } x \in (a, b).$$

(3) $M(z) = I + O(1/z)$ as $z \to \infty$.

To have uniqueness of a solution we also impose

(4) M has at most fourth-root singularities at the endpoints a and b.

The solution to this RH problem is given by

$$M(z) = \begin{pmatrix} \dfrac{\beta(z) + \beta^{-1}(z)}{2} & \dfrac{\beta(z) - \beta^{-1}(z)}{2i} \\ -\dfrac{\beta(z) - \beta^{-1}(z)}{2i} & \dfrac{\beta(z) + \beta^{-1}(z)}{2} \end{pmatrix}, \qquad z \in \mathbb{C} \setminus [a, b], \tag{6.4.14}$$

where

$$\beta(z) = \left(\frac{z-b}{z-a} \right)^{1/4}. \tag{6.4.15}$$

6.4.6 Local parametrix

The local parametrix is constructed in neighbourhoods $U_\delta(a) = \{z \mid |z-a| < \delta\}$ and $U_\delta(b) = \{z \mid |z-b| < \delta\}$ of the endpoints a and b, where $\delta > 0$ is small, but fixed.

In $U_\delta(a) \cup U_\delta(b)$ we want to have a 2×2 matrix valued function P satisfying the following (see also Figure 6.2 for the jumps in the neighbourhood of b):

(1) P is defined and analytic in $(U_\delta(a) \cup U_\delta(b)) \setminus \Sigma_S$ and has a continuous extension to $\left(\overline{U_\delta(a)} \cup \overline{U_\delta(b)}\right) \setminus \Sigma_S$.

(2) On $\Sigma_S \cap (U_\delta(a) \cup U_\delta(b))$ there is the jump

$$P_+ = P_- J_S$$

where J_S is the jump matrix (6.4.13) in the RH problem for S.

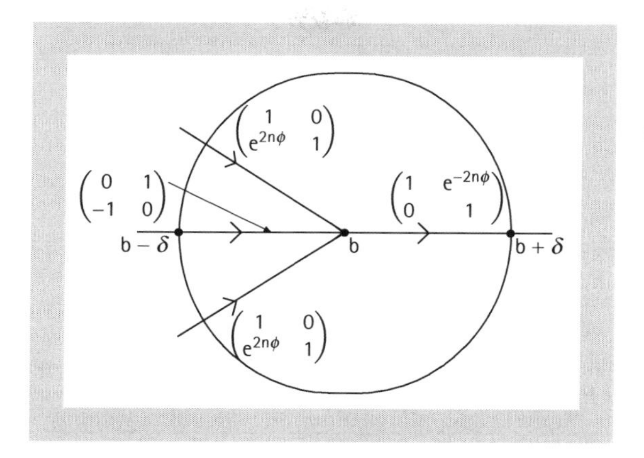

Fig. 6.2 Neighbourhood $U_\delta(b)$ of b and contours and jump matrices for the RH problem for P.

(3) P agrees with the global parametrix M on the boundaries of $U_\delta(a)$ and $U_\delta(b)$ in the sense that

$$P(z) = \left(I + O\left(\frac{1}{n}\right)\right) M(z) \tag{6.4.16}$$

as $n \to \infty$ uniformly for $z \in \partial U_\delta(a) \cup \partial U_\delta(b)$.

(4) $P(z)$ remains bounded as $z \to a$ or $z \to b$.

The solution of the RH problem for P is constructed out of the Airy function

$$y_0(z) = \mathrm{Ai}(z),$$

and its rotated versions

$$y_1(z) = \omega \, \mathrm{Ai}(\omega z), \qquad y_2(z) = \omega^2 \, \mathrm{Ai}(\omega^2 z), \qquad \omega = e^{2\pi i/3}.$$

These three solutions of the Airy differential equation $y''(z) = zy(z)$ are connected by the identity

$$y_0(z) + y_1(z) + y_2(z) = 0.$$

They are used in the solution of the following model RH problem posed on the contour Σ_A shown in Figure 6.3.

(1) $A : \mathbb{C} \setminus \Sigma_A \to \mathbb{C}^{2\times 2}$ is analytic.

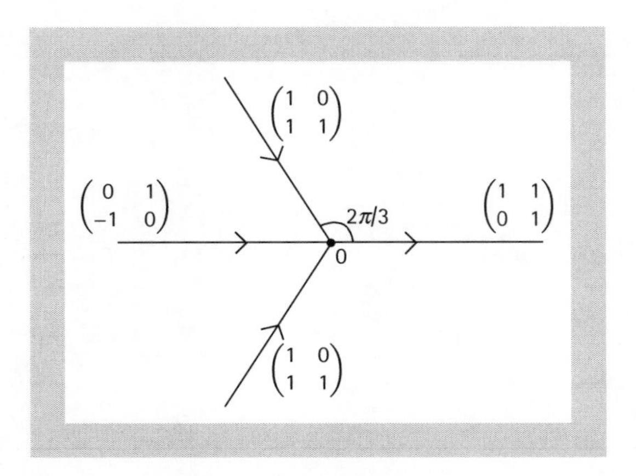

Fig. 6.3 Contour Σ_A and jump matrices for the Airy Riemann–Hilbert problem.

(2) A has jump $A_+ = A_- J_A$ on Σ_A with jump matrix J_A given by

$$J_A(z) = \begin{cases} \begin{pmatrix} 1 & 1 \\ 0 & 1 \end{pmatrix} & \text{for } \arg z = 0, \\ \begin{pmatrix} 1 & 0 \\ 1 & 1 \end{pmatrix} & \text{for } \arg z = \pm 2\pi i/3, \\ \begin{pmatrix} 0 & 1 \\ -1 & 0 \end{pmatrix} & \text{for } \arg z = \pi. \end{cases} \tag{6.4.17}$$

(3) As $z \to \infty$, we have

$$A(z) = \begin{pmatrix} z^{-1/4} & 0 \\ 0 & z^{1/4} \end{pmatrix} \frac{1}{\sqrt{2}} \begin{pmatrix} 1 & i \\ i & 1 \end{pmatrix} \left(I + O(z^{-3/2})\right) \begin{pmatrix} e^{-\frac{2}{3}z^{3/2}} & 0 \\ 0 & e^{\frac{2}{3}z^{3/2}} \end{pmatrix}. \tag{6.4.18}$$

(4) $A(z)$ remains bounded as $z \to 0$.

The rather complicated asymptotics in (6.4.18) corresponds to the asymptotic formulas

$$\begin{aligned} \mathrm{Ai}(z) &= \frac{1}{2\sqrt{\pi} z^{1/4}} e^{-\frac{2}{3}z^{3/2}} \left(1 + \mathcal{O}(z^{-3/2})\right), \\ \mathrm{Ai}'(z) &= -\frac{z^{1/4}}{2\sqrt{\pi}} e^{-\frac{2}{3}z^{3/2}} \left(1 + \mathcal{O}(z^{-3/2})\right), \end{aligned} \tag{6.4.19}$$

as $z \to 0$ with $-\pi < \arg z < \pi$, that are known for the Airy function and its derivative.

The solution of the Airy Riemann–Hilbert problem is as follows

$$A(z) = \sqrt{2\pi} \times \begin{cases} \begin{pmatrix} y_0(z) & -y_2(z) \\ -iy_0'(z) & iy_2'(z) \end{pmatrix}, & 0 < \arg z < 2\pi/3, \\ \begin{pmatrix} -y_1(z) & -y_2(z) \\ -iy_1'(z) & iy_2'(z) \end{pmatrix}, & 2\pi/3 < \arg z < \pi, \\ \begin{pmatrix} -y_2(z) & y_1(z) \\ iy_2'(z) & -iy_1'(z) \end{pmatrix}, & -\pi < \arg z < -2\pi/3, \\ \begin{pmatrix} y_0(z) & y_1(z) \\ -iy_0'(z) & -iy_1'(z) \end{pmatrix}, & -2\pi/3 < \arg z < 0. \end{cases} \tag{6.4.20}$$

The constants $\sqrt{2\pi}$ and $\pm i$ are such that $\det A(z) \equiv 1$ for $z \in \mathbb{C} \setminus \Sigma_A$.

To construct the local parametrix P in the neighbourhood $U_\delta(b)$ of b we also need the function

$$f(z) = \left[\frac{3}{2}\phi(z)\right]^{2/3}$$

which is a conformal map from $U_\delta(b)$ (provided δ is small enough) onto a neighbourhood of the origin. For this it is important that the density ρ_V vanishes as a square root at b. We may assume that the lens around (a, b) is opened in such a way that the parts of the lips of the lens within $U_\delta(b)$ are mapped by f into the rays $\arg z = \pm 2\pi/3$.

Then the local parametrix P is given in $U_\delta(b)$ by

$$P(z) = E_n(z)\, A(n^{2/3} f(z)) \begin{pmatrix} e^{n\phi(z)} & 0 \\ 0 & e^{-n\phi(z)} \end{pmatrix}, \qquad z \in U_\delta(b) \setminus \Sigma_S, \tag{6.4.21}$$

where the prefactor $E_n(z)$ is given explicitly by

$$\begin{aligned} E_n(z) = -\frac{1}{\sqrt{2}} \begin{pmatrix} 1 & -i \\ -i & 1 \end{pmatrix} \begin{pmatrix} n^{1/6}(z-a)^{1/4} & 0 \\ 0 & n^{-1/6}(z-a)^{-1/4} \end{pmatrix} \\ \times \begin{pmatrix} (f(z)/(z-b))^{1/4} & 0 \\ 0 & (f(z)/(z-b))^{-1/4} \end{pmatrix}. \end{aligned} \tag{6.4.22}$$

Then E_n is analytic in $U_\delta(b)$ and it does not change the jump conditions. It is needed in order to satisfy the matching condition (6.4.16) on the boundary $|z - b| = \delta$ of $U_\delta(b)$.

A similar construction gives the local parametrix P in the neighbourhood $U_\delta(a)$ of a.

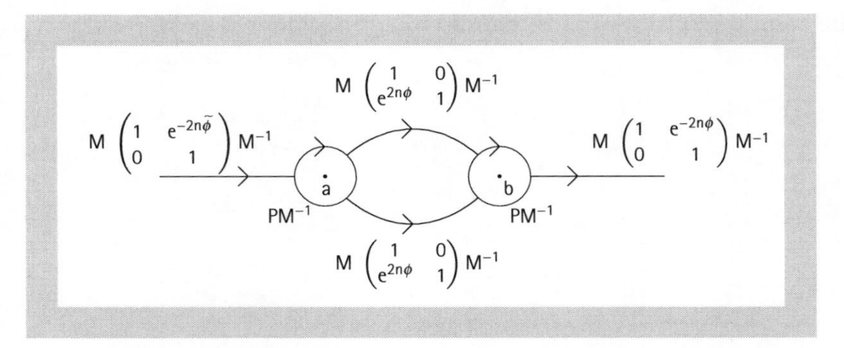

Fig. 6.4 Contour Σ_R and jump matrices for the RH problem for R.

6.4.7 Final transformation

In the final transformation we put

$$R(z) = \begin{cases} S(z)\, M(z)^{-1}, & z \in \mathbb{C} \setminus (\Sigma_S \cup \overline{U_\delta(a)} \cup \overline{U_\delta(b)}), \\ S(z)\, P(z)^{-1}, & z \in (U_\delta(a) \cup U_\delta(b)) \setminus \Sigma_S. \end{cases} \tag{6.4.23}$$

Then R has an analytic continuation to $\mathbb{C} \setminus \Sigma_R$ where Σ_R is shown in Figure 6.4.

R solves the following RH problem.

(1) R is analytic in $\mathbb{C} \setminus \Sigma_R$.
(2) R satisfies the jump conditions $R_+ = R_- J_R$ where J_R are the matrices given in Figure 6.4.
(3) $R(z) = I + O(z^{-1})$ as $z \to \infty$.

The jump matrices tend to the identity matrix as $n \to \infty$. Indeed, the jump matrix J_R on the boundaries of the disks satisfies

$$J_R(z) = P(z)\, M(z)^{-1} = I + O(n^{-1})$$

as $n \to \infty$, because of the matching condition (6.4.16). On the remaining parts of Σ_R we have that $J_R = I + O(e^{-cn})$ as $n \to \infty$ for some constant $c > 0$.

Technical estimates on solutions of RH problems, see [Dei99a], now guarantee that in this case

$$R(z) = I + O(1/n) \qquad \text{as } n \to \infty, \tag{6.4.24}$$

uniformly for $z \in \mathbb{C} \setminus \Sigma_R$.

6.4.8 Proof of bulk universality

Now we turn our attention again to the correlation kernel (6.4.3) which, since $n = N$, we denote by K_n instead of $K_{n,N}$. We follow what happens with K_n under the transformations $Y \mapsto T \mapsto S$. We assume that $x, y \in (a, b)$.

From the definition (6.4.8) we obtain

$$K_n(x, y) = \frac{\sqrt{e^{-nV(x)}}\sqrt{e^{-nV(y)}}}{2\pi i(x-y)} \begin{pmatrix} 0 & e^{n(g_+(y)+\ell/2)} \end{pmatrix} T_+^{-1}(y) T_+(x) \begin{pmatrix} e^{n(g_+(x)+\ell/2)} \\ 0 \end{pmatrix}$$

which by (non-trivial) properties of (6.4.7) and (6.4.9), based on the variational condition (6.4.5), can be rewritten as

$$K_n(x, y) = \frac{1}{2\pi i(x-y)} \begin{pmatrix} 0 & e^{-n\phi_+(y)} \end{pmatrix} T_+^{-1}(y) T_+(x) \begin{pmatrix} e^{-n\phi_+(x)} \\ 0 \end{pmatrix}. \tag{6.4.25}$$

Using the transformation (6.4.12) in the upper part of the lens, we see that (6.4.25) leads to

$$K_n(x, y) = \frac{1}{2\pi i(x-y)} \begin{pmatrix} -e^{n\phi_+(y)} & e^{-n\phi_+(y)} \end{pmatrix} S_+^{-1}(y) S_+(x) \begin{pmatrix} e^{-n\phi_+(x)} \\ e^{n\phi_+(x)} \end{pmatrix} \tag{6.4.26}$$

for $x, y \in (a, b)$. This is the basic formula for K_n in terms of S.

If $x, y \in (a+\delta, b-\delta)$, then by (6.4.23)

$$S_+^{-1}(y) S_+(x) = M_+^{-1}(y) R(y)^{-1} R(x) M_+(x).$$

The uniform estimate (6.4.24) on R can then be used to show that

$$S_+^{-1}(y) S_+(x) = I + O(x-y). \tag{6.4.27}$$

with an O-term that is uniform for $x, y \in (a+\delta, b-\delta)$. Using (6.4.27) in (6.4.26) we arrive at

$$K_n(x, y) = \frac{\sin(in(\phi_+(y) - \phi_+(x))}{\pi(x-y)} + O(1). \tag{6.4.28}$$

Take $x^* \in (a, b)$ fixed and let $c = \rho_V(x^*) > 0$. We may assume that $\delta > 0$ has been chosen so small that $x^* \in (a+\delta, b-\delta)$. Replace x and y in (6.4.28) by $x^* + \frac{x}{cn}$ and $x^* + \frac{y}{cn}$ respectively. Then after dividing through by cn, we have for fixed x and y,

$$\frac{1}{cn} K_n\left(x^* + \frac{x}{cn}, x^* + \frac{y}{cn}\right) = \frac{\sin\left(in\left(\phi_+\left(x^* + \frac{y}{cn}\right) - \phi_+\left(x^* + \frac{x}{cn}\right)\right)\right)}{\pi(x-y)} + O\left(\frac{1}{n}\right) \tag{6.4.29}$$

as $n \to \infty$, and the O-term is uniform for x and y in compact subsets of $\mathbb{R}$. Since $\phi_+'(s) = \pi i \rho_V(s)$ for $s \in (a, b)$, and $c = \rho_V(x^*)$, we have that

$$in\left(\phi_+\left(x^* + \frac{y}{cn}\right) - \phi_+\left(x^* + \frac{x}{cn}\right)\right) = \pi(x-y) + O\left(\frac{x-y}{n}\right)$$

and therefore the rescaled kernel (6.4.29) does indeed tend to the sine kernel (6.2.3) as $n \to \infty$. This proves the bulk universality.

Remark Letting $y \to x$ in (6.4.28) and dividing by n we also obtain

$$\begin{aligned}\frac{1}{n}K_n(x,x) &= \frac{1}{\pi i}\phi'_+(x) + O\left(\frac{1}{n}\right) \\ &= \frac{1}{\pi}h(x)\sqrt{(b-x)(x-a)} + O\left(\frac{1}{n}\right) && \text{(by (6.4.9))} \\ &= \rho_V(x) + O\left(\frac{1}{n}\right) && \text{(by (6.3.9))}\end{aligned}$$

as $n \to \infty$, uniformly for x in compact subsets of (a, b), which proves that ρ_V is the limiting means eigenvalue density.

6.4.9 Proof of edge universality

We will not give the proof of the edge universality in detail. The proof starts from the representation (6.4.26) of the correlation kernel in terms of S. In the neighbourhood of a and b we have by (6.4.23) that $S = RP$, where P is the local parametrix that is constructed out of Airy functions, as it involves the solution of the Airy Riemann–Hilbert problem.

From (6.4.20) it can be checked that the Airy kernel (6.2.4) is given in terms of the solution $A(z)$ by

$$K^{\mathrm{Ai}}(x,y) = \frac{1}{2\pi i(x-y)} \times \begin{cases} \begin{pmatrix} 0 & 1 \end{pmatrix} A_+^{-1}(y)A_+(x)\begin{pmatrix} 1 \\ 0 \end{pmatrix} & \text{if both } x, y > 0, \\ \begin{pmatrix} -1 & 1 \end{pmatrix} A_+^{-1}(y)A_+(x)\begin{pmatrix} 1 \\ 0 \end{pmatrix} & \text{if } x > 0 \text{ and } y < 0, \\ \begin{pmatrix} 0 & 1 \end{pmatrix} A_+^{-1}(y)A_+(x)\begin{pmatrix} 1 \\ 1 \end{pmatrix} & \text{if } x < 0 \text{ and } y > 0, \\ \begin{pmatrix} -1 & 1 \end{pmatrix} A_+^{-1}(y)A_+(x)\begin{pmatrix} 1 \\ 1 \end{pmatrix} & \text{if both } x, y < 0, \end{cases}$$

and this is exactly what comes out of the calculations for the scaling limit of the eigenvalue correlation kernels K_n near the edge point.

6.5 Non-standard universality classes

The standard universality classes (sine, Airy, and Bessel) describe the local eigenvalue statistics around regular points.

In the unitary ensemble (6.3.1) there are three types of singular eigenvalue behaviour. They all depend on the behaviour of the global eigenvalue density ρ_V. The three types of singular behaviour are:

- The density ρ_V vanishes at an interior point of the support.
- The density ρ_V vanishes to higher order at an edge point of the support (higher than square root).
- Equality holds in the variational inequality (6.4.6) at a point outside $\operatorname{supp}(\mu_V)$.

In each of these cases there exists a family of limiting correlation kernels that arise in a double scaling limit. In (6.3.1) and (6.3.3) one lets $n, N \to \infty$ with $n/N \to 1$ at a critical rate so that for some exponent γ, the limit

$$\lim_{n\to\infty} n^{\gamma}\left(\frac{n}{N} - 1\right) \tag{6.5.1}$$

exists. The family of limiting kernels is parameterized by the value of the limit (6.5.1).

Most rigorous results that have been obtained in this direction are based on the RH problem, and use an extension of the steepest descent analysis that was discussed in the previous section. Non-standard universality classes are also discussed in Chapters 12 and 13.

6.5.1 Interior singular point

An interior singular point is a point x^* where ρ_V vanishes. Varying the parameters in V may then either lead to a gap in the support around x^*, or to the closing of the gap. The local scaling limits at x^* depend on the order of vanishing at x^*, that is, on the positive integer k so that

$$\rho_V(x) = c(x - x^*)^{2k}(1 + o(1)) \qquad \text{as } x \to x^*$$

with $c > 0$.

The case of quadratic vanishing (i.e. $k = 1$) was considered in [Ble03] for the critical quartic potential

$$V(x) = \frac{1}{4}x^4 - x^2$$

and in [Cla06, Shc08] for more general V. For $k = 1$ one takes $\gamma = 2/3$ in (6.5.1). The limiting kernels are parameterized by a parameter s which is proportional to the limit (6.5.1)

$$s = c \lim_{n\to\infty} n^{2/3}\left(\frac{n}{N} - 1\right),$$

where the proportionality constant $c > 0$ is (for the case $\operatorname{supp}(\mu_V) = [a, b]$)

$$c = \frac{2}{(\pi \rho_V''(x^*))^{1/3}\sqrt{(b - x^*)(x^* - a)}}.$$

The s-dependence is then governed by the Hastings-Mcleod solution $q(s)$ of the Painlevé II equation

$$q'' = sq + 2q^3. \tag{6.5.2}$$

The limiting kernels are therefore called Painlevé II kernels and we denote them by $K^{\mathrm{PII}}(x, y; s)$, even though $q(s)$ itself does not appear in the formulas for the kernels. What does appear is a solution of the Lax pair equations

$$\begin{aligned}
\frac{\partial}{\partial x}\begin{pmatrix}\Phi_1(x;s)\\ \Phi_2(x;s)\end{pmatrix} &= \begin{pmatrix}-4ix^2 - i(s+2q^2) & 4xq + 2ir\\ 4xq - 2ir & 4ix^2 + i(s+2q^2)\end{pmatrix}\begin{pmatrix}\Phi_1(x;s)\\ \Phi_2(x;s)\end{pmatrix}\\
\frac{\partial}{\partial s}\begin{pmatrix}\Phi_1(x;s)\\ \Phi_2(x;s)\end{pmatrix} &= \begin{pmatrix}-ix & q\\ q & ix\end{pmatrix}\begin{pmatrix}\Phi_1(x;s)\\ \Phi_2(x;s)\end{pmatrix}
\end{aligned} \tag{6.5.3}$$

where $q = q(s)$ is the Hastings-Mcleod solution of (6.5.2) and $r = r(s) = q'(s)$.

There is a specific solution of (6.5.3) so that the family of limiting kernels takes the form

$$K^{\mathrm{PII}}(x, y; s) = \frac{-\Phi_1(x;s)\Phi_2(y;s) + \Phi_1(y;s)\Phi_1(y;s)}{2\pi i(x-y)}. \tag{6.5.4}$$

For $k \geq 2$, one takes $\gamma = 2k/(2k+1)$ in (6.5.1), and the limiting kernels can be described in a similar way by a special solution of the kth member of the Painlevé II hierarchy. The functions in the kernels itself are solutions of the associated Lax pair equations.

The connection with Painlevé II also holds for the unitary matrix model with a spectral singularity (6.3.18)

$$\frac{1}{\tilde{Z}_{n,N}}\left|\det M\right|^{2\alpha} e^{-N\mathrm{Tr}\,V(M)} dM, \qquad \alpha > -1/2.$$

In the multicritical case where ρ_V vanishes quadratically at $x = 0$, the limiting kernels are associated with a special solution of the general form of the Painlevé II equation

$$q'' = sq + 2q^3 - \alpha$$

with parameter α, see [Ake98, Cla08b].

6.5.2 Singular edge point

A singular edge point is a point x^* where ρ_V vanishes to higher order than square root. The local scaling limits depend again on the order of vanishing. If x^* is a right-edge point of an interval in the support, then there is an even integer k such that

$$\rho_V(x) = c(x^* - x)^{k+1/2}(1 + o(1)) \qquad \text{as } x \nearrow x^* \tag{6.5.5}$$

with $c > 0$. Here one takes $\gamma = (2k+2)/(2k+3)$ in (6.5.1). The limiting kernels are described by Lax pair solutions associated with a special solution of the kth member of the Painlevé I hierarchy. For $k = 2$ this is worked out in detail in

[Cla07]. For general k, and assuming x^* is the right-most point in the support, the largest eigenvalue distributions are studied in [Cla10]. These generalizations of the Tracy–Widom distribution are expressed in terms of members of a Painlevé II hierarchy.

The case $k = 1$ in (6.5.5) cannot occur in the unitary matrix model (6.3.1), although it frequently appears in the physics literature, see e.g. [DiF95], where, for example, it is associated with the potential

$$V(x) = -\frac{1}{48}x^4 + \frac{1}{2}x^2.$$

This potential does not satisfy (6.3.2) and the model (6.3.1) is not well-defined as a probability measure on Hermitian matrices, although it can be studied in a formal sense, see e.g. in Chapter 16.

In [Fok92] and [Dui06] the polynomials that are orthogonal on certain contours in the complex plane with a weight $e^{-NV(x)}$, where $V(x) = t\frac{x^4}{4} + \frac{x^2}{2}$ and $t \approx -\frac{1}{12}$, are studied and the connection with special solutions of the Painlevé I equation

$$q'' = 6q^2 + s$$

is rigorously established.

6.5.3 Exterior singular point

An exterior point x^* is singular if there is equality in the variational inequality (6.4.6). At such a point a new interval in the support may arise when perturbing the potential V. The limiting kernels at such a point depend on the order of vanishing of

$$2\int \log \frac{1}{|x-y|} d\mu_V(y) + V(x) - \ell$$

at $x = x^*$. In this situation the appropriate scaling is so that

$$\lim_{n\to\infty} \frac{n}{\log n}\left(\frac{n}{N} - 1\right) \tag{6.5.6}$$

exists, instead of (6.5.1). This special kind of scaling was discussed in [Ake97a, Eyn06]. Maybe surprisingly, there is no connection with Painlevé equations in this case. In the simplest case of quadratic vanishing at $x = x^*$ the possible limiting kernels are finite size GUE kernels and certain interpolants, see [Ber09, Cla08a, Mo08].

6.5.4 Pearcey kernels

The Painlevé II kernels (6.5.4) are the canonical kernels that arise at the closing of a gap in unitary ensembles. Brézin and Hikami [Bre98] were the first to identify a second one-parameter family of kernels that may arise at the closing of a gap. This is the family of Pearcey kernels

$$K^{\text{Pear}}(x, y; s) = \frac{p(x)q''(y) - p'(x)q'(y) + p''(x)q(y) - sp(x)q(y)}{x - y} \tag{6.5.7}$$

with $s \in \mathbb{R}$, where p and q are solutions of the Pearcey differential equations $p'''(x) = xp(x) - sp'(x)$ and $q'''(y) = yq(y) + sq'(y)$. The kernel is also given by the double integral

$$K^{\text{Pear}}(x, y; s) = \frac{1}{(2\pi i)^2} \int_C \int_{-i\infty}^{i\infty} e^{-\frac{1}{4}\eta^4 + \frac{s}{2}\eta^2 - \eta y + \frac{1}{4}\xi^4 - \frac{s}{2}\xi^2 + \xi x} \frac{d\eta\, d\xi}{\eta - \xi} \tag{6.5.8}$$

where the contour C consists of the two rays from $\pm\infty e^{i\pi/4}$ to 0 together with the two rays from 0 to $\pm\infty e^{-i\pi/4}$.

The Pearcey kernels appear at the closing of the gap in the Hermitian matrix model with external source

$$\frac{1}{\tilde{Z}_n} e^{-n\text{Tr}(V(M) - AM)} dM \tag{6.5.9}$$

where the external source A is a given Hermitian $n \times n$ matrix. This was proved in [Bre98] for the case where $V(x) = \frac{1}{2}x^2$ and A is a diagonal matrix

$$A = \text{diag}(\underbrace{a, \ldots, a}_{n/2 \text{ times}}, \underbrace{-a, \ldots, -a}_{n/2 \text{ times}}) \tag{6.5.10}$$

depending on the parameter $a > 0$. For $a > 1$ the eigenvalues of M accumulate on two intervals as $n \to \infty$ and for $0 < a < 1$ on one interval. The Pearcey kernels (6.5.8) appear in a double scaling limit around the critical value $a = 1$, see also [Ble07] for an analysis of an associated 3×3 matrix valued RH problem. The matrix model with external source (6.5.9) with quadratic potential has an interesting interpretation in terms of non-intersecting Brownian motions [Apt05].

The case of a quartic polynomial potential $V(x) = \frac{1}{4}x^4 - \frac{t}{2}x^2$ was analysed recently in [Ble11]. Here it was found that the closing of the gap can be either of the Pearcey type or of the Painlevé II type, depending on the value of $t \in \mathbb{R}$.

Acknowledgements

The author is supported in part by FWO-Flanders projects G.0427.09 and G.0641.11, by K.U. Leuven research grant OT/08/33, by the Belgian Interuniversity Attraction Pole P06/02, and by a grant from the Ministry of Education and Science of Spain, project code MTM2005-08648-C02-01.

References

[Ake97a] G. Akemann, Universal correlators for multi-arc complex matrix models, Nucl.Phys. B 507 (1997) 475–500.

[Ake97b] G. Akemann, P.H. Damgaard, U. Magnea, and S. Nishigaki, Universality of random matrices in the microscopic limit and the Dirac operator spectrum, Nucl. Phys. B 487,(1997), 721–738.

[Ake98] G. Akemann, P. H. Damgaard, U. Magnea, S. M. Nishigaki, Multicritical microscopic spectral correlators of hermitian and complex matrices, Nucl.Phys. B 519 (1998), 682–714.

[And10] G.W. Anderson, A. Guionnet, and O. Zeitouni, *An Introduction to Random Matrices*, Cambridge Univ. Press, Cambridge, 2010.

[Apt05] A.I. Aptekarev, P.M. Bleher, and A.B.J. Kuijlaars, Large n limit of Gaussian random matrices with external source II, Comm. Math. Phys. 259 (2005), 367–389.

[Ben97] G. Ben Arous and A. Guionnet, Large deviations for Wigner's law and Voiculescu's non-commutative entropy, Probab. Theory Related Fields 108 (1997), 517–542.

[Ber09] M. Bertola and S.Y. Lee, First colonization of a spectral outpost in random matrix theory, Constr. Approx. 30 (2009), 225–263.

[Ble11] P. Bleher, S. Delvaux, and A.B.J. Kuijlaars, Random matrix model with external source and a constrained vector equilibrium problem, Comm. Pure Appl. Math. 64 (2011), 116–160.

[Ble99] P. Bleher and A. Its, Semiclassical asymptotics of orthogonal polynomials, Riemann–Hilbert problem, and universality in the matrix model, Ann. of Math. 150 (1999), 185–266.

[Ble03] P. Bleher and A. Its, Double scaling limit in the random matrix model: the Riemann–Hilbert approach, Comm. Pure Appl. Math. 56 (2003), 433–516.

[Ble07] P.M. Bleher and A.B.J. Kuijlaars, Large n limit of Gaussian random matrices with external source III: double scaling limit, Comm. Math. Phys. 270 (2007), 481–517.

[Bow91] M.J. Bowick and E. Brézin, Universal scaling of the tail of the density of eigenvalues in random matrix models, Phys. Lett. B 268 (1991), 21–28.

[Bre98] E. Brézin and S. Hikami, Universal singularity at the closure of a gap in a random matrix theory, Phys. Rev. E 57 (1998), 4140–4149.

[Bre93] E. Brézin and A. Zee, Universality of the correlations between eigenvalues of large random matrices, Nuclear Physics B 402 (1993), 613–627.

[Cla08a] T. Claeys, Birth of a cut in unitary random matrix ensembles. Int. Math. Res. Notices 2008, Art. ID rnm166, 40 pp.

[Cla10] T. Claeys, A.R. Its, and I. Krasovsky, Higher order analogues of the Tracy–Widom distribution and the Painlevé II hierarchy, Comm. Pure Appl. Math. 63 (2010), 362–412.

[Cla06] T. Claeys and A.B.J. Kuijlaars, Universality of the double scaling limit in random matrix models, Comm. Pure Appl. Math. 59 (2006), 1573–1603.

[Cla08b] T. Claeys, A.B.J. Kuijlaars, and M. Vanlessen, Multi-critical unitary random matrix ensembles and the general Painlevé II equation, Ann. of Math. 168 (2008), 601–641.

[Cla07] T. Claeys and M. Vanlessen, Universality of a double scaling limit near singular edge points in random matrix models, Comm. Math. Phys. 273 (2007), 499–532.

[Dei99a] P. Deift, *Orthogonal Polynomials and Random Matrices: a Riemann–Hilbert Approach*, Amer. Math. Soc., Providence, RI, 1999.

[Dei07d] P. Deift, Universality for mathematical and physical systems, in: International Congress of Mathematicians, Vol. I, 125–152, Eur. Math. Soc., Zürich, 2007.

[Dei07a] P. Deift and D. Gioev, Universality in random matrix theory for orthogonal and symplectic ensembles, Int. Math. Res. Papers 2007, Art. ID rpm004, 116 pp.

[Dei07b] P. Deift and D. Gioev, Universality at the edge of the spectrum for unitary, orthogonal, and symplectic ensembles of random matrices, Comm. Pure Appl. Math. 60 (2007), 867–910.

[Dei09] P. Deift and D. Gioev, *Random Matrix Theory: Invariant Ensembles and Universality*, Courant Lecture Notes in Mathematics, 18, Amer. Math. Soc., Providence, RI, 2009.

[Dei07c] P. Deift, D. Gioev, T. Kriecherbauer, and M. Vanlessen, Universality for orthogonal and symplectic Laguerre-type ensembles, J. Stat. Phys. 129 (2007), 949–1053.

[Dei98] P. Deift, T. Kriecherbauer, and K.T-R McLaughlin, New results on the equilibrium measure for logarithmic potentials in the presence of an external field, J. Approx. Theory 95 (1998), 388–475.

[Dei99b] P. Deift, T. Kriecherbauer, K.T-R McLaughlin, S. Venakides, and X. Zhou, Strong asymptotics of orthogonal polynomials with respect to exponential weights, Comm. Pure Appl. Math. 52 (1999), 1491–1552.

[Dei99c] P. Deift, T. Kriecherbauer, K.T-R McLaughlin, S. Venakides, and X. Zhou, Uniform asymptotics for polynomials orthogonal with respect to varying exponential weights and applications to universality questions in random matrix theory, Comm. Pure Appl. Math. 52 (1999), 1335–1425.

[Dei93] P. Deift and X. Zhou, A steepest descent method for oscillatory Riemann–Hilbert problems. Asymptotics for the MKdV equation, Ann. of Math. 137 (1993), 295–368.

[DiF95] P. Di Francesco, P. Ginsparg, and J. Zinn-Justin, 2D gravity and random matrices, Phys. Rep. 254 (1995), 133 pp.

[Dui06] M. Duits and A.B.J. Kuijlaars, Painlevé I asymptotics for orthogonal polynomials with respect to a varying quartic weight, Nonlinearity 19 (2006), 2211–2245.

[Erd10a] L. Erdős, S. Péché, J. Ramírez, B. Schlein and H-T. Yau, Bulk universality for Wigner matrices, Comm. Pure Appl. Math. 63 (2010), 895–925.

[Erd10b] L. Erdős, J. Ramírez, B. Schlein, T. Tao, V. Vu, and H-T. Yau, Bulk universality for Wigner hermitian matrices with subexponential decay, Math. Res. Lett. 17 (2010), 667–674.

[Erd11] L. Erdős, Universality of Wigner random matrices: a survey of recent results, Russian Math. Surveys 66 (2011) 507–626.

[Eyn06] B. Eynard, Universal distribution of random matrix eigenvalues near the "birth of a cut" transition, J. Stat. Mech. Th. Exp. 2006, P07005, 33 pp.

[Fok92] A.S. Fokas, A.R. Its, and A.V. Kitaev, The isomonodromy approach to matrix models in 2D quantum gravity, Commun. Math. Phys. 147 (1992), 395–430.

[For93] P. J. Forrester, The spectrum edge of random matrix ensembles, Nucl. Phys. B 402 (1993), 709–728.

[For10] P.J. Forrester, *Log-Gases and Random Matrices*, LMS Mongraphs Series Vol. 34, Princeton Univ. Press, Princeton N.J., 2010.

[Fou11] A. Foulquié Moreno, A. Martínez-Finkelshtein and V.L. Sousa, Asymptotics of orthogonal polynomials for a weight with a jump on $[-1, 1]$, Constr. Approx. 33 (2011), 219–263.

[Hac95] G. Hackenbroich and H. A. Weidenmller, Universality of random-matrix results for non-Gaussian ensembles, Phys. Rev. Lett. 74 (1995), 4118–4121.

[Its08] A. Its and I. Krasovsky, Hankel determinant and orthogonal polynomials for the Gaussian weight with a jump, in: "Integrable Systems and Random Matrices" (Baik *et al.* eds.), Contemp. Math. 458, Amer. Math. Soc., Providence RI, 2008, pp. 215–247.

[Joh98] K. Johansson, On fluctuations of eigenvalues of random Hermitian matrices. Duke Math. J. 91 (1998), 151–204.

[Kan97] E. Kanzieper and V. Freilikher, Universality in invariant random-matrix models: existence near the soft edge, Phys. Rev. E 55 (1997), 3712–3715.

[Kan98] E. Kanzieper and V. Freilikher, Random matrix models with log-singular level confinement: method of fictitious fermions, Philos. Magazine B 77 (1998), 1161–1172.

[Kle00] B. Klein and J. Verbaarschot, Spectral universality of real chiral random matrix ensembles, Nucl. Phys. B 588 (2000), 483–507.

[Kui00] A.B.J. Kuijlaars and K.T-R McLaughlin, Generic behavior of the density of states in random matrix theory and equilibrium problems in the presence of real analytic external fields, Comm. Pure Appl. Math. 53 (2000), 736–785.

[Kui09] A.B.J. Kuijlaars and P. Tibboel, The asymptotic behaviour of recurrence coefficients for orthogonal polynomials with varying exponential weights, J. Comput. Appl. Math. 233 (2009), 775–785.

[Kui02] A.B.J. Kuijlaars and M. Vanlessen, Universality for eigenvalue correlations from the modified Jacobi unitary ensemble, Int. Math. Res. Notices 2002 (2002), 1575–1600.

[Kui03] A.B.J. Kuijlaars and M. Vanlessen, Universality for eigenvalue correlations at the origin of the spectrum, Comm. Math. Phys. 243 (2003), 163–191.

[Lev08] E. Levin and D.S. Lubinsky, Universality limits in the bulk for varying measures, Adv. Math. 219 (2008), 743–779.

[Lev09] E. Levin and D.S. Lubinsky, Universality limits for exponential weights, Constr. Approx. 29 (2009), 247–275.

[Lub08a] D.S. Lubinsky, Universality limits in the bulk for arbitrary measures with compact support, J. d'Analyse Math. 106 (2008), 373–394.

[Lub08b] D.S. Lubinsky, Universality limits at the hard edge of the spectrum for measures with compact support, Int. Math. Res. Notices 2008, Art. ID rnn 099, 39 pp.

[Lub09a] D.S. Lubinsky, A new approach to universality limits involving orthogonal polynomials, Ann. of Math. 170 (2009), 915–939.

[Lub09b] D.S. Lubinsky, Universality limits for random matrices and de Branges spaces of entire functions, J. Funct. Anal. 256 (2009), 3688–3729.

[Meh04] M.L. Mehta, *Random Matrices*, third edition, Elsevier/Academic Press, Amsterdam, 2004.

[McL08] K. T-R. McLaughlin and P.D. Miller, The $\overline{\partial}$ steepest descent method for orthogonal polynomials on the real line with varying weights, Int. Math. Res. Notices 2008, Art. ID rnn 075, 66 pp.

[Mo08] M.Y. Mo, The Riemann–Hilbert approach to double scaling limit of random matrix eigenvalues near the "birth of a cut" transition, Int. Math. Res. Notices 2008, Art. ID rnn042, 51 pp.

[Nag93] T. Nagao and M. Wadati, Eigenvalue distribution of random matrices at the spectrum edge, J. Phys. Soc. Japan 62 (1993), 3845–3856.

[Pas97] L. Pastur and M. Shcherbina, Universality of the local eigenvalue statistics for a class of unitary invariant random matrix ensembles, J. Statist. Phys. 86 (1997), 109–147.

[Pas08] L. Pastur, M. Shcherbina, Bulk universality and related properties of Hermitian matrix models, J. Stat. Phys. 130 (2008), 205–250.

[Saf97] E.B. Saff and V. Totik, *Logarithmic Potentials with External Fields*, Springer-Verlag, Berlin, 1997.

[Shc08] M. Shcherbina, Double scaling limit for matrix models with non analytic potentials, J. Math. Phys. 49 (2008) 033501, 34 pp.

[Shc09a] M. Shcherbina, On universality for orthogonal ensembles of random matrices, Comm. Math. Phys. 285 (2009), 957–974.

[Shc09b] M. Shcherbina, Edge universality for orthogonal ensembles of random matrices, J. Stat. Phys. 136 (2009), 35–50.

[Sos00] A. Soshnikov, Determinantal random point fields, Russian Math. Surveys 55 (2000), 923–975.

[Sto00] A. Stojanovic, Universality in orthogonal and symplectic invariant matrix models with quartic potential, Math. Phys. Anal. Geom. 3 (2000), 339–373.

[Tao11] T. Tao and V. Vu, Random matrices: Universality of local eigenvalue statistics, Acta Math. 206 (2011), 127–204.

[Tra94] C. Tracy and H. Widom, Level-spacing distributions and the Airy kernel, Comm. Math. Phys. 159 (1994), 151–174.

[Tra98] C. Tracy and H. Widom, Correlation functions, cluster functions, and spacing distributions for random matrices, J. Stat. Phys. 92 (1998), 809–835.

·7·

Supersymmetry

Thomas Guhr

Abstract

Supersymmetry is nowadays indispensable for many problems in random matrix theory. It is presented here with an emphasis on conceptual and structural issues. An introduction to supermathematics is given. The Hubbard–Stratonovich transformation as well as its generalization and superbosonization are explained. The supersymmetric non-linear σ model, Brownian motion in superspace and the colour-flavour transformation are discussed.

7.1 Generating functions

We consider $N \times N$ matrices H in the three symmetry classes [Dys62] real symmetric, Hermitian or quaternion real, that is, self-dual Hermitian. The Dyson index β takes the values $\beta = 1, 2, 4$, respectively. For $\beta = 4$, the $N \times N$ matrix H has 2×2 quaternion entries and all its eigenvalues are doubly degenerate. For a given symmetry, an ensemble of random matrices is specified by choosing a probability density function $P(H)$ of the matrix H. The ensemble is referred to as invariant or rotation invariant if

$$P(V^{-1}HV) = P(H) \tag{7.1.1}$$

where V is a fixed element in the group diagonalizing H, that is, in $\mathrm{SO}(N)$, $\mathrm{SU}(N)$ or $\mathrm{USp}(2N)$ for $\beta = 1, 2, 4$, respectively. Equation (7.1.1) implies that the probability density function only depends on the eigenvalues,

$$P(H) = P(X) = P(x_1, \ldots, x_N). \tag{7.1.2}$$

Here, we write the diagonalization of the random matrix as $H = U^{-1}XU$ with $X = \mathrm{diag}\,(x_1, \ldots, x_N)$ for $\beta = 1, 2$ and $X = \mathrm{diag}\,(x_1, x_1, \ldots, x_N, x_N)$ for $\beta = 4$. The k-point correlation function $R_k(x_1, \ldots, x_k)$ measures the probability density of finding a level around each of the positions $x_1, \ldots, x_k$, the remaining levels not being observed. One has [Dys62, Meh04]

$$R_k(x_1, \ldots, x_k) = \frac{N!}{(N-k)!} \int_{-\infty}^{+\infty} dx_{k+1} \cdots \int_{-\infty}^{+\infty} dx_N \, |\Delta_N(X)|^{\beta} P(X), \tag{7.1.3}$$

where $\Delta_N(X)$ is the Vandermonde determinant. If the probability density function factorizes,

$$P(X) = \prod_{n=1}^{N} P^{(E)}(x_n), \tag{7.1.4}$$

with a probability density function $P^{(E)}(x_n)$ for each of the eigenvalues, the correlation functions (7.1.3) can be evaluated with the Mehta–Mahoux theorem [Meh04]. They are $k \times k$ determinants for $\beta = 2$ and $2k \times 2k$ quaternion determinants for $\beta = 1, 4$ whose entries, the kernels, depend on only two of the eigenvalues $x_1, \ldots, x_k$.

Formula (7.1.3) cannot serve as the starting point for the supersymmetry method. A reformulation employing determinants is called for, because these can be expressed as Gaussian integrals over commuting or anticommuting variables, respectively. The key object is the resolvent, that is, the matrix $(x_p^- - H)^{-1}$ where the argument is given a small imaginary increment, $x_p^- = x_p - i\varepsilon$. The k-point correlation functions are then defined as the ensemble averaged imaginary parts of the traces of the resolvents at arguments $x_1, \ldots, x_k$,

$$R_k(x_1, \ldots, x_k) = \frac{1}{\pi^k} \int P(H) \prod_{p=1}^{k} \operatorname{Im} \operatorname{tr} \frac{1}{x_p^- - H} d[H]. \tag{7.1.5}$$

The necessary limit $\varepsilon \to 0$ is suppressed throughout in our notation. We write $d[\cdot]$ for the volume element of the quantity in square brackets, that is, for the product of the differentials of all independent variables. The definitions (7.1.3) and (7.1.5) are equivalent, but not fully identical. Formula (7.1.5) yields a sum of terms, only one coincides with the definition (7.1.3), all others contain at least one δ function of the form $\delta(x_p - x_q)$, see [Guh98].

Better suited for the supersymmetry method than the correlation functions (7.1.5) are the correlation functions

$$\widehat{R}_k(x_1, \ldots, x_k) = \frac{1}{\pi^k} \int d[H]\, P(H) \prod_{p=1}^{k} \operatorname{tr} \frac{1}{x_p - iL_p\varepsilon - H} \tag{7.1.6}$$

which also contain the real parts of the resolvents. The correlation functions (7.1.5) can always be reconstructed, but the way that this is done conveniently differs for different variants of the supersymmetry method. In Eq. (7.1.5), all imaginary increments are on the same side of the real axis. In Eq. (7.1.6), however, we introduced quantities L_p which determine the side of the real axis on which the imaginary increment is placed. They are either $+1$ or -1 and define a metric L. Hence, depending on L, there is an overall sign in Eq. (7.1.6) which we suppress. We use the shorthand notations $x_p^{\pm} = x_p - iL_{p\varepsilon}$ in the sequel. In some variants of the supersymmetry method, it is not important where the imaginary increments are, in the supersymmetric non-linear σ model, however, it is of crucial importance. We will return to this point.

To prepare the application of supersymmetry, one expresses the correlation functions (7.1.6) as derivatives

$$\widehat{R}_k(x_1, \ldots, x_k) = \frac{1}{(2\pi)^k} \frac{\partial^k}{\prod_{p=1}^k \partial J_p} Z_k(x+J)\bigg|_{J_p=0} \tag{7.1.7}$$

of the generating function

$$Z_k(x+J) = \int d[H]\, P(H) \prod_{p=1}^{k} \left(\frac{\det(H - x_p + i\,L_p\varepsilon - J_p)}{\det(H - x_p + i\,L_p\varepsilon + J_p)} \right)^{\gamma} \tag{7.1.8}$$

with respect to source variables $J_p,\ p = 1, \ldots, k$. For $\beta = 1, 2$ one has $\gamma = 1$ whereas $\gamma = 2$ for $\beta = 4$. For later purposes, we introduce the $2k \times 2k$ matrices $x = \mathrm{diag}\,(x_1, x_1, \ldots, x_k, x_k)$ and $J = \mathrm{diag}\,(-J_1, +J_1, \ldots, -J_k, +J_k)$ for $\beta = 2$ as well as the $4k \times 4k$ matrices $x = \mathrm{diag}\,(x_1, x_1, x_1, x_1 \ldots, x_k, x_k, x_k, x_k)$ and $J = \mathrm{diag}\,(-J_1, -J_1, +J_1, +J_1, \ldots, -J_k, -J_k, +J_k, +J_k)$ for $\beta = 1, 4$, which appear in the argument of Z_k. We write $x^{\pm} = x - iL\varepsilon$. Importantly, the generating function is normalized at $J = 0$, that is, $Z_k(x) = 1$.

7.2 Supermathematics

Martin [Mar59] appears to have written the first paper on anticommuting variables in 1959. Two years later, Berezin introduced integrals over anticommuting variables when studying second quantization. His posthumously published book [Ber87] is still the standard reference on supermathematics.

7.2.1 Anticommuting variables

We introduce Grassmann or anticommuting variables $\zeta_p,\ p = 1, \ldots, k$ by requiring the relation

$$\zeta_p \zeta_q = -\zeta_q \zeta_p, \qquad p, q = 1, \ldots, k. \tag{7.2.1}$$

In particular, this implies $\zeta_p^2 = 0$. These variables are purely formal objects. In contrast to commuting variables, they do not have a representation as numbers. The inverse of an anticommuting variable cannot be introduced in a meaningful way. Commuting and anticommuting variables commute. The product of an even number of anticommuting variables is commuting,

$$(\zeta_p \zeta_q)\zeta_r = \zeta_p \zeta_q \zeta_r = -\zeta_p \zeta_r \zeta_q = +\zeta_r \zeta_p \zeta_q = \zeta_r(\zeta_p \zeta_q). \tag{7.2.2}$$

We view the anticommuting variables as complex and define a complex conjugation, ζ_p^* is the complex conjugate of ζ_p. The variables ζ_p and ζ_p^* are independent in the same sense in which an ordinary complex variable and its conjugate are independent. The property (7.2.1) also holds for the complex conjugates as well as for mixtures, $\zeta_p\zeta_q^* = -\zeta_q^*\zeta_p$. There are two different but equivalent ways to interpret $(\zeta_p^*)^*$. The usual choice in physics is

$$(\zeta_p^*)^* = \zeta_p^{**} = -\zeta_p, \qquad p = 1, \ldots, k, \tag{7.2.3}$$

which has to be supplemented by the rule

$$(\zeta_p\zeta_q\cdots\zeta_r)^* = \zeta_p^*\zeta_q^*\cdots\zeta_r^*. \tag{7.2.4}$$

There is a concept of reality, since we have

$$(\zeta_p^*\zeta_p)^* = \zeta_p^{**}\zeta_p^* = -\zeta_p\zeta_p^* = \zeta_p^*\zeta_p. \tag{7.2.5}$$

Hence, we may interpret $\zeta_p^*\zeta_p$ as the modulus squared of the complex anticommuting variable ζ_p. Alternatively, one can use the plus sign in Eq. (7.2.3) and reverse the order of the anticommuting variables on the right-hand side of Eq. (7.2.4). In particular, this also preserves the property (7.2.5).

Because of $\zeta_p^2 = 0$ and since inverse anticommuting variables do not exist, functions of anticommuting variables can only be finite polynomials,

$$f(\zeta_1, \ldots, \zeta_k, \zeta_1^*, \ldots, \zeta_k^*) = \sum_{\substack{m_p=0,1 \\ l_p=0,1}} f_{m_1\cdots m_k l_1\cdots l_k}\zeta_1^{m_1}\cdots\zeta_k^{m_k}(\zeta_1^*)^{l_1}\cdots(\zeta_k^*)^{l_k} \tag{7.2.6}$$

with commuting coefficients $f_{m_1\cdots m_k l_1\cdots l_k}$. Thus, just like functions of matrices, functions of anticommuting variables are power series. For example, we have

$$\exp\left(a\zeta_p^*\zeta_p\right) = 1 + a\zeta_p^*\zeta_p = \frac{1}{1 - a\zeta_p^*\zeta_p}. \tag{7.2.7}$$

where a is a commuting variable.

7.2.2 Vectors and matrices

A supermatrix σ is defined via block construction,

$$\sigma = \begin{bmatrix} a & \mu \\ \nu & b \end{bmatrix}, \tag{7.2.8}$$

where a and b are matrices with ordinary complex commuting entries while the matrices μ and ν have complex anticommuting entries. Apart from the restriction that the blocks must match, all dimensions of the matrices are possible. Of particular interest are quadratic $k_1/k_2 \times k_1/k_2$ supermatrices, that is, a and b have dimensions $k_1 \times k_1$ and $k_2 \times k_2$, respectively, μ and ν have dimensions $k_1 \times k_2$ and $k_2 \times k_1$. A quadratic supermatrix σ can have an inverse σ^{-1}. Equally important are supervectors, which are defined as special supermatrices consisting of only one column. As seen in Eq. (7.2.8), there are two possibilities

$$\psi = \begin{bmatrix} z \\ \zeta \end{bmatrix} \quad \text{and} \quad \psi = \begin{bmatrix} \zeta \\ z \end{bmatrix}, \tag{7.2.9}$$

where z is a k_1 component vector of ordinary complex commuting entries z_p, and ζ is a k_2 component vector of complex anticommuting entries ζ_p. In the sequel we work with the first possibility, but everything to be said is accordingly valid for the second one. The standard rules of matrix addition and multiplication apply, if everything in Section 7.2.1 is taken into account. Consider for example the supervector ψ' given by

$$\psi' = \sigma\psi = \begin{bmatrix} a & \mu \\ \nu & b \end{bmatrix} \begin{bmatrix} z \\ \zeta \end{bmatrix} = \begin{bmatrix} az + \mu\zeta \\ \nu z + b\zeta \end{bmatrix}, \tag{7.2.10}$$

which has the same form as ψ. Hence the linear map (7.2.10) transforms commuting into anticommuting degrees of freedom and vice versa.

The transpose σ^T and the Hermitian conjugate $\sigma^\dagger$ are defined as

$$\sigma^T = \begin{bmatrix} a^T & \nu^T \\ -\mu^T & b^T \end{bmatrix} \quad \text{and} \quad \sigma^\dagger = (\sigma^T)^*. \tag{7.2.11}$$

The minus sign in front of ν^T ensures that $(\sigma_1\sigma_2)^T = \sigma_2^T\sigma_1^T$ carries over to supermatrices σ_1 and σ_2. Importantly, $(\sigma^\dagger)^\dagger = \sigma$ always holds, but $(\sigma^T)^T$ is in general not equal to σ. As a special application, we define the scalar product $\psi^\dagger\chi$ where each of the supervectors ψ and χ has either the first or the second of the forms (7.2.9). Because of the reality property (7.2.3), the scalar product $\psi^\dagger\psi$ is real and can be viewed as the length squared of the supervector ψ.

To have cyclic invariance, the supertrace is defined as

$$\operatorname{str} \sigma = \operatorname{tr} a - \operatorname{tr} b \tag{7.2.12}$$

such that $\operatorname{str} \sigma_1\sigma_2 = \operatorname{str} \sigma_2\sigma_1$ for two different supermatrices σ_1 and σ_2. Correspondingly, the superdeterminant is multiplicative owing to the definition

$$\operatorname{sdet} \sigma = \frac{\det\left(a - \mu b^{-1}\nu\right)}{\det b} = \frac{\det a}{\det\left(b - \nu a^{-1}\mu\right)} \tag{7.2.13}$$

for $\det b \neq 0$ such that $\operatorname{sdet} \sigma_1\sigma_2 = \operatorname{sdet} \sigma_1 \operatorname{sdet} \sigma_2$.

7.2.3 Groups and symmetric spaces

For an introduction to this topic see Chapter 3. Here we only present the salient features in the context of the supersymmetry method. The theory of Lie superalgebras was pioneered by Kac [Kac77]. Although the notion of supergroups, particularly Lie supergroups, seems to be debated in mathematics, a consistent definition from a physics viewpoint is possible and – as will become clear later on – urgently called for. All supermatrices u which leave the length of the supervector ψ invariant form the unitary supergroup $\mathrm{U}(k_1/k_2)$. With $\psi' = u\psi$ we require $(\psi')^\dagger\psi' = \psi^\dagger u^\dagger u\psi = \psi^\dagger\psi$ and the corresponding equation for $\psi' = u^\dagger\psi$. Hence we conclude

$$u^\dagger u = 1, \quad uu^\dagger = 1 \qquad \text{and thus} \qquad u^\dagger = u^{-1}. \tag{7.2.14}$$

The direct product $\mathrm{U}(k_1) \times \mathrm{U}(k_2)$ of ordinary unitary groups is a trivial subgroup of $\mathrm{U}(k_1/k_2)$, found by simply putting all anticommuting variables in u to zero. Non-trivial subgroups of the unitary supergroup exist as well. Consider commuting variables, real w_p, $p = 1, \ldots, k_1$ and complex z_{pj}, $p = 1, \ldots, k_1$, $j = 1, 2$. We introduce the real and quaternion-real supervectors

$$\psi = \begin{bmatrix} w_1 \\ \vdots \\ w_{k_1} \\ \zeta_1 \\ \zeta_1^* \\ \vdots \\ \zeta_{k_2} \\ \zeta_{k_2}^* \end{bmatrix} \qquad \text{and} \qquad \psi = \begin{bmatrix} z_{11} & -z_{12}^* \\ z_{12} & z_{11}^* \\ \vdots & \vdots \\ z_{k_1 1} & -z_{k_1 2}^* \\ z_{k_1 2} & z_{k_1 1}^* \\ \zeta_1^* & -\zeta_1 \\ \vdots & \vdots \\ \zeta_{k_2}^* & -\zeta_{k_2} \end{bmatrix}. \tag{7.2.15}$$

The unitary-ortho-symplectic subgroup of the unitary supergroup leaves the lengths of ψ invariant: $\mathrm{UOSp}(k_1/2k_2)$ the length of the first and $\mathrm{UOSp}(2k_1/k_2)$ the length of the second supervector in Eq. (7.2.15). Because of the quaternion structure in the commuting entries of the second supervector, the proper scalar product reads $\mathrm{tr}\,\psi^\dagger\psi$. The trivial ordinary subgroups are $\mathrm{O}(k_1) \times \mathrm{USp}(2k_2) \subset \mathrm{UOSp}(k_1/2k_2)$ and $\mathrm{USp}(2k_1) \times \mathrm{O}(k_2) \subset \mathrm{UOSp}(2k_1/k_2)$, respectively.

As in the ordinary case, non-compact supergroups result from the requirement that the bilinear form $\psi^\dagger L\psi$ remains invariant. The metric L is without loss of generality diagonal and only contains ± 1. We then have $u^\dagger L u = L$.

A Hermitian supermatrix σ is diagonalized by a supermatrix $u \in \mathrm{U}(k_1/k_2)$,

$$\sigma = u^{-1} s u \qquad \text{with} \qquad s = \mathrm{diag}\,(s_{11}, \ldots, s_{k_1 1}, s_{12}, \ldots, s_{k_2 2}). \tag{7.2.16}$$

All eigenvalues s_{pj} are real commuting. Zirnbauer [Zir96a] gave a classification of the Riemannian symmetric superspaces. The Hermitian symmetric superspace is denoted A|A. Of interest are also the symmetric superspaces AI|AII and AII|AI. The former consists of the $k_1/2k_2 \times k_1/2k_2$ supermatrices $\sigma = u^{-1}su$ with $u \in \mathrm{UOSp}(k_1/2k_2)$ and with $s = \mathrm{diag}\,(s_{11}, \ldots, s_{k_1 1}, s_{12}, s_{12}, \ldots, s_{k_2 2}, s_{k_2 2})$, the latter of the $2k_1/k_2 \times 2k_1/k_2$ supermatrices σ with $u \in \mathrm{UOSp}(2k_1/k_2)$ and with $s = \mathrm{diag}\,(s_{11}, s_{11}, \ldots, s_{k_1 1}, s_{k_1 1}, s_{12}, \ldots, s_{k_2 2})$.

7.2.4 Derivatives and integrals

Since anticommuting variables cannot be represented by numbers, there is nothing like a Riemannian integral over anticommuting variables either. The Berezin integral [Ber87] is formally defined by

$$\int d\zeta_p = 0 \qquad \text{and} \qquad \int \zeta_p d\zeta_p = \frac{1}{\sqrt{2\pi}}, \tag{7.2.17}$$

and accordingly for the complex conjugates ζ_p^*. The normalization involving $\sqrt{2\pi}$ is a common, but not the only convention used. The differentials $d\zeta_p$ have all the properties of anticommuting variables collected in Section 7.2.1. Thus, the Berezin integral of the function (7.2.6) is essentially the highest order coefficient, more precisely $f_{1\cdots11\cdots1}/(2\pi)^k$ apart from an overall sign determined by the chosen order of integration. For example, we have

$$\iint \exp\left(a\zeta_p^*\zeta_p\right) d\zeta_p d\zeta_p^* = \frac{a}{2\pi}. \tag{7.2.18}$$

This innocent-looking formula is at the heart of the supersymmetry method. Anticipating the later discussion, we notice that we would have found the inverse of the right-hand side for commuting integration variables z_p instead of ζ_p.

One can also define a derivative as the discrete operation $\partial\zeta_p/\partial\zeta_q = \delta_{pq}$. To avoid ambiguities with signs, one should distinguish left and right derivatives. Obviously, derivative and integral coincide apart from factors. Mathematicians often prefer to think of the Berezin integral as a derivation. In physics, however, the interpretation as integral is highly useful as seen when changing variables. We first consider the k_2 vectors ζ and $\eta = a\zeta$ of anticommuting variables where a is an ordinary complex $k_2 \times k_2$ matrix. From the definition (7.2.17) we conclude $d[\eta] = \det^{-1} a\, d[\zeta]$. This makes it plausible that the change of variables in ordinary space generalizes for an arbitrary transformation $\chi = \chi(\psi)$ of supervectors in the following manner: Let y be the vector of commuting and η be the vector of anticommuting variables in χ, then we have

$$d[\chi] = \operatorname{sdet}\frac{\partial\chi}{\partial\psi^T}\, d[\psi] = \operatorname{sdet}\begin{bmatrix} \partial y/\partial z^T & \partial y/\partial\zeta^T \\ \partial\eta/\partial z^T & \partial\eta/\partial\zeta^T \end{bmatrix} d[\psi] \tag{7.2.19}$$

with $d[\chi] = d[y]d[\eta]$ and $d[\psi] = d[z]d[\zeta]$. The Jacobian in superspace is referred to as Berezinian. Absolute value signs are not needed if we agree to only transform right-handed into right-handed coordinate systems. Changes of variables in superspace can lead to boundary contributions which have no analogue in ordinary analysis. In physics, they are referred to as Efetov-Wegner terms, see [Rot87] for a mathematical discussion.

Importantly, the concept of the δ function has a meaningful generalization in superspace. An anticommuting variable ζ_p acts formally as a δ function when integrating it with any function $f(\zeta_p)$, hence $\delta(\zeta_p) = \sqrt{2\pi}\zeta_p$. More complicated are expressions of the form $\delta(y - \zeta^\dagger\zeta)$ with an ordinary commuting variable y and a k component vector of complex anticommuting variables ζ. To make sense out of it, it has to be interpreted as

$$\delta(y - \zeta^\dagger\zeta) = \sum_{\kappa=0}^{k} \frac{(-1)^\kappa}{\kappa!} \delta^{(\kappa)}(y)(\zeta^\dagger\zeta)^\kappa. \qquad (7.2.20)$$

This is a terminating power series, because $(\zeta^\dagger\zeta)^\kappa = 0$ for $\kappa > k$.

7.3 Supersymmetric representation

Several problems in particle physics would be solved if each boson had a fermionic and each fermion had a bosonic partner. A review of this supersymmetry can be found in [Mar05]. Although mathematically the same, supersymmetry in condensed matter physics and random matrix theory have completely different interpretations: the commuting and anticommuting variables do not represent bosons or fermions, that is, physical particles. Rather, they are highly convenient bookkeeping devices making it possible to drastically reduce the number of degrees of freedom in the statistical model. Since as many commuting as anticommuting variables are involved, one refers to it as Super-"symmetry" – purely formally just like in particle physics. In 1979, Parisi and Sourlas [Par79] introduced superspace concepts to condensed matter physics. Three years later, Efetov [Efe82] constructed the supersymmetric non-linear σ model for the field theory describing electron transport in disordered systems. Efetov and his coworkers developed many of the tools and contributed a large body of work on supersymmetry [Efe83]. The first applications of supersymmetry to random matrices, that is, in the language of condensed matter physics, to the zero-dimensional limit of a field theory, were given by Verbaarschot and Zirnbauer [Ver85a] and by Verbaarschot, Zirnbauer, and Weidenmüller [Ver85b]. Reviews can be found in [Efe97, Guh98, Mir00], see also the chapters on chiral random matrix theory and on scattering.

7.3.1 Ensemble average

Using supersymmetry, the ensemble average in the generating function (7.1.8) is straightforward. We begin with the unitary case $\beta = 2$ and express the determinants as Gaussian integrals

$$\begin{aligned}
\frac{(2\pi)^N}{\det(H - x_p^\pm + J_p)} &= \int d[z_p] \exp\left(i L_p z_p^\dagger (H - x_p^\pm + J_p) z_p\right) \\
\frac{\det(H - x_p^\pm + J_p)}{(2\pi)^N} &= \int d[\zeta_p] \exp\left(i \zeta_p^\dagger (H - x_p^\pm - J_p) \zeta_p\right) \qquad (7.3.1)
\end{aligned}$$

over altogether k vectors $z_p,\ p = 1, \ldots, k$ with N complex commuting entries and k vectors $\zeta_p,\ p = 1, \ldots, k$ with N complex anticommuting entries. When integrating over the commuting variables, the imaginary increment is needed for convergence, for the integrals over anticommuting variables, convergence is never a problem. Hence we may write the metric tensor in the form $L = \text{diag}(L_1, \ldots, L_k, 1, \ldots, 1)$. Collecting all H dependences, the ensemble average in Eq. (7.1.8) amounts to calculating

$$\Phi(K) = \int d[H] P(H) \exp(i \text{tr}\, HK). \tag{7.3.2}$$

where the $N \times N$ matrix K assembles dyadic products of the vectors z_p and ζ_p,

$$K = \sum_{p=1}^{k} \left(L_p z_p z_p^\dagger - \zeta_p \zeta_p^\dagger \right). \tag{7.3.3}$$

For all L, this is a Hermitian matrix $K^\dagger = K$.

We now turn to the orthogonal case $\beta = 1$. At first sight it seems irrelevant whether H is Hermitian or real-symmetric in the previous steps. However, the Fourier transform (7.3.2) only affects the real part of K, because the imaginary part of K drops out in tr HK if H is real-symmetric. Thus, instead of the Gaussian integrals (7.3.1), we rather use

$$\begin{aligned}
\frac{\pi^N}{\det(H - x_p^\pm + J_p)} &= \int d[w_p^{(1)}] \exp\left(i L_p w_p^{(1)T} (H - x_p^\pm + J_p) w_p^{(1)}\right) \\
&\quad \int d[w_p^{(2)}] \exp\left(i L_p w^{(2)T} (H - x_p^\pm + J_p) w_p^{(2)}\right) \\
\frac{\det(H - x_p^\pm + J_p)}{\pi^N} &= \int d[\zeta_p] \exp\left(i \zeta_p^\dagger (H - x_p^\pm - J_p) \zeta_p\right) \\
&\quad \exp\left(-i \zeta_p^T (H - x_p^\pm - J_p) \zeta_p^*\right),
\end{aligned} \tag{7.3.4}$$

where the N component vectors $w_p^{(1)}$ and $w_p^{(2)}$ have real entries. For each p, we can construct a $4N$ component supervector out of $w_p^{(1)}$, $w_p^{(2)}$, ζ_p, and ζ_p^* whose structure resembles the one of the first of the supervectors (7.2.15), but with a different number of components. Reordering terms, we arrive at the Fourier transform (7.3.2), but now for real-symmetric H and with

$$K = \sum_{p=1}^{k} \left(L_p w_p^{(1)} w_p^{(1)T} + L_p w_p^{(2)} w_p^{(2)T} - \zeta_p \zeta_p^\dagger + \zeta_p^* \zeta_p^T \right), \tag{7.3.5}$$

which is $N \times N$ real-symmetric as well. For $\beta = 4$, one has to reformulate the steps in such a way that the corresponding K becomes self-dual Hermitian.

7.3.2 Hubbard–Stratonovich transformation

Because of universality, it suffices to assume a Gaussian probability density function $P(H) \sim \exp(-\beta \mathrm{tr}\, H^2/2)$ in almost all applications in condensed matter and many-body physics as well as in quantum chaos. Hence the random matrices are drawn from the Gaussian orthogonal (GOE), unitary (GUE), or symplectic ensemble (GSE). The Fourier transform (7.3.2) is then elementary and yields a Gaussian. The crucial property

$$\Phi(K) = \exp\left(-\frac{1}{2\beta}\mathrm{tr}\, K^2\right) = \exp\left(-\frac{1}{2\beta}\mathrm{str}\, B^2\right) \tag{7.3.6}$$

holds, where B is a supermatrix containing all scalar products of the vectors to be integrated over. The second equality sign has a purely algebraic origin. For $\beta = 2$, B has dimension $k/k \times k/k$ and reads

$$B = L^{1/2} \begin{bmatrix} z_1^\dagger z_1 & \cdots & z_1^\dagger z_k & z_1^\dagger \zeta_1 & \cdots & z_1^\dagger \zeta_k \\ \vdots & & \vdots & \vdots & & \vdots \\ z_k^\dagger z_1 & \cdots & z_k^\dagger z_k & z_k^\dagger \zeta_1 & \cdots & z_k^\dagger \zeta_k \\ -\zeta_1^\dagger z_1 & \cdots & -\zeta_1^\dagger z_k & -\zeta_1^\dagger \zeta_1 & \cdots & -\zeta_1^\dagger \zeta_k \\ \vdots & & \vdots & \vdots & & \vdots \\ -\zeta_k^\dagger z_1 & \cdots & -\zeta_k^\dagger z_k & -\zeta_k^\dagger \zeta_1 & \cdots & -\zeta_k^\dagger \zeta_k \end{bmatrix} L^{1/2}. \tag{7.3.7}$$

While K is Hermitian, the square roots $L^{1/2}$ destroy this property for B, since $L^{1/2}$ can have imaginary units i as entries, B is Hermitian only for $L = 1$. In general, B is in a deformed (non-compact) form of the symmetric superspace A|A. For $\beta = 1, 4$, the supermatrix B has dimension $2k/2k \times 2k/2k$ and it is in deformed (non-compact) forms of the symmetric superspaces AI|AII and AII|AI, respectively. We give the explicit forms later on. The identity (7.3.6) states the keystone of the supersymmetry method. The original model in the space of ordinary $N \times N$ matrices is mapped onto a model in space of supermatrices whose dimension is proportional to k, which is the number of arguments in the k-point correlation function.

The Gaussians (7.3.6) contain the vectors, that is, their building blocks to fourth order. To make analytical progress, a Hubbard–Stratonovich transformation in superspace is used,

$$\exp\left(-\frac{1}{2\beta}\mathrm{str}\, B^2\right) = c^{(\beta)} \int \exp\left(-\frac{\beta}{2}\mathrm{str}\,(L\sigma)^2\right) \exp\left(i\mathrm{str}\, L^{1/2}\sigma L^{1/2} B\right) d[\sigma], \tag{7.3.8}$$

where $c^{(2)} = 2^{k(k-1)}$ and $c^{(\beta)} = 2^{k(4k-3)/2}$ for $\beta = 1, 4$. We notice the appearance of the matrices L and $L^{1/2}$ in (7.3.8). For $L = 1$, the supermatrices σ and B have the same symmetries. However, as already observed in the early eighties for models in ordinary space [Sch80, Pru82], this choice is impossible for

$L \neq 1$, because it would render the integrals divergent. There are two ways out of this problem. One either constructs a proper explicit parameterization of σ or one inserts the matrices L and $L^{1/2}$ according to (7.3.8). A mathematically satisfactory understanding of these issues was put forward only recently in [Fyo08].

Another important remark is called for. Because of the minus sign in the supertrace (7.2.12), a Wick rotation is needed to make the integral convergent. This formally amounts to replacing the lower right block of σ, that is, b in Eq. (7.2.8), with ib. Apart from that, the metric L is also needed for convergence reasons. Now the vectors appear in second order. They can be ordered in one large supervector Ψ. For $\beta = 2$ it has the form (7.2.9) with $k_1 = k_2 = kN$, for $\beta = 1$ it has the first of the forms (7.2.15) with $k_1 = 2kN$, $k_2 = kN$ and for $\beta = 4$ it has the second of the forms (7.2.15) with $k_1 = kN$, $k_2 = 2kN$. The integral to be done is then seen to be the Gaussian integral in superspace

$$\int \exp\left(i\Psi^\dagger \left(L^{1/2}(L^{1/2}\sigma L^{1/2} - x^\pm - J)L^{1/2} \otimes 1_N\right)\Psi\right) d[\Psi] \\ = \mathrm{sdet}^{-N\beta/2\gamma}(\sigma L - x^\pm - J), \quad (7.3.9)$$

where the power N is due to the direct product structure. We eventually find

$$Z_k(x + J) = c^{(\beta)} \int \exp\left(-\frac{\beta}{2}\mathrm{str}\,(L\sigma)^2\right) \mathrm{sdet}^{-N\beta/2\gamma}(\sigma L - x^\pm - J)d[\sigma] \quad (7.3.10)$$

as supersymmetric representation of the generating function. The average over the $N \times N$ ordinary matrix H has been traded for an average over the matrix σ whose dimension is proportional to k, that is, independent of N.

7.3.3 Matrix δ functions and an alternative representation

In [Leh95, Hac95], an alternative route to the one outlined in Section 7.3.2 was taken. These authors used matrix δ functions in superspace and their Fourier representation to express functions $f(B)$ of the supermatrix B in the form

$$f(B) = \int f(\rho)\delta(\rho - B)d[\rho] = c^{(\beta)2} \int d[\rho] f(\rho) \int d[\sigma] \exp\left(-i\,\mathrm{str}\,\sigma(\rho - B)\right), \quad (7.3.11)$$

where auxiliary integrals over supermatrices ρ and σ are introduced. For simplicity, we only consider $L = 1$ here. The function $\delta(\rho - B)$ is the product of the δ functions of all independent variables. As discussed in Section 7.2.4, it is well-defined. For all functions f, formula (7.3.11) renders the integration over the supervector Ψ Gaussian. When studying Gaussian averages of ratios of characteristic polynomials, Fyodorov [Fyo02] built upon such insights to construct an alternative representation for the generating function. He employs a standard Hubbard–Stratonovich transformation for the lower right block of the supermatrix B in Eq. (7.3.7) which contains the scalar products $\zeta_p^\dagger \zeta_q$. He then inserts a δ function in the space of ordinary matrices to carry out the integrals over the vectors z_p. Although supersymmetry is used, the generating function is finally written as an integral over two ordinary matrices with commuting entries. In this derivation, the Ingham-Siegel integral

$$I^{(\mathrm{ord})}(R) = \int\limits_{S>0} \exp\left(-\mathrm{tr}\, RS\right) \det{}^m S d[S] \sim \frac{1}{\det{}^{m+N} R} \tag{7.3.12}$$

for ordinary Hermitian $N \times N$ matrices R and S appears, where $m \geq 0$.

7.3.4 Generalized Hubbard–Stratonovich transformation and superbosonization

Is supersymmetry only applicable to Gaussian probability density functions $P(H)$? In [Hac95], supersymmetry and asymptotic expansions were used to prove universality for arbitrary $P(H)$. The concept of superbosonization was put forward in [Efe04] and applied in [Bun07] to a generalized Gaussian model comprising a variety of correlations between the matrix elements. Extending the concept of superbosonization, a full answer to the question posed above was given in two different but related approaches in [Guh06, Kie09a], and [Lit08]. An exact supersymmetric representation exists for arbitrary, well-behaved $P(H)$. As the equivalence of the two approaches was proven in [Kie09b], we follow the line of argument in [Guh06, Kie09a]. For $\beta = 2$, we define the $N \times 2k$ rectangular supermatrix

$$A = [z_1 \cdots z_k \; \zeta_1 \cdots \zeta_k], \tag{7.3.13}$$

where the $z_p,\ p = 1, \ldots, k$ and $\zeta_p,\ p = 1, \ldots, k$ are N component vectors with complex commuting and anticommuting entries, respectively. We also define the $N \times 4k$ supermatrix

$$A = \left[z_1 \; z_1^* \cdots z_k \; z_k^* \; \zeta_1 \; \zeta_1^* \cdots \zeta_k \; \zeta_k^*\right] \tag{7.3.14}$$

for $\beta = 1$ and eventually the $2N \times 4k$ supermatrix

$$A = \begin{bmatrix} z_1 & -z_1^* & \ldots & z_k & -z_k^* & \zeta_1 & -\zeta_1^* & \ldots & \zeta_k & -\zeta_k^* \\ z_1 & z_1^* & & z_k & z_k^* & \zeta_1 & \zeta_1^* & & \zeta_k & \zeta_k^* \end{bmatrix} \tag{7.3.15}$$

for $\beta = 4$. This enables us to write the ordinary matrix K introduced in Section 7.3.1 and the supermatrix B introduced in Section 7.3.2 for all β in the form

$$\begin{aligned} K &= A L A^\dagger = (A L^{1/2})(L^{1/2} A^\dagger) \\ B &= (L^{1/2} A^\dagger)(A L^{1/2}) = L^{1/2} A^\dagger A L^{1/2}. \end{aligned} \tag{7.3.16}$$

For $\beta = 2$, we recover Eq. (7.3.7). This algebraic duality between ordinary and superspace has far-reaching consequences. One realizes [Guh91, Guh06, Lit08] that the integral (7.3.2) is the Fourier transform in matrix space of every, arbitrary probability density function $P(H)$ and that $\Phi(K)$ is the corresponding characteristic function. Since we assume that $P(H)$ is rotation invariant, the same must hold for $\Phi(K)$. Hence, $\Phi(K)$ only depends on the invariants $\mathrm{tr}\, K^m$, $m = 1, 2, 3, \ldots$. Due to cyclic invariance of the trace, the duality (7.3.16) implies for all m the crucial identity

$$\mathrm{tr}\, K^m = \mathrm{str}\, B^m, \qquad \text{such that} \qquad \Phi(K) = \Phi(B). \tag{7.3.17}$$

Hence, viewed as a function of the matrix invariants, Φ is a function in ordinary and in superspace. We now employ formula (7.3.11) for $\Phi(K) = \Phi(B)$, do the Gaussian Ψ integrals as usual find for the generating function

$$Z_k(x + J) = c^{(\beta)2} \int \exp\left(-i\mathrm{tr}\,(x + J)\, L\rho\right) \Phi(\rho)\, I(\rho) d[\rho] \tag{7.3.18}$$

with $I(\rho)$ being a supersymmetric version of the Ingham-Siegel integral. The supermatrices ρ and σ have the same sizes and symmetries as B. A convolution theorem in superspace yields the second form

$$Z_k(x + J) = \int \Pi(\sigma) \mathrm{sdet}^{-N\beta/2\gamma} \left(\sigma L - x^{\pm} - J\right) d[\sigma], \tag{7.3.19}$$

where $\Pi(\sigma)$ is the superspace Fourier backtransform of the characteristic function $\Phi(\rho)$. It plays the rôle of the probability density function for the supersymmetric representation. To apply these general results for exact calculations, explicit knowledge of either $\Phi(\rho)$ or $\Pi(\sigma)$ is necessary.

7.3.5 More complicated models

Most advantageously, supersymmetry allows one to make progress in important and technically challenging problems beyond the invariant and factorizing ensembles, for example:

- Invariant, but non-factorizing ensembles. The probability density function $P(H)$ has the property (7.1.1), but not the property (7.1.4). They can, in principle, be treated with the results of Section 7.3.4.
- Sparse or banded random matrices [Fyo91, Mir91], see Chapter 23. The probability density function $P(H)$ lacks the invariance property (7.1.1).
- Crossover transitions or external field models. One is interested in the eigenvalue correlations of the matrix $H(\alpha) = H^{(0)} + \alpha H$, where H is a random matrix as before and where $H^{(0)}$ is either a random matrix with symmetries different from H or a fixed matrix. The parameter α measures the relative strength. As the resolvent in question is now $(x_p^- - H(\alpha))^{-1} =$

$(x_p^- - (H^{(0)} + \alpha H))^{-1}$, we have to replace H by $H(\alpha)$ in the determinants in Eq. (7.1.8), but not in the probability density function $P(H)$ which usually is chosen invariant, see [Guh96b].

- Scattering theory and other problems, where matrix elements of the resolvents enter [Ver85b], see Chapter 2 on history and Chapter 34 on scattering. In the Heidelberg formalism [Mah69], scattering is modelled by coupling an effective Hamiltonian which describes the interaction zone to the scattering channels. The resolvent is then $(x_p^- - H + iW)^{-1}$ where the $N \times N$ matrix W contains information about the channels. One has to calculate averages of products of matrix elements $[(x_p^- - H + iW)^{-1}]_{nm}$. To make that feasible, the source variables J_p have to be replaced by $N \times N$ source matrices $\tilde{J}_p$ and instead of the derivatives (7.1.7), one must calculate derivatives with respect to the matrix elements $\tilde{J}_{p,nm}$. The probability density function $P(H)$ is unchanged.
- Field theories for disordered systems, see [Efe83, Efe97].

Of course, these and other non-invariant problems can not only be studied with the supersymmetry method, other techniques ranging from perturbative expansions, asymptotic analysis to orthogonal polynomials supplemented with group integrals are applied as well, see Chapters 4, 5, and 6. Nevertheless, the drastic reduction in the numbers of degrees of freedom, which is the key feature of supersymmetry, often yields precious structural insights into the problem.

7.4 Evaluation and structural insights

To evaluate the supersymmetric representation, a large N expansion, the celebrated non-linear σ model, is used in the vast majority of applications. We also sketch a method of exact evaluation which amounts to a diffusion process in superspace. Throughout, we focus on the structural aspects. A survey of the numerous results for specific systems is beyond the scope of this contribution, we refer the reader to reviews in [Efe97, Guh98, Mir00].

7.4.1 Non-linear σ model

The reduction in the numbers of degrees of freedom is borne out in the fact that the dimension N of the original random matrix H is an explicit parameter in Eqs. (7.3.10) and (7.3.19). Hence we can obtain an asymptotic expansion in $1/N$ by means of a saddle point approximation [Efe83, Efe97, Ver85a, Ver85b]. This suffices because one usually is interested in correlations on the local scale of the mean level spacing. Hence, the saddle point approximation goes hand in hand with the unfolding. The result of this procedure is the supersymmetric non-linear σ model. We consider the two-point function $k = 2$. The integrand in Eq. (7.3.10) is written as $\exp(-F(x+J))$ with the free energy

$$F(x+J) = \frac{\beta}{2}\text{str}\,(L\sigma)^2 + \frac{N\beta}{2\gamma}\text{str}\,\ln(\sigma L - x^{\pm} - J), \tag{7.4.1}$$

which is also referred to as Lagrangean. We introduce the centre $\bar{x} = (x_1 + x_2)/2$ and difference $\Delta x = x_2 - x_1$ of the arguments. In the large N limit, $\xi = \Delta x/D$ has to be held fixed where $D \sim 1/\sqrt{N}$ is the local mean level spacing. Hence, when determining the saddle points, we may set $\Delta x = 0$ such that $x = \bar{x}1$. Moreover, as we may choose the source variables arbitrarily small, we set $J = 0$ as well. Since all symmetry-breaking terms are gone, the free energy $F(x)$ with $x = \bar{x}1$ is invariant under rotations of σ which obey the metric L. Thus, variation of $F(x)$ with respect to σ yields the scalar equation

$$s_0(\bar{x} - s_0) = \frac{N}{2\gamma}, \qquad \text{such that} \qquad s_0 = \frac{1}{2}\left(\bar{x} \pm i\sqrt{\frac{2N}{\gamma} - \bar{x}^2}\right) \tag{7.4.2}$$

inside the spectrum, $|\bar{x}| \leq \sqrt{2N/\gamma}$. This is the famous Pastur equation and its solution s_0 [Pas72]. The latter is proportional to the large N one-point function whose imaginary part is the Wigner semicircle. We arrive at the important insight that the one-point function provides the stable points of the supersymmetric representation, the correlations on the local scale are the fluctuations around it. To make this more precise, we recall the result of Schäfer and Wegner [Sch80] for the non-linear σ model in ordinary space: when doing the large N limit as sketched above, the imaginary increments of the arguments x_1 and x_2 must lie on different sides of the real axis. Otherwise, the connected part of the two-point function cannot be obtained as seen from a contour-integral argument. Hence, the metric L must not be proportional to the unit matrix, the groups involved are non-compact and a hyperbolic symmetry is present. This carries over to the commuting degrees of freedom in superspace [Efe82], the groups are $\text{U}(1,1/2)$ for $\beta = 2$ and $\text{UOSp}(2,2/4)$ for $\beta = 1,4$. The full saddle point manifold is found to be given by all non-compact rotations of $\sigma_0 = \bar{x}/2 + i\sqrt{2N/\gamma - \bar{x}^2}L/2$ which leave $F(\bar{x}1)$ invariant. One parameterizes the group as $u = u_0 v$ with u_0 in the direct product $\text{U}(1/1) \times \text{U}(1/1)$ for $\beta = 1$ and in $\text{UOSp}(2/2) \times \text{UOSp}(2/2)$ for $\beta = 1,4$ and with v in the coset

$$\frac{\text{U}(1,1/2)}{\text{U}(1/1) \times \text{U}(1/1)} \quad \text{for} \quad \beta = 2\,,$$

$$\frac{\text{UOSp}(2,2/4)}{\text{UOSp}(2/2) \times \text{UOSp}(2/2)} \quad \text{for} \quad \beta = 1,4. \tag{7.4.3}$$

As u_0 and L commute, the saddle point manifold is $v^{-1}\sigma_0 v$, that is, essentially $Q = v^{-1}Lv$ with the crucial property $Q^2 = 1$. To calculate the correlations, we must re-insert the symmetry breaking terms Δx and J into the free energy. We put $\sigma = v^{-1}\sigma_0 v + \delta\sigma$ and expand to second order in the variables $\delta\sigma$ which are referred to as massive modes. They are integrated out in the generating function (7.3.10) as Gaussian integrals. One is left with integrals over the coset manifold, that is, over the Goldstone modes. On the unfolded scale, the two-point correlation functions (7.1.3) acquire the form $1 - Y_2(\xi)$. The two–level cluster functions read

$$Y_2(\xi) = -\mathrm{Re} \int \exp\left(i\xi \mathrm{str}\, QL\right) \mathrm{str}\, M_1 QL \, \mathrm{str}\, M_2 QL \, d\mu(Q), \tag{7.4.4}$$

where $d\mu(Q)$ is the invariant measure on the saddle point manifold. The matrices $M_i,\ i = 1, 2$ result from the derivatives with respect to the source variables, M_i is found by formally setting $J_i = 1$ and $J_l = 0,\ l \neq i$, in the matrix J. The expressions (7.4.4) can be reduced to two radial integrals on the coset manifold for the GUE and to three such integrals for GOE and GSE. Efetov [Efe83] discovered Eq. (7.4.4) when taking the zero-dimensional limit of his supersymmetric non-linear σ model for electron transport in disordered mesoscopic systems. He thereby established a most fruitful link between random matrix theory and mesoscopic physics.

The non-linear σ model, particularly its mathematical aspects, was recently reviewed in [Zir06].

7.4.2 Eigenvalues and diffusion in superspace

The supersymmetric representation and Fyodorov's alternative representation of Section 7.3 are exact for finite N. In some situations, it is indeed possible and advantageous to evaluate them without using the non-linear σ model. It has been shown for the supersymmetric representation [Guh06, Kie09a] that the imaginary increments of the arguments may then all lie on the same side of the real axis. We have $L = 1$ and all groups are compact. As we aim at structural aspects, we consider the crossover transitions involving $H(\alpha) = H^{(0)} + \alpha H$ as discussed in Section 7.3.5. We introduce the fictitious time $t = \alpha^2/2$. Dyson [Dys62, Dys72] showed that the eigenvalues of $H(t)$ follow a Brownian motion in t which implies that their probability density function is propagated by a diffusion equation. Without any loss of information, Supersymmetry reduces this stochastic process to a Brownian motion in a much smaller space [Guh96b]. This is precisely the radial part of the Riemannian symmetric superspaces discussed in Section 7.2.3. The quantity propagated is then the generating function $Z_k(x + J, t)$ of the correlations. The initial condition

$$Z_k^{(0)}(s) = \int P^{(0)}(H^{(0)}) \mathrm{sdet}^{-1}(s \otimes 1 + 1 \otimes H^{(0)}) d[H^{(0)}] \tag{7.4.5}$$

is arbitrary, it includes ensembles, but also a fixed matrix $H^{(0)}$ if the probability density $P^{(0)}(H^{(0)})$ is chosen accordingly. Because of the direct product structure, $Z_k^{(0)}(s)$ is rotation invariant. The diagonal matrix s is in the above mentioned radial space, such that $\sigma = u^{-1}su$, see Section 7.2.3. For $\beta = 2$, this space coincides with $x + J$, for $\beta = 1, 4$, it is slightly larger. The generating function is then a convolution in the radial space,

$$Z_k(r,t) = \int \Gamma_k(s,r,t) Z_k^{(0)}(s) B_k(s) d[s]. \tag{7.4.6}$$

When going to eigenvalue-angle coordinates $\sigma = u^{-1}su$ Berezinians $B_k(s)$ occur analogous to $|\Delta_N(X)|^\beta$ in ordinary space. The propagator is the supergroup integral

$$\Gamma_k(s,r,t) = c^{(\beta)} \exp\left(-\frac{\beta}{4t}\mathrm{str}\,(s^2 + r^2)\right) \int \exp\left(-\frac{\beta}{2t}\mathrm{str}\,u^{-1}sur\right) d\mu(u). \tag{7.4.7}$$

For $\beta = 2$ and all k, this integral is known explicitly [Guh91, Guh96a]. Unfortunately, for $\beta = 1, 4$ the available result [Guh02] is handy only for $k = 1$ but cumbersome for $k = 2$.

It is a remarkable inherent feature of supersymmetry that the propagator and thus the diffusion process of $Z_k(r,t)$ on the original scale in t carries over unchanged to the unfolded scale when introducing the proper time $\tau = t/D^2$. The initial condition is the unfolded large N limit of $Z_k^{(0)}(s)$. Moreover and in contrast to the hierarchical equations for the correlation functions [Fre88], the Brownian motion in superspace for the generating functions is diagonal in k.

7.5 Circular ensembles and Colour-Flavour transformation

In many physics applications, the random matrices H model Hamiltonians, implying that they are either real symmetric, Hermitian, or quaternion real. Mathematically speaking, they are in the non-compact forms of the corresponding symmetric spaces. However, if one aims at modelling scattering, it is often useful to work with random unitary matrices S, taken from the compact forms of these symmetric spaces [Dys62]. These are $\mathrm{U}(N)/\mathrm{O}(N)$, $\mathrm{U}(N)$ and $\mathrm{U}(2N)/\mathrm{Sp}(2N)$, respectively, leading to the circular orthogonal, unitary and symplectic ensembles COE, CUE, and CSE which are labelled $\beta = 1, 2, 4$. The phase angles of S play the same rôle as the eigenvalues of H. Because of the compactness, no Gaussian or other confining function is needed and the probability density function is just the invariant measure on the symmetric space in question. On the local scale of the mean level spacing, the correlations coincide with those of the Gaussian ensembles [Dys62, Meh04].

Zirnbauer [Zir96b] showed how to apply supersymmetry to circular ensembles. His approach works for all three symmetry classes, but for simplicity we only discuss the CUE which consists of the unitary matrices $S = U \in \mathrm{U}(N)$. Consider the generating function

$$Z_{k_+k_-}(\vartheta, \varphi) = \int d\mu(U) \prod_{p=1}^{k_+} \frac{\det\left(1 - \exp(i\varphi_{+p})U\right)}{\det\left(1 - \exp(i\vartheta_{+p})U\right)} \prod_{q=1}^{k_-} \frac{\det\left(1 - \exp(i\varphi_{-q})U^\dagger\right)}{\det\left(1 - \exp(i\vartheta_{-q})U^\dagger\right)}, \tag{7.5.1}$$

where $d\mu(U)$ is the invariant measure on $\mathrm{U}(N)$. To derive the correlation functions $R_k(\vartheta_1, \ldots, \vartheta_k)$ one sets $k_+ = k_- = k$, takes derivatives with respect to the variables $\varphi_{\pm p}$, and puts certain combinations of variables $\varphi_{\pm p}$ and $\vartheta_{\pm p}$ equal. The variables $\vartheta_{\pm p}$ in Eq. (7.5.1) have small imaginary increments to prevent $Z_{k_+k_-}(\vartheta, \phi)$ from becoming singular.

Since the Hubbard–Stratonovich transformation of Section 7.3.2 cannot be employed to construct a supersymmetric representation of $Z_{k_+k_-}(\vartheta, \phi)$, Zirnbauer [Zir96b] developed the colour-flavour transformation based on the identity

$$\int d\mu(U) \exp\left(\Psi^{n*}_{+pj} U^{nm} \Psi^{m}_{+pj} + \Psi^{n*}_{+qj} U^{nm*} \Psi^{m}_{+qj}\right) =$$
$$\int d[\Lambda] d[\tilde{\Lambda}] \mathrm{sdet}^{N}\left(1 - \tilde{\Lambda}\Lambda\right) \exp\left(\Psi^{n*}_{+pj} \Lambda_{pjql} \Psi^{n}_{-ql} + \Psi^{m*}_{-ql} \tilde{\Lambda}_{qlpj} \Psi^{m}_{+pj}\right) \tag{7.5.2}$$

which transforms an integral over the ordinary group $\mathrm{U}(N)$ into an integral over $k_+/k_+ \times k_-/k_-$ rectangular supermatrices Λ and $\tilde{\Lambda}$ parameterizing the coset space $\mathrm{U}(k_+ + k_-/k_+ + k_-)/\mathrm{U}(k_+/k_+) \times \mathrm{U}(k_-/k_-)$. The integrals depend on the supertensor Ψ with components Ψ^{n}_{+pj}. The indices $n, m = 1, \ldots, N$ label the elements of U in ordinary space. The indices pj and ql are superspace indices with $p = 1, \ldots, k_+$, $q = 1, \ldots, k_-$ and with $j, l = 1, 2$ labelling the four blocks of the supermatrices, see Eq. (7.2.8). Summation convention applies. The superfields Ψ are used to express the determinants in Eq. (7.5.1) as Gaussian integrals. After the colour-flavour transformation, they are integrated out again.

As the name indicates, the colour-flavour transformation is naturally suited for applications in lattice gauge theories where U is in the colour gauge group, see [Nag01]. In [Wei05], the colour-flavour transformation was derived for the gauge group $\mathrm{SU}(N)$ relevant in lattice quantum chromodynamics.

7.6 Concluding remarks

Apart from the wealth of results for specific physics systems which had to be left out, some important conceptual issues could not be discussed here either due to lack of space: we only mentioned applications of the other Riemannian symmetric superspaces [Zir96a] to Andreev scattering and chiral random matrix theory. As random matrix approaches are now ubiquitous in physics and beyond, one may also expect that the supersymmetry method spreads out accordingly. From a mathematical viewpoint, various aspects deserve further clarifying studies, in the present context most noticeably the theory of supergroups and harmonic analysis on superspaces.

Acknowledgements

I thank Mario Kieburg and Heiner Kohler for helpful discussions. I acknowledge support from Deutsche Forschungsgemeinschaft within Sonderforschungsbereich Transregio 12 "Symmetries and Universality in Mesoscopic Systems".

References

[Ber87] F.A. Berezin, *Introduction to Supermanifolds*, Reidel, Dordrecht 1987.

[Bun07] J.E. Bunder, K.B. Efetov, V.E. Kravtsov, O.M. Yevtushenko and M.R. Zirnbauer, J. Stat. Phys. **129** (2007) 809.

[Dys62] F.J. Dyson, J. Math. Phys. **3** (1962) 140; 157; 166; 1199.

[Dys72] F.J. Dyson, J. Math. Phys. **13** (1972) 90.

[Efe82] K.B. Efetov, Zh. Eksp. Teor. Fiz. **82** (1982) 872 [Sov. Phys. JETP **55** (1982) 514].

[Efe83] K.B. Efetov, Adv. in Phys. **32** (1983) 53.

[Efe97] K. B. Efetov, *Supersymmetry in Disorder and Chaos*, Cambridge University Press, Cambridge 1997.

[Efe04] K.B. Efetov, G. Schwiete and K. Takahashi, Phys. Rev. Lett. **92** (2004) 026807.

[Fre88] J.B. French, V.K.B. Kota, A. Pandey and S. Tomsovic, Ann. Phys. (NY) **181** (1988) 198.

[Fyo91] Y.V. Fyodorov and A.D. Mirlin, Phys. Rev. Lett. **67** (1991) 2405.

[Fyo02] Y.V. Fyodorov, Nucl. Phys. **B621** (2002) 643.

[Fyo08] Y.V. Fyodorov, Y. Wei and M.R. Zirnbauer, J. Math. Phys. **49** (2008) 053507.

[Guh91] T. Guhr, J. Math. Phys. **32** (1991) 336.

[Guh96a] T. Guhr, Commun. Math. Phys. **176** (1996) 555.

[Guh96b] T. Guhr, Ann. Phys. (NY) **250** (1996) 145.

[Guh98] T. Guhr, A. Müller–Groeling and H.A. Weidenmüller, Phys. Rep. **299** (1998) 189.

[Guh02] T. Guhr and H. Kohler, J. Math. Phys. **43** (2002) 2741.

[Guh06] T. Guhr, J. Phys. **A39** (2006) 13191.

[Hac95] G. Hackenbroich and H.A. Weidenmüller, Phys. Rev. Lett. **74** (1995) 4418.

[Kac77] V.G. Kac, Commun. Math. Phys. **53** (1977) 31.

[Kie09a] M. Kieburg, J. Grönqvist and T. Guhr, J. Phys. **A42** (2009) 275205.

[Kie09b] M. Kieburg, H.J. Sommers and T. Guhr, J. Phys. **A42** (2009) 275206.

[Leh95] N. Lehmann, D. Saher, V.V. Sokolov and H.J. Sommers, Nucl. Phys. **A582** (1995) 223.

[Lit08] P. Littlemann, H.J. Sommers and M.R. Zirnbauer, Commun. Math. Phys. **283** (2008) 343.

[Mah69] C. Mahaux and H.A. Weidenmüller, *Shell–Model Approach to Nuclear Reactions,* North–Holland, Amsterdam 1969.

[Mar59] I.L. Martin, Proc. Roy. Soc. **A251** (1959) 536.

[Mar05] S.P. Martin, *A Supersymmetry Primer*, in *Perspectives in Supersymmetry II*, G.L. Kane (Ed.) Advances Series on Directions in High Energy Physics, Vol. 21, World Scientific (2010) [arXive:hep-ph/9709356].

[Meh04] M.L. Mehta, *Random Matrices*, Academic Press, 3rd Edition, London 2004.

[Mir00] A. Mirlin, Phys. Rep. **326** (2000) 259.

[Mir91] A.D. Mirlin and Y. Fyodorov, J. Phys. **A24** (1991) 2273.

[Nag01] T. Nagao and S.M. Nishigaki, Phys. Rev. **D64** (2001) 014507.

[Par79] G. Parisi and N. Sourlas, Phys. Rev. Lett. **43** (1979) 744.

[Pas72] L. Pastur, Theor. Mat. Phys. **10** (1972) 67.

[Pru82] A.M.M. Pruisken and L. Schäfer, Nucl. Phys. **B200** (1982) 20.

[Rot87] M.J. Rothstein, Trans. Am. Math. Soc. **299** (1987) 387.

[Sch80] L. Schäfer and F. Wegner, Z. Phys. **B38** (1980) 113.

[Ver85a] J.J.M. Verbaarschot and M.R. Zirnbauer, J. Phys. **A17** (1985) 1093.

[Ver85b] J.J.M. Verbaarschot, H.A. Weidenmüller and M.R. Zirnbauer, Phys. Rep. **129** (1985) 367.

[Wei05] Y. Wei and T. Wettig, J. Math. Phys. **46** (2005) 072306.

[Zir96a] M.R. Zirnbauer, J. Math. Phys. **37** (1996) 4986.

[Zir96b] M.R. Zirnbauer, J. Phys. **A29** (1996) 7113.

[Zir06] M.R. Zirnbauer, in: *Encyclopedia of Mathematical. Physics*, vol. 5, pp. 151, Elsevier, Amsterdam 2006.

·8·

Replica approach in random matrix theory

Eugene Kanzieper

Abstract

This chapter outlines the replica approach in random matrix theory. Both fermionic and bosonic versions of the replica limit are introduced and its trickery is discussed. A brief overview of early heuristic treatments of zero-dimensional replica field theories is given to advocate an exact approach to replicas. The latter is presented in two elaborations: by viewing the $\beta = 2$ replica partition function as the Toda lattice and by embedding the replica partition function into a more general theory of τ functions.

8.1 Introduction

8.1.1 Resolvent as a field integral

In physics of disorder, all observables depend in highly non-linear fashion on a stochastic Hamiltonian hereby making a nonperturbative calculation of their ensemble averages very difficult. To determine the average quantities in an interactionless system, one has to know the spectral statistical properties of a single particle Hamiltonian $\mathcal{H}$ contained in the mean product of resolvents $G(z) = \mathrm{tr}\,(z - \mathcal{H})^{-1}$ defined for a generic, complex valued argument $z \in \mathbb{C} \setminus \mathbb{R}$. Each of the resolvents can exactly be represented as a ratio of two functional integrals running over an auxiliary vector field ψ which may consist of *either* commuting (bosonic, $\psi = s$) *or* anticommuting (fermionic, $\psi = \chi$) entries.

In the random matrix theory (RMT) limit, when a system Hamiltonian (or a scattering matrix) is modelled by an $N \times N$ random matrix $\mathcal{H}$ of prescribed symmetry, the resolvent $G(z)$ equals

$$G(z) = i\eta\,\mathfrak{s}_z \int \mathcal{D}[\bar{\psi}, \psi]\bar{\psi}_\ell \psi_\ell\, e^{i\mathfrak{s}_z \mathcal{S}_{\mathcal{H}}(z;\bar{\psi},\psi)} \left(\int \mathcal{D}[\bar{\psi}, \psi]\, e^{i\mathfrak{s}_z \mathcal{S}_{\mathcal{H}}(z;\bar{\psi},\psi)} \right)^{-1}. \qquad (8.1.1)$$

Here, $\mathcal{S}_{\mathcal{H}} = \bar{\psi}(z - \mathcal{H})\psi = \bar{\psi}_j \left(z\delta_{jk} - \mathcal{H}_{jk}\right)\psi_k$ (summation over repeated Latin indices is assumed throughout this section); vector ψ is defined by $\psi = (\psi_1, \cdots, \psi_N)^{\mathrm{T}}$, whilst $\bar{\psi} = (\bar{\psi}_1, \cdots, \bar{\psi}_N)$ is its proper conjugate. The parameter η accounts for the nature of both vectors: it is set to $+1$ for fermions ($\psi = \chi$) and to -1 for bosons ($\psi = s$). In the latter case, convergence of field integrals is ensured by the regularizer $\mathfrak{s}_z = \mathrm{sgn}(\Im\mathfrak{m}\, z)$. The notation $\mathcal{D}[\bar{\psi}, \psi]$ stands for the integration measure[1] $\mathcal{D}[\bar{\psi}, \psi] = (2\pi)^{-N} \prod_{j=1}^{N} d\bar{\psi}_j d\psi_j$.

Equation (8.1.1) may conveniently be viewed as a consequence of the identity

$$G(z) = \eta \frac{\partial}{\partial z} \log \det{}^{\eta}(z - \mathcal{H}), \tag{8.1.2}$$

combined with the field integral representations of the determinant ($\eta = +1$, $\psi = \chi$) and/or its inverse ($\eta = -1$, $\psi = s$):

$$\det{}^{\eta}(z - \mathcal{H}) = (i\eta\,\mathfrak{s}_z)^N \int \mathcal{D}[\bar{\psi}, \psi]\, e^{i\mathfrak{s}_z \mathcal{S}_{\mathcal{H}}(z;\bar{\psi},\psi)}. \tag{8.1.3}$$

Although exact, both bosonic and fermionic versions of Eq. (8.1.1) are a little too inconvenient for a nonperturbative ensemble averaging due to the awkward random denominator.

To get rid of this, several field theoretic frameworks have been devised by theoretical physicists: the replica trick [Weg79, Sch80, Efe80], the supersymmetry method [Efe82], and the dynamic (Keldysh) approach [Hor90, Kam99a]. Leaving aside the supersymmetry and the Keldysh techniques,[2] this chapter aims to provide an elementary introduction to the 'notorious' replica approach whose legitimacy has been a point of controversy [Ver85b, Kam99b, Zir99] for over two decades. Recently discovered integrability [Kan02, Kan05, Osi07] of zero-dimensional replica field theories will be a dominant motif of this contribution.

8.1.2 How replicas arise and why they are tricky

Both fermionic and bosonic replicas are based on the identity[3]

$$\log \mathfrak{X} = \lim_{n \to 0} \frac{\mathfrak{X}^n - 1}{n} \tag{8.1.4}$$

which can be very useful in evaluating the average of a logarithm $\langle \log \mathfrak{X} \rangle$ of the random variable $\mathfrak{X}$. Indeed, identifying $\mathfrak{X}$ with $\det^{\eta}(z - \mathcal{H})$ in Eq. (8.1.2), and further combining Eq. (8.1.4) with the field integral representation Eq. (8.1.3) of the (inverse) determinant, we *formally* relate the resolvent

[1] Integrals over anticommuting (Grassmann) variables χ and $\bar{\chi}$ are normalized according to

$$\int d\bar{\chi}\, \chi = \int d\chi\, \bar{\chi} = (2\pi)^{1/2}.$$

[2] An introductory exposition of the supersymmetry method can be found in Chapter 7 of this handbook as well as in the earlier review papers [Efe83, Ver85a] and the monograph [Efe97]. For a review of the Keldysh approach, the reader is referred to [Kam09]; see also the paper [Alt00] where the Keldysh technique is discussed in the RMT context.

[3] Mostly known in the condensed matter physics community since the paper by Edwards and Anderson [Edw75] on spin glasses (see also [Eme75]), the recipe for calculating the average of a logarithm based on Eq. (8.1.4) dates back at least as far as 1934, see the book [Har34] by Hardy, Littlewood, and Pólya.

$$G(z) = \eta \lim_{n\to 0} \frac{1}{n}\frac{\partial}{\partial z} \mathcal{Z}_n^{(\eta)}(z) \tag{8.1.5}$$

to the 'partition function'

$$\mathcal{Z}_n^{(\eta)}(z) = \prod_{a=1}^{n} \int \mathcal{D}[\bar{\psi}^{(a)}, \psi^{(a)}] \exp\left[i\mathfrak{s}_z \bar{\psi}^{(a)}(z - \mathcal{H})\psi^{(a)}\right] \tag{8.1.6}$$

of n copies, or replicas, of the initial random system. The pair of formulas Eqs. (8.1.5) and (8.1.6), known as the *replica trick,*[4] achieve our goal of removing a random denominator from Eq. (8.1.1) and hint that a nonperturbative calculation of the mean product of resolvents may become feasible.

In order to keep the discussion concrete and set up notation, we further assume that the matrix Hamiltonian $\mathcal{H}$ is drawn from the paradigmatic Gaussian unitary ensemble (GUE) associated with the probability measure [Meh04][5]

$$P_N[\mathcal{H}]\,\mathcal{D}\mathcal{H} = \pi^{-N^2/2} \exp(-\mathrm{tr}\,\mathcal{H}^2) \prod_{j=1}^{N} d\mathcal{H}_{jj} \prod_{1\le j<k}^{N} d\mathcal{H}_{jk} d\bar{\mathcal{H}}_{jk}. \tag{8.1.7}$$

Then, the resolvent $g(z) = \langle G(z)\rangle$ averaged with respect to the GUE probability measure should be furnished by the limiting procedure

$$g(z) = \eta \lim_{n\to 0} \frac{1}{n}\frac{\partial}{\partial z} \langle \mathcal{Z}_n^{(\eta)}(z)\rangle, \tag{8.1.8}$$

where the (average) *replica partition function* equals

$$\langle \mathcal{Z}_n^{(\eta)}(z)\rangle = \prod_{a=1}^{n} \int \mathcal{D}[\bar{\psi}^{(a)}, \psi^{(a)}] \exp\left(-i\eta\, \mathfrak{s}_z\, z\, \mathrm{tr}\,\sigma + \frac{\eta}{4}\mathrm{tr}\,\sigma^2\right). \tag{8.1.9}$$

Here, the Hermitian matrix $\sigma = \sigma^\dagger$ acting in the replica space is defined as

$$\sigma_{\alpha\beta} = \left(\psi_j \otimes \bar{\psi}_j\right)_{\alpha\beta} = \psi_j^{(\alpha)}\bar{\psi}_j^{(\beta)}, \quad \alpha, \beta \in (1, \ldots, n). \tag{8.1.10}$$

Notice that ensemble averaging of $\mathcal{Z}_n^{(\eta)}(z)$ [Eq. (8.1.6)] has induced an effective ψ^4 interaction between n replicated random systems as described by the interaction term $\mathrm{tr}\,\sigma^2 = \mathrm{tr}\,[(\psi_j \otimes \bar{\psi}_j)^2]$ in the action. To make further progress in evaluating $\langle \mathcal{Z}_n^{(\eta)}(z)\rangle$, the interaction term may routinely be decoupled by means of the Hubbard–Stratonovich transformation [Efe97, Ver85a].

[4] The alert reader might already detect some trickery behind Eqs. (8.1.5) and (8.1.6). Their mathematical status will be clarified below.

[5] See also Chapter 4 of this handbook which outlines the method of orthogonal polynomials for unitary invariant random matrix models.

Fermionic replicas

In the case of fermionic fields ($\psi = \chi$, $\eta = +1$), the χ^4 interaction can be decoupled via $n \times n$ Hermitian matrix field $\mathcal{Q}_n$ as

$$\exp\left(+\frac{1}{4}\mathrm{tr}\,\sigma^2\right) = \pi^{-n^2/2}\int_{\mathcal{Q}_n^\dagger=\mathcal{Q}_n} \mathcal{D}\mathcal{Q}_n \exp\left[-\mathrm{tr}\left(\mathcal{Q}_n^2 + \sigma\mathcal{Q}_n\right)\right]. \quad (8.1.11)$$

Performing the integration over fermionic fields in Eq. (8.1.9) with the help of Eq. (8.1.3),

$$\prod_{a=1}^{n}\int \mathcal{D}[\bar{\chi}^{(a)}, \chi^{(a)}]\, \exp\left(-i\mathfrak{s}_z\, z\,\mathrm{tr}\,\sigma + \frac{1}{4}\mathrm{tr}\,\sigma^2\right)$$
$$= \pi^{-n^2/2}(-i\mathfrak{s}_z)^{nN}\int_{\mathcal{Q}_n^\dagger=\mathcal{Q}_n} \mathcal{D}\mathcal{Q}_n\, e^{-\mathrm{tr}\,\mathcal{Q}_n^2}\,\det{}^N(z - i\mathfrak{s}_z\mathcal{Q}_n), \quad (8.1.12)$$

we arrive at the *fermionic* replica limit

$$g(z) = \lim_{n\to 0}\frac{1}{n}\frac{\partial}{\partial z}\langle \mathcal{Z}_n^{(+)}(z)\rangle \quad (8.1.13)$$

which relates the average resolvent $g(z)$ to the *fermionic* replica partition function

$$\langle \mathcal{Z}_n^{(+)}(z)\rangle = \int_{\mathcal{Q}_n^\dagger=\mathcal{Q}_n} \mathcal{D}\mathcal{Q}_n\; e^{-\mathrm{tr}\,\mathcal{Q}_n^2}\,\det{}^N\,(z - i\mathfrak{s}_z\mathcal{Q}_n). \quad (8.1.14)$$

Before discussing this result, let us turn to the bosonic version of the replica trick.

Bosonic replicas

In the case of bosonic fields ($\psi = s$, $\eta = -1$), decoupling of the s^4 interaction can be carried out in a similar fashion. Making use of yet another variant of the Hubbard–Stratonovich transformation

$$\exp\left(-\frac{1}{4}\mathrm{tr}\,\sigma^2\right) = \pi^{-n^2/2}\int_{\mathcal{Q}_n^\dagger=\mathcal{Q}_n} \mathcal{D}\mathcal{Q}_n \exp\left[-\mathrm{tr}\left(\mathcal{Q}_n^2 + i\sigma\mathcal{Q}_n\right)\right], \quad (8.1.15)$$

and integrating out bosonic fields in Eq. (8.1.9) with the help of Eq. (8.1.3),

$$\prod_{a=1}^{n}\int \mathcal{D}[\bar{s}^{(a)}, s^{(a)}]\, \exp\left(+i\mathfrak{s}_z\, z\,\mathrm{tr}\,\sigma - \frac{1}{4}\mathrm{tr}\,\sigma^2\right)$$
$$= \pi^{-n^2/2}(+i\mathfrak{s}_z)^{nN}\int_{\mathcal{Q}_n^\dagger=\mathcal{Q}_n} \mathcal{D}\mathcal{Q}_n\, e^{-\mathrm{tr}\,\mathcal{Q}_n^2}\,\det{}^{-N}(z - \mathfrak{s}_z\mathcal{Q}_n), \quad (8.1.16)$$

we express the average resolvent

$$g(z) = -\lim_{n\to 0}\frac{1}{n}\frac{\partial}{\partial z}\langle \mathcal{Z}_n^{(-)}(z)\rangle \quad (8.1.17)$$

through the *bosonic* replica partition function

$$\langle \mathcal{Z}_n^{(-)}(z) \rangle = \int_{\mathcal{Q}_n^\dagger = \mathcal{Q}_n} \mathcal{D}\mathcal{Q}_n \, e^{-\mathrm{tr}\,\mathcal{Q}_n^2} \det{}^{-N}(z - \mathfrak{s}_z \mathcal{Q}_n). \tag{8.1.18}$$

Subtleties of the replica limit

Seemingly innocent at first glance, both fermionic [Eqs. (8.1.13) and (8.1.14)] and bosonic [Eqs. (8.1.17) and (8.1.18)] replica prescriptions appear to be counterintuitive and raising fundamental mathematical questions [Par03]. Indeed, because of a particular integration measure which makes no sense for n other than positive integers ($n \in \mathbb{Z}^+$), the matrix-integral representation of average replica partition functions

$$\langle \mathcal{Z}_n^{(\eta)}(z) \rangle = \int_{\mathcal{Q}_n^\dagger = \mathcal{Q}_n} \mathcal{D}\mathcal{Q}_n \, e^{-\mathrm{tr}\,\mathcal{Q}_n^2} \det{}^{\eta N}\left(z - \sqrt{i}^{\,1+\eta} \mathfrak{s}_z \mathcal{Q}_n\right) \tag{8.1.19}$$

cannot be used directly to implement the replica limit Eqs. (8.1.8) as the latter is determined by the behaviour of $\langle \mathcal{Z}_n^{(\eta)}(z) \rangle$ in a close *vicinity* of $n = 0$. This mismatch between the 'available' ($n \in \mathbb{Z}^+$) and the 'needed' ($n \in \mathbb{R}$) is at the heart of the trickery the replica field theories have often been charged with [Ver85b, Zir99].

The canonical way to bridge this gap is to determine the average replica partition functions $\langle \mathcal{Z}_n^{(\eta)}(z) \rangle$ for $n \in \mathbb{Z}^+$, and then to attempt to analytically continue them to $n \in \mathbb{R}$, in general, and to a vicinity of $n = 0$, in particular. This is a non-trivial task for two major reasons:

(i) The analytic continuation of the replica partition function away from n integers should not necessarily be unique.[6]

(ii) To retain control over the analytic continuation, the latter must rest on an exact calculation of the average replica partition function for $n \in \mathbb{Z}^+$. Early approaches to replica field theories seem to underestimate these two points bringing a number of pathological results even in the RMT setting (see Section 8.2).

8.2 Early studies: heuristic approach to replicas

Exact evaluation of replica partition functions, whilst welcomed, is quite a challenge. At the same time, their approximate calculation is often feasible in a certain region of parameter space where a saddle point procedure can be justified. In doing so, one is naturally led to *'replica symmetric'* and *'replica asymmetric'* saddle point manifolds as discussed below.

[6] For entire functions, the uniqueness is guaranteed by a boundedness property as formulated by Carlson's theorem [Tit32].

8.2.1 Density of eigenvalues in the GUE

In the RMT context, a saddle point evaluation of the replica partition function Eq. (8.1.19) makes sense if the dimension N of the random matrix $\mathcal{H}$ is large enough. For not too large replica parameter $n \in \mathbb{Z}^+$ (in particular, n should not scale with N), the dominating contribution to $\langle \mathcal{Z}_n^{(\eta)}(z) \rangle$ is expected to come from the configurations $\mathcal{Q}_n^{(\mathrm{sp})}$ determined by the saddle point equation

$$\frac{\delta}{\delta \mathcal{Q}_n} \mathrm{tr}\Big[\mathcal{Q}_n^2 - \eta N \log(z - \sqrt{i}^{\,1+\eta} \mathfrak{s}_z \mathcal{Q}_n) \Big] = 0. \tag{8.2.1}$$

Its solutions form 2^n saddle point manifolds

$$\mathcal{Q}_n^{(\mathrm{sp})} = \frac{\mathfrak{s}_z}{\sqrt{i}^{\,1+\eta}} \sqrt{\frac{N}{2}}\, \mathrm{diag}(e^{i\kappa_1 \theta}, \dots, e^{i\kappa_n \theta}) \tag{8.2.2}$$

with κ_ℓ taking on the values ± 1 independently of each other. Here,

$$e^{i\theta} = z_s + i\sqrt{1 - z_s^2}, \tag{8.2.3}$$

where z_s stands for the scaled energy $z_s = z/\mathcal{D}_{\mathrm{edge}}$ with $\mathcal{D}_{\mathrm{edge}} = \sqrt{2N}$ being the endpoint of the spectrum support. (Hence, the spectrum bulk is situated within the segment $|\mathfrak{Re}\, z_s| < 1$.)

Bosonic replicas

Out of the plethora of saddles Eq. (8.2.2), only the distinguished *replica symmetric* manifold

$$\mathcal{Q}_n^{(\mathrm{sp})}\Big|_{\mathrm{sym}} = \mathfrak{s}_z \sqrt{\frac{N}{2}}\, e^{i\theta} \otimes \mathbb{1}_n \tag{8.2.4}$$

contributes the bosonic replica partition function Eq. (8.1.18). This is so because Eq. (8.2.4) is the only saddle [Zir99]

(i) reachable by continuous deformation of the integration contour in Eq. (8.1.18) without crossing the hypersurface defined by the singularities of $\det^{-N}(z - \mathfrak{s}_z \mathcal{Q}_n)$ and

(ii) compatible with the analyticity of the average resolvent Eq. (8.1.17) at infinity.

In the leading order in the large parameter N, the bosonic replica partition function is then approximated by

$$\begin{aligned} \langle \mathcal{Z}_n^{(-)}(z = \epsilon - i0) \rangle \;\simeq\; & \left(\frac{N}{2}\right)^{nN/2} (2 \sin\theta)^{-n^2/2} \\ & \times \exp\left[\frac{nN}{2} \left(e^{2i\theta} - 2i\theta\right) - i\,\frac{n^2}{2} \left(\theta - \frac{\pi}{2}\right) \right]. \end{aligned} \tag{8.2.5}$$

By derivation, Eq. (8.2.5) holds for $n \in \mathbb{Z}^+$. To retrieve the average density of eigenlevels

$$\varrho(\epsilon) = \frac{1}{\pi} \mathfrak{Im}\, g(\epsilon - i0) \tag{8.2.6}$$

through the replica limit Eq. (8.1.17), one should analytically continue $\langle \mathcal{Z}_n^{(-)}(z) \rangle$ away from $n \in \mathbb{Z}^+$. To be on the safe side, such an analytic continuation must rest on an exact integer-n result for $\langle \mathcal{Z}_n^{(-)}(z) \rangle$. The latter is sadly unavailable. Not being spoilt for choice, one could try to analytically continue the bosonic replica partition function to the domain $n \in \mathbb{R}^+$ taking the approximate result Eq. (8.2.5) as a starting point and merely assuming it to hold, as it stands, in the right vicinity of $n = 0$. Then, the replica limit Eq. (8.1.17) can be taken to yield the Wigner semicircle [Edw80]

$$\varrho(\epsilon_s) = \frac{2}{\pi}\sqrt{1 - \epsilon_s^2}, \quad |\epsilon_s| \leq 1. \tag{8.2.7}$$

Here, $\epsilon_s = \epsilon/\mathcal{D}_{\text{edge}}$. The $1/N$ correction to Eq. (8.2.7) can be obtained in a similar fashion and is known to vanish within the replica symmetric ansatz.

Similarly, replica symmetric saddle point calculations performed for the Gaussian orthogonal ensemble (GOE) and the Gaussian symplectic ensemble (GSE) of random matrices yield [Edw76, Ver84, Dhe90, Ito97]

$$\varrho(\epsilon_s) = \frac{2}{\pi}\sqrt{1 - \epsilon_s^2} + \frac{1}{2\pi N}\left(1 - \frac{2}{\beta}\right)\left[\frac{1}{\sqrt{1 - \epsilon_s^2}} - \pi\delta(\epsilon_s^2 - 1)\right] \tag{8.2.8}$$

with $|\epsilon_s| \leq 1$. Here, β is the Dyson symmetry index [Meh04] taking the values $\beta = 1$ for GOE, $\beta = 2$ for GUE, and $\beta = 4$ for GSE.

Importantly, both leading and subleading in $1/N$ terms in Eq. (8.2.8) for the average densities of eigenlevels contain no terms oscillating on the scale of the mean level spacing. Since the replica asymmetric saddles are inaccessible as explained on general grounds [Zir99] in (i) and (ii) below Eq. (8.2.4), we are led to conclude that the saddle point approach to bosonic replicas fails to reproduce truly non-perturbative features of the eigenlevel density.

Fermionic replicas

Looking for some insight into a possible rôle played by *replica asymmetric* saddles, we turn to the approximate performance of fermionic replicas Eq. (8.1.14). In this case, all 2^n saddle point manifolds

$$\mathcal{Q}_n^{(\text{sp})} = -i\mathfrak{s}_z\sqrt{\frac{N}{2}}\,\text{diag}(e^{i\kappa_1\theta}, \ldots, e^{i\kappa_n\theta}), \quad \kappa_\ell = \pm 1, \tag{8.2.9}$$

are accessible adding up, for $n = 2m$, to

$$\langle \mathcal{Z}_{2m}^{(+)}(z)\rangle \simeq \left(\frac{N}{2}\right)^{mN} e^{mN\cos 2\theta} \sum_{q=-m}^{q=+m} \mathcal{V}_{m,q} \left(\frac{N}{2\pi}\right)^{m^2-q^2} (2\sin\theta)^{m^2-3q^2}$$
$$\times \exp\left[iq\left(N(2\theta - \sin 2\theta) + 2m\left(\theta - \frac{\pi}{2}\right)\right)\right]. \quad (8.2.10)$$

Here, $\mathcal{V}_{m,q}$ denotes the volume of Grassmanian $\mathrm{U}(2m)/\mathrm{U}(m-q)\times \mathrm{U}(m+q)$ [Kam99b, Zir99]

$$\mathcal{V}_{m,q} = (2\pi)^{m^2-q^2} \frac{\prod_{j=1}^{m+q}\Gamma(j)\prod_{j=1}^{m-q}\Gamma(j)}{\prod_{j=1}^{2m}\Gamma(j)}. \quad (8.2.11)$$

The summation index q in Eq. (8.2.10) counts $(2m+1)$ families of saddle point manifolds [Eq. (8.2.9)], the qth family being represented by the configuration

$$\mathcal{Q}_n^{(\mathrm{sp})}[q] = -i\mathfrak{s}_z\sqrt{\frac{N}{2}}\,\mathrm{diag}(e^{i\theta}\otimes \mathbb{1}_{m-q},\ e^{-i\theta}\otimes \mathbb{1}_{m+q}) \quad (8.2.12)$$

taken with the obvious combinatorial weight

$$\binom{2m}{m+q} = \frac{\Gamma(2m+1)}{\Gamma(m+1-q)\Gamma(m+1+q)}.$$

Obtained in the large-N limit, Eq. (8.2.10) holds for $m\in\mathbb{Z}^+$ which do not scale with N.

Aimed at deriving the density of eigenlevels through the replica limit, one should first analytically continue Eq. (8.2.10) away from $m\in\mathbb{Z}^+$. Even though making an analytic continuation based on an *approximate* result is a dangerous ploy and is, with certainty, a mathematically questionable procedure, we embark on the proposal due to [Kam99b] who spotted that the volume of the Grassmanian [Eq. (8.2.11)] vanishes for all integers $|q|\geq m+1$. This observation makes it tempting to extend the summation over q in Eq. (8.2.10) to (minus and plus) infinities to end up with the following trial function for an 'analytically continued' fermionic replica partition function

$$\langle \mathcal{Z}_{2m}^{(+)}(z)\rangle \overset{?}{\simeq} \left(\frac{N}{2}\right)^{mN} e^{mN\cos 2\theta} \sum_{q=-\infty}^{q=+\infty} \mathcal{V}_{m,q} \left(\frac{N}{2\pi}\right)^{m^2-q^2} (2\sin\theta)^{m^2-3q^2}$$
$$\times \exp\left[iq\left(N(2\theta - \sin 2\theta) + 2m\left(\theta - \frac{\pi}{2}\right)\right)\right]. \quad (8.2.13)$$

A close inspection of this result reveals that it is flawed:

(i) the so-continued replica partition function $\langle \mathcal{Z}_{2m}^{(+)}(z)\rangle$ diverges[7] [Kam99b, Zir99] in the vicinity of $m=+0$, the region crucially important for retrieving the density of eigenlevels through the replica limit;

[7] The group volume $\mathcal{V}_{m,q}$ grows too fast with q for the series $\sum_{q=-\infty}^{q=+\infty}(\cdots)$ to converge.

(ii) because of the $q \mapsto -q$ symmetry of the summand, $\langle \mathcal{Z}_{2m}^{(+)}(z) \rangle$ must be *real* thus leaving no room for a non-zero density of states. Indeed, a formally derived small-m expansion of Eq. (8.2.13)

$$\langle \mathcal{Z}_{2m}^{(+)}(z) \rangle \overset{?}{\simeq} 1 + m\left[N\left(\cos 2\theta + \log \frac{N}{2} \right) + \frac{1}{4N(\sin\theta)^3} \cos\left[N(2\theta - \sin 2\theta) \right] \right] + \mathcal{O}(m^2) \tag{8.2.14}$$

considered together with Eq. (8.2.6) leads us to conclude that the replica limit [Eq. (8.1.13)] in the above elaboration fails to reproduce even the smooth part of the average density of eigenlevels yielding $\varrho(\epsilon) = 0$. Notice that the derivation of Eq. (8.2.14) boldly ignores the divergence of the infinite series Eq. (8.2.13).

This unphysical result is at odds with the work [Kam99b] where both the Wigner semicircle and the $1/N$ oscillating correction to it,

$$\varrho(\epsilon) \simeq \frac{2}{\pi}\sqrt{1-\epsilon_s^2}\left[1 - \frac{1}{4N} \frac{\cos\left(N(2\theta - \sin 2\theta)\right)}{\sin^3\theta} \right], \tag{8.2.15}$$

were reproduced out of fermionic replicas in almost the same elaboration. The only difference between the above treatment and that in [Kam99b] is that its authors used an *alternative enumeration* of saddle point manifolds Eq. (8.2.12) contributing the fermionic replica partition function.

To meet the parameterization of [Kam99b], we introduce a new summation index, $p = q + m$, in Eq. (8.2.10). This amounts to counting saddle point manifolds starting with the 'replica symmetric' one,

$$\tilde{\mathcal{Q}}_n^{(\mathrm{sp})}[p] = -i\mathfrak{s}_z \sqrt{\frac{N}{2}}\, \mathrm{diag}(e^{i\theta} \otimes \mathbb{1}_{2m-p},\ e^{-i\theta} \otimes \mathbb{1}_p) \tag{8.2.16}$$

as p varies from 0 to $2m$, so that

$$\langle \tilde{\mathcal{Z}}_{2m}^{(+)}(z) \rangle \simeq \left(\frac{N}{2}\right)^{mN} \frac{e^{mN\cos 2\theta}}{(2\sin\theta)^{2m^2}} \sum_{p=0}^{2m} \mathcal{V}_{m,\,p-m} \left(\frac{N}{2\pi}\right)^{p(2m-p)} (2\sin\theta)^{3p(2m-p)} \times \exp\left[i(p-m)\left(N(2\theta - \sin 2\theta) + 2m\left(\theta - \frac{\pi}{2}\right)\right)\right]. \tag{8.2.17}$$

For $m \in \mathbb{Z}^+$, Eqs. (8.2.10) and (8.2.17) are trivially identical. However, this is *not* the case for generic real valued m after Eq. (8.2.17) is 'analytically continued' by extending the summation over p to (minus and plus) infinities:

$$\langle \tilde{\mathcal{Z}}_{2m}^{(+)}(z) \rangle \overset{?}{\simeq} \left(\frac{N}{2}\right)^{mN} \frac{e^{mN\cos 2\theta}}{(2\sin\theta)^{2m^2}} \sum_{p=-\infty}^{p=+\infty} \mathcal{V}_{m,\,p-m} \left(\frac{N}{2\pi}\right)^{p(2m-p)} \times (2\sin\theta)^{3p(2m-p)} \exp\left[i(p-m)\left(N(2\theta - \sin 2\theta) + 2m\left(\theta - \frac{\pi}{2}\right)\right)\right]. \tag{8.2.18}$$

Now, contrary to the previous result [Eq. (8.2.14)], the small-m expansion of the fermionic replica partition function appears to contain a mysterious *imaginary* component,[8]

$$\langle \tilde{\mathcal{Z}}_{2m}^{(+)}(z) \rangle \stackrel{?}{\simeq} 1 + m \left[N \left(e^{2i\theta} - 2i\theta + \log \frac{N}{2} \right) + \frac{1}{4N(\sin\theta)^3}\, e^{iN(2\theta - \sin 2\theta)} \right] + \mathcal{O}(m^2). \tag{8.2.19}$$

This expansion coincides with the one in [Kam99b] and does reproduce, through the replica limit, the correct result [Eq. (8.2.15)] for the average density of eigenvalues in GUE, in both leading and subleading orders in $1/N$. As the latter term captures oscillatory behaviour of the eigenlevel density, it was assumed in the literature [Kam99b, Kam99c, Yur99] that the replica asymmetric saddles may describe a nonperturbative sector of replica field theories.

Let us stress that neither of the above two treatments of fermionic replicas [resulting in Eqs. (8.2.14) and (8.2.19), respectively] can be considered as mathematically satisfactory because they both rely on a nonexisting 'analytic continuation' of the replica partition function [Eq. (8.2.10)] to the vicinity of $m = +0$, as explained below Eq. (8.2.13).

8.2.2 Brief summary

A brief tour d'horizon on the saddle point approach to replica field theories indicates that the bosonic variation of the replica trick is restricted to the perturbative sector of the theory accounted for by the (only reachable) replica symmetric saddle point manifold. In the fermionic version of the replica trick, both replica symmetric and replica asymmetric saddle point manifolds contribute the replica partition function; however, its analytic continuation to a vicinity of $n = 0$ is ill defined, both in terms of convergence and uniqueness.

These drawbacks of the saddle point approach to replicas are not specific to the GUE description. Similar difficulties arise in replica studies of other random matrix ensembles [Dal01, Nis02] and in the analysis of physical systems (notably one-dimensional impenetrable bosons) admitting effective RMT description [Gan01, Nis03, Gan04].

Are the problems revealed in the above calculation indicative of *internal* difficulties of the replica method *itself* or should they be attributed to a particular computational framework? A little thought shows that the *approximate* evaluation of both bosonic and fermionic replica partition functions is the key point to blame for the inconsistencies encountered in the above elaboration of the

[8] It was argued in [Zir99] that the procedure of 'analytic continuation' that led to Eq. (8.2.18) and further to Eq. (8.2.19) favours so-called *causal* saddle points over their conjugate counterparts called *acausal*. Such selectivity is eventually responsible for the correct result for the density of eigenlevels as discussed below Eq. (8.2.19).

replica trick. In such a situation, leaning towards exact calculational schemes in replica field theories is a natural move.

8.3 Integrable theory of replicas

In this section, we outline an alternative way of treating replica partition functions. A connection [Kan02, Kan05] between zero-dimensional replica field theories and the theory of integrable hierarchies is central to our formalism.[9]

8.3.1 Density of eigenvalues in the GUE revisited (easy way)

For illustration purposes,[10] we choose the very same problem of calculating the average density of eigenlevels in the GUE specified by the probability measure Eq. (8.1.7). For the lack of space, only fermionic replicas will be considered; a bosonic replica treatment can be found in the tutorial paper [Osi10].

Replica partition function as Toda lattice

Our claim of exact solvability of the replica model Eq. (8.1.14) and models of the same ilk rests on *two* observations. To make the *first*, we routinely reduce the average fermionic partition function Eq. (8.1.14) to the n-fold integral

$$\langle \mathcal{Z}_n^{(+)}(z)\rangle = \int_{\mathbb{R}^n} \prod_{\ell=1}^{n} d\lambda_\ell \, e^{-\lambda_\ell^2} \, (\lambda_\ell - iz)^N \, \Delta_n^2(\lambda) \tag{8.3.1}$$

after diagonalizing the Hermitian matrix $\mathcal{Q}_n = \mathcal{U}_n \lambda \mathcal{U}_n^\dagger$ by unitary rotation $\mathcal{U}_n \in \mathrm{U}(n)$; here λ is a diagonal matrix $\lambda = \mathrm{diag}(\lambda_1, \dots, \lambda_n)$ composed of eigenvalues of $\mathcal{Q}_n$, and $\Delta_n(\lambda)$ is the Vandermonde determinant

$$\Delta_n(\lambda) = \det[\lambda_\ell^{k-1}] = \prod_{\ell>k}(\lambda_\ell - \lambda_k) \tag{8.3.2}$$

induced by the Jacobian of the transformation $\mathcal{Q}_n \mapsto (\mathcal{U}_n, \lambda)$. Further, making a proper shift of the integration variables and applying the Andréief-de Bruijn integration formula [And83, deB55]

$$\int_{\mathbb{R}^n} \prod_{\ell=1}^{n} d\mu(\lambda_\ell) \, \det[f_k(\lambda_\ell)] \det[g_k(\lambda_\ell)] = n! \, \det\left[\int_{\mathbb{R}} d\mu(\lambda) f_k(\lambda) \, g_\ell(\lambda)\right] \tag{8.3.3}$$

[9] An introductory exposition of integrability arising in the RMT context can be found in Chapter 10 of this handbook.

[10] Since the framework to be presented here rests solely on the *symmetry* underlying the matrix model, it can be readily adopted to other spectral statistics for random matrix ensembles falling into the same $\beta = 2$ Dyson's symmetry class [Meh04]. The reader interested in a general formulation of the integrable theory of replicas is referred directly to Section 8.3.2.

which holds for benign integration measure $d\mu(\lambda)$, one derives:

$$\langle \mathcal{Z}_n^{(+)}(z)\rangle = \exp\left[nz^2\right] \det_{k,\ell}\left[\int_{\mathbb{R}} d\lambda\, \lambda^{N+k+\ell} \exp\left(-\lambda^2 - 2iz\lambda\right)\right]. \tag{8.3.4}$$

The latter is equivalent to the remarkable representation

$$\langle \mathcal{Z}_n^{(+)}(z)\rangle = \exp\left[nz^2\right]\, \tau_n^{(+)}(z) \tag{8.3.5}$$

involving the *Hankel determinant*

$$\tau_n^{(+)}(z) = \det\left[\partial_z^{k+\ell}\, \tau_1^{(+)}(z)\right]_{k,\ell=0,\cdots,n-1} \tag{8.3.6}$$

with $\tau_1^{(+)}(z) = e^{-z^2} H_N(z)$ being related to the Hermite polynomial $H_N(z)$. In the above equations, no account was taken of prefactors, which tend to unity in the replica limit.

Consequences of the Hankel-determinant-like representation Eq. (8.3.5) of the fermionic replica partition function $\langle \mathcal{Z}_n^{(+)}(z)\rangle$ are far reaching. As was first shown by Darboux [Dar72] a century ago, any set of Hankel determinants meeting the 'initial condition' $\tau_0^{(+)}(z) = 1$ (which is indeed the case for Eq. (8.3.6) due to Eq. (8.3.5) and the normalization $\langle \mathcal{Z}_0^{(+)}(z)\rangle = 1$) satisfies the equation

$$\tau_n^{(+)}(z)\, \frac{\partial^2}{\partial z^2}\tau_n^{(+)}(z) - \left(\frac{\partial}{\partial z}\tau_n^{(+)}(z)\right)^2 = \tau_{n-1}^{(+)}(z)\tau_{n+1}^{(+)}(z), \quad n \in \mathbb{Z}^+. \tag{8.3.7}$$

Equations (8.3.5) and (8.3.7) taken together with the known initial conditions for $\tau_0^{(+)}(z)$ and $\tau_1^{(+)}(z)$ establish a differential recursive hierarchy between *non-perturbative* fermionic replica partition functions $\langle \mathcal{Z}_n^{(+)}(z)\rangle$ with different replica indices n. This is an *exact* alternative to the *approximate* solution Eq. (8.2.10) presented in the previous section.

Equation (8.3.7), known as the *positive Toda lattice* equation [Tod67] in the theory of integrable hierarchies [Mor94],[11] is the first indication of exact solvability of replica field theories. Importantly, the emergence of the Toda lattice hierarchy is eventually due to the $\beta = 2$ symmetry of the fermionic replica field theory encoded into the squared Vandermonde determinant in Eq. (8.3.1).

From Toda Lattice to Painlevé transcendent

While important from a conceptual point of view, the positive Toda lattice equation for the *fermionic* replica partition function $\langle \mathcal{Z}_n^{(+)}(z)\rangle$, if taken alone, is not much help in performing the replica limit.

Fortunately, here the *second observation* borrowed from [For01] comes in. Miraculously, the same Toda lattice equation governs the behaviour of so-called τ-functions arising in the Hamiltonian formulation [Oka86] of the six Painlevé

[11] See also Chapter 10 of this handbook.

equations [Cla03, Nou04],[12] which are yet another fundamental object in the theory of non-linear integrable systems. The Painlevé equations contain the hierarchy (or replica) index n as a *parameter*. For this reason, they serve as a proper starting point [Kan02] for building a consistent analytic continuation of replica partition functions away from n integers.

The aforementioned Painlevé reduction [Oka86, For01] of the Toda lattice equation Eq. (8.3.7) materializes in the exact representation [For01]

$$\langle \mathcal{Z}_n^{(+)}(z) \rangle = \langle \mathcal{Z}_n^{(+)}(0) \rangle \exp\left(\int_0^{i\epsilon} dt\, \sigma_{\text{IV}}(t) \right) \tag{8.3.8}$$

which holds as soon as $n \in \mathbb{Z}^+$. It involves the fourth Painlevé transcendent $\sigma_{\text{IV}}(t)$ satisfying the Painlevé IV equation[13] in the Jimbo-Miwa-Okamoto form [Jim81, Oka86]

$$(\sigma_{\text{IV}}'')^2 - 4(t\sigma_{\text{IV}}' - \sigma_{\text{IV}})^2 + 4\sigma_{\text{IV}}'(\sigma_{\text{IV}}' + 2n)(\sigma_{\text{IV}}' - 2N) = 0. \tag{8.3.9}$$

The boundary condition is $\sigma_{\text{IV}}(t) \sim (nN/t)\left(1 + \mathcal{O}(t^{-1})\right)$ as $t \to +\infty$. Note that Eq. (8.3.9), and therefore Eq. (8.3.8), contain the replica index n as a *parameter*.

By derivation, Eq. (8.3.8) holds for n positive integers only and, strictly speaking, there is no *a priori* reason to expect it to stay valid away from $n \in \mathbb{Z}^+$. It can be shown, however, that it *is* legitimate to extend Eq. (8.3.8), as it stands, beyond $n \in \mathbb{Z}^+$ and consider this extension as a *proper analytic continuation* to $n \in \mathbb{R}^+$ we are looking for (the reader is referred to [Kan02] for a detailed discussion).

As a result, the fermionic replica limit Eq. (8.1.13) can now be safely implemented to bring, via Eq. (8.2.6), the average density of eigenlevels [Meh60]

$$\varrho(\epsilon) = \frac{1}{2^N \Gamma(N)\sqrt{\pi}}\, e^{-\epsilon^2} \left[H_N'(\epsilon) H_{N-1}(\epsilon) - H_N(\epsilon) H_{N-1}'(\epsilon) \right] \tag{8.3.10}$$

expressed in terms of Hermite polynomials (see also Chapter 4 of this handbook). This is the famous GUE result firmly established by other methods [Meh04, Guh91]. Technically, the derivation of Eq. (8.3.10) is based on the small-n expansion of the Hamiltonian representation [Nou04] of the fourth Painlevé transcendent. The details of this somewhat cumbersome calculation can be found in [Osi10] where the nonperturbative result Eq. (8.3.10) is also re-derived within the bosonic variation of the replica trick.

Brief summary

The above treatment was largely based on a wealth of 'ready-for-use' results (Andréief-de Bruijn formula, Darboux theorem, and a connection between the

[12] See also Chapter 9 of this handbook.

[13] In the original paper [Kan02], Eq. (8.3.9) appears to have incorrect signs in front of n and N. I thank Nicholas Witte for bringing this fact to my attention.

Toda lattice and Painlevé transcendents) which, surprisingly, fitted well our goal of a nonperturbative evaluation of the particular replica partition function Eq. (8.3.1). Since existence of such an 'easy way' is clearly the exception rather than the rule, a regular yet flexible formalism is needed for a nonperturbative description of a general class of replica partition functions.

8.3.2 The τ function theory of replicas ($\beta = 2$)

In this section, we outline such a regular formalism [Osi07, Osi10] tailor-made for an exact analysis of both fermionic and bosonic zero-dimensional replica field theories belonging to the broadly interpreted $\beta = 2$ Dyson symmetry class.

From replica partition function to τ function

Let us concentrate on the fermionic and/or bosonic zero-dimensional replica field theories whose partition functions admit the eigenvalue representation

$$\langle \mathcal{Z}_n^{(\pm)}(\varsigma)\rangle = \int_{\mathcal{D}^n} \prod_{\ell=1}^{n} d\lambda_\ell\, \Gamma(\varsigma;\lambda_\ell)\, e^{-V_n(\lambda_\ell)} \Delta_n^2(\lambda), \quad n \in \mathbb{Z}^+. \tag{8.3.11}$$

Here $V_n(\lambda)$ is a 'confinement potential' which may depend on the replica index $\pm n$; $\Gamma(\varsigma;\lambda)$ is a function accommodating relevant physical parameters $\varsigma = (\varsigma_1, \varsigma_2, \dots)$ of the theory (e.g. energies in the multi-point spectral correlation functions). In order to treat the fermionic and bosonic replicas on the same footing, the integration domain $\mathcal{D}$ was chosen to be $\mathcal{D} = \bigcup_{j=1}^{r}[c_{2j-1}, c_{2j}]$.[14]

To determine the replica partition function $\langle \mathcal{Z}_n^{(\pm)}(\varsigma)\rangle$ nonperturbatively, we adopt the 'deform-and-study' approach, a standard string theory method of revealing hidden structures. Its main idea consists of 'embedding' $\langle \mathcal{Z}_n^{(\pm)}(\varsigma)\rangle$ into a more general theory of τ functions

$$\tau_n^{(s)}(\varsigma;\boldsymbol{t}) = \frac{1}{n!}\int_{\mathcal{D}^n} \prod_{\ell=1}^{n} d\lambda_\ell\, \Gamma(\varsigma;\lambda_\ell)\, e^{-V_{n-s}(\lambda_\ell)}\, e^{v(\boldsymbol{t};\lambda_k)}\, \Delta_n^2(\lambda) \tag{8.3.12}$$

which possess the infinite-dimensional parameter space $\boldsymbol{t} = (t_1, t_2, \cdots)$ arising as the result of the $\boldsymbol{t}$-deformation $v(\boldsymbol{t};\lambda) = \sum_{j=1}^{\infty} t_j \lambda^j$ of the confinement potential. The auxiliary parameter s is assumed to be an integer, $s \in \mathbb{Z}$. Studying the evolution of τ functions in the extended $(n, s, \boldsymbol{t}, \varsigma)$ space allows us to identify the highly non-trivial, non-linear differential hierarchical relations between them. Projection of these relations, taken at $s = 0$, onto the hyperplane $\boldsymbol{t} = \boldsymbol{0}$,

$$\langle \mathcal{Z}_n^{(\pm)}(\varsigma)\rangle = n!\, \tau_n^{(s)}(\varsigma;\boldsymbol{t})\Big|_{s=0,\, \boldsymbol{t}=\boldsymbol{0}}, \tag{8.3.13}$$

[14] Notice that $\mathcal{D} = [-1, +1]$ for (compact) fermionic replicas, and $\mathcal{D} = [0, +\infty)$ for (noncompact) bosonic replicas. A more general setting $\mathcal{D} = \bigcup_{j=1}^{r}[c_{2j-1}, c_{2j}]$ does not complicate the theory.

will generate, among others, a closed non-linear differential equation for the replica partition function $\langle \mathcal{Z}_n^{(\pm)}(\varsigma)\rangle$. Since this *nonperturbative* equation appears to contain the replica (or hierarchy) index n as a parameter, it is expected [Kan02] to serve as a proper starting point for building a consistent analytic continuation of $\langle \mathcal{Z}_n^{(\pm)}(\varsigma)\rangle$ away from n integers.

Having formulated the crux of the method, let us turn to its exposition. The two key ingredients of the exact theory of τ functions are (i) the bilinear identity [Adl95] and (ii) the (linear) Virasoro constraints [Mir90].

Bilinear identity and integrable hierarchies

The bilinear identity encodes an infinite set of hierarchically structured non-linear differential equations in the variables $\{t_j\}$. For the model introduced in Eq. (8.3.12), the bilinear identity reads [Adl95, Tu96, Osi07, Osi10]:

$$\oint_{\mathcal{C}_\infty} dz e^{a\, v(\boldsymbol{t}-\boldsymbol{t}';z)} \left(\tau_n^{(s)}(\boldsymbol{t}-[z^{-1}])\, \frac{\tau_{m+1}^{(m+1+s-n)}(\boldsymbol{t}'+[z^{-1}])}{z^{m+1-n}}\, e^{v(\boldsymbol{t}-\boldsymbol{t}';z)} \right.$$
$$\left. - \tau_m^{(m+s-n)}(\boldsymbol{t}'-[z^{-1}])\frac{\tau_{n+1}^{(s+1)}(\boldsymbol{t}+[z^{-1}])}{z^{n+1-m}} \right) = 0. \qquad (8.3.14)$$

Here, $a \in \mathbb{R}$ is a free parameter; the integration contour $\mathcal{C}_\infty$ encompasses the point $z = \infty$; the notation $\boldsymbol{t} \pm [z^{-1}]$ stands for the infinite set of parameters $\{t_j \pm z^{-j}/j\}$; for brevity, the physical parameters ς were dropped from the arguments of τ functions.

Being expanded in terms of $\boldsymbol{t}' - \boldsymbol{t}$ and a, Eq. (8.3.14) generates various integrable hierarchies. One of them, the Kadomtsev-Petviashvili (KP) hierarchy in the Hirota form[15]

$$\frac{1}{2} D_1 D_k\, \tau_n^{(s)}(\boldsymbol{t}) \circ \tau_n^{(s)}(\boldsymbol{t}) = s_{k+1}([\boldsymbol{D}])\, \tau_n^{(s)}(\boldsymbol{t}) \circ \tau_n^{(s)}(\boldsymbol{t}) \qquad (8.3.15)$$

($k \geq 3$) is of primary importance for the exact theory of replicas. The first non-trivial member of the KP hierarchy reads

$$\left(\frac{\partial^4}{\partial t_1^4} + 3\frac{\partial^2}{\partial t_2^2} - 4\frac{\partial^2}{\partial t_1 \partial t_3} \right) \log \tau_n^{(s)}(\varsigma;\boldsymbol{t}) + 6\left(\frac{\partial^2}{\partial t_1^2} \log \tau_n^{(s)}(\varsigma;\boldsymbol{t}) \right)^2 = 0. \quad (8.3.16)$$

[15] In Eq. (8.3.15), the jth component of the infinite-dimensional vector $[\boldsymbol{D}]$ equals $j^{-1} D_j$; the functions $s_k(\boldsymbol{t})$ are the Schur polynomials [Mac98] defined by the expansion

$$\exp\left(\sum_{j=1}^{\infty} t_j x^j \right) = \sum_{k=0}^{\infty} x^k s_k(\boldsymbol{t}).$$

The operator symbol $D_j\, f(\boldsymbol{t}) \circ g(\boldsymbol{t})$ stands for the Hirota derivative $\partial_{x_j} f(\boldsymbol{t}+\boldsymbol{x})\, g(\boldsymbol{t}-\boldsymbol{x})\,|_{\boldsymbol{x}=\boldsymbol{0}}$.

In what follows, it will be shown that its projection onto $s = 0$ and $\boldsymbol{t} = \boldsymbol{0}$ [Eq. (8.3.13)] gives rise to a non-linear differential equation for the replica partition function $\langle \mathcal{Z}_n^{(\pm)}(\varsigma)\rangle$.

Virasoro constraints

Since we are interested in deriving a differential equation for $\langle \mathcal{Z}_n^{(\pm)}(\varsigma)\rangle$ in terms of the derivatives over *physical parameters* $\{\varsigma_j\}$, we have to seek an additional block of the theory that would make a link between $\{t_j\}$-derivatives in Eq. (8.3.16) taken at $\boldsymbol{t} = \boldsymbol{0}$ and the derivatives over physical parameters $\{\varsigma_j\}$. The study [Adl95] by Adler, Shiota, and van Moerbeke suggests that the missing block is the *Virasoro constraints*[16] which reflect the invariance of the τ function [Eq. (8.3.12)] under a change of integration variables.

In the present context, it is useful to demand the invariance under an infinite set of transformations

$$\lambda_j \to \mu_j + \epsilon \mu_j^{q+1} f(\mu_j) \prod_{k=1}^{\mathfrak{m}} (\mu_j - c'_k), \quad q \geq -1, \tag{8.3.17}$$

labelled by integers q. Here, $\epsilon > 0$, the vector $\boldsymbol{c}'$ is $\boldsymbol{c}' = \{c_1, \cdots, c_{2r}\} \setminus \{\pm\infty, \aleph\}$ with $\aleph$ denoting a set of zeros of $f(\lambda)$, and $\mathfrak{m} = \dim(\boldsymbol{c}')$. The function $f(\lambda)$ is, in turn, related to the confinement potential $V_{n-s}(\lambda)$ through the parameterization

$$\frac{d V_{n-s}}{d\lambda} = \frac{g(\lambda)}{f(\lambda)}, \quad g(\lambda) = \sum_{k=0}^{\infty} b_k \lambda^k, \quad f(\lambda) = \sum_{k=0}^{\infty} a_k \lambda^k \tag{8.3.18}$$

in which both $g(\lambda)$ and $f(\lambda)$ depend on $n - s$ as do the coefficients b_k and a_k in the above expansions. The transformation Eq. (8.3.17) induces the Virasoro constraints [Osi07]

$$\left[\hat{\mathcal{L}}_q^V(\boldsymbol{t}) + \hat{\mathcal{L}}_q^\Gamma(\varsigma; \boldsymbol{t})\right] \tau_n^{(s)}(\varsigma; \boldsymbol{t}) = 0, \tag{8.3.19}$$

where the differential operator

$$\hat{\mathcal{L}}_q^V(\boldsymbol{t}) = \sum_{\ell=0}^{\infty} \sum_{k=0}^{\mathfrak{m}} s_{\mathfrak{m}-k}(-\boldsymbol{p}_{\mathfrak{m}}(\boldsymbol{c}')) \left(a_\ell \hat{\mathcal{L}}_{q+k+\ell}(\boldsymbol{t}) - b_\ell \frac{\partial}{\partial t_{q+k+\ell+1}}\right), \tag{8.3.20}$$

acting in the $\boldsymbol{t}$-space, is expressed in terms of the Virasoro operators[17]

$$\hat{\mathcal{L}}_q(\boldsymbol{t}) = \sum_{j=1}^{\infty} j t_j \frac{\partial}{\partial t_{q+j}} + \sum_{j=0}^{q} \frac{\partial^2}{\partial t_j \partial t_{q-j}}, \tag{8.3.21}$$

[16] See also Chapter 10 of this handbook.

[17] Equation (8.3.21) assumes that $\partial/\partial t_0$ is identified with the multiplicity of the matrix integral in Eq. (8.3.12), $\partial/\partial t_0 \equiv n$.

obeying the Virasoro algebra $[\hat{\mathcal{L}}_p, \hat{\mathcal{L}}_q] = (p-q)\hat{\mathcal{L}}_{p+q}$ for $p, q \geq -1$. The notation $s_k(-\boldsymbol{p}_{\mathrm{m}}(\boldsymbol{c}'))$ stands for the Schur polynomial, and $\boldsymbol{p}_{\mathrm{m}}(\boldsymbol{c}')$ is an infinite dimensional vector

$$\boldsymbol{p}_{\mathrm{m}}(\boldsymbol{c}') = \left(\mathrm{tr}_{\mathrm{m}}(\boldsymbol{c}'), \frac{1}{2}\mathrm{tr}_{\mathrm{m}}(\boldsymbol{c}')^2, \cdots, \frac{1}{k}\mathrm{tr}_{\mathrm{m}}(\boldsymbol{c}')^k, \cdots\right) \tag{8.3.22}$$

with $\mathrm{tr}_{\mathrm{m}}(\boldsymbol{c}')^k = \sum_{j=1}^{\mathrm{m}} (c_j')^k$.

While very similar in spirit, the calculation of $\hat{\mathcal{L}}_q^{\Gamma}(\boldsymbol{t})$, the second ingredient in Eq. (8.3.19), is more of an art since the function $\Gamma(\varsigma; \lambda)$ in Eq. (8.3.12) may significantly vary from one replica model to the other.

From τ function to replica partition function

Remarkably, for $\boldsymbol{t} = \boldsymbol{0}$, the two equations [Eqs. (8.3.16) and (8.3.19)] can be solved jointly to bring a closed non-linear differential equation for $\langle \mathcal{Z}_n^{(\pm)}(\varsigma)\rangle$. It is this equation which, being supplemented by appropriate boundary conditions, provides a truly nonperturbative description of the replica partition functions and facilitates performing the replica limit.

8.3.3 Exact (bosonic) replicas at work: density of eigenlevels in the chiral GUE

Definitions

To see the above formalism at work, let us consider the chiral Gaussian unitary ensemble (chGUE) of $N \times N$ random matrices

$$\mathcal{H}_{\mathcal{D}} = \begin{pmatrix} 0 & \mathcal{W} \\ \mathcal{W}^{\dagger} & 0 \end{pmatrix} \tag{8.3.23}$$

known to describe the low-energy sector of $\mathrm{SU}(N_c \geq 3)$ QCD in the fundamental representation [Ver93], see also Chapter 32 of this handbook. Composed of rectangular $n_L \times n_R$ random matrices $\mathcal{W}$ with the Gaussian distributed complex valued entries

$$P_{n_L, n_R}(\mathcal{W}) = \left(\frac{2\pi}{N\Sigma^2}\right)^{n_L n_R} \exp\left[-\frac{N\Sigma^2}{2}\mathrm{tr}\,\mathcal{W}^{\dagger}\mathcal{W}\right], \tag{8.3.24}$$

where $N = n_L + n_R$, the matrix $\mathcal{H}_{\mathcal{D}}$ has exactly $\nu = |n_R - n_L|$ zero eigenvalues identified with the topological charge ν; the remaining eigenvalues occur in pairs $\{\pm\lambda_j\}$; the parameter Σ denotes the chiral condensate.

Nonperturbative calculation of bosonic replica partition function

To determine the (microscopic) spectral density from the bosonic replicas, we define the replica partition function

$$\langle \mathcal{Z}_n^{(-)}(\varsigma)\rangle = \left\langle \det{}^{-n}(\varsigma + i\mathcal{H}_{\mathcal{D}})\right\rangle_{\mathcal{W}} \tag{8.3.25}$$

(the angular brackets denote averaging with respect to the probability density Eq. (8.3.24)) and map it onto a bosonic field theory. In the half-plane $\mathfrak{Re}\, \varsigma > 0$, the partition function $\langle \mathcal{Z}_n^{(-)}(\varsigma)\rangle$ reduces to [Dal01, Fyo02]

$$\mathcal{Z}_n^{(-)}(\omega) = \int_{\mathcal{S}_n} \mathcal{D}\mathcal{Q}_n \det{}^{\nu-n}\mathcal{Q}_n \exp\left[-\frac{\omega}{2}\mathrm{Tr}(\mathcal{Q}_n + \mathcal{Q}_n^{-1})\right], \tag{8.3.26}$$

where the integration domain $\mathcal{S}_n$ spans all $n \times n$ positive definite Hermitian matrices $\mathcal{Q}_n$. Equation (8.3.26) was derived in the thermodynamic limit $N \to \infty$ with the spectral parameter $\omega = \varsigma N\Sigma$ being kept fixed ($\mathfrak{Re}\,\omega > 0$).

Spotting the invariance of the integrand in Eq. (8.3.26) under the unitary rotation of the matrix $\mathcal{Q}_n$, one readily realizes that $\mathcal{Z}_n^{(-)}(\omega)$ belongs to the class of τ functions specified by Eq. (8.3.12) where $\mathcal{D}$ is set to $\mathbb{R}^+$, the potential V_{n-s} is $V_{n-s}(\lambda) = (n - s - \nu)\log\lambda$, and $\Gamma(\varsigma;\lambda)$ is replaced with $\Gamma(\omega;\lambda) = \exp\left[-(\omega/2)(\lambda + \lambda^{-1})\right]$. This observation implies that the associated τ function $\tau_n^{(s)}(\omega;\boldsymbol{t})$ satisfies both the first KP equation (8.3.16) and the Virasoro constraints Eq. (8.3.19) with [Osi07]

$$\hat{\mathcal{L}}_q^V(\boldsymbol{t}) = \hat{\mathcal{L}}_{q+1}(\boldsymbol{t}) + (\nu - n + s)\frac{\partial}{\partial t_{q+1}}, \tag{8.3.27}$$

$$\hat{\mathcal{L}}_q^\Gamma(\omega;\boldsymbol{t}) = -\frac{\omega}{2}\frac{\partial}{\partial t_{q+2}} - \delta_{q,-1}\left(\omega\frac{\partial}{\partial\omega} + \frac{\omega}{2}\frac{\partial}{\partial t_1}\right) + \left[1 - \delta_{q,-1}\right]\frac{\omega}{2}\frac{\partial}{\partial t_q}. \tag{8.3.28}$$

Projecting Eq. (8.3.16) taken at $s = 0$ onto $\boldsymbol{t} = \boldsymbol{0}$, and expressing the partial derivatives therein via the derivatives over ω with the help of Eqs. (8.3.19), (8.3.27), and (8.3.28), we conclude that $h_n(\omega) = (\partial/\partial\omega)\log \mathcal{Z}_n^{(-)}(\omega)$ obeys the differential equation [Osi07]

$$\begin{aligned} h_n''' &+ \frac{2}{\omega}h_n'' - \left(4 + \frac{1 + 4(n^2+\nu^2)}{\omega^2}\right)h_n' + 6(h_n')^2 \\ &+ \frac{1 - 4(n^2+\nu^2)}{\omega^3}h_n - \frac{2}{\omega^2}(h_n)^2 + \frac{4}{\omega}h_n h_n' + \frac{4n^2}{\omega^2} = 0 \end{aligned} \tag{8.3.29}$$

that can be reduced to the Painlevé III.

Considered together with the boundary conditions $h_n(\omega \to 0) \simeq -n\nu/\omega$ and $h_n(\omega \to \infty) \simeq -n - n^2/(2\omega)$, following from Eq. (8.3.26), the non-linear differential equation (8.3.29) provides a nonperturbative characterization of the *bosonic* replica partition function[18] $\mathcal{Z}_n^{(-)}(\omega)$ for all $n \in \mathbb{Z}^+$.

[18] Interestingly, the *fermionic* replica partition function [Leu92, Smi95, Dal01]

$$\mathcal{Z}_n^{(+)}(\omega) = \int_{\mathcal{U}_n\mathcal{U}_n^\dagger = \mathbb{1}_n} \mathcal{D}\mathcal{U}_n \det{}^{\nu}\mathcal{U}_n \exp\left[\frac{\omega}{2}\mathrm{Tr}(\mathcal{U}_n + \mathcal{U}_n^\dagger)\right],$$

admits a similar representation in terms of the Painlevé III. Specifically, the function $f_n(\omega) = (\partial/\partial\omega)\log \mathcal{Z}_n^{(+)}(\omega)$ can be shown [Kan02] to satisfy the very same Eq. (8.3.29) supplemented by the boundary conditions $f_n(\omega \to \omega_0) = h_{-n}(\omega \to \omega_0)$, where $\omega_0 = \{0, \infty\}$.

Implementing the replica limit

To pave the way for the replica calculation of the resolvent $g(\omega)$ determined by the replica limit

$$g(\omega) = -\lim_{n\to 0} \frac{1}{n}\frac{\partial}{\partial \omega} \log \mathcal{Z}_n^{(-)}(\omega) = -\lim_{n\to 0} \frac{1}{n} h_n(\omega), \qquad (8.3.30)$$

one has to analytically continue $h_n(\omega)$ away from n integers. Previous studies [Kan02, Kan05] suggest that the sought after analytic continuation is given by the very same Eq. (8.3.29) where the replica parameter n is allowed to explore the entire real axis. This leap makes the rest of the calculation straightforward. Representing $h_n(\omega)$ in the vicinity of $n = 0$ as $h_n(\omega) = \sum_{p=1}^{\infty} n^p a_p(\omega)$, we conclude that $g(\omega) = -a_1(\omega)$ satisfies the equation

$$\omega^3 g''' + 2\omega^2 g'' - \left(1 + 4\nu^2 + 4\omega^2\right) \omega g' + (1 - 4\nu^2)\, g = 0. \qquad (8.3.31)$$

Its solution, subject to the boundary conditions consistent with those specified below Eq. (8.3.29),

$$g(\omega) = \frac{\nu}{\omega} + \omega \left[I_\nu(\omega)\, K_\nu(\omega) + I_{\nu+1}(\omega)\, K_{\nu-1}(\omega) \right], \qquad (8.3.32)$$

brings the microscopic spectral density $\varrho(\omega) = \pi^{-1}\mathfrak{Re}\, g(i\omega + 0)$ in the form

$$\varrho(\omega) = \nu\delta(\omega) + \frac{\omega}{2}\left[J_\nu^2(\omega) - J_{\nu-1}(\omega) J_{\nu+1}(\omega) \right]. \qquad (8.3.33)$$

In the above formula (which is one of the celebrated RMT results originally obtained in [Ver93] within the orthogonal polynomial technique), the function J_ν denotes the Bessel function of the first kind, whilst I_ν and K_ν are the modified Bessel function of the first and second kind, respectively. Let us stress that the approximate, saddle point approach to bosonic replicas [Dal01] fails to produce Eq. (8.3.33).

8.4 Concluding remarks

Concluding this brief excursion into integrable theory of replica field theories, we wish to mention yet another important development due to Splittorff and Verbaarschot [Spl03, Spl04], not reviewed here due to lack of space. These authors showed that nonperturbative results for various RMT correlation functions at $\beta = 2$ can be derived by taking the replica limit of the *graded* Toda lattice equation whose positive ($n \in \mathbb{Z}^+$) and negative ($n \in \mathbb{Z}^-$) branches describe fermionic and bosonic replica partition functions, respectively. Being supersymmetric in nature, this approach greatly simplifies calculations of spectral correlation functions through a remarkable fermionic-bosonic factorization. For further details, the reader is referred to the original papers

[Spl03, Spl04], the lecture notes [Ver05], Chapter 32 of this handbook, and the tutorial paper [Osi10].

Certainly, more effort is needed to accomplish integrable theory of zero-dimensional replica field theories. In particular, its extension to the $\beta = 1$ and $\beta = 4$ Dyson symmetry classes is very much called for.

Acknowledgements

I am indebted to Vladimir Al. Osipov for collaboration on the 'replica project'. This work was supported by the Israel Science Foundation through Grant Nos. 286/04 and 414/08.

References

[Adl95] M. Adler, T. Shiota, and P. van Moerbeke, Phys. Lett. A **208**, 67 (1995).
[Alt00] A. Altland and A. Kamenev, Phys. Rev. Lett. **85**, 5615 (2000).
[And83] C. Andréief, Mém. de la Soc. Sci., Bordeaux **2**, 1 (1883).
[Cla03] P. A. Clarkson, J. Comp. Appl. Math. **153**, 127 (2003).
[Dal01] D. Dalmazi and J. J. M. Verbaarschot, Nucl. Phys. B **592** [FS], 419 (2001).
[Dar72] G. Darboux, *Lecons sur la Theorie Generale des Surfaces et les Applications Geometriques du Calcul Infinitesimal*, Vol. II: XIX (Chelsea, New York, 1972).
[deB55] N. G. de Bruijn, J. Indian Math. Soc. **19**, 133 (1955).
[Dhe90] G. S. Dhesi and R. C. Jones, J. Phys. A: Math. Gen. **23**, 5577 (1990).
[Edw75] S. F. Edwards and P. W. Anderson, J. Phys. F: Met. Phys. **5**, 965 (1975).
[Edw76] S. F. Edwards and R. C. Jones, J. Phys. A: Math. Gen. **9**, 1595 (1976).
[Edw80] S. F. Edwards and M. Warner, J. Phys. A: Math. Gen. **13**, 381 (1980).
[Efe80] K. B. Efetov, A. I. Larkin, and D. E. Khmelnitskii, Sov. Phys. JETP **52**, 568 (1980).
[Efe82] K. B. Efetov, Sov. Phys. JETP **55**, 514 (1982); Sov. Phys. JETP **56**, 467 (1982).
[Efe83] K. B. Efetov, Adv. Phys. **32**, 53 (1983).
[Efe97] K. B. Efetov, *Supersymmetry in Disorder and Chaos* (Cambridge University Press, Cambridge, 1997).
[Eme75] V. J. Emery, Phys. Rev. B **11**, 239 (1975).
[For01] P. J. Forrester and N. S. Witte, Commun. Math. Phys. **219**, 357 (2001).
[Fyo02] Y. V. Fyodorov, Nucl. Phys. B **621** [PM], 643 (2002).
[Gan01] D. M. Gangardt and A. Kamenev, Nucl. Phys. B **610** [PM], 578 (2001).
[Gan04] D. M. Gangardt, J. Phys. A: Math. and Gen. **37**, 9335 (2004).
[Guh91] T. Guhr, J. Math. Phys. **32**, 336 (1991).
[Har34] G. H. Hardy, J. E. Littlewood, and G. Pólya, *Inequalities* (Cambridge University Press, Cambridge, 1934).
[Hor90] M. Horbach and G. Schön, Physica A **167**, 93 (1990).
[Ito97] C. Itoi, H. Mukaida, and Y. Sakamoto, J. Phys. A.: Math. Gen. **30**, 5709 (1997).
[Jim81] M. Jimbo and T. Miwa, Physica D **2**, 407 (1981).
[Kam99a] A. Kamenev and A. Andreev, Phys. Rev. B **60**, 3944 (1999).
[Kam99b] A. Kamenev and M. Mézard, J. Phys. A: Math. and Gen. **32**, 4373 (1999).
[Kam99c] A. Kamenev and M. Mézard, Phys. Rev. B **60**, 3944 (1999).
[Kam09] A. Kamenev and A. Levchenko, Adv. Phys. **58**, 197 (2009).

[Kan02] E. Kanzieper, Phys. Rev. Lett. **89**, 250201 (2002).

[Kan05] E. Kanzieper, in: O. Kovras (ed.) *Frontiers in Field Theory*, p. 23 (Nova Science Publishers, New York, 2005).

[Leu92] H. Leutwyler and A. Smilga, Phys. Rev. D **46**, 5607 (1992).

[Mac98] I. G. Macdonald, *Symmetric Functions and Hall Polynomials* (Oxford University Press, Oxford, 1998).

[Meh60] M. L. Mehta and M. Gaudin, Nucl. Phys. **18**, 420 (1960).

[Meh04] M. L. Mehta, *Random Matrices* (Elsevier, Amsterdam, 2004).

[Mir90] A. Mironov and A. Morozov, Phys. Lett. B **252**, 47 (1990).

[Mor94] A. Morozov, Physics-Uspekhi (UK) **37**, 1 (1994).

[Nis02] S. M. Nishigaki and A. Kamenev, J. Phys. A: Math. and Gen. **35**, 4571 (2002).

[Nis03] S. M. Nishigaki, D. M. Gangardt, and A. Kamenev, J. Phys. A: Math. and Gen. **36**, 3137 (2003).

[Nou04] M. Noumi, *Painlevé Equations through Symmetry* (AMS, Providence, 2004).

[Oka86] K. Okamoto, Math. Ann. **275**, 221 (1986).

[Osi07] V. Al. Osipov and E. Kanzieper, Phys. Rev. Lett. **99**, 050602 (2007).

[Osi10] V. Al. Osipov and E. Kanzieper, Ann. Phys. **325**, 2251 (2010).

[Par03] G. Parisi, Bull. Symbolic Logic **9**, 181 (2003).

[Sch80] L. Schäfer and F. Wegner, Z. Phys. B **38**, 113 (1980).

[Smi95] A. Smilga and J. J. M. Verbaarschot, Phys. Rev. D **51**, 829 (1995).

[Spl03] K. Splittorff and J. J. M. Verbaarschot, Phys. Rev. Lett. **90**, 041601 (2003).

[Spl04] K. Splittorff and J. J. M. Verbaarschot, Nucl. Phys. B **683** [FS], 467 (2004).

[Tit32] E. C. Titchmarsh, *The Theory of Functions* (Oxford University Press, Oxford, 1932).

[Tod67] M. Toda, J. Phys. Soc. Japan **22**, 431 (1967).

[Tu96] M. H. Tu, J. C. Shaw, and H. C. Yen, Chinese J. Phys. **34**, 1211 (1996).

[Ver84] J. J. M. Verbaarschot and M. R. Zirnbauer, Ann. Phys. (N. Y.) **158**, 78 (1984).

[Ver85a] J. J. M. Verbaarschot, H. A. Weidenmüller, and M. R. Zirnbauer, Phys. Rep. **129**, 367 (1985).

[Ver85b] J. J. M. Verbaarschot and M. R. Zirnbauer, J. Phys. A: Math. and Gen. **17**, 1093 (1985).

[Ver93] J. J. M. Verbaarschot and I. Zahed, Phys. Rev. Lett. **70**, 3852 (1993).

[Ver05] J. J. M. Verbaarschot, AIP Conf. Proc. **744**, 277 (2005).

[Weg79] F. Wegner, Z. Phys. B **35**, 207 (1979).

[Yur99] I. V. Yurkevich and I. V. Lerner, Phys. Rev. B **60**, 3955 (1999).

[Zir99] M. R. Zirnbauer, arXiv: cond-mat/9903338 (1999).

·9·

Painlevé transcendents

Alexander R. Its

Abstract

We will briefly review the history and modern theory of Painlevé transcendents with emphasis on the Riemann–Hilbert method. The appearance of Painlevé functions in the theory of random matrices will be demonstrated.

9.1 Introduction

The six classical Painlevé transcendents were introduced at the turn of the twentieth century by Paul Painlevé and his school, as a solution for a specific classification problem of second order ODEs of the type

$$u_{xx} = F(x, u, u_x),$$

where F is a function meromorphic in x and rational in u and u_x. The problem was to find all equations of this form, which have the property that their solutions are free from movable critical points, i.e. the locations of possible branch points and essential singularities of the solution do *not* depend on the initial data. The motivation for posing this problem is quite clear: the absence of movable critical points means that every solution of the equation can be meromorphically extended to the entire universal covering of a punctured complex sphere, determined only by the equation. This implies that such equations share one of the fundamental properties of linear equations.

It was shown by Painlevé and Gambier (1900, 1910), that within a Möbius transformation,

$$u \mapsto \frac{\alpha(x)u + \beta(x)}{\gamma(x)u + \delta(x)}, \quad x \mapsto \phi(x), \tag{9.1.1}$$

$$\alpha, \beta, \gamma, \delta, \phi \quad - \quad \text{are meromorphic in } x,$$

there exist only fifty such equations (see [Inc56]). Each of them can either be integrated in terms of known functions, or can be mapped to a set of six equations, which cannot be integrated in terms of known functions. These six equations are called Painlevé equations, and their general solutions are called Painlevé functions or Painlevé transcendents. The canonical forms for the Painlevé equations are:

1. $u_{xx} = 6u^2 + x$
2. $u_{xx} = xu + 2u^3 - \alpha$
3. $u_{xx} = \frac{1}{u}{u_x}^2 - \frac{u_x}{x} + \frac{1}{x}(\alpha u^2 + \beta) + \gamma u^3 + \frac{\delta}{u}$
4. $u_{xx} = \frac{1}{2u}{u_x}^2 + \frac{3}{2}u^3 + 4xu^2 + 2(x^2 - \alpha)u + \frac{\beta}{u}$
5. $u_{xx} = \frac{3u-1}{2u(u-1)}{u_x}^2 - \frac{1}{x}u_x + \frac{(u-1)^2}{x^2}(\alpha u + \frac{\beta}{u}) + \frac{\gamma u}{x} + \frac{\delta u(u+1)}{u-1}$
6. $u_{xx} = \frac{1}{2}\left(\frac{1}{u} + \frac{1}{u-1} + \frac{1}{u-x}\right){u_x}^2 - \left(\frac{1}{x} + \frac{1}{x-1} + \frac{1}{u-x}\right)u_x$
$\quad + \frac{u(u-1)(u-x)}{x^2(x-1)^2}\left(\alpha + \beta\frac{x}{u^2} + \gamma\frac{x-1}{(u-1)^2} + \delta\frac{x(x-1)}{(u-x)^2}\right)$

Here, $\alpha, \beta, \gamma, \delta$ are complex parameters. During the rest of the twentieth century a great deal of facts about these equations were discovered: the structure of movable singularities, families of explicit solutions, their transformation properties, etc. We refer to the monograph [Gro02] for a comprehensive presentation of this part of the theory of Painlevé equations.

A new surge of interest in Painlevé functions occurred in the late 1970s after the pioneering work of M. J. Ablowitz and H. Segur [Abl77], and of B. M. McCoy, C. A. Tracy, and T. T. Wu [McC77] devoted to the self-similar reduction of the KdV equation and to the scaling theory of the 2D Ising model, respectively. Since that time the Painlevé equations have begun to appear in a wide range of physical applications and it is now becoming clear that the Painlevé transcendents play the same role in non-linear mathematical physics that the classical special functions, such as Airy functions, Bessel functions, etc., play in linear physics. Perhaps the most impressive showing of Painlevé functions is demonstrated in random matrix theory.

In random matrix theory, the Painlevé equations appear in two distinct contexts. They describe either the eigenvalue distribution functions in the classical ensembles for finite N or the universal eigenvalue distribution functions in the large N limit. In this review we will focus on the latter. For the full story concerning Painlevé functions in the finite N classical ensembles we refer the reader to the works [Adl95], [Tra94a] and [For01] (the "universal" Painlevé transcendents in the large N limit are considered in these papers as well). It is also worth mentioning that the finite N gap probabilities represent special solutions of the relevant Painlevé equations which are given in terms of classical special functions. The genuine "Painlevé transcendents" appear in the large N limit only. The first appearance of Painlevé transcendents in the theory of random matrices was in the 1990 works of Brézin and Kazakov, Gross and Migdal, and Douglas and Shenker devoted to the matrix model for 2D quantum gravity.

In this survey we will follow a Riemann–Hilbert point of view on the Painlevé equations. This point of view has been developed within the frameworks of the *Isomonodromy method* introduced in 1980 by H. Flaschka and A.C. Newell [Fla80], and by M. Jimbo, T. Miwa, and K. Ueno [Jim80b]. The isomonodromy method is based on the intrinsic relation of the Painlevé functions to monodromy theory of systems of linear ODE with rational coefficients. This relation had already been known to R. Garnier, R. Fuchs, and to L. Schlesinger. What apparently was missed in the classical papers is the possibility of using the isomonodromy interpretation of Painlevé functions for their global asymptotic analysis. We refer to the monograph [Fok06] for a detail exposition of the Riemann–Hilbert approach in the theory of Painlevé equations and for a historical review of the subject. It should also be mentioned that the first connection formulas were found in the above mentioned papers [Abl77] and [McC77].

Painlevé functions exhibit very rich algebraic and algebrageometric features, which we won't be able to discuss in this short review. We refer the reader to the monograph [Nou04] on this matter.

We also want to point to the chapter written by P. Clarkson in the up-coming *Handbook on Mathematical Functions* (to be published by Cambridge University Press) as the most complete collection of results concerning all aspects of the modern theory of Painlevé equations.

This chapter is organized as follows. In Sections 9.2 and 9.3, we will outline the main features of the Riemann–Hilbert method in the theory of Painlevé equations using the second Painlevé equation as a case study. Section 9.4 is devoted to the two most celebrated universal distribution functions of random matrix theory, that is, the sine-kernel and the Airy-kernel determinants. We will present their evaluation in terms of the Painlevé transcendents using the theory of integrable Fredholm operators and, once again, the Riemann–Hilbert technique.

9.2 Riemann–Hilbert representation of the Painlevé functions

Let

$$\Gamma_k = \left\{\lambda \in \mathbb{C} : \arg\lambda = \frac{\pi}{6} + \frac{\pi}{3}(k-1)\right\}, \quad k = 1, \ldots, 6, \tag{9.2.1}$$

be the six rays on the λ-plane, oriented from zero to infinity. With each ray Γ_k, we associate the 2×2 matrix S_k defined by the equations

$$S_{2l-1} = \begin{pmatrix} 1 & 0 \\ s_{2l-1} & 1 \end{pmatrix}, \quad S_{2l} = \begin{pmatrix} 1 & s_{2l} \\ 0 & 1 \end{pmatrix}, \quad l = 1, 2, 3, \tag{9.2.2}$$

where s_k are complex parameters. We also assume that the six matrices S_k satisfy the following constraints.

$$S_1 S_2 \dots S_6 = I, \quad S_{k+3} = \sigma_2 S_k \sigma_2, \quad k = 1, 2, 3, \tag{9.2.3}$$

where σ_2 is the second Pauli matrix,[1] These constraints, in terms of the scalar parameters s_k read,

$$s_1 - s_2 + s_3 + s_1 s_2 s_3 = 0, \quad s_{k+3} = -s_k, \quad k = 1, 2, 3. \tag{9.2.4}$$

The Riemann–Hilbert (RH) problem corresponding to the second Painlevé equation,

$$u_{xx} = xu + 2u^3, \tag{9.2.5}$$

consists in finding a 2×2 matrix function $Y(\lambda)$ satisfying the following three conditions.

- $Y(\lambda)$ is analytic in $\mathbb{C} \setminus \cup_{k=1}^{6} \Gamma_k$
- $Y_+(\lambda) = Y_-(\lambda) e^{-i\theta(\lambda)\sigma_3} S_k e^{i\theta(\lambda)\sigma_3}, \quad \lambda \in \Gamma_k$
- At $\lambda = \infty$, the function $Y(\lambda)$ satisfies the normalization equation,

$$Y(\lambda) = I + O\left(\frac{1}{\lambda}\right), \quad \lambda \to \infty \tag{9.2.6}$$

Here, $Y_\pm(\lambda)$ denote the boundary values of the function $Y(\lambda)$ on the contour Γ_k taken from the $\pm$ side, the exponent $\theta(\lambda)$ is given by the equation,

$$\theta(\lambda) = \frac{4}{3}\lambda^3 + x\lambda, \quad x \in \mathbb{C}, \tag{9.2.7}$$

and σ_3 stands for the third Pauli matrix. We will call this Riemann–Hilbert problem the *P2-RH problem*.

Theorem 9.2.1 *[Bol04] Let* $\mathbf{s} = (s_1, s_2, \dots, s_6)$ *be the set of complex numbers satisfying conditions (9.2.4). Then, the corresponding P2-RH problem is uniquely and meromorphically, with respect to x, solvable. Estimate (9.2.6) is extended to the full, differentiable in* λ *and x, asymptotic series,*

$$Y(\lambda) \sim I + \sum_{j=1}^{\infty} \frac{m_j}{\lambda^j}, \quad \lambda \to \infty, \tag{9.2.8}$$

with the matrix coefficients $m_j \equiv m_j(x)$ *being meromorphic functions of* x.

[1] We use the usual notations for the Pauli matrices,

$$\sigma_1 = \begin{pmatrix} 0 & 1 \\ 1 & 0 \end{pmatrix}, \quad \sigma_2 = \begin{pmatrix} 0 & -i \\ i & 0 \end{pmatrix}, \quad \sigma_3 = \begin{pmatrix} 1 & 0 \\ 0 & -1 \end{pmatrix}$$

The proof of this theorem, which is given in [Bol04], uses Malgrange's generalization of the classical Birkhoff–Grothendieck theorem to the case with the parameter. Alternatively, the proof can be obtained via application of the Fredholm theory of Riemann–Hilbert factorization problems developed in [Zho89]. The relation of the P2-RH problem to the second Painlevé equation (9.2.5) is given by the next theorem.

Theorem 9.2.2 *[Fla80] Let $m_1 \equiv m_1(x; \mathbf{s})$ be the first coefficient of series (9.2.8). Then formula,*

$$u \equiv u(x; \mathbf{s}) := 2(m_1)_{12}, \tag{9.2.9}$$

defines a solution of the second Painlevé equation (9.2.5). Moreover, every solution of (9.2.5) is given by equation (9.2.9) with the proper choice of the data $\mathbf{s}$ *in the P2-RH problem.*

We shall outline the proof of this theorem since it highlights all the basic ingredients of the algebraic side of the Riemann–Hilbert method (for more details see e.g. [Fok06]). The proof is based on the following two key facts.

Proposition 9.2.1 *Put*

$$\Psi(\lambda) := Y(\lambda)e^{-\theta(\lambda)\sigma_3}. \tag{9.2.10}$$

Then, the function $\Psi(\lambda) \equiv \Psi(\lambda, x)$ satisfies the following system of linear differential equations,

$$\begin{cases} \Psi_\lambda &= A(\lambda)\Psi \\ \\ \Psi_x &= U(\lambda)\Psi, \end{cases} \tag{9.2.11}$$

where the coefficient matrices $A(\lambda)$, $U(\lambda)$ are polynomial functions of λ, given by the formulae,

$$A(\lambda) = -(4i\lambda^2 + ix + 2iu^2)\sigma_3 - 4\lambda u\sigma_2 - 2u_x\sigma_1, \quad U(\lambda) = -i\lambda\sigma_3 - u\sigma_2, \tag{9.2.12}$$

with u as defined in (9.2.9).

Proposition 9.2.2 *The second Painlevé equation (9.2.5) is equivalent to the matrix commutation relation,*

$$U_\lambda(\lambda) - A_x(\lambda) = [A(\lambda), U(\lambda)], \quad \textit{identically in } \lambda, \tag{9.2.13}$$

where $A(\lambda)$ and $U(\lambda)$ as in (9.2.12).

In order to see how Propositions 9.2.1 and 9.2.2 yield the proof of Theorem 9.2.2, we first notice that the commutation relation (9.2.13) is the compatibility condition, $\Psi_{\lambda x} = \Psi_{x\lambda}$, of the overdetermined linear system (9.2.11). Therefore, the first statement of Theorem 9.2.2 is an immediate corollary of Proposition 9.2.1. The second statement of the theorem is the corollary of the principal observation that the commutation relation (9.2.13) describes the *isomonodromy deformation* of the first equation in system (9.2.11). In more detail, let

$$S_k \equiv S_k(x, u, u_x), \quad k = 1, \ldots, 6, \tag{9.2.14}$$

be the *Stokes* matrices corresponding to the first equation in (9.2.11) in the framework of the classical theory of systems of linear ordinary differential equations with rational coefficients (see e.g. [Sib90]). Then, relation (9.2.13) yields the x-independence of the Stokes matrices, i.e.

$$\frac{d}{dx} S_k(u, u_x, x) \equiv 0, \quad k = 1, \ldots, 6. \tag{9.2.15}$$

In other words, the Stokes matrices of the first equation in (9.2.11) are the *first integrals* of the second Painlevé equation (9.2.5).

Regarding the proofs of Propositions 9.2.1 and 9.2.2 themselves: the jump matrices of the function $\Psi(\lambda)$ are constant matrices. Therefore, the logarithmic derivatives, $\Psi_\lambda \Psi^{-1}(\lambda) \equiv A(\lambda)$ and $\Psi_x \Psi^{-1}(\lambda) \equiv U(\lambda)$ are entire functions of λ.[2] The polynomial structure of $A(\lambda)$ and $U(\lambda)$ and the formulae for their coefficients in terms of u and u_x, which are indicated in (9.2.12), follow from the asymptotic expansion of the function $\Psi(\lambda)$ at $\lambda = \infty$ and the symmetry $\sigma_2 \Psi(-\lambda) \sigma_2 = \Psi(\lambda)$ inherited from the symmetry condition (9.2.3) satisfied by the jump matrices S_k. This proves Proposition 9.2.1. The proof of Proposition 9.2.2 is straightforward.

Remark 9.2.1 *The Stokes matrices (9.2.14) and the matrices S_k participating in the setting of the P2-RH problem are the same objects.*

Remark 9.2.2 *According to the terminology of integrable systems, the linear system (9.2.11) and the non-linear equation (9.2.13) are the Lax pair and the zero-curvature (or Lax) representation of the second Painlevé equation (9.2.5). This Lax pair for (9.2.5) was found by Flaschka and Newell [Fla80].*

We call equation (9.2.9) the *Riemann–Hilbert representation* of the solutions of the second Painlevé equation (9.2.5). Let $\mathcal{S}$ be the algebraic manifold defined by the equation,

$$\mathcal{S} = \left\{ s = (s_1, s_2, s_3) \in \mathbb{C}^3 : s_1 - s_2 + s_3 + s_1 s_2 s_3 = 0 \right\}. \tag{9.2.16}$$

A direct corollary of Theorems 9.2.1 and 9.2.2 is that the map defined by equation (9.2.9) is a bijection of the manifold $\mathcal{S}$ onto the set of solutions of the second Painlevé equation (9.2.5). Hence the notation,

$$u(x) \equiv u(x; \mathbf{s}), \tag{9.2.17}$$

[2] The unimodularity of the jump matrices of the P2-RH problem implies that $\det\Psi(\lambda) \equiv 1$.

for the second Painlevé transcendents is justified. This fact and the meromorphicity of the function in $m_1(x)$ in (9.2.9) implies, in particular, that every solution of (9.2.5) is a meromorphic function of x. In other words, we arrive at the Painlevé property for the second Painlevé equation.[3]

In the next section we will outline the use of the Riemann–Hilbert representation of the Painlevé functions for their global asymptotic analysis.

9.3 Asymptotic analysis of the Painlevé functions

In order to give a flavour of the asymptotic results available in the modern theory of Painlevé equations within the framework of the Riemann–Hilbert method we shall consider the special one-parameter family of real for real x solutions of the second Painlevé equation (9.2.5) characterized by the following condition on the corresponding RH data $\mathbf{s}$,

$$s_2 = 0, \quad s_1 = -s_3 \equiv -ia, \quad a \in \mathbb{R}. \tag{9.3.1}$$

This family plays a prominent role in the theory of random matrices. The global facts concerning the asymptotic behaviour of the solutions from this family are collected in the following theorem.

Theorem 9.3.1 *Let $u(x) \equiv u(x;a)$ be the second Painlevé transcendent from family (9.3.1). Then, the behaviour of the solution $u(x;a)$ as $x \to +\infty$ is described by the formula,*

$$u(x;a) = \frac{a}{2\sqrt{\pi}} x^{-1/4} e^{-\frac{2}{3}x^{3/2}} \left(1 + o(1)\right), \quad x \to +\infty. \tag{9.3.2}$$

Moreover, the asymptotic condition (9.3.2) determines the solution of equation (9.2.5) uniquely. The behaviour of the solution $u(x;a)$ as $x \to -\infty$ depends on the value of $|a|$.

1. *If $|a| < 1$, then*

$$u(x;a) = (-x)^{-1/4} d \cos\left(\frac{2}{3}(-x)^{3/2} - \frac{3d^2}{4}\ln(-x) + \phi\right) + o(x^{-1/4}), \tag{9.3.3}$$

where

$$d^2 = -\frac{1}{\pi}\ln\left(1 - a^2\right), \ d > 0, \ \phi = -\frac{3d^2}{2}\ln 2 - \frac{\pi}{4} - \arg\Gamma\left(-\frac{id^2}{2}\right) + \frac{\pi}{2}\mathrm{sign}(a).$$

2. *If $|a| = 1$, then*

$$u(x;a) = \mathrm{sign}\,(a)\sqrt{-\frac{x}{2}}\left(1 + \frac{1}{8x^3} + O\left(x^{-6}\right)\right). \tag{9.3.4}$$

[3] This is, of course, not the only way to prove the Painlevé property. For a more direct approach, see e.g. [Jos90].

3. *if $|a| > 1$, then the function $u(x; a)$ has infinitely many poles on the negative real line (accumulating to $-\infty$).*

The case $|a| < 1$ is due to Ablowitz and Segur and the case $|a| = 1$ is due to Hastings and McLeod. In the case $|a| > 1$, the (singular) asymptotic behaviour as $x \to -\infty$ can also be written in terms of trigonometric functions (see e.g. [Fok06], Chapter 10), and this was done by Kapaev. In what follows, we outline the proof of the asymptotic formulae (9.3.2) using the RH representation (9.2.9) of the second Painlevé transcendents and the non-linear steepest descent method of Deift and Zhou [Dei93].

Proof of (9.3.2). We start with noticing that restriction $s_2 = 0$ on the RH data means, in particular, that the solution of the RH problem $Y(\lambda)$ has no jumps across the rays Γ_2 and Γ_5, and hence they can be excluded from the setting of the Riemann–Hilbert problem. In the next move, we perform the natural scaling transformation,

$$\lambda \to z = \lambda x^{-1/2}, \quad X(z) := Y(x^{1/2} z). \tag{9.3.5}$$

In terms of the function $X(z)$, the P2-RH problem reads as follows.

- $X(z)$ is analytic in $\mathbb{C} \setminus (\Gamma_1 \cup \Gamma_3 \cup \Gamma_4 \cup \Gamma_6)$
- $X_+(z) = X_-(z) e^{-t\vartheta(z)\sigma_3} S_k e^{t\vartheta(z)\sigma_3}, \quad z \in \Gamma_k, \quad k = 1, 3, 4, 6\,.$
- At $z = \infty$, the function $X(z)$ admits the asymptotic expansion,

$$X(z) \sim I + \sum_{j+1}^{\infty} \frac{m_j^X}{z^j}, \quad z \to \infty, \tag{9.3.6}$$

where,

$$\vartheta(z) = \frac{4i}{3} z^3 + iz, \quad \text{and} \quad t = x^{3/2}. \tag{9.3.7}$$

Simultaneously, formula (9.2.9) is replaced by the equation,

$$u(x) = 2\sqrt{x}\left(m_1^X\right)_{12}. \tag{9.3.8}$$

The saddle points of the exponent $\vartheta(z)$ are $z_\pm = \pm\frac{i}{2}$. Denote by $\gamma_\pm$ the stationary contours passing through the respective points $z_\pm$. On each of these contours, $\mathrm{Im}\vartheta(z) = 0$. The rays $\Gamma_{1,3}$ and $\Gamma_{4,6}$ are the asymptotes of the contour γ_+ and γ_-, respectively. It is an easy exercise to check that the rays $\Gamma_{1,3}$ and the contour γ_+ lie in the domain where $\mathrm{Re}\vartheta(z) < 0$, while the rays $\Gamma_{4,6}$ and the contour γ_- lie in the domain where $\mathrm{Re}\vartheta(z) > 0$. (In fact, one can easily derive the exact equations for the contours $\gamma_\pm$ as well as for the curves $\mathrm{Re}\vartheta(z) = 0$.) These observations, together with the fact that the Stokes matrices S_1 and S_3 are lower triangular while the Stokes matrices S_4 and S_6 are upper triangular, allow us to transform the X-RH problem to the following Riemann–Hilbert problem posed on the union of the contours γ_+ and γ_- (we keep the same notation X for the solution of the deformed problem).

- $X(z)$ is analytic in $\mathbb{C} \setminus \left(\gamma_+ \cup \gamma_-\right)$
- $X_+(z) = X_-(z)G_X(z) \quad z \in \gamma_+ \cup \gamma_-$.
- At $z = \infty$, the new function $X(z)$ admits exactly the same asymptotic expansion (9.3.6) and with the exactly the same matrix coefficients m_j^X as it had before the deformation.

The contours $\gamma_\pm$ are assumed to be oriented from the left to right. The jump matrix $G_X(z)$ is defined by the equations,

$$G_X(z) = \begin{cases} e^{-t\vartheta(z)\sigma_3} S_1 e^{t\vartheta(z)\sigma_3} \equiv \begin{pmatrix} 1 & 0 \\ se^{2t\vartheta(z)} & 1 \end{pmatrix} & \text{if} \quad z \in \gamma_+ \\ \\ e^{-t\vartheta(z)\sigma_3} S_6 e^{t\vartheta(z)\sigma_3} \equiv \begin{pmatrix} 1 & se^{-2t\vartheta(z)} \\ 0 & 1 \end{pmatrix} & \text{if} \quad z \in \gamma_-, \end{cases} \tag{9.3.9}$$

here $s = s_1 = -ia$. We have also taken into account that, under assumption (9.3.1), $S_3 = S_1^{-1}$ and $S_4 = S_6^{-1}$.

The contours γ_+ and γ_- are the steepest descent contours for the exponents $e^{2t\vartheta(z)}$ and $e^{-2t\vartheta(z)}$, respectively. The maximum values of these exponents are reached at the respective saddle points, and they are both are equal to $e^{-\frac{2}{3}x^{3/2}}$. Therefore, with the help of the standard arguments of the classical Laplace's method, we deduce from the equations (9.3.9) the following estimates for the jump matrix $G_X(z)$:

$$||I - G_X||_{L_\infty} \le Ce^{-\frac{2}{3}x^{3/2}}, \quad ||I - G_X||_{L_2} \le Cx^{-3/8}e^{-\frac{2}{3}x^{3/2}}. \tag{9.3.10}$$

Solution of the X-RH problem admits the following integral representation, (see e.g. [Dei93] or [Fok06], Chapter 3)

$$X(z) = I + \frac{1}{2\pi i}\int_{\gamma_+\cup\gamma_-} \rho(z')\left(G_X(z') - I\right)\frac{dz'}{z'-z}, \tag{9.3.11}$$

where $\rho(z)$ is the solution of the integral equation,

$$\rho = I + C_-\left[\rho(G_X - I)\right]. \tag{9.3.12}$$

Here C_- is the Cauchy operator corresponding to the contour $\gamma = \gamma_+ \cup \gamma_-$,

$$C_-[f](z) = \frac{1}{2\pi i}\lim_{z_1 \to z_-}\int_\gamma f(z')\frac{dz'}{z'-z_1}, \tag{9.3.13}$$

and $z_1 \to z_\pm$ denotes the non-tangential limit from $\pm$ side of γ. It is a simple, but still one of the key points of the non-linear steepest descent method of Deift and Zhou that the estimates (9.3.10), together with the L_2 - boundedness of the Cauchy operator C_-, yield, for sufficiently large x, the (unique) solvability of equation (9.3.12) and the estimate,

$$||I - \rho||_{L_2} \leq Cx^{-3/8}e^{-\frac{2}{3}x^{3/2}}, \tag{9.3.14}$$

for its solution.

From representation (9.3.11) it follows that

$$m_1^X = \frac{i}{2\pi}\int_\gamma \rho(z)\left(G_X(z) - I\right) dz, \tag{9.3.15}$$

which together with formula (9.3.8) produces the integral representation for the solution $u(x)$,

$$u(x) = \frac{i}{\pi}x^{1/2}\int_\gamma \Big[\rho(z)\left(G_X(z) - I\right)\Big]_{12} dz. \tag{9.3.16}$$

Using in this representation estimate (9.3.14) and the second estimate in (9.3.10), we have that,

$$u(x) = \frac{i}{\pi}x^{1/2}\int_\gamma \left(G_X(z) - I\right)_{12} dz + O\left(\frac{e^{-4/3x^{3/2}}}{x^{1/4}}\right)$$

$$= \frac{i}{\pi}x^{1/2}s\int_{\gamma_-} e^{-2t\vartheta(z)}dz + O\left(\frac{e^{-4/3x^{3/2}}}{x^{1/4}}\right) = is\mathrm{Ai}(x) + O\left(\frac{e^{-4/3x^{3/2}}}{x^{1/4}}\right), \quad x \to +\infty.$$

The last estimate can be also written in the form (9.3.2) which completes the proof of this estimate. □

The analysis of the behaviour of solutions of (9.2.5) as $x \to -\infty$ needs more work. The deformation of the original jump contour is more complicated. The rescaling is now $\lambda \to z = \lambda(-x)^{-1/2}$, so that the saddle points become real, $\pm 1/2$ and the jump matrix is not close to the identity near the saddle points. Hence the necessity to consider the non-trivial parametrices of the solution of the RH problem near these points. For a detailed derivation of formulae (9.3.3) and (9.3.4), as well as for the asymptotic formulae for general two-parameter solutions of the second Painlevé equation (9.2.5), we refer the reader to monograph [Fok06].

9.4 The Airy and the Sine kernels and the Painlevé functions

A remarkable fact about random matrix theory is that its basic correlation functions admit the Fredholm determinant representations. Moreover, the integral operators involved in these representations are of a special “integrable form”.

Therefore we shall start this section with an outline of the theory of integrable Fredholm operators [Its90], [Tra94a], [Dei97].

9.4.1 Integrable Fredholm operators

Let Γ be an oriented contour in the complex λ-plane. Consider the integral operator K in $L_2(\Gamma; d\lambda)$,

$$K(\phi)(\lambda) = \int_\Gamma K(\lambda, \mu)\phi(\mu)d\mu.$$

We shall call the operator K *integrable* if its kernel admits the representation,

$$K(\lambda, \mu) = \frac{\sum_{j=1}^m f_j(\lambda)h_j(\mu)}{\lambda - \mu} \equiv \frac{f^T(\lambda)h(\mu)}{\lambda - \mu}, \tag{9.4.1}$$

for some smooth vector functions $f(\lambda) = \Big(f_1(\lambda), \ldots, f_m(\lambda)\Big)^T$, $h(\lambda) = \Big(h_1(\lambda), \ldots, h_m(\lambda)\Big)^T$. We shall also assume that

$$f^T(\lambda)h(\lambda) = \sum_{j=1}^m f_j(\lambda)h_j(\lambda) = 0,$$

so that the kernel is continuous.

Let R denote the resolvent of the operator $1 - K$; that is,

$$1 + R = (1 - K)^{-1}, \tag{9.4.2}$$

assuming that the inverse exists. A key algebraic property of an integrable kernel is that the corresponding resolvent kernel has the same integrable form (9.4.1). Indeed,

$$R(\lambda, \mu) = \frac{F^T(\lambda)H(\mu)}{\lambda - \mu}, \tag{9.4.3}$$

where the components $F_j(\lambda)$ and $H_j(\lambda)$ of the vector functions $F(\lambda)$ and $H(\lambda)$ are defined by the relations,

$$F_j(\lambda) = \Big[(1 - K)^{-1} f_j\Big](\lambda), \quad \text{and} \quad H_j(\lambda) = \Big[(1 - K^T)^{-1} h_j\Big](\lambda). \tag{9.4.4}$$

A key analytic property of an integrable kernel is the existence of the alternative formulae for the functions $F_j(\lambda)$ and $H_j(\lambda)$ in terms of the solution $Y(\lambda)$ of the following $m \times m$ matrix Riemann–Hilbert problem.

- $Y(\lambda)$ is analytic in $\mathbb{C} \setminus \Gamma$
- $Y_+(\lambda) = Y_-(\lambda)G(\lambda), \quad \lambda \in \Gamma$, where the $m \times m$ jump matrix $G(\lambda)$ is defined by the equation,

$$G(\lambda) = I - 2\pi i f(\lambda)h^T(\lambda). \tag{9.4.5}$$

- If the contour Γ has an end point a, the singularity that the function $Y(\lambda)$ might have at this point is no more than logarithmic,

$$Y(\lambda) = O\Big(\ln(\lambda - a)\Big), \quad \lambda \to a. \tag{9.4.6}$$

- At $\lambda = \infty$, the function $Y(\lambda)$ satisfies the normalization condition,

$$Y(\lambda) = I + O\left(\frac{1}{\lambda}\right), \quad \lambda \to \infty \tag{9.4.7}$$

The (unique) solution $Y(\lambda)$ of this Riemann–Hilbert problem determines the functions $F(\lambda)$ and $H(\lambda)$ via the relations,

$$F(\lambda) = Y(\lambda)f(\lambda), \quad \text{and} \quad H(\lambda) = \Big[Y^T(\lambda)\Big]^{-1} h(\lambda). \tag{9.4.8}$$

Conversely, the solution $Y(\lambda)$ of the Riemann–Hilbert problem can be expressed in terms of the function $F(\lambda)$ via the Cauchy integral,

$$Y(\lambda) = I - \int_\Gamma F(\mu)h^T(\mu)\frac{d\mu}{\mu - \lambda}. \tag{9.4.9}$$

The proof of the above facts can be found in the above mentioned original papers.

We shall now apply the theory of integrable operators to the two basic distribution functions of the random matrix theory - to the edge and gap probabilities of the eigenvalues of the GUE matrices.

9.4.2 The Airy-kernel determinant and the second Painlevé transcendent

Let K_x be the trace-class operator in $L_2\Big((0, \infty); d\lambda\Big)$ with kernel

$$K_x(\lambda, \mu) = \frac{\mathrm{Ai}(\lambda + x)\mathrm{Ai}'(\mu + x) - \mathrm{Ai}'(\lambda + x)\mathrm{Ai}(\mu + x)}{\lambda - \mu}, \tag{9.4.10}$$

where $\mathrm{Ai}(x)$ is the Airy function,

$$\mathrm{Ai}(x) = \frac{1}{2\pi}\int_{-\infty}^{\infty} e^{-\frac{i}{3}s^3 - ixs} ds.$$

The *Airy-kernel determinant* $\det(I - K_x)$ describes the edge scaling limit for the largest eigenvalue of a random $n \times n$ Hermitian matrix H from the Gaussian unitary ensemble (GUE) as $n \to \infty$ (see [Meh04, Tra94b]). This determinant also appears in a number of other statistical problems related to random matrices.

The key property of the determinant $\det(I - K_x)$ is that it can be explicitly evaluated in terms of the Hastings-McLeod solution of the second Painlevé equation.

Theorem 9.4.1 *[Tra94b] The following formula is valid for the Fredholm determinant* $\det(1 - K_x)$,

$$\det(1 - K_x) = \exp\left\{-\int_x^\infty (y - x)u^2(y)dy\right\}, \tag{9.4.11}$$

where $u(x)$ is the Hastings-McLeod solution of the second Painlevé equations, i.e. $u(x)$ is the solution of (9.2.5) uniquely determined by the asymptotic condition (compare Theorem 9.3.1),

$$u(x) = \frac{1}{2\sqrt{\pi}} x^{-1/4} e^{-\frac{2}{3}x^{3/2}} (1 + o(1)), \quad x \to +\infty. \tag{9.4.12}$$

We shall now give a proof of the Tracy–Widom formula (9.4.11) which is slightly different from the original one given in [Tra94b]. As in [Tra94b], the proof presented below is based on the "integrable" form of the kernel (10.2), but it makes more use of the Riemann–Hilbert aspect of the theory of Painlevé functions.

Proof of Theorem 9.4.1. Our starting point is the observation that the integral operator K_x belongs to the integrable class discussed in the previous section. Indeed, the relevant vector functions $f(\lambda)$ and $h(\lambda)$ are

$$f(\lambda) = \Big(f_1(\lambda), f_2(\lambda)\Big)^T, \quad \text{and} \quad h(\lambda) = \Big(h_1(\lambda), h_2(\lambda)\Big)^T, \tag{9.4.13}$$

where

$$f_1(\lambda) = -h_2(\lambda) = \mathrm{Ai}(\lambda + x), \quad f_2(\lambda) = h_1(\lambda) = \mathrm{Ai}'(\lambda + x), \tag{9.4.14}$$

and hence the corresponding Riemann–Hilbert problem reads,

- $Y(\lambda)$ is analytic in $\mathbb{C} \setminus [0, \infty)$,
- $Y_+(\lambda) = Y_-(\lambda)G_{TW}(\lambda), \quad \lambda > 0$, where

$$G_{TW}(\lambda) = \begin{pmatrix} 1 - 2\pi i f_1(\lambda) f_2(\lambda) & 2\pi i f_1^2(\lambda) \\ -2\pi i f_2^2(\lambda) & 1 + 2\pi i f_1(\lambda) f_2(\lambda) \end{pmatrix}, \tag{9.4.15}$$

  and $f_{1,2}(\lambda)$ are given in (9.4.14),
- $Y(\lambda) = O\Big(\ln \lambda\Big), \quad \lambda \to 0,$
- $Y(\lambda) = I + O\left(\frac{1}{\lambda}\right), \quad \lambda \to \infty.$

In order to express the Fredholm determinant $\det(1 - K_x)$ in terms of the solution $Y(\lambda)$ of the Riemann–Hilbert problem (9.4.15) consider the logarithmic derivative of $\det(1 - K_x)$ with respect to the variable x. Taking into account that the Airy function satisfies the equation,

$$\mathrm{Ai}''(\lambda + x) = (\lambda + x)\mathrm{Ai}(\lambda + x),$$

we have that

$$\frac{d}{dx}K_x(\lambda,\mu) = -\mathrm{Ai}(\lambda + x)\mathrm{Ai}(\mu + x) = f_1(\lambda)h_2(\mu). \tag{9.4.16}$$

Therefore,

$$\frac{d}{dx}\ln\det(1 - K_x) = -\mathrm{Tr}(1 - K_x)^{-1}\frac{d}{dx}K_x = -\int_0^\infty \left[(1 - K_x)^{-1}f_1\right](\lambda)h_2(\lambda)\,d\lambda$$
$$= -\int_0^\infty F_1(\lambda)h_2(\lambda)\,d\lambda. \tag{9.4.17}$$

On the other hand, if we denote $m_1 \equiv m_1(x)$ the first matrix coefficient in the expansion,

$$Y(\lambda) \sim I + \frac{m_1}{\lambda} + \frac{m_2}{\lambda^2} + \ldots, \qquad \lambda \to \infty, \tag{9.4.18}$$

then from (9.4.9) it will follow that

$$m_1 = \int_0^\infty F(\lambda)h^T(\lambda)\,d\lambda. \tag{9.4.19}$$

Comparing this equation and equation (9.4.17), we arrive at the formula,

$$\frac{d}{dx}\ln\det(1 - K_x) = -(m_1)_{12}, \tag{9.4.20}$$

which connects the Airy determinant to the Riemann–Hilbert problem (9.4.15). Our next task is to link the Riemann–Hilbert problem (9.4.15) to the second Painlevé equation.

Observe that the jump matrix $G_{TW}(\lambda)$ can be written as the following product,

$$G_{TW}(\lambda) = \Phi(\lambda)\begin{pmatrix}1 & 2\pi i\\ 0 & 1\end{pmatrix}\Phi^{-1}(\lambda), \tag{9.4.21}$$

where

$$\Phi(\lambda) = \begin{pmatrix} f_1(\lambda) & g_1(\lambda)\\ f_2(\lambda) & g_2(\lambda)\end{pmatrix}, \tag{9.4.22}$$

where the only condition on $g_1(\lambda)$ and $g_2(\lambda)$ is that they ensure the identity,

$$\det\Phi(\lambda) \equiv 1. \tag{9.4.23}$$

For instance, we can choose, $g_1(\lambda) = 2\pi e^{-\frac{\pi i}{6}}\mathrm{Ai}(e^{-\frac{2\pi i}{3}}(\lambda + x))$, $g_2(\lambda) = g_1'(\lambda)$. Put

$$\Psi(\lambda) = Y(\lambda)\Phi(\lambda). \tag{9.4.24}$$

In terms of the function $\Psi(\lambda)$ the master Riemann–Hilbert problem (9.4.15) transforms into the following Riemann–Hilbert problem.

- $\Psi(\lambda)$ is analytic in $\mathbb{C} \setminus [0, \infty)$,
- $\Psi_+(\lambda) = \Psi_-(\lambda) G_\Psi(\lambda), \quad \lambda > 0$, where

$$G_\Psi(\lambda) = \begin{pmatrix} 1 & 2\pi i \\ 0 & 1 \end{pmatrix} \tag{9.4.25}$$

- In the neighbourhood of $\lambda = 0$,

$$\Psi(\lambda) = \hat{\Psi}(\lambda)\Big(I - \ln\lambda \begin{pmatrix} 0 & 1 \\ 0 & 0 \end{pmatrix}\Big) \tag{9.4.26}$$

where $\hat{\Psi}(\lambda)$ is analytic at $\lambda = 0$, and the branch of the logarithm is fixed by the condition,[4] $0 < \arg\lambda < 2\pi$
- In the neighbourhood of $\lambda = \infty$,

$$\Psi(\lambda) \sim \Big(I + \frac{m_1}{\lambda} + \frac{m_2}{\lambda^2} + \ldots\Big)\Phi(\lambda), \quad \lambda \to \infty. \tag{9.4.27}$$

We shall now show that this Riemann–Hilbert problem yields an alternative Lax pair for the second Painlevé equation.

The piecewise analytic function $\Psi(\lambda)$, similar to the Ψ-function from Proposition 9.2.1, has constant jumps (see (9.4.25)). Therefore, we can use the same arguments as we used when we discussed proof of Proposition 9.2.1 and arrive at the system of linear differential equations of the type (9.2.11) for function $\Psi(\lambda)$ defined by (9.4.24). However, because of the singularity (9.4.26) at $\lambda = 0$, the matrix function $A(\lambda)$, this time, is a rational function with the simple pole at $\lambda = 0$ while the matrix function $U(\lambda)$ is still a polynomial (of degree one). Indeed, we have that,

$$A(\lambda) = \lambda A_1 + A_0 + \frac{1}{\lambda} A_{-1} \quad \text{and} \quad U(\lambda) = \lambda U_1 + U_0, \tag{9.4.28}$$

where the coefficients $A_{1,0}$ and $U_{1,0}$ are given by the equations[5]

$$A_1 = U_1 = \begin{pmatrix} 0 & 0 \\ 1 & 0 \end{pmatrix}, \tag{9.4.29}$$

and

$$A_0 = U_0 = \begin{pmatrix} w & 1 \\ p & -w \end{pmatrix}, \quad A_{-1} = \begin{pmatrix} uv & -u^2 \\ v^2 & -uv \end{pmatrix}. \tag{9.4.30}$$

[4] Representation (9.4.26) is the direct corollary of the jump condition (9.4.25) and *a priori* estimate $Y(\lambda) = O(\ln\lambda)$.

[5] When deriving formulae (9.4.29)-(9.4.32), one has to take into account that,

$$\Phi_\lambda \Phi^{-1} = \Phi_x \Phi^{-1} = \begin{pmatrix} 0 & 1 \\ \lambda + x & 0 \end{pmatrix}.$$

The functional parameters w, p, u, and v in these formulae are determined by the matrix coefficients m_1 and $\hat{\Psi}(0)$ via the relations,

$$w = (m_1)_{12}, \quad p = x + (m_1)_{22} - (m_1)_{11}, \tag{9.4.31}$$

and

$$u = \left(\hat{\Psi}(0)\right)_{11} = \left(\Psi(0)\right)_{11}, \quad v = \left(\hat{\Psi}(0)\right)_{21} = \left(\Psi(0)\right)_{21}. \tag{9.4.32}$$

It is worth noticing that by virtue of the definition of the function $\Psi(\lambda)$ and due to the general equation (9.4.8), the quantities u and v coincide with the components of the vector $F(0)$; in fact,

$$u = F_1(0), \quad v = F_2(0). \tag{9.4.33}$$

The matrices $A(\lambda)$ and $U(\lambda)$ satisfy the zero-curvature equation (9.2.13). The latter, by a direct substitution, is equivalent to the following non-linear system on the functions w, v, p and q,

$$\begin{cases} w_x = u^2 \\ p_x = 1 + 2uv \\ u_x = uw + v \\ v_x = -vw + up. \end{cases} \tag{9.4.34}$$

This system has the first integral,

$$p + w^2 - x - u^2 \equiv \text{constant} \equiv c,$$

and it also yields the single second order ODE for the function u,

$$u_{xx} = (x + c)u + 2u^3, \tag{9.4.35}$$

which is the Painlevé II up to the shift of the argument x. In view of the first equation in (9.4.34), equation (9.4.20), and definition (9.4.31) of the function w, the theorem will be proven if we show that the function $u(x)$ has asymptotic behaviour (9.4.12) as $x \to +\infty$. (Note that, simultaneously, it will be shown that $c = 0$ in (9.4.35).) The fastest way to see this is to notice that for large positive x the kernel $K_x(\lambda, \mu)$ is exponentially small and hence the first equation in (9.4.4) would provide us with the estimate,

$$F_1(0) \sim f_1(0) = \mathrm{Ai}(x), \quad x \to +\infty.$$

The needed asymptotics for $u(x)$ would follow then from (9.4.33). □

In view of (9.3.4), the asymptotics of the Airy-determinant as $x \to -\infty$ is given by the equation,

$$\ln \det(1 - K_x) = \frac{x^3}{12} - \frac{1}{8}\ln(-x) + c_0 + o(1), \quad x \to -\infty. \tag{9.4.36}$$

The value of the constant c_0 was conjectured by Tracy and Widom in [Tra94b] to be given by the equation,

$$c_0 = \frac{1}{24}\ln 2 + \zeta'(-1), \tag{9.4.37}$$

where $\zeta(s)$ is the Riemann zeta function. This conjecture was independently proven in [Dei08] and [Bai08].

The function,

$$F_2(x) \equiv \det(1 - K_x) = \exp\left\{-\int_x^\infty (y - x)u^2(y)dy\right\}, \tag{9.4.38}$$

is called the *GUE Tracy–Widom distribution function*. The analogues of this distribution function for two the other classical β -ensembles, i.e. for Gaussian orthogonal ($\beta = 1$) and Gaussian symplectic ($\beta = 4$) ensembles are denoted $F_1(x)$ and $F_4(x)$, respectively. The GOE and GSE Tracy–Widom distribution functions describe the edge scaling limits at the respective matrix ensembles. Similarly to the GUE function $F_2(x)$, they admit explicit representations in terms of the Hastings-McLeod solution $u(x)$ of the second Painlevé equation (9.2.5). Indeed, as is shown in [Tra96],

$$F_1(x) = F(x)E(x), \quad F_4(x) = \frac{1}{2}\left\{E(x) + \frac{1}{E(x)}\right\}F(x), \tag{9.4.39}$$

where

$$F(x) = \exp\left\{-\frac{1}{2}\int_x^\infty (y - x)u^2(y)dy\right\} \equiv F_2^{1/2}(x),$$

and

$$E(x) = \exp\left\{-\frac{1}{2}\int_x^\infty u(y)dy\right\}.$$

The asymptotic expansions, including constant terms, for the distribution functions $F_1(x)$ and $F_4(x)$ as $x \to -\infty$, i.e. the GOE and GSE analogues of equations (9.4.36) and (9.4.37), have been found in [Bai08].

It is worth noticing that asymptotics (9.4.36) in conjunction with equation[6] (9.4.11) yield the following integral identity for the Hastings - McLeod Painlevé transcendent $u(x)$,

$$\text{v.p.}\int_{-\infty}^\infty yu^2(y)\,dy = -\frac{1}{8} - \frac{1}{24}\ln 2 - \zeta'(-1), \tag{9.4.40}$$

[6] One has also to take into account that $u^2 = -(u_x^2 - xu^2 - u^4)_x$.

where the principal value of the integral in the left-hand side is defined in the natural way, i.e.

$$\text{v.p.}\int_{-\infty}^{\infty} yu^2(y)\,dy \equiv \int_b^{\infty} yu^2(y)dy + \int_{-\infty}^{b}\left(yu^2(y)+\frac{y^2}{2}+\frac{1}{8y}\right)dy - \frac{b^3}{6} - \frac{\log|b|}{8}, \quad b<0.$$

The F_1 analogue of the asymptotic expansion (9.4.36), which was found in [Bai08] yields an addition integral formula involving $u(x)$,

$$\text{v.p.}\int_{-\infty}^{\infty} u(y)\,dy = \frac{1}{2}\ln 2, \tag{9.4.41}$$

with

$$\text{v.p.}\int_{-\infty}^{\infty} u(y)\,dy \equiv \int_b^{\infty} u(y)\,dy + \int_{-\infty}^{b}\left(u(y)-\sqrt{\frac{|y|}{2}}\right)dy + \frac{\sqrt{2}}{3}|b|^{3/2}, \quad b<0.$$

For more on the total integrals of the Painlevé functions, see [Bai09].

9.4.3 The Sine-kernel determinant and the fifth Painlevé transcendent

Let K_x be the trace-class operator in $L_2\big((-1,1);d\lambda\big)$ with kernel

$$K_x(\lambda,\mu) = \frac{\sin x(\lambda-\mu)}{\pi(\lambda-\mu)}. \tag{9.4.42}$$

The *Sine-kernel determinant* $\det(I-K_x)$ describes the probability of finding no eigenvalues in the interval $(-x/\pi, x/\pi)$ for a random Hermitian matrix H chosen from the Gaussian unitary ensemble (GUE), in the bulk scaling limit with mean spacing 1 (see [Meh04, Tra94a]). Similarly to the Airy-kernel determinant, the sine-kernel determinant appears to be one of the universal distribution functions in many statistical problems related to random matrices (see [Dei98] and references therein). Also similar to the Airy-kernel determinant, the sine-kernel determinant admits the Painlevé-type representation - this time, in terms of a particular solution of the fifth Painlevé equation.

Theorem 9.4.2 *[Jim80a] The following formula is valid for the Fredholm determinant* $\det(1-K_x)$,

$$\det(1-K_x) = \exp\left(\int_0^x \frac{\sigma(t)}{t}dt\right) \tag{9.4.43}$$

where $\sigma(x)$ *is the solution of the second-order differential equation,*

$$(x\sigma_{xx})^2 + 4(4\sigma - 4x\sigma_x - \sigma_x^2)(\sigma - x\sigma_x) = 0, \tag{9.4.44}$$

characterized by the following behaviour at $x = 0$,

$$\sigma(x) \sim -\frac{2}{\pi}x, \quad x \to 0. \tag{9.4.45}$$

Remark 9.4.1 *Equation (9.4.44) is the so-called Hirota σ-form of the fifth Painlevé equation whose canonical u-form is presented at the beginning of the chapter. The relevant parameters are: $\alpha = \beta = 0$, $\gamma = 2i$, and $\delta = 2$. The relation between the σ and u functions is given by the equation,*

$$\frac{1}{x}\sigma_x = \frac{8\hat{u} + 4i\hat{u}_x}{(\hat{u}-1)^2}, \quad \hat{u}(x) \equiv u(2x). \tag{9.4.46}$$

Proof of Theorem 9.4.2. The proof proceeds along the same lines of argument as in the proof of Theorem 9.4.1. Kernel (9.4.42) obviously belongs to the integrable class with

$$f(\lambda) = \left(e^{ix\lambda}, e^{-ix\lambda}\right)^T, \quad h(\lambda) = \left(\frac{e^{-ix\lambda}}{2\pi i}, -\frac{e^{ix\lambda}}{2\pi i}\right)^T. \tag{9.4.47}$$

The jump matrix $G_S(\lambda)$ of the corresponding Y - Riemann–Hilbert problem is then

$$G_S(\lambda) = \begin{pmatrix} 0 & e^{2ix\lambda} \\ -e^{-2ix\lambda} & 2 \end{pmatrix}, \tag{9.4.48}$$

and the setting of the Riemann–Hilbert problem includes two logarithmic singularities - one at $\lambda = -1$ and another at $\lambda = 1$.

Performing calculations very similar to the Airy-case, one arrives at the following analogue of formula (9.4.20),

$$\frac{d}{dx}\ln\det(1 - K_x) = i\Big((m_1)_{22} - (m_1)_{11}\Big), \tag{9.4.49}$$

where $m_1 \equiv m_1(x)$ is defined as the first coefficient of the asymptotic expansion (9.4.18) where $Y(\lambda)$ is now the solution of the Riemann–Hilbert problem posed on the interval $(-1, 1)$ and with jump matrix (9.4.48).

In order to connect the sine-RH problem with the fifth Painlevé equation, we should again take advantage of the fact that the problem's jump matrix is conjugated to the constant matrix (compare (9.4.21)),

$$G_S(\lambda) = e^{ix\lambda\sigma_3}\begin{pmatrix} 0 & 1 \\ -1 & 2 \end{pmatrix}e^{-ix\lambda\sigma_3}. \tag{9.4.50}$$

This factorization suggests introduction of the function $\Psi(\lambda)$ according to equation (compare (9.2.10), (9.4.24)),

$$\Psi(\lambda) = Y(\lambda)e^{ix\lambda\sigma_3}, \tag{9.4.51}$$

so that its logarithmic derivatives with respect to λ and x are again rational functions in λ. Indeed, we see that,

$$\Psi_\lambda(\lambda)\Psi^{-1}(\lambda) = ix\sigma_3 + \frac{B}{\lambda+1} + \frac{C}{\lambda-1} \equiv A(\lambda), \tag{9.4.52}$$

and

$$\Psi_x(\lambda)\Psi^{-1}(\lambda) = i\lambda\sigma_3 + U_0 \equiv U(\lambda). \tag{9.4.53}$$

Evaluating the principal parts of these logarithmic derivatives at the relevant singular points and taking into account the symmetry property of the function $\Psi(\lambda)$,

$$\sigma_1\Psi(-\lambda)\sigma_1 = \Psi(\lambda),$$

we would arrive at the following representations for the matrix coefficients B, C, and U_0 (compare (9.4.30)),

$$B = \sigma_1 C\sigma_1 = \begin{pmatrix} v & -v\gamma^{-1} \\ v\gamma & -v \end{pmatrix}, \quad \text{and} \quad U_0 = w\sigma_1. \tag{9.4.54}$$

Here the scalar functional parameters v, γ, and w are given in terms of the first matrix coefficient m_1 of the series (9.4.18) by the formulae,[7]

$$\begin{cases} v = \frac{1}{2}(m_1)_{11} + ix(m_1)_{12}^2 \\ v(\gamma - \gamma^{-1}) = 2ix(m_1)_{21} = -2ix(m_1)_{12} \\ w = 2i(m_1)_{21} = -2i(m_1)_{12} \end{cases} \tag{9.4.55}$$

In addition, by equating the terms of order λ^{-1} in equation (9.4.53), we obtain the differential identity,

$$\frac{d}{dx}(m_1)_{11} = -\frac{d}{dx}(m_1)_{22} = 2i(m_1)_{12}^2 = 2i(m_1)_{21}^2. \tag{9.4.56}$$

Direct substitution of the matrices $A(\lambda)$ and $U(\lambda)$ defined by (9.4.52) – (9.4.54) into the zero curvature equation (9.2.13) yields[8] the fifth Painlevé equation and the σ - form (9.4.44) of the fifth Painlevé equation for the functions,

$$u(x) := \gamma^2(x/2) \quad \text{and} \quad \sigma(x) := -4ixv + x^2w^2,$$

respectively. Simultaneously, from the first relation in (9.4.55) it follows that $\sigma = -2ix(m_1)_{11}$, which together with (9.4.49) implies (also taking into account the symmetry, $(m_1)_{11} = -(m_1)_{22}$),

$$\sigma(x) = x\frac{d}{dx}\det(1 - K_x). \tag{9.4.57}$$

The statement of the theorem follows directly from this equation. □

[7] The derivation of (9.4.54) and (9.4.55) is rather straightforward though somewhat tedious. The reader can find the details of this derivation in [Dei97].

[8] For details, we refer again to [Dei97].

The large x behaviour of the solution $\sigma(x)$ is described by the asymptotic formula ([Dys76], [Sul86]),

$$\sigma(x) \sim -x^2 - \frac{1}{4} + \sum_{k=1}^{\infty} \frac{c_k}{x^k}, \quad x \to \infty, \tag{9.4.58}$$

with all the coefficients c_k determined, in principal, recursively with the help of substitution of the series (9.4.58) into equation (9.4.44). This in turn yields the asymptotics of the sine determinant,

$$\ln \det(1 - K_x) \sim -\frac{x^2}{2} - \frac{1}{4} \ln x + c_0 - \sum_{k=1}^{\infty} \frac{c_k}{k x^k}, \quad x \to \infty, \tag{9.4.59}$$

The value of the constant term c_0 can not be deduced from series (9.4.58), and it was conjectured by Dyson in [Dys76] to be given by the equation (compare (9.4.37)),

$$c_0 = \frac{1}{12} \ln 2 + 3\zeta'(-1). \tag{9.4.60}$$

This conjecture was independently proven in [Kra04] and [Ehr06].

Similarly to the Airy-kernel case, equations (9.4.43) and (9.4.59) yield the integral identity for the fifth Painlevé function $\sigma(x)$ (compare (9.4.40) and (9.4.41)),

$$\text{v.p.} \int_0^{\infty} \frac{\sigma(t)}{t} dt = \frac{1}{12} \ln 2 + 3\zeta'(-1), \tag{9.4.61}$$

with the left-hand side defined as

$$\text{v.p.} \int_0^{\infty} \frac{\sigma(t)}{t} dt \equiv \int_0^{b} \frac{\sigma(t)}{t} dt + \int_b^{\infty} \left(\frac{\sigma(t)}{t} + t + \frac{1}{4t} \right) dt + \frac{b^2}{2} + \frac{1}{4} \ln b, \quad b > 0.$$

Acknowledgements

This work was partially supported by the NSF grant DMS-0701768.

References

[Abl77] M.J. Ablowitz and H. Segur, Stud. Appl. Math., **57** No 1 (1977) 13.
[Adl95] M. Adler, T. Shiota, and P. van Moerbeke, Phys. Lett. A. **208** (1995) 67.
[Bai08] J. Baik, R. Buckingham, and J. DiFranco, Commun. Math. Phys. **280** (2008) 463.
[Bai09] J. Baik, R. Buckingham, J. DiFranco, and A. Its, Nonlinearity, **22** (2009) 1021–1061.
[Bol04] A. A. Bolibruch, A. R. Its, and A. A. Kapaev, Algebra and Analysis, **16** (2004) 121.
[Dei98] P. Deift, *Orthogonal polynomials and random matrices: a Riemann–Hilbert approach,* Courant Lecture Notes in Math, 1998.
[Dei93] P.A. Deift and X. Zhou, Ann. of Math. **137** (1993) 295.
[Dei08] P. Deift, A. Its, and I. Krasovsky, Commun. Math. Phys. **278** (2008) 643.

[Dei97] P. A. Deift, A .R. Its, and X. Zhou, Ann. of Math. **146** (1997) 149.

[Dys76] F. Dyson, Commun. Math. Phys. **47** (1976) 171.

[Ehr06] T. Ehrhardt, Commun. Math. Phys. **262** (2006) 317.

[Fla80] H. Flaschka and A.C. Newell, Commun. Math. Phys. **76** (1980) 65.

[Fok06] A. Fokas, A. Its, A. Kapaev, and V. Novokshenov, *Painlevé Transcendents: The Riemann–Hilbert Approach,* AMS Mathematical Surveys and Monographs **128**, 2006.

[For01] P. J. Forrester and N. S. Witte, Commun. Math. Phys. **219** (2001) 357.

[Gro02] V. I. Gromak, I. Laine, and S. Shimomura, *Painlevé Differential Equations in the Complex Plane,* Walter de Gruyter, 2002.

[Its90] A .R. Its, A. G. Izergin, V .E. Korepin, and N. A. Slavnov, J. Mod. Phys. B, **4** (1990) 1003.

[Inc56] E. L. Ince, *Ordinary Differential Equations,* Dover, New York, 1956.

[Jim80a] M. Jimbo, T. Miwa, Y. Môri, and M. Sato, Physica D **1** (1980) 80.

[Jim80b] M. Jimbo, T. Miwa, and K. Ueno, Physica D **2** (1980) 306.

[Kra04] I. V. Krasovsky, Int. Math. Res. Not. **2004** (2004) 1249.

[Sib90] Y. Sibuya, *Linear differential equations in the complex domain: problems of analytic continuation,* Translations of mathematical monographs, **82**, Providence 1990.

[Sul86] B. I. Suleimanov, in the book: Lect. Notes in Math., Springer Verlag, **1191** (1986) 230.

[Jos90] N. Joshi and M.D. Kruskal, Preprint no. CMA-R06-90, Centre for Math. Anal., Australian Nat. Univ., 1990.

[McC77] B. M. McCoy, C .A. Tracy and T. T. Wu, J. Math. Phys. **18** No 5 (1977) 1058.

[Meh04] M.L. Mehta, *Random Matrices,* Academic Press, 3rd Edition, London 2004.

[Nou04] M. Noumi, *Painlevé equations through symmetry,* Translations of Math. Monographs, **223**, AMS, Providence, Rhode Island, 2004.

[Tra94a] C. Tracy and H. Widom, Commun. Math. Phys. **163** (1994) 33.

[Tra94b] C. Tracy and H. Widom, Commun. Math. Phys. **159** (1994) 151

[Tra96] C. Tracy and H. Widom, Commun. Math. Phys. **177** (1996) 727.

[Zho89] X. Zhou, SIAM J. Math. Anal. **20** No 4 (1989) 966.

·10·

Random matrix theory and integrable systems

Pierre van Moerbeke

Abstract

This is a brief introduction to the rich interaction between Random Matrix and Integrable Theory, leading to ordinary and partial differential equations for the eigenvalue distribution of random matrix models of size n and the transition probabilities of non-intersecting Brownian motion models, for finite n and for $n \to \infty$. This also provides a tool for asymptotic results.

10.1 Matrix models, orthogonal polynomials, and Kadomtsev-Petviashvili (KP)

10.1.1 Orthogonal polynomials and the KP-hierarchy

The theory of orthogonal polynomials is intimately connected with the KP-hierarchy in integrable systems. This section gives an outline of this connection; see e.g. [Adl99], [Van01], [Van10] and references within.

Given a weight $\rho(z)$ on $\mathbb{R}$ decaying (fast enough) at infinity, introduce a formal deformation

$$\rho(z) \mapsto \rho_t(z) := \rho(z) e^{\sum_1^\infty t_k z^k} \tag{10.1.1}$$

by means of an exponential involving time parameters $t := (t_1, t_2, \ldots) \in \mathbb{C}^\infty$. Then given a symmetric inner product $\int_{\mathbb{R}} f(z) g(z) \rho_t(z)\, dz$ between functions f and g, denote by $\tau_n(t)$ the determinant of the following t-dependent (Hankel) moment matrix:

$$\tau_n(t) := \det\Big(\int_{\mathbb{R}} z^{i+j} e^{\sum_1^\infty t_k z^k} \rho(z) dz \Big)_{0 \le i,j \le n-1}. \tag{10.1.2}$$

Theorem 10.1.1 *The functions $\tau_n(t)$ satisfy the following KP bilinear identities;*[1] *they amount to the vanishing of a residue about $z = \infty$:*

[1] Introduce the notation $[\alpha] := (\alpha, \frac{\alpha^2}{2}, \frac{\alpha^3}{3}, \ldots) \in \mathbb{C}^\infty$ for $\alpha \in \mathbb{C}$. For a given polynomial $p(t_1, t_2, \ldots)$, the Hirota symbol between functions $f = f(t_1, t_2, \ldots)$ and $g = g(t_1, t_2, \ldots)$ is defined by: $p(\frac{\partial}{\partial t_1}, \frac{\partial}{\partial t_2}, \ldots) f \circ g := p(\frac{\partial}{\partial y_1}, \frac{\partial}{\partial y_2}, \ldots) f(t+y) g(t-y)\Big|_{y=0}$. We also need the elementary Schur polynomials $\mathbf{s}_\ell$, defined by $e^{\sum_1^\infty t_k z^k} := \sum_{k \ge 0} \mathbf{s}_k(t) z^k$ for $\ell \ge 0$ and $\mathbf{s}_\ell(t) = 0$ for $\ell < 0$; moreover, set $\mathbf{s}_\ell(\tilde{\partial}_t) := \mathbf{s}_\ell(\frac{\partial}{\partial t_1}, \frac{1}{2}\frac{\partial}{\partial t_2}, \frac{1}{3}\frac{\partial}{\partial t_3}, \ldots)$. Setting $k = 3$ in formula (10.1.5), one obtains, using $\mathbf{s}_4(t) := \frac{t_1^4}{4!} + \frac{1}{2} t_2 {t_1}^2 + t_3 t_1 + \frac{1}{2} {t_2}^2 + t_4$, the classical KP equation

$$\oint_{z=\infty} \tau_n(t-[z^{-1}])\tau_{m+1}(t'+[z^{-1}])e^{\sum_1^\infty (t_i-t_i')z^i} z^{n-m-1}dz = 0, \quad \textit{for } n \geq m+1. \tag{10.1.4}$$

These bilinear identities imply, in particular, that the functions τ_n *satisfy the KP-hierarchy Hirota bilinear relations*

$$\left(\frac{\partial^2}{\partial t_k \partial t_1} - 2\mathbf{s}_{k+1}(\tilde{\partial}_t)\right)\tau_n \circ \tau_n = 0 \ \textit{for } k = 3, 4, \ldots. \tag{10.1.5}$$

Moreover, the symmetric tridiagonal matrix

$$L(t) := \begin{pmatrix} \frac{\partial}{\partial t_1}\log\frac{\tau_1}{\tau_0} & \left(\frac{\tau_0\tau_2}{\tau_1^2}\right)^{1/2} & 0 & \\ \left(\frac{\tau_0\tau_2}{\tau_1^2}\right)^{1/2} & \frac{\partial}{\partial t_1}\log\frac{\tau_2}{\tau_1} & \left(\frac{\tau_1\tau_3}{\tau_2^2}\right)^{1/2} & \\ 0 & \left(\frac{\tau_1\tau_3}{\tau_2^2}\right)^{1/2} & \frac{\partial}{\partial t_1}\log\frac{\tau_3}{\tau_2} & \\ & & & \ddots \end{pmatrix} \tag{10.1.6}$$

satisfies the commuting standard Toda lattice equations in terms of the Lie algebra splitting $\mathfrak{gl}(n) = \mathfrak{s} \oplus \mathfrak{b}$, *into skew-symmetric matrices and (lower) Borel matrices,*[2] *namely*

$$\frac{\partial L}{\partial t_k} = \left[\frac{1}{2}(L^k)_{\mathfrak{s}}, L\right] = -\left[\frac{1}{2}(L^k)_{\mathfrak{b}}, L\right]. \tag{10.1.7}$$

Proof Shifting t backwards by the $\mathbb{C}^\infty$-vector $[z^{-1}]$ in the function $\tau(t)$, defined in (10.1.2), yields t-dependent monic orthogonal polynomials in z, with regard to this weight $\rho_t(z)$, defined in (10.1.1), and shifting forward by $[z^{-1}]$ their Cauchy transform,

$$z^n\frac{\tau_n(t-[z^{-1}])}{\tau_n(t)} = p_n(z) \quad \text{and} \quad z^{-n-1}\frac{\tau_{n+1}(t+[z^{-1}])}{\tau_n(t)} = \int_{\mathbb{R}} \frac{p_n(x)}{z-x}\rho_t(x)dx. \tag{10.1.8}$$

Then the proof that the functions τ_n satisfy the bilinear identity (10.1.4) is based on the following computation for arbitrary $t, t' \in \mathbb{C}^\infty$,[3]

$$\left(\left(\frac{\partial}{\partial t_1}\right)^4 + 3\left(\frac{\partial}{\partial t_2}\right)^2 - 4\frac{\partial^2}{\partial t_1\partial t_3}\right)\log\tau_n + 6\left(\frac{\partial^2}{\partial t_1^2}\log\tau_n\right)^2 = 0. \tag{10.1.3}$$

[2] As an illustration, for a 2×2 matrix

$$\begin{pmatrix} a & b \\ c & d \end{pmatrix} = \begin{pmatrix} a & b \\ c & d \end{pmatrix}_{\mathfrak{b}} + \begin{pmatrix} a & b \\ c & d \end{pmatrix}_{\mathfrak{s}} := \begin{pmatrix} a & 0 \\ b+c & d \end{pmatrix} + \begin{pmatrix} 0 & b \\ -b & 0 \end{pmatrix}.$$

[3] The equality $\stackrel{*}{=}$ in (10.1.9) below is based on the following property, due to [Adl99]: if $\oint_\infty$ denotes the integral along a small circle about ∞, the following formal power series identity holds, for holomorphic

$$
\begin{aligned}
&\frac{1}{\tau_n(t)\tau_m(t')}\oint_{z=\infty}\tau_n(t-[z^{-1}])\tau_{m+1}(t'+[z^{-1}])e^{\sum_1^\infty(t_i-t_i')z^i}z^{n-m-1}dz\\
&=\oint_{z=\infty}dz\,e^{\sum_1^\infty(t_i-t_i')z^i}p_n(t;z)\int_{\mathbb{R}}\frac{p_m(t';u)}{z-u}e^{\sum_1^\infty t_i'u^i}\rho(u)du,\ \text{using (10.1.8)},\\
&\stackrel{*}{=}2\pi i\int_{\mathbb{R}}e^{\sum_1^\infty(t_i-t_i')z^i}p_n(t;z)\,p_m(t';z)e^{\sum_1^\infty t_i'z^i}\rho(z)dz,\ \text{using Footnote 3},\\
&=2\pi i\int_{\mathbb{R}}p_n(t;z)\,p_m(t';z)e^{\sum_1^\infty t_iz^i}\rho(z)dz=0,\ \text{when } m\le n-1, \qquad (10.1.9)
\end{aligned}
$$

by orthogonality of the polynomials $p_n(t;z)$, establishing (10.1.4). In particular for $m=n-1$, using (10.1.9) and expanding the integrand in $\stackrel{*}{=}$ below in a Laurent series in z, it is a standard computation that (see [Dat83] or [Van10])

$$
\begin{aligned}
0 &= \tau_n(t)\tau_{n-1}(t')\int_{\mathbb{R}}p_n(t;z)\,p_{n-1}(t';z)\rho_t(z)dz\Big|_{\substack{t\mapsto t-y\\ t'\mapsto t+y}}\\
&=\frac{1}{2\pi i}\oint_{z=\infty}dz\,e^{-\sum_1^\infty 2y_iz^i}\tau_n(t-y-[z^{-1}])\tau_n(t+y+[z^{-1}])\\
&\stackrel{*}{=}\frac{1}{2\pi i}\oint dz\,e^{-\sum_1^\infty 2y_iz^i}e^{\sum_1^\infty\frac{z^{-i}}{i}\frac{\partial}{\partial t_i}}e^{\sum_1^\infty y_k\frac{\partial}{\partial t_k}}\tau_n(t)\circ\tau_n(t)\\
&=e^{\sum_1^\infty y_k\frac{\partial}{\partial t_k}}\sum_0^\infty\mathbf{s}_i(-2y)\mathbf{s}_{i+1}(\tilde{\partial}_t)\tau_n\circ\tau_n,\ \text{upon evaluating the residue},\\
&=\left(\frac{\partial}{\partial t_1}+\sum_1^\infty y_k\left[\frac{\partial}{\partial t_k}\frac{\partial}{\partial t_1}-2\mathbf{s}_{k+1}(\tilde{\partial}_t)\right]\right)\tau_n\circ\tau_n+O(y^2), \qquad (10.1.10)
\end{aligned}
$$

for arbitrary y_k; this shows $\tau_n(t)$ satisfies in particular the hierarchy (10.1.5). □

This point of view is very robust: it will be generalized to moment matrices for several weights in Section 10.2.2.

10.1.2 Virasoro constraints

Virasoro constraints are linear PDEs satisfied by τ-functions, obtained by deforming certain matrix integrals. Consider weights $\rho(z)dz=e^{-V(z)}dz$ with a *rational logarithmic derivative,* defined on a fixed subset $F=[A,B]\subset\mathbb{R}$ and a disjoint union of intervals; namely:

$$
-\frac{\rho'}{\rho}=V'(z)=\frac{g}{f}:=\frac{\sum_0^\infty b_iz^i}{\sum_0^\infty a_iz^i}\quad\text{and}\quad E=\bigcup_1^r[c_{2i-1},c_{2i}]\subset F\subseteq\mathbb{R}, \quad (10.1.11)
$$

$f(z)=\sum_{i\ge0}a_iz^i$ and functions $g(z)$, assumed to have all its moments,

$$
\int_{\mathbb{R}}f(u)g(u)du=\frac{1}{2\pi i}\oint_\infty dzf(z)\int_{\mathbb{R}}\frac{g(u)}{z-u}du.
$$

such that $\lim_{z\to A,B} f(z)\rho(z)z^k = 0$ for all $k \geq 0$. Consider an integral for $n > 0$,

$$\int_{E^n} I_n(z)\prod_{k=1}^{n} dz_k \text{ with } I_n(z) =: |\Delta_n(z)|^{2\beta}\prod_{k=1}^{n} e^{-V(z_k)},\ \beta > 0, \tag{10.1.12}$$

with $\Delta_n(z)$ the Vandermonde of $z_1, \ldots, z_n$. Then the following holds:

Theorem 10.1.2 (Adler-van Moerbeke [Adl01c]) *The deformed integrals* $\tau_n(t, E)$

$$\tau_n(t, E) := \int_{E^n} \tilde{I}_n(z)\prod_{k=1}^{n} dz_k \ \textit{with}\ \tilde{I}_n(z) = I_n(z)\prod_{k=1}^{n} e^{\sum_{i=1}^{\infty} t_i z_k^i} \tag{10.1.13}$$

with $\tau_0 = 1$, *satisfy the following Virasoro constraints for all* $k \geq -1$,[4] *with* a_i, b_i, c_i *and* f *as in (10.1.11):*

$$\Big(-\sum_1^{2r} c_i^{k+1} f(c_i)\frac{\partial}{\partial c_i} + \sum_{i\geq 0}\Big(a_i\ \mathbb{J}^{(2)}_{k+i,n}(t,n) - b_i\ \mathbb{J}^{(1)}_{k+i+1,n}(t,n)\Big)\Big)\tau_n(t, E) = 0, \tag{10.1.14}$$

where $\mathbb{J}^{(2)}_{k,n}(t,n)$ *and* $\mathbb{J}^{(1)}_{k,n}(t,n)$ *are combined differential and multiplication (linear) operators. For each* $n \in \mathbb{Z}$, *the operators* $\mathbb{J}^{(2)}_{k,n}(t,n)$ *form a Virasoro algebra, with "central charge"* $c = 1 - 6\left(\beta^{1/2} - \beta^{-1/2}\right)^2$ *and* $\mathbb{J}^{(1)}_{k,n}(t,n)$ *a Heisenberg algebra, with a commutator involving the parameter* $c' = \frac{1}{2}(\beta^{-1} - 1)$,

$$\begin{aligned}\left[\mathbb{J}^{(2)}_{k,n}, \mathbb{J}^{(2)}_{\ell,n}\right] &= (k-\ell)\,\mathbb{J}^{(2)}_{k+\ell,n} + c\left(\frac{k^3-k}{12}\right)\delta_{k,-\ell} \\ \left[\mathbb{J}^{(2)}_{k,n}, \mathbb{J}^{(1)}_{\ell,n}\right] &= -\ell\ \mathbb{J}^{(1)}_{k+\ell,n} + c'k(k+1)\delta_{k,-\ell}, \quad \left[\mathbb{J}^{(1)}_{k,n}, \mathbb{J}^{(1)}_{\ell,n}\right] = \frac{k}{2\beta}\delta_{k,-\ell}.\end{aligned} \tag{10.1.15}$$

Remark 1: The operators $\mathbb{J}^{(2)}_{k,n} = \mathbb{J}^{(2)}_{k,n}(t,n)$, depending explicitly on n and the exponent β of the Vandermonde in (10.1.12), are defined as follows:

$$\begin{aligned}\mathbb{J}^{(2)}_{k,n}(t,n) &= \beta J^{(2)}_k + \Big(2n\beta + (k+1)(1-\beta)\Big)J^{(1)}_k + n\Big(n\beta + 1 - \beta\Big)J^{(0)}_k, \\ \mathbb{J}^{(1)}_{k,n}(t,n) &= J^{(1)}_k + nJ^{(0)}_k,\end{aligned}$$

$$\begin{aligned}\text{where}\quad J^{(0)}_k &= \delta_{k0}, \qquad J^{(1)}_k = \frac{\partial}{\partial t_k} + \frac{1}{2\beta}(-k)t_{-k} \\ J^{(2)}_k &= \sum_{i+j=k}\frac{\partial^2}{\partial t_i\partial t_j} + \frac{1}{\beta}\sum_{-i+j=k} it_i\frac{\partial}{\partial t_j} + \frac{1}{4\beta^2}\sum_{-i-j=k} it_i jt_j.\end{aligned} \tag{10.1.16}$$

Proof From the fundamental theorem of integration, one observes

$$\sum_1^{2r} c_i^{k+1}\frac{\partial}{\partial c_i}\int_{E^n}\tilde{I}_n(z)\prod_1^n dz_i = \int_{E^n}\left(\sum_{i=1}^{n}\frac{\partial}{\partial z_i}z_i^{k+1}\tilde{I}_n(z)\right)\prod_1^n dz_i. \tag{10.1.17}$$

[4] When E equals the whole range F, then the the first term, containing the partials with respect to the c_is, is absent in the formulae (10.1.14).

The sum of z_i-partials in the integrand above can further be expressed[5] as sums of derivatives in the deformation parameters t_i, implying (10.1.14). □

10.1.3 The spectrum of random matrices: combining the KP equation with Virasoro constraints

General method For $\beta = 1$, the time-deformed integral (10.1.13) can be expressed as the determinant of the corresponding moment matrix:

$$\int_{E^n} \tilde{I}_n(z) \prod_{k=1}^{n} dz_k = n! \det\Big(\int_{\mathbb{R}} z^{i+j} e^{\sum_1^\infty t_k z^k} \rho(z) dz\Big)_{0 \le i,j \le n-1},$$

which, by Theorems 10.1.1 and 10.1.2, satisfies the KP equation from the moment matrix representation, and Virasoro constraints from the integral representation. Upon division by $\tau(t, E)$, the Virasoro constraints (10.1.14) take on the form $\partial_k \log \tau = \mathcal{T}_k \log \tau$, where ∂_k denotes the differential operators involving the boundary points of the set E and $\mathcal{T}_k$ the t_i-differential operators in (10.1.14). Then $\partial_\ell \partial_k \log \tau = \partial_\ell \mathcal{T}_k \log \tau = \mathcal{T}_k \partial_\ell \log \tau = \mathcal{T}_k \mathcal{T}_\ell \log \tau$, using the fact that $\mathcal{T}_k$ and ∂_ℓ commute, since they involve differentiations with respect to different variables. One then proceeds with $\partial_m \partial_\ell \partial_k \log \tau$, etc. . . Then for *appropriate choices* of $V(z)$, the following miracle occurs: setting all $t_i = 0$ enables one to express all t-partials appearing in the KP equation (10.1.5) in terms of $\partial_m \log \tau$, which in the end implies a PDE for $\log \tau(0, E)$ in terms of the boundary points of E; when $E = (-\infty, c)$, the PDE will become an ODE in c. In all practical applications, as seen below, one is interested in the *log of the probability* $\log(\tau(0, E)/\tau(0, \mathbb{R})) = \log \tau(0, E) - \log \tau(0, \mathbb{R})$. Since $\log \tau(0, \mathbb{R})$ is independent of E, this method yields a PDE for $\log(\tau(0, E)/\tau(0, \mathbb{R}))$. Consider the following examples:

Gaussian ensemble (GUE) The probability that the n eigenvalues z_i of the $n \times n$ Hermitian matrices belong to $E = \bigcup_1^r [c_{2i-1}, c_{2i}] \subset \mathbb{R}$ for the probability measure $Z_n^{-1} e^{-\mathrm{Tr} M^2} dM$ and Haar measure dM,

$$\mathbb{P}_n := \mathbb{P}_n(\text{all } z_i \in E) = \frac{1}{Z_n} \int_{E^n} \Delta_n^2(z) \prod_1^n e^{-z_i^2} dz_i$$

satisfies a PDE, using the operators $\partial_k := \sum_1^{2r} c_i^{k+1} \frac{\partial}{\partial c_i}$ for $E = \bigcup_1^r [c_{2i-1}, c_{2i}]$,

$$(\partial_{-1}^4 + 8n\partial_{-1}^2 + 12\partial_0^2 + 24\partial_0 - 16\partial_{-1}\partial_1) \log \mathbb{P}_n + 6(\partial_{-1}^2 \log \mathbb{P}_n)^2 = 0; \qquad (10.1.18)$$

see [Adl98] and [Adl01c]. This follows from Theorems 10.1.1 and 10.1.2, applied to $V(z) = z^2$. Thus, in particular, $f(x) := \frac{d}{dx} \log \mathbb{P}_n(\max_i z_i \le x)$ satisfies a third order ODE of the Chazy form, which can be transformed to Painlevé IV; see [Tra94], [Adl01c], [Van10].

[5] Also using $\sum_{i=1}^n \frac{\partial}{\partial z_i}(z_i f) = nf + \sum_{i=1}^n z_i \frac{\partial f}{\partial z_i}$, together with the Vandermonde determinant properties, one checks that $\sum_{i=1}^n \frac{\partial}{\partial z_i} \Delta_n(z) = 0$ and $\sum_{i=1}^n z_i \frac{\partial}{\partial z_i} \Delta_n(z) = \frac{n(n-1)}{2} \Delta_n(z)$, for $n \ge 1$.

Laguerre ensemble The probability $Z_n^{-1}(\det M)^a e^{-\mathrm{Tr} M} dM$ on Hermitian matrices leads to the following probability for the eigenvalues (here $e^{-V} = z^a e^{-z}$):

$$\mathbb{P}_n := \mathbb{P}_n(\text{all } z_i \in E) = \frac{1}{Z_n} \int_{E^n} \Delta_n^2(z) \prod_1^n z_i^a e^{-z_i} dz_i,$$

which satisfies a fourth order PDE ([Adl01c] and [Van10]), upon defining a slightly different operator $\partial_k := \sum_1^{2r} c_i^{k+2} \frac{\partial}{\partial c_i}$:

$$\begin{pmatrix} \partial_{-1}^4 - 2\partial_{-1}^3 + (1-a^2)\partial_{-1}^2 - 4\partial_1\partial_{-1} + 3\partial_0^2 \\ +2(2n+a)\partial_0\partial_{-1} - 2\partial_1 - (2n+a)\partial_0 \end{pmatrix} \log \mathbb{P}_n$$
$$+6(\partial_{-1}^2 \log \mathbb{P}_n)^2 - 4(\partial_{-1}^2 \log \mathbb{P}_n)(\partial_{-1} \log \mathbb{P}_n) = 0. \tag{10.1.19}$$

In particular, $f(x) := x\frac{d}{dx} \log \mathbb{P}_n(\max_i z_i \leq x)$ satisfies a third order ODE of the Chazy type, which can be transformed to Painlevé V. ([Tra94], [Adl01c] and [Van10])

Jacobi ensemble Here the integral

$$\mathbb{P}_n := \mathbb{P}_n(\text{all } z_i \in E) = \frac{1}{Z_n} \int_{E^n} \Delta_n^2(z) \prod_1^n (1-z_i)^a (1+z_i)^b dz_i$$

satisfies a fourth order PDE, as well; then $f(x) = (1-x^2)\frac{d}{dx} \log \mathbb{P}_n(\max_i z_i \leq x)$ satisfies a third order Chazy ODE transformable to Painlevé VI. (see Haine-Sémengué [Hai99] and also [Adl01c, Van10]). For Chazy equations transformable to Painlevé, see [Cos00].

Further applications These methods and results also apply to the $\beta = 1/2$ and 2 ensembles, as also done in Adler-van Moerbeke [Adl01c]. These methods have further been extended, by Adler-van Moerbeke [Adl01b], Adler-Borodin-van Moerbeke [Adl07a], to weights $\rho(z)$ on an interval $F \subseteq \mathbb{R}$, with rational logarithmic derivative of the form

$$-\frac{\rho'(z)}{\rho(z)} = \frac{g}{f} := \frac{b_0 + b_1 z + b_2 z^2}{a_0 + a_1 z + a_2 z^2} =: \frac{B(z)}{A(z)}, \tag{10.1.20}$$

with boundary condition $f(z)\rho(z)z^k\big|_{\partial F} = 0$, for all $k = 0, 1, 2, \ldots$. The expression[6]

$$G(x) := \frac{\partial}{\partial x} \log \int_{F^n} \Delta^2(z) \prod_1^n e^{xz_k} \rho(z_k) dz_k \tag{10.1.21}$$

satisfies, as a function of x, a third order ordinary differential equation, which can be transformed into Painlevé V. The latter also comes up for integrals over the unitary group, integrals of the Grassmannian $\mathrm{Gr}(p, \mathbb{C}^n)$ of p-planes in $\mathbb{C}^n$

[6] When E is a finite interval, the integral always converges, but infinite intervals require some care.

$$\int_{\mathrm{U}(n)} dM\, e^{\mathrm{Tr}\sqrt{x}(M+\bar{M})}, \quad \int_{\mathrm{Gr}(p,\,\mathbb{C}^n)} e^{x\mathrm{Tr}(I+M^\dagger M)^{-1}} \det(M^\dagger M)^{p-q} dM,$$

and others; here dM is the Haar measure on $\mathrm{U}(n)$ and $\mathrm{Gr}(p, \mathbb{C}^n)$ respectively; see also Haine-Vanderstichelen [Hai10]. All such integrals make their appearance in combinatorial problems; see [Adl01a], [Adl01b] and [Adl07a]. Also, as a special case of (10.1.21), certain integrals over the space $\mathcal{H}_n$ of Hermitian matrices with a spectrum in $(-\infty, x]$ or $[0, x]$ relate to Painlevé IV and V equations: ([Adl07a] and [For02])

$$\int_{\mathcal{H}_n(-\infty,x]} \det(M - xI)^a e^{-\mathrm{Tr}M^2} dM, \quad \int_{\mathcal{H}_n[0,x]} \det(xI - M)^b \det M^a e^{-\mathrm{Tr}M} dM.$$

In relation to this, see the work of Igor Rumanov [Rum09, Rum08] and also the recent work of Osipov and Kanzieper [Osi10] on RMT characteristic polynomials.

10.2 Multiple orthogonal polynomials

10.2.1 Non-intersecting Brownian motions on $\mathbb{R}$

Consider N non-intersecting Brownian motions $x_1(t) < x_2(t) < \ldots < x_N(t)$ on $\mathbb{R}$ (Dyson's Brownian motions), all starting at source points $\gamma_1 < \gamma_2 < \cdots < \gamma_N$ at time $t = 0$ and forced to target points $\delta_1 < \delta_2 < \cdots < \delta_N$ at $t = 1$. According to the Karlin-McGregor formula [Kar59], the probability[7] $\mathbb{P}_{\mathrm{Br}}(t, \tilde{\mathbb{E}})$, that the N particles pass through the subsets $\tilde{E}_1, \tilde{E}_2, \ldots, \tilde{E}_m \subset \mathbb{R}$ respectively at times $0 < t_1 < t_2 < \cdots < t_m < 1$ is given by a multiple integral of a product of m determinants, involving the standard Brownian transition probability $p(t, x, y) := \frac{1}{\sqrt{\pi t}} e^{-\frac{(y-x)^2}{t}}$. When some source points and some target points coincide, one must take limits; see [Joh01, Ble07, Adl07b]. In this section, we consider the situation where *the source points all coincide with* 0, while some target points may coincide.

Consider thus $N = n_1 + n_2 + \cdots + n_p$ non-intersecting Brownian motions starting from the origin at $t = 0$, with n_i particles forced to reach p distinct target points β_i at time $t = 1$, with $\beta_1 < \beta_2 < \cdots < \beta_p$ in $\mathbb{R}$, as in Figure 10.1 above. As is well known, (see [Zin98, Joh01, Ble07, Dae07, Tra06, Adl07b]), the probability $\mathbb{P}_{\mathrm{Br}}(t, \tilde{\mathbb{E}})$ can also be viewed, after some change of variables, as the probability for the eigenvalues of a chain of m coupled Hermitian random matrices:

[7] $\tilde{\mathbb{E}} = \tilde{E}_1 \times \ldots \times \tilde{E}_m$.

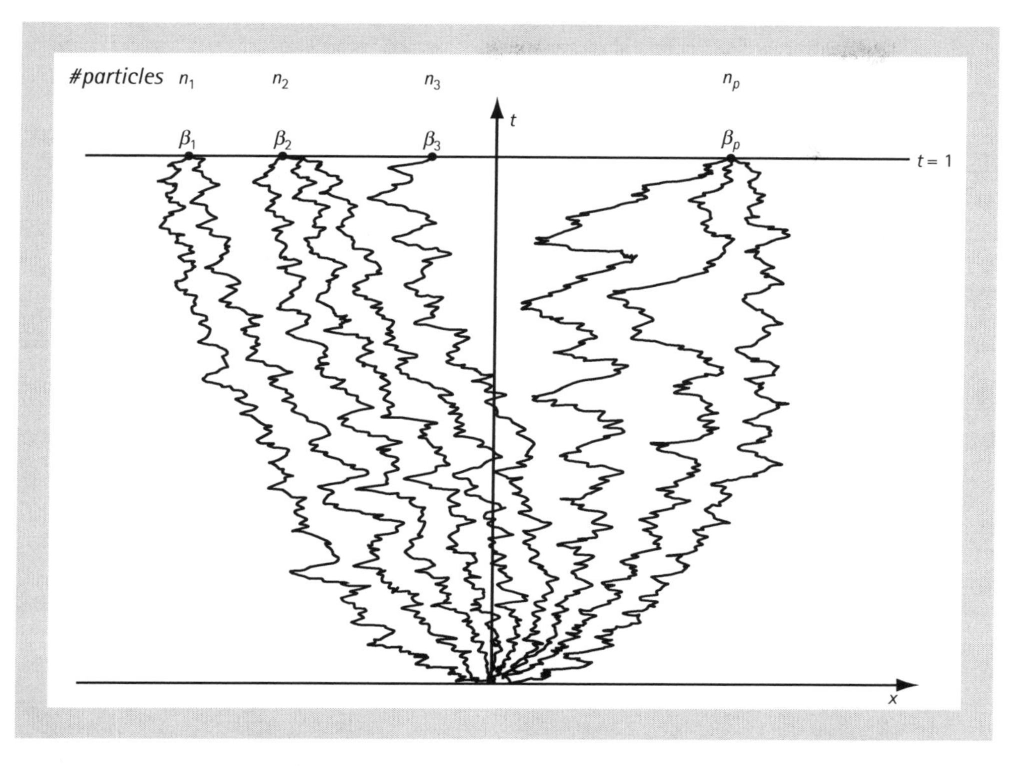

Fig. 10.1 Non-intersecting Brownian motions.

$$
\begin{aligned}
\mathbb{P}_{\text{Br}}(t, \tilde{\mathbb{E}}) &:= \mathbb{P}\left(\bigcap_{k=1}^{m} \left\{ \text{all } x_i(t_k) \in \tilde{E}_k \right\} \middle| \begin{array}{l} \text{all } x_i(0) = 0; \\ n_j \text{ paths end up at } \beta_j \text{ at } t = 1, \\ \text{for } 1 \leqslant j \leqslant p \end{array} \right) \\
&= \frac{1}{Z_n'''} \int_{\mathbb{E}^N} I_n(v) \prod_{k=1}^{m} dv_k \\
&= \frac{1}{Z_n'''} \det \begin{pmatrix} \left(\langle y^i e^{-\frac{1}{2}y^2 + b_1 y} \mid x^j e^{-\frac{1}{2}x^2} \rangle \right)_{\substack{0 \leqslant i < n_1 \\ 0 \leqslant j < N}} \\ \vdots \\ \left(\langle y^i e^{-\frac{1}{2}y^2 + b_p y} \mid x^j e^{-\frac{1}{2}x^2} \rangle \right)_{\substack{0 \leqslant i < n_p \\ 0 \leqslant j < N}} \end{pmatrix} \quad \begin{pmatrix} \text{matrix of} \\ p \text{ blocks} \end{pmatrix} \\
&= \frac{1}{\tilde{Z}_n} \int_{\operatorname{spec}(M_k) \in E_k} e^{-\frac{1}{2}\operatorname{Tr}\left(\sum_{k=1}^{m} M_k^2 - 2 \sum_{k=1}^{m-1} c_k M_k M_{k+1} - 2 A M_m \right)} \prod_{k=1}^{m} dM_k \\
&=: \mathbb{P}_n^A(c, \mathbb{E}),
\end{aligned}
\tag{10.2.22}
$$

where A is a diagonal matrix, defined in (10.2.25) below, and where

$$I_n(v) := \Delta_N(v_1)\prod_{\ell=1}^{p}\left(\Delta_{n_\ell}(v_m^{(\ell)})\prod_{i=1}^{n_\ell} e^{-\sum_{k=1}^{m}\frac{1}{2}v_{k;i}^{(\ell)2}+\sum_{k=1}^{m-1}c_k v_{k;i}^{(\ell)}v_{k+1;i}^{(\ell)}+b_\ell v_{m;i}^{(\ell)}}\right). \tag{10.2.23}$$

The inner-product[8] in (10.2.22) is given by (upon setting $w_1 := x$ and $w_m := y$):

$$\langle f \mid g\rangle := \iint_{E_1\times E_m} dx\, dy f(y)g(x)\left(\int_{\prod_{k=2}^{m-1}E_k}\prod_{k=1}^{m-1}e^{c_k w_k w_{k+1}}\prod_{k=2}^{m-1}e^{-\frac{1}{2}w_k^2}dw_k\right). \tag{10.2.24}$$

The integrand (10.2.23) above uses the following notation: for $k = 1, \ldots, m$ and for $\ell = 1, \ldots, p$, one defines the N-vector v_k and renames the variables according to the groups of size n_1, n_2,..., etc.

$$v_k := (v_{k;1}, \ldots, v_{k;N}) \;:=\; (v_{k;1}^{(1)}, \ldots, v_{k;n_1}^{(1)}, v_{k;1}^{(2)}, \ldots, v_{k;n_2}^{(2)}, \ldots, v_{k;1}^{(p)}, \ldots, v_{k;n_p}^{(p)}).$$

The change of variables in (10.2.22) is given by the following formulae[9], setting $t_0 := 0$ and $t_{m+1} := 1$,

$$A := \mathrm{diag}(\overbrace{b_1, \ldots, b_1}^{n_1}, \overbrace{b_2, \ldots, b_2}^{n_2}, \ldots, \overbrace{b_p, \ldots, b_p}^{n_p}), \text{ with } b_\ell = \sqrt{\frac{2(t_m - t_{m-1})}{(1-t_m)(1-t_{m-1})}}\beta_\ell$$

$$E_k := \tilde{E}_k\sqrt{\frac{2(t_{k+1} - t_{k-1})}{(t_k - t_{k-1})(t_{k+1} - t_k)}}, \quad c_k^2 := \frac{(t_{k+2} - t_{k+1})(t_k - t_{k-1})}{(t_{k+2} - t_k)(t_{k+1} - t_{k-1})}, \tag{10.2.25}$$

for $\ell = 1, \ldots, p$ and $k = 1, \ldots, m$. Finally, it is quite natural, as seen in the examples, to impose a *linear constraint on the target points* $\beta_1, \ldots, \beta_p$, namely

$$\sum_{\ell=1}^{p}\kappa_\ell\beta_\ell = 0, \text{ with } \sum_{\ell=0}^{p}\kappa_\ell = 0, \text{ setting } \kappa_0 := -1. \tag{10.2.26}$$

10.2.2 A moment matrix for several weights

Define two sets of weights $\psi_1(x), \ldots, \psi_q(x)$ and $\varphi_1(y), \ldots, \varphi_p(y)$, with $x, y \in \mathbb{R}$, and deform each of the weights $\psi_\alpha(x)$ by means of (formal) *time* parameters $s_\alpha = (s_{\alpha 1}, s_{\alpha 2}, \ldots)$ $(1 \le \alpha \le q)$ and each of the weights $\varphi_\beta(y)$ by means of the parameters $t_\beta = (t_{\beta 1}, t_{\beta 2}, \ldots)$ $(1 \le \beta \le p)$. That is, each weight goes with its own set of time-deformation variables. For each set of positive integers[10]

[8] For $m = 2$ the measure in brackets in formula (10.2.24) should be interpreted as $= 1$, and for $m = 1$ that measure $:= \delta(x - y)e^{x^2/2}$ (the delta distribution).

[9] For $m = 1$, the matrix integral above becomes a one-matrix integral with external potential. The change of variables below then becomes quite simple: $b_\ell = \sqrt{\frac{2t}{1-t}}\beta_\ell$, $E = \tilde{E}\sqrt{\frac{2}{t(1-t)}}$.

[10] $|m| = \sum_{\alpha=1}^{q} m_\alpha$ and $|n| = \sum_{\beta=1}^{p} n_\beta$.

$m = (m_1, \ldots, m_q)$, $\quad n = (n_1, \ldots, n_p)$ with $|m| = |n|$, consider the $m_\alpha \times n_\beta$ moment matrices with regard to a (not necessarily symmetric) inner product[11] $\langle \cdot \mid \cdot \rangle$

$$T^{\alpha\beta}_{mn} := \left(\left\langle x^i \psi_\alpha(x) e^{-\sum_{k=1}^{\infty} s_{\alpha k} x^k} \;\middle|\; y^j \varphi_\beta(y) e^{\sum_{k=1}^{\infty} t_{\beta k} y^k} \right\rangle \right)_{\substack{0 \le i < m_\alpha \\ 0 \le j < n_\beta}}.$$

Consider now the analogue of (10.1.2), namely the determinant of a moment matrix T_{mn} of size $|m| = |n|$, composed of the pq blocks $T^{\alpha\beta}_{mn}$ of sizes (m_α, n_β), defined above:

$$\tau_{mn}(s_1, \ldots, s_q; t_1, \ldots, t_p) := \det T_{mn} := \det \begin{pmatrix} T^{11}_{mn} & \ldots & T^{1p}_{mn} \\ \vdots & & \vdots \\ T^{q1}_{mn} & \ldots & T^{qp}_{mn} \end{pmatrix}. \tag{10.2.27}$$

Then the functions $\tau_{mn}(s_1, \ldots, s_q; t_1, \ldots, t_p)$ satisfy identities, *which characterize the τ-functions of the $(p+q)$-KP hierarchy* and which generalize (10.1.4).

Proposition 10.2.1 [Adl09b] ***($p+q$-component KP-hierarchy)*** *For arbitrary sets of times t_β, t^*_β, s_α, s^*_α, $(1 \le \beta \le p$ and $1 \le \alpha \le q)$ and for arbitrary sets of positive integers m, n, m^*, n^* such that $|m^*| = |n^*| + 1$ and $|m| = |n| - 1$, the functions $\tau_{mn}(s_1, \ldots, s_q; t_1, \ldots, t_p)$ of (10.2.27) satisfy the bilinear relations*[12]*:*

$$\begin{aligned} &\sum_{\beta=1}^{p} \oint_\infty (-1)^{\sigma_\beta(n)}\, \tau_{m, n-e_\beta}(t_\beta - [z^{-1}])\, \tau_{m^*, n^*+e_\beta}(t^*_\beta + [z^{-1}])\, e^{\sum_1^\infty (t_{\beta k} - t^*_{\beta k}) z^k} z^{n_\beta - n^*_\beta - 2}\, dz \\ &= \sum_{\alpha=1}^{q} \oint_\infty (-1)^{\sigma_\alpha(m)}\, \tau_{m+e_\alpha, n}(s_\alpha - [z^{-1}])\, \tau_{m^*-e_\alpha, n^*}(s^*_\alpha + [z^{-1}])\, e^{\sum_1^\infty (s_{\alpha k} - s^*_{\alpha k}) z^k} z^{m^*_\alpha - m_\alpha - 2} dz. \end{aligned} \tag{10.2.28}$$

The integrals above are taken along a small circle about $z = \infty$ and thus must be interpreted as residue evaluations about $z = \infty$.

Much in analogy with (10.1.5), the functions τ_{mn}, with $|m| = |n|$, satisfy the following PDEs, thus generalizing the KP hierarchy (10.1.5): (remembering Footnote 1)

$$\begin{aligned} \tau^2_{mn} \frac{\partial^2}{\partial t_{k,\ell+1} \partial t_{k',1}} \log \tau_{mn} &= \mathbf{s}_{\ell + 2\delta_{kk'}} (\tilde{\partial}_{t_k}) \tau_{m, n+e_k - e_{k'}} \circ \tau_{m, n+e_{k'} - e_k} \\ \tau^2_{mn} \frac{\partial^2}{\partial s_{k,\ell+1} \partial s_{k',1}} \log \tau_{mn} &= \mathbf{s}_{\ell + 2\delta_{kk'}} (\tilde{\partial}_{s_k}) \tau_{m+e_{k'} - e_k, n} \circ \tau_{m+e_k - e_{k'}, n} \end{aligned} \tag{10.2.29}$$

[11] A typical inner product to keep in mind is $\langle f(x) \mid g(y) \rangle = \iint_{\mathbb{R}^2} f(x) g(y)\, d\mu(x, y)$, where $\mu = \mu(x, y)$ is a fixed measure on $\mathbb{R}^2$, perhaps having support on a line or curve.

[12] Remember the notation $[\alpha] := (\alpha, \frac{\alpha^2}{2}, \frac{\alpha^3}{3}, \ldots)$ for $\alpha \in \mathbb{C}$. So, the function $\tau_{mn}(t_\ell - [z^{-1}])$ means that τ_{mn} still depends on all time parameters, but the variable t_ℓ only gets shifted. Throughout this section, we use the standard notation $e_1 = (1, 0, 0, \ldots)$, $e_2 = (0, 1, 0, \ldots)$, ... Also $\sigma_\alpha(m) = \sum_{\alpha'=1}^{\alpha} (m_{\alpha'} - m^*_{\alpha'})$ and $\sigma_\beta(n) = \sum_{\beta'=1}^{\beta} (n_{\beta'} - n^*_{\beta'})$.

and similar relations with mixed derivatives $\frac{\partial^2}{\partial s_{k,1}\partial t_{k',\ell+1}}\log\tau_{mn}$ *and* $\frac{\partial^2}{\partial t_{k',1}\partial s_{k,\ell+1}}\log\tau_{mn}$. *Setting* $k=k'$ *yields the KP equation for each* τ_{mn}, *the same as (10.1.5). Combining several such relations (10.2.29) leads to identities for a single function* τ_{mn}:

$$\frac{\partial}{\partial t_{k',1}}\left(\frac{\frac{\partial^2}{\partial t_{k,2}\partial t_{k',1}}\log\tau_{mn}}{\frac{\partial^2}{\partial t_{k,1}\partial t_{k',1}}\log\tau_{mn}}\right)+\frac{\partial}{\partial t_{k,1}}\left(\frac{\frac{\partial^2}{\partial t_{k',2}\partial t_{k,1}}\log\tau_{mn}}{\frac{\partial^2}{\partial t_{k',1}\partial t_{k,1}}\log\tau_{mn}}\right)=0. \qquad (10.2.30)$$

Proof The proof is a vast generalization of the proof of Theorem 10.1.1. Here one uses multiple orthogonal polynomials and their Cauchy transforms; multiple orthogonal polynomials are orthogonal polynomials with regard to several weights; see [Adl99, Adl09b, Apt03, Van01, Ble07]. In the identity (10.2.28), there appear *four kinds* of shifted τ-functions: the expressions $z^{n_\beta}\tau_{m,n}(t_\beta-[z^{-1}])$ are actually polynomials in z and provide sequences of multiple orthogonal polynomials by modifying the integers in the indices $n=(n_1,\dots,n_p)$; the expressions $z^{m_\alpha}\tau_{m,n}(s_\alpha+[z^{-1}])$ are polynomial in z as well and provide another sequence of multiple orthogonal polynomials. For precise statements; see [Ald09b]. Then $z^{-n_\beta}\tau_{m,n}(t_\beta+[z^{-1}])$ is a Cauchy transform,

$$z^{-n_\beta}\frac{\tau_{m,n}(t_\beta+[z^{-1}])}{\tau_{m,n}}=\left\langle\sum_{\alpha=1}^{q}\Box_\alpha\,\psi_\alpha(x)e^{-\sum_{k=1}^{\infty}s_{\alpha k}x^k}\;\middle|\;\frac{\varphi_\beta(y)e^{\sum_{k=1}^{\infty}t_{\beta k}y^k}}{z-y}\right\rangle$$

with $\Box_\alpha$ referring to polynomials in x, essentially given by $x^{m_\alpha}t_{m,n}(s_\alpha+[x^{-1}])$ above and where $\psi_\alpha(x)$ and $\varphi_\beta(y)$ are the weights above; similarly for $\tau_{m,n}(s_\alpha-[z^{-1}])$. This is to say that the integrals in (10.2.28) are essentially residues of products of polynomials with the Cauchy transform of other polynomials; all this much in analogy with (10.1.8) and (10.1.9). A procedure, similar to (10.1.9), but applied to (10.2.28), leads to (10.2.29). Then making the ratio of the two relations obtained leads to PDEs (10.2.30) in a single τ_{mn}. □

It is important to point out that the identities (10.2.28) can be encapsulated as matrix identities, involving square matrices $W_{mn}(z)$ and $W^*_{m^*n^*}(z)$ of size $p+q$, which satisfy the bilinear identities ($W^*_{m^*n^*}(z)$ turns out to be the inverse transpose of $W_{mn}(z)$),

$$\oint_\infty W_{mn}(z;s,t)\,W^*_{m^*n^*}(z;s^*,t^*)^\top dz=0, \qquad (10.2.31)$$

for all m,n,m^*,n^* such that $|m|=|n|$, $|m^*|=|n^*|$ and all $s,t,s^*,t^*\in\mathbb{C}^\infty$. The two left blocks of W_{mn} and the two right blocks of W^*_{mn} are mixed multiple orthogonal polynomials, and the remaining blocks are Cauchy transforms of such polynomials. It is remarkable that, upon setting all t and s parameters equal to zero, the matrix $W_{mn}(z)$ is precisely the *Riemann–Hilbert* matrix characterizing the multiple orthogonal polynomials! The Riemann–Hilbert matrix for the multiple-orthogonal polynomials has been defined in Daems-Kuijlaars-Veys

[Dae07], which is a generalization of the Riemann–Hilbert matrix of Fokas-Its-Kitaev [Fok92] and Deift-Its-Zhou [Dei97].

Return now to the **Dyson Brownian motions** governed by transition probability (10.2.22), and in particular focus on its block moment matrix representation. Upon adding variables in the inner-products $\langle y^i e^{-\frac{1}{2}y^2+b_\ell y} \mid x^j e^{-\frac{1}{2}x^2}\rangle$ in the following way:

$$\begin{pmatrix} \left(\left\langle y^i \exp(-\frac{y^2}{2} + b_1 y + b_1^{(2)} y^2 - \sum_{r\geq 1} s_r^{(1)} y^r) \middle| x^j \exp(-\frac{x^2}{2} + \sum_{r\geq 1} s_r^{(0)} x^r)\right\rangle\right)_{\substack{0\leq i<n_1\\ 0\leq j<N}} \\ \vdots \\ \left(\left\langle y^i \exp(-\frac{y^2}{2} + b_p y + b_p^{(2)} y^2 - \sum_{r\geq 1} s_r^{(p)} y^r) \middle| x^j \exp(-\frac{x^2}{2} + \sum_{r\geq 1} s_r^{(0)} x^r)\right\rangle\right)_{\substack{0\leq i<n_p\\ 0\leq j<N}} \end{pmatrix}, \tag{10.2.32}$$

it follows from Proposition 10.2.1 that the determinant of this matrix satisfies the $(p+1)$-KP hierarchy in the variables $s_r^{(\ell)}$ and $s_r^{(0)}$, and, in particular, the equations (10.2.29); this means that $\tau_n := \tau_{n_1,\ldots,n_p} :=$ det(matrix (10.2.32)) satisfies:

$$\frac{\partial}{\partial s_1^{(0)}} \log \frac{\tau_{n+e_\ell}}{\tau_{n-e_\ell}} = \frac{\frac{\partial^2}{\partial s_2^{(0)}\partial s_1^{(\ell)}} \log \tau_n}{\frac{\partial^2}{\partial s_1^{(0)}\partial s_1^{(\ell)}} \log \tau_n}, \quad \frac{\partial}{\partial s_1^{(\ell)}} \log \frac{\tau_{n+e_\ell}}{\tau_{n-e_\ell}} = -\frac{\frac{\partial^2}{\partial s_1^{(0)}\partial s_2^{(\ell)}} \log \tau_n}{\frac{\partial^2}{\partial s_1^{(0)}\partial s_1^{(\ell)}} \log \tau_n}, \tag{10.2.33}$$

where $n \pm e_\ell = (n_1, \ldots, n_p) \pm e_\ell := (n_1, \ldots, n_{\ell-1}, n_\ell \pm 1, n_{\ell+1}, \ldots, n_p)$.

10.2.3 Virasoro constraints

In (10.2.22), one considered the probability $\mathbb{P}_n^A(c, \mathbb{E})$ describing a chain of m Hermitian matrix integrals, with coupling terms $c_1, \ldots, c_{m-1}$ and with external potential, given by a diagonal matrix

$$A := \operatorname{diag}(\overbrace{b_1, \ldots, b_1}^{n_1}, \overbrace{b_2, \ldots, b_2}^{n_2}, \ldots, \overbrace{b_p, \ldots, b_p}^{n_p}), \text{ with } \begin{cases} \sum_1^p \kappa_\ell b_\ell = 0 \\ \sum_1^p \kappa_\ell = 1 \end{cases}, \tag{10.2.34}$$

and integrated over m subsets $E_1, \ldots, E_m \subset \mathbb{R}$, with $\mathbb{E} := E_1 \times \cdots \times E_m$. Replace the integrand $I_n(v)$ in the integral representation (10.2.22) by $\tilde{I}_n(v)$, given by

$$\tilde{I}_n(v) = I_n(v) \times$$

$$\prod_{\ell=1}^{p} \prod_{i=1}^{n_\ell} e^{b_\ell^{(2)} {v_{m;i}^{(\ell)}}^2 + \sum_{r\geqslant 1}(s_r^{(0)}(v_{1;i}^{(\ell)})^r - s_r^{(\ell)}(v_{m;i}^{(\ell)})^r) + \sum_{k=1}^{m-1} \sum_{(r,q)>(1,1)} c_{rq}^{(k)} (v_{k;i}^{(\ell)})^r (v_{k+1;i}^{(\ell)})^q + \sum_{k=2}^{m-1} \sum_{r\geqslant 1} \gamma_r^{(k)} (v_{k;i}^{(\ell)})^r},$$

much in tune with the extra-variables in (10.2.32); it is convenient here to add further additional variables $c_{rq}^{(k)}$ and $\gamma_r^{(k)}$. Then (define $b_\ell^{(1)} = b_\ell$)

$$\tau_n(\mathbb{E}) := \int_{\mathbb{E}^N} \tilde{I}_n(v)\Big|_{\sum_1^p \kappa_\ell b_\ell^{(1,2)}=0} \prod_{k=1}^m dv_k \tag{10.2.35}$$

satisfies Virasoro constraints, structurally analogous to (10.1.14), but only more complicated; for details, see [Adl10a].

The method to obtain those Virasoro constraints is quite analogous to formulae (10.1.17) upon using the constraints $\sum_1^p \kappa_\ell b_\ell^{(1,2)} = 0$ and the identities in Footnote 5. The point is that just enough deformation parameters have been added in order to express the partials in the deformation parameters in terms of partials in the end points of the intervals, the couplings $c_1, \ldots, c_{m-1}$ and the entries b_ℓ of the external potential A.

10.2.4 A PDE for non-intersecting Brownian motions

Consider a disjoint union of intervals $E = \cup_{i=1}^r [z_{2i-1}, z_{2i}]$, the diagonal matrix A as in (10.2.34) and couplings $c_1, \ldots, c_{m-1}$. Then define the following:

(i) The inverse of the following Jacobi matrix will play an important role:

$$\mathcal{J} := \begin{pmatrix} -1 & c_1 & & & \\ & & \ddots & 0 & \\ c_1 & -1 & & \ddots & \\ & \ddots & \ddots & \ddots & \\ & \ddots & -1 & & c_{m-1} \\ 0 & & \ddots & & \\ & & c_{m-1} & & -1 \end{pmatrix}^{-1}. \tag{10.2.36}$$

(ii) For any given vector $u = (u_1, \ldots, u_a)$, set

$$\partial_u := \sum_{i=1}^a \frac{\partial}{\partial u_i}, \quad \varepsilon_u := \sum_{i=1}^a u_i \frac{\partial}{\partial u_i}. \tag{10.2.37}$$

In particular, given the union of intervals E as above, denote by

$$\begin{aligned} \partial_E &:= \left\{ \begin{matrix} \text{sum of partials in the} \\ \text{boundary points of } E \end{matrix} \right\} = \textstyle\sum_{i=1}^{2r} \frac{\partial}{\partial z_i} \\ \varepsilon_E &:= \left\{ \begin{matrix} \text{Euler operator in the} \\ \text{boundary points of } E \end{matrix} \right\} = \textstyle\sum_{i=1}^{2r} z_i \frac{\partial}{\partial z_i}. \end{aligned} \tag{10.2.38}$$

(iii) In view of Theorem 10.2.2 below, given the entries $b = (b_1, \dots, b_{p-1}, b_p)$ of the matrix A, as in (10.2.34), subjected to the linear relations $\sum_1^p \kappa_\ell b_\ell = 0$, $\sum_1^p \kappa_\ell = 1$ and subsets E_i, define the linear differential operators:

$$
\begin{aligned}
\partial_b^{(\ell)} &:= \sum_{i=1}^{p-1} (\kappa_\ell - \delta_{\ell,i}) \frac{\partial}{\partial b_i}, \quad \partial_b^{(0)} := 0, \quad \text{implying} \sum_{\ell=1}^{p} \partial_b^{(\ell)} = 0, \\
\partial_\ell &:= \partial_b^{(\ell)} - \kappa_\ell \sum_{i=1}^{m} \partial_{E_i} \times \begin{cases} \mathcal{J}_{1i} & \text{for } \ell = 0, \\ \mathcal{J}_{mi} & \text{for } 1 \leqslant \ell \leqslant p, \end{cases} \\
\varepsilon_b &:= \sum_{1}^{p-1} b_i \frac{\partial}{\partial b_i}, \quad \varepsilon_0 := \varepsilon_{E_1} - \delta_{1,m}\varepsilon_b - c_1 \frac{\partial}{\partial c_1}, \\
\varepsilon_m &:= \varepsilon_{E_m} - \varepsilon_b - c_{m-1} \frac{\partial}{\partial c_{m-1}}.
\end{aligned}
\tag{10.2.39}
$$

Theorem 10.2.2 *[Adl10a] The probability, as in (10.2.22),*

$$
\mathbb{P}_n := \mathbb{P}_n^A(c, \mathbb{E}) = \frac{1}{Z_n} \int_{\operatorname{spec}(M_k) \in E_k} e^{-\frac{1}{2}\operatorname{Tr}\left(\sum_{k=1}^{m} M_k^2 - 2\sum_{k=1}^{m-1} c_k M_k M_{k+1} - 2AM_m\right)} \prod_{k=1}^{m} dM_k,
$$

with the linear constraint $\sum_1^p \kappa_\ell b_\ell = 0$ on the entries of the diagonal matrix A, satisfies a non-linear PDE in the boundary points of the subsets $E_1, \dots, E_m$ and in the target points $b_1, \dots, b_p$; it is given by the determinant of a $(p+1) \times (p+1)$ matrix, nearly a Wronskian, for the operator $' := \partial_0 = \sum_1^m \mathcal{J}_{1i} \partial_{E_i}$, for J as in (10.2.36)

$$
\det \begin{pmatrix}
F_1 & F_2 & F_3 & \dots & F_p & G_0 \\
F_1' & F_2' & F_3' & \dots & F_p' & G_1 \\
F_1'' & F_2'' & F_3'' & \dots & F_p'' & G_2 \\
\vdots & \vdots & \vdots & & \vdots & \vdots \\
F_1^{(p)} & F_2^{(p)} & F_3^{(p)} & \dots & F_p^{(p)} & G_p
\end{pmatrix} = 0,
\tag{10.2.40}
$$

where the F_ℓ and G_ℓ are given by $(1 \le \ell \le p)$

$$
\begin{aligned}
F_\ell &= -\partial_0 \partial_\ell \ln \mathbb{P}_n - n_\ell \mathcal{J}_{1m}, \quad H_\ell^{(2)} = (\delta_{1,m} - \varepsilon_0 + 2\mathcal{J}_{1m} b_\ell \partial_0) \partial_\ell \ln \mathbb{P}_n \\
H_\ell^{(1)} &:= (\kappa_\ell(\delta_{1,m} - \varepsilon_m)\partial_0 + 2\mathcal{J}_{1m}\partial_b^{(\ell)}) \ln \mathbb{P}_n + 2n_\ell \mathcal{J}_{1m} \Big(\mathcal{J}_{mm} b_\ell - \sum_{i \neq \ell} \frac{n_i}{b_\ell - b_i} \Big), \\
G_{\ell+1} &:= \partial_0 G_\ell + \sum_{i=1}^{p} (\partial_0)^\ell F_i \left(\partial_0 \frac{H_i^{(1)}}{F_i} - \partial_i \frac{H_i^{(2)}}{F_i} \right), \quad G_0 := 0.
\end{aligned}
\tag{10.2.41}
$$

Proof It is quite analogous to the method explained in Section 10.1.3. Upon composing boundary operators in the Virasoro representations and then setting all the additional variables equal to 0, one expresses first and second partials in the $s_i^{(\ell)}$ (appearing in (10.2.33)) of the function τ_n as in (10.2.33),

$$\frac{\partial}{\partial s_h^{(\ell)}}\ln\tau_n,\quad \frac{\partial^2}{\partial s_1^{(0)}\partial s_1^{(\ell)}}\ln\tau_n,\quad \frac{\partial^2}{\partial s_2^{(0)}\partial s_1^{(\ell)}}\ln\tau_n,\quad \frac{\partial^2}{\partial s_1^{(0)}\partial s_2^{(\ell)}}\ln\tau_n,\quad \text{for}\ \begin{matrix}\ell=1,\dots,p\\ h=1,2\end{matrix}$$

in terms of the boundary operators. Further lengthy manipulations, involving Wronskian identities, then yield (10.2.40). □

As a special case, we now consider the *probability* $\mathbb{P}_n^{(\beta)}(t,\tilde{E})$ for $m=1$ and $0<t_1=t<1$. For this single-time case, (10.2.22) becomes a one-matrix model with external potential $\mathbb{P}_n := \mathbb{P}_n^A(E)$, thus with no coupling. The expressions for (10.2.41) can be replaced by simpler expressions; note that the $\bar{H}_\ell^{(1)}$ below are not obtained from the $H_\ell^{(1)}$ above, by setting $m=1$; in fact, a further simplification occurs in the equations.

Corollary 10.2.1 *[Adl10a] When $m=1$, then $\ln\mathbb{P}_n=\ln\mathbb{P}_n^A(E)$ satisfies the same non-linear PDE (10.2.40), but with simpler expressions F_ℓ and $H_\ell^{(1)}$ and with $'=\partial_{\scriptscriptstyle E}$ and $\varepsilon := \varepsilon_E-\varepsilon_b$, with $\varepsilon_b=\sum_1^{p-1} b_i\frac{\partial}{\partial b_i}$.*

$$\begin{aligned}
F_\ell &:= \left(\partial_b^{(\ell)}+\kappa_\ell\partial_{\scriptscriptstyle E}\right)\partial_{\scriptscriptstyle E}\ln\mathbb{P}_n+n_\ell,\quad H_\ell^{(2)} := (1-\varepsilon+2b_\ell\partial_{\scriptscriptstyle E})\left(\partial_b^{(\ell)}+\kappa_\ell\partial_{\scriptscriptstyle E}\right)\ln\mathbb{P}_n\\
\bar{H}_\ell^{(1)} &:= \left(-\kappa_\ell\partial_{\scriptscriptstyle E}\varepsilon+(\kappa_\ell(\varepsilon-1)+2)(\partial_b^{(\ell)}+\kappa_\ell\partial_{\scriptscriptstyle E})\right)\ln\mathbb{P}_n+\bar{C}_\ell,\\
G_{\ell+1} &:= \partial_{\scriptscriptstyle E}G_\ell+\sum_{i=1}^{p}(\partial_{\scriptscriptstyle E})^\ell F_i\left(\partial_{\scriptscriptstyle E}\frac{\bar{H}_i^{(1)}}{F_i}-\partial_b^{(i)}\frac{H_i^{(2)}}{F_i}\right),\qquad G_0=0,\\
\bar{C}_\ell &:= -2n_\ell\Big((1-\kappa_\ell)b_\ell+\sum_{j\neq\ell}\frac{n_j}{b_\ell-b_j}\Big).
\end{aligned}\tag{10.2.42}$$

10.2.5 Examples

A couple of applications of Theorem 10.2.2 and Corollary 10.2.3, used later in Section 10.3, will be sketched here for different values of the spectrum of the matrix A, defined in (10.2.34).

The diagonal matrix $A=0$ and $m=2$ (coupled matrix case)

One checks $p=1$, $n_1=n$ and $\mathcal{J}=\frac{1}{1-c^2}\begin{pmatrix}-1 & -c\\ -c & -1\end{pmatrix}$ and

$$\kappa_0=-1,\ b_1=0,\ \kappa_1=1,\ \partial_b^{(0)}=\partial_b^{(1)}=\varepsilon_b=0,\ \varepsilon_0=\varepsilon_{E_1}-c\frac{\partial}{\partial c},\ \varepsilon_2=\varepsilon_{E_2}-c\frac{\partial}{\partial c},$$

$$\partial_0=-\frac{1}{1-c^2}\left(\partial_{E_1}+c\partial_{E_2}\right),\qquad \partial_1=\frac{1}{1-c^2}\left(c\partial_{E_1}+\partial_{E_2}\right).\tag{10.2.43}$$

So, for $\ell=1$, one has, from Theorem 10.2.2,

$$F_1=-\partial_0\partial_1\log\mathbb{P}_n+\frac{nc}{1-c^2},\quad H_1^{(1)}=-\varepsilon_2\partial_0\log\mathbb{P}_n,\quad H_1^{(2)}=-\varepsilon_0\partial_1\log\mathbb{P}_n\tag{10.2.44}$$

and thus

$$G_0 = 0, \quad G_1 = F_1 \left(\partial_0 \frac{H_1^{(1)}}{F_1} - \partial_1 \frac{H_1^{(2)}}{F_1} \right) = \frac{1}{F_1} \left(\left\{ H_1^{(1)}, F_1 \right\}_{\partial_0} - \left\{ H_1^{(2)}, F_1 \right\}_{\partial_1} \right)$$

leading to the PDE, $0 = \det \begin{pmatrix} F_1 & G_0 \\ \partial_0 F_1 & G_1 \end{pmatrix} = \left\{ H_1^{(1)}, F_1 \right\}_{\partial_0} - \left\{ H_1^{(2)}, F_1 \right\}_{\partial_1}$, with ∂_0 and ∂_1 as in (10.2.43) and $H_1^{(j)}$ and F_i as in (10.2.44); also $F_1 \neq 0$. Thus the probability

$$\mathbb{P}_n := \mathbb{P}_n^A(c, E_1 \times E_2) = \frac{1}{\tilde{Z}_n} \int_{\text{spec}(M_k) \in E_k} e^{-\frac{1}{2}\text{tr}\left(M_1^2 + M_2^2 - 2c M_1 M_2\right)} \prod_{k=1}^{m} d M_k \quad (10.2.45)$$

satisfies

$$\left\{ \varepsilon_2 \partial_0 \log \mathbb{P}_n, \partial_0 \partial_1 \log \mathbb{P}_n - \frac{nc}{1-c^2} \right\}_{\partial_0} - \left\{ \varepsilon_0 \partial_1 \log \mathbb{P}_n, \partial_0 \partial_1 \log \mathbb{P}_n - \frac{nc}{1-c^2} \right\}_{\partial_1} = 0. \quad (10.2.46)$$

The matrix A has two eigenvalues, with $m = 1$ (one-matrix case with external potential)

For $A = \text{diag}(a, \ldots, a, 0, \ldots, 0)$ with n_1 entries $= a$ and n_2 entries $= 0$, one has $p = 2$, $\kappa_0 = -1$, $\kappa_1 = 0$, $\kappa_2 = 1$, and $\varepsilon = \varepsilon_E - a \frac{\partial}{\partial a}$. From Corollary 10.2.3, one checks that

$$\mathbb{P}_n := \mathbb{P}_n^A(E) = \frac{1}{\tilde{Z}_n} \int_{\text{spec}(M) \in E} d M \, e^{-\frac{1}{2}\text{Tr}\left(M^2 - 2AM\right)}, \quad (10.2.47)$$

satisfies the (near-Wronskian) PDE (10.2.40), i.e.

$$\det \begin{pmatrix} F_1 & F_2 & 0 \\ F_1' & F_2' & \frac{H_1}{F_1} + \frac{H_2}{F_2} \\ F_1'' & F_2'' & \frac{H_1'}{F_1} + \frac{H_2'}{F_2} \end{pmatrix} = 0, \quad (10.2.48)$$

where $'$ is shorthand for ∂_E and

$$F_1 = -\frac{\partial}{\partial a} \partial_E \log \mathbb{P} + n_1, \qquad F_2 = \left(\frac{\partial}{\partial a} + \partial_E \right) \partial_E \log \mathbb{P} + n_2$$

$$H_1 = \left\{ H_1^{(1)}, F_1 \right\}_{\partial_E} + \left\{ H_1^{(2)}, F_1 \right\}_a, \quad H_2 = \left\{ H_2^{(1)}, F_2 \right\}_{\partial_E} - \left\{ H_2^{(2)}, F_2 \right\}_a,$$

with

$$H_1^{(2)} = \left(1 - \varepsilon_E + a \frac{\partial}{\partial a} + 2a \partial_E \right)\left(-\frac{\partial}{\partial a}\right) \log \mathbb{P}, \; H_2^{(2)} = \left(1 - \varepsilon_E + a \frac{\partial}{\partial a}\right)\left(\frac{\partial}{\partial a} + \partial_E\right) \log \mathbb{P}$$

$$H_1^{(1)} = -2 \frac{\partial}{\partial a} \log \mathbb{P} - 2n_1 \left(a + \frac{n_2}{a} \right), \; H_2^{(1)} = \left(1 + \varepsilon_E - a \frac{\partial}{\partial a} \right) \frac{\partial}{\partial a} \log \mathbb{P} + 2 \frac{n_1 n_2}{a}.$$

10.3 Critical diffusions

10.3.1 The Airy process

Consider n non-intersecting Brownian motions $x_1(t) < \ldots < x_n(t)$ on $\mathbb{R}$, all leaving from the origin and forced to return to the origin. For very large n, the average mean density of particles has its support, for each $0 < t < 1$, on the interval $(-\sqrt{2nt(1-t)}, \sqrt{2nt(1-t)})$. As sketched in Figure 10.2 below, the **Airy process** $\mathcal{A}(\tau)$ is defined in [Pra02] as the fluctuations of these non-intersecting Brownian particles for large n, but viewed by an observer through an "*Airy microscope*" belonging to the (right hand) edge-curve $\mathcal{C}$ (the microscope is indicated by a small circle about σ in Figure 10.2)

$$\mathcal{C}: \quad \{y = \sqrt{2nt(1-t)} > 0\} = \left\{ (y, t) = \Big(\frac{\sqrt{2n}}{2\cosh\sigma}, \frac{e^{\sigma}}{2\cosh\sigma}\Big), \quad -\infty < \sigma < \infty \right\}.$$

The microscope amounts to the following scaling: space is stretched by the customary GUE-edge rescaling $n^{1/6}$ and the new time σ is rescaled by a factor $n^{1/3}$ in accordance with the Brownian space-time rescaling; that is to say that in this new scale, which slows down the motion microscopically, the left-most particles appear infinitely far and the time horizon $t = 1$ lies in the very remote future. One now has the following statement, due to [Pra02, Joh05, Adl05, Adl09a]:

Theorem 10.3.1 *The following limit exists near every point of the curve $\mathcal{C}$ and defines the Airy process distribution:*

$$\begin{aligned}
&\mathbb{P}^{\mathcal{A}}\left(\bigcap_{k=1}^{m}\{\mathcal{A}(\tau_k) \cap E_k = \emptyset\}\right) \\
&:= \lim_{n\to\infty} \mathbb{P}^{(n)}_{Br}\left(\bigcap_{k=1}^{m}\left\{ \text{all } x_i \left(\frac{e^{\sigma+\frac{\tau_k}{n^{1/3}}}}{2\cosh(\sigma+\frac{\tau_k}{n^{1/3}})}\right) \in \frac{\sqrt{2n}+\frac{E_k^c}{\sqrt{2}n^{1/6}}}{2\cosh(\sigma+\frac{\tau_k}{n^{1/3}})}\right\}\right). \qquad (10.3.1)
\end{aligned}$$

It is independent of the point σ along the curve $\mathcal{C}$ (universality!) and is given by the matrix Fredholm determinant of the extended Airy kernel:

$$K^{A} = 2\tau\frac{\partial^3\mathbb{Q}}{\partial\tau\partial u\partial v} = \left(\frac{\tau^2}{2}\frac{\partial}{\partial u} - u\frac{\partial}{\partial v}\right)\left(\frac{\partial^2\mathbb{Q}}{\partial u^2} - \frac{\partial^2\mathbb{Q}}{\partial v^2}\right) + \left\{\frac{\partial^2\mathbb{Q}}{\partial u\partial v}, \frac{\partial^2\mathbb{Q}}{\partial v^2}\right\}_v,$$
$$\text{for } \tau = \tau_2 - \tau_1.$$

with $\Gamma_- \subset \mathcal{C}$ (resp. Γ_+) consisting of two rays making an angle of $\pi/3$ with the negative real axis (resp. positive real axis), with bottom to top orientation for Γ_- and top to bottom orientation for Γ_+. The Airy process is stationary, and for $m = 1$, the probability $\mathbb{P}(\mathcal{A}(\tau_1) \leq u)$ is the Fredholm determinant of the Airy kernel (10.5.6) (Tracy–Widom distribution). For $m = 2$, the log of the probability $\mathbb{Q}(\tau; u, v) := \ln\mathbb{P}(\mathcal{A}(\tau_1) \leq v + u, \mathcal{A}(\tau_2) \leq v - u)$, satisfies the PDE:

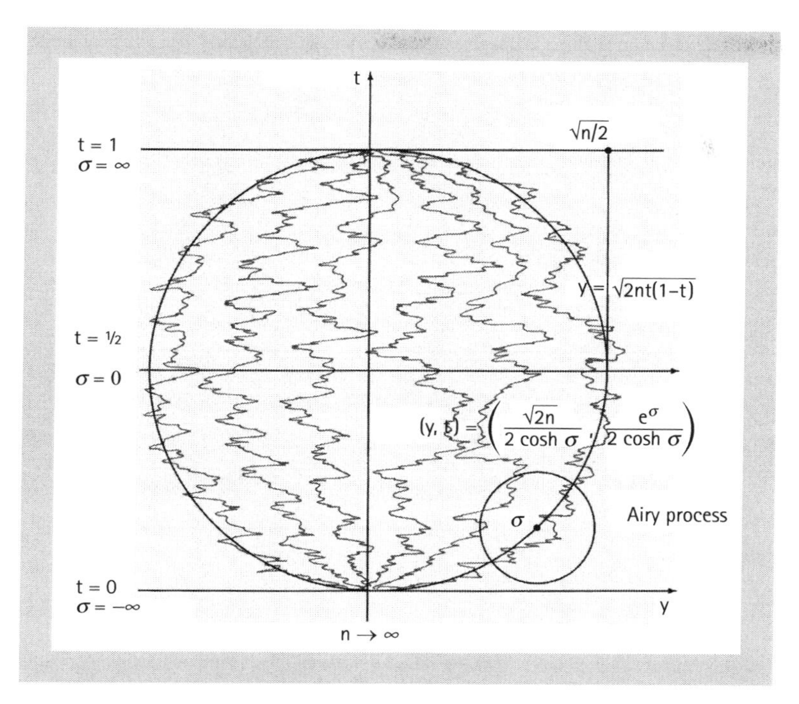

Fig. 10.2 Airy process, with extended kernel and with $\Gamma_- \subset \mathbb{C}$ (resp. Γ_+) making an angle of $\pi/3$ with the negative real axis (resp. positive real axis):

$$K^A(\tau_1, x_1; \tau_2, x_2) = \frac{1}{(2\pi i)^2}\int_{\Gamma_-} du \int_{\Gamma_+} dv \frac{e^{-\frac{u^3}{3}+x_2 u}}{e^{-\frac{v^3}{3}+x_1 v}} \frac{1}{(u+\tau_2)-(v+\tau_1)}$$
$$-\frac{\mathbf{I}(\tau_2 > \tau_1)}{\sqrt{4\pi(\tau_2-\tau_1)}} e^{-\frac{(x_2-x_1)^2}{4(\tau_2-\tau_1)} - \frac{1}{2}(\tau_2-\tau_1)(x_2+x_1) + \frac{1}{12}(\tau_2-\tau_1)^3} \tag{10.3.2}$$

Proof For one-time ($m = 1$), because of the stationarity, the non-intersecting Brownian motions reduce to the spectrum of the Gaussian Hermitian matrix ensemble (GUE), whose probability is given by Equation (10.1.18); one then applies the scaling given by (10.3.1) to this equation yielding a PDE for the transition probability in the end points of the disjoint union $E = \cup[x_{2i-1}, x_{2i}]$ ($\nabla_E := \sum_i \frac{\partial}{\partial x_i}$ and $\mathcal{E}_E := \sum_i x_i \frac{\partial}{\partial x_i}$)

$$0 = (\partial_{-1}^4 + 8n\partial_{-1}^2 + 12\partial_0^2 + 24\partial_0 - 16\partial_{-1}\partial_1) \ln \mathbb{P}_n + 6(\partial_{-1}^2 \ln \mathbb{P}_n)^2 \Big|_{c_i = \sqrt{2n} + \frac{x_i}{\sqrt{2}n^{1/6}}}$$
$$= 4n^{2/3}\left[(\nabla_E^3 - 4(\mathcal{E}_E - \frac{1}{2}))\nabla_E \ln \mathbb{P} + 6(\nabla_E^2 \ln \mathbb{P})^2\right] + o(n^{2/3}). \tag{10.3.3}$$

For an interval $E = (x, \infty)$, this PDE becomes a third order ODE for $\frac{\partial}{\partial x} \ln \mathbb{P}$, which can be transformed into Painlevé II, thus leading to the Tracy–Widom distribution. In Section 10.5, a different method, not using the scaling limit, will be presented in order to obtain Equation (10.3.3). For $m = 2$, it follows from (10.2.22) and (10.2.25) that

$$\mathbb{P}^{(n)}_{\mathrm{Br}}\,(\text{all } x_i(t_1) \in E_1, \text{all } x_i(t_2) \in E_2,\ )$$

$$= \mathbb{P}^A_n\left(\sqrt{\frac{t_1(1-t_2)}{t_2(1-t_1)}};\, E_1\sqrt{\frac{2t_2}{(t_2-t_1)t_1}} \;\times\; E_2\sqrt{\frac{2(1-t_1)}{(1-t_2)(t_2-t_1)}}\right)\Bigg|_{A=0}, \qquad (10.3.4)$$

where (see (10.2.22))

$$\mathbb{P}^A_n(c;\, E_1',\, E_2')\Big|_{A=0} := \frac{1}{Z_n}\iint_{\mathcal{H}(E_1')\times\mathcal{H}(E_2')} dM_1 dM_2\, e^{-\frac{1}{2}\mathrm{Tr}(M_1^2+M_2^2-2cM_1M_2)}. \qquad (10.3.5)$$

From (10.3.1), one uses the time scale $t_i = \frac{1}{1+e^{-2\tau_i n^{-1/3}}}$ and the space scale $E_{\frac{1}{2}} = (-\infty, \frac{\sqrt{2n}+\frac{v\pm u}{\sqrt{2}n^{1/6}}}{2\cosh(\tau_2 n^{-1/3})})$, yielding:[13] (setting $\tau = \tau_2 - \tau_1$, $\mathbb{P}^0_n := \mathbb{P}^A_n|_{A=0}$ and assuming $\sigma = 0$, without loss of generality)

$$\mathbb{Q}_n(\tau_2-\tau_1, u, v) \;:=\; \ln \mathbb{P}^{(n)}_{\mathrm{Br}}\begin{pmatrix} \text{all } x_i\left(\dfrac{1}{1+e^{-2\tau_1 n^{-1/3}}}\right) \le \dfrac{\sqrt{2n}+\dfrac{v+u}{\sqrt{2}n^{1/6}}}{2\cosh(\tau_1 n^{-1/3})} \\ \text{all } x_i\left(\dfrac{1}{1+e^{-2\tau_2 n^{-1/3}}}\right) \le \dfrac{\sqrt{2n}+\dfrac{v-u}{\sqrt{2}n^{1/6}}}{2\cosh(\tau_2 n^{-1/3})} \end{pmatrix}$$

$$= \ln \mathbb{P}^0_n\left(e^{-\frac{\tau}{n^{1/3}}};\, \frac{\left(2\sqrt{n}+\dfrac{v+u}{n^{1/6}}\right)}{\sqrt{1-e^{-2\tau n^{-1/3}}}},\, \frac{\left(2\sqrt{n}+\dfrac{v-u}{n^{1/6}}\right)}{\sqrt{1-e^{-2\tau n^{-1/3}}}}\right)$$

$$= \ln \mathbb{P}^0_n(c; a, b).$$

This gives a map $(\tau, u, v) \mapsto (c, a, b)$, with inverse given below, setting $z = n^{-1/6}$, replacing τ, u, v by functions of a, b, c,

$$\ln \mathbb{P}^0_n(c; a, b) = \mathbb{Q}_n\left(-\frac{\ln c}{z^2};\, \frac{a-b}{2z}\sqrt{1-c^2},\, \frac{a+b}{2z}\sqrt{1-c^2} - \frac{2}{z^4}\right) = \mathbb{Q}_n(\tau, u, v).$$

The PDE (10.2.46) for $\ln \mathbb{P}^0_n(c; a, b)$, in Section 10.2.5, involves operators $\partial_i, \varepsilon_i$ acting on $\ln \mathbb{P}_n(c; a, b)$, which by the chain rule implies operators acting on $\mathbb{Q}(\tau, u, v)$. One expands those operators in z, for $z \to 0$. Upon expanding $\mathbb{Q}_n(\tau_2-\tau_1, u, v) = \mathbb{Q}(\tau_2-\tau_1, u, v) + O(z)$ for large n and using the scaling, one checks the Equation (10.2.46) has a z^{-12} leading term, whose coefficient must be identically zero; i.e.

[13] For simplicity of notation, one replaces the set $(-\infty, x)$ simply by x in the probability $\mathbb{P}_n$ below.

$$
\begin{aligned}
0 &= \left\{\varepsilon_2\partial_0 \ln \mathbb{P}_n^0, \partial_0\partial_1 \ln \mathbb{P}_n^0 - \frac{nc}{1-c^2}\right\}_{\partial_0} - \left\{\varepsilon_0\partial_1 \ln \mathbb{P}_n^0, \partial_0\partial_1 \ln \mathbb{P}_n^0 - \frac{nc}{1-c^2}\right\}_{\partial_1} \\
&= -\frac{z^{-12}}{4\tau^2}\left(2\tau\frac{\partial^3\mathbb{Q}}{\partial\tau\partial u\partial v} - \left(\frac{\tau^2}{2}\frac{\partial}{\partial u} - u\frac{\partial}{\partial v}\right)\left(\frac{\partial^2\mathbb{Q}}{\partial u^2} - \frac{\partial^2\mathbb{Q}}{\partial v^2}\right) - \left\{\frac{\partial^2\mathbb{Q}}{\partial u\partial v}, \frac{\partial^2\mathbb{Q}}{\partial v^2}\right\}_v\right) \\
&\quad + O(z^{-8})
\end{aligned}
$$

establishing Theorem 10.3.1. □

10.3.2 The Pearcey process

Consider n nonintersecting Brownian motions leaving from the origin and forced to two points, with $n_1 = [np]$ particles forced to $a\sqrt{n}$ and $n_2 = [(1-p)n]$ particles forced to another point, say, 0, at time $t = 1$. When $n \to \infty$, the mean density of Brownian particles has its support on one interval for t near 0 and on two intervals for t near 1, so that a bifurcation appears for some intermediate time t_0, where one interval splits into two intervals; see Figure 10.3 below. At this point the boundary of the support of the mean density has a cusp, located at, say, $(x_0\sqrt{n}, t_0)$. Near this cusp appears a new "*universal*" process, upon looking through the "*Pearcey microscope*" near the point of bifurcation, different from the Airy microscope; the new process is independent of the values of a and p and is called the **Pearcey process**. Tracy and Widom [Tra06] showed the appearance of the Pearcey process in the symmetric case; namely when the target points are symmetric with respect to the origin and half of the particles go to either target point. Brézin and Hikami [Bre98a, Bre98b] first considered this kernel; see also Bleher-Kuijlaars [Ble04b]. The statement below is due to Adler-van Moerbeke [Adl07b] and Adler-Orantin-van Moerbeke [Adl10e] with an asymptotic result due to Adler–Cafasso–van Moerbeke [Adl11a]

Theorem 10.3.2 *It is convenient to reparameterize $p = \frac{1}{1+q^3}$ with $0 < q < \infty$. Let $E \in \mathbb{R}$ be a compact interval. The distribution of the Pearcey process is defined by the following limit near the cusp at $(x_0\sqrt{n}, t_0)$; it is given by an extended Pearcey kernel, as defined in Figure 10.3 below, and is independent of p and a:*

$$
\begin{aligned}
&\mathbb{P}^{\mathcal{P}}\left(\bigcap_{1\le k\le m}\{\mathcal{P}(\tau_k)\cap E = \emptyset\}\right) \\
&:= \lim_{n\to\infty}\mathbb{P}_{Br}^{(n)}\left(\bigcap_{1\le k\le m}\left\{all\ x_j\left(t_0 + \left(\frac{c_0\mu}{n^{1/4}}\right)^2 2\tau_k\right) \in x_0 n^{1/2} + c_0 A\tau_k + \frac{c_0\mu}{n^{1/4}}E^c\right\}\right),
\end{aligned}
\tag{10.3.6}
$$

with t_0, x_0, c_0, A and μ defined by ($r := \sqrt{q^2-q+1}$)

$$
\sqrt{\frac{2t_0}{1-t_0}} = \frac{q+1}{ar},\quad x_0 = \frac{2q-1}{q+1}at_0,\quad c_0 = \frac{at_0 r}{q+1},\quad A = \frac{q-(2q-1)t_0}{\sqrt{q}},\quad \mu = \frac{\sqrt{r}}{q^{1/4}}.
$$

The log of the transition probability for the Pearcey process $\mathbb{Q}(t, E) := \log \mathbb{P}^{\mathcal{P}}(\mathcal{P}(t) \cap E = \emptyset) = \log\det(I - K^{\mathcal{P}}(t, x_1; t, x_2))_E$, *with kernel (the contour* $\Gamma_\times$ *is a cross* $\subset \mathcal{C}$, *making angles of* $\pi/4$ *with the real axis):*

$$K_p = K^{\mathcal{P}}(\tau_1, x_1; \tau_2, x_2) = \frac{1}{(2\pi i)^2}\int_{-i\infty}^{i\infty} dz \int_{\Gamma_\times} d\tilde{z}\frac{1}{z - \tilde{z}}\frac{e^{-\frac{z^4}{4}+\tau_2\frac{z^2}{2}-x_2 z}}{e^{-\frac{\tilde{z}^4}{4}+\tau_1\frac{\tilde{z}^2}{2}-x_1\tilde{z}}} - \frac{\mathbf{I}(\tau_2 > \tau_1)}{\sqrt{2\pi(\tau_2 - \tau_1)}} e^{-\frac{(x_2-x_1)^2}{2(\tau_2-\tau_1)}} \tag{10.3.7}$$

satisfies the following third order non-linear PDE in $t = \tau_1 = \tau_2$ *and the boundary points of* E, *(for the notation* ∇_E *and* $\mathcal{E}_E$, *see formulae before (10.3.3))*

$$\frac{\partial^3 \mathbb{Q}}{\partial t^3} + \frac{1}{8}\left(\mathcal{E}_E - 2t\frac{\partial}{\partial t} - 2\right)\nabla_E^2 \mathbb{Q} - \frac{1}{2}\left\{\nabla_E^2 \mathbb{Q}, \nabla_E \frac{\partial \mathbb{Q}}{\partial t}\right\}_{\nabla_E} = 0. \tag{10.3.8}$$

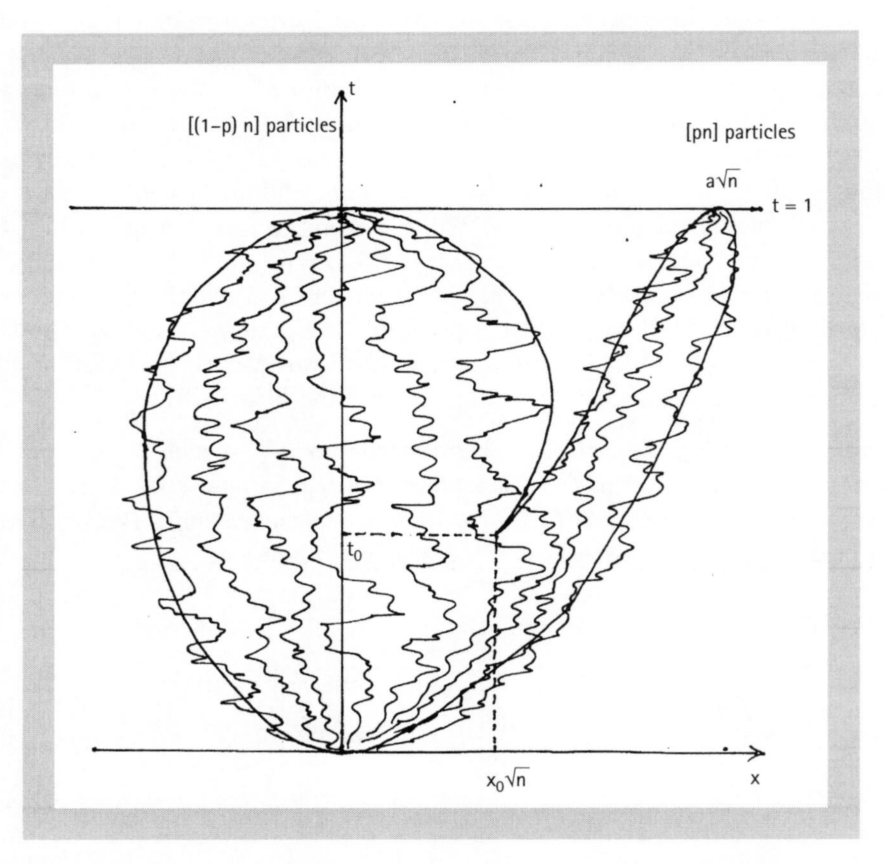

Fig. 10.3 Pearcey process, with extended kernel, where $\Gamma_\times$ is a cross $\subset \mathbb{C}$, making angles of $\pi/4$ with the real axis

For the multi-time probability, one finds a multi-time version of equation (10.3.8); see [Adl10a] . From this PDE, one deduces the following approximation of Airy by Pearcey: Given the finite parameters $t_1 < t_2$*, let both* $T_1, T_2 \to \infty$*, such that* $T_2 - T_1 \to \infty$ *behaves in the following way:*

$$\frac{T_2 - T_1}{2(t_2 - t_1)} = (3T_1)^{1/3} + \frac{t_2 - t_1}{(3T_1)^{1/3}} + \frac{2t_1 t_2}{3T_1} + O(\frac{1}{T^{5/3}});$$

this specifies two new times t_1 *and* t_2*. The parameters* t_1 *and* t_2 *provide the Airy times in the following approximation of the Airy process by the Pearcey process: :*

$$\mathbb{P}\left(\bigcap_{i=1}^{2}\left\{\frac{\mathcal{P}(T_i) - \frac{2}{27}(3T_i)^{3/2}}{(3T_i)^{1/6}} \cap (-E_i) = \emptyset\right\}\right) = \mathbb{P}\left(\bigcap_{i=1}^{2}\{\mathcal{A}(t_i) \cap (-E_i) = \emptyset\}\right) \times \left(1 + O(\frac{1}{T_1^{4/3}})\right). \quad (10.3.9)$$

Proof Equation (10.3.8) follows from doing asymptotics on the PDE (10.2.48). In Section 10.5, Equation (10.3.8) will be obtained without scaling limit. The kernels and PDEs for the Pearcey process with inliers have been considered in [Adl10a]; "inliers" here refers to a finite number of particles forced to a target point $an^{1/4}$, in between the target points of the Pearcey process. □

10.3.3 Airy process with wanderers

Consider now n non-intersecting Brownian particles leaving from 0 at time $t = -1$ and returning to 0 at time $t = 1$, as in Section 10.3.1; note the different origin of time, merely for convenience. The curve $\mathcal{C}$ of Section 10.3.1 becomes now $\mathcal{C}: \ y = \sqrt{2n(1-t^2)}$. Given a fixed integer $m > 0$, let m Brownian particles (wanderers or outliers) leave from the point $y_0^-\sqrt{2n}$ at time $t = -1$, with target $y_0^+\sqrt{2n}$ at $t = 1$. There are several possibilities:

(1) Either the line connecting the points $(y_0^\pm\sqrt{2n}, \pm 1)$ in the (y, t)-plane does not intersect the ellipse $\mathcal{C}$. Then the local fluctuations of the particles near the edge (along the curve $\mathcal{C}$) are given by the Airy process, as in Section 10.3.1; i.e., the wanderers have no influence on the edge behaviour of the bulk of the particles. This situation is depicted in the first figure of Figure 10.4 below.

(2) Or the line connecting the points $(y_0^\pm\sqrt{2n}, \pm 1)$ cuts through the ellipsoidal region $\mathcal{C}$; then the geodesic connecting the two points $(y_0^\pm\sqrt{2n}, \pm 1)$ in the region $\{(y, t)\text{-plane} \setminus \text{ellipsoidal region}\}$ will have two tangency points. At those points and only there will the statistical fluctuations be different from the Airy process: **two Airy processes with m one-sided outliers** (second figure of Figure 10.4).

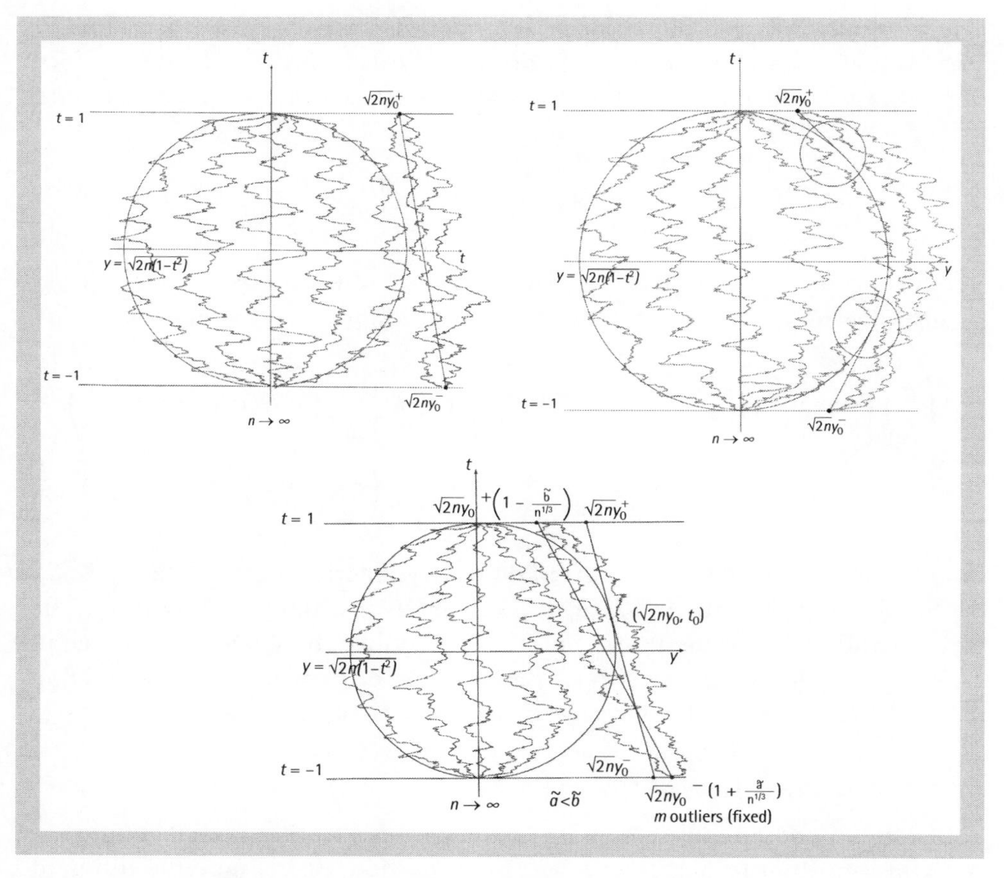

Fig. 10.4 The Airy process with outliers (one-sided and two-sided) describe the statistical fluctuations of the non-intersecting Brownian motions near the point(s) of tangency of the geodesic connecting $\sqrt{2ny_0^{\pm}}$ in ($\mathbb{R}^2\backslash$ellipsoidal region), when $n \to \infty$.

(3) Or the line connecting the points $(y_0^{\pm}\sqrt{2n}, \pm 1)$ is tangent to the ellipse C at the point, say, $(y_0\sqrt{2n}, t_0)$. Then new statistical fluctuations will appear, when the outliers leave from a point $y_0^-\sqrt{2n}(1 + \tilde a/n^{1/3})$ and will be forced to a point $y_0^+\sqrt{2n}(1 - \tilde b/n^{1/3})$, both of which are at a $n^{1/6}$-distance from the two points $(y_0^{\pm}\sqrt{2n}, \pm 1)$, but such that the line connecting them passes to the left of the tangency point $(y_0\sqrt{2n}, t_0)$; this amounts to the condition $\tilde a < \tilde b$. Looking at this tangency point with the "*Airy microscope*" reveals the existence of a new critical infinite-dimensional diffusion (also a universal process, but different from the Airy process) $\mathcal{A}_m(\tau)$, which defines an **Airy process with m two-sided outliers**. This can be generalized to m outliers leaving from and forced to m distinct points, subjected to $\tilde a_m \leq \ldots \leq \tilde a_1 < \tilde b_1 \leq \ldots \leq \tilde b_m$.

The next statement is due to Adler, Delépine, van Moerbeke [Adl09a] and Adler, Ferrari, van Moerbeke [Adl10b]. Theorem 10.3.3 below deals with the situation described in (3); see also Footnote 14.

Theorem 10.3.3 *[Adl09a, Adl10b]* **(Universal distribution)** *For n large, for ℓ distinct times $\tau_1, \tau_2, \ldots, \tau_\ell$ and compact sets $E_1, \ldots, E_\ell \subset \mathbb{R}$, pick ℓ points $(y_k, t_k) \in \mathcal{C}$ in a $n^{-1/3}$-neighbourhood of (y_0, t_0), with $t_k := t_0 + (1 - t_0^2)\tau_k n^{-1/3}$. Then, the following limit defines a new universal process $\mathcal{A}_m(\tau)$ (with m two-sided outliers depending on $\tilde{a}_i$ and $\tilde{b}_i$) describing the fluctuations about (y_0, t_0):*

$$\mathbb{P}^{\mathcal{A}_m}\left(\bigcap_{k=1}^{\ell}\{\mathcal{A}_m(\tau_k)\cap E_k=\emptyset\}\right)=\lim_{n\to\infty}\mathbb{P}_{Br}^{(n)}\left(\bigcap_{k=1}^{\ell}\left\{all\ x_i\ (t_k)\in\left(1+\frac{E_k^c}{2n^{2/3}}\right)y_k\right\}\right). \tag{10.3.10}$$

It is defined by the Fredholm determinant of the extended kernel, with contours $\Gamma_\pm$ as in (10.3.2),

$$\begin{aligned} K_m^{\tilde{a},\tilde{b}}(\tau_1, x_1; \tau_2, x_2) &= \frac{1}{(2\pi i)^2}\int_{\Gamma_-} du \int_{\Gamma_+} dv \frac{e^{-u^3/3+x_2u}}{e^{-v^3/3+x_1v}} \frac{\prod_{k=1}^m \left(\frac{v-\tilde{a}_k+\tau_1}{u-\tilde{a}_k+\tau_2}\right)\left(\frac{u-\tilde{b}_k+\tau_2}{v-\tilde{b}_k+\tau_1}\right)}{(u+\tau_2)-(v+\tau_1)} \\ &\quad - \frac{\mathbf{I}(\tau_2>\tau_1)}{\sqrt{4\pi(\tau_2-\tau_1)}} \times e^{-\frac{(x_2-x_1)^2}{4(\tau_2-\tau_1)}-\frac{1}{2}(\tau_2-\tau_1)(x_2+x_1)+\frac{1}{12}(\tau_2-\tau_1)^3} \end{aligned} \tag{10.3.11}$$

When the $\tilde{a}_j = \tilde{a} \to -\infty$ and all $\tilde{b}_j = \tilde{b}$, then the transition probability, for one time (i.e. $\ell = 1$), $Q(\tau, x) := \log \mathbb{P}^{\mathcal{A}_m}(\sup \mathcal{A}_m(\tau) \le x)$ satisfies the non-linear PDE, depending explicitly on the number m of wanderers:

$$\begin{aligned} 0 &= 2\left(m - \frac{\partial^2 Q}{\partial\tau\partial x}\right)^2 \left\{\frac{\partial^3 Q}{\partial\tau\,\partial x^2}, \frac{\partial^3 Q}{\partial x^3}\right\}_x \\ &\quad +2\left(m - \frac{\partial^2 Q}{\partial\tau\partial x}\right)\left\{\frac{\partial^3 Q}{\partial\tau\partial x^2}, \frac{\partial}{\partial\tau}\left(\frac{\partial}{\partial\tau}\left(\frac{\partial Q}{\partial\tau}+\tau\frac{\partial Q}{\partial x}\right)-x\frac{\partial^2 Q}{\partial x^2}\right)\right\}_x \\ &\quad +\left\{\frac{\partial^3 Q}{\partial\tau\,\partial x^2}, \frac{\partial^3 Q}{\partial\tau^2\,\partial x}\left(2m\tau+\frac{\partial^2 Q}{\partial\tau^2}\right)\right\}_x + \left(\frac{\partial^3 Q}{\partial\tau\partial x^2}\right)^2 \frac{\partial}{\partial\tau}\left(\frac{\partial^2 Q}{\partial\tau^2}+2\frac{\partial Q}{\partial x}\right). \end{aligned} \tag{10.3.12}$$

From the PDE one obtains the approximation of the Airy process by the Airy process with one-sided outliers, upon moving away from the tangency points along the curve $\mathcal{C}$.

Proof The kernel (10.3.11) and the PDE (10.3.12) are obtained by performing a scaling limit on the Brownian motion kernel and on Equation (10.2.48). The

[14] This and related kernels have also appeared in the work of Baik-BenArous-Péché [Ber04, Bai06] in the context of multivariate statistics and also of Borodin-Péché [Bor07] in the context of directed percolation.

two-time version for this PDE can be obtained as well. However, the analogue of PDE (10.3.12) for two-sided outliers remains unknown. □

10.4 The Tacnode process

Consider now non-intersecting Brownian motions leaving from two different points at time $t = 0$ and forced to two different points at $t = 1$. Letting the number of particles tend to infinity, the two ellipses will merely touch, upon tuning the starting and target points appropriately, creating a tacnode; i.e., the singularity created by letting two circles touch. The statistical fluctuations near the tacnode, after appropriate scaling, are expected to be of a different nature; see Figure 10.5.

In order to study this problem, first consider a continuous-time random walk on $\mathbb{Z}$ with jumps ± 1, which occur independently with rate 1; i.e. the waiting times of the left- and right-jumps are independent and exponentially distributed with mean 1. The transition probability of going from x to y during a time interval t is then given by $p_t(x, y) = e^{-2t} I_{|x-y|}(2t)$, where I_n is the modified Bessel function of degree n.

Consider now an infinite number of continuous time random walks starting from $\{\ldots, -m-2, -m-1\} \cup \{m+1, m+2, \ldots\}$ at time $\tau = -t$, returning

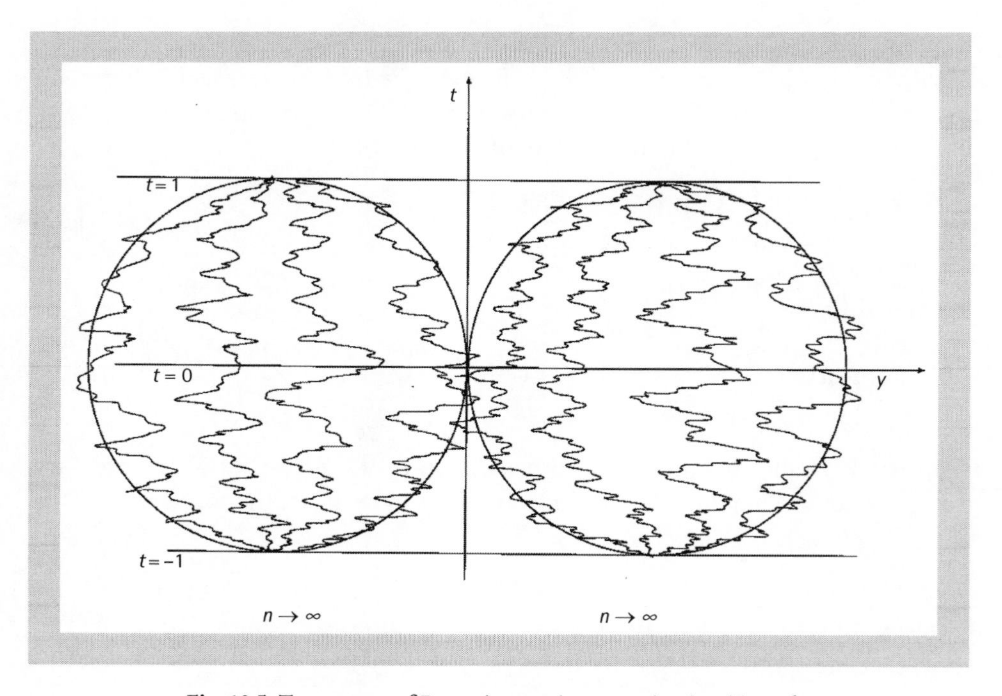

Fig. 10.5 Two groups of Brownian motions meeting in a Tacnode.

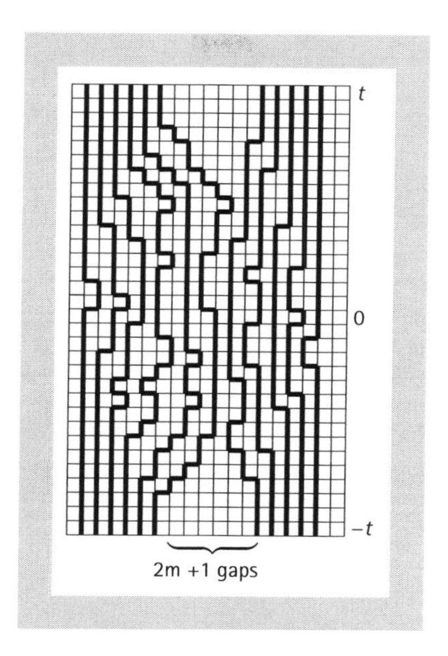

Fig. 10.6 Two groups of non-intersecting random walkers $\mathbf{x}(t)$ on $\mathbb{Z}$, with exponential holding time.

to the starting positions at time $\tau = t$, and *conditioned not to intersect*, see Figure 10.6. Denote by $x_k(\tau)$ the position of the walker who starts and ends at k. Then, the point process on $\mathbb{Z}$ defined by these walkers is determinantal, i.e. there exists a kernel $\mathbb{K}_m$, such that the n-point correlation function is given by $\det(\mathbb{K}_m(y_i, y_j))_{1\leq i,j\leq n}$. Then the probability that the process has a gap along a set E at time $-t \leq \tau \leq t$, is given by the Fredholm determinant of the kernel $\mathbb{K}_m$ projected onto E. This is done in [Adl10c]. For a Riemann–Hilbert approach to the Brownian motions problem, see [Del10] and, yet for another approach, see [Joh10].

Then, setting the scaling $m = 2t + \sigma t^{1/3}$, $x_i = \xi_i t^{1/3}$ and $\tau_i = s_i t^{2/3}$, the limit of the kernel $\mathbb{K}_m$ for $t \to \infty$, is then given by the following kernel, depending on a parameter σ (which appears in the scaling), with arbitrary $\delta > 0$. This kernel, due to [Adl10c], describes the space-time dynamics of the motion of the walkers near the tacnode:

$$\mathcal{K}(s_1, \xi_1; s_2, \xi_2) = -\frac{\mathbf{I}_{[s_2<s_1]}}{\sqrt{4\pi(s_1 - s_2)}} \exp\left(-\frac{(\xi_1 - \xi_2)^2}{4(s_1 - s_2)}\right) + \mathcal{C}(s_1 - s_2, \xi_1 - \xi_2)$$

$$+ \frac{1}{(2\pi i)^2} \int_{\delta+i\mathbb{R}} du \int_{-\delta+i\mathbb{R}} dv \, \frac{e^{\frac{u^3}{3}-\sigma u}}{e^{\frac{v^3}{3}-\sigma v}} \frac{e^{s_1 u^2}}{e^{s_2 v^2}} \left(\frac{e^{\xi_1 u}}{e^{\xi_2 v}} + \frac{e^{-\xi_1 u}}{e^{-\xi_2 v}}\right) \frac{(1 - \hat{\mathcal{P}}(u))(1 - \hat{\mathcal{P}}(-v))}{u - v}$$

$$- \frac{1}{(2\pi i)^2} \int_{2\delta+i\mathbb{R}} du \int_{\delta+i\mathbb{R}} dv \, \frac{e^{\frac{u^3}{3}-\sigma u}}{e^{-\frac{v^3}{3}-\sigma v}} \frac{e^{s_1 u^2}}{e^{s_2 v^2}} \left(\frac{e^{\xi_1 u}}{e^{\xi_2 v}} + \frac{e^{-\xi_1 u}}{e^{-\xi_2 v}}\right) \frac{(1 - \hat{\mathcal{P}}(u))\hat{\mathcal{Q}}(-v)}{u - v}$$

$$
-\frac{1}{(2\pi i)^2}\int_{-\delta+i\mathbb{R}} du \int_{-2\delta+i\mathbb{R}} dv \, \frac{e^{-\frac{u^3}{3}-\sigma u}}{e^{\frac{v^3}{3}-\sigma v}} \frac{e^{s_1u^2}}{e^{s_2v^2}} \left(\frac{e^{\xi_1 u}}{e^{\xi_2 v}} + \frac{e^{-\xi_1 u}}{e^{-\xi_2 v}}\right) \frac{(1-\hat{\mathcal{P}}(-v))\hat{\mathcal{Q}}(u)}{u-v}
$$

$$
+\frac{1}{(2\pi i)^2}\int_{-\delta+i\mathbb{R}} du \int_{\delta+i\mathbb{R}} dv \, \frac{e^{-\frac{u^3}{3}-\sigma u}}{e^{-\frac{v^3}{3}-\sigma v}} \frac{e^{s_1u^2}}{e^{s_2v^2}} \left(\frac{e^{\xi_1 u}}{e^{\xi_2 v}} + \frac{e^{-\xi_1 u}}{e^{-\xi_2 v}}\right) \frac{\hat{\mathcal{Q}}(u)\hat{\mathcal{Q}}(-v)}{u-v},
$$

with

$$
\hat{\mathcal{Q}}(u) := \int_{\tilde{\sigma}}^{\infty} d\kappa \, \mathcal{Q}(\kappa) e^{\kappa u 2^{1/3}},
$$

$$
\hat{\mathcal{P}}(u) := -\int_0^{\infty} d\kappa \, e^{-\kappa u 2^{1/3}} \int_{\tilde{\sigma}}^{\infty} d\mu \, \mathcal{Q}(\mu) \, \mathrm{Ai}(\mu+\kappa),
$$

and where Ai is the Airy function and $K^{\mathcal{A}}$ the Airy kernel (10.3.2) $\tau_1 = \tau_2$. Also

$$
\mathcal{Q}(\kappa) := [(\mathbf{I} - \chi_{\tilde{\sigma}} K^{\mathcal{A}} \chi_{\tilde{\sigma}})^{-1} \chi_{\tilde{\sigma}} \mathrm{Ai}](\kappa), \quad \text{with } \tilde{\sigma} := 2^{2/3}\sigma \text{ and } \chi_{\tilde{\sigma}}(x) = \mathbf{I}_{[x>\tilde{\sigma}]}, \tag{10.4.1}
$$

and $C(s, \xi)$ is an integral involving the Airy function and $\mathcal{Q}(\kappa)$. The integrable system underlying this kernel is unknown!

10.5 Kernels and p-reduced KP

The following integrals, along appropriate rays $\Gamma_\pm$ so as to guarantee convergence,

$$
\Psi_p^{\pm}(x, z) := \frac{\sqrt{p}}{\sqrt{2\pi}} z^{\frac{p-1}{2}} e^{\mp\frac{p}{p+1} z^{p+1}} \int_{\Gamma_\pm} du \, e^{\mp\frac{u^{p+1}}{p+1} \pm (x+z^p)u}, \tag{10.5.1}
$$

are eigenfunctions of the eigenvalue problem ($D := \partial/\partial x$))

$$
((\pm D)^p - x)\Psi_p^{\pm}(x, z) = z^p \Psi_p^{\pm}(x, z), \text{ with asymptotics } \Psi_p^{\pm}(0, z) = 1 + \sum_{i\geq 1} \frac{a_i^{\pm}}{z^i}.
$$

Flowing off the initial condition, $\mathcal{L}_\pm(x, t)\big|_{t=0} = \mathcal{L}_\pm := (\pm D)^p - x$, and $\Psi^\pm(x, t; z)\big|_{t=0} = \Psi_p^{\pm}(x, z)$, by means of the KP-hierarchy, while maintaining the eigenvalue problem,[15]

$$
\mathcal{L}_\pm(x, t)\Psi^\pm(x, t; z) = z^p \Psi^\pm(x, t; z) \quad \begin{cases} \frac{\partial \mathcal{L}_\pm(x,t)}{\partial t_n} = \left[(\mathcal{L}_\pm(x, t)^{\frac{n}{p}})_+, \mathcal{L}_\pm(x, t)\right] \\ \frac{\partial}{\partial t_n}\Psi^\pm(x, t; z) = \pm(\mathcal{L}_\pm(x, t)^{\frac{n}{p}})_+ \Psi^\pm(x, t; z) \end{cases}
$$

yields wave functions $\Psi^\pm(x, t; z)$, expressible by Sato's theory in terms of a τ-function as $\Psi^\pm(x, t; z) = e^{\pm\sum_1^\infty t_i z^i} \frac{\tau(t \mp [z^{-1}])}{\tau(t)}$, where $\tau(t)$ satisfies the KP-hierarchy

[15] $(\mathcal{L}_\pm(x, t)^{\frac{n}{p}})_+$ denotes the differential part of the pseudo-differential operator $\mathcal{L}_\pm(x, t)^{\frac{n}{p}}$

(10.1.5); see Footnote 1. The fact that $\mathcal{L}_\pm$ is a differential operator is maintained by the flow above, which implies that for all $n \geq 1$ one has $\frac{\partial \mathcal{L}_\pm}{\partial t_{np}} = 0$; thus $\tau(t)$ is independent of $t_p, t_{2p}, t_{3p}, \ldots$; this is the **$p$-reduced KP hierarchy** (Gel'fand-Dickey hierarchy). The following Proposition will be crucial.

Theorem 10.5.1 *[Adl98, Adl11b]* **(i)** *Integrating the product of the wave functions $\Psi^\pm(x,t;z)$ above in x defines a kernel, which restricted to the locus $t_p = t_{p+1} = \ldots = 0$, admits a representation as a double integral (for rays Γ_p and Γ_p^* in appropriate sectors in $\mathbb{C}$)*

$$
\begin{aligned}
&K^{(p)}_{x,t}(z^p, z'^p) && (10.5.2)\\
&:= e^{-\frac{p}{p+1}z^{p+1}} \frac{\frac{1}{2\pi p} D_x^{-1}\, \Psi^-(x,t;z)\Psi^+(x,t;z')}{z^{\frac{p-1}{2}} z'^{\frac{p-1}{2}}} e^{\frac{p}{p+1}z'^{p+1}}\\
&= \frac{1}{(2\pi i)^2}\int_{\Gamma_p} du \int_{\Gamma_p^*} \frac{dv}{u-v}\, \frac{e^{-V_p(u)+(z'^p+x)u}}{e^{-V_p(v)+(z^p+x)v}}, \quad \text{setting } t_p = t_{p+1} = \ldots = 0.
\end{aligned}
$$

The exponentials above contain the polynomial $V_p(u) := \frac{u^{p+1}}{p+1} + \sum_0^{p-2} \theta_i \frac{u^{i+1}}{i+1}$, with polynomial coefficients $\theta_i := \theta_i(t_1, \ldots, t_{p-1})$ in $t_1, \ldots, t_{p-1}$, obtained by solving the equation $w = V_p'(u)$ for u in terms of (large) w and by identifying the series obtained,[16] *namely the series*

$$
\begin{aligned}
u &= w^{\frac{1}{p}} - \frac{1}{p}\theta_{p-2} w^{-\frac{1}{p}} - \frac{1}{p}\theta_{p-3} w^{-\frac{2}{p}} - \frac{1}{p}\left(\theta_{p-4} - \frac{p-3}{2p}\theta_{p-2}^2\right) w^{-\frac{3}{p}}\\
&\quad + \cdots + O(w^{-1-\frac{1}{p}}), \qquad (10.5.3)
\end{aligned}
$$

and the series $u = w^{\frac{1}{p}} + \frac{1}{p}\sum_1^{p-1} i t_i w^{\frac{i-p}{p}} + O(w^{-1-\frac{1}{p}})$.
(ii) *The Fredholm determinant of the kernel (10.5.2) over a disjoint union $E = \bigcup_{i=1}^r [a_{2i-1}, a_{2i}] \subset \mathbb{R}^+$ is a ratio of two* **p-reduced KP τ-functions,** *$\tau_E^{(p)}(t)$ and $\tau^{(p)}(t)$. The latter is the so-called topological τ-function $\tau^{(p)}(t)$*[17] *and the former the exponential of the integral of the vertex operator $\mathbb{X}(t, \omega z, \omega' z)$*[18] *acting on $\tau^{(p)}(t)$, for pth-roots of unity $\omega \neq \omega'$, to wit, given the kernel $K^{(p)}_{x,t}(u,v)$ as in (10.5.2),*

$$
\det\left(I - 2\pi\lambda K^{(p)}_{x,t}\right)_E = \frac{\tau_E^{(p)}(t)}{\tau^{(p)}(t)} := \frac{1}{\tau^{(p)}(t)} \exp\left(-\lambda \int_{E^{\frac{1}{p}}} dz\, \mathbb{X}(t; \omega z, \omega' z)\right)\tau^{(p)}(t),
\tag{10.5.4}
$$

where $E^{\frac{1}{p}} := \{x \in \mathbb{R}^+ | x^p \in E\}$. Besides, both τ-functions satisfy the following **Virasoro constraints**[19] *with boundary differential operator for τ_E and without for τ*

[16] For example, $\frac{1}{p-1}\theta_{p-2} = -t_{p-1}$, $\frac{1}{p-2}\theta_{p-3} = -t_{p-2}$, $\frac{1}{p-3}\theta_{p-4} = -t_{p-3} + \frac{1}{2p}(p-1)^2 t_{p-1}^2, \ldots$.

[17] A few examples: $\log \tau^{(2)} = \frac{1}{12}t_1^3$, $\log \tau^{(3)} = -\frac{1}{3}t_1^2 t_2 - \frac{2}{27}t_2^4$, $\log \tau^{(4)} = -\frac{3}{8}t_1^2 t_3 - \frac{1}{2}t_1 t_2^2 - \frac{9}{16}t_2^2 t_3^2 - \frac{81}{1280}t_3^3$.

[18] Vertex operators $\mathbb{X}(t, y, z) := \frac{1}{z-y} e^{\sum_1^\infty (z^i - y^i)t_i} e^{\sum_1^\infty (y^{-i} - z^{-i})\frac{\partial}{\partial t_i}}$ typically generate Darboux transformations.

[19] where $W_n^{(2)} + (n+1)W_n^{(1)} = \sum_{i+j=n} : W_i^{(1)} W_j^{(1)} :$, with $W_i^{(1)} = \begin{cases} \frac{\partial}{\partial t_i} & \text{if } i > 0 \\ (-i)t_{-i} & \text{if } i \leq 0 \end{cases}$.

(topological τ-function) for all $j \geq 0$,

$$\left(-\sum_1^{2r} a_i^j \frac{\partial}{\partial a_i} + \left(\frac{1}{2p} W^{(2)}_{(j-1)p} + W^{(1)}_{jp+1} - \frac{p-1}{2p} W^{(1)}_{(j-1)p} + \frac{p^2-1}{12p^2}\right)\right) \tau_{_E}^{(p)} = 0. \tag{10.5.5}$$

Note that the kernels (10.3.2), (10.3.11), and (10.3.7) are all special cases of the kernel (10.5.2) of Theorem 10.5.1 for $p = 2$ and $p = 3$, except for the kernel (10.3.11) has an additional rational piece. In [Adl11b], other reductions (like k-vector constrained reductions) have been considered which also capture kernels with a rational piece of the type (10.3.11).

As an application of Theorem 10.5.1, equations of the KP-hierarchy are given by the coefficients in the y-Taylor series of (10.1.10); denote these equations by $\{y_k\}$. Also take into account that in each of these equations, the variables $t_p, t_{2p}, \ldots$ are not active. For instance, for $p = 2$, one finds the KdV hierarchy, for $p = 3$, the Boussinesq hierarchy, etc.... With this in mind, combining these equations with the Virasoro equations, one checks that, along the locus $L = \{x = t_1 = t_p = t_{p+1} = \ldots = 0\}$,

$$\mathbb{Q} := \ln \det(I - K^{(p)}_{x,t}(y,y')\chi_{_E}(y'))\big|_L = \ln \tau_{_E}^{(p)} - \ln \tau^{(p)}\big|_L$$

satisfies the following PDEs : (notation: $\partial_i := \partial/\partial t_i$, and for $E = \cup_1^r [x_{2i-1}, x_{2i}]$, one has $\partial_{_E} := \sum_i \frac{\partial}{\partial x_i}$ and $\varepsilon_{_E} := \sum_i x_i \frac{\partial}{\partial x_i}$)

$$p=2: \quad K^{(2)}_{x,t}(y,z)\Big|_L = \frac{1}{(2\pi i)^2}\int_{\Gamma_+} du \int_{\Gamma_-} \frac{dv}{u-v} \frac{e^{-\frac{u^3}{3}+zu}}{e^{-\frac{v^3}{3}+yv}} = \text{Airy kernel.}$$

$$\{y_3\}: \quad (\partial_{_E}^3 + 2 - 4\varepsilon_{_E})\partial_{_E}\mathbb{Q} + 6(\partial_{_E}^2\mathbb{Q})^2 = 0 \tag{10.5.6}$$

$$p=3: \quad K^{(3)}_{x,t}(y,z)\Big|_L = \frac{1}{(2\pi i)^2}\int_{\Gamma_+} du \int_{\Gamma_-} \frac{dv}{u-v} \frac{e^{-\frac{u^4}{4}+t_2u^2+zu}}{e^{-\frac{v^4}{4}+t_2v^2+yv}} = \left\{\begin{array}{l}\text{Pearcey}\\ \text{kernel}\end{array}\right.$$

$$\{y_3\}: \quad \partial_{_E}^4\mathbb{Q} + 6(\partial_{_E}^2\mathbb{Q})^2 - 8t_2\partial_{_E}^2\mathbb{Q} + 3\partial_2^2\mathbb{Q} = 0$$

$$\{y_4\}: \quad (\partial_2\partial_{_E}^2 - 2t_2\partial_2 - 3\varepsilon_{_E} + 1)\partial_{_E}\mathbb{Q} + 6(\partial_{_E}^2\mathbb{Q})(\partial_2\partial_{_E}\mathbb{Q}) = 0$$

$$\partial_2\{y_3\} - \partial_{_E}\{y_4\}: \quad \frac{1}{2}\partial_2^3\mathbb{Q} + (\frac{1}{2}\varepsilon_{_E} - t_2\partial_2 - 1)\partial_{_E}^2\mathbb{Q} + \{\partial_2\partial_{_E}\mathbb{Q}, \partial_{_E}^2\mathbb{Q}\}_{\partial_{_E}} = 0 \tag{10.5.7}$$

$$p=4: \quad K^{(4)}_{x,t}(y,z)\Big|_L = \frac{1}{(2\pi i)^2}\int_{\Gamma_+} du \int_{\Gamma_-} \frac{dv}{u-v} \frac{e^{-\frac{u^5}{5}+t_3u^3+t_2u^2+zu}}{e^{-\frac{v^5}{5}+t_3v^3+t_2v^2+yv}} = \left\{\begin{array}{l}\text{Quintic}\\ \text{kernel}\end{array}\right. \tag{10.5.8}$$

$$\{\gamma_3\}: \qquad \partial_E^4\mathbb{Q} + 6(\partial_E^2\mathbb{Q})^2 - 9t_3\partial_E^2\mathbb{Q} + 3\partial_2^2\mathbb{Q} - 4\partial_3\partial_E\mathbb{Q} = 0$$

$$\{\gamma_4\}: \qquad (\partial_2\partial_E^2\mathbb{Q} - 6t_2\partial_E - \frac{9}{2}t_3\partial_2)\partial_E\mathbb{Q} + 6(\partial_E^2\mathbb{Q})(\partial_2\partial_E\mathbb{Q}) + 2\partial_2\partial_3\mathbb{Q} = 0$$

$$\partial_2\{\gamma_3\} - \partial_E\{\gamma_4\}: \frac{1}{2}\partial_2^3\mathbb{Q} - \left(\frac{3t_3}{4}\partial_2\partial_E + \partial_2\partial_3 - t_2\partial_E^2\right)\partial_E\mathbb{Q} + \left\{\partial_2\partial_E\mathbb{Q}, \partial_E^2\mathbb{Q}\right\}_{\partial_E} = 0$$

Notice that for $p = 2$ equation $\{\gamma_3\}$ is nothing but the PDE (10.3.3) governing the GUE-spectrum, leading to Painlevé II and the Tracy–Widom distribution. For $p = 3$, equation $\partial_2\{\gamma_3\} - \partial_E\{\gamma_4\}$ coincides indeed with the Pearcey equation (10.3.8), after a time change $t_2 = t/2$. The Fredholm determinant of other kernels, like the ones governing the Airy process with outliers (10.5.11), will be related to the other reductions of the KP-hierarchy, mentioned above.

Acknowledgements

The support of a National Science Foundation grant # DMS-07-04271, a European Science Foundation grant (MISGAM), a Marie Curie Grant (ENIGMA), FNRS, and "Inter-University Attraction Pole (Belgium)" grants is gratefully acknowledged.

References

[Adl01a] M. Adler and P. van Moerbeke. *Integrals over classical groups, random permutations, Toda and Toeplitz lattices,* Comm. Pure Appl. Math., **54**(2):153–205, 2001. (arXiv: math.CO/9912143).

[Adl01b] M. Adler and P. van Moerbeke. *Integrals over Grassmannians and Random Permutations,* Adv. in Math., **181** 190–249, 2001.

[Adl01c] M. Adler and P. van Moerbeke, *Hermitian, symmetric and symplectic random ensembles: PDE's for the distribution of the spectrum,* Annals of Mathematics, **153**, 149–189, (2001).

[AdL03] M. Adler and P. van Moerbeke. *Recursion relations for Unitary integrals of Combinatorial nature and the Toeplitz lattice,* Comm. Math. Phys. **237**, 397–440 (2003) (arXiv: math-ph/0201063).

[Adl05] M. Adler and P. van Moerbeke. *PDEs for the joint distributions of the Dyson, Airy and sine processes,* Ann. of Prob., **33**(4):1326–1361, 2005.

[Adl07a] M. Adler, A. Borodin and P. van Moerbeke, *Expectations of hook products on large partitions and the chi-square distribution, Forum Math,* **19**, 159–175 *(2007)*, (arXiv:math/0409554).

[Adl07b] M. Adler and P. van Moerbeke. *PDE's for the Gaussian ensemble with external source and the Pearcey distribution,* Comm. Pure and Appl. Math, **60** 1–32 (2007) (arXiv:math.PR/0509047).

[Adl09a] M. Adler, J. Delépine, P. van Moerbeke, *Dyson's nonintersecting Brownian motions with a few outliers,* Comm. Pure Appl. Math., **62**, 334–395 (2009). (arXiv:0707.0442).

[Adl09b] M. Adler, P. van Moerbeke and P. Vanhaecke, *Moment matrices and multicomponent KP, with applications to random matrix theory*, Comm. Math Phys., **286**, p 1–38 (2009), (arXiv:math-ph/0612064).

[Adl10a] M. Adler, J. Delépine, P. Vanhaecke, P. van Moerbeke, *A PDE for Nonintersecting Brownian Motions and Applications*, Adv. in Math., **226**, 1715–1755 (2011).

[Adl10b] M. Adler, P. Ferrari and P. van Moerbeke, *Airy processes with wanderers and new universality classes*, Ann. of Prob. **38**, 714–769 (2010) (arXiv:0811.1863).

[Adl10c] M. Adler, P. Ferrari and P. van Moerbeke, *Nonintersecting random walks in the neighborhood of a symmetric tacnode*, Annals of Probability **41**, 2599–2647 (2013) (arXiv:1007.1163 [math-ph]).

[Adl10d] M. Adler, P. van Moerbeke, D. Vanderstichelen, *Non-intersecting Brownian motions leaving from and going to several points*, Physica D **241**, 443–460 (2012) (arXiv:1005.1303).

[Adl10e] M. Adler, N. Orantin and P. van Moerbeke, *Universality for the Pearcey process*, Physica D, **239** p 924–941 (2010).

[Adl11a] M. Adler, M. Cafasso, P. van Moerbeke, *From the Pearcey to the Airy process*, Electron. J. Probab. **16** (2011) 1048–1064 (arXiv:1009.0683 [math.PR]).

[Adl11b] M. Adler, M. Cafasso, and P. van Moerbeke, *Fredholm determinants of general* $(1, p)$*-kernels and reductions of non-linear integrable PDE* (2011).

[Adl98] M. Adler, T. Shiota and P. van Moerbeke, *Random matrices, Virasoro algebras, and noncommutative KP*, Duke Math. J. **94**, 379–431 (1998).

[Adl99] M. Adler and P. van Moerbeke, *Generalized orthogonal polynomials, discrete KP and Riemann–Hilbert problems*, Comm. Math. Phys. **207**, 589–620 (1999).

[Apt03] A. Aptekarev, A. Branquinho and W. Van Assche, *Multiple orthogonal polynomials for classical weights*, Trans. Amer. Math. Soc., **355**, 3887–3914 (2003).

[Apt05] A. Aptekarev, P. Bleher and A. Kuijlaars, *Large n limit of Gaussian random matrices with external source*. II. Comm. Math. Phys. **259** 367–389 (2005) (arXiv: math-ph/0408041).

[Ber04] J. Baik, G. Ben Arous, and S. Péché, *Phase transition of the largest eigenvalue for non-null complex sample covariance matrices*, Ann. Probab. **33**, no. 5, 1643–1697 (2005) (arXiv:math/0403022).

[Bai06] J. Baik, *Painlevé formulas of the limiting distributions for non-null complex sample covariance matrices*, Duke Math. J. **133**, no. 2, 205–235 (2006) (arXiv:math/0504606).

[Ble04a] P. Bleher and A. Kuijlaars, *Random matrices with external source and multiple orthogonal polynomials*, Internat. Math. Research Notices **3**, 109–129 (2004) (arXiv:math-ph/0307055).

[Ble04b] P. Bleher and A. Kuijlaars, *Large n limit of Gaussian random matrices with external source, Part I*, Comm. Math. Phys., **252**, 43–76 (2004).

[Ble07] P. Bleher and A. Kuijlaars, *Large n limit of Gaussian random matrices with external source. III. Double scaling limit*, Comm. Math. Phys. **270** 481–517 (2007).

[Bor07] A. Borodin, S. Péché, *Airy kernel with two sets of parameters in directed percolation and random matrix theory*, J. Stat. Phys. **132** (2008) 275–290 (arXiv:0712.1086 [math-ph]).

[Bre98a] E. Brézin and S. Hikami, *Universal singularity at the closure of a gap in a random matrix theory*, Phys. Rev., **E 57**, 4140–4149 (1998).

[Bre98b] E. Brézin and S. Hikami, *Level spacing of random matrices in an external source*, Phys. Rev., **E 58**, 7176–7185 (1998).

[Cos00] C. M. Cosgrove, *Chazy classes IX–XII of third-order differential equations*, Stud. Appl. Math. **104**, 3, 171–228 (2000).

[Dae07] E. Daems, A. Kuijlaars and W. Veys, *Multiple orthogonal polynomials of mixed type and non-intersecting Brownian motions*, J. Approx. Theory **146**, 91–114 (2007) (arXiv:math.CA/0511470).

[Dat83] E. Date, M. Jimbo, M. Kashiwara, T. Miwa, *Transformation groups for soliton equations*, In: Proc. RIMS Symp. Nonlinear integrable systems — Classical and quantum theory (Kyoto 1981), pp. 39–119. Singapore : World Scientific 1983.

[Dei97] P. Deift, A. Its, and X. Zhou, *A Riemann–Hilbert approach to asymptotic problems arising in the theory of random matrix models, and also in the theory of integrable statistical mechanics*, Ann. Math. **146**, 149–235 (1997).

[Del10] S. Delvaux, A. Kuijlaars, and L. Zhang, *Critical behavior of non-intersecting Brownian motions at a tacnode*, arXiv:1009.2457 (2010).

[Fok92] A. Fokas, A. Its and A. Kitaev, *The Isomonodromic approach to matrix models in 2D quantum gravity*, Comm. Math. Phys. **147**, 395–430 (1992).

[For93] P.J. Forrester, *The spectrum edge of random matrix ensembles* , Nucl. Phys. B, **402**, 709–728 (1993).

[For02] P.J. Forrester and N. Witte, *Application of the τ-function theory of Painlevé equations to random matrices: rmP_{rmVI}, the JUE, CyUE, cJUE and scaled limits*, Nagoya Math. J. **174** (2004) 29–114 (arXiv:math-ph/0204008).

[Hai99] L. Haine and J.-P. Semengue, *The Jacobi polynomial ensemble and the Painlev VI equation* J. Math. Phys. **40**, 2117–2134 (1999).

[Hai10] L. Haine and D. Vanderstichelen, *A centerless representation of the Virasoro algebra associated with the unitary circular ensemble*, Journal of Comp. and Appl. Math. (2010). (arXiv:1001.4244).

[Joh01] K. Johansson, *Universality of the Local Spacing distribution in certain ensembles of Hermitian Wigner Matrices*, Comm. Math. Phys. **215**, 683–705 (2001).

[Joh05] K. Johansson, *The Arctic circle boundary and the Airy process*, Ann. Probab. **33**, no. 1, 1–30 (2005) (arXiv: Math. PR/0306216).

[Joh10] K. Johansson, *Two groups of non-colliding Brownian motions*, 2010.

[Kar59] S. Karlin and J. McGregor, *Coincidence probabilities*, Pacific J. Math. **9**, 1141–1164 (1959).

[Osi10] V. Al. Osipov, E. Kanzieper, *Correlations of RMT Characteristic Polynomials and Integrability: Hermitean Matrices*, Annals of Physics **325** (2010) 2251–2306 (arXiv:1003.0757 [math-ph]).

[Pra02] M. Prähofer and H. Spohn, *Scale Invariance of the PNG Droplet and the Airy Process*, J. Stat. Phys. **108**, 1071–1106 (2002).

[Rum09] Igor Rumanov, *Universal Structure and Universal PDE for Unitary Ensembles* (arXiv:0910.4417).

[Rum08] Igor Rumanov, *The correspondence between Tracy–Widom (TW) and Adler–Shiota–van Moerbeke (ASvM) approaches in random matrix theory: the Gaussian case*, J. Math. Phys. **49** (2008) 043503 (arXiv:0712.0862 [math-ph]).

[Tra94] C. Tracy and H. Widom, *Level-spacing distributions and the Airy kernel*, Comm. Math. Phys. **159**, 151–174 (1994).

[Tra01] C. Tracy and H. Widom, *On the distributions of the lengths of the longest monotone subsequences in random words*, Probab. Theory Related Fields **119**, no. 3, 350–380 (2001).

[Tra02] C. Tracy and H. Widom, *Distribution functions for largest eigenvalues and their applications*, Proceedings of the International Congress of Mathematicians, Vol. I , 587–596, Higher Ed. Press, Beijing, 2002.

[Tra04] C.A. Tracy and H. Widom, *Differential equations for Dyson processes*, Comm. Math. Phys. **252**, no. 1–3, 7–41 (2004).

[Tra06] C. A. Tracy and H. Widom, *The Pearcey Process*, Comm. Math. Phys. **263**, no. 2, 381–400 (2006). (arXiv:math. PR /0412005).

[Van01] W. Van Assche, J.S. Geronimo and A. Kuijlaars, Riemann–Hilbert problems for multiple orthogonal polynomials, *Special functions* 2000*: Current Perspectives and Future directions* (J. Bustoz et al., eds) Kluwer, Dordrecht, 23–59 (2001).

[Van01] P. van Moerbeke, *Integrable lattices: random matrices and random permutations*, "Random matrices and their applications", MSRI Publications #40, Cambridge Univ. Press, pp 321–406, (2001) (http://www.msri.org/publications/books/Book40/files/moerbeke.pdf).

[Van10] P. van Moerbeke, *Random and integrable models in Mathematics and Physics*, CRM-lectures, Springer 2010. (arXiv:0712.3847).

[Zin98] P. Zinn-Justin, *Universality of correlation functions in Hermitian random matrices in an external field*, Comm. Math. Phys. **194**, 631–650 (1998).

·11·

Determinantal point processes

Alexei Borodin

Abstract

We present a list of algebraic, combinatorial, and analytic mechanisms that give rise to determinantal point processes.

11.1 Introduction

Let $\mathfrak{X}$ be a discrete space. A (simple) random point process $\mathcal{P}$ on $\mathfrak{X}$ is a probability measure on the set $2^{\mathfrak{X}}$ of all subsets of $\mathfrak{X}$. $\mathcal{P}$ is called *determinantal* if there exists a $|\mathfrak{X}| \times |\mathfrak{X}|$ matrix K with rows and columns marked by elements of $\mathfrak{X}$, such that for any finite $Y = (y_1, \ldots, y_n) \subset \mathfrak{X}$ one has

$$\Pr\{X \in 2^{\mathfrak{X}} \mid Y \subset X\} = \det[K(y_i, y_j)]_{i,j=1}^{n}.$$

A similar definition can be given for $\mathfrak{X}$ being any reasonable space; then the measure lives on locally finite subsets of $\mathfrak{X}$.

Determinantal point processes (with $\mathfrak{X} = \mathbb{R}$) have been used in random matrix theory since the early 1960s. As a separate class determinantal processes were first singled out in [Mac75] to model fermions in thermal equilibrium (compare [Ben73]) and the term 'fermion' point processes was used. The term 'determinantal' was introduced in [Bor00a], for the reason that the particles of the process studied there were of two kinds; particles of the same kind repelled, while particles of different kinds attracted. Nowadays, the expression 'determinantal point process (or field)' is standard.

There are several excellent surveys of the subject available, see [Sos00], [Lyo03], [Joh05], [Kön05], [Hou06], [Sos06]. The reader will find there a detailed discussion of probabilistic properties of determinantal processes as well as a wide array of their applications; many applications are also described in various chapters of this volume.

The goal of the present chapter is to bring together all known algebraic, combinatorial, and analytic mechanisms that produce determinantal processes. Many of the well-known determinantal processes fit into more than one class described below. However, none of the classes is superseded by any other.

11.2 Generalities

Let $\mathfrak{X}$ be a locally compact separable topological space. A *point configuration* X in $\mathfrak{X}$ is a locally finite collection of points of the space $\mathfrak{X}$. Any such point configuration is either finite or infinite. For our purposes it suffices to assume that the points of X are always pairwise distinct. The set of all point configurations in $\mathfrak{X}$ will be denoted as $\mathrm{Conf}(\mathfrak{X})$.

A relatively compact Borel subset $A \subset \mathfrak{X}$ is called *a window*. For a window A and $X \in \mathrm{Conf}(\mathfrak{X})$, set $N_A(X) = |A \cap X|$ (number of points of X in the window). Thus, N_A can be viewed as a function on $\mathrm{Conf}(\mathfrak{X})$. We equip $\mathrm{Conf}(\mathfrak{X})$ with the Borel structure generated by functions N_A for all windows A.

A *random point process* on $\mathfrak{X}$ is a probability measure on $\mathrm{Conf}(\mathfrak{X})$.

Given a random point process, one can usually define a sequence $\{\rho_n\}_{n=1}^{\infty}$, where ρ_n is a symmetric measure on $\mathfrak{X}^n$ called the nth *correlation measure*. Under mild conditions on the point process, the correlation measures exist and determine the process uniquely (compare [Len73]).

The correlation measures are characterized by the following property: for any $n \geq 1$ and a compactly supported bounded Borel function f on $\mathfrak{X}^n$ one has

$$\int_{\mathfrak{X}^n} f\rho_n = \left\langle \sum_{x_{i_1},\dots,x_{i_n}\in X} f(x_{i_1},\dots,x_{i_n}) \right\rangle_{X\in\mathrm{Conf}(\mathfrak{X})} \tag{11.2.1}$$

where the sum on the right is taken over all n-tuples of pairwise distinct points of the random point configuration X.

Often one has a natural measure μ on $\mathfrak{X}$ (called the *reference measure*) such that the correlation measures have densities with respect to $\mu^{\otimes n}$, $n = 1, 2, \dots$. Then the density of ρ_n is called the nth *correlation function* and it is usually denoted by the same symbol 'ρ_n'.

If $\mathfrak{X} \subset \mathbb{R}$ and μ is absolutely continuous with respect to the Lebesgue measure, then the probabilistic meaning of the nth correlation function is that of the density of probability of finding an eigenvalue in each of the infinitesimal intervals around points $x_1, x_2, \dots x_n$:

$$\rho_n(x_1, x_2, \dots x_n)\mu(dx_1)\cdots\mu(dx_n) \\ = \Pr\{\text{there is a particle in each interval } (x_i, x_i + dx_i)\}.$$

On the other hand, if μ is supported by a discrete set of points, then

$$\rho_n(x_1, x_2, \dots x_n)\mu(x_1)\cdots\mu(x_n) \\ = \Pr\{\text{there is a particle at each of the points } x_i\}.$$

Assume that we are given a point process $\mathcal{P}$ and a reference measure such that all correlation functions exist. The process $\mathcal{P}$ is called *determinantal* if there exists a function $K : \mathfrak{X} \times \mathfrak{X} \to \mathbb{C}$ such that

$$\rho_n(x_1, \ldots, x_n) = \det[K(x_i, x_j)]_{i,j=1}^n, \qquad n = 1, 2, \ldots. \tag{11.2.2}$$

The function K is called a *correlation kernel* of $\mathcal{P}$.

The determinantal form of the correlation functions (11.2.2) implies that many natural observables for $\mathcal{P}$ can be expressed via the kernel K. We mention a few of them. For the sake of simplicity, we assume that the state space $\mathfrak{X}$ is discrete and μ is the counting measure; under appropriate assumptions, the statements are easily carried over to more general state spaces.

- Let I be a (possibly infinite) subset of $\mathfrak{X}$. Denote by K_I the operator in $\ell^2(I)$ obtained by restricting the kernel K to I. Assume that K_I is a trace class operator.[1] Then the intersection of the random configuration X with I is almost surely finite and

$$\Pr\{|X \cap I| = N\} = \frac{(-1)^N}{N!} \frac{d^N}{dz^N} \det\Big(\mathbf{1} - zK_I\Big)\Big|_{z=1}.$$

  In particular, the probability that $X \cap I$ is empty is equal to

$$\Pr\{X \cap I = \varnothing\} = \det\Big(\mathbf{1} - K_I\Big).$$

  More generally, if $I_1, \ldots, I_m$ is a finite family of pairwise nonintersecting intervals such that the operators $K_{I_1}, \ldots, K_{I_m}$ are trace class then

$$\Pr\{|X \cap I_1| = N_1, \ldots, |X \cap I_m| = N_m\} = \frac{(-1)^{\sum_{i=1}^m N_i}}{\prod_{i=1}^m N_i!} \frac{\partial^{N_1+\cdots+N_m}}{\partial z_1^{N_1} \ldots \partial z_m^{N_m}} \det\Big(\mathbf{1} - z_1 K_{I_1} - \cdots - z_m K_{I_m}\Big)\Big|_{z_1=\cdots=z_m=1}. \tag{11.2.3}$$

- Slightly more generally, let ϕ be a function on $\mathfrak{X}$ such that the kernel $(1 - \phi(x))K(x, y)$ defines a trace class operator $(1 - \phi)K$ in $\ell^2(\mathfrak{X})$. Then

$$\mathbb{E}\left(\prod_{x_i \in X} \phi(x_i)\right) = \det(\mathbf{1} - (1 - \phi)K). \tag{11.2.4}$$

  Specifying $\phi = \sum_{j=1}^m (1 - z_j)\mathbf{1}_{I_j}$ leads to (11.2.3).
- For $I \subset \mathfrak{X}$ such that K_I is trace class and $\det(\mathbf{1} - K_I) \neq 0$, and arbitrary pairwise distinct locations $\{x_1, \ldots, x_n\} \subset I$, $n = 1, 2, \ldots$, set

$$\mathcal{J}_{I,n}(x_1, \ldots, x_n) = \Pr\{\text{there is a particle at each of the points } x_i \text{ and there are no other particles in } I\}.$$

  These are sometimes called *Janossy measures*. One has

$$\mathcal{J}_{I,n}(x_1, \ldots, x_n) = \det(\mathbf{1} - K_I) \cdot \det[L_I(x_i, x_j)]_{i,j=1}^n, \tag{11.2.5}$$

[1] For discrete $\mathfrak{X}$, a convenient sufficient condition for K_I to be of trace class is $\sum_{x,y \in I} |K(x, y)| < \infty$.

where L_I is the matrix of the operator $K_I(\mathbf{1} - K_I)^{-1}$.

Simple linear-algebraic proofs of (11.2.4) and (11.2.5) can be extracted from the proof of Proposition A.6 in [Bor00b]. We also refer to Chapter 4 in this volume for a detailed discussion of (11.2.3)–(11.2.5) and many related identities.

11.3 Loop-free Markov chains

Let $\mathfrak{X}$ be a discrete space, and let $P = [P_{xy}]_{x,y\in\mathfrak{X}}$ be the matrix of transition probabilities for a discrete time Markov chain on $\mathfrak{X}$. That is, $P_{xy} \geq 0$ for all $x, y \in \mathfrak{X}$ and

$$\sum_{y\in\mathfrak{X}} P_{xy} = 1 \quad \text{for any} \quad x \in \mathfrak{X}.$$

Let us assume that our Markov chain is *loop-free*, i.e. the trajectories of the Markov chain do not pass through the same point twice almost surely. In other words, we assume that

$$(P^k)_{xx} = 0 \qquad \text{for any} \quad k > 0 \quad \text{and} \quad x \in \mathfrak{X}.$$

This condition guarantees the finiteness of the matrix elements of the matrix

$$Q = P + P^2 + P^3 + \dots .$$

Indeed, $(P^k)_{xy}$ is the probability that the trajectory started at x is at y after the kth step. Hence, Q_{xy} is the probability that the trajectory started at x passes through $y \neq x$, and since there are no loops we have $Q_{xy} \leq 1$. Clearly, $Q_{xx} \equiv 0$.

The following (simple) fact was proved in [Bor08a].

Theorem 11.3.1 *For any probability measure $\pi = [\pi_x]_{x\in\mathfrak{X}}$ on $\mathfrak{X}$, consider the Markov chain with initial distribution π and transition matrix P as a probability measure on trajectories viewed as subsets of $\mathfrak{X}$. Then this measure on $2^{\mathfrak{X}}$ is a determinantal point process on $\mathfrak{X}$ with correlation kernel*

$$K(x, y) = \pi_x + (\pi Q)_x - Q_{yx}.$$

Note that the correlation kernel is usually not self-adjoint,[2] and self-adjoint examples should be viewed as 'exotic'. One such example goes back to [Mac75], and see also §2.4 of [Sos00]: it is a 2-parameter family of renewal processes – processes on $\mathbb{Z}$ or $\mathbb{R}$ with positive i.i.d. increments. Theorem 11.3.1 implies that if we do not insist on self-adjointness then any process with positive i.i.d. increments is determinantal.

[2] In fact, it can be written as the sum of a nilpotent matrix and a matrix of rank 1.

11.4 Measures given by products of determinants

Let $\mathfrak{X}$ be a finite set and N be any natural number no greater than $|\mathfrak{X}|$. Let Φ_n and Ψ_n, $n = 1, 2, \ldots, N$, be arbitrary complex-valued functions on $\mathfrak{X}$. To any point configuration $X \in \text{Conf}(\mathfrak{X})$ we assign its weight $W(X)$ as follows: If the number of points in X is not N then $W(X) = 0$. Otherwise, using the notation $X = \{x_1, \ldots, x_N\}$, we have

$$W(X) = \det\left[\Phi_i(x_j)\right]_{i,j=1}^{N} \det\left[\Psi_i(x_j)\right]_{i,j=1}^{N}.$$

Assume that the partition function of our weights does not vanish

$$Z := \sum_{X \in \text{Conf}(\mathfrak{X})} W(X) \neq 0.$$

Then the normalized weights $\widetilde{W}(X) = W(X)/Z$ define a (generally speaking, complex valued) measure on $\text{Conf}(\mathfrak{X})$ of total mass 1. Such measures are called *biorthogonal ensembles.*[3] For complex-valued point processes we use (11.2.1) to define their correlation functions.

An especially important subclass of biorthogonal ensembles consists of *orthogonal polynomial ensembles,* for which $\mathfrak{X}$ must be a subset of $\mathbb{C}$, and

$$W(X) = \prod_{1 \le i < j \le N} |x_i - x_j|^2 \cdot \prod_{i=1}^{N} w(x_i)$$

for a function $w : \mathfrak{X} \to \mathbb{R}_+$, see e.g. [Kön05] and Chapter 4 of this volume.

Theorem 11.4.1 *Any biorthogonal ensemble is a determinantal point process. Its correlation kernel has the form*

$$K(x, y) = \sum_{i,j=1}^{N} \left[G^{-t}\right]_{ij} \Phi_i(x)\Psi_j(y),$$

where $G = [G_{ij}]_{i,j=1}^{N}$ *is the Gram matrix:* $G_{ij} = \sum_{x \in \mathfrak{X}} \Phi_i(x)\Psi_j(x)$.[4]

The statement immediately carries over to $\mathfrak{X}$ being an arbitrary state space with reference measure μ; then one has $G_{ij} = \int_{\mathfrak{X}} \Phi_i(x)\Psi_j(x)\mu(dx)$.

Probably the first appearance of Theorem 11.4.1 is in the seminal work of F. J. Dyson [Dys62a], where it was used to evaluate the correlation functions of the eigenvalues of the Haar-distributed $N \times N$ unitary matrix. In that case, $\mathfrak{X}$ is the unit circle, μ is the Lebesgue measure on it,

$$\Phi_i(z) = z^{i-1}, \qquad \Psi_i(z) = \bar{z}^{i-1}, \qquad |z| = 1, \quad i = 1, \ldots, N,$$

[3] This term was introduced in [Bor99] and is now widely used.

[4] The invertibility of the Gram matrix is implied by the assumption $Z \neq 0$.

and the Gram matrix G coincides with the identity matrix.

In the same volume, Dyson [Dys62b] introduced a Brownian motion model for the eigenvalues of random matrices (currently known as the *Dyson Brownian motion*), and it took more than three decades to find a determinantal formula for the time-dependent correlations of eigenvalues in the unitarily invariant case. The corresponding claim has a variety of applications; let us state them. Again, for simplicity of notation, we work with finite state spaces.

Let $\mathfrak{X}^{(1)}, \ldots, \mathfrak{X}^{(k)}$ be finite sets. Set $\mathfrak{X} = \mathfrak{X}^{(1)} \sqcup \cdots \sqcup \mathfrak{X}^{(k)}$. Fix a natural number N. Let

$$\begin{aligned} \Phi_i : \mathfrak{X}^{(1)} \to \mathbb{C}, \qquad \Psi_i : \mathfrak{X}^{(k)} \to \mathbb{C}, \qquad i = 1, \ldots, N \\ \mathcal{T}_{j,j+1} : \mathfrak{X}^{(j)} \times \mathfrak{X}^{(j+1)} \to \mathbb{C}, \qquad j = 1, \ldots, k-1, \end{aligned}$$

be arbitrary functions. To any $X \in \mathrm{Conf}(\mathfrak{X})$ assign its weight $W(X)$ as follows. If X has exactly N points in each $X^{(j)}$, $j = 1, \ldots, k$ then denoting $X \cap \mathfrak{X}^{(j)} = \{x_1^{(j)}, \ldots, x_N^{(j)}\}$ we have

$$\begin{aligned} W(X) = \det\left[\Phi_i(x_j^{(1)})\right]_{i,j=1}^N \det\left[\mathcal{T}_{1,2}(x_i^{(1)}, x_j^{(2)})\right]_{i,j=1}^N \cdots \\ \times \det\left[\mathcal{T}_{k-1,k}(x_i^{(k-1)}, x_j^{(k)})\right]_{i,j=1}^N \det\left[\Psi_i(x_j^{(1)})\right]_{i,j=1}^N; \qquad (11.4.1) \end{aligned}$$

otherwise $W(X) = 0$.

As for biorthogonal ensembles above, we assume that the partition function of these weights is nonzero and define the corresponding normalized set of weights. This gives a (generally speaking, complex valued) random point process on $\mathfrak{X}$.

In what follows we use the notation

$$\begin{aligned} (f * g)(x, y) = \sum_z f(x, z)g(z, y), \quad h_1 * h_2 = \sum_x h_1(x)h_2(x), \\ (h_1 * f)(y) = \sum_x h_1(x)f(x, y), \quad (g * h_2)(x) = \sum_y g(x, y)h_2(y) \end{aligned}$$

for arbitrary functions $f(x, y)$, $g(x, y)$, $h_1(x)$, $h_2(x)$, where the sums are taken over all possible values of the summation variables.

Theorem 11.4.2 *The random point process defined by (11.4.1) is determinantal. The correlation kernel on $\mathfrak{X}^{(p)} \times \mathfrak{X}^{(q)}$, $p, q = 1, \ldots, N$, can be written in the form*

$$K(x^{(p)}, y^{(q)}) = -\mathbf{1}_{p>q} \cdot (\mathcal{T}_{q,q+1} * \cdots * \mathcal{T}_{p-1,p})(y^{(q)}, x^{(p)})$$
$$+ \sum_{i,j=1}^{N} \left[G^{-t}\right]_{ij} \left(\Phi_i * \mathcal{T}_{1,2} * \cdots * \mathcal{T}_{p-1,p}\right)(x^{(p)}) \left(\mathcal{T}_{q,q+1} * \cdots * \mathcal{T}_{k-1,k} * \Psi_j\right)(y^{(q)}), \tag{11.4.2}$$

where the Gram matrix $G = \left[G_{ij}\right]_{i,j=1}^{N}$ *is defined by*

$$G_{ij} = \Phi_i * \mathcal{T}_{1,2} * \cdots * \mathcal{T}_{k-1,k} * \Psi_j, \qquad i, j = 1, \ldots, N.$$

Similarly to Theorem 11.4.1, the statement is easily carried over to general state spaces $\mathfrak{X}^{(j)}$.

Theorem 11.4.2 is often referred to as the *Eynard–Mehta theorem*, it was proved in [Eyn98] and also independently in [Nag98]. Other proofs can be found in [Joh03], [Tra04], [Bor05].

The algebraically 'nice' case of the Eynard–Mehta theorem, which for example takes place for the Dyson Brownian motion, consists of the existence of an orthonormal basis $\{\Xi_i^{(j)}\}_{i\geq 1}$ in each $L^2(\mathfrak{X}^{(j)})$, $j = 1, \ldots, k$, such that

$$T_{j,j+1}(x, y) = \sum_{i\geq 1} c_{j,j+1;i}\, \Xi_i^{(j)}(x)\Xi_i^{(j+1)}(y), \qquad j = 1, 2, \ldots, k-1,$$

for some constants $c_{j,j+1;i}$, and

$$\mathrm{Span}\{\Xi_1^{(1)}, \ldots, \Xi_N^{(1)}\} = \mathrm{Span}\{\Phi_1, \ldots, \Phi_N\},$$
$$\mathrm{Span}\{\Xi_1^{(k)}, \ldots, \Xi_N^{(k)}\} = \mathrm{Span}\{\Psi_1, \ldots, \Psi_N\}.$$

Then, with the notation $c_{k,l;i} = c_{k,k+1;i}c_{k+1,k+2;i} \cdots c_{l-1,l;i}$, (11.4.2) reads

$$K(x^{(p)}, y^{(q)}) = \begin{cases} \displaystyle\sum_{i=1}^{N} \frac{1}{c_{p,q;i}}\, \Xi_i^{(p)}(x^{(p)})\, \Xi_i^{(q)}(y^{(q)}), & p \leq q, \\ \displaystyle -\sum_{i>N} c_{q,p;i}\, \Xi_i^{(p)}(x^{(p)})\, \Xi_i^{(q)}(y^{(q)}), & p > q. \end{cases}$$

The ubiquity of the Eynard–Mehta theorem in applications is explained by the combinatorial statement known as the Lindström–Gessel–Viennot (LGV) theorem, see [Ste90] and references therein, that we now describe.

Consider a finite directed acyclic graph and denote by V and E the sets of its vertices and edges.[5] Let $w : E \to \mathbb{C}$ be an arbitrary weight function. For any path π denote by $w(\pi)$ the product of weights over the edges in the path: $w(\pi) = \prod_{e\in\pi} w(e)$. Define the weight of a collection of paths as the product of weights of the paths in the collection (we will use the same letter w to denote this). We say that two paths π_1 and π_2 do not intersect (notation $\pi_1 \cap \pi_2 = \varnothing$) if they have no common vertices.

[5] The assumption of finiteness is not necessary as long as the sums in (11.4.3) converge.

For any $u, v \in V$, let $\Pi(u, v)$ be the set of all (directed) paths from u to v. Set

$$\mathcal{T}(u, v) = \sum_{\pi \in \Pi(u,v)} w(\pi). \tag{11.4.3}$$

Theorem 11.4.3 *Let $(u_1, \dots, u_n)$ and $(v_1, \dots, v_n)$ be two n-tuples of vertices of our graph, and assume that for any nonidentical permutation $\sigma \in S(n)$,*

$$\left\{(\pi_1, \dots, \pi_n) \mid \pi_i \in \Pi\left(u_i, v_{\sigma(i)}\right), \ \pi_i \cap \pi_j = \varnothing, \ i, j = 1, \dots, n\right\} = \varnothing.$$

Then

$$\sum_{\substack{\pi_1 \in \Pi(u_1,v_1), \dots, \pi_n \in \Pi(u_n,v_n) \\ \pi_i \cap \pi_j = \varnothing, \ i,j=1,\dots,n}} w(\pi_1, \dots, \pi_n) = \det\left[\mathcal{T}(u_i, v_j)\right]_{i,j=1}^{n}.$$

Theorem 11.4.3 means that if, in a suitable weighted oriented graph, we have nonintersecting paths with fixed starting and ending vertices, then the distributions of the intersection points of these paths with any chosen 'sections' have the same structure as (11.4.1), and thus by Theorem 11.4.2 we obtain a determinantal point process.

A continuous time analogue of Theorem 11.4.3 goes back to [Kar59], which in particular proved the following statement (the next paragraph is essentially a quotation).

Consider a stationary stochastic process whose state space is an interval on the extended real line. Assume that the process has a strong Markov property and that its paths are continuous everywhere. Take n points $x_1 < \cdots < x_n$ and n Borel sets $E_1 < \cdots < E_n$, and suppose n labelled particles start at $x_1, \dots, x_n$ and execute the process simultaneously and independently. Then the determinant $\det\left[P_t(x_i, E_j)\right]_{i,j=1}^{n}$, with $P_t(x, E)$ being the transition probability of the process, is equal to the probability that at time t the particles will be found in sets $E_1, \dots, E_n$ respectively without any of them ever having been coincident in the intervening time.

Similarly to Theorem 11.4.3, this statement coupled with Theorem 11.4.2 leads to determinantal processes, and this is exactly the approach that allows one to compute the time-dependent eigenvalue correlations of the Dyson Brownian motion.

We conclude this section with a generalization of the Eynard–Mehta theorem which allows the number of particles to vary.

Let $\mathfrak{X}_1, \dots, \mathfrak{X}_N$ be finite sets, and

$$\begin{aligned}
\phi_n(\cdot, \cdot) : \mathfrak{X}_{n-1} \times \mathfrak{X}_n \to \mathbb{C}, &\qquad n = 2, \dots, N, \\
\phi_n(\mathsf{virt}, \cdot) : \mathfrak{X}_n \to \mathbb{C}, &\qquad n = 1, \dots, N, \\
\Psi_j(\cdot) : \mathfrak{X}_N \to \mathbb{C}, &\qquad j = 1, \dots, N,
\end{aligned}$$

be arbitrary functions on the corresponding sets. Here the symbol virt stands for a 'virtual' variable, which is convenient to introduce for notational purposes. In applications, virt can sometimes be replaced by $+\infty$ or $-\infty$.

Let $c(1), \ldots, c(N)$ be arbitrary nonnegative integers, and let

$$t_0^N \le \cdots \le t_{c(N)}^N = t_0^{N-1} \le \cdots \le t_{c(N-1)}^{N-1} = t_0^{N-2} \le \cdots \le t_{c(2)}^2 = t_0^1 \le \cdots \le t_{c(1)}^1$$

be real numbers. In applications, these numbers may refer to time moments of an associated Markov process. Finally, let

$$\mathcal{T}_{t_a^n, t_{a-1}^n}(\cdot\,,\,\cdot) : \mathfrak{X}_n \times \mathfrak{X}_n \to \mathbb{C}, \qquad n = 1, \ldots, N, \quad a = 1, \ldots, c(n),$$

be arbitrary functions.

Set $\mathfrak{X} = (\mathfrak{X}_1 \sqcup \cdots \sqcup \mathfrak{X}_1) \sqcup \cdots \sqcup (\mathfrak{X}_N \sqcup \cdots \sqcup \mathfrak{X}_N)$ with $c(n) + 1$ copies of each $\mathfrak{X}_n$,[6] and to any $X \in \mathrm{Conf}(\mathfrak{X})$ assign its weight $W(X)$ as follows.

The weight $W(X)$ is zero unless X has exactly n points in each copy of $\mathfrak{X}_n$, $n = 1, \ldots, N$. In the latter case, denote the points of X in the mth copy of $\mathfrak{X}_n$ by $x_k^n(t_m^n)$, $k = 1, \ldots, n$, and set

$$\begin{aligned} W(X) = \prod_{n=1}^{N} \Bigg[&\det\Big[\phi_n\big(x_k^{n-1}(t_0^{n-1}),\, x_l^n(t_{c(n)}^n)\big)\Big]_{k,l=1}^{n} \\ &\times \prod_{a=1}^{c(n)} \det\Big[\mathcal{T}_{t_a^n, t_{a-1}^n}\big(x_k^n(t_a^n),\, x_l^n(t_{a-1}^n)\big)\Big]_{k,l=1}^{n}\Bigg] \cdot \det\Big[\Psi_l\big(x_k^N(t_0^N)\big)\Big]_{k,l=1}^{N}, \end{aligned} \tag{11.4.4}$$

where $x_n^{n-1}(\cdot) = \mathsf{virt}$ for all $n = 1, \ldots, N$.

Once again, we assume that the partition function does not vanish, and normalizing the weights we obtain a (generally speaking, complex valued) point process on $\mathfrak{X}$.

We need more notation. For any $n = 1, \ldots, N$ and two time moments $t_a^n > t_b^n$ we define

$$\mathcal{T}_{t_a^n, t_b^n} = \mathcal{T}_{t_a^n, t_{a-1}^n} * \mathcal{T}_{t_{a-1}^n, t_{a-2}^n} * \cdots * \mathcal{T}_{t_{b+1}^n, t_b^n}, \qquad \mathcal{T}^n = \mathcal{T}_{t_{c(n)}^n, t_0^n}.$$

For any time moments $t_{a_1}^{n_1} \ge t_{a_2}^{n_2}$ with $(a_1, n_1) \ne (a_2, n_2)$, we denote the convolution over all the transitions between them by $\phi^{(t_{a_1}^{n_1}, t_{a_2}^{n_2})}$:

$$\phi^{(t_{a_1}^{n_1}, t_{a_2}^{n_2})} = \mathcal{T}_{t_{a_1}^{n_1}, t_0^{n_1}} * \phi_{n_1+1} * \mathcal{T}^{n_1+1} * \cdots * \phi_{n_2} * \mathcal{T}_{t_{c(n_2)}^{n_2}, t_{a_2}^{n_2}}.$$

If there are no such transitions, i.e. if $t_{a_1}^{n_1} < t_{a_2}^{n_2}$ or $(a_1, n_1) = (a_2, n_2)$, we set $\phi^{(t_{a_1}^{n_1}, t_{a_2}^{n_2})} = 0$.

Furthermore, define the "Gram matrix" $G = [G_{kl}]_{k,l=1}^N$ by

$$G_{kl} = \big(\phi_k * \mathcal{T}^k * \cdots * \phi_N * \mathcal{T}^N * \Psi_l\big)(\mathsf{virt}), \qquad k, l = 1, \ldots, N,$$

[6] Instead of $c(n) + 1$ copies of $\mathfrak{X}_n$ one can take the same number of different spaces, and a similar result will hold. We decided not to do this in order to not clutter the notation any further.

and set

$$\Psi_l^{t_a^n} = \phi^{(t_a^n, t_0^N)} * \Psi_l, \qquad l = 1, \ldots, N.$$

Theorem 11.4.4 *The random point process on $\mathfrak{X}$ defined by (11.4.4) is determinantal. Its correlation kernel can be written in the form*

$$\begin{aligned} K(t_{a_1}^{n_1}, x_1; t_{a_2}^{n_2}, x_2) &= -\phi^{(t_{a_2}^{n_2}, t_{a_1}^{n_1})}(x_2, x_1) \\ &\quad + \sum_{i=1}^{n_1} \sum_{j=1}^{N} \left[G^{-t}\right]_{ij} (\phi_i * \phi^{(t_{c(i)}^{i}, t_{a_1}^{n_1})})(\mathsf{virt}, x_1)\, \Psi_j^{t_{a_2}^{n_2}}(x_2). \end{aligned}$$

One proof of Theorem 11.4.4 was given in [Bor08b]; another proof can be found in Section 4.4 of [For08]. Although we stated Theorem 11.4.4 for the case when all sets $\mathfrak{X}_n$ are finite, one can easily extend it to a more general setting.

11.5 L-ensembles

The definition of L-ensembles is closely related to (11.2.5).

Let $\mathfrak{X}$ be a finite set. Let L be a $|\mathfrak{X}| \times |\mathfrak{X}|$ matrix whose rows and column are parameterized by points of $\mathfrak{X}$. For any subset $X \subset \mathfrak{X}$ we will denote by L_X the symmetric submatrix of L corresponding to X: $L_X = \left[L(x_i, x_j)\right]_{x_i, x_j \in X}$. If determinants of all such submatrices are nonnegative (e.g. if L is positive definite), one can define a random point process on $\mathfrak{X}$ by

$$\Pr\{X\} = \frac{\det L_X}{\det(\mathbf{1} + L)}, \qquad X \subset \mathfrak{X}.$$

This process is called the *L-ensemble.*

The following statement goes back to [Mac75].

Theorem 11.5.1 *The L-ensemble as defined above is a determinantal point process with the correlation kernel K given by $K = L(\mathbf{1} + L)^{-1}$.*

Take a nonempty subset $\mathfrak{Y}$ of $\mathfrak{X}$ and, given an L-ensemble on $\mathfrak{X}$, define a new random point process on $\mathfrak{Y}$ by considering the intersections of the random point configurations $X \subset \mathfrak{X}$ of the L-ensemble with $\mathfrak{Y}$, provided that these point configurations contain the complement $\overline{\mathfrak{Y}}$ of $\mathfrak{Y}$ in $\mathfrak{X}$. It is not hard to see that this new process can be defined by

$$\Pr\{Y\} = \frac{\det L_{Y \cup \overline{\mathfrak{Y}}}}{\det(\mathbf{1}_{\mathfrak{Y}} + L)}, \qquad Y \in \mathrm{Conf}(\mathfrak{Y}).$$

Here $\mathbf{1}_{\mathfrak{Y}}$ is the block matrix $\begin{bmatrix} \mathbf{1} & 0 \\ 0 & 0 \end{bmatrix}$ where the blocks correspond to the splitting $\mathfrak{X} = \mathfrak{Y} \sqcup \overline{\mathfrak{Y}}$. This new process is called the *conditional L-ensemble*. The next statement was proved in [Bor05].

Theorem 11.5.2 *The conditional L-ensemble is a determinantal point process with the correlation kernel given by*

$$K = \mathbf{1}_{\mathfrak{Y}} - (\mathbf{1}_{\mathfrak{Y}} + L)^{-1}\big|_{\mathfrak{Y}\times\mathfrak{Y}}.$$

Note that for $\mathfrak{Y} = \mathfrak{X}$, Theorem 11.5.2 coincides with Theorem 11.5.1.

Not every determinantal process is an L-ensemble; for example, the processes afforded by Theorem 11.4.1 have exactly N particles, which is not possible for an L-ensemble. However, as shown in [Bor05], every determinantal process (on a finite set) is a conditional L-ensemble.

The definition of L-ensembles and Theorems 11.5.1, 11.5.2 can be carried over to infinite state spaces $\mathfrak{X}$, given that L satisfies appropriate conditions. In particular, the Fredholm determinant $\det(\mathbf{1}_{\mathfrak{Y}} + L)$ needs to be well defined.

Although L-ensembles do arise naturally, see e.g. [Bor00b], they also constitute a convenient computation tool. For example, proofs of Theorems 11.4.2 and 11.4.4 given in [Bor05] and [Bor08b] represent the processes in question as conditional L-ensembles and employ Theorem 11.5.2.

Here is another application of Theorem 11.5.2.

A random point process on (a segment of) $\mathbb{Z}$ is called *one-dependent* if for any two finite sets $A, B \subset \mathbb{Z}$ with $dist(A, B) \geq 2$, the correlation function factorizes: $\rho_{|A|+|B|}(A \cup B) = \rho_{|A|}(A)\rho_{|B|}(B)$.

Theorem 11.5.3 *Any one-dependent point process on (a segment of) $\mathbb{Z}$ is determinantal. Its correlation kernel can be written in the form*

$$K(x, y) = \begin{cases} 0, & x - y \geq 2, \\ -1, & x - y = 1, \\ \sum\limits_{r=1}^{y-x+1} (-1)^{r-1} \sum\limits_{x=l_0<l_1<\cdots<l_r=y+1} R_{l_0,l_1} R_{l_1,l_2} \cdots R_{l_{r-1},l_r} & x \leq y, \end{cases}$$

where $R_{a,b} = \rho_{b-a}(a, a+1, \ldots, b-1)$.

Details and applications can be found in [Bor09].

11.6 Fock space

A general construction of determinantal point processes via the Fock space formalism can be quite technical, see e.g. [Lyt02], so we will consider a much simpler (however non-trivial) example instead.

Recall that a *partition* $\lambda = (\lambda_1 \geq \lambda_2 \geq \cdots \geq 0)$ is a weakly decreasing sequence of nonnegative integers with finitely many nonzero terms. We will use standard notations $|\lambda| = \lambda_1 + \lambda_2 + \ldots$ for the size of the partition and $\ell(\lambda)$ for the number of its nonzero parts.

The *poissonized Plancherel measure* on partitions is defined by

$$\Pr\{\lambda\} = e^{-\theta^2} \left(\frac{\prod_{1\leq i<j\leq L}(\lambda_i - i - \lambda_j + j)}{\prod_{i=1}^{L}(\lambda_i - i + L)!} \theta^{|\lambda|} \right)^2, \tag{11.6.1}$$

where $\theta > 0$ is a parameter, and L is an arbitrary integer $\geq \ell(\lambda)$. We refer to [Bor00b], [Joh01] and references therein for details.

It is convenient to parameterize partitions by subsets of $\mathbb{Z}' := \mathbb{Z} + \frac{1}{2}$:

$$\lambda \mapsto \mathcal{L}(\lambda) = \left\{\lambda_i - i + \tfrac{1}{2}\right\}_{i\geq 1} \subset \mathbb{Z}'.$$

The pushforward of (11.6.1) via $\mathcal{L}$ defines a point process on $\mathbb{Z}'$, and we aim to show that it is determinantal. We follow [Oko01]; other proofs of this fact can be found in [Bor00b], [Joh01].

Let V be a linear space with basis $\left\{\underline{k} \mid k \in \mathbb{Z}'\right\}$. The linear space $\Lambda^{\frac{\infty}{2}} V$ is, by definition, spanned by vectors

$$v_S = \underline{s_1} \wedge \underline{s_2} \wedge \underline{s_3} \wedge \ldots,$$

where $S = \{s_1 > s_2 > \ldots\} \subset \mathbb{Z}'$ is such that both sets $S_+ = S \setminus \mathbb{Z}'_{<0}$ and $S_- = \mathbb{Z}'_{<0} \setminus S$ are finite. We equip $\Lambda^{\frac{\infty}{2}} V$ with the inner product in which the basis $\{v_S\}$ is orthonormal.

Creation and *annihilation* operators in $\Lambda^{\frac{\infty}{2}} V$ are introduced as follows. The creation operator ψ_k is the exterior multiplication by $\underline{k}$: $\psi_k(f) = \underline{k} \wedge f$. The annihilation operator ψ_k^* is its adjoint. These operators satisfy the canonical anti-commutation relations

$$\psi_k \psi_l^* + \psi_l^* \psi_k = \delta_{k,l}, \qquad k, l \in \mathbb{Z}'.$$

Observe that

$$\psi_k \psi_k^* \, v_S = \begin{cases} v_S, & k \in S, \\ 0, & k \notin S. \end{cases} \tag{11.6.2}$$

Let C be the *charge operator*: $C v_S = (|S_+| - |S_-|) v_S$. One easily sees that the zero-charge subspace $\ker C \subset \Lambda^{\frac{\infty}{2}} V$ is spanned by the vectors $v_{\mathcal{L}(\lambda)}$ with λ varying over all partitions. The *vacuum vector*

$$v_{\text{vac}} = \underline{-\tfrac{1}{2}} \wedge \underline{-\tfrac{3}{2}} \wedge \underline{-\tfrac{5}{2}} \wedge \ldots$$

corresponds to the partition with no nonzero parts.

Define the operators $\alpha_n = \sum_{k\in\mathbb{Z}'} \psi_{k-n} \psi_k^*$, $n \in \mathbb{Z} \setminus \{0\}$. Although the sums are infinite, the application of α_n to any v_S yields a finite linear combination of basis vectors. These operators satisfy the Heisenberg commutation relations

$$a_m a_n - a_n a_m = m\,\delta_{n,-m}, \qquad m, n \in \mathbb{Z} \setminus \{0\}.$$

For any $\theta > 0$, define $\Gamma_\pm(\theta) = \exp(\theta a_{\pm 1})$. It is not difficult to show that

$$\Gamma_\pm^*(\theta) = \Gamma_\mp(\theta), \qquad \Gamma_+(\theta)\Gamma_-(\theta') = e^{\theta\theta'} \cdot \Gamma_-(\theta')\Gamma_+(\theta), \qquad \Gamma_+(\theta)v_{\mathrm{vac}} = v_{\mathrm{vac}}. \tag{11.6.3}$$

One also proves that

$$\Gamma_-(\theta)v_{\mathrm{vac}} = \sum_\lambda \left(\frac{\prod_{1\le i<j\le L}(\lambda_i - i - \lambda_j + j)}{\prod_{i=1}^{L}(\lambda_i - i + L)!}\,\theta^{|\lambda|} \right) v_{\mathcal{L}(\lambda)},$$

where the sum is taken over all partitions, compare (11.6.1). This implies, together with (11.6.2), that for any $n \ge 1$ and $x_1, \ldots, x_n \in \mathbb{Z}'$, the correlation function of our point process can be written as a matrix element

$$\rho_n(x_1, \ldots, x_n) = e^{-\theta^2} \left(\left(\prod_{i=1}^{n} \psi_{x_i}\psi_{x_i}^* \right) \Gamma_-(\theta)\, v_{\mathrm{vac}}, \Gamma_-(\theta)v_{\mathrm{vac}} \right).$$

Using (11.6.3) we obtain

$$\rho_n(x_1, \ldots, x_n) = \left(\prod_{i=1}^{n} \Psi_{x_i}\,\Psi_{x_i}^*\, v_{\mathrm{vac}}, v_{\mathrm{vac}} \right), \tag{11.6.4}$$

where

$$\Psi_k = G\,\psi_k\,G^{-1}, \quad \Psi_k^* = G\,\psi_k^*\,G^{-1}, \quad G = \Gamma_+(\theta)\,\Gamma_-(\theta)^{-1}.$$

Theorem 11.6.1 *We have*

$$\rho_n(x_1, \ldots, x_n) = \det\left[K(x_i, x_j)\right]_{i,j=1}^{n}, \tag{11.6.5}$$

where $K(x, y) = \left(\Psi_x\Psi_y^*\, v_{\mathrm{vac}}, v_{\mathrm{vac}}\right)$.

The passage from (11.6.4) to (11.6.5) is an instance of the *fermionic Wick theorem*; it uses the fact that Ψ_x and Ψ_y^* are linear combinations of ψ_ks and ψ_l^*s respectively, together with the canonical anti-commutation relations.

A further computation gives an explicit formula for the correlation kernel:

$$K(x, y) = \theta\,\frac{J_{x-\frac{1}{2}}J_{y+\frac{1}{2}} - J_{x+\frac{1}{2}}J_{y-\frac{1}{2}}}{x - y} = \sum_{k\in\mathbb{Z}'_{>0}} J_{x+k}J_{y+k},$$

where $J_k = J_k(2\theta)$ are the J-Bessel functions. This is the so-called discrete Bessel kernel that was first obtained in [Bor00b], [Joh01].

We refer to [Oko01] and [Oko03] for far-reaching generalizations of Theorem 11.6.1, and to [Lyt02] for a general construction of determinantal processes via representations of the canonical anti-commutation relations corresponding to the *quasi-free states*.

11.7 Dimer models

Consider a finite planar graph $\mathcal{G}$. Let us assume that the graph is *bipartite*, i.e. its vertices can be coloured black and white so that each edge connects vertices of different colours. Let us fix such a colouring and denote by B and W the sets of black and white vertices.

A *dimer covering* or a *domino tiling* or a *perfect matching* of a graph is a subset of edges that covers every vertex exactly once. Clearly, in order for the set of dimer coverings of $\mathcal{G}$ to be nonempty, we must have $|B| = |W|$.

A *Kasteleyn weighting* of $\mathcal{G}$ is a choice of sign for each edge with the property that each face with 0 mod 4 edges has an odd number of minus signs, and each face with 2 mod 4 edges has an even number of minus signs. It is not hard to show that a Kasteleyn weighting of $\mathcal{G}$ always exists (here it is essential that the graph is planar), and that any two Kasteleyn weightings can be obtained one from the other by a sequence of multiplications of all edges at a vertex by -1.

A *Kasteleyn matrix* of $\mathcal{G}$ is a signed adjacency matrix of $\mathcal{G}$. More exactly, given a Kasteleyn weighting of $\mathcal{G}$, define a $|B| \times |W|$ matrix $\mathfrak{K}$ with rows marked by elements of B and columns marked by elements of W, by setting $\mathfrak{K}(b, w) = 0$ if b and w are not joined by an edge, and $\mathfrak{K}(b, w) = \pm 1$ otherwise, where $\pm$ is chosen according to the weighting.

It is a result of [Tem61], [Kas67] that the number of dimer coverings of $\mathcal{G}$ equals $|\det\mathfrak{K}|$. Thus, if there is at least one perfect matching, the matrix $\mathfrak{K}$ is invertible.

Assume that $\det\mathfrak{K} \neq 0$. Define a matrix K with rows and columns parameterized by the edges of $\mathcal{G}$ as follows: $K(e, e') = \mathfrak{K}^{-1}(w, b')$, where w is the white vertex on the edge e, and b' is the black vertex on the edge e'. The next claim follows from the results of [Ken97].

Theorem 11.7.1 *Consider the uniform measure on the dimer covers of $\mathcal{G}$ as a random point process on the set $\mathfrak{X}$ of edges of $\mathcal{G}$. Then this process is determinantal, and its correlation kernel is the matrix K introduced above.*

The theory of random dimer covers is a deep and beautiful subject that has been actively developing over the last 15 years. We refer the reader to [Ken09] and references therein for further details.

11.8 Uniform spanning trees

Let $\mathcal{G}$ be a finite connected graph. A *spanning tree* of $\mathcal{G}$ is a subset of edges of $\mathcal{G}$ that has no loops, and such that every two vertices of $\mathcal{G}$ can be connected within this subset.

Clearly, the set of spanning trees is nonempty and finite. Any probability measure on this set can be viewed as a random point process on the set $\mathfrak{X}$ of edges of $\mathcal{G}$. We are interested in the uniform measure on the set of spanning trees, and we denote the corresponding process on $\mathfrak{X}$ by $\mathcal{P}$.

Let us fix an orientation of all the edges of $\mathcal{G}$. For any two edges $e = \vec{xy}$ and f denote by $K(e, f)$ the expected number of passages through f, counted with a sign, of a random walk started at x and stopped when it hits y.

The quantity $K(e, f)$ also has an interpretation in terms of electric networks. Consider $\mathcal{G}$ with the fixed orientation of the edges as an electric network with each edge having unit conductance. Then $K(e, f)$ is the amount of current flowing through the edge f when a battery is connected to the endpoints x and y of e, and the voltage is such that unit current is flowing from y to x. For this reason, K is called the *transfer current matrix*.

This matrix also has a linear-algebraic definition. To any vertex v of $\mathcal{G}$ we associate a vector $a(v) \in \ell^2(\mathfrak{X})$ (recall that $\mathfrak{X}$ is the set of edges) as follows:

$$a(v) = \sum_{e \in \mathfrak{X}} a_e(v)\delta_e, \qquad a_x(e) = \begin{cases} 1, & \text{if } v \text{ is the tail of } e, \\ -1, & \text{if } v \text{ is the head of } e, \\ 0, & \text{otherwise.} \end{cases}$$

Then $[K(e, f)]_{e, f \in \mathfrak{X}}$ is the matrix of the orthogonal projection operator with image $\operatorname{Span}\{a(v)\}$, where v varies over all vertices of $\mathcal{G}$, see e.g. [Ben01].

Theorem 11.8.1 *$\mathcal{P}$ is a determinantal point process with correlation kernel K.*

Theorem 11.8.1 was proved in [Bur93]; another proof can be found in [Ben01]. The formulae for the first and second correlation functions via K go back to [Kir1847] and [Bro40], respectively. We refer to [Lyo03] and references therein for further developments of the subject.

Note that for planar graphs, the study of the uniform spanning trees may be reduced to that of dimer models on related graphs and *vice versa*, see [Bur93], [Ken00], [Ken04].

11.9 Hermitian correlation kernels

Let $\mathfrak{X}$ be $\mathbb{R}^d$ or $\mathbb{Z}^d$ with the Lebesgue or the counting measure as the reference measure. Let K be a nonnegative operator in $L^2(\mathfrak{X})$.[7] In the case $\mathfrak{X} = \mathbb{R}^d$, we also require K to be locally trace class, i. e. for any compact $B \subset \mathfrak{X}$, the operator $K \cdot \mathbf{1}_B$ is trace class. Then Lemma 2 in [Sos00] shows that one can choose an integral kernel $K(x, y)$ of K so that

[7] An Hermitian K must be nonnegative as we want $\det[K(x_i, x_j)] \geq 0$.

$$\mathrm{Trace}(K\cdot \mathbf{1}_B)^k = \int_{B^k} K(x_1,x_2)K(x_2,x_3)\cdots K(x_k,x_1)dx_1\cdots dx_k, \qquad k=1,2,\ldots$$

For $\mathfrak{X}=\mathbb{Z}^d$, $K(x,y)$ is just the matrix of K.

Theorem 11.9.1 *There exists a determinantal point process on $\mathfrak{X}$ with the correlation kernel $K(x,y)$ if and only if $0\le K\le 1$, i.e. both K and $\mathbf{1}-K$ are nonnegative.*

Theorem 11.9.1 was proved in [Sos00]; an incomplete argument was also given in [Mac75]. Remarkably, this remains the only known characterization of a broad class of kernels that yield determinantal point processes.

Although only Theorem 11.4.1 with $\Phi_i=\overline{\Psi}_i$, Theorem 11.6.1, and Theorem 11.8.1 from the previous sections yield manifestly nonnegative kernels, determinantal processes with such kernels are extremely important, and they are also the easiest to analyse asymptotically, compare [Hou06].

Let us write down the correlation kernels for the two most widely known determinantal point processes; they both fall into the class afforded by Theorem 11.9.1.

The *sine process* on $\mathbb{R}$ corresponds to the *sine kernel*

$$K^{\mathrm{sine}}(x,y)=\frac{\sin\pi(x-y)}{\pi(x-y)}=\int_{-\frac{1}{2}}^{\frac{1}{2}} e^{2i\pi\tau x}e^{-2i\pi\tau y}d\tau.$$

The Fourier transform of the corresponding integral operator K^{sine} in $L^2(\mathbb{R})$ is the operator of multiplication by an indicator function of an interval; hence K^{sine} is a self-adjoint projection operator.

The *Airy point process*[8] on $\mathbb{R}$ is defined by the *Airy kernel*

$$K^{\mathrm{Airy}}(x,y)=\frac{Ai(x)Ai'(y)-Ai'(x)Ai(y)}{x-y}=\int_0^{+\infty} Ai(x+\tau)Ai(y+\tau)d\tau,$$

where $Ai(x)$ stands for the classical Airy function. The integral operator K^{Airy} can be viewed as a spectral projection operator for the differential operator $\frac{d^2}{dx^2}-x$ that has the shifted Airy functions $\{Ai(x+\tau)\}_{\tau\in\mathbb{R}}$ as the (generalized) eigenfunctions.

11.10 Pfaffian point processes

A random point process on $\mathfrak{X}$ is called *Pfaffian* if there exists a 2×2 matrix valued skew-symmetric kernel K on $\mathfrak{X}$ such that the correlation functions of the process have the form

$$\rho_n(x_1,\ldots,x_n)=\mathrm{Pf}\big[K(x_i,x_j)\big]_{i,j=1}^n, \qquad x_1,\ldots,x_n\in\mathfrak{X}, \quad n=1,2,\ldots$$

[8] Not to be confused with the Airy process that describes the time evolution of the top particle of the Airy point process, see Chapter 37 of the present volume.

The notation Pf in the right-hand side stands for the Pfaffian, and we refer to [deB55] for a concise introduction to Pffafians and to [Dei09] for a more detailed exposition.

Pfaffian processes are significantly harder to study than determinantal ones. Let us list some Pffafian analogues of the statements from the previous sections.

- A Pfaffian analogue of the Fredholm determinant formula (11.2.4) for the generating functional can be found in Section 8 of [Rai00].
- A Pfaffian analogue of the Eynard–Mehta theorem is available in [Bor05]. One well-known corollary is that measures defined by weights of the form $\det[\Phi_i(x_j)]_{i,j=1}^N$ on $x_1 < \cdots < x_N$ give rise to Pfaffian processes.
- Pfaffians can be used to enumerate nonintersecting paths with free end-points, see [Ste90]. This leads to combinatorial examples for the Pfaffian Eynard–Mehta theorem.
- Pfaffian L-ensembles and conditional L-ensembles are treated in [Bor05].
- Fermionic Fock space computations leading to a Pfaffian point process were performed in [Fer04] and [Vul07].
- Pfaffians arise in the enumeration of dimer covers of planar graphs that are not necessarily bipartite, see [Tem61], [Kas67].

Acknowledgements

This work was supported by NSF grant DMS-0707163.

References

[Ben73] C. Benard and O. Macchi, *Detection and emission processes of quantum particles in a chaotic state*, J. Math, Phys. **14** (1973) 155–167.

[Ben01] I. Benjamini, R. Lyons, Y. Peres, and O. Schramm, *Uniform spanning forests*, Ann. Probab. **29** (2001) 1–65.

[Bor99] A. Borodin, *Biorthogonal ensembles*, Nuclear Phys. B **536** (1999) 704–732, [arXiv:math/9804027].

[Bor00a] A. Borodin and G. Olshanski, *Distributions on partitions, point processes and the hypergeometric kernel*, Comm. Math. Phys. **211** (2000) 335–358, [arXiv:math/9904010].

[Bor00b] A. Borodin, A. Okounkov and G. Olshanski, *Asymptotics of Plancherel measures for symmetric groups*, J. Amer. Math. Soc. **13** (2000) 491–515, [arXiv:math/9905032].

[Bor05] A. Borodin and E. M. Rains, *Eynard–Mehta theorem, Schur process, and their Pfaffian analogs*, Jour. Stat. Phys. **121** (2005), 291–317 [arXiv:math-ph/0409059].

[Bor08a] A. Borodin, *Loop-free Markov chains as determinantal point processes*, Ann. Ins. Henri Poincaré - Prob. et Stat. **44** (2008) 19–28, [arXiv:math/0605168].

[Bor08b] A. Borodin and P. L. Ferrari, *Large time asymptotics of growth models on space-like paths I: PushASEP*, Electr. Jour. Prob. **13** (2008) 1380–1418, [arXiv:0707.2813].

[Bor09] A. Borodin, P. Diaconis, J. Fulman, *On adding a list of numbers (and other one-dependent determinantal processes)*, Bull. Amer. Math. Soc. **47** (2010) 639–670 [arXiv:0904.3740 [math.PR]].

[Bro40] R. L. Brooks, C. A. B. Smith, A. H. Stone and W. T. Tutte, *The dissection of rectangles into squares*, Duke Math. Jour. **7** (1940) 312–340.

[Bur93] R. M. Burton and R. Pemantle, *Local characteristics, entropy and limit theorems for spanning trees and domino tilings via transfer-impedances*, Ann. Prob. **21** (1993) 1329–1371, [arXiv:math/0404048].

[deB55] N. G. de Bruijn, *On some multiple integrals involving determinants*, J. Indian Math. Soc. **19** (1955) 133–151.

[Dei09] P. Deift and D. Gioev, *Random Matrix Theory: Invariant Ensembles and Universality*, Courant Lecture Notes **18**, Amer. Math. Soc. 2009.

[Dys62a] F. J. Dyson, *Statistical theory of energy levels of complex systems III*, Jour. Math. Phys. **3** (1962), 166–175.

[Dys62b] F. J. Dyson, *A Brownian motion model for the eigenvalues of a random matrix*, Jour. Math. Phys. **3** (1962), 1191–1198.

[Eyn98] B. Eynard and M. L. Mehta, *Matrices coupled in a chain. I. Eigenvalue correlations*, J. Phys. A: Math. Gen. **31** (1998), 4449–4456 [arXiv:cond-mat/9710230].

[Fer04] P. L. Ferrari, *Polynuclear growth on a flat substrate and edge scaling of GOE eigenvalues*, Comm. Math. Phys. **252** (2004) 77–109.

[For08] P.J. Forrester, E. Nordenstam, *The Anti-Symmetric GUE Minor Process*, Mosc. Math. J. **9** (2009) 749–774 [arXiv:0804.3293 [math.PR]].

[Hou06] J. B. Hough, M. Krishnapur, Y. Peres and B. Virág, *Determinantal processes and independence*, Probab. Surv. **3** (2006) 206–229, [arXiv:math/0503110].

[Joh01] K. Johansson, *Discrete orthogonal polynomial ensembles and the Plancherel measure*, Ann. of Math. (2) **153** (2001) 259–296, [arXiv:math/9906120].

[Joh03] K. Johansson, *Discrete polynuclear growth and determinantal processes*, Comm. Math. Phys. **242** (2003) 277–329, [arXiv:math/0206208].

[Joh05] K. Johansson, *Random matrices and determinantal processes*, Lectures Les Houches summer school 2005, [arXiv:math-ph/0510038].

[Kar59] S. Karlin and J. McGregor, *Coincidence probabilities*, Pacific J. Math. **9** (1959) 1141–1164.

[Kas67] P. Kasteleyn, *Graph theory and crystal physics*, Graph Theory and Theoretical Physics 43–110, Academic Press, London 1967.

[Ken97] R. W. Kenyon, *Local statistics of lattice dimers*, Ann. Inst. H. Poincaré, Probabilités **33** (1997), 591–618, [arXiv:math/0105054].

[Ken00] R. W. Kenyon, J. G. Propp and D. B. Wilson, *Trees and matchings*, Electron. J. Combin. **7** (2000), Research Paper 25, [arXiv:math/9903025].

[Ken04] R. W. Kenyon and S. Sheffield, *Dimers, tilings and trees*, J. Combin. Theory Ser. B **92** (2004) 295–317, [arXiv:math/0310195].

[Ken09] R. W. Kenyon, *Lectures on dimers*, in: Statistical mechanics, pp. 191–230, IAS/Park City Math. Ser. **16**, Amer. Math. Soc., Providence, RI, 2009, [arXiv:0910.3129].

[Kir1847] G. Kirchhoff, *Über die Auflösung der Gleichungen, auf welche man bei der Untersuchung der linearen Verteilung galvanischer Ströme geführt wird*, Ann. Phys. Chem. 72 (1847) 497–508.

[Kön05] W. König, *Orthogonal polynomial ensembles in probability theory*, Probab. Surveys 2 (2005) 385–447, [arXiv:math/0403090].

[Len73] A. Lenard, *Correlation functions and the uniqueness of the state in classical statistical mechanics*, Comm. Math. Phys. **30** (1973) 35–44.

[Lyo03] R. Lyons, *Determinantal probability measures*, Publ. Math. Inst. Hautes Etudes Sci. **98** (2003) 167–212, [arXiv:math/0204325].

[Lyt02] E. Lytvynov, *Fermion and boson random point processes as particle distributions of infinite free Fermi and Bose gases of finite density*, Rev. Math. Phys. **14** (2002) 1073–1098, [arXiv:math-ph/0112006].

[Mac75] O. Macchi, *The coincidence approach to stochastic point processes*, Adv. Appl. Prob. **7** (1975) 83–122.

[Nag98] T. Nagao and P. J. Forrester, *Multilevel Dynamical Correlation Function for Dyson's Brownian Motion Model of Random Matrices*, Phys Lett. **A247** (1998), 42–46.

[Oko01] A. Okounkov, *Infinite wedge and random partitions*, Selecta Math. (N.S.) **7** (2001) 57–81, [arXiv:math/9907127].

[Oko03] A. Okounkov and N. Reshetikhin, *Correlation function of Schur process with application to local geometry of a random 3-dimensional Young diagram*, Jour. Amer. Math. Soc. **16** (2003) 581–603, [arXiv:math/0107056].

[Rai00] E. M. Rains, *Correlation functions for symmetrized increasing subsequences*, [arXiv:math/0006097].

[Sos00] A. Soshnikov, *Determinantal random point fields*, Russian Math. Surveys **55** (2000) 923–975, [arXiv: math/0002099].

[Sos06] A. Soshnikov, *Determinantal Random Fields*, in: Encyclopedia of Mathematical Physics, pp. 47–53, Oxford: Elsevier, 2006.

[Ste90] J. R. Stembridge, *Nonintersecting Paths, Pfaffians, and Plane Partitions*, Adv. in Math. **83** (1990), 96–131.

[Tem61] W. Temperley and M. Fisher, *Dimer problem in statistical mechanics an exact result*, Philos. Mag. **6** (1961) 1061–1063.

[Tra04] C. A. Tracy and H. Widom, *Differential equations for Dyson processes*, Comm. Math. Phys. **252** (2004) 7–41, [arXiv:math/0309082].

[Vul07] M. Vuletić, *The shifted Schur process and asymptotics of large random strict plane partitions*, Int. Math. Res. Not. (2007) Art. ID rnm043, [arXiv:math-ph/0702068].

·12·

Random matrix representations of critical statistics

V. E. Kravtsov

Abstract

We consider two random matrix ensembles that are relevant for describing critical spectral statistics in systems with multifractal eigenfunction statistics. One of them is the Gaussian non-invariant ensemble whose eigenfunction statistics is multifractal, while the other is the invariant random matrix ensemble with a shallow, log-square confinement potential. We demonstrate a close correspondence between the spectral as well as eigenfuncton statistics of these random matrix ensembles and those of the random tight-binding Hamiltonian in the point of the Anderson localization transition in three dimensions. Finally we present a simple field theory in 1+1 dimensions which reproduces level statistics of both of these random matrix models and the classical Wigner-Dyson spectral statistics in the framework of the unified formalism of Luttinger liquid. We show that the (equal-time) density correlations in both random matrix models correspond to the finite-temperature density correlations of the Luttinger liquid. We show that the spectral correlations in the invariant ensemble with log-square confinement correspond to the Luttinger liquid in the curved space-time with the event horizon and are similar to the phonon density correlations in the $1+1$ dimensional sonic analogue of Hawking radiation in black holes.

12.1 Introduction

It has been known since the pioneering work by F. Wegner [Weg80, Weg85] that the eigenfunctions $\psi_i(\mathbf{r})$ of the random Schrödinger operator $\hat{H}\psi_i(\mathbf{r}) = E_i\,\psi_i(\mathbf{r})$ at the mobility edge $E_i = E_m$ corresponding to the critical point of the Anderson localization transition, possess the property of multifractality. In particular, at $E = E_m$ the moments of the inverse participation ratio,

$$P_n(E) = \sum_{\mathbf{r}}\sum_{i}\langle|\psi_i(\mathbf{r})|^{2n}\,\delta(E_i - E)\rangle \propto L^{-d_n(n-1)}, \tag{12.1.1}$$

scale as a certain power-law with the total size L of the system. This power is a critical exponent which depends only on the basic symmetry (see [Meh04] and Chapter 3 of this book) of the Hamiltonian $\hat{H}$ and on the dimensionality of space d. The true extended states of the Schrodinger operator are characterized by all $d_n = d$. This allows us to interpret $d_n < d$ as a certain *fractal dimension* which depends on the order n of the moment. As a matter of fact the statistics of critical eigenfunctions are described by a *set* of fractal dimensions d_n which justifies the notion of *multi*fractality [Weg85].

Another aspect of criticality is the scaling with the energy difference $\omega = |E_i - E_j|$ between *two* eigenvalues. It is similar to *dynamical scaling* and is relevant for the correlation functions of *different* eigenfunctions $\psi_i(\mathbf{r})$ and $\psi_j(\mathbf{r})$. The most important of these is the local density of states correlation function (DoS),

$$C(\omega, R) = \sum_{\mathbf{r}} \sum_{i \neq j} \langle |\psi_i(\mathbf{r})|^2 \, |\psi_j(\mathbf{r} + \mathbf{R})|^2 \, \delta(E - E_i)\delta(E + \omega - E_j)\rangle. \qquad (12.1.2)$$

This correlation function is relevant to the matrix elements of the two-body interaction. The dynamical scaling connects the power law behaviour of $C(\omega, 0) \propto \omega^{-\mu}$ with that of $C(0, R) \propto R^{-(d-d_2)}$ by a conjecture [Cha90] on the dynamical exponent,

$$R^d \to \omega, \qquad \mu = 1 - \frac{d_2}{d}. \qquad (12.1.3)$$

Although there is extensive numerical evidence in favour of conjecture Eq. (12.1.3) its rigorous proof has been lacking so far. In a very recent work [Kra10] it was shown analytically that the dynamical scaling Eq. (12.1.3) holds true for the power-law banded random matrices Eq. (12.2.1) in the limit of strong fractality.

Last but not least, there is a growing interest in the spectral (level) statistics in quantum systems whose classical counterparts are in between chaos and integrability [Bog04], the simplest of them being the *two-level spectral correlation function* (TLSCF) $R(\omega) \propto \sum_{\mathbf{R}} C(\omega, R)$. Some of them apparently share the characteristic features of spectral statistics at the Anderson localization transition, e.g. the finite level compressibility $0 < \chi < 1$ [Alt88, Cha96, Aro95], and the Poisson tail of the level spacing distribution function $P(s) \propto e^{-s/2\chi}$ combined with the Wigner-Dyson level repulsion $P(s) \propto s^\beta$ at small level separation $s \ll 1$ [Shk93]. There is conjecture that these features are also related with the multifractality of critical eigenfunctions, however its exact formulation has not been developed beyond the limit of weak multifractality $\chi \ll 1$, where one can show that $\chi = \frac{1}{2}(1 - d_2/d)$ [Cha96]. For weak multifractality $d_2/d - 1 \propto n$ this result is equivalent to $\chi = \frac{1}{n}\left(1 - \frac{d_n}{d}\right)$ with an arbitrary n. Recently, new arguments have been presented [Bog11] that for an arbitrary strength of multifractality the results of Ref. [Cha96] can be generalized as follows: $\chi = \left(1 - \frac{d_1}{d}\right)$.

The main reason for the scarcity of rigorous knowledge about multifractality of critical eigenfunctions is a lack of exactly solvable models of sufficient generality. The most popular three-dimensional (3D) Anderson model of localization [And58] is quite efficient for numerical simulations but has so far evaded any rigorous analytical treatment in the critical region. More perspective seemed to be offered by Chalker's network model for the quantum Hall transition [Cha88] and its generalizations. However, numerous proposals for the critical field theory have not been successful so far [Eve08b]. In this situation the random matrix theory may prove to be the simplest universal and representative tool to obtain a rigorous knowledge about the critical eigenfunction and spectral statistics [Kra97].

Other invariant random matrix ensembles that show aspects of criticality have been introduced, e.g. in [Mosh94], and in [Gar00] in the context of applications to quantum chromodynamics (see Chapter 32 of this book). In fact [Mosh94] was shown to be equivalent in a certain limit [Kra97] to the non-invariant model we will study in detail in the next section.

For other aspects of random matrices with critical scaling see also Chapter 13 on ensembles with heavy tails, and Chapter 23 on banded matrices in this book.

12.2 Non-invariant Gaussian random matrix theory with multifractal eigenvectors

Guided by the idea that multifractality of eigenstates is the hallmark of criticality, we introduce the Gaussian random matrix ensemble [Mir96, Kra97] whose eigenvectors obey Eq. (12.1.1) with L being replaced by the matrix size N. This random matrix theory and its modifications describe very well not only the critical eigenfunction statistics at the Anderson localization transition in a three-dimensional (3D) Anderson model but also the off-critical states close to the transition [Kra07]. The critical random matrix ensemble (CRMT) suggested in [Mir96, Kra97] is *manifest non-invariant*, and is defined as follows:

$$\langle H_{nm}\rangle = 0, \qquad \langle |H_{nm}|^2\rangle = \begin{cases} \beta^{-1}, & n=m \\ \frac{1}{2}\left[1+\frac{(n-m)^2}{b^2}\right]^{-1}, & n\neq m \end{cases} \tag{12.2.1}$$

where H_{nm} is the Hermitian $N\times N$ random matrix with entries H_{nm} ($n>m$) being independent Gaussian random variables; $\beta=1,2,4$ for the Dyson orthogonal, unitary, and symplectic symmetry classes, and $b>0$ is the control parameter. This CRMT can be considered as a *particular deformation* of the Wigner-Dyson RMT which corresponds to $b=\infty$.

As is clear from the definition Eq. (12.2.1) the variance $\langle |H_{nm}|^2\rangle$ is non-invariant under unitary transformation $\hat{H}\to U\,\hat{H}\,U^{\dagger}$. The existence of the preferential basis is natural, as this RMT mimics the properties of the Anderson model of *localization* which happens in the *coordinate space*, and not necessarily e.g. in the momentum space.

The critical nature of the CRMT is encoded in the *power-law* decay of the variance matrix, the typical off-diagonal entry being proportional to $|n-m|^{-1}$ in the absolute value. This is exactly the decay with the power equal to the dimensionality of space ($d=1$ in the case of matrices). In contrast to the Wigner-Dyson RMT whose probability distribution is parameter-free, the CRMT is a *one-parameter family*. The parameter b controls the spectrum of fractal dimensions $d_n = d_n(b)$. One can show [Mir96, Mir00] that both at $b \gg 1$ and $b \ll 1$ the scaling relation Eq. (12.1.1) holds true, and the basic fractal dimension is equal to,

$$d_2 = \begin{cases} 1 - c_\beta\, B^{-1}, & B \gg 1 \\ c_\beta\, B, & B \ll 1 \end{cases} \tag{12.2.2}$$

where $c_\beta = \frac{\pi^{\frac{\beta}{2}-1}}{\beta} 2^{\frac{1}{4}}$, and $B = b\, \pi^{\frac{\beta}{2}}\, 2^{\frac{1}{4}}$.

Equation (12.2.2) can be cast in a form of the *duality relationship*,

$$d_2(B) + d_2(B^{-1}) = 1. \tag{12.2.3}$$

The duality relation has been checked numerically [Kra10] for the unitary CRMT Eq. (12.2.1). The results are presented in Figure 12.1. In particular, it was found that $2d_2(B=1) = 1.003 \pm 0.004$. The fact that the deviation of the sum $d_2(B) + d_2(1/B)$ from 1 does not exceed 1% in the entire region of $B \in [0, 1]$ looks extremely interesting. However, it is not yet known whether an exact

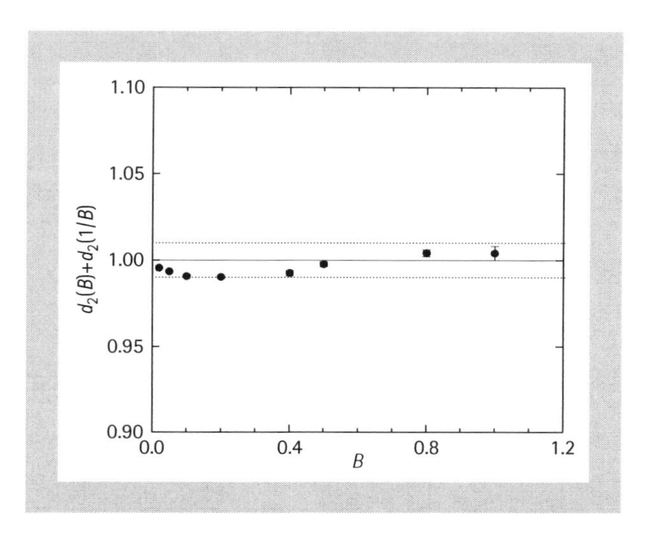

Fig. 12.1 Numerical verification of the duality relation Eq. (12.2.3), done by the box counting method [Kra10] for the unitary CRMT ensemble Eq. (12.2.1). The deviations from 1 in Eq. (12.2.3) do not exceed 1% in the entire region of $B \in [0, 1]$.

function $d_2(B)$ obeys this remarkable duality relationship which suggests that $d_2 = \frac{1}{2}$ at $b = \frac{1}{\sqrt{\sqrt{2}\pi\beta}}$. In the recent paper [Kra10] the correction of order B^2 to d_2 in the unitary ensemble was found analytically. It appeared to be numerically small $\approx 0.06\,B^2$ but non-zero. At the same time the large-B correction of order $1/B^2$ is reported to be strictly zero [Ras11] in the unitary ensemble (in contrast to the orthogonal one).

There was vast numerical evidence that the CRMT Eq. (12.2.1) reproduces the main qualitative features of the critical eigenfunction and spectral statistics (see detailed discussion in Section 1.1.7), including the power-law behaviour of the DoS correlation function $C(\omega, 0)$ (see [Kra07] and references therein), distributions of moments of inverse participation ratio P_q [Mir00], the role of rare realizations and multifractality spectrum close to termination point [Eve08a], and the hybrid Wigner-Dyson and Poisson level spacing distribution [Shk93].

The very fact that by a choice of control parameter B one can fit quite accurately the critical statistics of both eigenvalues and eigenfunctions of the Anderson localization model, is extremely encouraging.

12.3 Invariant random matrix theory (RMT) with log-square confinement

Quite remarkably, there is an *invariant* (but non-Gaussian) RMT whose *two-level spectral correlation function* $R_N(s, s')$ is closely related to that of the CRMT discussed in the previous section. Namely, in the unfolded energy variables s in which the mean spectral density ('global density of states') is equal to unity, and in the large B limit one finds:[1]

$$R_\infty(s - s')\,|_{\text{inv.}} = R_\infty(s - s')\,|_{\text{non-inv.}}, \tag{12.3.1}$$

where $R_\infty(s - s') = \lim_{s\to\infty}(\lim_{N\to\infty} R_N(s, s'))$.

For the unitary symmetry class one can show [Kra97] that

$$R_\infty(s - s')\,|_{\text{non-inv}} = 1 - \pi^2\kappa^2 \frac{\sin^2(\pi(s - s'))}{\sinh^2[\pi^2\kappa(s - s')]}, \qquad \kappa = \frac{\beta}{2}\chi(b), \tag{12.3.2}$$

where κ is related to the level compressibility $\chi(b)$ (see Eqs. (12.8.3), (12.8.6) below). This suggests [Meh04] the form of the kernel $K(s - s')$ of the invariant RMT which is the only input one needs to compute all many-point spectral correlation functions,

$$K(s - s') = \pi\kappa \frac{\sin(\pi(s - s'))}{\sinh[\pi^2\kappa(s - s')]}. \tag{12.3.3}$$

[1] In the opposite limit of small B and large κ this correspondence breaks down.

Thus a remarkable correspondence between an invariant and a non-invariant RMT can be conjectured [Kra97], which allows to use the full power of the unitary invariance for calculation of spectral statistics.

Now it is time to specify the invariant RMT whose counterpart in Eq. (12.3.1) is the critical RMT with multifractal eigenstates. The probability distribution for the random matrix Hamiltonian $\hat{H}$ is [Mut93, Kra97]

$$P(\hat{H}) \propto \exp\left[-\beta \operatorname{tr} V(\hat{H})\right], \quad V(x) = \sum_{n=0}^{\infty} \ln\left[1 + 2q^{n+1}(1+2x^2) + q^{2(n+1)}\right], \tag{12.3.4}$$

where $1 < q < 0$ is a control parameter.

It is extremely important that the 'confinement potential' $V(x)$ at large $|x|$ behaves as,

$$V(x) \to A \ln^2 |x|, \qquad A = \frac{2}{\ln(q^{-1})}. \tag{12.3.5}$$

thus being an example of a 'shallow' confinement. The correspondence Eq. (12.3.1) holds for $\ln q^{-1} << 1$, while for $\ln q^{-1} > \pi^2$ an interesting phenomenon of 'crystallization' of eigenvalues happens with the TLSCF taking a triangular form [Mut93, Kra97, Bog97].

The form of the confinement potential in Eq. (12.3.4) is quite specific, even within the class of shallow potentials with a double-logarithmic asymptotic behavior Eq. (12.3.5). It has been chosen in [Mut93] in order to enable an exact solution in terms of the q-deformed Hermite polynomials.[2] However, there exists another, more simple argument why this particular form leads to an exact solution [Sed05]. The point is that the measure Eq. (12.3.4) can be considered as a *generalized Cauchy distribution*:

$$e^{-\beta V(E_i)} = \prod_{n=1}^{\infty} \frac{1}{4q^{n+1}} \prod_{i=1}^{N} \frac{1}{E_i^2 + \Gamma_n^2}, \qquad \Gamma_n = \frac{1+q^{n+1}}{2q^{\frac{n+1}{2}}}. \tag{12.3.6}$$

12.4 Self-unfolding and not self-unfolding in invariant RMT

As we will see below, the cause of all the peculiarities of the RMT with double-logarithmic confinement is the fact that the mean density of states $\rho_\infty(E) = \lim_{N\to\infty} \sum_{i=1}^{N} \langle \delta(E - E_i) \rangle$ approaches some stable *non-trivial* form in the bulk of the spectrum as the size of matrix N tends to infinity. This is in contrast with

[2] Note that the measure Eq. (12.3.4) is not the unique one leading to the q-deformed Hermite (Ismail-Masson) polynomials. This is perhaps the main reason for all peculiarities of the double-logarithmic confinement, that there is no one-to-one correspondence between the confinement potential $V(x)$ and the set of orthogonal polynomials. In particular, the confinement potential $V(x) = [\sinh^{-1}(x)]^2/\ln(1/q)$ leads to the same set of orthogonal polynomials [Mut01].

the Wigner-Dyson classical RMT where $\lim_{N\to\infty} \rho_N(E)/\sqrt{2N} = 1$ is independent of E. This is the reason why the Wigner-Dyson RMT can be referred to as *self-unfolding*, while the RMT with the double-logarithmic confinement is *not self-unfolding*.

In order to find a criterion for an invariant RMT to be self-unfolding, let us apply the Wigner-Dyson plasma analogy [Meh04]. According to this approximation the mean density of states $\rho_N(E)$ obeys the integral equation of the equilibrium classical plasma with logarithmic interaction subject to the confining force $-dV/dE$,

$$v.p. \int_{-\infty}^{+\infty} \rho_N(E') \frac{dE'}{E - E'} = \frac{dV}{dE} \equiv f(E), \tag{12.4.1}$$

where $v.p.$ denotes the principle value of the integral. The solution to this equation in the case of symmetric confining potential is,

$$\rho_N(E) = \frac{1}{\pi^2} \sqrt{D_N^2 - E^2}\; v.p. \int_{-D_N}^{D_N} \frac{f(E')}{\sqrt{D_N^2 - E'^2}} \frac{dE'}{E' - E}, \tag{12.4.2}$$

where the band-edge D_N should be determined from the normalization condition $\int_{-D_N}^{D_N} \rho_N(E')\, dE' = N$.

One can see that if $|f(E')|$ increases slower than $|E'|$ as $|E'| \to \infty$, the integral converges to a non-trivial function as N and $D_N \to \infty$,

$$\rho_\infty(E) = \frac{1}{\pi^2}\; v.p. \int_{-\infty}^{+\infty} dE' \frac{f(E')}{E' - E}. \tag{12.4.3}$$

Otherwise, the large values of $|E'| \sim D_N$ dominate the integral in Eq. (12.4.2), so that at a fixed E and $N \to \infty$ (when $D_N \gg |E|$) the dependence of $\rho_N(E)$ on E disappears. We conclude therefore [Can95, Can96] that the criterion of a *non-self-unfolding* RMT is,

$$\lim_{|x|\to\infty} \frac{V(x)}{|x|} = 0. \tag{12.4.4}$$

Let us consider the *shallow* confining potential with the power-law large-x asymptotic behaviour,

$$V(x) = A|x|^\alpha, \quad (\alpha < 1). \tag{12.4.5}$$

Then at large $|E| \gg 1$ we have [Can95, Can96],

$$\rho_\infty(E) = \frac{A\alpha}{\pi} \tan\left(\frac{\pi\alpha}{2}\right) \frac{1}{|E|^{1-\alpha}}. \tag{12.4.6}$$

In particular for the log-square confining potential,

$$V(x) = A\ln^2|x| = A\lim_{\alpha\to 0}\left(\frac{|x|^\alpha - 1}{\alpha}\right)^2 = A\lim_{\alpha\to 0}\frac{|x|^{2\alpha} - 2|x|^\alpha + 1}{\alpha^2}, \qquad (12.4.7)$$

one finds using the linear dependence of Eq. (12.4.3) $\rho_\infty(E)$ on $f(E)$ and Eq. (12.4.5):[3]

$$\rho_\infty(E) = \frac{A}{|E|}. \qquad (12.4.8)$$

There is a qualitative and far-reaching difference between the shallow power-law confinement potential with $0 < \alpha < 1$, and the log-square confinement. Although both lead to a non-self-unfolding RMT, the case of finite α can be called *weakly non-self-unfolding*, because for large enough E the variation of the mean density at a scale of the mean level spacing $\Delta = \rho_\infty^{-1}$ is much smaller than the density itself,

$$\frac{\rho_\infty(E+\Delta) - \rho_\infty(E)}{\rho_\infty(E)} = \frac{\rho'_\infty}{\rho^2_\infty} \propto |E|^{-\alpha} \to 0. \qquad (12.4.9)$$

In contrast to that, the log-square confinement is *strongly non-self-unfolding*, since the ratio in Eq. (12.4.9) is always finite.[4]

We conclude that, depending on the steepness of the confinement potential at large E, there are three qualitatively different cases [Can95, Can96]:

- *self-unfolding RMT* for $\lim_{x\to\infty} V(x)/|x| > 0$
- *weakly non-self-unfolding RMT* for $\lim_{x\to\infty} V(x)/|x| = 0$ but $\exists\alpha > 0$ such that $\lim_{x\to\infty} V(x)/|x|^\alpha > 0$
- *strongly non-self unfolding RMT* if $\forall\alpha > 0$ holds $\lim_{x\to\infty} V(x)/|x|^\alpha = 0$

The first case is characterized by the Wigner-Dyson universality of the spectral correlation functions. In the second case this universality holds only approximately for sufficiently large distance from the origin, while if one or two energies are close to the origin, the Wigner-Dyson universality is no longer valid, even after unfolding. In the third case, the Wigner-Dyson universality is not valid also in the bulk of the spectrum.

[3] This case of log-square confinement is very special [Cho09]. For $V(x) = A\ln^n|x|$ with $n \geq 2$ consideration similar to Eq. (12.4.7) gives the leading term at large E of the form $\rho_\infty(E) = C\,\frac{\ln^{n-2}|E|}{|E|}$ with $C = \frac{An(n-1)}{2}$.

[4] We put the confinement potentials $\ln^n|E|$ with $n > 2$ to the class of strongly non-self unfolding potentials even though the ratio Eq. (12.4.9) vanishes logarithmically at large E.

12.5 Unfolding and the spectral correlations

Let us consider the power-law confinement Eq. (12.4.5) and find the corresponding *unfolding coordinates* $s(E)$ in which the spectral density is 1:

$$s(E) = \mathrm{sgn}(E) \int_0^E \rho_\infty(E')\, dE' = \frac{A}{\pi} \tan\left(\frac{\pi\alpha}{2}\right) \mathrm{sgn}(E)\, |E|^\alpha. \qquad (12.5.1)$$

Note that for large enough E the unfolding coordinates are given by Eq. (12.5.1) even if $V(E)$ is not a pure power-law at small E but is rather deformed to have a regular behaviour at the origin.

The corresponding unfolding coordinates for the log-square potential Eq. (12.4.7) are,

$$s(E) = A\, \mathrm{sgn}(E)\, \ln(c|E|), \quad c\, E(s) = \mathrm{sgn}(s)\, e^{\frac{|s|}{A}}, \qquad (12.5.2)$$

where c is a constant that depends on the regularization of the log-square potential close to the origin.

As we will see, the exponential change of coordinates Eq. (12.5.2) leads to dramatic consequences for two-level spectral correlations, and even has a far-reaching analogy in the physics of black holes [Can95, Fra09].

In order to show how a non-trivial unfolding changes the form of the spectral correlations consider the spectral kernel $K_N(E, E')$ given in terms of the orthogonal polynomials by the Christoffel–Darboux formula [Meh04]:

$$K_N(E, E') = \left| \frac{\varphi_{N-1}(E)\varphi_N(E') - \varphi_{N-1}(E')\varphi_N(E)}{E - E'} \right|, \qquad (12.5.3)$$

where $\varphi_N(E)$ is the (properly normalized) 'wave-function' which is related with the orthogonal polynomials p_n,

$$\varphi_N(E) = p_n(E)\, e^{-\beta V(E)/2}, \quad \int_{-\infty}^{+\infty} p_n(E')\, p_m(E')\, e^{-\beta V(E')}\, dE' = \delta_{nm}. \qquad (12.5.4)$$

Any spectral correlation function can be expressed [Meh04] in terms of the kernel Eq. (12.5.3). In particular, the n-point density of states correlation function at $\beta = 2$ takes the form [Meh04],

$$R_\infty(E_1, \ldots E_n) = \det\left[K_\infty(E_i, E_j)\right] \prod_{i=1}^{n} K^{-1}(E_i, E_i). \qquad (12.5.5)$$

Below we derive a *semi-classical* form of the kernel which is valid in the $N \to \infty$ limit provided that the coefficient A in Eqs. (12.4.5), (12.4.7) is large. In particular for the RMT with log-square confinement given by Eq. (12.3.4) the condition of applicability of the analysis done below is $\ln q^{-1} < 2\pi$ [Mut93].

In this semiclassical limit the 'wave functions' take the form (we assume that the confinement potential is an even function of E),

$$\varphi_{N-1}(E) = \sin(\pi s(E)), \qquad \varphi_N = \cos(\pi s(E)), \tag{12.5.6}$$

if N is even and $\cos \to \sin$ if N is odd. Then one finds in the unfolding coordinates,

$$K_\infty(E, E') = \frac{\sin(\pi(s - s'))}{E(s) - E(s')}. \tag{12.5.7}$$

Equation Eq. (12.5.7) demonstrates that as long as the semiclassical approach applies, the non-trivial unfolding $E = E(s)$ is the only source of deformation of the kernel and its deviation from the universal Wigner-Dyson form.

The main conclusion one can draw from Eq. (12.5.7) is that the translational invariance is lost in the $N \to \infty$ limit for all non-self-unfolding invariant RMT. It is not sufficient to have the mean density $\rho_\infty(s)$ equal to unity for all values of s in order to have all the correlation functions of the universal Wigner-Dyson form.

12.6 Ghost correlation dip in RMT and Hawking radiation

The breakdown of translational invariance takes especially dramatic form in the case of log-square confinement. We will show below that in this case a *ghost correlation dip* appears in the two-level spectral correlation function which position at $s \approx -s'$ is *mirror reflected* relative to the position of the *translational-invariant correlation dip* at $s \approx s'$.

Indeed let us consider the semiclassical kernel Eq. (12.5.7) with exponential unfolding Eq. (12.5.2) for $s \gg 1$ and $|s - s'| << |s|$. Then after a simple algebra we obtain for $s\,s' > 0$ the Two-level spectral correlation function given by Eq. (12.3.2) ($\beta = 2$):

$$R_\infty(s\,s' > 0) = 1 - \pi^2\kappa^2\,\frac{\sin^2(\pi(s - s'))}{\sinh^2[\pi^2\kappa(s - s')]}, \qquad \kappa = \frac{1}{2\pi^2\,A} = \frac{\ln q^{-1}}{4\pi^2} \ll 1. \tag{12.6.1}$$

This correlation functions exhibits a dip of anti-correlation (level repulsion) at small $s - s'$ and approaches exponentially fast the uncorrelated asymptotic $R_\infty(s, s') = 1$ for $|s - s'| > 1/\pi^2\kappa$. An amazing fact is that the anti-correlation revives again when $s + s'$ becomes small. Considering $s\,s' < 0$ and plugging the exponential unfolding Eq. (12.5.2) into Eq. (12.5.7) we obtain for the unitary case $\beta = 2$,

$$R_{\infty}(s\,s' < 0) = 1 - \pi^2\kappa^2\,\frac{\sin^2(\pi(s-s'))}{\cosh^2(\pi^2\kappa\,(s+s'))}. \tag{12.6.2}$$

This is the *ghost anti-correlation peak* discovered in [Can95] but also present in the exact solution [Mut93].

The existence of such a ghost correlation dip is requested by the normalization sum rule [Can95] which for the invariant RMT survives taking the limit $N \to \infty$:

$$\int_{-\infty}^{+\infty} (1 - R_{\infty}(s,s'))\,ds' = 1. \tag{12.6.3}$$

Substituting in Eq. (12.6.3) the sum of the translational invariant and the ghost peak found from Eqs. (12.6.1), (12.6.2) and doing the integral one obtains,

$$\coth(\kappa^{-1}) - \frac{1}{\sinh(\kappa^{-1})}\cos(4\pi\,s) \approx 1. \tag{12.6.4}$$

This expression is equal to 1 with the exponential accuracy in $e^{-\frac{1}{\kappa}} \ll 1$, which is exactly the accuracy of the semiclassical approximation Eqs. (12.6.1), (12.6.2).

It appears that the situation with the ghost correlation dip has an important physical realization [Fra09]. Let us consider the *sonic* analogue of the black hole [Unr81, Bal08] described in Figure 12.3. The stream of cold atoms moves with

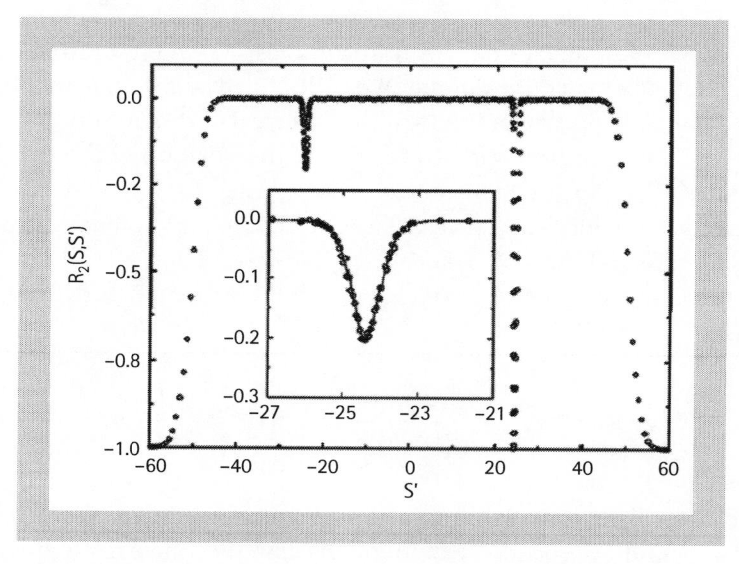

Fig. 12.2 The irreducible part $R_{\infty}(s,s') - 1$ of the TLSCF obtained [Can96] by the classical Monte-Carlo simulations at a temperature $T = \frac{1}{2}$ on the one-dimensional plasma with logarithmic interaction in the log-squared confinement potential Eq. (12.4.7) with $A = 0.5$. Two dips of the anti-correlations are clearly seen with the smaller *ghost dip* shown in detail in the insert. The edges of the correlation function correspond to the finite number of particles (energy levels) $N = 101$.

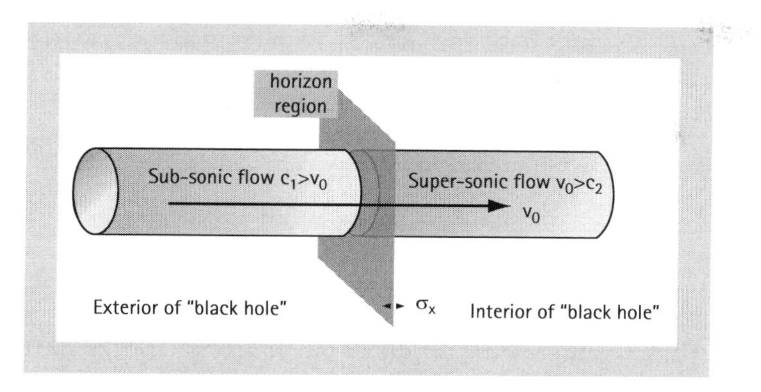

Fig. 12.3 The sonic analogue of a black hole (courtesy of I. Carusotto): the stream of cold atoms with velocity v_0 and the phonon velocity $c(x)$ changing near the origin from the value $c_1 > v_0$ to the value $c_2 < v_0$. This region of the super-sonic flow is analogous to the interior of the black hole from where light cannot escape. This figure is reproduced in colour in the colour plates section.

a constant velocity v_0, while the interaction between atoms is tuned so that the sound velocity is larger than the stream velocity for $x < -\delta$ and smaller than the stream velocity for $x > \delta$. The latter region from where no phonon may escape is analogous to the interior of a black hole. The point in space where $c(x) = v_0$ is analogous to the *horizon* in black hole physics and the phonon radiation emerging when the horizon is formed is analogous to the Hawking radiation [Haw74].

As was shown in [Bal08], Hawking radiation possesses a peculiar correlation of photons (phonons): despite being bosons, they statistically *repel* each other forming a dip in the density correlation function not only at $x \approx x'$ but also near the mirror point $x \approx -x'$, with the envelopes of the normal and ghost dips proportional to $\sinh^{-2}$ and $\cosh^{-2}$, respectively. As in the case of RMT with log-square confinement, the ghost correlation dip appears in the absence of any symmetry that would require the eigenvalues E_i or the Hawking bosons to appear in pairs simultaneously at points $\pm E_i$ ($\pm x_i$). Closer inspection shows that the mechanisms of its appearance in quantum gravity and in random matrix theory are very similar: it is the exponential change of variables similar to Eq. (12.5.2) which is aimed to make flat the mean DoS (unfolding) in RMT or to make flat the metric of space-time in quantum gravity (exponential red-shift) [Fra09]. Thus formation of the strong non-self-unfolding RMT with log-square confinement appears to be analogous to formation of a horizon in general relativity.

12.7 Invariant-noninvariant correspondence

Note that the translational-invariant part of spectral correlations described by Eq. (12.6.1) is *well separated* from the translational non-invariant ghost dip if $|s| \gg |s-s'|$. As correlations decrease exponentially at $|s-s'| > (\pi^2\kappa)^{-1}$, the scale separation takes place when $|s| \gg (\pi^2\kappa)^{-1}$. Essentially what happens because of the scale separation is that (as in a black hole) the world is divided in two parts, and in each part local spectral correlations (which are approximately translational-invariant) are not affected by the presence of a 'parallel world'. Moreover, the local correlations in the *invariant* RMT Eq. (12.3.5) with log-square confinement and $\kappa \ll 1$ were conjectured (for the two-level correlations this conjecture has been proven in [Kra97], see Eq. (12.3.1)) to be the same as in the *non-invariant* critical RMT [Kra97, Kra10]. This conjecture is related to the idea of the *spontaneous breaking of unitary invariance* in RMT with shallow confinement [Can95]. Although this idea has never been proven or even convincingly demonstrated numerically, it seems to be the only physically reasonable cause of invariant-noninvariant correspondence as given by Eq. (12.3.1).

Invariant-noninvariant correspondence allows us to use the invariant RMT with log-square confinement for computing spectral correlations of the *non-invariant* RMT considered in section 1.2. The idea of such calculations is to use the kernel in the form Eq. (12.3.3) to be plugged into the conventional machinery of invariant RMT [Meh04]. Here we present some results of such calculations taken from [Nis99]. One can see from Figure 12.4 that the choice of one single parameter A in the invariant RMT with log-square confinement allows us to fit well both the *body* and the *exponential tail* of the level spacing distribution function obtained by numerical diagonalization of the 3D Anderson localization model at criticality. The corresponding parameters b of the non-invariant CRMT Eq. (12.2.1) should be chosen so that the parameter κ in the kernel Eq. (12.3.3) expressed through A and b is the same. Comparing expressions for κ in Eqs. (12.3.2), (12.6.1) we conclude that b is a solution to the equation,

$$\frac{1}{\pi^2 A} = \beta\chi(b). \tag{12.7.1}$$

For the *unitary* ensemble the function $\chi(b)$ in the range $\chi = (2\pi^2 A)^{-1} \approx 0.36$ is well described [Nda03] by the large-b asymptotic formula [Mir00], $\chi(b) = (4\pi b)^{-1}$. This gives an estimate $b = 0.22$. Numerical diagonalization of the CRMT shows that $b = 0.22$ corresponds to $d_2 \approx 0.46 \pm 0.01$. To establish correspondence with the Anderson localization model of unitary symmetry at critical disorder we note that the fractal dimension in the 1d CRMT corresponds to the *reduced* dimension $d_2/3$ of the three-dimensional Anderson model. Thus the fractal dimension of the Anderson model should be compared with $0.46 \times 3 = 1.38$. This appears to be in an excellent agreement with the direct diagonalization of the Anderson model of unitary symmetry which gives [Cue02] $d_2 = 1.37$. This example demonstrates that the values of parameters A and b found from fitting the *spectral* statistics of RMT to that of the 3D Anderson localization model automatically give an excellent fit of the *eigenfunction* statistics.

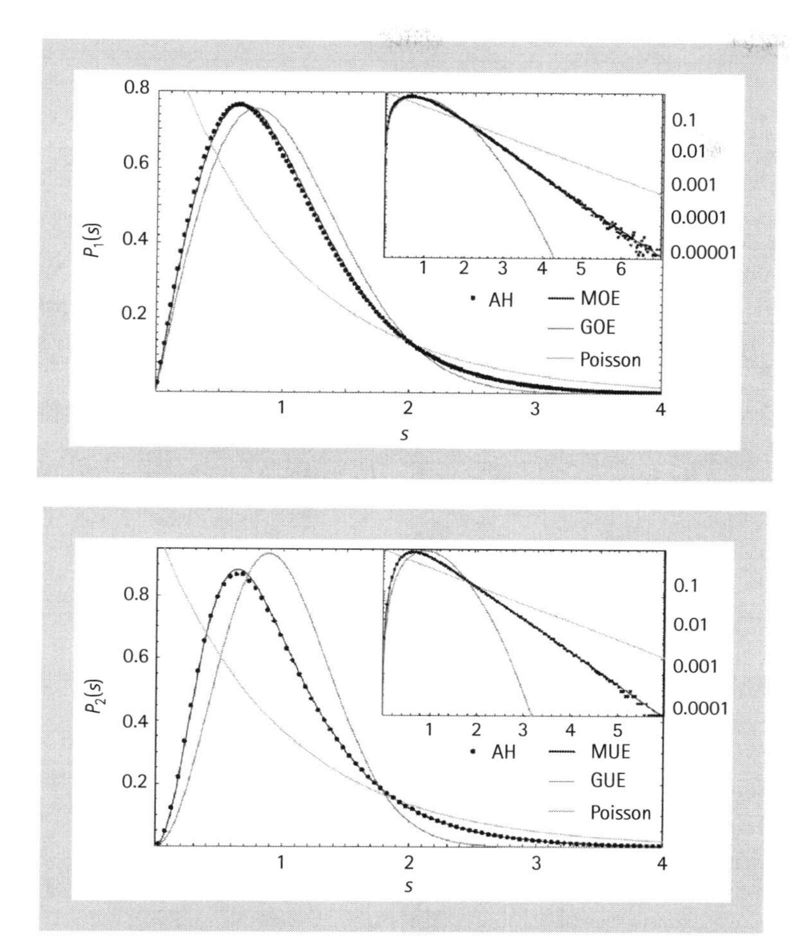

Fig. 12.4 Level spacing distribution function (after Nishigaki [Nis99]) for the orthogonal and unitary ensembles with log-square confinement (black solid lines) and the corresponding distributions for the 3D Anderson localization model with critical disorder (data points). Grey solid lines are the Wigner-Dyson and Poisson distributions. The same values of $A = 0.34$ (orthogonal ensemble) and $A = 0.14$ (unitary ensemble) allow us to fit well both the body and the far tail of the distribution shown in the inserts.

12.8 Normalization anomaly, Luttinger liquid analogy and the Hawking temperature

As has been already mentioned, the local spectral correlations with $|s - s'| \ll |s|$ are well described by the translational-invariant kernel Eq. (12.3.3). An important difference between this kernel and that of the conventional Wigner-Dyson theory is that it contains a *second energy scale* κ^{-1} in addition to the mean level spacing $\Delta = 1$. Moreover, the way this scale appears through $s - s' \to \sinh(\pi^2 \kappa\, (s - s'))$ suggests an analogy with the system of one-dimensional fermions at a *finite temperature*. Indeed, the density correlations of non-interacting one-dimensional fermions (whose ground state correlations are equivalent to the Wigner-Dyson spectral statistics at $\beta = 2$) is described by Eq. (12.3.2), where

$$T = \pi\kappa. \tag{12.8.1}$$

Given the kernel Eq. (12.3.3) it is not difficult to obtain the level density correlation functions for $\beta = 1, 4$ using the standard formulae of the Wigner-Dyson RMT [Meh04]. It appears that both in the region of $|s - s'| \gg 1$ and in the region of $|s - s'| \ll \kappa^{-1}$ they coincide with the generalization of Eq. (12.3.2) to $\beta = 1, 4$ obtained [Kra10] directly from the *non-invariant* CRMT Eq. (12.2.1). For $\kappa \ll 1$ these regions overlap on a parametrical large interval, and one can again demonstrate the invariant-noninvariant correspondence.

It is important to note that breaking the unitary invariance explicitly by a finite b in Eq. (12.2.1) leads to a *normalization anomaly*. Namely, the normalization sum rule [Can95] which holds exactly for the finite size of matrix N,

$$2\int_{s>0}^{+\infty} (1 - R_N(s' - s))\, ds' = 1, \tag{12.8.2}$$

is violated if the limit $N \to \infty$ is taken *prior* to doing the integral. For instance at $\beta = 2$ we have with exponential accuracy $e^{-1/\kappa} \ll 1$,

$$\eta = 2\int_{s>0}^{\infty} (1 - R_\infty(s' - s))\, ds' = \coth(\kappa^{-1}) - \kappa \approx 1 - \kappa. \tag{12.8.3}$$

For $\beta = 1$ the corresponding expression is,

$$\eta = 2\cot(\kappa^{-1}) - \tanh(1/2\kappa) - 2\kappa \approx 1 - 2\kappa. \tag{12.8.4}$$

This leads to the finite *level compressibility* [Cha96],

$$\chi \equiv \frac{d}{d\bar{n}}\langle (n - \bar{n})^2\rangle = 1 - \eta, \qquad N \gg \bar{n} = \langle n\rangle \gg 1, \tag{12.8.5}$$

and the exponential (instead of the Gaussian in the Wigner-Dyson RMT) tail of the level spacing distribution function,

$$\ln P(s) \approx -\frac{s}{2\chi}. \tag{12.8.6}$$

Both properties Eq. (12.8.5) and Eq. (12.8.6) are the signatures of criticality [Shk93]. Because of invariant-noninvariant correspondence the normalization anomaly Eq. (12.8.3) holds also for the invariant RMT with log-square confinement thus again raising a question on the *spontaneous breaking of unitary invariance*.

In the absence of a formal proof of this conjecture, we present here a simple theory which may *unify* both of the ensembles. The idea of such a theory stems from the fact [Sim93] that the classical Wigner-Dyson RMT [Meh04] is equivalent to the *ground state* of the one-dimensional Calogero–Sutherland model [Sut95] of *fermions* with inverse square interaction of the strength $\frac{\beta}{2}\left(\frac{\beta}{2}-1\right)$ in a harmonic confinement potential. The large-scale properties of such a model are described by the Luttinger liquid phenomenology [Hal81]. Its simplest finite-temperature formulation [Gog98] is in terms of the free-*bosonic* field in a $1+1$ space-time $\Phi(s,\tau)=\frac{1}{2}\left[\Phi_R(s,\tau)+\Phi_L(s,\tau)\right]$ with the quadratic action,

$$S_T[\Phi]=\frac{1}{2\pi K}\int_0^{1/T}d\tau\int_{-\infty}^{+\infty}ds\,[(\partial_s\Phi)^2+(\partial_\tau\Phi)^2],\qquad \Phi(s,\tau)=\Phi(s,\tau+1/T),\tag{12.8.7}$$

where $K=\frac{2}{\beta}$ is the Luttinger parameter.

The density correlation functions are given by the *functional averages* of the density operator,

$$\rho(s,\tau)=\rho_0+\frac{1}{\pi}\,\partial_s\Phi(s,\tau)+A_1\,\cos[2\pi s+2\Phi(s,\tau)]+A_2\,\cos[4\pi s+4\Phi(s,\tau)]+\ldots,\tag{12.8.8}$$

where A_k are structural constants which are determined by details of interaction at small distances and take some fixed values for the Calogero–Sutherland model corresponding to the symmetry class β. Using Eq. (12.8.8) one may express the density correlation functions in terms of the fundamental Green's function,

$$\langle\Phi(s,\tau)\,\Phi(s',\tau)\rangle_S-\frac{1}{2}\langle\Phi(s,\tau)\,\Phi(s,\tau)\rangle_S-\frac{1}{2}\langle\Phi(s',\tau))\,\Phi(s',\tau)\rangle_S=\frac{\pi}{\beta}\,G(s,s'),\tag{12.8.9}$$

where $\langle\ldots\rangle_S$ is the functional average with the action $S[\Phi]$. For example, the two-level spectral correlation functions are equal to,

$$R_\infty(s,s')=1+\frac{2}{\pi\beta}\,\partial_s\partial_{s'}\,G(s,s')+\frac{2}{\beta\,(2\pi^2)^{\frac{2}{\beta}}}\cos(2\pi s)\,e^{8\pi\beta^{-1}G(s,s')}.\tag{12.8.10}$$

For $T=0$ the Green's function is $G(s-s')=-\frac{1}{4\pi}\ln(s-s')^2$ at $|s-s'|\gg 1$ and one reproduces the asymptotic form of the Wigner-Dyson correlations.

In order to reproduce by the same token the corresponding correlation functions [Kra10] for the *non-invariant* CRMT Eq. (12.2.1) one merely substitutes in Eq. (12.8.7) the finite T given by Eq. (12.8.1) and replaces the zero-temperature Green's function by the finite-temperature one,

$$G(s-s')=\frac{1}{2\pi}\ln\left(\frac{\pi T}{\sinh(\pi T\,|s-s'|)}\right).\tag{12.8.11}$$

Thus we conclude that the deformation of the Wigner-Dyson RMT given by Eq. (12.2.1) with large but finite b, retains the analogy with the Calogero–Sutherland model. However, instead of the ground state, the non-invariant

CRMT with $\kappa \ll 1$ is equivalent to the Calogero–Sutherland model at a small but finite temperature Eq. (12.8.1).

It is remarkable that there exists a deformation of the free-boson functional Eq. (12.8.7) capable of reproducing the two-level correlation function for the *invariant RMT with log-square confinement*, including the ghost correlation dip. All one has to do for that is to replace the action Eq. (12.8.7) defined on a cylinder by that defined on a curved space-time with a horizon,

$$\int_0^\infty d\tau \int_{-\infty}^{+\infty} ds\,[(\partial_s \Phi)^2 + (\partial_\tau \Phi)^2] \to \int d^2\xi\,\sqrt{-g(\xi)}\,g^{\mu\nu}\,\partial_\mu \Phi\,\partial_\nu \Phi, \qquad (12.8.12)$$

where $g \equiv \det g_{\mu\nu}$ with $g_{\mu\nu}$ being the metric, i.e. $ds^2 = g_{\mu\nu}\,d\xi^\mu\,d\xi^\nu$ with $x^1 = x$ and $x^0 = t = -i\tau$. The main requirement for the metric is that the transformation of co-ordinates to the frame $(\bar{x}, \bar{t})$ where the metric is flat ('Minkovski space' with $ds^2 = d\bar{x}^2 - d\bar{t}^2$), is exponential for large enough x with a factorized x and t dependence, e.g.

$$\bar{x} = \varphi(x)\,\cosh(t/A), \quad \bar{t} = \varphi(x)\,\sinh(t/A), \qquad (12.8.13)$$

$$\varphi(x) \sim \mathrm{sgn}(x)\,e^{\frac{|x|}{A}} \quad \text{at} \quad |x| \gg A = 2\pi\,T. \qquad (12.8.14)$$

In order to represent the invariant RMT with an *even* confinement potential, the function $\varphi(x)$ should be *odd*. This last requirement plus the continuity of the function at $x = 0$ necessarily implies $g_{00}(x = 0) = 0$, that $x = 0$ is the horizon. One example of such a metric is,

$$\bar{x} = A\,\sinh(x/A)\,\cosh(t/A), \quad \bar{t} = A\,\sinh(x/A)\,\sinh(t/A) \qquad (12.8.15)$$

$$ds^2 = \cosh^2(x/A)\,dx^2 - \sinh^2(x/A)\,dt^2 \qquad (12.8.16)$$

$$\sqrt{-g(\xi)}\,g^{\mu\nu}\,\partial_\mu \Phi\,\partial_\nu \Phi = |\tanh(x/A)|\,(\partial_x \Phi)^2 + |\coth(x/A)|\,(\partial_\tau \Phi)^2. \qquad (12.8.17)$$

It maps the strip $x \geq 0$, $0 < \tau < 2\pi\,A$ and the strip $x \leq 0$, $0 < \tau < 2\pi\,A$ *separately* onto the entire plane $(\bar{x}, \bar{\tau})$.

In order to compute the Green's function on the curved space-time corresponding to Eq. (12.8.13) we use the well-known formula [Tsv03],

$$G(z, z') = -\frac{1}{2\pi}\ln|\bar{z}(z) - \bar{z}(z')| - \frac{1}{4\pi}\ln[|\partial_z \bar{z}(z)\,\partial_{z'}\bar{z}(z')|], \qquad (12.8.18)$$

where $\bar{z} = \bar{x} \pm i\bar{\tau}$, $z = x \pm i\tau$ and $\partial_z = \partial_x \mp i\partial_\tau$ with $\pm = \mathrm{sgn}(x)$. Then from Eq. (12.8.13) for x, x' sufficiently far from the origin one easily obtains,

$$G(x, x') = -\frac{1}{4\pi}\ln\left[\frac{(\varphi(x) - \varphi(x'))^2}{4|\varphi(x)\,\varphi(x')|}\right]$$

$$= -\frac{1}{2\pi}\begin{cases}\ln\left[2A\sinh[(x - x')/2A]\right], & x\,x' > 0\\ \ln\left[2A\cosh[(x + x')/2A]\right], & x\,x' < 0\end{cases} \qquad (12.8.19)$$

The origin of sinh and cosh in Eq. (12.8.19) is very much the same as that in Eqs. (12.6.1), (12.6.2). The exponential transformation of coordinates Eq. (12.8.13) plays the same role as the exponential unfolding Eq. (12.5.2). Finally substituting Eq. (12.8.19) into the expression for the two-level correlation function Eq. (12.8.10) one reproduces Eqs. (12.6.1), (12.6.2).

Note that the new scale $A = (2\pi T)^{-1}$ sets the temperature scale T given by Eq. (12.8.1) which means the *Hawking temperature* in the black hole analogy. While the finite temperature arises because of the periodicity of transformations Eq. (12.8.15) in the imaginary time $\tau = i\,t$, this compactification is different from the standard one: each of the semi-strips $x \geq 0$ and $x \leq 0$ are mapped on the entire plane $(\bar{x}, \bar{\tau})$ independently of each other. This is essentially the effect of the horizon.

12.9 Conclusions

I would like to close this chapter with some concluding remarks. First of all, it is by no means a comprehensive review but rather an introduction to the subject written by a physicist motivated by the physics applications and not by the formal rigour. My goal was to show that the subject is rich and poorly explored and that efforts are likely to be rewarded by non-trivial discoveries. Let me formulate finally the (highly subjective) list of open problems as I see them.

- *further study of the non-invariant Gaussian critical RMT*
  Some progress has been made in this direction by development of the regular expansion in $b \ll 1$, the so-called *virial expansion method* [Yev03, Kra06, Yev07], which extends the ideas of [Lev90, Mir00]. This approach has made it possible to compute analytically the level compressibility up to b^2 terms and to find an extremely good and simple approximation [Yev07, Kra07] to the DoS correlation function Eq. (12.1.2). These works are the basis for a perturbative proof of the dynamical scaling Eq. (12.1.3)[Cha90] and the duality relation Eq. (12.2.3).
- *Non-perturbative solution to the non-invariant critical RMT*
  Some nice relationships such as the duality relation Eq. (12.2.3), raise a question about the possibility of exact solution to the CRMT. In our opinion this possibility does exist.
- *Invariant-noninvariant correspondence*
  This is a very interesting issue with lots of applications in the case where the origin of this correspondence is understood. It is also important to invest some effort in studying the issue of correspondence between spectral and eigenvector statistics, in particular the possible spontaneous breaking of unitary invariance and emergence of a preferential basis [Can95].

- *Level statistics in weakly non-self-unfolding RMT*
  This is a broad class of invariant RMT where the Wigner-Dyson universality is broken. The spectral statistics in such RMT exhibits unusual features like super-strong level repulsion near the origin.
- *Crystallization of eigenvalues*
  This phenomenon takes place in the non-perturbative regime $\kappa > 1$ of RMT with log-square confinement and has its counterpart in the 'crystallization' of roots of orthogonal polynomials [Bog97]. So far it does not have physical realization, with the Calogero–Sutherland model being a clear candidate [Ber08].

Acknowledgements

I am grateful to my coworkers C.M.Canali, J.T. Chalker, E. Cuevas, F. Franchini, I.V. Lerner, M. Ndawana, K.A. Muttalib, A. Ossipov, A.M. Tsvelik and O. Yevtushenko for extremely insightful and enjoyable collaboration.

References

[Alt88] B.L. Altshuler, I.K. Zharekeshev, S.A. Kotochigova, and B. I. Shklovskii, Zh. Eksp. Teor. Fiz. **94**, 343 (1988) [Sov. Phys. JETP **63**, 625 (1988).

[And58] P.W. Anderson, Phys. Rev. **109**, 492 (1958).

[Aro95] A.G. Aronov and A.D. Mirlin, Phys. Rev. B 51 (1995) 6131.

[Bal08] R. Balbinot, A. Fabbri, S. Fagnocchi, A. Recati, and I. Carusotto, Phys. Rev. A **78**, 021603(R) (2008).

[Ber08] B.A. Bernevig, and F.D.M. Haldane, Phys. Rev. B **77**, 184502 (2008).

[Bog97] E. Bogomolny, O. Bohigas, and M.P. Pato, Phys. Rev. E **55**, 6707 (1997).

[Bog04] E. Bogomolny and C. Schmit, Phys. Rev. Lett. **92**, 244102 (2004); B.L. Altshuler and L.S. Levitov, Phys. Rep. **288**, 487 (1997).

[Bog11] E. Bogomolny and O. Giraud, Phys. Rev. Lett. **106**, 044101 (2011).

[Can95] C.M. Canali and V.E. Kravtsov, Phys. Rev. E **51**, R5185 (1995).

[Can96] C.M. Canali, Phys. Rev. B **53**, 3713 (1996).

[Cha88] J.T. Chalker and P.D. Coddington, J. Phys. C, **21**, 2665 (1988).

[Cha90] J.T. Chalker, Physica A, **167**, 253 (1990).

[Cha96] J.T. Chalker, V.E. Kravtsov and I.V. Lerner, JETP Lett. **64**, 386 (1996).

[Cho09] J. Choi and K.A. Muttalib, J. Phys. A: Math. Gen. **42**, 152001 (2009).

[Cue02] E. Cuevas, Phys. Rev. B **66**, 233103 (2002).

[Eve08a] F. Evers and A.D. Mirlin, Rev. Mod. Phys. **80**, 1355 (2008).

[Eve08b] F. Evers, A. Mildenberger, and A.D. Mirlin, Phys. Rev. Lett. **101**, 116803 (2008); A.R. Subramaniam, I.A. Gruzberg, and A.W.W. Ludwig, Phys. Rev. B **78**, 245105 (2008).

[Fra09] F. Franchini and V.E. Kravtsov, Phys. Rev. Lett. **103**, 166401 (2009).

[Gar00] A.M. Garcia-Garcia and J.J.M. Verbaarschot, Nucl. Phys. **B 586**, 668 (2000).

[Gog98] A.O. Gogolin, A.A. Nersesyan, and A.M. Tsvelik, *Bosonization and strongly correlated systems*, Cambridge, University Press, 1998.

[Hal81] F.D.M. Haldane, J. Phys. C **14**, 2585 (1981).

[Haw74] S.W. Hawking, Nature **248**, 30 (1974).

[Kra97] V.E. Kravtsov and K. Muttalib, Phys. Rev. Lett. **79**, 1913 (1997).

[Kra06] V.E. Kravtsov, O. Yevtushenko and E. Cuevas, J. Phys. A:Math. Gen. **39**, 2021(2006).

[Kra07] V.E. Kravtsov and E. Cuevas, Phys. Rev. B **76**, 235119 (2007).

[Kra10] V.E. Kravtsov, A. Ossipov, O.M. Yevtushenko, and E. Cuevas, Phys. Rev. B **82**, 161102(R) (2010).

[Lev90] L.S. Levitov, Phys. Rev. Lett. **64**, 547 (1990).

[Meh04] M.L. Mehta, *Random Matrices*, Academic Press, 3rd Edition, London 2004.

[Mir96] A.D. Mirlin, Y.V. Fyodorov, F.M. Dittes, J. Quezada, and T.H. Seligman, Phys. Rev. E **54**, 3221 (1996).

[Mir00] A.D. Mirlin and F. Evers, Phys. Rev. B **62**, 7920 (2000).

[Mosh94] M. Moshe, H. Neuberger, and B. Shapiro, Phys. Rev. Lett. **73**, 1497 (1994).

[Mut93] K.A. Muttalib, Y. Chen, M.E.H. Ismail, and V.N. Nicopoulos, Phys. Rev. Lett. **71**, 471 (1993).

[Mut01] K.A. Muttalib, Y. Chen and M.E.H. Ismail in: *Symbolic Computation, Number Theory, Special Functions, Physics and Combinatorics*, Eds. F.G, Garvan and M.E.H. Ismail, Kluwer Academic (2001).

[Nda03] M.L. Ndawana and V.E. Kravtsov, J. Phys. A:Math. Gen. **36**, 3639 (2003).

[Nis99] S.M. Nishigaki, Phys. Rev. E **59**, 2853 (1999).

[Ras11] I. Rushkin, A. Ossipov, and Y.V. Fyodorov, J. Stat. Mech. L03001 (2011).

[Sed05] T.A. Sedrakyan, Nucl. Phys. B **729**, 526 (2005).

[Sim93] B.D. Simons, P.A. Lee, and B.L. Altshuler, Phys. Rev. Lett. **70**, 4122 (1993).

[Shk93] B.I. Shklovskii, B. Shapiro, B.R. Sears, P. Lambrianides, and H. B. Shore, Phys. Rev. B, **47**, 11487 (1993).

[Sut95] B. Sutherland in: *Lecture notes in physics*, **242**, Springer Verlag, Berlin 1985.

[Tsv03] A.M. Tsvelik, *Quantum field theory in condensed matter physics*, Cambridge University Press, Cambridge, 2003, p.219.

[Unr81] W.G. Unruh, Phys. Rev. Lett. **46**, 1351 (1981).

[Weg80] F. Wegner, Z. Phys. **B36** (1980) 209.

[Weg85] F. Wegner, in: *Localisation and Metal Insulator transitions*, ed. by H. Fritzsche and D. Adler (Plenum, N.Y., 1985) , 337.

[Yev03] O. Yevtushenko and V.E. Kravtsov, J. Phys. A: Math. Gen. **30**, 8265 (2003).

[Yev07] O. Yevtushenko and A. Ossipov, J. Phys. A:Math. Gen. **40**, 4691(2007).

·13·
Heavy-tailed random matrices

Z. Burda and J. Jurkiewicz

Abstract

We discuss non-Gaussian random matrices whose elements are random variables with heavy-tailed probability distributions. In probability theory heavy tails of distributions describe rare but violent events which usually have a dominant influence on the statistics. They also completely change the universal properties of eigenvalues and eigenvectors of random matrices. We concentrate here on the universal macroscopic properties of (1) Wigner matrices belonging to the Lévy basin of attraction, (2) matrices representing stable free random variables, and (3) a class of heavy-tailed matrices obtained by parametric deformations of standard ensembles.

13.1 Introduction

Gaussian random matrices have been studied over many decades and are well known by now [Meh04]. Much less is known about matrices whose elements display strong fluctuations described by probability distributions with heavy tails.

Probably the simplest example of a matrix from this class is a real symmetric ($A_{ij} = A_{ji}$) random matrix A_N with elements A_{ij}, $1 \leq i \leq j \leq N$, being independent identically distributed (i.i.d.) centred real random variables with a probability density function (p.d.f.) falling off as a power

$$p(x) \sim |x|^{-\alpha-1} \tag{13.1.1}$$

for $|x| \to \infty$. The smaller the value of α the heavier is the tail. Higher moments of this distribution do not exist. For $\alpha \in (0, 1]$ the tail is extremely heavy and the mean value does not exist since the corresponding integral $\int xp(x)dx$ is divergent. For $\alpha \in (1, 2]$ the mean value does exist but the variance does not. The influence of heavy tails on the statistical properties of a random matrix is enormous. It is particularly apparent for $\alpha < 1$. In this case the elements A_{ij} assume values scattered over a wide range which itself increases quickly when N goes to infinity. The largest element $|a_{max}|$ of the matrix is of the order $|a_{max}| \sim N^{2/\alpha}$ and its value fluctuates strongly from matrix to matrix. The distribution of the normalized value of the maximal element $x = |a_{max}|/N^{2/\alpha}$ is given by a Fréchet distribution which itself has a heavy tail. The largest element of the matrix may be larger than the sum of all remaining elements. The values of the elements change so dramatically from matrix to matrix that one cannot speak about a typical matrix or about self-averaging for large N. In the limit $N \to \infty$ matrices A_N may look effectively very sparse.[1] Indeed if one considers a rescaled matrix $A_N/|a_{max}|$ one will find that only a finite fraction of all elements of this matrix will be significantly different from zero. This effective sparseness is quantified by the inverse participation ratio Y_2 constructed from normalized weights $w_{ij} = |a_{ij}|/\sum_{ij}|a_{ij}|$, which sum up to unity $\sum_{ij} w_{ij} = 1$,

$$Y_2 = \frac{2}{N(N+1)}\overline{\sum_{i\leq j} w_{ij}^2}. \tag{13.1.2}$$

The bar denotes the average over matrices A_N. In the limit $N \to \infty$ the participation ratio is $Y_2 = 1 - \alpha > 0$ for $\alpha \in (0, 1)$ [Bou97]. This means that only a finite fraction of matrix elements is relevant in a given realization of the matrix. This is a completely different behaviour from the one known from considerations of Gaussian random matrices. For $\alpha \leq [1, 2)$, although $Y_2 = 0$ in the limit $N \to \infty$, one still observes very large fluctuations of individual matrix elements which in particular lead to a localization of eigenvectors which will be discussed shortly, towards the end of the next section. Only for $\alpha > 2$ does the behaviour of matrices resemble that known for Gaussian matrices.[2] In this case, the variance σ^2 of (13.1.1) is finite and the eigenvalue density of the matrix $A_N/\sqrt{N}$ converges for $N \to \infty$ to the Wigner semicircle law $\rho(\lambda) = \sqrt{4\sigma^2 - \lambda^2}/(2\pi\sigma^2)$, independently of the details of the probability distribution. Random matrices having the same limiting eigenvalue density for $N \to \infty$ are said to belong to the same macroscopic universality class. For $\alpha > 2$ it is called a Gaussian universality. This class is very broad and comprises a whole variety of random matrices. In particular one can prove [Pas72] that if A_N is a symmetric random matrix with independent (but not necessarily identically distributed) centred entries with the same variance σ^2, then the condition for the eigenvalue distribution of $A_N/\sqrt{N}$ to converge to the Wigner semicircle law is

$$\lim_{N\to\infty} \frac{1}{N^2} \sum_{i\leq j} \int_{|x|>\epsilon\sqrt{N}} x^2 p_{ij}(x)\, dx = 0, \tag{13.1.3}$$

where ϵ is any positive number and $p_{ij}(x)$ is the p.d.f. for the ijth element of the matrix. As a matter of fact this condition is almost identical to the Lindeberg condition known from the central limit theorem for the distribution of a sum of random numbers to converge to a normal distribution [Fel71]. The fastest

[1] Sparse random matrices are discussed in Chapter 23 of this book.

[2] The convergence to the limiting semicircle law is generically very slow in the presence of power-law tails and moreover for $\alpha \leq 4$ microscopic properties are significantly different from those for generic Gaussian matrices, as we will mention later.

convergence to the limiting semicircle law is achieved for matrices whose elements are independent Gaussian random variables. A prominent place in this macroscopic universality class is taken by the ensemble of symmetric Gaussian matrices whose diagonal elements have a twice larger variance than the off-diagonal ones $\mathcal{N}(0, \sigma^2(1+\delta_{ij}))$. Clearly, such matrices fulfill the Lindeberg condition. The probability measure in the ensemble of such matrices can be written as

$$d\mu(A) = DA \exp -\frac{1}{2\sigma^2} \mathrm{tr}\, A^2, \tag{13.1.4}$$

where DA is a flat measure $DA = \prod_{1\leq i\leq j\leq N} dA_{ij}$. The measure $d\mu(A)$ is manifestly invariant with respect to the orthogonal transformations: $A \to OAO^T$, where O is an orthogonal matrix. This is the GOE ensemble. In the limit $N \to \infty$ the eigenvalue density of $A_N/\sqrt{N}$ approaches a well-known semicircle distribution. In a similar way one can construct GUE and GSE ensembles which are extensively discussed in other chapters of the book. In this chapter we will be mostly interested in matrices for which the variance does not exist.

13.2 Wigner–Lévy matrices

In this section we will discuss properties of heavy-tailed symmetric matrices with i.i.d. elements (13.1.1) for $\alpha \in (0, 2)$. We call them Wigner–Lévy matrices since the Lévy distribution is the corresponding stable law for $0 < \alpha < 2$ which plays an analogous role from the point of view of the central limit theorem to the Gaussian distribution for $\alpha \geq 2$ [Gne68]. The Lévy distributions are sometimes called α-stable laws.

Before we discuss Wigner–Lévy matrices let us briefly recall basic facts about Lévy distributions. In addition to a stability index α these laws are characterized by an asymmetry parameter $\beta \in [-1, 1]$, which will be discussed below, and a scale parameter $R > 0$, called the range, which plays a similar role to the standard deviation σ for the normal law. The statement that a Lévy distribution is a stable law (with respect to addition) means that a sum of two independent Lévy random variables $x = x_1 + x_2$ with a given index α is again a Lévy random variable with the same α. The range R and asymmetry β of the resulting distribution can be calculated as

$$R^\alpha = R_1^\alpha + R_2^\alpha, \quad \beta = \frac{\beta_1 R_1^\alpha + \beta_2 R_2^\alpha}{R_1^\alpha + R_2^\alpha}. \tag{13.2.1}$$

For $\beta_1 = \beta_2$ the asymmetry is preserved and the relation for the effective range is a generalization of the corresponding one for independent Gaussian random variables, where the sum is also a Gaussian random variable with variance

$\sigma^2 = \sigma_1^2 + \sigma_2^2$. Actually for $\alpha = 2$ the range and the standard deviation are related as $\sigma = \sqrt{2}R$.

The p.d.f. $L_\alpha^{R,\beta}(x)$ of the Lévy distribution with stability index α, asymmetry β and range R is conventionally written as a Fourier transform of the characteristic function, since its form is known explicitly. There are several definitions used in the literature, here we quote one, which seems to be the most common [Gne68, Nol10]

$$L_\alpha^{R,\beta}(x) = \frac{1}{2\pi}\int_{-\infty}^{+\infty} dk \widehat{L}_\alpha^{R,\beta}(k) e^{-ikx} \tag{13.2.2}$$

where[3]

$$c(k) = \ln \widehat{L}_\alpha^{R,\beta}(k) = -R^\alpha |k|^\alpha \left(1 + i\beta \mathrm{sgn}(k) \tan(\pi\alpha/2)\right). \tag{13.2.3}$$

The logarithm of the characteristic function $c(k)$ is usually called a cumulant-generating function. This name is slightly misleading since for Lévy distributions only the first cumulant exists for $\alpha \in (1, 2)$ or none for $\alpha \in (0, 1]$. Therefore we would rather call it the c-transform. The characteristic function is known explicitly in contrast to the corresponding p.d.f. $L_\alpha^{R,\beta}(x)$ which can be expressed in terms of simple functions only for $\alpha = 1$, $\beta = 0$ (Cauchy distribution) $\alpha = 3/2$, $\beta = \pm 1$ (Smirnoff distribution). For $\alpha = 2$, the characteristic function (13.2.3) becomes a Gaussian function independent of β. A stability of the Lévy distribution can be easily verified. A p.d.f. for the sum of two independent variables $x_{1+2} = x_1 + x_2$ is a convolution of the p.d.f.s for individual components x_1 and x_2 and thus the corresponding characteristic function is a product of two characteristic functions. It is easy to check by inspection of (13.2.3) that the relations (13.2.1) are indeed satisfied. It is less trivial to demonstrate that the Fourier transform of the characteristic function (13.2.2) is a non-negative function. Actually this is only the case for $0 < \alpha \leq 2$. Lévy distributions $L_\alpha^{R,\beta}(x)$ for $\alpha \in (0, 2]$ are the only stable laws. The asymmetry parameter β controls the skewness of the distribution. For $\beta = 0$, the characteristic function (13.2.3) is even and so is the p.d.f. $L_\alpha^{R,0}(x) = L_\alpha^{R,0}(-x)$. For $\beta \neq 0$ the p.d.f. is skew and has a different left and right asymptotic behaviour,

$$L_\alpha^{R,\beta}(x) \overset{x\to\pm\infty}{\longrightarrow} (1 \pm \beta)\frac{\gamma_\alpha R^\alpha}{|x|^{\alpha+1}}, \tag{13.2.4}$$

where $\gamma_\alpha = \Gamma(1+\alpha)\sin(\pi\alpha/2)/\pi$. In the extreme cases $\beta = \pm 1$ one of the tails is suppressed and for $\alpha < 1$ the distribution becomes fully asymmetric with a support only on the positive (resp. negative) semiaxis. One should note that the mean value of the Lévy distribution for $\alpha > 1$ equals zero, independently of the asymmetry β. For $\alpha = 2$ the dependence on β disappears. If one sets $R = 1$ in (13.2.3) one obtains a standardized Lévy distribution. A Lévy random variable x

[3] For $\alpha = 1$ it assumes a slightly different form: $c(k) = -R|k| \left(1 + i(2\beta/\pi)\mathrm{sgn}(k)\ln(Rk)\right)$.

with an arbitrary R can be obtained from the corresponding standardized one x_* by a rescaling $x = Rx_*$, hence $L_a^{R,\beta}(x) = L_a^{1,\beta}(x/R)/R$.

Let us now consider a symmetric matrix A_N with elements A_{ij} for $1 \leq i \leq j \leq N$ being i.i.d. Lévy random variables with the p.d.f. $L_a^{R,\beta}(x)$. We are now interested in the eigenvalue density of such a matrix in the limit $N \to \infty$. As we shall see below the eigenvalue density of the matrix $A_N/N^{1/a}$ converges to a limiting density $\rho(\lambda)$ which is completely different from the Wigner semicircle law and has an infinite support and heavy tails. The choice of the scaling factor $N^{1/a}$ is related to the universal scaling properties of the Lévy distribution and for this reason can be viewed as a generalization of the scaling for Gaussian random matrices ($a=2$). The problem of the determination of the limiting eigenvalue distribution of Wigner–Lévy matrices is not simple because the standard methods used for Gaussian matrices or matrices from invariant ensembles do not apply here. A special method tailored to the specific universal features of heavy-tailed distributions was necessary to attack this problem. Such a method, being a beautiful adaptation of the cavity method which maximally exploits universal properties of a-stable distributions, was invented in [Ciz94]. The description of the method is beyond the scope of this chapter and we refer the interested reader to the original paper [Ciz94] and to the paper [Bur07] where the derivation was explained step by step and some details were corrected. Here we only quote the final result. The eigenvalue density $\rho(\lambda)$ of a Wigner–Lévy matrix is given by a Lévy function with an index $a/2$, being a half of the index a of the p.d.f. $L_a^{R,\beta}(x)$ used to generate the matrix elements, and an effective "running" range $\widehat{R}(\lambda)$ and asymmetry parameter $\widehat{\beta}(\lambda)$

$$\rho(\lambda) = \frac{1}{\Lambda}\widehat{\rho}\left(\frac{\lambda}{\Lambda}\right) \quad \text{where} \quad \widehat{\rho}(\lambda) = L_{a/2}^{\widehat{R}(\lambda),\widehat{\beta}(\lambda)}(\lambda) \tag{13.2.5}$$

and

$$\Lambda = R\left(\frac{\Gamma(1+a)\cos(\pi a/4)}{\Gamma(1+a/2)}\right)^{1/a}. \tag{13.2.6}$$

The scale parameter Λ is proportional to the original range R. The functions $\widehat{R}(\lambda), \widehat{\beta}(\lambda)$ satisfy a set of integral equations

$$\widehat{R}^{\frac{a}{2}}(\lambda) = \int_{-\infty}^{+\infty} dx\, |x|^{-\frac{a}{2}} L_{a/2}^{\widehat{R}(x),\widehat{\beta}(x)}(\lambda - x) \tag{13.2.7}$$

$$\widehat{\beta}(\lambda) = \frac{\int_{-\infty}^{+\infty} dx\, \mathrm{sign}(x)|x|^{-\frac{a}{2}} L_{a/2}^{\widehat{R}(x),\widehat{\beta}(x)}(\lambda - x)}{\int_{-\infty}^{+\infty} dx\, |x|^{-\frac{a}{2}} L_{a/2}^{\widehat{R}(x),\widehat{\beta}(x)}(\lambda - x)} \tag{13.2.8}$$

where the integrals should be interpreted as Cauchy principal values. The equation (13.2.8), which was derived in [Bur07], is a corrected version of this equation given in [Ciz94].

We conclude this section with some comments.

The dependence on λ on the right-hand side of (13.2.5) appears not only through the main argument of the function but also through the dependence of the effective parameters $\widehat{R}(\lambda)$ and $\widehat{\beta}(\lambda)$ on λ. This makes the resulting expression very complex. The equations for $\widehat{\beta}(\lambda)$ and $\widehat{R}(\lambda)$ cannot be solved analytically. It is also not easy to solve them numerically because the computation of the function $L_a^{R,\beta}(x)$, which is the main building block of the above integral equations, is numerically unstable if one does the Fourier integral (13.2.2) in a straightforward way since its integrand is a strongly oscillating function. It is necessary to apply a non-trivial transformation of the integration contour in the complex plane to assure that the integration becomes numerically stable [Nol10]. We skip the details here and again refer the interested reader to [Bur07] for further details.

The eigenvalue density (13.2.5) holds for any $\alpha \in (1, 2)$ and is independent of the asymmetry β of the p.d.f. for matrix elements. The independence of β might be surprising but it can easily be checked by inspection. Indeed, equations (13.2.7) and (13.2.8) depend only on α and do not involve any dependence on the parameters β or R of the p.d.f. for matrix elements, $L_a^{R,\beta}(x)$. Numerically the solution (13.2.5) seems to extend also to the range of $\alpha \in (0, 1]$ but for symmetric distributions ($\beta = 0$) only.

The limiting eigenvalue density $\rho(\lambda)$ is a unimodal even function. The height of the maximum at $\lambda = 0$ is

$$\rho(0) = \frac{\Gamma(1 + 2/\alpha)}{\pi R} \left(\frac{\Gamma^2(1 + \alpha/2)}{\Gamma(1 + \alpha)} \right)^{1/\alpha}. \tag{13.2.9}$$

For large $|\lambda| \to \infty$ the eigenvalue density has heavy tails

$$\rho(\lambda) \sim \frac{1}{\pi} \Gamma(1 + \alpha) \sin\left(\frac{\alpha\pi}{2}\right) \frac{R^\alpha}{|\lambda|^{\alpha+1}} \tag{13.2.10}$$

with the same power as the p.d.f. for the matrix elements, although at first sight one might have the impression that the power should rather be $\alpha/2$ for it is the index of the Lévy function on the right-hand side of (13.2.5). This is not the case because also the dependence of the running parameters $\widehat{R}(\lambda)$ and $\widehat{\beta}(\lambda)$ on λ contributes to a net asymptotic behaviour. In Figure 13.1 we show limiting distributions $\rho(\lambda)$ of the Wigner-Lévy matrices with $R = 1$ and different values of α and compare them to eigenvalue histograms obtained numerically by Monte-Carlo generation of symmetric 200×200 matrices with i.i.d. entries with the p.d.f. $L_\alpha^{1,0}(x)$. The agreement is very good and finite size effects are small.

Another feature of Wigner–Lévy matrices which distinguishes them from Wigner matrices from the Gaussian universality class is a localization of the eigenvectors [Ciz94]. The degree of localization is measured by the inverse participation ratio

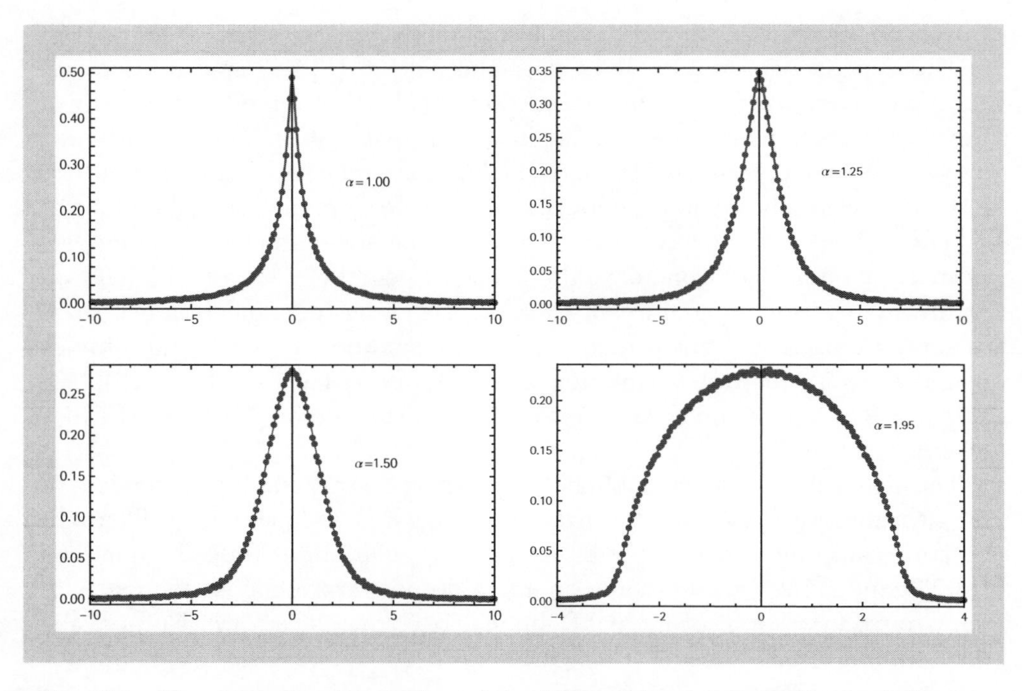

Fig. 13.1 The eigenvalue density for infinite Wigner–Lévy matrices with the index $\alpha = 1.0, 1.25, 1.5, 1.95$ and the range $R = 1$ (solid line) and the corresponding numerical densities obtained by Monte-Carlo generated matrices of size 200×200 with entries distributed according to $L^{1,0}_{\alpha}(x)$.

$$y_2 = \sum_{i=1}^{N} \psi_i^4 \tag{13.2.11}$$

calculated for the elements ψ_is of a normalized eigenvector $\sum_{i=1}^{N} \psi_i^2 = 1$. For Wigner matrices from the Gaussian universality class $y_2 = 0$ in the limit $N \to \infty$ and it is independent of the corresponding eigenvalue. For Wigner–Lévy matrices with the index $\alpha \in (1, 2)$, the average inverse participation ratio $y_2 = y_2(\lambda)$ depends on the eigenvalue λ. For λ smaller than a certain critical value, $\lambda \leq \lambda_{cr}$, $y_2(\lambda) = 0$, however for $\lambda > \lambda_{cr}$ above this value $y_2(\lambda)$ is positive and it grows monotonically to one when λ increases [Ciz94]. This means that only a finite number of elements ψ_i are significantly different from zero or, phrasing it differently, that the vector is localized on a subset of cardinality of order $1/y_2(\lambda)$. When λ goes to infinity $y_2(\lambda)$ tends to one. This means that for extremely large eigenvalues all but one elements of the corresponding vector are equal to zero and the eigenvector is localized on exactly one state. The critical value λ_{cr} depends on α and grows monotonically from zero for $\alpha = 1$ to infinity for $\alpha = 2$. The fact that $\lambda_{cr} = 0$ for $\alpha = 1$ means that in this case (and also for $\alpha < 1$) all eigenvectors are localized.

The eigenvalue density (13.2.5) defines a whole universality class of Wigner–Lévy matrices. It is a counterpart of the Wigner semicircle law. The eigenvalue density of a matrix $A_N/N^{1/\alpha}$ with i.i.d. entries with a p.d.f. $p(x)$ will converge to the same limiting law if the function $p(x)$ has the same asymptotic behaviour as $L_\alpha^{R,\beta}(x)$. More precisely, if the p.d.f. $p(x)$ is centred ($\int_{-\infty}^{+\infty} dx x p(x) = 0$) for $\alpha \in (1, 2)$ (or even for $\alpha \in (0, 1]$) and has the following asymptotic behaviour:

$$p(x) \overset{x \to \pm\infty}{\longrightarrow} \frac{C_\pm}{|x|^{1+\alpha}}, \tag{13.2.12}$$

then the eigenvalue distribution of the matrix $A_N/N^{1/\alpha}$ converges to the same limiting density $\rho(x)$ (13.2.5) as for the corresponding matrix with the p.d.f. $L_\alpha^{R,\beta}(x)$ with the same index α and

$$R = \left(\frac{C_+ + C_-}{2\gamma_\alpha}\right)^{1/\alpha}, \quad \beta = \frac{C_+ - C_-}{C_+ + C_-}, \tag{13.2.13}$$

where as before $\gamma_\alpha = \Gamma(1+\alpha)\sin(\pi\alpha/2)/\pi$. One can probably extend this macroscopic universality to a class of matrices with independent but not necessarily identically distributed entries, however having distributions with the same asymptotic behaviour.

We finish this section with a comment on microscopic properties of Wigner matrices with heavy tails (13.1.1). We have seen so far that macroscopic properties of Wigner matrices change at $\alpha = 2$. For $\alpha > 2$ Wigner matrices belong to the Gaussian universality class and their eigenvalue density converges for large N to the Wigner semicircle law while for $\alpha < 2$ they belong to the Lévy universality class and their eigenvalue density converges to the limiting distribution given by (13.2.5). One may ask if $\alpha = 2$ is also a critical value for microscopic properties like for instance eigenvalue correlations or the statistics of the largest eigenvalue λ_{max}. This question has already been partially studied. It was found that in this case the critical value of the exponent α is rather $\alpha = 4$ [Bir07a]. For $\alpha > 4$ the largest eigenvalue of the matrix $A_N/\sqrt{N}$, for $N \to \infty$, fluctuates around the upper edge of the support of the Wigner semicircle distribution and the fluctuations are of order $N^{-2/3}$. A rescaled quantity $x = (\lambda_{max} - \lambda_{edge})N^{2/3}$ obeys the Tracy–Widom statistics [Tra94, Tra96] although the convergence to this limiting distribution is rather slow. For $\alpha < 4$ the largest eigenvalue is of the order $N^{(4-\alpha)/(2\alpha)}$ and a rescaled quantity $y = \lambda_{max}/N^{(4-\alpha)/(2\alpha)}$ is distributed according to a modified Fréchet law. For $\alpha = 4$ which is a marginal case the two regimes are mixed in the proportions depending on details of the p.d.f. for matrix elements, in particular on the amplitude of the tail. Roughly speaking for $\alpha > 4$ the eigenvalue repulsion shapes the microscopic properties of the matrix and leads to the Tracy–Widom statistics while for $\alpha < 4$ the repulsion plays a secondary role. The dominant effect is in this case related to fluctuations in the tail of the distribution which are so large that the repulsion can be neglected and the largest eigenvalues can be treated as independent of each other. Actually it has been known for some time [Sos99] that indeed the largest eigenvalues are given by a Poisson point process with a Fréchet intensity related to the statistics of the largest elements in the random matrix coming from the tail of the distribution (13.1.1) for $\alpha \leq 2$. In the paper [Bir07a] an argument was given that basically the same picture holds also for $2 < \alpha < 4$.

Wigner matrices are discussed in Chapter 21. The interested reader can also find there a discussion of other aspects of this class of random matrices.

13.3 Free random variables and free Lévy matrices

It is sometimes convenient to think of whole matrices as entities and to formulate for them probabilistic laws. One can ask for example if one can calculate the eigenvalue density $\rho_{1+2}(\lambda)$ of a sum of two symmetric (or Hermitian) $N \times N$ random matrices

$$A_{1+2} = A_1 + A_2 \tag{13.3.1}$$

given the densities $\rho_1(\lambda)$ and $\rho_2(\lambda)$ of A_1 and A_2. In general the answer to this question is negative since the resulting distribution depends on many other factors. The situation becomes less hopeless for large matrices. It turns out that for $N \to \infty$, $\rho_{1+2}(\lambda)$ depends only on $\rho_1(\lambda)$ and $\rho_2(\lambda)$ if A_1 and A_2 are independent random matrices or, more precisely, if they are free. The freeness is a concept closely related to the independence. The independence itself is not sufficient. The freeness additionally requires a complete lack of angular correlations. Such correlations may appear even for independent matrices if they are generated from a matrix-ensemble which hides some characteristic angular pattern. An example is just the ensemble of Wigner–Lévy matrices whose probability measure is not rotationally invariant. This measure favours some specific angular directions. In effect, even if one picks at random two matrices from this ensemble, they will both prefer some angular directions and thus will have some sort of mutual correlations. One can remove the correlations by a uniform angular randomization of the matrices A_1 and A_2

$$A_{1\boxplus 2} = O_1 A_1 O_1^T + O_2 A_2 O_2^T \tag{13.3.2}$$

where O_is are random orthogonal matrices with a uniform probability measure in the group of orthogonal matrices.[4] The matrix $O_i A_i O_i^T$ has exactly the same eigenvalue content as A_i. Actually to achieve the effect of angular decorrelation it is sufficient to rotate only one of the A_is. This type of addition is called a free addition and we denote it by $\boxplus$. Of course if A_is are generated from

[4] For Hermitian matrices A_i one uses unitary rotations $A_{1+2} = U_1 A_1 U_1^\dagger + U_2 A_2 U_2^\dagger$.

an ensemble with a rotationally invariant probability measure, as for instance $DA \exp -\mathrm{tr}\, V(A)$, then the two types of additions (13.3.1) and (13.3.2) are identical and $\rho_{1+2}(\lambda) = \rho_{1\boxplus 2}(\lambda)$. Otherwise they are different and $\rho_{1+2}(\lambda) \neq \rho_{1\boxplus 2}(\lambda)$. For example a sum of independent identically distributed Wigner–Lévy matrices (13.3.1) is a Wigner–Lévy matrix which is not rotationally invariant while a free sum of Wigner–Lévy matrices is a rotationally invariant matrix, which has a different eigenvalue density, so in this case $\rho_{1+2}(\lambda) \neq \rho_{1\boxplus 2}(\lambda)$. Generally, for large matrices in the limit $N \to \infty$ the eigenvalue density $\rho_{1\boxplus 2}(\lambda)$ depends only on the individual densities $\rho_1(\lambda)$ and $\rho_2(\lambda)$ and it can be uniquely determined from them [Voi92].

A theory of free addition was actually developed in probability theory of non-commutative objects (operators) long before random matrices entered the scene [Voi85]. It was originally formulated in terms of von Neumann algebras equipped with a trace-like normal state τ, which was introduced to generalize the concept of uncorrelated variables known from a classical probability. More specifically in classical probability two real random variables x_1 and x_2 are uncorrelated if the correlation function, calculated as the expectation value $E(\hat{x}_1\hat{x}_2) = 0$ for centred variables, $\hat{x}_i = x_i - E(x_i)$, vanishes. In free probability, by analogy, elements X_i are free if $\tau(\hat{x}_{\pi(1)}\, \hat{x}_{\pi(2)} \ldots \hat{x}_{\pi(m)}) = 0$ for any permutation π of the corresponding centered elements $\hat{X}_i = X_i - \tau(X_i) \cdot 1$, where 1 is the identity matrix. A link to large matrices was discovered later [Voi91] when it was realized that a non-commutative probability space with free elements X_is can be mapped onto a set of large $N \times N$ matrices, $N \to \infty$, of the form $X_i = U_i D_i U_i^{\dagger}$, where D_i are diagonal real matrices and U_i random unitary (or orthogonal, O_i) matrices distributed with a uniform probability measure on the group. In this mapping the state function τ corresponds to the standard normalized trace operation $\tau(\ldots) \leftrightarrow \frac{1}{N}\mathrm{tr}(\ldots)$. This observation turned out to be very fruitful both for free probability and for the theory of large random matrices since one could successfully apply results of free probability to random matrices and vice versa [Voi92]. Free probability and its relation to random matrices and planar combinatorics for the case when all moments of the probability distribution exist [Spe94, Nic06] is discussed in detail in Chapter 22. Here we concentrate on heavy-tailed free distributions for which higher moments do not exist. In particular we shall discuss free stable laws and their matrix realizations. Before we do that we recall basic concepts to make the discussion of this section self-contained.

Actually the law of addition of free random variables is in many respects analogous to the law of addition of independent real random variables. We will therefore begin by briefly recalling the law of addition in classical probability and by describing the corresponding steps in free probability. The p.d.f. $p_{1+2}(x)$ of a sum $x = x_1 + x_2$ of independent real random variables x_1 and x_2 is given by a convolution of the individual p.d.f.s $p_{1+2}(x) = (p_1 * p_2)(x)$. Thus, the characteristic function $\widehat{p}(k) = \int dk e^{ikx} p(x)$ for the sum is a product of the corresponding characteristic functions $\widehat{p}_{1+2}(k) = \widehat{p}_1(k)\widehat{p}_2(k)$. This is more conveniently expressed in terms of a cumulant-generating function (c-transform) defined as $c(k) = \ln \widehat{p}(k)$, as a simple additive law

$$c_{1+2}(k) = c_1(k) + c_2(k). \tag{13.3.3}$$

It turns out that one can find a corresponding object in free probability [Voi92], a free cumulant-generating function alternatively called R-transform, for which the free addition (13.3.2) leads to the corresponding additive rule,

$$R_{1\boxplus 2}(z) = R_1(z) + R_2(z). \tag{13.3.4}$$

For a given eigenvalue density $\rho(\lambda)$ one defines a moment-generating function as the Cauchy transform of the eigenvalue density [Voi92],

$$G(z) = \int_{-\infty}^{+\infty} \frac{\rho(\lambda) d\lambda}{z - \lambda}, \tag{13.3.5}$$

also known as the resolvent or Green function. The free-cumulant-generating function (R-transform) is related to the Green function as

$$z = G\left(R(z) + \frac{1}{z}\right). \tag{13.3.6}$$

The last equation can be inverted for $G(z)$,

$$z = R(G(z)) + \frac{1}{G(z)}. \tag{13.3.7}$$

In short $z \to R(z) + 1/z$ is the inverse function of $z \to G(z)$ and thus for a given resolvent one can determine the R-transform and vice versa. Using the addition law (13.3.4) one can now give a step-by-step algorithm to calculate the eigenvalue density $\rho_{1\boxplus 2}(\lambda)$ of the free sum (13.3.2) from $\rho_1(\lambda)$ and $\rho_2(\lambda)$. First one determines the resolvents $G_1(z)$ and $G_2(z)$ using (13.3.5), then the corresponding R-transforms $R_1(z)$ and $R_2(z)$ using (13.3.6), and finally $R_{1\boxplus 2}(z)$ using the addition law (13.3.4). Having found $R_{1\boxplus 2}(z)$ one proceeds in the opposite order. One reconstructs the corresponding resolvent $G_{1\boxplus 2}(z)$ (13.3.7) and then the density $\rho_{1\boxplus 2}(\lambda)$ by the inverse of (13.3.5),

$$\rho(\lambda) = -\frac{1}{\pi} \text{Im} G(\lambda + i0^+), \tag{13.3.8}$$

which follows from the relation $(x + 0^+)^{-1} = P.V.x^{-1} - i\pi\delta(x)$. This completes the task of calculating the eigenvalue density of a free sum $\rho_{1\boxplus 2}(\lambda)$ from $\rho_1(\lambda)$ and $\rho_2(\lambda)$. This procedure can be fully automated and it has actually been implemented for a certain class of matrices [Rao06].

The correspondence between the laws of addition (13.3.3) and (13.3.4) and between the logical structures behind these laws in classical and free probability also has profound theoretical implications. One of these is a bijection between infinitely divisible laws of classical probability and the laws in free probability

[Ber93]. Using this bijection one can derive the R-transform for stable laws in free probability [Ber99][5]

$$R(z) = \begin{cases} bz^{\alpha-1} & \text{for } \alpha \in (0,2] \text{ and } \alpha \neq 1 \\ -i(1+\beta) - (2\beta/\pi)\ln z & \text{for } \alpha = 1 \end{cases} \tag{13.3.9}$$

where $b = -e^{i\alpha(1+\beta)\pi/2}$ for $\alpha \in (0,1)$ and $b = e^{i(\alpha-2)(1+\beta)\pi/2}$ for $\alpha \in (1,2]$. The stable R-transforms (13.3.9) are in one-to-one correspondence with the c-transforms of Lévy distributions (13.2.3) and they fully classify all free stable laws and allow one to determine the free probability densities for these stable laws. These densities are equal to the eigenvalue densities of free Lévy matrices, being matrix realizations of the stable free random variables.

Let us illustrate how this procedure works for a stable law with the stability index $\alpha = 2$ and the unit range $R = 1$. Using (13.3.9) we have $R(z) = z$. Inserting this into (13.3.7) we obtain an equation for the resolvent $G(z) = z - 1/G(z)$ which gives in the upper complex half-plane $G(z) = (z - \sqrt{z^2-4})/2$. Finally using (13.3.8) we find a Wigner law $\rho(\lambda) = \sqrt{4-\lambda^2}/2\pi$ with $\sigma = 1$. We see that the Wigner law is equivalent in free probability to the normal law in classical probability. Secondly consider the case for $\alpha = 1, \beta = 0$. Proceeding in the same way as above we find respectively,

$$R(z) = -i, \quad G(z) = \frac{1}{z+i}, \quad \rho(\lambda) = \frac{1}{\pi}\frac{1}{1+\lambda^2}. \tag{13.3.10}$$

This case is special because the stable density is identical in classical and free probability. For other values of α one can find the free stable laws by applying Equation (13.3.7) to the R-transform (13.3.9) which gives the following equation for the resolvent,

$$bG^{\alpha}(z) - zG(z) + 1 = 0. \tag{13.3.11}$$

This equation can be solved analytically for a couple of values of the parameter α for which it is just a quadratic, cubic, or quartic equation. The solution can then be used to calculate the eigenvalue density (13.3.8). For other values the eigenvalue density can be determined numerically.

Equation (13.3.11) may be used to extract the asymptotic behaviour of the corresponding eigenvalue density (13.3.8). We will only give the result for the symmetric case ($\beta = 0$) and for the range $R = 1$. For small eigenvalues $|\lambda| \to 0$ it reads [Bur02]

$$\rho(\lambda) = \frac{1}{\pi}\left(1 - \frac{3-\alpha}{2\alpha^2}\lambda^2 + \ldots\right) \tag{13.3.12}$$

[5] We give only the standardized version which corresponds to the unit range. An R-transform with a range r can be obtained from the standardized one by a rescaling $R_r(z) = rR(rz)$.

while for large ones, $|\lambda| \to \infty$

$$\rho(\lambda) \sim \frac{1}{\pi} \sin\left(\frac{\alpha\pi}{2}\right) |\lambda|^{-\alpha-1}. \tag{13.3.13}$$

The distribution has a smooth quadratic maximum at $\lambda = 0$ while for large λ it has heavy power-law tails with the same power as the corresponding stable law for real variables. For $\alpha = 2$ the tails disappear.

The procedure described above to derive the free probability density $\rho(\lambda)$ of free stable laws is based solely on relations between the R-transform, the resolvent, and the density. It does not involve any matrix calculations. Free random matrices appear as a representation of these free random variables and correspond to infinitely large random matrices with a given eigenvalue density $\rho(\lambda)$ and with a rotationally invariant probability measure. Actually there are many different matrix realizations of the same free random variable. We shall discuss below two of the simplest ones which have completely different microscopic properties. The most natural realization is a matrix generated by the uniform angular randomization ODO^T of a large diagonal matrix D of size $N \to \infty$ which has i.i.d. random variables on the diagonal. If we choose the p.d.f. of the diagonal elements as $\rho(\lambda)$, which corresponds to a free stable law, we obtain a matrix realization of a free random variable which is stable under addition. Clearly, the eigenvalues of this matrix are by construction uncorrelated. We can now independently generate many such matrices $O_i D_i O_i^T$, $i = 1, \ldots, K$. Because of the stability the following sum,

$$A_{\boxplus K} = \frac{1}{K^{1/\alpha}} \sum_{i=1}^{K} O_i D_i O_i^T \tag{13.3.14}$$

also has exactly the same eigenvalue density $\rho(\lambda)$ as each term in the sum, so it is a representation of the same free random variable. However the sum $A_{\boxplus K}$ belongs to a different microscopic class since its eigenvalues repel each other, contrary to eigenvalues of each matrix in the sum which are uncorrelated by construction. Already for $K = 2$ one observes a standard repulsion characteristic for invariant ensembles as illustrated in Figure 13.2. It is tempting to conjecture that for $K \to \infty$ the sum $A_{\boxplus K}$ becomes a random matrix which maximizes entropy for the stable eigenvalue density $\rho(\lambda)$. The probability measure for such matrices is known to be rotationally invariant $d\mu(A) = DA \exp - N \mathrm{tr} V(A)$ and to have a potential $V(\lambda)$ which is related to $\rho(\lambda)$ as [Bal68]

$$V'(\lambda) = 2P.V. \int d\zeta \frac{\rho(\zeta)}{\lambda - \zeta} = G(\lambda + i0^+) + G(\lambda - i0^-). \tag{13.3.15}$$

For free stable laws the resolvent $G(\lambda)$ is given by a solution of (13.3.11). In particular, for $\alpha = 2$ the last equation gives a quadratic potential $V(\lambda) = \lambda^2/2$, while for $\alpha = 1$ and $\beta = 0$ a logarithmic one $V(\lambda) = \ln(\lambda^2 + 1)$. The potential

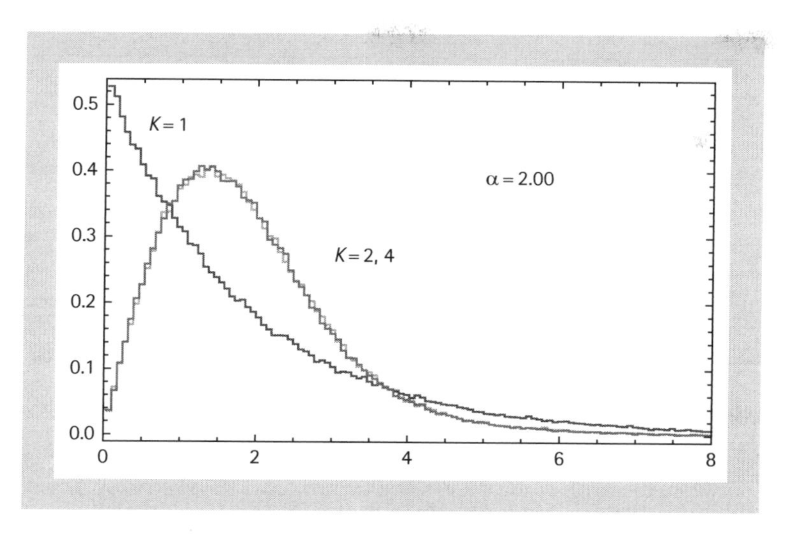

Fig. 13.2 Level spacing histograms of the matrices $A_{\boxplus K}$ (13.3.14) of size 200 × 200 obtained from diagonal matrices D_i whose eigenvalues were generated independently from a Wigner semicircle. For $K = 1$ the histogram properly reflects the Poissonian nature of eigenvalues of a single matrix D_i. For $K > 1$ (already for $K = 2$) the histogram has a shape of the Wigner surmise characteristic for a typical eigenvalue repulsion.

can also be determined for other values of α. Generally, for large $|\lambda| \to \infty$ it behaves as,

$$V(\lambda) = 2\ln\lambda - 2\alpha^{-1}\text{Re}\,\left(b\lambda^{-\alpha}\right) + \ldots. \tag{13.3.16}$$

For maximal entropy random matrices one can also find the joint probability which takes a standard form,

$$\rho(\lambda_1, \ldots, \lambda_N) = Ce^{-N\sum_i V(\lambda_i)} \prod_{i<j} (\lambda_i - \lambda_j)^\beta, \tag{13.3.17}$$

where C is a normalization and β as usual is equal to 1 or 2 for orthogonal or unitary invariant ensemble respectively. For free stable laws the potential $V(\lambda)$ assumes, however, a highly non-standard non-polynomial form [Bur02].

The stable laws in free probability play a similar role to the corresponding stable laws in classical probability where a sum of i.i.d. centred random variables $s_n = (x_1 + \ldots + x_n)/n^{1/\alpha}$ is known to become an α-stable random variable for $n \to \infty$. We expect a similar effect for random matrices [Hia00]. For example, if one generates a sequence of K independent Wigner–Lévy matrices $(A_1, A_2, \ldots, A_K)$ and a sequence of independent random orthogonal matrices $(O_1, O_2, \ldots, O_K)$ and one forms a sum,

$$A_{\boxplus K} = \frac{1}{K^{1/\alpha}} \sum_{i=1}^{K} O_i A_i O_i^T \tag{13.3.18}$$

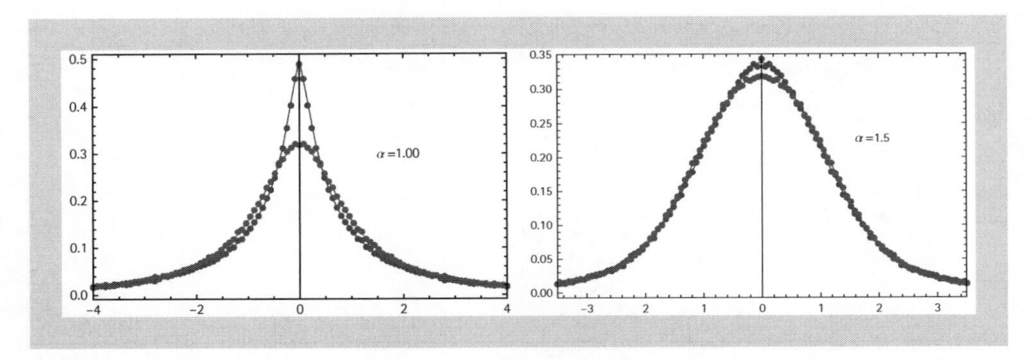

Fig. 13.3 Eigenvalue density for Wigner–Lévy matrices and corresponding free Lévy matrices for $\alpha = 1.00$ (left) and $\alpha = 1.50$ (right) are represented by the solid line. The corresponding Monte-Carlo histogram for a free sum (13.3.18) of $K = 32$ Wigner–Lévy matrices of size $N = 200$ is shown. For comparison we chose the range of the Wigner–Lévy matrix to be $R = (\Gamma(1+\alpha))^{-1/\alpha}$ since then the asymptotic behaviour of the eigenvalue density for Wigner–Lévy matrices (13.2.10) is identical to that for standardized free random variables (13.3.13).

in analogy to (13.3.14) one expects that for large K the sum will become a free random matrix with an eigenvalue density given by a free α-stable law [Bur07]. Actually in practice the convergence to the limiting distribution is very fast. We observe that already for K of order 10 and N of order 100 the eigenvalue density of the matrix $A_{\boxplus K}$ does not significantly differ from the stable density $\rho(\lambda)$ for free random variables (see Figure 13.3). If we skipped random rotations in (13.3.18) and just added many Wigner–Lévy matrices $(A_1 + \ldots + A_K)/K^{1/\alpha}$ we would obtain a Wigner–Lévy matrix. As mentioned already, the independence of the matrices A_i is not sufficient in itself to make the addition free and only the rotational randomization brings the sum to the universality class of free random variables. The comparison of the eigenvalue density of the Wigner–Lévy and the corresponding free Lévy matrices is shown in Figure 13.3.

13.4 Heavy-tailed deformations

In this section we discuss ensembles of random matrices obtained from standard ensembles by a reweighting of the probability measure. We will begin by sketching the idea for real random variables where the procedure is simple and well known and then we will generalize it to random matrices. We will in particular concentrate on heavy-tailed deformations of the probability measure of the Wishart ensemble, which are applicable to the statistical multivariate analysis of heavy-tailed data.

Consider a random variable x constructed as a product $x = \sigma\xi$ where ξ is a normally distributed variable $\mathcal{N}(0, 1)$ and $\sigma \in (0, \infty)$ is an independent random variable representing a fluctuating scale factor having a p.d.f. $f(\sigma)$. One can think of x as a Gaussian variable $\mathcal{N}(0, \sigma^2)$ for which the variance itself is a random variable. Obviously the p.d.f. of x can be calculated as

$$p(x) = \int_0^\infty d\sigma f(\sigma) \frac{1}{\sqrt{2\pi}\sigma} e^{-\frac{x^2}{2\sigma^2}}. \tag{13.4.1}$$

Choosing appropriately the frequency function $f(\sigma)$ one can thus model the p.d.f. of x. For example for [Tos04, Ber04, Abu05]

$$f(\sigma) = \frac{1}{\sigma} \frac{2}{\Gamma\left(\frac{a}{2}\right)} \left(\frac{a^2}{2\sigma^2}\right)^{\frac{a}{2}} e^{-\frac{a^2}{2\sigma^2}} \tag{13.4.2}$$

where a is a constant, the integral (13.4.1) takes the following form,

$$p(x) = \frac{1}{a\sqrt{\pi}} \frac{1}{\Gamma\left(\frac{a}{2}\right)} \int_0^\infty d\zeta\, \zeta^{\frac{a-1}{2}} e^{-\zeta} e^{-\zeta\left(\frac{x}{a}\right)^2} = \frac{1}{a\sqrt{\pi}} \frac{\Gamma\left(\frac{a+1}{2}\right)}{\Gamma\left(\frac{a}{2}\right)} \left(1 + \left(\frac{x}{a}\right)^2\right)^{-\frac{a+1}{2}}. \tag{13.4.3}$$

In doing the integral we changed the integration variable[6] to $\zeta = a^2/(2\sigma^2)$. For large $|x|$ the p.d.f. has a power-law tail $p(x) \sim |x|^{-\alpha-1}$. The variance of the distribution exists only for $\alpha > 2$ and is equal to $a^2/(\alpha - 2)$.

If one applies this procedure to each matrix element independently $A_{ij} = \sigma_{ij}\xi_{ij}$, $i \leq j$, one obtains a Wigner matrix, discussed in the first section of this chapter. One can however construct a slightly different matrix $A_N(\sigma)$ whose elements have the same common random scale factor $A_{ij} = \sigma\xi_{ij}$ and differ only by ξ_{ij} which are i.i.d. Gaussian random variables $\mathcal{N}(0, 1)$. In the limit $N \to \infty$ the eigenvalue density of the matrix $A_N(\sigma)/\sqrt{N}$ converges to the Wigner semicircle law $\rho_\sigma(\lambda) = \sqrt{4\sigma^2 - \lambda^2}/(2\pi\sigma^2)$. The scale factor σ changes however from matrix to matrix with frequency $f(\sigma)$ so in analogy to (13.4.1) the effective eigenvalue density in the ensemble of matrices is given by the average of the semicircle law over σ [Boh08],

$$\rho(\lambda) = \int d\sigma f(\sigma) \rho_\sigma(\lambda) = \int_{\frac{\lambda}{2}}^\infty d\sigma f(\sigma) \frac{1}{2\pi\sigma^2} \sqrt{4\sigma^2 - \lambda^2}. \tag{13.4.4}$$

In particular, for the frequency function (13.4.2) we obtain the following eigenvalue density [Ber04],

$$\rho(\lambda) = \frac{1}{a} \frac{\sqrt{2}}{\pi\Gamma\left(\frac{a}{2}\right)} \int_0^{\frac{2a^2}{\lambda^2}} d\zeta\, \zeta^{\frac{a-1}{2}} e^{-\zeta} \sqrt{1 - \frac{\zeta\lambda^2}{2a^2}} \tag{13.4.5}$$

which has power-law tails, $\rho(\lambda) \sim |\lambda|^{-\alpha-1}$, with the same power as the p.d.f. (13.4.3) for matrix elements. One can generalize the result to other frequency functions [Mut05].

[6] The variable ζ has a χ^2 distribution.

In fact, exactly the same strategy can be applied to calculate the joint probability since fluctuations of matrix elements are modified by a common scale factor $\rho(\lambda_1, \ldots, \lambda_N) = \int d\sigma f(\sigma)\rho_\sigma(\lambda_1, \ldots, \lambda_N)$, where ρ_σ is given by (13.3.17) with $V(\lambda) = \lambda^2/(2\sigma^2)$. From the joint probability one can then derive microscopic properties of matrices, including the microscopic correlation functions and the distribution of the largest eigenvalue [Boh09].

One can use this procedure to deform probability measures of other matrix ensembles as well. In what follows we concentrate on deformations of the Wishart ensemble. The probability measure for the standardized Wishart ensemble of real matrices is given by[7]

$$d\mu_*(\xi) = (2\pi)^{-\frac{NT}{2}} e^{-\frac{1}{2}\mathrm{tr}\, \xi\xi^T} D\xi \tag{13.4.6}$$

where ξ is a rectangular matrix ξ_{it}, $i = 1, \ldots, N$, $t = 1, \ldots, T$ and $D\xi = \prod_{i,t}^{N,T} d\xi_{it}$. The eigenvalue density of the matrix $(1/T)\xi\xi^T$ is known to converge to the Marčenko–Pastur law $\rho_*(\lambda) = \sqrt{(\lambda_+ - \lambda)(\lambda - \lambda_-)}/(2\pi r\lambda)$, where $\lambda_\pm = (1 \pm r)^2$ and $r = N/T \leq 1$ in the limit $N \to \infty$, $r = \mathrm{const}$ [Mar67]. The elements of the matrix ξ represent normally distributed fluctuations of uncorrelated random numbers with unit variance. Now using the reweighting method we can consider a matrix $A_{it} = \sigma\xi_{it}$ with a common fluctuating scale factor being an independent random variable with a p.d.f. $f(\sigma)$. The effective probability measure can easily be derived from (13.4.6) and reads,

$$d\mu(A) = DA \int d\sigma f(\sigma)\sigma^{-NT} e^{-\frac{1}{2\sigma^2}\mathrm{tr} AA^T} \tag{13.4.7}$$

where the factor σ^{-NT} comes from the change of variables in the measure $DA = \sigma^{NT} D\xi$. In particular for the frequency function (13.4.2) this gives the measure of a multivariate Student's distribution[8]

$$d\mu(A) = DA \frac{\Gamma\left(\frac{a+NT}{2}\right)}{(a\sqrt{\pi})^{NT}\Gamma\left(\frac{a}{2}\right)} \left(1 + \frac{\mathrm{tr}\, AA^T}{a^2}\right)^{-\frac{a+NT}{2}}. \tag{13.4.8}$$

The eigenvalue density of the matrix $(1/T)XX^T$, where X is generated with the probability measure given above can be obtained by the same reweighting method as before. The eigenvalue density of the Gaussian part in (13.4.7) is $\rho_\sigma(\lambda) = \rho_*(\lambda/\sigma^2)/\sigma^2 = \sqrt{(\sigma^2\lambda_+ - \lambda)(\lambda - \sigma^2\lambda_-)}/(2\pi r\sigma^2\lambda)$. It has to be reweighted with the frequency function (13.4.2) $\rho(\lambda) = \int d\sigma f(\sigma)\rho_\sigma(\lambda)$. The result reads [Bur06],

[7] For complex matrices the corresponding measure reads $d\mu_*(\xi) = \pi^{-NT} e^{-\mathrm{tr}\xi\xi^\dagger} D\xi$ where $D\xi = \prod_{it}^{NT} d\mathrm{Re}\xi_{it}\, d\mathrm{Im}\xi_{it}$.

[8] An analogous expression [Tos04, Ber04] can be obtained for GOE and GUE Wigner matrices reweighted with the frequency function (13.4.2).

$$\rho(\lambda) = \frac{\left(\frac{\alpha}{2}\right)^{\alpha/2}}{2\pi r \Gamma\left(\frac{\alpha}{2}\right)} \lambda^{-\alpha/2-1} \int_{\lambda_-}^{\lambda_+} \sqrt{(\lambda_+ - \zeta)(\zeta - \lambda_-)}\, e^{-\frac{\alpha\zeta}{2\lambda}} \zeta^{\alpha/2-1} d\zeta. \tag{13.4.9}$$

The support of this eigenvalue distribution is infinite. The exponent $\alpha/2$ of the tail is a half of the exponent α of p.d.f. for the matrix elements X that one can expect for a matrix $(1/T)XX^T$ which is a 'square' of X. The reweighting can be applied to derive the corresponding joint probability function and to determine the microscopic correlations of the deformed Wishart ensembles [Ake08].

This result can be generalized in a couple of ways. One can change the frequency function $f(\sigma)$ [Abu09] but one can also change the relation between the A and ξ matrix from $A_{it} = \sigma\xi_{it}$ to, for instance, $A_{it} = \sigma \sum_j S_i O_{ij} \xi_{jt}$ where O_{ij} is a rotation matrix and S_i is a vector of positive numbers. The matrix O and the vector S are fixed in this construction and the only fluctuating elements are ξ_{it} which are i.i.d. $\mathcal{N}(0, 1)$ and σ which is an independent random variable with a given p.d.f. $f(\sigma)$, as before. The interpretation of the construction is clear. The factors S_is change the scale of fluctuations and the matrix O rotates the main axes. If one applies it to (13.4.6) one will obtain a deformed measure (13.4.8) where $\mathrm{trAA^T}$ will be substituted by $\mathrm{trAC^{-1}A^T}$. The matrix $C_{ij} = \sum_k O_{ik} S_k^2 O_{kj}$ introduces explicit correlations between the degrees of freedom. In a similar way one can introduce correlations $C_{tt'}$ between A_{it} and $A_{it'}$ at different times t and t' [Bur06].

Another interesting generalization of the reweighting procedure is to consider a matrix $A_{it} = \sigma_i \xi_{it}$ where now the scale factors σ_i are independent random variables for each row [Bir07b]. This case corresponds to a Wishart ensemble with a fluctuating covariance matrix $C_{ij} = \delta_{ij}\sigma_i^2$ where σ_i are i.i.d. random variables with a given p.d.f. $f(\sigma)$. This problem can be solved analytically thanks to an explicit relation between the Greens functions and eigenvalue densities of the matrices C and $(1/T)XX^T$ [Bur04].

The idea of reweighting is quite general. So far we have described reweighting through the scale parameter σ but one can also use other quantities as a basis for the reweighting scheme. For example, one can use the trace $t = \mathrm{tr}XX^T$ of the whole matrix. In this scheme the idea is to calculate quantities for the ensemble with a probability measure[9] $d\mu_t(X) = DX\delta(t - \mathrm{tr}XX^T)$ and then to reweight them using a frequency function $g(t)$ to obtain the corresponding values for the ensemble with the measure $d\mu(X) = DXg(\mathrm{tr}XX^T)$ [Ake99, Abu05]. It turns out that the first step, that is the calculations for the fixed trace ensemble, can be done analytically so this procedure also gives a practical recipe for handling 'non-standard' ensembles. Of course it works only for ensembles for

[9] Alternatively one can use $d\mu_t'(X) = \theta(\mathrm{tr}XX^T - t)\,DX$, where $\theta(x)$ is the Heaviside step function [Ber04].

which the measure depends on $\mathrm{tr}XX^T$ as for instance (13.4.8). In particular it can be applied to the multivariate Wishart-Student ensembles [Bur06].

13.5 Summary

Heavy-tailed random matrices is a relatively new branch of random matrix theory. In this chapter we have discussed several matrix models belonging to this class and presented methods for integrating them. We believe that the models, methods and concepts can be applied to many statistical problems where non-Gaussian effects play an important role.

Acknowledgements

We thank G. Akemann, J.P. Bouchaud, A. Görlich, R.A. Janik, A. Jarosz, M.A. Nowak, G. Papp, P. Vivo, B. Waclaw and I. Zahed for many interesting discussions. This work was supported by the Marie Curie ToK project "COCOS", No. MTKD-CT-2004-517186, the EC-RTN Network "ENRAGE", No. MRTN-CT-2004-005616 and the Polish Ministry of Science Grant No. N N202 229137 (2009-2012).

References

[Abu05] A. Y. Abul-Magd, Phys. Rev. E **71** (2005) 066207.
[Abu09] A.Y. Abul-Magd, G. Akemann, P. Vivo, J. Phys. A: Math. Theor. **42** (2009) 175207.
[Ake99] G. Akemann, G.M. Cicuta, L. Molinari, G. Vernizzi, Phys. Rev. E **59** (1999) 1489; Phys. Rev. E **60** (1999) 5287.
[Ake08] G. Akemann and P. Vivo, J. Stat. Mech. (2008) P09002.
[Bal68] R. Balian, Nuovo Cimento B **57** (1968) 183.
[Ber93] H. Bercovici and D. Voiculescu, Ind. Univ. Math. J. **42** (1993) 733.
[Ber99] H. Bercovici and V. Pata, Annals of Mathematics **149** (1999) 1023; Appendix by P. Biane.
[Ber04] A.C. Bertuola, O. Bohigas, and M.P. Pato, Phys. Rev E **70** (2004) 065102(R).
[Bir07a] G. Biroli, J.-P. Bouchaud, and M. Potters, Europhys. Lett. **78** (2007) 10001.
[Bir07b] G. Biroli, J.-P. Bouchaud, and M. Potters, Acta Phys. Pol. B **38** (2007) 4009.
[Boh08] O. Bohigas, J.X. de Carvalho and M.P. Pato, Phys. Rev. E **77** (2008) 011122.
[Boh09] O. Bohigas, J.X. de Carvalho and M.P. Pato, Phys. Rev. E **79** (2009) 031117.
[Bou97] J.-P. Bouchaud and M. Mézard, J. Phys. A, Math. Gen. **30** (1997) 7997.
[Bur02] Z. Burda, R.A. Janik, J. Jurkiewicz, M.A. Nowak, G. Papp and I. Zahed, Phys. Rev. E **65** (2002) 021106.
[Bur04] Z. Burda, A. Görlich, A. Jarosz, J. Jurkiewicz, Physica A **343** (2004) 295.
[Bur06] Z. Burda, A. Görlich, B. Waclaw, Phys. Rev. E **74** (2006) 041129.
[Bur07] Z. Burda, J. Jurkiewicz, M. A. Nowak, G. Papp and I. Zahed, Phys. Rev. E **75** (2007) 051126; arXiv:cond-mat/0602087.
[Ciz94] P. Cizeau and J.P. Bouchaud, Phys. Rev. E **50** (1994) 1810.

[Fel71] W. Feller, *An Introduction to Probability Theory and Its Applications*, Wiley, 3rd Edition, New York 1971.

[Gne68] B.V. Gnedenko, A. N. Kolmogorov, *Limit distributions for sums of independent random variables*, Revised Edition Addison-Wesley, Cambridge 1968.

[Hia00] F. Hiai and D. Petz, *The Semicircle Law, Free Random Variables and Entropy*, Am. Math. Soc., Providence 1992.

[Mar67] V.A. Marčenko and L. A. Pastur, Math. USSR-Sb, **1**, (1967) 457.

[Meh04] M.L. Mehta, *Random Matrices*, Academic Press, 3rd Edition, London 2004.

[Mut05] K.A. Muttalib and J.R. Klauder, Phys. Rev. E **71**, (2005) 055101(R).

[Nic06] A. Nica and R. Speicher, *Lectures on the Combinatorics of Free Probability*, London Mathematical Society Lecture Note Series, vol. 335, Cambridge University Press, 2006.

[Nol10] J.P. Nolan, *Stable Distributions - Models for Heavy Tailed Data*, Birkhäuser, Boston 2010; http://academic2.american.edu/~jpnolan.

[Pas72] L.A. Pastur: Teor. Mat. Fiz., **10** (1972) 102.

[Rao06] N.R. Rao, *RMTool: A random matrix and free probability calculator in MAT-LAB*; http://www.mit.edu/raj/rmtool/.

[Sos99] A. Soshnikov, Elect. Comm. in Probab. **9** (2004) 82.

[Spe94] R. Speicher, Math. Ann. **298** (1994) 611.

[Tos04] F. Toscano, R.O. Vallejos and C. Tsallis, Phys. Rev. E **69** (2004) 066131.

[Tra94] C.A. Tracy and H. Widom, Commun. Math. Phys. **159** (1994) 151.

[Tra96] C.A. Tracy and H. Widom, Commun. Math. Phys. **177** (1996) 724.

[Voi85] D.V. Voiculescu, in *Operator algebras and their connections with topology and ergodic theory*, (Busteni, 1983), Lecture Notes in Math. Series, vol. **1132**, 556, Springer, New York 1985.

[Voi91] D.V. Voiculescu, Invent. Math. **104** (1991), 201220.

[Voi92] D.V. Voiculescu, K.J. Dykema and A. Nica, *Free Random Variables*, Am. Math. Soc., Providence 1992.

·14·

Phase transitions

G. M. Cicuta and L. G. Molinari

Abstract

A review is presented to orientate the reader in the vast and complex literature concerning phase transitions in matrix models that are invariant under a symmetry group. These phase transitions often have relevant applications in physics. We also mention phase transitions that occur in some matrix ensembles with preferred basis, like the Anderson transition.

14.1 Introduction

Phase transitions were discovered almost thirty years ago in some random matrix ensembles which had a definite interpretation in theoretical physics. Perhaps the most relevant ones are the Gross-Witten phase transition in two dimensional Yang-Mills theory and the models of matter coupled to two-dimensional gravity. The following decades witnessed a vast increase of interest in the study of random matrices in fields of pure and applied mathematics and in theoretical physics, together with a development of accurate mathematical techniques. Phase transitions were found in several models, with universal properties which are often the main subject of interest.

The simplest setting is the ensemble of $n \times n$ Hermitian matrices with a probability density that is invariant under the action of the unitary group. The partition function $Z_n = \int dH e^{-n \operatorname{tr} V(H)}$ can be expressed in terms of the eigenvalues $\{\lambda_i\}$ of the random matrix H and defines the equilibrium statistical mechanics of a Dyson gas of n particles with positions λ_i in the line, in the potential $V(\lambda)$. Neglecting irrelevant constants,

$$Z_n = \int \prod_{j>k} (\lambda_j - \lambda_k)^2 \prod_{j=1}^{n} e^{-nV(\lambda_j)} d\lambda_j = \int e^{-E_n(\vec{\lambda})} d\lambda_1 \cdots d\lambda_n. \qquad (14.1.1)$$

The particles interact by the repulsive electrostatic potential of 2D world and are bounded by $V(\lambda)$. The energy of a configuration is $E_n(\vec{\lambda}) = -\sum_{j>k} \log(\lambda_j - \lambda_k)^2 + n\sum_j V(\lambda_j)$. For large n, the partition function can be reformulated as a functional integral on normalized particle densities ρ_n, $Z_n = \int \mathcal{D}\rho_n e^{-n^2 E_n}$, with Boltzmann weight,

$$E_n[\rho_n] = \int \rho_n(\lambda)\, V(\lambda)\, d\lambda + \frac{1}{n}\int \rho_n(\lambda)\, \log \rho_n(\lambda)\, d\lambda$$
$$- \int\int d\lambda\, d\mu\, \rho_n(\lambda)\, \rho_n(\mu)\, \log|\lambda - \mu| - 2\gamma\left(\int \rho_n(\lambda)\, d\lambda - 1\right).$$

Two new terms appear: an entropic one resulting from the Jacobian (negligible in the large n limit) and the Lagrange multiplier γ enforcing normalization. The large n limit is both the thermodynamic and the zero temperature limit of the model, and allows for a saddle point evaluation of the partition function. Under certain conditions on V, there is a unique limiting spectral density $\rho(\lambda)$. It is the solution of the limit of the saddle-point equation,

$$\frac{1}{2}V(\lambda) - \int_\sigma d\mu\, \log|\lambda - \mu|\, \rho(\mu) = \gamma, \quad \lambda \in \sigma, \tag{14.1.2}$$

which also minimizes the limit free energy,

$$\mathcal{F} = \lim_{n\to\infty} \frac{1}{n^2} \log Z_n = -\int d\lambda \rho(\lambda)\, V(\lambda) - \iint d\lambda\, d\mu\, \rho(\lambda)\rho(\mu) \log|\lambda - \mu|. \tag{14.1.3}$$

Generically $\mathcal{F}$ is an analytic function of the parameters of the potential, except for possible critical points or lines. When they occur, they divide the parameter space into different phases of the model.[1]

In Section 14.2 we summarize the results for the simplest model with a non-trivial set of phases, the one-matrix Hermitian model with polynomial potential. We refer the reader to the beautiful lecture notes by P. Bleher [Ble08] for proofs and references to the mathematical literature.

We present a view of the several solutions of the saddle point equation, which simplifies the current analysis of the phases of the model. Generically, the limit eigenvalue density ρ has support on different numbers of intervals, in different phases of the model. Its behaviour near an edge of the support is typically a square root. Parameters can be adjusted to soften this edge singularity and, in the one-cut phase, the continuum limit of the model is approached. The universal distributions that describe various scalings of the density in the bulk or close to an edge are affected by phase transitions; we refer the reader to Chapter 6 of this book. In the orthogonal polynomial approach, a phase transition manifests in the doubling phenomenon of recurrence equations.

For several matrix models that are invariant under a continuous group it is possible to obtain a partition function for the eigenvalues only. We call

[1] It may be useful and important to also consider complex potentials. Early papers on the subject are [Dav91][Moo90] [Fok91].

In the paper by Eynard [Eyn07] many important results are summarized.

We have chosen to limit ourselves to real potentials to stress the analogy with statistical mechanics and the Boltzmann weight.

them eigenvalue matrix models. Quite often they display phase transitions. In Section 14.3 we review *circular* models and their Cayley transform to Hermitian models, and *fixed trace* models. Models with *normal, chiral, Wishart, and rectangular* matrices are discussed briefly, because they are subjects of chapters in this book. We do not include results about matrix models in nonzero spacetime.

Though non-Hermitian random matrices are discussed in Chapter 18, it is appropriate to present here, in Section 14.4, the curious single-ring theorem, restricting the phase transitions of complex spectra with rotational symmetry.

Multi-matrix models are recalled in Section 14.5, with their spectacular success in describing phase transitions of classical statistical models on fluctuating two-dimensional surfaces. We refer the reader to Chapters 15 and 16.

A large and important number of matrix models have a preferred basis; their analysis is usually performed with specific tools. In Section 14.6 the delocalization transition is summarized for the Anderson, Hatano-Nelson, and Euclidean random matrix models.

14.2 One-matrix models with polynomial potential

In our view the occurrence of phase transitions is best understood in the steepest descent solution of matrix models. As happens with many discoveries, the two-cut and other one-cut asymmetric solutions of the saddle point equation were discovered at least twice, first by Shimamune in $D = 0, 1$, next by Cicuta *et al.* [Shi82][Cic86]. Soon after, two puzzling features emerged: 1) Multi-cut solutions of the (derivative of the) saddle point equation

$$\frac{1}{2}V'(\lambda) = P\int_{\sigma} d\mu\,\frac{\rho(\mu)}{\lambda-\mu}, \qquad \lambda \in \sigma \tag{14.2.1}$$

seemed to have an insufficient number of constraints to fully determine them, 2) the recurrence relations which determine the large n behaviour of orthogonal polynomials were found to need two interpolating functions in correspondence of the two-cut phase [Mol88]. The intriguing feature was that the recurrence relations seemed to exhibit unstable behaviour [Jur91] [Sen92].
The following decades witnessed a rigorous derivation of the limit spectral density as the solution of a variational problem [Joh98], and the development of Riemann–Hilbert and resolvent-based approaches [Dei99],[Pas06]. The latter seem crucial for the present understanding of the quasi-periodic asymptotics of recurrence coefficients. The rigorous derivations confirmed much of the previous heuristic work of theoretical physicists, including the set of missing equations (14.2.7), which were predicted [Jur90],[Lec91].

We outline the multi-cut solutions of the saddle point equation for polynomial potentials. The interest in this model was revived by recent work on the 'birth of a cut' [Ble03],[Eyn06],[Cla08],[Mo07].[2] Consider the Hermitian one-matrix model with a polynomial potential of even degree p, and positive leading coefficient. The limiting density $\rho(\lambda)$ with its support σ can be evaluated from the Green function,

$$F(z) = \int_\sigma d\lambda \frac{\rho(\lambda)}{z-\lambda}, \qquad \rho(\lambda) = \frac{1}{\pi}\text{Im } F(\lambda - i\epsilon). \tag{14.2.2}$$

The saddle point approximation [Bre78] or the loop equations [Wad81] provide

$$F(z) = \frac{1}{2}V'(z) - \frac{1}{2}\sqrt{V'(z)^2 - 4Q(z)}, \tag{14.2.3}$$

$$Q(z) = \int_\sigma d\lambda\, \rho(\lambda) \frac{V'(z) - V'(\lambda)}{z-\lambda}. \tag{14.2.4}$$

$Q(z)$ is a polynomial of degree $p-2$, that contains unknown parameters $\langle x\rangle$, ..., $\langle x^{p-2}\rangle$, which are moments of the density. Since $F(z) \approx Q(z)/V'(z) = 1/z$ for $|z| \to \infty$, normalization of ρ is ensured. The relevant question arises about the polynomial $V'(z)^2 - 4Q(z)$: *how many are the pairs of simple real zeros?* The pairs of zeros are the endpoints of cuts of the function $F(z)$ that become intervals in the support of ρ: $\sigma = \bigcup_{j=1}^q [a_j, b_j]$. The remaining $2(p-q-1)$ zeros must be such to factor a squared polynomial $M(z)^2$:

$$V'(z)^2 - 4Q(z) = M(z)^2 \prod_{j=1}^q (z-a_j)(z-b_j). \tag{14.2.5}$$

The density is then evaluated,

$$\rho(\lambda) = \frac{1}{2\pi}|M(\lambda)| \sqrt{\prod_{j=1}^q |(\lambda - a_j)(b_j - \lambda)|}, \quad \lambda \in \sigma. \tag{14.2.6}$$

The polynomial identity (14.2.5) provides $2p-1$ equations for $2q + (p-q) = p+q$ unknowns in the r.h.s. (the endpoints and the coefficients of M). However, the l.h.s. is fully determined only in its monomials $z^{2p-2}, \ldots, z^{p-2}$. Therefore, only $p+1$ equations are useful to fix the $p+q$ parameters of the density. The remaining $p-2$ are equations for the moments of the density (they are not self-consistency relations, and add nothing to the knowledge of the density).

The 1-cut density ($q=1$) is then completely (but not uniquely!) determined, while the q-cut density ($q>1$) still depends on $q-1$ unknown parameters. It is a continuum set of normalized solutions of the *derivative* of the saddle point equation (14.1.2). This equation, (14.2.1), ensures that the chemical potential

[2] The merging of the extrema of a cut, leading to its disappearance, and the reverse phenomenon, had been known to some experts for a long time. See for instance Section 6.2 in [Ake97].

γ is constant for arbitrary λ in a single interval. But the saddle point equation requires γ to be the same for λ in the whole support σ: this gives extra $q-1$ conditions. They can be chosen as the vanishing of the integrals of the density on the gaps between the cuts [Jur90][Lec91]:

$$\int_{b_j}^{a_{j+1}} d\lambda\, \rho(\lambda) = 0, \quad j = 1, \ldots, q-1. \tag{14.2.7}$$

Necessarily $M(z)$ must have at least one real zero in each gap. Hint: the equations can be obtained by integrating on a gap Eq. (14.2.3), with the input of Eq. (14.2.5).

As the parameters of the potential V are changed, the zeros of ρ move continuously, and eventually two of them collide on the real axis; collisions provide mechanisms for phase transitions:

- a pair of complex zeros coalesce into a double real zero *inside* an interval. The density ceases to be positive, and an extra cut must be considered in (14.2.3). The reverse is the closure of a gap, with the ends turning into a complex conjugate pair.
- a pair of complex zeros coalesce into a double real zero *outside* an interval and an extra interval is born. The reverse is the closure of an interval, with the ends turning into a complex conjugate pair.
- a zero other that an endpoint coalesces with an endpoint, thus changing the edge's singular behaviour.

The first two are *multicut transitions*, the third is an *edge singularity transition.*

With the conditions (14.2.7) the number of equations equals the number of parameters that determine the density. However:

1) the decoupling of the algebraic equations produces equations for the single unknowns which may have several solutions with the same q;
2) the requirement that ρ must be non-negative on its support eliminates solutions.
   In general one remains with more than one q-cut solution, and selects *the* saddle point solution by comparing their free energies.

We have recently redone the analysis of the phase diagram for the quartic model.[3] We summarize the results of the new analysis, which suggest that multiple solutions with different numbers of cuts are trivially related.

[3] We studied this same problem long ago [Cic87]. At the time we did not identify the correct missing equation (14.2.7), which led us to conclusions quite different from those presented here.

The quartic potential

The general quartic potential can be rescaled and shifted to $V(\lambda) = hz + \frac{1}{2}a\lambda^2 + \frac{1}{4}\lambda^4$, which depends on two parameters h and a. Let us first summarize the simple case $h = 0$: for $a \geq -2$ there is the BIPZ 1-cut solution [Bre78]. At $a = -2$ the density has a zero in the middle of its support. For $a < -2$ one must consider a two-cut solution $\rho(\lambda) \approx |\lambda|\sqrt{(b^2 - \lambda^2)(\lambda^2 - a^2)}$.

In the general case, the plane (h, a) is partitioned in three phases (I,II,III). In I, which includes the half-plane $a \geq 0$, the solution is one-cut; in II only the two-cut solution exists. In III, three solutions coexist: the two-cut solution and *two* one-cut solutions. As we show below, the two-cut solution has lower free energy; III has the line $h = 0$ in its interior ($a < -\sqrt{15}$).

In III, before the condition (14.2.7) is imposed, one has a one-parameter family of two-cut solutions of (14.2.1):

$$\rho(\lambda) = \frac{1}{2\pi}|\lambda - R|\sqrt{\prod_{j=1,2}|(\lambda - a_j)(\lambda - b_j)|}, \qquad \lambda \in [a_1, b_1] \cup [a_2, b_2] \quad (14.2.8)$$

If R is chosen as a free parameter, it is fixed by eq.(14.2.7).[4] Another choice is x, which gives the filling fractions of the intervals: $x = \int_{a_1}^{b_1} \rho(\lambda)\, d\lambda$ and $1 - x = \int_{a_2}^{b_2} \rho(\lambda)\, d\lambda$ [Bon00]. One checks that the two one-cut solutions which also exist in III correspond to the limit values $x = 0$ and $x = 1$ where an interval degenerates to a point. The corresponding free energies are then higher. Since a two-cut solution is analytic in h for small h, there is no spontaneous breaking of the Z_2 symmetry.

The three different phases are separated by lines of third-order phase transitions. Their occurrence finds application in the study of the Homolumo gap in a one-fermion spectrum [And07], structural changes in glasses [Deo02] and gluons in baryons (with addition of log-term in potential) [Kri06].

14.2.1 Orthogonal polynomials

Multicut phase transitions arise when the matrix potential is endowed by two or more minima, and the large n limit suppresses tunnelling among the wells. In the saddle point solution this is manifest in multicut support of the eigenvalue density. In the approach with orthogonal polynomials, see Chapter 4 (and Chapter 5), the phenomenon appears in a different guise.

The partition function, the spectral density and correlators can be evaluated formally for all n, by expanding the Vandermonde determinant in monic polynomials $P_r(x)$ constrained by orthogonality: $\int dx e^{-nV} P_r P_s = h_r \delta_{rs}$. Starting with $P_0(x) = 1$, the relation allows us in principle to evaluate all coefficients and the partition function $Z_n = n! h_0 \cdots h_{n-1}$. In practice, one looks for asymptotics

[4] The determination of ρ in the gap (b_1, a_2) has the factor $\lambda - R$ without modulus.

in n as follows. Because of orthogonality the polynomials are linked by a three-term relation $x P_r = P_{r+1} + A_r P_r + R_r P_{r-1}$. The coefficients A_r and R_r themselves solve finite order recurrency equations, resulting from the identities

$$\frac{k}{n} = \frac{1}{h_{k-1}} \int dx e^{-nV} V' P_{k-1} P_k, \quad 0 = \int dx e^{-nV} V'(x) P_k^2. \tag{14.2.9}$$

The initial conditions $R_0 = 0$, $A_0, \ldots$ must be computed explicitly. Since $R_r = h_{r+1}/h_r > 0$, the free energy is evaluated:

$$\mathcal{F}_n = -\frac{1}{n^2} \log Z_n = -\frac{1}{n^2} \log(n!) - \frac{1}{n} \log h_0 - \frac{1}{n} \sum_{r=1}^{n-1} (1 - \frac{r}{n}) \log R_r. \tag{14.2.10}$$

For large n, the *string equations* (14.2.9) become equations for interpolating functions $R(x)$ and $A(x)$, with $R(k/n) = R_k$ and $A(k/n) = A_k$: $x = w_1(R(x), A(x); R'(x), \ldots)$ and $0 = w_2(R(x), A(x); R'(x), \ldots)$. The leading order in n is purely algebraic in R and A, and does not involve derivatives. The solutions must comply with the initial conditions $R(0) = 0 = R_1 = \ldots$ and $A(0) = A_0 = A_1 = \ldots$. However, the finite sets of initial conditions $\{R_k\}$ and $\{A_k\}$ may not collapse in the large n limit to unique values $R(0)$ and $A(0)$. This is the signal of a *multicut phase*: coefficients have to be interpolated by as many functions $R_\ell(x)$ and $A_\ell(x)$ as the initial conditions.

The quartic potential

For the symmetric quartic potential $V(x) = \frac{1}{2} a x^2 + \frac{1}{4} x^4$ the orthogonal polynomials have definite parity ($A_k = 0$) and only the first string equation is used,

$$\frac{k}{n} = a R_k + R_k(R_{k+1} + R_k + R_{k-1}) \tag{14.2.11}$$

with initial conditions $R_0 = 0$ and $R_1 = \langle x^2 \rangle$. If $a > 0$ it is $R_1 \to 0$ and a unique function $R(x)$ is needed, with $R(0) = 0$. The recurrence equation becomes the quadratic equation $x = a R(x) + 3 R(x)^2$ which leads to the one-cut solution [Bre78]. If $a < 0$ it is $R_1 \to |a|$ and two interpolating functions $R_1(x)$ and $R_2(x)$ are needed: $R_1(2k/n) = R_{2k}$ with $R_1(0) = 0$ and $R_2(2k/n + 1/n) = R_{2k+1}$ with $R_2(0) = |a|$. They solve: $x = a R_1 + R_1^2 + 2 R_1 R_2$ and $x = a R_2 + R_2^2 + 2 R_1 R_2$. One recovers the solution $R_1 = R_2$ but also the new ones $R_2(x) = -R_1(x) - a$ and $x = g R_1(x) R_2(x)$. In the latter case the equation for $R_1(x)$ yields a real positive solution only for $a < -2$. This is the two-cut phase.
The asymmetric quartic potential is more complicated because of $A_k \neq 0$. One reobtains regions I,II,III of the previous discussion.

14.2.2 The edge singularity limit

A special status is owned by the edge singularity limit in the one-cut phase. The critical line marks the boundary of analyticity of the perturbative phase where the partition function can be expanded in all couplings above quadratic. The perturbative expansion provides numbers that enumerate Feynman diagrams with fixed numbers of vertices. This is a useful tool in the theory of graphs and in statistical mechanics, where each graph may correspond to a configuration in the partition function of some model on random graphs [Amb97][Gro92] (see Chapters 30 and 31).

In the *topological expansion* ('t Hooft) the perturbative series is rearranged

$$\mathcal{F}_n = \sum_{h=0}^{\infty} \frac{1}{n^{2h-2}} \mathcal{F}_h(g), \qquad \mathcal{F}_h(g) = \sum_V g^V \mathcal{F}_{h,V} \tag{14.2.12}$$

to enumerate the vacuum graphs of given genus h and number V of vertices. The leading term $\mathcal{F}_0$ is the *planar* free energy. A remarkable feature of the expansion is that $\mathcal{F}_h(g)$ has a finite radius of convergence g_c. This singularity is where two zeros of the eigenvalue density collide, and in the single-well phase it corresponds to an edge-singularity limit.

The *area* of a graph is defined as the number of vertices (faces of the dual graph). The average area at fixed genus diverges near the critical point, thus providing the continuum limit,

$$\langle area \rangle = \frac{1}{\mathcal{F}_h} \sum_V V g^V \mathcal{F}_{h,V} = \frac{1}{\mathcal{F}_h} g \frac{\partial}{\partial g} \mathcal{F}_h \approx const. \frac{g_c}{g_c - g}. \tag{14.2.13}$$

The double scaling is a prescription to reach the critical point, in order to enhance subleading orders of the topological expansion.

14.3 Eigenvalue matrix models

Several matrix models with continuous symmetry, after integration of angular degrees of freedom, produce a partition function that generalizes Eq. (14.1.1) and describes only the eigenvalues or the singular values. Many models, but not all, derive from classification schemes of the symmetries (Chapter 3). Depending on the eigenvalue measure, phase transitions may appear. The analytic methods are almost the same as in the previous section.

14.3.1 Unitary circular ensembles

Perhaps the first and most influential phase transition was found by Gross and Witten in the study of one plaquette in QCD [Gro80], see Chapter 17,

$$Z_n = \int dU \, e^{\frac{1}{g^2} \mathrm{tr}(U + U^\dagger)}, \qquad U \in U(n). \tag{14.3.1}$$

Let $\{e^{i\alpha_k}\}$ be the n eigenvalues of the random unitary matrix U, then

$$Z_n = \int_{-\pi}^{\pi} d\alpha_1 \cdots d\alpha_n \, e^{\sum_{i\neq j} \log|\sin \frac{\alpha_i - \alpha_j}{2}| + \frac{2}{g^2}\sum_j \cos\alpha_j}. \tag{14.3.2}$$

In the large-n limit the spectral density $\rho(\alpha)$ which makes the energy stationary is the solution of the integral equation

$$\frac{1}{g^2 n}\cos\alpha + \int \rho(\beta)\, \log|\sin\frac{\alpha-\beta}{2}|\, d\beta = 0. \tag{14.3.3}$$

As is well known, there is a strong and a weak coupling solution:

$$\rho(\alpha) = \frac{1}{2\pi} \quad (g^2 n \geq 2), \qquad \rho(\alpha) = \frac{1}{2\pi}\left[1 + \frac{2}{g^2 n}\cos\alpha\right] \quad (g^2 n \leq 2),$$

with support on the whole circle or an arc. The Cayley map provides a one-to-one correspondence between unitary and Hermitian matrices:

$$U = \frac{i - H}{i + H}, \qquad dU = \frac{dH}{\det(1 + H^2)}.$$

Accordingly, the model (14.3.2) can be mapped into a Hermitian model. The corresponding partition function is written in terms of the eigenvalues λ_j of the Hermitian matrix,

$$Z_n = \int_{-\infty}^{\infty} \prod_j d\lambda_j \, e^{-n\sum_j \log(1+\lambda_j^2) + \sum_{i\neq j}\log|\lambda_i - \lambda_j| + \frac{2}{g^2}\sum_j \frac{1-\lambda_j^2}{1+\lambda_j^2}}.$$

Instead of Eq. (14.3.3) one has the integral equation

$$\log(1+\lambda^2) - \frac{2}{g^2 n}\frac{1-\lambda^2}{1+\lambda^2} = 2\int d\mu\, \rho(\mu)\, \log|\lambda - \mu| \tag{14.3.4}$$

One finds a strong coupling solution with support on the whole real axis,

$$\rho(\lambda) = \frac{1}{\pi}\left[\frac{1}{1+\lambda^2} + \frac{2}{g^2 n}\frac{1-\lambda^2}{(1+\lambda^2)^2}\right], \quad 2 \leq g^2 n. \tag{14.3.5}$$

Mizoguchi [Miz05] recently studied the weak coupling solution. It has support on a finite interval,

$$\rho(\lambda) = \frac{2}{\pi}\frac{\sqrt{1+b^2}}{b^2}\frac{\sqrt{b^2-\lambda^2}}{(1+\lambda^2)^2}, \quad b^2 = \frac{g^2 n}{2 - g^2 n} \tag{14.3.6}$$

Therefore, the Cayley map takes the Gross-Witten phase transition to a phase transition in a Hermitian matrix model, where it separates a compact support phase (typical of confining potentials) and an infinite support phase (typical of logarithmic external potentials). The Gross-Witten model was solved for polynomial potentials [Man90] [Dem91], $V(U) = \sum_k c_k U^k + c_k^* U^{\dagger k}$, and phases with support on several arcs of the unit circle were found. The Cayley map takes them to multi-cut solutions of Hermitian models.

14.3.2 Restricted trace ensembles

Restricted trace ensembles of Hermitian matrices have joint probability density

$$p(\lambda_1, \cdots, \lambda_n) = const. \quad \Phi\left(r^2 - \frac{1}{n}\sum_{j=1}^{n} V(\lambda_j)\right) \prod_{1 \leq i < j \leq n} |\lambda_i - \lambda_j|^2$$

with $\Phi(x) = \delta(x)$ for the fixed trace ensembles or $\Phi(x) = \theta(x)$ for the bounded trace ensembles. The constraint replaces the standard Boltzmann factor of matrix models. These invariant matrix ensembles bear the same relation to unrestricted ensembles as microcanonical ensembles do to canonical ones in statistical physics. They were studied at the early stages of random matrix theory [Ros63], with quadratic potential. It was interesting to generalize them to polynomial potential, as phase transitions in these models provide good examples for the limited equivalence of ensembles in the thermodynamic limit. [Ake99]. Further recent works include [Del00],[Got08], later extended to the fixed trace beta ensembles [LeC07],[Liu09]. See also Chapter 37 for fixed trace ensembles of Wishart type.

14.3.3 External field

Ensembles of unitary or Hermitian matrices coupled to a fixed matrix source, called external field, had been investigated for a long time [Bre80][Gro91]. The limiting eigenvalue density of the ensemble depends on the eigenvalue density of the fixed matrix. Several important tools of one-matrix models may be generalized to the external field problem to prove that the short distance behaviour of correlation functions is not affected by the source [Zin98], [Ble08]. We refer the reader to Chapter 16.

If the eigenvalue density of the external source has a gap which may be tuned to vanish, a new class of universality appears [BrH98]. This phase transition has a relevant role in several problems, including the spectral statistics of low-lying eigenvalues of QCD Dirac operators, see Chapter 32.

14.3.4 Other models

Several models are described in chapters of this book, and display various critical behaviours. We list some here.

1) Normal matrices have a complex spectrum. With the potential $XX^\dagger + V(X) + V(X)^\dagger$ the boundary of the support of eigenvalues describes the growth of a 2D fluid droplet, as the parameters change. At singular points the boundary develops cusp-like singularities. [Teo06] (Ch.38).

2) Chiral matrix ensembles were introduced by Verbaarschot to study spectral properties of the QCD Dirac operator (Chapter 32) linked to the formation of a condensate, with beautiful accordance with lattice calculations. A natural application of chiral random matrix ensemble is the study of single particle excitations in bulk type-II superconductors [Bah96].
3) Wishart matrices have the form $W = R^\dagger R$, where R is a rectangular matrix. They occur in multivariate statistics (Chapter 28 and 39), in transport (as suggested by Buttiker's Landauer formula for conductance) and in the study of rare events (Chapter 36).
4) Transfer matrix ensembles were introduced to reproduce the statistics of conductance (Chapter 35).
5) Models of rectangular matrices $n \times nL$ with $O(L)$ symmetry can be viewed as $L-$matrix models with square matrices of size n, and the large n limit is the planar limit of a vector model [Cic87b][And91][Fei97b]. The singular values undergo a phase transition in the double-well potential, where a gap opens near the origin. Multicritical behaviour and underlying random surfaces are studied. An ample introduction and references to rectangular matrices are in a paper on coloured graphs [DiF03].
6) Non-polynomial potentials may originate different universality classes for correlation functions (see also Chapter 12). Some of these models may be seen as arising from integration over additional bosonic or fermionic fields. Early relevant examples include [Kaz90][Kon92]. Recent examples include [Ake02][Jan02].

14.4 Complex matrix ensembles

The powerful spectral methods available for Hermitian matrices, based on the resolvent, do not apply to complex matrices. With the exclusion of normal matrices, it is very difficult to extract a measure for the eigenvalues. By regarding the eigenvalues as point charges in the complex plane, the search of the density can be formulated as an electrostatic problem [Cri88]. The ensemble-averaged logarithmic potential,

$$U(z, z^*) = \langle \frac{1}{n} \log[\det(zI_n - X)\det(z^* I_n - X^\dagger)] \rangle \tag{14.4.1}$$

and the average eigenvalue density $\rho(x, y) = \langle \frac{1}{n} \sum_a \delta(x - \mathrm{Re} z_a)\delta(y - \mathrm{Im} z_a) \rangle$ are linked by the Poisson equation $\partial\partial^* U = \pi\rho$, where $z = x + iy$ and $\partial = \frac{1}{2}(\partial_x - i\partial_y)$. The problem is simplified by introducing the Green function,

$$G(z, z^*) = \langle \frac{1}{n} \mathrm{tr} \frac{1}{z - M} \rangle = \int d^2 w \frac{\rho(w, w^*)}{z - w}, \tag{14.4.2}$$

then: $\rho(z, z^*) = \frac{1}{\pi}\partial^* G(z, z^*)$. By noting that, up to a constant, the potential U contains the determinant of the $2n \times 2n$ Hermitian matrix,

$$H(z, z^*) = \begin{bmatrix} 0 & X - z \\ X^\dagger - z^* & 0 \end{bmatrix} \tag{14.4.3}$$

Feinberg and Zee proposed the *Method of Hermitian reduction* [Fei97]. This allows us to evaluate the Green function $G(z, z^*)$ from the resolvent matrix of $H(z, z^*)$: $F(\eta)_{ab} = \langle (\eta - H)^{-1}_{ab} \rangle$. For the Ginibre ensemble [Gin65] the resolvent is $G(z, z^*) = 1/z$ for $|z| > 1$ and z^* for $|z| < 1$. Then the eigenvalue distribution is uniform in the disk $|z| < 1$. Other tools are the extension to non-Hermitian matrices of *Blue functions* [Jar06] (the functional inverse of the Green function $G(B(z)) = B(G(z)) = z$), or the fermionic replica trick [Nis02].

14.4.1 The single ring theorem

The method of Hermitian reduction can be worked out for the probability distribution

$$p(X) = Z^{-1} e^{-n \operatorname{tr} V(XX^\dagger)} \tag{14.4.4}$$

where V is a polynomial. The eigenvalue density $\rho(x, y)$ for X only depends on $r = \sqrt{x^2 + y^2}$. By resummation of the planar diagrams in the perturbative expansion of the resolvent F of the hermitized model, Feinberg, Zee, and Scalettar showed that the fraction $\gamma(r)$ of eigenvalues with modulus less than r solves in the large n limit the algebraic equation

$$r^2 \frac{\gamma(r)}{1 - \gamma(r)} F\left(r^2 \frac{\gamma(r)}{1 - \gamma(r)} \right) = 1. \tag{14.4.5}$$

It admits various solutions, but only two are r-independent, namely $\gamma = 0$ and $\gamma = 1$. The actual solution is obtained by a smooth matching of solutions such that $\gamma(r)$ is nondecreasing, with b.c. $\gamma(0) = 0$ and $\gamma(\infty) = 1$. Surprisingly, the support of the density ρ can only be a disk or a single annulus. This is the single ring theorem [Fei97][Jan97]. The proof is simple: two or more annuli would imply a gap, hence a solution $\gamma(r)$ which is constant on the gap interval of r, and different from the only two allowed values 0, 1.

Although the eigenvalues of the positive matrix $XX^\dagger$ may distribute with a multicut density, the complex eigenvalues of X are only allowed to coalesce in a disk or in an annulus. As the parameters of the potential V are changed, one observes phase transitions between the two configurations. For the double-well potential $V = \operatorname{tr}(2aXX^\dagger + g(XX^\dagger)^2)$, the phase transition takes place at $a = -\sqrt{2g}$. An interesting insight was provided in [Fyo07], where Eq. (14.4.5) was recovered with the hypothesis that, for large n, the log can be taken out of the ensemble average (14.4.1), thus simplifying the evaluation.

A disk–annulus transition was observed in other ensembles. Complex tridiagonal random matrices have exponentially localized eigenvectors, with inverse localization length given by the Thouless formula,

$$\gamma(z) = \int d^2w \log|z - w|\rho(w, w^*) + \text{const.} \qquad (14.4.6)$$

If boundary conditions (b.c.) $u_{n+1} = e^{n\xi}u_1$ and $u_0 = e^{-n\xi}u_n$ are used, then for $\xi > \xi_c$ a hole opens in the support of the spectrum [Mol09b]. The eigenvalues that are removed from the hole belong to eigenvectors that are delocalized by the b.c., i.e. $\gamma(z) < \xi$. For large ξ the complex spectrum becomes circular.

The transition was also observed in the ensemble $A + H_0$, where H_0 is a matrix with a highly unstable zero eigenvalue, and A is random and asymmetric [Kho96].

14.5 Multi-matrix models

The free energy of a group-invariant matrix model has a topological large-n expansion (14.2.12), where the terms associated with inverse powers of n^2 (unitary ensembles) or n (orthogonal and symplectic ensembles) are the generating functions of Feynman graphs embeddable on orientable (or non-orientable, in the second case) surfaces of different genus.

With two or more matrix variables, each graph has both the meaning of a random triangulation and fixes a 'configuration' of two or more variables associated with its vertices or links. The matrix integrations simultaneously perform the summation on configurations of variables on a graph (a random triangulation) and the summation on inequivalent triangulations. Therefore, multi-matrix models may be considered as a definition for classical statistical mechanics on random lattices. By tuning the couplings of the potential to critical values it is possible to interpret the triangulations provided by the Feynman graphs, or their duals, as becoming dense and reaching a continuum limit.

In chain models, the potential of two (or more) random matrices has the form $V(A, B) = V_1(A) + V_2(B) + AB$. The analysis of such chain-linked ensembles was possible after two major discoveries:

1) Integration of angular variables by Itzykson and Zuber [Itz80] through the Harish–Chandra–Itzykson–Zuber formula, which reduces the partition function to integrals on the eigenvalues of the two (or more) random matrices.
2) Bi-orthogonal polynomials by Mehta [Meh81], which allow the formal exact evaluation of the free energy and asymptotic behaviour.

It is well known that near the critical singularities of statistical mechanical models on fixed, regular 2D lattices, the correlation lengths become much bigger than the lattice spacing. The critical behaviour of thermodynamic functions are then described by classes of universality of 2D field theories in the continuum. Analogous universality classes occur for phase transitions of statistical models on random surfaces.

If the critical singularities of a model on a fixed lattice are described by a conformal field theory with central charge c, the Knizhnik-Polyakov-Zamolodchikov relation predicts the conformal dimensions of operators of the corresponding model on inequivalent random triangulations (i.e. dressed by gravity). This relation was checked in several multi-matrix models: Ising, Potts, $O(n)$, 8-vertex, edge-colouring, ... These impressive accomplishments are described in classical reviews [Amb97][DiF95]. More recent ones by Di Francesco [DiF02] focus on the structure of random lattices. Recent results on the Riemann–Hilbert problem for the two matrix model are found in [Dui08],[Mo08].

14.6 Matrix ensembles with preferred basis

14.6.1 Lattice Anderson models

Anderson models describe the dynamics of a particle in a lattice, in a random potential; the hopping amplitudes may also be random. They arise as tight-binding descriptions of a particle in a crystal with impurities, disordered materials, random alloys.[5]

The Hamiltonian is $H = T + V$ with kinetic term $(Tu)_k = \sum_\mu u_{k+\mu}$ and potential $(Vu)_k = v_k u_k$, where k labels the lattice sites. The sum runs on all sites linked to k, the numbers v_k are i.i.d. random variables with uniform distribution in the interval $[-w/2, w/2]$ (w is the disorder parameter) or else. For the Cauchy distribution, $p(v) = (w/\pi)(v^2 + w^2)^{-1}$, the density of states can be computed analytically. The lattice is usually taken as Z^D, but also other structures have been studied. In direct space the Hamiltonian has a matrix representation where T is fixed (a Laplacian matrix, up to a diagonal shift) and V is diagonal and random. The matrix size is equal to the number of sites. The matrix H is block-tridiagonal, with the number of blocks being equal to the number of sites in one direction ($n = n_z$), and the size of blocks being equal to the number of sites in a section ($m = n_x n_y$). The eigenvalue equation in terms of the diagonal blocks h_k and off diagonal ones (unit matrices if hopping amplitude is one) is

$$h_k u_k + u_{k+1} + u_{k-1} = \lambda u_k, \quad u_k \in \mathbb{C}^m. \tag{14.6.1}$$

[5] For physics we address the reader to the reviews [Kra93],[Vol92],[Jan98]. For the random matrix theory approach see Chapter 35 and references therein, [Bee97],[Eve08]. The mathematically orientated reader can consult the review [Bel04], the books [Pas92],[Car90] or (as a start) [Fro87]. Material can be found at the websites of the Newton and Poincaré Institutes [NI][IHP].

The components of u_k are the occupation amplitudes of the m sites having longitudinal coordinate k, λ is the energy of the particle. A fundamental analytic tool is the transfer matrix, of size $2m \times 2m$, that links the wave-vector components at the ends of the sample of length n:

$$\begin{bmatrix} u_{n+1} \\ u_n \end{bmatrix} = T(\lambda) \begin{bmatrix} u_1 \\ u_0 \end{bmatrix}, \quad T(\lambda) = \prod_k \begin{bmatrix} \lambda - h_k & -I_m \\ I_m & 0 \end{bmatrix}. \tag{14.6.2}$$

With Dirichlet b.c. it is $u_{n+1} = u_0 = 0$. The transfer matrix is the product of n random matrices; its eigenvalues describe the long range behaviour of eigenstates of the Hamiltonian. Oseledec's theorem [Pas92][Cri93] states that the matrix $TT^\dagger$ in the limit $n \to \infty$ converges to a nonrandom matrix $e^{-n\Gamma(\lambda)}$ and the eigenvalues of Γ come in pairs $\pm\gamma(\lambda)$ (Lyapunov spectrum). The smallest positive one is the inverse localization length, that controls transport properties. So far no analytic expression is known for the Lyapunov spectrum (in $D = 1$ it is the Thouless formula). However, for any n and m, the eigenvalues of the transfer matrix can be linked to the spectrum of the Hamiltonian matrix, but with non-Hermitian boundary conditions, via an algebraic spectral duality [Mol09a]:

$$z^{nm}\det[\lambda - H(z^n)] = (-1)^m \det[T(\lambda) - z^n]. \tag{14.6.3}$$

The main feature of Anderson models is a phase transition that occurs in 3D in the infinite-size limit: for $w < w_c$ the spectrum of H is a.c. while for $w > w_c$ the spectrum is p.p. The eigenvectors are, respectively, extended or exponentially localized. Anderson's transition (also named *metal insulator transition*, MIT) and Mott's transition (due to interaction effects on the band filling) are cornerstones of the theory of electronic transport (both physicists earned the Nobel prize in 1977 with van Vleck). The transition is observed in various experimental situations [Kra93]. Numerically, it can be detected in various ways: scaling in transverse dimensions of the smallest Lyapunov exponent of the transfer matrix [Mac81], spacing distribution of the energy levels [Shk93], response of energy levels to changes of boundary conditions [Zyc94] (Thouless' approach to conductance). In the metallic phase the Thouless conductance is well described by the distribution of level curvatures of the random matrix ensemble $H(\varphi) = A\cos\varphi + B\sin\varphi$, A and B in GOE or GUE, which was conjectured in [Zak93] and analytically proved in [vOp94].

The two phases are characterized by order parameters: the localization length (localized phase) and the dimensionless conductance g (delocalized phase) that diverge approaching the transition. A finite size one-parameter scaling theory was established by Abrahams *et al.* which implies the phase transition in $D = 3$, but no transition in $D = 1, 2$. In $D = 1, 2$ such as wires or electron layers in heterostructures, the eigenstates are localized. This is crucial for explaining the occurrence of plateaux in the integer quantum Hall effect. Anderson localization is caused by destructive interference on random scatterers, and has been observed in diverse wave phenomena as electrons, sound, light, and Bose-Einstein condensates.

Mathematical proofs of localization in 1D became available in the late seventies; Molchanov proved localization in the 1D continuous case and the Poisson law for energy levels. Theorems for D>1 appeared in the early eighties (Frohlich, Spencer, Martinelli, and Scoppola [Mar86]) and established localization at large disorder. Minami proved the Poisson law for energy levels of lattice models. Multilevel correlators are studied by various groups (see [Aiz08]). The extended phase still lacks rigorous results. A Wigner-Dyson statistics is expected and seen numerically in the transport regime. The level statistics at the Anderson transition is of a new type [Kra94][Zha97] (Chapter 12).

The quantization of time-dependent classically chaotic systems brings in the phenomenon of dynamical localization, which is analogous to disorder localization. In model systems such as kicked rotator the quantum system does not increase its energy as classically, but reaches a stationary state [Cas89][Haa]. This has been observed experimentally in optical systems, or in microwave ionization of Rydberg atoms.

Quantum chaos eventually revived, in the early eighties, the interest for *banded random matrices*, for the transition in level statistics from Wigner-Dyson to Poissonian, or the semiclassical limit of quantum mechanics of chaotic systems [Fei89]. The study of quantum maps suggested that banded random matrices have localized eigenstates and level statistics with scaling laws governed by the ratio b^2/n [Cas90][Haa], where $2b+1$ is the bandwidth. Analytic results became accessible by supersymmetric methods after the papers by Fyodorov and Mirlin [Fyo91]. Banded random matrices are now on stage to study the Anderson transition (Chapter 23).

14.6.2 Hatano-Nelson transition

Hatano and Nelson [Hat96] proposed a model for the depinning of flux tubes in type II superconductors, which turned out to be useful for studying the delocalization transition, as well as phase transitions [Nak06]. In essence, it tests localization through eigenvalue sensitivity to b.c. that drive the model off Hermiticity. The Hatano-Nelson deformation of a 1D Anderson model with periodic b.c. is

$$e^{\xi}u_{k+1} + e^{-\xi}u_{k-1} + v_k u_k = \lambda u_k \qquad (14.6.4)$$

By similarity it is equivalent to $\tilde{u}_{k+1} + \tilde{u}_{k-1} + v_k\tilde{u}_k = \lambda\tilde{u}_k$ with b.c. $\tilde{u}_{n+1} = e^{n\xi}\tilde{u}_1$ and $\tilde{u}_0 = e^{-n\xi}\tilde{u}_n$. For $\xi = i\varphi$ (Bloch phase), eigenvalues sweep n non-intersecting real bands; for ξ real they all enter the gaps. If $\xi > \xi_c$ they start to collide and enter the complex plane, where they fill a closed curve [Gol98] which encloses the depleted segment of real eigenvalues. The extrema of the segment are 'mobility edges' $\pm\lambda_c(\xi)$ beyond which the eigenvalues are real and (numerically) unmodified by ξ; their eigenstates are localized enough not to feel the b.c. The equation of the spectral curve is $\xi = \gamma(\lambda)$, where γ is the Lyapunov exponent. The Hatano-Nelson deformation has been applied to complex tridiagonal matrices [Mol09b].

14.6.3 Euclidean random matrix models

Euclidean random matrices were introduced by Mézard, Parisi, and Zee[Mez99] to describe statistical properties of disordered systems such as harmonic vibrations in fluids, glasses, or electron hopping in amorphous semiconductors. Given n points $\{\vec{x}_i\}$ in R^D and a real function $F : R^D \to R$, one constructs the matrix $E_{ij} = F(\vec{x}_i - \vec{x}_j)$. As the n points are chosen randomly, an ensemble of Euclidean random matrices is constructed. In distance matrices, F is just the distance of points [Bog03], with interesting connections between spectral properties and geometry. Another example are the Hessian matrices of some pair potential. For Lennard-Jones potential a mobility edge was found, i.e. a threshold frequency value between regimes of low energy localized modes (phonon-like) and a delocalized regime, with the critical exponents of Anderson's transition [Gri03][Hua09]. The transition may explain the excess in the density of vibrational states (Boson peak) with respect to Debye's ν^2 law observed in glasses.

Several new papers appeared which are relevant to the subject of phase transitions in random matrix models. Very surprising are the papers by K. Zarembo and collaborators exhibiting an infinite number of phase transitions. From the recent preprint [Zar14] the reader may find an introduction and references to a series of works.

References

[Ake97] G. Akemann, Nucl. Phys. B **507** (1997) 475.

[Ake99] G. Akemann, G. M. Cicuta, L. Molinari and G. Vernizzi, Phys. Rev. E **69** (1999) 1489, Phys. Rev. E **60** (1999) 5287;
G. Akemann and G. Vernizzi, Nucl. Phys. B **583** (2000) 739.

[Ake02] G. Akemann and G. Vernizzi, Nucl. Phys. B **631** (2002) 471.

[Aiz08] M. Aizenman and S. Warzel, arXiv:0804.4231 [math-ph]

[Amb97] J. Ambjorn, B. Durhuus, T. Jonsson, *Quantum Geometry*, Cambridge University Press (1997).

[And91] A. Anderson, R.C. Myers and V. Periwal, Nucl. Phys. B **360** (1991) 463; R.C. Myers and V. Periwal, Nucl. Phys. B **390** (1993) 716.

[And07] I. Andrić, L. Jonke and D. Jurman, Phys. Rev. D **77** (2008) 127701.

[Bac00] C. Bachas and P.M.S. Petropoulos, Phys. Lett. B **247** (1990) 363.

[Bah96] S. R. Bahcall, Phys. Rev. Lett. **77** (1996) 5276.

[Bee97] C. W. J. Beenakker, Rev. Mod. Phys. **69** (1997) 731.

[Bel04] J. Bellissard, J. Stat. Phys. **116** (2004) 739.

[Ble03] P. M. Bleher and B. Eynard, J. Phys. A: Math. Gen. **36** (2003) 3085.

[Ble08] P. M. Bleher, *Lectures on random Matrix Models. The Riemann–Hilbert Approach*, arXiv:0801-1858 [math-ph].

[Bog03] E. Bogomolny, O. Bohigas and C. Schmit, J. Phys. A: Math. Gen. **36** (2003) 3595.

[Bon00] G. Bonnet, F. David and B. Eynard, J. Phys. A: Math. Gen. **33** (2000) 6739.

[Bre78] E. Brézin, C. Itzykson, G. Parisi and J.-B. Zuber, Comm. Math. Phys. **59** (1978) 35.

[Bre80] E. Brézin and D. J. Gross, Phys. Lett. B **97** (1980) 120.

[BrH98] E. Brézin and S. Hikami, Phys. Rev. E **57** (1998) 4140.

[Car90] R. Carmona and J. Lacroix, *Spectral theory of random Schroedinger operators*, Birkhauser (1990).

[Cas89] G. Casati, I. Guarneri and D. L. Shepelyansky, Phys. Rev. Lett. **62** (1989) 345.

[Cas90] G. Casati, F. Izrailev and L. Molinari, Phys. Rev. Lett. **64** (1990) 1851.

[Cic86] G. M. Cicuta, L. Molinari and E. Montaldi, Mod. Phys. Lett. A **1** (1986) 125.

[Cic87] G. M. Cicuta, L. Molinari and E. Montaldi, J. Phys. A: Math. Gen. **20** (1987) L67.

[Cic87b] G. M. Cicuta, L. Molinari, E. Montaldi and F. Riva, J. Math. Phys. **28** (1987) 1716.

[Cla08] T. Claeys, Int. Math. Res. Notices 2008, ID rnm166.

[Cri88] H. J. Sommers, A. Crisanti, H. Sompolinsky and Y. Stein, Phys. Rev. Lett. **60** (1988) 1895.

[Cri93] A. Crisanti, G. Paladin, A. Vulpiani, *Products of random matrices*, Springer series in solid state sciences 104 (1993).

[Dav91] F. David, Nucl. Phys. B **348** (1991) 507; Phys. Lett. B **302** (1993) 403.

[Dei99] P. Deift, *Orthogonal Polynomials and Random Matrices : A Riemann–Hilbert Approach*, Courant Institute of Mathematical Sciences (1999).

[Del00] R. Delannay, G. Le Caer, J. Physics A: Math. Gen. **33** (2000) 2611; G. Le Caer and R. Delannay, J. Physics A: Math. Gen. **36** (2003) 9885.

[Dem91] K. Demeterfi, C. I. Tan, Phys. Rev. D **43** (1991) 2622.

[Deo02] N. Deo, Phys. Rev. E **65** (2002) 056115.

[DiF95] P. Di Francesco, P. Ginsparg, J. Zinn-Justin, Phys. Rep. **254** (1995) 1.

[DiF02] cond-mat/0211591; Applications of Random Matrices in Physics, NATO Science Series II: Mathematics, Physics and Chemistry **221** (2006) 33–88.

[DiF03] P. Di Francesco, Nucl. Phys. B **648** (2003) 461.

[Dui08] M. Duits and A. B. J. Kuijlaars, Comm. Pure Appl. Math. **62** (2009) 1076.

[Eve08] F. Evers and A. D. Mirlin, Rev. Mod. Phys. **80** (2008) 1355.

[Eyn06] B. Eynard, J. Stat. Mech. (2006) P07005.

[Eyn07] B. Eynard and N. Orantin, Comm. Number Th. and Phys. **1** (2007) 347, arXiv:0702045 [math-ph].

[Fei97] J. Feinberg and A. Zee, Nucl. Phys. B **504** [FS] (1997) 579;
J. Feinberg, R. Scalettar and A. Zee, J. Math. Phys. **42** (2001) 5718;
J. Feinberg, J. Phys. A: Math. Gen. **39** (2006) 10029.

[Fei97b] J. Feinberg and A. Zee, J. Stat. Phys. **87** (1997) 473.

[Fei89] M. Feingold, D. M. Leitner and O. Piro, Phys. Rev. A **39** (1989) 6507.

[Fok91] A. S. Fokas, A. R. Its and A. V. Kitaev, Comm. Math. Phys. **142** (1991) 313; Comm. Math. Phys. **147** (1992) 395.

[Fro87] H. Froese, Cycon and B. Simon, *Schroedinger operators*, Texts and Monographs in Physics, Springer (1987).

[Fyo91] Y. V. Fyodorov and A. D. Mirlin, Phys. Rev. Lett. **67** (1991) 2405.

[Fyo07] Y. Fyodorov and B. Khoruzhenko, Acta Phys. Pol. B **38** (2007) 4067.

[Gin65] J. Ginibre, J. Math. Phys. **6** (1965) 440.

[Gol98] I. Goldsheid and B. Khoruzhenko, Phys. Rev. Lett. **80** (1998) 2897.
[Got08] F. Gotze and M. Gordin, Comm. Math. Phys. **281** (2008) 203.
[Gri03] T. S. Grigera, V. Martin-Mayor, G. Parisi and P. Verrocchio, Nature **422** (2003) 289.
[Gro80] D. J. Gross and E. Witten, Phys. Rev. D **21** (1980) 446.
[Gro91] D. J. Gross and M. J. Newman, Phys. Lett. B **266** (1991) 291.
[Gro92] D. J. Gross, T. Piran, S. Weinberg editors, *Two dimensional quantum gravity and random surfaces*, World Scientific (1992).
[Haa] F. Haake, *Quantum signatures of chaos*, 2nd ed. Springer (2000).
[Hat96] N. Hatano and D. R. Nelson, Phys. Rev. Lett. **77** (1996) 570.
[Hua09] B. J. Huang and T-M. Wu, Phys. Rev. E **79** (2009) 041105.
[Itz80] C. Itzykson and J. B. Zuber, J. Math. Phys. **21** (1980) 411.
[IHP] http://www.andersonlocalization.com/canonal50.php
[Jan97] R. A. Janik, M. A. Nowak, G. Papp and I. Zahed, Acta Phys. Pol. B **28** (1997) 2949; Phys. Rev. E **55** (1997) 4100; Nucl. Phys. B **501** (1997) 603.
[Jan02] R. A. Janik, Nucl. Phys. B **635** (2002) 492.
[Jan98] M. Janssen, Phys. Rep. **295** (1998) 1.
[Jar06] A. Jarosz and M. A. Nowak, J. Phys. A: Math. Gen. **39** (2006) 10107.
[Joh98] K. Johansson, Duke Math.J. **91** (1998) 151.
[Jur90] J. Jurkiewicz, Phys. Lett. B **245** (1990) 178.
[Jur91] J. Jurkiewicz, Phys. Lett. B **261** (1991) 260.
[Kaz90] V. A. Kazakov, Phys. Lett. B **237** (1990) 212.
[Kho96] B. Khoruzhenko, J. Phys. A: Math. Gen. **29** (1996) L165.
[Kon92] M. Kontsevich, Comm. Math. Phys. **147** (1992) 1.
[Kra93] B. Kramer and A. MacKinnon, Rep. Progr. Phys. **56** (1993) 1469.
[Kra94] V. E. Kravtsov, I. V. Lerner, B. L. Altshuler and A. G. Aronov, Phys. Rev. Lett. **72** (1994) 888.
[Kri06] G. S. Krishnaswami, JHEP 03 (2006) 067.
[LeC07] G. Le Caer and R. Delannay, J. Physics A: Math. Gen. **40** (2007) 1561.
[Lec91] O. Lechtenfeld, Int. J. Mod. Phys A **9** (1992) 2335.
[Liu09] Dang-Zheng Liu and Da-Sheng Zhou, arXiv:0905.4932.
[Mac81] A. MacKinnon and B. Kramer, Phys. Rev. Lett. **47** (1981) 1546.
[Man90] G. Mandal, Mod. Phys. Lett. A **5** (1990) 1147.
[Mar86] F. Martinelli, *A rigorous analysis of Anderson localization*, Lecture Notes in Physics 262, Springer (1986).
[Meh81] M. L. Mehta, Comm. Math. Phys. **70** (1981) 327.
[Mez99] M. Mézard, G. Parisi and A. Zee, Nucl. Phys. B **559** [FS] (1999) 689.
[Miz05] S. Mizoguchi, Nucl. Phys. B **716** (2005) 462.
[Mo07] M. Y. Mo, Int. Math. Res. Notices 2008, ID rnn042.
[Mo08] M. Y. Mo, arXiv:0811.0620 [math-ph].
[Mol88] L. Molinari, J. Phys. A: Math. Gen. **21** (1988) 1.
[Mol09a] L. G. Molinari, J. Phys. A: Math. Theor. **42** (2009) 265204.
[Mol09b] L. G. Molinari and G. Lacagnina, J. Phys. A: Math. Theor. **42** (2009) 395204.
[Moo90] G. Moore, Comm. Math. Phys. **133** (1990) 261; Progr. Theor. Phys. Supp. **102** (1990) 255.
[Nak06] Y. Nakamura and N. Hatano, J. Phys. Soc. Jpn. **75** (2006) 114001.
[NI] http://www.newton.ac.uk/webseminars/pg+ws/2008/mpa/mpaw01/
[Nis02] S. Nishigaki and A. Kamenev, J. Phys. A: Math. Gen. **35** (2002) 4571.
[Pas92] L. Pastur and A. Figotin, *Spectra of Random and Almost-Periodic Operators*, Springer (1992).

[Pas06] L. Pastur, J. Math. Phys. **47** (2006) 103303; L. Pastur and M. Shcherbina, J. Stat. Phys. **130** (2008) 205.

[Ros63] N. Rosenzweig, *Statistical Physics*, Brandeis Summer Inst. 1962, ed. by G. Uhlenbeck et al., Benjamin, New York (1963);
B. V. Bronk, Ph.D Thesis, Princeton Univ. 1964.

[Sen92] D. Sénéchal, Int. J. Mod. Phys. A **7** (1992) 1491.

[Shk93] B. I. Shklovskii, B. Shapiro, B. R. Sears, P. Lambrianides and H. B. Shore, Phys. Rev. B **47** (1993) 11487.

[Shi82] Y. Shimamune, Phys. Lett. B **108** (1982) 407.

[Teo06] R. Teodorescu, J. Phys. A: Math. Gen. **39** (2006) 8921.

[Vol92] D. Vollhardt and P. Wolfle, *Self consistent theory of Anderson localization*, in Electronic Phase Transitions, Ed. by W. Hanke and Yu. V. Kopaev, Elsevier (1992).

[vOp94] F. von Oppen, Phys. Rev. Lett. **73** (1994) 798.

[Wad81] S. R. Wadia, Phys. Rev. D **24** (1981) 970.

[Zak93] J. Zakrzewski and D.Delande, Phys. Rev. E **47** (1993) 1650.

[Zha97] I. K. Zharekeshev and B. Kramer, Phys. Rev. Lett. **79** (1997) 717.

[Zar14] K. Zarembo, Strong-Coupling Phases of Planar $N = 2^*$ Super–Yang–Mills Theory, Theor. Math. Phys. **181** (2014) 1522.

[Zin98] P. Zinn-Justin, Comm. Math. Phys. **194** (1998) 631.

[Zyc94] K. Zyczkowski, L. Molinari and F. Izrailev, J. Physique I France **4** (1994) 1469.

·15·

Two-matrix models and biorthogonal polynomials

M. Bertola

Abstract

This chapter deals with two instances of two-matrix models with Itzykson-Zuber and Cauchy interactions, that are reducible to biorthogonal polynomials and points out some features and outstanding problems.

15.1 Introduction: chain-matrix models

The term of (Hermitian) 'multi-matrix-models' is quite generic and hence we prefer here to use the term 'chain'; these models consist of probability measures on collection of (Hermitian) matrices $M_1, \ldots, M_K$ with a measure of the form

$$d\mu(M_1, \ldots, M_K) = e^{-\Lambda \operatorname{tr} \sum_{\ell=1}^{K} V_\ell(M_\ell)} I_1(M_1, M_2) \cdots I_{K-1}(M_{K-1}, M_K) \prod_{\ell=1}^{K} dM_\ell \tag{15.1.1}$$

where dM_ℓ denote the Lebesgue measures on the vector space of Hermitian $N \times N$ matrices, and V_ℓ (the **potentials**) are some scalar functions (typically polynomials). The measure is not factorized but the 'interaction' is only between the neighbours in a chain of matrices, whence the terminology. The interaction functions $I_\ell(A, B)$ are invariant under diagonal $U(N)$ conjugations $I_\ell(A, B) = I_\ell(UAU^\dagger, UBU^\dagger)$. The cases that are known to be amenable to biorthogonal polynomials are

- **Itzykson-Zuber interaction (IZ)**: $I_\ell(A, B) = e^{c_\ell \operatorname{tr}(AB)}$ where the constants c_ℓ are usually taken as the same.
- **Cauchy interaction (C)**: $I_\ell(A, B) = \det(A + B)^{-N_\ell}$ where the matrices of the chain are supposed to be positive definite and N_ℓ is usually taken as $N_\ell = N$.

In either case the crucial ingredient for the reduction to a biorthogonal polynomial ensemble is the existence of some sort of generalized Harish-Chandra formula in the sense that there exists a function $F(x, y)$ such that for any diagonal matrices $X = \operatorname{diag}(x_1, \ldots, x_N)$, $Y = \operatorname{diag}(y_1, \ldots, y_N)$,

$$\int_{U(N)} I(X, UYU^\dagger)\,\mathrm{d}U = C\frac{\det[F(x_i, y_j)]_{i,j}}{\Delta(X)\Delta(Y)}, \qquad \Delta(Z) = \prod_{i<j}(z_i - z_j). \quad (15.1.2)$$

In the chapter we will consider mostly the two-matrix model with IZ interaction and briefly mention the two-matrix model with Cauchy interaction, pointing out the features and applications of the biorthogonal polynomials relevant to either case.

15.2 The Itzykson-Zuber Hermitian two-matrix model

We will focus on the following (unnormalized) density of probability over the space of pairs (M_1, M_2) of Hermitian matrices of size $N \times N$

$$\mathrm{d}\mu(M_1, M_2) = \mathrm{d}M_1\,\mathrm{d}M_2 \mathrm{e}^{-\Lambda \mathrm{tr}(V_1(M_1)+V_2(M_2)-\tau M_1 M_2)}, \quad \Lambda > 0 \qquad (15.2.1)$$

In order to ensure the normalizability of the measure, the *potentials* $V_1(x)$, $V_2(y)$ (two scalar functions) must be bounded below on the support of the eigenvalues of M_1 and M_2 (respectively). In general this is taken to be $\mathbb{R}$ for both, but one may want to consider a model where the Hermitian matrices M_1, M_2 have their spectrum restricted to some interval (or union thereof). This situation is usually referred to as the introduction of 'hard-edges' in the spectra. The constant Λ could be absorbed into a rescaling of the potentials, but it is used in the scaling limit when $N \to \infty$: in this case it is usually scaled as $\Lambda = \frac{N}{T}$ for some fixed $T > 0$ (sometimes referred to as *temperature* or *total charge*). The case on which we will focus for most of the chapter is based on the assumption:

Assumption 15.2.1 *The potentials are* **polynomials**

$$V_1(x) = \sum_{J=1}^{d_1+1} \frac{u_J}{J} x^J, \qquad V_2(y) = \sum_{K=1}^{d_2+1} \frac{v_K}{K} y^K \qquad (15.2.2)$$

Remark 15.2.1 *If we want to deal with a* bona fide *probability density we must require that the potential be real and bounded below, which implies also that their degrees are even $d_1 + 1 \in 2\mathbb{N} \ni d_2 + 1$ and the leading coefficient is positive. Much of the formalism does not depend on these requirements. However, failing those, the model has to be defined on* normal *matrices with their spectrum lying on contours in the complex plane instead, see [Ber03b]).*

15.2.1 Reduction to eigenvalues

The expectation values of functions $F(M_1, M_2)$ which are invariant under conjugations in both variables separately $F(M_1, M_2) = F(UM_1U^\dagger, VM_2V^\dagger)$, $U, V \in U(N)$ and hence are functions only of the eigenvalues of M_1, M_2 can be computed from the knowledge of the joint probability distributions for the eigenvalues of M_1, M_2. The latter can be expressed effectively in terms of biorthogonal polynomials as we briefly sketch now.

Since both matrices are Hermitian, we can write them in diagonalized form (up to permutations of the eigenvalues)

$$M_1 = UXU^\dagger,\ M_2 = VYV^\dagger,\ U, V \in U(N),$$
$$X = \mathrm{diag}(x_1, \dots, x_N),\ Y = \mathrm{diag}(y_1, \dots, y_N) \tag{15.2.3}$$

This decomposition defines a map $Diag \times U(N) \to \mathcal{H}$ of degree $N!$. The pull-back of the measure (15.2.1) to the space of eigenvalues/eigenvectors (X and U) requires the computation of the Jacobian; in [Meh04] it is shown that this Jacobian is nothing but the square of the Vandermonde determinant $\Delta(Z)$ of the eigenvalues, so that we obtain

$$\mathrm{d}M_1 = \mathrm{d}X\,\mathrm{d}U\Delta(X)^2,\ \mathrm{d}M_2 = \mathrm{d}Y\,\mathrm{d}V\Delta(Y)^2,\ \Delta(Z) = \prod_{i<j}(z_i - z_j) \tag{15.2.4}$$

$$\mathrm{d}\mu(X, U, Y, V) \propto \mathrm{d}X\,\mathrm{d}Y\,\mathrm{d}U\,\mathrm{d}V e^{-\Lambda \mathrm{tr}(V_1(X)+V_2(Y)-\tau UXU^\dagger VYV^\dagger)} \tag{15.2.5}$$

where $\mathrm{d}U$, $\mathrm{d}V$ stand for the Haar measures on the two copies of $U(N)$. In order to obtain the joint probability distribution function (jpdf) for the eigenvalues (and realize the model as a determinantal point process) one needs to perform the 'angular' integration over U, V; the left/right invariance of the Haar measure allows us to re-define $U \mapsto UV^\dagger$ so that we are led to the evaluation of the so-called Itzykson-Zuber-Harish-Chandra integral [Itz80, Har57]

$$\int_{U(N)} \mathrm{d}U e^{\Lambda\tau \mathrm{tr}(XUYU^\dagger)} = C\frac{\det\left[e^{\Lambda\tau x_i y_j}\right]_{i,j\le N}}{\Delta(X)\Delta(Y)} \tag{15.2.6}$$

where the constant C depends only on N and on the normalization of the Haar measure. Using (15.2.6) into (15.2.5) one obtains the jpdf of the $2N$ eigenvalues (as ordered sets)

$$\mathcal{R}^{(N)}(x_1, \dots, x_N; y_1, \dots y_N) = \frac{N!^2}{Z_N}\Delta(X)\Delta(Y)\det\left[e^{\Lambda\tau x_i y_j}\right]_{i,j\le N} e^{-\Lambda\sum_{j=1}^N V_1(x_j)+V_2(y_j)} \tag{15.2.7}$$

$$Z_N := \int_{\mathbb{R}^{2N}} \Delta(X)\Delta(Y)\det\left[e^{\Lambda\tau x_i y_j}\right]_{i,j\le N} e^{-\Lambda\sum_{j=1}^N V_1(x_j)+V_2(y_j)}\mathrm{d}X\mathrm{d}Y \tag{15.2.8}$$

To have the jpdf of k eigenvalues of M_1 and ℓ eigenvalues of M_2 one needs to integrate $\mathcal{R}^{(N)}$ given above over the remaining variables

$$\mathcal{R}^{(N)}_{k,\ell}(x_1, \dots, x_k; y_1, \dots, y_\ell) = \frac{\int \prod_{r=k+1}^N \mathrm{d}x_r \prod_{h=\ell+1}^N \mathrm{d}y_h}{(N-k)!(N-\ell)!}\mathcal{R}^{(N)}(x_1, \dots, x_N; y_1, \dots, y_N) \tag{15.2.9}$$

Determinantal point process and biorthogonal polynomials

Suppose we can construct two sequences of monic polynomials $\{p_n(x)\}_{n\in\mathbb{N}}$, $\{q_n(y)\}_{n\in N}$ uniquely characterized by the requirement of **biorthogonality**

$$\int_{\mathbb{R}}\int_{\mathbb{R}} p_n(x)q_m(y)\mathrm{e}^{-\Lambda(V_1(x)+V_2(y)-\tau xy)}\,\mathrm{d}x\,\mathrm{d}y = h_n\delta_{nm} \tag{15.2.10}$$

If we introduce the matrix of bimoments

$$\mathcal{M} = [\mu_{ij}], \qquad \mu_{ij} := \iint_{\mathbb{R}^2} x^i y^j \mathrm{e}^{-\Lambda(V_1(x)+V_2(y)-\tau xy}\,\mathrm{d}x\,\mathrm{d}y, \quad i,j\geq 0 \tag{15.2.11}$$

then one has the following representation for $p_n(x)$, $q_n(y)$:

$$\begin{aligned}
p_n(x) &= \frac{1}{\Delta_n}\det\begin{bmatrix} \mu_{00} & \cdots & \mu_{n0} \\ \vdots & & \\ \mu_{0n-1} & \cdots & \mu_{n\,n-1} \\ 1\ x & \cdots & x^n \end{bmatrix} \\
q_n(y) &= \frac{1}{\Delta_n}\det\begin{bmatrix} \mu_{00} & \cdots \mu_{n-1 0} & 1 \\ \mu_{01} & \cdots & y \\ \vdots & & \\ \mu_{0n} & \cdots & y^n \end{bmatrix} \\
\Delta_n &:= \det\left[\mu_{ij}\right]_{0\leq i,j\leq n-1}
\end{aligned} \tag{15.2.12}$$

It can be shown that $h_n = \frac{\Delta_{n+1}}{\Delta_n} > 0$ and that for any potentials the zeros of the biorthogonal polynomials are real and simple [Meh02]. Now define

$$\begin{aligned}
K_{12}^{(N)}(x,y) &:= \sum_{\ell=0}^{N-1} \frac{p_\ell(x)q_\ell(y)}{h_\ell}\mathrm{e}^{-\Lambda(V_1(x)+V_2(y))} \\
K_{11}^{(N)}(x,x') &:= \mathrm{e}^{-\Lambda V_1(x')}\int_{\mathbb{R}} \mathrm{e}^{\Lambda\tau x' y} K_{12}(x,y)\,\mathrm{d}y \\
K_{22}^{(N)}(y',y) &:= \mathrm{e}^{-\Lambda V_2(y')}\int_{\mathbb{R}} \mathrm{e}^{\Lambda\tau x y'} K_{12}(x,y)\,\mathrm{d}x \\
K_{21}^{(N)}(y',x') &:= \mathrm{e}^{-\Lambda(V_1(x')+V_2(y'))}\iint_{\mathbb{R}^2} \mathrm{e}^{\Lambda\tau(x'y+xy')} K_{12}(x,y)\,\mathrm{d}x\,\mathrm{d}y + \\
&\quad - \mathrm{e}^{-\Lambda(V_1(x')+V_2(y')-\tau x'y')}
\end{aligned} \tag{15.2.13}$$

The following theorem realizes that the eigenvalues of the two-matrix model constitute a determinantal point process

Theorem 15.2.1 (main theorem in [Eyn98]) *The jpdf $\mathcal{R}_{k,\ell}$ is expressed in terms of the kernels (15.2.13) as follows*

$$\mathcal{R}^{(N)}_{k,\ell}(x_1...x_k; y_1...y_\ell) = \det\left[\begin{array}{c|c} \left[K^{(N)}_{11}(x_i, x_j)\right]_{i,j=1..k} & \left[K^{(N)}_{12}(x_i, y_j)\right]_{\substack{i=1..k\\ j=1..\ell}} \\ \hline \left[K^{(N)}_{21}(y_i, x_j)\right]_{\substack{i=1..\ell\\ j=1..k}} & \left[K^{(N)}_{22}(y_i, y_j)\right]_{i,j=1..\ell} \end{array}\right] \tag{15.2.14}$$

We note that in [Eyn98] a more general situation of p matrices coupled in a chain was considered, and the theorem above is the specialization thereof to the case $p = 2$.

Remark 15.2.2 *It should be pointed out that the Eynard–Mehta proof holds in general for any multi-level determinantal point process as long as the coupling is in the determinantal form* $\det[F(x_i, y_j)]_{i,j}$ *for some function* $F(x, y)$ *(instead of* $\det[e^{\Lambda_T x_i y_j}]$*). This was exploited and generalized in [Har04].*

15.3 Biorthogonal polynomials: Christoffel–Darboux identities

In view of an eventual asymptotic analysis $N \to \infty$, $N/\Lambda = T = \mathcal{O}(1) > 0$ (*scaling limit*) one would like to achieve a control over $K_{12}(x, y)$ and its integral transforms as per (15.2.13) in the scaling limit. The situation is conceptually similar to the one-matrix model (Chapter 4 in this handbook) where the kernel $K(x, x')$ can be reduced from its expression as a sum over N terms to a finite expression involving only two consecutive orthogonal polynomials via the Christoffel–Darboux identity.

In the present setting there are Christoffel–Darboux identities that can help in simplifying the analysis, however the situation is not as ideal as in the case of the one-matrix model, as we will see.

We first point out that the kernel $K_{12}(x, y)$ can be expressed simply in terms of the matrix of bimoments (15.2.11). Indeed if $\mathcal{M}_N$ is the principal block of $\mathcal{M}$ of size N then one can also write (as follows from direct manipulations using the LDU decomposition of $\mathcal{M}_N$)

$$K^{(N)}_{12}(x, y) = e^{-\Lambda(V_1(x)+V_2(y))} \sum_{i,j=0}^{N-1} y^i x^j ({\mathcal{M}_N}^{-1})_{ij} \tag{15.3.1}$$

This formula is however ineffective in deriving asymptotic results (and we point out that it is also very unstable from the point of view of numerical approximations). This is the main reason why a great deal of effort in deriving a Riemann–Hilbert approach to the model has been spent over the years, although some information can be derived from semi-formal manipulations based on the invariance of the integral under reparametrizations (*loop equations*) [Eyn03a, Eyn03b].

It is convenient to define the biorthogonal 'wave functions' and 'wave vectors'

$$\psi_n(x) := \frac{1}{\sqrt{h_n}} p_n(x) \mathrm{e}^{-\Lambda V_1(x)}, \qquad \phi_n(y) := \frac{1}{\sqrt{h_n}} q_n(y) \mathrm{e}^{-\Lambda V_2(y)}$$

$$\Psi(x) := [\psi_0, \ldots, \psi_n, \ldots]^t, \quad \Phi(y) := [\phi_0, \ldots, \phi_n, \ldots]^t$$

$$\int_{\mathbb{R}} \int_{\mathbb{R}} \Psi(x) \mathrm{e}^{\Lambda \tau x y} \Phi^t(y) \, \mathrm{d}x \, \mathrm{d}y = \mathbf{1} \qquad (15.3.2)$$

We then have

Proposition 15.3.1 ([Ber02]) *Under Assumption 15.2.1 the sequences ψ_n, ϕ_n satisfy the following multiplicative recurrence relations*

$$x\psi_n(x) = \gamma_n \psi_{n+1}(x) + \alpha_0(n)\psi_n(x) + \cdots + \alpha_{d_2}(n)\psi_{n-d_2}(x) \qquad (15.3.3)$$

$$y\phi_n(y) = \gamma_n \phi_{n+1}(y) + \beta_0(n)\phi_n(y) + \cdots + \beta_{d_1}(n)\phi_{n-d_1}(y)$$

$$\gamma_n \neq 0, \quad \alpha_{d_2}(n) \neq 0 \neq \beta_{d_1}(n). \qquad (15.3.4)$$

which can be written in matrix form as

$$x\Psi(x) = Q\Psi(x), \qquad y\Phi(y) = P\Phi(y) \qquad (15.3.5)$$

and both Q, P are matrices of a finite band *with nonzero elements $\gamma_0, \ldots$ on the supra-diagonal and d_2 (d_1 respectively) non-trivial sub-diagonals.*

The proof of Prop.15.3.1 relies upon integration by parts. It appears from this proposition that the length of the recurrence relation is directly related to the degrees of the potentials; this suggests that if both potentials are just some arbitrary functions, no particular form of the recurrence relations will follow. A notable exception (which falls outside the scope of this chapter) is the so–called 'semiclassical' case; the term was used in [Mar98] for ordinary orthogonal polynomials to describe those polynomials whose weight is of the form $\mathrm{e}^{-R(x)}\,\mathrm{d}x$ with $R'(x)$ a rational function and also restricted to an arbitrary number of intervals (the endpoints of which are called in the random-matrix literature, 'hard-edges'). In [Ber07a] some recurrence relations (and Christoffel–Darboux identities thereof) were found in the case where *both* weights are semiclassical in the above sense. However, the recurrence relations have the form of a *generalized eigenvalue problem*, namely $x(\mathbf{1} + L)\Psi(x) = A\Psi(x)$ with both L, A finite-band matrices, and L strictly lower triangular.

Corollary 15.3.1 ([Ber02]) *The wave vectors satisfy the differential recurrence relations*

$$-\frac{1}{\Lambda\tau}\partial_x \Psi(x) = P^t \Psi(x), \qquad -\frac{1}{\Lambda\tau}\partial_y \Phi(y) = Q^t \Phi(y) \qquad (15.3.6)$$

The proof of Cor. 15.3.1 is a consequence of the orthogonality (15.3.2).

15.3.1 Christoffel–Darboux identities

A simple manipulation of the recurrence relations yields several identities of Christoffel–Darboux type; we notice, in passing, that this is possibly not the most effective way of proceeding, although it is possibly the simplest and shortest to explain.

Define the projector $\Pi_N := \mathrm{diag}(1, \dots 1, 0, \dots)$ (with N 'ones' on the diagonal); simple algebra shows that

$$K_{12}^{(N)}(x, y) = \Phi^t(y)\Pi_N\Psi(x) \tag{15.3.7}$$

and the recurrence relations in Prop. 15.3.1 and Cor. 15.3.1 yield

$$\left(x + \frac{1}{\Lambda\tau}\frac{\partial}{\partial y}\right) K_{12}^{(N)}(x, y) = \Phi^t(y)\Big[\Pi_N, Q\Big]\Psi(x) \tag{15.3.8}$$

$$\left(y + \frac{1}{\Lambda\tau}\frac{\partial}{\partial x}\right) K_{12}^{(N)}(x, y) = \Psi^t(x)\Big[\Pi_N, P\Big]\Phi(y) \tag{15.3.9}$$

Since the matrices P, Q have a finite-band shape, the commutators in (15.3.8, 15.3.9) only involve a finite number of entries of the wave vectors; we will henceforth focus only on the relations for the Ψs, since the two wave vectors play a completely dual and interchangeable rôle.

We then have

$$\Phi^t(y)\Big[\Pi_N, Q\Big]\Psi(x) = \Phi^N(y)\mathbb{A}_N\Psi_N(x) \tag{15.3.10}$$

$$\Psi_N(x) := [\psi_N, \dots, \psi_{N-d_2}]^t, \quad \Phi^N(y) := [\phi_{N-1}, \dots, \phi_{N+d_2-1}] \tag{15.3.11}$$

$$\mathbb{A}_N := \left[\begin{array}{cccc|c} & & & & -\gamma_N \\ \hline \alpha_{d_2}(N) & \dots & & \alpha_1(N) & \\ & \ddots & & & \\ & & & \alpha_{d_2}(N+d_2-1) & \end{array}\right] \tag{15.3.12}$$

where empty entries are understood as zeros; note that Φ^N is a row vector, whereas Ψ_N is a column vector.

The identity (15.3.8) is *almost* a Christoffel–Darboux identity, if we were able to 'divide' both sides by the differential operator appearing in the left-hand side. This becomes possible for the Laplace transform, namely, the kernel $K_{11}^{(N)}$ (or $K_{22}^{(N)}$ starting from (15.3.9)). Indeed

$$\begin{aligned}(x - x')K_{11}(x, x') &= \int_{\mathbb{R}} \left(x + \frac{1}{\Lambda\tau}\frac{\partial}{\partial y}\right) K_{12}^{(N)}(x, y)\mathrm{e}^{-\Lambda\tau x' y}\,\mathrm{d}y \\ &= \left(\int_{\mathbb{R}} \mathrm{e}^{-\Lambda\tau x' y}\Phi^N(y)\,\mathrm{d}y\right)\mathbb{A}_N\Psi_N(x)\end{aligned} \tag{15.3.13}$$

This shows that

Proposition 15.3.2 *The kernels $K_{11}^{(N)}(x, x')$, $K_{22}^{(N)}(y, y')$ are* integrable kernels *in the sense of Its-Izergin-Korepin-Slavnov [Its94, Its93a, Its93b, Its90], and of rank $d_2 + 1, d_1 + 1$ respectively.*

We recall that the notion of *integrable kernel of rank r* consists of those integral operators on some (collection of) contour(s) with kernels of the form

$$K(x, x') = \frac{\mathbf{f}(x)\mathbf{g}^t(x')}{x - x'}, \quad \mathbf{f}(x), \ \mathbf{g}(x) \in \mathbb{C}^r, \quad \mathbf{f}(x)\mathbf{g}(x) \equiv 0. \tag{15.3.14}$$

The importance of the above statement relies on the applications to spacing distributions in terms of Fredholm determinants and their relationship with a Riemann–Hilbert problem. It is a notable obstacle to the biorthogonal polynomial approach for this model that the kernels K_{12} and K_{21} do *not* fall in this class, and hence the methods for studying Fredholm determinants of matrix-symbols K_{ij} must be enlarged in a suitable way. This is an open problem.

15.3.2 Differential equations

Since the recurrence relations in Prop.15.3.1 have a finite number of terms, one can use them to recursively express any $\psi_m(x)$ (or $\phi_m(y)$) in terms of $d_2 + 1$ $(d_1 + 1)$ consecutive $\psi_N, \psi_{N-1}, \ldots, \psi_{N-d_2}$ (and analogous statement for the ϕ_ns wave functions) with polynomial coefficients. This process, named *folding* in [Ber02], allows us to re-express the infinite differential equation in Cor. 15.3.1 as a closed ODE of rank $d_2 + 1$ and degree d_1 for the ψ_n's.

Proposition 15.3.3 *Let $\Psi_N(x) := [\psi_N, \ldots, \psi_{N-d_2}]^t$ and $\underline{\Phi}^N(x) := \int_{\mathbb{R}} e^{\Lambda_{\mathcal{T}} xy} \Phi^N(y)$; then there are matrices $D_N(x), D^N(x) \in gl(d_2 + 1, \mathbb{C}[x])$, $\deg D_N = \deg D^N = d_1$ such that*

$$-\frac{1}{\Lambda_{\mathcal{T}}}\frac{\mathrm{d}}{\mathrm{d}x}\Psi_N(x) = D_N(x)\Psi_N(x), \qquad \frac{1}{\Lambda_{\mathcal{T}}}\frac{\mathrm{d}}{\mathrm{d}x}\underline{\Phi}^N(x) = \underline{\Phi}^N(x)\, D^N(x) \tag{15.3.15}$$

The coefficients of the matrices D^N, D_N can be effectively and explicitly expressed in terms of the recurrence coefficient matrices P, Q (15.3.5) and we refer to [Ber07b] for the expressions.

Integral representation of the solutions: Riemann–Hilbert problem

The two systems (15.3.15) are put in **duality** by the Christoffel–Darboux pairing $\mathbb{A}_N$ in the sense that

$$0 \equiv D_N(x)\mathbb{A}_N - D^N(x)\mathbb{A}_N \ \Leftrightarrow \ 0 \equiv \partial_x \left(\underline{\Phi}^N(x)\mathbb{A}_N\Psi_N(x)\right) \tag{15.3.16}$$

The question arises as to how to find the remaining d_2 solutions of the pair of ODEs; this was accomplished in the unpublished work [Unp] and expanded upon in [Ber07a]. It was also a crucial ingredient in the steepest descent recent analysis by [Dui09].

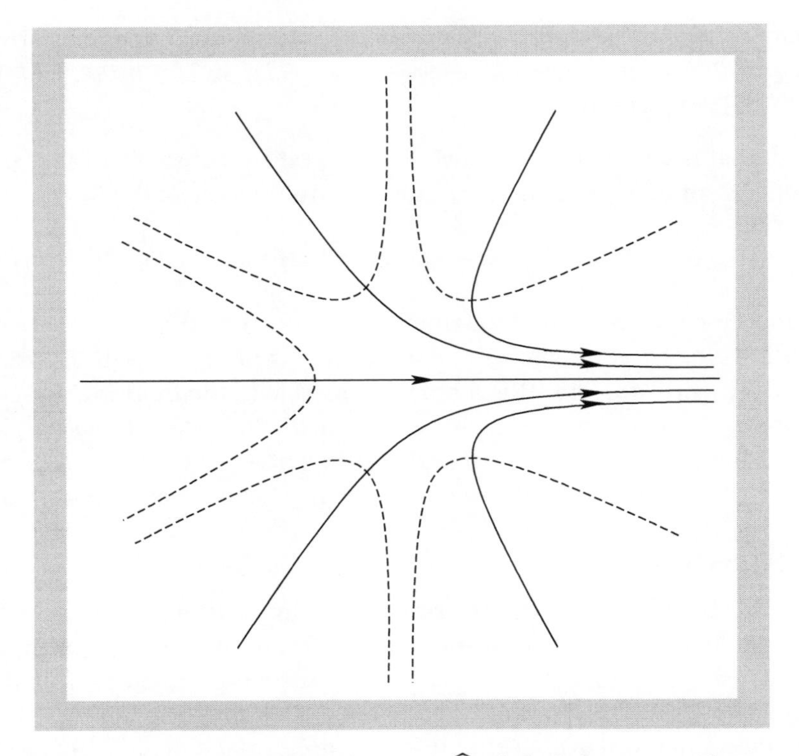

Fig. 15.1 The contours Γ_μ (solid) and dual contours $\widehat{\Gamma}_\nu$ (dashed) for a potential $V_2(y)$ of degree $d_2+1=6$. Note that each contour intersects precisely one of the dual contours, and vice versa. For pictorial reasons only, the contours are chosen as smooth arcs, but one can take them simply as a union of straight rays from the origin.

The quasipolynomial[1] solution $\Psi_N^{(0)}(x) := \Psi_N(x)$ of the ODE (15.3.15) can be complemented by a d_2 non-polynomial solution of the same equation

$$\Psi_N^{(\mu)}(x) = \frac{1}{2i\pi}\int_{\widehat{\Gamma}_\mu}\mathrm{d}s\iint_{\mathbb{R}^2}\mathrm{d}z\,\mathrm{d}w\,\frac{\Psi_N(z)}{x-z}\,\frac{V_2'(s)-V_2'(w)}{s-w}\,\mathrm{e}^{-\Lambda(V_2(w)-V_2(s)-\tau zw+\tau xs)} \tag{15.3.17}$$

The contours of integration $\widehat{\Gamma}_\mu$ are d_2 contours extending from ∞ to ∞ and connecting sectors where $\mathrm{Re}\,V_2(s)\searrow -\infty$ (simple arguments based upon Cauchy theorem show that there are precisely d_2 'homologically' independent such contours: see Fig. 15.1 for an example with $d_2=5$).

The formula (15.3.17), for each fixed μ defines *two* analytic expressions for $\mathrm{Im}(x)>0$ and $\mathrm{Im}(x)<0$, both of which can be extended to entire functions by deformation of the integration contours $\mathbb{R}$ in the z-variable. The proof of

[1] Here 'quasipolynomial' means that it is a polynomial up to multiplication by a scalar function, in this case $\mathrm{e}^{-\Lambda V_1(x)}$.

this (non-trivial) fact is the analogue of similar statements for the Cauchy transformed orthogonal polynomials and can be found in a more general setting of semiclassical weights in [Ber07a].

The square matrix $Y_N := [\Psi_N, \Psi_N^{(1)}, \ldots, \Psi_N^{(d_2)}]$ solves a Riemann–Hilbert problem with constant jump on $\mathbb{R}$ which plays an essential role in the steepest descent analysis in [Dui09], namely

$$Y_N(x)_+ = Y_N(x)_- \begin{bmatrix} 1 & 1 & 0\ldots \\ 0 & 1 & 0\ldots \\ & & \ddots \end{bmatrix} \tag{15.3.18}$$

Dual systems

Along similar lines the other d_2 solutions of the system for $\Phi^N(x)$ in (15.3.15) are written as follows; there are $d_2 - 1$ independent choices of contours Γ_ν where $\mathrm{Re}\, V_2(y) \nearrow +\infty$ (beside the real axis) and hence we can define

$$\Phi^N_{(\nu)}(x) = \int_{\Gamma_\nu} \mathrm{e}^{\Lambda\tau x y} \Phi^N(y) \frac{\mathrm{d}y}{2i\pi}, \qquad \Gamma_1 = \mathbb{R}. \tag{15.3.19}$$

The last row is given by

$$\Phi^N_{(0)}(x) = \mathrm{e}^{\Lambda V_1(x)} \iint_{\mathbb{R}^2} \frac{\Phi^N(y)}{x - \zeta} \mathrm{e}^{-\Lambda V_1(\zeta) + y\zeta} \frac{\mathrm{d}y\, \mathrm{d}\zeta}{2i\pi} \tag{15.3.20}$$

Similarly the matrix

$$Y^N := \begin{bmatrix} \Phi^N_{(0)}(x) \\ \Phi^N_{(1)}(x) \\ \vdots \\ \Phi^N_{(d_2)}(x) \end{bmatrix} \tag{15.3.21}$$

solves another Riemann–Hilbert problem with jump matrix on $\mathbb{R}$ inverse to (15.3.18).

A direct, although not completely trivial computation involving explicit representations allows us to prove that (choosing the contours Γ_ν and $\widehat{\Gamma}_\mu$ appropriately so that they intersect only for $\nu = \mu$)

$$Y^N(x) \mathbb{A}_N Y_N(x) = \mathbf{1} \tag{15.3.22}$$

which is an extension of (15.3.16) to the full fundamental systems for the pair of ODEs 15.3.15. We refer to [Ber07a] for the details and generalizations.

15.4 The spectral curve

Since $\mathbb{A}_N$ is x-independent (and can be shown to be invertible), the duality relation (15.3.16) immediately implies that the characteristic polynomials of $D_N(x)$, $D^N(x)$ are identical.

If we repeated the construction reversing the rôles of V_1, V_2, etc., we would have found ODEs for

$$\Phi_N(y) := [\phi_N, \ldots, \phi_{N-d_2}]^t \tag{15.4.1}$$

$$\Psi^N(y) = \int_{\mathbb{R}} e^{\Lambda_T xy}[\psi_{N-1}, \ldots, \psi_{N+d_1-1}](x)\, dx \tag{15.4.2}$$

$$\frac{d}{dy}\Phi_N(y) = \widetilde{D}_N(y)\Phi_N(y) \tag{15.4.3}$$

$$\frac{d}{dy}\Psi^N(y) = -\Psi^N(y)\,\widetilde{D}^N(y) \tag{15.4.4}$$

It was shown in [Ber02] that the characteristic polynomials of both ODEs (15.3.15) coincide, or more precisely

Proposition 15.4.1 (Prop. 4.1 in [Ber02]) *The following identities hold*

$$u_{d_1+1}\det\Big(y\mathbf{1}_{d_2+1} - D_N(x)\Big) = u_{d_1+1}\det\Big(y\mathbf{1}_{d_2+1} - D^N(x)\Big) = v_{d_2+1}\det\Big(x\mathbf{1}_{d_1+1} - \widetilde{D}^N(y)\Big)$$
$$= v_{d_2+1}\det\Big(x\mathbf{1}_{d_1+1} - \widetilde{D}_N(y)\Big) =: E_N(x, y) \tag{15.4.5}$$

This exact relation can be interpreted as follows. If we think of the algebraic relation $E_N(x, y)$ as defining an algebraic curve $\mathcal{C}$ in $\mathbb{C}\times\mathbb{C}$ (and suitably compactified) we can then view $x, y : \mathcal{C} \to \mathbb{C}$ as functions on $\mathcal{C}$. If we look at the first line of (15.4.5), we then view $y = Y(x)$ as an eigenvalue of $D_N(x)$; looking at the second line we are thinking of $x = X(y)$ as an eigenvalue of $\widetilde{D}_N(y)$. Then, the fact that they are solving the *same* algebraic relation can be regarded as an exact, finite-N version of the original *Matytsin duality*; indeed, using saddle-point heuristic arguments in the scaling $N \to \infty$ limit Matytsin showed that the two functions ($T = N/\Lambda$)

$$X(y) := V_2'(y) - T\int \frac{\rho_2(\eta)\, d\eta}{y - \eta}, \quad Y(x) := V_1'(x) - T\int \frac{\rho_1(\xi)\, d\eta}{x - \xi} \tag{15.4.6}$$

are functionally inverse [Mat94]: $X \circ Y = Id = Y \circ X$.

Eynard-Bergère formula for $E_N(x, y)$ in terms of expectation values

A surprising fact, that mimics a similar occurrence for orthogonal polynomials [Ber06] is that the coefficients of the polynomial $E_N(x, y)$ can be directly related to the expectation values of the matrix integrals. Specifically it can be proved

Proposition 15.4.2 ([Eyn06a]) *The spectral curve* $E_N(x, y)$ *is given by*

$$E_N(x, y) = (x - V_2'(y))(y - V_1'(x)) - \frac{N}{\Lambda_T} + \frac{1}{\Lambda}\left\langle \operatorname{tr}\left(\frac{V_1'(M_1) - V_1'(x)}{M_1 - x}\frac{V_2'(M_2) - V_2'(y)}{M_2 - y}\right)\right\rangle_N \quad (15.4.7)$$

where the angle brackets stand for the matrix expectation value

$$\langle F(M_1, M_2)\rangle_N := \frac{1}{Z_N}\iint F(M_1, M_2)\,\mathrm{d}\mu(M_1, M_2) \quad (15.4.8)$$

The proof of this highly non-trivial statement was written only for polynomial potentials, albeit of arbitrary degree and was first conjectured as a finite-N exact version of a similar asymptotic formula derived using the master-loop equation [Eyn03a, Eyn03b]. It is not known whether the formula can be extended to more general potentials; the most general class of potentials for which a Riemann–Hilbert problem and a finite-rank rational ODE can be set up for the biorthogonal polynomials was considered in [Ber07a]. In principle one may argue by taking limits of polynomial potentials of arbitrary high degree that the same formula (15.4.7) should hold true for more general potentials. On the other hand, if one does not have a corresponding ODE like 15.3.15, it is not clear what the limiting expression (15.4.7) should represent in the limit. Even in the setting of [Ber07a] the ODEs of the biorthogonal polynomials do *not* share the same spectral curve, and hence some modification to the right-hand-side of (15.4.7) may be required. This is an open problem.

15.4.1 Mixed correlation functions and biorthogonal polynomials

The jpdf are sufficient to compute all expectation values of functions $f(M_1, M_2)$ that are invariant under *independent* conjugations of the two matrices, namely, that depend only on the eigenvalues of M_1 and M_2. In the combinatorial applications of the two-matrix model, however, remarkable interest is directed to the so-called *mixed correlation functions*; these are required to compute the expectation values of, for example,

$$\left\langle \operatorname{tr}\left(M_1^{j_1} M_2^{k_1} \cdots M_1^{j_\ell} M_2^{k_\ell}\right)\cdots\right\rangle_N. \quad (15.4.9)$$

It is not at all obvious that these computations can be reduced to a computation involving only eigenvalues and biorthogonal polynomials, but in fact this is the case; in [Ber03a] it was shown that the generating function of the simplest amongst such objects, namely, the expectation values of $\operatorname{tr}(M_1^k M_2^\ell)$, can be expressed in terms of the recurrence coefficients of the biorthogonal polynomials

$$1+\left\langle \mathrm{tr}\left(\frac{1}{x-M_1}\frac{1}{y-M_2}\right)\right\rangle_N = \det\left(\mathbf{1}_N + \left((y-P^t)^{-1}(x-Q)^{-1}\right)_N\right) \quad (15.4.10)$$

where the subscript $()_N$ in the right-hand side denotes the principal submatrix of the indicated size. This $N \times N$ determinant can also expressed as a $(d_2+1)\times(d_2+1)$ (or $(d_1+1)\times(d_1+1)$) determinant (Thm.7.2 in [Eyn06a]) but the formula would require too much notation to be set up and hence we refer directly to the source; we point out that the *existence* of such a formula is potentially very important since it may allow us to derive universality results solely from the asymptotics for large degrees of the biorthogonal polynomials.

The problem of finding *all* mixed expectation values was settled in [Eyn06b]; the authors provided an algorithm to compute first any generalized Itzykson-Zuber integral

$$\int_{U(N)} F(X, UYU^\dagger)\mathrm{e}^{\Lambda \tau \mathrm{tr}(XUYU^\dagger)}\,\mathrm{d}U \quad (15.4.11)$$

where $F(A, B)$ is an arbitrary (polynomial) function of two $N \times N$ matrices A, B that is invariant under the *diagonal* action

$$F(A, B) = F(UAU^\dagger, UBU^\dagger). \quad (15.4.12)$$

This includes all words in $\mathrm{tr}\,A^k$, $\mathrm{tr}\,B^j$ (which are invariant under the independent action) but also arbitrary words in $\mathrm{tr}\,(A^{j_1}B^{k_1}\cdots A^{j_\ell}B^{k_\ell})$. Subsequently they also managed to express the results in terms of certain determinants that involve directly the recurrence-coefficient matrices P, Q. The result is summarized below (in an unfairly reduced form, due to size constraints) from [Eyn06b].

The formula for mixed correlation functions

The statement of the formula requires a certain number of definitions.

The goal: given a polynomial function $F(A, B)$ satisfying (15.4.12) compute the expectation value

$$< F > := \langle F(M_1, M_2)\rangle \quad (15.4.13)$$

in terms of the recurrence coefficients of the biorthogonal polynomials. The computation is best described for the *generating functions* of such invariants.

Definition 15.4.1 (The $\mathcal{M}$ matrix) *Fix $R \in \mathbb{N}$, $\vec{z}, \vec{w} \in \mathbb{C}^R$ and two permutations π, ρ in $\mathfrak{S}_R$. The $\mathcal{M}$-matrix is the following function of ξ, η with values on the matrix algebra of the permutation group, namely, a matrix indexed by two permutations $\pi, \rho \in \mathfrak{S}_R$ as follows*

$$\left[\mathcal{M}^{(R)}_{\vec{z},\vec{w}}\right]_{\pi,\rho}(\xi,\eta) := \prod_{j=1}^{R}\left(\delta_{\pi(j),\rho(j)} + \frac{1}{(z_j-\xi)(w_{\pi(j)}-\eta)}\right) \quad (15.4.14)$$

with the eigenvalues depending rationally on ξ, η and on the components of $\vec{z}, \vec{w}$.

It can be shown that

Proposition 15.4.3 *The family of $\mathcal{M}$-matrices as functions of ξ, η are a commutative family and they are symmetric. Hence they can be simultaneously diagonalized by (ξ, η)–independent* orthogonal *matrices*

$$\mathcal{M}^{(R)}_{\vec{z},\vec{w}}(\xi, \eta) = \mathcal{U}^{(R)^t}_{\vec{z},\vec{w}} \Lambda^{(R)}_{\vec{z},\vec{w}}(\xi, \eta)\mathcal{U}^{(R)}_{\vec{z},\vec{w}} \tag{15.4.15}$$

The formula should be understood in the sense of generating functions, expanding it in inverse powers of the components of $\vec{z}, \vec{w}$. Indeed the coefficient of one monomial in such inverse powers is a polynomial in the variables ξ, η.

Definition 15.4.2 (The $\mathcal{M}$-determinant) *Let* $X := \mathrm{diag}(x_1, \dots, x_N)$, $Y := \mathrm{diag}(y_1, \dots, y_N)$. *The $\mathcal{M}$-determinant is defined as the matrix with indices $\pi, \rho \in \mathfrak{S}_R$ and given by*

$$\mathcal{M}\mathrm{det}\left[\mathcal{M}^{(R)}_{\vec{z},\vec{w}}(x_i, y_j)\right]_{i,j\le N} := \frac{1}{N!}\sum_{\sigma,\tau\in\mathfrak{S}_N}(-1)^{\sigma\tau}\prod_{j=1}^{N}\mathcal{M}^{(R)}_{\vec{z},\vec{w}}\left(x_{\sigma(j)}, y_{\tau(j)}\right) \tag{15.4.16}$$

Note that the order of the products is irrelevant because of the commutativity of the $\mathcal{M}$-family.

Definition 15.4.3 (Generating functions for polynomial invariant functions) *Let $\vec{x}, \vec{y} \in \mathbb{C}^R$, and let $\pi, \pi' \in \mathfrak{S}_R$ be two permutations. Let $\mathfrak{c}_k$, $k = 1 \dots p$ be the decomposition of $\sigma := \pi\pi'^{-1}$ into p disjoint cycles, each of which of length r_k and containing the index j_k, namely*

$$\mathfrak{c}_k = \left(j_k,\ \pi(j_k),\ \sigma(j_k),\ \sigma\pi(j_k),\ \sigma^2(j_k),\ \dots\ \sigma^{r_k-1}\pi(j_k)\right) \tag{15.4.17}$$

Then set

$$F_{\pi,\pi'}(\vec{z}, \vec{w}; A, B) := \prod_{k=1}^{p}\left(\delta_{r_k,1} + \mathrm{tr}\left(\prod_{\ell=1}^{r_k}\frac{1}{z_{\sigma^{\ell-1}(j_k)} - A}\frac{1}{w_{\sigma^{\ell-1}\pi(j_k)} - B}\right)\right) \tag{15.4.18}$$

This formula is to be understood as a generating function: the expansion in inverse powers of the components of $\vec{z}, \vec{w}$ has coefficients some invariant polynomials consisting of the products of p traces, each of which contains a word $A^{t_1}B^{s_1}\cdots A^{t_{r_k}}B^{s_{r_k}}$ of length $2r_k$.

With these definitions in place we can state the generalization of the Harish-Chandra Itzykson Zuber integral formula.

Theorem 15.4.1 *Let X, Y be two diagonal matrices of size N and let $F_{\pi,\pi'}(\vec{z}, \vec{w}; X, Y)$ be as in (15.4.18). Then*

$$\int_{U(N)} \mathrm{d}U F_{\pi,\pi'}(\vec{z}, \vec{w}; X, UYU^\dagger) \mathrm{e}^{-\mathrm{tr}\,(XUYU^\dagger)} =$$

$$= \left(\prod_{\ell=0}^{N-1} 2^\ell \ell!\right) \left[\frac{\mathcal{M}\det\left[\mathrm{e}^{-x_i y_j} \mathcal{M}^{(R)}_{\vec{z},\vec{w}}(x_i, y_j)\right]_{i,j\le N}}{\Delta(X)\Delta(Y)}\right]_{\pi,\pi'} \tag{15.4.19}$$

Example 15.4.1 *$R = 2$ The group $\mathfrak{S}_2 = \{Id, (12)\}$ and F and $\mathcal{M}$ are thus 2×2 matrices*

$$F_{Id\ Id}(\vec{z}, \vec{w}; A, B) = \left(1 + \mathrm{tr}\left(\frac{1}{z_1 - A}\frac{1}{w_1 - B}\right)\right)\left(1 + \mathrm{tr}\frac{1}{z_2 - A}\frac{1}{w_2 - B}\right)\right)$$

$$F_{Id\ (12)}(\vec{z}, \vec{w}; A, B) = \mathrm{tr}\frac{1}{z_1 - A}\frac{1}{w_1 - B}\frac{1}{z_2 - A}\frac{1}{w_2 - B}$$

$$F_{(12)\ Id}(\vec{z}, \vec{w}; A, B) = \mathrm{tr}\frac{1}{z_1 - A}\frac{1}{w_2 - B}\frac{1}{z_2 - A}\frac{1}{w_1 - B}$$

$$F_{(12)\ (12)}(\vec{z}, \vec{w}; A, B) = \left(1 + \mathrm{tr}\frac{1}{z_1 - A}\frac{1}{w_2 - B}\right)\left(1 + \mathrm{tr}\frac{1}{z_2 - A}\frac{1}{w_1 - B}\right)$$

$$\mathcal{M}^{(2)}_{\vec{z},\vec{w}} = \begin{bmatrix} \left(1 + \frac{1}{(z_1 - A)(w_1 - B)}\right)\left(1 + \frac{1}{(z_2 - A)(w_2 - B)}\right) & \frac{1}{(z_1 - A)(w_1 - B)(z_2 - A)(w_2 - B)} \\ \frac{1}{(z_1 - A)(w_2 - B)(z_2 - A)(w_1 - B)} & \left(1 + \frac{1}{(z_1 - A)(w_2 - B)}\right)\left(1 + \frac{1}{(z_2 - A)(w_1 - B)}\right) \end{bmatrix} \tag{15.4.20}$$

Theorem 15.4.2 *The expectation value of $F_{\pi,\pi'}(\vec{z}, \vec{w})$ is given by*

$$\langle F_{\pi,\pi'}(\vec{z}, \vec{w})\rangle = \sum_{\rho\in\mathfrak{S}_R} \mathcal{U}^{(R)}_{\pi,\rho}(\vec{z}, \vec{w})\mathcal{U}^{(R)}_{\pi',\rho}(\vec{z}, \vec{w})\det\left(:\Lambda_\rho(\vec{z}, \vec{w}; Q, P^t) :_{(N)}\right) \tag{15.4.21}$$

where the notation $: :_{(N)}$ means that the matrix Q is written to the right of the matrix P^t in the rational function $\Lambda_\rho(\vec{z}, \vec{w}; \xi, \eta)$ and then the result is truncated to the principal minor of size N.

Example 15.4.2 ($R = 1$, $\pi = (1) = \pi'$)

$$\mathcal{M}^{(1)}_{z,w}(\xi, \eta) = 1 + \frac{1}{(z - \xi)(w - \eta)} \tag{15.4.22}$$

$$\left\langle 1 + \mathrm{tr}\left(\frac{1}{z - M_1}\frac{1}{w - M_2}\right)\right\rangle = \det\left(1 + \frac{1}{w - P^t}\frac{1}{z - Q}\right)_N \tag{15.4.23}$$

15.5 Cauchy two-matrix models

In a recent work [Ber09] a different model for pairs of *positive* Hermitian matrices $M_1, M_2 > 0$ has been put forward. This consist of an unnormalized probability measure

$$\mathrm{d}\mu(M_1, M_2) = \frac{\mathrm{d}M_1\,\mathrm{d}M_2}{\det(M_1 + M_2)^N} \mathrm{e}^{-\Lambda \mathrm{tr}\,(V_1(M_1)+V_2(M_2))} \tag{15.5.1}$$

Here the coupling between the two matrices is given by $\det(M_1 + M_2)^{-N}$ instead of $\mathrm{e}^{\Lambda\tau \mathrm{tr}\, M_1 M_2}$; the treatment of the model follows similar general lines as for the Itzykson-Zuber interaction, where the IZ formula (15.2.6) has been replaced by the Harnad-Orlov formula [Har06]

$$\int_{U(N)} \frac{\mathrm{d}U}{\det(X + UYU^\dagger)^N} = \frac{\det\left[\frac{1}{x_i+y_j}\right]}{\Delta(X)\Delta(Y)} \tag{15.5.2}$$

Following the same steps as before, we are led to considering the *Cauchy biorthogonal polynomials*

$$\iint_{\mathbb{R}_+^2} p_n(x)q_m(y)\frac{\mathrm{e}^{-\Lambda(V_1(x)+V_2(y))}}{x+y}\,\mathrm{d}x\,\mathrm{d}y = h_n\delta_{nm} \tag{15.5.3}$$

Thanks to Remark 15.2.2 the correlation functions and jpdfs are built out of the same kernels, which (after the due modifications) read

$$\begin{aligned}
K_{12}^{(N)}(x, y) &:= \sum_{\ell=0}^{N-1} \frac{p_\ell(x)q_\ell(y)}{h_\ell} \mathrm{e}^{-\Lambda(V_1(x)+V_2(y))} \\
K_{11}^{(N)}(x, x') &:= \mathrm{e}^{-\Lambda V_1(x')} \int_{\mathbb{R}_+} \frac{\mathrm{d}y}{x'+y} K_{12}(x, y) \\
K_{22}^{(N)}(y', y) &:= \mathrm{e}^{-\Lambda V_2(y')} \int_{\mathbb{R}_+} \frac{\mathrm{d}x}{x+y'} K_{12}(x, y) \\
K_{21}^{(N)}(y', x') &:= \mathrm{e}^{-\Lambda(V_1(x')+V_2(y'))} \iint_{\mathbb{R}_+^2} \frac{K_{12}(x, y)\,\mathrm{d}x\,\mathrm{d}y}{(x+y')(y+x')} + \\
&\quad - \frac{\mathrm{e}^{-\Lambda(V_1(x')+V_2(y'))}}{x'+y'}
\end{aligned} \tag{15.5.4}$$

Properties of the polynomials

In a 'complication scale' the Cauchy model turns out to lie in between the one-matrix model and the previously described IZ two-matrix model. The first hint of this lies in the algebraic properties of the Cauchy biorthogonal polynomials (CBOPs). Indeed the sequence of $p_n(x)$ satisfies (with parallel statements for the other sequence of polynomials)

- they *always* solve a four-term recurrence relation of the form

$$x\Big(p_n(x) - \pi_n p_{n-1}(x)\Big) = p_{n+1}(x) + \alpha_n p_n + \beta_n p_{n-1} + \gamma_n p_{n-2} \tag{15.5.5}$$

  for some sequences of constants $\pi_n, \alpha_n, \beta_n, \gamma_n$;

- the roots of the polynomials $p_n(x)$ are *positive and simple*;
- the roots of the polynomial $p_n(x)$ are *interlaced* with the roots of p_{n+1} and p_{n-1}.

All these properties are independent of the choice of the potentials (as long as the measures $\mathrm{e}^{-\Lambda V_j(x)}\,\mathrm{d}x$ are positive); each of the above properties is similar to a parallel one in the context of orthogonal polynomials (related to the Unitary ensemble).

A (non-immediate) consequence of the four-term recurrence relation is that the polynomials can be characterized by a *Riemann–Hilbert problem* [Ber09]:

$$\Gamma_+(x) = \Gamma_-(x)\begin{bmatrix} 1 & \mathrm{e}^{-\Lambda V_1(x)}\chi_{\mathbb{R}_+} & 0 \\ 0 & 1 & \mathrm{e}^{-\Lambda V_2(-x)}\chi_{\mathbb{R}_-} \\ 0 & 0 & 1 \end{bmatrix}$$

$$\Gamma(x) \sim (\mathbf{1} + \mathcal{O}(x^{-1}))\begin{bmatrix} x^n & 0 & 0 \\ 0 & 1 & 0 \\ 0 & 0 & x^{-n} \end{bmatrix} \tag{15.5.6}$$

The sequence $p_n(x)$ appears as the $(1, 1)$ entry of the solution $\Gamma(x) = \Gamma_n(x)$ of the sequences of problems sketchily summarized in (15.5.6). Besides the potential usefulness for a non-linear steepest-descent analysis, the solution $\Gamma(x)$ provides directly the four kernels (15.5.4).

Proposition 15.5.1 (Prop. 3.2 in [Ber09]) *The four kernels (15.5.4) are given by*

$$K_{12}(x, y) = \mathrm{e}^{-\Lambda(V_1(x)+V_2(y))}\frac{\left[\Gamma^{-1}(-y)\Gamma(x)\right]_{3,1}}{(2i\pi)^2(x+y)}$$

$$K_{11}(x, y) = \mathrm{e}^{-\Lambda(V_1(x)+V_1(x'))}\frac{\left[\Gamma^{-1}(x')\Gamma(x)\right]_{2,1}}{(2i\pi)(x'-x)}$$

$$K_{22}(y', y) = \mathrm{e}^{-\Lambda(V_2(y')+V_2(y))}\frac{\left[\Gamma^{-1}(-y)\Gamma(-y')\right]_{3,2}}{(2i\pi)(y'-y)}$$

$$K_{21}(x, y) = \mathrm{e}^{-\Lambda(V_1(x)+V_2(y))}\frac{\left[\Gamma^{-1}(x)\Gamma(-y)\right]_{3,1}}{x+y} \tag{15.5.7}$$

Remark 15.5.1 *If both potentials are, say, polynomials, then one can associate to the biorthogonal polynomials a 3×3 ODE that plays the same rôle as the ODE for the IZ-BOPs. In general this ODE has a simple pole at the origin.*

References

[Ber09] M. Bertola, M. Gekhtman and J. Szmigielski. The Cauchy two–matrix model. *Comm. Math. Phys.*, 287(3):983–1014, 2009.

[Ber07a] M. Bertola. Biorthogonal polynomials for two-matrix models with semiclassical potentials. *J. Approx. Theory*, 144(2):162–212, 2007.

[Ber07b] M. Bertola and B. Eynard. The PDEs of biorthogonal polynomials arising in the two-matrix model. *Math. Phys. Anal. Geom.* 9(1):23–52, 2007.

[Ber06] M. Bertola, B. Eynard, and J. Harnad. Semiclassical orthogonal polynomials, matrix models and isomonodromic tau functions. *Comm. Math. Phys.*, 263(2):401–437, 2006.

[Ber03a] M. Bertola and B. Eynard. Mixed correlation functions of the two–matrix model. *J. Phys. A* 36(28):7733–7750, 2003.

[Ber03b] M. Bertola, B. Eynard, and J. Harnad. Differential systems for biorthogonal polynomials appearing in 2-matrix models and the associated Riemann-Hilbert problem. *Comm. Math. Phys.*, 243(2):193–240, 2003.

[Unp] M. Bertola, J. Harnad, A. Its. unbublished 2005.

[Ber02] M. Bertola, B. Eynard, and J. Harnad. Duality, biorthogonal polynomials and multi-matrix models. *Comm. Math. Phys.*, 229(1):73–120, 2002.

[Dui09] Duits, Maurice; Kuijlaars, Arno B. J. Universality in the two-matrix model: a Riemann-Hilbert steepest-descent analysis. Comm. Pure Appl. Math. 62 (2009), no. 8, 1076–1153.

[Eyn06a] B. Eynard and M. Bergère. Mixed correlation function and spectral curve for the 2–matrix model. *J. Phys. A* 39(49):15091–15134, 2006.

[Eyn06b] B. Eynard and A. Prats Ferrer. 2-matrix versus complex matrix model, integrals over the unitary group as triangular integrals. *Comm. Math. Phys.*, 264(1):115–144, 2006. See also Erratum *Comm. Math. Phys.*, 277(3):861–863, 2008.

[Eyn03a] B. Eynard. Large-N expansion of the 2-matrix model. *J. High Energy Phys.*, (1):051, 38, 2003.

[Eyn03b] B. Eynard. Master loop equations, free energy and correlations for the chain of matrices. *J. High Energy Phys.*, (11):018, 45 pp. (electronic), 2003.

[Eyn98] B. Eynard and M. L. Mehta. Matrices coupled in a chain. I. Eigenvalue correlations. *J. Phys. A*, 31(19):4449–4456, 1998.

[Har57] Harish-Chandra. Differential operators on a semisimple Lie algebra. *Amer. J. Math.*, 79:87–120, 1957.

[Har06] J. Harnad and A. Yu. Orlov. Fermionic construction of partition functions for two-matrix models and perturbative Schur function expansions. *J. Phys. A*, 39(28): 8783–8809, 2006.

[Har04] J. Harnad. Janossy densities, multimatrix spacing distributions and Fredholm resolvents. *Int. Math. Res. Not.*, (48):2599–2609, 2004.

[Its94] A. G. Izergin, A. R. Its, V. E. Korepin, and N. A. Slavnov. The matrix Riemann-Hilbert problem and differential equations for correlation functions of the XXO Heisenberg chain. *Algebra i Analiz*, 6(2):138–151, 1994.

[Its93a] A. R. Its, A. G. Izergin, V. E. Korepin, and N. A. Slavnov. The quantum correlation function as the τ function of classical differential equations. In *Important developments in soliton theory*, Springer Ser. Nonlinear Dynam., pages 407–417. Springer, Berlin, 1993.

[Its93b] A. G. Izergin, A. R. Its, V. E. Korepin, and N. A. Slavnov. Integrable differential equations for temperature correlation functions of the Heisenberg XXO chain. *Zap.*

Nauchn. Sem. S.-Peterburg. Otdel. Mat. Inst. Steklov. (POMI), 205 (Differentsialnaya Geom. Gruppy Li i Mekh. 13):6–20, 179, 1993.

[Its90] A. R. Its, A. G. Izergin, V. E. Korepin, and N. A. Slavnov. Differential equations for quantum correlation functions. In *Proceedings of the Conference on Yang-Baxter Equations, Conformal Invariance and Integrability in Statistical Mechanics and Field Theory*, volume 4, pages 1003–1037, 1990.

[Itz80] C. Itzykson and J.B. Zuber, "The planar approximation (II)", *J. Math. Phys.* **21**, 411 (1980).

[Mar98] F. Marcellán and I. A. Rocha. Complex Path Integral Representation for Semiclassical Linear Functionals. *J. Appr. Theory* 94:107–127, 1998.

[Mat94] V. D. Matytsin. On the large N–limit of the Itzykson–Zuber integral. *Nuclear Phys. B* 411(2-3):805–820, 1994.

[Meh04] M. L. Mehta. *Random matrices*, volume 142 of *Pure and Applied Mathematics (Amsterdam)*. Elsevier/Academic Press, Amsterdam, third edition, 2004.

[Meh02] M. L. Mehta. Zeros of some bi-orthogonal polynomials. *J. Phys. A* 35(3):517–525, 2002.

[Meh94] M. L. Mehta and P. Shukla. Two coupled matrices: eigenvalue correlations and spacing functions. *J. Phys. A* 27(23):7793–7803, 1994.

·16·

Chain of matrices, loop equations, and topological recursion

N. Orantin

Abstract

Random matrices are used in fields as different as the study of multi-orthogonal polynomials or the enumeration of discrete surfaces. Both of these are based on the study of a matrix integral. However, this term can be confusing since the definition of a matrix integral in these two applications is not the same. These two definitions, perturbative and non-perturbative, are discussed in this chapter as well as their relationship. The so-called loop equations satisfied by integrals over random matrices coupled in a chain is discussed as well as their recursive solution in the perturbative case when the matrices are Hermitian.

16.1 Introduction: what is a matrix integral?

The diversity of aspects of mathematics and physics exposed in the present volume witnesses how rich the theory of random matrices can be. This large spectrum of applications of random matrices does not only come from the numerous possible ways of solving them but it is also intrinsically due to the existence, and use, of different definitions of the matrix integral giving rise to the partition function of the theory under study.

Back to the original work of Dyson [Dys62], the study of random matrices is aimed at computing integrals over some given set of matrices with respect to some probability measure on this set of matrices. In order to be computed, these integrals are obviously expected to be convergent. Nevertheless, one of the main applications of random matrices in modern physics follows from a slightly different definition. Following the work of [Bre78], the matrix integral can be considered, through its expansion around a saddle point of the integrand, as a formal power series seen as the generating function of random maps, i.e. random surfaces composed of polygons glued by their sides.[1] Whether this formal series has a non-vanishing radius of convergency or not does not make any difference: only its coefficients, which take finite values, are meaningful.

[1] See Chapter 26 for an introduction to this topic and [Eyn06] and references therein for the generalization to multi-matrix integrals.

The issue as to whether these two definitions coincide or not was not addressed for a long time and led to confusion. In particular it led to a puzzling non-coincidence of some results in the literature [Ake96b, Bre99, Kan98]. Their computations of the same quantity, even if proved to be right, did not match. This puzzle was solved by Bonnet, David, and Eynard [Bon00] who were able to show that the mismatch between the two solutions is a consequence of the discrepancy between the definitions of the matrix integrals taken as partition functions.

Since some of the topics discussed in the present chapter do depend on the definition one considers for the partition function, whereas some other issues do not, Section 16.2 is devoted to the precise definition of these different matrix integrals. In Section 16.3, we present the loop equations which can be used to compute the partition function and correlation functions of a large family of matrix models. Section 16.4 is devoted to a review of one of the solutions of the one Hermitian matrix model through the use of so-called loop equations. Section 16.5 generalizes this method to an arbitrary number of Hermitian matrices coupled in chain. Finally, Section 16.6 gives a short overview of generalizations and applications of this very universal method.

16.2 Convergent versus formal matrix integral

One of the most interesting features in the study of random matrices is the behaviour of the statistic of eigenvalues, or correlation functions, as the random matrices become arbitrarily large. This limit is not only very interesting for its applications in physics (study of heavy nuclei, condensed matter...) but also in mathematics: the knowledge of the large size limit allows us to access the asymptotics of a large set of multi-orthogonal polynomials.[2]

Most of the usual techniques used in random matrix theory fail in the study of formal matrix integrals in the large matrix limit. However, one possible way to address this problem is to try to use naively some saddle point analysis. Let us consider the example of a Hermitian one-matrix with polynomial potential to illustrate this procedure. The partition function is given by the matrix integral:

$$\mathcal{Z}(V) = \int_{\mathcal{H}_N} dM e^{-\frac{N}{t}\operatorname{Tr} V(M)},$$

where one integrates over the group $\mathcal{H}_N$ of Hermitian matrices of size N with respect to the measure

[2] See [Meh04] for a nice review of these applications and all the other chapters of the present volume.

$$dM := \frac{\prod_{i=1}^{N} k!}{\pi^{\frac{N(N-1)}{2}}} \prod_{i=1}^{N} dM_{ii} \prod_{i<j} d\mathrm{Re}\left(M_{ij}\right) \, d\mathrm{Im}\left(M_{ij}\right)$$

defined as the product of the Lebesgues measures of the real components of the matrix M divided by the volume of the unitary group of size N. For the sake of simplicity, one assumes that the potential $V(x) = \sum_{k=0}^{d} \frac{t_k}{k+1} x^{k+1}$ is a polynomial. *Notice that the direct saddle point analysis of this integral does not make sense in general.*

In order to fix this, let us consider a more general problem. Instead of considering Hermitian matrices, we consider normal matrices of size N whose eigenvalues lie on some arbitrary path γ in the complex plane: $H_N(\gamma)$ is the set of matrices M of size $N \times N$ such that there exists $U \in U(N)$ and $X = \mathrm{diag}(x_1, \ldots, x_N)$ with $x_i \in \gamma$ satisfying $M = UXU^{\dagger}$.

With this notation, the set of Hermitian matrices is $H_N(\mathbb{R})$. Given a fixed potential $V(x)$, one considers the family of matrix integrals over formal matrices on arbitrary contours γ:

$$\mathcal{Z}(V, \gamma) = \int_{\mathcal{H}_N(\gamma)} dM e^{-\frac{N}{t} \mathrm{Tr} V(M)}.$$

As in the Hermitian case, one can integrate out the unitary group to turn this partition function into an integral over the eigenvalues of the random matrix:[3]

$$\mathcal{Z}(V, \gamma) = \int_{\gamma} \cdots \int_{\gamma} \prod_{i=1}^{N} dx_i \prod_{i<j} (x_i - x_j)^2 \, e^{-\frac{N}{t} \sum_{i=1}^{N} V(x_i)}.$$

However, given a polynomial potential of degree $d+1$, not every path γ is admissible. Indeed, there are only $d+1$ directions going to infinity where $\mathrm{Re}\left[V(x)\right] > 0$ as $x \to \infty$ and where the integrand decreases rapidly enough for the integral to converge. Thus there exist d homologically independent paths on which the integral $\int dx e^{-\frac{N}{t} V(x)}$ is convergent. Let us choose a basis $\{\gamma_i\}_{i=1}^{d}$ of such paths. Every admissible path γ for the eigenvalues of the random matrix can thus be decomposed in this basis: $\gamma = \sum_{i=1}^{d} \kappa_i \gamma_i$.

[3] This procedure can be generalized to multi-matrix models using the HCIZ formula [Itz80, Har57] presented in Chapter 17.

Using this decomposition, for any admissible path γ, the partition function reduces to

$$\mathcal{Z}(V, \{\gamma_i\} \,|\, \{\kappa_i\}) = N! \sum_{\{n_i\}} \prod_{i=1}^{d} \frac{\kappa_i^{n_i}}{n_i!} \int_{\gamma_1^{n_1} \times \cdots \times \gamma_d^{n_d}} \prod_{i=1}^{N} dx_i \prod_{i<j} (x_i - x_j)^2 \, e^{-\frac{N}{t} \sum_{i=1}^{N} V(x_i)}, \tag{16.2.1}$$

where one sums over all integer d-partitions $(n_1, \ldots, n_d)$ of N, i.e. the sets of d integers $\{n_i\}_{i=1}^{d}$ satisfying $n_1 + \cdots + n_d = N$.

The requirement of convergence of the integral only fixes the asymptotic directions of the paths γ_is. We still have the freedom to choose their behaviour away from their asymptotic directions. Does there exist one choice that is better than the others? One is interested in performing a saddle point analysis of the matrix integral. One thus has to look for the singular points of the action, i.e. the solutions of $V'(x) = 0$. There exist d such solutions ξ_i, $i = 1, \ldots, d$, i.e. as many as the number of paths γ_i in one basis. In the case of the one-matrix model with polynomial potential exposed in the present section, it was proved following [Ber07] that there exists a good basis in the sense that every path γ_i is a steepest descent contour.[4] More precisely, along any path γ_i, the effective potential felt by an eigenvalue x, $V_{eff}(x) = V(x) - \frac{t}{N} \langle \ln(\det x - M) \rangle$, behaves as follows: its real part decreases then stays constant on some interval and then increases, whereas its imaginary part is constant then increases and finally is constant.

Such a steepest descent path can thus be seen as a possible vacuum for one eigenvalue. Each d-partition of N hence corresponds to one vacuum for the theory, or one saddle configuration for the random matrix. The formula Eq. 16.2.1 can be understood as a sum over all possible vacua of the theory:

$$\mathcal{Z}(V, \gamma) = \sum_{n_1 + \cdots + n_d = N} N! \prod_{i=1}^{d} \frac{\kappa_i^{n_i}}{n_i!} \mathcal{Z}(V, \{\gamma_i\} \,|\, n_1, \ldots, n_d)$$

where the partition function with fixed filling fractions $\epsilon_i = \frac{n_i}{N}$

$$\mathcal{Z}(V, \{\gamma_i\} \,|\, n_1, \ldots, n_d) := \int_{\gamma_1^{n_1} \times \cdots \times \gamma_d^{n_d}} \prod_{i=1}^{N} dx_i \prod_{i<j} (x_i - x_j)^2 \, e^{-\frac{N}{t} \sum_{i=1}^{N} V(x_i)}$$

is the weight of a fixed configuration of eigenvalues, or the partition function of the theory with a fixed vacuum labelled by a partition $(n_1, \ldots, n_d)$.

[4] The existence of a good path is conjectured to hold for all other matrix models discussed in this chapter. However, at the time of writing, the proof is known, only in the one-matrix model case.

Assuming that the paths γ_i are good steepest descent paths, the partition functions with fixed filling fractions can be computed by saddle point approximation, i.e. perturbative expansion of the integral around a saddle as $t \to 0$. Further *assuming that one can commute the integral and the power series expansion*, the result is a formal power series in t whose coefficients are Gaussian matrix integrals:

$$\mathcal{Z}(V, \{\gamma_i\}|n_1, \ldots, n_d) \sim \mathcal{Z}_{formal}(V, \{\gamma_i\}|n_1, \ldots, n_d) \qquad \text{when} \qquad t \to 0$$

with

$$\mathcal{Z}_{formal} := e^{-\frac{N}{t}\sum_i n_i V(\xi_i)} \sum_{k=0}^{\infty} \frac{(-1)^k N^k}{t^k\, k!} \prod_{i=1}^{d} \left(\int_{H_{n_i}(\gamma_i)} dM_i \right) \left(\sum_{i=1}^{d} \mathrm{Tr}\,\delta V_i(M_i) \right)^k$$
$$\times\ e^{-\frac{N}{2t}\sum_i V''(\xi_i)\mathrm{Tr}(M_i - \xi_i \mathbf{1}_{n_i})^2} \prod_{j<i} \det\left(M_i \otimes \mathbf{1}_{n_j} - \mathbf{1}_{n_i} \otimes M_j\right)^2, \quad (16.2.2)$$

where $\{\xi_i\}_{i=1..d}$ denote the d solutions of the saddle point equation $V'(\xi_i) = 0$, $\delta V_i(x) := V(x) - V(\xi_i) - \frac{V''(\xi_i)}{2}(x - \xi_i)^2$ denotes the non-Gaussian part of the Taylor expansion of the potential around the saddle ξ_i, and the notation $\prod_{i=1}^{d} \left(\int_{H_{n_i}(\gamma_i)} dM_i \right)$ stands for the multiple integral $\int_{H_{n_1}(\gamma_1)} dM_1 \ldots \int_{H_{n_d}(\gamma_d)} dM_d$. This formal series in t is referred to as a *formal matrix integral* even though **it is not a matrix integral** but a formal power series in t.

This construction can be thought of as a perturbation theory: the matrix integral $\mathcal{Z}(V)$ is the non-perturbative partition function of the theory whereas the formal matrix integral $\mathcal{Z}_{formal}(V, \mathbb{R}|n_1, \ldots, n_d)$ is a perturbative partition function corresponding to the expansion around a fixed vacuum $(n_1, \ldots, n_d)$ in the basis $(\gamma_1, \ldots, \gamma_d)$.

Since these two possible definitions of the partition function might be confused, let us emphasize the main differences, concerning their properties as well as their applications:

- The convergent matrix integral is fixed by a choice of potential V together with an admissible path γ. The formal matrix integral depends on a potential V of degree d, a basis of admissible paths $\{\gamma_i\}_{i=1}^d$ and a d partition of N, $\{n_i\}_{i=1}^d$.
- By definition, the non-perturbative partition function is a convergent matrix integral for arbitrary potential, provided the paths γ_i are chosen consistently. The perturbative integral is a power series defined for arbitrary potentials, integration paths, and filling fractions. It might be non-convergent, and will be for most combinatorial applications.

- The logarithm of the perturbative partition function always has a $\frac{1}{N^2}$ expansion, whereas the non-perturbative one does not most of time (see Section 16.3.3).
- The formal matrix integral is typically used to solve problems of enumerative geometry such as enumeration of maps or topological string theory. The convergent matrix integral is related for example to the study of multi-orthogonal polynomials.

16.3 Loop equations

Even if the perturbative and non-perturbative partition functions do not coincide in general, they share some common properties. One of the most useful properties is the existence of a set of equations linking the correlation functions of the theory: the loop equations. These equations, introduced by Migdal [Mig83], are simply the Schwinger-Dyson equations applied to the matrix model setup. They have proved to be an efficient tool for the computation of formal matrix integrals as the explicit computation of one class of one Hermitian formal matrix integral by Ambjorn *et al.* [Amb93] illustrates.

16.3.1 Free energy and correlation functions

One of the main quantities used in the study of matrix integrals is the *free energy* which is defined as the logarithm of the partition function:

$$\mathcal{F} := -\frac{1}{N^2}\mathcal{Z}.$$

In the formal case, where $\mathcal{Z}$ is the generating function of closed discrete surfaces, the free energy enumerates only connected such surfaces.

In order to be able to compute the free energy, but also for their own interpretation in combinatorics of maps or string theory,[5] it is convenient to introduce the following correlation functions:

$$W_k(x_1, \dots, x_k) := \left\langle \mathrm{Tr}\frac{1}{x_1 - M}\mathrm{Tr}\frac{1}{x_2 - M}\dots\mathrm{Tr}\frac{1}{x_k - M}\right\rangle_c$$

where the index c denotes the connected part and

$$\frac{1}{x-M} = \sum_{i=1}^{d}\sum_{k=0}^{\infty}\frac{\left(M_i - \xi_i \mathbb{I}_{n_i}\right)^k}{(x-\xi_i)^{k+1}} \tag{16.3.1}$$

[5] They are generating functions of open surfaces, as opposed to the free energy which generates surfaces without boundaries. The interested reader is referred to Chapter 31 of the present book or to the review [Eyn06] for details of this interpretation.

has to be understood as a formal power series. It is also useful to introduce the polynomial of degree $d-1$ in x,

$$P_k(x, x_1, \ldots, x_k) := \left\langle \mathrm{Tr}\frac{V'(x)-V'(M)}{x-M}\prod_{i=1}^{k}\mathrm{Tr}\frac{1}{x_i-M}\right\rangle_c .$$

16.3.2 Loop equations

The non-perturbative partition function is given by a convergent matrix integral. It should thus be invariant under change of the integration variable M (or its entries). The name loop equation refers to any equation obtained from the invariance to first order in $\epsilon \to 0$ of the partition function under a change of variable of the form $M \to M + \epsilon\delta(M)$:[6]

$$\int_{\mathcal{H}_N(\gamma)} dM e^{-\frac{N}{t}\mathrm{Tr}V(M)} = \int_{\mathcal{H}_N(\gamma)} d(M+\epsilon\delta(M))e^{-\frac{N}{t}\mathrm{Tr}V(M+\epsilon\delta(M))}.$$

To first order in ϵ, this means that the variation of the action should be compensated by the Jacobian $\mathcal{J}(M)$ of the change of variables:

$$\frac{N}{t}\left\langle \mathrm{Tr}\,V'(M)\delta(M)\right\rangle = \left\langle \mathcal{J}(M)\right\rangle .$$

Actually, the form of the change of variables considered is limited to two main families of $\delta(M)$. This allows us to give a recipe to compute the Jacobian rather easily as follows.

- Leibniz rule:

$$\mathcal{J}\left[A(M)\,B(M)\right] = \left\{\mathcal{J}\left[A(M)\,B(m)\right]\right\}_{m\to M} + \left\{\mathcal{J}\left[A(m)\,B(M)\right]\right\}_{m\to M};$$

- Split rule:

$$\mathcal{J}\left[A(m)\;M^l\;B(m)\right] = \sum_{j=0}^{l-1}\mathrm{Tr}\left[A(m)\,M^j\right]\mathrm{Tr}\left[M^{l-j-1}\,B(m)\right];$$

- Merge rule:

$$\mathcal{J}\left[A(m)\mathrm{Tr}\left(M^l\,B(m)\right)\right] = \sum_{j=0}^{l-1}\mathrm{Tr}\left[A(m)\,M^j\,B(m)\,M^{l-j-1}\right];$$

- if there is no M:

$$\mathcal{J}\left[A(m)\right] = 0.$$

The formal matrix integral is obtained from Gaussian convergent integrals by algebraic computations which commute with the loop equations. It thus follows

[6] It can be equivalently seen as as an integration by parts.

Theorem 16.3.1 *The formal matrix integrals satisfy the same loop equations as the convergent matrix integrals.*

16.3.3 Topological expansion

The loop equations are a wonderful tool for the study of formal matrix integrals. From now on, we restrict our study to these formal power series, leaving aside convergent matrix integrals.

Following an observation originally made by t'Hooft in the study of Feynman graphs of QCD [tHo74], one can see that the exponent of N in the free energy $\mathcal{F}$ is the Euler characteristic of the surface enumerated by this partition function. Thus, $\mathcal{F}$ admits a $\frac{1}{N^2}$ expansion

$$\mathcal{F} = \sum_{g=0}^{\infty} N^{-2g} F^{(g)}$$

commonly called *topological expansion* since the terms $F^{(g)}$ of this expansion are generating functions of connected closed surfaces of fixed genus g.

As for the free energy, one can collect together coefficients with the same power of N in the correation function and get

$$W_k(x_1, \ldots, x_k) = \sum_{h=0}^{\infty} \left(\frac{N}{t}\right)^{2-2h-k} W_k^{(h)}(x_1, \ldots, x_k)$$

as well as

$$P_k(x, x_1, \ldots, x_k) = \sum_{h=0}^{\infty} \left(\frac{N}{t}\right)^{1-2h-k} P_k^{(h)}(x, x_1, \ldots, x_k)$$

where the coefficients are formal power series in t independent of N.

Both the convergent (non-perturbative) partition function and the formal (perturbative) matrix integral are a solution of the loop equations. Nevertheless they do not coincide in general, considering that the loop equations have no unique solution. Indeed, in order to make the solution of these equations unique, one has to fix some 'initial conditions' satisfied by the sought for solution, and the convergent and formal matrix integrals are not constrained by the same kind of conditions.

On the one hand, the formal matrix integral has well-defined constraints: it has a $\frac{1}{N^2}$ expansion and the small-t and large-x limit of any correlation function is fixed by the choice of filling fractions. In other words, by fixing the filling fractions, one prevents the eigenvalues of the random matrix from tunnelling from one saddle to another, i.e. from one steepest descent path to another. There is no instanton contribution.

On the other hand, the convergent matrix integral does not admit, in general, a $\frac{1}{N^2}$ expansion. Moreover, its resolvent is not normalized by an arbitrarily fixed choice of filling fraction: it is rather normalized by some equilibrium conditions on the configuration of the eigenvalues, which, thanks to tunnelling, give instanton corrections to the classical partition function around the true vacuum of the theory. This means that the eigenvalues of the matrix distribute on the different paths of the basis in such a way that they are in equilibrium with respect to the action of the potential and their mutual logarithmic repulsion.

In the formal case, one of the main properties of the correlation functions is the existence of a topological expansion. Let us plug these topological expansions into one set of equations obtained by considering the change of variable of type $\delta M = \frac{1}{x-M}\prod_{i=1}^{k}\mathrm{Tr}\frac{1}{x_i-M}$. They read, order by order in N^{-2}:

$$V'(x)\,W_{n+1}^{(h)}(x,J) = W_{n+1}^{(h-1)}(x,x,J) + \sum_{m=0}^{h}\sum_{I\subset J} W_{1+|I|}^{(m)}(x,I)\,W_{1+n-|I|}^{(h-m)}(x,J\backslash I)$$
$$+\, P_n^{(h)}(x,J) + \sum_{j=1}^{n}\frac{\partial}{\partial x_j}\frac{W_n^{(h)}(x,J\backslash\{x_j\}) - W_n^{(h)}(J)}{x-x_j} \qquad (16.3.2)$$

where J stands for $\{x_1,\ldots,x_n\}$. This is the hierarchy of equations which is solved in the following section.

Remark 16.3.1 Remember that the correlation functions can be seen as the generating functions of discrete surfaces of a given topology. In this picture, the loop equations get a combinatorial interpretation: they summarize all the possible ways of erasing one edge from surfaces of a given topology. This gives a recursive relation among generating functions of surfaces with different Euler characteristics. This inductive method was introduced in the case of triangulated surfaces by Tutte [Tut62] without any matrix model representation of the considered generating functions.

16.4 Solution of the loop equations in the one-matrix model

The solution of the loop equations in their topological expansion has been under intensive study since their introduction by Migdal [Mig83]. In particular, [Amb93] proposed a general solution of these equations in the one-matrix model case for the so-called one cut case, i.e. the case where only one of the filling fractions ϵ_i doesn't vanish. The first steps in the study of the two-cut case were then performed by Akemann in [Ake96a].

Later, in 2004, Eynard [Eyn04] solved the loop equations Eq. 16.3.2 for the formal integral for an arbitrary number of cuts, i.e. compute all the terms in the topological expansion of any correlation function as well as the free energy's $\frac{1}{N^2}$-expansion for arbitrary filling fractions. This solution relies heavily on the existence of an algebraic curve encoding all the properties of the considered matrix model: the spectral curve. Let us first remind ourselves how the latter can be derived.

16.4.1 Spectral curve

Consider Eq. 16.3.2 for $(h, n) = (0, 0)$: it is a quadratic equation satisfied by the genus 0 one-point function:

$$W_1^{(0)}(x)^2 - V'(x)\, W_1^{(0)}(x) = -\, P^{(0)}(x) \tag{16.4.1}$$

called the master loop equation. This can be written as an algebraic equation

$$H_{1MM}(x, W_1^{(0)}) = 0$$

where $H_{1MM}(x, y)$ is a polynomial of degree d in x and 2 in y. The algebraic equation $H_{1MM}(x, y) = 0$ is the basis of the solution presented in this section and will be referred to as the *spectral curve* of the considered matrix model.

A first corollary of this equation is the multi-valuedness of $W_1^{(0)}(x)$ as a function of x. Indeed, for a known $P(x)$ one can solve this equation and get:

$$W_1^{(0)}(x) = \frac{V'(x) \pm \sqrt{V'(x)^2 - 4\, P^{(0)}(x)}}{2}. \tag{16.4.2}$$

A priori, for any value of the complex variable x, there exist two values of $W_1^{(0)}(x)$. Since, from the definition 16.4.2, its large x behaviour is known to be

$$W_1^{(0)}(x) \sim \frac{t}{x} \sum_{i=1}^{d} \frac{n_i}{N} = \frac{t}{x} \quad \text{as } x \to \infty, \tag{16.4.3}$$

one has to select the minus sign in order to get the physically meaningful correlation function.

If one can fix this ambiguity by hand for the genus zero one-point function, the computation of the complete topological expansion of all the correlation functions would imply such a choice at each step.

On the other hand, one can totally get rid of this problem by understanding where it originates. Any correlation function is defined as a formal power series both in $t \to 0$ and in $x \to \infty$. The coefficients of the $\frac{1}{N^2}$-expansion of $W_1(x)$ are thus well defined only around $x \to \infty$, as this series might have a finite radius of convergency: it is not an analytic function in x. In order to get a single-valued function, one has to extend this series further than this radius. The master loop equation tells us how one can proceed: instead of considering the correlation function $W_1(x)$ as a function of the complex variable x, one should consider it as a function defined on the spectral curve. That is to say that one should not consider $W_1^{(h)}(x)$ as functions of a complex variable x but rather as functions of a complex variable x together with a $+$ or $-$ sign corresponding to the choice of one branch of solution of the master loop equation. The tools of algebraic geometry are built to be able to deal with such situations and we present this in the next section.

16.4.2 Algebraic geometry

Consider an algebraic equation $\mathcal{E}(x, y) = 0$ in y and x of respective degrees degree $d_y + 1$ and $d_x + 1$.

A classical result of algebraic geometry states that there exists a compact Riemann surface $\mathcal{L}$ and two meromorphic functions $x(p)$ and $y(p)$ on it such that:

$$\forall p \in \mathcal{L}, \ \mathcal{E}(x(p), y(p)) = 0.$$

By abuse of language, we shall use the term spectral curve to denote the Riemann surface $\mathcal{L}$, the triple $(\mathcal{L}, x, y)$, and the equation $\mathcal{E}(x, y) = 0$ in the following, when no ambiguity can occur.

Let us detail some general properties of this spectral curve useful for the solution of the matrix model.[7]

Sheeted structure and branch points

For a generic fixed value of x, there exist $d_y + 1$ functions of x, $y^i(x)$, $i = 0, \dots, d_y$ solutions of the equation $\mathcal{E}(x, y^i(x)) = 0$. In other words, a given complex number $x(p)$ has $d_y + 1$ preimages p^i, $i = 0, \dots, d_y$ on the surface $\mathcal{L}$ corresponding to different values of $y(p^i)$: one can thus view the Riemann surface $\mathcal{L}$ as $d_y + 1$ copies of the Riemann sphere, denoted as x-sheets, glued together, the function x being injective on each copy.[8]

How are these sheets glued together to form the Riemann surface $\mathcal{E}$? Two sheets merge when two branches of solution in y coincide: $y(p^i) = y(p^j)$ for $i \neq j$. These critical points a_i, called *branch points*, are characterized by the vanishing of the differential dx, i.e. the branch points are solutions of the equation $dx(a_i) = 0$. From now on, we suppose that all the branch points are simple zeros of the one-form dx. This means that only two sheets merge at these points.

This last assumption implies that, around a branch point a, one has

$$y(p) \sim y(a) + \sqrt{x(p) - x(a)}.$$

[7] Most of the properties needed for the study of matrix models can be found in [Far92, Fay73] as well as in Chapter 29 of this volume.

[8] Each copy of the Riemann sphere corresponds to a branch of the solution in y of the equation $\mathcal{E}(x, y) = 0$.

This assumption also implies that, for any branch point a_i and any point z close to a_i, there exists a unique point $\bar{z}$ such that $x(\bar{z}) = x(z)$ and $\bar{z} \to a_i$ as $z \to a_i$.[9] We call $\bar{z}$ the point conjugated to z around a_i.

The spectral curve $\mathcal{L}$ is thus a $d_y + 1$ covering of the Riemann sphere with simple ramification points, the solutions of $dx(a_i) = 0$.

Example: hyperelliptic curve

Let us consider a hyperelliptic spectral curve, i.e. a curve given by a quadratic equation $d_y = 1$, as in the one Hermitian matrix model case:

$$\mathcal{E}(x, y) = y^2 - \prod_{i=1}^{d} (x - x(a_i))(x - x(b_i)) = y^2 - \sigma(x)$$

where $d_x := 2d$ is to match the notations of the previous section. The corresponding Riemann surface can be seen as a two-sheeted cover of the Riemann sphere: one sheet corresponding to the branch $y(x) = \sqrt{\sigma(x)}$, and the other one to the other branch $y(x) = -\sqrt{\sigma(x)}$. These two sheets merge when $y(x)$ takes the same value on both sheets, i.e. when $y(x)$ vanishes. The branch points are thus the preimages of the points $x(a_i)$ and $x(b_i)$ on the spectral curve. The latter can thus be described as two copies of $\mathbb{CP}^1$ glued by d cuts $[a_i, b_i]$.

Genus and cycles

Generically, the compact Riemann surface $\mathcal{L}$ associated to an algebraic equation may have non-vanishing genus g, and this will be the case in most of the applications of the present chapter.

The Riemann–Hurwitz theorem allows us to get this genus out of the branched covering picture of the spectral curve. For example, if there are only simple ramification points, it states that

$$g = -d_y + \frac{\text{number of branch points}}{2}$$

where $d_y + 1$ is the number of x-sheets.

If the Riemann surface has non-vanishing genus, i.e. it is not conformally equivalent to $\mathbb{CP}^1$, there exist non-contractible cycles on it. In order to deal with these, it will be useful to choose a canonical homology basis of cycles $(\mathcal{A}_1, \ldots, \mathcal{A}_g, \mathcal{B}_1, \ldots, \mathcal{B}_g)$ satisfying the intersection conditions

$$\forall i, j = 1 \ldots, g\,, \ \mathcal{A}_i \bigcap \mathcal{A}_j = \mathcal{B}_i \bigcap \mathcal{B}_j = 0, \qquad \mathcal{A}_i \bigcap \mathcal{B}_j = \delta_{i,j}.$$

[9] The application $z \to \bar{z}$ is defined only locally around the branch points and depends on the branch point considered and the notation $\bar{z}$ is abusive. Nevertheless, this application will always be used in the vicinity of a branch point in such a way that no ambiguity will occur.

Examples of a hyperelliptic curve

Let us keep on considering the example of an hyperelliptic curve with $2d$ branch points. From the Riemann–Hurwitz theorem, it has genus $g = d - 1$. This follows the intuitive picture of two Riemann spheres glued by d segments, giving rise to a genus $d - 1$ surface.

One can also make a canonical homology basis explicit as follows. First choose one cut, for example $[a_1, b_1]$ and one sheet, e.g. the sheet corresponding to the minus sign of the square root. Then define the $\mathcal{A}_i$-cycle as the cycles on the chosen sheet around the cut $[a_{i+1}, b_{i+1}]$. Finally, define the $\mathcal{B}_i$-cycle as the composition of the segment $[a_1, a_{i+1}]$ in the chosen sheet and $[a_{i+1}, a_1]$ followed in the opposite direction in the other sheet (see Fig.16.1 for the simplest example of a genus 1 surface).

Differentials

The meromorphic differentials on the Riemann surface $\mathcal{L}$ and their properties will play a crucial role in the following. In particular, let us remember a fundamental result concerning meromorphic differentials: a meromorphic differential df on Riemann surface $\mathcal{L}$ of genus g equipped with a basis of cycles $\{\mathcal{A}_i, \mathcal{B}_i\}_{i=1}^{g}$, is defined uniquely by its $\mathcal{A}$-cycles $\int_{\mathcal{A}_i} df$ on the one hand, and by its singular behaviour, i.e. the position of its poles and the divergent part of its Laurent expansion around the latter, on the other hand.

For example, one introduces one of the main building blocks of the solution of loop equations as follows:

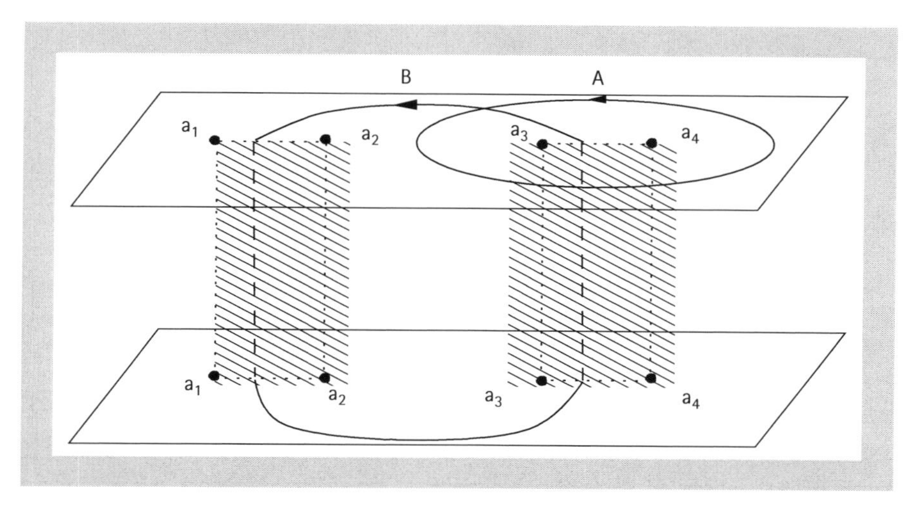

Fig. 16.1 Genus 1 hyperelliptic curve: it is built as two copies of the Riemann sphere glued by two cuts $[a_1, a_2]$ and $[a_3, a_4]$. The unique $\mathcal{A}$-cycle encircles $[a_3, a_4]$ while the $\mathcal{B}$-cycles goes through both cuts.

Definition 16.4.1 *Let the Bergman kernel $B(p, q)$ be the unique bi-differential in p and q on $\mathcal{L}$ defined by the constraints as a differential in p:*

- *it has a unique pole located at $p \to q$ which is double without residue. In local coordinates, it reads*

$$B(p, q) \sim \frac{dp\, dq}{(p-q)^2} + \textit{regular} \qquad \textit{when} \qquad p \to q;$$

- *it has vanishing $\mathcal{A}$-cycle integrals:*

$$\forall i = 1, \dots, g, \quad \oint_{\mathcal{A}_i} B(p, q) = 0.$$

It is also useful to define the primitive of the Bergman kernel:

$$d\, S_{p_1, p_2}(q) = \int_{z=p_1}^{p_2} B(z, q)$$

which is a one-form in q with simple poles in $q \to p_1$ and $q \to p_2$ with respective residues -1 and $+1$.

16.4.3 The one-point function and the spectral curve

With these few elements of algebraic geometry on hand, let us complete our study of the spectral curve of the Hermitian one-matrix model. Up to now, one has obtained that $W_1^{(0)}(x)$ is a solution of a quadratic equation which depends on a polynomial $P^{(0)}(x)$ of degree $d-1$, i.e. d variables remain to be fixed.

From the definition Eq. 16.3.1 of the correlation functions, considering the $\mathcal{A}_i$-cycles as a circle, independent of t around ξ_i,[10] one gets d constraints

$$\forall i = 1, \dots, d\ , \quad \frac{1}{2i\pi} \oint_{\mathcal{A}_i} W_1^{(0)}(x) dx = \frac{n_i t}{N}$$

allowing us to fix the coefficients of the polynomial $P^{(0)}(x)$, since the contour integral $\oint_{\mathcal{A}_i}$ picks only one residue at ξ_i. One thus gets all the parameters of the spectral curve as well as the one-point function $W_1^{(0)}(x)$.

Properties of the one-matrix model's spectral curve

The polynomial $H_{1MM}(x, y)$ has degree 2 in y. This means that the embedding of $\mathcal{L}_{1MM}$ is composed by two copies of the Riemann sphere glued by $g+1$ cuts so that the resulting Riemann surface $\mathcal{L}_{1MM}$ has genus g. Each copy of the Riemann sphere corresponds to one particular branch of the solutions of the equation $H_{1MM}(x, y) = 0$. Since there are only two sheets in involution, this spectral curve is said to be hyperelliptic. It also means that the application $z \to$

[10] Indeed, when $t \to 0$, the cuts are reduced to double points at ξ_is. As t grows, these double points give rise to cuts of length of order $\frac{n_i t}{N}$.

$\bar{z}$ is globally defined since it is the map that exchanges both sheets, i.e. that exchanges the two branches of the square root in Eq. 16.4.2.

The Riemann surface $\mathcal{L}_{1MM}$ has genus g lower than $d - 1$.[11]

The function $x(z)$ on the Riemann surface $\mathcal{L}_{1MM}$ has two simple poles (call them α_+ and α_-), one on each sheet. Near $\alpha_\pm$, $y(z)$ behaves like:

$$y(z) \underset{z\to\alpha_+}{\sim} \frac{t}{x(z)} + O(1/x(z)^2)$$

and

$$y(z) \underset{z\to\alpha_-}{\sim} t_d x^d(z) + O(x^{d-1}(z)).$$

16.4.4 Two-point function

Let us go one step further and consider the loop equation (16.3.2) for $k = 2$ and $h = 0$. This allows us to obtain a formula for $W_2^{(0)}(x, x_1)$:

$$W_2^{(0)}(x, x_1) = \frac{\frac{\partial}{\partial x_1} \frac{W_1^{(0)}(x) - W_1^{(0)}(x_1)}{x - x_1} + P_2^{(0)}(x; x_1)}{2(V'(x) - W_1^{(0)}(x))}.$$

A first look at this expression allows us to see that this function is multivalued in terms of the complex variables x and x_1. However, one can lift it to a single-valued function, actually a two-form, on the spectral curve by defining

$$\widehat{\omega}_2^{(0)}(z, z_1) := W_2^{(0)}(x(z), x(z_1))dx(z)dx(z_1).$$

Thus $\widehat{\omega}_2^{(0)}\omega(z, z_1)$ is a meromorphic bi-differential on $\mathcal{L}$. One can then study all possible singularities of this formula and see that $\widehat{\omega}_2^{(0)}(z, z_1)$ has poles only at $z \to \bar{z}_1$. On the other hand, the normalization of the two-point function around the $\mathcal{A}$-cycles reads $\oint_{\mathcal{A}_i} \widehat{\omega}_2^{(0)}(z, z_1) = 0$ for $i = 1, \ldots, d$. These two conditions imply that $\widehat{\omega}_2^{(0)}(z, z_1)$ is given by the Bergman kernel (see for example Section 5.2.3 of [Eyn09])

$$\widehat{\omega}_2^{(0)}(z, z_1) = -B(z, \bar{z}_1) = B(z, z_1) - \frac{dx(z)dx(z_1)}{(x(z) - x(z_1))^2}.$$

16.4.5 Correlation functions

We now have everything at hand to compute any correlation function by solving the loop equations. First of all, the study of the one- and two-point functions

[11] Notice that $d - 1$ is an upper bound. It might happen that two branch points coincide resulting in the closing of one cut and decreasing of the genus by one. For some special value of the coefficients of the polynomial H_{1MM} one can even get a genus zero spectral curve. For application of matrix models to enumeration of surfaces, this very non-generic constraint is almost always satisfied (see [Eyn06] for further considerations on this point).

proved that it is more convenient to promote the multivalued functions $W_n^{(h)}$ on the complex plane to single-valued meromorphic forms on $\mathcal{L}$:[12]

$$\omega_n^{(h)}(z_1,\ldots,z_n) := W_n^{(h)}(z_1,\ldots,z_n)\prod_{i=1}^{n} dx(z_i) + \delta_{n,2}\delta_{h,0}\frac{dx(z_1)dx(z_2)}{(x(z_1)-x(z_2))^2}$$

and

$$y(z)dx(z) := W_1^{(0)}(z)dz.$$

It is important to remember that the physical quantities encoded in the correlation functions are obtained as terms of the expansion of the latter, when their variables approach the physical pole α_+ of the spectral curve.

From the loop equations (16.3.2), one can prove by induction that $\omega_n^{(h)}(z_1,\ldots,z_n)$ with $2h+n\geq 3$ can neither have poles at coinciding points $x(z_i)=x(z_j)$, nor at the poles of x, nor at the double points. It may only have poles at the branch points.

Let us now write down the Cauchy formula on the spectral curve:

$$\omega_{n+1}^{(h)}(z,z_1,\ldots,z_n) = \operatorname*{Res}_{z'\to z} d\,S_{z',o}(z)\omega_{n+1}^{(h)}(z',z_1,\ldots,z_n)$$

where o is an arbitrary point of $\mathcal{L}$. Since $\omega_{n+1}^{(h)}(z',z_1,\ldots,z_n)$ only has poles at the branch point a_i, moving the integration contours on $\mathcal{L}$ (and not $\mathbb{C}$!), one gets contributions from the latter and the boundaries of the fundamental domain of $\mathcal{L}$ according to the Riemann bilinear formula [Far92]

$$\begin{aligned}\omega_{n+1}^{(h)}(z,z_1,\ldots,z_n) &= -\sum_i \operatorname*{Res}_{z'\to a_i} d\,S_{z',o}(z)\omega_{n+1}^{(h)}(z',z_1,\ldots,z_n)\\ &+\sum_{i=1}^{g}\left[\oint_{z'\in\mathcal{A}_i} B(z,z')\oint_{z'\in\mathcal{B}_i}\omega_{n+1}^{(h)}(z',z_1,\ldots,z_n)\right.\\ &\left.+\oint_{z'\in\mathcal{B}_i} B(z,z')\oint_{z'\in\mathcal{A}_i}\omega_{n+1}^{(h)}(z',z_1,\ldots,z_n)\right].\end{aligned}$$

Since the correlation functions and the Bergman kernel have vanishing $\mathcal{A}$-cycle integrals, the second and third lines vanish. One can then plug the expression for $\omega_{n+1}^{(h)}(z',z_1,\ldots,z_n)$ coming from the loop equations (16.3.2) into this formula. Since the polynomial $P_{n+1}^{(g)}(x(z'),z_1,\ldots,z_n)$ is regular at the branch points, it does not give any contribution and one gets the recursion formula

[12] The single valuedness of the differential form $\omega_n^{(h)}$ on the spectral curve is obtained by induction on the Euler characteristic $2h+n-2$ through the use of the loop equations (16.3.2).

$$\omega_{n+1}^{(h)}(z, z_1, \ldots, z_n) = \sum_i \text{Res}_{z' \to a_i} K(z, z')\Big[\omega_{n+2}^{(h-1)}(z', \overline{z'}, z_1, \ldots, z_n)$$

$$+ \sum_{j=0}^{h} \sum_{I \subset \{z_1, \ldots, z_n\}}{}' \omega_{|I|+1}^{(j)}(z', I)\omega_{n-|I|+1}^{(h-j)}(\overline{z'}, \{z_1, \ldots, z_n\} \setminus I)\Big]$$

where the sign $\sum'$ means that the sum does not involve the terms with $(j, |I|) = (0, 0)$ or $(j, |I|) = (h, n)$ and the recursion kernel is

$$K(z, z') := \frac{d S_{z', \overline{z'}}(z)}{2(y(z') - y(\overline{z'}))dx(z')}.$$

It is easy to see that, provided $\omega_2^{(0)}(z, z_1) = B(z, z_1)$ is known, this recursive relation in $2h + n - 2$ determines all the other correlation functions through their topological expansion.

Remark 16.4.1 This recursion can be represented graphically in such a way that it becomes very easy to remember, and allows us to recover some of the properties of the correlation functions using only diagrammatic proofs. Details on this diagrammatic representation can be found in [Eyn09].

16.4.6 Free energies

In the preceding section, we have been able to compute the topological expansion of any correlation function W_n for $n > 0$. Let us now address the case $n = 0$, that is to say, the computation of the topological expansion of the free energy.

For this purpose, one can build an operator acting from the space of $n+1$-differentials on $\mathcal{L}$ into the space of n-differentials, mapping the $n+1$-point function to the n-point function

Theorem 16.4.1 *For any h and n satisfying $2 - 2h - n < 0$ and any primitive Φ of ydx, one has*

$$\omega_n^{(h)}(z_1, \ldots, z_n) = \frac{1}{2 - 2h - n} \sum_i \mathop{\text{Res}}_{z \to a_i} \Phi(z)\omega_{n+1}^{(h)}(z, z_1, \ldots, z_n).$$

One can guess that this definition can be extended to $n = 0$ in order to get the topological expansion of the free energies as follows:

Theorem 16.4.2 *The terms of the topological expansion of the free energy of the Hermitian one-matrix model are given by:*

$$F^{(h)} = \frac{1}{2 - 2h} \sum_i \mathop{\text{Res}}_{z \to a_i} \Phi(z)\omega_1^{(h)}(z)$$

for $h \geq 2$.

This guess can be proved to be right by looking at the derivative of the result with respect to all the moduli of the formal integral, i.e. the coefficient of the potential and the filling fractions. Indeed, they match the expected variations of the free energies when varying these moduli [Che06a].

16.5 Matrices coupled in a chain plus external field

It is remarkable that the recursive formula giving the topological expansion of the free energy and the correlation functions depends on the moduli of the model only through the spectral curve. One can thus wonder whether the same procedure can be applied to solve other matrix models that are known to be related to a spectral curve. This is indeed the case for the model of two matrices coupled in chain [Che06b], but also for the an arbitrary long chain of matrices in an external field [Eyn08b].

In order to deal with a large family of Hermitian matrix models at once, let us consider an arbitrarily long sequence of matrices coupled in chain and submitted to the action of an external field.

The partition function is given by the chain of matrices formal matrix integral:

$$Z_{\text{chain}} = \int_{\text{formal}} e^{-\frac{N}{t}\text{Tr}\left(\sum_{k=1}^{m} V_k(M_k) - \sum_{k=1}^{m} c_{k,k+1}\, M_k M_{k+1}\right)} d\,M_1 \dots d\,M_m$$

where the integral is a formal integral in the sense of the preceding section,[13] M_{m+1} is a constant given diagonal matrix $M_{m+1} = \Lambda$ with s distinct eigenvalues λ_i with multiplicities l_i:

$$M_{m+1} = \Lambda = \text{diag}\left(\overbrace{\lambda_1, \dots, \lambda_1}^{l_1}, \dots, \overbrace{\lambda_i, \dots, \lambda_i}^{l_i}, \dots, \overbrace{\lambda_s, \dots, \lambda_s}^{l_s}\right)$$

with $\sum_i l_i = N$ and one considers the m polynomial potentials[14]

$$V_k(x) = -\sum_{j=2}^{d_k+1} \frac{t_{k,j}}{j}\, x^j .$$

As in the one-matrix model case, the definition of the formal integral requires one to choose around which saddle point one expands. Saddle points are solutions of the set of equations

[13] The formal integral is a power series in t whose coefficients are Gaussian integrals. See [Eyn06] for a review on this topic.

[14] It is possible to generalize all this section to potentials whose derivatives are arbitrary rational functions without any significant modification of the present procedure.

$$\forall k = 1, \ldots, m, \qquad V_k'(\xi_k) = c_{k-1,k}\xi_{k-1} + c_{k,k+1}\xi_{k+1}, \qquad \exists j,\ \xi_{m+1} = \lambda_j$$

which can be reduced to an algebraic equation with $D = sd_1d_2\ldots d_m$ solutions.

This choice is thus equivalent to the choice of a D-partition $(n_1, \ldots, n_D)$ of N giving rise to the filling fractions:

$$\epsilon_i = t\frac{n_i}{N}$$

for $i = 1, \ldots, D$ with $D = d_1d_2\ldots d_m s$ and n_i arbitrary integers satisfying

$$\sum_i n_i = N.$$

Definition of the correlation functions

The loop equations of the chain of matrices were derived in [Eyn03, Eyn08b], and they require the definition of several quantities.

For convenience, we introduce $G_i(x_i) := \frac{1}{x_i - M_i} = \sum_{k=0}^{\infty} \frac{M_i^k}{x_i^{k+1}}$ as a formal power series in $x_i \to \infty$ as well as a polynomial in x, $Q(x) = \frac{1}{c_{n,n+1}} \frac{S(x)-S(\Lambda)}{x-\Lambda}$, where $S(x)$ is the minimal polynomial of Λ, $S(x) = \prod_{i=1}^{s}(x - \lambda_i)$. We also define the polynomials $f_{i,j}(x_i, \ldots, x_j)$ by $f_{i,j} = 0$ if $j < i - 1$, $f_{i,i-1} = 1$, and

$$f_{i,j}(x_i, \ldots, x_j) = \det \begin{pmatrix} V_i'(x_i) & -c_{i,i+1}x_{i+1} & & 0 \\ -c_{i,i+1}x_i & V_{i+1}'(x_{i+1}) & \ddots & \\ & \ddots & \ddots & -c_{j-1,j}x_j \\ 0 & & -c_{j-1,j}x_{j-1} & V_j'(x_j) \end{pmatrix}$$

if $j \geq i$. The latter satisfy the recursion relations

$$c_{i-1,i}f_{i,j}(x_i, \ldots, x_j) = V_i'(x_i)f_{i+1,j}(x_{i+1}, \ldots, x_j) - c_{i,i+1}\, x_i\, x_{i+1}\, f_{i+2}(x_{i+2}, \ldots, x_j).$$

Let us finally define the correlation functions and some useful auxiliary functions. In the following $\mathrm{Pol}_x f(x)$ refers to the polynomial part of $f(x)$ as $x \to \infty$. For $i = 2, \ldots, m$, we define

$$\begin{aligned} &W_i(x_1, x_i, \ldots, x_m, z) := \\ &\mathrm{Pol}_{x_i,\ldots,x_m} f_{i,m}(x_i, \ldots, x_m) \left\langle \mathrm{Tr}\,(G_1(x_1)G_i(x_i)\ldots G_m(x_m)\,Q(z))\right\rangle, \end{aligned}$$

which is a polynomial in variables $x_i, \ldots, x_m, z$, but not in x_1; for $i = 1$,

$$\begin{aligned} &W_1(x_1, x_2, \ldots, x_m, z) := \\ &\mathrm{Pol}_{x_1,\ldots,x_m} f_{1,m}(x_1, \ldots, x_m) \left\langle \mathrm{Tr}\,(G_1(x_1)G_2(x_2)\ldots G_m(x_m)\,Q(z))\right\rangle \end{aligned}$$

which is a polynomial in all variables and, for $i = 0$, $W_0(x) = \langle \mathrm{Tr} G_1(x) \rangle$. We also define:

$$W_{i;1}(x_1, x_i, \ldots, x_m, z; x_1') :=$$
$$\mathrm{Pol}_{x_i,\ldots,x_m} f_{i,m}(x_i, \ldots, x_m) \left\langle \mathrm{Tr}\left(G_1(x_1')\right) \mathrm{Tr}\left(G_1(x_1) G_i(x_i) \ldots G_m(x_m) Q(z)\right) \right\rangle_c .$$

All these functions admit a topological expansion:

$$W_i = \sum_g (N/t)^{1-2g}\, W_i^{(g)} \qquad \text{and} \qquad W_{i;1} = \sum_g (N/t)^{-2g}\, W_{i;1}^{(g)}.$$

Loop equations and spectral curve

In this model, the master loop equation reads [Eyn03, Eyn08b]:

$$\begin{aligned} &W_{2;1}(x_1, \ldots, x_{m+1}; x_1) + \frac{t}{N} W_1(x_1, \ldots, x_{m+1}) - \left(V_1'(x_1) - c_{1,2}x_2\right) S(x_{m+1}) \\ &+ (c_{1,2}x_2 - V_1'(x_1) + \frac{t}{N} W_0(x_1))\Big(\frac{t}{N} W_2(x_1, \ldots, x_{m+1}) - S(x_{m+1})\Big) \\ &= \frac{t}{N} \sum_{i=2}^{m} \left(V_i'(x_i) - c_{i-1,i}x_{i-1} - c_{i,i+1}x_{i+1}\right) W_{i+1}(x_1, x_i, \ldots, x_{m+1}). \end{aligned} \tag{16.5.1}$$

Let us now consider specific values for the variables x_i in order to turn it into an equation involving only x_1 and x_2. One defines $\left\{\hat{x}_i(x_1, x_2)\right\}_{i=3}^{m+1}$ as functions of the two first variables x_1 and x_2 by

$$c_{i,i+1}\hat{x}_{i+1}(x_1, x_2) = V_i'(\hat{x}_i(x_1, x_2)) - c_{i-1,i}\hat{x}_{i-1}(x_1, x_2). \tag{16.5.2}$$

for $i = 2, \ldots, m$ with the initial conditions $\hat{x}_1(x_1, x_2) = x_1$ and $\hat{x}_2(x_1, x_2) = x_2$.

Choosing $x_i = \hat{x}_i(x_1, x_2)$, reduces the master loop equation to an equation in x_1 and x_2:

$$\widehat{W}_{2;1}(x_1, x_2; x_1) + \frac{t}{N} (c_{1,2}x_2 - Y(x_1))\, \widehat{U}(x_1, x_2) = \widehat{E}(x_1, x_2)$$

where $Y(x) = V_1'(x) - \frac{t}{N} W_0(x)$; the hat means that the functions are considered at the value $x_i = \hat{x}_i(x_1, x_2)$, i.e. $\widehat{f}(x_1, x_2) := f(x_1, x_2, \hat{x}_3, \hat{x}_4, \ldots, \hat{x}_n)$ for an arbitrary function f, and one has defined

$$\widehat{U}(x_1, x_2) = W_2(x_1, x_2, \hat{x}_3, \ldots, \hat{x}_{m+1}) - \frac{N}{t} S(\hat{x}_{m+1}),$$

and

$$\widehat{E}(x_1, x_2) = -\frac{t}{N} \widehat{W}_1(x_1, x_2) + \left(V_1'(x_1) - c_{1,2}x_2\right) \widehat{S}(x_1, x_2).$$

Finally, the leading order in the topological expansion gives

$$\widehat{E}^{(0)}(x_1, x_2) = \left(c_{1,2}x_2 - Y^{(0)}(x_1)\right) \widehat{U}^{(0)}(x_1, x_2) \tag{16.5.3}$$

where one should notice that $\widehat{W}_1(x_1, x_2)$, and thus $\widehat{E}(x_1, x_2)$, is a polynomial in both x_1 and x_2.

Again, this equation is valid for any x_1 and x_2, and, if we choose x_2 such that $c_{1,2}x_2 = Y^{(0)}(x_1)$, we get:

$$H_{chain}(x_1, x_2) := \widehat{E}^{(0)}(x_1, x_2) = 0. \tag{16.5.4}$$

This algebraic equation is the spectral curve of our model.

Study of the spectral curve

The algebraic plane curve $H_{chain}(x_1, x_2) = 0$, can be parameterized by a variable z living on a compact Riemann surface $\mathcal{L}_{chain}$ of some genus g, and two meromorphic functions $x_1(z)$ and $x_2(z)$ on it. Let us study this in greater detail.

The polynomial $H_{chain}(x_1, x_2)$ has degree $1 + \frac{D}{d_1}$ (resp. $d_1 + \frac{D}{d_1 d_2}$) in x_2 (resp. x_1). This means that the embedding of $\mathcal{L}_{chain}$ is composed by $1 + \frac{D}{d_1}$ (resp. $d_1 + \frac{D}{d_1 d_2}$) copies of the Riemann sphere, called x_1-sheets (resp. x_2-sheets), glued by cuts so that the resulting Riemann surface $\mathcal{L}_{chain}$ has genus g. Each copy of the Riemann sphere corresponds to one particular branch of the solutions of the equation $H_{chain}(x_1, x_2) = 0$ in x_2 (resp. x_1).

The Riemann surface $\mathcal{L}_{chain}$ has genus g lower than $D - s$ with $D = s\, d_1 \ldots d_m$.

One can consider all the variables $x_i(p) := \hat{x}_i(x_1(p), x_2(p))$ as meromorphic functions on $\mathcal{L}_{chain}$ as opposed to only x and y in the one-matrix model case. Their negative divisors are given by

$$[x_k(p)]_- = -r_k \infty - s_k \sum_{i=1}^{s} \hat{\lambda}_i$$

where ∞ is the only point of $\mathcal{L}_{chain}$ where x_1 has a simple pole, the $\hat{\lambda}_i$ are the preimages of λ_i under the map $x_{m+1}(p)$, $x_{m+1}(\hat{\lambda}_i) = \lambda_i$, and the degrees r_k and s_k are integers given by $r_1 := 1$, $r_k := d_1 d_2 \ldots d_{k-1}$, $s_{m+1} := 0$, $s_m := 1$, and $s_k := d_{k+1} d_{k+2} \ldots d_m\, s$.

Note that the presence of an external matrix creates as many poles as the number of distinct eigenvalues of this external matrix $M_{m+1} = \Lambda$.[15]

Solution of the loop equations

The procedure used to solve the loop equations in the one-matrix model cannot be generalized in this setup, mainly because the involution $z \to \bar{z}$ is not globally defined on the spectral curve. However, the loop equations can be solved using a detour [Eyn08b]. This resolution proceeds in three steps. One first shows that

[15] The cases of matrix models without external field correspond to a totally degenerate external matrix $\Lambda = c\,\mathrm{Id}$ with only one eigenvalue. There are thus two poles as in the one-matrix model studied earlier.

the loop equations Eq. 16.5.1 have a unique solution admitting a topological expansion. One then finds a solution of these equations:

Lemma 16.5.1

$$E(x(z), y) = -K \text{`}\left\langle \prod_{i=0}^{d_2} (y - V_1'(x(z^i)) + \frac{t}{N}\mathrm{Tr}\frac{1}{x(z^i) - M_1}) \right\rangle_c \text{'} \qquad (16.5.5)$$

where K is a constant and the inverted commas '< . >' mean that every time one encounters a two-point function $\left\langle \mathrm{Tr}\frac{1}{x(z^i)-M_1}\mathrm{Tr}\frac{1}{x(z^j)-M_1}\right\rangle_c$*, one replaces it by* $W_{0;1}^{(0)}(z_i; z_j) := \left\langle \mathrm{Tr}\frac{1}{x(z^i)-M_1}\mathrm{Tr}\frac{1}{x(z^j)-M_1}\right\rangle_c + \frac{1}{(x(z_i)-x(z_j))^2}$.

The matching of the coefficients of the polynomials in y on the left- and right-hand sides of Eq. 16.5.5, and a few algebro-geometrical computations allow us to solve the loop equations to get[16]

Theorem 16.5.1 *The correlation functions of the formal integral of a chain of matrices are recursively obtained by computing residues on $\mathcal{L}_{chain}$:*

$$\omega_{n+1}^{(h)}(z, z_1, \ldots, z_n) = \sum_i \operatorname*{Res}_{z' \to a_i} K(z, z')\big[\omega_{n+2}^{(h-1)}(z', \overline{z}', z_1, \ldots, z_n) + \sum_{j=0}^{h} \sum_{I \subset \{z_1,\ldots,z_n\}}{}' \omega_{|I|+1}^{(j)}(z', I)\omega_{n-|I|+1}^{(h-j)}(\overline{z}', \{z_1, \ldots, z_n\} \setminus I)\big] \qquad (16.5.6)$$

where, as in the preceding section,

$$\omega_n^{(h)}(z_1, \ldots, z_n) = \operatorname*{Res}_{N \to \infty} N^{n+2h-3} \left\langle \prod_{i=1}^{n} \mathrm{Tr}G_1(x_i) \right\rangle_c dx(z_1)\ldots dx(z_n) + \delta_{n,2}\delta_{g,0}\frac{dx(z_1)dx(z_2)}{(x(z_1) - x(z_2))^2}.$$

The two-point function $\omega_2^{(0)}$ is the Bergman kernel of the spectral curve $\mathcal{L}_{chain}$, the recursion kernel is

$$K(z, z') := \frac{dS_{z',\overline{z}'}(z)}{2(y(z') - y(\overline{z}'))dx(z')},$$

x and y are two meromorphic functions on $\mathcal{L}_{chain}$ such that $H_{chain}(x(z), y(z)) = 0$ for any point $z \in \mathcal{L}_{chain}$, and the a_i are the x-branch points, i.e. solutions to $dx(a_i) = 0$.

Remember that H_{chain} was defined in Eq. 16.5.4 by $H_{chain}(x_1, c_{12}x_2) = 0$ for $c_{12}x_2 = Y^{(0)}(x_1)$. Thus the functions x and y can be thought of as continuation to the whole spectral curve of x_1 and $c_{12}x_2$ respectively.

[16] See [Che06b, Eyn08b] for the detailed proof.

The free energy is also obtained by using the same formula as in the one-matrix case:

Theorem 16.5.2 *For $h > 1$ and any primitive Φ of $y\,dx$, one has*

$$F^{(h)} = \frac{1}{2-2h} \sum_i \operatorname*{Res}_{z \to a_i} \Phi(z) \omega_1^{(h)}(z). \tag{16.5.7}$$

Thus the solution of any chain-matrix model with an external field is obtained by applying exactly the same formula as for the solution of the one-matrix model: the only difference is the spectral curve to be applied in this recursion.

16.6 Generalization: topological recursion

We have seen that the loop equation method gives a unique solution for a large family of formal matrix models. The only input of this solution is the spectral curve of the considered model. In [Eyn07], it has been proposed to use Equations 16.5.6 and 16.5.7 to associate infinite sets of correlation functions and free energies to any spectral curve $(\mathcal{L}, x, y)$ where $\mathcal{L}$ is a compact Riemann surface and x and y two functions analytic in some open domain of $\mathcal{L}$.

The free energies and correlation functions built from this recursive procedure show many interesting properties such as invariance under a large set of transformations of the spectral curve, special geometry relations, modular invariance, or integrable properties. In particular, it is a very convenient tool to study critical regimes and to get the universal properties of the matrix integrals described in Chapter 6 of this volume. It is also very useful to compare different matrix integrals. This procedure eventually proved to be efficient in the resolution of many problems of enumerative geometry or statistical physics such as string theory, Gromov-Witten invariants theory, Hurwitz theory, or exclusion processes such as TASEP or PASEP. Most of the results proposed by this approach are still conjecture up to now but the numerous checks passed so far tend to indicate that this generalization of the loop equation method is a very promising field.[17]

The inductive procedure presented in this chapter only allows us to compute one particular set of observables of multi-matrix models. It does not compute correlation functions involving more than one type of matrix inside the same trace. These more complicated objects are very important for their application to quantum gravity or conformal field theories where they correspond to the insertion of boundary operators [Sta93]. In the two-matrix model, the loop equation method allowed us to compute the topological expansion of any of these operators [Eyn08a]. In the chain of matrices case, only a few of them were

[17] For a review on this subject, see [Eyn09] and references therein.

computed in their large N limit [Eyn03]. The computation of any observable of the chain of matrices is still an open problem which is very likely to be solved by the use of loop equations.

Acknowledgements

It is a pleasure to thank Bertrand Eynard who developed most of the material exposed in this chapter and patiently taught me all I know about these topics.

References

[Ake96a] G. Akemann, Nucl. Phys. **B482** (1996) 403 [arXiv:hep-th/9606004]
[Ake96b] G. Akemann and J. Ambjørn, J. Phys. **A29** (1996) 555 [arXiv:cond-mat/9606129]
[Amb93] J. Ambjørn, L. Chekhov, C. F. Kristjansen and Yu. Makeenko, Nucl. Phys. **B404** (1993) 127; Erratum ibid. **B449** (1995) 681 [arXiv:hep-th/9302014]
[Ber07] M. Bertola, Analysis and Mathematical Physics **1** (2011) 167–211 [arXiv:0705.3062 [nlin.SI]].
[Bon00] G. Bonnet, F. David, and B. Eynard, J. Phys. **A33** (2000) 6739 [arXiv:cond-mat/0003324]
[Bre78] E. Brézin, C. Itzykson, G. Parisi, and J.B. Zuber, Comm. Math. Phys. **59**, 35 (1978)
[Bre99] E. Brézin and N. Deo, Phys. Rev. **E59** (1999) 3901 [arXiv:cond-mat/9805096]
[Che06a] L. Chekhov and B. Eynard, J. High Energy Phys. **03** (2006) 014 [arXiv:hep-th/0504116]
[Che06b] L. Chekhov, B. Eynard and N. Orantin, J. High Energy Phys. **12** (2006) 053 [arXiv:math-ph/0603003]
[Dys62] F.J. Dyson, J. Math. Phys. **3** (1962) 1191
[Eyn04] B. Eynard, J. High Energy Phys. **11** (2004) 31 [arXiv:hep-th/0407261]
[Eyn03] B. Eynard, J. High Energy Phys. **11** (2003) 018 [arXiv:hep-th/0309036]
[Eyn08b] B. Eynard, A.P. Ferrer, JHEP **07** (2009) 096 [arXiv:0805.1368 [math-ph]].
[Eyn06] B. Eynard, Random Matrices, Random Processes and Integrable Systems, CRM Series in Mathematical Physics 2011, pp 415–442 [arXiv:math-ph/0611087].
[Eyn07] B. Eynard and N. Orantin, Communications in Number Theory and Physics **1.2** (2007) 347 [arXiv:math-ph/0702045]
[Eyn08a] B. Eynard and N. Orantin, J. High Energy Phys. **06** (2008) 037 [arXiv:0710.0223]
[Eyn09] B. Eynard and N. Orantin, J. Phys. **A42** (2009) 293001 [arXiv:0811.3531]
[Far92] H.M. Farkas and I. Kra, *Riemann surfaces*, Springer Verlag, 2nd edition, 1992
[Fay73] J.D. Fay, *Theta functions on Riemann surfaces*, Springer Verlag, 1973
[Har57] Harish-Chandra, Amer. J. Math. **79** (1957) 87
[Itz80] C. Itzykson and J.B. Zuber, J. Math. Phys. **21** (1980) 411
[Kan98] E. Kanzieper and V. Freilikher, Phys. Rev. **E57** (1998) 6604 [arXiv:cond-mat/9709309]
[Meh04] M.L. Mehta, *Random Matrices*, Academic Press, 3rd Edition, London 2004
[Mig83] A.A. Migdal, Phys. Rep. **102**(1983) 199
[Sta93] M. Staudacher, Phys. Lett. **B305** (1993) 332
[tHo74] G. 't Hooft, Nucl. Phys. **B72** (1974) 461
[Tut62] W.T. Tutte, Can. J. Math. **14** (1962) 21

·17·

Unitary integrals and related matrix models

A. Morozov

Abstract

A concise review of the basic properties of unitary matrix integrals is given. They are studied with the help of the following three matrix models: the ordinary unitary model, the Brézin–Gross–Witten model and the Harish-Chandra-Itzykson-Zuber model. Special attention is paid to the tricky sides of the story, from the de Wit-'t Hooft anomaly in unitary integrals to the problem of correlators with Itzykson-Zuber measure. A technical tool emphasized is the method of character expansions. The subject of unitary integrals remains highly under-investigated and many results are expected in this field when it attracts sufficient attention.

17.1 Introduction

Integrals over unitary and other non-Hermitian matrices have attracted far less attention during the entire history of matrix model theory. This is unjust, both because they are no less important in applications than Hermitian models, and because they can effectively be studied by the same methods and are naturally included into the unifying M-theory of matrix models [Ale07]. Still, in the recent reincarnation of matrix model theory non-Hermitian models are once again under-investigated, and neither their non-trivial phase structure *à la* [Ale04], nor string-field-theory-like reformulation *à la* [Eyn04] are explicitly analysed – despite the fact that these and other subjects are clearly within reach of the newly developed approaches.

The present text is also too short to address these important issues. Instead it concentrates on the *complement* of these *universal* subjects: on methods and results *specific* to one particular model – the most important of all non-Hermitian models – Gaussian *unitary* ensembles (for a review of complex, symplectic, and real asymmetric matrix models see Chapter 18). The importance of unitary integrals is obvious from the Yang-Mills-theory perspective: they describe angular – colour – degrees of freedom, the complement of diagonal – colourless or hadronic – components, which are adequately captured by the Hermitian eigenvalue models [Mor94], and by their Generalized Kontsevich Model (GKM) 'duals' [Kon92, Kha92, Mor94]. As already mentioned, this does not mean that various kinds of eigenvalue techniques – from orthogonal polynomials to Virasoro constraints and their iterative or quantum-field-theory-style solutions – can not be applied to unitary integrals – they can and they are pretty effective, as usual. This means that the main *questions* of interest here are of another type, concerning the correlators of *non-diagonal* matrix elements rather than those of traces or determinants.

This paper is no more than a brief introduction into these aspects of unitary integrals, for its fuller version see [Mor09a]. Our presentation is mostly about the following three basic unitary matrix models:

- the original unitary model [Bow91], describing correlators of traces of unitary matrices by standard matrix-model methods, thus we touch on it only briefly;
- the Brézin–Gross–Witten (BGW) model [Bre80], describing the correlators of arbitrary matrix elements, which was made – perhaps, surprisingly at the time – a part of the GKM theory in [Kha92, Mir96], and was finally incorporated in [Ale09] into M-theory of matrix models [Ale07];
- the Harish-Chandra-Itzykson-Zuber (HCIZ) model [Itz80], a simplified single-plaquette version of the Kazakov-Migdal theory [Kaz93], describing arbitrary correlators with non-trivial but exactly solvable weight (Itzykson-Zuber (IZ) action), an old and difficult subject [Cas92, Kog93, Mor92a, Sha93, Kha95], that has attracted some fresh attention recently [Ber03, Pra09].

A lot of new – and much more profound – results are expected in the study of unitary integrals in the foreseeable future, which – unlike that of Hermitian integrals – is in a rather early stage of development. This makes the presentation of this paper essentially incomplete and temporal – it is no more than a preliminary introduction. Still, we have tried to concentrate on facts and formulae, which have a good chance of remaining in the core of the subject when it becomes more complete and self-contained. Instead, some currently important – but temporal – results, i.e. those which will supposedly be significantly improved and represented in more adequate terms, are discussed in less detail and even unjustly ignored because of space limitations.

A number of very significant results are available in the literature on closely related orthogonal and symplectic models: group integrals [Bre01, Guh03, Ber03, Pra09], character expansions [Bal02], and large-N limit [Wei80, Zub08, Col09]. Interesting applications are further discussed in other chapters of this book: $2d$-quantum gravity (Chapter 30), random folding (Chapter 42), and finance and economics (Chapter 40).

17.2 Unitary integrals and the Brézin–Gross–Witten model

17.2.1 The measure

Unitary integrals are those over unitary $N \times N$ matrices U, $UU^\dagger = U^\dagger U = I$ with invariant Haar measure $[dU]$. In contrast to the case of Hermitian matrices $H = H^\dagger$, where $dH = \wedge_{i,j} dH_{ij}$, the Haar measure $[dU]$ is non-trivial and non-linear in U. If unitary matrices are expressed through Hermitian ones, one can express $[dU]$ through dH. For example, in the vicinity of diagonal H

$$U = e^{iH} \Longrightarrow ||\delta U||^2 = \mathrm{tr}\,(\delta U^\dagger \delta U)\big|_{H\text{ diagonal}} = \mathrm{tr}\,(\delta H)^2 \Longrightarrow [dU] = dH,$$

$$U = \frac{I+iH}{I-iH} \Rightarrow \underset{H\text{ diag}}{||\delta U||^2 =} 4\,\mathrm{tr}\left(\frac{I}{I+H^2}\,\delta H \frac{I}{I+H^2}\,\delta H\right) \overset{[\text{Bow91}]}{\Rightarrow} [dU] \sim \frac{dH}{\det^N(1+H^2)}$$

and so on. The restriction to diagonal H (but not δH) is enough because of the $U(N)$ invariance of the measure. An alternative representation is given through a generic complex matrix M [Kha95, Mir96, Zir98, Ber03, Pra09]:

$$[dU] = \delta(MM^\dagger - I)d^2M = d^2M \int dS \exp\left(i\,\mathrm{tr}\,S(MM^\dagger - I)\right) \quad (17.2.1)$$

where S is an auxiliary Hermitian matrix, used as a Lagrange multiplier to impose the unitarity constraint on M. Following [Zir98], in QCD applications the Hubbard-Stratonovich-style reformulation (17.2.1) (see also Chapter 7, Section 7.3) is often called the color-flavour transform. Below we often absorb i into S, which makes it anti-Hermitian, i.e. it changes the contour of the S-integration. More than that, sometimes one can absorb a non-trivial *matrix* factor into S, which completely breaks its Hermiticity; still for the integral this is often no more than an innocent change of integration contour. In what follows we normalize the measure $[dU]$ so that $\int [dU] = 1$, i.e. we divide $[dU]$ by the volume of the unitary group.

17.2.2 Elementary unitary correlators in fundamental representation

With this normalization agreement and the symmetry properties one immediately gets:

$$\langle U_{ij} U^\dagger_{kl} \rangle = \int U_{ij} U^\dagger_{kl} [dU] = \frac{1}{N}\delta_{il}\delta_{jk} \quad (17.2.2)$$

$$\int U_{ij} U_{i'j'} U^\dagger_{kl} U^\dagger_{k'l'} [dU] =$$
$$\frac{N^2\left(\delta_{il}\delta_{i'l'}\delta_{jk}\delta_{j'k'} + \delta_{il'}\delta_{i'l}\delta_{jk'}\delta_{j'k}\right) - N\left(\delta_{il'}\delta_{i'l}\delta_{jk}\delta_{j'k'} + \delta_{il}\delta_{i'l'}\delta_{jk'}\delta_{j'k}\right)}{N^2(N^2-1)}. \quad (17.2.3)$$

It is clear from these examples that the correlators with more entries become increasingly complicated and one needs a systematic approach to evaluate them. This approach is provided by the theory of the BGW model [Mir96].

Before proceeding to this subject in Section 17.2.4, we note that these correlators are significantly simplified if we contract indices, i.e. consider the correlators of *traces* instead of particular matrix elements. Indeed, from Eq. (17.2.3)

$$\int \mathrm{tr}\, U^2 \mathrm{tr}\,(U^\dagger)^2\, [dU] = 2 - \delta_{N1},$$
$$\int \mathrm{tr}\, U^2 (\mathrm{tr}\, U^\dagger)^2\, [dU] = \delta_{N1},$$
$$\int (\mathrm{tr}\, U)^2 (\mathrm{tr}\, U^\dagger)^2\, [dU] = 2 - \delta_{N1} \tag{17.2.4}$$

and so on (the case of $N = 1$ in these formulae requires additional consideration, see below). Such *colourless* correlators are studied by the ordinary unitary-matrix model [Bow91].

Note that unitary correlators in (17.2.3) possess a non-trivial N-dependence, with spurious poles at some integer values of N. This phenomenon is known as the de Wit-'t Hooft (DWH) anomaly [DeW77]. In correlators of colourless quantities (17.2.4) these poles turn into peculiar non-singular corrections at the same values of N.

17.2.3 On the theory of the unitary model [Bow91]

The partition function of the unitary matrix model is the generating function of correlators of colourless trace operators:

$$Z_U(t) = \int \exp\left(\sum_{k=-\infty}^{\infty} t_k \mathrm{tr}\, U^k \right) [DU] \tag{17.2.5}$$

This was also considered in [Mar90]. Note that $U^\dagger = U^{-1}$ and conjugate U-matrices are represented by negative powers of U.

In matrix model theory partition functions are best defined from a complete set of the Ward identities, also known as *Virasoro constraints*, because they are often made from differential operators, which form Virasoro or some other closely related loop algebras. Such a description is independent of a particular choice of integration contours and therefore can be used to provide various analytical continuations of partition functions. For details of this approach and further references see [Fuk91, Mor94, Ale04, Eyn04, Ale07]. In the case of $Z_U(t)$ the Virasoro constraints have the form [Bow91]

$$\left\{\sum_{k=-\infty}^{\infty} kt_k\left(\frac{\partial}{\partial t_{k+n}}-\frac{\partial}{\partial t_{k-n}}\right)+\sum_{1\le k\le n}\left(\frac{\partial^2}{\partial t_k\partial t_{n-k}}+\frac{\partial^2}{\partial t_{-k}\partial t_{k-n}}\right)\right\} Z_U(t)=0,$$

$$\left\{\sum_{k=-\infty}^{\infty} kt_k\left(\frac{\partial}{\partial t_{k+n}}+\frac{\partial}{\partial t_{k-n}}\right)+\sum_{1\le k\le n}\left(\frac{\partial^2}{\partial t_k\partial t_{n-k}}-\frac{\partial^2}{\partial t_{-k}\partial t_{k-n}}\right)\right\} Z_U(t)=0,$$

and in the first line $n \geq 1$, while in the second line $n \geq 0$. The standard *discrete-Virasoro* generators

$$\hat{L}_n = \sum_{k=-\infty}^{\infty} kt_k \frac{\partial}{\partial t_{k+n}} + \sum_{1\le k\le n} \frac{\partial^2}{\partial t_k \partial t_{n-k}} \tag{17.2.6}$$

appear as building blocks of these equations. The use of these Virasoro relations is a little bit unusual: there is no shift of any time variable, and the formal series solution is given directly in terms of ts and not their ratio, with some background values of time-variables, which label the choice of phase, say, in the Hermitian model [Ale04]. The analogue of the $\hat{L}_0$ constraint states that

$$\sum_{k=-\infty}^{\infty} kt_k \frac{\partial Z_U(t)}{\partial t_k} = 0 \tag{17.2.7}$$

and requests that monomial items in $Z_U(t)$ should have the form $\prod_k t_k^{n_k}$ with $\sum_{k\geq 1} kn_k = \sum_{k\geq 1} kn_{-k}$, i.e. $t_k t_{-k}$, $t_k t_l t_{-k-l}$ are allowed, and so on, but not t_k or $t_k t_{-k-l}$ with $l \neq 0$. With this selection rule the first terms of the series are

$$Z_U(t) = e^{Nt_0}\Big(1 + a_{1|1}t_1t_{-1} + a_{2|2}t_2t_{-2} + a_{11|2}(t_1^2t_{-2} + t_{-1}^2t_2) + \dots\Big) \tag{17.2.8}$$

where the coefficients $a_{Y|\bar{Y}}$, labelled by pairs of Young diagrams, still need to be found. They describe the correlators (17.2.4), and are dictated by the lowest Virasoro constraints with $n = \pm 1, \pm 2$, but in a somewhat non-trivial manner:

$$0 = e^{-Nt_0}\left(\sum_{k=-\infty}^{\infty} kt_k \frac{\partial}{\partial t_{k+1}} + \frac{\partial^2}{\partial t_0 \partial t_1}\right) Z_U = t_{-1}\Big(-N + Na_{1|1}\Big) +$$
$$+t_1t_{-2}\Big(-2a_{1|1} + a_{2|2} + 2Na_{2|11}\Big) + t_1t_{-1}^2\Big(-Na_{1|1} + a_{2|11} + 2Na_{11|11}\Big) + \dots$$

Similarly,

$$\begin{aligned} 0 &= e^{-Nt_0}\left(\sum_{k=-\infty}^{\infty} kt_k \frac{\partial}{\partial t_{k+2}} + \frac{\partial^2}{\partial t_0 \partial t_2} + \frac{\partial^2}{\partial t_1^2}\right) Z_U(t) \\ &= t_{-2}\Big(-2N + Na_{2|2} + 2a_{2|11}\Big) + t_{-1}^2\Big(-a_{1|1} + Na_{2|11} + 2a_{11|11}\Big) + \dots \end{aligned} \tag{17.2.9}$$

Virasoro constraints require that each bracket on the right-hand side vanishes, so that

$$
\begin{array}{rcl}
& & \alpha_{1|1} = 1, \\
(N^2-1)\alpha_{2|2} = 2(N^2-\alpha_{1|1}) = 2(N^2-1) & \Rightarrow & \alpha_{2|2} = 2 - s\delta_{N,1}, \\
(N^2-1)\alpha_{2|11} = 0 & \Rightarrow & \alpha_{2|11} = s\delta_{N,1} \\
2(N^2-1)\alpha_{11|11} = (N^2-1)\alpha_{1|1} = N^2-1 & \Rightarrow & \alpha_{11|11} = \frac{1}{4}(2-s\delta_{N,1}) \\
& \ldots &
\end{array}
$$

Thus Virasoro constraints do not define coefficients $\alpha_{2|2}$, $\alpha_{2|11}$ and $\alpha_{11|11}$ unambiguously – at $N = 1$. In fact there is a single ambiguous parameter s in these three coefficients, and it can be fixed by direct evaluation of the correlators (17.2.4) at $N = 1$. Similarly, coefficients with higher subscript label $l = \sum_{k\geq 0} kn_k > 2$ are not fully defined by Virasoro constraints at $N = 1, \ldots, l$.

17.2.4 On the theory of the BGW model [Mir96]

The partition function of the BGW matrix model [Bre80] is the generating function of all unitary correlators:

$$
\begin{aligned}
Z_{BGW}(t) &= \int e^{\operatorname{tr}(J^\dagger U + J U^\dagger)}[dU] = \iint e^{iS(MM^\dagger - I)} e^{\operatorname{tr} J^\dagger M + J M^\dagger} d^2M dS \\
&\overset{(17.2.1)}{=} \int \frac{dS}{\det S^N} e^{\operatorname{tr}(JJ^\dagger/S) - \operatorname{tr} S} \overset{S\to 1/S}{=} \int \frac{dS}{\det S^N} e^{\operatorname{tr}(JJ^\dagger S) - \operatorname{tr} 1/S}
\end{aligned} \tag{17.2.10}
$$

Obviously, Z_{BGW} depends only on the eigenvalues of the matrix $JJ^\dagger$, i.e. on the time variables $t_k = \frac{1}{k}\operatorname{tr}(JJ^\dagger)^k$.

The model has two essentially different phases, named Kontsevich and character phase in [Mir96]. Here we consider only the character phase, when Z_{BGW} is a formal series in *positive* powers of t_k (in the Kontsevich phase the expansion is in negative powers of the external fields J and $J^\dagger$). The Virasoro constraints in the character phase can be derived in the usual way: as Ward identities *à la* [Fuk91], but in this particular case they can alternatively be considered as a direct corollary of the identity $UU^\dagger = I$ for unitary matrices, expressed in terms of t-variables. This identity implies that

$$
0 = \left(\frac{\partial}{\partial J^*}\frac{\partial}{\partial J} - I\right) Z_{BGW}(t) = \sum_{n=1}^{\infty} (JJ^\dagger)^{n-1} \hat{L}_n Z_{BGW}(t) - Z_{BGW}(t) \tag{17.2.11}
$$

with * denoting complex conjugation, i.e.

$$
\hat{L}_n Z_{BGW}(t) \overset{(17.2.6)}{=} \left(\sum_{k=1}^{\infty} kt_k \frac{\partial}{\partial t_{k+n}} + \sum_{\substack{a+b=n \\ a,b\geq 0}} \frac{\partial^2}{\partial t_a \partial t_b} \right) Z_{BGW}(t) = \delta_{n,1} \tag{17.2.12}
$$

are the usual *discrete-Virasoro* constraints of [Ger91], familiar from the theory of the Hermitian model, with $n \geq 1$ instead of $n \geq -1$ and with a modified right-hand side.

Substituting a formal-series ansatz for $Z_{BGW}(t)$ into these Virasoro constraints, one can iteratively reconstruct any particular coefficient, see [Mir96]:

$$Z_{BGW}(t) = 1 + \sum_{M\geq 1}\left(\sum_{k_1\geq\ldots\geq k_M>0} c_N(\{k_a\})\frac{k_1 t_{k_1}\ldots k_M t_{k_M}}{(k_1+\ldots+k_M)!}\right) \tag{17.2.13}$$

where

$$\begin{gathered}
c_N(\{k_a\}) = \hat{c}_N(\{k_a\}) \prod_{l=0}^{k_1+\ldots+k_M-1} \frac{1}{N^2-l^2} \\
\hat{c}_N(1) = N, \\
\hat{c}_N(2) = -N, \quad \hat{c}_N(1,1) = N^2, \\
\hat{c}_N(3) = 4N, \quad \hat{c}_N(2,1) = -3N^2, \quad \hat{c}_N(1,1,1) = N(N^2-2), \\
\hat{c}_N(4) = -30N, \quad \hat{c}_N(3,1) = 8(2N^2-1), \quad \hat{c}_N(2,2) = 3(N^2+6), \\
\hat{c}_N(2,1,1) = -6N(N^2-4), \quad \hat{c}_N(1,1,1,1) = N^4-8N^2+6, \\
\ldots
\end{gathered} \tag{17.2.14}$$

The first two lines of this list reproduce Eqs. (17.2.2) and (17.2.3).

17.2.5 De Wit-'t Hooft anomaly [DeW77, Mir96]

A remarkable feature of Eq. (17.2.4) is the occurrence of poles at integer values of the matrix size N. It is worth noting that this is a property of the character phase: nothing like this happens in the Kontsevich phase of the BGW model – as usual, if properly defined, the Kontsevich partition function is actually independent of N [Kha92, Mir96]. As we saw in Section 17.2.3, these poles are associated with the *ambiguity* of a solution to the Virasoro equations in the ordinary unitary matrix model.

In fact, DWH poles do not show up in correlators: numerators vanish at the same time as denominators, and the singularity can be resolved by the L'Hospital rule. However, some care is needed. For example, in (17.2.3) one can consider

$$\int U_{11}^2 (U_{11}^\dagger)^2 [dU] = \frac{2(N^2-N)}{N^2(N^2-1)} = \frac{2}{N(N+1)} \xrightarrow{N=1} 1$$

in accordance with (17.2.4). However, if one first takes traces in (17.2.3), and then puts $N=1$, then one does not obtain (17.2.4):

$$\int \mathrm{tr}\, U^2 \mathrm{tr}\,(U^\dagger)^2\,[dU] \overset{(17.2.3)}{=} \frac{2(N^2\cdot N^2 - N\cdot N)}{N^2(N^2-1)} = 2 \neq 2 - \delta_{N1}$$

An anomaly of this kind is familiar from experience with dimensional regularizations, where taking traces and putting the dimension equal to a particular value are also non-commuting operations. For more details about the DWH anomaly and its possible interpretation see [Bar80].

17.2.6 BGW model in M-theory of matrix models [Ale09]

As explained in [Ale07] (see also [Giv00] and [Ale04] for preliminary results), the partition function of every matrix model in every phase can be decomposed into elementary constituents, belonging to the Kontsevich family of [Kha92]. The most important building block is the original cubic Kontsevich tau-function τ_K, while the BGW tau-function, made from $Z_{BGW}(t)$ in the Kontsevich phase, appears to be the next most important one (for example, it arises, along with τ_K, in the decomposition of the complex matrix model). Both τ_K and τ_{BGW} satisfy the same *continuous-Virasoro* constraints of [Fuk91], only with $n \geq -1$ in the case of τ_K and with $n \geq 0$ in the case of τ_{BGW}. Another difference is that instead of a *shift* in the second time variable in the case of τ_K, for τ_{BGW} the first time variable is *shifted*. This allows a smaller set of constraints to unambiguously define the partition function. See [Ale09] for many more details of this important construction.

17.2.7 Other correlators and representations

A slight modification of the BGW model is to make the power of $\det S$ in Eq. (17.2.10) an additional variable. In this way we get the Leutwyler-Smilga integral [Leu92],

$$Z_{BGW}(\nu; t) = \int (\det U)^{\nu} e^{\mathrm{tr}(J^{\dagger}U + JU^{\dagger})}[dU] \tag{17.2.15}$$

which plays a role in applications, see Chapter 32, Section 32.2.4, and is widely discussed in the literature.

Another modification involves the change of representation: in (17.2.10) the partition function is defined as generating the integrals over unitary matrices in the fundamental representation of dimension N of $GL(N)$, but one can instead consider an arbitrary representation R of dimension $D_R(N)$. All correlators will, of course, be related to those in the fundamental representation, but in a somewhat non-trivial way, and they will be different. For example, instead of (17.2.2) in a generic representation we have

$$\int \mathcal{U}_{ab}\mathcal{U}^{\dagger}_{cd}\,[dU] = \frac{\delta_{ad}\delta_{bc}}{D_R(N)}. \tag{17.2.16}$$

Moreover, different representations are orthogonal in the sense that

$$\int \mathcal{U}_R\mathcal{U}^{\dagger}_{R'}\,[dU] \sim \delta_{RR'} \tag{17.2.17}$$

An important way to handle this kind of generalization is through the comprehensive theory of character expansions, see Section 3 of [Mor09a] and references therein. We will come back to this in Section 17.3.2.

17.3 Theory of the Harish–Chandra–Itzykson–Zuber integral

A very important application of the character expansion appears in the Harish–Chandra–Itzykson–Zuber integral. [Har57, Itz80]. As mentioned in the introduction, it is the simplified version – or the main building block – of the Kazakov-Migdal model of lattice gluodynamics and, actually, of many other matrix-model-based approaches to non-perturbative dynamics of gauge theories with propagating (particle-like) degrees of freedom. It is also a non-trivial example of a group integral, which satisfies the requirements of the Duistermaat–Heckman theorem [Sem76, Sza00] and can be treated exactly by methods now known as the *localization technique* (with Nekrasov's instanton calculus in Seiberg-Witten theory as one of its remarkable achievements).

HCIZ integrals can be handled by a variety of methods, starting from the Ward identities of [Itz80]. We present here the character-based approach [Kha95, Bal00] and a closely related GKM-based technique of [Kha92, Kog93, Mir96], recently revived in [Ber03, Pra09] and applied to the old (and still unsolved) problem of arbitrary correlators with the IZ weight [Mor92a, Sha93].

17.3.1 HCIZ integral and duality formula

The HCIZ integral is a unitary integral with non-trivial weight of the special form:

$$J_{HCIZ}(X,Y) = \int e^{\operatorname{tr} XUYU^{\dagger}}[dU]. \tag{17.3.1}$$

It depends on the eigenvalues $\{x_i\}$ and $\{y_j\}$ of two Hermitian $N \times N$ matrices X and Y through the celebrated determinant formula [Itz80]

$$J_{HCIZ}(X,Y) \sim \frac{\det_{ij}[\exp(x_i y_j)]}{\Delta(X)\Delta(Y)} = \frac{1}{\Delta(X)\Delta(Y)} \sum_{P \in S_N} (-)^P \exp\left(x_i y_{P(i)}\right), \tag{17.3.2}$$

where $\Delta(X) = \prod_{i<j}^{N}(x_i - x_j) = (-)^{N(N-1)/2}\det_{ij}[x_i^{j-1}]$ and $\Delta(Y)$ are the two Vandermonde determinants, and we sum over all P in the group of permutations S_N of N elements.

The standard way to derive Eq. (17.3.2) is to observe that J_{HCIZ} depends only on the eigenvalues of X and Y and substitute a function of eigenvalues into the obvious Ward identities: for all $k \geq 0$

$$\left\{ \mathrm{tr}\left(\frac{\partial}{\partial X^T}\right)^k - \mathrm{tr}\, Y^k \right\} J_{HCIZ}(X, Y) = 0, \quad \left\{ \mathrm{tr}\left(\frac{\partial}{\partial Y^T}\right)^k - \mathrm{tr}\, X^k \right\} J_{HCIZ}(X, Y) = 0,$$

which reduces the problem to Calogero-Dunkl-like equations. The derivatives are defined as

$$\mathrm{tr}\left(\frac{\partial}{\partial X^T}\right)^2 = \sum_i \frac{\partial^2}{\partial x_i^2} + \sum_{i \neq j} \frac{1}{x_i - x_j}\left(\frac{\partial}{\partial x_i} - \frac{\partial}{\partial x_j}\right) \tag{17.3.3}$$

when acting on a function of eigenvalues. One can then check that (17.3.2) is indeed a solution. This method is rather tedious and the answer is in no way obvious, at least for people unfamiliar with Calogero-Dunkl equations and the associated theory of 'zonal spherical functions'. It has non-trivial generalizations, a recent one [Mor09b] being the theory of so-called *integral discriminants* – an important branch of *non-linear algebra* [Gel94, Dol07].

A direct way from Eq. (17.3.1) to the answer (17.3.2) is provided by the Duistermaat–Heckman approach. This claims that an integral invariant under the action of a compact group that satisfies an appropriate relation between the classical action and quantum measure – conditions, which are trivially satisfied by the group-theoretical integral (17.3.1) – is exactly given by its *full* quasiclassical approximation, the sum over *all* extrema of the classical action:

$$\int d\phi e^{-S(\phi)} \sim \sum_{\phi_0:\ \delta S(\phi_0)=0} \left(\det \frac{\partial^2 S(\phi_0)}{\partial \phi^2}\right)^{-1/2} e^{-S(\phi_0)}\,. \tag{17.3.4}$$

As explained in [Mor95, Ber03, Pra09], the equations of motion for the action $\mathrm{tr}\, XUYU^\dagger$,

$$\left[X, UYU^\dagger\right] = 0, \tag{17.3.5}$$

have any permutation matrix $P \in S_N$ as their solution: $U = P$, after performing a unitary transformation which makes X and Y diagonal (more solutions arise when some eigenvalues of X or Y coincide, but this does not affect the answer). Then the Duistermaat–Heckman theorem (17.3.4) implies that

$$J_{HCIZ}(X, Y) \sim \sum_{P \in S_N} \frac{(-)^P}{\Delta(X)\Delta(Y)} \exp\left(\sum_i x_i y_{P(i)}\right) \tag{17.3.6}$$

where the Vandermonde determinants arise from the pre-exponential factor and the sign factor $(-)^P$ naturally arises from the square root and distinguishes the contributions of minima and maxima to the quasiclassical sum. The Duistermaat–Heckman method, also known as the *localization* technique, has been applied to a variety of applications since it was first discovered in the study of HCIZ integrals.

Of certain importance also is the following representation of (17.3.1) as a Hermitian-matrix integral [Kog93, Mir96, Ber03]. Immediately from the definition we get:

$$J_{HCIZ}(X, Y) \overset{(17.3.1)}{=} \int e^{\mathrm{tr}\, XUYU^\dagger}[dU] \overset{(17.2.1)}{=} \int d^2 M dS e^{\mathrm{tr}\, S(I-MM^\dagger)+\mathrm{tr}\, XMYM^\dagger}$$
$$= \int \frac{e^{\mathrm{tr}\, S} dS}{\det_{N^2\times N^2}(S\otimes I - X\otimes Y)} \tag{17.3.7}$$

One can easily convert to the basis where Y is diagonal, $Y = \mathrm{diag}\{y_k\}$. Then

$$J_{HCIZ}(X, Y) \sim \int \frac{e^{\mathrm{tr}\, S} dS}{\prod_k \det(S - y_k X)} \longrightarrow \int \frac{e^{\mathrm{tr}\, SX} dS}{\prod_k \det(S - y_k I)}$$
$$\sim \int J_{HCIZ}(X, S) \frac{\prod_{i\neq j}(s_i - s_j)\prod_i ds_i}{\prod_{i,k}(s_i - y_k)} \tag{17.3.8}$$

where determinants and traces are now over $N \times N$ matrices. The step denoted by an arrow results from the change of variables $S \to SX$ (which implies a change of integration contour over S). The last step is a transition to eigenvalues of S, with the angular integration again given by the same HCIZ integral. According to [Ber03, Pra09] this duality relation provides an effective recursion in the size N of the unitary matrix.

Of certain interest also is a different representation of $J_{HCIZ}(X, Y)$, the one in terms of *characters*. In order to describe these kinds of formulae in Section 17.3.3 we first recall some basic definitions from the theory of characters.

17.3.2 Basics of character calculus

A character depends on the representation R and on the conjugation class of a group element, which can be represented through time-variables $t_1, t_2, \ldots$ or through an auxiliary matrix (Miwa) variable X, accordingly we use two notations $\chi_R(t) = \chi_R[X]$.

Representations R of $GL(\infty)$ are labelled by Young diagrams, i.e. by ordered integer partitions $R: \lambda_1 \geq \lambda_2 \geq \ldots \geq 0$, of $|R|$ – the number of boxes in the diagram – such that $\sum_j \lambda_j = |R|$ and all λ_j are integers.

Time variables form an infinite chain, which – if one likes – can be considered as eigenvalues of the infinite matrix from $GL(\infty)$. Miwa variables define an N-dimensional subspace in the space of time variables, parameterized by the eigenvalues x_i of an $N \times N$ matrix X, in the following way: $t_k = \frac{1}{k}\mathrm{tr}\, X^k = \frac{1}{k}\sum_i x_i^k$. Here and below tr without additional indices denotes a trace of an $N \times N$ dimensional matrix, i.e. a trace in the first fundamental representation of $GL(N)$, associated with a single-box Young diagram $R = 1 = \Box$. The same matrix X be can converted into an arbitrary representation R of $SL(N)$, then we denote it by $\mathcal{X}^{(R)}$. While X be can actually considered as independent of

N, i.e. defined as an element of $GL(\infty)$, with N just the number of non-vanishing eigenvalues, its conversion into $\mathcal{X}^{(R)}$ actually depends on N, thus one should be a little more accurate with formulae, which involve $\mathcal{X}^{(R)}$. However, for many purposes the N-dependence enters only through dimensions $D_R(N)$ of representations R of $SL(N)$, see Eq. (17.3.16) below. In other words, while $\mathcal{X}^{(R)}$ and $D_R(N)$ essentially depend on N, if formulae are written in terms of these objects, N often does not show up. Accordingly we suppress label N in $\mathcal{X}^{(R)}$ and D_R.

There are three important definitions of characters:

- the first Weyl determinant formula

$$\chi_R(t) = \det_{ij}\Big(s_{\lambda_i - i + j}(t)\Big), \tag{17.3.9}$$

expressing $\chi_R(t)$ through a determinant of Schur polynomials, which are defined by

$$\exp\left(\sum_k t_k z^k\right) = \sum_k s_k(t) z^k \tag{17.3.10}$$

- the second Weyl determinant formula

$$\chi_R(t) = \chi_R[X] = \frac{\det_{ij} x_i^{\lambda_j + N - j}}{\Delta(X)} = \frac{\det_{ij} x_i^{\lambda_j - j}}{\det_{ij} x_i^{-j}}, \tag{17.3.11}$$

where the Vandermonde determinant satisfies $\Delta(X) = \det_{ij} x_i^{N-j}$,

- the trace formula

$$\chi_R(t) = \chi_R[X] = \mathrm{Tr}_R\, \mathcal{X}^{(R)}. \tag{17.3.12}$$

Like Janus, these characters have two faces. They are actually characters of two very different universal groups: $GL(\infty)$ and $S(\infty)$, where S_N is a group of permutations of N elements. An intimate relation between these two groups is already reflected in the definition of Miwa variables, expressing time variables as symmetric functions of X-eigenvalues.

An important role in character calculus is played by two additional functions: the dimension of representations of $SL(N)$ and S_N, associated with the Young diagram R:

$$D_R = \dim_R\Big(GL(N)\Big) = \chi_R\left(t_k = \frac{N}{k}\right) = \chi_R[I], \quad N = |R| \tag{17.3.13}$$

and

$$d_R = \frac{1}{N!}\dim_R\Big(S(N)\Big) = \chi_R\Big(t_k = \delta_{k1}\Big), \quad N = |R| \tag{17.3.14}$$

The parameter d_R is given by the hook formula:

$$d_R = \prod_{\substack{\text{over all boxes of} \\ \text{Young diagrams R}}} \frac{1}{\text{hook length}} = \det_{1 \le i,j \le |R|} \frac{1}{(\lambda_j + i - j)!} \tag{17.3.15}$$

and the ratio

$$\frac{D_R}{d_R} = \prod_{i=1} \frac{(\lambda_i + N - i)!}{(N-i)!} \tag{17.3.16}$$

Note that d_R, obtained by division of dimension by $N! = |R|!$, depends only on parameters λ_i, while D_R contains an additional explicit dependence on N.

One more important fact is that the determinant of a matrix, i.e. a product of all its eigenvalues, is also equal to a character in the totally antisymmetric representation $R = [\underbrace{1, 1, \ldots, 1}_{N}]$,

$$\det X = \chi_{11\ldots1}[X], \quad \text{rank}(X) = N \tag{17.3.17}$$

Moreover, integer powers of a determinant are also characters:

$$(\det X)^\nu = \chi_{\nu\nu\ldots\nu}[X], \quad \text{rank}(X) = N \tag{17.3.18}$$

Note that when $\text{rank}(X) = N$ the time variables t_k with $k > N$ are algebraic functions of those with $k \le N$, and, when restricted to this subspace, the characters with $|R| > N$ also become algebraic functions of characters with $|R| \le N$.

Characters satisfy a number of 'sum rules': certain sums over *all* representations (i.e. over all Young diagrams or over all ordered partitions) are equal to some distinguished quantity. So far the most important sum rules in applications have been the following three:

$$\sum_R d_R \chi_R(t) = e^{t_1} = e^{\text{tr}\, X} \tag{17.3.19}$$

$$\sum_R \chi_R(t)\chi_R(t') = \exp\left(\sum_{k=1}^{\infty} k t_k t'_k\right) = \frac{1}{\text{Det}\,(I \otimes I - X \otimes X')} \tag{17.3.20}$$

$$\sum_R \frac{d_R}{D_R} \chi_R[X]\chi_R[X'] = \int [dU] \exp(\text{tr}\, XUX'U^\dagger). \tag{17.3.21}$$

Equation (17.3.19) is a particular case of Eq. (17.3.20), because $d_R = \chi_R(\delta_{k1})$. Equation (17.3.21) is actually a character expansion of the HCIZ integral, but it can effectively be used as a sum rule as well.

A very important role in the formalism of character theory is played by the orthogonality relation:

$$\chi_R(\tilde{\partial})\chi_{R'}(t) = \delta_{R,R'} \quad \text{for} \quad |R| = |R'| \tag{17.3.22}$$

Here one substitutes time derivatives $\tilde{\partial}_k \equiv \frac{\partial}{\partial p_k} = \frac{1}{k}\frac{\partial}{\partial t_k}$ instead of t_k in the arguments of $\chi_R(t)$ – a polynomial of its variables. For example,

$$\begin{aligned}
\chi_1(t) &= t_1, \quad \Longrightarrow \quad \chi_1(\tilde{\partial}) = \frac{\partial}{\partial t_1}, \\
\chi_2(t) &= t_2 + \frac{t_1^2}{2}, \quad \Longrightarrow \quad \chi_2(\tilde{\partial}) = \frac{1}{2}\left(\frac{\partial}{\partial t_2} + \frac{\partial^2}{\partial t_1^2}\right), \\
\chi_{11}(t) &= -t_2 + \frac{t_1^2}{2}, \quad \Longrightarrow \quad \chi_{11}(\tilde{\partial}) = \frac{1}{2}\left(-\frac{\partial}{\partial t_2} + \frac{\partial^2}{\partial t_1^2}\right)
\end{aligned} \tag{17.3.23}$$

and one can easily check that Eq. (17.3.22) works in these simple cases of $|R| = 1, 2$. It is also clear that the orthogonality relation fails for $|R| \neq |R'|$.

17.3.3 Character expansion of the HCIZ integral

From character calculus we can immediately derive an alternative formula, called character expansion: one can express $e^{\mathrm{tr}\,\Psi}$ with $\Psi \equiv XUYU^\dagger$ as a linear combination of characters, $e^{\mathrm{tr}\,\Psi} = \sum_R d_R \chi_R(\Psi)$, and after that substitute the average of characters over the unitary group:

$$\int \chi_R[\Psi]\,[dU] = \int \mathrm{Tr}_R\Big(\mathcal{X}_R \mathcal{U} \mathcal{Y}_R \mathcal{U}^\dagger\Big)[dU] = \frac{1}{D_R}\chi_R[X]\chi_R[Y] \tag{17.3.24}$$

Thus

$$J_{HCIZ}(X, Y) = \int e^{\mathrm{tr}\,\Psi}[dU] = \sum_R d_R \int \chi_R[\Psi][dU] \overset{(17.3.24)}{=} \sum_R \frac{d_R \chi_R[X]\chi_R[Y]}{D_R} \tag{17.3.25}$$

This formula can be found, for example, in [Bal00, Kaz00]. Similarly one obtains a character expansion in a more sophisticated case, with two terms in the exponent, known as (generalized) Berezin-Karpelevich integral [Ber58, Bal00]:

$$\begin{aligned}
J_{BK} &= \int [dU] \int [dV] \exp\left(XUYV^\dagger + VYU^\dagger X\right) \\
&= \sum_{R, R'} d_R d_{R'} \int\!\!\int \chi_R(XUYV^\dagger)\chi_{R'}(VYU^\dagger X)[dU][dV] \\
&\overset{(17.2.17)}{=} \sum_R d_R^2 \int\!\!\int \mathrm{Tr}_R\Big(\mathcal{X}_R \mathcal{U} \mathcal{Y}_R \mathcal{V}^\dagger\Big)\mathrm{Tr}_R\Big(\mathcal{X}_R \mathcal{V} \mathcal{Y}_R \mathcal{U}^\dagger\Big)[dU][dV] \\
&\overset{(17.2.16)}{=} \sum_R \frac{d_R^2}{D_R^2}\chi_R[X^2]\chi_R[Y^2].
\end{aligned} \tag{17.3.26}$$

because, as a consequence of (17.2.17) we have

$$\int \chi_R[U]\chi_{R'}[U]\,[dU] = \delta_{R,R'} \tag{17.3.27}$$

For applications to two-matrix models see, for example, [Kaz00], as well as Chapters 15 and 16 in this book.

Representation (17.3.2) follows from Eq. (17.3.25) if one expresses the characters on the right-hand side through the eigenvalues with the help of the second Weyl formula:

$$J_{HCIZ}(X,Y) \overset{(17.3.25)}{=} \sum_R \frac{d_R \chi_R[X]\chi_R[Y]}{D_R} \tag{17.3.28}$$

$$= \frac{1}{\Delta(X)\Delta(Y)} \sum_R \frac{d_R}{D_R} \det_{ij} x_i^{N-j+\lambda_j} \det_{ij} y_i^{N-j+\lambda_j}$$

$$= \frac{1}{\Delta(X)\Delta(Y)} \sum_{\lambda_1 \geq \lambda_2 \geq \dots} \prod_{i=1}^{N} \frac{(N-i)!}{(\lambda_i + N - i)!} \det_{ij} x_i^{N-j+\lambda_j} \det_{ij} y_i^{N-j+\lambda_j}$$

$$\overset{\lambda_i+N-i=k_i}{=} \frac{\prod_{i=1}^{N}(N-i)!}{\Delta(X)\Delta(Y)} \sum_{k_1>k_2>\dots} \frac{\det_{ij} x_i^{k_j} \det_{ij} y_i^{k_j}}{\prod_j k_j!} = \left(\prod_{j=0}^{N-1} j!\right) \frac{\det_{ij} e^{x_i y_j}}{\Delta(X)\Delta(Y)}$$

since

$$\sum_{k_1>k_2>\dots} \left(\prod_{j=1}^{N} a_j\right) \det_{ij} x_i^{k_j} \det_{ij} y_i^{k_j} = \frac{1}{N!} \sum_{P,P'} \sum_{k_j} \left(\phi_{P(j)} \psi_{P'(j)}\right)^{k_j} = \det_{ij} A(\phi_i \psi_j) \tag{17.3.29}$$

for $A(x) = \sum_k a_k x^k$, see [Bal00].

Similarly from (17.3.26) we obtain another celebrated formula [Ber58, Bal00]:

$$J_{BK} \overset{(17.3.26)}{=} \sum_R \frac{d_R^2 \chi_R[X^2]\chi_R[Y^2]}{D_R^2} = \frac{\left(\prod_{j=0}^{N-1} j!\right)^2}{\Delta(X^2)\Delta(Y^2)} \sum_{k_1>k_2>\dots} \frac{\det_{ij} x_i^{2k_j} \det_{ij} y_i^{2k_j}}{\left(\prod_j k_j!\right)^2}$$

$$\overset{(17.3.29)}{=} \left(\prod_{j=0}^{N-1} j!\right)^2 \frac{\det_{ij} \mathcal{I}_0(2x_i y_j)}{\Delta(X^2)\Delta(Y^2)}, \tag{17.3.30}$$

with Bessel function $\mathcal{I}_0(2x) = \sum_k \frac{x^{2k}}{(k!)^2}$. This is a kind of a transcendental generalization of the HCIZ integral. A further generalization was found in [Sch03]:

$$\int [dU] \int [dV] \det{}^{\nu}(UV) \exp\left(AUBV^\dagger + VCU^\dagger D\right) =$$

$$= \left(\prod_{j=1}^{N-1} j!\right)^2 \left(\frac{\det(CD)}{\det(AB)}\right)^{\frac{\nu}{2}} \frac{\det \mathcal{I}_\nu(2x_i y_j)}{\Delta(X^2)\Delta(Y^2)} \tag{17.3.31}$$

where this time x_i^2 and y_i^2 are the eigenvalues of AD and CB. One can even take matrices U and V of different sizes N and M [Sch03]. Then matrices A, B, C, D are rectangular, there is no ν-parameter and for $M \leq N$

$$\int_{N\times N}[dU]\int_{M\times M}[dV]\,\exp\left(AUBV^{\dagger}+VCU^{\dagger}D\right)=$$
$$=\left(\prod_{j=N-M}^{N-1}j!\prod_{l=N-M}^{N-1}l!\right)\frac{\det\mathcal{I}_{N-M}(2x_i y_j)}{\Delta(X^2)\Delta(Y^2)\prod_{i=1}^{M}(x_i y_i)^{N-M}} \tag{17.3.32}$$

and x_i^2, y_i^2 are eigenvalues of the smaller $M\times M$ matrices AD and CB. For a further extension to super-matrices see [Leh08].

17.3.4 Character expansion for the BGW model and Leutwyler-Smilga integral

Similar character expansions exist, of course, in the simpler case of the BGW model. However, since the integral (17.2.10) contains two items in the exponent, it is in fact closer to the transcendental Berezin-Karpelevich integral, than to the HCIZ integral. Instead of (17.3.26) we get for $Z_{BGW}(t)$ in Eq. (17.2.10):

$$\int e^{\mathrm{tr}\,(J^{\dagger}U+JU^{\dagger})}[dU]=\sum_{R,R'}d_R d_{R'}\int\chi_R[J^{\dagger}U]\chi_{R'}[JU^{\dagger}]\,[dU]\overset{(17.2.17)}{=}\sum_R\frac{d_R^2}{D_R}\chi_R[JJ^{\dagger}]$$

At the next step also the hook formula Eq. (17.3.15) is used [Mir96, Bal00]:

$$Z_{BGW}(t)=\sum_R\frac{d_R^2}{D_R}\chi_R[JJ^{\dagger}]=\frac{\prod_{j=0}^{N-1}j!}{\Delta(JJ^{\dagger})}\sum_{k_1>k_2>\dots}\frac{\det_{ij}J_i^{2k_j}}{\left(\prod_j k_j!\right)}\det_{ij}\frac{1}{(k_j-N+i)!}$$
$$=\left(\prod_{j=0}^{N-1}j!\right)\frac{\det_{ij}\left(J_j^{N-i}\mathcal{I}_{i-N}(2J_j)\right)}{\Delta(JJ^{\dagger})}=\left(\prod_{j=0}^{N-1}j!\right)\frac{\det_{ij}\left(J_i^{j-1}\mathcal{I}_{j-1}(2J_i)\right)}{\Delta(JJ^{\dagger})}$$

where J_i^2 are the eigenvalues of the Hermitian matrix $JJ^{\dagger}$ and the Bessel function satisfies

$$\mathcal{I}_s(2x)=\sum_{k=0}^{\infty}\frac{x^{2k+s}}{k!(k+s)!}=\mathcal{I}_{-s}(2x) \tag{17.3.33}$$

The Leutwyler-Smilga integral (17.2.15) can be handled in the same way:

$$(\det U)^{\nu}e^{\mathrm{tr}\,U}=\sum_{\lambda_1\geq\lambda_2\geq\dots\geq 0}\chi_{\lambda_1\dots\lambda_N}[U]\det_{ij}\frac{1}{(\lambda_j-\nu+i-j)!} \tag{17.3.34}$$

From this formula we get for $Z_{BGW}(\nu;t)$ [Leu92, Bal00, Sch03]

$$\int(\det U)^{\nu}e^{\mathrm{tr}\,(J^{\dagger}U+JU^{\dagger})}[dU]=\frac{\prod_{j=0}^{N-1}j!}{\Delta(JJ^{\dagger})}\sum_{k_1>k_2>\dots}\frac{\det_{ij}J_i^{2k_j}}{\left(\prod_j k_j!\right)}\det_{ij}\frac{1}{(k_j-N-\nu+i)!}$$
$$=\left(\prod_{j=0}^{N-1}j!\right)\frac{\det_{ij}\left(J_i^{j-1}\mathcal{I}_{\nu+j-1}(2J_i)\right)}{\Delta(JJ^{\dagger})}. \tag{17.3.35}$$

17.3.5 Correlators of Ψ-variable in the HCIZ integral [Miro9]

The character expansion allows one to evaluate arbitrary correlators of traces of $\Psi = XUYU^\dagger$ and its powers in the HCIZ integral.

For the set $\Delta = \Big\{\delta_1 \geq \delta_2 \geq \dots\Big\} = \Big\{\underbrace{m,\dots,m}_{a_m},\ \dots\ ,\underbrace{1,\dots,1}_{a_1}\Big\}$ of the size $|\Delta| = \sum_k \delta_k = \sum_m m a_m$ and $z_\Delta = \left(\prod_j a_j!\, j^{a_j}\right)^{-1}$ we have:

$$\Psi^\Delta \equiv (\mathrm{tr}\,\Psi)^{a_1}(\mathrm{tr}\,\Psi^2)^{a_2}\dots(\mathrm{tr}\,\Psi^m)^{a_m} = z_\Delta \sum_{|R|=|\Delta|} d_R \chi_R(\Psi)\varphi_R(\Delta) \tag{17.3.36}$$

where the symmetric-group characters $\varphi_R(\Delta)$ are defined as expansion coefficients of the characters $\chi_R(p)$ expanded in time variables:

$$\chi_R(p) = \sum_{|\Delta|=|R|} d_R \varphi_R(\Delta) p^\Delta, \tag{17.3.37}$$

where we recall $p_k = kt_k$. Instead of (17.3.25) we can now write

$$\begin{aligned} \int \Psi^\Delta e^{\mathrm{tr}\,\Psi}[dU] &= \sum_{n=0}^{\infty} \frac{1}{n!} \Psi^{\Delta+n\cdot 1}[dU] \\ &\overset{(17.3.36)}{=} \sum_n \frac{z_{\Delta+n\cdot 1}}{n!} \sum_{|R|=|\Delta|+n} d_R\varphi_R(\Delta+n\cdot 1)\int \chi_R(\Psi)[dU] \\ &\overset{(17.3.24)}{=} \sum_R \frac{z_{\Delta+(|R|-|\Delta|)\cdot 1}}{(|R|-|\Delta|)!}\frac{d_R}{D_R}\varphi_R\Big(\Delta+(|R|-|\Delta|)\cdot 1\Big)\chi_R[X]\chi_R[Y]. \end{aligned} \tag{17.3.38}$$

Note that sum over n is actually traded for the sum over *all* R, then only the contribution with $n = |R| - |\Delta|$ is picked up from the sum over n. If Δ does not contain any unit entries, then $z_{\Delta+(|R|-|\Delta|)\cdot 1} = (|R|-|\Delta|)!\, z_\Delta$ and this sophisticated formula simplifies a little: if $1 \notin \Delta$, then

$$\int \Psi^\Delta e^{\mathrm{tr}\,\Psi}[dU] \sim z_\Delta \sum_R \frac{d_R}{D_R}\varphi_R\Big(\Delta+(|R|-|\Delta|)\cdot 1\Big)\chi_R[X]\chi_R[Y] \tag{17.3.39}$$

Coming back to Eq. (17.3.38), this provides a character expansion for generic correlators of products of traces Ψ. Of course, this is only an implicit formula: an analogue of the transformation (17.3.28) is still needed to convert it into a more explicit expression. Unfortunately, in any case the correlators of traces Ψ are not the most general ones needed in HCIZ theory, and an evaluation of other correlators remains an unsolved, underestimated, and under-investigated problem to which we devote a separate short section.

17.3.6 Pair correlator in HCIZ theory

The evaluation of unitary integrals with IZ measure has become an important scientific problem since the appearance of the Kazakov-Migdal model [Kaz93] of lattice gluodynamics and it was originally addressed in [Kog93] and [Mor92a, Sha93]. Given the Duistermaat–Heckman nature of the HCIZ integral Eq. (17.3.6) [Mor95, Ber03, Pra09], this problem is also a natural chapter in the development of 'localization' techniques and BRST-like approaches. Unfortunately, the subject of IZ correlators is still underestimated and remains poorly investigated – not a big surprise, given the lack of knowledge about the ordinary unitary integrals, overviewed in Section 17.2 above. Still, the interest is returning to this field, in particular the relatively recent papers [Ber03, Pra09] provide a new confirmation and serious extension of the old result of [Mor92a]. This section provides nothing more than a sketchy review of the present state of affairs.

Correlators can, in principle, be evaluated directly – by choosing one's favorite parametrization of unitary matrices and explicitly evaluating all the N^2 integrals. A clever and convenient parametrization [Sha93, Guh03] makes use of Gelfand-Zeitlin variables [Gel50], which are unconstrained (free), it also arises naturally in the recursive procedure in N of [Ber03], which starts from the duality relation (17.3.8). Though seemingly natural, this approach does not provide any general formulae, at least in any straightforward way. As usual, a more practical approach is the 'stringy' one of [Mor92b] – that of generating functions.

The generating function of interest in HCIZ theory is

$$
\begin{aligned}
Z_{HCIZ} &= \int e^{\mathrm{tr}\, XUYU^{\dagger}+\mathrm{tr}\, J^{\dagger}U+\mathrm{tr}\, J\, U^{\dagger}}[dU] \overset{(17.2.1)}{\sim} \\
&\sim \int d^2 M dS \exp\left(\mathrm{tr}\, S(I - MM^{\dagger})\right) \exp\left(\mathrm{tr}\, XMYM^{\dagger} + \mathrm{tr}\, J^{\dagger}M + JM^{\dagger}\right) \\
&= \int \frac{dS\, e^{\mathrm{tr}\, S}}{\det_{N^2\times N^2}(S\otimes I - X\otimes Y)} \exp\left(\mathrm{tr}_{\,N^2\times N^2} J^{\dagger}\left(S\otimes I - X\otimes Y\right)^{-1} J\right) \\
&= \int \frac{dS}{\prod_k \det(S - y_k X)} \exp\left\{\sum_k \left(S_{kk} + J^{\dagger}_{ki}(S - Xy_k)^{-1}_{ij} J_{jk}\right)\right\},
\end{aligned}
$$

as a direct generalization of Eqs. (17.3.7) and (17.3.8). The integral over M is Gaussian and Wick's theorem is applicable, the non-trivial part is the evaluation of the S integral. In can be interesting to include also the Leutwyler-Smilga factor $(\det U)^{\nu}$ into the integrand, but – as we saw in Section 17.3.4 – this leads to non-trivial modifications. Unfortunately not much is known about this function even at $\nu = 0$. There are no serious obstacles for going beyond the first non-trivial correlator of $\langle U_{ij}U^{\dagger}_{kl}\rangle_{IZ}$, nevertheless it remains the only well-studied example.

This pair correlator simplifies considerably if we consider it in the basis where X and Y are both *diagonal* – then [Kog93]

$$\int U_{ij}\,U^{\dagger}_{kl}\,[dU] = \mathcal{M}_{ij}\,\delta_{kj}\,\delta_{il} \tag{17.3.40}$$

where $\mathcal{M}_{ij}$ still needs to be found. According to [Mor92a], *in this basis* the generating function

$$\sum_{i,j=1}^{N} a_i\mathcal{M}_{ij}\,b_j = \sum_{i,j=1}^{N} a_i b_j \int |U_{ij}|^2 \exp\left(\sum_{k,l} x_k y_l |U_{kl}|^2\right)[dU] =$$

$$= \frac{1}{\Delta(X)\Delta(Y)}\sum_{P\in S_N}(-)^P e^{x_i y_{P(i)}}\sum_{n=0}^{N-1}(-)^n \tag{17.3.41}$$

$$\times \sum_{1\le i_1<\ldots<i_{n+1}\le N} \frac{\det\begin{pmatrix} 1 & \ldots & 1\\ x_{i_1} & \ldots & x_{i_{n+1}}\\ \vdots & & \vdots\\ x_{i_1}^{n-1} & \ldots & x_{i_{n+1}}^{n-1}\\ a_{i_1} & \ldots & a_{i_{n+1}}\end{pmatrix}\det\begin{pmatrix} 1 & \ldots & 1\\ y_{P(i_1)} & \ldots & y_{P(i_{n+1})}\\ \vdots & & \vdots\\ y_{P(i_1)}^{n-1} & \ldots & y_{P(i_{n+1})}^{n-1}\\ b_{P(i_1)} & \ldots & b_{P(i_{n+1})}\end{pmatrix}}{\det\begin{pmatrix} 1 & \ldots & 1\\ x_{i_1} & \ldots & x_{i_{n+1}}\\ \vdots & & \vdots\\ x_{i_1}^{n} & \ldots & x_{i_{n+1}}^{n}\end{pmatrix}\det\begin{pmatrix} 1 & \ldots & 1\\ y_{P(i_1)} & \ldots & y_{P(i_{n+1})}\\ \vdots & & \vdots\\ y_{P(i_1)}^{n} & \ldots & y_{P(i_{n+1})}^{n}\end{pmatrix}}.$$

This sophisticated formula was transformed in [Ber03, Pra09] into a more elegant form:

$$\mathcal{M}_{ij} = \frac{1}{\Delta(X)\Delta(Y)}\,\mathrm{Res}\left\{\det E - \det\left(E - \frac{1}{X-u}\,E\,\frac{1}{Y-v}\right)\right\}\Bigg|_{\substack{u=x_i\\ v=y_j}} \tag{17.3.42}$$

where the matrix E is defined as $E_{ij} = e^{x_i y_j}$. According to this residue formula one should pick up the coefficients in front of $\Big((u-x_i)(v-y_j)\Big)^{-1}$ in the sum

$$-\frac{1}{\Delta(X)\Delta(Y)}\sum_{P\in S_N}(-)^P\prod_k e^{x_k y_{P(k)}}\left(1-\frac{1}{(u-x_k)(v-y_{P(k)})}\right) \tag{17.3.43}$$

over permutations. The first term $\det E$ in (17.3.42) does not contribute to the residue. There will be two different kinds of distributions: one for permutations P with the property $j = P(i)$ and another – for all other permutations:

$$\mathcal{M}_{ij} = \frac{1}{\Delta(X)\Delta(Y)} \sum_{P\in S_N} \sum_{P} (-)^P \left\{ \delta_{P(i),j} \prod_{k\neq i}^{N} \left(1 - \frac{1}{(x_i - x_k)(y_j - y_{P(k)})}\right) + \right.$$

$$\left. - \frac{\left(1 - \delta_{P(i),j}\right)}{(x_i - x_{P(j)})(y_j - y_{P(i)})} \prod_{k\neq i,\ P^{-1}(j)}^{N} \left(1 - \frac{1}{(x_i - x_k)(y_j - y_{P(k)})}\right) \right\} \prod_{k=1}^{N} e^{x_k y_{P(k)}}$$

For $N = 1$ we have a single term of the first kind and $\mathcal{M}_{11} = e^{x_1 y_1}$.
For $N = 2$ the terms of both types contribute:

$$\mathcal{M}_{11} = \frac{1}{x_{12}y_{12}} \left\{ \left(1 - \frac{1}{x_{12}y_{12}}\right) e^{x_1y_1+x_2y_2} + \frac{1}{x_{12}y_{12}} e^{x_1y_2+x_2y_1} \right\} = \mathcal{M}_{22},$$

$$\mathcal{M}_{12} = \frac{1}{x_{12}y_{12}} \left\{ \frac{1}{x_{12}y_{12}} e^{x_1y_1+x_2y_2} - \left(1 + \frac{1}{x_{12}y_{12}}\right) e^{x_1y_2+x_2y_1} \right\} \tag{17.3.44}$$

which is exactly the same as Eq. (17.3.41), where just the first two terms survive on the right-hand side for $N = 2$. Of course, $\mathcal{M}_{11} + \mathcal{M}_{12} = J_{HCIZ}$.

A character decomposition of pair correlators can easily be obtained similarly to Eq. (17.3.24), provided one knows an appropriate generalization of Eqs. (17.2.16) and (17.2.17), namely

$$\int \mathcal{U}_R \mathcal{U}_R^\dagger \mathcal{U}_{R'} \mathcal{U}_{R'}^\dagger [dU]\,, \tag{17.3.45}$$

a generalization of Eq. (17.2.3) to arbitrary representations. Actually, the case when R' is the fundamental representation would be enough for decomposition of Eq. (17.3.41).

Like the HCIZ integral itself, the pair correlator satisfies a kind of duality relation – a direct generalization of Eq. (17.3.8): from

$$\int U_{ij} U_{kl}^\dagger [dU] \overset{(17.3.40)}{\sim} \delta_{jk} \int \left(\frac{1}{S - Xy_k}\right)^{-1}_{il} \frac{e^{\operatorname{tr} S} dS}{\prod_k \det(S - Xy_k)} \tag{17.3.46}$$

it follows, after the change of integration matrix-variable $S \longrightarrow XS$ and after diagonalization of X and S that

$$\mathcal{M}_{ij}(X, Y) \sim \int \frac{\mathcal{M}_{ij}(X, S)\, \Delta^2(S) \prod_p ds_p}{(s_i - y_j)\, x_j \prod_{p,q} (s_p - y_q)}\,. \tag{17.3.47}$$

Similar relations can be straightforwardly deduced from Eq. (17.3.40) for higher correlators in HCIZ theory.

For generalizations to other simple Lie groups see [Bre01, Guh03] and [Ber03, Pra09].

Acknowledgements

I am indebted to A. Alexandrov, A. Mironov, A. Popolitov and Sh. Shakirov for collaboration and help. This work is partly supported by Russian Federal Nuclear Energy Agency, by the joint grants 09-02-90493-Ukr, 09-02-91005-ANF, 09-01-92440-CE, 09-02-93105-CNRSL and 10-02-92109-Yaf-a, by CNRS and by RFBR grant 10-02-00499.

References

[Ale04] A. Alexandrov, A. Mironov and A. Morozov, Int. J. Mod. Phys. **A19** (2004) 4127, Theor. Math. Phys. **142** (2005) 349, hep-th/0310113; Int. J. Mod. Phys. **A21** (2006) 2481, hep-th/0412099; Fortsch. Phys. **53** (2005) 512, hep-th/0412205;
A. Alexandrov, A. Mironov, A. Morozov and P. Putrov, Int. J. Mod. Phys. **A** (2009), arXiv:0811.2825

[Ale07] A. Alexandrov, A. Mironov and A. Morozov, Theor. Math. Phys. **150** (2007) 179, hep-th/0605171; Physica **D 235** (2007) 126;
N. Orantin, arXiv:0808.0635

[Ale09] A. Alexandrov, A. Mironov, and A. Morozov, JHEP 0912 (2009) 053

[Bal00] A. Balantekin, Phys. Rev. **D62** (2000) 085017, hep-th/0007161

[Bal02] A.B. Balantekin and P. Cassak, J. Math. Phys. **43** (2002) 604–620

[Bar80] I. Bars, J. Math. Phys. 21 (1980) 2678

[Ber58] F. Berezin and F. Karpelevich, Dokl.Acad.Nauk SSSR, **118** (1958) 9;
T. Guhr and T. Wettig, J. Math. Phys. **37** (1996) 6395;
A.Jackson, M.Sener and J.Verbaarschot, Phys.Lett. **B387** (1996) 355

[Ber03] M. Bertola and B. Eynard, J. Phys, **A36** (2003) 7733, hep-th/0303161;
B. Eynard, math-ph/0406063;
B. Eynard and A. Prats Ferrer, Comm. Math. Phys. **264** (2005) 115–144, hep-th/0502041

[Bre80] E. Brézin and D. Gross, Phys. Lett. **B97** (1980) 120;
D. Gross and E. Witten, Phys. Rev. **D21** (1980) 446

[Bre01] E. Brézin and S. Hikami, Comm. Math. Phys. **223** (2001) 363

[Bow91] M. Bowick, A. Morozov, D. Shevitz, Nucl. Phys. **B354** (1991) 496–530

[Cas92] M. Caselle, A. D'Adda and S. Panzer, Phys. Lett. **B293** (1992) 161–167, hep-th/9207086; Phys. Lett. **B302** (1993) 80–86, hep-th/9212074

[Col09] B. Collins, A. Guionnet and E. Maurel-Segala, Adv. Math. **222** (2009) 172–215

[DeW77] B. de Wit and G.'t Hooft, Phys. Lett. **B69** (1977) 61

[Dol07] V. Dolotin and A. Morozov, *Introduction to Non-Linear Algebra*, World Scientific, 2007, hep-th/0609022

[Eyn04] B. Eynard, JHEP **0411** (2004) 031, hep-th/0407261;
B. Eynard and N.Orantin, JHEP **0612** (2006) 026; math-phys/0702045;
L. Chekhov and B. Eynard, JHEP **0603** (2006) 014, hep-th/0504116; JHEP **0612** (2006) 026, math-ph/0604014;
B. Eynard, M. Mariño and N.Orantin, JHEP **0706** (2007) 058;
N. Orantin, PhD thesis, arXiv:0709.2992; arXiv:0803.0705

[Fuk91] M. Fukuma, H. Kawai and R. Nakayama, Int. J. Mod. Phys. **A6** (1991) 1385;
R. Dijkgraaf, E. Verlinde and H. Verlinde, Nucl. Phys. **B348** (1991) 565;

A. Mironov and A. Morozov, Phys. Lett. **B252** (1990) 47–52;
F. David, Mod. Phys. Lett. **A5** (1990) 1019;
J. Ambjorn and Yu. Makeenko, Mod. Phys. Lett. **A5** (1990) 1753;
H. Itoyama and Y. Matsuo, Phys. Lett. **B255** (1991) 202

[Gel50] I. Gelfand and M. Zeitlin, Dokl. Acad. Nauk SSSR **71** (1950) 825–829; 1017–1020

[Gel94] I. Gelfand, M. Kapranov and A. Zelevinsky, *Discriminants, Resultants and Multidimensional Determinants*, Birkhäuser, 1994

[Ger91] A. Gerasimov, A. Marshakov, A. Mironov, A. Morozov, and A. Orlov, Nucl. Phys. **B357** (1991) 565–618;
Yu. Makeenko, A. Marshakov, A. Mironov and A. Morozov, Nucl. Phys. **B356** (1991) 574;
S. Kharchev, A. Marshakov, A. Mironov, A. Morozov and S. Pakuliak, Nucl. Phys. **B404** (1993) 717–750, hep-th/9208044

[Giv00] A. Givental, math.AG/0008067

[Guh03] T. Guhr and H. Kohler, J. Math. Phys. **44** (2003) 4267, math-ph/0212059; J. Math. Phys. **45** (2004) 3636, math-ph/0212060

[Har57] Harish-Chandra, Am. J. Math. **79** (1957) 87; **80** (1958) 241

[Itz80] C. Itzykson and J. Zuber, J. Math. Phys. **21** (1980) 411;
P. Zinn-Justin and J.-B. Zuber, J. Phys. **A 36** (2003) 3173–3193

[Kaz00] V. Kazakov, *Solvable Matrix Models*, hep-th/0003064

[Kaz93] V. Kazakov and A. Migdal, Nucl. Phys. **B397** (1993) 214–238, 1993, hep-th/9206015

[Kha95] S. Kharchev, A. Marshakov, A. Mironov and A. Morozov, Int. J. Mod. Phys. **A10** (1995) 2015, hep-th/9312210

[Kha92] S. Kharchev, A. Marshakov, A. Mironov, A. Morozov and A. Zabrodin, Phys. Lett. **B275** (1992) 311–314, hep-th/9111037; Nucl. Phys. **B380** (1992) 181–240, hep-th/9201013
A. Marshakov, A. Mironov and A. Morozov, Mod. Phys. Lett. **A7** (1992) 1345–1360, hep-th/9201010; Phys. Lett. **274B** (1992) 280–288;
P. Di Francesco, C. Itzykson and J.-B. Zuber, Comm. Math. Phys. **151** (1993) 193–219, hep-th/9206090; M. Adler and P. van Moerbeke, Comm. Math. Phys. **147** (1992) 25;
S. Kharchev, A. Marshakov, A. Mironov and A. Morozov, Nucl. Phys. **B397** (1993) 339–378, hep-th/9203043; Mod. Phys. Lett. **A8** (1993) 1047–1062, hep-th/9208046

[Kog93] I. Kogan, A. Morozov, G. Semenoff and H. Weiss, Nucl. Phys. **B395** (1993) 547–580, hep-th/9208012; Int. J. Mod. Phys. **A8** (1993) 1411–1436, hep-th/9208054

[Kon92] M. Kontsevich, Funk. Anal. Prilozh., **25:2** (1991) 50–57; Comm. Math. Phys. **147** (1992) 1–23

[Leh08] C. Lehner, T. Wettig, T. Guhr and Y. Wei, J. Math. Phys. **49** (2008) 063510, arXiv:0801.1226

[Leu92] H. Leutwyler and A. Smilga, Phys. Rev. **D 46** (1992) 5607–5632;
J. Verbaarschot, hep-th/9710114;
R. Brower, P. Rossi and C.-I. Tan, Nucl. Phys. **B190** (1981) 699;
T. Akuzawa and M. Wadati, J. Phys. Soc. Jap. **67** (1998) 2151

[Mar90] E. Martinec, Commun. Math. Phys. **138** (1990) 437–450,1991;
V. Periwal and D. Shevitz, Phys. Rev. Lett. **64** (1990) 1326; Nucl. Phys. **B344** (1990) 731;
K. Demeterfi and C. Tan, Mod. Phys. Lett. **A5** (1990) 1563

[Mir96] A. Mironov, A. Morozov and G. Semenoff, Int. J. Mod. Phys. **A11** (1996) 5031–5080, hep-th/9404005

[Mir09] A. Mironov, A. Morozov, and S. Natanzon, Theor. Math. Phys. **166** (2011) 1–22.
[Mor92a] A. Morozov, Mod. Phys. Lett. **A7** (1992) 3503–3508, hep-th/9209074
[Mor92b] A. Morozov, Sov. Phys. Usp. **35** (1992) 671–714
[Mor94] A. Morozov, Phys. Usp. **37** (1994) 1–55, hep-th/9303139
[Mor95] A. Morozov, *Matrix Models as Integrable Systems*, hep-th/9502091
[Mor09a] A. Morozov, Theor. Math. Phys. **162** (2010) 1–33.
[Mor09b] A. Morozov and Sh. Shakirov, JHEP 0912 (2009)002; arXiv:0911.5278 [math-ph]
K. Fujii, arXiv:0905.1363 [math-ph]; SIGMA **7** (2011) 022.
[Pra09] A. Prats Ferrer, B. Eynard, P. Di Francesco and J.-B. Zuber, J. Stat. Phys. **129** (2009) 885–935, math-ph/0610049;
M. Bertola and A. Prats Ferrer, Int. Math. Res. Notices (2008) 2008;
Michel Bergère and B. Eynard, J. Phys. A **42** (2009) 265201.
[Sem76] M. Semenov, Tyan-Shansky, Izv. AN SSSR, Physics, **40** (1976) 562;
J. J. Duistermaat and G. I. Heckman, Invent. Math. **69** (1982) 259–268; **72** (1983) 153–158;
A. Alekseev, L. Faddeev and S. Shatashvili, Geometry and Physics, **3** (1989);
M. Blau, E. Keski-Vakkuri and A. Niemi, Phys. Lett. **B246** (1990) 92;
E. Keski-Vakkuri, A. Niemi, G. Semenoff and O. Tirkkonen, Phys. Rev. **D 44** (1991) 3899;
A. Hietamaki, A. Morozov, A. Niemi and K. Palo, Phys. Lett. **B263** (1991) 417–424;
A. Morozov, A. Niemi and K. Palo, Phys. Lett. **B271** (1991) 365–371; Nucl. Phys. **B377** (1992) 295–338;
R. Szabo, *Equivariant Localization of Path Integrals*, hep-th/9608068;
Y. Karshon, *Lecture Notes on Group Action on Manifolds*, 1996–97;
M. Stone, Nucl. Phys. **B314** (1989) 557–586
[Sha93] S. Shatashvili, Comm. Math. Phys.**154** (1993) 421–432
[Sch03] B. Schlittgen and T. Wettig, J. Phys. **A36** (2003) 3195–3202
[Sza00] R.-J. Szabo, Lect. Notes in Phys. **M63** (2000) 1
[Wei80] D. Weingarten, Phys. Lett. **B90** (1980) 285; J. Math. Phys. **19** (1998) 999
[Zir98] M. Zirnbauer, chao-dyn/9810016
[Zub08] J.-B. Zuber, J. Phys. A: Math. Theor. **41** (2008) 382001

·18·

Non-Hermitian ensembles

Boris A. Khoruzhenko and Hans-Jürgen Sommers

Abstract

This is a concise review of the complex, real, and quaternion real Ginibre random matrix ensembles and their elliptic deformations. Eigenvalue correlations are exactly reduced to two-point kernels and discussed in the strongly and weakly non-Hermitian limits of large matrix size.

18.1 Introduction

The study of eigenvalue statistics in the complex plane was initiated in 1965 by Ginibre [Gin65] who introduced a three-fold family of Gaussian random matrices (complex, real, and quaternion real) as a mathematical extension of Hermitian random matrix theory. Although no physical applications of the theory were in sight at that time, Ginibre expressed the hope that 'the methods and results will provide further insight in the cases of physical interest or suggest as yet lacking applications'. Nowadays, statistics of complex eigenvalues have many interesting applications in modelling of a wide range of physical phenomena. They appeared in the studies of quantum chromodynamics (see Chapter 32), dissipative quantum maps [Gro88] and scattering in chaotic quantum systems (see Chapter 34), growth processes (see Chapter 38), fractional quantum-Hall effect [DiF94] and Coulomb plasma [For97], stability of complex biological [May72] and neural networks [Som88], directed quantum chaos in randomly pinned superconducting vortices [Efe97], delayed time series in financial markets [Kwa06], random operations in quantum information theory [Bru09], and others.

This chapter gives an overview of the three Ginibre ensembles and their elliptic deformations. We have tried to keep our exposition self-contained providing hints of derivations. Some of the derivations included are new. We do not have space to cover the chiral extensions of the Ginibre ensembles, only mentioning briefly the chiral companion of the complex Ginibre ensemble [Osb04]. This topic is partly covered in Chapter 32. The real and quaternion real (qu-r) companions were solved only recently, see [Ake10] and [Ake05a]. There are also important non-Hermitian ensembles of random matrices relevant in the context of quantum chaotic scattering which are only mentioned here (but see Chapter 34 for a summary of results and the survey paper [Fyo03] for details). Also, we will not discuss the complete classification of non-Hermitian matrix ensembles depending on the action of a few number of involutions [Ber02, Mag08].

On a macroscopic scale, all three Ginibre ensembles exhibit similar patterns of behaviour with a uniform distribution of eigenvalues and sharp (Gaussian) fall in the eigenvalue density when one transverses the boundary of the eigenvalue support. The similarities extend to the microscopic scale as well but only away from the real line where all three ensembles exhibit a cubic repulsion of eigenvalues. In the vicinity of the real line and on the real line their behaviour is very different due to the differences in symmetries. The eigenvalue correlation functions have either determinantal (complex) or pfaffian (real and qu-r) form with the kernel being almost identical far away from the real line; again the differences coming from a pre-exponential factor describing the transition from the real line into the bulk of the spectrum.

One expects the eigenvalue statistics provided by the Ginibre ensembles to be universal within their symmetry classes. Establishing such universality is an open and challenging problem, especially for the real and qu-r ensembles. There is some evidence for universality in the complex case where every solved model leads to Ginibre's form of correlations, including complex normal matrices where the universality of Ginibre's correlations has been proved in a general class of matrix distributions [Ame08]. This result can be applied to complex matrices (18.2.4) as for this class of ensembles the induced eigenvalue distribution does not differ from the one for normal matrices [Oas97], see also Chapter 38.

18.2 Complex Ginibre ensemble

Measure, change of variables The complex Ginibre ensemble is defined on the space of complex $N \times N$ matrices by the probability measure

$$d\mu(J) = \exp\left(-\operatorname{tr} J J^{\dagger}\right)|DJ|. \tag{18.2.1}$$

Here $DJ = \prod_{i,j=1}^{N}(dJ_{ij}dJ_{ij}^{*}/2\pi)$ is the (exterior) product of the one-forms in matrix entries and $|DJ|$ is the corresponding Cartesian volume element. With probability one, the matrix J has N distinct eigenvalues z_j. On ordering the eigenvalues in an arbitrary but fixed way, one can think of them as random variables. The corresponding joint probability distribution function (jpdf) can be obtained by changing variables in (18.2.1). This can be done in several ways, we outline here a calculation due to Dyson [Meh04].

On making use of the Schur decomposition, J can be brought to triangular form $J = U(\Lambda + \Delta)U^{-1}$. Here U is unitary, $\Lambda = \mathrm{diag}(z_1, \ldots, z_N)$ and Δ is strictly upper-triangular. It is apparent that U can be restricted to the space of right cosets $\mathrm{U}(N)/\mathrm{U}(1)^N$. The variations in J are related to those in Λ, Δ and U by the formula $U^{-1}\mathrm{d}J\,U = \mathrm{d}\Lambda + \mathrm{d}\Delta + \mathrm{d}M$, where

$$(\mathrm{d}M)_{ij} = (U^{-1}\mathrm{d}U)_{ij}(z_j - z_i) + \sum_{k<j}(U^{-1}\mathrm{d}U)_{ik}\Delta_{kj} - \sum_{l>i}\Delta_{il}(U^{-1}\mathrm{d}U)_{lj}.$$

The volume form $\mathrm{D}J$ does not change on conjugation by unitary matrices. Hence $\mathrm{D}J = \mathrm{D}(\Lambda + \Delta + M)$ and one gets the Jacobian of the coordinate transformation by multiplying through the matrix entries of $\mathrm{d}\Lambda + \mathrm{d}\Delta + \mathrm{d}M$ using the calculus of alternating differential forms. This leads to the important relation

$$\mathrm{d}\mu(J) = C_N\,\mathrm{e}^{-\mathrm{tr}\,(\Delta\Delta^\dagger + \Lambda\Lambda^\dagger)} \prod_{1\le i<j\le N} |z_i - z_j|^2\,|\mathrm{D}U|\,|\mathrm{D}\Lambda|\,|\mathrm{D}\Delta|, \qquad (18.2.2)$$

where $\mathrm{D}U = \prod_{i<j}(U^{-1}\mathrm{d}U)_{ij}(U^{-1}\mathrm{d}U)_{ji}$ is a volume form on the coset space. The density function on the right-hand side is symmetric in the eigenvalues of J. It does not depend on U and is Gaussian in Δ, and these two variables can easily be integrated out. Thus, if $f(z_1, \ldots, z_N)$ is symmetric in eigenvalues of J then [Gin65] $\int f\mathrm{d}\mu = \int \mathrm{d}^2 z_1 \cdots \int \mathrm{d}^2 z_N\, P(z_1, \ldots, z_N) f(z_1, \ldots, z_N)$ where $\mathrm{d}^2 z = |\mathrm{d}z\mathrm{d}z^*|/2$ is the element of area in the complex plane and

$$P(z_1, \ldots, z_N) = \frac{1}{\pi^N \prod_{j=0}^N j!}\,\mathrm{e}^{-\sum_{j=1}^N |z_j|^2} \prod_{1\le i<j\le N} |z_i - z_j|^2 \qquad (18.2.3)$$

is the *symmetrized* jpdf of the eigenvalues of J.

In contrast to the Hermitian matrices, the above calculation of the jpdf is not easily extended to other invariant distributions. For example, it breaks down if $JJ^\dagger$ in (18.2.1) is replaced by a higher-order polynomial in $JJ^\dagger$ [Fei97] as U, Δ and Λ do not decouple in that case, although it still works well for (see Chapter 38)

$$\mathrm{d}\mu(J) \propto \exp\big(-\mathrm{tr}\,JJ^\dagger - \mathrm{Re}\,\mathrm{tr}\,\Phi(J)\big)|\mathrm{D}J|, \qquad (18.2.4)$$

with $\Phi(J)$ being a potential that ensures existence of the normalization integral. We will discuss in detail the special case of (18.2.4) with $\Phi(J) = J^2$ later.

Correlation functions and orthogonal polynomials The eigenvalue correlation functions are marginals of the symmetrized jpdf,

$$R_n(z_1, \ldots, z_n) = \frac{N!}{(N-n)!}\int \mathrm{d}^2 z_{n+1} \ldots \int \mathrm{d}^2 z_N\, P(z_1, \ldots, z_N), \qquad (18.2.5)$$

normalized to $\int d^2z_1 \ldots \int d^2z_n \; R_n(z_1, \ldots, z_n) = N(N-1)...(N-n+1)$. The one-point correlation function is just the density of eigenvalues $\rho(z) = \sum_j \delta^{(2)}(z - z_j)$ averaged over the ensemble distribution, $R_1(z) = \langle \rho(z) \rangle$, so that if n_D is the number of eigenvalues in D then $\langle n_D \rangle = \int_D d^2z R_1(z)$.

The eigenvalue correlation functions for the complex Ginibre ensemble can be found in a closed form. The corresponding calculation is almost identical to that for Hermitian ensembles. For the purpose of future reference we shall outline it in a slightly more general setting.

Let $p_m(z)$ be the monic orthogonal polynomials associated with weight function $w(z) \geq 0$ in the complex plane, i.e. $\int d^2z \; w(z) \, p_m(z) p_n(z)^* = h_n \delta_{m,n}$ and $p_m(z) = z^m + \ldots$. Then

$$P(z_1, \ldots, z_N) = \frac{1}{N! \prod_{l=0}^{N-1} h_l} \prod_{j=1}^{N} w(z_j) \prod_{i<j} |z_i - z_j|^2 \qquad (18.2.6)$$

is a probability density in $\mathbb{C}^N$ symmetric with respect to permutations of z_j. Recalling the Vandermonde determinant $\prod_{1 \leq i < j \leq N}(z_i - z_j) = \det(z_i^{N-j})_{1 \leq i,j \leq N} = \det(p_{N-j}(z_i))_{1 \leq i,j \leq N}$, one obtains the important determinantal representation of the jpdf:

$$P(z_1, \ldots, z_N) = \frac{1}{N!} \det(K(z_i, z_j))_{i,j=1}^{N} \qquad (18.2.7)$$

with

$$K(z_1, z_2) = \sqrt{w(z_1)} \sqrt{w(z_2)} \sum_{n=0}^{N-1} \frac{p_n(z_1) p_n(z_2)^*}{h_n}. \qquad (18.2.8)$$

The kernel K is Hermitian, $K(z_1, z_2) = K(z_2, z_1)^*$, and $\int d^2z \; K(z, z) = N$, $\int d^2z \; K(z_1, z) K(z, z_2) = K(z_1, z_2)$. Hence by Mehta's 'integrating out' lemma [Meh04] (see also the relevant section in Chapter 4)

$$R_n(z_1, \ldots, z_n) = \det(K(z_i, z_j))_{i,j=1}^{n}. \qquad (18.2.9)$$

In particular, the one- and two-point correlation functions are given by $R_1(z) = K(z, z)$ and $R_2(z_1, z_2) = K(z_1, z_1) K(z_2, z_2) - |K(z_1, z_2)|^2$.

In the complex Ginibre ensemble $w(z) = e^{-|z|^2}$. In view of the rotational symmetry the power functions are orthogonal. Thus $p_n(z) = z^n$, $h_n = \pi n!$ and

$$K(z_1, z_2) = \frac{1}{\pi} e^{-\frac{1}{2}|z_1|^2 - \frac{1}{2}|z_2|^2} \sum_{n=0}^{N-1} \frac{(z_1 z_2^*)^n}{n!}. \qquad (18.2.10)$$

The sum on the right-hand side is the truncated exponential series and can be expressed in terms of the incomplete Gamma function: $\sum_{l=0}^{N-1} \frac{x^l}{l!} = e^x \Gamma(N, x)/\Gamma(N)$. The saddle-point integral $\Gamma(N, x) = \int_x^\infty dt \; e^{-t} t^{N-1}$ comes in handy for asymptotic analysis. We have

$$R_1(z) = \frac{1}{\pi}\frac{\Gamma(N, |z|^2)}{\Gamma(N)} \simeq \frac{1}{\pi}\Theta(\sqrt{N} - |z|), \quad N \to \infty, \tag{18.2.11}$$

where Θ is the Heaviside function, meaning that on average most of the eigenvalues are distributed within the disk $|z| < \sqrt{N}$ with constant density in agreement with the Circular Law [Gir85, Tao08]. The expected number of eigenvalues outside this disk $\simeq \sqrt{\frac{N}{2\pi}}$. The eigenvalue density in the transitional region around the circular boundary, as found from the integral $\Gamma(N, x)$, is given in terms of the complementary error function

$$R_1\big((\sqrt{N} + x)\,\mathrm{e}^{i\varphi}\big) \simeq \frac{1}{2\pi}\mathrm{erfc}\,(\sqrt{2}\,x), \quad \text{as } N \to \infty, \tag{18.2.12}$$

[For99, Kan05a]. Recalling that $\mathrm{erfc}\,(x) \simeq 2\,\Theta(-x) + \mathrm{e}^{-x^2}/(\sqrt{\pi}x)$ for large real $|x|$, one concludes that the density vanishes at a Gaussian rate at the edge.

The kernel in (18.2.10) has a well-defined limit as $N \to \infty$ and $|z_{1,2}| = O(1)$, leading to a simple expression for the correlation functions in this limit [Gin65]

$$R_n(z_1, \ldots, z_n) \simeq \frac{1}{\pi^n}\,\mathrm{e}^{-\sum_j |z_j|^2}\det(\mathrm{e}^{z_i z_j^*})_{i,j=1}^{n}. \tag{18.2.13}$$

Equation (18.2.13) describes eigenvalue correlations at the origin $z_0 = 0$ and on the (local) scale when the mean separation between eigenvalues is of the order of unity. It also holds true (locally) at any other reference point inside the disk $|z| < \sqrt{N}$, see, e.g. [Bor09]. In particular, $R_2(z_1, z_2) \simeq \frac{1}{\pi^2}(1 - \mathrm{e}^{-|z_1 - z_2|^2})$. Note that the dependence of R_n on the reference point disappears in the limit $N \to \infty$ and complete homogeneity arises. The eigenvalue correlation functions at the edge of the eigenvalue support can also be found, see [For99, Bor09].

Gap probability and nearest neighbour distance Consider the disk D of radius s centred at z_0 and denote by χ_D its characteristic function. Define $H(s; z_0)$ to be the conditional probability that given that one eigenvalue lies at z_0 all others are outside D. With χ_D being the characteristic function of D,

$$H(s; z_0) = \frac{N}{R_1(z_0)}\int \mathrm{d}^2 z_2 \cdots \int \mathrm{d}^2 z_N\, P(z_0, z_2, \ldots, z_N)\prod_{k=2}^{N}(1 - \chi_D(z_k)). \tag{18.2.14}$$

When s gets infinitesimal increment δs, the decrement in H, $H(s, z_0) - H(s + \delta s, z_0)$, is the probability for the distance between the eigenvalue at z_0 and its nearest neighbour to lie in the interval $(s, s + \delta s)$. Therefore $p(s; z_0) = -\frac{\mathrm{d}}{\mathrm{d}s}H(s; z_0)$ is the density of nearest neighbour distances at z_0.

For the complex Ginibre ensemble $H(s; 0)$ can easily be computed with the help of the Vandermonde determinant and Andréief-de Bruijn integration formula (Eq. (8.3.41) in Chapter 8). The rotational invariance of the weight function $\mathrm{e}^{-|z|^2}$ ensures that the monomials z^m stay orthogonal when integrated over the disk D, leading to [Gro88]

$$H(s;0) = \prod_{n=1}^{N-1} \frac{\Gamma(n+1,s^2)}{\Gamma(n+1)} = \prod_{n=1}^{N-1}\left(1 - \mathrm{e}^{-s^2}\sum_{l=n+1}^{\infty}\frac{s^{2l}}{l!}\right). \tag{18.2.15}$$

The product on the right-hand side converges quite rapidly in the limit $N \to \infty$ and $H(s; z_0)$ has a well-defined limit. In this limit $H(s, 0) = 1 - s^4/2 + s^6/6 - s^8/24 + O(s^{10})$ for small s and [Gro88] $H(s, 0) = \exp[-s^4/4 - s^2(\ln s + O(1))]$ for large s. Taking the derivative, one gets the cubic law of the eigenvalue repulsion: $p(s; 0) = 2s^3 + O(s^5)$ for small s.

The small-s behaviour of $H(s, z_0)$ can also be obtained by expanding the product in (18.2.14). For example, on retaining the first two terms $H(s, z_0) = 1 - \frac{1}{R_1(z_0)}\int_{|z-z_0|\le s} \mathrm{d}^2z\, R_2(z_0, z) + O(s^4)$. This approach is general and does not rely on the rotational invariance. The universality of Ginibre's correlations then implies universality of the cubic law of the eigenvalue repulsion, which can also be verified directly for complex ensembles with known jpdf [Oas97]. Retaining all terms leads to a Fredholm determinant, see Chapter 4, giving access to various gap probabilities beyond the Ginibre ensemble [Ake09b].

18.3 Random contractions

Let U be a unitary matrix taken at random from the unitary group $\mathrm{U}(M+L)$. Denote by J its top left corner of size $M \times M$. The matrix J is a random contraction: with probability one, it has all of its eigenvalues inside the unit disk $|z| < 1$. The jpdf of eigenvalues of J was computed in [Zyc00] and is given by (18.2.6) with $w_L(z) = (1-|z|^2)^{L-1}\Theta(1-|z|)$. Since $w_L(z)$ is rotation invariant, the associated orthogonal polynomials are again powers $p_m(z) = z^m$ and the normalization constants are easily computed in terms of the Beta function, $h_m = \pi B(m+1, L)$. Applying the orthogonal polynomials formalism, one obtains the correlation functions in the determinantal form (18.2.9) with

$$K(z_1, z_2) = \frac{L}{\pi}(1-|z_1|^2)^{\frac{L-1}{2}}(1-|z_2|^2)^{\frac{L-1}{2}}\sum_{m=0}^{M-1}\binom{L+m}{m}(z_1z_2^*)^m.$$

The truncated binomial series for $(1-x)^{-(L+1)}$ on the right-hand side can be expressed in terms of the incomplete Beta function $I_x(a,b) = \frac{1}{B(a,b)}\int_0^x t^{a-1}(1-t)^{b-1}\,\mathrm{d}t$ via the relation $I_x(M, L+1) = 1 - (1-x)^{L+1}\sum_{m=0}^{M-1}\binom{L+m}{m}x^m$. This leads to a useful representation

$$K(z_1, z_2) = \frac{L}{\pi}\frac{(1-|z_1|^2)^{\frac{L-1}{2}}(1-|z_2|^2)^{\frac{L-1}{2}}}{(1-z_1z_2^*)^{L+1}}\Big(1 - I_{z_1z_2^*}(M, L+1)\Big). \tag{18.3.1}$$

There are several asymptotic regimes to be considered in the context of $M \times M$ truncations of random unitary matrices of an increasing dimension $M + L$.[1] The simplest one is the limit when M stays finite and $L \to \infty$. In this regime the jpdf becomes Gaussian and one immediately recovers the complex Ginibre ensemble. In fact one can allow M to grow with L and still recover the Ginibre ensemble provided that $M \ll L$; see [Jia06] and references therein for bounds on the rate of growth of M for the Gaussian Law to hold. In the opposite case when the size of the truncation increases at the same rate as the overall dimension $M + L$, one has two distinct regimes: (i) $M, L \to \infty$, $M/L = \alpha > 0$, and (ii) $M \to \infty$, L is finite.

We start with (i) which is the limit of strong non-unitarity. In this limit the eigenvalues are distributed inside the disk $|z|^2 \leq \frac{\alpha}{(1+\alpha)}$ with density [Zyc00]

$$R_1(z) = \frac{L}{\pi} \frac{1 - I_{|z|^2}(M, L+1)}{(1 - |z|^2)^2} \simeq \frac{M}{\pi\alpha} \frac{1}{(1 - |z|^2)^2} \Theta\Big(\frac{\alpha}{1+\alpha} - |z|^2\Big)$$

in the bulk and $R_1\big(\sqrt{\frac{\alpha}{1+\alpha}} + \frac{x}{\sqrt{M}}\big) \simeq \frac{M}{2\pi} \frac{(1+\alpha)^2}{\alpha} \operatorname{erfc}\big(\sqrt{2}\, \frac{1+\alpha}{\sqrt{\alpha}}\, x\big)$ at the edge. Modulo a simple rescaling, the edge density is the same as in the complex Ginibre ensemble. It can also be seen that the average number of eigenvalues outside the boundary $\simeq \sqrt{M(1+\alpha)/(2\pi)}$.

One can also find the eigenvalue correlation functions in the bulk. At the origin this task is especially simple. Scaling z by $\sqrt{\pi L}$, which is the mean distance between the eigenvalues at the origin, and noticing that for u in a bounded region in the complex plane $\sum_{m=0}^{M-1} \binom{L+m}{m} \left(\frac{u}{L}\right)^m \simeq e^u$ in the strong non-unitarity limit, one concludes that the rescaled correlation functions are given by Ginibre's expression (18.2.13). Although the eigenvalue distribution is not now homogeneous, after appropriate rescaling (18.2.13) describes the eigenvalue correlations at any point in the bulk [Zyc00].

The gap probability at the origin, $H(s;0)$, can be computed by exploiting the rotational symmetry, in the same way as for the Ginibre ensemble, $H(s;0) = \prod_{m=1}^{M-1} \big(1 - I_{s^2}(m+1, L)\big)$. After appropriate rescaling one obtains the same expression as in the Ginibre ensemble (compare (18.2.15)), $H\Big(\frac{s}{\sqrt{L}};0\Big) \simeq \prod_{m=1}^{\infty} \Gamma(m+1, s^2)/\Gamma(m+1)$ in the limit $M, L \to \infty$, $M/L = \alpha$, leading to the cubic law of eigenvalue repulsion in the bulk.

Now, consider the limit of weak non-unitarity $M \to \infty$, L is finite. In this limit the eigenvalues of J lie close to the unit circle, with the magnitude of the

[1] The integral $I_x(a, b)$ is convenient for finding the relevant limits by the Laplace method. For example, one finds that $I_x(a, b) \simeq \Theta(x - \frac{\alpha}{1+\alpha})$ as $a, b \to \infty$ and $\alpha = a/b$. The remainder term here is exponentially small when x is away from the transitional point $x_0 = \frac{\alpha}{(1+\alpha)}$. In the transitional region $I_{x_0 + \frac{t}{\sqrt{a}}}(a, b) \simeq 1 - \frac{1}{2}\operatorname{erfc}\big(\frac{(1+\alpha)^{3/2}}{\sqrt{2}\,\alpha}\, t\big)$. Another asymptotic relation of interest is $I_{1-\frac{y}{a}}(a, b) \simeq \Gamma(b, y)/\Gamma(b)$ which holds for positive y in the limit when $a \to \infty$, and b is fixed.

typical deviation being of the order $1/M$. Scaling z accordingly, one finds the eigenvalue density in the limit of weak non-unitarity [Zyc00]:

$$R_1\left(\left(1-\frac{y}{M}\right)\mathrm{e}^{\mathrm{i}\phi}\right) \simeq \frac{M^2}{\pi}\frac{(2y)^{L-1}}{(L-1)!}\int_0^1 \mathrm{e}^{-2yt}t^L\,\mathrm{d}t\,, \quad M\to\infty \text{ and } L \text{ is fixed.}$$

Setting $z_j = \left(1-\frac{y_j}{M}\right)\mathrm{e}^{\mathrm{i}\varphi_0+\mathrm{i}\frac{\varphi_j}{M}}$, one finds the correlations in this limit:

$$R_n(z_1,\ldots,z_n)\simeq\left(\frac{M^2}{\pi}\right)^n\prod_{j=1}^n\frac{(2y_j)^{L-1}}{(L-1)!}\det\left(\int_0^1 \mathrm{e}^{-(y_i+y_j+\mathrm{i}(\varphi_i-\varphi_j))t}\,t^L\,\mathrm{d}t\right)^n_{i,j=1}.$$

Interestingly, a different random matrix ensemble, $J = H + \mathrm{i}\gamma W$, leads to the same form of the correlation functions (after appropriate rescaling) as on the right-hand side above [Fyo99]. Here H is drawn from the GUE, $\gamma > 0$ and W is a diagonal matrix with L 1s and M zeros on the diagonal, with $M \gg 1$ and finite L. The above equation is a particular case of a universal formula describing correlations in more general ensembles of random contractions and non-Hermitian finite rank deviations from the GUE, see [Fyo03] and Chapter 34 for discussion and results.

18.4 Complex elliptic ensemble

Let H_1, H_2 be two independent samples from distribution $\mathrm{d}\mu(H) = \mathrm{e}^{-\mathrm{tr}\,H^2}|DH|$ on the space of Hermitian $N\times N$ matrices. Then $J = \sqrt{1+\tau}H_1 + i\sqrt{1-\tau}H_2$ is a random matrix ensemble interpolating between the (circular) Ginibre ensemble ($\tau = 0$) and the GUE ($\tau = 1$). We only consider the interval $0 \le \tau \le 1$. The matrix J is complex Gaussian,

$$\mathrm{d}\mu(J) \;\propto\; \exp\left\{-\frac{1}{1-\tau^2}\mathrm{tr}\left[JJ^\dagger - \frac{\tau}{2}(J^2+J^{\dagger 2})\right]\right\}\,|\mathrm{D}J|. \qquad (18.4.1)$$

The jpdf, $P(z_1,\ldots,z_N)$, can be obtained by bringing J to triangular form, as in Section 18.2. The resulting expression is given by (18.2.6) with weight function $w_\tau(z) = \frac{1}{\pi\sqrt{1-\tau^2}}\exp\{-\frac{|z|^2}{1-\tau^2}+\frac{\tau(z^2+z^{*2})}{2(1-\tau^2)}\}$. The associated orthogonal polynomials are scaled Hermite polynomials [DiF94]. Indeed, by making use of the integral representation $H_n(z) = \frac{n!}{2\pi\mathrm{i}}\oint \mathrm{e}^{2zt-t^2}t^{-(n+1)}\,dt$, for the Hermite polynomials $H_n(z)$ one can easily verify that

$$\int_{\mathbb{C}} H_n\Big(\frac{z}{\sqrt{2\tau}}\Big)H_m\Big(\frac{z^*}{\sqrt{2\tau}}\Big)w_\tau(z)\,\mathrm{d}^2z = \delta_{m,n}\,n!\Big(\frac{2}{\tau}\Big)^n. \qquad (18.4.2)$$

The monic polynomials $p_n(z)$ are easily found, $p_n(z) = (\tau/2)^{n/2}H_n(z/\sqrt{2\tau})$, $h_n = n!$, and on applying the orthogonal polynomial formalism, one obtains the correlation functions in the determinantal form (18.2.9) with

$$K(z_1, z_2) = w_\tau^{1/2}(z_1) w_\tau^{1/2}(z_2^*) \sum_{n=0}^{N-1} \frac{\tau^n}{2^n n!} H_n\Big(\frac{z_1}{\sqrt{2\tau}}\Big) H_n\Big(\frac{z_2^*}{\sqrt{2\tau}}\Big). \tag{18.4.3}$$

The sum on the right is the truncated exponential series in Mehler's formula

$$\sum_{n=0}^{\infty} \frac{\tau^n}{2^n n!} H_n\Big(\frac{z_1}{\sqrt{2\tau}}\Big) H_n\Big(\frac{z_2^*}{\sqrt{2\tau}}\Big) = \frac{1}{\sqrt{1-\tau^2}} \exp\left\{\frac{z_1 z_2^*}{1-\tau^2} - \frac{\tau(z_1^2 + z_2^{*2})}{2(1-\tau^2)}\right\}.$$

A quick comparison of (18.4.3) with Mehler's formula convinces us that the density $K(z, z)$ is constant in the limit $N \to \infty$. More care is needed to determine the boundary of the eigenvalue support. To this end, another integral for the Hermite polynomials comes in handy, $H_n(z) = (\pm 2\mathrm{i})^n \int_{-\infty}^{+\infty} \mathrm{e}^{-(t\pm \mathrm{i}z)^2} t^n \, \mathrm{d}t/\sqrt{\pi}$. On substituting this into (18.4.3) and after a simple change of variables, one writes the kernel in a form suitable for asymptotic analysis

$$K(z_1, z_2) = \frac{1}{\pi(1-\tau^2)} \exp\left\{-\frac{|z_1|^2 + |z_2|^2 - 2z_1 z_2^*}{2(1-\tau^2)}\right\} \tag{18.4.4}$$

$$\times \int_{-\infty}^{+\infty} \mathrm{d}u \int_{-\infty}^{+\infty} \mathrm{d}v \, f^-(u) f^+(v) \, \frac{\Gamma(N, N\tau(u^2 - v^2))}{\Gamma(N)}, \tag{18.4.5}$$

with $f^\pm(q) = \sqrt{\frac{N(1\pm\tau)}{\pi}} \exp\left\{-\left(q\sqrt{N(1\pm\tau)} - \frac{\mathrm{i}(z_1 \pm z_2^*)}{2\sqrt{\tau(1\pm\tau)}}\right)^2\right\}$.

For N large with $1-\tau > 0$ uniformly in N and $x, y = O(\sqrt{N})$, the functions $f^\pm$ can formally be replaced by delta functions, $f^+(v) = \delta\big(v - \frac{\mathrm{i}x}{(1+\tau)\sqrt{N\tau}}\big)$ and $f^-(u) = \delta\big(u + \frac{y}{(1-\tau)\sqrt{N\tau}}\big)$. Such a replacement can be justified by deforming the v-integral into the complex plane to pick up the sharp peak of $f^+(v)$ along the imaginary axis. This gives a simpler expression for the eigenvalue density $R_1(z) \simeq (\pi(1-\tau^2))^{-1} \Gamma\big(N, \frac{x^2}{(1+\tau)^2} + \frac{y^2}{(1-\tau)^2}\big)/\Gamma(N)$, cf (18.2.11). Hence, for large N the eigenvalues density is $1/\pi(1-\tau^2)$ inside the ellipse with half-axes $\sqrt{N}(1\pm\tau)$ along the x and y directions, in agreement with Girko's elliptic law [Gir86]. It falls to zero exponentially fast when one transverses the boundary of the ellipse.

For z_1, z_2 inside the ellipse and such that $|z_1 - z_2| = O(1)$ the integral in (18.4.5) converges to 1 as $N \to \infty$, and after a trivial rescaling one obtains the same expression (18.2.13) for the correlations as in the circular case (at the origin this readily follows from Mehler's formula).

The limit $N \to \infty$, $1-\tau > 0$ is the limit of strong non-Hermiticity. There is another important limit, $N \to \infty$ and $1-\tau = \alpha^2/N$, the so-called limit of weak non-Hermiticity [Fyo97] describing the crossover from Wigner-Dyson to Ginibre eigenvalue statistics. In this limit the eigenvalues of J can be thought of as those of $\sqrt{2}H_1$ (the Hermitian part of J) displaced from the real axis by the 'perturbation' term $\mathrm{i}\frac{\alpha}{\sqrt{N}} H_2$ (the skew-Hermitian part of J). The eigenvalues

x_j of $\sqrt{2}H_1$ fill the interval $(-2\sqrt{N}, 2\sqrt{N})$ with density $\rho_{sc}(x) = \frac{1}{\pi}\sqrt{N - x^2/4}$ (Wigner's semicircle law). A simple perturbation theory calculation [Fyo97] gives the density of eigenvalues $z = x + iy$ of J in the factorized form $\rho_{sc}(x)\rho(y)$, with the Gaussian distribution $\rho(y) = \sqrt{N/2\pi\alpha^2}\,\exp(-Ny^2/2\alpha^2)$ of the displacements. For such a calculation to be well defined, the width of this distribution should be much smaller than the (mean) eigenvalue spacing $1/\rho_{sc}(x)$ of the unperturbed eigenvalues, making it natural introducing the control parameter $a(x) = \alpha\rho_{sc}(x)/\sqrt{N}$. Formulae (18.4.4)–(18.4.5) make it possible to go beyond the perturbation theory and compute the eigenvalue density and correlations in the limit of weak non-Hermiticity exactly. In this limit the function $f^+(v)$ is singular and the same as in the limit of strong non-Hermiticity, however the function $f^-(u)$ is not. Let $z_j = x + \zeta_j/\rho_{sc}(x)$. Then, to the leading order, $f^-(u) \simeq \pi^{-1/2}a\exp\left(ua - \frac{i(\zeta_1-\zeta_2^*)}{2a\rho_{sc}(x)}\right)^2$ and the integral in (18.4.5) $\simeq$ $2\sqrt{\pi}\,a\,\exp\{\frac{(\zeta_1-\zeta_2^*)^2}{4a^2}\}\int_0^1 e^{-\pi^2a^2u^2}\cos(\pi u(\zeta_1 - \zeta_2^*))\,du$, where $a \equiv a(x)$ is the control parameter defined above. One then obtains the scaled correlation functions $\tilde{R}_n(\zeta_1, \ldots, \zeta_n) = (\rho_{sc}^2(x))^{-n} R_n(z_1, \ldots, z_n)$ in the form [Fyo97]:

$$\tilde{R}_n(\zeta_1, \ldots, \zeta_n) \simeq \left(\frac{1}{\sqrt{\pi}a}\right)^n e^{-\frac{1}{a^2}\sum_j(\operatorname{Im}\zeta_j)^2}\det\left(\int_0^1 e^{-\pi^2a^2u^2}\cos(\pi u(\zeta_i - \zeta_j^*))\,du\right). \tag{18.4.6}$$

It is easy to check that $\tilde{R}_n$ interpolates between the Wigner-Dyson correlations ($a \to 0$) and Ginibre's ($a \to \infty$), and the above equation allows one to study the crossover from one set of eigenvalue statistics to the other, see [Fyo98]. The eigenvalue density is

$$\tilde{R}_1(\zeta) \simeq \frac{1}{\sqrt{\pi}a}\,e^{-\frac{(\operatorname{Im}\zeta)^2}{a^2}}\int_0^1 e^{-\pi^2a^2u^2}\cosh(2\pi u\operatorname{Im}\zeta)\,du, \tag{18.4.7}$$

correcting the perturbation theory result. Interestingly, the distribution of the scaled imaginary parts of eigenvalues as given in (18.4.7) appears to be universal. A supersymmetry calculation shows [Fyo98] that it does not depend on the details of the Hermitian and skew-Hermitian parts of J and extends to non-invariant matrix distributions (like the semicircular and circular laws). It is an open challenging problem to find a proof of this universality satisfying the rigour of pure mathematics. Staying within the class of invariant distributions, the eigenvalue density and correlations also appear to be universal: it was argued in [Ake03] that the weakly non-Hermitian limit in the ensemble (18.2.4) at the origin is also described by (18.4.6).

In conclusion, we would like to mention briefly two topics related to the complex elliptic ensemble. One is the recent studies of edge scaling limits [Gar02, Ben09]. By scaling τ with N so that $1 - \tau = \alpha/N^{1/3}$ [Ben09] one gets access to the crossover from Airy ($\alpha \ll 1$, GUE) to Poisson edge statistics

($a \gg 1$, Ginibre). And the other is the chiral extension $J = \begin{pmatrix} 0 & iA+\mu B \\ iA^\dagger+\mu B^\dagger & 0 \end{pmatrix}$ of the complex elliptic ensemble. Here A and B are two independent samples from the Gaussian measure with density $e^{-\mathrm{tr}\, A^\dagger A}$ on the space of complex $(N+\nu)\times N$ matrices, $\nu \geq 0$, $0 \leq \mu \leq 1$. This ensemble can be studied along the same lines as (18.4.1). Computation of the jpdf of eigenvalues [Osb04] is though more involved and leads to (18.2.6) with weight function $w(z) = |z|^{2\nu+2} e^{-\frac{1-\mu^2}{4\mu^2}(z^2+z^{*2})} K_\nu(\frac{1+\mu^2}{2\mu^2}|z|^2)$ where K_ν is a modified Bessel function with the associated orthogonal polynomials being scaled Laguerre polynomials [Osb04, Ake05a]. At $\mu = 0$ we recover the Wishart (Laguerre) ensemble of Hermitian matrices and the corresponding weakly non-Hermitian limit $N \to \infty$ and $a = N\mu^2 = O(1)$ describes its neighbourhood. The eigenvalue correlations in this limit were computed in [Osb04], with the answer being somewhat different from (18.4.6). This is not surprising given that the two ensembles belong to different symmetry classes. By letting $a \to \infty$ one obtains the correlations in the chiral ensemble in the regime of strong non-Hermiticity, see, e.g. [Ake05b].

18.5 Real and quaternion-real Ginibre ensembles

Measure, change of variables Restricting to even dimensions N, we will treat the real and quaternion real (qu-r) ensembles in a unifying way. Consider the normalized measure

$$d\mu(J) = \exp(-\frac{1}{2}\mathrm{tr}\, J J^\dagger)\, |DJ| \tag{18.5.1}$$

on the space of $N \times N$ matrices subject to the constraints

$$\begin{aligned} J^T &= J^\dagger && \text{in the real case} \\ ZJ^T Z^T &= J^\dagger && \text{in the quaternion-real (qu-r) case,} \end{aligned} \tag{18.5.2}$$

where $Z = \bigoplus_{j=1}^{N/2} \begin{pmatrix} 0 & 1 \\ -1 & 0 \end{pmatrix}$ is the symplectic unit. Here $DJ = \prod_{i,j=1}^{N} (dJ_{ij}/\sqrt{2\pi})$ is the (exterior) product of the one-forms in matrix entries and $|DJ|$ is the corresponding Cartesian volume element.

As follows from (18.5.2), both real and qu-r matrices have non-real eigenvalues occurring in pairs z and z^*, and any real eigenvalue of a qu-r matrix has multiplicity ≥ 2. Hence, the probability for a qu-r matrix drawn from the distribution (18.5.1) to have a real eigenvalue is zero.

As with complex matrices, the jpdf of eigenvalues can be obtained from a Schur decomposition. Because of the symmetries, it is convenient to work with matrices partitioned into 2×2 blocks. Ignoring multiple eigenvalues, we can bring J to block-triangular form, $J = U(\Delta + \Lambda)U^{-1}$ where Λ is block-diagonal and Δ has nonzero blocks only above Λ. The matrix U is orthogonal in the real case and unitary symplectic in the qu-r case and can be restricted to the space of right cosets, $O(N)/O(2)^{N/2}$ and $USp(N)/USp(2)^{N/2}$ respectively.

The variations in J are related to those in U, Λ, Δ by

$$U^{-1}\mathrm{d}J\,U = \mathrm{d}\Lambda + \mathrm{d}\Delta + [U^{-1}\mathrm{d}U, \Delta] + [U^{-1}\mathrm{d}U, \Lambda], \tag{18.5.3}$$

with $[A, B] = AB - BA$. The matrix $U^{-1}\mathrm{d}U$ is skew-symmetric in the real case, $(U^{-1}\mathrm{d}U)^T = -(U^{-1}\mathrm{d}U)$, and anti self-dual in the qu-r case, $Z(U^{-1}\mathrm{d}U)^T Z^T = -(U^{-1}\mathrm{d}U)$.

To find the Jacobian associated with the change of variables from J to U, Λ, Δ, we multiply entries of the matrix on the right-hand side in (18.5.3) using the calculus of alternating differential forms. The matrices $\mathrm{d}\Lambda$ and $\mathrm{d}\Delta$ yield the volume forms for Λ and Δ, respectively. Because of the triangular structure, the matrix $[U^{-1}\mathrm{d}U, \Delta]$ does not contribute, and the matrix $[U^{-1}\mathrm{d}U, \Lambda]$ yields the coset volume form times a factor depending on the eigenvalues λ_j of Λ,

$$\prod{}'(U^{-1}\mathrm{d}U\Lambda - \Lambda U^{-1}\mathrm{d}U)_{ij} = \prod{}'(U^{-1}\mathrm{d}U)_{ij}(\lambda_j - \lambda_i),$$

with the dashed product running over nonzero entries in the lower triangle of $U^{-1}\mathrm{d}U$. On gathering all terms, we arrive at

$$\mathrm{d}\mu(J) = \mathrm{e}^{-\frac{1}{2}\mathrm{tr}\,(\Delta\Delta^\dagger + \Lambda\Lambda^\dagger)}\,|D\Delta|\,|D\Lambda|\,|\prod{}'(U^{-1}\mathrm{d}U)_{ij}(\lambda_j - \lambda_i)/\sqrt{2\pi}| \tag{18.5.4}$$

with $D\Lambda = \prod\,(\mathrm{d}\Lambda_{ij}/\sqrt{2\pi})$ and $D\Delta = \prod\,(\mathrm{d}\Delta_{ij}/\sqrt{2\pi})$, the products running over nonzero entries. The jpdf of eigenvalues follows from (18.5.4) on integrating out all auxiliary variables. Since the density function does not depend on U, the U-integral gives the volume of the coset space. The integral in Δ is Gaussian, and this variable can be integrated out with ease as well. Thus we are left with the problem of integrating over the $N/2$ (2×2)-blocks appearing in Λ keeping the eigenvalues in each block fixed. These are precisely the eigenvalues of the matrix J and our problem reduces to 2×2 matrices.

Dimension $\mathbf{N = 2}$ and the jpdf The generic form of a qu-r 2×2 matrix is $J = \left(\begin{smallmatrix} a & b \\ -b^* & a^* \end{smallmatrix}\right)$ with a and b complex. The eigenvalues are complex conjugate, $\lambda_{1,2} = \mathrm{Re}\ a \pm \mathrm{i}\sqrt{(\mathrm{Im}\ a)^2 + |b|^2}$, and we choose $\mathrm{Im}\,\lambda_1 = -\,\mathrm{Im}\,\lambda_2 > 0$. Thus, $\int \mathrm{d}\mu(J)\,\delta^2(\lambda_1 - \mathrm{Re}\,a - \mathrm{i}\sqrt{(\mathrm{Im}\,a)^2 + |b|^2}) = \frac{|\lambda_1 - \lambda_2|}{\pi} f^q(\lambda_1)^2$ with $f^q(\lambda)$ as in (18.5.6). The right-hand side is normalized wrt integrating over λ_1 in the half-plane $\mathrm{Im}\,\lambda_1 > 0$.

Turning to real 2×2 matrices $J = \left(\begin{smallmatrix} a & b \\ c & d \end{smallmatrix}\right)$ with a, b, c, d real, the eigenvalues are $\lambda_{1,2} = \frac{1}{2}(a + d \pm \sqrt{(a-b)^2 - 4bc})$, and we now have two possibilities (i) $\lambda_1 = \lambda_2^*$ (choosing $\mathrm{Im}\,\lambda_1 = -\,\mathrm{Im}\,\lambda_2 > 0$), or (ii) both λ_1 and λ_2 are real (choosing $\lambda_1 > \lambda_2$). In the first case we obtain [Leh91, Som08] $\int \mathrm{d}\mu(J)\,\delta^2(\lambda_1 - \frac{1}{2}[a + d - \mathrm{i}\sqrt{4bc - (a-b)^2}]) = \frac{|\lambda_1 - \lambda_2|}{\sqrt{2\pi}} f^r(\lambda_1)^2$ with $f^r(\lambda)$ as in (18.5.6). In the second case

($\lambda_{1,2}$ real) the corresponding average over the product of two δ-constraints results in $\frac{|\lambda_1-\lambda_2|}{2\sqrt{2\pi}} f^r(\lambda_1) f^r(\lambda_2)$ [Som08].

Consequently, the joint distribution of eigenvalues for both ensembles real and qu-r can be written in the form [Gin65, Meh04, Som08]

$$\mathrm{d}\mu(\lambda_1, \lambda_2, \ldots, \lambda_N) = C_N \prod_{1 \le i < j \le N} (\lambda_i - \lambda_j) \prod_{i=1}^{N} f(\lambda_i)\, \mathrm{d}\lambda_1 \mathrm{d}\lambda_2 \cdots \mathrm{d}\lambda_N, \tag{18.5.5}$$

where

$$f(\lambda)^2 = \begin{cases} f^r(\lambda)^2 = \operatorname{erfc}\left(\frac{|\lambda-\lambda^*|}{\sqrt{2}}\right) \mathrm{e}^{-\frac{1}{2}(\lambda^2+\lambda^{*2})} & \text{in the real case} \\ f^q(\lambda)^2 = |\lambda-\lambda^*|\, \mathrm{e}^{-|\lambda|^2} & \text{in the qu-r case,} \end{cases} \tag{18.5.6}$$

with $f(\lambda) = f(\lambda^*) \ge 0$ in both cases. We have to put the eigenvalues in such an order that $\mathrm{d}\mu \ge 0$. Thus we consider each case with m complex conjugate pairs of eigenvalues ($m = 0, 1, 2, \ldots, N/2$) separately, arranging the λ_is in (18.5.5) in the following order: for $m = 0$ (all real) $\lambda_1 > \lambda_2 > \ldots > \lambda_N$, for $m = 1$ (one complex conjugate pair $\lambda_1 = \lambda_2^*$) $\operatorname{Im}\lambda_1 > \operatorname{Im}\lambda_2$, $\lambda_3 > \lambda_4 > \ldots > \lambda_N$, for $m = 2$ (two complex conjugate pairs $\lambda_1 = \lambda_2^*$ and $\lambda_3 = \lambda_4^*$) $\operatorname{Im}\lambda_1 > \operatorname{Im}\lambda_2$, $\operatorname{Re}\lambda_2 > \operatorname{Re}\lambda_3$, $\operatorname{Im}\lambda_3 > \operatorname{Im}\lambda_4$, $\lambda_5 > \ldots > \lambda_N$, etc.. Summing and integrating over all cases and ranges yields total probability 1. In the qu-r case we only have $m = N/2$ complex conjugate pairs, and in the real case the probabilities of finding m complex conjugate pairs have been calculated in [Kan05b, For07].

The jpdf (18.5.5) immediately shows the repulsion behaviour of the eigenvalues. Due to the Vandermonde it contains for two eigenvalues in the upper half-plane the factor $|\lambda_1 - \lambda_2|^2$ which means cubic repulsion in the distance, the additional power coming from the two-dimensional volume element [Gro88]. For two eigenvalues exactly on the real axis one has the factor $|\lambda_1 - \lambda_2|$ and thus linear repulsion as in GOE. This applies only in the real case. In this case due to the factor $|\lambda_1 - \lambda_1^*|$ there is linear repulsion of complex eigenvalues with distance from the real axis, while in the qu-r case there is quadratic repulsion from the real axis due to an additional factor $|\lambda_1 - \lambda_1^*|$ coming from $f^q(\lambda_1)^2$.

Correlation functions as Pfaffians It is convenient to abandon the λ_js in favour of the real two-dimensional vectors $z_j = \lambda_j$ with $z_j = x_j + \mathrm{i}y_j$ and the two-dimensional volume element $\mathrm{d}^2z = \mathrm{d}x\,\mathrm{d}y$ (in contrast, in (18.5.5) $\mathrm{d}\lambda\mathrm{d}\lambda^* = -2\mathrm{i}\mathrm{d}x\,\mathrm{d}y$). Equation (18.5.5) defines the symmetrized jpdf of eigenvalues $P(z_1, \ldots, z_N)$ in the obvious way (however, care must be taken to account for complex conjugate pairs and the ordering of eigenvalues before symmetrizing the density). $P(z_1, \ldots, z_N)$ is a formal density, as it contains delta functions accounting for complex conjugate pairs and also for eigenvalues on the real line.

The eigenvalue correlation functions (18.2.5) (more precisely they are measures, since they contain delta-function contributions) can be obtained by functional derivatives from the generating functional [Tra98, Kan02, Som08]

$$Z[g] = \int d^2 z_1 \dots \int d^2 z_N \, g(z_1) \dots g(z_N) \, P(z_1, z_2, ..., z_N) \qquad (18.5.7)$$

as $R_n(z_1, z_2, \dots, z_n) = \frac{\delta^n}{\delta g(z_1)\dots\delta g(z_n)} Z[g]\big|_{g(z)\equiv 1}$. They can also be obtained by integration from the symmetrized jpdf via (18.2.5). This implies a simple integration theorem for the full correlations R_n (a Pfaffian analogue of Mehta's 'integrating out' lemma, see (18.2.9)). Restricting to the smooth (in the upper half-plane) and singular (on the real line) parts of the R_n leads to a more sophisticated integration theorem [Kan05b] connecting these quantities recursively. We find it more convenient to work with the generating function $Z[g]$ which can be calculated in the form of a Pfaffian with the help of the Grassmann integral representation of the Vandermonde determinant [Som08] $\prod_{1\le i<j\le N}(z_i - z_j) = \int d\chi_1 \dots d\chi_N \prod_{i=1}^{N}(\sum_{k=1}^{N} z_i^{k-1}\chi_k)$ with $\chi_k\chi_l = -\chi_l\chi_k$. Since the integrand $g(z_1)\dots g(z_N)$ in (18.5.7) is symmetric in the z_is, one can integrate it against the joint distribution of ordered eigenvalues as in (18.5.5) in order to obtain $Z[g]$. This helps to avoid the absolute values in the Vandermonde determinant, making it possible to apply the Grassmann integral above. Recalling the Grassmann integral representation $\mathrm{Pfaff}(A_{kl}) = \int d\chi_1 \dots d\chi_N \exp(-\frac{1}{2}\sum_{kl}\chi_k A_{kl}\chi_l)$ for the Pfaffian of an antisymmetric matrix (A_{kl}) one finds immediately

$$Z[g] = C_N \, \mathrm{Pfaff}(\tilde{A}_{kl}), \qquad (18.5.8)$$

with $\tilde{A}_{kl} = \int\int d^2z_1 d^2z_2 \, \mathcal{F}(z_1, z_2)\, z_1^{k-1} z_2^{l-1} \, g(z_1) g(z_2)$ and

$$\mathcal{F}(z_1, z_2) = f(z_1) f(z_2) (\, 2\mathrm{i}\, \delta^2(z_1 - z_2^*) \,\mathrm{sgn}(y_1) + \delta(y_1)\delta(y_2)\,\mathrm{sgn}(x_2 - x_1)\,). \qquad (18.5.9)$$

This simple form of $\mathcal{F}$ is a consequence of the fact that in (18.5.5) the normalization constant does not depend on the chosen number m of complex conjugate pairs. The first and second terms on the right-hand side account for eigenvalues coming in complex conjugate pairs and eigenvalues on the real axis, respectively. The second term vanishes in the qu-r case since $f^q = 0$ on the real axis.

Putting $g(z) = 1 + u(z)$ and expanding $Z[1+u]$ in powers of u one again obtains a series of Pfaffians [Som08]. This method goes back to Mehta's alternate variables [Meh04] and Tracy and Widom's paper [Tra98]. With $\tilde{A}_{kl}\big|_{g\equiv 1} = A_{kl}$, $A = (A_{kl})$ and defining the kernel

$$\mathcal{K}_N(z_1, z_2) = \sum_{k=1}^{N}\sum_{l=1}^{N} A_{kl}^{-1} z_1^{k-1} z_2^{l-1} \qquad (18.5.10)$$

the n-point densities are given by

$$R_n(z_1, \dots, z_n) = \mathrm{Pfaff}(Q_{kl}) \quad \text{with } Q_{kl} = \begin{pmatrix} K_{kl} & G_{kl} \\ -G_{lk} & W_{kl} \end{pmatrix}. \qquad (18.5.11)$$

This means the Pfaffian of the $2n \times 2n$ matrix built of the n^2 quaternions (2×2 matrices) Q_{kl}, $k, l = 1, 2, \ldots, n$ with entries $G_{kl} = \int d^2 z \mathcal{K}_N(z_k, z)\mathcal{F}(z, z_l)$, $K_{kl} = \mathcal{K}_N(z_k, z_l)$, and $W_{kl} = \int d^2 z \int d^2 z' \mathcal{F}(z_k, z)\mathcal{K}_N(z, z')\mathcal{F}(z', z_l) - \mathcal{F}(z_k, z_l)$ and symmetries $K_{kl} = -K_{lk}$, $W_{kl} = -W_{lk}$. With the above expressions one finds, e.g., the one-point density:

$$R_1(z_1) = \int d^2 z_2\, \mathcal{F}(z_1, z_2)\mathcal{K}_N(z_2, z_1) = R_1^C(z) + \delta(y)\, R_1^R(x) \tag{18.5.12}$$

and the two-point density, see Eq. (21) in [Som07]. Note that $\mathcal{F}(z_1, z_2)$ is composed of two parts. Correspondingly, $R_1(z)$ contains two parts, a smooth part $R_1^C(z)$ which describes the density of complex eigenvalues and a singular part $R_1^R(x)$ that describes the density of real eigenvalues. In the qu-r case $R_1^R(x) = 0$ and only the complex part remains. Restricting oneself to points $z_1, z_2, \ldots, z_n$ in the upper half-plane no terms containing delta functions like $\delta^2(z_i - z_j^*)$ appear in the correlation functions and (18.5.11) gives the correlations directly in terms of Pfaffians involving the kernel $\mathcal{K}_N$. Also, the real correlations can be obtained by considering only the terms concentrated on the real axis [Som07, For07]. These however still involve some real integrations [Som07].

Kernel, characteristic polynomials In order to use (18.5.11) one needs to know the kernel $\mathcal{K}_N(z_1, z_2)$ which we are now going to find. It follows from (18.5.12) and (18.5.9) that $R^C(z) = 2f(z)f(z^*)|\mathcal{K}_N(z, z^*)|$. On the other hand one can find $R_1^C(z)$ (and $R_1(z)$) directly from the jpdf (18.5.5) by integrating out $N-1$ variables. Since the eigenvalues are real or come in complex conjugate pairs this leads to $R_1^C(z) \propto |z - z^*|\langle \det(J - z)\det(J - z^*)\rangle_{N-2}$, where $\langle \ldots \rangle_{N-2}$ means averaging over the ensemble (18.5.1) in $N-2$ dimensions [Ede97]. On comparing the two expressions for $R_1^C(z)$ one arrives at the important relation:[2]

$$\mathcal{K}_N(z, z^*) \propto (z - z^*)\langle \det(J - z)\det(J - z^*)\rangle_{N-2}. \tag{18.5.13}$$

Thus, we are left with the task of calculating $\langle \det(J - u)\det(J - v)\rangle_N$. Writing the determinants as Grassmann integrals, $\det(J - u) = \int D\eta\, e^{-\eta^\dagger(J-u)\eta}$,

$$P_N(u, v) = \langle \det(J - u)\det(J - v)\rangle_N = \int D\eta D\zeta\, e^{u\eta^\dagger\eta + v\zeta^\dagger\zeta}\, e^{\frac{1}{2}\langle(\mathrm{tr}\, J(\eta\eta^\dagger + \zeta\zeta^\dagger)^2\rangle_N} \tag{18.5.14}$$

where we have used the Gaussian property. The lowest cumulants of J are $\langle J_{ij}\rangle_N = 0$ and $\langle J_{ij}J_{kl}\rangle_N = \Delta_{ik}\Delta_{jl}$, where in the real case $\Delta_{ik} = \delta_{ik}$, while in the qu-r case $\Delta_{ik} = Z_{ik}$ with Z being the symplectic unit.

[2] One can also express the kernel and/or correlation functions via averages of the characteristic polynomials in the qu-r and complex Ginibre ensembles and in some ensembles beyond Gaussian. The averages as on the right-hand side in (18.5.13) can be computed in a variety of ways and this gives an alternative way of calculating the eigenvalue correlation functions in the complex plane in a variety of random matrix ensembles, see [Ede94, Fyo99, Ake03, Ake07, Fyo07, Ake09a].

In the real case, after a complex Hubbard–Stratonovich (HS) transformation,

$$P_N(u,v) = \int \frac{\mathrm{d}^2 a}{\pi}\, \mathrm{e}^{-|a|^2} \left(\int \mathrm{d}\eta^* \mathrm{d}\eta \mathrm{d}\zeta^* \mathrm{d}\zeta \mathrm{e}^{u\eta^*\eta + v\zeta^*\zeta + ia\eta^\dagger\zeta^\dagger + i\bar{a}\eta\zeta} \right)^N . \tag{18.5.15}$$

On evaluating the simple Grassmann integration, one arrives at

$$P_N(u,v) = \int \frac{\mathrm{d}^2 a}{\pi}\, \mathrm{e}^{-|a|^2} (uv + |a|^2)^N = N! \sum_{n=0}^{N} \frac{(uv)^n}{n!}. \tag{18.5.16}$$

After restoring the normalization one finds the kernel

$$\mathcal{K}_N(z_1, z_2) = \frac{z_1 - z_2}{2\sqrt{2\pi}} \sum_{n=0}^{N-2} \frac{(z_1 z_2)^n}{n!} = \frac{z_1 - z_2}{2\sqrt{2\pi}}\, \mathrm{e}^{z_1 z_2}\, \Gamma(N-1,\, z_1 z_2). \tag{18.5.17}$$

This immediately gives the density of complex eigenvalues [Ede97], compare to (18.2.11),

$$R_1^C(z) = \frac{2|y|}{\sqrt{2\pi}}\, \mathrm{e}^{2y^2} \mathrm{erfc}\,(\sqrt{2}|y|)\, \frac{\Gamma(N-1, |z|^2)}{\Gamma(N-1)}, \qquad z = x + \mathrm{i}y. \tag{18.5.18}$$

and, in view of (18.5.12), the density of real eigenvalues [Ede94]

$$R_1^R(x) = \frac{\Gamma(N-1, x^2)}{\sqrt{2\pi}\Gamma(N-1)} + \frac{\mathrm{e}^{-x^2/2} x^{2N-2}}{2^{N-1/2}\Gamma(N/2)} \gamma^*\left(\frac{N-1}{2}, \frac{x^2}{2} \right), \tag{18.5.19}$$

with $\gamma^*(N,x) = x^{-N}(1 - \Gamma(N,x)/\Gamma(N))$. The right-hand side in (18.5.18) and (18.5.19) is analytic in N and these equations hold for odd N as well [Som08, For09a].

One can easily analyse (18.5.18) and (18.5.19) in the limit $N \to \infty$. For example, the average number $\langle n_R \rangle_N$ of real eigenvalues can be found by integration, recovering the result by Edelmann, Kostlan and Shub [Ede94]

$$\langle n_R \rangle_N = 1 + \frac{\sqrt{2}}{\pi} \int_0^1 \frac{\mathrm{d}t\; t^{1/2}(1 - t^{N-1})}{(1-t)^{3/2}(1+t)} \simeq \sqrt{\frac{2N}{\pi}} \tag{18.5.20}$$

and verifying the conjecture $\langle n_R \rangle_N \propto \sqrt{N}$ made in [Leh91]. Interestingly, the variance of n_R is also proportional to $\sqrt{N}$ [For07]. For large N and away from the real line ($|y| \gg 1$) the density of complex eigenvalues obeys the circular law (18.2.11) with the same edge profile (18.2.12) as in the complex Ginibre ensemble. Inside the circle $|z| < \sqrt{N}$ and close to the real line ($y = O(1)$) $R_1^C(z) \simeq \sqrt{2/\pi}\, |y|\, \mathrm{e}^{2y^2} \mathrm{erfc}\,(\sqrt{2}|y|)$. One also finds a constant density of real eigenvalues inside the same circle[Ede94] $R_1^R(x) \simeq \frac{1}{\sqrt{2\pi}} \Theta(\sqrt{N} - |x|)$ which is consistent with the asymptotics in (18.5.20). This constant density is in contrast with the Wigner semicircle law for Gaussian Hermitian or real symmetric matrices. In the transitional region around the end point $x = \sqrt{N}$ [For07, Bor09]

$$R_1^R(\sqrt{N}+u) \simeq \frac{1}{2\sqrt{2\pi}}\text{erfc}\,(\sqrt{2}u) + \frac{1}{4\sqrt{\pi}}\,\exp(-u^2)\text{erfc}\,(-u).$$

Although in the qu-r case $\mathcal{K}_N(z_1, z_2)$ can be obtained as a saddle point integral similarly to (18.5.16), it is convenient to derive it directly from the integral A_{kl}

$$A_{kl} = \int d^2z\, 2(z-z^*)e^{-|z|^2} z^{k-1} z^{*l-1} = 2\pi(k!\,\delta_{k,l-1} - l!\,\delta_{l,k-1}). \qquad (18.5.21)$$

We see that in the qu-r case the matrix (A_{kl}) has a simple tridiagonal structure, while in the real case its inverse is tridiagonal (as evident from (18.5.17)). Due to this structure we were able to find a beautiful formula for all Schur function averages (moments) in the real Ginibre ensemble [Som09]. With this formula one can calculate the moments of symmetric functions in eigenvalues by making use of the Schur function expansion. A similar formula exists for the qu-r Ginibre ensemble [For09b].

Using the duplication formula for $\Gamma(z)$ and introducing $a_k = 2^{k/2}\Gamma(k/2)$,

$$A_{kl} = -\sqrt{\pi/2}\, a_{k+1}a_{l+1}\,\epsilon_{kl} \text{ in the qu-r case,} \qquad A_{kl} = a_k\, a_l\, \epsilon_{kl}^{-1} \text{ in the real case,}$$

where $(\epsilon_{kl}) = \text{tridiag}(1, 0, -1)$ is the antisymmetric tridiagonal matrix with 1s below the main diagonal and -1's above and its inverse (also antisymmetric) $\epsilon_{kl}^{-1} = 1$ for k odd and j even in the upper triangle $k < j$ and the remaining entries being zero in that triangle. From these formulae and (18.5.10) one finds the kernel $\mathcal{K}_N(z_1, z_2)$ in the qu-r case thus recovering Mehta's result [Meh04]

$$\mathcal{K}_N(z_1, z_2) = -\frac{1}{\sqrt{2\pi}} \sum_{k=1}^{N}\sum_{l=1}^{N} \frac{2^{-(k+l)/2}}{\Gamma((k+1)/2))\,\Gamma((l+1)/2)}\,\epsilon_{kl}^{-1}\, z_1^{k-1} z_2^{l-1}.$$

Also, the normalization constant C_N in (18.5.5) follows via (18.5.8)

$$\frac{1}{C_N} = \begin{cases} \left(\frac{\pi}{2}\right)^{N/4} \prod_{k=1}^{N} a_{k+1} = (2\pi)^{N/2}\; 1!\; 3! \ldots (N-1)! & \text{in the qu-r case} \\ \prod_{k=1}^{N} a_k = (2\sqrt{2\pi})^{N/2}\; 0!\; 2! \ldots (N-2)! & \text{in the real case.} \end{cases} \qquad (18.5.22)$$

We see that the problem with all correlations and moments is to calculate the matrix A_{kl} and its inverse A_{kl}^{-1}. In the qu-r case the calculation of A_{kl} is simple and leads then to A_{kl}^{-1}. In the real case the calculation of A_{kl}^{-1} via characteristic polynomials is simple and a direct calculation of A_{kl} and taking then the inverse is much more involved but possible with the help of the method of skew orthogonal polynomials wrt to the form $\mathcal{F}(z_1, z_2)$, as was demonstrated in [For07, Bor09]. The method of skew-orthogonal polynomials can also be used to derive the correlation functions in the qu-r ensemble [Kan02]. We will quote these results in connection with more general elliptic ensembles.

18.6 Real and quaternion-real elliptic ensembles

Consider now the two families of normalized measures

$$d\mu_\tau(J) \;=\; B_\tau\, \mathrm{e}^{-\frac{1}{2(1-\tau^2)}\mathrm{tr}\,(JJ^\dagger - \frac{\tau}{2}(J^2+J^{\dagger 2}))}\, |\mathrm{D}J| \tag{18.6.1}$$

on the space of $N \times N$ matrices subject to the symmetry constraints (18.5.2). The normalization constant B_τ is $(1-\tau)^{-N(N-1)/2}(1+\tau)^{-N(N+1)/2}$ in the real case and $(1+\tau)^{-N(N-1)/2}(1-\tau)^{-N(N+1)/2}$ in the qu-r case. When τ varies between -1 and $+1$ $d\mu_\tau$ interpolates between anti-Hermitian (antisymmetric or antiselfdual, $\tau = -1$) and Hermitian (symmetric or selfdual, $\tau = +1$) ensembles. On comparing to (18.5.1), one concludes that the jpdf of eigenvalues in ensembles (18.6.1) can simply be obtained by a rescaling of the term $\mathrm{tr}\, JJ^\dagger$ and otherwise multiplying with the factor $\exp(\lambda^2 + \lambda^{*2})\tau/4(1-\tau^2))$, thus leading to the same expression (18.5.5) with $f(\lambda)$ replaced by $f_\tau(\lambda)$, with $f_\tau^2(\lambda) = f^2(\lambda/\sqrt{1-\tau^2}) \exp\{\frac{\tau}{2(1-\tau^2)}(\lambda^2 + \lambda^{*2})\}$ and a new normalization constant $C_{N,\tau}$. Correspondingly, the skew-symmetric form is given by (18.5.9), again with $f(z)$ replaced by $f_\tau(z)$. Hereby the matrices A_{kl}, A^{-1}_{kl}, the kernel $\mathcal{K}_N(z_1, z_2)$ and the correlations can be obtained.

Kernel in the elliptic case The kernel is again related to the average of two characteristic polynomials (18.5.13)–(18.5.14), but now the second cumulants of J are $\langle J_{ij} J_{kl}\rangle_N = \Delta_{ik}\Delta_{jl} + \tau\, \Delta_{il}\Delta_{jk}$ with Δ_{ik} as before.

In the real case one can evaluate the corresponding Grassmann integral by introducing two complex and two real HS transformations. This gives

$$P_N(u, v) = \int \frac{d^2a}{\pi}\frac{d^2b}{\pi}\mathrm{e}^{-|a|^2-|b|^2} \int \frac{dxdy}{2\pi}\mathrm{e}^{-(x^2+y^2)/2} P^N_{r,\tau} \tag{18.6.2}$$

where $P_{r,\tau}$ is the Pfaffian of a four-dimensional antisymmetric matrix equivalent to a fourfold Gaussian Grassmann integral with the value $P_{r,\tau} = (u + ix\sqrt{\tau})(v + iy\sqrt{\tau}) + |a|^2 + \tau|b|^2$. From (18.6.2) one can obtain [Ake09a] the kernel

$$\mathcal{K}_N(z_1, z_2) = \sum_{k,l=1}^{N} P_{k-1}(z_1)\frac{Z^{-1}_{kl}}{r_{l-1}} P_{l-1}(z_2) \tag{18.6.3}$$

expanded in terms of skew orthogonal monic polynomials $P_{k-1}(z)$ defined by

$$\int d^2z_1\, d^2z_2 \mathcal{F}(z_1, z_2) P_{k-1}(z_1) P_{l-1}(z_2) = Z_{kl}\, r_{l-1} = r_{k-1}\, Z_{kl}. \tag{18.6.4}$$

For $\tau > 0$, (18.6.3) was also obtained directly from the jpdf in [For08]. In this case the skew-orthogonal polynomials can be expressed [For08] in terms of the Hermite polynomials $P_{2n}(z) = p_{2n}(z)$ and $P_{2n+1} = p_{2n+1}(z) - 2np_{2n-1}(z)$ with $p_n(z) = (\tau/2)^{n/2} H_n(z/\sqrt{2\tau})$ being the scaled Hermite polynomials as in Section 18.4, and with $r_n = r_{n+1} = 2\sqrt{2\pi}\, n!\, (1+\tau)$ for n even.

In the qu-r case the saddle-point integral representation of the kernel following from its relation to the average of characteristic polynomials is more involved, however the kernel can again be expanded in skew orthogonal polynomials (for $\tau > 0$) [Kan02]. The resulting expression is the same as in (18.6.3) only now $P_{2n}(z) = \sum_{l=0}^{n} \frac{2^n n!}{2^l l!} p_{2l}(z)$ and $P_{2n+1} = p_{2n+1}(z)$, with the polynomials p_n as before and the constants r_n in (18.6.4) given by $r_n = r_{n-1} = 2\pi\, n!\,(1-\tau)$ for n odd. Note that the expressions for r_n immediately give the normalization constant in (18.5.5): $C_{N,\tau} = (1+\tau)^{-N/2} C_N$ in the real case and and $C_{N,\tau} = (1-\tau)^{-N/2} C_N$ in the qu-r case with C_N as in (18.5.22).

Having an explicit form for the kernel, all correlations, including in the real case the real-real and real-complex ones, can be found [For08, Bor09, Som08].

Strongly non-Hermitian limit In this limit u, v are assumed to be of order $\sqrt{N}$ as $N \to \infty$. We are going to evaluate the kernel $\mathcal{K}_N(u, v)$ (18.5.13) by the saddle-point analysis of $P_{N-2}(u, v)$. Let us start with the circular real case (18.5.16) which is the simplest. Then the saddle-point equations are $a^* = \frac{Na^*}{uv+aa^*}$ and $a = \frac{Na}{uv+aa^*}$ with two solutions $a = a^* = 0$ and $aa^* = N - uv$. The first saddle point yields $P_{N-2} \simeq (uv)^{N-2}/(1 - N/uv)$ and the second, which is actually a manifold, yields $P_{N-2} \simeq N^{N-2}\sqrt{2\pi N}\, \mathrm{e}^{uv-N}$. Thus the second is dominating for $\mathrm{Re}(uv) < N$ and otherwise the first is (excluding the neighbourhood of $N = uv$). Hence, the one-point density $R_1(z)$ is asymptotically $R_1(z) = 1/\pi$ for $|z| < \sqrt{N}$ and is exponentially small outside this circle. Also, this immediately gives the asymptotic form of the kernel in the circular case in the bulk ($\mathrm{Re}(z_1 z_2) < N$):

$$\mathcal{K}_N(z_1, z_2) \simeq (2\sqrt{2\pi})^{-1}(z_1 - z_2)\,\mathrm{e}^{z_1 z_2} \quad \text{as } N \to \infty. \tag{18.6.5}$$

Obviously, (18.6.5) could have been obtained directly from (18.5.17). Here we have determined in addition the region of validity. The correlations (n-point densities) are then given with this formula in the region $|z_j| < \sqrt{N}$, $j = 1, 2, \ldots, n$. Hence we call it the circular real case.

By similar reasoning we find in the circular qu-r case

$$\mathcal{K}_N(z_1, z_2) \simeq (2\pi(z_2 - z_1))^{-1}\,\mathrm{e}^{z_1 z_2} \quad \text{as } N \to \infty, \tag{18.6.6}$$

for $\mathrm{Re}(z_1 z_2) < N$ with both z_1 and z_2 (and also $z_1 - z_2$) being of order $\sqrt{N}$. This implies the circular law in the qu-r case: $R_1(z) \simeq 1/\pi$ inside the circle $|z| = \sqrt{N}$ and away from the real line ($\mathrm{Im}\, z \sim \sqrt{N}$) and $R_1(z) \simeq 0$ outside.

For the elliptic real case the relevant saddle point leads to the condition for the elliptic support of $R_1(z)$. This has already been found numerically and analytically using the replica trick in the early paper [Som88]. Similarly, for the elliptic qu-r case we expect it can be shown by a saddle-point analysis that in the large N-limit $R_1(z)$ is constant inside the ellipse with main half-axes $\sqrt{N}(1 \pm \tau)$ along the x-and y- directions, and zero outside.

Having the expressions (18.6.3) for the kernel in terms of skew orthogonal polynomials, one can find the scaling limit of the kernel and the correlations for N to infinity considering $z_{1,2}$ as being of order 1 and letting the edge going to infinity. Since the one-point density is constant it is not necessary to unfold. It turns out that in the real case for $\tau = 0$ this limit $\mathcal{K}_\infty(z_1, z_2)$ is already given by expression (18.6.5). For $\tau > 0$ $\mathcal{K}_\infty(z_1, z_2)$ has been calculated in [For08]

$$\mathcal{K}_\infty(z_1, z_2) = \frac{z_1 - z_2}{2\sqrt{2\pi}(1-\tau^2)^{3/2}} \exp\left(\frac{z_1 z_2}{1-\tau^2} - \frac{\tau(z_1^2 + z_2^2)}{2(1-\tau^2)}\right) \tag{18.6.7}$$

with the normalization corrected. Obviously this is also valid for $-1 < \tau < 0$. In the circular qu-r case ($\tau = 0$) the asymptotic form of the kernel has been calculated in [Kan02] and is already found in [Meh04] in a somewhat different form. Scaling it appropriately one obtains for the elliptic qu-r case

$$\mathcal{K}_\infty(z_1, z_2) = \frac{1}{2\sqrt{2\pi}(1-\tau^2)} \exp\left(\frac{z_1^2 + z_2^2}{2(1+\tau)}\right) \operatorname{erf}\left(\frac{z_1 - z_2}{\sqrt{2(1-\tau^2)}}\right). \tag{18.6.8}$$

Both expressions (18.6.7, 18.6.8) yield the bulk density $R_1(z) \simeq 1/\pi(1-\tau^2)$ corresponding to an elliptic support with main half-axes $\sqrt{N}(1 \pm \tau)$. In the real case the density of real eigenvalues is again asymptotically constant $R_1^R(x) \simeq 1/\sqrt{2\pi(1-\tau^2)}$ which together with the support $|x| < \sqrt{N}(1+\tau)$ gives a number of real eigenvalues of $\sqrt{2N(1+\tau)/\pi(1-\tau)}$ (compare (18.5.20)) [For08].

Weakly non-Hermitian limit Finally we consider the limits of weak non-Hermiticity [Fyo97]. Here we put $\tau = 1 - a^2$ with $1 - \tau$ being of order $1/N$ and let N go to infinity assuming z_i to be in the neighbourhood of the origin. In the real case in this limit the kernel is given by [For08]

$$\mathcal{K}_N(z_1, z_2) \simeq \frac{N}{2\pi} \int_0^1 du\, u \exp(-Na^2u^2) \sin(\sqrt{N}\, u(z_1 - z_2)), \tag{18.6.9}$$

and $f_\tau^2(z) \simeq \operatorname{erfc}(\frac{|z-z^*|}{2a})$. In the qu-r case $f_\tau^2(z) \simeq e^{\frac{(z-z^*)^2}{4a^2}}$ and [Kan02]

$$\mathcal{K}_N(z_1, z_2) \simeq \frac{\pi^{3/2}}{4a^3 N^{3/2}} \int_0^1 \frac{du}{u} \exp(-Na^2u^2) \sin(\sqrt{N}\, u(z_1 - z_2)). \tag{18.6.10}$$

As in the complex elliptic ensemble, it is convenient to scale z_1, z_2, a and $\mathcal{K}_N$ with the local mean level spacing (here $= \pi/\sqrt{N}$ since the ensembles go in this limit to GOE/GSE with the semicircular density of eigenvalues $\rho_{sc}(x) = \frac{1}{\pi}\sqrt{N - x^2/4}$). Then one obtains a universal form of the correlations. For example, on the large x-scale ($z = x + iy$) the one-point density can be written as $R_1(z) \simeq \rho_{sc}(x)^2 P(\rho_{sc}(x)y, \rho_{sc}(x)a)$ with the probability density function

$$P(y, a) = \delta(y) \int_0^1 du\, e^{-\pi^2 a^2 u^2} + \pi \text{erfc}\left(\frac{|y|}{a}\right) \int_0^1 du\, u\, e^{-\pi^2 a^2 u^2} \sinh(2\pi u |y|)$$

in the real case (GOE limit) [Efe97] and in the qu-r case (GSE limit) [Kol99]

$$P(y, a) = \frac{y}{\pi^{3/2} a^3} \exp\left(-\frac{y^2}{a^2}\right) \int_0^1 \frac{du}{u} \exp(-\pi^2 a^2 u^2) \sinh(2\pi u y).$$

Note that in the qu-r case despite the fact that the complex eigenvalue density vanishes exactly on the real axis in the limit pairs of complex conjugate eigenvalues approach the real axis and collapse giving Kramers degeneracy for the Hermitian ensemble with symplectic symmetry. In the real case despite the fact that the density of real eigenvalues becomes constant, but very low, complex eigenvalues approach the real axis and in the limit give rise to the Wigner semicircle density which can be considered as a projection of the elliptic law in the complex plane onto the real axis.

Acknowledgements

H-JS acknowledges support by SFB/TR12 of the Deutsche Forschungsgemeinschaft. Gernot Akemann, Yan Fyodorov, Eugene Kanzieper and Dmitry Savin are thanked for helpful comments on earlier versions of this manuscript.

References

[Ake03] G. Akemann, G. Vernizzi, Nucl. Phys. B **660 [FS]** (2003) 532
[Ake05a] G. Akemann, Nucl. Phys. B **730** (2005) 253
[Ake05b] G. Akemann, J.C. Osborn, K. Splittorf, J.J.M. Verbaarschot, Nucl. Phys. B **712** (2005) 287
[Ake07] G. Akemann, F. Basile, Nucl. Phys. B **766** (2007) 766
[Ake09a] G. Akemann, M. J. Phillips, H.-J. Sommers, J. Phys. A **42** (2009) 012001
[Ake09b] G. Akemann, M.J. Phillips, L.Shifrin, J. Math. Phys. **50** (2009) 063504
[Ake10] G. Akemann, M.J. Phillips, H.-J. Sommers, J. Phys. **A 43** (2010) 085211
[Ame08] Y. Ameur, H. Hedenmalm, N. Makarov, Duke Math. J. Volume 159, Number 1 (2011), 31.
[Ben09] M. Bender, Probability Theory and Related Fields (2009) doi: 10.1007/s00440-009-0207-9 (arXiv:0808.2608 [math.PR])
[Ber02] D. Bernard, A. LeClair, J. Phys A **35** (2002) 2555; also arXiv:cond-mat/0110649
[Bor09] A. Borodin, C.D. Sinclair, Commun. Math. Phys. **291** (2009) 177
[Bru09] W. Bruzda, V. Cappelini, H.-J. Sommers, K. Życzkowski, Phys. Lett. A **373** (2009) 320
[DiF94] F. Di Francesco, M. Gaudin, C. Itzykson, F. Lesage, Int. J. Mod. Phys. A **9** (1994) 4257
[Ede97] A. Edelman, J. Multivar. Anal. **60** (1997), 203
[Ede94] A. Edelman, E. Kostlan, M. Shub, J. Amer. Math. Soc. **7** (1994), 247
[Efe97] K.B. Efetov, Phys. Rev. Lett. **79** (1997), 491

[Fei97] J. Feinberg, A. Zee, Nucl. Phys. B **501** (1997) 643;
J. Feinberg, J. Phys. A **39** (2006) 10029.

[For97] P.J. Forrester, B. Jancovici, Int. J. Mod. Phys. A **11** (1997) 941

[For99] P.J. Forrester, G. Honner, J. Phys. A **32**, (1999), 2961

[For07] P.J. Forrester, T. Nagao, Phys. Rev. Lett. **99**, (2007), 050603

[For08] P.J. Forrester, T. Nagao, J. Phys. A **41** (2008), 375003

[For09a] P.J. Forrester, A. Mays, J. Stat. Phys. **134** (2009), 443

[For09b] P.J. Forrester, E. M. Rains, J. Phys. A **42** (2009) 385205

[Fyo97] Y.V. Fyodorov, B.A. Khoruzhenko, H.-J. Sommers, Phys. Lett. **A226** (1997) 46; Phys. Rev. Lett. **79** (1997) 557

[Fyo98] Y.V. Fyodorov, B.A. Khoruzhenko, H.-J. Sommers, Ann. Inst. H. Poincaré Phys. Theór. **68** (1998) 449

[Fyo99] Y.V. Fyodorov, B.A. Khoruzhenko, Phys. Rev. Lett. **83** (1999) 65

[Fyo03] Y.V. Fyodorov, H.-J. Sommers, J. Phys. A **36** (2003) 3303

[Fyo07] Y.V. Fyodorov, B.A. Khoruzhenko, Commun. Math. Phys. **273** (2007) 561

[Gar02] A.M. Garćia-Garćia, S.M. Nishigaki, J.J. Verbaarschot, Phys. Rev. E **66** (2002) 016132

[Gin65] J. Ginibre, J. Math. Phys. **6** (1965) 440

[Gir85] Girko V.L., Theor. Prob. Appl. **29** (1985) 694

[Gir86] Girko V.L., Theor. Prob. Appl. **30** (1986) 677

[Gro88] R. Grobe, F. Haake, H.-J. Sommers, Phys. Rev. Lett. **61** (1988) 1899

[Jia06] T. Jiang, Ann. Prob. **34** (2006) 1497

[Kwa06] J.Kwapien, S. Drozdz, A.Z. Gorski, F. Oswiecimka, Acta Phys. Pol. B **37** (2006) 3039

[Leh91] N. Lehmann, H.-J. Sommers, Phys. Rev. Lett. **67** (1991) 941

[May72] R.M. May, Nature **298** (1972) 413

[Meh04] M.L. Mehta, *Random Matrices*, 3rd ed., Academic Press, 2004.

[Kan02] E. Kanzieper, J. Phys. A **35** (2002) 6631

[Kan05a] E. Kanzieper, In: Frontiers in Field Theory, ed. O. Kovras, 2005 Nova Science Publ. pp. 23–51.

[Kan05b] E. Kanzieper, G. Akemann Phys. Rev. Lett. **95** (2005) 230201;
G. Akemann, E. Kanzieper, J. Stat. Phys. **129** (2007), 1159

[Kol99] A.V. Kolesnikov, K.B. Efetov, Waves Ran. Media **9** (1999) 71

[Mag08] U. Magnea, J. Phys. A **41** (2008) 045203

[Oas97] G. Oas, Phys. Rev. E **55** (1997) 205

[Osb04] J. C. Osborn, Phys. Rev. Lett. **93** (2004) 222001

[Som88] H.-J. Sommers, A. Crisanti, H. Sompolinsky, Y. Stein, Phys. Rev. Lett. **60** (1988) 1895

[Som07] H.-J. Sommers J. Phys. A **40** (2007) F671

[Som08] H.-J. Sommers, W. Wieczorek, J. Phys. A **41** (2008) 405003

[Som09] H.-J. Sommers, B.A. Khoruzhenko, J. Phys. A **42** (2009) 222002

[Tao08] T. Tao, V. Vu, M. Krishnapur, Ann. Probab. Volume 38, Number 5 (2010), 2023.

[Tra98] C. Tracy, H. Widom, J. Stat. Phys. **92** (1998) 809

[Zyc00] K. Życzkowski, H.-J. Sommers, J. Phys. A **33** (2000) 2045

·19·

Characteristic polynomials

E. Brézin and S. Hikami

Abstract

The average of products of characteristic polynomials provides new correlation functions in random matrix theory. They may be expressed as determinants, and they exhibit a useful duality property which interchanges the size of the random matrices with the number of points of these new correlators. The relation to the standard eigenvalue correlation functions is discussed by the replica method, and by the supersymmetric method. Universal factors for the asymptotic moments and universal correlators are derived in the large N limit of the averaged characteristic polynomials. The topological invariants of Riemann surfaces, such as the intersection numbers of the moduli space of curves, may also be derived from averaged characteristic polynomials.

19.1 Introduction

Characteristic polynomials of random matrices turn out to be useful objects in random matrix theory and its numerous applications in physics and mathematics. Introduced by Keating and Snaith [Kea00] in their work on the circular unitary ensemble and its number theoretic connection, they led to a number of new results in the theory of random matrices and its mathematical implications for classical groups [Zir96] [Bump06] and number theory (see Chapter 24). Several studies for the classical ensembles (GUE, GOE and GSE) [Bor06][Str03] and the universality of the asymptotic resulting kernels [Str03][Ake03a] were followed by applications to the Ginibre ensembles of complex matrices [Ake03b], to studies in quantum chromodynamics [Fyo03] (see Chapter 32) and many other topics in mathematical physics.

This chapter is devoted to a review of a few useful results obtained in simple Gaussian models of random Hermitian matrices in the presence of an external matrix source. In spite of the simplicity of Gaussian models, the tuning of the external matrix source may be used to derive results of non-trivial models, such as the Kontsevich [Kon92] model and its implications for computing the intersection numbers of the moduli of curves on Riemann surfaces of arbitrary genera. We consider here the characteritic polynomials

$$p(\lambda) = \det(\lambda - M) \tag{19.1.1}$$

where M is an $N \times N$ Hermitian random matrix. The average with respect to a probability distribution Pr(M), is defined as

$$F(\lambda) = \int dM \mathrm{Pr}(\mathrm{M}) \det(\lambda - M). \quad (19.1.2)$$

The Gaussian distribution in the presence of an external matrix source A is defined with

$$\mathrm{Pr}(\mathrm{M}) = \frac{1}{Z_N} \exp(-\frac{1}{2} \mathrm{tr} M^2 + \mathrm{tr} M A) \quad (19.1.3)$$

in which A is a given (i.e. non-random) Hermitian matrix.

The zeros of the characteristic polynomial $p(\lambda)$ are random numbers on the real axis, and the connections with the zeros of Riemann's zeta function on the critical line have been extensively studied. The two-point correlation of the zeros of the Riemann zeta function is described by Dyson's sine kernel [Mon73]. The distribution of spacings between the zeros of the zeta function agrees with the level spacing correlation in random matrix theory [Odl87].

The products of the characteristic polynomials

$$F_{k,N}(\lambda_1, \ldots, \lambda_k) =< \prod_{a=1}^{k} \det(\lambda_a - M) > \quad (19.1.4)$$

are a new type of correlators for the eigenvalues λ_a. Those new correlation functions exhibit a universal behaviour in the large N limit, for the Dyson limit $((\lambda_a - \lambda_\beta)N$ is fixed) and for the multicritical scaling regions near the edge of the spectrum.

In the degenerate case, in which all the λ_a are equal to λ, the above product of characteristic polynomial reduces to the k-th moment of the characteristic polynomial,

$$F_{k,N}(\lambda) =< [\det(\lambda - M)]^k > . \quad (19.1.5)$$

In this chapter, characteristic polynomials are discussed in relation to the orthogonal polynomials and kernels. Several useful formulae for products and the ratios are given for the Gaussian unitary ansemble GUE and for the orthogonal enssemble GOE. Duality formulae for characteristic polynomials in the presence of an external source are given and the intersection numbers of the moduli space of p-spin curves are obtained as a consequence of this duality.

19.2 Products of characteristic polynomials

We consider a general probability distribution $Pr(M)$, with a potential $V(M)$ which is a polynomial in M,

$$Pr(M) = \frac{1}{Z} e^{-N\mathrm{tr} V(M)} \tag{19.2.1}$$

where

$$Z = \int dM e^{-N\mathrm{tr} V(M)}. \tag{19.2.2}$$

The product (19.1.4) has a simple determinantal expression [Bre00] in terms of the orthogonal polynomials $q_n(\lambda)$ with respect to the weight $e^{-NV(\lambda)}$,

Theorem 19.2.1

$$\begin{aligned} F_{k,N}(\lambda_1, \ldots, \lambda_k) &= < \prod_{a=1}^{k} \det(\lambda_a - M) > \\ &= \frac{1}{\Delta(\lambda)} \det \begin{vmatrix} q_N(\lambda_1) & q_{N+1}(\lambda_1) & \cdots & q_{N+k-1}(\lambda_1) \\ \cdots & \cdots & \cdots & \cdots \\ q_N(\lambda_k) & q_{N+1}(\lambda_k) & \cdots & q_{N+k-1}(\lambda_k) \end{vmatrix} \end{aligned} \tag{19.2.3}$$

where the orthogonal polynomial $q_n(x) = x^n +$ lower order, and $\Delta(\lambda)$ is the Vandermonde determinant, $\Delta(\lambda) = \prod_{i<j}(\lambda_i - \lambda_j)$.

The Vandermonde determinant $\Delta(\lambda)$ may be written as $\det[q_n(x_m)]$ $(0 \le n \le k-1, 1 \le m \le k)$. For the degenerate case, $\lambda_a = \lambda$, $(a = 1, \ldots, k)$, the above theorem reduces to

$$\begin{aligned} F_{k,N}(\lambda) &= < [\det(\lambda - M)]^k > \\ &= \det \begin{vmatrix} q_N & q_{N+1} & \cdots & q_{N+k-1} \\ q'_N & q'_{N+1} & \cdots & q'_{N+k-1} \\ \cdots & \cdots & \cdots & \cdots \\ q_N^{(k-1)} & q_{N+1}^{(k-1)} & \cdots & q_{N+k-1}^{(k-1)} \end{vmatrix} \end{aligned} \tag{19.2.4}$$

where $q_N^{(k-1)}$ is a $k-1$ th derivative of $q_N(\lambda)$ by λ. The determinant (19.2.4) is a Wronskian of the orthogonal polynomials, $W(q_N, \cdots, q_{N+k-1})$.

As a corollary of Th.19.2.1, this correlation function for $k = 2$ gives

$$< \det(\lambda_1 - M)\det(\lambda_2 - M) >= K_{N+1}(\lambda_1, \lambda_2) \tag{19.2.5}$$

which is a Szegö kernel which has a simple expression in terms of orthogonal polynomials after use of the Christoffel–Darboux formula.

The k-th moment of the characteristic polynomials for the simple GUE $(V(M) = 1/2M^2)$

$$F_{k,N}(\lambda) = \frac{1}{N!} \int \cdots \int \prod (x_i - x_j)^2 \prod_{l=1}^{N} (\lambda - x_l)^k e^{-\frac{1}{2}x_l^2} dx_l \tag{19.2.6}$$

is a Hankel determinant, sometimes called a generalized Hermite polynomial.

The derivative of the logarithm of the ratio of $F_{k,N}(\lambda)$, denoted a_N, turns out to be a rational solution of the Painlevé IV equation [Che06]:

$$a_N = -\frac{1}{2}\frac{\partial}{\partial\lambda}\log\frac{F_{k,N+1}(\lambda)}{F_{k,N}(\lambda)} \tag{19.2.7}$$

which satisfies

$$a''_N = \frac{(a'_N)^2}{2a_N} + 6a_N^3 - 4\lambda a_N^2 + (\frac{1}{2}\lambda^2 - k - 2N - 1)a_N - \frac{k^2}{8a_N}. \tag{19.2.8}$$

For instance, the polynomials $F_{k,N}(\lambda)$ reduce for $k = 2$ to

$$\begin{aligned} F_{2,2}(\lambda) &= \lambda^4 + 3, \\ F_{2,3}(\lambda) &= \lambda^6 - 3\lambda^4 + 9\lambda^2 + 9. \end{aligned} \tag{19.2.9}$$

and the derivative of the ratio of these two polynomials, a_N in (19.2.7), satisfies (19.2.8).

There is a duality relation when one interchanges the parameters k and N, The characteristic polynomial $F_{k,N}(\lambda)$, expressed by the Wronskian in (19.2.4), coincides with $F_{N,k}(i\lambda)$, which is given by a Wronskian $W(p_k(i\lambda), \ldots, p_{k+N-1}(i\lambda))$, with the exchange of k and N,

$$F_{k,N}(\lambda) = F_{N,k}(i\lambda) \tag{19.2.10}$$

This may be also derived from a duality theorem Th.19.4.1 given below. Such dualities, which are very powerful for the analysis of the large N-limit, and for applications to the intersection numbers of moduli of curves, are also present in other problems, such as a colour-flavour transformation in random matrix modes of quantum chromodynamics [Zir99]

The asymptotic analysis of the large N-limit gives [Bre00]

$$e^{-NkV(\lambda)}F_{2k}(\lambda, \ldots, \lambda) = (2\pi N\rho(\lambda))^{k^2}e^{-Nk}\prod_0^{k-1}\frac{l!}{(k+l)!} \tag{19.2.11}$$

The last combinatorial factor is a universal quantity, which coincides with the universal factor in the average of the 2k-th moment of the Riemann zeta function.

These ensembles may be extended to classical Lie algebras. The complex Hermitian matrix corresponds to the unitary group $U(N)$. For the orthogonal group $O(2N)$, $O(2N+1)$, and the symplectic group $Sp(N)$, the characteristic polynomials in the Gaussian case have been investigated in relation to the moment of L-functions [Bre00].

For the $o(2N)$ Lie algebra, the average of the product of characteristic polynomials is

$$F_{k,N}(\lambda_1,\ldots,\lambda_k) = \frac{1}{Z_N}\int \prod d\mu(x_i)\Delta^2(x_1^2,\ldots,x_N^2)\prod_{a=1}^{k}\prod_{i=1}^{N}(\lambda_a{}^2 - x_i^2) \quad (19.2.12)$$

in which $\prod d\mu(x_i) = \exp(-N\sum x_i^2)\prod_{i<j}(x_i - x_j)^2\prod dx_i$. By the transformation, $\mu_i = \lambda_i^2$ and $x_i^2 = y_i$, it becomes

$$F_{k,N}(\mu_1,\ldots,\mu_k) = \int_0^{\infty}\prod_{i=1}^{N} dy_i \prod y_i^{-\frac{1}{2}}\prod(y_i - y_j)^2\prod_{a=1}^{k}\prod_{i=1}^{N}(\mu_a - y_i)e^{-N\Sigma y_i} \quad (19.2.13)$$

This is expressed in a determinantal form by the use of the Laguerre polynomials. When all λ_a are degenerated, it is described by the Wronskian, similar to the unitary case.

For the $Sp(N)$ Lie algebra, the random matrix X is an Hermitian $2N \times 2N$ matrix, which satisfies the condition,

$$X^T J + JX = 0 \quad (19.2.14)$$

where

$$J = \begin{vmatrix} 0 & 1_N \\ -1_N & 0 \end{vmatrix} \quad (19.2.15)$$

The average of the characteristic polynomials is given by

$$F_{k,N}(\lambda_1,\ldots,\lambda_k) =< \prod_{a=1}^{k}\det(\lambda_a - X) >$$

$$= \frac{1}{N}\int \prod d\mu_i(x_i)\Delta^2(x_1^2,\ldots,x_N^2)\prod_{i=1}^{N}x_i^2\prod_{a=1}^{k}\prod_{i=1}^{N}(\lambda_a{}^2 - x_i^2) \quad (19.2.16)$$

where $\prod d\mu(x_i) = \exp(-N\sum x_i^2)\prod_{i<j}(x_i - x_j)^4\prod dx_i$

The case of $O(2N+1)$ group becomes similar to the $Sp(N)$ case.

In the RMT, Gaussian orthogonal ensemble (GOE) and Gaussian symplectic ensemble (GSE) are important classes for various applications. The characteristic polynomials for these cases provide new correlation functions [Bre01].

In the GOE case, the average of the characteristic polynomials becomes

$$F_{k,N}(\lambda_1,\ldots,\lambda_k) =< \prod_{a=1}^{k}\det(\lambda_a - M) > \quad (19.2.17)$$

where M is a real symmetric matrix.

$$F_{k,N}(\lambda_1,\ldots,\lambda_k) = \int dB dD e^{-N\mathrm{tr}(B^2 + D^\dagger D)}[-\mathrm{Pf}M]^N \quad (19.2.18)$$

where

$$M = \begin{vmatrix} \mathrm{D} & \Lambda - i\,B \\ -(\Lambda - i\,B) & D^{\dagger} \end{vmatrix}$$

$$F_{k,N}(\lambda_1, \ldots, \lambda_k) = \int d\,\tilde{M}_0(\mathrm{Qdet}\,\tilde{M})^N e^{-N\mathrm{tr}\,\tilde{M}_0{}^2} \tag{19.2.19}$$

where Qdet$M$ is a quaternion determinant, and $\tilde{M}$ is a quaternionic matrix for M [Bre01],

$$\mathrm{Qdet}\,\tilde{M} = -\mathrm{Pf}M. \tag{19.2.20}$$

The moment of the characteristic polynomials of GOE $F_{k,N}(\lambda) = F_{k,N}(\lambda, \ldots, \lambda)$ becomes

$$\exp(-2Nk\lambda^2)\,F_{2k,N}(\lambda) = \gamma_k\,N^{2k^2}(2\pi\rho(\lambda))^{2k^2+k} \tag{19.2.21}$$

$$\begin{aligned} \gamma_k &= \frac{(2k)!}{k!k!}\frac{[\prod_{l=1}^{k}(2l)!]^2}{\prod_{l=1}^{2k}(2l)!} \\ &= \prod_{l=1}^{k}\frac{(2l-1)!}{(2k+2l-1)!} \end{aligned} \tag{19.2.22}$$

The two-point correlation of the characteristic polynomial $F_{2,N}(\lambda_1, \lambda_2)$ becomes a universal form in the large N limit [Bre01],

$$F_{2,N}(\lambda_1, \lambda_2) = Ce^{-2N\lambda^2}[\frac{\cos x}{x^2} - \frac{\sin x}{x^3}] \tag{19.2.23}$$

where $x = \pi N(\lambda_1 - \lambda_2)\rho(\lambda)$ and $\lambda = (\lambda_1 + \lambda_2)/2$.

19.3 Ratio of characteristic polynomials

The average of the products of the ratios of the characteristic polynomials are expressed by

$$F_{k,N}(\lambda_1, \ldots, \lambda_k; \mu_1, \ldots, \mu_k) =< \prod_{a=1}^{k}\frac{\det(\lambda_a - M)}{\det(\mu_a - M)} > \tag{19.3.1}$$

The reason for considering expectation values of such characteristic polynomials is that they turn out to be simpler than the usual correlation functions of the type $< \prod \mathrm{tr}\frac{1}{\lambda_a - M} >$, but they may be used to recover the same information. For instance

$$\frac{\partial}{\partial\lambda} < \frac{\det(\lambda - M)}{\det(\mu - M)} > |_{\mu=\lambda} =< \mathrm{tr}\frac{1}{\lambda - M} > \tag{19.3.2}$$

The average of the single ratio of characteristic polynomials $F_N(\lambda;\mu)$ with Gaussian distribution $Pr(M) = Z^{-1}\exp[-(N/2)\mathrm{tr}M^2]$,

$$F_N(\lambda;\mu) =< \frac{\det(\lambda - M)}{\det(\mu - M)} > \tag{19.3.3}$$

is expressed by the integral of Grassmann variables $\bar{\theta}$ and θ, and by commuting variables z and z^* [Bre03].

$$F_N(\lambda;\mu) = \int \prod_{a=1}^{N} dz_a^* dz_a d\bar{\theta}_a d\theta_a < \exp i N[\bar{\theta}_a(\lambda\delta_{ab} - M_{ab})\theta_b + z_a^*(\mu\delta_{ab} - M_{ab})z_b] > \tag{19.3.4}$$

After the integral of M, by the introduction of auxiliary commuting and Grassmannian variables, b and η,

$$F_N(\lambda,\mu) = \frac{1}{2}\int dbdb'd\eta d\bar{\eta} e^{-\frac{N}{2}(b^2+b'^2+2\bar{\eta}\eta)}[\frac{\lambda + ib'}{\mu - b} + \frac{\bar{\eta}\eta}{(\mu - b)^2}]^N \tag{19.3.5}$$

This leads to

$$F_N(\lambda;\mu) = \frac{1}{2^N\sqrt{\pi}(N-1)!}\int_{-\infty}^{+\infty} db \frac{e^{b^2}}{(\bar{\mu} - b)}[H_N(\bar{\lambda})H_{N-1}(b) - H_{N-1}(\bar{\lambda})H_N(b)] \tag{19.3.6}$$

where $\bar{\lambda} = \lambda(N/2)^{1/2}$, $\bar{\mu} = \mu(N/2)^{1/2}$ and $H_N(x)$ is a Hermite polynomial. From the imaginary part $\mathrm{Im}\,F_N(\lambda;\mu)$, the density of state $\rho(\lambda)$ is obtained as

$$\mathrm{Im}\,F_N(\lambda;\mu) = -\frac{\sqrt{\pi}}{2^N(N-1)!}e^{-\bar{\mu}^2}[H_N(\bar{\lambda})H_{N-1}(\bar{\mu}) - H_{N-1}(\bar{\lambda})H_N(\bar{\mu})]$$

$$\rho(\lambda) = -\lim_{\mu\to\lambda}\frac{1}{\pi N}\frac{\partial}{\partial\mu}\mathrm{Im}\,F_N(\lambda;\mu) \tag{19.3.7}$$

which becomes the well-known finite N expression. Another method is the use of the super group symmetry, which is more powerful for the analysis of higher correlation functions.

For GOE, the well-known universal two-point correlation function is also derived from the ratio of the characteristic polynomials. For this analysis, extension of the Harish-Chandra, and Itzykson-Zuber formula for GUE to GOE is required. This is done by the heat kernel differential equation.

$$F_N(\lambda_1,\lambda_2;\mu_1,\mu_2) = \int [\mathrm{Sdet}Q]^{-\frac{N}{2}} e^{-\frac{1}{2}\mathrm{Str}Q^2 + \mathrm{Str}Q\Lambda} \tag{19.3.8}$$

where Q is an 8×8 supermatrix, which has Grassmannian numbers in the off-diagonal block, and Sdet is the superdeterminant, Str is a supertrace. By the diagonalization of this supermatrix Q, $\mathrm{diag}(u_1, u_2, u_3, u_4, it_1, it_2, it_3, it_4)$, it becomes

$$F_N(\lambda_1,\lambda_2;\mu_1,\mu_2) = \int dt_1 dt_2 du_1 du_2 du_3 du_4 \frac{[(it_1)(it_2)]^N}{(u_1u_2u_3u_4)^{N/2}} J \cdot I e^{-N\Sigma t_i^2 - N\Sigma u_j^2} \tag{19.3.9}$$

where J is Jacobian, and I is a group integral,

$$J = \frac{\prod_{i<j}^4 |u_i - u_j|(t_1 - t_2)^4}{\prod_{i=1}^2 \prod_{a=1}^4 (it_i - u_a)^2}$$

$$I = \int dg e^{N\mathrm{Strg}\, Qg^{-1}\Lambda} \tag{19.3.10}$$

where the group g is the orthogonal group $O(2N)$.

In the Dyson's scaling limit, it gives a well-known formula [Bre03]

$$\rho(\lambda_1,\lambda_2) = \rho^2(\lambda)[1 - (\frac{\sin x}{x})^2 - \frac{d}{dx}(\frac{\sin x}{x})\int_x^\infty \frac{\sin z}{z} dz] \tag{19.3.11}$$

where $x = \pi N\rho(\lambda)(\lambda_1 - \lambda_2)$. Thus the ratio of the characteristic polynomials gives the previous results derived by the orthogonal polynomial method.

19.4 Duality formula for an external source

The duality theorems for two different characteristic polynomials of Gaussian weights with external sources is useful. This duality formula is about $N-k$ exchange, where N is a size of the matrix M and k is a size of the matrix B. The external sources A and Λ are dual.

The Gaussian ensemble is extended to include the external source matrix A. The probability distribution becomes

$$\Pr(M) = e^{-\frac{1}{2}\mathrm{tr}M^2 + \mathrm{tr}MA} \tag{19.4.1}$$

where A is a fixed $N \times N$ matrix.

Theorem 19.4.1 *For the $N\times N$ Hermitian random matrix M and $k \times k$ Hermitian random matrix B, there exists a dual representation for the characteristic polynomials,*

$$\begin{aligned} F(\lambda_1,\ldots,\lambda_k) &= \frac{1}{Z_N}\int dM \prod_{a=1}^k \det(\lambda_a \cdot \mathrm{I} - M) e^{-\frac{1}{2}\mathrm{tr}(M-A)^2} \\ &= \frac{1}{Z_k}\int dB \prod_{j=1}^N \det(-a_j \cdot \mathrm{I} + iB) e^{-\frac{1}{2}\mathrm{tr}(B+i\Lambda)^2} \end{aligned} \tag{19.4.2}$$

where $Z_N = 2^{N/2}\pi^{N^2/2}$, $Z_k = 2^{k/2}\pi^{k^2/2}$ and A is an external source matrix with the eigenvalues $a_j (j = 1,\ldots,N)$ and Λ has eigenvalues $\lambda_i (i = 1,\ldots,k)$.

The proof is given by the use of the Grassmann variables which represent the determinants [Bre00],[Bre08a]. When the external source A is absent, and in the degenerated case, $\lambda_i = \lambda (i = 1, \ldots, N)$, this theorem reduces to (19.2.10).

Similar dualities exist for ensembles based on classical Lie algebras. For the orthogonal group $O(2N)$ [Bre07a, Bre08b],

Theorem 19.4.2

$$< \prod_{\alpha=1}^{k} \det(\lambda_\alpha \cdot \mathrm{I} - X) >_A = < \prod_{n=1}^{N} \det(a_n \cdot \mathrm{I} - Y) >_\Lambda \qquad (19.4.3)$$

where X is a $2N \times 2N$ real antisymmetric matrix ($X^t = -X$) and Y is a $2k \times 2k$ real antisymmetric matrix. A is also a $2N \times 2N$ antisymmetric matrix, and it couples to X as an external source matrix. Λ is $2k \times 2k$ antisymmetric matrix, coupled to Y.

We assume, without a loss of the generality, that A and Λ have canonical forms. It is expressed as

$$A = a_1 v \oplus \cdots \oplus a_N v, \qquad v = \begin{vmatrix} 0 & 1 \\ -1 & 0 \end{vmatrix} \qquad (19.4.4)$$

Λ is expressed also as

$$\Lambda = \lambda_1 v \oplus \cdots \oplus \lambda_k v \qquad (19.4.5)$$

The Gaussian averages of Th.19.4.2 are

$$\begin{aligned} < \cdots >_A &= \int dM e^{-\frac{1}{2}\mathrm{tr} M^2 + \mathrm{tr} MA} \\ < \cdots >_\Lambda &= \int dX e^{-\frac{1}{2}\mathrm{tr} X^2 + \mathrm{tr} X\Lambda} \end{aligned} \qquad (19.4.6)$$

19.5 Fourier transform $U(s_1, \ldots, s_k)$

The m-point correlation functions of the eigenvalues in the Gaussian unitary ensemble are conveniently deduced from their Fourier transforms $U(s_1, \ldots, s_m)$, defined as

$$\begin{aligned} U(s_1, s_2, \ldots, s_m) &= < \mathrm{tr} e^{s_1 M} \mathrm{tr} e^{s_2 M} \cdots \mathrm{tr} e^{s_m M} > \\ &= \int \prod_{l=1}^{m} d\lambda_l e^{\Sigma i t_l \lambda_l} < \prod_{1}^{m} \mathrm{tr}\delta(\lambda_j - M) > \end{aligned} \qquad (19.5.1)$$

where $s_l = i t_l$; M is an $N \times N$ Hermitian random matrix. The bracket stands for averages with the Gaussian probability measure

$$< X >_A = \int dMe^{-\frac{N}{2}\mathrm{tr}M^2 + N\mathrm{tr}MA} X(M), \tag{19.5.2}$$

A is an $N \times N$ external Hermitian source matrix. We may assume that this matrix A is diagonal with eigenvalues a_j.

An exact and useful representation for $U(s_1, \ldots, s_n)$ in the presence of an external matrix source A is given in the integral form [Bre97].

Theorem 19.5.1 *The Fourier transform of an n-point correlation function is expressed by the contour integral*

$$\begin{aligned} U(s_1, \cdots, s_n) &= \int \prod_{i=1}^{n} d\lambda_i e^{-i\Sigma t_i \lambda_i} \langle \mathrm{tr}\delta(\lambda_1 - M) \cdots \mathrm{tr}\delta(\lambda_n - M) > \\ &= \frac{1}{N} \langle \mathrm{tr}e^{s_1 M} \cdots \mathrm{tr}e^{s_n M} \rangle \\ &= e^{\Sigma_1^n s_i^2} \oint \prod_1^n \frac{du_i}{2\pi i} e^{\Sigma_1^n u_i s_i} \prod_{a=1}^{N} \prod_{i=1}^{n} (1 - \frac{s_i}{a_a - u_i}) \det \frac{1}{u_i - u_j + s_i} \end{aligned} \tag{19.5.3}$$

where the contours are around $u_i = a_a$, and $s_i = it_i$.

When $k = 1$, and $a_a = 0$,

$$U(t) = \frac{1}{it} \oint \frac{du}{2\pi i} (1 + \frac{it}{Nu})^N e^{-\frac{1}{2N}t^2 + itu} \tag{19.5.4}$$

In the large N limit, neglecting the $1/N$ terms,

$$\begin{aligned} U(t) &= \frac{1}{it} \oint \frac{du}{2\pi i} e^{it(u + \frac{1}{u})} \\ &= \frac{1}{t} J_1(2t) \end{aligned} \tag{19.5.5}$$

The inverse Fourier transformation gives the semicircle law for the density of state $\rho(x)$,

$$\begin{aligned} \rho(x) &= \int_{-\infty}^{\infty} \frac{dt}{2\pi i} U(t) e^{-ixt} \\ &= \begin{cases} \frac{1}{\pi}\sqrt{1 - (\frac{x}{2})^2}, & if\, |x| \le 4 \\ 0, & if\, |x| \ge 4 \end{cases} \end{aligned} \tag{19.5.6}$$

The integral representation of Th.19.5.1 is useful for the large N limit and the universal quantities are derived from this integral representation through the saddle-point method.

19.6 Replica method

The correlation function of the characteristic polynomials is related to the standard n-point correlation function $\rho(\lambda_1, \ldots, \lambda_k)$, which is defined by

$$\rho(\lambda_1, \ldots, \lambda_k) =< \mathrm{tr}\delta(\lambda_1 - M)\mathrm{tr}\delta(\lambda_2 - M)\cdots\mathrm{tr}\delta(\lambda_k - M) > \tag{19.6.1}$$

When k=1, above $\rho(\lambda)$ gives the density of state of the energy. From the expression of $F_k(\lambda, \ldots, \lambda)$, we have

$$< [\det(\lambda - M)]^k >=< e^{k\mathrm{trlog}(\lambda - M)} > \tag{19.6.2}$$

Therefore, in the limit $k \to 0$ (replica limit),

$$\lim_{k\to 0} \frac{1}{k}\frac{\partial}{\partial\lambda} < [\det(\lambda - M)]^k >=< \mathrm{tr}\frac{1}{\lambda - M} > \tag{19.6.3}$$

This is a Green function $G(\lambda)$. The density of state is obtained from the imaginary part of $G(\lambda)$, $\rho(\lambda) = -\frac{1}{\pi}\mathrm{Im}G(\lambda)$.

As a simple exercise, we derive the expression of $G(\lambda)$ from $F_k(\lambda, \ldots, \lambda)$ in the replica trick. From Th.19.4.1, we have

$$< [\det(\lambda - M)]^k >= \int dB[\det(B - i\lambda\cdot \mathrm{I})]^N e^{-\frac{1}{2}\mathrm{tr}B^2} \tag{19.6.4}$$

Expanding the exponentiated logarithmic term $\log(1 + iB/\lambda)$, in powers of λ^{-1},

$$< [\det(\lambda - M)]^k >= e^{Nk\log\lambda} \int_{k\times k} dB\exp\Big[-\frac{1}{2}\mathrm{tr}B^2 + \frac{iN}{\lambda}\mathrm{tr}B + \frac{N}{2\lambda^2}\mathrm{tr}B^2 + \cdots\Big] \tag{19.6.5}$$

In the limit $k \to 0$, this perturbational calculation can be done by the later theorem Th.19.6.1 of the replica formula, which leads to

$$\lim_{k\to 0}\frac{1}{Nk}\frac{\partial}{\partial\lambda} < [\det(\lambda - M)]^k >= \frac{1}{\lambda} + \frac{1}{\lambda^3} + \frac{2}{\lambda^5} + \frac{1}{N^2\lambda^5} + \cdots \tag{19.6.6}$$

which is the expansion of $G(\lambda) = \frac{1}{2}(\lambda - \sqrt{\lambda^2 - 4})$.

From the previous integral representation of $U(s_1, \ldots, s_k)$ in Th.19.5.1, when the external source is absent, the continuation to non-integer k is straightforward and leads to

$$\lim_{n\to 0} U(s_1, \cdots, s_k) = (-1)^{k(k-1)/2} e^{\Sigma_1^k \frac{s_i^2}{2}}$$
$$\times \oint \prod_1^k \frac{du_i}{2i\pi} e^{\Sigma_1^k u_i s_i} \sum_1^k \log(1 + \frac{s_i}{u_i})\det\frac{1}{u_i + s_i - u_j} \tag{19.6.7}$$

Calculation of the contour integrals can be done explicitly and gives the following theorem, which is a remarkably compact result [Bre08a].

Theorem 19.6.1

$$\lim_{n\to 0} U(s_1,\ldots,s_k) = \lim_{n\to 0} < \mathrm{tr} e^{s_1 B}\cdots \mathrm{tr} e^{s_k B} > = \frac{1}{\sigma^2}\prod_{i=1}^{k} 2\sinh\frac{\sigma s_i}{2} \tag{19.6.8}$$

with $\sigma = s_1 + \cdots + s_k$. and B is an $n \times n$ Hermitian matrix. The average is taken under the Gaussian probability distribution.

This theorem provides the replica limit $n \to 0$ for the average of arbitory products $< \mathrm{tr}\, B^{m_1} \mathrm{tr}\, B^{m_2} \cdots >$.

19.7 Intersection numbers of moduli space of curves

As a model for two-dimensional quantum gravity, a matrix model for the triangulation of the Riemann surfaces has been developed [Bre90][Dou90][Gro90]. $Z = \int \exp(\mathrm{tr}\, P(M)) d\,M$, These integrals were evaluated using orthogonal polynomials. This partition function of two-dimensional gravity is a series in an infinite number of variables and coincides with the logarithm of some τ-function for KdV-hierarchy. Another approach is to choose some specific action. Using supersymmetry the integral over the space of all metrics reduces to the integral over the finite-dimensional space of conformal structures. This integral has a cohomological description as an intersection theory on the compactified moduli space of curves. The proposed specific model is an Airy matrix model [Kon92],

$$Z = \int dM \exp[\frac{i}{3}\mathrm{tr}\, B^3 + i\,\mathrm{tr}\, B\Lambda] \tag{19.7.1}$$

and a more general Airy type,

$$Z = \int dM \exp[\frac{i}{p+1}\mathrm{tr}\, B^{p+1} + i\,\mathrm{tr}\, B\Lambda] \tag{19.7.2}$$

The partition function of the Airy matrix model in (19.7.1) is a generating function of the intersection numbers of the moduli space of curves,

$$\log Z = \sum < \tau_0^{k_0}\tau_1^{k_1}\cdots > \prod_{i=0}^{\infty}\frac{t_i^{k_i}}{k_i!} \tag{19.7.3}$$

The series of $F(t_0, t_1, \ldots) = \log Z$ obeys the KdV hierarchy [Wit93], in which the first equation is

$$\frac{\partial U}{\partial t_1} = U\frac{\partial}{\partial t_0} + \frac{1}{12}\frac{\partial^3 U}{\partial t_0^3}, \quad (U = \frac{\partial^2 F}{\partial t_0^2}). \tag{19.7.4}$$

The coefficient $< \tau_0^{k_0} \tau_1^{k_1} \cdots >$ in (19.7.3) denotes the intersection numbers for genus g and for the numbers of marked points s with the following condition,

$$3g = \sum_{i=1}^{s} d_i - s + 3 \tag{19.7.5}$$

for $< \tau_{d_1} \cdots \tau_{d_s} >$. For example, $< \tau_0 \tau_0 \tau_0 >= 1$, $< \tau_1 >= \frac{1}{24}$. The intersection numbers are defined by

$$< \tau_{d_1} \cdots \tau_{d_s} >= \int_{\overline{\mathcal{M}}_{g,s}} \prod_{i=1}^{s} c_1(\mathcal{L}_i)^{d_i} \tag{19.7.6}$$

where $\mathcal{L}_i$, $(i = 1, \ldots, s)$ are line bundles on the moduli space $\overline{\mathcal{M}}_{g,s}$, and $c_1(\mathcal{L}_i)$ is the first Chern class. The parameter t_i is given by

$$t_i = -(2i - 1)!! \mathrm{tr} \frac{1}{\Lambda^{i+\frac{1}{2}}} \tag{19.7.7}$$

The Airy matrix model in (19.7.1) is deduced from the duality theorem in the presence of the external sources Th.19.4.1. Let all $a_i = 1$, and expand about the matrix B, it becomes

$$\det(1 - i\,B)^N = \exp[-i\,N \mathrm{tr}\, B + \frac{N}{2} \mathrm{tr}\, B^2 + i \frac{N}{3} \mathrm{tr}\, B^3 + \cdots] \tag{19.7.8}$$

The linear term of B combined with the linear term of the exponent of Th.19.4.1, and the B^2 term cancel. In a scale in which the initials λ_a are close to one, or more precisely $N^{2/3}(\lambda_a - 1)$ is finite, the large N asymptotics is given by matrices B of order $N^{-1/3}$. Then the higher terms in (19.7.8) are negligible and this leads to

$$F_{k,N}(\lambda_1, \ldots, \lambda_k) = e^{\frac{N}{2} \mathrm{tr} \Lambda^2} \int d\,B e^{i \frac{N}{3} \mathrm{tr}\, B^3 + i\,N \mathrm{tr}\, B(\Lambda - 1)} \tag{19.7.9}$$

The limit $a_i \to 1$ corresponds to the edge of the semicircle spectrum of the matrix M, and the kernel is the Airy kernel.

For the generalized Airy matrix model, $F = \log Z$ is a generating function for the intersection numbers of the moduli space of p-spin curves [Wit93].

$$< \tau_{k_1}(U_{r_1}) \cdots \tau_{k_s}(U_{r_s}) >= \frac{1}{p^g} \int_{\bar{\mathcal{M}}_{g,s}} c_D(\mathcal{V}) \prod_{i=1}^{s} c_1(\mathcal{L}_i)^{d_i} \tag{19.7.10}$$

where $\mathcal{V}$ is a vector bundle ($\mathcal{V} = H^1(\Sigma, \mathcal{T})$, $\mathcal{L} \simeq \mathcal{T}^{\otimes p}$), and $c_D(\mathcal{V})$ is its top Chern class (Euler class). The line bundle $\mathcal{L}$ has p-th roots. r_i takes the values from 0 to $p - 2$, and it has a meaning of spin. For non-vanishing intersection numbers, we have a condition $\sum_{i=1}^{s} d_s + D = 3g - 3 + s$. The field U_{r_i} is a primary field. The case $p=2$ reduces to the previous Airy matrix model in (19.7.1). We write $\tau_{k_i}(U_{r_i})$ by τ_{k_i,r_i}.

$$\log Z = \sum < \prod \tau_{k_i,r_i}^{d_i} > \prod \frac{t_{k_i,r_i}^{d_i}}{d_i!} \tag{19.7.11}$$

where

$$t_{k_i,r_i} = (-p)^{\frac{r_i-p-k_i(p+2)}{2(p+1)}} \prod_{l=0}^{k_i-1}(1+r_i+lp)\mathrm{tr}\frac{1}{\Lambda^{k_i+\frac{r_i}{p}+\frac{1}{p}}} \tag{19.7.12}$$

The condition of (19.7.5) for p=2 becomes for arbitrary p as

$$(p+1)(2g-2+s) = \sum_{i=1}^{s}(pk_i+r_i+1) \tag{19.7.13}$$

This generalized Airy matrix model is derived by Th.19.4.1 from the characteristic polynomials when the eigenvalues a_i of the external source matrix A are tuned to a special set of values. For $p=3$, the eigenvalues of a_i are ± 1 [Bre98a][Bre98b].

The external source A with $(p-1)$ distinct eigenvalues, each of them being $\frac{N}{p-1}$ times degenerate: $A = \mathrm{diag}(a_1, \ldots, a_1, \ldots, a_{p-1}, \ldots, a_{p-1})$.

$$< \prod_{\alpha=1}^{p-1} \det(a_\alpha - iB)^{\frac{N}{p-1}} >=< \exp[\frac{N}{p-1}\sum_{\alpha=1}^{p-1}\mathrm{trlog}(1-\frac{iB}{a_\alpha}) + N\sum \log a_\alpha] > \tag{19.7.14}$$

The critical values of a_α

$$\sum_{\alpha}^{p-1}\frac{1}{a_\alpha^2} = p-1, \quad \sum_{\alpha}^{p-1}\frac{1}{a_\alpha^m} = 0, (m=3,4,\ldots)$$
$$\sum_{\alpha}^{p-1}\frac{1}{a_\alpha^{p+1}} \neq 0. \tag{19.7.15}$$

Then, this characteristic polynomial reduces to the generalized Airy matrix model of (19.7.2) in the scaling limit $N \to \infty$.

By the argument of the replica, the intersection numbers of k-marked points are shown to be related to the k-point correlation functions of the dual matrix model near the critical edge. More precisely, the Fourier transform of the k-point correlation function $U(t_1, \ldots, t_k)$ is a generating function of the intersection numbers of the moduli space of p-th spin curves [Bre07a][Bre07b][Bre08a][Bre09].

Theorem 19.7.1 *The Fourier transform of a k-point correlation function,*

$$U(s_1, \ldots, s_k) =< \mathrm{tr}e^{s_1 M}\mathrm{tr}e^{s_2 M}\cdots \mathrm{tr}e^{s_k M} > \tag{19.7.16}$$

is a generating function of the intersection numbers $< \prod_{i=1}^{k} \tau_{k_i,r_i} >$ *of the moduli space of p-th spin curves, and is evaluated by the integral representation of Th.19.5.1.*

For one marked point, the Fourier transform $U(s)$ is given from Th.19.5.1,

$$U(s) = \frac{1}{Ns} \int \frac{du}{2i\pi} e^{-\frac{c}{p+1}[(u+\frac{1}{2}s)^{p+1}-(u-\frac{1}{2}s)^{p+1}]} \tag{19.7.17}$$

This case, obtained by the tuning of the external source a_i, corresponds to the multicritical edge behaviours for the density of state; the gap closing in the density of state as $\rho(\lambda) \sim \lambda^{\frac{1}{p}}$ [Bre98a][Bre98b]. When $p = 3$, $U(s)$ becomes an Airy function. Expanding $U(s)$ in the power series, the intersection numbers $< \tau_{n,j} >$ for arbitrary p and g are obtained.

The intersection numbers for p-th spin curves become a polynomial of p with a factor of Γ-function from (19.7.17). For example, in the cases of $g = 1$ and $g = 2$,

$$\begin{aligned} < \tau_{1,0} >_{g=1} &= \frac{p-1}{24} \\ < \tau_{n,j} >_{g=2} &= \frac{(p-1)(p-3)(1+2p)}{p \cdot 5! \cdot 4^2 \cdot 3} \frac{\Gamma(1-\frac{3}{p})}{\Gamma(1-\frac{1+j}{p})} \end{aligned} \tag{19.7.18}$$

An interesting limit $p \to -1$ exists. Taking $p \to -1$ in (19.7.17), we obtain the Euler characteristics for one marked point $\chi(\overline{\mathcal{M}}_{g,1})$ [Har86]

$$\chi(\overline{\mathcal{M}}_{g,1}) = \zeta(1-2g) = -\frac{B_{2g}}{2g} \tag{19.7.19}$$

where ζ is a Riemann zeta function and B_{2g} are Bernoulli numbers. The intersection numbers, for example, of (19.7.18) become for the spin zero ($j = 0$) case, $-\frac{1}{12}$ and $\frac{1}{120}$, respectively. These numbers are $\chi(\overline{\mathcal{M}}_{g,1})$, since $B_2 = \frac{1}{6}$ and $B_4 = -\frac{1}{30}$.

In the case genus zero ($g = 0$), the ring structure of the primary fields U_{r_j}, is deduced from the values of the intersection numbers evaluated from Th.19.5.1. This shows that the random matrix theory with external source near critical edges, has a structure of a minimal $\mathcal{N} = 2$ superconformal field theory with Lie algebra $\mathcal{A}_{p-1}$ [Wit93].

References

[Ake03a] G. Akemann and Y.V. Fyodorov, Universal random matrix correlations of ratios of characteristic polynomials at the spectral edges. Nucl. Phys. **B 664** (2003) 457.

A.B.J. Kuijlaars and M. Vanlessen, Universality for eigenvalue correlations at the origin of the spectrum. Commun. Math. Phys. **243** (2003) 163.

M.Vanlessen, Universal behavior for averages of characteristic polynomials at the origin of the spectrum. Commun. Math. Phys. **253** (2005) 535.

[Ake03b] G. Akemann and G. Vernizzi, Characteristic polynomials of complex random matrix models. Nucl. Phys. **B660** (2003) 532.
G. Akemann, M.J. Phillips and H.J. Sommers, Characteristic polynomials in real Ginibre ensembles. Journ. Math. Phys. A, Mathematical and General, **42** (2009) 012001.

[Bor06] A. Borodin and E. Strahov, Averages of characteristic polynomials in random matrix theory. Commun. pure and applied Math. **59** (2006) 161.

[Bre96] E. Brézin and S. Hikami, Correlations of nearby levels induced by a random potential, Nucl. Phys. **479**, 697–706 (1996).

[Bre97] E. Brézin and S. Hikami, Extension of level-spacing universality. Phys. Rev. **E56**, 264 (1997).

[Bre98a] E. Brézin and S. Hikami, Universal singularity at the closure of a gap in a random matrix theory, Phys. Rev. **B57**, 4140 (1998).

[Bre98b] E. Brézin and S. Hikami, Level spacing of random matrices in an external source, Phys. Rev. **E58** (1998) 7176.

[Bre00] E. Brézin and S. Hikami, Characteristic polynomials of random matrices, Commun. Math. Phys. **214** (2000) 111.

[Bre01] E. Brézin and S. Hikami, Characteristic polynomials of real symmetric random matrices, Commun. Math. Phys. **223** (2001) 363–382.

[Bre03] E. Brézin and S. Hikami, New correlation functions for random matrices and integrals over supergroups. J. Phys. **A36**, (2003) 711.

[Bre07a] E. Brézin and S. Hikami, Vertices from replica in a random matrix theory, J. Phys. **A40**, (2007) 13545–13566.

[Bre07b] E. Brézin and S. Hikami, Intersection numbers of Riemann surfaces from Gaussian matrix models, JHEP **10** (2007) 096.

[Bre08a] E. Brézin and S. Hikami, Intersection theory from duality and replica. Commun. Math. Phys. **283**, (2008) 507–521.

[Bre08b] E. Brézin and S. Hikami, Intersection numbers from the antisymmetric Gaussian matrix model, JHEP **07** (2008) 050.

[Bre09] E. Brézin and S. Hikami, Computing topological invariants with one and two-matrix models, JHEP **04** (2009) 110.

[Bre90] E. Brézin and V. Kazakov, Exactly solvable field theories of closed strings, Phys. Lett. **B236**, (1990) 144.

[Bump06] D. Bump and A. Gamburd, On the averages of characteristic polynomials from classical groups Commun. Math. Phys. **265** (2006) 227.

[Che06] Y. Chen and M.V. Feigin, Painlevé IV and degenerate Gaussian unitary ensembles, J. Phys. A **39**, (2006) 12381–12393.

[Dou90] M.R. Douglas and S.H. Shenker, Strings less than one dimension, Nucl. Phys. **B335**, (1990) 635–654.

[Fyo03] Y.V. Fyodorov and G. Akemann, On the supersymmetric partition function in QCD-inspired random matrix models JETP Letters **77** (2003) 438.

[Gro90] D.J. Gross and A.A. Migdal, Nonperturbative two-dimensional quantum gravity, Phys. Rev. Lett. **64** (1990), 127.

[Har86] J. Harer and D. Zagier, The Euler characteristic of the moduli space of curves, Invent. Math. **85** (1986) 457.

[Kea00] J.P. Keating and N.C. Snaith, Random matrix theory and zeta(1/2+it). Commun. Math. Phys. **214** (2000) 57.

[Kon92] M. Kontsevich, Intersection theory on the moduli space of curves and the matrix Airy function, Commun. Math. Phys. **147**, 1 (1992).

[Mon73] H.L. Montgomery, The pair correlation of zeros of the zeta function, Proc. Symp. Pure. Math. (AMS). **24**, 181–193 (1973).

[Odl87] A.M. Odlyzko, On the distribution of spacings between the zeros of zeta function, Math. Comp. **48** (1987) 273–308.

[Str03] E. Strahov and Y.V. Fyodorov, Universal results for correlations of characteristic polynomials: Riemann–Hilbert approach. Commun. Math. Phys. **241** (2003) 343.
J. Baik, P. Deift and E. Strahov, Products and ratios of characteristic polynomials of random Hermitian matrices. Journ. Math. Phys. **44** (2003) 3657.

[Wit93] E. Witten, Algebraic geometry associated with matrix models of two dimensional gravity, in *Topological Methods in Modern Mathematics*, Publish or Perish, INC. 1993. P.235.

[Zir96] M. Zirnbauer, Supersymmetry for systems with unitary disorder: Circular ensembles. Journ. Phys. A, **29** (1996) 7113.

[Zir99] M. Zirnbauer, The color-flavor transformation and a new approach to quantum chaotic maps. XIIth International Congress of Mathematical Physics (ICMP 1997), 290, Int. Press, Cambridge, MA (1999).
Y.V. Fyodorov and B.A. Khoruzhenko, A few remarks on colour-flavour transformations, truncations of random unitary matrices, Berezin reproducing kernels and Selberg-type integrals Journ. Phys. A, Mathematical and Theoretical, **40** (2007) 669.
Z.M. Feng and J.P. Song, Integrals over the circular ensembles relating to classical domains Journ. Phys. A, Mathematical and Theoretical, **42** (2009) 325204.
Y. Wei, Moments of ratios of characteristic polynomials of a certain class of random matrices Journ. Math. Phys. **50** (2009) 043518.

·20·

Beta ensembles

Peter J. Forrester

Abstract

Classical random matrix ensembles contain a parameter β, taking on the values 1, 2, and 4. This relates to the underlying symmetry, and shows itself as a repulsion s^{β} between neighbouring eigenvalues for small s. Different viewpoints of the eigenvalue probability density function for the classical random matrix ensembles – as the Boltzmann factor for a log-gas, the squared ground state wave function of a quantum many-body system, or the eigenvalue probability density for certain random tridiagonal or unitary Hessenberg matrices – gives prominence to regarding β as a continuous positive parameter. Some of the most recent advances in random matrix theory have resulted from this perspective.

20.1 Log-gas systems

It has been known since the work of Dyson [Dys62a] that much insight into the statistical properties of the eigenvalues of random matrices can be gained from the log-gas analogy. A log-gas is a classical particle system in equilibrium, and thus the probability density function for the event that the particles are at positions $\vec{r}_1, \ldots, \vec{r}_N$ is proportional to the Boltzmann factor $e^{-\beta U(\vec{r}_1,\ldots,\vec{r}_N)}$. In the latter $U(\vec{r}_1, \ldots, \vec{r}_N)$ denotes the total potential energy of the system, and β is effectively the inverse temperature. The distinguishing feature of a log-gas is that the particles interact through a repulsive logarithmic pair potential $\Phi(\vec{r}, \vec{r}\,') = -\log|\vec{r} - \vec{r}\,'|$. Physically, this is the electrostatic potential at the point $\vec{r}$, due to a two-dimensional charge (long thin wire perpendicular to the plane) of unit strength at $\vec{r}\,'$. Thus a log-gas system consists of N two-dimensional charged particles, all of the same charge, confined to some region Ω of the plane. For such a system to be stable, a smeared out background charge density $-\rho_b(\vec{r})$ must be imposed, satisfying the neutrality condition $\int_\Omega \rho_b(\vec{r})\, d\vec{r} = N$. This in turn couples with the particles to create the one-body potential $V(\vec{r}) = \int_\Omega \log|\vec{r} - \vec{r}\,'|\rho_b(\vec{r})\, d\vec{r}\,'$, and so the Boltzmann factor for a log-gas system is proportional to

$$\prod_{l=1}^{N} e^{-\beta V(\vec{r}_l)} \prod_{1\le j<k\le N} |\vec{r}_k - \vec{r}_j|^{\beta}, \qquad (\vec{r}_l \in \Omega).$$

The functional form is familiar from the preceding chapters, but with one crucial extension: rather than β being restricted to one of the special values $\beta = 1, 2,$ or 4 (when Ω is a one-dimensional domain) or $\beta = 2$ (when Ω is $\mathbb{R}^2$), it is now naturally regarded as an arbitrary positive real valued coupling constant.

The input of a prescribed background charge density has some immediate (heuristic) consequence for the eigenvalue density in the corresponding random matrix ensemble. Thus to leading order the particle density in a log-gas system must be equal to the background density $\rho_b(x)$, for otherwise the charge imbalance would create an electric field and so put the system out of equilibrium. As an example, consider a log-gas confined to the interval $(-\sqrt{2N}, \sqrt{2N})$ with background density

$$\rho_b(x) = \frac{\sqrt{2N}}{\pi}\sqrt{1 - \frac{x^2}{2N}}. \tag{20.1.1}$$

This satisfies the neutrality condition, and (after a somewhat tricky calculation; see e.g. [For10, §1.4.2]) gives

$$V(x) = \int_{-\sqrt{2N}}^{\sqrt{2N}} \log|x - y|\rho_b(y)\,dy = \frac{x^2}{2}, \qquad x \in [-\sqrt{2N}, \sqrt{2N}].$$

Thus log-gas heuristics tell us that the particle density corresponding to the Boltzmann factor

$$\prod_{l=1}^{N} e^{-\beta x_l^2/2} \prod_{1\le j<k\le N} |x_k - x_j|^\beta \tag{20.1.2}$$

will to leading order have the profile (20.1.1) for $x \in [-\sqrt{2N}, \sqrt{2N}]$, and be zero outside. In random matrix theory, we recognize (20.1.2), for $\beta = 1, 2,$ and 4, as the eigenvalue probability density function for the GOE, GUE, and GSE respectively, and we recognize (20.1.1) as the Wigner semicircle law.

The truncated two-point bulk correlation $\rho_{(2)}^T(x, y)$ for (20.1.2) can also be probed using the log-gas analogy. First we recall that the bulk limit corresponds to scaling the coordinates $x_j \mapsto \pi x_j/\sqrt{2N}$ so that the eigenvalue density in the neighbourhood of the origin is unity, and taking the limit $N \to \infty$. Physically, a log-gas system must be a perfect conductor. Thus if an external charge is introduced as a perturbation, the system will respond to redistribute its charge so as to effectively neutralize the perturbation, and will furthermore similarly screen an oscillatory external charge in the long wavelength limit. Mathematically, the former property is captured by the sum-rule

$$\int_{-\infty}^{\infty} \rho_{(2)}^T(x, 0)\,dx = -1 \tag{20.1.3}$$

and the latter by

$$\int_{-\infty}^{\infty} \Big(\rho_{(2)}^{T}(x,0) + \delta(x)\Big) e^{ikx}\, dx \underset{k\to 0}{\sim} \frac{|k|}{\pi\beta}, \tag{20.1.4}$$

or equivalently

$$\rho_{(2)}^{T}(x,0) \underset{x\to\infty}{\dot{\sim}} -\frac{1}{\pi\beta x^2}, \tag{20.1.5}$$

where $\dot{\sim}$ indicates that all oscillatory terms have been ignored. These predictions are indeed properties of the truncated two-particle correlation for bulk correlations of the classical Gaussian ensembles (see e.g. [For10, Ch. 7]).

Generally, if a log-gas is perturbed in such a way so as to create a perturbation of the one-body potential $\delta U = \sum_{j=1}^{N} u(x_j)$, where $u(x)$ varies on a length scale much greater than the inter-particle spacing, the characterization of a log-gas being in a conductive phase tells us that the perturbed particle density $\delta\rho_{(1)}(x)$ will be such that

$$-\int_{\Omega} \log|x-y|\, \delta\rho_{(1)}(y)\, dy = u(x) + C. \tag{20.1.6}$$

This is an asymptotically exact linear relation between $\delta\rho_{(1)}$ and u. It can be used [Pol89, Jan95] to argue that for the log-gas system (20.1.2) in the infinite density limit, $x_j \mapsto x_j\sqrt{2N}$ and $N \to \infty$, so that the support is the finite interval $(-1, 1)$, the distribution of a linear statistic $A = \sum_{j=1}^{N} a(x_j)$ will to leading order be a Gaussian with the N independent variance

$$\frac{2}{\beta}\sum_{n=1}^{\infty} n c_n^2, \qquad a(\cos\theta) = c_0 + 2\sum_{n=1}^{\infty} c_n \cos n\theta \tag{20.1.7}$$

(see [For10, §14.4.1]). In fact, under certain technical conditions on $a(x)$, is has been proved in [Joh98] that for a class of log-gas systems with polynomial one-body potentials and thus including (20.1.2),

$$\Big\langle \prod_{l=1}^{N} e^{a(x_l)} \Big\rangle \underset{N\to\infty}{\sim} \exp\Big(N\int_I a(x)\rho_{(1)}(x)\, dx + \frac{2}{\beta}\sum_{n=1}^{\infty} n c_n^2 \Big), \tag{20.1.8}$$

where I is the eigenvalue support, in keeping with these predictions.

The variance of some linear statistics diverges in the infinite density limit. For log-gas systems corresponding to (20.1.8) the sum in (20.1.7) will then diverge. Thus the statistic must exhibit a strong enough singularity for this to happen. In some cases when the sum specifying the variance diverges only logarithmically, by an appropriate scaling a Gaussian fluctuation formula can still be exhibited. An explicit example, relating to the log-gas on a circle, is the linear statistic

$$A = \sum_{j=1}^{N} \chi_{\theta_j \in (0,\phi)}, \tag{20.1.9}$$

where $\chi_A = 1$ for A true and $\chi_a = 0$ otherwise. Making use of the expression for the variance in terms of the two-point correlation (see [For10, Ch. 14]), together with (20.1.4), one can show that for large N, Var $A \sim (2/\pi^2\beta) \log N$. The linear response relation (20.1.6) should then be applied to the linear statistic scaled by the standard deviation, implying that

$$(A - \langle A \rangle)/(\text{Var } A)^{1/2} \tag{20.1.10}$$

will be distributed as a standard Gaussian.

Dyson himself [Dys62a] applied the log-gas analogy to the problem of determining the large s asymptotic form of the gap probability $E_\beta(0; s)$ for the interval $(0, s)$ to be free of eigenvalues. The basic hypothesis is that for large s

$$E_\beta(0; s) \sim e^{-\beta\delta F},$$

where δF is the change in free energy resulting from constraining the interval $(0, s)$ to be free of mobile charges. The change δF is taken to consist of an electrostatic energy V_1 and a free energy V_2, given in terms of the particle density $\rho_{(1)}(x)$ by

$$V_1 = \frac{1}{2} \int_{-\infty}^{\infty} (\rho_{(1)}(x) - \rho)\phi(x)\, dx \tag{20.1.11}$$

and

$$V_2 = \int_{-\infty}^{\infty} \rho_{(1)}(x)(f_\beta[\rho_{(1)}(x)] - f_\beta[\rho])\, dx. \tag{20.1.12}$$

With ρ the constant background density, in (20.1.11)

$$\phi(x) = -\int_{-\infty}^{\infty} (\rho_{(1)}(y) - \rho) \log |x - y|\, dy,$$

while in (20.1.12) f_β is the free energy per particle for the log-gas system. The difference $f_\beta[\rho_{(1)}(x)] - f_\beta[\rho]$ is simple to compute for a circular domain, since then the volume scales from the partition function, allowing (20.1.12) to be simplified to read

$$V_2 = \left(\frac{1}{\beta} - \frac{1}{2}\right) \int_{-\infty}^{\infty} \rho_{(1)}(x) \log(\rho_{(1)}(x)/\rho)\, dx. \tag{20.1.13}$$

After some calculation, the particle density $\rho_{(1)}(x)$, and the energies V_1, V_2 can be computed explicitly, giving the prediction

$$E_\beta(0;s) \underset{s\to\infty}{\sim} e^{-\beta(\pi\rho s)^2/16+(\beta-1)\pi\rho s/2}. \tag{20.1.14}$$

Very recently [Val09], using the stochastic differential equation characterization of the bulk log-gas state (see §20.4 below), (20.1.14) has been proved and furthermore extended to the next order.

With $E_\beta(n;s)$ denoting the probability that the interval $(0,s)$ contains k eigenvalues, the log-gas argument can be extended [Dys95, Fog95] to give the prediction

$$\frac{E_\beta(n;2t)}{E_\beta(0;2t)} \underset{\substack{t,n\to\infty\\ t\gg n}}{\sim} c_{\beta,N}\frac{e^{\beta n\pi\rho t}}{(\pi\rho t)^{\beta n^2/4+(\beta/2-1)n/2}} \tag{20.1.15}$$

for some $c_{\beta,N}$. Exact results for $\beta=1,2$ and 4 indicate that this remains true for n fixed (see [For10, Ch. 9]).

20.2 Fokker–Planck equation and Calogero–Sutherland system

A member of the GUE can be constructed from an $N\times N$ Gaussian matrix X, each entry a standard complex Gaussian, according to $(X+X^\dagger)/2$. A generalization is for each entry of the matrix X to be specified by independent complex Brownian motions, confined by a harmonic potential, and thus with real and imaginary parts satisfying the Langevin equation

$$\frac{d}{d\tau}x(\tau) = -x(\tau) + B'(\tau)$$

where $B(\tau)$ is a standard Brownian motion. The GOE and GSE can be similarly generalized. In this setting it was shown by Dyson [Dys62b] that the corresponding eigenvalue probability density function $p_\tau = p_\tau(\lambda_1,\dots,\lambda_N)$ satisfies the multi-dimensional Fokker–Planck equation

$$\frac{\partial p_\tau}{\partial\tau} = \mathcal{L}p_\tau \tag{20.2.1}$$

where

$$\mathcal{L} := \frac{1}{\beta}\sum_{j=1}^N \frac{\partial^2}{\partial\lambda_j^2} + \sum_{j=1}^N \frac{\partial}{\partial\lambda_j}\Big(\lambda_j - \sum_{\substack{k=1\\k\neq j}}^N \frac{1}{\lambda_j-\lambda_k}\Big)$$

for $\beta = 1, 2$, or 4 as appropriate.

Writing

$$W = -\sum_{1\le j<k\le N}\log|\lambda_j-\lambda_k| + \frac{1}{2}\sum_{j=1}^N\lambda_j^2 \tag{20.2.2}$$

shows

$$\mathcal{L} = \sum_{j=1}^{N} \frac{\partial}{\partial \lambda_j}\Big(\frac{\partial W}{\partial \lambda_j} + \beta^{-1}\frac{\partial}{\partial \lambda_j}\Big). \tag{20.2.3}$$

This has immediate significance from the viewpoint of the log-gas analogy. Thus in general one has that an interacting particle system with a general potential energy W, executing overdamped Brownian motion in a fictitious viscous fluid with unit friction coefficient at inverse temperature β, evolves according to the Fokker–Planck equation (20.2.1) with $\mathcal{L}$ given by (20.2.3). The $\tau \to \infty$ steady-state solution of the Fokker–Planck equation is proportional to the Boltzmann factor $e^{-\beta W}$. So we have that the eigenvalues of the ensembles of Hermitian matrices with Brownian elements are statistically identical to the log-gas on a line, specified by the pair potential (20.2.2), undergoing Brownian motion.

It is well known [Ris92] that the Fokker–Planck operator (20.2.3) is under conjugation by $e^{-\beta W/2}$ equal to a Schrödinger operator. In the case that W is given by (20.2.2) one finds

$$-e^{\beta W/2}\mathcal{L}e^{-\beta W/2} = (H - E_0)/\beta$$

where $E_0 = N\beta/2 + (\beta^2/4)N(N-1)$ and

$$H = -\sum_{j=1}^{N}\frac{\partial^2}{\partial x_j^2} + \frac{\beta^2}{4}\sum_{j=1}^{N} x_j^2 + \beta(\beta/2-1)\sum_{1\le j<k\le N}\frac{1}{(x_j-x_k)^2}. \tag{20.2.4}$$

The Schrödinger operator H specifies a Calogero–Sutherland quantum many-body system [Cal69, Sut71]. Notice that when $\beta = 2$ (complex Hermitian matrices) the pair potential in (20.2.4) vanishes identically, and the quantum system corresponds to free fermions in an harmonic potential.

All the classical random matrix ensembles – Gaussian, circular, Laguerre (viewed in terms of chiral ensembles), and Jacobi (in circular variables) – allow for generalizations leading to the Fokker–Planck equation (20.2.1), with W the potential energy of the corresponding log-gas. However only for the Gaussian and chiral matrices (Laguerre ensemble) are the matrix elements literally Brownian variables. In the circular and Jacobi cases the Brownian motion is introduced implicitly (see [For10, Ch. 11]). In all cases the Fokker–Planck operator, after conjugation, gives a Schrödinger operator for a Calogero–Sutherland many-body system, characterized by the pair potential being proportional to the inverse square distance. In the circular ensemble case this reads

$$H^{(C)} := -\sum_{j=1}^{N}\frac{\partial^2}{\partial \theta_j^2} + \frac{\beta(\beta/2-1)}{4}\sum_{1\le j<k\le N}\frac{1}{\sin^2(\theta_j-\theta_k)/2}, \tag{20.2.5}$$

while for the Gaussian ensembles one obtains

$$H^{(H)} := -\sum_{j=1}^{N} \frac{\partial^2}{\partial x_j^2} + \frac{\beta^2}{4}\sum_{j=1}^{N} x_j^2 + \beta(\beta/2-1)\sum_{1\le j<k\le N} \frac{1}{(x_j-x_k)^2}. \quad (20.2.6)$$

The Calogero–Sutherland quantum many-body systems associated with the Fokker–Planck operator $\mathcal{L}$ have as their ground state wave function $e^{-\beta W/2}$. A remarkable feature is that the complete set of eigenfunctions factorize as a product of the ground state times a polynomial. In the case of (20.2.5) these polynomials (in the variables $z_j = e^{i\theta_j}$, and with $\alpha = 2/\beta$) are symmetric eigenfunctions of the differential operator

$$\sum_{j=1}^{N}\Big(z_j \frac{\partial}{\partial z_j}\Big)^2 + \frac{N-1}{\alpha}\sum_{j=1}^{N} z_j \frac{\partial}{\partial z_j} + \frac{2}{\alpha}\sum_{1\le j<k\le N} \frac{z_j z_k}{z_j - z_k}\Big(\frac{\partial}{\partial z_j} - \frac{\partial}{\partial z_k}\Big). \quad (20.2.7)$$

They are referred to as the symmetric Jack polynomials (see [For10, Ch. 12]) and are denoted $P_\kappa(z) = P_\kappa(z_1, \dots, z_N; \alpha)$ where the quantum number $\kappa := (\kappa_1, \dots, \kappa_N)$, $\kappa_1 \ge \cdots \ge \kappa_N \ge 0$ is a partition. Their explicit characterization is as the polynomial eigenfunctions of (20.2.7) with the structure

$$P_\kappa(z) = m_\kappa + \sum_{\sigma<\kappa} b_{\kappa\sigma} m_\sigma,$$

where m_κ denotes the monomial symmetric function in the variables $\{z_j\}$ associated with κ, $\sigma < \kappa$ refers to the dominance partial ordering on partitions and the coefficients $b_{\kappa\sigma}$ are independent of N.

In the case of (20.2.6), the polynomial factors of the wave functions are the symmetric eigenfunctions $P_\kappa^{(H)}$ of the differential operator

$$\sum_{j=1}^{N}\Big(\frac{\partial^2}{\partial y_j^2} - 2y_j\frac{\partial}{\partial y_j}\Big) + \frac{2}{\alpha}\sum_{j<k} \frac{1}{y_j - y_k}\Big(\frac{\partial}{\partial y_j} - \frac{\partial}{\partial y_k}\Big) \quad (20.2.8)$$

with the structure when expressed in terms of the Jack polynomials

$$P_\kappa^{(H)}(y) = P_\kappa(y) + \sum_{|\mu|<|\kappa|} a_{\kappa\mu} P_\mu(y).$$

They are multidimensional generalizations of the Hermite polynomials [Bak97].

The theory of Jack polynomials plays a crucial role in the calculation of correlation functions in the log-gas systems for general even β, and in the case of the two-point function for the circular ensemble in the bulk limit, for general rational β. The theory of generalized Hermite polynomials allows for an explicit evaluation of the one-point correlation at the soft edge for even β.

Space constraints restrict our presentation to statement of only a selection of results of this type. Details can be found in [For10, Ch. 13].

For the log-gas on a circle of circumference length L with $N+2$ particles and β even the two-point correlation can be written as the β-dimensional integral [For93]

$$\rho_{(2)}(r_1, r_2) = \frac{(N+2)(N+1)}{L^2} \frac{(\beta N/2)!((\beta/2)!)^{N+2}}{(\beta(N+2)/2)!} \frac{M_N(n\beta/2, n\beta/2, \beta/2)}{S_\beta(1-2/\beta, 1-2/\beta, 2/\beta)}$$
$$\times (2\sin \pi(r_1 - r_2)/L)^\beta e^{-\pi i\beta N(r_1-r_2)/L} \int_{[0,1]^\beta} du_1 \cdots du_\beta$$
$$\times \prod_{j=1}^{\beta}(1 - (1 - e^{2\pi i(r_1-r_2)/L})u_j)^N u_j^{-1+2/\beta}(1-u_j)^{-1+2/\beta} \prod_{j<k} |u_k - u_j|^{4/\beta}. \tag{20.2.9}$$

In the bulk limit $N, L \to \infty$ with $N/L = \rho$ fixed this reduces to

$$\rho_{(2)}^{\text{bulk}}(r_1, r_2) = \rho^2(\beta/2)^\beta \frac{((\beta/2)!)^3}{\beta!(3\beta/2)!} \frac{e^{-\pi i\beta\rho(r_1-r_2)}(2\pi\rho(r_1-r_2))^\beta}{S_\beta(-1+2/\beta, -1+2/\beta, 2/\beta)} \int_{[0,1]^\beta} du_1 \cdots du_\beta$$
$$\times \prod_{j=1}^{\beta} e^{2\pi i\rho(r_1-r_2)u_j} u_j^{-1+2/\beta}(1-u_j)^{-1+2/\beta} \prod_{j<k} |u_k - u_j|^{4/\beta}. \tag{20.2.10}$$

Here $M_n(a, b, c)$ and $S_n(a, b, c)$ are the Morris and Selberg integrals respectively, which permit evaluations in terms of gamma functions [For10, Ch. 4]. A striking feature of (20.2.9) and (20.2.10) is that the product of differences, which is a characteristic of a Boltzmann factor for a log-gas system, re-appears but with $\beta/2 \mapsto 2/\beta$. In the setting of the Calogero–Sutherland model, the exponent $\beta/2$ appears in the exponent of the phase change $(x_1 - x_2)^{\beta/2} \mapsto e^{\pi i\beta/2}(x_2 - x_1)^{\beta/2}$, when two-particles are interchanged in the ground state $e^{-\beta W/2}$, defined first with $x_1 > x_2 > \cdots > x_N$ so the absolute value signs around the product of differences can be removed. The exponent $2/\beta$ relates to the same property for the quasi-particles determining the excitations [Hal94].

For the Brownian motion dynamics of the log-gas on a circle, initially in its equilibrium state, and perturbed by fixing a particle at point x, the truncated two-point correlation $\rho_{(1,1)}^{T\text{bulk}}$ corresponding to the truncated density at point y after time τ is, for $\beta/2 := \lambda = p/q$ (p and q relatively prime) and in the bulk limit, given by [Ha95]

$$\rho_{(1,1)}^{T\text{bulk}}(x, y; \tau) = C_{p,q}(\lambda) \int_0^\infty dx_1 \cdots \int_0^\infty dx_q \int_0^1 dy_1 \cdots \int_0^1 dy_p \, Q_{p,q}^2$$
$$\times F(p, q, \lambda | \{x_i, y_j\}) \cos(Q_{p,q}(x-y)) \exp(-E_{p,q}\tau/2\lambda). \tag{20.2.11}$$

Here the momentum Q and the energy E variables are given by

$$Q_{p,q} := 2\pi\rho\Big(\sum_{i=1}^{q} x_i + \sum_{j=1}^{p} y_j\Big), \quad E_{p,q} := (2\pi\rho)^2\Big(\sum_{i=1}^{q} \epsilon_P(x_i) + \sum_{j=1}^{p} \epsilon_H(y_j)\Big)$$

with $\epsilon_P(x) = x(x+\lambda)$ and $\epsilon_H(y) = \lambda y(1-y)$, the so called form factor F is given by

$$F(p,q,\lambda|\{x_i, y_j\}) = \frac{\prod_{i<i'} |x_i - x_{i'}|^{2\lambda} \prod_{j<j'} |y_j - y_{j'}|^{2/\lambda}}{\prod_{i=1}^{q}(\epsilon_P(x_i))^{1-\lambda} \prod_{j=1}^{p}(\epsilon_H(y_j))^{1-1/\lambda} \prod_{i=1}^{q}\prod_{j=1}^{p}(x_i + \lambda y_j)^2} \tag{20.2.12}$$

and the normalization is given by

$$C_{p,q}(\lambda) = \frac{\lambda^{2p(q-1)}\Gamma^2(p)}{2\pi^2 p!q!} \frac{\Gamma^q(\lambda)\Gamma^p(1/\lambda)}{\prod_{i=1}^{q}\Gamma^2(p-\lambda(i-1))\prod_{j=1}^{p}\Gamma^2(1-(j-1)/\lambda)}. \tag{20.2.13}$$

Integrals involving (20.2.12) first appeared in the work of Dotesenko and Fateev on conformal field theory [Dot85].

We remark that according to the definitions $\rho_{(1,1)}^{T\text{bulk}}(x, y; 0) = \rho_{(2)}^{\text{bulk}}(x, y) - \rho^2 + \delta(x-y)$. This allows us to compute the structure function $\hat{S}(k;\beta)$ for the equilibrium log-gas according to [For00],

$$\begin{aligned}\hat{S}(k;\beta) &= \int_{-\infty}^{\infty} \rho_{(1,1)}^{T\text{bulk}}(x,0;0)e^{ikx}\,dx \\ &= \pi C_{p,q}(\lambda)\prod_{i=1}^{q}\int_0^\infty dx_i \prod_{j=1}^{p}\int_0^1 dy_j\; Q_{p,q}^2\, F(q,p,\lambda|\{x_i,y_j\})\,\delta(k - Q_{p,q}).\end{aligned} \tag{20.2.14}$$

It is easy to see from this that $\hat{S}(k;\beta) = (|k|/\pi\beta)f(|k|;\beta)$ where $f(k;\beta)$ is analytic in the interval $|k| < \min(2\pi, \pi\beta)$. The exact form of $f(k;\beta)$ can be deduced from (20.2.14), and from this one sees the exact relationship between couplings 2β and $2/\beta$,

$$f(k;\beta) = f\Big(-\frac{2k}{\beta};\frac{4}{\beta}\Big). \tag{20.2.15}$$

Expanding

$$f(k;\beta) = 1 + \sum_{j=1}^{\infty} A_j(\beta/2)\Big(\frac{|k|}{\pi\beta}\Big)^j, \tag{20.2.16}$$

and by postulating that $A_j(x)$ is a polynomial of degree j, (20.2.15) has been used to give the exact expansion of $\hat{S}(k;\beta)$ up to and including $O(k^{10})$. Without any assumptions, one can deduce from (20.2.14) the expansion

$$\hat{S}(k;\beta) \underset{k\to 0}{\sim} \frac{|k|}{\pi\beta} + \frac{1}{2\pi\rho}\Big(\frac{\beta-2}{\beta}\Big)k^2. \tag{20.2.17}$$

The leading-order term is (20.1.4), which was deduced from the physical principle that a log-gas must, in the long-wavelength limit, screen an external charge density. In fact the screening argument can be extended to second order (see [For10, Ch. 14]), allowing the $O(k^2)$ term to also be derived from physical principles (interpretation of the factor $\beta - 2$ is then from the pressure).

For the special couplings $\beta = 1$, 2, and 4 the structure function can be computed in terms of elementary functions, and one has [For10, Ch. 7]

$$f(k;\beta) = \begin{cases} 1 - (1/2)\log(1 + k/\pi), & \beta = 1 \\ 1, & \beta = 2 \\ 1 - (1/2)\log(1 - k/(2\pi)), & \beta = 4, \end{cases}$$

which are all consistent with (20.2.15).

For the log-gas system specified by (20.1.2), with $N \mapsto N+1$ for convenience and β even, the theory of generalized Hermite polynomials can be used to express the one-point correlation as a β-dimensional integral [Bak97],

$$\rho_{(1)}(r) = \frac{e^{-\beta r^2/2}}{C_{N,\beta}} \int_{-\infty}^{\infty} du_1 \cdots \int_{-\infty}^{\infty} du_\beta \prod_{j=1}^{\beta} (iu_j + r)^N e^{-u_j^2} \prod_{1\le j<k\le\beta} |u_k - u_j|^{4/\beta}. \tag{20.2.18}$$

Here $C_{N,\beta}$ is known explicitly in terms of products of gamma functions. We know the leading-order support is $(-\sqrt{2N}, \sqrt{2N})$. The neighbourhood of the edge of the support is referred to as the soft edge. Introducing the scaling $x \mapsto \sqrt{2N} + x/(\sqrt{2}N^{1/6})$, well-defined limiting correlations result. In particular, from (20.2.18) one obtains an integral form for the soft edge scaled one-point function with β even [Des06],

$$\begin{aligned} \rho_{(1)}^{\mathrm{soft},\beta}(x) &:= \lim_{N\to\infty} \frac{1}{\sqrt{2}N^{1/6}} \rho_{(1)}(\sqrt{2N} + x/(\sqrt{2}N^{1/6}) \\ &= \frac{\Gamma(1+\beta/2)}{2\pi}\Big(\frac{4\pi}{\beta}\Big)^{\beta/2} \prod_{j=1}^{\beta} \frac{\Gamma(1+2/\beta)}{\Gamma(1+2j/\beta)} K_{\beta,\beta}(x) \end{aligned} \tag{20.2.19}$$

where

$$K_{n,\beta}(x) := -\frac{1}{(2\pi i)^n} \int_{-i\infty}^{i\infty} dv_1 \cdots \int_{-i\infty}^{i\infty} dv_n \prod_{j=1}^{n} e^{v_j^3/3 - xv_j} \prod_{1\le k<l\le n} |v_k - v_l|^{4/\beta}. \tag{20.2.20}$$

Asymptotic analysis of (20.2.19) shows

$$\rho_{(1)}^{\text{soft},\beta}(x) \underset{x\to\infty}{\sim} \frac{1}{2\pi}\frac{\Gamma(1+\beta/2)}{(4\beta)^{\beta/2}}\frac{e^{-2\beta x^{3/2}/3}}{x^{3\beta/4-1/2}} + \mathrm{O}\Big(\frac{1}{x^{3\beta/4+1}}\Big)$$

$$\rho_{(1)}^{\text{soft},\beta}(x) \underset{x\to-\infty}{\sim} \frac{\sqrt{|x|}}{\pi} - \frac{\Gamma(1+\beta/2)}{2^{6/\beta-1}|x|^{3/\beta-1/2}}\cos\Big(\frac{4}{3}|x|^{3/2} - \frac{\pi}{2}\Big(1-\frac{2}{\beta}\Big)\Big)$$
$$+\,\mathrm{O}\Big(\frac{1}{|x|^{5/2}}, \frac{1}{|x|^{6/\beta-1/2}}\Big). \qquad (20.2.21)$$

Note how the second of these formulae connects to the boundary of the Wigner semicircle (20.1.1), and we remark that the first gives the leading-order form of the right-hand tail for the scaled distribution of the largest eigenvalue.

20.3 Matrix realization of β ensembles

The three fold way of Dyson [Dys62c], isolating for example Gaussian ensembles of Hermitian matrices with real, complex, or real quaternion elements, and giving the special values $\beta = 1, 2$, and 4 in (20.1.2), is rooted in a viewpoint in physics. Changing mindsets, and thinking of random matrices from the viewpoint of numerical analysis, it is natural to investigate the Householder reduction of the classical Gaussian Hermitian ensembles to their symmetric tridiagonal form

$$\begin{bmatrix} a_N & b_{N-1} & & & & \\ b_{N-1} & a_{N-1} & b_{N-2} & & & \\ & b_{N-2} & a_{N-2} & b_{N-3} & & \\ & \ddots & \ddots & \ddots & & \\ & & b_2 & a_2 & b_1 \\ & & & b_1 & a_1 \end{bmatrix}. \qquad (20.3.1)$$

In the case of real entries ($\beta = 1$), this was done in a paper by Trotter [Tro84], who showed that each entry a_i, b_j is again independently distributed, with $a_i \sim \mathrm{N}[0, 1]$ and $b_j^2 \sim \Gamma[j/2, 1]$. Here $\mathrm{N}[\mu, \sigma]$ denotes the Gaussian distribution, and $\Gamma[a, \sigma]$ denotes the gamma distribution. Dumitriu and Edelman [Dum02], in the case of the Gaussian unitary ensemble scaled by multiplication by $\sqrt{2}$, showed the tridiagonal form (20.3.1) with $a_i \sim \mathrm{N}[0, 1]$ and $b_j^2 \sim \Gamma[j, 1]$ results. One can now pose the question, starting with a random tridiagonal matrix of this specification, calculate directly the corresponding eigenvalue probability density function. The main technical task is to compute the Jacobian for the change of variables from $\{a_i\}$, $\{b_j\}$ to the eigenvalues $\{\lambda_i\}$ and the first component of the normalized eigenvectors $\{q_j\}_{j=1,\dots,N-1}$ (note that q_N is not independent by the normalization condition). In [Dum02] this was carried out indirectly using Trotter's result. Subsequently, a direct proof was given in [For06]. Either way, the Jacobian is computed as

$$\frac{1}{q_N}\frac{\prod_{i=1}^{N-1} b_i}{\prod_{i=1}^{N} q_i}. \tag{20.3.2}$$

From this, it was shown in [Dum02] that random tridiagonal matrices (20.3.1) with $a_i \sim \mathrm{N}[0, 1]$ and $b_j^2 \sim \Gamma[\beta j/2, 1]$, and scaled by multiplication by $1/\sqrt{\beta}$ have an eigenvalue probability density function proportional to (20.1.2). Thus, with this change of viewpoint, the values $\beta = 1, 2$, and 4 are no longer special and (20.1.2) can be realized as an eigenvalue probability density function of a symmetric random tridiagonal matrix for general $\beta > 0$.

There is another change of viewpoint on the classical Gaussian ensembles leading to (20.1.2) for general $\beta > 0$ [For05]. This is based on the observation that a member M_{N+1} of the $(N+1) \times (N+1)$ GOE can be constructed from a member M_N of the $N \times N$ GOE according to a bordering procedure,

$$M_{N+1} = \begin{bmatrix} M_N & \vec{b} \\ \vec{b}^T & a \end{bmatrix}. \tag{20.3.3}$$

Here $a \sim \mathrm{N}[0, 1]$ while $b_j^2 \sim \Gamma[1/2, 1]$ $(j = 1, \dots, N)$. Moreover, the distribution of $\{M_{N+1}\}$ is unchanged if we replace M_N in (20.3.3) by its diagonal form D_N, and so write

$$M_{N+1} = \begin{bmatrix} D_N & \vec{b} \\ \vec{b}^T & a \end{bmatrix}. \tag{20.3.4}$$

The same bordering procedure can be used to generate GUE distributed matrices, except that now $b_j^2 \sim \Gamma[1, 1]$, and also M_{N+1} should be multiplied by $1/\sqrt{2}$. This suggests that a β generalization of the classical Gaussian ensembles will result from matrices defined inductively according to (20.3.4) with $a \sim \mathrm{N}[0, 1]$ and $b_j^2 \sim \Gamma[\beta/2, 1]$. In fact we can show that the eigenvalue probability density of $\{M_N\}$, when multiplied by $1/\sqrt{\beta}$, is precisely (20.1.2).

To demonstrate this, denote $p_n(\lambda) = \det(\lambda \mathbb{I}_n - M_n)$ for the characteristic polynomial and $D_N = \mathrm{diag}[a_1, \dots, a_N]$ for the diagonal form. Then we see from (20.3.4) that

$$\frac{p_{N+1}(\lambda)}{p_N(\lambda)} = \lambda - a - \sum_{i=1}^{N} \frac{b_i^2}{\lambda - a_i}. \tag{20.3.5}$$

The eigenvalues $\{\lambda_j\}_{j=1,\dots,N+1}$ of M_{N+1} are given by the zeros of this random rational function. We would thus like to change variables from $\{a, b_1^2, \dots, b_N^2\}$ to these zeros. By noting from (20.3.5) that

$$-b_i^2 = \frac{\prod_{j=1}^{N+1}(a_i - \lambda_j)}{\prod_{l=1, l\neq i}^{N}(a_i - a_l)},$$

the Jacobian for the transformation can be computed, and one finds that the probability density function for the eigenvalues of M_{N+1}, with $\{a_i\}$ given, is equal to

$$\frac{1}{\sqrt{2\pi}} \frac{1}{(\Gamma(\beta/2))^N} \frac{\prod_{1\le j<k\le N+1}(\lambda_j - \lambda_k)}{\prod_{1\le j<k\le N}(a_j - a_k)^{\beta-1}} \prod_{j=1}^{N+1}\prod_{p=1}^{N} |\lambda_j - a_p|^{\beta/2-1}$$
$$\times \exp\Big(-\frac{1}{2}\Big(\sum_{j=1}^{N+1}\lambda_j^2 - \sum_{j=1}^{N} a_j^2\Big)\Big) \qquad (20.3.6)$$

supported on

$$\infty > \lambda_1 > a_1 > \cdots > \lambda_{N+1} > -\infty. \qquad (20.3.7)$$

Let the conditional probability density (20.3.6) be denoted $G_N(\{\lambda_j\}, \{a_k\})$, and denote the domain of its support (20.3.7) by R_N. Let $p_n(\lambda_1, \ldots, \lambda_n)$ denote the eigenvalue probability density of M_n $(n = 1, 2, \ldots)$ and set $p_0 := 1$. Then the above working tells us that $\{p_n\}$ is uniquely determined by the recurrence

$$p_{n+1}(\lambda_1, \ldots, \lambda_{n+1}) = \int_{R_n} da_1 \cdots da_n \, G_n(\{\lambda_j\}, \{a_k\}) p_n(a_1, \ldots, a_n). \qquad (20.3.8)$$

We know that for $\beta = 1$ and 2 the solution of this recurrence is proportional to (20.1.2) (after the scaling $\sqrt{\beta}x_j \mapsto x_j$). Seeking (20.1.2) so scaled as the solution for general β, the right-hand side reads

$$e^{-\sum_{j=1}^{n+1}\lambda_j^2/2} \prod_{1\le j<k\le n+1}(\lambda_j - \lambda_k) \int_{R_n} da_1 \cdots da_n$$
$$\times \prod_{1\le j<k\le n}(a_j - a_k) \prod_{1\le j<k\le n}(a_j - a_k) \prod_{j=1}^{n+1}\prod_{p=1}^{n} |\lambda_j - a_p|^{\beta/2-1}, \qquad (20.3.9)$$

up to proportionality. The integral appearing in (20.3.9) is a special case of the Dixon-Anderson integral from the theory of the Selberg integral [For10, Ch. 4]. Up to proportionality, it is equal to $\prod_{1\le j<k\le n+1}(\lambda_j - \lambda_k)^{\beta-1}$, thus demonstrating the correctness of our trial solution.

A feature of this analysis is that it gives not just the eigenvalue probability density of M_N itself, but also the joint eigenvalue probability density of $\{M_n\}_{n=1,2,\ldots}$ constructed by the bordering (20.3.4). Thus with the eigenvalues of M_n in this sequence denoted $\{x_j^{(n)}\}_{j=1,\ldots,n}$, the joint probability of $\cup_{n=1}^{N}\{x_j^{(n)}\}_{j=1,\ldots,n}$ is equal to

$$p_n(x_1^{(n)}, \ldots, x_n^{(n)}) \prod_{l=2}^{n} G(\{x_j^{(l)}\}_{j=1,\ldots,l}, \{x_j^{(l-1)}\}_{j=1,\ldots,l-1}). \tag{20.3.10}$$

The simplest case is $\beta = 2$, for which $\{x_j^{(n)}\}_{j=1,\ldots,n}$ corresponds to the eigenvalues of the top $n \times n$ sub-block of a GUE matrix (scaled by multiplication by $\sqrt{2}$) when (20.3.10) reads, up to proportionality

$$\prod_{l=1}^{n} e^{-(x_l^{(n)})^2/2} \prod_{1 \le j < k \le n} (x_j^{(n)} - x_k^{(n)}) \prod_{j=1}^{n-1} \chi(x^{(j+1)} > x^{(j)}).$$

Here,

$$\chi(x^{(j+1)} > x^{(j)}) := \chi_{x_1^{(j+1)} > x_1^{(j)} > \cdots > x_j^{(j+1)} > x_j^{(j)} > x_{j+1}^{(j+1)}}.$$

This specifies the so called GUE minor process [Bar01, Joh06].

In general the characteristic polynomial for the tridiagonal matrix (20.3.1) satisfies the three-term recurrence

$$p_{n+1}(\lambda) = (\lambda - a_{n+1}) p_n(\lambda) - b_n^2 p_{n-1}(\lambda) \qquad n = 0, 1, 2, \ldots \tag{20.3.11}$$

where $p_{-1} := 0$, $p_0 := 1$. With a_{n+1} and b_n^2 specified below (20.3.2), this gives a random recurrence for the characteristic polynomials associated with the β generalization of the Gaussian ensemble. One can show that the same random recurrence results for the characteristic polynomial of $\{M_n\}$ as generated by (20.3.4). For the other classical random matrix ensembles – Laguerre, Jacobi, and circular – it is also true that their β-generalizations, with eigenvalue probability density function specifying the Boltzmann factor of certain log-gases, are such that the corresponding characteristic polynomials satisfy random recurrences. These are given in [Dum02] for the Laguerre case and in [Kil04] in the circular case and Jacobi case. Alternative recurrences in the Laguerre and Jacobi cases are given in [For05], and an alternative recurrence in the circular case is given in [For06].

The strategy taken in [Kil04] to set up a β-generalization of the classical circular ensembles is to first seek the similarity reduction of a random unitary matrix to upper Hessenberg form. The latter matrix has all the entries below the first sub-diagonal equal to zero, and requires all the elements in the first sub-diagonal to be real and positive. Such a reduction can be carried out by conjugation by a sequence of complex Householder reflections and diagonal unitary matrices. The entries of the Hessenberg matrix can be written in terms of complex numbers $\{\alpha_j\}_{j=0,\ldots,N-2}$ with $|\alpha_j| < 1$ and a further complex number α_{N-1} of unit modulus, such that the characteristic polynomial $\chi_k(\lambda)$ for the top $k \times k$ sub-block satisfies the coupled recurrence

$$\chi_k(\lambda) = \lambda\chi_{k-1}(\lambda) - \bar{a}_{k-1}\tilde{\chi}_{k-1}(\lambda)$$
$$\tilde{\chi}_k(\lambda) = \tilde{\chi}_{k-1}(\lambda) - \lambda a_{k-1}\chi_{k-1}(\lambda) \tag{20.3.12}$$

($k = 1, \ldots, N$). Here $\chi_0(\lambda) = \tilde{\chi}_0(\lambda) = 1$ and $\tilde{\chi}_k(\lambda) = \lambda^k \bar{\chi}_k(1/\lambda)$ with $\bar{\chi}_k$ denoting the polynomial χ_k with its coefficients replaced by their complex conjugates. Moreover, with Θ_ν, $\nu > 1$ the distribution on complex numbers $|z| < 1$ with probability density

$$\frac{\nu - 1}{2\pi}(1 - |z|^2)^{(\nu-3)/2},$$

and Θ_1 denoting the uniform distribution on complex numbers $|z| = 1$, for the CUE the parameters $\{\alpha_j\}_{j=0,\ldots,N-1}$ are distributed according to $\alpha_{N-j-1} \sim \Theta_{2j+1}$ ($j = 0, \ldots, N-1$).

This result suggests that analogous to tridiagonal matrices, it is possible to compute the Jacobian for the change of variables from $\{\alpha_j\}_{j=0,\ldots,N-1}$ to the eigenvalues $\{\lambda_j\}_{j=1,\ldots,N}$ and first (positive) components $\{q_i\}_{i=1,\ldots,N-1}$ of the independent normalized eigenvectors. Indeed, as a corollary to the above CUE to Hessenberg reduction, the Jacobian was computed in [Kil04] to equal

$$\frac{\prod_{i=0}^{N-2}(1 - |\alpha_i|^2)}{q_N \prod_{i=1}^{n} q_i}. \tag{20.3.13}$$

A direct proof was later given in [For06]. Continuing to proceed by analogy with the tridiagonal matrix construction of the β-generalized Gaussian β-ensemble, one would like to now make use of (20.3.13) to show that an appropriate deformation of the distribution of $\{\alpha_j\}_{j=0,1,\ldots,N-1}$ corresponding the CUE gives a unitary Hessenberg realization of the β-generalization of the circular ensemble. This was done in [Kil04] where the eigenvalue probability density for unitary Hessenberg matrices with $\alpha_{n-j-1} \sim \Theta_{\beta j+1}$ ($j = 0, \ldots, N-1$) was computed to be proportional to

$$\prod_{1 \le j < k \le N} |e^{i\theta_k} - e^{i\theta_j}|^\beta,$$

thus giving a random matrix realization of the β-generalization of the circular ensemble.

20.4 Stochastic differential equations

A feature of the construction of the realizations of the β-ensembles is that the corresponding characteristic polynomials satisfy random three-term recurrences. Moreover, in all cases except (20.3.12) the zeros of the successive characteristic polynomials interlace. In the construction (20.3.4) this can be seen to be a consequence of a fundamental result from matrix theory regarding the interlacing of eigenvalues of a symmetric matrix before and after bordering by a single row and column. With $r_i = -p_i(\lambda)/p_{i-1}(\lambda)$ it follows that the number of negative values in the sequence $\{r_i\}_{i=1,\dots,N}$ is equal to the number $N(\lambda)$ of eigenvalues less than λ. Since, from (20.3.11),

$$r_i = \begin{cases} a_1 - \lambda, & i = 1 \\ (a_i - \lambda) - b_i^2/r_{i-1}, & i = 2, \dots, N \end{cases}$$

the number of sign changes and thus $N(\mu)$ can be computed very efficiently in numerical calculations [Alb09].

The recurrences also have implications at an analytic level. Introduce the Prüfer phase θ_j^λ by writing

$$\cot \theta_j^\lambda = \frac{1}{(b_{j-1})^2} \frac{p_{j-1}(\lambda)}{p_{j-2}(\lambda)} \qquad (j = 2, \dots, N+1). \tag{20.4.1}$$

Some of the fundamental properties of the Prüfer phase are that $d\theta_j^\lambda/d\lambda < 0$; the k-th largest eigenvalue λ_k of the $N \times N$ matrix is such that $\theta_{N+1}^{\lambda_k} = (\pi/2) + (k-1)$; and

$$\left| \frac{1}{\pi} \theta_{N+1}^\lambda - (N - N(\lambda)) \right| \le \frac{1}{2}.$$

In the case of the bulk scaling limit of the β-generalization of the Gaussian ensemble, an analysis of the Prüfer phase for the corresponding random three-term recurrence due to Valkó and Virág [Val10] has led to the stochastic differential equation

$$d\alpha_\lambda = (\beta\lambda/4)e^{-\beta t/4} + 2\sin(\alpha_\lambda/2)dB, \tag{20.4.2}$$

with the property that $\alpha_\lambda(\infty)/2\pi$ is distributed as the number of eigenvalues in the interval $(0, \lambda)$ of the limiting matrix ensemble (dB denotes standard Brownian motion). As an application, this has been used to prove the asymptotic formula (20.1.14), and furthermore extended to the next order [Val09].

The coupled recurrences (20.3.12) for the characteristic polynomial in the β-generalized circular ensemble can also be analysed from the viewpoint of a Prüfer phase [Kil09]. For this one introduces

$$B_k(\lambda) = \frac{\lambda \chi_{k-1}(\lambda)}{\tilde{\chi}_{k-1}(\lambda)}. \tag{20.4.3}$$

This has the property $|B_k(\lambda)| = 1$ for $|\lambda| = 1$ and so permits the parameterization

$$\frac{B_k(e^{i\theta})}{B_k(1)} = e^{i\psi_k(\theta)}. \tag{20.4.4}$$

Analogous to θ_j^μ in (20.4.1), $\psi_k(\theta)$ – referred to as a relative Prüfer phase– has the property that it is an increasing function of θ, and with $\bar{\alpha}_{N-1} = e^{-i\eta}$

the eigenvalues of the $N \times N$ matrix occur when $\psi_{N-1}(\theta) = 2\pi k + \eta$, $k \in \mathbb{Z}_{\geq 0}$. Moreover, it follows from (20.4.3), (20.4.4), and (20.3.12) that $\{\psi_k(\theta)\}_{k=0,1,\dots}$ satisfies the first-order recurrence

$$\psi_{k+1}(\theta) = \psi_k(\theta) + \theta + 2\,\mathrm{Im}\,\log\Big(\frac{1-\gamma_k}{1-\gamma_k e^{i\psi_k(\theta)}}\Big), \qquad \psi_0 = \theta \tag{20.4.5}$$

where $\gamma_k := B_k(1)\alpha_k$. Killip [Kil08] has used this to prove that the linear statistic (20.1.9), scaled according to (20.1.10), is in the limit $N \to \infty$ distributed as a standard Gaussian, in accordance with the log-gas prediction.

At the hard and soft edges there are two possible strategies to obtain stochastic characterizations of the eigenvalue distributions. The first has its origin in the work of Edelman and Sutton [Ede06]. The basic idea is that the tridiagonal and bidiagonal matrix constructions of the β-generalized Gaussian and Laguerre ensembles can be regarded as discretizations of differential operators with a noise term. At the soft edge, this leads to the result that the distribution of the scaled largest eigenvalues in the β-generalized Gaussian ensemble is identical to the smallest eigenvalues of the so-called stochastic Airy operator

$$-\frac{d^2}{dx^2} + x + \frac{2}{\sqrt{\beta}}B'(x), \tag{20.4.6}$$

where $B(x)$ defines a standard Brownian path. This operator is to act on functions defined on $[0, \infty)$, which vanish at $x = 0$ and decay to zero as $x \to \infty$. Convergence of random tridiagonal matrices to the stochastic Airy operator was made rigorous by Ramírez, Rider and Virág [Ram06] who also established an interpretation as a diffusion. Furthermore, in [Ram06], this characterization has been used to prove that the right-hand tail of the distribution of the largest eigenvalue has the leading-order decay given by the first formula in (20.2.21), and also to prove that the leading-order decay of the left-hand tail is $\exp(-\beta|s|^3/24)$. The other approach to the soft edge scalings is to again analyse the appropriate limit of the recurrences [Val10]. This also leads to (20.4.6).

The positive definite matrices LL^T – L bidiagonal – realizing the Laguerre β-ensemble have been proved in [Ram09] to converge in the hard edge limit to the stochastic operator $\mathcal{J}_{\tilde{a}}^{\beta}(\mathcal{J}_{\tilde{a}}^{\beta})^*$. Here $\mathcal{J}_{\tilde{a}}^{\beta}$ is the stochastic Bessel operator [Ede06]

$$\mathcal{J}_{\tilde{a}}^{\beta} = -2\sqrt{x}\frac{d}{dx} + \frac{\tilde{a}}{\sqrt{x}} + \frac{2}{\sqrt{\beta}}B'(x), \quad x \in [0, 1], \tag{20.4.7}$$

where, with a corresponding to the exponent in the Laguerre weight $x^a e^{-x}$, $\tilde{a} = a - 1 + 2/\beta$ and $(\mathcal{J}_{\tilde{a}}^{\beta})^*$ is the adjoint of $\mathcal{J}_{\tilde{a}}^{\beta}$. The operator (20.4.7) is to act on square integrable functions $v(x)$ on $[0, 1]$ with the properties that $v(1) = 0$ and $(\mathcal{J}_{\tilde{a}}^{\beta}v)(0) = 0$. As an application, the soft to hard edge transition known from [Bor03] for $\beta = 1, 2$, and 4, which takes place in the limit $a \to \infty$ with an appropriate scaling and re-centring of the eigenvalues, was established for general $\beta > 0$.

Acknowledgements

This work was supported by the Australian Research Council.

References

[Alb09] J.T. Albrecht, C. Chan and A. Edelman, Found. Comp. Math. **9** (2009) 461
[Bak97] T.H. Baker and P.J. Forrester, Comm. Math. Phys. **188** (1997) 175
[Bar01] Y. Baryshnikov, Prob. Th. Relat. Fields **119** (2001) 256
[Bor03] A. Borodin and P.J. Forrester, J. Phys. A **36** (2003) 2963
[Cal69] F. Calogero, J. Math. Phys. **10** (1969) 2191
[Des06] P. Desrosiers and P.J. Forrester, Nucl. Phys. B, **743** (2006) 307
[Dot85] V.S. Dotsenko and V.A. Fateev, Nucl. Phys. B **251** (1985) 691
[Dys62a] F.J. Dyson, J. Math. Phys. **3** (1962) 157
[Dys62b] F.J. Dyson, J. Math. Phys. **3** (1962) 1191
[Dys62c] F.J. Dyson, J. Math. Phys. **3** (1962) 1199
[Dys95] F.J. Dyson, in *Chen Ning Yang* ed. S.-T. Yau, International Press, Cambridge MA, (1995) 131
[Dum02] I. Dumitriu and A. Edelman, J. Math. Phys. **43** (2002) 5830
[Ede06] A. Edelman and B.D. Sutton, J. Stat. Phys. **127** (2006) 1121
[Fog95] M. Fogler and B.I. Shklovskii, Phys. Rev. Lett. **74** (1995) 3312
[For93] P.J. Forrester, Phys. Lett. A **179** (1993) 127
[For10] P.J. Forrester, *Log-gases and Random Matrices*, Princeton University Press, Princeton (2010)
[For05] P.J. Forrester and E.M. Rains, Prob. Th. Relat. Fields **131** (2005) 1
[For06] P.J. Forrester and E.M. Rains, Int. Math. Res. Not. **2006** (2006) 48306
[For00] P.J. Forrester, B. Jancovici and D.S. McAnally, J. Stat. Phys. **102** (2000) 737
[Ha95] Z.N.C. Ha, Nucl. Phys. B **435** (1995) 604
[Hal94] F.D.M. Haldane, in *Correlation effects in low-dimensional electron systems*, ed. A. Okiji and N. Kawakami, Springer, Berlin (1994) 3
[Jan95] B. Jancovici, J. Stat. Phys. **80** (1995) 445
[Joh98] K. Johansson, Duke Math. J. **91** (1998) 151
[Joh06] K. Johansson and E. Nordenstam, Elect. J. Probability **11** (2006) 1342
[Kil04] R. Killip and I. Nenciu, Int. Math. Res. Not. **50** (2004) 2665
[Kil08] R. Killip, Int. Math. Res. Not. **2008** (2008) rnn007
[Kil09] R. Killip and M. Stoiciu, Duke Math. J. **146** (2009) 361
[Pol89] H.D. Politzer, Phys. Rev. B **40** (1989) 11917
[Ris92] H. Risken, *The Fokker-Planck Equation*, Springer, Berlin, 1992.
[Ram09] J. Ramírez and B. Rider, Commun. Math. Phys. **288** (2009) 887
[Ram06] J. Ramírez, B. Rider, B. Virág, J. Amer. Math. Soc. **24** (2011) 919.
[Sut71] B. Sutherland, J. Math. Phys. **12** (1971) 246
[Tro84] H.F. Trotter, Adv. Math. **54** (1984) 67
[Val10] B. Valko and B. Virág, Ann. Prob. **38** (2010) 1263
[Val09] B. Valko and B. Virág, Inv. Math. **177** (2009) 463

·21·

Wigner matrices

G. Ben Arous and A. Guionnet

Abstract

This is a brief survey of some of the important results in the study of the eigenvalues and the eigenvectors of Wigner random matrices, i.e. random Hermitian (or real symmetric) matrices with iid entries. We review briefly the known universality results, which show how much the behaviour of the spectrum is insensitive to the distribution of the entries.

21.1 Introduction

In the fifties, Wigner introduced a very simple model of random matrices to approximate generic self-adjoint operators. It is given as follows. Consider a family of independent, zero mean, real-or complex-valued random variables $\{Z_{i,j}\}_{1\leq i<j}$, independent from a family $\{Y_i\}_{1\leq i}$ of iid centered real-valued random variables. Consider the (real-symmetric or Hermitian) $N \times N$ matrix X_N with entries

$$X_N(j,i) = \bar{X}_N(i,j) = \begin{cases} Z_{i,j}, & \text{if } i < j, \\ Y_i, & \text{if } i = j. \end{cases} \tag{21.1.1}$$

We call such a matrix a *Wigner matrix*, and if the random variables $Z_{i,j}$ and Y_i are Gaussian, we use the term *Gaussian Wigner matrix*. The case of Gaussian Wigner matrices in which $EY_1^2 = 2$ and $E|Z_{1,2}|^2 = 1$ is of particular importance, since their law is invariant under the action of the orthogonal, (resp. unitary) group if the entries are real (resp. complex), see e.g. Chapter 3 in this handbook. In this case, the eigenvalues of the matrix X_N are independent of the eigenvectors which are Haar distributed. The joint distribution $P_N^{(\beta)}$ of the eigenvalues $\lambda_1(X) \leq \cdots \leq \lambda_N(X)$ is given by

$$P_N^{(\beta)}(dx_1, \ldots, dx_N) = \bar{C}_N^{(\beta)} \mathbf{1}_{x_1 \leq \cdots \leq x_N} |\Delta(x)|^\beta \prod_{i=1}^N e^{-\beta x_i^2/4} dx_1 \cdots dx_N, \tag{21.1.2}$$

where

$$\bar{C}_N^{(\beta)} = \left(\int_{-\infty}^{\infty} \cdots \int_{-\infty}^{\infty} |\Delta(x)|^\beta \prod_{i=1}^N e^{-\beta x_i^2/4} dx_i \right)^{-1}.$$

In this formula, $\beta = 1$ (resp. 2) if the entries are real (resp. complex). The distributions of the eigenvalues of these random matrices are usually called the Gaussian orthogonal ensembles (GOE) or the Gaussian Unitary Ensemble (GUE) respectively. These and more general invariant ensembles are the main focus of the other chapters of this handbook. The present chapter is devoted to the study of general non-invariant Wigner ensembles and of their universal properties, that is the properties they share with the invariant Gaussian ensembles, at the global and at the local levels, in the bulk and at the edge of the spectrum. This is a very wide domain of research, impossible to summarize in any depth in the format of this review article. The study of the properties of the spectrum of Wigner ensembles has a very long and rich history, and a few recent expository works cover the results and their proof in much greater depth than can be achieved here ([Bai99], [And09]). Compounding the difficulty is the fact that the field has recently seen a burst of very important results. Indeed, the question of universality in the bulk at the local level for the spectrum of Wigner matrices, which had long been seen as one of the central open problems in random matrix theory, has been solved recently by two groups of mathematicians (L.Erdos, J.Ramirez, B.Schlein, S.Peche, H.T. Yau and T.Tao, and V.Vu). These new universality results open a very exciting period for the field of random matrix theory (see the Bourbaki seminar [Gui10]). We will try to cover these new developments succinctly, without any claim for completeness given the rapid pace of recent progress. We will give some of the universality results as well as some results known for the Gaussian ensembles which have not yet been proved to be universal, i.e. the open questions left after the recent wave of progress. We will also give some of the known limits of universality, i.e. some of the cases where it is known that the results differ from the Gaussian ensembles. Before proceeding, let us comment very briefly and informally about the methods of approach for general non-invariant Wigner ensembles. In fact there are very few such possible approaches, essentially three. Let us note first that the main tool used for invariant ensembles, i.e. the explicit computation of the distribution of the spectrum, is of course not possible for the non-invariant ensembles. The first general approach for non-invariant Wigner ensembles is the moment method, which consists in computing the asymptotic behaviour of the moments of the spectral measure L_N, that is the empirical measure of the renormalized eigenvalues of X_N, or equivalently the normalized trace of powers of the random matrix

$$\int x^k dL_N(x) = \frac{1}{N^{\frac{k}{2}+1}} \mathrm{Tr}[X_N^k].$$

The second one, the resolvent method, consists in computing the asymptotic behaviour of the normalized trace of the resolvent, i.e. the Stieltjes transform of the spectral measure,

$$\int \frac{1}{z-x} dL_N(x) = \frac{1}{N}\mathrm{Tr}[(zId - N^{-\frac{1}{2}}X_N)^{-1}].$$

For a long time these two methods were basically the only ones available. The survey [Bai99] shows how far one can go using these tools. Another approach to universality is based on explicit formulae and concerns a special case of non-invariant ensembles, the Gaussian divisible case. This is the case of random Wigner matrices, where the distribution of the entries is the convolution of an arbitrary distribution and of a centred Gaussian one. These matrices can be written as

$$X_N = \sqrt{\varepsilon} G_N + \sqrt{1-\varepsilon}\, V_N \tag{21.1.3}$$

with a matrix G_N taken from the GOE or the GUE, independent from a self-adjoint matrix V_N, and some $\varepsilon \in (0, 1)$. Note that when V_N is a Wigner matrix, so is X_N. K. Johansson [Joh01] studied such matrices when X_N is taken from the GUE based on rather explicit formulae for the joint law of the eigenvalues of X_N. Another point of view is based on the fact that the spectrum of such matrices is described by the Dyson Brownian motion, see e.g. [And09, Theorem 4.3.2], at time $t = -\log \varepsilon$ which is a weak solution to the system

$$d\lambda_i^N(t) = \frac{\sqrt{2}}{\sqrt{\beta N}} dW_i(t) + \frac{1}{N}\sum_{j:j\neq i} \frac{1}{\lambda_i^N(t) - \lambda_j^N(t)} dt, \quad i = 1, \ldots, N, \tag{21.1.4}$$

with the initial conditions given by the eigenvalues of V_N and W_i, $1 \leq i \leq N$ independent Brownian motions. This fourth perspective, based on Dyson's Brownian motion, has been considerably strengthened and proved to be very useful for far more general matrices, in the recent work of the group around Erdos, Schlein and Yau.

Universality can also be proved by using approximation arguments from the above models, see e.g. [R06, Tao09a, Tao09b].

21.2 Global properties

In this section, we describe the global properties of the spectrum of Wigner matrices. It turns out that, when the entries have a finite second moment, the eigenvalues of X_N are of order $\sqrt{N}$ with overwhelming probability, and the global properties of the spectrum of X_N shall be described by the spectral measure

$$L_N := \frac{1}{N}\sum_{i=1}^{N} \delta_{\frac{\lambda_i^N(X)}{\sqrt{N}}}$$

which is a probability measure on the real line. Note that in particular, for any $a < b$

$$L_N([a,b]) = \frac{1}{N}\sharp\{i : \lambda_i^N(X) \in \sqrt{N}[a,b]\}$$

is the proportion of normalized eigenvalues falling in the interval $[a, b]$.

The first result of RMT is, of course, Wigner's theorem which says that L_N converges towards the semicircle law. We will give some of the known results about the fluctuations around this limit. We will also mention concentration results, and see that the law of L_N concentrates under fairly general hypotheses. One first result which seems out of reach for general Wigner ensembles concerns the large deviations of the spectral measure. Whereas in the case of the Gaussian ensembles, one knows a full large deviations principle, it seems impossible at this time to get such a result for any non-invariant ensemble. We will recall this large deviations principle as well as the moderate deviations principle for Gaussian ensembles. A moderate deviation principle has been given for some non-invariant Gaussian divisible ensembles.

Recently, it was observed [Erd08] that the convergence of the spectral measure towards the semicircle law holds in a very local sense, that is it can be obtained on intervals with width going to zero just more slowly than the typical spacing between eigenvalues. This result thus interpolates between global and local properties of the spectrum. Moreover, this local convergence is intimately related to the fact that the eigenvectors are *delocalized* in the sense that all of their entries are at most of order $(\log N)^4/\sqrt{N}$ with overwhelming probability.

Finally, in Section 21.2.5, we will describe the global behaviour of the spectrum when the entries do not have a finite second moment but have heavy tails, provided they belong to the domain of attraction of an α-stable law.

21.2.1 Convergence to the semicircle law

The main result of this section goes back to Wigner [Wig55] and gives the weak convergence of L_N towards the semicircle law σ given by

$$d\sigma(x) = \frac{1}{2\pi}\mathbf{1}_{[-2,2]}\sqrt{4-x^2}dx\,.$$

Theorem 21.2.1 *Assume $E[|Z_{1,2}|^2] = 1$ and $E[Y_1^2] < \infty$. Then, for any continuous function f with polynomial growth,*

$$\lim_{N\to\infty}\int f(x)dL_N(x) = \int f(x)d\sigma(x)\;a.s.$$

This theorem was originally proved for polynomial functions and the convergence in probability, under the condition that the entries all have their moments finite (see [Wig55] or [And09, Theorem 2.1.1]). However, almost sure convergence can be obtained by using the Borel-Cantelli lemma and fine estimates of the covariances [And09, Exercise 2.1.16]. Finally, the assumptions that all moments are finite can be removed by approximation using the Hoffman-Wielandt inequality [And09, Lemma 2.1.19].

The next question is about the speed of convergence of the spectral measure to the semicircle law. It is not universal, and should depend on the law of the entries and the metric used to measure it. In [Bai93, Theorem 4.1], it was shown that

$$\sup_{x\in\mathbb{R}} |E[L_N((-\infty, x])] - \sigma([-2, x \vee -2])| = O(N^{-\frac{1}{4}})$$

under the assumption that the entries have a finite fourth moment. When the sixth moment is finite, it was shown [Bai09, Lemma 6.1] that the speed is at most of order $O(N^{-\frac{1}{2}})$. For complex Gaussian divisible ensembles, it was proved to be at most of order $O(N^{-\nu})$ for all $\nu < 2/3$ in [Got07, Theorem 1.2]. The conjecture is that the optimal speed is $O(N^{-1})$, as proved for the GUE case in [Got05]. A similar question may be asked for the expectation of the distance between L_N and σ. We refer to [Cha04] and references therein.

21.2.2 Fluctuations around the semicircle law

The standard fluctuations results of the spectral measure for invariant Wigner ensembles state that for any smooth enough function f the law of

$$\delta_N(f) = N(\int f(x)dL_N(x) - \int f(x)d\sigma(x))$$

converges towards a Gaussian variable (compare [Joh98] for the Gaussian ensembles). Note here the intriguing speed $1/N$, to be compared with the classical $1/\sqrt{N}$ speed of the classical central limit theorem. Such a result for non-invariant ensembles was first proved in the slightly different Wishart matrix model case by Jonsson [Jon82] for polynomial functions f. We refer the reader to [And09, Theorem 2.1.31] for similar techniques for Wigner matrices under the hypothesis that the entries all have their moments finite and the centring is done with respect to the mean instead of the limit. This last result was extended in [Lyt09] under only the finite fourth moment condition. An interesting point is that the covariance of the limiting Gaussian variable depends on the fourth moment of the distribution of the entries. The fluctuations result for $\delta_N(f)$ was generalized, under the assumption of a finite sixth moment, in [Bai09, Theorem 1.1] for functions f with four continuous derivatives. An interesting feature of [Bai09] is a generalization of the result of [Joh98] for the GOE which shows that the limiting Gaussian variable is not centred in general in the real case.

Such central limit theorems hold for many models of random matrices, see e.g. unitary matrices following the Haar measure [Dia94].

21.2.3 Deviations and concentration properties

Under stronger conditions about the distribution of the entries, more can be said about the convergence of the spectral measure. One can prove concentration results. Indeed, it turns out that the evaluation of the spectral measure against a smooth function is a smooth function of the entries. Therefore, the theory of concentration of measure applies to such random variables. For instance, we have the following concentration of measure property. To begin with, recall that a probability measure P on $\mathbb{R}$ is said to satisfy the *logarithmic Sobolev inequality* (LSI) with constant c if, for any differentiable function f,

$$\int f^2 \log \frac{f^2}{\int f^2 dP} dP \leq 2c \int |f'|^2 dP.$$

When the distribution of the entries satisfies the log-Sobolev inequality, not only does the spectral measure concentrate but also the eigenvalues themselves. More precisely we have the following statement.

Theorem 21.2.2 *Suppose that the laws of the independent entries* $\{X_N(i,j)\}_{1\leq i\leq j\leq N}$ *all satisfy the (LSI) with constant* c*. Then, for any Lipschitz function* f *on* $\mathbb{R}$*, for any* $\delta > 0$*,*

$$P\left(|\int f(x)dL_N(x) - E[\int f(x)dL_N(x)]| \geq \delta\right) \leq 2e^{-\frac{1}{4c|f|_{\mathcal{L}}^2}N^2\delta^2}. \quad (21.2.1)$$

Further, for any $k \in \{1, \ldots, N\}$*,*

$$P\left(|f(N^{-\frac{1}{2}}\lambda_k(X_N)) - Ef(N^{-\frac{1}{2}}\lambda_k(X_N))| \geq \delta\right) \leq 2e^{-\frac{1}{4c|f|_{\mathcal{L}}^2}N\delta^2}. \quad (21.2.2)$$

These results can also be generalized to the case when the distribution of the entries satisfies a Poincaré inequality or when they are simply bounded (but then the test function f has to be convex). We refer to [Gui00] or [And09] for precise statements and generalizations. Interestingly, concentration inequalities hold under the mere assumption of independence (in fact only of the vectors $((X_N(ij))_{j\leq i})_{1\leq j\leq N}$), but with a worst speed estimate [Bord10].

Theorem 21.2.3 *Assume* f *has finite variation norm*

$$\|f\|_{TV} := \sup_{\substack{k\in\mathbb{N} \\ x_0<x_1<x_2\cdots<x_k}} \sum_{\ell=1}^{k} |f(x_\ell) - f(x_{\ell-1})|.$$

Then, for any $\delta > 0$*,*

$$P\left(|\int f(x)dL_N(x) - E[\int f(x)dL_N(x)]| \geq \delta\|f\|_{TV}\right) \leq 2e^{-\frac{1}{2}N\delta^2}. \quad (21.2.3)$$

The proof follows from Martingale inequalities.

The advantage of concentration inequalities is that they hold for any fixed N. However, they do not provide in general the optimal asymptotic speed. Such optimal constants are provided by a moderate deviations principle, or by a large deviations principle. Recall that a sequence of laws $(P_N, N \geq 0)$ on a Polish space Ξ satisfies a large deviation with good rate function $I : \Xi \to \mathbb{R}^+$ and speed a_N going to infinity with N if and only if the level sets $\{x : I(x) \leq M\}$, $M \geq 0$, of I are compact and for all closed set F

$$\limsup_{N\to\infty} a_N^{-1} \log P_N(F) \leq -\inf_F I$$

whereas for all open set O

$$\liminf_{N\to\infty} a_N^{-1} \log P_N(O) \geq -\inf_O I.$$

Large deviation results for the spectral measure of Wigner ensembles are still only known for the Gaussian ensembles since their proof is based on the explicit joint law of the eigenvalues $P_N^{(\beta)}$. This question was studied in [Ben97], in relation to Voiculescu's non-commutative entropy. The latter is defined as the real-valued function on the set $M_1(\mathbb{R})$ of probability measures on the real line given by

$$\Sigma(\mu) = \begin{cases} \int\int \log|x-y| d\mu(x)d\mu(y) & \text{if } \int \log|x| d\mu(x) < \infty, \\ -\infty & \text{else.} \end{cases} \tag{21.2.4}$$

Theorem 21.2.4 *Let*

$$I_\beta(\mu) = \begin{cases} \frac{\beta}{2}\int x^2 d\mu(x) - \frac{\beta}{2}\Sigma(\mu) - c_\beta & \text{if } \int x^2 d\mu(x) < \infty, \\ \infty & \text{else,} \end{cases} \tag{21.2.5}$$
$$\text{with } c_\beta = \inf_{\nu\in M_1(\mathbb{R})}\{\tfrac{\beta}{2}\int x^2 d\nu(x) - \tfrac{\beta}{2}\Sigma(\nu)\}.$$

Then, the law of L_N under $P_N^{(\beta)}$, as an element of $M_1(\mathbb{R})$ equipped with the weak topology, satisfies a large deviation principle in the scale N^2, with good rate function I_β.

A moderate deviations principle for the spectral measure of the GUE or GOE is also known, giving a sharp rate for the decrease of deviations from the semicircle law in a smaller scale.

Theorem 21.2.5 *For any sequence $a_N \to 0$ so that $Na_N \to \infty$, the sequence $Na_N(L_N([x,\infty)) - \sigma([x,\infty)))$ in $L_c^1(\mathbb{R})$ equipped with Stieljes'-topology and the corresponding cylinder σ-field satisfies the large deviation principle in the scale $(Na_N)^2$ and with good rate function*

$$J(F) = \sup\{\int h'(x)F(x)dx - \frac{1}{2}\int_0^1 \int (h')^2(\sqrt{s}x)\sigma(dx)ds\}.$$

Here the supremum is taken over the complex vector space generated by the Stieltjes functions $f(x) = (z - x)^{-1}$, $z \in \mathbb{C}$ and the Stieltjes'-topology is the weak topology with respect to derivatives of such functions.

This moderate deviation result does not yet have a fully universal version for general Wigner ensembles. It has however been generalized to Gaussian divisible matrices (21.1.3) with a deterministic self-adjoint matrix V_N with converging spectral measure [Dem01] and to Bernouilli random matrices [Dor09].

21.2.4 The local semicircle law and delocalization of the eigenvectors

A crucial result has been obtained much more recently, by Erdös, Schlein, and Yau [Erd08, Theorem 3.1] proving that the convergence to the semicircle law also holds locally, namely on more or less any scale larger than the typical spacings between the normalized eigenvalues which is of order N^{-1}. More precisely, they consider the case where the distribution of the entries has sub-exponential tails, and in the case of complex entries when the real and the imaginary part are independent. They then showed the following theorem.

Theorem 21.2.6 *For an interval $I \subset]-2, 2[$, let $\mathcal{N}_I$ be the number of eigenvalues of $X_N/\sqrt{N}$ which belong to I. Then, there exist positive constants c, C so that for all $\kappa \in (0, 2)$, all $\delta \leq c\kappa$, any $\eta > \frac{(\log N)^4}{N}$ sufficiently small, we have*

$$P\left(\sup_{|E|\leq 2-\kappa}\left|\frac{\mathcal{N}_{[E-\eta, E+\eta]}}{2N\eta} - \rho_{sc}(E)\right| > \delta\right) \leq Ce^{-c\delta^2\sqrt{N\eta}}. \tag{21.2.6}$$

Amazingly, this local convergence, concentration inequalities, independence and equi-distribution of entries, entails delocalization of the eigenvectors of X_N, namely it was shown in [Erd08] that

Corollary 21.2.1 *Under the same hypotheses, for any $\kappa > 0$ and $K > 0$, there exist positive finite constants C, c such that for all $N \geq 2$,*

$$P\left(\exists v \text{ so that } X_N v = \sqrt{N}\mu v,\ \|v\|_2 = 1,\ \mu \in [-2+\kappa, 2-\kappa] \text{ et } \|v\|_\infty \geq \frac{(\log N)^{\frac{9}{2}}}{N^{\frac{1}{2}}}\right)$$

is bounded above by $Ce^{-c(\log N)^2}$.

21.2.5 The limits of universality: the spectral measure of heavy-tailed Wigner random matrices

This brief section shows that there are natural obvious limits to the universality properties of random matrix theory. When the second moment of the distribution of the entries is infinite, the global behaviour of the spectrum cannot be expected to be similar to Gaussian invariant cases, and indeed it changes dramatically. Let us assume more precisely that the entries $(Z_{i,j})_{i\leq j}$, $(Y_i)_{i\geq 0}$ are independent and have the same distribution P on $\mathbb{R}$ belonging to the domain of attraction of an α-stable law. This means that there exists a slowly varying function L so that

$$P(|x| \geq u) = \frac{L(u)}{u^{\alpha}}$$

with $\alpha < 2$. We then let

$$a_N := \inf\{u : P(|x| \geq u) \leq \frac{1}{N}\}$$

which is of order $N^{\frac{1}{\alpha}} \gg \sqrt{N}$ for $\alpha < 2$. Note that a_N is the order of magnitude of the largest entries on a row (or column) of the matrix X_N or of the sum of the entries on a row (centred if $\alpha < 1$). Then, it was stated in [Bou94] (see Chapter 13 of this handbook) and proved in [Ben08] (see [Bel09] for the proof of the almost sure convergence, and [Bord09] for another approach) that the eigenvalues of X_N are of order a_N and that the spectral measure of X_N correctly renormalized converges towards a probability measure which is different from the semicircle law. More precisely we have

Theorem 21.2.7 *Let $\alpha \in (0, 2)$ and put $L_N = N^{-1} \sum_{i=1}^{N} \delta_{a_N^{-1}\lambda_i}$. Then*

- *L_N converges weakly to a probability measure μ_α almost surely.*
- *μ_α is symmetric, has unbounded support, has a smooth density ρ_α which satisfies*

$$\rho_\alpha(x) \sim \frac{L_\alpha}{|x|^{\alpha+1}} \qquad |x| \to \infty.$$

21.3 Local properties in the bulk

Analysis of the local properties of the eigenvalues of X_N in the bulk of the spectrum goes back to Gaudin, Dyson, and Mehta, among others, for the Gaussian ensembles. The asymptotic behaviour of the probability that no eigenvalues belong to an interval of width N^{-1}, the asymptotic distribution of the typical spacing between two nearest eigenvalues, the asymptotic behaviour of the k-points correlation functions are, for instance, well understood [Meh04]. The generalization of these results to large classes of non-invariant Wigner ensembles has been a major challenge for a long time. (As anecdotal evidence of this, it could be noted that during the 2006 conference at the Courant Institute in honour of Percy Deift's 60th birthday, a panel of five experts all quoted this as the main open question of random matrix theory).

The first universality result in this direction was obtained by K. Johansson [Joh01] for the correlation functions of complex Gaussian divisible entries (21.1.3). Johansson's proof follows and expands on an idea of Brézin and Hikami [Bré97]. This universality result has recently been extended to real entries by using dynamics and Dyson Brownian motion (21.1.4) by Erdös, Ramirez, Schlein, Yau, and Yin [Erd09a, Erd09b].

It was only very recently generalized to a general case of non-invariant ensembles (with an assumption of sub-exponential moments and the same four first moments as Gaussian variables) by Tao and Vu [Tao09a]. Combining these two sets of results, one can prove the universality of the local statistics of the eigenvalues in the bulk provided the entries have sub-exponential tails (and at least three points in their support in the real case). We shall detail these results in the next section, about spacing distributions and refer to the original papers for correlation functions.

21.3.1 Spacings in the bulk for Gaussian Wigner ensembles

Let us first concentrate on a sample of local results for the Hermitian case. Based on the fact that the law $P_N^{(2)}$ is determinantal, the following asymptotics are well known, see [Tra94], with ρ_{sc} the density of the semicircle law.

Theorem 21.3.1

- *For any $x \in (-2, 2)$, any $k \in \mathbb{N}$, the joint law of k (unordered) rescaled eigenvalues $\eta_i = N^{\frac{1}{2}}\rho_{sc}(x)^{-1}(\lambda_i - N^{\frac{1}{2}}x)_{1\le i\le k}$ converges vaguely towards the measure which is absolutely continuous with respect to Lebesgue measure and with density*

$$\rho^{(k)}(\eta_1, \dots, \eta_k) = C_k \det\left((S(\eta_i, \eta_j))_{1\le i,j\le k}\right)$$

with S the Sine kernel

$$S(y_1, y_2) := \frac{\sin(y_1 - y_2)}{\pi(y_1 - y_2)}.$$

- *For any $x \in (-2, 2)$, for any compact set B,*

$$\lim_{N\to\infty} P_N^{(2)}\left(\lambda_i \notin N^{\frac{1}{2}}x + N^{-\frac{1}{2}}\rho_{sc}(x)B, i = 1, \dots, N\right) = \Delta(B, S)$$

with Δ the Fredholm determinant

$$\Delta(B, S) := 1 + \sum_{k=1}^{\infty} \frac{(-1)^k}{k!} \int_A \cdots \int_A \det\left((S(x_i, x_j))_{i,j=1}^k\right) \prod_{i=1}^k dx_i.$$

This result implies, see e.g. [Meh04, Appendix A.8] (or a more complete proof in [And09, Theorem 4.2.49] together with [Dei99] which gives the uniform asymptotics of the Hermite kernel towards the Sine kernel on intervals with width going to infinity with N), that the empirical spacing distribution between two eigenvalues, near a given point in the bulk, converges. More precisely, we give the statement for spacings near the origin. Define the Gaudin distribution by

$$P_{Gaudin}([2t, +\infty)) = -C\partial_t \Delta(S, \mathbf{1}_{(-t,t)^c})$$

and consider a sequence l_N increasing to infinity, such that $l(N) = o(N)$.

Theorem 21.3.2 *The number of eigenvalues of $N^{-\frac{1}{2}}X_N$ at a distance less than l_N/N whose nearest neighbours spacing is smaller than $N^{-1}\pi s$, divided by l_N, converges in probability towards $P_{Gaudin}([0,s])$.*

These results generalize to the real case, even though the law $P_N^{(1)}$ is not determinantal anymore. For instance one has, see e.g. [And09, Section 3.1.2]

Theorem 21.3.3 *There exists an increasing function $F_1 : \mathbb{R}^+ \to \mathbb{R}^+$, $F_1(0)=0$, $F_1(\infty) = 1$ so that*

$$1 - F_1(t) = \lim_{N\to\infty} P_N^{(1)}\left(\cap_{1\le i\le N}\{\lambda_i^N \in \frac{1}{\sqrt{N}}(-\frac{t}{2},\frac{t}{2})^c\}\right)$$

Explicit formulae for F_1 are known, see e.g. [And09, Theorem 3.1.6].

21.3.2 Universality for Gaussian divisible Wigner matrices

We will consider matrices given by (21.1.3) with V_N a Wigner matrix. We shall assume that the entries of V_N are independent, equidistributed, and have sub-exponential tails. In the complex case, we shall assume that the real and imaginary parts of the entries of V_N are independent and equidistributed.

Then, we have

Theorem 21.3.4 *For any $\varepsilon \in (0,1]$, the results of Theorem 25.3.4 extend to the eigenvalues of $X_N = \sqrt{\varepsilon}G_N + \sqrt{1-\varepsilon}V_N$.*

This result was proved by Johansson [Joh01] in the Hermitian case, based on the fact that the density of the eigenvalues of X_N is then explicitly given by the Harich-Chandra-Itzykson-Zuber integral. It was initially valid only in the middle of the bulk, i.e. very near the centre of the semicircle, but has been later extended to the whole bulk. He showed more recently [Joh10] that the existence of the second moment is sufficient to grant the result. In the real-symmetric case, such a formula does not exist and the result was only recently proved by Erdös, Schlein, and Yau by noticing that the entries of X_N can be seen as the evolution of Brownian motions starting from the entries of V_N and taken at time $-\log\varepsilon$, see (21.1.4). Based on this fact, Erdös, Schlein, and Yau developed techniques coming from hydrodynamics theory to show that a short time evolution of the Brownian motion is sufficient to guarantee that the correlation functions are close to equilibrium, given by the Gaussian matrix. In fact, they could show that this time can even be chosen going to zero with N. The universality result of Theorem 21.3.4 is weaker than those stated in Chapter 6 of this handbook, where convergence of the density of the joint law of the eigenvalues is required; for the time being such an averaging is needed to apply Erdös, Schlein, and Yau techniques.

21.3.3 Universality and the four moments theorem

The approach proposed by Tao and Vu to prove universality follows Lindenberg's replacement argument, which is classical for sums of iid random variables. The idea is to show that the eigenvalues of X_N are smooth functions of the entries so that one can replace one by one the entries of a Wigner matrix by the entries of another Wigner matrix (for which we can control the local properties of the spectrum because they are Gaussian or Gaussian divisible) and control the difference of the expectations by $o(N^{-2})$ if the four first moments are the same. The statement of the result is more precisely the following. Let us consider two Wigner matrices X_N and X'_N whose entries have sub-exponential tails (in fact a sufficiently large number of moments should be enough) and, in the complex case, with independent imaginary and real parts. We assume that the moments $C(\ell, p) = E[\Re(X_{ij})^\ell \Im(X_{ij})^p]$ are the same for X_N and X'_N for all $\ell + p \leq 4$. We denote by $\lambda_1(M_N) \leq \lambda_2(M_N) \cdots \leq \lambda_{N-1}(M_N) \leq \lambda_N(M_N)$ the ordered eigenvalues of a Hermitian $N \times N$ matrix M_N. Then, Theorem 15 of [Tao09a] states that

Theorem 21.3.5 *For any $c_0 > 0$ sufficiently small, any $\epsilon \in (0, 1)$ and $k \geq 1$, any function $F : \mathbb{R}^k \to \mathbb{R}$ such that*

$$|\nabla^j F(x)| \leq N^{c_0}$$

for all $0 \leq j \leq 5$ and $x \in \mathbb{R}^k$, we have for any $i_1, \ldots, i_k \in [\epsilon N, (1-\epsilon)N]$,

$$\left| E\left[F(\lambda_{i_j}(\sqrt{N}X_N), 1 \leq j \leq k)\right] - E\left[F(\lambda_{i_j}(\sqrt{N}X'_N), 1 \leq j \leq k)\right]\right| \leq N^{-c_0}.$$

One can then take X'_N to be a Gaussian or a Gaussian divisible matrix to deduce from the previous results on such matrices that any Wigner matrix having the same moments of order smaller than four have asymptotically the same spacing distributions in the bulk. It is not hard to see [Tao09a, Corollary 30] that one can always match the first four moments with those of a Gaussian divisible matrix provided the law of the entries has at least three points in its support. In the Hermitian case, using a finer analysis based on the explicit formula for the Harich-Chandra-Itzykson-Zuber integral, one can prove [Erd09b] that Theorem 21.3.4 holds with ε of order $N^{-\frac{3}{4}}$ in which case one can always approximate the first four moments of the entries with such a Gaussian divisible law to get rid of the 'three points in the support' hypothesis [Erd09c].

21.3.4 The extreme gaps and fluctuations of eigenvalues in the bulk

We saw in the preceding question that the typical spacing between normalized eigenvalues in the bulk is of order N^{-1}, and that the distribution of a typical spacing is known and universal. What about the size and distribution of the smallest and of the largest spacings?

For the GUE ensemble, it has recently been proved ([Ben10] after initial results in [Vin]) that the smallest spacing between normalized eigenvalues in the bulk, say in the interval $(-2+\epsilon, 2-\epsilon)$ for a positive ϵ, is of order $N^{-4/3}$. The distribution of this smallest spacing is known. Moreover the point process of the smallest spacings is asymptotically Poissonian.

We first consider the smallest gaps, studying the point process

$$\chi^N = \sum_{i=1}^{N-1} \delta_{(N^{4/3}(\lambda_{i+1}-\lambda_i),\lambda_i)} 1_{|\lambda_i|<2-\epsilon},$$

for any arbitrarily small fixed $\epsilon > 0$. Then, Theorem 1.4 of [Ben10] states that

Theorem 21.3.6 *As $N \to \infty$, the process χ^N converges to a Poisson point χ process with intensity*

$$E[\chi(A \times I)] = \left(\frac{1}{48\pi^2}\int_A u^2 du\right)\left(\int_I (4-x^2)^2 dx\right)$$

for any bounded Borel sets $A \subset \mathbb{R}_+$ and $I \subset (-2+\epsilon, 2-\epsilon)$.

In fact this result obtained in [Ben10] is the same one would get if one assumed (wrongly) that the spacings are independent. The correlations between the spacings are not felt across the macroscopic distance between the smallest gaps. The following corollary about the smallest gaps is an easy consequence of the previous theorem. Introduce $t_1^{(n)} < \cdots < t_k^{(n)}$ the k smallest spacings in I, i.e. of the form $\lambda_{i+1} - \lambda_i$, $1 \le i \le n-1$, with $\lambda_i \in I$, $I = [a, b]$, $-2 < a < b < 2$. Let

$$\tau_k^{(n)} = \left(\int_I (4-x^2)^2 dx/(144\pi^2)\right)^{1/3} t_k^{(n)}.$$

Corollary 21.3.1 *For any $0 \le x_1 < y_1 < \cdots < x_k < y_k$, with the above notations*

$$P\left(x_\ell < n^{4/3}\,\tilde{\tau}_\ell^{(n)} < y_\ell, 1 \le \ell \le k\right) \underset{n\to\infty}{\longrightarrow} \left(e^{-x_k^3} - e^{-y_k^3}\right)\prod_{\ell=1}^{k-1}\left(y_\ell^3 - x_\ell^3\right).$$

In particular, the kth smallest normalized space $N^{4/3}\tau_k^N$ converges in law to τ_k, with distribution

$$P(\tau_k \in dx) = \frac{3}{(k-1)!}\, x^{3k-1} e^{-x^3} dx.$$

For the largest spacings, the situation is less understood. Nevertheless it is proven that the largest gap normalized by $\frac{\sqrt{\log N}}{N}$ converges to a constant.

The universality question related to extreme gaps is still open. Can these results be generalized to non-invariant ensembles? The real-symmetric case is not understood even for the GOE.

The size $\frac{\sqrt{\log N}}{N}$ of the largest spacing is natural, it is the same as one would guess from the tail of the Gaudin distribution by making the ansatz that the spacings are independent. But it is also interestingly small. Indeed this size $\frac{\sqrt{\log N}}{N}$ is also the size of the standard deviation of the position of a given eigenvalue in the bulk [Gus05]. Thus the maximal spacing is not bigger than a typical fluctuation of one eigenvalue! This is of course linked to the fact that two adjacent eigenvalues have perfectly correlated Gaussian fluctuations.

We mention here that Gustavsson's result gives more than what is alluded to above. It proves that an eigenvalue in the bulk of GUE fluctuates around its mean as a Gaussian variable in the scale $\frac{\sqrt{\log N}}{N}$, it computes its asymptotic variance. It also gives a joint Gaussian fluctuation result for mesoscopically separated eigenvalues in the bulk. These results have been extended to the GOE case by O'Rourke [Rou10], who also proved them to be universal for Wigner matrices using the four moments theorem [Tao09a].

21.4 Local properties at the edge

It has long been observed that extreme eigenvalues of random matrices tend to stick to the bulk, and hence converge to the boundary of the limiting spectral measure. We shall describe this phenomenon and study the fluctuations of the extreme eigenvalues and their universality. For heavy-tailed distribution of the entries, and as mentioned above in the case of the spectral measure, this universality breaks down for the behaviour of extreme eigenvalues. The extreme eigenvalues are then Poissonian and the associated eigenvectors are very localized.

We will also mention an interesting universality question related to the sensitivity of the extreme eigenvalues to the addition of a finite rank matrix.

21.4.1 Convergence of the extreme eigenvalues

The convergence of the extremal eigenvalues of a Wigner matrix towards the boundary of the support of the semicircle law goes back to Füredi and Komlós [Fur81]. The following result has been proved by [Bai99]

Theorem 21.4.1 *Assume that the fourth moment of the distribution of the entries is finite. Then,* $\bar{\lambda}_N := \max_{1\le i\le N} \lambda_i^N$ *and* $\underline{\lambda_N} := \min_{1\le i\le N} \lambda_i^N$ *converge to* 2 *and* -2 *in probability.*

The proof relies on fine estimates of the moments of the averaged spectral measure, at powers going to infinity with N faster than logarithmically. It is not hard to see that the convergence in probability can be improved into an almost sure convergence. This result breaks down if the fourth moment is not finite.

21.4.2 Fluctuations of extreme eigenvalues and the Tracy–Widom law

For Gaussian ensembles, one can again rely on the explicit joint law of the eigenvalues to study precisely the fluctutations of extreme eigenvalues. This is again simpler for the GUE, based on the determinantal structure of the law $P_N^{(2)}$. After the work of Mehta [Meh04], a complete mathematical analysis was given by the works of P. Forrester [For93] and C. Tracy and H. Widom [Tra94, Tra00]. The main result goes as follows.

Theorem 21.4.2 *For $\beta = 1$ or 2, for any $t \in \mathbb{R}$, the largest eigenvalue $\bar{\lambda}_N$ of X_N is such that*

$$F_\beta(t) := \lim_{N\to\infty} P_N^{(\beta)}\left(N^{\frac{1}{6}}(\bar{\lambda}_N - 2\sqrt{N}) \le t\right)$$

with F_β the partition function for the Tracy–Widom law.

This result can be generalized to describe the joint convergence of the k-th largest eigenvalues.

21.4.3 Universality

Using very fine combinatorial arguments, A. Soshnikov [Sos99] showed that the moments of the spectral measure up to order $N^{\frac{2}{3}}$ are the same as those of the Gaussian ensembles. This allowed him to show that

Theorem 21.4.3 *Assume that the entries $(Z_{i,j})_{i\le j}$ and $(Y_i)_{i\ge 0}$ are iid with distribution which is symmetric and with a sub-Gausian tail. Then, the results of Theorem 21.4.2 extend to X_N.*

By approximation arguments, it was shown in [R06] that it is sufficient to have the first 36 moments finite, whereas new combinatorial arguments allowed in [Kho09] reducing the assumption to the twelve first moments finite. It is conjectured, and proved in the Gaussian divisible case [Joh10], that the optimal assumption is to have four moments finite. However, the hypothesis that the law is symmetric could not be completely removed using this approach, see [PS07] for an attempt in this direction.

21.4.4 Universality and the four moments theorem

T.Tao and V.Vu have generalized their four moments theorem to deal with the eigenvalues at the edge [Tao09b, Theorem 1.13].

Theorem 21.4.4 *Let X_N and X'_N be two Wigner matrices with entries with sub-exponential tails and moments which match up to order four. Assume in the case of*

complex entries that the real and the imaginary parts are independent. Then, there exists a small constant $c_0 > 0$ so that for any function $F : \mathbb{R}^k \to \mathbb{R}$ such that

$$|\nabla^j F(x)| \leq N^{c_0}$$

for all $0 \leq j \leq 5$ and $x \in \mathbb{R}^k$, we have for any $i_1, \ldots, i_k \in [1, N]$,

$$\left| E\left[F(\lambda_{i_j}(\sqrt{N}X_N), 1 \leq j \leq k)\right] - E\left[F(\lambda_{i_j}(\sqrt{N}X'_N), 1 \leq j \leq k)\right]\right| \leq N^{-c_0}.$$

As a corollary, we have [Tao09b, Theorem 1.16].

Corollary 21.4.1 *Under the assumptions of the previous theorem, let k be a fixed integer number and X_N a Wigner matrix with centred entries with sub-exponential tails. Assume that they have the same covariance matrix as the GUE (resp. GOE) and that all third moments vanish. Then the joint law of $((\lambda_N(X_N/\sqrt{N})-2) N^{\frac{2}{3}}, \ldots, (\lambda_{N-k}(X_N/\sqrt{N}) - 2) N^{\frac{2}{3}})$ converges weakly to the same limit as for the GUE (resp. GOE).*

Note that the advantage of this approach is that it does not require symmetry of the distribution of the entries as Soshnikov's result did.

21.4.5 Extreme eigenvalues of heavy-tailed Wigner matrices

Let us now consider what happens when the entries have no finite moments of order four but are in the domain of attraction of an α-stable law as in Section 21.2.5. Then, the behaviour of the extreme eigenvalues is dictated by the largest entries of the matrix, and therefore the point process of the extreme eigenvalues, once correctly normalized, converges to a point process, as in the extreme value theory of independent variables. More precisely, assume that the $(Z_{i,j}, i \leq j)$ and $(Y_i, i \geq 0)$ are iid with law P and let

$$b_N = \inf\{x : P(u : |u| \geq x) \leq \frac{2}{N(N+1)}\}$$

which is of order $N^{\frac{2}{\alpha}}$. Then it was shown in [Auf09], generalizing a result from [Sos04], that

Theorem 21.4.5 *Take $\alpha \in (0, 4)$ and for $\alpha \in [2, 4)$ assume that the entries are centred. Then, the point process*

$$\mathcal{P}_N = \sum_{i \geq 0} \mathbf{1}_{\lambda_i \geq 0} \delta_{b_N^{-1} \lambda_i}$$

converges in distribution to the Poisson point process on $(0, \infty)$ with intensity $\rho(x) = \frac{\alpha}{x^{1+\alpha}}$.

This result simply states that the largest eigenvalues behave as the largest entries of the matrix.

21.4.6 Non-universality and extreme eigenvalues of finite rank perturbation of Wigner matrices

In [Baik05], the authors consider the effect on the largest eigenvalue of adding a finite rank matrix to a Gaussian sample covariance matrix. The phenomena they observed, namely a phase transition in the asymptotic behaviour and fluctuations of the extreme eigenvalues, also happen when one considers Wigner matrices, see e.g. [Cap09a]. A finite rank perturbation can pull some eigenvalues away from the bulk.

We will only dwell on the simplest case, where the rank of the perturbation is one. But more is known, see e.g. [Fer07, Cap09a, Bai08]. Consider the deformed Wigner matrix

$$M_N = \frac{1}{\sqrt{N}} X_N + A_N$$

with X_N a Wigner matrix with independent entries with symmetric law μ satisfying Poincaré inequality and A_N a deterministic rank-one matrix. We will look at the two extreme cases.

First let $A_N = \frac{\theta}{N} J_N$, where $\theta > 0$ and J_N is the matrix whose entries are all ones, see e.g. [Fer07]. Obviously A_N is rank-one and its eigenvalues are 0 and θ. The only non-trivial eigenvector of A_N is maximally delocalized. When the parameter θ is small enough ($\theta < 1$), the perturbation has no influence on the top of the spectrum, the top eigenvalue λ_1 of M_N 'sticks to the bulk', i.e. λ_1 converges to 2, the edge of the bulk. If X_N is Gaussian it is proved that the fluctuations are also unaffected by the perturbation, i.e. that $(\lambda_1 - 2) N^{2/3}$ converges to the Tracy–Widom distribution. This result is expected to be universal. When the parameter θ is large enough ($\theta > 1$), the top eigenvalue λ_1 is pulled away from the bulk, it converges to $\rho_\theta = \theta + \frac{1}{\theta} > 2$. Moreover in this case the fluctuations of the top eigenvalue are in the scale $\sqrt{N}$ and Gaussian, i.e. $\sqrt{N}(\lambda_1 - \rho_\theta)$ converges to a Gaussian distribution. This result is universal.

But if one now chooses a rank-one perturbation with a very localized eigenvector, the situation is quite different. Let

$$A_N = diag(\frac{\theta}{N}, 0, \dots, 0)$$

Then again A_N is rank-one and its eigenvalues are 0 and θ. The only non-trivial eigenvector of A_N is maximally localized. In this case again when θ is small enough ($\theta < 1$), the perturbation has no influence on the top of the spectrum, the top eigenvalue λ_1 of M_N 'sticks to the bulk', i.e. λ_1 converges to 2, the edge of the bulk. Its fluctuations are Tracy–Widom in the invariant case and expected to also be in the non-invariant cases. When the parameter θ is large enough ($\theta > 1$), the top eigenvalue λ_1 is pulled away from the bulk, it converges to $\rho_\theta > 2$. But now the fluctuations are no longer Gaussian, $\sqrt{N}(\lambda_1 - \rho_\theta)$ converges in distribution to the convolution of a Gaussian law and of μ. This limiting law thus remembers the law of the entries and this result is definitely not universal. The reason is quite clear, and is due to the fact that, because of the localization of the top eigenvector of the perturbation A_N, the top eigenvalue of M_N remembers very much one entry of the matrix X_N.

This interesting non-universal behaviour has been studied in more general cases in [Cap09b].

Acknowledgements

The work of Gérard Ben Arous was supported in part by the National Science Foundation under grants DMS-0806180 and OISE-0730136. The work of Alice Guionnet was supported by grant ANR-08-BLAN-0311-01.

References

[And09] G. Anderson, A. Guionnet and O. Zeitouni, An introduction to Random matrices, Cambridge University Press, Cambridge 2009.
[Auf09] A. Auffinger, G. Ben Arous and S. Péché, Ann. Inst. Henri Poincaré Probab. Stat. **45** (2009) 589.
[Bai93] Z. D. Bai, Ann. Probab. **21** (1993) 625.
[Bai99] Z. D. Bai, Statistica Sinica **9** (1999) 611.
[Bai09] Z. D. Bai, X. Wang, and W. Zhou, Electron. J. Probab. **14** (2009) 2391.
[Bai08] Z. D. Bai and J-F Yao, Ann. Inst. Henri Poincaré Probab. Stat. **44** (2008) 447.
[Baik05] J. Baik, G. Ben Arous and S. Péché, Ann. Probab. **33** (2005) 1643.
[Bel09] S. Belinschi, A. Dembo and A. Guionnet, Comm. Math. Phys. **289** (2009) 1023.
[Ben97] G. Ben Arous and A. Guionnet, Probab. Theory Related Fields, **108** (1997) 517.
[Ben08] G. Ben Arous and A. Guionnet, Comm. Math. Phys. **278** (2008) 715.
[Ben10] G. Ben Arous and P. Bourgade, preprint (2010)
[Bord09] C. Bordenave, P. Caputo, D Chafaï, Annals of Probability **39** (2011) 1544–1590.
[Bord10] C. Bordenave, private communication (2010)
[Bou94] J. Bouchaud and P. Cizeau, Phys. Rev. E **50** (1994) 1810.
[Bré97] E. Brézin and S. Hikami, Phys. Rev. E (3) **55** (1997) 4067.
[Cap09a] M. Capitaine, C. Donati-Martin and D. Féral, Ann. Probab. **37** (2009) 1.
[Cap09b] M. Capitaine, C. Donati-Martin, D. Féral, Ann. Inst. H. Poincaré Probab. Statist. **48** (2012) 107–133.
[Cha04] S. Chatterjee and A. Bose, J. Theoret. Probab. **17** (2004) 1003.
[Dei99] P. Deift, T. Kriecherbauer, K. McLaughlin, S. Venakides and X. Zhou, Comm. Pure Appl. Math. **52** (1999) 1491.
[Dem01] A. Dembo, A. Guionnet and O. Zeitouni, Ann. Inst. H. Poincaré Probab. Statist. **39** (2003) 1013.
[Dia94] P. Diaconis and M. Shahshahani, Studies in applied probability **31A** (1994) 49.
[Dor09] H. Döring and P. Eichelsbacher, Electron. J. Probab. **14** (2009) 2636.
[Erd08] L. Erdös, B. Schlein, H.T. Yau, Comm. Math. Phys. **287** (2009) 641.
[Erd09a] L Erdös, J. Ramirez, B. Schlein, H.T. Yau, Electr. J. Prob. **15** (2010) 526–603.

[Erd09b] L. Erdös, B. Schlein, H.T. Yau, J. Yin, Ann. Inst. H. Poincaré Probab. Statist. **48** (2012) 1–46.
[Erd09c] L. Erdös, J. Ramirez, B. Schlein, T. Tao, V. Vu and H.T. Yau [arXive:0906.4400]
[Erd09d] L. Erdös, S. Péché, J. A. Ramírez, B. Schlein, and H.-T. Yau, Commun. Pure App. Math. **63** (2010) 895–925.
[Fer07] D. Féral and S. Péché, Comm. Math. Phys. **272** (2007) 185.
[For93] P. Forrester – Nuclear Phys. B, **402** (1993) 709.
[Fur81] Z. Füredi and J. Komlós, The eigenvalues of random symmetric matrices, Combinatorica **1** (1981) 233.
[Got05] F. Götze and A. Tikhomirov, Cent. Eur. J. Math. **3** (2005) 666.
[Got07] F. Götze, A. Tikhomirov, and D. Timushev, Cent. Eur. J. Math. **5** (2007) 305.
[Gui00] A. Guionnet and O. Zeitouni, Elec. Comm. in Probab. **5** (2000) 119.
[Gui02] A. Guionnet and O. Zeitouni, J. Funct. Anal. **188** (2002) 461.
[Gui10] A. Guionnet, Séminaire Bourbaki, (2010)
[Gus05] J. Gustavsson, Ann. Inst H. Poincaré. **41** (2005) 151.
[Joh98] K. Johansson, Duke J. Math. **91**, (1998) 151.
[Joh01] K. Johansson, Comm. Math. Phys. **215** (2001) 683.
[Joh10] K. Johansson, Ann. Inst. H. Poincaré Probab. Statist. **48** (2012) 47–79.
[Jon82] D. Jonsson, J. Mult. Anal. **12** (1982) 1.
[Koh09] O. Khorunzhiy, Random Operators and Stochastic Equations, **20** (2012) 25–68.
[Lej02] A.Lejay and L.Pastur, Seminaire de Probabilites XXXVI, Lecture Notes in Mathematics 1801, Springer, (2002).
[Lyt09] A. Lytova and L.Pastur, Ann. Probab. **37** (2009) 1778.
[Meh04] M.L. Mehta, *Random Matrices*, Academic Press, 3rd Edition, London 2004.
[Rou10] S. O'Rourke, J. Stat. Phys. (2010) 1045.
[PS07] S. Péché and S. Soshnikov J. Stat. Phys., **129** (2007) 857.
[R06] A. Ruzmaikina Comm. Math. Phys. **261** (2006) 277.
[Sos99] A. Soshnikov Comm. Math. Phys. **207** (1999) 697.
[Sos04] A. Soshnikov, Electronic Commun. in Probab. **9** (2004) 82.
[Tao09a] T. Tao, V. Vu, Acta Mathematica **206** (2011) 127–204.
[Tao09b] T. Tao, V. Vu, Commun. Math. Phys. **298** (2010) 549–572.
[Tra98] C. Tracy and H. Widom, J. Stat. Phys. **92** (1998) 809. [arXive:solv-int/9804004]
[Tra94] C. Tracy and H. Widom, Comm. Math. Phys. **159** (1994) 151.
[Tra00] C. Tracy and H. Widom, Calogero-Moser-Sutherland models (Montréal, QC, 1997), (2000) 461.
[Vin] J.Vinson, Ph.D. Thesis, Princeton University (2001).
[Wig55] E. P. Wigner, Ann. Math. **62** (1955) 548.

·22·

Free probability theory

Roland Speicher

Abstract

Free probability theory was created by Dan Voiculescu around 1985, motivated by his efforts to understand special classes of von Neumann algebras. His discovery in 1991 that random matrices also satisfy asymptotically the *freeness* relation transformed the theory dramatically. Not only did this yield spectacular results about the structure of operator algebras, but it also brought new concepts and tools into the realm of random matrix theory. In the following we will give, mostly from the random matrix point of view, a survey on some of the basic ideas and results of free probability theory.

22.1 Introduction

Free probability theory allows one to deal with asymptotic eigenvalue distributions in situations involving several matrices. Let us consider two sequences A_N and B_N of selfadjoint $N \times N$ matrices such that both sequences have an asymptotic eigenvalue distribution for $N \to \infty$. We are interested in the asymptotic eigenvalue distribution of the sequence $f(A_N, B_N)$ for some non-trivial selfadjoint function f. In general, this will depend on the relation between the eigenspaces of A_N and of B_N. However, by the concentration of measure phenomenon, we expect that for large N this relation between the eigenspaces concentrates on *typical* or *generic positions*, and then the asymptotic eigenvalue distribution of $f(A_N, B_N)$ depends in a deterministic way only on the asymptotic eigenvalue distribution of A_N and on the asymptotic eigenvalue distribution of B_N. Free probability theory replaces this vague notion of *generic position* by the mathematical precise concept of *freeness* and provides general tools for calculating the asymptotic distribution of $f(A_N, B_N)$ out of the asymptotic distribution of A_N and the asymptotic distribution of B_N.

22.2 The moment method for several random matrices and the concept of freeness

The *empirical eigenvalue distribution* of a selfadjoint $N \times N$ matrix A is the probability measure on $\mathbb{R}$ which puts mass $1/N$ on each of the N eigenvalues

λ_i of A, counted with multiplicity. If μ_A is determined by its moments then it can be recovered from the knowledge of all traces of powers of A,

$$\operatorname{tr}(A^k) = \frac{1}{N}(\lambda_1^k + \cdots + \lambda_N^k) = \int_{\mathbb{R}} t^k d\mu_A(t),$$

where by tr we denote the normalized trace on matrices (so that we have for the identity matrix 1 that $\operatorname{tr}(1) = 1$). This is the basis of the *moment method* which tries to understand the asymptotic eigenvalue distribution of a sequence of matrices by determination of the asymptotics of traces of powers.

Definition 22.2.1 *We say that a sequence $(A_N)_{N\in\mathbb{N}}$ of $N \times N$ matrices* has an asymptotic eigenvalue distribution *if the limit $\lim_{N\to\infty} \operatorname{tr}(A_N^k)$ exists for all $k \in \mathbb{N}$.*

Consider now our sequences A_N and B_N, each of which is assumed to have an asymptotic eigenvalue distribution. We want to understand, in the limit $N \to \infty$, the eigenvalue distribution of $f(A_N, B_N)$, not just for one f, but for a wide class of different functions. By the moment method, this asks for investigation of the limit $N \to \infty$ of $\operatorname{tr}\left(f(A_N, B_N)^k\right)$ for all $k \in \mathbb{N}$ and all f in our considered class of functions. If we choose for the latter all polynomials in non-commutative variables, then it is clear that the basic objects which we have to understand in this approach are the asymptotic *mixed moments*

$$\lim_{N\to\infty} \operatorname{tr}(A_N^{n_1} B_N^{m_1} \cdots A_N^{n_k} B_N^{m_k}) \qquad (k \in \mathbb{N};\, n_1, \ldots, n_k, m_1, \ldots, m_k \in \mathbb{N}). \quad (22.2.1)$$

Thus our fundamental problem is the following. If A_N and B_N each have an asymptotic eigenvalue distribution, and if A_N and B_N are in generic position, do the asymptotic mixed moments $\lim_{N\to\infty} \operatorname{tr}(A_N^{n_1} B_N^{m_1} \cdots A_N^{n_k} B_N^{m_k})$ exist? If so, can we express them in a deterministic way in terms of

$$\left(\lim_{N\to\infty} \operatorname{tr}(A_N^k)\right)_{k\in\mathbb{N}} \qquad \text{and} \qquad \left(\lim_{N\to\infty} \operatorname{tr}(B_N^k)\right)_{k\in\mathbb{N}}. \qquad (22.2.2)$$

Let us start by looking at the second part of the problem, namely by trying to find a possible relation between the mixed moments (22.2.1) and the moments (22.2.2). For this we need a simple example of matrix sequences A_N and B_N which we expect to be in generic position.

Whereas up to now we have only talked about sequences of matrices, we will now go over to random matrices. It is actually not clear how to produce two sequences of deterministic matrices whose eigenspaces are in generic position. However, it is much easier to produce two such sequences of random matrices for which we almost surely have a generic situation. Indeed, consider two independent random matrix ensembles A_N and B_N, each almost surely with a limiting eigenvalue distribution, and assume that one of them, say B_N, is a *unitarily invariant ensemble*, which means that the joint distribution of its entries does not change under unitary conjugation. This implies that taking $U_N B_N U_N^*$, for any unitary $N \times N$-matrix U_N, instead of B_N does not change anything. But then we can use this U_N to rotate the eigenspaces of B_N against those of A_N into a generic position, thus for typical realizations of A_N and B_N the eigenspaces should be in a generic position.

The simplest example of two such random matrix ensembles is two independent Gaussian random matrices A_N and B_N. In this case one can calculate everything concretely: in the limit $N \to \infty$, $\mathrm{tr}\,(A_N^{n_1} B_N^{m_1} \cdots A_N^{n_k} B_N^{m_k})$ is almost surely given by the number of non-crossing or planar pairings of the pattern

$$\underbrace{A \cdot A \cdots A}_{n_1\text{-times}} \cdot \underbrace{B \cdot B \cdots B}_{m_1\text{-times}} \cdots \underbrace{A \cdot A \cdots A}_{n_k\text{-times}} \cdot \underbrace{B \cdot B \cdots B}_{m_k\text{-times}},$$

which do not pair A with B. (A pairing is a decomposition of the pattern into pairs of letters; if we connect the two elements from each pair by a line, drawn in the half-plane below the pattern, then non-crossing means that we can do this without getting crossings between lines for different pairs.)

After some contemplation, it becomes obvious that this implies that the trace of a corresponding product of centred powers,

$$\lim_{N\to\infty} \mathrm{tr}\,\Big(\big(A_N^{n_1} - \lim_{M\to\infty} \mathrm{tr}\,(A_M^{n_1}) \cdot 1\big) \cdot \big(B_N^{m_1} - \lim_{M\to\infty} \mathrm{tr}\,(B_M^{m_1}) \cdot 1\big) \cdots \\ \cdots \big(A_M^{n_k} - \lim_{M\to\infty} \mathrm{tr}\,(A_M^{n_k}) \cdot 1\big) \cdot \big(B_N^{m_k} - \lim_{M\to\infty} \mathrm{tr}\,(B_M^{m_k}) \cdot 1\big)\Big) \quad (22.2.3)$$

is given by the number of non-crossing pairings which do not pair A with B and for which, in addition, each group of As and each group of Bs is connected with some other group. It is clear that if we want to connect the groups in this way we will get some crossing between the pairs, thus there are actually no pairings of the required form and we have that the term (22.2.3) is equal to zero.

One might wonder what advantage is gained by trading the explicit formula for mixed moments of independent Gaussian random matrices for the implicit relation (22.2.3)? The drawback to the explicit formula for mixed moments of independent Gaussian random matrices is that the asymptotic formula for $\mathrm{tr}\,(A_N^{n_1} B_N^{m_1} \cdots A_N^{n_k} B_N^{m_k})$ will be different for different random matrix ensembles (and in many cases an explicit formula fails to exist). However, the vanishing of (22.2.3) remains valid for many matrix ensembles. The vanishing of (22.2.3) gives a precise meaning to our idea that the random matrices should be in generic position; it constitutes Voiculescu's definition of asymptotic freeness.

Definition 22.2.2 *Two sequences of matrices $(A_N)_{N\in\mathbb{N}}$ and $(B_N)_{N\in\mathbb{N}}$ are* asymptotically free *if we have the vanishing of* (22.2.3) *for all $k \geq 1$ and all $n_1, m_1, \ldots, n_k, m_k \geq 1$.*

Provided with this definition, the intuition that unitarily invariant random matrices should give rise to generic situations now becomes a rigorous theorem. This basic observation was proved by Voiculescu [Voi91] in 1991.

Theorem 22.2.1 *Consider $N \times N$ random matrices A_N and B_N such that: both A_N and B_N almost surely have an asymptotic eigenvalue distribution for $N \to \infty$; A_N and B_N are independent; B_N is a unitarily invariant ensemble. Then, A_N and B_N are almost surely asymptotically free.*

In order to prove this, one can replace A_N and B_N by A_N and $U_N B_N U_N^*$, where U_N is a Haar unitary random matrix (i.e. from the ensemble of unitary matrices equipped with the normalized Haar measure as a probability measure); furthermore, one can then restrict to the case where A_N and B_N are deterministic matrices. In this form it reduces to showing almost sure asymptotic freeness between Haar unitary matrices and deterministic matrices. The proof of that statement proceeds then as follows. First one shows asymptotic freeness in the mean and then one strengthens this to almost certain convergence.

The original proof of Voiculescu [Voi91] for the first step reduced the asymptotic freeness for Haar unitary matrices to a corresponding statement for non-selfadjoint Gaussian random matrices; by realizing the Haar measure on the group of unitary matrices as the pushforward of the Gaussian measure under taking the phase. The asymptotic freeness result for Gaussian random matrices can be derived quite directly by using the genus expansion for their traces. Another more direct way to prove the averaged version of unitary freeness for Haar unitary matrices is due to Xu [Xu97] and relies on *Weingarten type formulae* for integrals over products of entries of Haar unitary matrices.

In the second step, in order to strengthen the above result to almost sure asymptotic freeness one can either [Voi91] invoke the concentration of measure results of Gromov and Milman (applied to the unitary group) or [Spe93, Hia00] more specific estimates for the variances of the considered sequence of random variables.

Though unitary invariance is the most intuitive reason for having asymptotic freeness among random matrices, it is not a necessary condition. For example, the above theorem includes the case where B_N are Gaussian random matrices. If we generalize those to Wigner matrices (where the entries above the diagonal are iid, but not necessarily Gaussian), then we lose the unitary invariance, but the conclusion of the above theorem still holds true. More precisely, we have the following theorem.

Theorem 22.2.2 *Let X_N be a selfadjoint Wigner matrix, such that the distribution of the entries is centred and has all moments, and let A_N be a random matrix which is independent from X_N. If A_N has almost surely an asymptotic eigenvalue distribution and if we have*

$$\sup_{N \in \mathbb{N}} \| A_N \| < \infty,$$

then A_N and X_N are almost surely asymptotically free.

The case where A_N consists of block diagonal matrices was treated by Dykema [Dyk93], for the general version see [Min10, And09].

22.3 Basic definitions

The freeness relation, which holds for many random matrices asymptotically, was actually discovered by Voiculescu in a quite different context; namely canonical generators in operator algebras given in terms of free groups satisfy the same relation with respect to a canonical state, see Section 22.9.1. Free probability theory investigates these freeness relations abstractly, inspired by the philosophy that freeness should be considered and treated as a kind of non-commutative analogue of the classical notion of independence.

Some of the main probabilistic notions used in free probability are the following.

Notation 22.3.1 *A pair $(\mathcal{A}, \varphi)$ consisting of a unital algebra $\mathcal{A}$ and a linear functional $\varphi : \mathcal{A} \to \mathbb{C}$ with $\varphi(1) = 1$ is called a* non-commutative probability space. *Often the adjective 'non-commutative' is just dropped. Elements from $\mathcal{A}$ are addressed as* (non-commutative) random variables, *the numbers $\varphi(a_{i(1)} \cdots a_{i(n)})$ for such random variables $a_1, \dots, a_k \in \mathcal{A}$ are called* moments, *the collection of all moments is called the* joint distribution of $a_1, \dots, a_k$.

Definition 22.3.1 *Let $(\mathcal{A}, \varphi)$ be a non-commutative probability space and let I be an index set.*

(1) *Let, for each $i \in I$, $\mathcal{A}_i \subset \mathcal{A}$, be a unital subalgebra. The subalgebras $(\mathcal{A}_i)_{i \in I}$ are called* free *or* freely independent, *if $\varphi(a_1 \cdots a_k) = 0$ whenever we have: k is a positive integer; $a_j \in \mathcal{A}_{i(j)}$ (with $i(j) \in I$) for all $j = 1, \dots, k$; $\varphi(a_j) = 0$ for all $j = 1, \dots, k$; and neighbouring elements are from different subalgebras, i.e. $i(1) \neq i(2), i(2) \neq i(3), \dots, i(k-1) \neq i(k)$.*

(2) *Let, for each $i \in I$, $a_i \in \mathcal{A}$. The elements $(a_i)_{i \in I}$ are called* free *or* freely independent, *if their generated unital subalgebras are free, i.e. if $(\mathcal{A}_i)_{i \in I}$ are free, where, for each $i \in I$, $\mathcal{A}_i$ is the unital subalgebra of $\mathcal{A}$ which is generated by a_i.*

Freeness, like classical independence, is a rule for calculating mixed moments from knowledge of the moments of individual variables. Indeed, one can easily show by induction that if $(\mathcal{A}_i)_{i \in I}$ are free with respect to φ, then φ restricted to the algebra generated by all $\mathcal{A}_i$, $i \in I$, is uniquely determined by $\varphi|_{\mathcal{A}_i}$ for all $i \in I$ and by the freeness condition. For example, if $\mathcal{A}$ and $\mathcal{B}$ are free, then one has for $a, a_1, a_2 \in \mathcal{A}$ and $b, b_1, b_2 \in \mathcal{B}$ that $\varphi(ab) = \varphi(a)\varphi(b)$, $\varphi(a_1 b a_2) = \varphi(a_1 a_2)\varphi(b)$, and $\varphi(a_1 b_1 a_2 b_2) = \varphi(a_1 a_2)\varphi(b_1)\varphi(b_2) + \varphi(a_1)\varphi(a_2)\varphi(b_1 b_2) - \varphi(a_1)\varphi(b_1)\varphi(a_2)\varphi(b_2)$. Whereas the first two factorizations are the same as for the expectation of independent random variables, the last one is different, and more complicated, from the classical situation. It is important to note that freeness plays a similar role in the non-commutative world as independence plays in the classical world, but that freeness is not a generalization of independence: independent random variables can be free only in very trivial situations. Freeness is a theory for non-commuting random variables.

22.4 Combinatorial theory of freeness

The defining relations for freeness from Def. 22.3.1 are quite implicit and not easy to handle directly. It has turned out that replacing moments by other quantities, so-called *free cumulants*, is advantageous for many questions. In particular, freeness is much easier to describe on the level of free cumulants. The relation between moments and cumulants is given by summing over non-crossing partitions. This combinatorial theory of freeness is due to Speicher [Spe94]; many consequences of this approach were worked out by Nica and Speicher, see [Nic06].

Definition 22.4.1 *For a unital linear functional* $\varphi : \mathcal{A} \to \mathbb{C}$ *on a unital algebra* $\mathcal{A}$ *we define* cumulant functionals $\kappa_n : \mathcal{A}^n \to \mathbb{C}$ *(for all* $n \geq 1$*) by the* moment-cumulant relations

$$\varphi(a_1 \cdots a_n) = \sum_{\pi \in NC(n)} \kappa_\pi[a_1, \ldots, a_n]. \tag{22.4.1}$$

In Equation (22.4.1) the summation is running over *non-crossing partitions* of the set $\{a_1, a_2, \ldots, a_n\}$; these are decompositions of that set into disjoint non-empty subsets, called *blocks*, such that there are no crossings between different blocks. In diagrammatic terms this means that if we draw the blocks of such a π below the points $a_1, a_2, \ldots, a_n$, then we can do this without having crossings in our picture. The contribution κ_π in (22.4.1) of such a non-crossing π is a product of cumulants corresponding to the block structure of π. For each block of π we have as a factor a cumulant which contains as arguments those a_i which are connected by that block.

An example of a non-crossing partition π for $n = 10$ is

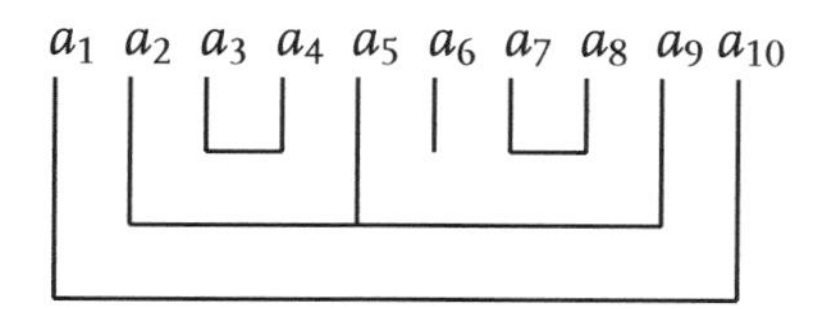

In this case the blocks are $\{a_1, a_{10}\}$, $\{a_2, a_5, a_9\}$, $\{a_3, a_4\}$, $\{a_6\}$, and $\{a_7, a_8\}$; and the corresponding contribution κ_π in (22.4.1) is given by

$$\kappa_\pi[a_1, \ldots, a_{10}] = \kappa_2(a_1, a_{10}) \cdot \kappa_3(a_2, a_5, a_9) \cdot \kappa_2(a_3, a_4) \cdot \kappa_1(a_6) \cdot \kappa_2(a_7, a_8).$$

Note that in general there is only one term in (22.4.1) involving the highest cumulant κ_n, thus the moment cumulant formulae can be inductively resolved for the κ_n in terms of the moments. More concretely, the set of non-crossing partitions forms a lattice with respect to refinement order and the κ_n are given by the *Möbius inversion* of the formula (22.4.1) with respect to this order.

For $n = 1$, we get the mean, $\kappa_1(a_1) = \varphi(a_1)$ and for $n = 2$ we have the covariance, $\kappa_2(a_1, a_2) = \varphi(a_1 a_2) - \varphi(a_1)\varphi(a_2)$.

The relevance of the κ_n in our context is given by the following characterization of freeness.

Theorem 22.4.1 *Freeness is equivalent to the vanishing of mixed cumulants. More precisely, the fact that $(a_i)_{i\in I}$ are free is equivalent to: $\kappa_n(a_{i(1)}, \ldots, a_{i(n)}) = 0$ whenever $n \geq 2$ and there are k, l such that $i(k) \neq i(l)$.*

This description of freeness in terms of free cumulants is related to the planar approximations in random matrix theory. In a sense some aspects of this theory of freeness were anticipated (but mostly neglected) in the physics community in the paper [Cvi82].

22.5 Free harmonic analysis

For a meaningful harmonic analysis one needs some positivity structure for the non-commutative probability space $(\mathcal{A}, \varphi)$. We will usually consider selfadjoint random variables and φ should be positive. Formally, a good frame for this is a *C^*-probability space*, where $\mathcal{A}$ is a C^*-algebra (i.e. a norm-closed $*$-subalgebra of the algebra of bounded operators on a Hilbert space) and φ is a state, i.e. it is positive in the sense $\varphi(aa^*) \geq 0$ for all $a \in \mathcal{A}$. Concretely this means that our random variables can be realized as bounded operators on a Hilbert space and φ can be written as a vector state $\varphi(a) = \langle a\xi, \xi\rangle$ for some unit vector ξ in the Hilbert space.

In such a situation the distribution of a selfadjoint random variable a can be identified with a compactly supported probability measure μ_a on $\mathbb{R}$, via

$$\varphi(a^n) = \int_{\mathbb{R}} t^n d\mu_a(t) \qquad \text{for all } n \in \mathbb{N}.$$

22.5.1 Sums of free variables: the $\mathcal{R}$-transform

Consider two selfadjoint random variables a and b which are free. Then, by freeness, the moments of $a + b$ are uniquely determined by the moments of a and the moments of b.

Notation 22.5.1 *We say the distribution of $a+b$ is the* free convolution, *denoted by* $\boxplus$, *of the distribution of a and the distribution of b,*

$$\mu_{a+b} = \mu_a \boxplus \mu_b.$$

Notation 22.5.2 *For a random variable a we define its* Cauchy transform G *and its* $\mathcal{R}$-transform $\mathcal{R}$ *by*

$$G(z) = \frac{1}{z} + \sum_{n=1}^{\infty} \frac{\varphi(a^n)}{z^{n+1}} \qquad \text{and} \qquad \mathcal{R}(z) = \sum_{n=1}^{\infty} \kappa_n(a, \ldots, a) z^{n-1}.$$

One can see quite easily that the moment-cumulant relations (22.4.1) are equivalent to the following functional relation

$$\frac{1}{G(z)} + \mathcal{R}(G(z)) = z. \tag{22.5.1}$$

Combined with the additivity of free cumulants under free convolution, which follows easily by the vanishing of mixed cumulants in free variables, this yields the following basic theorem of Voiculescu.

Theorem 22.5.1 *Let $G(z)$ be the Cauchy transform of a, as defined in Notation 22.5.2 and define its $\mathcal{R}$-transform by the relation* (22.5.1). *Then we have*

$$\mathcal{R}^{a+b}(z) = \mathcal{R}^a(z) + \mathcal{R}^b(z)$$

if a and b are free.

We have defined the Cauchy and the $\mathcal{R}$-transform here only as formal power series. Also (22.5.1) is proved first as a relation between formal power series. But if a is a selfadjoint element in a C^*-probability space, then G is also the analytic function

$$G : \mathbb{C}^+ \to \mathbb{C}^-; \qquad G(z) = \varphi\left(\frac{1}{z-a}\right) = \int_{\mathbb{R}} \frac{1}{z-t} d\mu_a(t);$$

and one can also show that (22.5.1) defines then $\mathcal{R}$ as an analytic function on a suitably chosen subset of $\mathbb{C}^+$. In this form Theorem 22.5.1 is amenable to analytic manipulations and so gives an effective algorithm for calculating free convolutions. This can be used to calculate the asymptotic eigenvalue distribution of sums of random matrices which are asymptotically free.

Furthermore, by using analytic tools around the Cauchy transform (which exist for any probability measure on $\mathbb{R}$) one can extend the definition of and most results on free convolution to all probability measures on $\mathbb{R}$. See [Ber93, Voi00] for more details.

We would like to remark that the machinery of free convolution was also found at around the same time, independently from Voiculescu and independently from each other, by different researchers in the context of random walks on the free product of groups: by Woess, by Cartwright and Soardi, and by McLaughlin; see, for example, [Woe86].

22.5.2 Products of free variables: the S-transform

Consider a, b free. Then, by freeness, the moments of ab are uniquely determined by the moments of a and the moments of b.

Notation 22.5.3 *We say the distribution of ab is the* free multiplicative convolution, *denoted by $\boxtimes$, of the distribution of a and the distribution of b,*

$$\mu_{ab} = \mu_a \boxtimes \mu_b.$$

Note: even if we start from selfadjoint a and b, their product ab is not selfadjoint, unless a and b commute (which is rarely the case, when a and b are free). Thus the above does not define an operation on probability measures on $\mathbb{R}$ in general. However, if one of the operators, say a, is positive (and thus μ_a supported on $\mathbb{R}_+$), then $a^{1/2}ba^{1/2}$ makes sense; since it has the same moments as ab (note for this that the relevant state is a trace, as the free product of traces is tracial) we can identify μ_{ab} then with the probability measure $\mu_{a^{1/2}ba^{1/2}}$.

Again, Voiculescu introduced an analytic object which allows us to deal effectively with this multiplicative free convolution.

Theorem 22.5.2 *Put $M_a(z) := \sum_{m=1}^{\infty} \varphi(a^m)z^m$ and define the* S-transform *of a by*

$$S_a(z) := \frac{1+z}{z} M_a^{<-1>}(z),$$

where $M^{<-1>}$ denotes the inverse of M under composition. Then we have

$$S_{ab}(z) = S_a(z) \cdot S_b(z)$$

if a and b are free.

As in the additive case, the moment generating series M and the S-transform are not just formal power series, but analytic functions on suitably chosen domains in the complex plane. For more details, see [Ber93, Hia00].

22.5.3 The free central limit theorem

One of the first theorems in free probability theory, proved by Voiculescu in 1985, was the free analogue of the central limit theorem. Surprisingly, it turned out that the analogue of the Gaussian distribution in free probability theory is the semicircular distribution.

Definition 22.5.1 *Let $(\mathcal{A}, \varphi)$ be a C^*-probability space. A selfadjoint element $s \in \mathcal{A}$ is called* semicircular *(of variance 1) if its distribution μ_s is given by the probability measure with density $\frac{1}{2\pi}\sqrt{4-t^2}$ on the interval $[-2, +2]$. Alternatively, the moments of s are given by the Catalan numbers,*

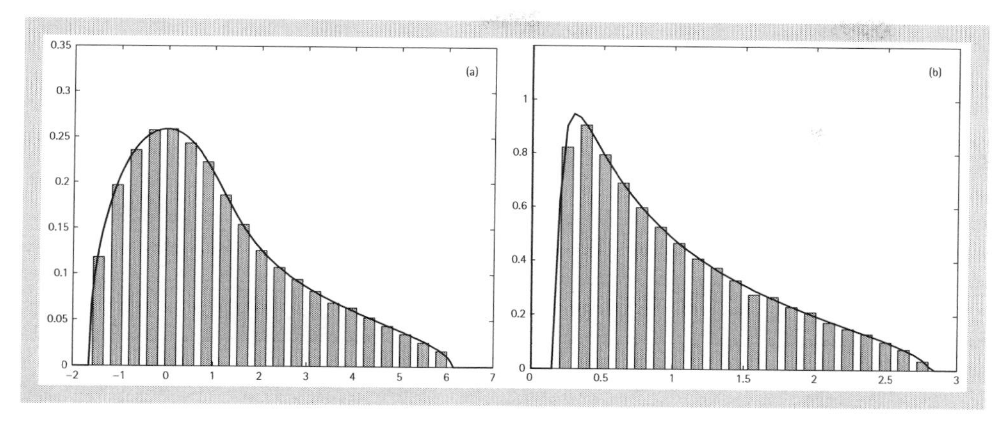

Fig. 22.1 Comparison of free probability result with histogram of eigenvalues of an $N \times N$ random matrix, for $N = 2000$: (a) histogram of the sum of independent Gaussian and Wishart matrices, compared with the free convolution of semicircular and free Poisson distribution (rate $\lambda = 1/2$), calculated by using the $\mathcal{R}$-transform; (b) histogram of the product of two independent Wishart matrices, compared with the free multiplicative convolution of two free Poisson distributions (both with rate $\lambda = 5$), calculated by using the S-transform.

$$\varphi(s^n) = \begin{cases} \frac{1}{k+1}\binom{2k}{k}, & \textit{if } n = 2k \textit{ even} \\ 0, & \textit{if } n \textit{ odd} \end{cases}$$

Theorem 22.5.3 *If ν is a compactly supported probability measure on $\mathbb{R}$ with vanishing mean and variance 1, then*

$$D_{1/\sqrt{N}}\nu^{\boxplus N} \Rightarrow \mu_s,$$

where D_α denotes the dilation of a measure by the factor α, and $\Rightarrow$ means weak convergence.

By using the analytic theory of $\boxplus$ for all, not necessarily compactly supported, probability measures on $\mathbb{R}$, the free central limit theorem can also be extended to this general situation.

The occurrence of the semicircular distribution as limit both in Wigner's semicircle law as well as in the free central limit theorem was the first hint of a relationship between free probability theory and random matrices. The subsequent development of this connection culminated in Voiculescu's discovery of asymptotic freeness between large random matrices, as exemplified in Theorem 22.2.1. When this contact was made between freeness and random matrices, the previously introduced $\mathcal{R}$- and S-transforms gave powerful new techniques for calculating asymptotic eigenvalue distributions of random matrices. For computational aspects of these techniques we refer to [Rao09], for applications in electrical engineering see [Tul04], and also Chapter 40.

22.5.4 Free Poisson distribution and Wishart matrices

There exists a very rich free parallel of classical probability theory, of which the free central limit theorem is just the starting point. In particular, one has the free analogue of infinitely divisible and of stable distributions and corresponding limit theorems. For more details and references see [Ber99, Voi00].

Let us present here only as another instance of this theory, the free Poisson distribution. As with the semicircle distribution, the free counterpart of the Poisson law, which is none other than the Marchenko-Pastur distribution, appears very naturally as the asymptotic eigenvalue distribution of an important class of random matrices, namely Wishart matrices.

As in the classical theory the Poisson distribution can be described by a limit theorem. The following statement deals directly with the more general notion of a compound free Poisson distribution.

Proposition 22.5.1 *Let $\lambda \geq 0$ and ν be a probability measure on $\mathbb{R}$ with compact support. Then the weak limit for $N \to \infty$ of*

$$\left(\left(1 - \frac{\lambda}{N}\right)\delta_0 + \frac{\lambda}{N}\nu\right)^{\boxplus N}$$

has free cumulants $(\kappa_n)_{n\geq 1}$ which are given by $\kappa_n = \lambda \cdot m_n(\nu)$ $(n \geq 1)$ (m_n denotes here the n-th moment) and thus an $\mathcal{R}$-transform of the form

$$\mathcal{R}(z) = \lambda \int_{\mathbb{R}} \frac{x}{1 - xz}\, d\nu(x).$$

Definition 22.5.2 *The probability measure appearing in the limit of Prop. 22.5.1 is called a* compound free Poisson distribution *with rate λ and jump distribution ν.*

Such compound free Poisson distributions show up in the random matrix context as follows. Consider rectangular Gaussian $M \times N$ random matrices $X_{M,N}$, where all entries are independent and identically distributed according to a normal distribution with mean zero and variance $1/N$; and a sequence of deterministic $N \times N$ matrices T_N such that the limiting eigenvalue distribution μ_T of T_N exists. Then almost surely, for $M, N \to \infty$ such that $N/M \to \lambda$, the limiting eigenvalue distribution of $X_{M,N}T_N X^*_{M,N}$ exists, too, and it is given by a compound free Poisson distribution with rate λ and jump distribution μ_T.

One notes that the above frame of rectangular matrices does not fit directly into the theory presented up to now (so it is, e.g. not clear what asymptotic freeness between $X_{M,N}$ and T_N should mean). However, rectangular matrices can be treated in free probability by either embedding them into bigger square matrices and applying some compressions at appropriate stages or, more directly, by using a generalization of free probability due to Benaych-Georges [Ben09] which is tailor-made to deal with rectangular random matrices.

22.6 Second-order freeness

Asymptotic freeness of random matrices shows that the mixed moments of two ensembles in generic position are deterministically calculable from the moments of each individual ensemble. The formula for calculation of the mixed moments is the essence of the concept of freeness. The same philosophy applies also to finer questions about random matrices, most notably to *global fluctuations of linear statistics.* With this we mean the following: for many examples (like Gaussian or Wishart) of random matrices A_N the magnified fluctuations of traces around the limiting value, $N\big(\mathrm{tr}\,(A_N^k) - \lim_{M\to\infty} \mathrm{tr}\,(A_M^k)\big)$, form asymptotically a Gaussian family. If we have two such ensembles in generic position (e.g. if they are independent and one of them is unitarily invariant), then this is also true for mixed traces and the covariance of mixed traces is determined in a deterministic way by the covariances for each of the two ensembles separately. The formula for calculation of the mixed covariances constitutes the definition of the concept of *second-order freeness.* There exist again cumulants and an R-transform on this level, which allow explicit calculations. For more details, see [Col07, Min10].

22.7 Operator-valued free probability theory

There exists a generalization of free probability theory to an *operator-valued* level, where the complex numbers $\mathbb{C}$ and the expectation state $\varphi : \mathcal{A} \to \mathbb{C}$ are replaced by an arbitrary algebra $\mathcal{B}$ and a conditional expectation $E : \mathcal{A} \to \mathcal{B}$. The formal structure of the theory is very much the same as in the scalar-valued case, one only has to take care of the fact that the 'scalars' from $\mathcal{B}$ do not commute with the random variables.

Definition 22.7.1

(1) *Let $\mathcal{A}$ be a unital algebra and consider a unital subalgebra $\mathcal{B} \subset \mathcal{A}$. A linear map $E : \mathcal{A} \to \mathcal{B}$ is a* conditional expectation *if $E[b] = b$ for all $b \in \mathcal{B}$ and $E[b_1 a b_2] = b_1 E[a] b_2$ for all $a \in \mathcal{A}$ and all $b_1, b_2 \in \mathcal{B}$. An* operator-valued probability space *consists of $\mathcal{B} \subset \mathcal{A}$ and a conditional expectation $E : \mathcal{A} \to \mathcal{B}$.*

(2) *Consider an operator-valued probability space $\mathcal{B} \subset \mathcal{A}$, $E : \mathcal{A} \to \mathcal{B}$. Random variables $(x_i)_{i\in I} \subset \mathcal{A}$ are* free with respect to E *or* free with amalgamation over $\mathcal{B}$ *if $E[p_1(x_{i(1)}) \cdots p_k(x_{i(k)})] = 0$, whenever $k \in \mathbb{N}$, $p_j(x_{i(j)})$ are elements from the algebra generated by $\mathcal{B}$ and $x_{i(j)}$, neighbouring elements are different, i.e. $i(1) \neq i(2) \neq \cdots \neq i(k)$, and we have $E[p_j(x_{i(j)})] = 0$ for all $j = 1, \dots, k$.*

Voiculescu introduced this operator-valued version of free probability theory in [Voi85] and also provided in [Voi95] a corresponding version of free convolution and $\mathcal{R}$-transform. A combinatorial treatment was given by Speicher [Spe98] who showed that the theory of free cumulants also has a nice counterpart in the operator-valued frame.

For $a \in \mathcal{A}$ we define its *(operator-valued) Cauchy transform* $G_a : \mathcal{B} \to \mathcal{B}$ by

$$G_a(b) := E[\frac{1}{b-a}] = \sum_{n\geq 0} E[b^{-1}(ab^{-1})^n].$$

The *operator-valued* $\mathcal{R}$*-transform* of a, $\mathcal{R}_a : \mathcal{B} \to \mathcal{B}$, can be defined as a power series in operator-valued free cumulants, or equivalently by the relation $bG(b) = 1 + \mathcal{R}(G(b)) \cdot G(b)$ or $G(b) = \big(b - \mathcal{R}(G(b))\big)^{-1}$. One then has as before: If x and y are free over $\mathcal{B}$, then $\mathcal{R}_{x+y}(b) = \mathcal{R}_x(b) + \mathcal{R}_y(b)$. Another form of this is the *subordination property* $G_{x+y}(b) = G_x\big[b - \mathcal{R}_y\big(G_{x+y}(b)\big)\big]$.

There also exists the notion of a *semicircular element* s in the operator-valued world. It is characterized by the fact that only its second-order free cumulants are different from zero, or equivalently that its $\mathcal{R}$-transform is of the form $\mathcal{R}_s(b) = \eta(b)$, where $\eta : \mathcal{B} \to \mathcal{B}$ is the linear map given by $\eta(b) = E[sbs]$. Note that in this case the equation for the Cauchy transform reduces to

$$bG(b) = 1 + \eta[G(b)] \cdot G(b); \tag{22.7.1}$$

more generally, if we add an $x \in \mathcal{B}$, for which we have $G_x(b) = E[(b-x)^{-1}] = (b-x)^{-1}$, we have for the Cauchy transform of $x + s$ the implicit equation

$$G_{x+s}(b) = G_x\big[b - \mathcal{R}_s\big(G_{x+s}(b)\big)\big] = \big(b - \eta[G_{x+s}(b)] - x\big)^{-1}. \tag{22.7.2}$$

It was observed by Shlyakhtenko [Shl96] that operator-valued free probability theory provides the right frame for dealing with a more general kind of random matrices. In particular, he showed that so-called *band matrices* become asymptotically operator-valued semicircular elements over the limit of the diagonal matrices.

Theorem 22.7.1 *Suppose that $A_N = A_N^*$ is an $N \times N$ random band matrix, i.e. $A_N = (a_{ij})_{i,j=1}^N$, where $\{a_{ij} \mid i \leq j\}$ are centred independent complex Gaussian random variables, with $E[a_{ij}\bar{a}_{ij}] = (1 + \delta_{ij}\sigma^2(i/N, j/N))/N$ for some $\sigma^2 \in L^\infty([0,1]^2)$. Let $\mathcal{B}_N$ be the diagonal $N \times N$ matrices, and embed $\mathcal{B}_N$ into $\mathcal{B} := L^\infty[0,1]$ as step functions. Let $B_N \in \mathcal{B}_N$ be selfadjoint diagonal matrices such that $B_N \to f \in L^\infty[0,1]$ in $\|\cdot\|_\infty$. Then the limit distribution of $B_N + A_N$ exists, and its Cauchy transform G is given by*

$$G(z) = \int_0^1 g(z,x)dx,$$

where $g(z,x)$ is analytic in z and satisfies

$$g(z,x) = \left[z - f(x) - \int_0^1 \sigma^2(y,x)g(z,y)dy\right]^{-1}. \tag{22.7.3}$$

Note that (22.7.3) is nothing but the general equation (22.7.2) specified to the situation $\mathcal{B} = L^\infty[0,1]$ and $\eta : L^\infty[0,1] \to L^\infty[0,1]$ acting as integration operator with kernel σ^2.

Moreover, Gaussian random matrices with a certain degree of correlation between the entries are also asymptotically semicircular elements over an appropriate subalgebra, see [Ras08].

22.8 Further free-probabilistic aspects of random matrices

Free probability theory also provides new ideas and techniques for investigating other aspects of random multi-matrix models. In particular, Haagerup and Thorbjornsen [Haa02, Haa05] obtained a generalization to several matrices for a number of results concerning the largest eigenvalue of a Gaussian random matrix.

Much work is also devoted to deriving rigorous results about the large N limit of random multi-matrix models given by densities of the type

$$c_N e^{-N^2 \operatorname{tr} P(A_1,\ldots,A_N)} d\lambda(A_1, \ldots, A_n),$$

where $d\lambda$ is Lebesgue measure, $A_1, \ldots, A_n$ are selfadjoint $N \times N$ matrices, and P a noncommutative selfadjoint polynomial. To prove the existence of that limit in sufficient generality is one of the big problems. For a rigorous mathematical treatment of such questions, see [Gui06].

Free Brownian motion is the large N limit of the Dyson Brownian motion model of random matrices (with independent Brownian motions as entries, compare Chapter 11). Free Brownian motion can be realized concretely in terms of creation and annihilation operators on a full Fock space (see Section 22.9.1). There also exists a corresponding *free stochastic calculus* [Bia98]; for applications of this to multi-matrix models, see [Gui09].

There is also a surprising connection with the representation theory of the symmetric groups S_n. For large n, representations of S_n are given by large matrices which behave in some respects like random matrices. This was made precise by Biane who showed that many operations on representations of the symmetric group can be asymptotically described by operations from free probability theory, see [Bia02].

22.9 Operator algebraic aspects of free probability

A survey on free probability without mentioning at least some of its operator algebraic aspects would be quite unbalanced and misleading. We will highlight some of these operator algebraic facets in this last section. For the sake of brevity, we will omit definitions of standard concepts from operator algebras, since these can be found elsewhere (we refer the reader to [Voi92, Voi05, Hia00] for more information on notions, as well as for references related to the following topics).

22.9.1 Operator algebraic models for freeness

Free group factors

Let $G = \star_{i\in I} G_i$ be the free product of groups G_i. Let $L(G)$ denote the group von Neumann algebra of G, and φ the associated trace state, corresponding to the neutral element of the group. Then $L(G_i)$ can be identified with a subalgebra of $L(G)$ and, with respect to φ, these subalgebras $(L(G_i))_{i\in I}$ are free. This freeness is nothing but the rewriting in terms of φ what is meant by the groups G_i being free as subgroups in G. The definition of freeness was modelled according to the situation occurring in this example. The *free* in free probability theory refers to this fact.

A special and most prominent case of these von Neumann algebras are the *free group factors* $L(\mathbb{F}_n)$, where $\mathbb{F}_n$ is the free group on n generators. One hopes to eventually be able to resolve the isomorphism problem: whether the free groups factors $L(\mathbb{F}_n)$ and $L(\mathbb{F}_m)$ are, for $n, m \geq 2$, isomorphic or not.

Creation and annihilation operators on full Fock spaces

Let $\mathcal{H}$ be a Hilbert space. The *full Fock space* over $\mathcal{H}$ is defined as $\mathcal{F}(\mathcal{H}) := \bigoplus_{n=0}^{\infty} \mathcal{H}^{\otimes n}$. The summand $\mathcal{H}^{\otimes 0}$ on the right-hand side of the last equation is a one-dimensional Hilbert space. It is customary to write it in the form $\mathbb{C}\Omega$ for a distinguished vector of norm one, which is called *the vacuum vector*. The vector state $\tau_{\mathcal{H}}$ on $B(\mathcal{F}(\mathcal{H}))$ given by the vacuum vector, $\tau_{\mathcal{H}}(T) := \langle T\Omega, \Omega\rangle$ $(T \in B(\mathcal{F}(\mathcal{H})))$, is called a *vacuum expectation state.*

For each $\xi \in \mathcal{H}$, the operator $l(\xi) \in B(\mathcal{F}(\mathcal{H}))$ determined by the formula $l(\xi)\Omega = \xi$ and $l(\xi)\xi_1 \otimes \cdots \otimes \xi_n = \xi \otimes \xi_1 \otimes \cdots \otimes \xi_n$ for all $n \geq 1, \xi_1, \ldots, \xi_n \in \mathcal{H}$, is called the *(left) creation operator* given by the vector ξ. As one can easily verify, the adjoint of $l(\xi)$ is described by the formula: $l(\xi)^*\Omega = 0$, $l(\xi)^*\xi_1 = \langle \xi_1, \xi\rangle\Omega$, and $l(\xi)^*\xi_1 \otimes \cdots \otimes \xi_n = \langle \xi_1, \xi\rangle \xi_2 \otimes \cdots \otimes \xi_n$ and is called the *(left) annihilation operator* given by the vector ξ.

The relevance of these operators comes from the fact that orthogonality of vectors translates into free independence of the corresponding creation and annihilation operators.

Proposition 22.9.1 *Let $\mathcal{H}$ be a Hilbert space and consider the probability space $(B(\mathcal{F}(\mathcal{H})), \tau_{\mathcal{H}})$. Let $\mathcal{H}_1, \ldots, \mathcal{H}_k$ be a family of linear subspaces of $\mathcal{H}$, such that $\mathcal{H}_i \perp \mathcal{H}_j$ for $i \neq j$ $(1 \leq i, j \leq k)$. For every $1 \leq i \leq k$ let $\mathcal{A}_i$ be the unital C^*-subalgebra of $B(\mathcal{F}(\mathcal{H}))$ generated by $\{l(\xi) : \xi \in \mathcal{H}_i\}$. Then $\mathcal{A}_1, \ldots, \mathcal{A}_k$ are freely independent in $(B(\mathcal{F}(\mathcal{H})), \tau_{\mathcal{H}})$.*

Also semicircular elements show up very canonically in this frame; namely, if we put $l := l(\xi)$ for a unit vector $\xi \in \mathcal{H}$, then $l + l^*$ is a semicircular element of variance 1. More generally, one has that $l + f(l^*)$ has $\mathcal{R}$-transform $\mathcal{R}(z) = f(z)$ (for f a polynomial, say). This, together with the above proposition, was the basis of Voiculescu's proof of Theorem 22.5.1. Similarly, a canonical realization for the S-transform is $(1 + l)g(l^*)$, for which one has $S(z) = 1/g(z)$. This representation is due to Haagerup who used it for a proof of Theorem 22.5.2.

22.9.2 Free entropy

Free entropy is, as the name suggests, the counterpart of entropy in free probability theory. The development of this concept is at present far from complete. The current state of affairs is that there are two distinct approaches to free entropy. These should give isomorphic theories, but at present we only know that they coincide in a limited number of situations. The first approach to a theory of free entropy is via *microstates*. This goes back to the statistical mechanics roots of entropy via the Boltzmann formula and is related to the theory of large deviations. The second approach is *microstates free*. This draws its inspiration from the statistical approach to classical entropy via the notion of Fisher information. We will in the following only consider the first approach via microstates, as this relates directly to random matrix questions.

Wigner's semicircle law states that as $N \to \infty$ the empirical eigenvalue distribution μ_{A_N} of an $N \times N$ Gaussian random matrix A_N converges almost surely to the semicircular distribution μ_W, i.e. the probability that μ_{A_N} is in any fixed neighbourhood of the semicircle converges to 1. We are now interested in the deviations from this: what is the rate of decay of the probability that μ_{A_N} is close to ν, where ν is an arbitrary probability measure? We expect that this probability behaves as $e^{-N^2 I(\nu)}$, for some *rate function* I vanishing at the semicircle distribution. By analogy with the classical theory of large deviations, I should correspond to a suitable notion of free entropy. This heuristics led Voiculescu to define in [Voi93] the *free entropy* χ in the case of one variable to be

$$\chi(\nu) = \iint \log|s - t| d\nu(s) d\nu(t) + \frac{3}{4} + \frac{1}{2}\log 2\pi.$$

Inspired by this, Ben-Arous and Guionnet proved in [Ben97] a rigorous version of a large deviation for Wigner's semicircle law, where the rate function $I(\nu)$ is, up to a constant, given by $-\chi(\nu) + \frac{1}{2}\int t^2 d\nu(t)$.

Consider now the case of several matrices. By Voiculescu's generalization of Wigner's theorem we know that n independent Gaussian random matrices $A_N^{(1)}, \ldots, A_N^{(n)}$ converge almost surely to a freely independent family $s_1, \ldots, s_n$ of semicircular elements. Similarly as for the case of one matrix, large deviations from this limit should be given by

$$\text{Prob}\left\{(A_N^{(1)}, \ldots, A_N^{(n)}) : \text{distr}((A_N^{(1)}, \ldots, A_N^{(n)}) \approx \text{distr}(a_1, \ldots, a_n)\right\}$$
$$\sim e^{-N^2 I(a_1,\ldots,a_n)},$$

where $I(a_1, \ldots, a_n)$ should be related to the *free entropy* of the random variables $a_1, \ldots, a_n$. Since the distribution $\text{distr}(a_1, \ldots, a_n)$ of several non-commuting random variables $a_1, \ldots, a_n$ is a mostly combinatorial object (consisting of the collection of all joint moments of these variables), it is much harder to deal with these questions and, in particular, to get an analytic formula for I. Essentially, the above heuristics led Voiculescu to the following definition [Voi94b] of a free entropy for several variables.

Definition 22.9.1 *Given a tracial W^*-probability space (M, τ) (i.e. M a von Neumann algebra and τ a faithful and normal trace), and an n-tuple $(a_1, \ldots, a_n)$ of selfadjoint elements in M, put*

$$\Gamma(a_1, \ldots, a_n; N, r, \epsilon) := \Big\{(A_1, \ldots, A_n) \in M_N(\mathbb{C})^n_{sa} : |\text{tr}\,(A_{i_1} \ldots A_{i_k}) - \tau(a_{i_1} \ldots a_{i_k})| \leq \epsilon$$
$$\textit{for all } 1 \leq i_1, \ldots, i_k \leq n, 1 \leq k \leq r\Big\}$$

In words, $\Gamma(a_1, \ldots, a_n; N, r, \epsilon)$ is the set of all n-tuples of $N \times N$ selfadjoint matrices which approximate the mixed moments of the selfadjoint elements $a_1, \ldots, a_n$ of length at most r to within ϵ.

Let Λ denote Lebesgue measure on $M_N(\mathbb{C})^n_{sa}$. Define

$$\chi(a_1, \ldots, a_n; r, \epsilon) := \limsup_{N\to\infty} \frac{1}{N^2} \log \Lambda(\Gamma(a_1, \ldots, a_n; N, r, \epsilon) + \frac{n}{2} \log N,$$

and

$$\chi(a_1, \ldots, a_n) := \lim_{\substack{r\to\infty \\ \epsilon\to 0}} \chi(a_1, \ldots, a_n; r, \epsilon).$$

The function χ is called the free entropy.

Many of the expected properties of this quantity χ have been established (in particular, it behaves additive with respect to free independence), and there have been striking applications to the solution of some old operator algebra problems. A celebrated application of free entropy was Voiculescu's proof of the fact that free group factors do not have Cartan subalgebras (thus settling a longstanding open question). This was followed by several results of the same nature; in particular, Ge showed that $L(\mathbb{F}_n)$ cannot be written as a tensor product of two II_1 factors. The rough idea of proving the absence of some property for the von Neumann algebra $L(\mathbb{F}_n)$ using free entropy is the following string of arguments: finite matrices approximating in distribution any set of generators of $L(\mathbb{F}_n)$ should also show an approximate version of the considered property; one then has to show that there are not many finite matrices with this approximate property; but for $L(\mathbb{F}_n)$ one has many matrices, given by independent Gaussian random matrices, which approximate its canonical generators.

However, many important problems pertaining to free entropy remain open. In particular, we only have partial results concerning the relation to large deviations for several Gaussian random matrices. For more information on those and other aspects of free entropy we refer to [Voi02, Bia03, Gui04].

22.9.3 Other operator algebraic applications of free probability theory

The fact that freeness occurs for von Neumann algebras as well as for random matrices means that the former can be modelled asymptotically by the latter and this insight resulted in the first progress on the free group factors since Murray and von Neumann. In particular, Voiculescu showed that a compression of some $L(\mathbb{F}_n)$ results in another free group factor; more precisely, one has $(L(\mathbb{F}_n))_{1/m} = L(\mathbb{F}_{1+m^2(n-1)})$. By introducing interpolated free group factors $L(\mathbb{F}_t)$ for all real $t > 1$, this formula could be extended by Dykema and Radulescu to any real $n, m > 1$, resulting in the following dichotomy: One has that either all free group factors $L(\mathbb{F}_n)$ $n \geq 2$ are isomorphic or that they are pairwise not isomorphic.

There exist also type III versions of the free group factors; these free analogues of the Araki-Woods factors were introduced and largely classified by Shlyakhtenko.

The study of free group factors via free probability techniques has also had an important application to subfactor theory. Not every set of data for a subfactor inclusion can be realized in the hyperfinite factor; however, work of Shlyakhtenko, Ueda, and Popa has shown that this is possible using free group factors.

By relying on free probability techniques and ideas, Haagerup also achieved a crucial breakthrough on the famous invariant subspace problem: every operator in a II_1 factor whose Brown measure (which is a generalization of the spectral measure composed with the trace to non-normal operators) is not concentrated in one point has non-trivial closed invariant subspaces affiliated with the factor.

Acknowledgements

This work was supported by a Discovery Grant from NSERC.

References

[And09] G. Anderson, A. Guionnet, and O. Zeitouni, *An Introduction to Random Matrices*, Cambridge University Press, Cambridge UK (2009)

[Ben97] G. Ben-Arous and A. Guionnet, Prob. Th. Rel. Fields **108** (1997) 517

[Ben09] F. Benaych-Georges, Prob. Th. Rel. Fields **144** (2009) 471

[Ber93] H. Bercovici and D. Voiculescu, Indiana Univ. Math. J. **42** (1993) 733
[Ber99] H. Bercovici and V. Pata (with an appendix by P. Biane), Ann. of Math. **149** (1999) 1023
[Bia98] P. Biane and R. Speicher, Prob. Th. Relat. Fields **112** (1998) 373
[Bia02] P. Biane, Proceedings of the International Congress of Mathematicians, Beijing 2002, Vol. 2 (2002) 765
[Bia03] P. Biane, M. Capitaine, and A. Guionnet, Invent. Math. **152** (2003) 433
[Cvi82] P. Cvitanovic, P.G. Lauwers, and P.N. Scharbach, Nucl. Phys. B **203** (1982) 385
[Col07] B. Collins, J. Mingo, P. Sniady, and R., Speicher, Documenta Math. **12** (2007) 1
[Dyk93] K. Dykema, J. Funct. Anal. **112** (1993) 31
[Gui04] A. Guionnet, Probab. Surv. **1** (2004) 72
[Gui06] A. Guionnet, Proceedings of the International Congress of Mathematicians, Madrid 2006, Vol. III (2006) 623
[Gui09] A. Guionnet and D. Shlyakhtenko, Geom. Funct. Anal. 18 (2009), 1875.
[Haa02] U. Haagerup, Proceedings of the International Congress of Mathematicians, Beijing 2002, Vol 1 (2002) 273
[Haa05] U. Haagerup, S. Thorbjørnsen, Ann. of Math. **162** (2005) 711
[Hia00] F. Hiai and D. Petz, *The Semicircle Law, Free Random Variables and Entropy*, Math. Surveys and Monogr. 77, AMS 2000
[Nic06] A. Nica and R. Speicher, *Lectures on the Combinatorics of Free Probability*, London Mathematical Society Lecture Note Series, vol. 335, Cambridge University Press, 2006
[Min10] J. Mingo and R. Speicher, *Free Probability and Random Matrices*, Fields Monograph Series (to appear)
[Rao09] N. R. Rao and A. Edelman, Foundations of Computational Mathematics (to appear)
[Ras08] R. Rashidi Far, T. Oraby, W. Bryc, and R. Speicher, IEEE Trans. Inf. Theory **54** (2008) 544
[Shl96] D. Shlyakhtenko, IMRN 1996 **1996** 1013
[Spe93] R. Speicher, Publ. RIMS **29** (1993) 731
[Spe94] R. Speicher, Math. Ann. **298** (1994) 611
[Spe98] R. Speicher, *Combinatorial theory fo the free product with amalgamtion and operator-valued free probability theory*, Memoirs of the AMS **627** 1998
[Tul04] A. Tulino, S. Verdu, *Random matrix theory and wirless communications*, Foundations and Trends in Communications and Information Theory **1** (2004)
[Voi85] D. Voiculescu, in *Operator Algebras and their Connections with Topology and Ergodic Theory*, Lecture Notes in Math. 1132 (1985), Springer Verlag, 556
[Voi91] D. Voiculescu, Invent. Math. **104** (1991) 201
[Voi92] D. Voiculescu, K. Dykema, and A. Nica, *Free Random Variables*, CRM Monograph Series, Vol. 1, AMS 1992
[Voi93] D. Voiculecu, Comm. Math. Phys. **155** (1993) 71
[Voi94a] D. Voiculescu, Proceedings of the International Congress of Mathematicians, Zürich 1994, 227
[Voi94b] D. Voiculescu, Invent. Math. **118** (1994) 411
[Voi95] D. Voiculescu, Asterisque **223** (1995) 243
[Voi00] D. Voiculescu, in *Lectures on Probabiltiy Theory and Statistics (Saint-Flour, 1998)*, Lecture Notes in Mathematics 1738, Springer, 2000, 279
[Voi02] D. Voiculescu, Bulletin of the London Mathematical Society **34** (2002) 257
[Voi05] D. Voiculescu, Reports on Mathematical Physics **55** (2005) 127
[Woe86] W. Woess, Bollettino Un. Mat. Ital. **5-B** (1986) 961
[Xu97] F. Xu, Commun. Math. Phys. **190** (1997) 287

·23·

Random banded and sparse matrices

Thomas Spencer

Abstract

The aim of this article is to review results and conjectures about the spectral properties of large random band matrices H. The band matrices H_{jk} described here are indexed by lattice sites $j, k \in \Lambda \subseteq \mathbb{Z}^d$ with $d = 1, 2$, or 3 and Λ a large box. Matrix elements are assumed to be small or 0 when $|j - k|$ is large. Thus these matrices reflect the geometry of the lattice. One of the motivations for studying such random band matrices is that they interpolate between the random Schrödinger operators on $\mathbb{Z}^d$ and the mean field GOE or GUE ensembles.

The second part of this article reviews the spectral theory of large random sparse matrices. Such matrices arise for example in the study of the adjacency matrix of large graphs whose vertices have degrees selected from a fixed distribution.

23.1 Introduction

Over the past few decades great progress has been made in understanding the spectral statistics of large matrix ensembles of mean field type. There are many classes of such matrices. One of the most studied class of N by N matrices has a distribution with a weight proportional to $\exp(-NTr\ V(H))$ where $V(x)$, $x \in \mathbb{R}$ is a real function bounded from below with suitable growth for large $|x|$. If the matrices H are Hermitian, then this ensemble is invariant under unitary transformations $U(N)$. Similarly, when H is symmetric the distribution is invariant under $O(N)$. For the special case when V is proportional to x^2 the ensemble has a Gaussian distribution and the distributions are referred as the Gaussian unitary ensemble (GUE) and Gaussian orthogonal ensemble (GOE). For a wide class of functions V, a detailed spectral analysis can be obtained by using the theory of orthogonal polynomials. One of the major achievements of this analysis is that the local eigenvalue spacing statistics in the limit of large N is independent of V (modulo a simple scaling) and depends only on the symmetry class. Since these matrix ensembles are invariant under $U(N)$ or $O(N)$, the eigenvectors share the same property and therefore their statistical properties are well understood. We refer the reader to Chapters 4–6 of this volume for an overview of these ensembles.

The main goal of this article is to describe some mathematical results and conjectures about random band matrix ensembles (RBM) as well as sparse matrix ensembles. Since these ensembles are not invariant under $O(N)$ or $U(N)$, many of the methods that have been applied so successfully to the unitary or orthogonal ensembles described above no longer work. In one dimension, random band matrices can be thought of as large N by N matrices with matrix elements concentrated in a band about the diagonal of width W. The matrix elements outside this band are assumed to be 0 or exponentially small. Random band matrices, of fixed width W, indexed by elements of $\mathbb{Z}^d$ are closely related to a discrete Schrödinger operator with a random potential on $\mathbb{Z}^d$.

In comparison to the orthogonal or unitary ensembles described above, our mathematical understanding of the spectral properties of RBM is relatively primitive. On the other hand, the theoretical physics literature makes rather detailed predictions about their spectral properties on the basis of scaling ideas and supersymmetric (SUSY) statistical mechanics. This review will attempt to present both mathematical and theoretical physics perspectives on RBM and sparse ensembles.

As early as 1955, Wigner investigated random band matrices in [Wig55, Wig57]. He called them bordered matrices and was interested in the case where the diagonal elements grew linearly. See [Fyo96] for more recent developments. The case in which the diagonal elements are strongly fluctuating is also interesting and arises in the study of two interacting particles in a random background, [She94, Fra95, Fyo95].

Modern interest in RBM can be traced back in the work of [Sel85, Cas90, Cas91] who studied RBM as a model of quantum chaos. These authors considered large N by N symmetric random matrices ($d = 1$) whose width W grows with N. Numerical simulations suggested that as W varied, local eigenvalue statistics changed from Poisson, for $W \ll N^{1/2}$, to GOE or GUE when $W \gg N^{1/2}$. These results were supported strongly by theoretical work of Fyodorov and Mirlin [Fyo91a], which studied RBM in terms of a related supersymmetric (SUSY) statistical mechanics model. See also [Kho02] for some partial mathematical results.

Spectral problems of RBM and sparse matrices can be expressed in terms of SUSY statistical mechanics. Averages of Green's functions are given by correlations in the field variables. Supersymmetric statistical mechanics provides a dual representation for disordered quantum systems. This representation offers important insights into nonperturbative aspects of the spectrum and eigenfunctions of RBM. The foundations of the SUSY statistical mechanics models lies in work of Wegner [Weg79a, Sch80] and Efetov [Efe83, Efe97]. We refer the reader to Guhr's contribution to this volume and to [Mir00a, Zir04] for expositions of SUSY.

The main topics discussed in this survey include the density of states, the behaviour of eigenvectors, and eigenvalue statistics for RBM and sparse random matrices. Connections with random Schrödinger (RS) and the role of the dimension of the lattice will be highlighted. The remainder of this article is organized as follows: the next section provides the basic definitions of RBM. Section 23.3 reviews results on the density of states. In Section 23.4 a connection between RBM and statistical mechanics is briefly described. Conjectures and theorems about eigenvectors and local spacing statistics are presented in Section 23.5. The final sections give a brief survey of sparse random matrices and of the random Schrödinger operator on the Cayley tree or Bethe lattice.

23.2 Definition of random banded matrix (RBM) ensembles

Let $\Lambda \subseteq \mathbb{Z}^d$ be a large box. This review will focus on Gaussian RBM ensembles whose matrix elements are indexed by vertices in Λ and whose average $\langle H_{ij} \rangle = 0$. Let $J_{ij} \geq 0$ be a symmetric function on Λ^2, (possibly depending on Λ). Consider Gaussian Hermitian and symmetric matrices H with variance given by

$$\langle H_{ij} \bar{H}_{i'j'} \rangle = \delta_{ii'}\delta_{jj'} J_{ij}, \quad \langle H_{ij} H_{i'j'} \rangle = (\delta_{jj'}\delta_{ii'} + \delta_{ij'}\delta_{ji'}) J_{ij} \tag{23.2.1}$$

respectively. Here $\bar{H}$ denotes the complex conjugate of H and $\langle \cdot \rangle$ is the Gaussian expectation. Thus the matrix elements $i \geq j$ are assumed to be independent. More generally, one can consider an ensemble of independent identically distributed random variables R_{ij} with $i \geq j$ and define the symmetric RBM by

$$H_{ij} = R_{ij}\sqrt{J_{ij}}. \tag{23.2.2}$$

In one dimension, GUE and GOE ensembles are obtained by setting $\Lambda = [1, 2, \ldots N]$ and $J_{ij} = 1/N$. Some results for the case when the H_{ij} are correlated, are described in [Jan99].

The band structure appears provided J_{ij} is small or vanishes when $|i - j| \geq W$ and $i, j \in \Lambda \subset \mathbb{Z}^d$. W is called the width of the band. When $d = 1$, the matrix elements can be thought of as lying in a band about the diagonal. It will be convenient to assume that

$$\sum_j J_{ij} \approx 1. \tag{23.2.3}$$

Note that random band matrices are closely related to Wegner's **m**-orbital model [Weg79b]. Here J is short range but the lattice index $j \in \mathbb{Z}^d$ is replaced by (j, α) with $\alpha = 1, 2 \ldots, \mathbf{m}$, denoting the orbital index. The parameters **m** and W are comparable. See [Opp79, Sch80] for an analysis of these models.

One is interested in the spectral statistics of H as Λ becomes infinite. Sometimes W will be fixed and in other cases it will grow with the volume of the box $|\Lambda|$. One especially convenient choice of J is given by the lattice Green's function

$$J_{jk} = (-W^2\Delta + 1)^{-1}(j,k) \tag{23.2.4}$$

where Δ is the discrete Laplacian on Λ with suitable boundary conditions

$$\Delta f(j) = \sum_{j' \sim j} (f(j') - f(j)) \tag{23.2.5}$$

and the sum is over j' adjacent to j. It will often be convenient to choose periodic boundary conditions on the boundary of Λ. If $W = 0$ then the matrix H is diagonal. Note that for large W, J decays as $W^{-1}\exp(-|j-k|/W)$ in one dimension and $\sum_j J_{jk} = 1$ for all dimensions in a periodic box. A remarkable feature of this particular choice of J is that in one dimension some spectral properties of H can be studied via a *nearest neighbour* transfer matrix. This is because although J has a range approximately W, its matrix inverse, J^{-1}, is local (nearest neighbour) and W now enters only as a parameter, [Fyo91a]. The inverse of J appears naturally in a dual statistical mechanics representation as is explained in Section 23.4.

The Anderson random Schrödinger operator (RS) on the lattice $\mathbb{Z}^d$ has the form,

$$H = -\Delta + g v_j \tag{23.2.6}$$

and $|g| \in \mathbb{R}^+$ measures the strength of the disorder. The v_j are usually assumed to be identically distributed random variables indexed by $j \in \Lambda \subseteq \mathbb{Z}^d$. Hence the disorder in (23.2.6) only appears in the diagonal elements. Although random band matrices and RS look quite different, one expects that for fixed W and large boxes they have similar spectral characteristics with g roughly corresponding to W^{-1}. The fact that RBM have randomness in all matrix elements and RS have deterministic off-diagonal matrix elements seems to make RBM easier to analyse mathematically.

23.3 Density of states

The density of states (DOS) is one of the basic objects in the study of random matrices and RS operators. If J is translation invariant and decays rapidly for $|i-j|$ large, condition (23.2.3) guarantees that most of the eigenvalues of H lie in the interval [−2,2]. Let $n_\Lambda(E) = \mathcal{N}_\Lambda(E)/|\Lambda|$ be the number of eigenvalues of H less than E per unit volume. The limit of $n_\Lambda(E)$ as N goes to infinity or $\Lambda \to \mathbb{Z}^d$ is the integrated density of states denoted by $n(E)$. For a wide class of random ergodic systems, this limit is known to exist and is non random with probability 1. Its formal derivative is the density of states $\rho(E)$. For GUE or GOE matrices the integrated density of states is given by Wigner's well-known semicircle law:

$$n(E) = \frac{1}{2\pi} \int_{-2}^{E} \sqrt{4 - u^2}\, du \tag{23.3.1}$$

for $|E| \leq 2$. The density of states is $\rho(E) = \frac{1}{2\pi}\sqrt{4 - E^2}$. For Gaussian RBM with fixed W the density of states cannot be calculated explicitly, but it is known to be everywhere positive. However, if both W, N, $\to \infty$, and $W/N \to 0$, the DOS coincides with the semicircle law, for a wide class of (RBM). See [Bog91, Mol92].

In [Gui02, And06] a central limit theorem was established for N by N band matrices with $J_{ij} = N^{-1} g(i/N, j/N)$ and g a positive, smooth, symmetric function on $[0, 1]^2$. Given a smooth function, f, the sum

$$\sum_{j}^{N} [f(\lambda_j) - \mathbb{E} f(\lambda_j)] \tag{23.3.2}$$

approaches a Gaussian distribution as $N \to \infty$. Here $\mathbb{E}$ denotes the expectation over the matrix ensemble and λ_j are the eigenvalues. Earlier results for the unbanded case are due to [Jon82, Cos95, Sin98, Joh98]. See also [Cha09].

23.3.1 Finite volume estimates on DOS

For many applications, the above results on the density of states are not sufficient. For example, in a finite box Λ, it is important to estimate the probability that there is an eigenvalue λ of H very close to E. Wegner's method [Weg81] shows that

$$\text{Prob}\ [|\lambda - E| \leq \delta,\ \lambda \in \text{spec}\ H] \leq C\delta\, |\Lambda|\, W^{1/2}. \tag{23.3.3}$$

The factor of $W^{1/2}$ on the right side should be replaced by the density of states. In the case of Wigner matrices, large deviation estimates on the density of states over intervals of order N^{-a} with $1 \geq a \geq 1/2$ have been proved in [Erd09]. These estimates play a central role in understanding the eigenvectors and eigenvalue spacing statistics.

To obtain more refined information about the spectrum we shall analyse the Green's function

$$G(E_\epsilon; j, k) = [E_\epsilon - H]^{-1}(j, k) \qquad E_\epsilon = E - i\epsilon \tag{23.3.4}$$

with $\epsilon > 0$. In a periodic box, Λ, the average Green's function

$$Im\langle G(E_\epsilon; j, k)\rangle = \pi \bar{\rho}_{\epsilon,\Lambda}(E)\delta(j, k) \tag{23.3.5}$$

is proportional to the identity matrix. Off-diagonal elements in (23.3.5) vanish because H_{ij} and $H_{ij}\, \sigma_i \sigma_j$ with $\sigma = \pm 1$ are assumed to have the same distribution. To see the relation to the density of states recall

$$Im\, G(E_\epsilon; j, j) = \frac{1}{|\Lambda|} Im\, tr\ G(E_\epsilon) = \frac{\epsilon}{|\Lambda|} \sum_j [(E - \lambda_j)^2 + \epsilon^2]^{-1}. \tag{23.3.6}$$

The right side of (23.3.6) is proportional to $\pi\, \Sigma \delta_\epsilon\, (E - \lambda_j)$. Note that for fixed $\epsilon > 0$, $\bar{\rho}_{\epsilon,\Lambda}(E)$ approaches the Stieltjes transform of Wigner's law as $\Lambda \to \mathbb{Z}^d$ and $W \to \infty$. However, if $\epsilon \leq 1/|\Lambda|$, the proof that $\bar{\rho}_{\epsilon,\Lambda}(E)$ is well approximated by Wigner's law, for large but fixed W, is more difficult. For this case rigorous results on the average Green's function have only been obtained for certain Gaussian RBM with J given by (23.2.4) by an exact mapping to SUSY statistical mechanics. In this case smoothness of the density of states for fixed W can be established. See [Con87, Dis02] for precise statements and details in the band case and see [Dis04] for an expository account of SUSY for the density of states for GUE. In addition it is proved that

$$|\langle G(E_\epsilon; 0, x)\, G(E_\epsilon; x, 0)\rangle| \;\leq Ce^{-m|x|} \tag{23.3.7}$$

where $x \in \mathbb{Z}^3$ and $m > 0$ is uniform in ϵ and Λ.

23.3.2 Perturbation theory

Next we make formal calculations in perturbation theory to explain the semicircle law and its correction when W is large. Consider Hermitian RBM with Gaussian distribution and J given by (23.2.4). The perturbation scheme described here is very closely related to the one used for random Schrödinger operators. The reader may consult [Sil97, Sod10] for a systematic exposition. To calculate the density of states write

$$G(E_\epsilon) = [E_\epsilon - H]^{-1} = [\tilde{E} - H + (E_\epsilon - \tilde{E})]^{-1} \tag{23.3.8}$$

where $\tilde{E} = \tilde{E}(E, \epsilon)\, I$ and I is the identity matrix. We shall perturb about $\tilde{E}^{-1} \equiv G_0$,

$$\begin{aligned}\langle G(E_\epsilon)(i, i)\rangle &= G_0 + \langle G_0\, H_{ii}\, G_0\rangle \\ &\quad + \sum_j \langle G_0\, H_{ij}\, G_0\, H_{ji}\, G_0\rangle - G_0(E_\epsilon - \tilde{E})G_0 + \ldots\end{aligned} \tag{23.3.9}$$

We shall define $\tilde{E}$ so that the third and fourth terms on the right side of (23.3.9) cancel. The second term vanishes because $\langle H\rangle = 0$. Since

$$\sum_j \langle H_{ij}\, H_{ji}\rangle = \sum_j J_{ij} = 1 \tag{23.3.10}$$

the third and fourth terms cancel when

$$G_0 = \tilde{E}^{-1} = E_\epsilon - \tilde{E} \qquad \text{hence,} \quad G_0 = [E_\epsilon + i\sqrt{4 - E_\epsilon^2}]/2. \tag{23.3.11}$$

The imaginary part of G_0 gives Wigner's semicircle law for the density of states. Of course, this expansion has been done only to second order in H. If we calculate to fourth order, we will see that the correction is $O(W^{-1})$. The reason that higher-order averages are smaller is that adjacent factors $H_{jk}\, G_0\, H_{kj'}$ appearing in the expansion about G_0 with $j = j'$ are cancelled by $E_\epsilon - \tilde{E}$ to leading order as above. Thus in the Hermitian case, $\sum_j H_{ij_1} H_{j_1j_2} H_{j_2j_3} H_{j_3i}$ has only terms with $(i, j_1) = (j_3, j_2)$ and $(i, j_3) = (j_1, j_2)$ which contributes after averaging. Hence we get a contribution $J^2_{i,i} \le W^{-2}$. Proceeding beyond fourth order one must re-sum classes of graphs. This can be done by grouping together diagrams according to their 'genus'. Roughly speaking diagrams contributing with higher genus are suppressed by powers of W^{-1}.

It is also instructive to calculate (23.3.7) to leading order. Assume that we are in a periodic box and J is defined by (23.2.4). Let us expand $G(E_\epsilon; 0, x)$, $G(E_\epsilon; x, 0)$ about G_0 as above. This expansion will have terms of the form

$$G_0\, H_{0,j_1}\, G_0\, H_{j_1,j_2} \dots H_{j_n,x}\, G_0, \qquad G_0\, H_{x,k_n}\, G_0\, H_{k_n,k_{n-1}} \dots H_{k_1,0}\, G_0. \tag{23.3.12}$$

The leading contribution to the average of this expression occurs when we pair $j_i = k_i$ producing the average

$$\sum_j G_0^2 J_{0,j_1} G_0^2 J_{j_1,j_2} \dots J_{j_n,x} G_0^2 = (G_0^2 J)^n_{0,x} G_0^2 \tag{23.3.13}$$

here J^n denotes the n-fold convolution of J. Summing over n we get

$$\int [G_0^{-2} - \hat{J}]^{-1}(p) e^{ix\cdot p} dp \tag{23.3.14}$$

where $\hat{J}(p)$ is the Fourier transform of J. Exponential decay in x can be seen explicitly from (23.2.4) and (23.3.11). Rigorous control of the above perturbation scheme for small ϵ seems to be difficult to achieve for E inside [−2,2] unless the SUSY approach is used.

23.4 Statistical mechanics and RBM

In this section we shall illustrate how statistical mechanics can be used to get information about random band matrices. To keep our discussion brief and self contained we shall only consider the so-called bosonic sector of a SUSY model. Let us first observe that in a finite box Λ, Cramer's rule tells us that the Green's function, $G(E_\epsilon)$, given by (23.3.8), is a ratio of determinants. The goal of the section is to express $\langle \det G(E_\epsilon) \rangle$ in terms of a statistical mechanics model which can be analysed. To average the ratio of determinants or $\langle G \rangle$ one needs to introduce Grassmann variables which will be omitted here for ease of exposition. See Guhr's chapter 7 on SUSY in this volume for more details and references.

Let $\phi_j = \phi_{1,j} + i\phi_{2,j},\ j \in \Lambda$, be a complex field and ϕ_j^* its complex conjugate. The quadratic form

$$Q = i \sum_{kj} \phi_k^* (E_\epsilon \delta_{kj} - H_{kj}) \phi_j \tag{23.4.1}$$

has a positive real part because $\epsilon > 0$, see (23.3.4). If we set

$$D\phi = \prod_{j \in \Lambda} \frac{1}{\pi} \, d\phi_{1,j} \, d\phi_{2,j} \tag{23.4.2}$$

then the identity

$$\int e^{-i \sum_{kj} \phi_k^* (E_\epsilon \delta_{kj} - H_{kj}) \phi_j} \, D\phi = \det G(E_\epsilon) \tag{23.4.3}$$

holds for any Hermitian matrix H. (Factors of i disappear assuming there are an even number of sites in Λ.) If H has a Gaussian distribution with covariance J, then we may explicitly compute the average of (23.4.3) over H:

$$\langle \det G(E_\epsilon) \rangle = \int e^{-i E_\epsilon \sum \phi_j^* \phi_j} \, e^{-\frac{1}{2} \sum J_{kj} \phi_k^* \phi_k \, \phi_j^* \phi_j} \, D\phi. \tag{23.4.4}$$

To simplify this nonlocal quartic expression we introduce another independent Gaussian field $a_j \in \mathbb{R}$ of mean 0 and covariance J_{ij}. Then (23.4.4) can be written as

$$\langle \det G(E_\epsilon) \rangle = \int \left\langle e^{-i \sum_j (E_\epsilon - a_j) \phi_j^* \phi_j} \right\rangle_a D\phi \tag{23.4.5}$$

where $\langle \cdot \rangle_a$ is the expectation over the Gaussian a variables. The integral over the ϕ fields can now be done explicitly and we are left with an integral over the a_j fields. The action of the a_j field is given by $\sum a_j \, J_{jk}^{-1} a_k$. If J is given by (23.2.4), J_{jk}^{-1} vanishes unless $|j - k| \leq 1$. Thus in one dimension a nearest neighbour transfer matrix can be applied.

We conclude with some comments about how to analyse the resulting statistical mechanics in the a_j field. First, note that after integration over the ϕ fields we get an expression $\Pi_j (E_\epsilon - a_j)^{-1}$ which is singular for small ϵ. This problem is easily fixed by deforming the contour of integration. We replace a_j by $a_j + i\sigma(E)$ with σ chosen so the contour of integration passes through a saddle point. This saddle point is a complex number independent of j which satisfies a quadratic equation essentially equivalent to (23.3.11). The saddle point gives the main contribution to the integral over the a_j fields. The main correction comes from the Hessian at the saddle. This is non-degenerate when $|E| \leq 1.5$ and is proportional to W^2. This enables one to use standard methods of statistical mechanics to control fluctuations about the saddle even as the box Λ gets large.

In this way one can get rigorous estimates on $\langle \det G \rangle$ which are asymptotic in W^{-1}.

A somewhat more complicated representation for $\langle Tr\, G \rangle$ involves both the Grassmann field ζ_j and the ϕ_j field. The Grassmann fields are needed to get $\det[E_\epsilon - H]$. The resulting statistical mechanics model is supersymmetric because the ϕ and ζ fields appear symmetrically. A mathematical analysis of the saddle points and fluctuations of this system was established in [Con87, Dis02]. Although the average Green's function for (23.2.6) can also be recast as a SUSY functional integral, this case is more difficult to analyse because the Laplacian in (23.2.6) produces strong oscillations which cannot be averaged out.

23.5 Eigenvectors of RBM

Although the average Green's function is a useful tool needed to understand certain spectral properties of RBM, it does not contain crucial information about the nature of the eigenfunctions. This information is contained in

$$\langle |G(E_\epsilon; j, k)|^2 \rangle \tag{23.5.1}$$

as $\epsilon \to 0$. In theoretical physics, (23.5.1) is also expressed in terms of statistical mechanics with a noncompact hyperbolic supersymmetry [Efe83, Mir00a, Zir04]. This noncompact symmetry arises because, unlike (23.3.7), we are averaging G times $\bar{G}$ and hence various phase cancellations are no longer present. Furthermore, divergences as the symmetry breaking term, $\epsilon \to 0$, may occur in (23.5.1) with $j = k$ indicating the presence of point spectrum or localized states.

Fyodorov and Mirlin applied SUSY to make precise predictions about the eigenvectors of RBM [Fyo91a], and sparse matrices [Mir91a]. For a more detailed exposition of the method see [Fyo94]. Although these arguments are convincing, the analysis of the associated SUSY statistical mechanics is much more complicated than that for $\langle G \rangle$ and remains open mathematically. However, there are some partial results. See [Spe04] for results on a 3D bosonic hyperbolic sigma model. In [Dis09a, Dis09b] a phase transition for a simplified version of Efetov's 3D SUSY model is established. This transition is analogous to the Anderson transition described below.

23.5.1 Localization

Let ψ_j, with $j \in \Lambda$, be an eigenvector of a random band matrix H, normalized so that $\sum_j |\psi_j|^2 = 1$. We shall say that ψ is localized with a localization length ℓ if for some $c \in \Lambda$,

$$|\psi_j| \leq C e^{-|j-c|/\ell} \tag{23.5.2}$$

holds with probability 1, for some random C whose large values are strongly suppressed. A weaker form of localization holds if $\sum |\psi_j|^4 \geq C_\ell^{-1}$, with $C_\ell > 0$ independent of Λ.

In the extreme case of random diagonal matrices, e.g. $W = 0$ in (23.2.4), all eigenvectors are localized at one site. If W is fixed, it is known that outside the interval $[-2 - \delta_W, 2 + \delta_W]$ all eigenvectors are localized with probability one as $\Lambda \to \mathbb{Z}^d$. Moreover, $\delta_W \to 0$ for large W. The proof uses the fact that the density of states is small outside this interval, for W large. The rigorous mathematical methods developed to establish localization for RS, [Fro83, Dre89, Aiz01], together with estimate (23.3.3) can be applied to prove this result. In one dimension, there is recent work of Schenker [Sch09] which proves that for a class of RBM of width W, all eigenvectors are localized with a localization length $\ell \leq W^8$. Theoretical work of [Fyo91b, Efe83] strongly suggest that $\ell = O(W^2)$ is the optimal bound. This result is obtained analysing (23.5.1) via a 1D SUSY transfer matrix. In two dimensions, renormalization group arguments due to Abrahams, Anderson, Licciardello, and Ramakrishnan [Abr79] predict that all eigenvectors are localized with ℓ exponentially large in W.

23.5.2 Extended states

For GUE and GOE ensembles, it is known that eigenvectors are not localized. In fact since the eigenvectors are invariant under $U(N)$ or $O(N)$ it is easy to see that, with probability approaching 1 as $N \to \infty$, the eigenvectors are strongly extended:

$$I_2 = \sum |\psi_j|^4 = O(|\Lambda|^{-1}) = O(N^{-1}) \tag{23.5.3}$$

This means that the typical eigenvector is uniformly spread out over $[1, 2, \ldots N]$.

For RBM or RS, an eigenstate of energy E is extended if we have uniform bounds on (23.5.1) with $j = k$ and $\epsilon \approx |\Lambda|^{-1}$. These eigenstates may be thought of as conducting states whereas the localized states are insulating. As mentioned in the introduction, when $W^2 \gg N$ both numerics and theory [Cas90, Fyo91a] show that eigenvectors of a 1D band matrix with eigenvalues strictly inside [−2,2] are extended. This is of course consistent with the prediction that the localization length is $\ell = O(W^2)$. Moreover, the eigenvalue spacing statistics coincide with GOE or GUE depending on whether the matrices are symmetric or Hermitian. While there is little doubt that these assertions are correct, a rigorous mathematical proof is still missing.

The eigenvalues of GUE or GOE matrices near the spectral edge ($E = \pm 2$) are well known to be governed by an Airy point process. In particular the largest eigenvalue has the classical Tracy–Widom distribution. Recent mathematical work by A. Sodin [Sod10] proves that the same results hold for a class of RBM with $W \gg N^{5/6}$. The eigenvectors corresponding to these eigenvalues near the

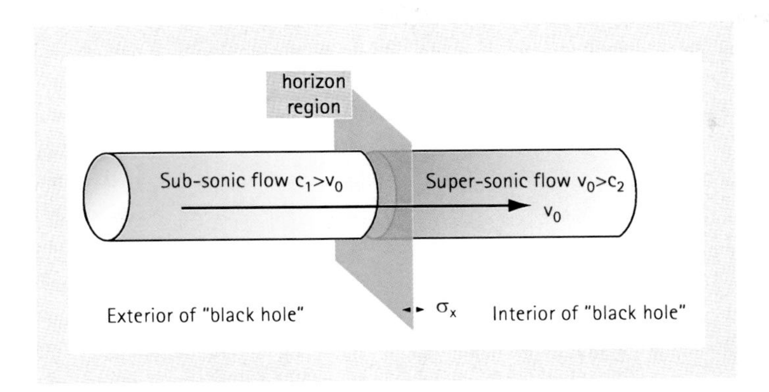

Fig. 12.3 The sonic analogue of a black hole (courtesy of I. Carusotto): the stream of cold atoms with velocity v_0 and the phonon velocity $c(x)$ changing near the origin from the value $c_1 > v_0$ to the value $c_2 < v_0$. This region of the super-sonic flow is analogous to the interior of the black hole from where light cannot escape.

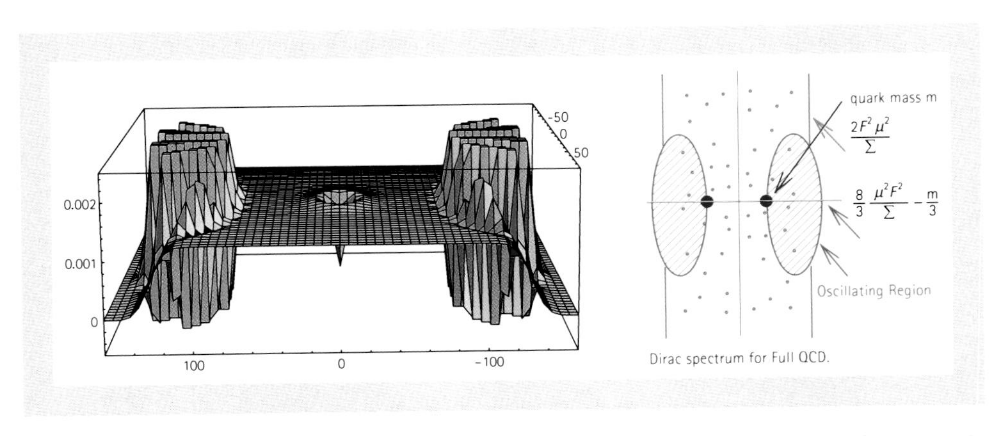

Fig. 32.4 Left: The real part of the spectral density of the QCD Dirac operator for one-flavour QCD at $\mu \neq 0$. For better illustration the z-axis has been clipped. Right: The phase diagram of the Dirac spectrum.

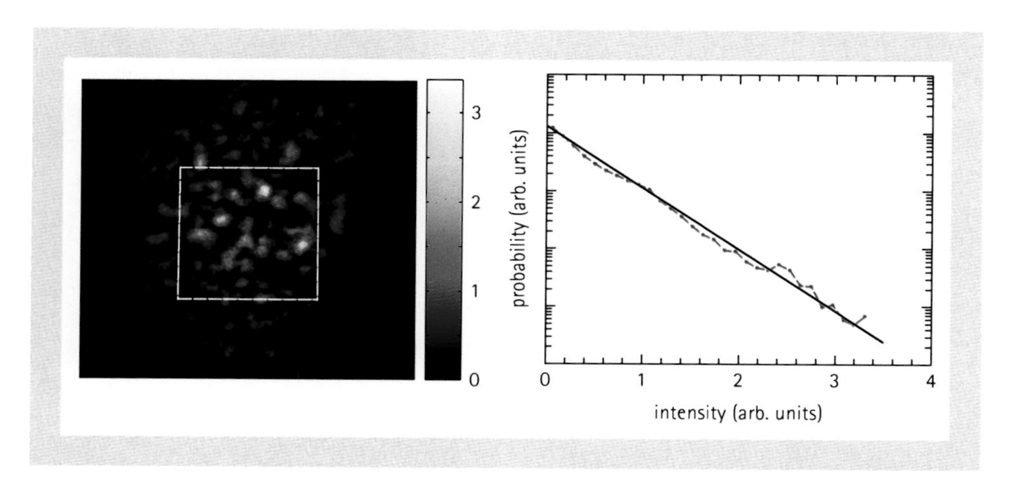

Fig. 36.2 *Left panel:* Speckle pattern produced by a laser beam behind a diffusor (full scale 45 mrad × 45 mrad). The vertical bar indicates the colour coding of the intensity, in arbitrary units. The average angular opening angle $\delta a \approx 1.3$ mrad of a bright or dark spot (a 'speckle') is equal to $\lambda/\pi R$, with $\lambda = 830$ nm the wave length and $R = 200\,\mu$m the radius of the illuminated area on the diffusor. The envelope of the intensity pattern reflects the 18 mrad opening angle of the directional scattering from this type of diffusor. The intensity distribution $P(I)$ of the speckle pattern measured inside the white square is plotted in the right panel, and compared with the exponential distribution (36.2.1) (straight line in the semi-logarithmic plot). Figure courtesy of M.P. van Exter.

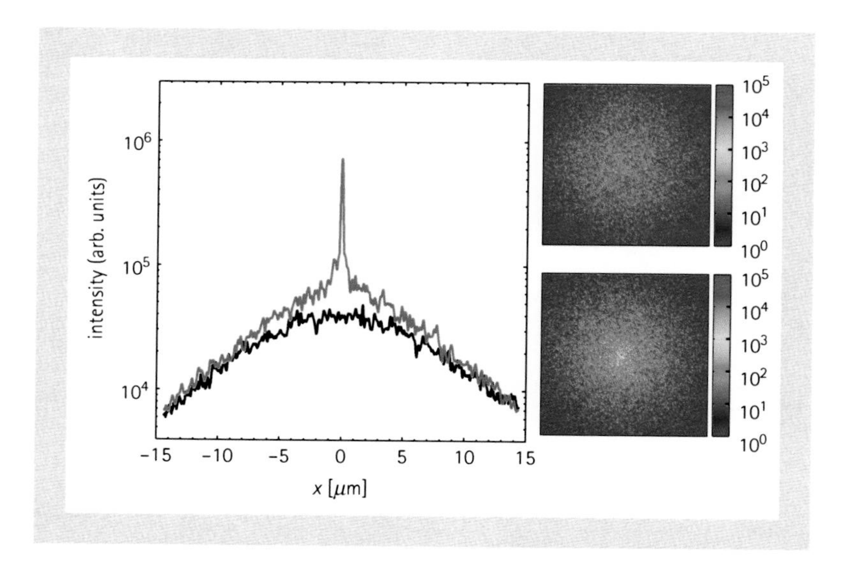

Fig. 36.6 *Right panels:* Speckle pattern (area $30\,\mu$m × $30\,\mu$m) behind a diffusor (a $11.3\,\mu$m layer of ZnO particles with mean free path $l = 0.85\,\mu$m), for a random incident wave front (top) and for a wave front optimized to couple to open transmission channels (bottom). The intensity of the bright speckle at the centre in the bottom panel is a factor of 750 greater than the background. *Left panel:* Intensity profile, integrated over the y-direction to average out the speckle pattern. The optimized wave front (red) has a peak, which the random wave front (black) lacks. Adapted from [Vel08].

edge are shown to be extended. If $W \ll N^{5/6}$ another distribution appears. This work is based on a detailed study of averages of the trace of Chebyshev polynomials in H. The advantage of using these polynomials has been noticed in [Bai03]. These results extend those of Soshnikov [Sos99] who studied Wigner matrices with W=N. Earlier related work on RBM was done by Silvestrov [Sil97]. A very recent paper of Erdös and Knowles [Erd11a] proves that most eigenstates in the bulk are extended for band matrices of width $W \gg N^{6/7}$.

In three dimensions, RBM with fixed large width W are expected to exhibit a phase transition as $\Lambda \to \mathbb{Z}^3$. Eigenvalues slightly outside $[-2, 2]$ are known to correspond to localized states while those inside the interval should be strongly extended (23.5.3). The existence of an interval of extended states has been long understood in theoretical physics, however, a mathematical proof remains a major open problem. The Anderson transition occurs at the mobility edge $E_m(W)$ which is characterized by the property that for $|E| > E_m$ all eigenstates are localized whereas in the range $|E| < E_m$ eigenvectors are extended and electrons are mobile. Here we have assumed that the probability distribution is symmetric as in the Gaussian case. Near the mobility edge, the behaviour of the eigenfunctions is predicted to be multi-fractal. See [Eve08] for an analysis of such states and further references. Although the qualitative features of the transition are believed to be universal (modulo symmetry), exponents of the transition such as the localization length exponent, mostly come from numerical simulation.

The Anderson transition is related through supersymmetry to phase transitions in statistical mechanics models with continuous symmetry. Roughly speaking, the ordered phase, or magnetized phase, corresponds to extended states and the disordered phase to the localized phase. However, the phase transition for SUSY models seem to be more difficult to analyse than conventional phase transitions. In fact there is no known upper critical dimension.

23.5.3 More perturbation theory and scaling theory

In this section we shall present perturbative and scaling approaches to understanding localized and extended eigenvectors for RBM. Although these approaches have only been made rigorous in special cases such as in [Sod10], they provide a valuable guide to the phenomena described above. Let us calculate the leading contribution to (23.5.1) in perturbation theory. Our approach is very similar to the way (23.3.14) was obtained. We shall assume that J is given by (23.2.4). However, in this section the two Green's functions are complex conjugates of each other. The leading contribution to the corresponding sum is given by (23.3.14) but G_0^2 is replaced by

$$|G_0|^2 = 1 - 2\epsilon(4 - E^2)^{-1/2}. \tag{23.5.4}$$

The Fourier transform of (23.5.1) is to leading order diffusive:

$$(-r(E)W^2\Delta(p)+\epsilon)^{-1}\approx(r(E)W^2p^2+\epsilon)^{-1} \tag{23.5.5}$$

where $r(E)=\sqrt{1-(E/2)^2}$ is proportional to the density of states. The parameter ϵ may be thought of as the inverse time scale and $r(E)W^2\equiv D(E)$ is the (bare) diffusion constant. This leading order diffusion arises because $1-\hat{J}(p)$ vanishes at $p=0$. See (23.2.4). The expression above comes from the sum of the so-called ladder graphs. The corrections to this leading order contribution are very important in one and two dimensions. They involve the integral over p of (23.5.5) which is divergent in 1 and 2D as $\epsilon\to 0$. This divergence should be the mechanism for the absence of diffusion (localization) in one and two dimensions for fixed W and $N\to\infty$. In three dimensions, there is no divergence for E inside $[-2,2]$ and a formal perturbation expansion in powers of $1/D(E)$ can be done. This expansion is quite complicated and involves numerous cancellations coming from Ward identities. Rigorous estimates on the remainder terms of such expansions are hard to get. Thus unlike (23.3.7) for which decay can be established in 3D, the corrections to (23.5.5) are mathematically not well understood for small ϵ due to the presence of diffusive slow modes.

Next we briefly review the Thouless scaling theory, [Edw72, Tho77], which approximately predicts when localized and extended states occur. Consider a random band matrix ensemble of width $W\gg 1$ in a box of side $N\gg W$ on a d-dimensional lattice. The average spacing between the eigenvalues near E is proportional to $\Delta(E,N)=1/[r(E)N^d]$, where $r(E)=r_N(E)$ is the 'finite volume' density of states. Let us imagine that a particle undergoes diffusion for some time with a diffusion constant $D=D(E,N)$. From (23.5.5) we see that to leading order $D\approx r(E)W^2$. Here we have neglected backscattering contributions and so this value of D may be thought of as semi-classical. Then the time such a particle needs to hit (or 'feel') the boundary is $t_{Th}\approx N^2/D(E)$. The Thouless energy is proportional to the inverse of this time and may be thought of as a measure of how much an eigenvalue near E changes if one alters the boundary condition from periodic to anti-periodic. Alternatively it tells us that Green's function $|G(E_\epsilon,0,x)|$ is small if x is on the boundary of the box and $\epsilon\gg t_{Th}^{-1}$. The conductance of our sample at energy E is defined to be the ratio of these two energies

$$\mathbf{C}=\frac{1}{t_{Th}\Delta(E,N)}\approx\frac{r(E)N^d}{N^2/(r(E)W^2)}. \tag{23.5.6}$$

If the conductance is small for large N then one should expect localized states, whereas if $\mathbf{C}$ is large as $N\to\infty$, extended states and diffusion are expected.

Before commenting on this picture notice that for $d=1$ when $r(E)\approx 1$ we see that (23.5.6) predicts that eigenstates near E are extended provided $W^2\gg N$. Next we follow Fyodorov's adaptation of Thouless scaling for energies near the spectral edge of RBM. When $E \approx \pm 2$, then $r(E) \approx N^{-1/3}$ is N dependent. Thus near the edge of the spectrum (23.5.6) predicts eigenvectors are extended provided that $W \gg N^{5/6}$ as proved in [Sod10]. Notice that in two dimensions for fixed W, $\mathbf{C}$ is independent of N. However, the corrections to diffusion coming from crossed ladder graphs together with a renormalization group argument indicate that the effective conductance tends to 0 for large N. Hence, localization is predicted [Abr79].

Although the Thouless scaling theory has been very influential, there are some implicit assumptions that are not easy to justify mathematically. The main assumption is that diffusion or sub-diffusion holds for some appropriate time scale. Localization in 1D can be established with this approach if one assumes that the transport is strictly sub-ballistic in a strong sense. Similarly, in 2D we need to assume that it is strictly sub-diffusive, see [Wan92, Ger04]. The idea is that for $\mathbf{C}$ small, the eigenstates or Green's functions do not feel the boundary. Strong insensitivity to boundary conditions in a suitable finite box is a sufficient condition to rigorously establish localization [Fro83, Spe88, Dre89, Aiz01]. On the other hand, there is as yet no rigorous mathematical theorem which provides us with a finite volume criterion for diffusion. Hence there is as yet no way to deduce extended states if $\mathbf{C}$ is large in three dimensions.

A mathematical version of the Thouless localization criterion may be briefly explained as follows. Consider a Green's function $|G(E_\epsilon; 0, j)|$ in a box of side N with 0 at the centre and $|j| \approx N$ at its boundary. The diffusion assumption tells us that $|G(E_\epsilon; 0, N)| \approx e^{-\sqrt{\epsilon D}\, N}$ is small when $Dt = D\epsilon^{-1} \ll N^2$ where D is the diffusion constant. Thus the boundary is not felt, i.e. $|G(E_\epsilon(0, N)|$ is small, if $\epsilon \gg 1/t_{Th} = D/N^2$. In order to obtain the finite volume criterion for localization we must get rid of ϵ in the Green's function. The idea is that ϵ can be dropped if it is less than $\Delta(E, N)$. To see this one expands $|G(E; 0, N)|$ about $|G(E_\epsilon; 0, N)|$. This can be controlled with the help of (23.3.3). The expansion converges provided that the norm of $\epsilon\, G(E_\epsilon)$ is small. Note that the norm of $G(E)$ is typically given by the inverse of the eigenvalue separation. Thus if $\mathbf{C}$ is small one can be argued that with very high probability $|G(E; 0, N)|$ is small. This gives us a rigorous finite volume criterion for localization. See Chapter 3 of W-M Wang's thesis [Wan92] and also [Ger04] for details of this discussion in the case of RS.

23.5.4 Power law RBM

In 1D, consider the case when J_{ij} decays like $(1 + |i - j|/W)^{-p}$ so W plays a somewhat different role than it did in earlier sections. The motivation for studying this class of matrices also comes from models of quantum chaos. The work of [Mir96, Mir00b] predicts that when $p = 2$ this system is critical in the sense that if $p > 2$ there is power law localization where as for $p < 2$ the eigenvectors are delocalized. At $p = 2$, multi-fractal exponents d_q are predicted for the eigenvectors:

$$I_q = \left\langle \sum |\psi_j|^{2q} \right\rangle \approx N^{-d_q(q-1)}. \tag{23.5.7}$$

The fractal exponents d_q depend on a one parameter family related to W. The eigenvalue spacing statistics in this model look like a hybrid of Poisson and GOE statistics. There is earlier work of Levitov [Lev90] who studied related problems in 3D using a real space renormalization analysis on pairs of resonances. Recently in [Fyo09], the Anderson transition has been studied for an ultrametric analogue of power law random matrices. We refer to [Eve08] for a review of multi-fractal statistics at the Anderson transition.

23.6 Random sparse matrices

This section will describe some spectral properties of a class of N by N, symmetric, sparse random matrices as N becomes large. These matrices will be indexed by $[1, 2, \dots N] \equiv I_N$. Unlike the random band case, the geometry of the lattice $\mathbb{Z}$ will not be reflected in these models but instead the geometry of the underlying graph is central. Sparse matrices have the property that for a fixed i, there are typically only a finite set of j for which $H_{ij} \neq 0$. We shall primarily consider ensembles associated with the adjacency matrix of random graphs whose vertices are labelled by I_N. For $i \geq j$ let R_{ij} denote independent identically distributed random variables and A_{ij} the adjacency matrix of a random graph and define H to be the symmetric matrix

$$H_{ij} = A_{ij} R_{ij}. \tag{23.6.1}$$

We focus on two families of random graphs:

(A) Unordered pairs i, j from I_N are selected independently with probability p/N. The resulting class of graphs with these edges, $G(N, p/N)$, is called Erdös-Rényi graphs and their connectivity properties have been extensively studied.

(B) For $k \in \mathbb{Z}^+$, let $RG_N(k)$ denote the set of all k-regular graphs on I_N. By k-regular we mean that each vertex has exactly k edges touching it. Each graph in $RG_N(k)$ is given equal weight. Regular graphs often arise as Cayley graphs of families of discrete groups such as $SL(2, \mathbb{Z}_q)$ with q prime. Given a k-regular graph, set $-\Delta = kI - A$ where A is its adjacency matrix. The second eigenvalue of $-\Delta$ plays an important role in computer science, [Hoo06].

If $p > 1$, then with probability 1, graphs in $G(N, p/N)$ have single large component of size proportional to N as $N \to \infty$. On the other hand if $p < 1$ components are of cardinality $O(\ln N)$. In this case eigenstates of H are clearly supported on a set of size $O(\ln N)$. The density of states for these ensembles was studied by Rodgers and Bray, [Rod88] as $N \to \infty$ via replica methods. These results were corroborated by [Fyo91b] using a SUSY saddle analysis. Both of these papers show that the DOS is the solution to a highly non-linear integral equation which cannot be explicitly solved. Numerical simulations as well as theoretical considerations show that for fixed p the DOS has spikes (especially near 0) as well as regular components. See [Kho04, Kuh08] for some recent developments. As p gets large, the DOS approaches the Wigner semicircle law. The behaviour of the eigenstates and the level spacing statistics were examined numerically in [Eva92] and via SUSY in [Mir91a]. If $p < p_q \approx 1.4$, all eigenvectors appear localized whereas for $p > p_q$ extended eigenvectors (23.5.3) appear. As noted above, classically the transition occurs at $p = 1$. Some localized eigenvectors may also coexist with extended states even for $p \geq 2$. The level statistics look like those of the GOE ensemble at energies where extended states are present.

For random k-regular graphs, the density of states of the adjacency matrix A is given by the Kesten-McKay formula [McK81]:

$$\rho(E) = \frac{k[4(k-1) - E^2]^{1/2}}{2\pi(k^2 - E^2)} \tag{23.6.2}$$

for $|E| \leq 2(k-1)^{1/2}$. Numerical simulations [Jak89] show that the eigenvalue statistics approximately match those of GOE. The correlations of a typical eigenvector ψ_j at different vertices j, has been investigated in [Elo08] and found to be approximately Gaussian. See [Ore09] for a study of spectral properties of random adjacency matrices via trace formulae and random walks. SUSY or replicas cannot be easily applied because of the complicated character of the randomness in $RG_N(k)$. Sodin [Sod09] has proved that for $3 \leq k \ll N^{2/3}$ and $R_{ij} = \pm 1$ the eigenvalues of H near the spectral edges are given statistically by an Airy point process with corresponding extended eigenvectors.

The sparse random graphs we have introduced have the property that locally these graphs have a tree-like structure. If v is a vertex then a ball centred at v of radius $\log N$ looks like a tree with high probability. This fact is used to calculate the Kesten-MacKay distribution. The non-linear equations appearing in [Mir91b, Eva92] are closely related to the study of a random potential on the Cayley tree.

Recent work of Erdős et al., [Erd11b], proves universality of the bulk eigenvalue spacing for the adjacency matrix of a random Erdős-Renyi graph $G(N, p/N)$ provided that $p(N)/N \gg N^{-1/3}$. Thus the local spacing coincides with that for GOE.

23.7 Random Schrödinger on the Bethe lattice

We conclude this review with a summary of mathematical results for the random Schrödinger operator $H = A + \lambda v_j$ on the Cayley tree or Bethe lattice $\mathbb{B}$. As before, A is the adjacency matrix. The Bethe lattice is an infinite tree graph for which every vertex has fixed degree $k \geq 3$. H acts on $\ell_2(\mathbb{B})$, the Hilbert space of square integrable functions on the Bethe lattice. This system was first investigated in [Abo73, Abo74] as a self-consistent model of localized and extended states. There are numerous investigations of the transition for this model in the physics literature, including [Efe85, Zir86, Mir91b]. These studies analyse an associated SUSY model. See also [Kun83]. For small λ localized states are predicted for $|E| > k$. This was later proved in [Aiz94]. A. Klein, [Kle98], proved the existence of an absolutely continuous spectrum for energies inside $[-2\sqrt{k-1}, 2\sqrt{k-1}]$, provided λ is small. See also [Aiz06b]. In particular, there are no localized states in this energy range. In [Kle96], Klein proves that the mean square displacement of a solution to the equation $-i\partial_t\psi(j,t) = H\psi(j,t)$ is ballistic:

$$\sum_j |\psi|^2(j,t)|j|^2 \approx Ct^2. \tag{23.7.1}$$

For RBM with W large or RS on $\mathbb{Z}^d$ with $d > 2$ and λ small, the motion is expected to be diffusive, i.e. the right side is replaced by Dt.

Recently, Aizenman and Warzel [Aiz06b] investigated the eigenvalue statistics of H on an increasing sequence of finite sub-trees the Bethe lattice. In a range of energies for which there are no localized states on $\ell_2(\mathbb{B})$, the eigenvalue statistics were proven to be Poisson. The explanation for this seemingly surprising result is that the associated operator on the canopy graph (the graph as seen from the boundary) has localized eigenvectors. Aizenman and Warzel, [Aiz11], have recently proved that for weak disorder, H has absolutely continuous spectrum in the interval $2\sqrt{k-1} < |E| < k$. Such energies lie in the Lifshitz tail where the density of states is very small.

Acknowledgments

I would like to thank Yan Fyodorov and Sasha Sodin for very helpful discussions and for their many comments on preliminary drafts of this review. Thanks also to Margherita Disertori and the referee for their suggestions and corrections.

References

[Abo73] R. Abou-Chacra, P. Anderson and D. Thouless, J. Phys. C **6** (1973):1734.
[Abo74] R. Abou-Chacra and D. Thouless, J. Phys. C **7** (1974):65.

[Abr79] E. Abrahams, P. W. Anderson, D. C. Licciardello, and T. V. Ramakrishnan, Phys. Rev. Lett. **42** (1979):673.
[Aiz94] M. Aizenman, Rev. Math. Phys. **6** (1994):1163.
[Aiz01] M. Aizenman, J.H. Schenker, R.H. Friedrich and D. Hundertmark: Commun. Math. Phys. **224** (2001):219.
[Aiz06a] M. Aizenman and S. Warzel, Math. Phys. Anal. Geom. **9** (2006):291.
[Aiz06b] M. Aizenman, R. Sims and S. Warzel, Probab. Theory Relat. Fields **136** (2006):363.
[Aiz11] M. Aizenman and S. Warzel, Phys. Rev. Lett. **106**, 136804(2011).
[And06] G. Anderson and O. Zeitouni, Probab. Theory Relat. Fields **134** (2006):283.
[Bai03] Z. Bai and Y. Yin, Annals of Probability, **21** (1993):1275.
[Bog91] L. Bogachev, S. Molchanov and L. Pastur, Matematicheskie Zametki **50** (1991):31.
[Cas90] G. Casati, F. Izrailev and L. Molinari, Phys. Rev. Lett. **64** (1990):1851.
[Cas91] G. Casati, F. Izrailev and L. Molinari, J. Phys. A: Math. Gen. (1991):4755.
[Cha09] S. Chatterjee, Probab. Theory Relat. Fields **143** (2009):1.
[Con87] F. Constantinescu, G. Felder, K. Gawedzki and A. Kupiainen, J. of Stat. Phys. **48** (1987):365.
[Cos95] O. Costin and J. Lebowitz, Phys. Rev. Lett. **75** (1995):69.
[Dis02] M. Disertori, H. Pinson and T. Spencer, Commun. Math. Phys. **232** (2002):83.
[Dis04] M. Disertori, Rev. in Math. Phys. **16** (2004):1191.
[Dis09a] M. Disertori, T. Spencer, M. R. Zirnbauer, Commun. Math. Phys. **300** (2010) 435–486.
[Dis09b] M. Disertori, T. Spencer, Commun. Math. Phys. **300** (2010) 659–671.
[Dre89] H. von Dreifus and A. Klein, Commun. Math. Phys. **124** (1989):285.
[Edw72] J.T. Edwards and D.J. Thouless, J. Phys. C **5** (1972):807.
[Efe83] K. B. Efetov, Adv. in Phys. **32** (1983):53.
[Efe85] K.B. Efetov Sov. Phys. JETP **61** (1985):606.
[Efe97] K. B. Efetov, *Supersymmetry in Disorder and Chaos*, Cambridge University Press, Cambridge 1997.
[Elo08] Y. Elon, Phys. A, **41** (2008):435203.
[Erd09] L. Erdös, B. Schlein, H-T Yau, Commun. Math. Phys. **287** (2009):641.
[Erd11a] L. Erdös and A. Knowles, Commun. Math. Phys. **303** (2011):59.
[Erd11b] L. Erdös, A. Knowles, H.T. Yau, J. Yin, Commun. Math. Phys., **314** (2012) 587–640.
[Eva92] S. Evangelou, J. Stat. Phys. **69** (1992):361.
[Eve00] F. Evers and A. Mirlin, Phys. Rev. Lett. **84** (2000):3690.
[Eve08] F. Evers and A. Mirlin, Rev. Mod. Phys. **80** (2008) 1355; arXiv:0707.4378 (2007).
[Fra95] K. Frahm and A. Müller-Groeling, Europhys. Lett. **32** (1995):385.
[Fro83] J. Fröhlich and T. Spencer, Commun. Math. Phys. **88** (1983):151.
[Fyo91a] Y. V. Fyodorov and A. D. Mirlin, Phys. Rev. Lett. **67** (1991):2405.
[Fyo91b] Y. V. Fyodorov and A. D. Mirlin, J. Phys. A: Math. Gen. **24** (1991):2219.
[Fyo94] Y. V. Fyodorov and A.D. Mirlin, Int. J. Mod. Phys. B, **8** (1994):3795.
[Fyo95] Y.V. Fyodorov and A.D. Mirlin, Phys. Rev. **B52** (1995):R11580.
[Fyo96] Y. V. Fyodorov, O. Chubykalo, F. Izrailev, and G. Casati, Phys. Rev. Lett. **76** (1996):1603.
[Fyo09] Y. V. Fyodorov, A. Ossipov and A. Rodriguez, J. Stat. Mech. (2009): L 12001.
[Ger04] F. Germinet and A. Klein, Duke Math. J. **124** (2004):309.
[Gui02] A. Guionnet, Ann. I. H. Poincaré, PR**38** (2002):341.
[Hoo06] S. Hoory, N. Linial and A. Wigderson, Bull. Am. Math. Soc. **43** (2006):439.
[Jak89] D. Jakobson, S.D. Miller, I. Rivin, Z. Rudnick, Emerging Applications of Number Theory, The IMA Volumes in Mathematics and its Applications **109** (1999) 317–327.
[Jan99] M. Janssen and K. Pracz, Phys. Rev. E **61** (2000):6278.
[Joh98] K. Johansson, Duke Math. J. **91** (1998):151.

[Jon82] D. Jonsson, J. Mult. Anal. **12** (1982):1.
[Kho02] A. Khorunzhy and W. Kirsch, Commun. Math. Phys. **231** (2002):223.
[Kho04] A. Khorunzhy [O. Khorunzhy], M. Shcherbina and V. Vengerovsky, J. Math. Phys. **45**, (2004):1648.
[Kho93] A. M. Khorunzhy and L. A. Pastur, Comm. Math. Phys. **153** (1993):605.
[Kle96] A. Klein, Commun. Math. Phys. **177** (1996):755.
[Kle98] A. Klein, Adv. in Math. **133** (1998):163.
[Kuh08] R. Kühn, J. Phys. A, **41** (2008):295002.
[Kun83] H. Kunz and B. Souillard, J.Phys (Paris) **44** (1983):L411.
[Lev90] L. Levitov, Phys. Rev. Lett. **64** (1990):547.
[McK81] B. McKay, Lin. Alg. and Its Appl. **40** (1981):203.
[Mir91a] A. D. Mirlin and Y. V. Fyodorov, J. Phys. A **24** (1991):2273.
[Mir91b] A. D. Mirlin and Y.V. Fyodorov, Nuclear Phys. B **366** (1991):507.
[Mir96] A. D. Mirlin, Y. V. Fyodorov, F.-M. Dittes, J. Quezada, T. H. Seligman, Phys. Rev. E **54** (1996):3221.
[Mir00a] A. D. Mirlin, 'New Directions in Quantum Chaos' Eds. G.Casati, I.Guarneri, and U.Smilansky (IOS Press, Amsterdam, 2000), pp.223; arXiv:0006421 (2000).
[Mir00b] A. D. Mirlin and F. Evers, Phys. Rev. B **62** (2000):7290.
[Mol92] S. Molchanov, L. Pastur, and A. Khorunzhy, Teor. Mat. Fiz. **90** (1992):108.
[Opp79] R. Opperman and F. Wegner, Z. Phys. B **34** (1979):327.
[Ore09] I. Oren, A. Godel, and U. Smilansky, ArXiv:0908.3944
[Rod88] G. Rodgers and A. Bray, Phys. Rev. B **37** (1998):3557.
[Sch80] L. Schaefer and F. Wegner, Z. Phys. B **38** (1980):113.
[Sch09] J. Schenker, Comm. Math. Phys. **290** (2009):1065, arXiv:0809.4405 (2008).
[Sel85] T. Seligman, J. Verbaarschot and M. Zirnbauer, J. Phys. A: Math. Gen. **18** (1985):2751.
[She94] D.L. Shepelyansky, Phys.Rev.Lett. **73** (1994):2607.
[Sil97] P.G. Silvestrov, Phys. Rev. E, **55** (1997):6419.
[Sin98] Y. Sinai and A. Soshnikov, Funct. Anal. Appl. **32** (1998):114.
[Sod10] S. Sodin, Annals of Math. 172 (2010):2223.
[Sod09] S. Sodin, J. Stat. Phys. 136 (2009):834.
[Sos99] A. Soshnikov, Commun. Math. Phys., **207** (1999):697.
[Spe88] T. Spencer, J. Stat. Phys. **51** (1988):1009.
[Spe04] T. Spencer and M.R. Zirnbauer, Comm. Math. Phys. **252** (2004):167.
[Tho77] D.J. Thouless, Phys. Rev. Lett. **39** (1977):1167.
[Wan92] W-M Wang, Princeton University PhD. Thesis 1992.
[Weg79a] F. Wegner, Z. Phys. B **35** (1979):209.
[Weg79b] F. Wegner, Phys. Rev. B **19** (1979):783.
[Weg81] F. Wegner, Z. Phys. B **44** (1981):9.
[Wig55] E. Wigner, Ann. Math. **62** (1955):548.
[Wig57] E. Wigner, Ann. Math. **65** (1957):203.
[Zir86] M.R. Zirnbauer, Phys. Rev. B **34** (1986):6394.
[Zir04] M.R. Zirnbauer, arXiv:0404057 (2004).

PART III
Applications of random matrix theory

·24·

Number theory

J. P. Keating and N. C. Snaith

Abstract

We review some of the connections between random matrix theory and number theory. These include modelling the value distributions of the Riemann zeta function and other L-functions, and the statistical distribution of their zeros.

24.1 Introduction

One of the more surprising recent applications of random matrix theory (RMT) has been to problems in number theory. There it has been used to address seemingly disparate questions, ranging from modelling mean and extreme values of the Riemann zeta function to counting points on curves. In order to cover the many aspects of this rapidly developing area, we will focus on a selection of key ideas and list specific references where more information can be found. In particular, we direct readers who seek a more in-depth overview to the book of lecture notes edited by Mezzadri and Snaith [Me05], or previous reviews [Con01, KS03, Kea05].

24.2 The number theoretical context

Although the applications of random matrix theory (RMT) to number theory appear very diverse, they all have one thing in common: L-functions. The statistics of the critical zeros of these functions are believed to be related to those of the eigenvalues of random matrices.

The most well-known L-function is the Riemann zeta function

$$\zeta(s) = \sum_{n=1}^{\infty} \frac{1}{n^s} = \prod_p (1 - 1/p^s)^{-1}, \quad \mathrm{Re}(s) > 1. \tag{24.2.1}$$

The above expressions allow for continuation to a meromorphic function on the whole complex plane, with a pole at $s = 1$ and a symmetry defined by the functional equation

$$\zeta(s) = \pi^{s-\frac{1}{2}} \frac{\Gamma(\frac{1}{2} - \frac{1}{2}s)}{\Gamma(\frac{1}{2}s)} \zeta(1-s), \tag{24.2.2}$$

or, equivalently,

$$\tilde{\zeta}(s) = \pi^{-s/2}\Gamma(s/2)\zeta(s) = \tilde{\zeta}(1-s). \tag{24.2.3}$$

The critical line $\mathrm{Re}(s) = 1/2$, about which this formula effectively reflects, is where the Riemann hypothesis places all the complex zeros of $\zeta(s)$.

The L-functions we will discuss share similar properties with the Riemann zeta function. They have a more general Dirichlet series representation,

$$L(s) = \sum_{n=1}^{\infty} \frac{a_n}{n^s}, \tag{24.2.4}$$

and also a representation as an Euler product over the primes. Each L-function then has an analytic continuation that extends it to a meromorphic function in the whole complex plane, a functional equation, and a generalized Riemann hypothesis that puts the non-trivial zeros of the L-function on a critical (vertical) line in the complex plane. (See, for example, [Con05] for a short introduction to these properties.)

We shall, in the remainder of this article, assume the truth of the generalized Riemann hypothesis. This is not strictly necessary, but it makes the connections with random matrix theory more transparent.

24.3 Zero statistics

Denoting the nth zero up the critical line by $\rho_n = 1/2 + i\gamma_n$, there is a great deal of interest in studying the statistical distribution of the heights γ_n. The Riemann zeros grow logarithmically more dense with height up the critical line, so defining

$$w_n = \gamma_n \frac{1}{2\pi} \log \tfrac{|\gamma_n|}{2\pi} \tag{24.3.1}$$

we have

$$\lim_{W\to\infty} \frac{1}{W} \# \{w_n < W\} = 1. \tag{24.3.2}$$

Montgomery [Mon73] studied the pair correlation, or two-point correlation function, of these scaled zeros and conjectured that

$$\lim_{N\to\infty} \frac{1}{N} \sum_{1\le n,m\le N} f(w_n - w_m) = \int_{-\infty}^{\infty} f(x)\big(R_2(x) + \delta(x)\big)dx, \tag{24.3.3}$$

where

$$R_2(x) = 1 - \left(\frac{\sin(\pi x)}{\pi x}\right)^2. \qquad (24.3.4)$$

He proved that (24.3.3) is true for $f(x)$ such that

$$\hat{f}(\tau) = \int_{-\infty}^{\infty} f(x)e^{2\pi i x\tau}dx \qquad (24.3.5)$$

has support in $[-1, 1]$.

Dyson pointed out that (24.3.4) coincides with the two-point correlation function of random Hermitian (GUE) and unitary (CUE) matrices in the limit as the matrix size tends to infinity; that is, it coincides with the two-point correlation function of the eigenvalues of matrices averaged with respect to Haar measure on $U(N)$ in the limit $N \to \infty$. Odlyzko [Odl89] later produced striking numerical support for Montgomery's conjecture, and for its generalization to other local statistics of the Riemann zeros.

Montgomery's theorem, that (24.3.3) is true for $\hat{f}(\tau)$ with support in $[-1, 1]$, follows directly from the prime number theorem, via the connection, known as the *explicit formula*, between the Riemann zeros and the primes. (The explicit formula is a consequence of the Euler product and the functional equation.) Montgomery's theorem generalizes to all n-point correlations between the zeros [RS96]. The extension to values of τ outside $[-1, 1]$ requires information about correlations between the primes, as embodied in the Hardy-Littlewood conjecture concerning the number of integers $n < X$ such that both n and $n + h$ are prime. See, for example [BK95, BK96b], where the the Hardy-Littlewood conjecture is used to show heuristically that all n-point correlations of the Riemann zeros coincide with the corresponding CUE/GUE expressions in the large-matrix-size limit.

It is not just in the zero statistics high on the critical line of an individual L-function, such as the Riemann zeta function, that one sees random matrix statistics. Katz and Sarnak [KS99a, KS99b] proposed that the zero statistics of L-functions averaged over naturally defined families behave like the eigenvalues of matrices from $U(N)$, $O(N)$ or $USp(2N)$. Which ensemble of matrices models a given family depends on details of that family.

In many statistics, the eigenvalues of matrices from each of the three compact groups above show the same behaviour in the large matrix limit. For example, the limiting two-point correlation function (24.3.4) is common to all three. However, the orthogonal and unitary symplectic matrices have eigenvalues appearing in complex conjugate pairs, making the points 1 and –1 symmetry points of their spectra. Katz and Sarnak showed that the distribution of the first eigenvalue (or more generally the first few eigenvalues) near to this symmetry point is ensemble specific (i.e. group specific). If we define the distribution of the k-th eigenvalue $e^{i\theta_k}$ of a matrix A varying over $G(N) = U(N),\ O(N)$ or $USp(2N)$,

$$\nu_k(G(N))[a, b] = \text{meas}\{A \in G(N) : \tfrac{\theta_k N}{2\pi} \in [a, b]\}, \tag{24.3.6}$$

then Katz and Sarnak show that the limit

$$\lim_{N\to\infty} \nu_k(G(N)) = \nu_k(G) \tag{24.3.7}$$

exists, but, in contrast to the two-point correlation function, depends on G.

To be more explicit, let $\mathcal{F}$ denote a family of L-functions. An individual L-function within the family is associated to a particular $f \in \mathcal{F}$ – we can write $L(s, f)$ – and has conductor c_f (the conductor often orders the L-functions within the family and it also features in the density of the zeros). If γ_f^j denotes the height up the critical line of the jth zero, then the zeros near $s = 1/2$ (the point corresponding to 1 on the unit circle in the RMT analogy) are normalized to have constant mean spacing by scaling them in the following way:

$$\frac{\gamma_f^j \log c_f}{2\pi}. \tag{24.3.8}$$

Let $\mathcal{F}_X$ denote the members of the family $\mathcal{F}$ with conductor less than X. Then Katz and Sarnak define the distribution of the jth zero above $s = 1/2$ as

$$\nu_j(X, \mathcal{F})[a, b] = \frac{\#\left\{f \in \mathcal{F}_X : \frac{\gamma_f^{(j)} \log c_f}{2\pi} \in [a, b]\right\}}{\#\mathcal{F}_X}. \tag{24.3.9}$$

It is then expected, and Katz and Sarnak provide analytical and numerical evidence for this, that $\nu_j(X, \mathcal{F})$ converges, as $X \to \infty$, to $\nu_j(G)$, where G represents the symmetry type of the family: unitary, orthogonal, or symplectic. Similarly, other statistics of the lowest zeros are also expected, upon averaging over the family, to tend to the corresponding random matrix statistics of the appropriate symmetry type in this limit.

It should be noted that the influence of the symmetry points on non-local statistics may extend much further into the spectrum, even in the large-matrix-size limit. The nature of the transition in various statistics, both local and non-local, has been explored in [KO08].

The question of determining the symmetry type of a given family *a priori* is, in general, a difficult one. The method used by Katz and Sarnak is that for some families of L-functions a related family of zeta functions on finite fields can be defined (see Section 24.6 for further details). In the case of these zeta functions the definition of families is straightforward, the Riemann hypothesis has been proven (in that all zeros lie on a circle) and the symmetry type is determined by the monodromy of the family (see [KS99a]). The symmetry type of the related family of L-functions is then assumed to be the same.

24.4 Values of the Riemann zeta function

In many instances in number theory questions arise involving the values of L-functions on or near the critical line. There is now a substantial literature, starting with [KS00b, KS00a], providing evidence that these values can be modelled using characteristic polynomials of random matrices. The characteristic polynomial is zero at the eigenvalues of the matrix, so for an $N \times N$ unitary matrix A with eigenvalues $e^{i\theta_1}, \ldots, e^{i\theta_N}$ we define

$$\Lambda(s) = \Lambda_A(s) = \det(I - A^* s) = \prod_{n=1}^{N}(1 - se^{-i\theta_n}). \tag{24.4.1}$$

Averaging functions of this characteristic polynomial over one of the classical compact groups is a straightforward calculation using Selberg's integral (see [Meh04], Chapter 17). For example, averaging over the unitary group with respect to Haar measure, moments of the characteristic polynomial [KS00b] are

$$\begin{aligned} M_N(\lambda) := \int_{U(N)} |\Lambda_A(e^{i\theta})|^{2\lambda} dA_{Haar} &= \prod_{j=1}^{N} \frac{\Gamma(j)\Gamma(2\lambda + j)}{(\Gamma(j+\lambda))^2} \\ &= \frac{G(N+1)G(2\lambda + N + 1)G^2(\lambda+1)}{G(2\lambda+1)G^2(\lambda+N+1)}, \end{aligned} \tag{24.4.2}$$

where in the last line the Barnes double gamma function [Bar00] is used to express more compactly products over gamma functions by invoking

$$\begin{aligned} G(1) &= 1, \\ G(z+1) &= \Gamma(z)\, G(z). \end{aligned} \tag{24.4.3}$$

For large N (24.4.2) is asymptotic to

$$M_N(\lambda) \sim \frac{(G(1+\lambda))^2}{G(1+2\lambda)} N^{\lambda^2}. \tag{24.4.4}$$

We compare this to the moments of the Riemann zeta function averaged along the critical line. Number theorists have conjectured that these grow asymptotically like

$$\frac{1}{T}\int_0^T |\zeta(1/2 + it)|^{2\lambda} dt \sim c_\lambda (\log T)^{\lambda^2} \tag{24.4.5}$$

for large T. Numerical evidence [KS00b, CFK$^+$05] supports the conjecture that

$$c_\lambda = a_\lambda \frac{(G(1+\lambda))^2}{G(1+2\lambda)}, \tag{24.4.6}$$

where a_λ is a product over primes

$$a_\lambda = \prod_p \left(1 - \tfrac{1}{p}\right)^{\lambda^2} \sum_{m=0}^{\infty} \left(\frac{\Gamma(m+\lambda)}{m!\Gamma(\lambda)} \right)^2 p^{-m}. \tag{24.4.7}$$

This agrees with the known values, $c_1 = a_1$ and $c_2 = a_2 \frac{2}{4!}$, and those conjectured independently based on number theoretical heuristics, $c_3 = a_3 \frac{42}{9!}$ and $c_4 = a_4 \frac{24024}{16!}$.

In the previous section it was seen in the comparison of Montgomery's pair correlation of the Riemann zeros and Dyson's two-point correlation of eigenvalues of matrices from $U(N)$ that the limit of large height up the critical line, $T \to \infty$, is related to the limit $N \to \infty$. Here we see that in the range where they are large but finite, N plays the role of $\log T$. This is a correspondence that is observed throughout this subject. It has been tested numerically in many situations and is motivated by equating the density of zeros with the density of eigenvalues on the unit circle: $\frac{1}{2\pi} \log \frac{T}{2\pi} = \frac{N}{2\pi}$.

One striking feature of the leading order moment conjecture (24.4.6) is the factorization into an arithmetical component, a_λ, and a random-matrix component coming from (24.4.4). The arithmetical component may be understood by noting that it arises essentially from substituting in (24.4.5) the formula for the zeta function as a prime product, ignoring the fact that this product diverges on the line $\mathrm{Re}s = 1/2$, and interchanging the t-average and the product (i.e. assuming independence of the primes). However, this fails to capture the random-matrix component. Alternatively, the random-matrix component may be understood by substituting for the zeta function its (Hadamard) representation as a product over its zeros and assuming that the zeros are distributed like CUE eigenvalues *over the range of scales that contribute to this product*. This fails to capture the arithmetical component, essentially because the moments are not local statistics.

The resolution of this problem comes from using a 'hybrid' representation of the zeta function:

$$\zeta(s) = P_X(s) Z_X(s)(1 + o(1)), \tag{24.4.8}$$

where $P_X(s)$ behaves like the prime product for the zeta function truncated smoothly so that only primes $p < X$ contribute to it, and $Z_X(s)$ behaves like the Hadamard product over the zeros of the zeta function, truncated smoothly so that only zeros within a distance of the order of $1/\log X$ from s contribute [GHK07]. If X is not too large compared to T, the moments of $P_X(1/2 + it)$ can be determined rigorously in terms of a_λ. The moments of $Z_X(1/2 + it)$ may be modelled by averaging correspondingly truncated characteristic polynomials over $U(N)$ as $N \to \infty$. The RMT average can be computed in this limit using the asymptotic properties of Toeplitz determinants [Bas78]. Assuming that the moments of $\zeta(1/2+it)$ split into a product of the moments of $P_X(1/2+it)$ and those of $Z_X(1/2+it)$, the dependence on X drops out, leading directly to (24.4.6) [GHK07].

There is a specific interest in the moments when λ is a positive integer, k. On the random matrix theory side, the $2k$th moment is then a polynomial in N of degree k^2, as can be seen from (24.4.2). This polynomial can be represented as a contour integral (the equality follows by evaluating the integral using Cauchy's theorem):

$$\begin{aligned}\int_{U(N)} |\Lambda_A(e^{i\theta})|^{2k} dA_{Haar} &= \prod_{j=0}^{k-1}\left(\frac{j!}{(k+j)!}\prod_{i=1}^{k}(N+i+j)\right)\\ &= \frac{(-1)^k}{k!^2}\frac{1}{(2\pi i)^{2k}}\oint\cdots\oint \frac{G(z_1,\ldots,z_{2k})\Delta^2(z_1,\ldots,z_{2k})}{\prod_{j=1}^{2k} z_j^{2k}}\\ &\quad\times e^{\frac{1}{2}N\sum_{j=1}^{k} z_j - z_{k+j}}\,dz_1\cdots dz_{2k},\end{aligned} \tag{24.4.9}$$

where the Vandermonde determinant is defined as

$$\Delta(x_1,\ldots,x_n) = \prod_{1\le j<\ell\le N}(x_\ell - x_j), \tag{24.4.10}$$

we define

$$G(z_1,\ldots,z_{2k}) = \prod_{i=1}^{k}\prod_{j=1}^{k}(1-e^{-z_i+z_{j+k}})^{-1}, \tag{24.4.11}$$

and the integration is on small contours around the origin. See [BH00] for earlier examples of such contour integrals for Hermitian matrices.

The final equality in (24.4.9) is directly analogous to a similar contour integral which is conjectured to give the full main term of the corresponding moment of the Riemann zeta function:

$$\begin{aligned}\frac{1}{T}\int_0^T |\zeta(1/2+it)|^{2k}dt &= \frac{1}{T}\int_0^T \frac{(-1)^k}{k!^2}\frac{1}{(2\pi i)^{2k}}\\ &\quad\times\oint\cdots\oint\frac{G_\zeta(z_1,\ldots,z_{2k})\Delta^2(z_1,\ldots,z_{2k})}{\prod_{j=1}^{2k} z_j^{2k}}\\ &\quad\times e^{\frac{1}{2}\log\frac{t}{2\pi}\sum_{j=1}^{k} z_j - z_{k+j}}\,dz_1\ldots dz_{2k}\,dt + o(1),\end{aligned} \tag{24.4.12}$$

where

$$G_\zeta(z_1,\ldots,z_{2k}) = A_k(z_1,\ldots,z_{2k})\prod_{i=1}^{k}\prod_{j=1}^{k}\zeta(1+z_i-z_{k+j}), \tag{24.4.13}$$

and A_k is another Euler product which is analytic in the regions we are interested in (see [CFK+08] for more on the development of these integral formulae to moments with non-integer values of λ). A close inspection of the residue structure of the contour integral in (24.4.12) shows that it is a polynomial of degree k^2 in $\log \frac{t}{2\pi}$, in analogy with the random matrix case.

The integral expressions in (24.4.9) and (24.4.12) are remarkably similar. We note the identical role played by N in the random matrix formulae and $\log T$ in the number theory expression. Also, to a large extent the behaviour of (24.4.9) and (24.4.12) are dominated by the poles of G and G_ζ, which are the same due to $(1-e^{-x})^{-1}$ having a simple pole with residue 1 at $x=0$ exactly as does $\zeta(1+x)$.

These contour integral formulae for moments generalize to averages of the forms

$$\int_0^T \prod_{\alpha\in A}\zeta(\tfrac{1}{2}+it+\alpha)\prod_{\beta\in B}\zeta(\tfrac{1}{2}-it+\beta)dt, \tag{24.4.14}$$

[CFK+05] and

$$\int_0^T \frac{\prod_{\alpha\in A}\zeta(\frac{1}{2}+it+\alpha)\prod_{\beta\in B}\zeta(\frac{1}{2}-it+\beta)}{\prod_{\gamma\in C}\zeta(\frac{1}{2}+it+\gamma)\prod_{\delta\in D}\zeta(\frac{1}{2}-it+\delta)}dt, \tag{24.4.15}$$

with $\mathrm{Re}\gamma, \mathrm{Re}\delta > 0$ [CFZ08]. Again there is a striking similarity with corresponding formulae for the characteristic polynomials of random matrices [CFK+03, CFZ]. A remarkable amount of information can be extracted from the resulting formulae about statistics of zeros and values of the zeta function and its derivatives [CS07]. For example, the ratios conjecture with two zeta functions in the numerator and two in the denominator yields a formula, first written down by Bogomolny and Keating [BK96a], that includes all the significant lower-order terms for the two-point correlation function for which Montgomery conjectured the limiting ($T\to\infty$) form:

$$\begin{aligned}\sum_{\gamma,\gamma'\le T} f(\gamma-\gamma') = \frac{1}{(2\pi)^2}\int_0^T \Bigg(2\pi f(0)\log\frac{t}{2\pi} + \int_{-T}^{T} f(r)\Bigg(\log^2\frac{t}{2\pi} \\ +2\Bigg(\left(\frac{\zeta'}{\zeta}\right)'(1+ir) + \left(\frac{t}{2\pi}\right)^{-ir}\zeta(1-ir)\zeta(1+ir)A(ir) \\ - B(ir)\Bigg)\Bigg)dr\Bigg)dt + O(T^{1/2+\epsilon});\end{aligned} \tag{24.4.16}$$

here the integral is to be regarded as a principal value near $r=0$,

$$A(\eta) = \prod_p \frac{(1-\frac{1}{p^{1+\eta}})(1-\frac{2}{p}+\frac{1}{p^{1+\eta}})}{(1-\frac{1}{p})^2}, \tag{24.4.17}$$

and

$$B(\eta) = \sum_p \left(\frac{\log p}{(p^{1+\eta} - 1)} \right)^2 . \qquad (24.4.18)$$

The limit coincides with random matrix theory (taking $T \to \infty$ in (24.4.16) with $f(x)$ replaced by $g\big(x(\log \frac{T}{2\pi})/(2\pi)\big)\big(\frac{T}{2\pi} \log \frac{T}{2\pi}\big)$ reproduces Montgomery's conjecture (24.3.3)), but the lower-order contributions contain important arithmetic information specific to the Riemann zeta function. See [BK99] for further details and numerical illustrations of the importance of these lower-order terms and [CS08] for a generalization to all n-point correlations.

24.5 Values of L-functions

Many applications in number theory require one to consider mean values of L-functions evaluated at the critical point (the centre of the critical strip – where the critical line crosses the real axis) and averaged over a suitable family. Two examples of such families are given below and more details can be found in, for example, [CF00, KS00a, CFK$^+$05].

The Dirichlet L-functions associated to real, quadratic characters form a family with symplectic symmetry:

$$L(s, \chi_d) = \sum_{n=1}^{\infty} \frac{\chi_d(n)}{n^s} = \prod_p \left[1 - \chi_d(p) p^{-s}\right]^{-1}, \qquad (24.5.1)$$

where $\chi_d(n) = \left(\frac{d}{n}\right)$ is Kronecker's extension of Legendre's symbol which is defined for p prime,

$$\left(\frac{d}{p}\right) = \begin{cases} +1 & \text{if } p \nmid d \text{ and } x^2 \equiv d (\operatorname{mod} p) \text{ is soluble} \\ \ \ 0 & \text{if } p | d \\ -1 & \text{if } p \nmid d \text{ and } x^2 \equiv d (\operatorname{mod} p) \text{ is not soluble.} \end{cases} \qquad (24.5.2)$$

The character χ_d exists for all fundamental discriminants d, and the L-functions attached to these characters are said to form a family as we vary d. The family can be partially ordered by the *conductor* $|d|$. The symplectic symmetry type can be seen in the statistics of the zeros, studied numerically by Rubinstein [Rub01] and theoretically in the limit of large d by Özlük and Snyder [ÖS99].

The symplectic symmetry type also determines the behaviour of the moments of these L-functions. A moment conjecture similar to that leading to (24.4.12) can be applied here to obtain a conjectural contour integral formula [CFK$^+$05] which implies:

$$\sum_{|d|\leq D}^{*} L(\tfrac{1}{2},\chi_d)^k = \frac{6}{\pi^2} D\, \mathcal{Q}_k(\log D) + O(D^{\frac{1}{2}+\varepsilon}), \tag{24.5.3}$$

where $\sum^*$ is over fundamental discriminants and so the sum is over all real, primitive Dirichlet characters of conductor up to D. Here $\mathcal{Q}_k$ is a polynomial of degree $k(k+1)/2$, with leading coefficient $f_k a_k$, where

$$a_k = \prod_p \frac{(1-\frac{1}{p})^{\frac{k(k+1)}{2}}}{1+\frac{1}{p}} \left(\frac{(1-\frac{1}{\sqrt{p}})^{-k} + (1+\frac{1}{\sqrt{p}})^{-k}}{2} + \frac{1}{p} \right) \tag{24.5.4}$$

and

$$f_k = \prod_{j=1}^{k} \frac{j!}{(2j)!}. \tag{24.5.5}$$

The crucial point to note is the degree of the polynomial $\mathcal{Q}_k$, which is $k(k+1)/2$, mirroring RMT calculations with matrices from $USp(2N)$ [KS00a]:

$$\int_{USp(2N)} \Lambda_A(1)^k dA_{Haar} = \left(2^{k(k+1)/2} \prod_{j=1}^{k} \frac{j!}{(2j)!} \right) \prod_{1\leq i\leq j\leq k} (N + \tfrac{i+j}{2}). \tag{24.5.6}$$

This is to be compared to the corresponding degree, k^2, for the unitary group. Also, in this case equating the density of zeros near the point 1/2 with the density of eigenvalues gives an equivalence $2N = \log|d|$, for large $|d|$ and N. The main term of conjecture (24.5.3) for $k = 1, 2, 3$ has been proved, see [GV79, Jut81, Sou00].

As for the Riemann zeta function, there is a contour integral expression for shifted moments and averages of ratios. To give a specific example [CS07] (where the two terms below result from residues of the contour integral formula):

$$\begin{aligned}\sum_{d\leq X} \frac{L(1/2+\alpha,\chi_d)}{L(1/2+\gamma,\chi_d)} = \sum_{d\leq X} \Bigg(& \frac{\zeta(1+2\alpha)}{\zeta(1+\alpha+\gamma)} A_D(\alpha;\gamma) \\ & + \left(\frac{d}{\pi}\right)^{-\alpha} \frac{\Gamma(1/4-\alpha/2)}{\Gamma(1/4+\alpha/2)} \frac{\zeta(1-2\alpha)}{\zeta(1-\alpha+\gamma)} A_D(-\alpha;\gamma) \Bigg) + O(X^{1/2+\epsilon}),\end{aligned} \tag{24.5.7}$$

where

$$A_D(\alpha;\gamma) = \prod_p \left(1 - \frac{1}{p^{1+\alpha+\gamma}}\right)^{-1} \left(1 - \frac{1}{(p+1)p^{1+2\alpha}} - \frac{1}{(p+1)p^{\alpha+\gamma}}\right) \tag{24.5.8}$$

From this conjecture one can determine the one-level density of the zeros of these Dirichlet L-functions and see the dependence of the lower-order terms on arithmetical quantities (the Riemann zeta function and products over primes) in exactly the same way as for the two-point correlation function (24.4.16):

$$\sum_{d\leq X}\sum_{\gamma_d} f(\gamma_d) = \frac{1}{2\pi}\int_{-\infty}^{\infty} f(t)\sum_{d\leq X}\left(\log\frac{d}{\pi} + \frac{1}{2}\frac{\Gamma'}{\Gamma}(1/4+it/2)\right.$$
$$+\frac{1}{2}\frac{\Gamma'}{\Gamma}(1/4-it/2) + 2\left(\frac{\zeta'(1+2it)}{\zeta(1+2it)} + A'_D(it;it)\right.$$
$$\left.\left.-\left(\frac{d}{\pi}\right)^{-it}\frac{\Gamma(1/4-it/2)}{\Gamma(1/4+it/2)}\zeta(1-2it)A_D(-it;it)\right)\right)dt + O(X^{1/2+\epsilon}).$$
(24.5.9)

Here $A'_D(it;it) = \frac{\partial}{\partial\alpha}A_D(\alpha;\gamma)\big|_{\alpha=\gamma=it}$.

If the zeros γ_d in (24.5.9) are scaled by their mean density then in the limit as $X\to\infty$ the expression multiplying $f(t)$ in the integrand reduces to $1-[\sin(2\pi t)/(2\pi t)]^2$, which is the limiting one-level density of the group of unitary symplectic matrices $USp(2N)$. This is the limiting behaviour predicted by Katz and Sarnak (see Section 24.3). However, one can see from (24.5.9) that the approach to the limit is determined by arithmetical structure.

A family with zero statistics corresponding to an orthogonal family of matrices is that of elliptic curve L-functions. An elliptic curve is represented by an equation of the form $y^2 = x^3 + ax + b$, where a and b are integers and $-16(4a^3+27b^2)\neq 0$ which ensures that the curve is nonsingular. Let

$$L_E(s) = \sum_{n=1}^{\infty}\frac{a(n)}{n^s}, \tag{24.5.10}$$

where $a(p) = p+1-\#E(\mathbb{F}_p)$ ($\#E(\mathbb{F}_p)$ being the number of points on E counted over $\mathbb{F}_p$), be the L-function associated with an elliptic curve E. A family can be created by 'twisting' this L-function by χ_d and varying d:

$$L_E(s,\chi_d) = \sum_{n=1}^{\infty}\frac{a(n)\chi_d(n)}{n^s}. \tag{24.5.11}$$

This family is then ordered by d in exactly the same way as the symplectic family above (which, in fact, could be thought of as 'twisting' the Riemann zeta function by the real, quadratic characters). Similar calculations to those described for the family of Dirichlet L-functions and for the Riemann zeta function can also be carried out in this case. In particular, one sees once again the symmetry type, in this case orthogonal, dictating the form of the moments. The moments of characteristic polynomials of matrices from $SO(2N)$ are [KS00a]

$$\int_{SO(2N)}\Lambda_A(1)^k dA_{Haar} = \left(2^{k(k-1)/2}\,2^k\prod_{j=1}^{k-1}\frac{j!}{(2j)!}\right)\prod_{0\leq i<j\leq k-1}(N+\tfrac{i+j}{2}),$$
(24.5.12)

which is a polynomial in N of degree $k(k-1)/2$. From the elliptic curve family one can select the L-functions which have an even functional equation (forcing even symmetry of the zeros on the critical line around the point $1/2$ – in analogy with selecting from $O(2N)$ the matrices $SO(2N)$ which have even symmetry of their eigenvalues around the point 1 on the unit circle). The average value of the kth power of these L-functions, with $|d| < D$, approximates a polynomial in $\log D$ of degree $k(k-1)/2$.

In this case the moment conjectures inspired by RMT have very significant applications to ranks of elliptic curves; that is, to the long-standing and important question of how likely it is that a given elliptic curve has a finite as opposed to an infinite number of rational solutions (points) [CKRS02, CKRS06]. The connection comes via the conjecture of Birch and Swinnerton-Dyer, which relates this number to the value of L_E at the centre of the critical strip. Random matrix theory predicts the value distribution of L_E at this point and hence the asymptotic fraction of curves with $d < X$ having an infinite number of solutions.

24.6 Further areas of interest

Many of the connections between RMT and number theory are still somewhat speculative, being a mixture of theorems proved in a limited class of cases, numerical computations, and heuristic arguments. The area in which they have the most solid foundations is that relating to zeta functions of curves over finite fields. Let $\mathbb{F}_q$ denote a finite field with q elements and $\mathbb{F}_q[x]$ polynomials $f(x)$ with coefficients in $\mathbb{F}_q$. A polynomial $f(x)$ in $\mathbb{F}_q[x]$ is *reducible* if $f(x) = g(x)h(x)$ with $\deg(g) > 0$ and $\deg(h) > 0$, and *irreducible* otherwise. One defines an extension field $\mathbb{E}$ of $\mathbb{F}_q$, of degree n, by adjoining to $\mathbb{F}_q$ a root of an irreducible polynomial of degree n. Polynomials in $\mathbb{F}_q$ play the role of integers, and monic irreducible polynomials play the role of the primes. One can define a zeta function for $\mathbb{F}_q[x]$ in terms of a product over the monic irreducible polynomials in direct analogy to the prime product for the Riemann zeta function. More generally, one can associate a zeta function to a smooth, geometrically connected, proper curve C defined over $\mathbb{F}_q$ in the following way: let N_n denote the number of points of C over an extension field of degree n, then

$$Z_C(u) = \exp\Big(\sum_{n=1}^{\infty} \frac{N_n u^n}{n}\Big). \tag{24.6.1}$$

It turns out that $Z_C(u)$ is a rational function of u, of the form

$$Z_C(u) = \frac{P_C(u)}{(1-u)(1-qu)} \tag{24.6.2}$$

where $P_{\mathcal{C}}(u)$ is a monic integer polynomial of degree $2g$, where g is the genus of $\mathcal{C}$. $Z_{\mathcal{C}}(u)$ satisfies a functional equation connecting u with $1/(qu)$, and importantly, can be *proved* to obey a Riemann hypothesis, namely all of the inverse roots of $P_{\mathcal{C}}(u)$ have absolute value $\sqrt{q}$. Of particular relevance is the fact that $P_{\mathcal{C}}(u)$ turns out to be the characteristic polynomial of a symplectic matrix of size $2g$.

The connection with RMT comes when the curve $\mathcal{C}$ is allowed to vary. Consider, for example, the moduli space of hyperelliptic curves $\mathcal{C}_Q$ of genus g over $\mathbb{F}_q$, namely curves of the form $y^2 = Q(x)$, where $Q(x)$ is a square-free monic polynomial of degree $2g + 1$. Averaging $Q(x)$ uniformly over all possible square-free monic polynomials generates an ensemble of $(q - 1)q^{2g}$ curves. The central result, which follows from work of Deligne, is that, when g is fixed, the matrices associated with $P_{\mathcal{C}}(u)$ become equidistributed with respect to Haar measure on $USp(2g)$ in the limit $q \to \infty$ [KS99a, KS99b]. The zeros of the zeta functions $Z_{\mathcal{C}}(u)$ are therefore distributed like the eigenvalues of random matrices associated with $USp(2g)$. For other families of curves, equidistribution with respect to the unitary or orthogonal groups can similarly be established. Examples have also been found relating to the exceptional group G_2 [KLR03]. It is believed that when q is fixed and $g \to \infty$ the local statistics of the eigenvalues of the matrices associated with $P_{\mathcal{C}}(u)$ have a limit which coincides with the corresponding limit for averages of $USp(2g)$ [KS99a, KS99b], but this has been established in only a few cases, principally relating to the value distribution of the traces of matrices in question [KR03].

A number of further results have followed the initial work linking RMT and number theory. Random matrix theory calculations play various roles in these examples. Often rigorous calculations with random matrices suggest conjectures in number theory or that theorems hold more generally than can be proven number theoretically, as in the example of Hughes and Rudnick's mock Gaussian statistics [HR03b, HR02, HR03a]. They prove [HR02] the following theorem about linear statistics of the zeros of the Riemann zeta function:

Theorem 24.6.1 *Define*

$$N_f(\tau) := \sum_{j=\pm1,\pm2,\ldots} f\Big(\frac{\log T}{2\pi}(\gamma_j - \tau)\Big), \tag{24.6.3}$$

and consider τ as varying near T in an interval of size $H = T^a, 0 < a \leq 1$. With f a real-valued, even test function having Fourier transform $\hat{f} := \int_{-\infty}^{\infty} f(x)e^{-2\pi i x u}dx \in C_C^{\infty}(\mathbb{R})$ and $\operatorname{supp}\hat{f} \subseteq (-2a/m, 2a/m)$, then the first m moments of N_f converge as $T \to \infty$ to those of a Gaussian random variable with expectation $\int_{-\infty}^{\infty} f(x)dx$ and variance

$$\sigma_f^2 = \int_{-\infty}^{\infty} \min(|u|, 1)\hat{f}(u)^2 du. \tag{24.6.4}$$

Interestingly, the higher moments for test functions with this same range of support are not Gaussian.

Hughes and Rudnick illustrate exactly the same behaviour for eigenvalues $e^{i\theta_1}, \ldots, e^{i\theta_N}$ of an $N \times N$ random unitary matrix U. They consider the 2π-periodic function

$$F_N(\theta) := \sum_{j=-\infty}^{\infty} f\left(\tfrac{N}{2\pi}(\theta + 2\pi j)\right) \tag{24.6.5}$$

and model N_f with

$$Z_f(U) := \sum_{j=1}^{N} F_N(\theta_j). \tag{24.6.6}$$

They find the same variance as (24.6.4), although, significantly, there is no restriction on the support of the Fourier transform of the test function in order for the variance to have this form. In the RMT calculation, they see the same mock-Gaussian behaviour for the moments up to the mth, defined as $\lim_{N\to\infty} \mathbb{E}\{(Z_f - \mathbb{E}\{Z_f\})^m\}$ however in the RMT setting they can prove that the mth moment is Gaussian if $\operatorname{supp}\hat{f} \subseteq [-2/m, -2/m]$. This leads them to conjecture that the more restrictive condition on the support is not necessary in the number theory case.

Random matrix theory has also played a part in the study of average values and the statistics of zeros of the derivative of the Riemann zeta function. Methods for proving a lower bound for the fraction of zeros of the Riemann zeta function lying on the critical line could be improved if more was known about the positions of the zeros of the derivative of the zeta function near the critical line. In fact, if the Riemann hypothesis is true then all the zeros of the derivative will have a real part greater than or equal to 1/2. More delicate information about their horizontal distribution would allow for more accurate counting of zeros of zeta on the critical line. Inspired by this, Mezzadri [Mez03] conjectured and the team [DFF$^+$] confirmed the form for the leading-order term near the unit circle of the radial distribution of the zeros of the derivative of the characteristic polynomial of a random unitary matrix. (Note that all the zeros of the derivative of a polynomial with roots on the unit circle lie *inside* the unit circle.) This led to a similar calculation for the distribution of the zeros of the derivative of the Riemann zeta function near the critical line, and the prediction compares very well with the distribution computed numerically. The theorem of [DFF$^+$] is

Theorem 24.6.2 *Let $\Lambda(z)$ be the characteristic polynomial of a random matrix in $U(N)$ distributed with respect to Haar measure, and let $Q(s; N)$ be the probability density function of $S = N(1 - |z'|)$ for z' a root of $\Lambda'(z)$. Then*

$$Q(s) = \lim_{N\to\infty} Q(s;N) \sim \frac{4}{3\pi} s^{1/2}. \tag{24.6.7}$$

Various moments of the derivative of characteristic polynomials have also been considered, with a view to providing insight into moments of the derivative of $\zeta(s)$, but this problem seems much less tractable than the moments of characteristic polynomials themselves. In particular, for averages over $U(N)$ at an arbitrary point on the unit circle of even integer powers of the derivative of the characteristic polynomial, only the leading order results, for large matrix size, have been obtained explicitly. This is to be compared with the exact formula for the moments of $\Lambda(s)$ itself (24.4.2). In his thesis [Hug01] (Chapter 6), Hughes calculated the averages

$$\int_{U(N)} |\Lambda_A(1)|^{2k-2h} |\Lambda'_A(1)|^{2h} dA_{Haar} \sim F(h,k)\, N^{k^2+2h}, \tag{24.6.8}$$

and so conjectured the leading-order form of joint moments of the Riemann zeta function and its derivative:

$$\frac{1}{T}\int_0^T |\zeta(\tfrac{1}{2}+it)|^{2k-2h} |\zeta'(\tfrac{1}{2}+it)|^{2h} \sim F(h,k) a_k \left(\log\frac{T}{2\pi}\right)^{k^2+2h}. \tag{24.6.9}$$

Here a_k is given in (24.4.7). Hughes gives $F(h,k)$ as a combinatorial sum, but this problem was recently revisited by Dehaye [Deh08] who gives more insight into its structure. There is also a similar result for the $2k$th moment of the absolute value of the derivative of the characteristic polynomial [CRS06]. This case is covered by Hughes' work, but the latter paper gives a different form for the coefficient, the analogue of $F(h,k)$, in terms of $k \times k$ determinants of Bessel functions. Looking to the future, a goal would be to derive explicit lower-order contributions to (24.6.9) and to be able to calculate non-integer moments.

Another moment calculation involving the derivative of the characteristic polynomial is the discrete moment, where the derivative is evaluated at the zeros of $\zeta(s)$. This was first modelled using RMT in [HKO00]. As can be seen below, in this case an exact formula for the RMT moment can be obtained, leading to an asymptotic conjecture for the analogous number theoretical quantity. Here the eigenvalues of the unitary matrix A are $e^{i\theta_1}, \ldots, e^{i\theta_N}$, $G(z)$ is the Barnes double gamma function, and we require $k > -3/2$:

$$\begin{aligned}\int_{U(N)} \frac{1}{N}\sum_{n=1}^{N} |\Lambda'_A(e^{i\theta_n})|^{2k} dA_{Haar} &= \frac{G^2(k+2)}{G(2k+3)} \frac{G(N+2k+2)G(N)}{N\, G^2(N+k+1)} \\ &\sim \frac{G^2(k+2)}{G(2k+3)} N^{k(k+2)}, \quad as\ N\to\infty.\end{aligned} \tag{24.6.10}$$

This result led to the following conjecture for large T, where $N(T)$ is the number of zeros of the Riemann zeta function with height γ_n between 0 and T,

$$\frac{1}{N(T)} \sum_{0<\gamma_n \le T} |\zeta'(\tfrac{1}{2} + i\gamma_n)|^{2k} \sim \frac{G^2(k+2)}{G(2k+3)} a_k \left(\log \frac{T}{2\pi} \right)^{k(k+2)}. \tag{24.6.11}$$

Lower-order terms for the moments with $k = 1$ and $k = 2$ were determined in [CS07] (Sections 7.1 and 7.2) using the ratio conjectures mentioned at (24.4.15).

A significant recent development in the subject is that Bourgade, Hughes, Nikeghbali, and Yor [BHNY08] have shown that averages of characteristic polynomials behave in the same way as products of independent beta variables:

Theorem 24.6.3 *Let $V_N \in U(N)$ be distributed with the Haar measure $\mu_{U(N)}$. Then for all $\theta \in \mathbb{R}$,*

$$\det(I - e^{i\theta} V_N) \stackrel{\text{law}}{=} \prod_{k=1}^{N} (1 + e^{i\theta_k} \sqrt{\beta_{1,k-1}}) \tag{24.6.12}$$

with $\theta_1, \ldots, \theta_n, \beta_{1,0}, \ldots, \beta_{1,n-1}$ independent random variables, the θ_ks uniformly distributed on $[0, 2\pi]$, and the $\beta_{1,j}$s $(0 \le j \le N-1)$ being beta-distributed with parameters 1 and j. (By convention, $\beta_{1,0}$ is the Dirac distribution on 1.)

The probability density of a beta-distributed random variable, defined on $[0, 1]$, with parameters α and β is

$$\frac{\Gamma(\alpha+\beta)}{\Gamma(\alpha)\Gamma(\beta)} x^{\alpha-1}(1-x)^{\beta-1}. \tag{24.6.13}$$

This leads to an alternative proof of (24.4.2) and a probabilistic proof of the Selberg integral formula. It also provides an explanation for the fact, which may be proved by analysing the moment generating function for $\log \Lambda_A$ (given by a formula generalizing (24.4.2)), that the logarithm of the characteristic polynomial of a random $U(N)$ matrix satisfies a central limit theorem when $N \to \infty$ [BF97, KS00b]. The limiting Gaussian behaviour follows naturally from the probabilistic model because the logarithm of the characteristic polynomial may be decomposed into sums of independent random variables and then classical central limit theorems can be applied. This approach also leads to new estimates on the rate of convergence to the Gaussian limit. That $\log \Lambda_A$ satisfies a central limit theorem is relevant to number theory because of a theorem due to Selberg [Tit86, Odl89]:

Theorem 24.6.4 (Selberg)
For any rectangle $B \in \mathbb{C}$,

$$\lim_{T\to\infty} \frac{1}{T} \left| \left\{ t : T \le t \le 2T, \frac{\log \zeta(1/2 + it)}{\sqrt{\frac{1}{2} \log\log T}} \in B \right\} \right| = \frac{1}{2\pi} \int\int_B e^{-(x^2+y^2)/2} dx\, dy.$$

That is, in the limit as T, the height up the critical line, tends to infinity, the value distributions of the real and imaginary parts of $\log \zeta(1/2 + iT)/\sqrt{(1/2)\log\log T}$ each tend, independently, to a Gaussian with unit variance and zero mean. This coincides exactly with the central limit theorem for $\log \Lambda_A$, if, as in other calculations, N and $\log(T/2\pi)$ are identified.

The fact that $\log \Lambda_A$ models the value distribution of $\log \zeta(1/2 + it)$ extends to the large deviations regime too. See [HKO01], where also the ergodicity of the Gaussian limit of the value distribution of $\log \Lambda_A$ is established.

Acknowledgements

Both authors were supported by EPSRC Research Fellowships when this article was written.

References

[Bar00] E.W. Barnes. The theory of the G-function. *Q. J. Math.*, 31:264–314, 1900.

[Bas78] E. Basor. Asymptotic formulas for Toeplitz determinants. *Trans. Amer. Math. Soc.*, 239:33–65, 1978.

[BF97] T.H. Baker and P.J. Forrester. Finite-N fluctuation formulas for random matrices. *J. Stat. Phys.*, 88:1371–1385, 1997.

[BH00] E. Brézin and S. Hikami. Characteristic polynomials of random matrices. *Comm. Math. Phys.*, 214:111–135, 2000. arXiv:math-ph/9910005.

[BHNY08] P. Bourgade, C.P. Hughes, A. Nikeghbali, and M. Yor. The characteristic polynomial of a random unitary matrix: A probabilistic approach. *Duke Math. J.*, 145(1):45–69, 2008.

[BK95] E.B. Bogomolny and J.P. Keating. Random matrix theory and the Riemann zeros I: three- and four-point correlations. *Nonlinearity*, 8:1115–1131, 1995.

[BK96a] E.B. Bogomolny and J.P. Keating. Gutzwiller's trace formula and spectral statistics: beyond the diagonal approximation. *Phys. Rev. Lett.*, 77(8):1472–1475, 1996.

[BK96b] E.B. Bogomolny and J.P. Keating. Random matrix theory and the Riemann zeros II:n-point correlations. *Nonlinearity*, 9:911–935, 1996.

[BK99] M.V. Berry and J.P. Keating. The Riemann zeros and eigenvalue asymptotics. *SIAM Rev.*, 41(2):236–266, 1999.

[CF00] J.B. Conrey and D.W. Farmer. Mean values of L-functions and symmetry. *Int. Math. Res. Notices*, 17:883–908, 2000. arXiv:math.nt/9912107.

[CFK+03] J.B. Conrey, D.W. Farmer, J.P. Keating, M.O. Rubinstein, and N.C. Snaith. Autocorrelation of random matrix polynomials. *Comm. Math. Phys.*, 237(3):365–395, 2003. arXiv:math-ph/0208007.

[CFK+05] J.B. Conrey, D.W. Farmer, J.P. Keating, M.O. Rubinstein, and N.C. Snaith. Integral moments of L-functions. *Proc. London Math. Soc.*, 91(1):33–104, 2005. arXiv:math.nt/0206018.

[CFK+08] J.B. Conrey, D.W. Farmer, J.P. Keating, M.O. Rubinstein, and N.C. Snaith. Lower order terms in the full moment conjecture for the Riemann zeta function. *J. Number Theory*, 128(6):1516–54, 2008. arXiv:math/0612843.

[CFZ] J.B. Conrey, D.W. Farmer, and M.R. Zirnbauer. Howe pairs, supersymmetry, and ratios of random characteristic polynomials for the unitary groups $U(N)$. *preprint.* arXiv:math-ph/0511024.

[CFZ08] J.B. Conrey, D.W. Farmer, and M.R. Zirnbauer. Autocorrelation of ratios of L-functions. *Comm. Number Theory and Physics*, 2(3):593–636, 2008. arXiv:0711.0718.

[CKRS02] J.B. Conrey, J.P. Keating, M.O. Rubinstein, and N.C. Snaith. On the frequency of vanishing of quadratic twists of modular L-functions. In M.A. Bennett et al., editor, *Number Theory for the Millennium I: Proceedings of the Millennial Conference on Number Theory*, pages 301–315. A K Peters, Ltd, Natick, 2002. arXiv:math.nt/0012043.

[CKRS06] J.B. Conrey, J.P. Keating, M.O. Rubinstein, and N.C. Snaith. Random matrix theory and the Fourier coefficients of half-integral weight forms. *Experiment. Math.*, 15(1):67–82, 2006. arXiv:math.nt/0412083.

[Con01] J.B. Conrey. L-functions and random matrices. In B. Enquist and W. Schmid, editors, *Mathematics Unlimited* 2001 *and Beyond*, pages 331–352. Springer-Verlag, Berlin, 2001. arXiv:math.nt/0005300.

[Con05] J.B. Conrey. Families of L-functions and 1-level densities. In *Recent perspectives on random matrix theory and number theory, LMS Lecture Note Series* 322, pages 225–49. Cambridge University Press, Cambridge, 2005.

[CRS06] J.B. Conrey, M.O. Rubinstein, and N.C. Snaith. Moments of the derivative of characteristic polynomials with an application to the Riemann zeta-function. *Comm. Math. Phys.*, 267(3):611–629, 2006. arXiv:math.NT/0508378.

[CS07] J.B. Conrey and N.C. Snaith. Applications of the L-functions ratios conjectures. *Proc. London Math. Soc.*, 94(3):594–646, 2007. arXiv:math.NT/0509480.

[CS08] J.B. Conrey and N.C. Snaith. Correlations of eigenvalues and Riemann zeros. *Comm. Number Theory and Phyics*, 2(3):477–536, 2008. arXiv:0803.2795.

[Deh08] P.-O. Dehaye. Joint moments of derivatives of characteristic polynomials. *Alg. Number Theory*, 2(1):31–68, 2008. arXiv:math/0703440.

[DFF$^+$] E. Dueñez, D. W. Farmer, S. Froehlich, C. P. Hughes, F. Mezzadri and T. Phan. Roots of the derivative of the Riemann zeta function and of characteristic polynomials. Nonlinearity **23** (2010) 2599.

[GHK07] S.M. Gonek, C.P. Hughes, and J.P. Keating. A hybrid Euler-Hadamard product formula for the Riemann zeta function. *Duke Math. J.*, 136(3):507–549, 2007.

[GV79] D. Goldfeld and C. Viola. Mean values of L-functions associated to elliptic, Fermat and other curves at the centre of the critical strip. *S. Chowla Anniversary Issue, J. Number Theory*, 11:305–320, 1979.

[HKO00] C.P. Hughes, J.P. Keating, and N. O'Connell. Random matrix theory and the derivative of the Riemann zeta function. *Proc. R. Soc. Lond. A*, 456:2611–2627, 2000.

[HKO01] C.P. Hughes, J.P. Keating, and N. O'Connell. On the characteristic polynomial of a random unitary matrix. *Comm. Math. Phys.*, 220(2):429–451, 2001.

[HR02] C.P. Hughes and Z. Rudnick. Linear statistics for zeros of Riemann's zeta function. *Comptes Rendus Math.*, 335(8):667–670, 2002. arXiv:math.nt/0208220.

[HR03a] C.P. Hughes and Z. Rudnick. Linear statistics of low-lying zeros of L-functions. *Q. J. Math.*, 54(3):309–333, 2003. arXiv:math.nt/0208230.

[HR03b] C.P. Hughes and Z. Rudnick. Mock-Gaussian behaviour for linear statistics of classical compact groups. *J. Phys. A*, 36(12):2919–32, 2003. arXiv:math.pr/0206289.

[Hug01] C.P. Hughes. *On the characteristic polynomial of a random unitary matrix and the Riemann zeta function.* PhD thesis, University of Bristol, 2001.

[Jut81] M. Jutila. On the mean value of $L(1/2, \chi)$ for real characters. *Analysis*, 1:149–161, 1981.

[Kea05] J.P. Keating. L-functions and the characteristic polynomials of random matrices. In *Recent perspectives on random matrix theory and number theory, LMS Lecture Note Series 322*, pages 251–78. Cambridge University Press, Cambridge, 2005.

[KLR03] J.P. Keating, N. Linden, and Z. Rudnick. Random matrix theory, the exceptional Lie groups, and L-functions. *J. Phys. A: Math. Gen.*, 36(12):2933–44, 2003.

[KO08] J.P. Keating and B.E. Odgers. Symmetry transitions in random matrix theory and L-functions. *Commun. Math. Phys.*, 281:499–528, 2008.

[KR03] P. Kurlberg and Z. Rudnick. The fluctuations in the number of points on a hyperelliptic curve over a finite field. *J. Number Theory*, 129:2933–44, 2003.

[KS99a] N.M. Katz and P. Sarnak. *Random Matrices, Frobenius Eigenvalues and Monodromy*. American Mathematical Society Colloquium Publications, 45. American Mathematical Society, Providence, Rhode Island, 1999.

[KS99b] N.M. Katz and P. Sarnak. Zeros of zeta functions and symmetry. *Bull. Amer. Math. Soc.*, 36:1–26, 1999.

[KS00a] J.P. Keating and N.C. Snaith. Random matrix theory and L-functions at $s = 1/2$. *Comm. Math. Phys*, 214:91–110, 2000.

[KS00b] J.P. Keating and N.C. Snaith. Random matrix theory and $\zeta(1/2 + it)$. *Comm. Math. Phys.*, 214:57–89, 2000.

[KS03] J.P. Keating and N.C. Snaith. Random matrices and L-functions. *J. Phys. A: Math. Gen.*, 36(12):2859–81, 2003.

[Me05] F. Mezzadri and N.C. Snaith (editors). *Recent perspectives on random matrix theory and number theory, LMS Lecture Note Series 322.* Cambridge University Press, Cambridge, 2005.

[Meh04] M.L. Mehta. *Random Matrices*. Elsevier, Amsterdam, third edition, 2004.

[Mez03] F. Mezzadri. Random matrix theory and the zeros of $\zeta'(s)$. *J. Phys. A*, 36(12): 2945–62, 2003. arXiv:math-ph/0207044.

[Mon73] H.L. Montgomery. The pair correlation of the zeta function. *Proc. Symp. Pure Math*, 24:181–93, 1973.

[Odl89] A.M. Odlyzko. The 10^{20}th zero of the Riemann zeta function and 70 million of its neighbors. *preprint*, 1989. http://www.dtc.umn.edu/~ odlyzko/unpublished/index.html.

[ÖS99] A. E. Özlük and C. Snyder. On the distribution of the nontrivial zeros of quadratic L-functions close to the real axis. *Acta Arith.*, 91(3):209–228, 1999.

[RS96] Z. Rudnick and P. Sarnak. Zeros of principal L-functions and random matrix theory. *Duke Math. J.*, 81(2):269–322, 1996.

[Rub01] M.O. Rubinstein. Low-lying zeros of L-functions and random matrix theory. *Duke Math. J.*, 109(1):147–181, 2001.

[Sou00] K. Soundararajan. Non-vanishing of quadratic Dirichlet L-functions at $s = \frac{1}{2}$. *Ann. of Math.*, 152(2):447–488, 2000.

[Tit86] E.C. Titchmarsh. *The Theory of the Riemann Zeta Function.* Clarendon Press, Oxford, second edition, 1986.

·25·

Random permutations and related topics

Grigori Olshanski

Abstract

We present an overview of selected topics in random permutations and random partitions highlighting analogies with random matrix theory.

25.1 Introduction

An ensemble of random permutations is determined by a probability distribution on S_n, the set of permutations of $[n] := \{1, 2, \ldots, n\}$. Even the simplest instance of a uniform probability distribution is already very interesting and leads to deep theory. The symmetric group S_n is in many ways linked to classical matrix groups, and ensembles of random permutations should be treated on an equal footing with random matrix ensembles, such as the ensembles of classical compact groups and symmetric spaces of compact type with normalized invariant measure. The role of matrix eigenvalues is then played by partitions of n that parameterize the conjugacy classes in S_n. The parallel with random matrices becomes especially striking in applications to constructing representations of 'big groups' – inductive limits of symmetric or classical compact groups.

The theses stated above are developed in Section 25.2. The two main themes of this section are the space $\mathfrak{S}$ of virtual permutations ($\mathfrak{S}$ is a counterpart of the space of Hermitian matrices of infinite size) and the Poisson–Dirichlet distributions (a remarkable family of infinite-dimensional probability distributions). We focus on a special family of probability distributions on S_n with nice properties, the so-called Ewens measures (they contain uniform distributions as a particular case). It turns out that the large-n limits of the Ewens measures can be interpreted as probability measures on $\mathfrak{S}$. On the other hand, the Ewens measures give rise to ensembles of random partitions, from which one gets, in a limit transition, the Poisson–Dirichlet distributions.

A remarkable discovery of Frobenius, the founder of representation theory, was that partitions of n not only parameterize conjugacy classes in S_n but also serve as natural coordinates in the dual space $\widehat{S_n}$ – the set of equivalence classes of irreducible representations of S_n. This fact forms the basis of a 'dual' theory of random partitions, which turns out to have many intersections with random matrix theory. This is the subject of Sections 25.3–25.4. Here we survey results related to the Plancherel measure on $\widehat{S}_n$ and its consecutive generalizations: the z-measures and the Schur measures.

Thus, the two-faced nature of partitions gives rise to two kinds of probabilistic models. At first glance, they seem to be weakly related, but under a more general approach one sees a bridge between them. The idea is that the probability measures in the 'dual picture' can be further generalized by introducing an additional parameter, which is an exact counterpart of the β parameter in random matrix theory. This parameter interpolates between the group level and the dual space level, in the sense that in the limit as $\beta \to 0$, the 'beta z-measures' degenerate to the measures on partitions derived from the Ewens measures, see Section 25.4.4 below.

25.2 The Ewens measures, virtual permutations, and the Poisson–Dirichlet distributions

25.2.1 The Ewens measures

Permutations $s \in S_n$ can be represented as $n \times n$ unitary matrices $[s_{ij}]$, where s_{ij} equals 1 if $s(j) = i$, and 0 otherwise. This makes it possible to view random permutations as a very special case of random unitary matrices.

Given a probability distribution on each set S_n, $n = 1, 2, \dots$, one may speak of a sequence of ensembles of *random permutations* and study their asymptotic properties as $n \to \infty$. The simplest yet fundamental example of a probability distribution on S_n is the *uniform distribution* $P^{(n)}$, which gives equal weight $1/n!$ to all permutations $s \in S_n$; this is also the normalized Haar measure on the symmetric group. However, a more complete picture is achieved by considering a one-parameter family of distributions, $\{P_\theta^{(n)}\}_{\theta>0}$, forming a deformation of $P^{(n)}$:

$$P_\theta^{(n)}(s) = \frac{\theta^{\ell(s)}}{\theta(\theta+1)\dots(\theta+n-1)}, \qquad s \in S_n, \tag{25.2.1}$$

where $\ell(s)$ denotes the number of cycles in s (the uniform distribution corresponds to $\theta = 1$).

For reasons explained below we call $P_\theta^{(n)}$ the *Ewens measure* on the symmetric group S_n with parameter θ. Obviously, the Ewens measures are invariant under the action of S_n on itself by conjugations.

We propose to consider $(S_n, P^{(n)})$ as an analogue of CUE_N, Dyson's circular unitary ensemble [For10a] formed by the unitary group $U(N)$ endowed with the normalized Haar measure $P^{U(N)}$. More generally, the Ewens family $\{P_\theta^{(n)}\}$ should be viewed as a counterpart of a family of probability distributions on the unitary group $U(N)$ forming a deformation of the Haar measure:

$$P_t^{U(N)}(dU) = \text{const}\ |\det((1+U)^t)|^2\, P^{U(N)}(dU), \qquad U \in U(N), \tag{25.2.2}$$

where t is a complex parameter, $\operatorname{Re} t > -\frac{1}{2}$ (the Haar measure corresponds to $t = 0$). Using the Cayley transform one can identify the manifold $U(N)$, within a negligible subset, with the flat space $H(N)$ of $N \times N$ Hermitian matrices; then $P_t^{U(N)}$ turns into the so-called *Hua–Pickrell measure* on $H(N)$:

$$P_t^{H(N)}(dX) = \text{const}\ |\det(1+iX)^{-t-N}|^2 dX, \tag{25.2.3}$$

where 'dX' in the right-hand side denotes the Lebesgue measure. For more detail, see [Bor01b], [Ner02], [Ols03a]. A similarity between $P_\theta^{(n)}$ and $P_t^{U(N)}$ (or $P_t^{H(N)}$) is exploited in [Bou07].

A fundamental property of the Ewens measures is their consistency with respect to some natural projections $S_n \to S_{n-1}$ which we are going to describe.

Given a permutation $s \in S_n$, the *derived permutation* $s' \in S_{n-1}$ sends each $i = 1, \ldots, n-1$ either to $s(i)$ or to $s(n)$, depending on whether $s(i) \neq n$ or $s(i) = n$. In other words, s' is obtained by removing n from the cycle of s that contains n. For instance, if $s = (153)(24)$ (meaning that one cycle in s is $1 \to 5 \to 3 \to 1$ and the other is $2 \to 4 \to 2$), then $s' = (13)(24)$. The map $S_n \to S_{n-1}$ defined in this way is called the *canonical projection* [Ker04] and denoted as $p_{n-1,n}$. For $n \geq 5$, it can be characterized as the only map $S_n \to S_{n-1}$ commuting with the two-sided action of the subgroup S_{n-1}.

The next assertion [Ker04] is readily verified:

Lemma 25.2.1 *For any $n = 2, 3, \ldots$, the push-forward of $P_\theta^{(n)}$ under $p_{n-1,n}$ coincides with $P_\theta^{(n-1)}$.*

The Hua-Pickrell measures enjoy a similar consistency property with respect to natural projections $p^H_{N-1,N}\colon H(N) \to H(N-1)$ (removal of the Nth row and column from an $N \times N$ matrix) [Bor01b], [Ols03a]. This fact is hidden in the old book by Hua Lokeng [Hua58], in his computation of the matrix integral

$$\int_{H(N)} \det(1+X^2)^{-t} dX \tag{25.2.4}$$

by induction on N. (Note that [Hua58] contains a lot of masterly computations of matrix integrals.) Much later, the consistency property was rediscovered and applied to constructing measures on infinite-dimensional spaces by Shimomura [Shi75], Pickrell [Pic87], and Neretin [Ner02]. Note that analogues of the projections $p^H_{N-1,N}$ can be defined for other matrix spaces including the three series of compact classical groups and, more generally, the ten series of classical compact symmetric spaces [Ner02].

25.2.2 Virtual permutations, central measures, and Kingman's theorem

As we will see, the consistency property of the Ewens measures $P_\theta^{(n)}$ makes it possible to build an '$n = \infty$' version of these measures. In the particular case $\theta = 1$, this leads to a concept of 'uniformly distributed infinite permutations'.

Let $\mathfrak{S} = \varprojlim S_n$ be the projective limit of the sets S_n taken with respect to the canonical projections. By the very definition, each element $\sigma \in \mathfrak{S}$ is a sequence $\{\sigma_n \in S_n\}_{n=1,2,\dots}$ such that $\sigma_{n-1} = p_{n-1,n}(\sigma_n)$ for any $n = 2, 3, \dots$. We call $\mathfrak{S}$ the *space of virtual permutations* [Ker04]. It is a compact topological space with respect to the projective limit topology.

By classical Kolmogorov's theorem, any family $\{\mathcal{P}^{(n)}\}$ of probability measures on the groups S_n, consistent with the canonical projections, gives rise to a probability measure $\mathcal{P} = \varprojlim \mathcal{P}^{(n)}$ on the space $\mathfrak{S}$. Taking $\mathcal{P}^{(n)} = P_\theta^{(n)}$, $\theta > 0$, we get some measures P_θ on $\mathfrak{S}$ with nice properties; we still call them the Ewens measures.

A parallel construction exists in the context of matrix spaces. In particular, by making use of the projections $p^H_{N-1,N}$, one can define a projective limit space $\mathfrak{H} := \varprojlim H(N)$, which is simply the space of all Hermitian matrices of infinite size. This space carries the measures $P_t^{\mathfrak{H}} := \varprojlim P_t^{H(N)}$, $\mathrm{Re}\, t > -\frac{1}{2}$. Using the Cayley transform, the measures $P_t^{\mathfrak{H}}$ can be transformed to some measures $P_t^{\mathfrak{U}}$ on a projective limit space $\mathfrak{U} := \varprojlim U(N)$.

As is argued in [Bor01b], the $t = 0$ case of the probability space $(\mathfrak{H}, P_t^{\mathfrak{H}}) \simeq (\mathfrak{U}, P_t^{\mathfrak{U}})$ may be viewed as an '$N = \infty$' version of CUE_N. Likewise, we regard $(\mathfrak{S}, P_1)$ as an '$n = \infty$' version of the finite uniform measure space $(S_n, P_1^{(n)})$.

A crucial property of CUE_N is its invariance under the action of the unitary group $U(N)$ on itself by conjugation. This property is shared by the deformed ensemble with parameter t as well. In the infinite-dimensional case, the analogous property is $U(\infty)$-invariance of the measures $P_t^{\mathfrak{H}}$; here by $U(\infty)$ we mean the inductive limit group $\varinjlim U(N)$, which acts on the space $\mathfrak{H}$ by conjugations.

Likewise, we define the *infinite symmetric group* S_∞ as the inductive limit $\varinjlim S_n$, or simply as a union of the finite symmetric groups S_n, where each group S_n is identified with the subgroup of S_{n+1} fixing the point $n+1$ of the set $[n+1]$. The group S_∞ is countable and can be realized as the group of all permutations of the set $\mathbb{N} = \{1, 2, \dots\}$ moving only finitely many points. The space $\mathfrak{S}$ contains S_∞ as a dense subset, and the action of the group S_∞ on itself by conjugation extends by continuity to the space $\mathfrak{S}$; we will still call the latter action the *conjugation action* of S_∞.

It is convenient to give a name to measures that are invariant under the conjugation action of S_n or S_∞; let us call such measures *central*. Now, a simple lemma says:

Lemma 25.2.2 *Let, as above, $\mathcal{P} = \varprojlim \mathcal{P}^{(n)}$ be a projective limit probability measure on $\mathfrak{S}$. If $\mathcal{P}^{(n)}$ is central for each n, then so is $\mathcal{P}$.*

As a consequence we get that the measures P_θ are central.

In a variety of random matrix problems, the invariance property under an appropriate group action makes it possible to pass from matrices to their eigenvalues or singular values. For random permutations from S_n directed by a central measure, a natural substitute of eigenvalues is another invariant – partitions of n parameterizing the cycle structure of permutations.

In combinatorics, by a *partition* one means a sequence $\rho = (\rho_1, \rho_2, \dots)$ of weakly decreasing non-negative integers with infinitely many terms and finite sum $|\rho| := \sum \rho_i$. Of course, the number of non-zero terms ρ_i in ρ is always finite; it is denoted as $\ell(\rho)$. The finite set of all partitions ρ with $|\rho| = n$ will be denoted as $\mathscr{P}(n)$. To a permutation $s \in S_n$ we assign a partition $\rho \in \mathscr{P}(n)$ (in words, a partition of n) comprised of the cycle-lengths of s written in weakly decreasing order. Obviously, ρ is a full invariant of the conjugacy class of s. The projection $s \mapsto \rho$ takes any probability measure on S_n to a probability measure on $\mathscr{P}(n)$. Now, the point is that this establishes a one-to-one correspondence $\mathcal{P}^{(n)} \leftrightarrow \Pi^{(n)}$ between arbitrary central probability measures $\mathcal{P}^{(n)}$ on S_n and arbitrary probability measures $\Pi^{(n)}$ on $\mathscr{P}(n)$. In this sense, random permutations $s \in S_n$ (directed by a central measure) may be replaced by *random partitions* $\rho \in \mathscr{P}(n)$.

The link between $\mathcal{P}^{(n)}$ and $\Pi^{(n)}$ is simple: given $\rho \in \mathscr{P}(n)$, let $C(\rho) \subset S_n$ denote the corresponding conjugacy class in S_n. Then $\Pi^{(n)}(\rho) = |C(\rho)|\mathcal{P}^{(n)}(s)$ for any $s \in C(\rho)$. Further, there is an explicit expression for $|C(\rho|$: it is equal to $n!/z_\rho$, where $z_\rho = \prod k^{m_k} m_k!$ and m_k stands for the multiplicity of $k = 1, 2, \dots$ in ρ.

In this notation, the measure $\Pi_\theta^{(n)}$ corresponding to the Ewens measure $\mathcal{P}^{(n)} = P_\theta^{(n)}$ is given by the expression

$$\Pi_\theta^{(n)}(\rho) = \frac{\theta^{\ell(\rho)}}{\theta(\theta+1)\dots(\theta+n-1)} \prod_k \frac{1}{k^{m_k} m_k!}, \qquad \rho \in \mathscr{P}(n), \tag{25.2.5}$$

widely known under the name of *Ewens sampling formula* [Ewe98]. This justifies the name given to the measure (25.2.1).

The following result provides a highly non-trivial '$n = \infty$' version of the evident correspondence $\mathcal{P}^{(n)} \leftrightarrow \Pi^{(n)}$:

Theorem 25.2.1 *There exists a natural one-to-one correspondence $\mathcal{P} \leftrightarrow \Pi$ between arbitrary central probability measures $\mathcal{P}$ on the space $\mathfrak{S}$ of virtual permutations and arbitrary probability measures Π on the space*

$$\overline{\nabla}_\infty := \{(x_1, x_2, \dots) \in [0,1]^\infty : x_1 \ge x_2 \ge \dots, \quad \sum x_i \le 1\}. \tag{25.2.6}$$

In other words, each central measure $\mathcal{P}$ on $\mathfrak{S}$ is uniquely representable as a mixture of *indecomposable* (or *ergodic*) central measures, which in turn are parameterized by the points of $\overline{\nabla}_\infty$. The measure Π assigned to $\mathcal{P}$ is just the mixing measure.

Idea of proof. The theorem is a reformulation of the celebrated *Kingman's theorem* [Kin78b]. Kingman did not deal with virtual permutations but worked with some sequences of random permutations that he called *partition structures*. Represent $\mathcal{P}$ as a projective limit measure, $\mathcal{P} = \varprojlim \mathcal{P}^{(n)}$. By Lemma 25.2.2, all measures $\mathcal{P}^{(n)}$ are central. Pass to the corresponding measures $\Pi^{(n)}$ on partitions. The consistency of the family $\{\mathcal{P}^{(n)}\}$ with canonical projections $S_n \to S_{n-1}$ then translates as the consistency of the family $\{\Pi^{(n)}\}$ with some canonical Markov transition kernels $\mathscr{P}(n) \to \mathscr{P}(n-1)$. In Kingman's language this just means that $\{\mathcal{P}^{(n)}\}$ is a partition structure. Kingman's theorem provides a kind of Poisson integral representation of partition structures via probability measures on $\overline{\nabla}_\infty$, which is equivalent to the claim of Theorem 25.2.1.

Other proofs of Kingman's theorem can be found in [Ker03], [Ker98], where this result is placed in the broader context of potential theory for branching graphs. For our purpose it is worth emphasizing that the claim of Theorem 25.2.1 has a counterpart in the random matrix context – description of $U(\infty)$-invariant probability measures on $\mathfrak{H}$, which in turn is equivalent to the classical Schoenberg's theorem on totally positive functions [Sch51], [Pic91], [Ols96].

From the proof of Kingman's theorem it is seen that the space $\overline{\nabla}_\infty$ arises as a large-n limit of finite sets $\mathscr{P}(n)$, and every measure Π can be interpreted as a limit of the corresponding measures $\Pi^{(n)}$. In this picture, the ergodic measures $\Pi_{(x_1,x_2,\dots)}$ that are parameterized by points $(x_1, x_2, \dots) \in \overline{\nabla}_\infty$ arise as limits of uniform distributions on conjugacy classes $C(\rho)$ in growing finite symmetric groups. Thus, it is tempting to regard the $\Pi_{(x_1,x_2,\dots)}$s as a substitute of those uniform measures.

25.2.3 Application to representation theory

Pickrell's pioneer work [Pic87] demonstrated how some non-Gaussian measures on infinite-dimensional matrix spaces can be employed in representation theory. The idea of [Pic87] was further developed in [Ols03a]. As shown there, the measures $P_t^{\mathfrak{H}}$ on $\mathfrak{H}$ (or, equivalently, the measures $P_t^{\mathfrak{U}}$ on $\mathfrak{U}$) meet a lack of the Haar measure in the infinite-dimensional situation and can be applied to the construction of some 'generalized (bi)regular representations' of the group $U(\infty) \times U(\infty)$.

Pickrell's work was also a starting point for a parallel theory for the group S_∞, [Ker93c], [Ker04]. The key point is that the Ewens measures P_θ on $\mathfrak{S}$ have good transformation properties with respect to a natural action of the group $S_\infty \times S_\infty$ on $\mathfrak{S}$ extending the two-sided action of S_∞ on itself. Namely, the measure P_1 is $S_\infty \times S_\infty$-invariant and is the only probability measure on $\mathfrak{S}$ with such a property, so that it may be viewed as a substitute for the uniform distribution on S_n. As for the Ewens measures P_θ with general $\theta > 0$, these turn out to be quasi-invariant with respect to the action of $S_\infty \times S_\infty$. The quasi-invariance property forms the basis of the construction of some 'generalized (bi)regular representations' T_z of the group $S_\infty \times S_\infty$. Here z is a parameter ranging over $\mathbb{C}$, and the Hilbert space of T_z is $L^2(\mathfrak{S}, P_{|z|^2})$. We refer to [Ker04] and [Ols03b] for details.

25.2.4 Poisson–Dirichlet distributions

The probability measures on $\overline{\nabla}_\infty$ assigned by Theorem 25.2.1 to the Ewens measures P_θ are known under the name of *Poisson–Dirichlet distributions*; we denote these by $PD(\theta)$. Continuing our juxtaposition of the Ewens measures and the Hua-Pickrell measures one may say that the Poisson–Dirichlet distributions are counterparts of the determinantal point processes directing the decomposition of the measures $P_t^{\mathfrak{H}}$ into ergodic components (those processes involve, in a slightly disguised form, the sine-kernel process, see [Bor01b]). Although the Poisson–Dirichlet distributions $PD(\theta)$ seem to be much simpler than the sine-kernel process, they are still very interesting objects with a rich structure. Below we list a few equivalent descriptions of the $PD(\theta)$s:

(a) *Projection of a Poisson process.* Let $P(\theta)$ denote the inhomogeneous Poisson process on the half-line $\mathbb{R}_{>0} := \{\tau \in \mathbb{R} \mid \tau > 0\}$ with intensity $\theta\tau^{-1}e^{-\tau}$, and let $y = \{y_i\}$ be the random point configuration on $\mathbb{R}_{>0}$ with law $P(\theta)$. Due to the fast decay of the intensity at ∞, the configuration y is almost surely bounded from above, so that we may arrange the y_is in weakly decreasing order: $y_1 \geq y_2 \geq \cdots > 0$. Furthermore, the sum $r := \sum y_i$ is almost surely finite. Finally, it turns out that r and the normalized vector $x := (y_1/r, y_2/r, \dots) \in \nabla_\infty$ are independent of each other, the random variable r has the gamma distribution on $\mathbb{R}_{>0}$ with density $(\Gamma(\theta))^{-1}t^{\theta-1}\exp(-t)$, and the random vector x is distributed according to $PD(\theta)$.

 This means, in particular, that $PD(\theta)$ arises as the pushforward of the Poisson process $P(\theta)$ under the projection $y \mapsto x$.

(b) *Limit of Dirichlet distributions* [Kin75]. Let $D_n(\theta)$ denote the probability distribution on the $(n-1)$-dimensional simplex

$$\Delta_n := \{(x_1, \dots, x_n) \in \mathbb{R}^n \mid x_1, \dots, x_n \geq 0, \quad \sum x_i = 1\}$$

 with the density proportional to $\prod x_i^{n^{-1}\theta-1}$ (with respect to the Lebesgue measure on Δ_n). Note that $D_n(\theta)$ enters the family of the Dirichlet distributions. Rearranging the coordinates x_i in weakly decreasing order and adding infinitely many 0s one gets a map $\Delta_n \to \overline{\nabla}_\infty$; let $\widetilde{D}_n(\theta)$ stand for the pushforward of $D_n(\theta)$ under this map. Then $PD(\theta)$ appears as the weak limit of the measures $\widetilde{D}_n(\theta)$ as $n \to \infty$.

(c) *Projection of a product measure* [Ver77a], [Arr03, §4.11]. Consider the infinite-dimensional simplex

$$\overline{\Delta}_\infty := \left\{(u_1, u_2, \dots) \in [0,1]^\infty \mid u_1 + u_2 + \dots \le 1\right\}.$$

The triangular transformation of coordinates $v = (v_1, v_2, \dots) \mapsto u = (u_1, u_2, \dots)$ given by

$$u_1 = v_1; \qquad u_n = v_n(1 - v_1)\dots(1 - v_{n-1}), \quad n \ge 2,$$

maps the cube $[0,1]^\infty$ onto the simplex $\overline{\Delta}_\infty$. This map is almost one-to-one: it admits the inversion $u \mapsto v$,

$$v_1 = u_1; \qquad v_n = \frac{u_n}{1 - u_1 - \dots - u_{n-1}}, \quad n \ge 2, \tag{25.2.7}$$

which is well defined provided that all the partial sums of the series $u_1 + u_2 + \dots$ are strictly less than 1.

Next, rearrangement of the coordinates in weakly decreasing order determines a projection $\overline{\Delta}_\infty \to \overline{\nabla}_\infty$.

Denoting by $B(\theta)$ the probability measure on $[0,1]^\infty$ obtained as the product of infinitely many copies of the measure $\theta(1-t)^{\theta-1}dt$ on $[0,1]$, the Poisson–Dirichlet distribution $PD(\theta)$ coincides with the pushforward of $B(\theta)$ under the composition map $[0,1]^\infty \to \overline{\Delta}_\infty \to \overline{\nabla}_\infty$.

(d) *Characterization via correlation functions* [Wat76, §3]. Removing possible 0s from a sequence $x \in \overline{\nabla}_\infty$ one may interpret it as a locally finite point configuration on the semi-open interval $(0,1]$. This allows one to interpret any probability measure on $\overline{\nabla}_\infty$ as a random point process on $(0,1]$ (see [Bor10] for basic definitions). It turns out that the correlation functions of the point process associated with $PD(\theta)$ have a very simple form:

$$\rho_n(u_1, \dots, u_n) = \begin{cases} \dfrac{\theta^n(1 - u_1 - \dots - u_n)^{\theta-1}}{u_1 \dots u_n}, & \sum_{i=1}^n u_i < 1; \\ 0, & \text{otherwise.} \end{cases}$$

This provides one more characterization of $PD(\theta)$.

The literature devoted to the Poisson–Dirichlet distributions and their various connections and applications is very large. The interested reader will find rich material in [Arr03], [Ver72], [Wat76], [Ver77a], [Ver78], [Ign82], [Pit97], [Hol01] [Kin75].

Note that $PD(\theta)$ describes the asymptotics of the *large* cycle-lengths of random permutations with law $P_\theta^{(n)}$ (namely, the ith coordinate x_i on $\overline{\nabla}_\infty$ corresponds to the ith largest cycle-length scaled by the factor of $1/n$). The literature also contains results concerning the asymptotics of other statistics on random permutations, for instance, small cycle-lengths and the number of cycles [Arr03].

25.3 The Plancherel measure

25.3.1 Definition of the Plancherel measure

Partitions parameterize not only the conjugacy classes in the symmetric groups but also their irreducible representations. So far we have focused on the conjugacy classes, but now we will exploit the connection with representations. It is convenient to identify partitions of n with Young diagrams containing n boxes. The set of such diagrams will be denoted as $\mathbb{Y}_n$. Given a diagram $\lambda \in \mathbb{Y}_n$, let V_λ denote the corresponding irreducible representation of S_n and $\dim \lambda$ its dimension. In particular, the one-row diagram $\lambda = (n)$ and the one-column diagram $\lambda = (1^n)$ correspond to the only one-dimensional representations, the trivial and the sign ones. Note that the symmetry map $\mathbb{Y}_n \to \mathbb{Y}_n$ given by transposition $\lambda \mapsto \lambda'$ amounts to tensoring V_λ with the sign representation, so that $\dim \lambda' = \dim \lambda$.[1]

By virtue of Burnside's theorem,

$$\sum_{\lambda \in \mathbb{Y}_n} (\dim \lambda)^2 = n!. \qquad (25.3.1)$$

This suggests the definition of a probability distribution $M^{(n)}$ on $\mathbb{Y}_n$:

$$M^{(n)}(\lambda) := \frac{(\dim \lambda)^2}{n!}, \qquad \lambda \in \mathbb{Y}_n . \qquad (25.3.2)$$

Following [Ver77b], one calls $M^{(n)}$ the *Plancherel measure* on $\mathbb{Y}_n$.

In purely combinatorial terms, $\dim \lambda$ equals the number of *standard tableaux* of shape λ [Sag01, Section 2.5]. Several explicit expressions for $\dim \lambda$ are known, see [Sag01, Sections 3.10, 3.11].

25.3.2 Limit shape and Gaussian fluctuations

We view each $\lambda \in \mathbb{Y}_n$ as a plane shape, of area n, in the (r, s) plane, where r is the row coordinate and s the column coordinate. In new coordinates $x = s - r$, $y = r + s$, the boundary $\partial\lambda$ of the shape $\lambda \subset \mathbb{R}^2_+$ may be viewed as the graph of a continuous piecewise linear function, which we denote as $y = \lambda(x)$. Note that $\lambda'(x) = \pm 1$, and $\lambda(x)$ coincides with $|x|$ for sufficiently large values of $|x|$. The area of the shape $|x| \le y \le \lambda(x)$ equals $2n$. (See Figure 25.1.)

Assuming λ to be the random diagram distributed according to the Plancherel measure $M^{(n)}$, we get a random ensemble $\{\lambda(\,\cdot\,)\}$ of polygonal lines. We will describe the behaviour of this ensemble as $n \to \infty$.

[1] In the context of conjugacy classes the operation of transposition has no natural interpretation.

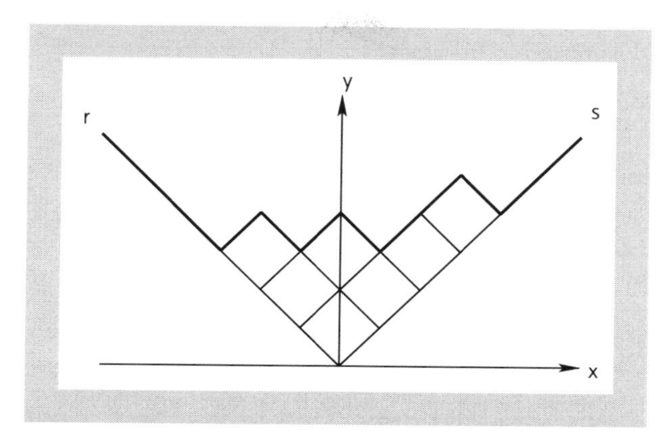

Fig. 25.1 The function $y = \lambda(x)$ for the Young diagram $\lambda = (4, 2, 1)$.

Informally, the result can be stated as follows: Let $y = \bar{\lambda}(x)$ be obtained from $y = \lambda(x)$ by shrinking along both the x and y axes with coefficient $\sqrt{n}$,

$$\bar{\lambda}(x) = \frac{1}{\sqrt{n}} \lambda(\sqrt{n}\, x);$$

then we have

$$\bar{\lambda}(x) \quad \approx \quad \Omega(x) + \frac{2}{\sqrt{n}} \Delta(x), \qquad n \to \infty, \tag{25.3.3}$$

where $y = \Omega(x)$ is a certain nonrandom curve coinciding with $y = |x|$ outside $[-2, 2] \subset \mathbb{R}$, and $\Delta(x)$ is a generalized Gaussian process. Let us explain the exact meaning of (25.3.3).

First of all, the purpose of the scaling $\lambda(\,\cdot\,) \to \bar{\lambda}(\,\cdot\,)$ is to put the random ensembles with varying n on the same scale: note that the area of the shape $|x| \leq y \leq \bar{\lambda}(x)$ equals 2 for any n.

The function $y = \Omega(x)$ is given by two different expressions depending on whether or not x belongs to the interval $[-2, 2] \subset \mathbb{R}$:

$$\Omega(x) = \begin{cases} \frac{2}{\pi}(x \arcsin \frac{x}{2} + \sqrt{4 - x^2}), & |x| \leq 2 \\ |x|, & |x| \geq 2 \end{cases}$$

In the first approximation, the asymptotic relation (25.3.3) means concentration of the random polygonal lines $y = \bar{\lambda}(x)$ near a limit curve. The exact statement (see [Log77], [Ver77b], and also [Iva02]) is:

Theorem 25.3.1 (Law of large numbers) *For each $n = 1, 2, \ldots$, let $\lambda \in \mathbb{Y}_n$ be the random Plancherel diagram and $\bar{\lambda}(\,\cdot\,)$ be the corresponding random curve, as defined above. As $n \to \infty$, the distance in the uniform metric between $\bar{\lambda}(\,\cdot\,)$ and the curve $\Omega(x)$ tends to 0 in probability:*

$$\lim_{n\to\infty} M^{(n)}\{\lambda \in \mathbb{Y}_n \mid \sup_{x\in\mathbb{R}} |\bar{\lambda}(x) - \Omega(x)| \le \varepsilon\} = 1, \qquad \forall\varepsilon > 0.$$

The second term in the right-hand side of (25.3.3) describes the fluctuations around the limit curve. The Gaussian process $\Delta(x)$ can be defined by a random trigonometric series on the interval $[-2, 2] \subset \mathbb{R}$, as follows. Let $\xi_2, \xi_3, \ldots$ be independent Gaussian random variables with mean 0 and variance 1, and set $x = 2\cos\theta$, where $0 \le \theta \le \pi$. Then

$$\Delta(x) = \Delta(2\cos\theta) = \frac{1}{\pi}\sum_{k=2}^{\infty}\frac{\xi_k}{\sqrt{k}}\sin(k\theta), \qquad x \in [-2, 2].$$

This is a *generalized process*, meaning that its trajectories are not ordinary functions but generalized ones (i.e. distributions). In other words, it is a Gaussian measure on the space of distributions supported by $[-2, 2]$. For any smooth test function φ on $\mathbb{R}$, the smoothed series

$$\frac{1}{\pi}\sum_{k=2}^{\infty}\frac{\xi_k}{\sqrt{k}}\int_{-2}^{2}\sin(k\theta)\varphi(x)dx, \qquad \theta = \arccos(x/2),$$

converges and represents a Gaussian random variable. However, the value of $\Delta(x)$ at a point x is not defined.

More precisely, the result about the Gaussian fluctuations appears as follows:

Theorem 25.3.2 (Central limit theorem for global fluctuations) *Let, as above, $\{\bar{\lambda}(\,\cdot\,)\}$ be the random ensemble governed by the Plancherel measure $M^{(n)}$, and set*

$$\Delta_n(x) = \frac{\sqrt{n}}{2}(\bar{\lambda}(x) - \Omega(x)), \qquad x \in \mathbb{R}.$$

For any finite collection of polynomials $\varphi_1(x), \ldots, \varphi_m(x)$, the joint distribution of the random variables

$$\int_{\mathbb{R}} \varphi_i(x)\Delta_n(x)dx, \qquad 1 \le i \le m \tag{25.3.4}$$

converges, as $n \to \infty$, to that of the Gaussian random variables

$$\int_{\mathbb{R}} \varphi_i(x)\Delta(x)dx, \qquad 1 \le i \le m.$$

This result is due to Kerov [Ker93a]; a detailed exposition is given in [Iva02].

Note that for any diagram λ, the function $\bar{\lambda}(x) - \Omega(x)$ vanishes for $|x|$ large enough, so that the integral in (25.3.4) makes sense.

The theorem implies that the normalized fluctuations $\Delta_n(x)$, when appropriately smoothed, are of finite order. This can be rephrased by saying that, in the (r, s) coordinates, the global fluctuations of the boundary $\partial\lambda$ of the random Plancherel diagram $\lambda \in \mathbb{Y}_n$ in the direction parallel to the diagonal $r = s$ have finite order.

A different central limit theorem is stated in [Bog07]: that result describes fluctuations at points (so that there is no smoothing); then an additional scaling of order $\sqrt{\log n}$ along the y-axis is required.

Theorem 25.3.1 should be compared to a similar concentration result for spectra of random matrices (convergence to Wigner's semicircle law). A similarity between the two pictures becomes especially convincing in view of the fact (discovered in [Ker93b]) that there is a natural transform relating the curve Ω to the semicircle law. As for Theorem 25.3.2, it has a strong resemblance to the central limit theorems for random matrix ensembles, established in [Dia94], [Joh98].

Biane [Bia01] considered a modification of the Plancherel measures $M^{(n)}$ related to the Schur-Weyl duality and found a one-parameter family of limit curves forming a deformation of Ω.

25.3.3 The poissonized Plancherel measure as a determinantal process

Let $\mathbb{Y} = \mathbb{Y}_0 \cup \mathbb{Y}_1 \cup \dots$ be the countable set of all Young diagrams including the empty diagram $\varnothing$. To each $\lambda \in \mathbb{Y}$ we assign an infinite subset $\mathcal{L}(\lambda)$ on the lattice $\mathbb{Z}' := \mathbb{Z} + \frac{1}{2}$ of half-integers, as follows

$$\mathcal{L}(\lambda) = \left\{\lambda_i - i + \tfrac{1}{2} \mid i = 1, 2, \dots\right\}. \tag{25.3.5}$$

We interpret $\mathcal{L}(\lambda)$ as a particle configuration on the nodes of the lattice $\mathbb{Z}'$ and regard the unoccupied nodes $\mathbb{Z}' \setminus \mathcal{L}(\lambda)$ as *holes*. In particular, the configuration $\mathcal{L}(\varnothing)$ is $\mathbb{Z}'_- := \{\dots, -\frac{3}{2}, -\frac{1}{2}\}$ and the corresponding holes occupy $\mathbb{Z}'_+ := \{\frac{1}{2}, \frac{3}{2}, \dots\}$. In this picture, appending a box to a diagram λ results in moving a particle from $\mathcal{L}(\lambda)$ to the neighbouring position on the right. Thus, growing λ from the empty diagram $\varnothing$ by consecutively appending a box can be interpreted as a passage from the configuration $\mathbb{Z}'_-$ to the configuration $\mathcal{L}(\lambda)$ by moving at each step one of the particles to the right by 1.

The configurations $\mathcal{L}(\lambda)$ are precisely those configurations for which the number of particles on $\mathbb{Z}'_+$ is finite and equal to the number of holes on $\mathbb{Z}'_-$. Note also that transposition $\lambda \to \lambda'$ translates as replacing particles by holes and vice versa, combined with the reflection map $x \to -x$ on $\mathbb{Z}'$.

The *poissonized* Plancherel measure with parameter $\nu > 0$ [Bai99] is a probability measure M_ν on $\mathbb{Y}$, which is obtained by mixing together the measures $M^{(n)}$ (see (25.3.2)), $n = 0, 1, 2, \dots$, by means of a Poisson distribution on the set of indices n:

$$M_\nu(\lambda) = e^{-\nu} \frac{\nu^{|\lambda|}}{|\lambda|!} M^{(|\lambda|)}(\lambda) = e^{-\nu} \nu^{|\lambda|} \left(\frac{\dim \lambda}{|\lambda|!}\right)^2, \qquad \lambda \in \mathbb{Y}$$

(see also [Bor10, Sect. 1.6]).

Theorem 25.3.3 *Under the correspondence $\lambda \to \mathcal{L}(\lambda)$ defined by (25.3.5), the poissonized Plancherel measure M_ν turns into the determinantal point process on the lattice $\mathbb{Z}'$ whose correlation kernel is the discrete Bessel kernel.*

About determinantal point processes in general, see [Bor10], the discrete Bessel kernel is written down in [Bor10, Sect. 11.6] (there, replace θ by $\sqrt{\nu}$). Note that it is a projection kernel. Theorem 25.3.3 was obtained in [Joh01a] and (in a slightly different form) in [Bor00b]. Johansson's approach [Joh01a] is also discussed in his note [Joh01b] and the expository paper [Joh05].

25.3.4 The bulk limit [Bor00b]

Fix $a \in (-2, 2)$. Recall that the point $(a, \Omega(a))$ on the limit curve $y = \Omega(x)$ (see Section 25.3.2) corresponds to the intersection of the boundary $\partial\lambda$ of the typical large Plancherel diagram $\lambda \in \mathbb{Y}_n$ with the line $j - i = a\sqrt{n}$. The next result describes the asymptotic behaviour of the boundary $\partial\lambda$ near this point.

Theorem 25.3.4 *Assume that $n \to \infty$ and $x(n) \in \mathbb{Z}'$ varies together with n in such a way that $x(n)/\sqrt{n} \to a \in (-2, 2)$. Let $\lambda \in \mathbb{Y}_n$ be the random diagram with law $M^{(n)}$ given by (25.3.2) and let X_n be the random particle configuration on $\mathbb{Z}$ obtained from the configuration $\mathcal{L}(\lambda)$ defined by (25.3.5) under the shift $x \mapsto x - x(n)$ mapping $\mathbb{Z}'$ onto $\mathbb{Z}$. Then X_n converges to a translation invariant point process on $\mathbb{Z}$, with the correlation kernel*

$$S^a(k,l) = \begin{cases} \dfrac{\sin(\arccos(a/2)(k-l))}{\pi(k-l)}, & k,l \in \mathbb{Z}, \quad k \neq l; \\ \dfrac{\arccos(a/2)}{\pi}, & k = l. \end{cases}$$

The kernel $S^a(k,l)$ is called the *discrete sine kernel*. It is a projection kernel and should be viewed as a lattice analogue of the famous sine kernel on $\mathbb{R}$ originated in random matrix theory. Like the sine kernel, the discrete sine kernel possesses a universality property [Bai07].

Theorem 25.3.4 is derived from Theorem 25.3.3: Let $J^\nu(x,y)$ denote the discrete Bessel kernel; one shows that

$$\lim_{\nu \to \infty} J^\nu(x(\nu)+k,\ x(\nu)+l) = S^a(k,l), \qquad x(\nu) \in \mathbb{Z}', \quad x(\nu) \sim a\sqrt{\nu},$$

and then one applies a depoissonization argument to check that the large-n limit and the large-ν limit are equivalent.

25.3.5 The edge limit

Theorem 25.3.5 *Let $\lambda = (\lambda_1, \lambda_2, \dots) \in \mathbb{Y}_n$ be distributed according to the nth Plancherel measure $M^{(n)}$ given by (25.3.2). For any fixed $k = 1, 2, \dots,$ introduce real-valued random variables $u_1, \dots, u_k$ by setting*

$$\lambda_i = 2n^{1/2} + u_i n^{1/6}, \qquad i = 1, \dots, k. \tag{25.3.6}$$

Then, as $n \to \infty$, the joint distribution of $u_1, \dots, u_k$ converges to that of the first k particles in the Airy point process.

Recall ([Bor10, Sect. 1.9]) that the *Airy point process* is a determinantal process on $\mathbb{R}$ living on point configurations $(u_1 > u_2 > \dots)$ bounded from above; it is determined by the *Airy correlation kernel*, which is a projection kernel on $\mathbb{R}$.

The Airy point process arises in the edge limit transition from a large class of random matrix ensembles. It turns out that it also describes the limit distribution of a few (appropriately scaled) largest rows of the random Plancherel diagram.

Due to symmetry of $M^{(n)}$ under transposition $\lambda \to \lambda'$, the same result holds for the largest column lengths as well.

Already the simplest case $k = 1$ of Theorem 25.3.5 is very interesting, especially because of its connection to longest increasing subsequences in random permutations (see Section 25.3.6 below). The claim for $k = 1$ was first established by Baik, Deift, and Johansson [Bai99]; they then proved the claim for $k = 2$, [Bai00]. Their work completed a long series of investigations and at the same time opened the way to generalizations. The general case of Theorem 25.3.5 is due to Okounkov [Oko00]; note that his approach is very different from that of [Bai99], [Bai00]. Shortly afterwards, the theorem was obtained by yet another method in independent papers [Bor00b] and [Joh01a], by using Theorem 25.3.3 as an intermediate step. Note that once one knows Theorem 25.3.3, the precise form of the scaling (25.3.6) can be guessed by a simple argument, see [Ols08].

25.3.6 Longest increasing subsequences

Given a permutation $s \in S_n$, let $L_n(s)$ stand for the length of the *longest increasing subsequence* in the permutation word $\widetilde{s} := s(1)s(2)\dots s(n)$.[2] Under the uniform distribution on S_n, L_n becomes a random variable. In the sixties, S. Ulam raised the question about its asymptotic properties as $n \to \infty$. This seemingly rather particular problem turned out to be surprisingly deep (about the history of the problem and many related results, see [Bai99] and the survey papers [Ald99], [Dei00], [Sta07]). The next claim relates L_n to the Plancherel measure $M^{(n)}$:

Theorem 25.3.6 *The distribution of L_n under the uniform measure on S_n coincides with the distribution of λ_1, the first row length of the random Young diagram $\lambda \in \mathbb{Y}_n$ with law $M^{(n)}$.*

[2] An increasing subsequence in $\widetilde{s}$ is a subword $s(i_1)\dots s(i_k)$ such that $i_1 < \dots < i_k$ and $s(i_1) < \dots < s(i_k)$.

This result is obtained with the help of the *Robinson-Schensted correspondence*, which establishes an explicit bijection $RS : s \leftrightarrow (\mathcal{P}, \mathcal{Q})$ between permutations $s \in S_n$ and couples $(\mathcal{P}, \mathcal{Q})$ of standard tableaux of one and the same shape $\lambda \in \mathbb{Y}_n$. The bijection RS is described in detail in many textbooks, e.g. [Ful97] and [Sag01]. The latter book also contains an elegant geometric interpretation of RS due to Viennot. By the very definitions, the pushforward under RS of the uniform measure on S_n is $M^{(n)}$. A non-trivial fact is that under this bijection, $L_n(s) = \lambda_1$.

By virtue of Theorem 25.3.6, the Ulam problem is completely solved by the $k = 1$ case of Theorem 25.3.5 discussed above: the limit distribution of the scaled random variable $(L_n - 2\sqrt{n})n^{-1/6}$ is the GUE Tracy–Widom distribution F_2 [Tra94].

Given a subset $S_n^* \subset S_n$, denote by L_n^* the random variable $L_n(\,\cdot\,)$ directed by the uniform measure on S_n^*. A modification of the Ulam problem consists in studying the limit distribution of L_n^* (suitably centred and scaled) for subsets S_n^* determined by certain symmetry conditions imposed on the matrix $[s_{ij}]$ of a permutation $s \in S_n$. Baik and Rains (see [Bai01] and references therein) showed that in this way one can get two other Tracy–Widom distributions [Tra96], F_1 and F_4, as well as a large family of allied probability distributions including an interpolation between F_1 and F_4. These results demonstrate once again a similarity in asymptotic properties of random permutations and random matrices. Here is the simplest example from [Bai01], which shows that involutions $s = s^{-1}$ in S_n (i.e. symmetric permutation matrices) model real symmetric matrices:

Theorem 25.3.7 *Take as S_n^* the subset of involutions in S_n, and let L_n^* be the corresponding random variable. Then the limit distribution of $(L_n^* - 2\sqrt{n})n^{-1/6}$ is the GOE Tracy–Widom distribution F_1.*

25.4 The z-measures and Schur measures

25.4.1 The z-measures

The identity (25.3.1) admits an extension depending on two parameters $z, z' \in \mathbb{C}$:

$$\sum_{\lambda \in \mathbb{Y}_n} (z)_\lambda (z')_\lambda (\dim \lambda)^2 = (zz')_n n!,$$

where $(x)_n := x(x+1)\dots(x+n-1)$ is the Pochhammer symbol and $(x)_\lambda$ is its generalization,

$$(x)_\lambda := \prod_{(i,j)\in\lambda} (x + j - i),$$

the product taken over the boxes (i, j) belonging to λ, where i and j stand for the row and column number of a box. The (complex-valued) *z-measure* $M^{(n)}_{z,z'}$ on $\mathbb{Y}_n$ assigns weights

$$M^{(n)}_{z,z'}(\lambda) = \frac{(z)_\lambda(z')_\lambda}{(zz')_n} M^{(n)}(\lambda) = \frac{(z)_\lambda(z')_\lambda}{(zz')_n} \frac{(\dim\lambda)^2}{n!}$$

to diagrams $\lambda \in \mathbb{Y}_n$. This is a deformation of the Plancherel measure $M^{(n)}$ in the sense that $M^{(n)}_{z,z'}(\lambda) \to M^{(n)}(\lambda)$ as $z, z' \to \infty$. In what follows we assume that the parameters take *admissible values* meaning that $(z)_\lambda(z')_\lambda \geq 0$ for any $\lambda \in \mathbb{Y}$ and $zz' > 0$ (for instance, one may assume $z' = \bar{z} \in \mathbb{C} \setminus \{0\}$). Then $M^{n}_{z,z'}$ is a probability measure for every n.

The z-measures first emerged in [Ker93c]; they play an important role in the representation theory of the infinite symmetric group S_∞: Recall that in Section 25.2.3 we have mentioned generalized regular representations T_z; it turns out that when $z' = \bar{z}$, a suitably defined large-n scaled limit of the z-measures governs the spectral decomposition of T_z into irreducibles: [Bor01a, §3], [Ols03b].

The *mixed* z-measure $M_{z,z',\xi}$ on $\mathbb{Y}$ with admissible parameters (z, z') and an additional parameter $\xi \in (0, 1)$ is obtained by mixing up the z-measures with varying superscript n by means of a negative binomial distribution on $\mathbb{Z}_+$:

$$M_{z,z',\xi}(\lambda) = (1-\xi)^{zz'} \frac{(zz')_{|\lambda|}\xi^{|\lambda|}}{|\lambda|!} M^{(|\lambda|)}_{z,z'}(\lambda) = (1-\xi)^{zz'}\xi^{|\lambda|}(z)_{|\lambda|}(z')_{|\lambda|}\left(\frac{\dim\lambda}{|\lambda|!}\right)^2,$$

where λ ranges over $\mathbb{Y}$. This procedure is similar to poissonization of the Plancherel measure and serves the same purpose of facilitating the study of limit transitions. Note that the poissonized Plancherel measure M_ν is a degeneration of $M_{z,z',\xi}$ when $z, z' \to \infty$ and $\xi \to 0$ in such a way that $zz'\xi \to \nu$.

Theorem 25.4.1 *Under the correspondence $\lambda \to \mathcal{L}(\lambda)$ defined by* (25.3.5), *the mixed z-measure $M_{z,z',\xi}$ turns into a determinantal point process on the lattice $\mathbb{Z}'$ whose correlation kernel can be explicitly expressed through the Gauss hypergeometric function.*

This is a generalization of Theorem 25.3.3. Various proofs have been given in [Bor00a], [Bor00c], [Oko01b], [Bor06].

For the lattice determinantal process from Theorem 25.4.1 there are three interesting limit regimes, as $\xi \to 1$, leading to continuous and discrete determinantal processes:

(1) Split $\mathbb{Z}'$ into positive and negative parts, $\mathbb{Z}' = \mathbb{Z}'_+ \sqcup \mathbb{Z}'_-$. Given $\lambda \in \mathbb{Y}$, let $\mathcal{L}^\circ(\lambda) \subset \mathbb{Z}'$ be obtained from $\mathcal{L}(\lambda)$ by switching from particles to holes on $\mathbb{Z}'_-$; then $\mathcal{L}^\circ(\lambda)$ is finite and contains equally many particles in $\mathbb{Z}'_+$ and in $\mathbb{Z}'_-$. Note that this *particle/hole involution* does not affect the determinantal

property. Next, scale the lattice $\mathbb{Z}'$ making its mesh equal to small parameter $\varepsilon = 1 - \xi$. Letting $\xi \to 1$, one gets in this way from $M_{z,z',\xi}$ a determinantal process living on the punctured real line $\mathbb{R} \setminus \{0\}$. The corresponding correlation kernel is called the *Whittaker kernel*, because it is expressed through the classical Whittaker function. This limit process is of great interest for harmonic analysis on the infinite symmetric groups. For more detail, see [Bor00a], [Ols03b].

(2) No scaling, we remain on the lattice. The limit determinantal process is directed by a diffuse measure on the space $\{0, 1\}^{\mathbb{Z}'}$ of all lattice point configurations, and the limit correlation kernel is expressed through Euler's gamma function, see [Bor05c].

(3) An 'intermediate' limit regime assuming a scaling. It leads to a stationary limit process whose correlation kernel is expressed through trigonometric functions and is a deformation of the sine kernel, see [Bor05c].

These three different regimes describe the asymptotics of the largest, smallest, and intermediate Frobenius coordinates of random Young diagrams, respectively.

Remark 25.4.1 *Note a special role of the quantity* $\dim \lambda / |\lambda|!$ *in the expression for* $M_{z,z',\xi}$*: this is a Vandermonde-like object, which creates a kind of log-gas pair interaction between particles from the random configuration* $\mathcal{L}(\lambda)$ *(about log-gas systems, see [For10a], [For10b]). The particle/hole involution* $\mathcal{L}(\lambda) \to \mathcal{L}^{\circ}(\lambda)$ *changes the sign of interaction between particles on the different sides from 0, so that we get two kinds of particles which are oppositely charged. Note that in the first regime, the particle/hole involution is necessary for existence of a limiting point process. The Whittaker kernel is an instance of a correlation kernel which is symmetric with respect to an indefinite inner product.*

25.4.2 Special instances of z-measures

(a) *Meixner and Laguerre ensembles.* Assume $z = N = 1, 2, \ldots$ and $z' = N + b - 1$ with $b > 0$; these are admissible values. Then $M_{z,z',\xi}$ is supported by the subset $\mathbb{Y}(N) \subset \mathbb{Y}$ of Young diagrams with at most N non-zero rows. Under the correspondence

$$\mathbb{Y}(N) \ni \lambda \mapsto (l_1, l_2, \ldots, l_N) = (\lambda_1 + N - 1, \lambda_2 + N - 2, \ldots, \lambda_N) \subset \mathbb{Z}_+,$$

the measure turns into a random-matrix-type object: the N-particle Meixner orthogonal polynomial ensemble with discrete weight function $(b)_l \xi^l / l!$, where the argument l ranges over $\mathbb{Z}_+$ (for generalities about orthogonal ensembles, see [Kon05]). It follows that for general values of (z, z'), the measure $M_{z,z',\xi}$ may be viewed as the result of *analytic continuation* of the Meixner ensembles with respect to parameters N and b. This observation is exploited in [Bor06]. In a scaling limit regime as

$\xi \to 1$, the N-particle Meixner ensemble turns into the N-point Laguerre ensemble; the correlation kernel for the latter ensemble is a degeneration of the Whittaker kernel, see [Bor00a].

(b) *Generalized permutations.* Recall that the Plancherel measure $M^{(n)}$ on $\mathbb{Y}_n$ coincides with the pushforward of the uniform measure on S_n under the projection $S_n \to \mathbb{Y}_n$ afforded by the Robinson–Schensted correspondence RS (Section 25.3.6). Here is a generalization:

Fix natural numbers $N \le N'$ and replace S_n by the finite set $S^{(n)}_{N,N'}$ consisting of all $N \times N'$ matrices with entries in $\mathbb{Z}_+$ such that the sum of all entries equals n. Elements of $S^{(n)}_{N,N'}$ are called *generalized permutations.* Knuth's generalization of the Robinson-Schensted correspondence (the RSK correspondence, see, e.g. [Ful97, Section 4.1]) provides a projection of $S^{(n)}_{N,N'}$ onto $\mathbb{Y}_n(N) := \mathbb{Y}_n \cap \mathbb{Y}(N)$, the set of Young diagram with n boxes and at most N non-zero rows. It turns out that the pushforward of the uniform distribution on $S^{(n)}_{N,N'}$ coincides with the z-measure $M^{(n)}_{N,N'}$, see [Bor01a].

(c) *A variation.* In the same way one can get the mixed z-measure $M_{N,N',\xi}$ if instead of $S^{(n)}_{N,N'}$ one takes $N \times N'$ matrices whose entries are iid random variables, the law being the geometric distribution with parameter ξ.

(d) *Random words.* Denote by $S^{(n)}_{N,\infty}$ the set of words of length n in the alphabet $[N] := \{1, \dots, N\}$. Endowing $S^{(n)}_{N,\infty}$ with the uniform measure we get a model of random words. This model may be viewed as a degeneration of the model of random generalized permutations (item (b) above) in the limit $N' \to \infty$ (this explains the notation $S^{(n)}_{N,\infty}$). The RSK correspondence (or rather its simpler version due to Schensted) provides a projection $S^{(n)}_{N,\infty} \to \mathbb{Y}_n(N)$ taking random words to random Young diagrams $\lambda \in \mathbb{Y}_n(N)$ with distribution $M^{(n)}_{N,\infty} := \lim_{N' \to \infty} M^{(n)}_{N,N'}$. Asymptotic properties of random words are studied in [Tra01] and [Joh01a]. The model of random words can be generalized by allowing non-uniform probability distributions on the alphabet (see [Its01] and references therein). As explained in [Its01], this more general model is connected to the Schur measure discussed in Section 25.4.3 below.

(e) *The Charlier ensemble and the Plancherel degeneration.* Poissonization of the measure $M^{(n)}_{N,\infty}$ with respect to parameter n leads to the N-particle Charlier ensemble [Bor01a, §9]. Alternatively, it can be obtained as a limit case of the mixed z-measures $M_{N,N',\xi}$. The poissonized Plancherel measure M_ν appears as the limit of the mixed z-measures $M_{z,z',\xi}$ when $z, z' \to \infty$ and $\xi \to 0$ in such a way that $zz'\xi \to \nu$. This fact prompted derivation of the discrete Bessel kernel (Theorem 25.3.3) in [Bor00b]. Alternatively, M_ν can be obtained through a limit transition from the Charlier or Meixner ensembles; this leads to another derivation of the discrete Bessel kernel: [Joh01a], [Joh01b].

25.4.3 The Schur measures

Let Λ denote the graded algebra of symmetric functions. The Schur functions s_λ, indexed by arbitrary partitions $\lambda \in \mathbb{Y}$, form a distinguished homogeneous basis in Λ. As a graded algebra, Λ is isomorphic to the algebra of polynomials in countably many generators; as these generators, one can take, for instance the complete homogeneous symmetric functions $h_1, h_2, \dots$ where $\deg h_k = k$. One has $s_\lambda = \det[h_{\lambda_i - i + j}]$ with the understanding that $h_0 = 1$ and $h_k = 0$ for $k < 0$ (the Jacobi-Trudi formula); here the order of the determinant can be chosen arbitrarily provided it is large enough. For more detail, see, e.g. [Sag01].

Given two multiplicative functionals $\varphi, \psi\colon \Lambda \to \mathbb{C}$, the corresponding (complex-valued) *Schur measure* $M_{\varphi,\psi}$ on $\mathbb{Y}$ is defined by

$$M_{\varphi,\psi}(\lambda) = \text{const}^{-1}\, \varphi(s_\lambda)\psi(s_\lambda), \quad \lambda \in \mathbb{Y}, \qquad \text{const} = \sum_{\lambda\in\mathbb{Y}} \varphi(s_\lambda)\psi(s_\lambda),$$

provided that the sum is absolutely convergent (which is a necessary condition on φ, ψ). This notion, due to Okounkov [Oko01a], provides a broad generalization of the mixed z-measures. Since a multiplicative functional is uniquely determined by its values on the generators h_k, the Schur measure has a doubly-infinite collection of parameters $\{\varphi(h_k), \psi(h_k); k = 1, 2, \dots\}$. In this picture, the z-measures correspond to a very special collection of parameters

$$\varphi(h_k) = \xi^{k/2}(z)_k/k!, \quad \psi(h_k) = \xi^{k/2}(z')_k/k!, \qquad k = 1, 2, \dots,$$

and the poissonized Plancherel measure M_ν appears when $\varphi(h_k) = \psi(h_k) = \nu^{k/2}/k!$.

As shown in [Oko01a], Theorem 25.4.1 extends to Schur measures: if the parameters are such that the measure $M_{\varphi,\psi}$ is non-negative (and hence is a probability measure), then it gives rise to a lattice determinantal point process. Moreover, for the corresponding correlation kernel one can write down an explicit contour integral representation [Bor00c]. Such a representation is well suited for asymptotic analysis.

If φ and ψ are evaluations of symmetric functions at finitely many positive variables, the first row λ_1 can be interpreted as the last passage percolation time in a suitable directed percolation model on the plane, see [Joh05].

25.4.4 Some generalizations

Kerov [Ker00] generalized the construction of the z-measures $M^{(n)}_{z,z'}$ by introducing an additional parameter related to Jack polynomials. This new parameter is similar to the β parameter in random matrix ensembles [For10b]. In particular, the Plancherel measure $M^{(n)}$, which is a limit case of the z-measures, also allows a β-deformation [Ker00], [Oko05], [Oko06]. The ordinary z-measures correspond to the special value $\beta = 2$, and in the limit $\beta \to 0$ the beta z-measures degenerate to the measures (25.2.5) derived from the Ewens measures, see [Ols10, Section 1.2]. Thus, the β parameter interpolates between the models of Section 25.2 and those of Sections 25.3–25.4, as has been pointed out in the end of Section 25.1.1. Note also that replacing the Schur functions by the Jack symmetric functions leads to a natural β-deformation of the Schur measures.

As in random matrix theory, the value $\beta = 2$ is a distinguished one, while in the general case $\beta > 0$ the situation is much more complex. Some results for $\beta \neq 2$ can be found in [Bor05b], [Ful04], [Ols10], [Str10a], [Str10b].

In a somewhat different direction, one can define natural analogues of the Plancherel measure and Schur measures for shifted Young diagrams (equivalently, strict partitions): [Tra04], [Mat05]. This theory is related to Schur's Q-functions (a special case of Hall-Littlewood symmetric functions that appears in the theory of projective representations of the symmetric group). Surprisingly enough, a natural analogue of the z-measures for shifted diagrams, discovered by Borodin and recently studied in [Pet10], seems not to be related to Schur's Q-functions.

Finally, note that there are many points of contact between the results described in this chapter and Fulman's work on 'random matrix theory over finite fields', see his survey [Ful01] and references therein.

Acknowledgements

This work was supported by the RFBR grant 08-01-00110. I am grateful to Jinho Baik, Alexei Borodin, Alexander Gnedin, and the referee for helpful comments.

References

[Ald99] D. Aldous and P. Diaconis, *Longest increasing subsequences: From patience sorting to the Baik–Deift–Johansson theorem.* Bull. Amer. Math. Soc. **36** (1999), 413–432.

[Arr97] R. Arratia, A. D. Barbour, and S. Tavaré, *Random combinatorial structures and prime factorizations.* Notices Amer. Math. Soc. **44**, no. 8 (1997), 903–910.

[Arr03] R. Arratia, A. D. Barbour, and S. Tavaré, *Logarithmic combinatorial structures: A probabilistic approach.* EMS Monographs in Mathematics. Zürich, Europ. Math. Soc., 2003.

[Bai99] J. Baik, P. Deift, and K. Johansson, *On the distribution of the length of the longest increasing subsequence of random permutations.* J. Amer. Math. Soc. **12** (1999), 1119–1178 [arXiv:math/9810105].

[Bai00] J. Baik, P. Deift, and K. Johansson, *On the distribution of the length of the second row of a Young diagram under Plancherel measure.* Geom. Funct. Anal. **10**(2000), 702–731 [arXiv:math/9901118].

[Bai01] J. Baik and E. M. Rains, *Symmetrized random permutations*. In: "Random matrix models and their applications", MSRI Publ. **40**, Cambridge Univ. Press, Cambridge, 2001, pp. 1–19 [arXiv:math/9910019].

[Bai07] J. Baik, T. Kriecherbauer, K. T.-R. McLaughlin, and P. D. Miller, *Discrete orthogonal polynomials. Asymptotics and applications*. Annals of Mathematics Studies, **164**. Princeton University Press, Princeton, NJ, 2007.

[Bia01] P. Biane, *Approximate factorization and concentration for characters of symmetric groups*. Intern. Math. Res. Notices **2001** (2001), no. 4, 179–192 [arXiv:math/0006111].

[Bog07] L. V. Bogachev and Z. Su, *Gaussian fluctuations of Young diagrams under the Plancherel measure*. Proc. Roy. Soc. Lond. Ser. A Math. Phys. Eng. Sci. **463** (2007), no. 2080, 1069–1080 [arXiv:math/0607635].

[Bor00a] A. Borodin and G. Olshanski, *Distributions on partitions, point processes and the hypergeometric kernel*. Comm. Math. Phys. **211** (2000), 335–358 [arXiv:math/9904010].

[Bor00b] A. Borodin, A. Okounkov, and G. Olshanski, *Asymptotics of Plancherel measures for symmetric groups*. J. Amer. Math. Soc. **13** (2000), 491–515 [arXiv:math/9905032].

[Bor00c] A. Borodin and A. Okounkov, *A Fredholm determinant formula for Toeplitz determinants*. Integral Equations Oper. Theory **37** (2000), 386–396 [arXiv:math/9907165].

[Bor01a] A. Borodin and G. Olshanski, *Z-Measures on partitions, Robinson-Schensted-Knuth correspondence, and $\beta = 2$ ensembles*. In: " Random matrix models and their applications" (P. M. Bleher and A. R. Its, eds). MSRI Publications **40**, Cambridge Univ. Press, 2001, pp. 71–94 [arXiv:math/9905189].

[Bor01b] A. Borodin and G. Olshanski, *Infinite random matrices and ergodic measures*. Commun. Math. Phys. **223** (2001), 87–123 [arXiv:math-ph/0010015].

[Bor05a] A. Borodin and G. Olshanski, *Harmonic analysis on the infinite–dimensional unitary group and determinantal point processes*. Ann. Math. **161** (2005), 1319–1422 [arXiv:math/0109194].

[Bor05b] A. Borodin and G. Olshanski, *Z-measures on partitions and their scaling limits*. European J. Combin. **26** (2005), no. 6, 795–834 [arXiv:math-ph/0210048].

[Boro5c] A. Borodin and G. Olshanski, *Random partitions and the Gamma kernel*. Adv. Math. **194** (2005), 141–202 [arXiv:math-ph/0305043].

[Bor06] A. Borodin and G. Olshanski, *Meixner polynomials and random partitions*. Moscow Math. J. **6** (2006), 629–655 [arXiv:math/0609806].

[Bor10] A. Borodin, *Determinantal point processes*. Chapter 11 of the present Handbook.

[Bou07] P. Bourgade, A. Nikeghbali, and A. Rouault, *Ewens measures on compact groups and hypergeometric kernels*, Séminaire de Probabilités XLIII Lecture Notes in Mathematics **2006** (2011) 351–377 Springer, 2011 [arXiv:0712.0848].

[Dei00] P. Deift, *Integrable systems and combinatorial theory*. Notices Amer. Math. Soc. 47 (2000), no. 6, 631–640.

[Dia94] P. Diaconis and M. Shahshahani, *On the eigenvalues of random matrices*. In: "Studies in applied probability: Essays in honor of Lajos Takács", J. Appl. Probab., special volume **31A** (1994), 49–62.

[Ewe98] W. J. Ewens and S. Tavaré, *The Ewens sampling formula*. In: "Encyclopedia of Statistical Science", Vol. 2 (S. Kotz, C. B. Read, and D. L. Banks, eds.), pp. 230–234, Wiley, New York, 1998.

[For10a] P. J. Forrester, *Log-gases and random matrices*. Princeton Univ. Press, 2010.

[For10b] P. J. Forrester, *Beta ensembles*, Chapter 20 of the present Handbook.

[Ful97] W. Fulton, *Young tableaux, with applications to representation theory and geometry*, Cambridge Univ. Press, 1997.

[Ful01] J. Fulman, *Random matrix theory over finite fields*. Bull. Amer. Math. Soc. (New Series) **39** (2001), no. 1, 51–85 [arXiv:math/0003195].

[Ful04] J. Fulman, *Stein's method, Jack measure, and the Metropolis algorithm.* J. Comb. Theory, Ser. A **108** (2004), 275–296 [arXiv:math/0311290]

[Hol01] L. Holst, *The Poisson–Dirichlet distribution and its relatives revisited.* KTH preprint. Stockholm, 2001. Available from http://www.math.kth.se/matstat/fofu/reports/PoiDir.pdf

[Hua58] L. K. Hua, *Harmonic analysis of functions of several complex variables in the classical domains.* Chinese edition: Science Press, Peking, 1958; Russian edition: IL, Moscow, 1959; English edition: Transl. Math. Monographs **6**, Amer. Math. Soc., 1963.

[Ign82] T. Ignatov, *On a constant arising in the theory of symmetric groups and on Poisson-Dirichlet measures.* Theory Probab. Appl. **27**, 136–147.

[Iva02] V. Ivanov and G. Olshanski, *Kerov's central limit theorem for the Plancherel measure on Young diagrams.* In: "Symmetric functions 2001. Surveys of developments and perspectives". Proc. NATO Advanced Study Institute (S. Fomin, ed.), Kluwer, 2002, pp. 93–151 [arXiv:math/0304010].

[Its01] A. R. Its, C. A. Tracy and H. Widom, *Random words, Toeplitz determinants and integrable systems.* II. Physica D **152–153** (2001), 199–224 [arXiv:nlin/0004018].

[Joh98] K. Johansson, *On fluctuations of eigenvalues of random Hermitian matrices.* Duke Math. J. **91** (1998), 151–204.

[Joh01a] K. Johansson, *Discrete orthogonal polynomial ensembles and the Plancherel measure.* Ann. Math. (2) **153** (2001) 259–296 [arXiv:math/9906120].

[Joh01b] K. Johansson, Random permutations and the discrete Bessel kernel. In: *Random matrix models and their applications.* (P. M. Bleher and A. R. Its, eds). MSRI Publications **40**, Cambridge Univ. Press, 2001, pp. 259–269.

[Joh05] K. Johansson, *Random matrices and determinantal processes.* In: A. Bovier et al. editors, Mathematical Statistical Physics, Session LXXXIII: Lecture Notes of the Les Houches Summer School 2005, pages 1–56. Elsevier Science, 2006 [arXiv:math-ph/0510038].

[Ker93a] S. V. Kerov, *Gaussian limit for the Plancherel measure of the symmetric group.* Comptes Rendus Acad. Sci. Paris, Série I, **316** (1993), 303–308.

[Ker93b] S. V. Kerov, *Transition probabilities of continual Young diagrams and Markov moment problem.* Funct. Anal. Appl. **27** (1993), 104–117.

[Ker93c] S. Kerov, G. Olshanski, and A. Vershik, *Harmonic analysis on the infinite symmetric group. A deformation of the regular representation.* Comptes Rendus Acad. Sci. Paris. Sér. I, **316** (1993), 773–778.

[Ker97] S. V. Kerov and N. V. Tsilevich, *Stick breaking process generated by virtual permutations with Ewens distribution.* J. Math. Sci. (New York) **87** (1997), no. 6, 4082–4093.

[Ker98] S. Kerov, A. Okounkov, G. Olshanski, *The boundary of the Young graph with Jack edge multiplicities.* Internat. Math. Res. Notices **1998** (1998), no. 4, 173–199 [arXiv:q-alg/9703037].

[Ker00] S. V. Kerov, *Anisotropic Young diagrams and Jack symmetric functions.* Funct. Anal. Appl. **34** (2000), no. 1, 45–51, [arXiv:math/9712267].

[Ker03] S. V. Kerov, *Asymptotic representation theory of the symmetric group and its applications in analysis.* Translations of Mathematical Monographs, **219**. American Mathematical Society, Providence, RI, 2003.

[Ker04] S. Kerov, G. Olshanski, and A. Vershik, *Harmonic analysis on the infinite symmetric group.* Invent. Math. **158** (2004), 551–642 [arXiv:math/0312270].

[Kin75] J. F. C. Kingman, *Random discrete distributions.* J. Roy. Statist. Soc. Ser. B **37** (1975), 1–22.

[Kin78a] J. F. C. Kingman, *Random partitions in population genetics.* Proc. Roy. Soc. London Ser. A **361** (1978), no. 1704, 1–20.

[Kin78b] J. F. C. Kingman, *The representation of partition structures*. J. London Math. Soc. (2) **18** (1978), no. 2, 374–380.

[Kin93] J. F. C. Kingman, *Poisson Processes*. Oxford University Press, 1993.

[Kon05] W. König, *Orthogonal polynomial ensembles in probability theory*. Prob. Surveys **2** (2005), 385–447.

[Log77] B. F. Logan and L. A. Shepp, *A variational problem for random Young tableaux*. Advances in Math. **26** (1977), no. 2, 206–222.

[Mat05] Sho Matsumoto, *Correlation functions of the shifted Schur measure*. J. Math. Soc. Japan 57(2005), 619–637 [arXiv:math/0312373].

[Ner02] Yu. A. Neretin, *Hua type integrals over unitary groups and over projective limits of unitary groups*. Duke Math. J. **114** (2002), 239–266 [arXiv:math-ph/0010014].

[Oko00] A. Okounkov, *Random matrices and random permutations*. Intern. Mathem. Research Notices **2000** (2000), no. 20, 1043–1095 [arXiv:math/9903176].

[Oko01a] A. Okounkov, *Infinite wedge and random partitions*. Selecta Math. (N.S.) **7** (2001), 57–81 [arXiv:math/9907127].

[Oko01b] A. Okounkov, *$SL(2)$ and z–measures*, in: Random matrix models and their applications (P. M. Bleher and A. R. Its, eds). Mathematical Sciences Research Institute Publications **40**, Cambridge Univ. Press, 2001,407–420 [arXiv:math/0002136].

[Oko05] A. Okounkov, *The uses of random partitions*. In: "XIVth International Congress on Mathematical Physics", World Sci. Publ., 2005, pp. 379–403 [arXiv:math-ph/0309015].

[Oko06] A. Okounkov, *Random partitions*, Encyclopedia of Mathematical Physics, vol. 4, Elsevier, 2006, p. 347.

[Ols96] G. Olshanski and A. Vershik, *Ergodic unitarily invariant measures on the space of infinite Hermitian matrices*. In "Contemporary Mathematical Physics. F. A. Berezin's memorial volume" (R. L. Dobrushin et al., eds). Amer. Math. Soc. Transl. Ser. 2, **175**, 1996, pp. 137–175 [arXiv:math/9601215].

[Ols03a] G. Olshanski, *The problem of harmonic analysis on the infinite-dimensional unitary group*. J. Funct. Anal. **205** (2003), no. 2, 464–524 [arXiv:math/0109193].

[Ols03b] G. Olshanski, *An introduction to harmonic analysis on the infinite symmetric group*. In: Asymptotic Combinatorics with Applications to Mathematical Physics (A.M.Vershik, ed.), Springer Lect. Notes Math. **1815** (2003), 127–160 [arXiv:math/0311369].

[Ols08] G. Olshanski, *Difference operators and determinantal point processes*. Funct. Anal. Appl **42** (2008), no. 4, 317–329 [arXiv:0810.3751].

[Ols10] G. Olshanski, *Anisotropic Young diagrams and infinite-dimensional diffusion processes with the Jack parameter*. Intern. Research Math. Notices, **2010** (2010), no. 6, 1102–1166 [arXiv:0902.3395].

[Pet10] L. Petrov, *Random strict partitions and determinantal point processes*, Electr. Commun. Probab. **15** (2010), 162–175. [arXiv:1002.2714].

[Pic87] D. Pickrell, *Measures on infinite dimensional Grassmann manifold*, J. Funct. Anal. **70** (1987), 323–356.

[Pic91] D. Pickrell, *Mackey analysis of infinite classical motion groups*. Pacific J. Math. **150** (1991), 139–166.

[Pit97] J. Pitman and M. Yor, *The two-parameter Poisson–Dirichlet distribution derived from a stable subordinator*. Ann. Probab. **25** (1997), 855–900.

[Sag01] B. Sagan, *The Symmetric Group* (second ed.) Graduate Texts in Math. **203**, Springer-Verlag, New York, 2001.

[Sch51] I. J. Schoenberg, *On Pólya frequency functions. I. The totally positive functions and their Laplace transforms*. Journal d'Analyse Mathématique **1** (1951), 331–374.

[Shi75] H. Shimomura, *On the construction of invariant measure over the orthogonal group on the Hilbert space by the method of Cayley transformation*, Publ. RIMS, Kyoto Univ. **10** (1975), 413–424.

[Sta07] R. P. Stanley, *Increasing and decreasing subsequences and their variants*. In: "International Congress of Mathematicians". Vol. I, 545–579, Europ. Math. Soc., Zürich, 2007 [arXiv:math/0512035].

[Str10a] E. Strahov, *Z-measures on partitions related to the infinite Gelfand pair* $(S(2\infty), H(\infty))$. J. Alg. **323** (2010), 349–370 [arXiv:0904.1719].

[Str10b] E. Strahov, *The z-measures on partitions, Pfaffian point processes, and the matrix hypergeometric kernel*. Adv. Math. **224** (2010), no. 1, 130–168 [arXiv:0905.1994].

[Tra94] C. A. Tracy and H. Widom, *Level-spacing distributions and the Airy kernel*. Commun. Math. Phys. **159** (1994), 151–174 [arXiv:hep-th/9211141].

[Tra96] C. A. Tracy and H.Widom, *On orthogonal and symplectic matrix ensembles*. Commun. Math. Phys. **177** (1996), 727–754 [arXiv:solv-int/9509007].

[Tra01] C. A. Tracy and H. Widom, *On the distributions of the lengths of the longest monotone subsequences in random words*. Probab. Theory Relat. Fields **119** (2001), 350–380 [arXiv:math/9904042].

[Tra04] C. A. Tracy and H. Widom, *A limit theorem for shifted Schur measures*. Duke Math. J. **123** (2004), 171–208 [arXiv:math/0210255].

[Ver72] A. M. Vershik and A. A. Shmidt, *Symmetric groups of high degree*. Soviet Math. Dokl. **13** (1972), 1190–1194.

[Ver77a] A. M. Vershik and A. A. Shmidt, *Limit measures arising in the asymptotic theory of symmetric groups, I*. Probab. Theory Appl. **22** (1977), 70–85.

[Ver77b] A. M. Vershik, and S. V. Kerov, *Asymptotics of the Plancherel measure of the symmetric group and the limiting form of Young tableaux*. Soviet Math. Dokl. **18** (1977), 527–531.

[Ver78] A. M. Vershik and A. A. Shmidt, *Limit measures arising in the asymptotic theory of symmetric groups, II*. Probab. Theory Appl. **23** (1978), 36–49.

[Wat76] G. A. Watterson, *The stationary distribution of the infinitely-many neutral alleles diffusion model*. J. Appl. Prob. **13** (1976), 639–651.

·26·

Enumeration of maps

J. Bouttier

Abstract

This chapter is devoted to the connection between random matrices and maps, i.e. graphs drawn on surfaces. We concentrate on the one-matrix model and explain how it encodes and allows us to solve a map enumeration problem.

26.1 Introduction

Maps are fundamental objects in combinatorics and graph theory, originally introduced by Tutte in his series of 'census' papers [Tut62a, Tut62b, Tut62c, Tut63]. Their connection to random matrix theory was pioneered in the seminal paper entitled 'Planar diagrams' by Brézin, Itzykson, Parisi, and Zuber [Bre78] building on an original idea of 't Hooft [Hoo74]. In parallel with the idea that planar diagrams (i.e. maps) form a natural discretization for the random surfaces appearing in 2D quantum gravity (see chapter 30), this led to huge developments in physics among which some of the highlights are the solution by Kazakov [Kaz86] of the Ising model on dynamical random planar lattices (i.e. maps) then used as a check for the Knizhnik-Polyakov-Zamolodchikov relations [Kni88].

Matrix models, or more precisely *matrix integrals*, are efficient tools to address map enumeration problems, and complement other techniques which have their roots in combinatorics, such as Tutte's original recursive decomposition or the bijective approach. In spite of the fact that these techniques originate from different communities, they are undoubtedly intimately connected. For instance, it was recognized that the loop equations for matrix models correspond to Tutte's equations [Eyn06], while knowledge of the matrix model result proved instrumental in finding a bijective proof [Bou02b] for a fundamental result of map enumeration [Ben94].

In this chapter, we present an introduction to the method of matrix integrals for map enumeration. To keep a clear example in mind, we concentrate on the problem considered in [Ben94, Bou02b] of counting the number of planar maps with a prescribed degree distribution, though we mention a number of possible generalizations of the results encountered on the way. Of course many other problems can be addressed, see for instance Chapter 30 for a review of models of statistical physics on discrete random surfaces (i.e. maps) which can be solved either exactly or asymptotically using matrix integrals.

This chapter is organized as follows. In Section 26.2, we introduce maps and related objects. In section 26.3, we discuss the connection between matrix integrals and maps itself: we focus on the Hermitian one-matrix model with a polynomial potential, and explain how the formal expansion of its free energy around a Gaussian point (quadratic potential) can be represented by diagrams identifiable with maps. In Section 26.4, we show how techniques of random matrix theory introduced in previous chapters allow us to deduce the solution of the map enumeration problem under consideration. We conclude in Section 26.5 by explaining how to translate the matrix model result into a bijective proof.

26.2 Maps: definitions

Colloquially, maps are 'graphs drawn on a surface'. The purpose of this section is to make this definition more precise. We recall the basic definitions related to graphs in Section 26.2.1, before defining maps as graphs embedded into surfaces in Section 26.2.2. Section 26.2.3 is devoted to a coding of maps by pairs of permutations, which will be useful for the following section.

26.2.1 Graphs

A *graph* is defined by the data of its *vertices*, its *edges*, and their *incidence relations.* Saying that a vertex is incident to an edge simply means that it is one of its extremities, hence each edge is incident to two vertices. Actually, we allow for the two extremities of an edge to be the same vertex, in which case the edge is called a loop. Moreover several edges may have the same two extremities, in which case they are called multiple edges. A graph without loops and multiple edges is said to be *simple.* (In other terminologies, graphs are always simple and multigraphs refer to the ones containing loops and multiple edges.) A graph is *connected* if it is not possible to split its set of edges into two or more non-empty subsets in such a way that no edge is incident to vertices from different subsets. The *degree of a vertex* is the number of edges incident to it, counted with multiplicity (i.e. a loop is counted twice).

26.2.2 Maps as embedded graphs

The surfaces we consider here are supposed to be compact, connected, orientable[1] and without boundary. It is well known that such surfaces are characterized, up to homeomorphism, by a unique non-negative integer called the genus, the sphere having genus 0.

[1] It is however possible to extend our definition to non-orientable surfaces.

The *embedding* of a graph into a surface is a function associating each vertex with a point of the surface and each edge with a simple open arc, in such a way that the images of distinct graph elements (vertices and edges) are disjoint and incidence relations are preserved (i.e. the extremities of the arc associated with an edge are the points associated with its incident vertices). The embedding is *cellular* if the connected components of the complementary of the embedded graph are simply connected. In that case, these are called the *faces* and we extend the incidence relations by saying that a face is incident to the vertices and edges on its boundary. Each edge is incident to two faces (which might be equal) and the *degree of a face* is the number of edges incident to it (counted with multiplicity). It is not difficult to see that the existence of cellular embedding requires the graph to be connected (for otherwise, graph components would be separated by non-contractible loops).

Maps are cellular embeddings of graphs considered up to continuous deformation, i.e. up to an homeomorphism of the target surface. Clearly such a continuous deformation preserves the incidence relations between vertices, edges, and faces. It also preserves the genus of the target surface, hence it makes sense to speak about the genus of a map. A *planar map* is a map of genus 0. In all rigour, this notion should not be confused with that of a planar graph,[2] which is a graph having at least one embedding into the sphere. Figure 26.1 shows a few cellular embeddings of the same (planar) graph into a sphere (drawn in the plane by stereographic projection): the first three are equivalent to each other, but not to the fourth (which might immediately be seen by comparing the face degrees). Let us conclude this section by stating the following well-known result.

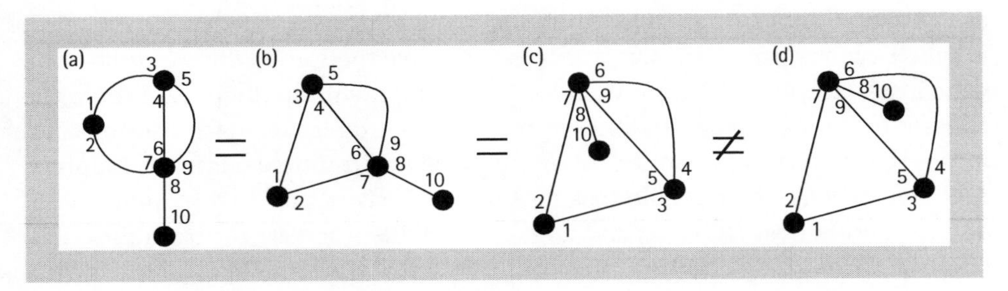

Fig. 26.1 Several cellular embeddings of the same graph into a sphere (drawn in the plane by stereographic projection), where half-edges are labelled from 1 to 10. (a) and (b) differ only by a simple deformation. (c) is obtained by choosing another pole for the projection. Thus (a), (b), and (c) represent the same planar map encoded by the permutations $\sigma = (1\,2)(3\,4\,5)(6\,7\,8\,9)(10)$ and $\alpha = (1\,3)(2\,7)(4\,6)(5\,9)(8\,10)$. (d) represents a different map, encoded by the permutations $\sigma' = \sigma \circ (8\,9)$ and $\alpha' = \alpha$. Their inequivalence may be checked by comparing the face degrees or noting that $\sigma \circ \alpha$ is not conjugate to $\sigma' \circ \alpha'$.

[2] Unfortunately the literature is often not that careful. To illustrate the distinction let us mention that, while the first enumerative results for maps date back to Tutte in the 1960s, the mere asymptotic counting of planar graphs (in our present definition) is a fairly recent result [Gim05].

Theorem 26.2.1 (Euler characteristic formula) *In a map of genus g, we have*

$$\#\{\text{vertices}\} - \#\{\text{edges}\} + \#\{\text{faces}\} = 2 - 2g. \tag{26.2.1}$$

This quantity is the Euler characteristic χ *of the map.*

In the planar case $g = 0$, the formula is also called the Euler relation.

26.2.3 Combinatorial maps

As discussed above, a map contains more data than a graph, and one may wonder what the missing data is. It turns out that the embedding into an oriented surface amounts to defining a cyclic order on the edges incident to a vertex.[3] More precisely, as we allow for loops, it is better to consider *half-edges*, each of them being incident to exactly one vertex. Let us label the half-edges by distinct consecutive integers $1, \dots, 2m$ (where m is the number of edges) in an arbitrary manner. Given a half-edge i, let $\alpha(i)$ be the other half of the same edge, and let $\sigma(i)$ be the half-edge encountered after i when turning counterclockwise around its incident vertex (see again Figure 26.1 for an example with $m = 5$). It is easily seen that σ and α are permutations of $\{1, \dots, 2m\}$, which furthermore satisfy the following properties:

(A) α is an involution without fixed point,
(B) the subgroup of the permutation group generated by σ and α acts transitively on $\{1, \dots, 2m\}$.

The latter property simply expresses that the underlying graph is necessarily connected. The subgroup generated by α and σ is called the cartographic group [Zvo95].

These permutations fully characterize the map. Actually, a pair of permutations (σ, α) satisfying properties (A) and (B) is called a *labelled combinatorial map* and there is a one-to-one correspondence between maps (as defined in the previous section) with labelled half-edges and labelled combinatorial maps. In this correspondence, the vertices are naturally associated with the cycles of σ, the edges with the cycles of α, and the faces with the cycles of $\sigma \circ \alpha$. At this stage one may have noticed that vertices and faces play a symmetric role, indeed $(\sigma \circ \alpha, \alpha)$ is also a labelled combinatorial map corresponding to the *dual map*. Degrees are given the length of the corresponding cycles, and the Euler characteristic is given by

$$\chi(\sigma, \alpha) = c(\sigma) - c(\alpha) + c(\sigma \circ \alpha) \tag{26.2.2}$$

where $c(\cdot)$ denotes the number of cycles.

[3] This observation is generally attributed to Edmonds [Edm60]. For a comprehensive graph-theoretical treatment of embeddings into surfaces, we refer to the book of Mohar and Thomassen [Moh01].

Our coding of maps via pairs of permutations depends on an arbitrary labelling of the half-edges, which might seem unsatisfactory. We shall identify configurations differing by a relabelling, and it is easily seen that this amounts to identifying (σ, α) to all $(\rho \circ \sigma \circ \rho^{-1}, \rho \circ \alpha \circ \rho^{-1})$ where ρ is an arbitrary permutation of $\{1, \ldots, 2m\}$. These equivalence classes are in one-to-one correspondence with unlabelled maps. This distinction has some consequences for enumeration: for instance the number of labelled maps with m edges is *not* equal to $(2m)!$ times the number of unlabelled maps, because some equivalence classes have fewer than $(2m)!$ elements (due to symmetries). By the orbit-stabilizer theorem, the number of elements in the class of (σ, α) is $(2m)!/\,\Gamma(\sigma, \alpha)$ where $\Gamma(\sigma, \alpha)$ is the number of permutations ρ such that $\sigma = \rho \circ \sigma \circ \rho^{-1}$ and $\alpha = \rho \circ \alpha \circ \rho^{-1}$ (such ρ is an *automorphism*). $\Gamma(\sigma, \alpha)$ is often called the 'symmetry factor' in the literature, though it is seldom properly defined.

As we shall see in the next section, matrix integrals are naturally related to labelled maps. Enumerating unlabelled maps is a harder problem as it requires classifying their possible symmetries, which is beyond the scope of this text.[4] The distinction is circumvented when considering *rooted* maps i.e. maps with a distinguished half-edge (often represented as a marked oriented edge): such maps have no non-trivial automorphism hence the enumeration problem is equivalent in the labelled and unlabelled case. Most enumeration results in the literature deal with rooted maps. Furthermore maps of large size are 'almost surely' asymmetric, so the distinction is irrelevant in this context.

26.3 From matrix integrals to maps

In this section, we return to random matrices in order to present their connection with maps. We will concentrate on the so-called one-matrix model (though we will allude to its generalizations): maps appear as diagrams representing the expansion of its partition function around a Gaussian point. Our goal is to explain this construction in some detail, as this might also be useful for the comprehension of other chapters.

This section is organized as follows. Section 26.3.1 provides the definitions and the main statement (Theorem 26.3.1) of the topological expansion in the one-matrix model. The following sections are devoted to its derivation and generalizations. Section 26.3.2 discusses Wick's theorem for Gaussian matrix models. Section 26.3.3 introduces *ab initio* the diagrammatic expansion of the one-matrix model. Section 26.3.4 formalizes this computation and shows its natural relation with combinatorial maps. Section 26.3.5 finally presents a few generalizations of the one-matrix model.

[4] For more on the topic of combinatorial maps and their automorphisms, we refer the reader to [Cor75, Cor92] and references therein. [Cor92] also discusses hypermaps, which are the natural generalization of combinatorial maps obtained when relaxing constraint (A): these are actually bipartite maps in disguise, associated with the two-matrix model of equation (26.3.26).

26.3.1 The one-matrix model

We consider the model of a Hermitian random matrix in a polynomial potential, often simply called the *one-matrix model*, which has already been discussed in previous chapters. Different notations and conventions exist, for the purposes of this chapter we define its partition function as

$$\Xi_N(t, V) = \frac{\int e^{N\mathrm{Tr}(-M^2/(2t)+V(M))}\, dM}{\int e^{-N\mathrm{Tr}\, M^2/(2t)}\, dM} \tag{26.3.1}$$

where dM is the Lebesgue measure over the space of Hermitian matrices, and V stands for a 'perturbation' of the form

$$V(x) \equiv \sum_{n=1}^{\infty} \frac{v_n}{n} x^n. \tag{26.3.2}$$

Here we shall consider the coefficients v_n as formal variables hence, rather than a polynomial, V is a formal power series in x and v_n. In this sense, a more *proper definition of the partition function* $\Xi_N(t, V)$ is

$$\Xi_N(t, V) = \left\langle e^{N\mathrm{Tr}\, V(M)} \right\rangle \tag{26.3.3}$$

where $\langle \cdot \rangle$ denotes the expectation value with respect to the Gaussian measure proportional to $e^{-N\mathrm{Tr}\, M^2/(2t)}\, dM$, acting coefficient-wise on $e^{N\mathrm{Tr}\, V(M)}$ viewed as a formal power series in v_n whose coefficients are polynomials in the matrix elements. In other words, the matrix integral in (26.3.1) must be understood in the 'formal' sense of Chapter 16. We define furthermore the *free energy* by

$$F_N(t, V) = \log \Xi_N(t, V). \tag{26.3.4}$$

The main purpose of this section is to establish the following theorem, which is essentially a formalization of ideas presented in [Bre78, Hoo74].

Theorem 26.3.1 (Topological expansion) *The free energy of the one-matrix model has the 'topological' expansion*

$$F_N(t, V) = \sum_{g=0}^{\infty} N^{2-2g} F^{(g)}(t, V) \tag{26.3.5}$$

where $F^{(g)}(t, V)$ is equal to the exponential generating function for labelled maps of genus g with a weight t per edge and, for all $n \geq 1$, a weight v_n per vertex of degree n.

Corollary 26.3.1 *The quantity*

$$E^{(g)}(t, V) = 2t \frac{\partial F^{(g)}(t, V)}{\partial t} \tag{26.3.6}$$

is the generating function for rooted maps (i.e. maps with a distinguished half-edge) of genus g. Similarly $n v_n \frac{\partial F^{(g)}(t, V)}{\partial v_n}$ corresponds to rooted maps of genus g whose root *vertex (i.e. the vertex incident to the distinguished half-edge) has degree n. Maps with several marked edges or vertices are obtained by taking multiple derivatives.*

We recall that, from the discussion of Section 26.2.3, a labelled map is a map whose half-edges are labelled $\{1, \ldots, 2m\}$ where m is the number of edges. Hence by the *exponential generating function for labelled maps* we mean

$$F^{(g)}(t, V) = \sum_{m=0}^{\infty} \frac{t^m}{(2m)!} F^{(g,m)}(V) \tag{26.3.7}$$

where $F^{(g,m)}(V)$ is the (finite) sum over labelled maps of genus g with m edges of the product of vertex weights. If we want to reduce $F^{(g,m)}(V)$ to a sum over unlabelled maps instead, then the multiplicity of an individual unlabelled map is $(2m)!/\Gamma$, where Γ is its number of automorphisms. The $(2m)!$ cancels the denominator in (26.3.7) leading to an ordinary generating function where, however, the weight $1/\Gamma$ has to be kept. It differs from the 'true' generating function for unlabelled maps where this weight is absent. For rooted maps there is no difference since $\Gamma = 1$ for all of them: we simply refer to *the generating function for rooted maps* without specifying between exponential/labelled and ordinary/unlabelled.

$F^{(0)}(t, V)$ is called the *planar free energy*. Informally, it is 'dominant' in the large N limit. Actually, Equation (26.3.5) makes sense as a sum of formal power series (at a given order in t, only a finite number of terms contribute).

26.3.2 Gaussian model, Wick theorem

In order to derive Theorem 26.3.1, we first consider the Gaussian measure

$$\left(\frac{N}{2\pi t}\right)^{N^2/2} e^{-N\mathrm{Tr}M^2/(2t)}\, dM \tag{26.3.8}$$

where dM is the Lebesgue (translation-invariant) measure over the set of Hermitian matrices of size N. It is easily seen that the matrix elements are centred jointly Gaussian random variables, with covariance

$$\langle M_{ij} M_{kl} \rangle = \delta_{il}\delta_{jk}\frac{t}{N}. \tag{26.3.9}$$

More generally, the expectation value of the product of arbitrarily many matrix elements is given via Wick's theorem (generally valid for any Gaussian measure). This classical result can be stated as follows.

Theorem 26.3.2 (Wick's theorem for matrix integrals) *The expectation value of the product of an arbitrary number of matrix elements is equal to the sum, over all possible pairwise matchings of the matrix elements, of the product of pairwise covariance.*

For instance, for four matrix elements we have

$$\begin{aligned}\langle M_{i_1 j_1} M_{i_2 j_2} M_{i_3 j_3} M_{i_4 j_4}\rangle &= \langle M_{i_1 j_1} M_{i_2 j_2}\rangle\langle M_{i_3 j_3} M_{i_4 j_4}\rangle \\ &+ \langle M_{i_1 j_1} M_{i_3 j_3}\rangle\langle M_{i_2 j_2} M_{i_4 j_4}\rangle \\ &+ \langle M_{i_1 j_1} M_{i_4 j_4}\rangle\langle M_{i_2 j_2} M_{i_3 j_3}\rangle.\end{aligned} \tag{26.3.10}$$

Clearly, for an odd number of elements the expectation value vanishes by parity, while for an even number $2n$ of elements the sum involves $(2n-1)!! = (2n-1)(2n-3)\cdots 5\cdot 3\cdot 1$ terms.

Let us immediately note that these results extend easily to a Gaussian model of K random Hermitian matrices $M^{(1)}, \ldots, M^{(K)}$ of same size N, with measure

$$\left(\frac{N^K \det Q}{2^K \pi^K}\right)^{N^2/2} \exp\left(-\frac{N}{2}\sum_{a,b=1}^{K} Q_{ab} \mathrm{Tr}\, M^{(a)} M^{(b)}\right) dM^{(1)} \cdots dM^{(K)} \tag{26.3.11}$$

where Q is a $K \times K$ real symmetric matrix. The covariance of two matrix elements is

$$\langle M_{ij}^{(a)} M_{kl}^{(b)}\rangle = \delta_{il}\delta_{jk}\frac{(Q^{-1})_{ab}}{N}. \tag{26.3.12}$$

and, taking the extra index into account, Wick's theorem still applies, for instance

$$\begin{aligned}\langle M_{i_1 j_1}^{(a_1)} M_{i_2 j_2}^{(a_2)} M_{i_3 j_3}^{(a_3)} M_{i_4 j_4}^{(a_4)}\rangle &= \langle M_{i_1 j_1}^{(a_1)} M_{i_2 j_2}^{(a_2)}\rangle\langle M_{i_3 j_3}^{(a_3)} M_{i_4 j_4}^{(a_4)}\rangle \\ &+ \langle M_{i_1 j_1}^{(a_1)} M_{i_3 j_3}^{(a_3)}\rangle\langle M_{i_2 j_2}^{(a_2)} M_{i_4 j_4}^{(a_4)}\rangle \\ &+ \langle M_{i_1 j_1}^{(a_1)} M_{i_4 j_4}^{(a_4)}\rangle\langle M_{i_2 j_2}^{(a_2)} M_{i_3 j_3}^{(a_3)}\rangle.\end{aligned} \tag{26.3.13}$$

26.3.3 Diagrammatics of the one-matrix model: a first approach

We now return to the partition function (26.3.3), which is a formal power series in the variables v_n. If $\boldsymbol{k} = (k_n)_{n\geq 1}$ denotes a family of non-negative integers with finite support, the coefficient of the monomial $\boldsymbol{v}^{\boldsymbol{k}} = \prod_{n=1}^{\infty} v_n^{k_n}$ in $\Xi_N(t, V)$ reads[5]

$$\left[\boldsymbol{v}^{\boldsymbol{k}}\right] \Xi_N(t, V) = \left\langle \prod_{n=1}^{\infty} \frac{(N \mathrm{Tr}\, M^n)^{k_n}}{n^{k_n} k_n!}\right\rangle. \tag{26.3.14}$$

Inside this expression, each trace $\mathrm{Tr}\, M^n$ can be rewritten as a sum of the product of n elements of M, for instance

$$\mathrm{Tr}\, M^3 = \sum_{i,j,k=1}^{N} M_{ij} M_{jk} M_{ki}. \tag{26.3.15}$$

[5] A similar expression appears in [Pen88], albeit with slightly different conventions.

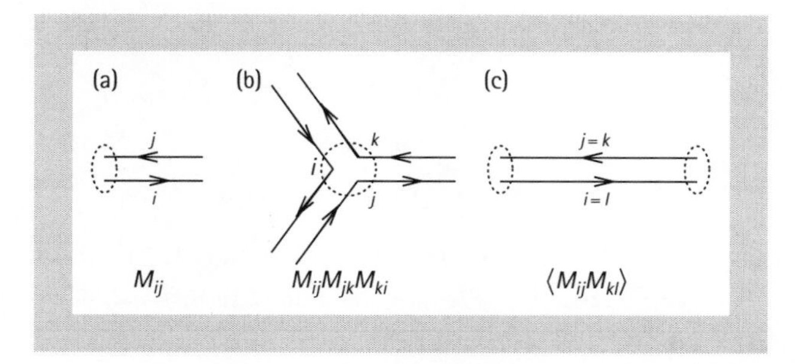

Fig. 26.2 Elements constituting the Feynman diagrams for the Hermitian one-matrix model: (a) a leg representing a matrix element, (b) a three-leg vertex representing a term in $\mathrm{Tr}\, M^3$, (c) two paired elements forming an edge.

hence (26.3.14) may itself be rewritten as the expectation value of a finite linear combination of products of elements of M, to be evaluated via Wick's theorem.

We may represent graphically the factors appearing in this decomposition as follows (see also Figure 26.2):

(a) Each matrix element M_{ij} is represented as a double line originating from a point, forming a *leg*. The lines are oriented in opposite directions (*incoming* and *outgoing*) and 'carry' respectively an index i and j.
(b) Each product of matrix elements appearing in the expansion of $\mathrm{Tr}\, M^n$ is represented as n legs incident to the same point forming a *vertex*. The legs are cyclically ordered around the vertex so that each incoming line is connected to the outgoing line of the consecutive leg, and carries the same index. This directly translates the pattern of indices obtained when writing a trace as product of matrix elements. For instance, for $n = 3$ the decomposition (26.3.15) yields the vertex shown on Figure 26.2(b), where each index i, j, or k takes N possible values.
(c) Wick's theorem states that the expectation value of a product of matrix elements is obtained by matching them pairwise in all possible manners, and taking the corresponding product of covariances. A pair of matched elements is represented by linking the corresponding legs, forming an *edge*. More precisely, the incoming line of the one leg is connected to the outgoing line of the other leg, and carries the same index. This translates relation (26.3.9): if the connected lines do not carry the same index, then the covariance vanishes hence the matching does not contribute to the expectation value.

Globally, the factors appearing in (26.3.14) form a collection of $k = \sum k_n$ vertices, consisting of k_n n-leg vertices for all $n \geq 1$. The expectation value is obtained by matching the $\sum nk_n$ legs, i.e. merging them pairwise into edges, in all possible manners. Hence, the quantity (26.3.14) is expressed as a sum over all diagrams built out of these vertices and edges, which are sometimes called *fatgraphs* or *ribbon graphs* in the literature. By the rules discussed above, the index lines that are merged together must carry the same index, and they form closed oriented cycles. There is clearly a finite number of diagrams, since there are $(\sum nk_n)!!$ ways to merge the legs (in particular, if the number of legs is odd, there are no such diagrams, and correspondingly the expectation value vanishes). It remains to determine what is the contribution of an individual diagram. Because of relation (26.3.9), each of the $m = \sum nk_n/2$ edges produces a factor $1/\lambda = t/N$. Hence all diagrams will have the same contribution $t^m N^{k-m}/\prod_{n=1}^{\infty} n^{k_n} k_n!$, taking into account the extra factors present in (26.3.14). Evaluating this matrix integral amounts to counting the number of such diagrams.

However, the sum involves many 'equivalent' diagrams, i.e. diagrams which differ only by the choice of line indices or which have the same 'shape'. For instance, Figure 26.3 displays the three possible types of diagram obtained when expanding $\langle \mathrm{Tr}\, M^3 \cdot \mathrm{Tr}\, M^3 \rangle$ (in this example, all three diagrams are connected but this is not true in general). Clearly, we may forget about the line indices by counting each index-less diagram with a multiplicity N^ℓ, where ℓ is the number of cycles of index lines. In our example, we have $\ell = 3$ for the first two diagrams while $\ell = 1$ for the third. Evaluating the shape multiplicity is slightly more subtle, and we leave its general discussion to the next section. It is not difficult to do the computation for the diagrams of Figure 26.3, which yields the respective shape multiplicities 9, 3, and 3, and gathering the various factors we arrive at

$$\langle \mathrm{Tr}\, M^3 \cdot \mathrm{Tr}\, M^3 \rangle = (9N^3 + 3N^3 + 3N)(t/N)^3 = \left(12 + \frac{3}{N^2}\right) t^3. \tag{26.3.16}$$

Generally, as mentioned above, the coefficient (26.3.14) will correspond to a sum over all (not necessarily connected) diagrams made out of k_n vertices with n legs for all n. The partition function $\Xi_N(t, V)$ is then obtained by summing over all k_n's, attaching a weight v_n per n-leg vertex, leading to the generating function for all diagrams (possibly empty or disconnected). Then $F_N(t, V) = \log \Xi_N(t, V)$ is the generating function for connected diagrams, as

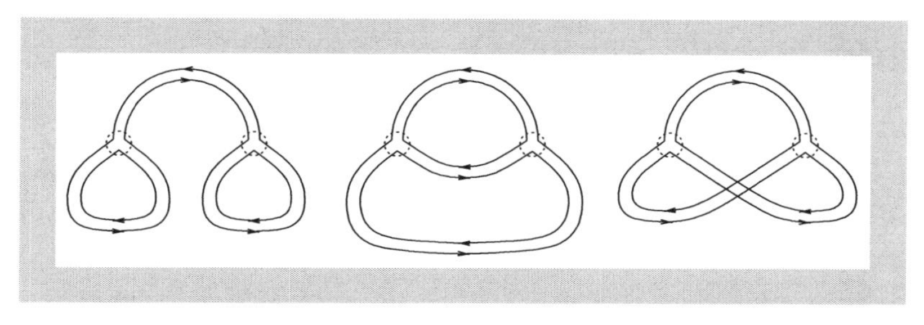

Fig. 26.3 The three possible types of diagram appearing in the expansion of $\langle \mathrm{Tr}\, M^3 \cdot \mathrm{Tr}\, M^3 \rangle$.

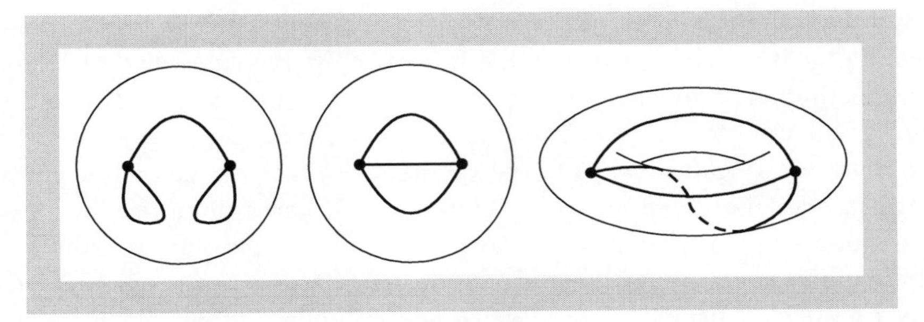

Fig. 26.4 The maps corresponding to the diagrams of Figure 26.3: the first two have genus 0, the third genus 1.

is well known. In $\Xi_N(t, V)$ as well as in $F_N(t, V)$, the exponent of N in the contribution of a diagram is equal to the number of vertices minus the number of edges plus the number of index lines.

At this stage, it may be rather clear that our (connected) diagrams are nothing but maps in disguise (see Figure 26.4). Indeed they are graphs endowed with a cyclic order of half-edges (legs) around the vertices, which is a characterization of maps as discussed in Section 26.2. The cycles of index lines correspond to faces, hence the exponent of N in the contribution of a diagram to $F_N(t, V)$ is equal to the Euler characteristic of the corresponding map. This essentially establishes Theorem 26.3.1.

26.3.4 One-matrix model and combinatorial maps

In this section, we revisit the calculation done above in a more formal manner. The purpose is to show that it naturally involves labelled combinatorial maps, i.e. pairs of permutations, as defined in Section 26.2.3.

We start from a 'classical' formula for enumerating permutations with prescribed cycle lengths. Let $\mathcal{S}_p$ denote the set of permutations of $\{1, \dots, p\}$ and $c_n(\sigma)$ denote the number of n-cycles in the permutation σ, $c(\sigma)$ being the total number of cycles. Then the numbers of permutations σ in $\mathcal{S}_p$ with prescribed values of $c_n(\sigma)$ for all $n \geq 1$ are encoded into the exponential multivariate generating function

$$\exp\left(\sum_{n=1}^{\infty} \frac{A_n}{n}\right) = \sum_{p=0}^{\infty} \frac{1}{p!} \sum_{\sigma \in \mathcal{S}_p} \prod_{n=1}^{\infty} A_n^{c_n(\sigma)}. \tag{26.3.17}$$

Here $(A_n)_{n\geq 1}$ is a family of formal variables. Establishing this formula is a simple exercise in combinatorics [Fla08]: it simply translates the decomposition of a permutation into cycles. We recover $\Xi_N(t, V)$ of (26.3.1) on the left-hand side by the substitution

$$A_n = N v_n \mathrm{Tr} M^n \tag{26.3.18}$$

and taking the expectation value over the Gaussian random matrix M. By this substitution, the σ term on the right-hand side is, up to a factor independent of M, equal to a product of traces which we may rewrite as

$$\prod_{n=1}^{\infty} (\mathrm{Tr}\, M^n)^{c_n(\sigma)} = \sum_{(i_1,\dots,i_p)\in\{1,\dots,N\}^p} \prod_{q=1}^{p} M_{i_q\, i_{\sigma(q)}}. \tag{26.3.19}$$

Wick's theorem and relation (26.3.9) yield

$$\left\langle \prod_{q=1}^{p} M_{i_q\, i_{\sigma(q)}} \right\rangle = \left(\frac{t}{N}\right)^{p/2} \sum_{\alpha\in\mathcal{I}_p} \prod_{q=1}^{p} \delta_{i_q\, i_{\sigma(\alpha(q))}} \tag{26.3.20}$$

where $\mathcal{I}_p \subset \mathcal{S}_p$ is the set of involutions without fixed point (also known as pairwise matchings) of $\{1, \dots, p\}$, which is empty for p odd. We then observe that the product on the right-hand side of (26.3.20) is equal to 1 if the index i_q is constant over the cycles of the permutation $\sigma \circ \alpha$, otherwise it is 0. Therefore, when summing over all values of $(i_1, \dots, i_p)$, we find

$$\left\langle \prod_{n=1}^{\infty} (\mathrm{Tr}\, M^n)^{c_n(\sigma)} \right\rangle = \left(\frac{t}{N}\right)^{p/2} \sum_{\alpha\in\mathcal{I}_p} N^{c(\sigma\circ\alpha)}. \tag{26.3.21}$$

Plugging into (26.3.17) and (26.3.18) and writing $p = 2m = 2c(\alpha)$, we arrive at

$$\Xi_N(t, V) = \sum_{m=0}^{\infty} \frac{t^m}{(2m)!} \sum_{(\sigma,\alpha)\in\mathcal{S}_{2m}\times\mathcal{I}_{2m}} N^{c(\sigma)-c(\alpha)+c(\sigma\circ\alpha)} \prod_{n=1}^{\infty} v_n^{c_n(\sigma)}. \tag{26.3.22}$$

We are very close to recognizing a sum over combinatorial maps, with the Euler characteristic (26.2.2) appearing as the exponent of N, but we lack the requirement that σ and α generate a transitive subgroup. Again, this is implemented by taking the logarithm (in combinatorial terms [Fla08], $\mathcal{S}_{2m} \times \mathcal{I}_{2m}$ is equal to the labelled set construction applied to the class of combinatorial maps – as seen by decomposing $\{1, \dots, 2m\}$ into orbits – and all parameters are inherited, hence $\Xi_N(t, V)$ is the exponential of the generating function for combinatorial maps), leading to

$$F_N(t, V) = \log \Xi_N(t, V) = \sum_{m=0}^{\infty} \frac{t^m}{(2m)!} \sum_{(\sigma,\alpha)\in\mathcal{M}_m} N^{\chi(\sigma,\alpha)} \prod_{n=1}^{\infty} v_n^{c_n(\sigma)} \tag{26.3.23}$$

where $\mathcal{M}_m$ is the set of labelled combinatorial maps with m edges. Upon regrouping the maps according to their genus, this yields relation (26.3.7) and formally establishes Theorem 26.3.1.

26.3.5 Generalization to multi-matrix models

Let us now briefly discuss multi-matrix models. Informally, we consider a 'perturbation' of the multi-matrix Gaussian model (26.3.11) by a $U(N)$-invariant potential $N\mathrm{Tr}\, V(M^{(1)}, M^{(2)}, \dots, M^{(K)})$, where $V(x_1, x_2, \dots, x_K)$ is a 'polynomial' in K non-commutative variables. Actually, V shall be viewed as a formal linear combination of monomials in p non-commutative variables $x_1, x_2, \dots, x_K$ namely

$$V(x_1, x_2, \dots, x_K) = \sum_{n=0}^{\infty} \sum_{(a_1,\dots,a_n)\in\{1,\dots,K\}^n} \frac{v(a_1, \dots, a_n)}{n} x_{a_1} \cdots x_{a_n} \qquad (26.3.24)$$

where the $v(a_1, \dots, a_n)$ are formal variables with the identification $v(a_2, \dots, a_n, a_1) = v(a_1, a_2, \dots, a_n)$ (since the trace is invariant by cyclic shifts). The partition function is defined as the expectation value of $e^{N\mathrm{Tr}\, V(M^{(1)}, M^{(2)}, \dots, M^{(K)})}$ under the Gaussian measure (26.3.11), and the free energy as its logarithm. By a simple extension of the arguments of Section 26.3.3 (using Wick's theorem for multi-matrix integrals), the free energy may be written as a sum over maps, whose half-edges now carry one of K 'colours' corresponding to the extra index $a = 1, \dots, K$. The formal variable $v(a_1, \dots, a_n)$ is the weight for vertices around which the half-edge colours are $(a_1, \dots, a_n)$ in cyclic order, and Q_{ab}^{-1} is the weight per edge whose halves are coloured a and b. Furthermore the topological expansion (26.3.5) still holds.

In the particular case

$$V(x_1, x_2, \dots, x_K) = \sum_{n=1}^{\infty} \sum_{a=1}^{K} \frac{v_n^{(a)}}{n} x_a^n, \qquad (26.3.25)$$

all legs incident to a same vertex carry the same colour. The free energy yields the generating function for (labelled) maps of a given genus whose vertices are coloured in K colours, with a weight $v_n^{(a)}$ per vertex with colour a and degree n, and a weight $Q_{a,b}^{-1}$ per edge linking of vertex of colour a to a vertex of colour b.

Instances of these models appears in several other chapters. Those corresponding to the form (26.3.25) include the chain matrix model in Chapter 16, the Ising and Potts models in Chapter 30. Models of the form (26.3.24) but not (26.3.25) include the complex matrix model in Chapter 27 (for the diagrammatic expansion, M and $M^{\dagger}$ may be treated as two independent Hermitian matrices, which are represented with outgoing and incoming arrows), the $O(n)$, six-vertex and SOS/ADE models in Chapter 30. We particularly emphasize the two-matrix model with free energy

$$\log \frac{\int \exp\left(-\frac{N}{t}\mathrm{Tr}\, M_1 M_2 + \sum_n \frac{N v_n^{(1)}}{n} \mathrm{Tr}\, M_1^n + \sum_n \frac{N v_n^{(2)}}{n} \mathrm{Tr}\, M_2^n\right) d M_1\, d M_2}{\int \exp\left(\frac{N}{t}\mathrm{Tr}\, M_1 M_2\right) d M_1\, d M_2} \qquad (26.3.26)$$

which yields generating functions for bipartite maps with a weight per vertex depending on degree and colour. It is the most natural generalization of the counting problem addressed with the one-matrix model (recovered by setting $v_n^{(2)} = \delta_{n,2}/t$).

Let us finally mention that the diagrammatic expansion for real symmetric matrices corresponds to maps on unoriented surfaces.

26.4 The vertex degree distribution of planar maps

This section is devoted to the enumeration of rooted planar maps with a prescribed vertex degree distribution, i.e. with a given number of vertices of each degree. This is equivalent to deriving the generating function for rooted planar maps with a weight t per edge and, for all $n \geq 1$, a weight v_n per vertex of degree n. By Theorem 26.3.1 and its corollary, this in turn amounts to computing the derivative with respect to t of the planar free energy of the one-matrix model.

Let us first state the result, in the form given in [Bou02b]. A different form was obtained independently in [Ben94] (without matrices), and a check of their agreement can be found in [Bou06].

Theorem 26.4.1 *Let R, S be the formal power series in t and $(v_n)_{n\geq 1}$ satisfying*

$$\begin{aligned} R &= t + t\sum_{n=1}^{\infty} v_n \sum_{j=1}^{\lfloor \frac{n}{2} \rfloor} \frac{(n-1)!}{j!(j-1)!(n-2j)!} R^j S^{n-2j} \\ S &= t\sum_{n=1}^{\infty} v_n \sum_{j=0}^{\lfloor \frac{n-1}{2} \rfloor} \frac{(n-1)!}{(j!)^2(n-2j-1)!} R^j S^{n-2j-1}. \end{aligned} \tag{26.4.1}$$

Then the generating function of rooted planar maps with a weight t per edge and, for all n, a weight v_n per vertex of degree n is given by

$$E^{(0)} = \frac{1}{t}\left(R + S^2 - \sum_{n=1}^{\infty} v_n \sum_{j=2}^{\lfloor \frac{n+2}{2} \rfloor} \frac{(2n-3j+2)(n-1)!}{j!(j-2)!(n-2j+2)!} R^j S^{n-2j+2} - t \right). \tag{26.4.2}$$

Remark 26.4.1 *Clearly, R, S are uniquely determined from the requirement that $R = S = 0$ for $t = 0$. They have a direct combinatorial interpretation: R is the generating function for planar maps with two distinguished vertices of degree 1 (without weights), S is the generating function for planar maps with one distinguished vertex (without weight) of degree 1 and one distinguished face.*

Let us mention that the planar free energy $F^{(0)}$ itself has a more complicated expression, involving logarithms.

This section is organized as follows. In Section 26.4.1 we discuss the main equation describing the planar limit. In Section 26.4.2 we explain how to solve this equation and derive Theorem 26.4.1. Finally in Section 26.4.3 we present a few instances with explicit counting formulae.

26.4.1 Saddle-point, loop, Tutte's equations

Our goal is to 'solve' the one-matrix model in the large N limit, i.e. extract the genus 0 contribution in (26.3.5). This problem has already been approached several times in this book (particularly in Chapters 14 and 16), let us recall what the 'master' equation is. It is an equation is for a quantity called the resolvent in the context of matrix integrals. For our purposes, it is nothing but a generating function for rooted maps involving an extra variable attached to the degree of the root vertex.

More precisely, we define here the planar resolvent as

$$W(z) = \sum_{n=0}^{\infty} \frac{W_n}{z^{n+1}} = \sum_{n=0}^{\infty} \frac{n}{z^{n+1}} \frac{\partial F_n^{(0)}}{\partial v_n} \tag{26.4.3}$$

where z is a new formal variable and W_n is the generating function for rooted (i.e. with a distinguished half-edge) planar maps whose root vertex (i.e. the vertex incident to the root) has degree n, with a weight t per edge and, for all n, a weight v_n per non-root vertex. By convention we set $W_0 = 1$. In comparison with the corollary of Theorem 26.3.1 we do not attach a weight to the root vertex.

Then, the planar resolvent satisfies the master equation

$$W(z)^2 - \left(\frac{z}{t} - V'(z)\right) W(z) + P(z) = 0 \tag{26.4.4}$$

which is a quadratic equation for $W(z)$ immediately solved into

$$W(z) = \frac{1}{2}\left(\frac{z}{t} - V'(z) \pm \sqrt{\left(\frac{z}{t} - V'(z)\right)^2 - 4\,P(z)}\right). \tag{26.4.5}$$

At this stage $P(z)$ is still an unknown quantity but all derivations of (26.4.4) show that, unlike $W(z)$, *$P(z)$ contains only non-negative powers of z.* Actually, if $V(z)$ is a polynomial of degree d (i.e. we set $v_n = 0$ for $n > d$), then $P(z)$ is a polynomial of degree $d - 2$. These remarks are instrumental in solving the equation, as discussed in the next section. Let us first briefly review some methods for deriving the master equation.

Saddle-point approximation. The saddle-point approximation is the original 'physical' method used in [Bre78]. It consists in treating the partition function (26.3.1) as a 'genuine' matrix integral and extracting its analytical large N asymptotics. This is done classically by reducing to an integral over the eigenvalues, then determining the dominant eigenvalue distribution. See Chapter 14, Sections 14.1 and 14.2, for a general discussion of this method. Equation (14.2.6) is nothing but Equation (26.4.5) in different notation: the quantities denoted by

$F(z)$, $Q(z)$ and $V'(z)$ in Chapter 14 correspond respectively to $W(z)$, $P(z)$ and $\frac{z}{t} - V'(z)$ here.

Loop equations. Loop equations correspond to the Schwinger-Dyson equations of quantum field theory applied in the context of matrix models [Wad81, Mig83], see Chapter 16 for a general discussion and application in the context of the one-matrix model. An interesting feature of loop equations is that they provide an easier access to higher genus contributions than the saddle-point approximation. However we are here interested in the planar case for which they are essentially equivalent. Again, Equation (16.4.1) is (26.4.4) in different notation: the quantities denoted by $W_1^{(0)}(x)$, $P^{(0)}(z)$ and $V'(z)$ in Chapter 16 correspond respectively to $W(z)$, $P(z)$ and $\frac{z}{t} - V'(z)$ here.

Tutte's recursive decomposition. Tutte's original approach consists in recursively decomposing rooted maps by 'removing' (contracting or deleting) the root edge. It translates into an equation determining their generating function, upon introducing an extra 'catalytic' variable in order to make the decomposition bijective. It is now recognized that Tutte's equations are essentially equivalent to loop equations [Eyn06] despite their very different origin.

Let us explain the recursive decomposition in our setting [Tut68, Bou06]. We consider a rooted planar map whose root degree (i.e. the degree of the root vertex) is n, and we decompose it as follows.

- If the root edge is a loop (i.e. connects the root vertex to itself), then it naturally 'splits' the map into two parts, which may be viewed as two rooted planar maps. If there are i half-edges incident to the root vertex on one side (excluding those of the loop), then there are $n-2-i$ on the other side. These are the respective root degrees of the corresponding maps.
- If the root edge is not a loop, then we contract it (and we may canonically pick a new root). If m denotes the degree of the other vertex incident to the root edge in the original map, then the root degree of the contracted map is $n+m-2$.

This decomposition is clearly reversible. Taking into account the weights, it leads to the equation

$$W_n = t \sum_{i=0}^{n-2} W_i \, W_{n-2-j} + t \sum_{m=1}^{\infty} v_m \, W_{n+m-2} \tag{26.4.6}$$

valid for all $n \geq 1$ with the convention $W_0 = 1$. Equation (26.4.4) is deduced using (26.4.3), in particular $P(z)$ is given by

$$P(z) = -\sum_{n=0}^{\infty} \left(\sum_{m=n+2}^{\infty} v_m \, W_{m-2-n} \right) z^n. \tag{26.4.7}$$

26.4.2 One-cut solution

We now turn to the solution of Equation (26.4.4). In the context of matrix models, it gives the 'one-cut solution' discussed for instance in Chapter 14. Here we concentrate on expressing it in combinatorial form. Let us first suppose that $V(z)$ (hence $P(z)$) is a polynomial in z. The one-cut solution is obtained by assuming that the polynomial $\Delta(z) = (z/t - V'(z))^2 - 4P(z)$ appearing under the square root in (26.4.5) has exactly two simple zeros, say in a and b, and only double zeros elsewhere. This leads to

$$W(z) = \frac{1}{2}\left(\frac{z}{t} - V'(z) + G(z)\sqrt{(z-a)(z-b)}\right) \qquad (26.4.8)$$

where $G(z)$ is a polynomial. This assumption is physically justified in the saddle-point picture by saying that the dominant eigenvalue distribution has a support made of a single interval $[a, b]$ (corresponding to the cut of $W(z)$), as a perturbation of Wigner's semicircle distribution. Alternatively, a rigorous proof comes via Brown's lemma [Bro65], which translates the fact that $W(z)$ hence $\sqrt{\Delta(z)}$ are power series without fractional powers, we refer to [Bou06, Section 10] for details in the current context. $\sqrt{(z-a)(z-b)}$ must be understood as a Laurent series in $1/z$, i.e. $\sqrt{(z-a)(z-b)} = z +$ (lower powers in z).

Now, it turns out that a, b, and $G(z)$ in (26.4.8) may be fully determined from the mere condition that $W(z) = 1/z +$ (lower powers in z). Indeed, let us rewrite (26.4.8) as

$$\frac{W(z)}{\sqrt{(z-a)(z-b)}} = \frac{1}{2}\left(\frac{\frac{z}{t} - V'(z)}{\sqrt{(z-a)(z-b)}} + G(z)\right). \qquad (26.4.9)$$

Then, we first extract the coefficients of z^{-1} and z^{-2} on both sides. On the left-hand side, we obtain respectively 0 and 1 by the above condition. On the right-hand side, $G(z)$ does not contribute since it is a polynomial in z. Therefore we arrive at

$$\left.\frac{\frac{z}{t} - V'(z)}{\sqrt{(z-a)(z-b)}}\right|_{z^{-1}} = 0 \qquad \left.\frac{\frac{z}{t} - V'(z)}{\sqrt{(z-a)(z-b)}}\right|_{z^{-2}} = 2. \qquad (26.4.10)$$

These equations determine a and b in terms of t and $V(z)$. The coefficients may be extracted via a contour integration around $z = \infty$, but a nicer form is obtained by performing a change of variable $z \to u$ given by

$$z = u + S + \frac{R}{u} \qquad (26.4.11)$$

also known as Joukowsky's transform. S and R are chosen such that $(z-a)(z-b)$ becomes a perfect square namely

$$S = \frac{a+b}{2} \qquad R = \frac{(b-a)^2}{16} \qquad (z-a)(z-b) = \left(u - \frac{R}{u}\right)^2. \tag{26.4.12}$$

Then, by this change of variable, relations (26.4.10) yield

$$S = t\ V'\left(u + S + \frac{R}{u}\right)\bigg|_{u^0} \qquad R = t + t\ V'\left(u + S + \frac{R}{u}\right)\bigg|_{u^{-1}}. \tag{26.4.13}$$

Upon expanding $V'(z) = \sum v_n z^n$, then extracting the respective coefficients of u^0 and u^{-1} in $(u + S + R/u)^k$ via the multinomial formula, we obtain Equations (26.4.1).

Extracting further coefficients $z^{-3}, z^{-4}, \ldots$ in (26.4.9) , we may determine step by step the first few W_n. In particular, W_2 is, up to a factor, the generating function $E^{(0)}$ for rooted planar maps (without condition on the root vertex), since marking a bivalent vertex is tantamount to marking an edge. A slightly tedious computation yields

$$E^{(0)} = \frac{R + S^2 - V'\left(u + S + \frac{R}{u}\right)\big|_{u^{-3}} - 2S\ V'\left(u + S + \frac{R}{u}\right)\big|_{u^{-2}} - t}{t}. \tag{26.4.14}$$

which can then be put into the form (26.4.2). This establishes Theorem 26.4.1, upon noting the restriction that $V(z)$ is a polynomial may now be 'lifted': at a given order in t, the coefficient of $E^{(0)}$ and the corresponding sum over maps both depend on finitely many v_n, therefore by suitably truncating $V(z)$ we may establish their equality.

26.4.3 Examples

Tetravalent maps. In the case of a quartic potential $V(x) = x^4/4$, the diagrammatic expansion involves only vertices of degree four, forming 4-regular or *tetravalent* maps. Equations (26.4.1) and (26.4.2) reduce to

$$E_4^{(0)} = \frac{R - R^3}{t} \qquad S = 0 \qquad R = t + 3t\,R^2. \tag{26.4.15}$$

R is given by a particularly simple quadratic equation solved as

$$R = \frac{1 - \sqrt{1 - 12t^2}}{6t} = \sum_{k=0}^{\infty} \frac{1}{k+1}\binom{2k}{k} 3^k\, t^{2k+1} \tag{26.4.16}$$

where we recognize the celebrated Catalan numbers. Substituting into $E_4^{(0)}$ we obtain the series expansion

$$E_4^{(0)} = \sum_{k=0}^{\infty} \frac{2(2k)!}{k!(k+2)!} 3^k\, t^{2k} \tag{26.4.17}$$

where we identify the number of rooted planar tetravalent maps with $2k$ edges (hence k vertices). This is the same number as that of (general) rooted planar maps with k edges [Tut63], as seen by Tutte's equivalence, and that of rooted planar quadrangulations (i.e. maps with only faces of degree 4) with k faces, as seen by duality.

Trivalent maps. In the case of a cubic potential $V(x) = x^3/3$, the diagrammatic expansion involves only vertices of degree three, forming 3-regular or *trivalent* maps. Equations (26.4.1) and (26.4.2) reduce to

$$E_3^{(0)} = \frac{R + S^2 - 2R^2 S - t}{t} \qquad S = t(S^2 + 2R) \qquad R = t + 2tRS \quad (26.4.18)$$

and we may eliminate R, yielding a cubic equation for S namely

$$t^3 = \frac{tS(1 - tS)(1 - 2tS)}{2}. \quad (26.4.19)$$

(hence tS will be a power series in t^3). Substituting into $E_3^{(0)}$, we may use the Lagrange inversion formula to compute explicitly its expansion as

$$E_3^{(0)} = \sum_{k=0}^{\infty} \frac{2^{2k+1}(3k)!!}{(k+2)!k!!} t^{3k}. \quad (26.4.20)$$

where we recognize the number of rooted planar trivalent maps with $3k$ edges (hence $2k$ vertices) [Mul70].

Eulerian maps. We now consider the case of a general even potential ($v_n = 0$ for n odd). This corresponds to counting maps with vertices of even degree, which are called *Eulerian* (these maps admit a Eulerian path, i.e. a path visiting each edge exactly once). A drastic simplification occurs in (26.4.1), namely that $S = 0$ and R is given by

$$R = t + t\sum_{n=1}^{\infty} v_{2n} \binom{2n-1}{n} R^n. \quad (26.4.21)$$

Hence the generating function $E^{(0)}$ for rooted planar Eulerian maps depends on a single function R satisfying an algebraic equation. This paves the way for an application of the Lagrange inversion formula, allowing us to compute the general term in the series expansion of $E^{(0)}$ namely the number of rooted planar Eulerian maps having a prescribed number k_n of vertices of degree $2n$ for all n, given by

$$2\frac{(\sum_{n=1}^{\infty} nk_n)!}{(\sum_{n=1}^{\infty}(n-1)k_n + 2)!} \prod_{k=1}^{\infty} \frac{1}{k_n!}\binom{2k-1}{k}^{k_n}. \quad (26.4.22)$$

This formula was first derived combinatorially by Tutte [Tut62c].

26.5 From matrix models to bijections

To conclude this chapter, we move a little away from matrices and explain how to rederive the enumeration result of Theorem 26.4.1 through a bijective approach. Such an approach consists in counting objects (here maps) by transforming them into other objects easier to enumerate. We have already encountered several bijections in this chapter: between maps and some pairs of permutations, between fatgraphs and maps, in Tutte's recursive decomposition. However they do not directly yield an enumeration formula, as a non-bijective step is needed.

The 'easier' objects we shall look for are *rooted plane trees* (which we may view as rooted planar maps with one face). This might not come as a surprise ever since the appearance of Catalan numbers at Equation (26.4.16). In general, trees are indeed easy to enumerate by recursive decomposition: removing the root cuts the tree into subtrees (forming an ordered sequence due to planarity) and this is often immediately translated into an algebraic equation for their generating function. Here, we will perform the inverse translation: we will *construct* the trees corresponding to a given equation, namely (26.4.1) or equivalently (26.4.13).

Let us indeed define two classes of rooted trees $\mathcal{R}$ and $\mathcal{S}$ recursively as follows. An $\mathcal{R}$-tree (i.e. a tree in $\mathcal{R}$) is either reduced to a 'white' leaf, or it consists of a root vertex to which are attached a sequence of subtrees that can be $\mathcal{R}$-trees, $\mathcal{S}$-trees, or single 'black' leaves, with the condition that the number of such black leaves is equal to the number of $\mathcal{R}$-subtrees minus one. An $\mathcal{S}$-tree consists of a root vertex to which are attached a sequence of the same possible subtrees, with the condition that the number of black leaves is equal to the number of $\mathcal{R}$-subtrees. It is straightforward to check that this is a well-defined recursive construction, which translates into Equations (26.4.1) or (26.4.13) for the corresponding generating functions R and S, provided that we attach a weight t per vertex or white leaf, and a weight v_n per vertex with $n-1$ subtrees. Figure 26.5 displays a tree in $\mathcal{S}$.

We may then wonder how such trees are related to maps. A natural idea is to 'match' the black and white leaves together, creating new edges. Consider for instance an $\mathcal{S}$-tree: it has the same number of black and white leaves, and given an orientation there is a 'canonical' matching procedure, see again Figure 26.5. This creates a planar map out of an $\mathcal{S}$-tree, and we further observe that the vertex degrees are preserved. It is possible to show that this defines a one-to-one correspondence between $\mathcal{S}$ and the second class of maps mentioned in the remark below Theorem 26.4.1, see [Bou02b] for details. We proceed similarly with $\mathcal{R}$, and then a few more steps allow us to establish bijectively Theorem 26.4.1. The knowledge of the matrix model solution was instrumental in 'guessing' the suitable family of trees, which encompasses the one found in [Sch97] based

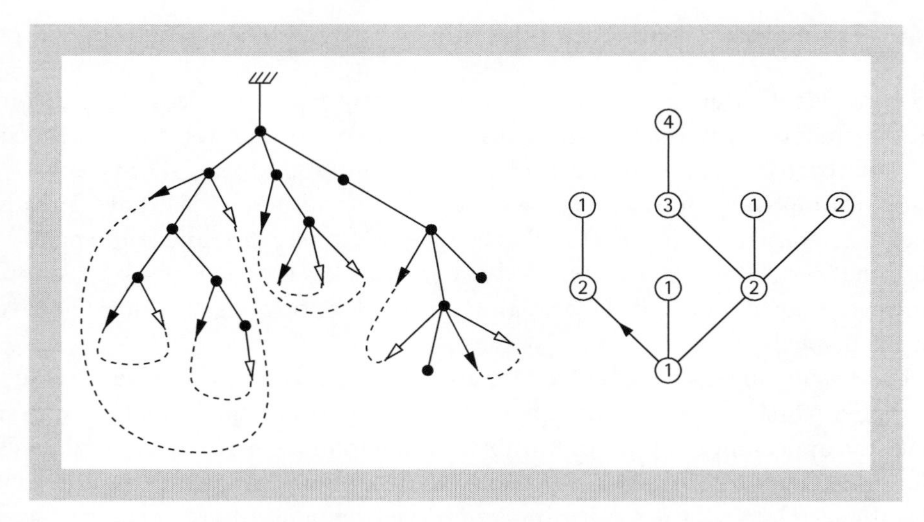

Fig. 26.5 Left: an $\mathcal{S}$-tree, together with the canonical matching of black and white leaves (dashed lines between arrows). Right: a well-labelled tree.

on Tutte's formula (26.4.22). The present construction was further extended to bipartite maps corresponding to the two-matrix model (26.3.26) [Bou02a], and maps corresponding to a chain-matrix model [Bou05].

The bijective approach has a great virtue. It was indeed realized that it is intimately connected with the 'geodesic' distance in maps [Bou03]. Actually, there is another 'dual' family of bijections with so-called well-labelled trees or mobiles [Cor81, Mar01, Bou04, Bou07] for which the connection is even more apparent. In the simplest instance [Cor81], a well-labelled tree is a rooted plane tree whose vertices carry a positive integer label, in such a way that labels on adjacent vertices differ by at most 1 (see again Figure 26.5 for an example). It encodes bijectively a rooted quadrangulation (i.e. a map whose faces have degree 4), a vertex with label ℓ in the tree corresponding to a vertex at *distance* ℓ from the root vertex in the quadrangulation [Mar01] (where the distance is the graph distance, i.e. the minimal number of consecutive edges connecting two vertices). It is easily seen that the generating function for well-labelled trees with root label $\ell \geq 1$ satisfies

$$R_\ell = t + t\, R_\ell \left(R_{\ell-1} + R_\ell + R_{\ell+1} \right) \tag{26.5.1}$$

together with the boundary condition $R_0 = 0$, t being a weight per edge or vertex. Through the bijection, R_ℓ yields the generating function for quadrangulations with two marked points at a distance at most ℓ, related to the so-called two-point function [Amb95]. Equation (26.5.1) is nothing but a refinement of the third relation of (26.4.15) (and we have $E_0^{(4)} = R_1/t$). Remarkably, it has an explicit solution [Bou03, Section 4.1]. Furthermore it looks surprisingly analogous to (yet different from) the 'first string equation' (14.2.14). This still mysterious analogy is much more general, as one may refine equations (26.4.1) into discrete recurrence equations involving the distance, similar to the string equations for the one-matrix model, and having again explicit solutions [Bou03, DiF05].

The correspondence between maps and trees has sparked an active field of research between physics, combinatorics, and probability theory, devoted to the study of the geometry of large random maps, see for instance [Mie09] and references therein.

References

[Amb95] J. Ambjørn and Y. Watabiki, "Scaling in quantum gravity", Nucl. Phys. **B445** (1995) 129–144, arXiv:hep-th/9501049.

[Ben94] E.A. Bender and E.R. Canfield, "The number of degree-restricted rooted maps on the sphere", SIAM J. Discrete math. **7** (1994) 9–15.

[Bou02a] M. Bousquet-Mélou and G. Schaeffer, "The degree distribution in bipartite planar maps: applications to the Ising model", arXiv:math/0211070.

[Bou06] M. Bousquet-Mélou and A. Jehanne, "Polynomial equations with one catalytic variable, algebraic series, and map enumeration", Journal of Combinatorial Theory Series B **96** (2006) 623–672, arXiv:math/0504018.

[Bou02b] J. Bouttier, P. Di Francesco and E. Guitter, "Census of planar maps: from the one-matrix model solution to a combinatorial proof", Nucl. Phys. **B645** [PM] (2002) 477–499, arXiv:cond-mat/0207682.

[Bou03] J. Bouttier, P. Di Francesco and E. Guitter, "Geodesic Distance in Planar Graphs", Nucl. Phys. **B663** (2003) 535–567, arXiv:cond-mat/0303272.

[Bou04] J. Bouttier, P. Di Francesco and E. Guitter, "Planar maps as labeled mobiles", Elec. Jour. of Combinatorics **11** (2004) R69, arXiv:math/0405099.

[Bou05] J. Bouttier, P. Di Francesco and E. Guitter, "Combinatorics of bicubic maps with hard particles", J. Phys. A **38** (2005) 4529–4559, arXiv:math/0501344.

[Bou07] J. Bouttier, P. Di Francesco and E. Guitter, "Blocked edges on Eulerian maps and mobiles: Application to spanning trees, hard particles and the Ising model", J. Phys. A: Math. Theor. **40** (2007) 7411–7440, arXiv:math/0702097.

[Bre78] E. Brézin, C. Itzykson, G. Parisi and J.-B. Zuber, "Planar diagrams", Commun. Math. Phys. **59** (1978) 35–51.

[Bro65] W.G. Brown, "On the existence of square roots in certain rings of power series", Math. Ann. **158** (1965) 82–89.

[Cor75] R. Cori, *Un code pour les graphes planaires et ses applications*, Société Mathématique de France, Paris 1975. With an English abstract, Astérisque **27**.

[Cor81] R. Cori and B. Vauquelin, "Planar maps are well labeled trees", Canad. J. Math. **33** (1981) 1023–1042.

[Cor92] R. Cori and A. Machì, "Maps, hypermaps and thier automorphisms: a survey, I, II, III" Exposition. Math. **10** (1992) 403–427, 429–447, 449–467.

[DiF05] P. Di Francesco, "Geodesic Distance in Planar Graphs: An Integrable Approach", Ramanujan J. **10** (2005) 153–186, arXiv:math/0506543.

[Edm60] J.R. Edmonds, "A combinatorial representation for polyhedral surfaces", Notices Amer. Math. Soc. **7** (1960) 646.

[Eyn06] B. Eynard, Random Matrices, Random Processes and Integrable Systems, CRM Series in Mathematical Physics 2011, pp 415–442 [arXiv:math-ph/0611087].

[Fla08] P. Flajolet and R. Sedgewick, *Analytic Combinatorics*, Cambridge University Press, 2008.

[Gim05] O. Giménez and M. Noy, "Asymptotic enumeration and limit laws of planar graphs", J. Amer. Math. Soc. **22** (2009) 309–329, arXiv:math/0501269.

[Hoo74] G. 't Hooft, "A planar diagram theory for strong interactions", Nucl. Phys. **B72** (1974) 461–473.

[Kaz86] V.A. Kazakov, "Ising model on a dynamical planar random lattice: Exact solution", Phys. Lett. **A119** (1986) 140–144.

[Kni88] V.G. Knizhnik, A.M. Polyakov and A.B. Zamolodchikov, "Fractal structure of 2D quantum gravity", Mod. Phys. Lett. **A3** (1988) 819–826.

[Mar01] M. Marcus and G. Schaeffer, "Une bijection simple pour les cartes orientables", manuscript (2001), available online at http://www.lix.polytechnique.fr/~schaeffe/Biblio/MaSc01.ps; see also G. Chapuy, M. Marcus, G. Schaeffer, "A bijection for rooted maps on orientable surfaces", SIAM Journal on Discrete Mathematics, 23(3) (2009) 1587–1611, arXiv:0712.3649.

[Mie09] G. Miermont, "Random maps and their scaling limits", Progress in Probability **61** (2009) 197–224.

[Mig83] A.A. Migdal, "Loop equations and 1/N expansion", Phys. Rep. **102** (1983) 199–290.

[Moh01] B. Mohar and C. Thomassen, *Graphs on surfaces*, The Johns Hopkins University Press, Baltimore 2001.

[Mul70] R.C. Mullin, E. Nemeth and P.J. Schellenberg, "The enumeration of almost cubic maps", Proceedings of the Louisiana Conference on Combinatorics, Graph Theory and Computer Science **1** (1970) 281–295.

[Pen88] R.C. Penner, "Perturbative series and the moduli space of Riemann surfaces", J. Differential Geom. **27**(1) (1988) 35–53.

[Sch97] G. Schaeffer, "Bijective census and random generation of eulerian planar maps", Elec. Jour. of Combinatorics **4** (1997) R20.

[Tut62a] W. Tutte, "A census of planar triangulations", Canad. J. Math. **14** (1962) 21–38.

[Tut62b] W. Tutte, "A census of Hamiltonian polygons", Canad. J. Math. **14** (1962) 402–417.

[Tut62c] W. Tutte, "A census of slicings", Canad. J. Math. **14** (1962) 708–722.

[Tut63] W. Tutte, "A census of planar maps", Canad. J. Math. **15** (1963) 249–271.

[Tut68] W. Tutte, "On the enumeration of planar maps", Bull. Amer. Math. Soc. **74** (1968) 64–74.

[Wad81] S. R. Wadia, "Dyson–Schwinger equations approach to the large-N limit: Model systems and string representation of Yang-Mills theory", Phys. Rev. **D24** (1981) 970–978.

[Zvo95] A. Zvonkin, "How to Draw a Group", Discrete Mathematics **180** (1998) 403–413.

·27·

Knot theory and matrix integrals

Paul Zinn-Justin and Jean-Bernard Zuber

Abstract

The large size limit of matrix integrals with quartic potential may be used to count alternating links and tangles. The removal of redundancies amounts to renormalization of the potential. This extends into two directions: higher genus and the counting of 'virtual' links and tangles; and the counting of 'coloured' alternating links and tangles. We discuss the asymptotic behaviour of the number of tangles as the number of crossings goes to infinity.

27.1 Introduction and basic definitions

This chapter is devoted to some enumeration problems in knot theory. For a general review of the subject, see [Fin03]. Here we are interested in the application of matrix integral techniques. We start with basic definitions of knot theory.

27.1.1 Knots, links, and tangles

We first recall the definitions of the knotted objects under consideration. A *knot* is a closed loop embedded in three-dimensional space. A *link* is made of several entangled knots. An *n-tangle* is a knotted pattern with $2n$ open ends. We shall be interested in particular in 2-tangles, where it is conventional to attach the four outgoing strands to the four cardinal points SE, SW, NW, NE.

This figure depicts a knot, two links, and three 2-tangles.

All these objects are regarded as equivalent under isotopy, i.e. under deformations in which strands do not cross one another, and (for tangles) open ends are maintained fixed. Our problem is to count topologically inequivalent knots, links, and tangles.

It is usual to represent knots etc. by their *planar projection* with a minimal number of over/under-crossings. There is an important

Theorem 27.1.1 (Reidemeister) *Two projections represent the same knot, link or tangle if they may be transformed into one another by a sequence of Reidemeister moves:*

Also, in the classification or the counting of knots etc., one tries to avoid redundancies by keeping only *prime* links. A link is non-prime if cutting tranversely two strands may yield two disconnected non-trivial parts. Here is a non-prime link:

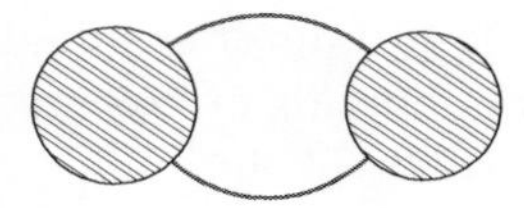

27.1.2 Alternating links and tangles

We shall now restrict ourselves to the subclass of *alternating* knots, links, and tangles, in which one meets alternatingly over- and under-crossings, when one follows any strand.

For low numbers of crossings, all knots, links, or tangles may be drawn in an alternating pattern, but for $n \geq 8$ (resp. 6) crossings, there are knots (links) which cannot be drawn in an alternating form. Here is an example of an 8-crossing non-alternating knot:

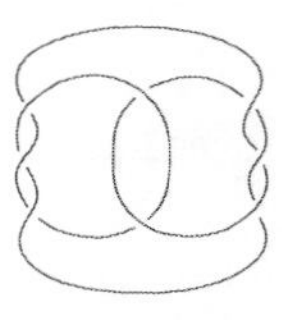

One may show that asymptotically, the alternating links and knots are subdominant. Still the tabulation and counting of this subclass is an important task, as a preliminary step in the general classification programme.

A major result conjectured by Tait (1898) and proved in [MT91, MT93] is

Theorem 27.1.2 (Menasco, Thistlethwaite) *Two alternating reduced knots or links represent the same object if they are related by a sequence of 'flypes', where a flype is a combination of Reidemeister moves respecting the alternating character of tangles:*

We shall thus restrict ourselves to the (manageable)

Problem 27.1.1 *Count alternating prime links and tangles.*

This problem was given a first substantial answer by Sundberg and Thistlethwaite in [ST98]. We will discuss in the rest of this text how the matrix integral approach has allowed us to make significant progress building on their work.

27.2 Matrix integrals, alternating links, and tangles

27.2.1 The basic integral

Consider the integral over complex (*non Hermitian*) $N \times N$ matrices

$$Z_C = \int dM \, \mathrm{e}^{N[-t\, \frac{\mathrm{tr}}{N} MM^\dagger + \frac{g}{2} \frac{\mathrm{tr}}{N}(MM^\dagger)^2]} \tag{27.2.1}$$

with $dM = \prod_{i,j} d\mathrm{Re} M_{ij} \, d\mathrm{Im} M_{ij}$. It was proposed in the context of knot enumeration in [ZJZ02].

According to the discussion of Chapters 26 and 30, its diagrammatic expansion involves oriented double-line propagators $M_j^i \rightleftarrows {}^l_k M^\dagger$, while its vertices may be drawn in a one-to-one correspondence with the previous link crossings, with, say, over-crossing associated with outgoing arrows.

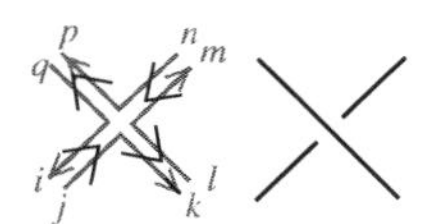

As usual, the perturbative (small g) expansion of the integral (27.2.1) or of the associated correlation functions involves only planar diagrams in the large N limit. Moreover the conservation of arrows implies that the diagrams are alternating:

It follows from the discussion of Chapter 26 that in the large N limit $\lim_{N\to\infty} \frac{1}{N^2} \log Z_C = \sum_{\substack{\text{planar connected alternating} \\ \text{diagrams } D \text{ with } n \text{ vertices}}} \frac{g^n}{|\mathrm{Aut}\, D|}$, where $|\mathrm{Aut}\, D|$ is the order of the automorphism group of D. But going from complex matrices to Hermitian matrices doesn't affect that 'planar limit', up to a global factor 2. We thus conclude that, provided we remove redundancies including flypes, the counting of Feynman diagrams of the following integral over $N \times N$ Hermitian matrices M, for $N \to \infty$,

$$Z = \int dM\, e^{N[-\frac{t}{2}\frac{\mathrm{tr}}{N} M^2 + \frac{g}{4}\frac{\mathrm{tr}}{N} M^4]} \tag{27.2.2}$$

with $dM = \prod_i dM_{ii} \prod_{i<j} d\mathrm{Ree}\, M_{ij}\, d\mathrm{Imm} M_{ij}$, yields the counting of alternating links and tangles.

27.2.2 Computing the integral

The large N limit of the integral (27.2.2) may be computed by the saddle-point method, by means of orthogonal polynomials or of the loop equations, as reviewed elsewhere in this book.

In that $N \to \infty$ limit, the eigenvalues λ form a continuous distribution with density $u(\lambda)$ of support $[-2a, 2a]$, forming a deformed semicircle law [BIPZ78]

$$u(\lambda) = \frac{1}{2\pi}(1 - 2\frac{g}{t^2}a^2 - \frac{g}{t^2}\lambda^2)\sqrt{4a^2 - \lambda^2} \tag{27.2.3}$$

with a^2 related to g and t by

$$3\frac{g}{t^2}a^4 - a^2 + 1 = 0 \tag{27.2.4}$$

and one finds that the large N limit of the 'free energy' F is

$$F(g,t) := \lim_{N\to\infty} \frac{1}{N^2} \log \frac{Z(g,t)}{Z(t,0)} = \frac{1}{2}\log a^2 - \frac{1}{24}(a^2-1)(9-a^2)$$

$$F(g,t) = \sum F_p \left(\frac{g}{t^2}\right)^p = \sum_{p=1} \left(\frac{3g}{t^2}\right)^p \frac{(2p-1)!}{p!(p+2)!}\,.$$

We recall that this formal power series of F, the 'perturbative expansion of F', is a generating function for the number of *connected* planar diagrams, (as usual, weighted by their inverse symmetry factor)

For future reference, we note that the asymptotic behaviour of F_p as $p \to \infty$ is

$$F_p \sim \mathrm{const}(12)^p p^{-7/2}. \tag{27.2.5}$$

Also, all the $2p$-point functions $\frac{1}{N}\langle\frac{\mathrm{tr}}{N} M^{2p}\rangle = \int \rho(\lambda)\lambda^{2p}$ may be computed. We only give here two expressions that we need below, the 2-point function

$$\Delta = \frac{1}{3t}a^2(4-a^2) = \text{—◯—} \tag{27.2.6}$$

and the connected 4-point function Γ

$$\Gamma(g,t) = \frac{1}{9t^2}a^4(1-a^2)(2a^2-5) = \quad , \tag{27.2.7}$$

whose diagrams, after removal of redundancies, will count 2-tangles. The p-th term γ_p in the g expansion of Γ behaves as $12^p p^{-5/2}$.

27.2.3 Removal of redundancies

The removal of redundancies for the counting links and tangles will be done in two steps. First 'nugatory' that are in fact irrelevant diagrams representing patterns that may be unknotted, = , and 'non-prime' diagrams ,

may be both removed by adjusting $t = t(g)$ in such a way that $\Delta = \quad = 1$. In the language of quantum field theory, this is a 'wave function renormalization'.

We then find $F(g) = F(g, t(g))$

$$F(g) = \frac{g^2}{4} + \frac{g^3}{3} + \frac{3g^4}{4} + \frac{11g^5}{5} + \frac{91g^6}{12} + \cdots \tag{27.2.8}$$

and $\Gamma(g) = \Gamma(g, t(g)) = \frac{(5-2a^2)(a^2-1)}{(4-a^2)^2} = 2\frac{\mathrm{d}F}{\mathrm{d}g}$.

In that way, one gets the correct counting of links up to six crossings and of 2-tangles up to three crossings. This is apparent on the following table where we have listed in (a) the first links with their traditional nomenclature; (b) the corresponding Feynman diagrams with their symmetry weight; (c) the first 2-tangles in Feynman diagram notation. It appears that the diagrams of each of the last two pairs in (c) are flype equivalent.

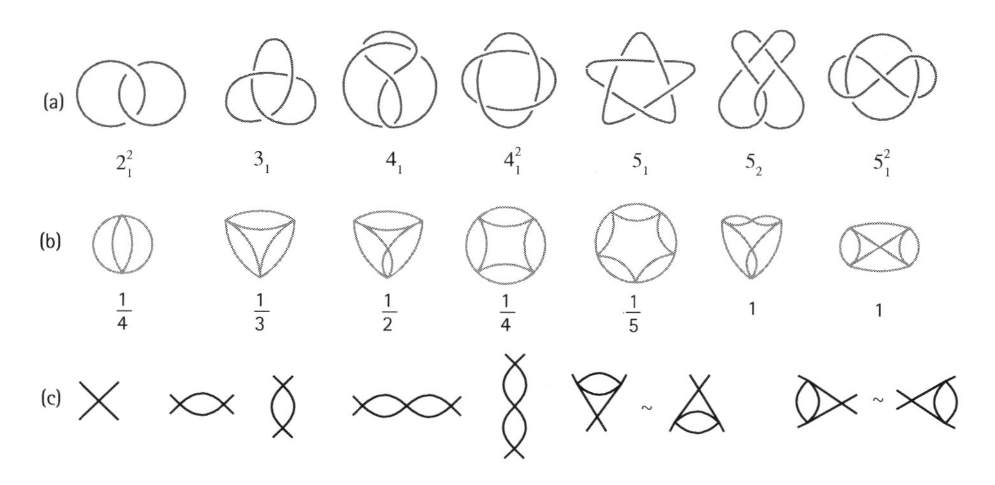

For links the first flype equivalence occurs at order 6:

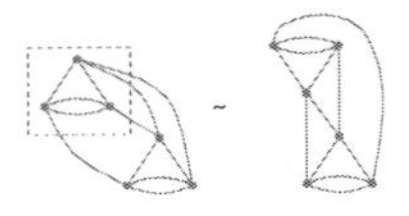

The asymptotic behaviour $F_p \sim \text{const}\,(27/4)^p\, p^{-7/2}$ exhibits the same 'critical exponent' $-7/2$ as in (27.2.5) but an increased radius of convergence, as expected. Likewise $\gamma_p \sim \text{const}\,(27/4)^p\, p^{-5/2}$.

In a second step we must take the quotient by the flype equivalence. Sundberg and Thistlethwaite [ST98] proved that the flype equivalence can be dealt with by a suitable combinatorial analysis. The net result of their rigourous analysis is that the connected four-point function $\tilde{\Gamma}$ can be deduced from $\Gamma(g)$ by a suitable change of variable: this final computation has been rephrased in [ZJ03] where it is shown that it can be elegantly presented as a coupling constant renormalization $g \to g_0$. In other words, start from $N\frac{\text{tr}}{N}\left(\frac{1}{2}tM^2 - \frac{g_0}{4}M^4\right)$, fix $t = t(g_0)$ as before. Then compute $\Gamma(g_0)$ and determine $g_0(g)$ as the solution of

$$g_0 = g\left(-1 + \frac{2}{(1-g)(1+\Gamma(g_0))}\right), \tag{27.2.9}$$

then the desired generating function is $\tilde{\Gamma}(g) = \Gamma(g_0)$.

To show this, we introduce $H(g)$, the generating function of 'horizontally two-particle-irreducible' (H2PI) 2-tangle diagrams, i.e. of diagrams whose left part cannot be separated from the right by cutting two lines. Its Feynman diagram expansion reads

$H =$ = + + + ...

Then the four-point function Γ is a geometric series of H

$H =$ = + + + ...

summing up to $\Gamma = H/(1-H)$.

Now under the flype equivalence ~ . Thus, with $\tilde{\Gamma}$, resp. $\tilde{H}$ denoting generating functions of *flype equivalence classes* of prime tangles, resp. of H2PI tangles and if $\tilde{H}'$ is the non-trivial part of $\tilde{H}$, $\tilde{H} = g + \tilde{H}'$, $\tilde{\Gamma}$ satisfies a simple recursive equation

$$\tilde{\Gamma} = g + g\tilde{\Gamma} + \frac{\tilde{H}'}{1-\tilde{H}'}, \tag{27.2.10}$$

both relations being depicted as

$\tilde{H} =$

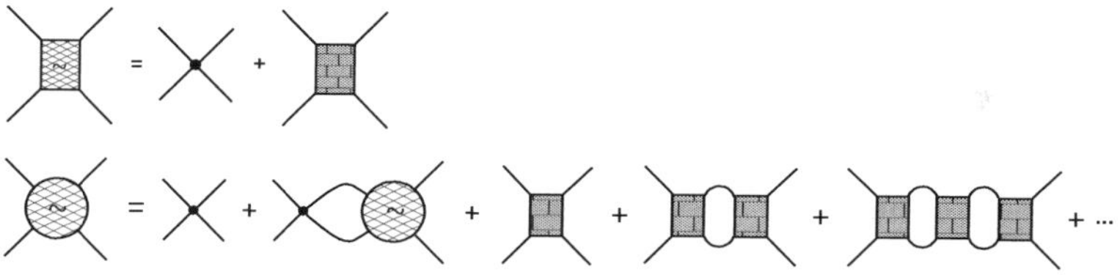

$\tilde{\Gamma} =$

Consider now the perturbative expansion of $\Gamma(g_0)$ computed for a new value g_0 of the coupling constant, depicted as an open circle

$\Gamma(g_0) =$

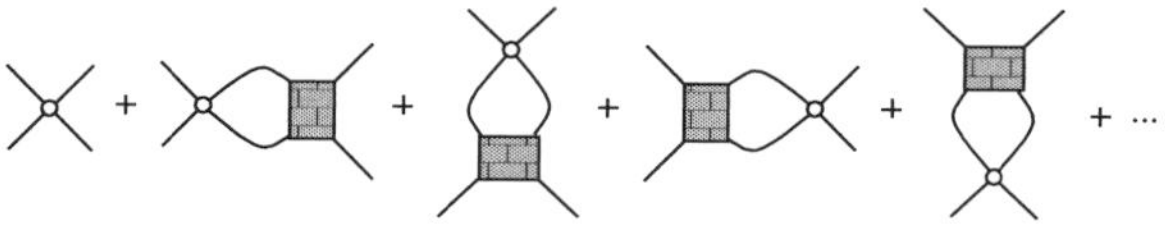

If we want to identify it to $\tilde{\Gamma}(g)$, it is suggested to determine $g_0 = g_0(g)$ by demanding that $g_0 = g - 2g\,\tilde{H}' - \ldots$

$g_0 =$

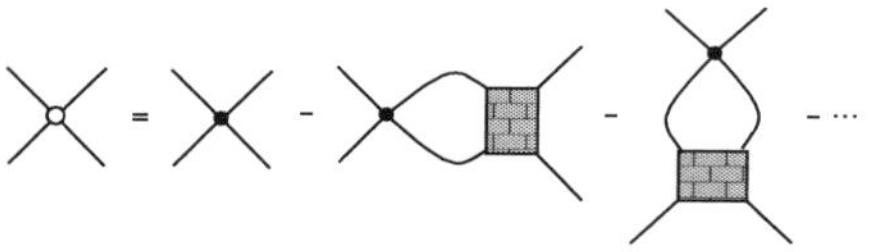

so as to remove the first flype redundancies, and the remarkable point is that the ellipsis may be omitted and that no further term is required. Indeed eliminating $\tilde{H}'$ between the two relations (27.2.10) and $g_0 = g - 2g\,\tilde{H}'$ gives

$$g_0 = g\Big(-1 + \frac{2}{(1-g)(\tilde{\Gamma}+1)}\Big) \tag{27.2.11}$$

which is equivalent to (27.2.9) and also to relations found in [ST98]. In the case of the matrix integral (27.2.2), it is convenient to parameterize things in terms of $A = \frac{6}{4-a^2}$. One finds

$$\tilde{\Gamma} = \frac{(A-2)(4-A)}{4} \tag{27.2.12}$$

$$g_0 = \frac{4(A-2)}{A^3}, \tag{27.2.13}$$

where $\tilde{\Gamma}$ is the wanted generating function of the number of flype-equivalence classes of prime alternating 2-tangles. Eliminating $\tilde{\Gamma}$ and g_0 between the three latter equations results in a degree five equation for A

$$A^5 g - 6A^4 g + \frac{4A^3\,(g^2-2g-1)}{g-1} - 32A^2 + 64A - 32 = 0 \tag{27.2.14}$$

of which we have to find the solution which goes to 2 as $g \to 0$

$$A = 2+2g+6g^2+20g^3+78g^4+334g^5+1532g^6+7372g^7+36734g^8+187902g^9+\ldots$$

This then gives for $\tilde{\Gamma}$ the following expansion (given up to order 50 in [ST98])

$$\tilde{\Gamma}(g) = g + 2g^2 + 4g^3 + 10g^4 + 29g^5 + 98g^6 + 372g^7 + 1538g^8 + 6755g^9 + \cdots \tag{27.2.15}$$

and the asymptotic behaviour of the p-th order of that expansion reads

$$\tilde{\gamma}_p \sim \text{const} \left(\frac{101 + \sqrt{21001}}{40} \right)^p p^{-5/2} \tag{27.2.16}$$

with again the same exponent $-5/2$ but a still increased radius of convergence.

At this stage, we have merely reproduced the results of [ST98]. Our matrix integral approach has however two merits. It simplifies the combinatorics and recasts the quotient by flype equivalence in the (physically) appealing language of renormalization. For example using the results of [BIPZ78], one may easily compute the connected 2ℓ-function which counts the number of flype-equivalence classes of prime alternating ℓ-tangles [ZJ03]

$$\begin{aligned} \Gamma_{2\ell} &= \frac{c_\ell}{\ell!}(A-2)^{\ell-1}(3\ell - 2 - (\ell-1)A) \\ c_{\ell+1} &= \frac{1}{3\ell+1} \sum_{\ell/2 \le q \le \ell} (-4)^{q-\ell} \frac{(\ell+q)!}{(2q-\ell)!(\ell-q)!} \end{aligned}$$

and the numbers of 3- and 4-tangles up to nine crossings are given by

$$\Gamma_6 = 3g^2 + 14g^3 + 51g^4 + 186g^5 + 708g^6 + 2850g^7 + 12099g^8 + 53756g^9 + \cdots \tag{27.2.17}$$

$$\Gamma_8 = 12g^3 + 90g^4 + 468g^5 + 2196g^6 + 10044g^7 + 46170g^8 + 215832g^9 + \cdots \tag{27.2.18}$$

Our approach also opens the route to generalizations in two directions:

- higher genus surfaces and 'virtual' links.
- counting of 'coloured' links, with a potential access to the still open problem of disentangling knots from links.

This is what we explore in the next two sections.

27.3 Virtual knots

27.3.1 Definition

The large N 'planar' limit of the matrix integral (27.2.1) has been shown to be directly related to the counting of links and tangles. It is thus a natural question to wonder what the subleading terms in the N^{-2} expansion of that integral, i.e. its higher genus contributions, correspond to from the knot theoretic standpoint. If one realizes that ordinary links and knots may always be deformed to live in a spherical shell $S^2 \times I$, where the interval I is homeomorphic to $[0, 1]$, one is ready to see that higher genus analogues exist. In fact, these objects may be defined in two alternative ways.

First, as just suggested, they are curves embedded in a 'thickened' Riemann surface $\Sigma := \Sigma \times [0, 1]$, modulo isotopy in Σ, *and* modulo orientation-preserving homeomorphisms of Σ, *and* modulo addition or subtraction of empty handles.

But one may also focus on the planar representations of these objects. This leads to the concept of virtual knot diagrams [Kau99, KM06]. In addition to the ordinary under- and over-crossings, one must introduce a new type of *virtual* crossing, which somehow represents the crossing of two different strands that belong to different sides of the surface but are seen as crossing in the planar projection. Thus virtual knots diagrams are made of , and virtual links and knots are equivalence classes of such diagrams with respect to the following generalized Reidemeister moves

That the two definitions are equivalent was proved in [CKS02, Kup03]. See [Gre04] for a table of virtual knots.

Virtual alternating links and tangles are defined in the same way as in Section 27.2: along each strand, one encounters alternatingly over- and under-crossings, paying no attention to possible virtual crossings.

Here is a virtual link depicted in several alternative ways:

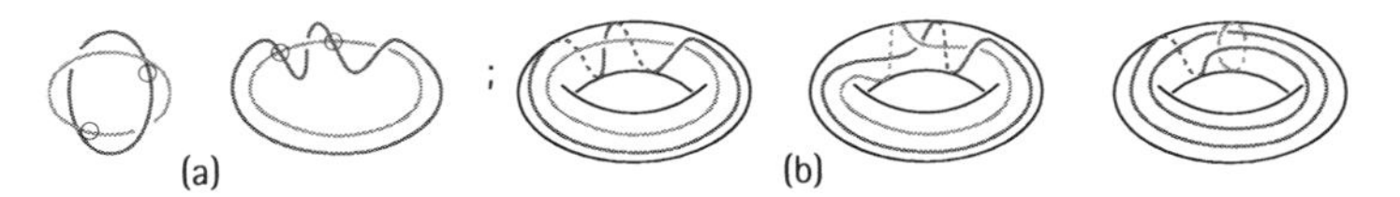

in (a), using ordinary and virtual crossings; in (b), three equivalent representations on a Riemann surface. As illustrated by this example, in the thickened Riemann surface picture, the counting should be done irrespective of the choice of homology basis or of the embedding of the link/knot. But this is precisely what higher genus Feynman diagrams of the matrix integral do for us!

Remark 27.3.1 *there is a notion of genus for knots (minimal genus of a Seifert surface) which is unrelated to the genus defined above (genus of the surface Σ). For the former notion in the context of knot enumeration, see [STV02, SV05].*

27.3.2 Higher genus contributions to integral (27.2.1)

We thus return to the integral (27.2.1) over *complex matrices*

$$Z(g, t, N) = \int dM \, e^{N[-t \frac{\mathrm{tr}}{N} MM^\dagger + \frac{g}{2} \frac{\mathrm{tr}}{N}(MM^\dagger)^2]}$$

and compute $F(g, t, N) = \frac{1}{N^2} \log Z(g, t, N)/Z(0, t, N)$ in an N^{-2} expansion

$$F(g, t, N) = \sum_{h=0}^{\infty} N^{2-2h} F^{(h)}(g, t)$$

$F^{(h)}(g, t)$ receives contributions from Feynman diagrams of genus h. $F^{(0)}$ is (up to a factor 2) what was called F in the previous section. $F^{(1)}$ was computed in [Mor91], $F^{(2)}$ and $F^{(3)}$ in [Ake97] and [Ada97]. From F one derives the expressions of $\Delta = \frac{1}{t} - \frac{\partial F}{\partial t}$ and $\Gamma = 2\frac{\partial F}{\partial g} - 2\Delta^2$. Moreover the first two terms in the g power series expansion of any $F^{(h)}(g)$ are easy to get [ZJZ04] and provide some additional information.

As before, we remove the non-prime diagrams by imposing that $\Delta(g, t(g, N), N) = 1$, which determines $t = t(g, N)$ as a double g and $1/N^2$ expansion. One then finds the generating function of prime 2-tangles of minimal genus h, $\Gamma^{(h)}(g)$, as the N^{2-2h} term in the $1/N^2$ expansion of $\Gamma(g, N)$:

$$\Gamma^{(0)}(g) = g+2g^2+6g^3+22g^4+91g^5+408g^6+1938g^7+9614g^8+49335g^9+260130g^{10}+\cdots$$

$$\Gamma^{(1)}(g) = g+8g^2+59g^3+420g^4+2940g^5+20384g^6+140479g^7+964184g^8+6598481g^9+45059872g^{10}+\cdots$$

$$\Gamma^{(2)}(g) = 17g^3+456g^4+7728g^5+104762g^6+1240518g^7+13406796g^8+135637190g^9+1305368592g^{10}+\cdots$$

$$\Gamma^{(3)}(g) = 1259g^5+62072g^6+1740158g^7+36316872g^8+627368680g^9+9484251920g^{10}+\cdots$$

$$\Gamma^{(4)}(g) = 200589\,g^7+14910216\,g^8+600547192\,g^9+17347802824g^{10}+\cdots$$

$$\Gamma^{(5)}(g) = 54766516\,g^9+5554165536g^{10}+\cdots$$

27.3.3 Table of genus 1, 2, and 3 virtual links with four crossings

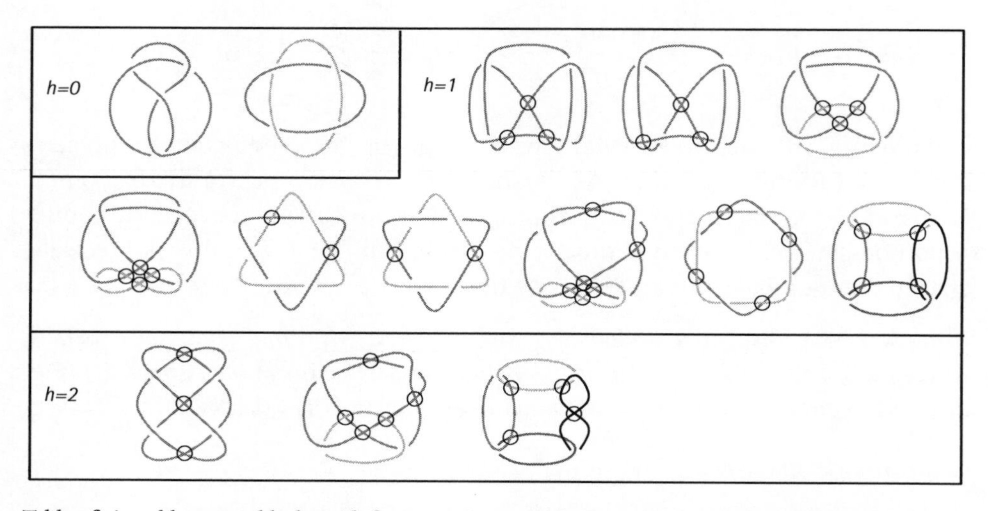

Table of virtual knots and links with four crossings. Objects are not distinguished from their mirror images, see [ZJZ04] for details.

27.3.4 Removing the flype redundancies.

The first occurences of flype equivalences occur in tangles with three crossings:

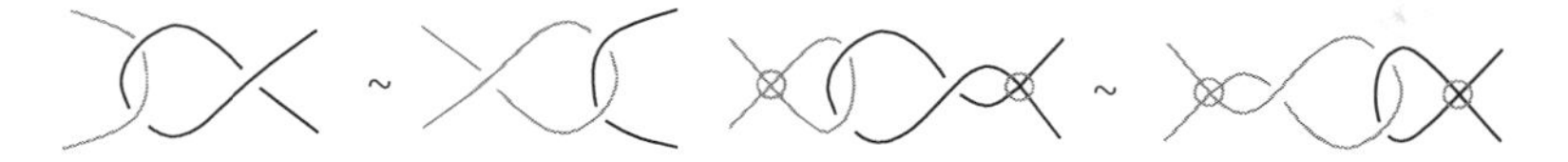

It has been suggested [ZJZ04] that it is (necessary and) sufficient to take the quotient by *planar* flypes, thus to perform the *same* renormalization $g \to g_0(g)$ as for genus 0. In other words, we have the

Generalized flype conjecture: *For a given (minimal) genus h, $\widetilde{\Gamma}^{(h)}(g) = \Gamma^{(h)}(g_0)$ is the generating function of flype-equivalence classes of virtual alternating tangles.*

Then denoting by $\widetilde{\Gamma}^{(h)}(g) = \Gamma^{(h)}(g_0)$ the generating function of the number of flype equivalence classes of prime virtual alternating 2-tangles of minimal genus h, $\widetilde{\Gamma}^{(0)}(g)$ is what was called $\widetilde{\Gamma}(g)$ in Section 27.2, Eq. (27.2.10), while

$$\widetilde{\Gamma}^{(1)}(g) = g+8\,g^2+57\,g^3+384\,g^4+2512\,g^5+16158\,g^6+102837\,g^7+649862\,g^8+4086137\,g^9+25597900\,g^{10}+\cdots$$

$$\widetilde{\Gamma}^{(2)}(g) = 17\,g^3+456\,g^4+7626\,g^5+100910\,g^6+1155636\,g^7+11987082\,g^8+115664638\,g^9+1056131412\,g^{10}+\cdots$$

$$\widetilde{\Gamma}^{(3)}(g) = 1259\,g^5+62072\,g^6+1727568\,g^7+35546828\,g^8+601504150\,g^9+8854470134\,g^{10}+\cdots$$

$$\widetilde{\Gamma}^{(4)}(g) = 200589\;g^7+14910216\;g^8+597738946\;g^9+17103622876\,g^{10}+\cdots$$

$$\widetilde{\Gamma}^{(5)}(g) = 54766516\;g^9+5554165536\,g^{10}+\cdots$$

The asymptotic behaviour of the number of inequivalent tangles of order p is

$$\tilde{\gamma}_p^{(h)} \sim \left(\frac{101+\sqrt{21001}}{40}\right)^p p^{\frac{5}{2}(h-1)}.$$

In [ZJZ04], this generalized flype conjecture was tested up to four crossings for links and five crossings for tangles by computing as many distinct invariants of virtual links as possible. We refer the reader to that reference for a detailed discussion. No counterexamples were found.

27.4 Coloured links

27.4.1 The bare matrix model

Let us first describe the 'bare' model that describes coloured link diagrams. Since we are only interested in the dominant order as the size of the matrices N goes to infinity, we can consider, as was argued in Section 27.2.1, a model of Hermitian matrices (as opposed to the complex matrices that were necessary in Section 27.3 for virtual tangles).

Let us fix a positive integer τ – the number of colours – and define the following measure on the space of τ Hermitian matrices M_a:

$$\prod_{a=1}^{\tau} dM_a \exp\left(N \operatorname{tr}\left(-\frac{1}{2}\sum_{a=1}^{\tau} M_a^2 + \frac{g}{4}\sum_{a,b=1}^{\tau}(M_a M_b)^2\right)\right) \tag{27.4.1}$$

This measure has an $O(\tau)$ symmetry where the matrices M_a are in the fundamental representation of $O(\tau)$.

Expansion in perturbation series of the constant g produces the following Feynman diagrams: they are fat graphs (planar maps) with vertices of valence 4, in which the colours cross each other at each vertex, see the figure. The summation over $O(\tau)$ indices produces a factor of τ for every colour loop.

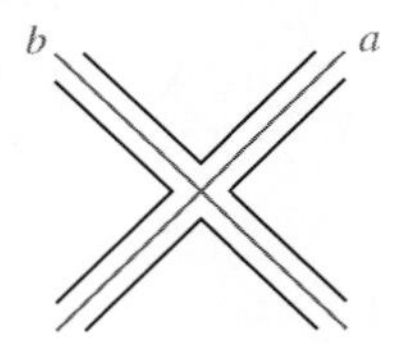

Thus, we have the following double expansion in g and τ:

$$F = \lim_{N\to\infty} \frac{\log Z}{N^2} = \sum_{\text{4-valent diagrams } D} \frac{1}{|\operatorname{Aut} D|} g^{\text{number of vertices}(D)} \tau^{\text{number of loops}(D)}$$

where it is understood that the number of loops is computed by considering that colour loops cross each other at vertices. In other words, the model of coloured links gives us more information than the one-matrix model because it allows for a 'refined' enumeration in which one distinguishes the number of components of the underlying link.

Note that F is at each order in g a polynomial in τ, so that we can formally continue it to arbitrary non-integer values of τ.

Observables

Let P_{2k} be the set of pairings of $2k$ points (sometimes called 'link patterns'), that is involutions of $\{1, \ldots, 2k\}$ without fixed points.

To each given link pattern π of $2k$ points one can associate the quantity I_π (I stands for 'internal connectivity') as follows. It is the generating series of the number of alternating $2k$-tangle diagrams (or simply, of 4-valent fat graphs with $2k$ external legs) with a weight of τ per closed loop and a weight of g per vertex, in such a way that the connectivity of the external legs, which are numbered say clockwise from 1 to $2k$, is represented by π (assuming as usual that colours cross at each vertex). See Fig. 27.1(a).

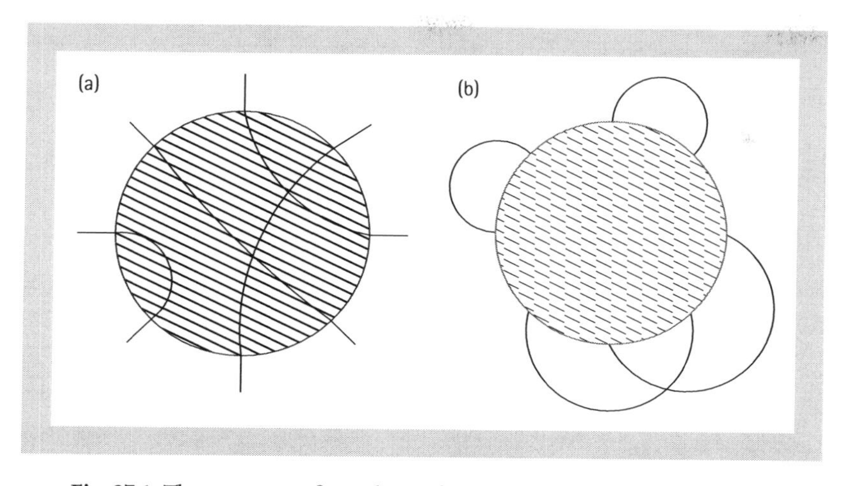

Fig. 27.1 The two types of correlation functions of the $O(\tau)$ matrix model.

From the point of view of the matrix model these observables I_π are not so natural. In principle one can define them as follows:

$$I_\pi = \lim_{N\to\infty} \left\langle \frac{\mathrm{tr}}{N} M_{a_1} \dots M_{a_{2k}} \right\rangle \qquad a_i = a_j \Leftrightarrow j = i \text{ or } j = \pi(i)$$

By $O(\tau)$ symmetry, the result is independent of the choice of the a_i as long as they satisfy the condition above, i.e. that indices occur exactly twice according to the link pattern π. However this formula only makes sense if $\tau \geq k$.

A more natural quantity in the matrix model is the 'external connectivity' correlation function E_π, which is defined in a very similar way:

$$E_\pi = \lim_{N\to\infty} \sum_{a_1=1}^{\tau} \cdots \sum_{a_{2k}=1}^{\tau} \prod_{i=1}^{2k} \delta_{a_i, a_{\pi(i)}} \left\langle \frac{\mathrm{tr}}{N} M_{a_1} \dots M_{a_{2k}} \right\rangle$$

The only difference is that this time one sums over all a_i (which might produce additional coincidences of indices, and in fact always will if $\tau < k$).

The graphical meaning of E_π is that it is the generating function of tangle diagrams with $2k$ external legs and prescribed connectivity *outside* the diagram, compare Fig. 27.1(b). Closing the external legs will produce closed loops which must be given a weight of τ. However, crossings outside the diagram should *not* be given a weight of g.

Noting that all the diagrams that contribute to E_π must have a certain internal connectivity, we can write

$$E_\pi = \sum_{\pi'} G_{\pi,\pi'} I_{\pi'} \tag{27.4.2}$$

The coefficients $G_{\pi,\pi'}$ are nothing but the natural scalar product on link patterns of same size $2k$, defined as follows:

$$G_{\pi,\pi'} = \tau^{\frac{1}{2}\text{number of cycles of } \pi\circ\pi'} \qquad \pi,\pi' \in P_{2k}$$

Graphically, this corresponds to gluing together the two pairings and giving a weight of τ to each closed loop that has been produced.

As a consequence of the formulae presented below, for positive integer τ and $k > \tau$, the matrix G has zero determinant and formula (27.4.2) cannot be inverted in the sense that the E_π are actually linearly dependent. For example at $\tau = 1$ there is really only one observable per k (with one colour one cannot distinguish connectivities). It is however convenient to introduce the pseudo-inverse W of G, that is the matrix that satisfies $WGW = W$ and $GWG = G$. The definition of G still makes sense for non-integer τ, in which case G is invertible and $W = G^{-1}$. We now sketch the computation of W following [CM09] (where it is called the Weingarten matrix, in reference to [Wei78]). See also [Zub08] for a recursive way to compute W for generic τ.

Matrix G is a $(2k-1)!! \times (2k-1)!!$ symmetric matrix, with the property that it is invariant by the action of the symmetric group, where the latter acts on involutions by conjugation ($\sigma \cdot \pi = \sigma\pi\sigma^{-1}$): $G_{\sigma\cdot\pi,\sigma\cdot\pi'} = G_{\pi,\pi'}$; or equivalently $\sigma G = G\sigma$ for all $\sigma \in \mathcal{S}_{2k}$. Furthermore, one easily finds that $\mathbb{C}[P_{2k}]$ contains exactly once every irreducible representation of $\mathcal{S}_{2k}$ associated to a Young diagram with *even lengths of rows*. Thus, G is a linear combination of projectors onto these irreducible subrepresentations, which are of the form

$$P^\lambda_{\pi,\pi'} = \frac{\chi^\lambda(1)}{|\mathcal{S}_{2k}|} \sum_{\sigma\in\mathcal{S}_{2k}:\sigma\cdot\pi'=\pi} \chi^\lambda(\sigma^{-1})$$

where λ is a Young diagram with $2k$ boxes ($\lambda = 2\mu$ for the projector P^λ to be non-zero) and χ^λ is the associated character of the symmetric group.

Finally one can write $G = \sum_\mu c_\mu P^{2\mu}$ where μ is a Young diagram with k boxes, and the coefficients c_μ can be computed [CM09, ZJ10]:

$$c_\mu = \prod_{(i,j)\in\mu} (\tau + 2j - i - 1) \tag{27.4.3}$$

Therefore, the pseudo-inverse W of G can be written as

$$W = \sum_{\mu:\, c_\mu\neq 0} c_\mu^{-1} P^{2\mu} \tag{27.4.4}$$

Loop equations

Loop equations are simply recursion relations satisfied by the correlation functions of our matrix model. They can in fact be derived graphically without any reference to the matrix model, in which case the parameter τ can be taken to be

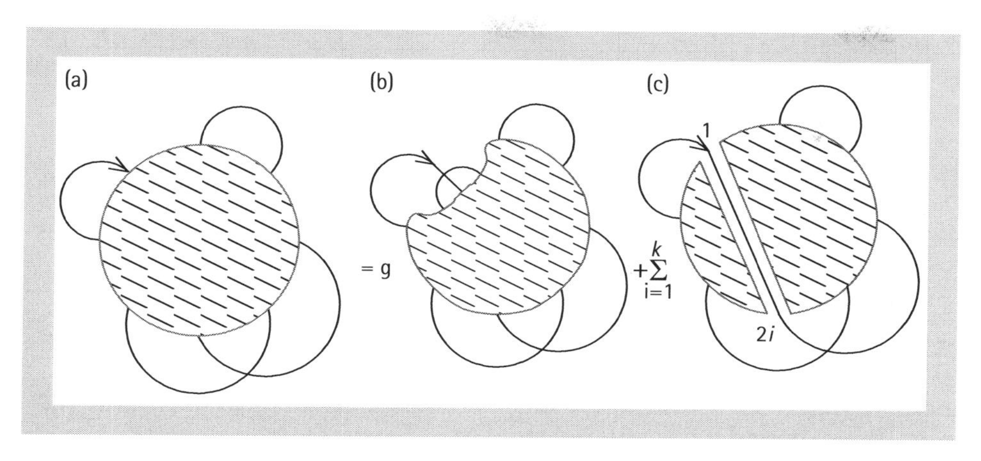

Fig. 27.2 Graphical decomposition for E_π.

arbitrary (not necessarily a positive integer). We recall that we limit ourselves to the dominant order as $N \to \infty$.

The recursion satisfied by E_π is illustrated on Fig. 27.2. Start with one of the external legs (say leg numbered one), and look at what happens to it once one moves inside the 'blob'. There are two possibilities: (i) it reaches a crossing, in which case one gets a factor of g and a new correlation function $E_{\pi'}$ where π' is obtained from π by adding one arch around the leg number one; or (ii) it goes out directly and connects to the external leg $2i$, $i = 1, \ldots, k$ (possibly creating a loop and therefore a factor of τ if $\pi(1) = 2i$). This second situation is more complex because, naively, the two blobs created by cutting the initial blob into two may still be connected by say $2\ell_i$ lines. Let us consider the two limiting cases. If $\ell_i = 0$ we simply have two disconnected blobs and the contribution is $E_{\pi_1} E_{\pi_2}$ where π_1 and π_2 are connectivities of size $2(i-1)$ and $2(k-i)$. On the contrary if $\ell_i = k-1$, the two blobs are fully connected to each other according to a certain permutation $\sigma \in \mathcal{S}_{2(k-1)}$ and it is clear that internal connectivity for one becomes external connectivity for the other, so that the contribution is of the form $\sum_{\pi_1 \in P_{2(k-1)}} E_{\pi_1} I_{\sigma\cdot\pi_1}$. The crucial remark is that one can rewrite this as $\sum_{\pi_1,\pi_2 \in P_{2(k-1)}} W_{\sigma\cdot\pi_1,\pi_2} E_{\pi_1} E_{\pi_2}$ even if G is non-invertible. Indeed the E_π, due to formula (27.4.2), live in the image of G and therefore one can ignore the zero modes of G (G being symmetric, its image and kernel are orthogonal). In the general case in which there are $2\ell_i$ connections between the two blobs with associated permutation $\sigma \in \mathcal{S}_{2\ell_i}$, one has to break these connections by using the matrix W for link patterns of size $2\ell_i$; calling $\pi_1(\rho_1)$ the connectivity of the first blob in which the $2\ell_i$ legs connecting it to the other blob have been replaced with the link pattern $\sigma^{-1}\cdot\rho_1$ of size $2\ell_i$, and similarly for $\pi_2(\rho_2)$ and link pattern ρ_2, we get an expression of the form

$$E_\pi = g\, E_{\pi'} + \sum_{i=1}^{k} \tau^{\delta_{\pi(1),2i}} \sum_{\rho_1,\rho_2 \in P_{2\ell_i}} W_{\rho_1,\rho_2}\, E_{\pi_1(\rho_1)}\, E_{\pi_2(\rho_2)} \tag{27.4.5}$$

(where $W_{\emptyset,\emptyset} = E_\emptyset = 1$). This equation allows us to calculate the E_π iteratively, in the sense that to compute the left-hand side at a given order, the E_π appearing in the right-hand side are either needed at a lower order in g, or at the same order in g but have fewer external legs than the E_π in the left-hand side.

27.4.2 Removal of redundancies and renormalized model

As in Section 27.2.3, we now discuss how to go from the counting of (coloured) alternating link diagrams to the counting of actual (coloured) alternating links, that is up to topological equivalences. We recall that the process involves two steps: removal of nugatory crossings and consideration of prime tangles only, which amounts to a wave function renormalization (i.e. renormalization of the quadratic term of the action); and inclusion of flypes, which amounts to a renormalization of the quartic term of the action. However, a crucial difference with the model discussed in Section 27.2.3 is that in the $O(\tau)$ model of coloured links one can introduce not just one, but two $O(\tau)$-invariant quartic terms: besides the already present term of the form $\mathrm{tr} \sum_{a,b} (M_a M_b)^2$, one can also have another term of the form $\mathrm{tr} \sum_{a,b} M_a^2 M_b^2$, and one expects that this term will be generated by the renormalization [ZJ01]. We now summarize the equations that we find. We start from the measure

$$\prod_{a=1}^{\tau} d M_a \exp\left(N \,\mathrm{tr}\left(-\frac{t}{2} \sum_{a=1}^{\tau} M_a^2 + \frac{g_1}{4} \sum_{a,b=1}^{\tau} (M_a M_b)^2 + \frac{g_2}{2} \sum_{a,b=1}^{\tau} M_a^2 M_b^2 \right)\right) \tag{27.4.6}$$

The Feynman rules of this model now allow loops of different colours to 'avoid' each other, which one can imagine as tangencies. The loop equations of this model generalize in an obvious way those of Chapter 16 and 29 and will not be written here.

Next we define the following correlation functions Δ and $\Gamma_{0,\pm}$:

$$\begin{aligned}
\Delta &= E_{(12)}/\tau \\
\Gamma_0 &= (E_{(12)(34)} - \tau(\tau+1)\Delta^2)/n \\
\Gamma_\pm &= I_{(12)(34)} - \Delta^2 \pm I_{(13)(24)}
\end{aligned}$$

The I_π are not directly defined in the matrix model, but the E_π are – in fact the four-point functions E_π are obtained by differentiating the free energy with respect to g_1 and g_2. But from the formulae of Section 27.4.1 one can check that the Weingarten function for four-point functions is invertible for $\tau \neq 1, -2$ (special cases which only require Γ_0, as discussed in detail in [ZJ03], and which we exclude from now on). Thus one can deduce the four-point I_π from the E_π.

Also define the auxiliary objects (generating series of horizontally two-particle irreducible diagrams)

$$H_0 = 1 - \frac{1}{(1-g)(1+\Gamma_0)}$$
$$H_\pm = 1 - \frac{1}{(1\mp g)(1+\Gamma_\pm)}$$

Then the equations to impose on the bare parameters g_1, g_2 and t as functions of the renormalized coupling constant g are

$$\Delta = 1 \tag{27.4.7}$$
$$g_1 = g(1 - H_+ - H_-) \tag{27.4.8}$$
$$g_2 = -g(H_0/\tau + (1/2 - 1/\tau)H_+ - H_-/2) \tag{27.4.9}$$

Up to order 8, we find

$$\begin{aligned}
g_1 &= g - (2g^4 + (2+2\tau)g^5 + (14+2\tau)g^6 + (26+16\tau+2\tau^2)g^7 + (134+56\tau+2\tau^2)g^8) + \cdots \\
g_2 &= -(g^3 + g^4 + 3g^5 + (5+2\tau)g^6 + (27+5\tau)g^7 + (89+32\tau+\tau^2)g^8) + \cdots \\
t &= 1 + 2g + \tau g^2 - 2\tau g^3 - 6g^4 - (8+10\tau)g^5 - (38+16\tau+3\tau^2)g^6 \\
&\quad - (104+86\tau+14\tau^2)g^7 - (410+338\tau+56\tau^2+2\tau^3)g^8 + \cdots
\end{aligned}$$

Composing these series with the correlation functions allows us to produce generating series for the number of coloured (prime) alternating tangles with arbitrary connectivity. For example, we find for 4- and 6-tangles (we only mention one pairing per class of rotationally equivalent pairings):

$$\begin{aligned}
I^c_{(12)(34)} &= g^2 + g^3 + (3+\tau)g^4 + (9+\tau)g^5 + (21+11\tau+\tau^2)g^6 + (101+32\tau+\tau^2)g^7 + \cdots \\
I^c_{(13)(24)} &= g + 2g^3 + 2g^4 + (6+3\tau)g^5 + (30+2\tau)g^6 + (62+40\tau+2\tau^2)g^7 + \cdots \\
I^c_{(14)(25)(36)} &= 2g^3 + 18g^5 + 18g^6 + (156+24\tau)g^7 + \cdots \\
I^c_{(14)(26)(35)} &= g^2 + 7g^4 + 6g^5 + (53+8\tau)g^6 + (154+6\tau)g^7 + \cdots \\
I^c_{(12)(35)(46)} &= 2g^3 + 2g^4 + (16+2\tau)g^5 + (42+2\tau)g^6 + (171+44\tau+2\tau^2)g^7 + \cdots \\
I^c_{(14)(23)(56)} &= 4g^4 + 8g^5 + (42+7\tau)g^6 + (156+14\tau)g^7 + \cdots \\
I^c_{(12)(34)(56)} &= 3g^4 + 9g^5 + (41+7\tau)g^6 + (168+21\tau)g^7 + \cdots
\end{aligned}$$

The superscript c means we are considering the *connected* generating series (corresponding to tangles which cannot be broken into several disentangled pieces) e.g. $I^c_{(12)(34)} = I_{(12)(34)} - 1$, $I^c_{(13)(24)} = I_{(13)(24)}$, etc.

In particular, note that $2I^c_{(12)(34)} + I^c_{(13)(24)}$ at $\tau = 1$ reproduces Eq. (27.2.15), and similarly $I^c_{(14)(25)(35)} + 3I^c_{(14)(26)(35)} + 6I^c_{(12)(35)(46)} + 3I^c_{(14)(23)(56)} + 2I^c_{(12)(34)(56)}$ at $\tau = 1$ reproduces Eq. (27.2.17).

27.4.3 Case of $\tau = 2$ or the counting of oriented tangles and links

In the case of the (renormalized) matrix model with $\tau = 2$ matrices, a subset of correlation functions can be computed exactly, including the two- and four-point functions which are necessary for our enumeration problem. Instead of giving two colours to each loop, one can equivalently give them two orientations: not only does this give a nice interpretation of the enumeration problem as the counting of *oriented* tangles, but it is also the first step towards the exact solution of the problem. Indeed, as shown in [ZJZ00], this reduces it to the solution of the *six-vertex model* on dynamical random lattices, which was studied in [ZJ00, Kos00].

The explicit generating series are given in terms of elliptic theta functions and will not be given here; even their asymptotic (large order) behaviour is somewhat non-trivial to extract, and we quote here the result of [ZJZ00]: if γ_p is the pth term of one of the four-point correlation functions,

$$\gamma_p \overset{p\to\infty}{\sim} \text{cst}\, g_c^{-p} p^{-2} (\log p)^{-2}$$

where g_c is the closest singularity to the origin of these generating series; $1/g_c \approx 6.28329764$. Though the latter number is non-universal, the subleading corrections are; they correspond to a $c = 1$ conformal field theory of a free boson coupled to quantum gravity.

27.4.4 Case of $\tau = 0$ or the counting of knots

A case of particular interest is the limit $\tau \to 0$ of the matrix model (27.4.1). This can be considered as a 'replica limit' where one sends the number of replicas to zero. Alternatively, the $\tau \to 0$ matrix model can be written explicitly using a supersymmetric combination of usual (commuting) and of Grassmannian (anticommuting) variables, see [ZJ03].

The observables are defined as follows:

$$\hat{E}_\pi = \lim_{\tau\to 0}\left(\frac{1}{\tau} E_\pi\right)$$

that is they correspond to tangles which, once closed from the outside, form exactly one loop (i.e. form knots as opposed to links).

The loop equations of the bare model become

$$\hat{E}_\pi = g\,\hat{E}_{\pi'} + \sum_{\substack{i=1,\dots,k\\ 2i\neq\pi(1)}} \sum_{\rho_1,\rho_2\in P_{2\ell_i}} \hat{W}_{\rho_1,\rho_2}\,\hat{E}_{\pi_1(\rho_1)}\,\hat{E}_{\pi_2(\rho_2)} \tag{27.4.10}$$

where $\hat{W}$ is the pseudo-inverse of $\hat{G} = \lim_{\tau\to 0}(\tau^{-1}G)$. Note that according to (27.4.3, 27.4.4), the factor τ^{-1} cancels the trivial zeros of G at $\tau = 0$. These zeros are simple for diagrams with $\lambda_3 \leq 1$; the remaining diagrams (in size $n \geq 6$) have higher zeros, making $\hat{G}$ non-invertible.

Although Eqs. (27.4.10) cannot be solved analytically, it is worth mentioning that they are easily amenable to an iterative solution by computer; in fact the resulting algorithm is notably better than the transfer matrix approach of [JZJ02, JZJ01], and one finds for example for the two point function $\Delta = \hat{E}_{(12)}$ the following power series in g: (using a PC with 8 Gb of memory and 24h of CPU)

$$
\begin{gathered}
1, 2, 8, 42, 260, 1796, 13396, 105706, 870772, 7420836, 65004584, 582521748, 5320936416, \\
49402687392, 465189744448, 4434492302426, 42731740126228, 415736458808868, \\
4079436831493480, 40338413922226212, 401652846850965808, 4024556509468827432, \\
40558226664529024000, 410887438338905738908, 4182776248940752113344, \\
42770152711524569532616, 439143340987014152920384, 4526179842103708969039296\dots
\end{gathered}
$$

The objects being counted by this formula are also known as self-intersecting plane curves or long curves, see e.g. [GZD98].

One can similarly take the limit $\tau \to 0$ in the renormalized model. However little is known beyond the general facts mentioned above for arbitrary τ.

27.4.5 Asymptotics

The most interesting unsolved question about the $O(\tau)$ matrix model of coloured links and tangles concerns the large-order behaviour of the generating series in the coupling constant g, i.e. the asymptotic number of coloured alternating tangles as the number of crossing is sent to infinity. If one considers, in the spirit of chapter 30, that the model represents a statistical model on random lattices, then it is expected that the model is critical for $|\tau| < 2$ and non-critical for $|\tau| > 2$. This should affect the universal subleading power-law corrections to the asymptotic behaviour.

In [SZJ04], the following conjecture was made. For $|\tau| < 2$, the model corresponds to a theory with central charge $c = \tau - 1$ (corresponding to the analytic continuation of a model of $\tau - 1$ free bosons). This implies the following behaviour for the series $\sum_p \gamma_p g^p$ counting coloured prime alternating tangles (with, say, four external legs):

$$
\gamma_p(\tau) \stackrel{p\to\infty}{\sim} \text{cst}\, g_c(\tau)^{-p}\, p^{\gamma(\tau)-2} \qquad \gamma(\tau) = \frac{\tau - 2 - \sqrt{(2-\tau)(26-\tau)}}{12} \qquad |\tau| < 2
$$

This was tested numerically in [SZJ04], but the results are not entirely conclusive (see also [JZJ01, ZJ05]).

In particular, as a corollary of the conjecture above, one would have the following asymptotic behaviour for the number of prime alternating knots:

$$
f_p \stackrel{p\to\infty}{\sim} \text{cst}\, g_c(0)^{-p}\, p^{-\frac{19+\sqrt{13}}{6}}
$$

It is most likely that one can remove the 'prime' property without changing the form of the asymptotic behaviour (only the non-universal coefficient of the exponential growth would be modified); one can speculate that removing the 'alternating' property will not change it either.

Note the similarity between our problem of counting knots with that of counting meanders [DFGG00]. There too, the problem can be rewritten as a matrix model and the asymptotic behaviour is dictated by 2D quantum gravity, leading to a non-rational critical exponent. A key difference is that in the case of meanders, corrections to the leading behaviour are expected to be power-law, making numerical checks reasonably easy. In contrast, if the conjecture above for knots is correct, the corrections are expected to be logarithmic (the theory being asymptotically free in the infrared), which would make numerical checks extremely hard.

Acknowledgements

This work was supported in part by EU Marie Curie Research Training Network "ENRAGE" MRTN-CT-2004-005616, ESF program "MISGAM", and ANR program "GRANMA" BLAN08-1-13695.

References

[Ada97] P. Adamietz. Kollektive Feldtheorie und Momentenmethode in Matrixmodellen, 1997. PhD thesis, internal report DESY T-97-01.

[Ake97] G. Akemann, 1997. unpublished notes, private communication.

[BIPZ78] E. Brézin, C. Itzykson, G. Parisi, and J.-B. Zuber. Planar diagrams. *Comm. Math. Phys.*, 59(1):35–51, 1978.

[CKS02] J. S. Carter, S. Kamada, and M. Saito. Stable equivalence of knots on surfaces and virtual knot cobordisms. *J. Knot Theory Ramifications*, 11(3):311–322, 2002. Knots 2000 Korea, Vol. 1 (Yongpyong).

[CM09] B. Collins and S. Matsumoto. On some properties of orthogonal Weingarten functions. *J. Math. Phys.*, 50(11):113516, 14, 2009.

[DFGG00] P. Di Francesco, O. Golinelli, and E. Guitter. Meanders: exact asymptotics. *Nuclear Phys. B*, 570(3):699–712, 2000.

[Fin03] S. Finch. Knots, links and tangles, 2003. http://algo.inria.fr/csolve/knots.pdf.

[Gre04] J. Green. A table of virtual knots, 2004. http://www.math.toronto.edu/~drorbn/Students/GreenJ/.

[GZD98] S. M. Guseĭn-Zade and F. S. Duzhin. On the number of topological types of plane curves. *Uspekhi Mat. Nauk*, 53(3(321)):197–198, 1998.

[JZJ01] J. Jacobsen and P. Zinn-Justin. A transfer matrix approach to the enumeration of colored links. *J. Knot Theory Ramifications*, 10(8):1233–1267, 2001.

[JZJ02] J. Jacobsen and P. Zinn-Justin. A transfer matrix approach to the enumeration of knots. *J. Knot Theory Ramifications*, 11(5):739–758, 2002.

[Kau99] L. Kauffman. Virtual knot theory. *European J. Combin.*, 20(7):663–690, 1999.

[KM06] L. Kauffman and V. Manturov. Virtual knots and links. *Tr. Mat. Inst. Steklova*, 252(Geom. Topol., Diskret. Geom. i Teor. Mnozh.):114–133, 2006.

[Kos00] I. Kostov. Exact solution of the six-vertex model on a random lattice. *Nuclear Phys. B*, 575(3):513–534, 2000.

[Kup03] G. Kuperberg. What is a virtual link? *Algebr. Geom. Topol.*, 3:587–591 (electronic), 2003.

[Mor91] T.R. Morris. Chequered surfaces and complex matrices. *Nucl. Phys. B*, 356:703–728, 1991.

[MT91] W. Menasco and M. Thistlethwaite. The Tait flyping conjecture. *Bull. Amer. Math. Soc. (N.S.)*, 25(2):403–412, 1991.

[MT93] W. Menasco and M. Thistlethwaite. The classification of alternating links. *Ann. of Math. (2)*, 138(1):113–171, 1993.

[ST98] C. Sundberg and M. Thistlethwaite. The rate of growth of the number of prime alternating links and tangles. *Pacific J. Math.*, 182(2):329–358, 1998.

[STV02] A. Stoimenow, V. Tchernov, and A. Vdovina. The canonical genus of a classical and virtual knot. In *Proceedings of the Conference on Geometric and Combinatorial Group Theory, Part II (Haifa, 2000)*, volume 95, pages 215–225, 2002.

[SV05] A. Stoimenow and A. Vdovina. Counting alternating knots by genus. *Math. Ann.*, 333(1):1–27, 2005.

[SZJ04] G. Schaeffer and P. Zinn-Justin. On the asymptotic number of plane curves and alternating knots. *Experiment. Math.*, 13(4):483–493, 2004.

[Wei78] D. Weingarten. Asymptotic behavior of group integrals in the limit of infinite rank. *J. Mathematical Phys.*, 19(5):999–1001, 1978.

[ZJ00] P. Zinn-Justin. The six-vertex model on random lattices. *Europhys. Lett.*, 50(1): 15–21, 2000.

[ZJ01] P. Zinn-Justin. Some matrix integrals related to knots and links. In *Random matrix models and their applications*, volume 40 of *Math. Sci. Res. Inst. Publ.*, pages 421–438. Cambridge Univ. Press, Cambridge, 2001. proceedings of the 1999 semester at the MSRI.

[ZJ03] P. Zinn-Justin. The general $O(n)$ quartic matrix model and its application to counting tangles and links. *Comm. Math. Phys.*, 238(1–2):287–304, 2003.

[ZJ05] P. Zinn-Justin. Conjectures on the enumeration of alternating links. In *Physical and numerical models in knot theory*, volume 36 of *Ser. Knots Everything*, pages 597–606. World Sci. Publ., Singapore, 2005.

[ZJ10] P. Zinn-Justin. Jucys–Murphy elements and Weingarten matrices. *Letters in Mathematical Physics*, 91(2), 2010.

[ZJZ00] P. Zinn-Justin and J.-B. Zuber. On the counting of colored tangles. *J. Knot Theory Ramifications*, 9(8):1127–1141, 2000.

[ZJZ02] P. Zinn-Justin and J.-B. Zuber. Matrix integrals and the counting of tangles and links. *Discrete Math.*, 246(1–3):343–360, 2002. proceedings of "Formal power series and algebraic combinatorics" (Barcelona, 1999).

[ZJZ04] P. Zinn-Justin and J.-B. Zuber. Matrix integrals and the generation and counting of virtual tangles and links. *J. Knot Theory Ramifications*, 13(3):325–355, 2004.

[Zub08] J.-B. Zuber. The large N limit of matrix integrals over the orthogonal group. *J. Phys. A*, 41:382001, 2008.

·28·

Multivariate statistics

Noureddine El Karoui

Abstract

This chapter discusses some classical and more modern results obtained in random matrix theory for applications in statistics.

28.1 Introduction

The classic paradigm of parametric statistics is that we observe data generated randomly according to a probability distribution indexed by parameters, and, from this data, which is by nature random, we try to infer properties of the *deterministic* (and unknown) parameters. Here is a simple matrix-related example: suppose that we observe n vectors $X_1, \ldots, X_n$ generated according to a normal distribution with mean 0 and covariance Σ – i.e. $X_i \sim \mathcal{N}(0, \Sigma)$ – what can we say about Σ? Can we estimate it? Or can we estimate for instance a few of its eigenvalues and eigenvectors?

Such questions are essential in multivariate statistical analysis, the sub-part of statistics dealing with vector-valued data. We will present some salient aspects of this and refer the reader to [And03], [Mui82], and [MKB79] for a thorough introduction.

Going back to our simple example, it is clear that our ability to infer properties of the unknown Σ (the *population covariance matrix*) will depend on the quality of our *estimator*, $\widehat{\Sigma}$, of Σ. Formally, $\widehat{\Sigma} = F(X_1, \ldots, X_n)$, where F is a deterministic function. $\widehat{\Sigma}$ is naturally a random matrix, computed from the data, so it is observable in practice. The standard estimator of Σ is the sample covariance matrix defined by

$$\widehat{\Sigma} = \frac{1}{n}\sum_{i=1}^{n} X_i X_i' - \bar{X}\bar{X}',$$

where X_i' is the transpose of X_i and $\bar{X} = \frac{1}{n}\sum_{i=1}^{n} X_i$. If X is an $n \times p$ data matrix whose i-th row is X_i, we have in matrix notation,

$$\widehat{\Sigma} = \frac{1}{n} X' H X\,, \tag{28.1.1}$$

where $H = \mathrm{Id}_n - \mathbf{e}\mathbf{e}'/n$ and $\mathbf{e}$ is the n-dimensional vector whose entries are all equal to 1. Another popular choice is $\widetilde{\Sigma} = X'HX/(n-1)$, which is unbiased when the X_is are iid with a second moment, i.e. $\mathbf{E}\left(\widetilde{\Sigma}\right) = \Sigma$. In what follows, we will refer to $\widehat{\Sigma}$ as the *sample covariance matrix*.

At this point, we can already ask several interesting statistical questions. For instance, can we test, based e.g. on $\widehat{\Sigma}$, that $\Sigma = \Sigma_0$, for a given deterministic Σ_0? In other words, can we find a procedure involving only the data such that the probability that we declare $\Sigma \neq \Sigma_0$ when $\Sigma = \Sigma_0$ is less than a certain prespecified threshold (or level) α? Can we test that Σ has a few 'separated' eigenvalues? For reasons that will be clear later, spectral methods, based on eigenvalues and eigenvectors of $\widehat{\Sigma}$, are popular in statistics. Answering the testing questions we just mention therefore requires us to develop a good understanding of the spectral properties of $\widehat{\Sigma}$.

Before we proceed, let us set some notations. We will use $\lambda_i(\Sigma)$ to denote the decreasingly ordered eigenvalues of Σ (i.e. $\lambda_1 \geq \lambda_2 \geq \ldots$). We will use l_i with the same ordering convention when we refer to eigenvalues computed from sample covariance matrices. Finally, the symbol $\Longrightarrow$ denotes weak convergence of distributions and X' is the transpose of X.

28.1.1 Two spectral statistical techniques: PCA and CCA

To motivate the theoretical discussions that will follow, we first discuss some techniques widely used in practice. The first one is a (linear) dimensionality reduction technique, principal components analysis (PCA). It is also known under various names such Karhunen-Loève decomposition and is literally used in tens of thousands of published research articles. Entire books have been written on it (see [Jol02]).

PCA PCA seeks to answer the following question: given p-dimensional data, what is the optimal k-dimensional (linear) representation of this data? Here optimality is understood in the sense of preserving the maximal possible amount of variance. In other words, if Y is a p-dimensional vector with covariance Σ, we seek a matrix A, which is $k \times p$ with unit length rows, such that the coordinates of AY are uncorrelated and $\mathrm{trace}\,(\mathrm{cov}\,(AY))$ is maximal. Because $\mathrm{cov}\,(AY) = A\Sigma A'$, and by construction we require this matrix to be diagonal, it is clear that the optimal A is the matrix containing the first k eigenvectors of Σ. In practice, since we do not know Σ, we replace this construction with the equivalent construction using the first k eigenvectors of $\widehat{\Sigma}$, our estimate of Σ (this is equivalent to minimizing the ℓ^2 error made by projecting the centred data on a k-dimensional hyperplane – see [HTF09], Section 14.5). PCA is a particularly useful technique when $\sum_{i=1}^{k} \lambda_i(\Sigma)/\mathrm{trace}\,(\Sigma)$ is large for k small, in which case a low-dimensional projection of the data contains most of the 'information' (i.e. variance) of the data. A natural question is of course: how should k be chosen? We get back to this particular question later in Section 28.3, but it should now be clear that inferential work about PCA will require understanding of the statistical behaviour of eigenvalues (and eigenvectors) of sample covariance matrices.

CCA Another widely used spectral method in statistics is canonical correlation analysis (CCA). By contrast to PCA, CCA operates on two data sets. Where PCA sought to find projections of the data that contain a maximal amount of variance, CCA seeks projections of the two data sets that are maximally correlated with one another. The population version of CCA is the following: suppose X (resp. Y) is a p-dimensional (resp. q-dimensional) vector. Call Σ the covariance matrix of $Z' = (X'Y')$, Σ_{11} the covariance of X, Σ_{22} the covariance of Y and $\Sigma_{12} = \mathbf{E}(XY')$ the cross-covariance between X and Y. The first canonical correlation coefficient is defined as, if a and b are vectors of dimension p and q,

$$\rho_1 = \max_{a,b} a'\Sigma_{12}b \text{ subject to } a'\Sigma_{11}a = b'\Sigma_{22}b = 1.$$

The vectors a and b realizing this maximum (which are clearly defined up to sign) are called the canonical correlation vectors. Other canonical correlation coefficients are defined sequentially by requiring that the subsequent canonical correlation variables (i.e. $a_i'X$ and $b_i'Y$) be uncorrelated with all the previous ones. Practically, canonical correlation coefficients and vectors are computed by solving a generalized eigenvalue problem, now involving $\widehat{\Sigma}$. Hence, understanding the behaviour of CCA in practice also requires that we understand the spectral properties of $\widehat{\Sigma}$ and functions of its sub-blocks.

28.1.2 Classical theory, modern questions

Random matrices and their spectra are an essential tool of multivariate statistical analysis and have therefore been studied in great detail by statisticians. The seminal paper of Wishart [Wis28] gave the distribution of sample covariance matrices computed from Gaussian data. Later papers [Fis39, Hsu39, Gir39] proved joint distribution results about the eigenvalues of these matrices. In the 1960s, Anderson [And63] obtained limiting distribution results for these eigenvalues, while A.T James [Jam64] and collaborators developed zonal polynomial techniques to obtain closed-form expressions for the distribution of some of these statistics. So why is there a renewed interest for random matrix theory in statistics?

One factor is that with data collection made easier and with increase in computing power making spectral analysis of large data sets practically feasible, the reach of statistical methods has extended to very high-dimensional data sets. In particular, while the important results of [And63] were obtained in an asymptotic setting where p, the number of variables in the data set, was held fixed, and n, the number of observations, was going to infinity, it is now often the case that statisticians work with data sets for which p and n are comparable to one another and, sometimes, p is even much larger than n (for instance in genomic studies, where one can get data about thousands of genes for a patient, but the number of patients is limited to fifty or a hundred). Understanding the behaviour and limitations of well-known and widely used techniques such as PCA and CCA in modern practice therefore leads to the study of the spectral properties of large-dimensional random matrices. Most of these modern efforts are centred on asymptotics where p and n both go to infinity – the 'large p, large n' setting, in the hope that they give more informative approximations for high-dimensional data sets than the classical 'fixed p, large n' asymptotics. (There is also interest in 'large p, fixed n' asymptotics, see e.g. [HMN05].)

This chapter is organized as follows: we discuss the Wishart distribution in 28.2 and then relatively recent results concerning extreme eigenvalues of large-dimensional sample covariance matrices (in 28.3). We then briefly discuss questions concerning limiting spectral distribution results (in 28.4) and conclude.

28.2 Wishart distribution and normal theory

A significant part of classical statistical random matrix theory has been concerned with the *Wishart* distribution. A Wishart matrix with (positive definite) covariance Σ and n degrees of freedom is defined as

$$W = \sum_{i=1}^{n} Z_i Z_i' \ , \ \text{where } Z_i \overset{iid}{\backsim} \mathcal{N}(0, \Sigma).$$

In matrix form, this is equivalent to saying that if $W \sim \mathcal{W}_p(\Sigma, n)$, $W = \Sigma^{1/2} Z' Z \Sigma^{1/2}$, where Z is a $n \times p$ matrix with iid $\mathcal{N}(0, 1)$ entries and Σ is $p \times p$. Note that the sample covariance matrix defined in Equation (28.1.1) can be written in the Gaussian case $n\widehat{\Sigma} = \Sigma^{1/2} Z' H Z \Sigma^{1/2}$ and at first glance might not appear to be Wishart, because of the presence of H. However, H is a rank $(n-1)$ orthogonal projection matrix, so after diagonalization and use of elementary properties of Gaussian random vectors, we see that

$$n\widehat{\Sigma} \sim \mathcal{W}_p(\Sigma, n-1) \ . \tag{28.2.1}$$

This simple result explains why much effort in the statistics literature has been devoted to understanding Wishart distributions. The Wishart distribution has many remarkable properties and numerous results involving it yield exact distributional results in finite dimension (see also Section 37.2). We present a non-trivial and recent one in Section 28.2.3.

28.2.1 Classical results, distribution of eigenvalues

The fundamental result of [Wis28] is the following theorem (see e.g. [Mui82], Theorem 3.2.1, or [And03], Theorem 7.2.2):

Theorem 28.2.1 *Suppose W is distributed according to a Wishart distribution $\mathcal{W}_p(\Sigma, n)$, where $n \geq p$ and Σ is $p \times p$. Then the density of W is*

$$\frac{(\det(W))^{(n-p-1)/2}}{2^{np/2}(\det(\Sigma))^{n/2}\Gamma_p(n/2)} \exp(-\frac{1}{2}\mathrm{trace}(\Sigma^{-1}W)), \qquad (28.2.2)$$

for $W \succ 0$ (i.e. positive definite) and 0 otherwise. Γ_p is the multivariate gamma function defined by $\Gamma_p(t) = \pi^{p(p-1)/4}\prod_{i=1}^{p}\Gamma(t-(i-1)/2)$.

In theoretical statistics, both this density result and the stochastic representation mentioned earlier are extremely useful. However, the density result is particularly convenient when one needs to deal with determinants and other functions of eigenvalues. In particular, we have the following result concerning the joint distribution of eigenvalues (see e.g. [And03], Theorem 13.3.2).

Theorem 28.2.2 *Suppose $W \sim \mathcal{W}_p(\mathrm{Id}_p, n)$ and $n \geq p$. Then the joint density of the eigenvalues $l_1 > l_2 > \ldots > l_p > 0$ of W is*

$$f(l_1, l_2, \ldots, l_p) = C\prod_{i<j}(l_i - l_j)\left(\prod_{i=1}^{p} l_i^{(n-p-1)/2}\right)\exp(-\sum_{i=1}^{p} l_i)\,, \qquad (28.2.3)$$

where

$$C = \frac{\pi^{p^2/2}}{2^{pn/2}\Gamma_p(n/2)\Gamma_p(p/2)}\,.$$

The joint density described in Equation (28.2.3) is amenable to analysis using modern tools of random matrix theory. Because the density is of the form $V((l_i)_{i=1}^{p})\prod_{i=1}^{p} l_i^{\frac{n-p-1}{2}} \mathrm{e}^{-l_i}$ (where V is the Vandermonde determinant appearing above) and because $l_i^k \mathrm{e}^{-l_i}$ is the Laguerre weight in classical orthogonal polynomial theory (see [Sze75], Chapter 5), the $\mathcal{W}_p(\mathrm{Id}, n)$ distribution is known as the Laguerre orthogonal ensemble (LOE) in random matrix literature.

28.2.2 Classical asymptotics: fixed **p**, large **n**

Classical asymptotics in statistics have been studied while keeping the number of variables in the problem, p, fixed and letting the number of samples (or observations), n, go to infinity. For applications in principal component analysis, two main cases were of interest: the case where all eigenvalues of Σ have multiplicity one and the case where $\Sigma = \mathrm{Id}_p$, the latter case corresponding to the 'null' case where the variance is the same in all directions, and hence all directions are equally informative.

The following was shown in [And63] (see also [And03], Chapter 13.5.1):

Theorem 28.2.3 *Suppose that all eigenvalues of the covariance matrix Σ have multiplicity 1. Then if $n\widehat{\Sigma} \sim \mathcal{W}_p(\Sigma, n)$, and $(l_i)_{i=1}^p$ denotes the vector of ordered eigenvalues of $\widehat{\Sigma}$, we have, when p is held fixed and n goes to infinity,*

$$\sqrt{n}(l_i(\widehat{\Sigma}) - \lambda_i(\Sigma)) \Longrightarrow \mathcal{N}(0, 2\lambda_i^2).$$

A similar result holds for the corresponding eigenvectors. Anderson's proof is perturbative in nature and can be obtained by using now classic results concerning perturbation of eigenvalues and eigenvectors of symmetric matrices (see e.g. [Kat95], p. 80) in connection with simple weak convergence results for the entries of Wishart matrices.

The case of $\Sigma = \mathrm{Id}_p$ is slightly more complicated because the fact that the population eigenvalues (the λ_is) are not isolated precludes the use of simple perturbative arguments. Nonetheless Anderson [And63] obtained fluctuation results here also. In modern random matrix parlance, his results read as follows (see also Theorem 13.5.2 in [And03]):

Theorem 28.2.4 *Suppose that $n\widehat{\Sigma} \sim \mathcal{W}_p(\mathrm{Id}_p, n)$, and call l_i the eigenvalues of $\widehat{\Sigma}$. Then, when p is held fixed and $n \to \infty$,*

$$\sqrt{n}\,(l_i - 1)_{i=1}^p \Longrightarrow \mathcal{L}(\textit{spectrum } GOE_{p\times p}).$$

In words, the vector of eigenvalues of $\widehat{\Sigma}$ now behaves (after recentring around the population value 1, and rescaling by $\sqrt{n}$) in the same way as the eigenvalues of a $p \times p$ GOE matrix (see Chapters 5 and 21). We note that under the assumptions of the previous theorem, the eigenvectors of $\widehat{\Sigma}$ are distributed uniformly on the $p \times p$ orthogonal group – after properly dealing with the sign indeterminacy. An interesting alternative approach to these questions was developed in [ET91].

28.2.3 An application in optimization and finance

The Wishart distribution has many remarkable analytic properties which sometimes allow us to obtain exact distributional results for non-trivial statistics in finite dimensions (for other applications to finance see Chapter 40). We now present a recent result at the intersection of statistics and optimization that illustrates this fact (the proof relies on classic statistical results (see [Eat72], [Eat07], Proposition 8.9)).

Consider the following quadratic program with linear equality constraints (see e.g [BV04] for definitions):

$$\begin{cases} \min_{w\in\mathbb{R}^p} \frac{1}{2}w'\Sigma w \\ w'v_i = u_i\,,\ 1 \leq i \leq k-1\ , \\ w'\mu = u_k \end{cases} \qquad \text{(QP-eqc-Pop)}$$

This problem arises for instance in financial applications where an investor tries to find an optimal investing strategy which guarantees a certain level of returns on the investments, while minimizing risk (measured here by the variance of the portfolio). The above problem requires us to know the population (or 'true') parameters Σ and μ – assuming, as we will, that the v_i are given constraints. The u_is are also assumed to be deterministic and known. In practice, we need to estimate these parameters and a natural first approach would be to estimate the solution of Problem (QP-eqc-Pop) by the solution of its sample version:

$$\begin{cases} \min_{w \in \mathbb{R}^p} \frac{1}{2} w' \widehat{\Sigma} w \\ w' v_i = u_i \,, \ 1 \leq i \leq k-1, \\ w' \widehat{\mu} = u_k \end{cases} \qquad \text{(QP-eqc-Emp)}$$

We call w_{emp} the solution of Problem (QP-eqc-Emp) and w_{theo} the solution of Problem (QP-eqc-Pop). It is very interesting to practitioners to try to relate the naive measure of risk $w'_{\text{emp}} \widehat{\Sigma} w_{\text{emp}}$ with the population (or true) version $w'_{\text{theo}} \Sigma w_{\text{theo}}$. The following theorem (see [El 09b]) gives an answer (exact distributional result in finite dimension) to this question.

Theorem 28.2.5 *Suppose that we observe iid data* $\{X_i\}_{i=1}^n$ *with* $X_i \sim \mathcal{N}(\mu, \Sigma)$ *and* Σ *positive definite. Suppose* $\widehat{\Sigma}$ *is defined as in 28.1.1, with* $n-1$ *scaling (instead of* n*) and* $\widehat{\mu}$ *is the sample mean. Call* w_{oracle} *the solution of Problem* (QP-eqc-Pop) *where* μ *is replaced by* $\widehat{\mu}$*. Finally, call* $\widehat{V}$ *the* $p \times k$ *matrix containing the vectors* $v_1, \ldots, v_k = \widehat{\mu}$*. Then we have, if* $n-1 \geq p$*, and* $\widehat{V}' \Sigma^{-1} \widehat{V}$ *is invertible,*

$$w'_{\text{emp}} \widehat{\Sigma} w_{\text{emp}} = w'_{\text{oracle}} \Sigma w_{\text{oracle}} \frac{\chi^2_{n-1-p+k}}{n-1} \,, \tag{28.2.4}$$

where $w'_{\text{oracle}} \Sigma w_{\text{oracle}}$ *is random (because* $\widehat{\mu}$ *is) but is statistically independent of* $\chi^2_{n-1-p+k}$ *(a* χ^2 *random variable with* $n-1-p+k$ *degrees of freedom).*

We do not discuss this result and its intuitive meaning any further but wish to mention that it is not robust asymptotically to 'interesting' modifications of the Gaussianity assumptions (see 28.4.2 for more on this topic).

28.3 Extreme eigenvalues, Tracy–Widom laws

As we have explained before, modern and high-dimensional statistics calls for new types of asymptotics. In this section, we will contrast results obtained in the asymptotic setting of 'large p, large n' to the classical results described earlier. Strikingly, the asymptotic predictions obtained from taking these doubly infinite limits often work very well in low dimensions (p and n sometimes as small as five for results concerning largest eigenvalues). In particular, they are sometimes closer to simulation results than 'classical' predictions, perhaps capturing the fact that they are sensitive to what would have been in lower dimensions a second-order effect (due to p), but turns out to be in high-dimension a first-order effect. Finally, let us mention that both scaling and limiting laws are often different in the 'large p, large n' setting and the 'small p, large n' settings.

28.3.1 Statistical motivation: largest root tests in PCA

An important problem in the practice of PCA is to determine the number of principal components to keep, i.e. to determine the dimensionality of the subspace on which to project the data. There are several empirical rules of thumb (see [MKB79], Chapter 8): for instance k could be chosen so that $\sum_{i=1}^{k} l_i(\widehat{\Sigma})/\text{trace}\,(\widehat{\Sigma}) \geq C$, where C is a given threshold. This is taken to mean that the fraction of variance explained by the first k principal components is greater than C (classical asymptotics validate this interpretation). A slightly more model-based approach can also be taken: suppose that we observe Gaussian data, and say that the covariance matrix is a small rank perturbation of Id_p. For ease of exposition, let us even assume that Σ is diagonal. Clearly, the principal components corresponding to the part of the covariance matrix that is identity are not terribly informative, since the variance is the same in all those directions – hence no directions should be preferred over others. So we would like to know (from testing) how many eigenvalues of Σ are equal to 1.

Let us now talk about two approaches to select k: one graphical (the scree plot) and one based on formal testing method. The *scree plot* [Cat66] seeks an 'elbow' in the graph of the ordered eigenvalues of $\widehat{\Sigma}$ versus their rank to decide on the number of components to keep. We illustrate graphically the effect of high dimensionality on this procedure with Figures 28.1 and 28.2. when $n = p = 50$, the 'high-dimensional' setting (because p/n is not small), the clear demarcation that exists when $n = 500$ and $p = 50$ between the 'signal' part (separated eigenvalues with multiplicity one) and the 'noise' part (equal eigenvalues) of the covariance matrix has vanished. Hence, there might be some question about the efficiency of the scree plot method in high dimension.

More formal testing approaches have been proposed to tackle this problem. One is a sequential application of Roy's test ([Roy57], Section 6.4, or [And03], Section 10.8.3) for the null hypothesis that $\Sigma = \text{Id}$. This test is based on the fluctuation behaviour of the extreme eigenvalues of $\widehat{\Sigma}$. Another one is a sequential application of the likelihood ratio test for sphericity (i.e. all eigenvalues equal; see [MKB79], pp.235–236); this is a χ^2 test involving the arithmetic and geometric means of the sample eigenvalues. In both cases, the tests can be applied by taking into account $p - k$ eigenvalues, with k starting at 0 and being increased until the null hypothesis is not rejected. (A correction for multiple testing should nonetheless be applied.)

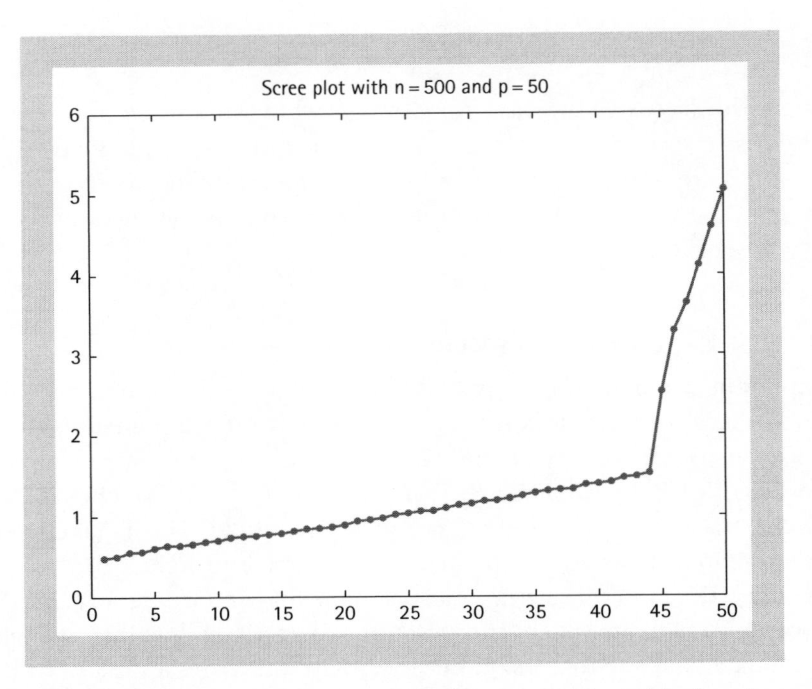

Fig. 28.1 Scree plot for $n = 500$, $p = 50$: X_i are iid $\mathcal{N}(0, \Sigma)$. Σ is diagonal, its first six eigenvalues being (5,4.5,4,3.5,3,2.5), all the other eigenvalues are equal to 1. The elbow after the sixth empirical eigenvalue is clear.

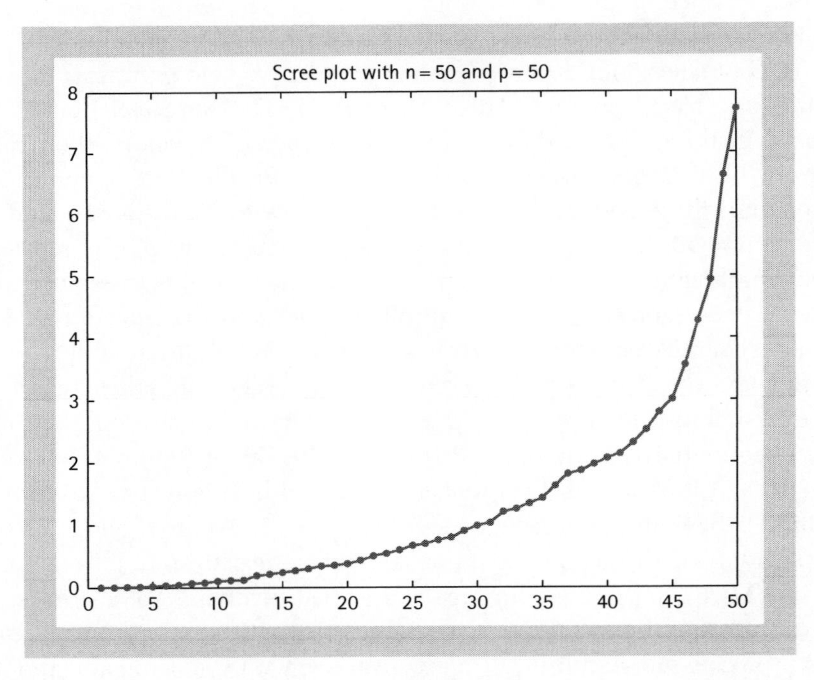

Fig. 28.2 Scree plot for $n = 50$, $p = 50$: X_i are iid $\mathcal{N}(0, \Sigma)$. Σ is diagonal, its first six eigenvalues being (5,4.5,4,3.5,3,2.5), all the other eigenvalues are equal to 1. The elbow after the sixth empirical eigenvalue has disappeared.

28.3.2 Tracy–Widom convergence

In light of the requirements for performing Roy's test in high dimension and also because of the sheer interest of the question for principal component analysis, it is natural to study the behaviour of the largest eigenvalues of sample covariance matrices in the 'large p, large n' setting. Before we proceed, we note that the marginal distribution of the extreme eigenvalues of a Wishart matrix (for general Σ) is known in closed form (see [Con63] and [Mui82], Section 9.7). However, evaluating the corresponding expressions is difficult, even numerically, because they involve hypergeometric functions of matrix arguments and hence series of zonal polynomials which converge slowly.

The mathematical groundwork for this type of asymptotic results concerning largest eigenvalues was done in work of Tracy and Widom [TW94b, TW94a, TW96]. A key and remarkable result in this area was obtained by Johnstone [Joh01]. It is the following:

Theorem 28.3.1 *Suppose $\widehat{\Sigma} \sim \mathcal{W}_p(\mathrm{Id}_p, n)/n$ and denote by l_1 its largest eigenvalue. Then as p and n tend to infinity, while $p/n \to \rho \in (0, \infty)$, we have*

$$(n-1)^{2/3} \frac{a_n l_1(\widehat{\Sigma}) - \mu_{n-1,p}}{\sigma_{n-1,p}} \Longrightarrow TW_1 , \qquad (28.3.1)$$

with

$$a_n = \frac{n}{n-1} , \ \mu_{n,p} = (1 + \sqrt{p/n})^2 \quad \textit{and} \ \sigma_{n,p} = \left(1 + \sqrt{\frac{n}{p}}\right)\left(1 + \sqrt{\frac{p}{n}}\right)^{1/3} .$$

Here TW_1 is the Tracy–Widom distribution appearing in the study of GOE.

We note that the work of Johnstone is connected to work of Forrester [For93] and Johansson [Joh00] on related but different problems. Later work of Soshnikov [Sos02] showed that the convergence results extended to the first k eigenvalues of a Wishart matrix, for any fixed k. Subsequently, it was shown in [El 03] that the previous results held when p/n tends to 0 or infinity at any rate.

This result is striking in comparison to the classical ones we described in Section 28.2. The largest eigenvalue is asymptotically close to $(1 + \sqrt{p/n})^2$ and does not converge to the largest eigenvalue of the population covariance, which is equal to 1. The scaling is $n^{2/3}$ compared to the classical $n^{1/2}$ appearing in Anderson's Theorem 28.2.4. The limit law is Tracy–Widom (this result was the first appearance of such laws in the statistics literature). Johnstone showed numerically that the Tracy–Widom approximation to the marginal of the largest eigenvalue worked well or very well in practice for n and p as small as 5, especially in the right tail of the distribution, which made the result also potentially useful for practitioners. The previous theorem also helped explain limitations of PCA and its classical asymptotic theory that had been observed by practitioners.

Two natural questions about Theorem 28.3.1 might need to be answered, at least at an intuitive level: why is the sample largest eigenvalue a biased estimator of the population largest eigenvalue? How does the previous result fit with classical theory? The first question will be partially answered below (Section 28.4), but a beginning of the explanation might come from looking at the joint density of eigenvalues (Equation (28.2.3)). The Vandermonde term shows that the eigenvalues repel each other, since the density is 0 if two eigenvalues are equal. Hence it is at least conceivable that the empirical distribution of sample eigenvalues might be overspread compared to the population distribution. And if the repulsion is strong enough, this will lead to the bias we observe. The second question is essentially answered by the result of [El 03]: if one were to take carelessly the doubly infinite limit that is studied in the papers mentioned above by first fixing p and letting n go to infinity, one would get according to Anderson's result a $\text{GOE}_{p\times p}$ limit. Now letting p go to infinity, one would get a Tracy–Widom limit, since the largest eigenvalue of GOE is asymptotically Tracy–Widom [TW96]. The (rigorous) results of [El 03] show in some sense that this invalid way of taking limits would yield the right prediction.

Beyond $\Sigma = \text{Id}$ Johnstone's theorem clarified the behaviour of the largest eigenvalue of a Wishart random matrix with identity covariance in high dimension. However, as should be clear from our discussion of PCA, PCA is not well suited for the case of $\Sigma_p = \text{Id}_p$. The case $\Sigma \neq \text{Id}_p$ has therefore also been investigated. Two routes have been taken: the study of 'spiked' models, where one or a few eigenvalues are not equal to 1, motivated partly by the fact that PCA 'works well' when a few directions contain most of the variance in the data. And a study of the general Σ case motivated in part by questions of power of tests based on largest eigenvalues.

Spiked models The first results on spiked models were obtained in [BBAP05] (for complex Gaussian data), [BS06] (for first-order results without Gaussianity assumptions) and [Pau07] (for the real Gaussian case). Though it does not do justice to these papers to try to summarize them in a few bullet points, let us point out some striking and noteworthy results (see also [BGR09] for very recent advances):

- [BBAP05] show that if $\lambda_1(\Sigma_p) < 1 + \sqrt{p/n}$, then the behaviour of $l_1(\widehat{\Sigma})$ is unaffected by the spikes, and in particular one cannot detect a spike below this threshold by considering the largest sample eigenvalue. If $\lambda_1(\Sigma_p) > 1 + \sqrt{p/n}$, the sample eigenvalues corresponding to the largest spiked population eigenvalue (with multiplicity possibly higher than 1) revert to Anderson-type fluctuation behaviour, with modified centring and scaling constants, but the same classical $n^{-1/2}$ scale. If $\lambda_1 = 1 + \sqrt{p/n}$, the largest sample eigenvalue fluctuates at scale $n^{-2/3}$, but the limit law is not Tracy–

Widom and depends on the multiplicity of λ_1. This is shown in the setting where $p \leq n$.

- [BS06] characterize the effect of multiple spikes (both large and small) of various multiplicities on the a.s convergence of largest and smallest eigenvalues without assuming normality. The threshold $1 + \sqrt{p/n}$ plays again a key role here and the results are derived also when $p > n$.
- [Pau07] considers mostly the case where each spike has multiplicity 1, in the case of real Gaussian data. Gaussian fluctuation results (at scale $n^{-1/2}$) are obtained for sample eigenvalues corresponding to spikes beyond the threshold $1 + \sqrt{p/n}$. Angles between sample and population eigenvectors are also considered. It is for instance shown that if a spike is below the threshold, the angle between population and sample eigenvectors goes to 0 a.s, while if a spike is above the threshold, the angle goes to a non-zero limit that depends on p/n and the size of the spike.

General covariance case There has been less work on the case of general covariance Σ. [El 07] characterized (for complex Gaussian data) a large class of population covariance matrices for which the largest eigenvalues have Tracy–Widom fluctuations at scale $n^{-2/3}$ (i.e. joint convergence of the k largest eigenvalues for any k). A similar 'detectability' threshold for spikes was shown to exist and explicit formulae for the threshold (depending in a complicated manner on the spectrum of Σ, p, and n) and the centring and scaling sequences were given. These results were derived in the case where $p \leq n$ and were later shown to hold for $p > n$ [Ona08].

Universality questions Some of these fluctuation results have been shown, in the case of identity covariance, to be valid beyond the Gaussian setting, by requiring that the data matrix X contain iid entries with mean 0 and subgaussian tails [Sos02, Péc09] (though 'only' for matrices of the type $X'X$ and not $X'HX$, the latter being the one used in applications). see also Chapter 21.

Rates of convergence In an attempt to explain the excellent agreement between Tracy–Widom approximations and simulations, and potentially even improve on it, efforts have been made to understand rates of convergence in Johnstone's theorem. It was shown in [El 06] (for the complex case) that by appropriately changing the centring and scaling sequences, one could get an asymptotic Berry-Esseen result at scale $\min(n, p)^{-2/3}$ and exponential control of the errors in the right tail, giving some theoretical justifications for the observation that the Tracy–Widom approximation is especially good in the right tail of the distribution.

CCA and Jacobi ensemble Questions concerning the asymptotic behaviour of the largest canonical correlation coefficients are closely related to the Jacobi ensemble of random matrices. Johnstone [Joh08] has recently shown Tracy–Widom fluctuation in this case again, with accompanying rates of convergence results.

28.3.3 Possible applications to model selection

It is appealing to try to apply these various Tracy–Widom fluctuation results to problems in statistics. We will raise some general robustness questions in 28.4.2, but let us briefly discuss various methods that could be used.

In [El 04], the following method was proposed to select the number of components to keep in PCA: sequentially test equality of the $p-k+1$ smallest eigenvalues by comparing $l_k(\widehat{\Sigma})/s_k^2$, where s_k is a measure of variance, to the quantiles of the Tracy–Widom distribution, the quantiles being possibly dependent on k to account for multiple testing problems. The procedure is stopped at the first k for which we cannot reject the null. s_k^2 can be naively estimated as $\sum_{j\geq k} l_j(\widehat{\Sigma})/(p-k)$, or in a more sophisticated fashion (suggested by Iain Johnstone) as an estimate of the slope of the Wachter plot [Wac76], a (qq)plot of the sample eigenvalues vs the quantiles of the Marčenko–Pastur law (see Equation (28.4.2)).

Another possibility is to use the joint convergence of eigenvalues to find scale-free statistics and do sequential testing based on those. This would obliterate the need to estimate parameters such as the s_k^2 appearing in the previous discussion. One could consider for instance statistics of the type $(l_1(\widehat{\Sigma})-l_2(\widehat{\Sigma}))/(l_1(\widehat{\Sigma})-l_3(\widehat{\Sigma}))$ (or other combinations of large indices) and compare the empirical values to the distribution of a similar statistic for the Tracy–Widom (also known as Airy) point process. Getting these quantiles might require simulations, but tridiagonal representations of random matrices [Tro84, Sil85, DE02] make such computations quite fast. Ideas similar in spirit to this one have already appeared in the literature [Ona08].

28.4 Limiting spectral distribution results

28.4.1 Limiting spectra of sample covariance matrices

In the context of high-dimensional statistics, it is important for questions concerning likelihood ratio tests or estimation of covariance matrices to understand the spectral distributions of sample covariance matrices, i.e. the probability distributions that put mass $1/p$ at each one of the p eigenvalues of $\widehat{\Sigma}$. This is the more 'classical' part of large-dimensional random matrix theory, the landmark paper [MP67] settling a number of important questions. Here is a version of a main result of [MP67] incorporating a strengthening due to Silverstein [Sil95] (see also [SB95] and Chapter 21 for details on Stieltjes transforms).

Theorem 28.4.1 *Consider the random matrix $\widehat{\Sigma}_p = \Sigma_p^{1/2} X' X \Sigma_p^{1/2}/n$, where X is an $n \times p$ matrix whose entries are iid, with $\mathbf{E}(X_{i,j}) = 0$ and $\text{var}(X_{i,j}) = 1$, and Σ_p is a deterministic covariance matrix. Call $m_p(z)$ the Stieltjes transform of the spectral distribution of $\widehat{\Sigma}_p$, i.e. $m_p(z) = \text{trace}\left((\widehat{\Sigma}_p - z\text{Id}_p)^{-1}\right)/p$. Suppose that the spectral distribution of Σ_p converges weakly to H and $p/n \to \rho > 0$. Then, for each fixed $z \in \mathbb{C}^+$, $m_p(z) \to m(z)$ a.s , where $m(z)$ is the Stieltjes transform of a probability distribution and satisfies the fixed point equation:*

$$m(z) = \int \frac{d\,H(\tau)}{\tau(1 - \rho - \rho z m(z)) - z} . \tag{28.4.1}$$

In particular, this implies that the spectral distribution of $\widehat{\Sigma}_p$ is asymptotically non-random and has Stieltjes transform $m(z)$.

This result extends immediately (for simple linear algebraic reasons) to sample covariance matrices of the type $\Sigma^{1/2} X' H X \Sigma^{1/2}/n$ which are used in statistical practice. An interpretation of Equation (28.4.1) is that the histogram of sample eigenvalues is a non-linear deformation of the histogram of population eigenvalues, Equation (28.4.1) encoding this deformation. Figure 28.3 illustrates the

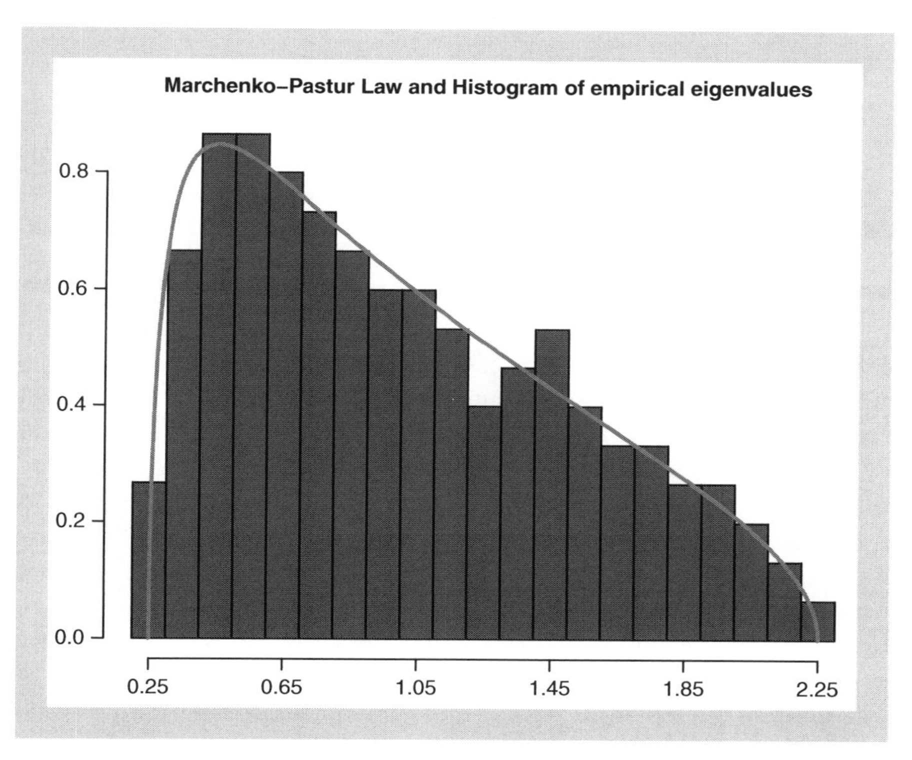

Fig. 28.3 Normalized histogram of eigenvalues and Marčenko–Pastur density (solid line), $n = 600$, $p = 150$, iid Gaussian data, $\mu = 0$, $\Sigma = \text{Id}_p$. The population (or true) eigenvalues are all equal to 1. The 'overspreading' of sample eigenvalues is striking.

situation when $\Sigma = \mathrm{Id}_p$ and the data is Gaussian, showing the dramatic impact that high-dimensionality has on eigenvalue estimation.

In the case where $\Sigma = \mathrm{Id}_p$ and $\rho > 0$, Equation (28.4.1) can be solved explicitly and it is known that the limiting spectral distribution has a density when $\rho \in (0, 1]$. When $0 < \rho \leq 1$, it can be shown that this limiting density is

$$f_\rho(x) = \frac{1}{2\pi\rho} \frac{\sqrt{b-x}\sqrt{x-a}}{x} 1_{a \leq x \leq b} \;, \tag{28.4.2}$$

where $b = (1 + \sqrt{\rho})^2$ and $a = (1 - \sqrt{\rho})^2$ (note that b is similar to the centring sequence appearing in Theorem 28.3.1); f_ρ is commonly known as the density of the Marčenko–Pastur law.

A number of these spectral distribution results can also be obtained through the tools of free probability (see [Voi00] and Chapter 22) and are starting to find interesting statistical applications (see e.g. [RMSE08] and [RE08] for software).

It should also be noted that recent results on convergence of linear spectral statistics [BS04] (i.e. $\sum_{i=1}^p h(l_i(\widehat{\Sigma}))/p$ for some function h) lead to a better understanding of classical likelihood ratio tests in the high-dimensional setting. Because the size of the fluctuations is $1/n$, they could prove useful in practice.

28.4.2 Robustness questions

We have mostly described so far results about Gaussian data models. There have been great efforts to show that these results (either about spectral distributions or extreme eigenvalues) hold for more general distributions, a phenomenon commonly called 'universality' in random matrix theory (see also Chapters 6 and 21). Very often, the model for the data matrix is changed from $X = Y\Sigma^{1/2}$, where Y is $n \times p$ and $Y_{i,j}$ are iid $\mathcal{N}(0, 1)$ to the same model with $Y_{i,j}$ iid with mean 0, variance 1 and some conditions on higher moments. We will refer to these models, for which the data is assumed to be a linear transformation of a vector with iid entries, as standard random matrix models.

It has recently been pointed out [El 09a] that standard random matrix models imply a very peculiar geometry for the data: asymptotically (under some mild technical conditions) $\{\|X_i\|/\sqrt{p}\}_{i=1}^n$ are assumed to be essentially constant, and $X_i' X_j / p$ is assumed to be very close to 0, for all $i \neq j$. This is easily seen in the Gaussian case and is an instance of the concentration of measure phenomenon more generally [Led01].

Some recent work on robustness of random matrix models [El 09a] has therefore focused on understanding how fundamental these implicit geometric assumptions are to universality for random matrix results. It has been found that they are extremely important. In particular, standard random matrix results are not robust when Gaussianity assumptions (and generalizations) are changed to elliptical assumptions, i.e. when considering $\widetilde{X}_i = \lambda_i X_i$ instead of

X_i, where λ_i are random, independent of X_i and $\mathbf{E}(\lambda_i^2) = 1$. Since $\text{cov}(\widetilde{X}_i) = \text{cov}(X_i)$, this means that two models with the same population covariance can generate very different limiting spectral distributions, an observation that raises questions for anyone who is trying to naively (or blindly) apply standard random matrix results to test hypotheses about population covariance. This also affects the convergence of extreme eigenvalues, since the largest and smallest eigenvalues are asymptotically at or beyond the edge of the limiting spectral distribution of a random matrix. On the other hand, it was shown in the same paper that standard random matrix results concerning spectral distributions could be extended to situations where only concentration properties of high-dimensional random vectors were assumed – extending their reach (and range of robustness) considerably.

This discussion suggests that while random matrix results and techniques are very helpful in understanding the behaviour of standard tools of multivariate analysis in high dimension, their use in practice should be tempered by careful checks of adequacy of the model with the data, in particular in terms of data geometry. It also suggests new avenues for statisticians for further theoretical investigations and the design of generically robust high-dimensional procedures.

28.5 Conclusion

Random matrix theory has historically been a central part of multivariate statistical analysis. New theoretical needs – arising from the recent availability of high-dimensional data sets – have made aspects of modern random matrix theory a crucial tool for our understanding of statistical techniques in high dimension. Furthermore, the accuracy of random matrix approximations – even in low dimension – have made them an applicable and useful tool for practitioners. Geometric limitations of standard models are currently raising questions about the (otherwise great) robustness of certain random matrix results and their direct applicability in practice.

Beyond the spectral analysis aspects of random matrix theory we have discussed here, a number of random matrix-related problems are currently the subject of active investigation in statistics, in particular in connection with covariance estimation. We refer the reader to [BJM+08] for a special issue of the Annals of Statistics on random matrix theory and Statistics and a glance at the variety of problems tackled there.

Acknowledgements

This work was supported in part by NSF grants DMS-0605169, DMS-0847647 (CAREER) and an Alfred P. Sloan Research Fellowship. The author thanks the editors for the opportunity to write this chapter, Peter Bickel, Iain Johnstone, and Debashis Paul for many interesting discussions on random matrices and Statistics, and an anonymous referee for insightful comments and references.

References

[And63] T. W. Anderson. Asymptotic theory for principal component analysis. *Ann. Math. Statist.*, 34:122–148, 1963.

[And03] T. W. Anderson. *An introduction to multivariate statistical analysis.* Wiley Series in Probability and Statistics. Wiley-Interscience [John Wiley & Sons], Hoboken, NJ, third edition, 2003.

[BBAP05] J. Baik, G. Ben Arous, and S. Péché. Phase transition of the largest eigenvalue for non-null complex sample covariance matrices. *Ann. Probab.*, 33(5):1643–1697, 2005.

[BGR09] Florent Benaych-Georges and Raj N. Rao. The eigenvalues and eigenvectors of finite, low rank perturbations of large random matrices. 2009.

[BJM+08] P.J. Bickel, I. M. Johnstone, H. Massam, D. Nychka, D. Richards, and C. Tracy. Special issue on random matrix theory. *Annals of Statistics*, 36(6), December 2008.

[BS04] Z. D. Bai and Jack W. Silverstein. CLT for linear spectral statistics of large-dimensional sample covariance matrices. *Ann. Probab.*, 32(1A):553–605, 2004.

[BS06] J. Baik and J. Silverstein. Eigenvalues of large sample covariance matrices of spiked population models. *Journal of Multivariate Analysis*, 97(6):1382–1408, 2006.

[BV04] Stephen Boyd and Lieven Vandenberghe. *Convex optimization.* Cambridge University Press, Cambridge, 2004.

[Cat66] R.B Cattell. The scree test for the number of factors. *Multivariate Behav. Res.*, 1:245–276, 1966.

[Con63] A. G. Constantine. Some non-central distribution problems in multivariate analysis. *Ann. Math. Statist.*, 34:1270–1285, 1963.

[DE02] Ioana Dumitriu and Alan Edelman. Matrix models for beta ensembles. *J. Math. Phys.*, 43(11):5830–5847, 2002.

[Eat72] Morris L. Eaton. *Multivariate Statistical Analysis.* Institute of Mathematical Statistics, University of Copenhagen, Copenhagen, Denmark, 1972.

[Eat07] Morris L. Eaton. *Multivariate statistics.* Institute of Mathematical Statistics Lecture Notes—Monograph Series, 53. Institute of Mathematical Statistics, 2007. Reprint of the 1983 original.

[El 03] Noureddine El Karoui. On the largest eigenvalue of Wishart matrices with identity covariance when n, p and $p/n \to \infty$. *arXiv:math.ST/0309355*, September 2003.

[El 04] Noureddine El Karoui. *New results about random covariance matrices and statistical applications.* PhD thesis, Stanford University, 2004.

[El 06] Noureddine El Karoui. A rate of convergence result for the largest eigenvalue of complex white Wishart matrices. *Ann. Probab.*, 34(6):2077–2117, 2006.

[El 07] Noureddine El Karoui. Tracy-Widom limit for the largest eigenvalue of a large class of complex sample covariance matrices. *The Annals of Probability*, 35(2):663–714, March 2007.

[El 09a] Noureddine El Karoui. Concentration of measure and spectra of random matrices: Applications to correlation matrices, elliptical distributions and beyond. *The Annals of Applied Probability*, 19(6):2362–2405, December 2009.

[El 09b] Noureddine El Karoui. High-dimensionality effects in the Markowitz problem and other quadratic programs with linear equality constraints: risk underestimation. Technical Report 781, Department of Statistics, UC Berkeley, 2009.

[ET91] Morris L. Eaton and David E. Tyler. On Wielandt's inequality and its application to the asymptotic distribution of the eigenvalues of a random symmetric matrix. *Ann. Statist.*, 19(1):260–271, 1991.

[Fis39] R.A Fisher. The sampling distribution of some statistics obtained from non-linear equations. *Ann. Eugenics*, 9:238–249, 1939.

[For93] P. J. Forrester. The spectrum edge of random matrix ensembles. *Nuclear Phys. B*, 402(3):709–728, 1993.

[Gir39] M.A Girshick. On the sampling theory of roots of determinantal equations. *Ann. Math. Statist.*, 10:203–224, 1939.

[HMN05] Peter Hall, J. S. Marron, and Amnon Neeman. Geometric representation of high dimension, low sample size data. *J. R. Stat. Soc. Ser. B Stat. Methodol.*, 67(3): 427–444, 2005.

[Hsu39] P.L Hsu. On the distribution of the roots of certain determinantal equations. *Ann. Eugenics*, 9:250–258, 1939.

[HTF09] T. Hastie, R. Tibshirani, and J. Friedman. *The Elements of Statistical Learning.* Springer Series in Statistics. Springer-Verlag, New York, 2nd ed. edition, 2009. Data mining, inference, and prediction.

[Jam64] A. T. James. Distributions of matrix variates and latent roots derived from normal samples. *Ann. Math. Statist.*, 35:475–501, 1964.

[Joh00] K. Johansson. Shape fluctuations and random matrices. *Comm. Math. Phys.*, 209(2):437–476, 2000.

[Joh01] I.M. Johnstone. On the distribution of the largest eigenvalue in principal component analysis. *Ann. Statist.*, 29(2):295–327, 2001.

[Joh08] Iain M. Johnstone. Multivariate analysis and Jacobi ensembles: largest eigenvalue, Tracy-Widom limits and rates of convergence. *Ann. Statist.*, 36(6):2638–2716, 2008.

[Jol02] I. T. Jolliffe. *Principal component analysis.* Springer Series in Statistics. Springer-Verlag, New York, second edition, 2002.

[Kat95] Tosio Kato. *Perturbation theory for linear operators.* Classics in Mathematics. Springer-Verlag, Berlin, 1995. Reprint of the 1980 edition.

[Led01] M. Ledoux. *The concentration of measure phenomenon*, volume 89 of *Mathematical Surveys and Monographs.* American Mathematical Society, Providence, RI, 2001.

[MKB79] Kantilal Varichand Mardia, John T. Kent, and John M. Bibby. *Multivariate analysis.* Academic Press, London, 1979.

[MP67] V. A. Marčenko and L. A. Pastur. Distribution of eigenvalues in certain sets of random matrices. *Mat. Sb. (N.S.)*, 72 (114):507–536, 1967.

[Mui82] R. J. Muirhead. *Aspects of multivariate statistical theory.* John Wiley & Sons Inc., New York, 1982. Wiley Series in Probability and Mathematical Statistics.

[Ona08] Alexei Onatski. The Tracy-Widom limit for the largest eigenvalues of singular complex Wishart matrices. *Ann. Appl. Probab.*, 18(2):470–490, 2008.

[Pau07] Debashis Paul. Asymptotics of sample eigenstructure for a large dimensional spiked covariance model. *Statistica Sinica*, 17(4):1617–1642, October 2007.

[Péc09] Sandrine Péché. Universality results for the largest eigenvalues of some sample covariance matrix ensembles. *Probab. Theory Related Fields*, 143(3–4):481–516, 2009.

[RE08] N.R. Rao and A. Edelman. The polynomial method for random matrices. *Foundations of Computational Mathematics*, 8(6):649–702, 2008.

[RMSE08] N. Raj Rao, James Mingo, Roland Speicher, and Alan Edelman. Statistical eigen-inference from large Wishart matrices. *The Annals of Statistics*, Volume 36(Number 6):2850–2885, 2008.

[Roy57] S. N. Roy. *Some aspects of multivariate analysis.* John Wiley and Sons Inc., New York, 1957.

[SB95] Jack W. Silverstein and Z. D. Bai. On the empirical distribution of eigenvalues of a class of large-dimensional random matrices. *J. Multivariate Anal.*, 54(2):175–192, 1995.

[Sil85] Jack W. Silverstein. The smallest eigenvalue of a large-dimensional Wishart matrix. *Ann. Probab.*, 13(4):1364–1368, 1985.

[Sil95] Jack W. Silverstein. Strong convergence of the empirical distribution of eigenvalues of large-dimensional random matrices. *J. Multivariate Anal.*, 55(2):331–339, 1995.

[Sos02] A. Soshnikov. A note on universality of the distribution of the largest eigenvalues in certain sample covariance matrices. *J. Statist. Phys.*, 108(5/6):1033–1056, September 2002.

[Sze75] Gábor Szegő. *Orthogonal polynomials.* American Mathematical Society, Providence, R.I., fourth edition, 1975. American Mathematical Society, Colloquium Publications, Vol. XXIII.

[Tro84] Hale F. Trotter. Eigenvalue distributions of large Hermitian matrices; Wigner's semicircle law and a theorem of Kac, Murdock, and Szegő. *Adv. in Math.*, 54(1):67–82, 1984.

[TW94a] C.A. Tracy and H. Widom. Fredholm determinants, differential equations and matrix models. *Comm. Math. Phys.*, 163(1):33–72, 1994.

[TW94b] C.A. Tracy and H. Widom. Level-spacing distribution and the Airy kernel. *Comm. Math. Phys.*, 159:151–174, 1994.

[TW96] C.A. Tracy and H. Widom. On orthogonal and symplectic matrix ensembles. *Comm. Math. Phys.*, 177:727–754, 1996.

[Voi00] Dan Voiculescu. Lectures on free probability theory. In *Lectures on probability theory and statistics (Saint-Flour, 1998)*, volume 1738 of *Lecture Notes in Math.*, pages 279–349. Springer, Berlin, 2000.

[Wac76] Kenneth W. Wachter. Probability plotting points for principal components. *Proceedings of the Computer Science and Statistics 9th Annual Symposium on the Interface*, pages 299–308, 1976.

[Wis28] John Wishart. The generalised product moment distribution in samples from a normal multivariate population. *Biometrika*, 20(A):32–52, 1928.

·29·

Algebraic geometry and matrix models

L. O. Chekhov

Abstract

We describe several applications of matrix models to algebraic geometry: the Kontsevich matrix model describing intersection indices on moduli spaces of curves with marked points, the Hermitian matrix model free energy at the leading expansion order as the prepotential of the Seiberg-Witten-Whitham-Krichever hierarchy, and the other orders of free energy and resolvent expansions as symplectic invariants and possibly amplitudes of open/closed strings.

29.1 Introduction

In this chapter, we describe relations between the matrix models and algebraic geometry. At the moment, this is one of the most actively developing fields of the application of matrix models. We restrict ourself to the following topics: in the first section, we review the seminal Kontsevich's paper in which the free energy of the special matrix model with external matrix field (the Kontsevich matrix model) was proved to be the generating function for the intersection indices (integrals of the products of the first Chern classes) on the moduli space $\mathcal{M}_{g,n}$ of (stable) Riemann surfaces (RS) of genus g with n marked points ($n > 0$, $2g - 2 + n > 0$). The partition function obtained is simultaneously the τ-function of the KdV hierarchy being one of examples of the deep connection between matrix models and integrable systems.

The second section is devoted to the relation between the so-called *formal matrix models* (solutions of the loop equation) and algebraic hierarchies of Dijkgraaf-Witten-Whitham-Krichever type. Namely, we demonstrate that already the leading order $\mathcal{F}_0$ of the free energy asymptotic expansion (the so-called planar graph approximation) is the prepotential (the τ-function) of the Seiberg-Witten-Whitham-Krichever hierarchy. The central part in this construction is played by the algebraic curve describing the corresponding RS $C(x, y) = 0$ and by the special 1-differential $dS = ydx$. We demonstrate the 'geometric engineering' equations that are finalized in the statement that the third derivatives of the free energy w.r.t. the canonical variables satisfy the Witten-Dijkgraaf-Verlinde-Verlinde (WDVV) equations; the (finite) dimension of the

WDVV system coincides then with the total number of dynamical (i.e. not rigidly fixed) branch points.

In the third section, we use the loop equation to extend the structure of $\mathcal{F}_0$ to higher expansion orders and to show that the obtained free energy terms $\mathcal{F}_h$ are generating functions for symplectic invariants of the corresponding algebraic curve.

29.2 Moduli spaces and matrix models

The link between algebraic geometry and physics was pushed forward by Witten who suggested a new class of topological theories describing objects on the *moduli spaces* [Wit90, Wit91]. The Witten's cohomological approach is based on the two-dimensional gravity phase space being the moduli space of complex curves.

More concretely, consider a linear holomorphic bundle $\mathcal{F}_i$ with a fibre being the tangent space (complex one-dimensional vector space) to the RS Σ at some fixed point x_i. Then, the correlation functions of topological theory are defined to be the *intersection indices* of such bundles. These indices are geometrical invariants, which are integrals of the products of the first Chern classes (closed 2-forms) over the compactified *a lá* Deligne and Mumford moduli (orbi)spaces $\overline{\mathcal{M}}_{g,n}$,

$$\langle \tau_{d_1} \cdots \tau_{d_n} \rangle_g = \int_{\overline{\mathcal{M}}_{g,n}} \prod_{i=1}^{n} \omega_i^{d_i}. \tag{29.2.1}$$

Here ω_i is the first Chern class associated with the ith marked point (puncture). As is the case in a topological theory, expression (29.2.1) is coordinate independent. Would $\overline{\mathcal{M}}_{g,n}$ be a standard manifold admitting a smooth coordinatization, such quantities would be non-negative *integers*. The orbifold nature of moduli spaces means that a proper closure $\overline{\mathcal{M}}_{g,n}$ of an orbispace $\mathcal{M}_{g,n}$ can be obtained from the covering manifold $\mathcal{T}_{g,n}$ by factoring over a discrete finitely generated symmetry group $\Gamma_{g,n}$: $\mathcal{M}_{g,n} = \mathcal{T}_{g,n}/\Gamma_{g,n}$. Stationary points of this group are singular (conical) points of the metric on $\mathcal{M}_{g,n}$. Because of such factoring, the corresponding indices become non-negative *rational* numbers. They are equal to zero unless $6g - 6 + 2n = 2\sum d_i$ and are strictly positive otherwise. These numbers satisfy a series of recurrent relations equivalent to the Virasoro conditions in the double scaling limit of the Hermitian one-matrix model [Dij91a].

The topology of the moduli spaces $\mathcal{M}_{g,n}$ of algebraic curves of genus g with n marked points (insertions of the matter fields) was described by Kontsevich [Kon91, Kon92], where the matrix model that yields the generating function for the intersection indices on the moduli spaces $\mathcal{M}_{g,n}$ was constructed. The generating function is given by the novel *Kontsevich matrix model* (KMM) with the *external* matrix field Λ. The standard times of the KdV hierarchy are related to this matrix by the *Miwa transformation* $t_n = \frac{1}{2n+1}\mathrm{tr}\,\Lambda^{-2n-1}$, whereas the KMM integral itself is a τ-function of the KdV hierarchy. Below we present the details of this construction.

29.2.1 The moduli space of algebraic curves and its parameterization via the Jenkins-Strebel differentials

In what follows, g and n are integers satisfying the conditions

$$g \geq 0, \qquad n > 0, \qquad 2 - 2g - n < 0.$$

We let $\mathcal{M}_{g,n}$ denote the moduli space of smooth complete complex curves C of genus g with n distinct marked points $x_1, \ldots, x_n$, whose closure $\overline{\mathcal{M}}_{g,n}$ is a smooth compactification of the Deligne-Mumford type (its form actually follows from the combinatorics of matrix model diagrams and is irrelevant here; it suffices to know that such a compactification exists). We let $\mathcal{L}_i$, $i = 1, \ldots, n$, denote the linear fibre bundles over $\overline{\mathcal{M}}_{g,n}$. A fibre of a bundle $\mathcal{L}_i$ at the point $(C, x_1, \ldots, x_n)$ is the cotangent space $T^*_{x_i}C$.

Let $d_1, \ldots, d_n$ be non-egative integers such that

$$d := \sum_{i=1}^{n} d_i = \dim_{\mathbb{C}} \overline{\mathcal{M}}_{g,n} = 3g - 3 + n.$$

We then let $\langle \tau_{d_1} \cdots \tau_{d_n} \rangle_g$ denote the *intersection index* determined by formula (29.2.1) in which $\omega_i \equiv c_1(\mathcal{L}_i)$ are the first Chern classes of the corresponding linear fibre bundles over the moduli space $\overline{\mathcal{M}}_{g,n}$.

To parameterize these line bundles, we need the *quadratic Strebel differential* φ on the RS C, which is a holomorphic section of the linear bundle $(T^\star)^{\otimes 2}$. In local coordinates, this section determines a *flat metric* on the complement to the discrete set of its zeros and poles:

$$ds^2 = |\varphi(z)| \cdot |dz|^2, \qquad \text{where} \quad \varphi = \varphi(z)\, dz^2. \tag{29.2.2}$$

All the poles of a Strebel differential are purely double poles placed at the puncture points. Moreover, the second-order residues p_i^2 at the double poles are confined to be strictly positive real numbers. Because the Riemann theorem for quadratic differentials implies that #zeros – #poles $= 4(g-1)$ (taking the multiplicities into account), then for a point in the general position in $\mathcal{M}_{g,n}$ (a RS with n punctures), we have exactly $4g - 4 + 2n$ simple zeros.

A *horizontal line* (a geodesic line) of a quadratic differential is a curve along which the differential $\varphi(z)\, dz^2$ is real and positive. In contrast, a *vertical line* (also a geodesic line) is a curve along which the differential $\varphi(z)\, dz^2$ is real and negative. We now consider the system of horizontal and vertical lines in the vicinity of a double pole where $\varphi(z) = p_i^2/\big((2\pi)^2(z - z_i)^2\big)$ (Fig. 29.1a).

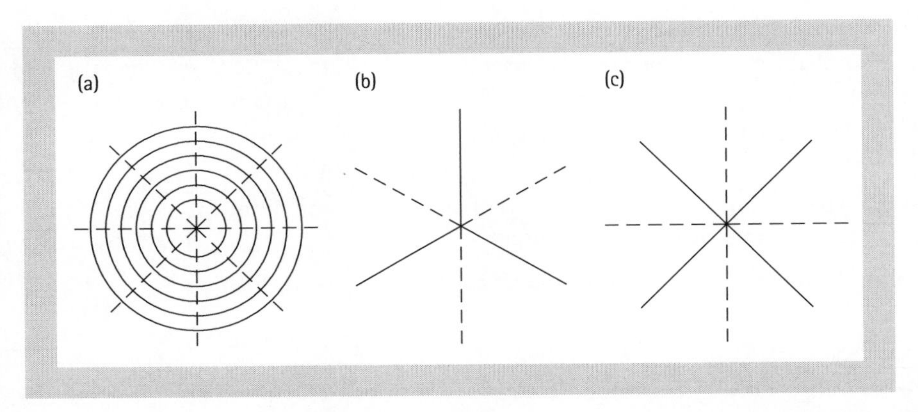

Fig. 29.1 Horizontal and vertical lines of the Strebel metric in the double pole (a), simple zero (b), and double zero (c).

Horizontal lines are then concentric circles (all having the same geodesic length p_i) around the point z_i, their maximum extension is called the maximum circle domain, and vertical lines are half-lines beginning at the same point, so every maximum circle domain is metrically a half-infinite cylinder. Taking a kth order zero of the function $\varphi(z)$, we obtain exactly $k+2$ horizontal and $k+2$ vertical geodesic half-lines issued from the point z_j (see Fig. 29.1b and 29.1c for the respective cases $k=1$ and $k=2$). Horizontal lines of a general differential φ are not closed, it is the *Strebel differentials* for which the union of nonclosed horizontal lines has measure zero and the complement of the RS to this set of nonclosed lines is exactly the union of maximum circle domains. These nonclosed geodesics are always those connecting zeros of the differential. In what follows, we call these maximum circle domains the *faces* of the RS, each face containing exactly one puncture.

The Strebel theorem [Str84] states that for a given RS C with n distinct points $x_1, \ldots, x_n \in C$ ($n > 0$ for $g > 0$ and $n > 2$ for $g = 0$) and for n positive real numbers $p_1, \ldots, p_n$, we have a unique quadratic *Jenkins–Strebel differential* on $C \backslash \{x_1, \ldots, x_n\}$, whose maximum circle domains are n punctured discs D_i, $x_i \in D_i$, with the perimeters p_i.

The union of all nonclosed horizontal lines is the fat, or ribbon, graph corresponding to the given RS. This means that the graph edges are labelled and can be thought of as (infinitely thin) stripes whose two sides belong to two faces of the surface separated by the given edge. In what follows, the index i runs along the face boundary in the matrix model technique (see Fig. 29.1).

In the Strebel metric, each face becomes a half-infinite cylinder with the boundary composed from edges of stripes of a ribbon graph, i.e. the ith face boundary is a polygon with the perimeter p_i.

We endow an oriented graph with an additional structure: to each edge we assign a positive number $l_s \in \mathbb{R}^+$ where s is the number of the edge. The

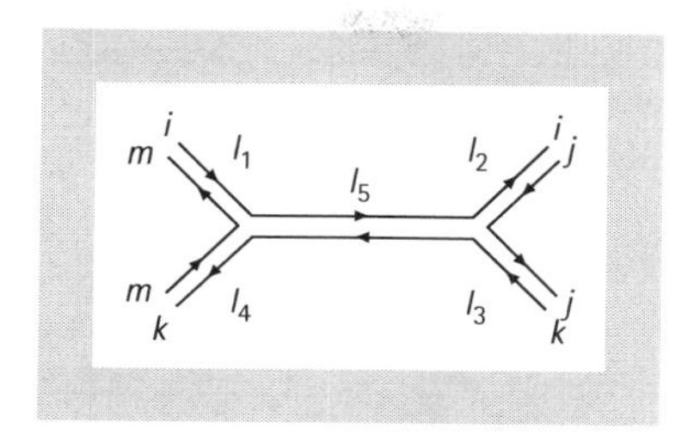

Fig. 29.2 A typical subgraph in a matrix model with three-valent vertices.

numbers l_s being just the edge lengths of a ribbon graph Γ_g of the genus g determine the coordinates on the $(6g-6+3n)$-dimensional linear space of the metrizable graphs (with valences of each vertex greater or equal three). The set of all metrizable graphs up to the symmetry groups of graphs constitutes the *combinatorial moduli space* $\mathcal{M}_{g,n}^{\text{comb}}$. The perimeters p_i are then the sums of lengths of edges incident to the ith cycle. The dimension of $\mathcal{M}_{g,n}^{\text{comb}}$ is equal to the real dimension of the moduli space $\mathcal{M}_{g,n}$ plus n additional dimensions associated with the perimeters of faces. Structures on the compactified moduli space $\overline{\mathcal{M}}_{g,n}$ itself must be independent of these additional parameters. Next, we can put the graph into correspondence to each Jenkins-Strebel differential. The converse statement is also true: given a metrizable graph we can construct a unique RS with the Jenkins-Strebel differential structure and with the second-order residues equal to the squared perimeters of the graph cycles. The theorem by Penner, Harer, and Zagier [Har86], [Pen88] claims that the mapping $\mathcal{M}_{g,n} \otimes \mathbb{R}_+^n \to \mathcal{M}_{g,n}^{\text{comb}}$, which sets the critical graph of the canonical Jenkins-Strebel differential into correspondence with the RS C and the set of positive numbers $p_1, \ldots, p_n$, is one-to-one.

Considering the set of all graphs with the given g and n and all possible (nonzero) edge lengths, we therefore obtain a *stratification* of the moduli space $\mathcal{M}_{g,n}^{\text{comb}}$ with the strata dimension equal to the number of edges of the corresponding graph. Open strata correspond to the combinatorial types of the three-valent graphs and have the dimension $6g-6+3n$.

29.2.2 The bundle geometry on $\overline{\mathcal{M}}_{g,n}$ and the matrix integral

Kontsevich [Kon91, Kon92] proposed a method for finding the intersection indices (or, equivalently, the integrals of the first Chern classes) on the moduli spaces. The perimeter of the kth boundary component is

$$p_k = \sum_{l_a \in I_k} l_a.$$

A convenient choice of the representative of the closed 2-form ω_k (the first Chern class $c_1(\mathcal{L}_k)$) on this component is then

$$\omega_k = \sum_{1\le i<j\le l_k} d\left(\frac{l_i}{p}\right)\wedge d\left(\frac{l_j}{p}\right) \tag{29.2.3}$$

if we assume the existence of moduli space compactification that is consistent with cell decomposition.

We let $\pi : \mathcal{M}_{g,n}^{\text{comb}} \to \mathbb{R}_+^n$ denote the projection operator to the perimeter space. The intersection indices are then determined by the formula

$$\langle\tau_{d_1}\cdots\tau_{d_n}\rangle = \int_{\pi^{-1}(p_*)} \prod_{i=1}^{n} \omega_i^{d_i}, \tag{29.2.4}$$

where $p_* = (p_1, \ldots, p_n)$ is an arbitrary set of positive numbers and $\pi^{-1}(p_*)$ is the fibre in $\mathcal{M}_{g,n}^{\text{comb}}$ isomorphic to $\overline{\mathcal{M}}_{g,n}$.

We now introduce the 2-form Ω on open strata of the space $\mathcal{M}_{g,n}^{\text{comb}}$,

$$\Omega = \sum_{k=1}^{n} p_k^2\omega_k, \tag{29.2.5}$$

whose restrictions to fibres of the projection π have constant coefficients in the coordinates l_e. We let $d = 3g - 3 + n$ denote the complex dimension of the space $\mathcal{M}_{g,n}$. The volume of a fibre of the projection operator π w.r.t. the form Ω is then

$$\begin{aligned}\mathrm{Vol}\left(\pi^{-1}(p_1,\ldots,p_n)\right) &= \int_{\pi^{-1}(p_*)} \frac{\Omega^d}{d!} = \\ &= \frac{1}{d!}\int_{\pi^{-1}(p_*)} \left(p_1^2c_1(\mathcal{L}_1)+\cdots+p_n^2c_1(\mathcal{L}_n)\right)^d = \\ &= \sum_{\sum d_i=d}\prod_{i=1}^{n}\frac{p_i^{2d_i}}{d_i!}\langle\tau_{d_1}\cdots\tau_{d_n}\rangle_g.\end{aligned} \tag{29.2.6}$$

This formula becomes rigorous only if we continue integrations to the Deligne-Mumford closure of the moduli space $\overline{\mathcal{M}}_{g,n}$ (i.e. we must consider a stable cohomological class of curves). But choosing actual compactification type does not affect the results because integrals in (29.2.4) with ω_k from (29.2.3) are nonsingular and bounded and only integrations over higher-dimensional cells are relevant, whereas all additional simplices added for compactifying $\mathcal{M}_{g,n}$ have lower dimensions. This consideration also shows that we must take into consideration only the graphs with three-valent vertices.

To introduce a matrix model, we perform the Laplace transformation w.r.t. the variables p_i of the volumes of the projection π for the quantities in the right-hand side of (29.2.6):

$$\int_0^\infty dp_i e^{-p_i\lambda_i} p_i^{2d_i} = (2d_i)!\,\lambda_i^{-2d_i-1}. \tag{29.2.7}$$

In the left-hard side., we then obtain

$$\int_0^\infty \cdots \int_0^\infty dp_1 \wedge \cdots \wedge dp_n e^{-\sum_i p_i\lambda_i} \int_{\overline{\mathcal{M}}_{g,n}} e^{\Omega} \tag{29.2.8}$$

and because all the factors p_i^2 in the denominators of the form Ω cancel, we obtain that

$$\left|e^{\Omega} dp_1 \wedge \cdots \wedge dp_n\right| = \rho \left|\prod_{e\in X_1} dl_e\right|. \tag{29.2.9}$$

We now let $\#X_q$ denote the total number of q-dimensional cells of a simplicial complex ($\#X_1$ is the number of edges, $\#X_0$ is the number of vertices, etc.). The function ρ is a positive-definite function defined on open cells:

$$\rho = \left|\prod_{i=1}^n dp_i \wedge \frac{\Omega^d}{d!}\right| \Big/ \prod_{e\in X_1} |dl_e|. \tag{29.2.10}$$

Lemma 29.2.1 *The constant ρ in (29.2.10) depends only on the Euler characteristic of the graph Γ,*

$$\rho = 2^{d+\#X_1-\#X_0}. \tag{29.2.11}$$

Proof The original proof by Kontsevich was by highly elaborated cohomology arguments. Below we present a purely combinatorial proof. For this, we consider the *flip transformation* w.r.t. the edge l_5

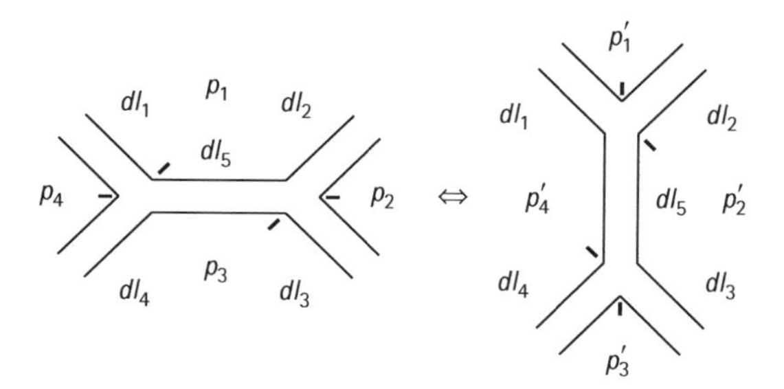

(the primes at corners denote the starting points for the counterclockwise oriented linear ordering of edges in expression (29.2.3) for ω_k). We now take the corresponding two forms, Ω and Ω'. Introducing the standard scalar product $\langle dl_i, dl_j\rangle = \delta_{i,j}$ on the linear space of 1-differentials and segregating in Ω and Ω' all the terms depending on $dl_1, \ldots, dl_5$ systematically excluding terms proportional to $dp_1, \ldots, dp_4$ in Ω and those proportional to $dp_1', \ldots, dp_4'$ in Ω', we obtain

$$\Omega \simeq dl_5 \wedge d(-l_1 + l_2 = l_3 + l_4) + d(l_1 - l_3) \wedge d(l_2 - l_4) + \widetilde{\Omega},$$
$$\Omega' \simeq -dl_5 \wedge d(-l_1 + l_2 = l_3 + l_4) + d(l_2 - l_4) \wedge d(l_1 - l_3) + \widetilde{\Omega},$$

where $\widetilde{\Omega}$ contains none of $dl_1, \ldots, dl_5$. So the forms Ω and Ω' differ only by the first term. Note that the vector $d(-l_1 + l_2 = l_3 + l_4)$ is orthogonal to all the vectors $dp_1, \ldots, dp_4$, $dp'_1, \ldots, dp'_4$, to dl_5, and to both $d(l_1 - l_3)$ and $d(l_2 - l_4)$ being therefore orthogonal to all the vectors in $\widehat{\Omega} := d(l_1 - l_3) \wedge d(l_2 - l_4) + \widetilde{\Omega}$. We then have that (all the products here are the wedge-products)

$$\Omega^d dp_1 \cdots dp_n = d \cdot \left(\widehat{\Omega}\right)^{d-1} dl_5\, d(-l_1 + l_2 - l_3 + l_4) dp_1 \cdots dp_n$$

and

$$\Omega'^d dp'_1 \cdots dp'_n = d \cdot \left(\widehat{\Omega}\right)^{d-1} dl_5\, d(l_1 - l_2 + l_3 - l_4) dp'_1 \cdots dp'_n,$$

but all dp'_i differ from dp_i by adding or subtracting dl_5 and since we have already segregated the multiplier dl_5, we can replace all the dp'_i by dp_i without changing the expression in the latter formula thus obtaining that these two expressions differ only by the sign. Since all the diagrams of the same topological type can be transformed one into another by a set of flip transformations, they all have the same factor ρ, and one can evaluate its actual value by choosing some selected diagrams corresponding to gluing a handle or adding a new hole. □

The integral

$$I_g(\lambda_*) := \int_{\mathcal{M}^{\mathrm{comb}}_{g,n}} e^{-\sum_i \lambda_i p_i} \prod_{e \in X_1} |dl_e| \tag{29.2.12}$$

equals the sum of integrals over open cells in $\mathcal{M}^{\mathrm{comb}}_{g,n}$. These open cells are in one-to-one correspondence with the complete set of three-valent graphs contributing to the given order of expansion in g and n. We must also take the internal automorphisms of the graphs (their number indicates how many copies of the moduli space domain are inside the given cell) into account. Eventually, the sum $\sum_i \lambda_i p_i$ can be expressed through l_e,

$$\sum_{i=1}^{n} \lambda_i p_i = \sum_{e \in X_1} l_e (\lambda_e^{(1)} + \lambda_e^{(2)}), \tag{29.2.13}$$

for each graph. Here, $\lambda_e^{(1)}$ and $\lambda_e^{(2)}$ are the variables of two cycles incident to the eth edge (see, e.g. Fig. 29.2). Now performing the Laplace transformation, we obtain

$$\sum_{d_1,\ldots,d_n=0}^{\infty} \langle \tau_{d_1} \ldots \tau_{d_n} \rangle_g \prod_{i=1}^{n} (2d_i - 1)!!\, \lambda_i^{-(2d_i+1)} = \sum_{\Gamma} \frac{2^{-\#X_0}}{\#\mathrm{Aut}\,(\Gamma)} \prod_{\{ij\}} \frac{2}{\lambda_i + \lambda_j}, \tag{29.2.14}$$

where the summation in the right-hand side ranges all the oriented connected three-valent ribbon graphs Γ with n marked boundary components, the product is taken over all edges of a graph, $\#X_0$ is the number of vertices of a graph Γ, and #Aut is the volume of the discrete symmetry group of a graph Γ.

The key observation of Kontsevich was that the quantity in the right-hand side of (29.2.14) is the free energy of the matrix model (KMM).

Theorem 29.2.1 (Kontsevich [Kon91, Kon92]). *Considering matrix integrals over the Hermitian $N \times N$ matrices as asymptotic expansions over the times*

$$t_n = \frac{1}{2n+1}\mathrm{tr}\,\frac{1}{\Lambda^{2n+1}},$$

$n = 0, 1, 2, \ldots,\ (-1)!! \equiv 1$, *we obtain*

$$\sum_{g=0}^{\infty}\sum_{\substack{n=1,g>0\\ n=3,g=0}}^{\infty}\frac{1}{(\alpha N)^{2g-2+n}}\sum_{\Sigma d_i=3g-3+n}\langle\tau_{d_1}\ldots\tau_{d_n}\rangle_g\frac{1}{n!}\prod_{i=1}^{n}(2d_i+1)!!\,t_{d_i} =$$

$$= \ln\frac{\int DX\,\exp\{-\alpha N\mathrm{tr}\left(\frac{X^2\Lambda}{2}+\frac{X^3}{6}\right)\}}{\int DX\,\exp\{-\alpha N\mathrm{tr}\,\frac{X^2\Lambda}{2}\}} =$$

$$= \mathcal{F}_{\mathrm{K}}(t_0, t_1, \ldots) \equiv \ln\mathcal{Z}_{\mathrm{K}}\big(\{t_n\}, \alpha N\big), \tag{29.2.15}$$

$$\Lambda = \mathrm{diag}\{\lambda_1, \ldots, \lambda_N\}.$$

The Feynman rules for the KMM are as follows: as in the standard matrix models, the ribbon graph propagators have two sides each carrying the corresponding index (see Fig. 29.2). The KMM differs from the one-matrix model by the additional variables λ_i associated with loops with indices i in a diagram, while the propagator is $2/(\lambda_i+\lambda_j)$, where λ_i and λ_j are the variables of two cycles (possibly, the same cycle) incident to the two sides of the ribbon graph edge. All vertices are three valent, which corresponds to the higher-dimensional cells in the moduli space stratification.

29.3 The planar term $\mathcal{F}_0$ and Witten-Dijkgraaf-Verlinde-Verlinde

Basic integrals under consideration in this and the next sections are the Hermitian one- and two-matrix model integrals

$$\int_{N\times N;\,E} DX\,e^{-\frac{1}{\hbar}\mathrm{tr}\,V(X)} = e^{\frac{1}{\hbar^2}\mathcal{F}} \equiv \mathcal{Z}, \tag{29.3.1}$$

$$\iint_{N\times N;\,E} DX\,DY e^{-\frac{1}{\hbar}\mathrm{tr}\,(V_1(X)+V_2(Y)+\kappa XY)} = e^{\frac{1}{\hbar^2}\mathcal{F}} \equiv \mathcal{Z}, \tag{29.3.2}$$

with $V_i(x) = \sum_{k=1}^{d_i+1} t_k^{(i)} x^k$, $t_k^{(i)}$ being the times of the corresponding potential. In the cases where structures related to these two integrals exhibit similar properties, we use the same notation for both of them.

29.3.1 Matrix integrals and resolvents

Consider the matrix-model integral (29.3.1) or (29.3.2) with $V(X) = \sum_{k\geq 1} t_k X^k$ the potential and $\hbar = \frac{t_0}{N}$ a formal expansion parameter (note that in the two-matrix model case, we assume the size of the matrices X and Y to be the same). The integration goes over the $N \times N$ matrices with the standard Haar measure $DX \propto \prod_{i<j} d\mathrm{Re}X_{ij} d\mathrm{Im}X_{ij} \prod_i d\mathrm{Re}X_{ii}$, and we also assume that the integration domain E for the eigenvalues of the matrix X comprises a number of (possibly half-infinite) intervals, $E = \cup_{\beta=1}^{q/2}[a_{2\beta-1}, a_{2\beta}]$. The topological expansion of the Feynman diagrams series is then equivalent to the expansion in even powers of $\hbar$ for the free energy,

$$\mathcal{F} = \sum_{h=0}^{\infty} \hbar^{2h-2}\mathcal{F}_h, \tag{29.3.3}$$

Customarily $t_0 = \hbar N$ is the scaled number of eigenvalues. We assume the potential V (or both V_1 and V_2) to be a polynomial of the fixed degree.

The averages, corresponding to the partition function (29.3.1), are defined as usual:

$$\langle f(X)\rangle = \frac{1}{\mathcal{Z}} \int_{N\times N;\,E} DX\, f(X)\, e^{-\frac{1}{\hbar}\mathrm{tr}\, V(X)} \tag{29.3.4}$$

We define the one-point resolvent to be a 1-*differential*

$$W(\lambda) = \hbar \sum_{k=0}^{\infty} \frac{\langle \mathrm{tr}\, X^k\rangle}{\lambda^{k+1}} d\lambda = \hbar \left\langle \mathrm{tr}\, \frac{1}{\lambda - X}\right\rangle, \tag{29.3.5}$$

and the s-point resolvents ($s \geq 2$) to be symmetric s-differentials

$$W(\lambda_1, \ldots, \lambda_s) == \hbar^{2-s} \left\langle \mathrm{tr}\, \frac{1}{\lambda_1 - X} \cdots \mathrm{tr}\, \frac{1}{\lambda_s - X}\right\rangle_{\mathrm{conn}} d\lambda_1 \cdots d\lambda_s \tag{29.3.6}$$

where the subscript 'conn' pertains to the connected part of the corresponding correlation function.

These resolvents are obtained from the free energy $\mathcal{F}$ through the action of the *loop insertion operator*

$$\frac{\partial}{\partial V(\lambda)} \equiv -\sum_{j=1}^{\infty} \frac{1}{\lambda^{j+1}} \frac{\partial}{\partial t_j} d\lambda \tag{29.3.7}$$

using the relations

$$W(\lambda_1,\ldots,\lambda_s) = \hbar^2 \frac{\partial}{\partial V(\lambda_s)} \frac{\partial}{\partial V(\lambda_{s-1})} \cdots \frac{\partial \mathcal{F}}{\partial V(\lambda_1)} = \\ = \frac{\partial}{\partial V(\lambda_s)} \frac{\partial}{\partial V(\lambda_{s-1})} \cdots \frac{\partial}{\partial V(\lambda_2)} W(\lambda_1). \tag{29.3.8}$$

Therefore, if one knows exactly the one-point resolvent for arbitrary potential, all multi-point resolvents can be calculated by induction. In the above normalization, the genus expansion has the form

$$W(\lambda_1,\ldots,\lambda_s) = \sum_{h=0}^{\infty} \hbar^{2h} W_h(\lambda_1,\ldots,\lambda_s), \quad s \geq 1, \tag{29.3.9}$$

which is analogous to genus expansion (29.3.3).

29.3.2 Master loop equation and the spectral curve

One-matrix model

We need one more quantity: the polynomial in x expectation value

$$P_n(x; x_1,\ldots,x_n) = \left\langle \mathrm{tr}\, \frac{V'(x) - V'(M)}{x - M} \mathrm{tr}\, \frac{1}{x_1 - M} \cdots \mathrm{tr}\, \frac{1}{x_n - M} \right\rangle_{\mathrm{conn}} dx\, dx_1 \cdots dx_n \tag{29.3.10}$$

with the corresponding expansion in $\hbar$. We also use the capital I to denote arbitrary sets of noncoinciding variables $\{x_{a_1},\ldots,x_{a_{|I|}}\}$, where we let $|\cdot|$ denote the cardinality of a finite set.

Performing the change of variables $X \to X + \varepsilon \frac{1}{X-x}$ in the matrix integral (29.3.1), exploiting the invariance and integrating by parts, we obtain the loop equation (see [Mak91]) for the Hermitian one-matrix model (in the presence of q hard edges a_β). Below, we write it already in the $\hbar$-expanded form:

$$W^{(h-1)}_{n+2}(x,x,J) + \sum_{\xi=0}^{h} \sum_{I \subseteq J} W^{(\xi)}_{|I|+1}(x,I)\, W^{(h-\xi)}_{n-|I|+1}(x, J\setminus I) \\ + \sum_{j=1}^{n} \frac{\partial}{\partial x_j} \frac{W^{(h)}_n(x, J\setminus\{x_j\}) - W^{(h)}_n(J)}{x - x_j} + \sum_{\beta=1}^{q} \frac{\partial}{\partial a_\beta} W^{(h)}_n(J) \frac{1}{x - a_\beta} dx\, dx = \\ = V'(x)\, W^{(h)}_{n+1}(x,J) dx - P^{(h)}_n(x;J) dx, \quad J = \{x_1,\cdots,x_n\}. \tag{29.3.11}$$

The particular case of this equation for $h = 0$ and $n = 0$ gives the equation commonly known as the master loop equation:

$$\left(W^{(0)}_1(x)\right)^2 = V'(x) dx\, W^{(0)}_1(x) - P^{(0)}_0(x) dx + \sum_{\beta=1}^{q} \frac{\partial}{\partial a_\beta} \mathcal{F}^{(h)} \frac{1}{x - a_\beta} dx\, dx \tag{29.3.12}$$

Introducing the *new variable*

$$y = \frac{V'(x)}{2} - W_1^{(0)}(x), \tag{29.3.13}$$

we present (29.3.12) in the form $\mathcal{C}(x, y) = 0$, where

$$\mathcal{C}(x, y) = y^2 - \frac{\left(V'(x)\right)^2}{4} + P_0^{(0)}(x) + \sum_{\beta=1}^{m} \frac{\partial \mathcal{F}^{(h)}}{\partial a_\beta} \frac{1}{x - a_\beta} \tag{29.3.14}$$

is the (hyperelliptic) *spectral curve* of the Hermitian one-matrix model. The branch points of $y = \sqrt{\frac{\left(V'(x)\right)^2}{4} + P_0^{(0)}(x) + \sum_{\beta=1}^{m} \frac{\partial \mathcal{F}^{(h)}}{\partial a_\beta} \frac{1}{x-a_\beta}}$ are q hard edges a_β and p 'dynamical' branch points μ_α (for actual models, their number is customarily lesser than n – the order of $V'(x)$ – because of the presence of double points (degenerate cuts).

Using the asymptotic condition (which follows from the definition of the matrix integral)

$$W_1^{(h)}(x)|_{x\to\infty} = \frac{t_0}{x}\delta_{h,0} + O(1/x^2), \tag{29.3.15}$$

we can manifestly solve (29.3.11) for genus zero. For this, note that the number of free parameters ($2n$ branch points and k double points) always exceeds by $n-1$ the number of constraints imposed by (29.3.13) and (29.3.15). So we must impose $n-1$ additional constraints to fix the solution unambiguously. We discuss these condition in the next section.

Two-matrix model

To simplify the presentation we present the loop equation in the case without hard edges; accounting for the loop equation in the presence of hard edges can be found in [Eyn05].

In the two-matrix model case, we can consider means of resolvents of the two matrices, X and Y. Because we are mostly interested in the free energy terms $\mathcal{F}^{(h)}$ and because, due to the symplectic invariance property to be described below, these terms are independent on which of the matrices we take the 'basic' one when considering correlation functions (another choice pertains to the change of variables $x \to y$, $y \to -x$), we let $W_n^{(h)}(J)$ denote the means of resolvents of the first matrix X. The loop equations for the 2-matrix model were first studied by Staudacher [Sta93] and they were then presented in a concise form by Eynard [Eyn03].

We now need the expectation value

$$P_n(x, y; x_1, \ldots, x_n) = \left\langle \mathrm{tr}\left(\frac{V_1'(x) - V_1'(X)}{x - X}\frac{V_2'(y) - V_2'(Y)}{y - Y}\right) \times \right.$$
$$\left. \times \mathrm{tr}\,\frac{1}{x_1 - X}\cdots \mathrm{tr}\,\frac{1}{x_n - X}\right\rangle_{\mathrm{conn}} dx\,dy\,dx_1\cdots dx_n, \quad (29.3.16)$$

which is a polynomial in x and y, and

$$U_n(x, y; x_1, \ldots, x_n) = \left\langle \mathrm{tr}\left(\frac{1}{x - X}\frac{V_2'(y) - V_2'(Y)}{y - Y}\right) \times \right.$$
$$\left. \times \mathrm{tr}\,\frac{1}{x_1 - X}\cdots \mathrm{tr}\,\frac{1}{x_n - X}\right\rangle_{\mathrm{conn}} dx\,dy\,dx_1\cdots dx_n, \quad (29.3.17)$$

which is a polynomial only in y. For all these quantities, we use their expansions in $\hbar$.

To obtain the loop equation, we now perform two changes of variables (see [Eyn03] for details): $Y \to Y + \varepsilon\frac{1}{x-X}$ and $X \to X + \varepsilon\left\{\frac{V_2'(y)-V_2'(Y)}{y-Y}, \frac{1}{x-X}\right\}$ in the matrix integral (29.3.2), exploiting the invariance and integrating by parts. The obtained loop equation for the Hermitian two-matrix model reads (before expanding in $\hbar$)

$$\hbar^{-1}(y - V'(x))U_n(x, y; J) + U_{n+1}(x, y; x, J) + \sum_{I\subseteq J} W_{|I|+1}(x, I)U_{n-|I|}(x, y; J\setminus I)$$
$$+\sum_{j=1}^{n}\frac{\partial}{\partial x_j}\frac{U_{n-1}(x, y; J\setminus\{x_j\}) - U_{n-1}(x_j, y; J\setminus\{x_j\})}{x - x_j} =$$
$$= -\hbar^{-1}P_n(x, y; J)dx + \hbar^{-2}\kappa, \quad J = \{x_1, \cdots, x_n\}. \quad (29.3.18)$$

The particular case of this equation for $h = 0$ and $n = 0$ has the form

$$(y - V_1'(x) + W_1^{(0)}(x))U_0^{(0)}(x, y) = (V_2'(y) - x)W_1^{(0)}(x) - P_0^{(0)}(x, y) + \kappa \quad (29.3.19)$$

valid for any y, and upon the substitution $y = V_1'(x) - W_1^{(0)}(x)$, we obtain the equation $C(x, y) = 0$, where

$$C(x, y) = (V_2'(y(x)) - x)(V_1'(x) - y) - P_0^{(0)}(x, y(x)) + \kappa \quad (29.3.20)$$

is the *spectral curve* of the Hermitian two-matrix model.

Above we introduced a very important object: the 1-differential $y(x)dx$ for the variables y and x related by the spectral curve equation. Note that in the both matrix models, ydx behaves differently at different branch points: $ydx \sim \sqrt{x - \mu_a}dx$ at the "dynamical" (movable) branch points, whereas $ydx \sim \dfrac{dx}{\sqrt{x - a_\beta}}$ at hard edges.

29.3.3 Second and third derivatives of $\mathcal{F}_0$

Eigenvalue pattern

From the interpretation of $W_1^{(0)}(x)dx = V'(x)dx - y(x)dx$ as the one-point resolvent, it follows immediately that if we restrict consideration to the *physical sheet* of the obtained RS with the appropriate cuts (chosen from physical considerations; say, belonging to the real axis), then the jump of ydx on edges of cuts determines the *eigenvalue density* of the corresponding matrix in the large-N limit. The different behaviour of ydx in the vicinity of branch points of different sort (dynamical and hard edges) mentioned at the end of the preceding section reflects the different type of eigenvalue distributions in the vicinity of these points: for the dynamical branch points, we observe locally the Wigner semicircular distribution law, whereas the distributions at hard edges satisfy the reciprocal law: eigenvalues have the tendency to concentrate when approaching the wall.

To define quantities on the RS, we must first choose the set of A- and B-cycles on the RS $\mathcal{C}(x, y)$ of genus g in such a way that their intersection indices are $A_i \circ B_j = \delta_{i,j}, i, j = 1, \dots, g$, an example of such a choice for a hyperelliptic surface is in Fig. 29.3.3. There the integrals of ydx over the A-cycles determine fractions of eigenvalues lying in the corresponding (separated) intervals of eigenvalue distribution. This interpretation of the A-cycle integrals is however lost outside the physical sheet, so in what follows we consider *formal matrix models* without referring to the physical nature of these cycle integrals.

We also segregate the point ∞_x that is the infinity point on the physical sheet with the local coordinate $1/x$ and introduce the set of variables

$$
\begin{aligned}
S_i &= \frac{1}{2\pi i}\oint_{A_i} ydx, \quad i = 1, \dots, g;\\
t_0 &= \operatorname*{res}_{x=\infty_x} ydx; \\
t_k &= \operatorname*{res}_{x=\infty_x} x^{-k} ydx, \quad k = 1, 2, \dots
\end{aligned}
\tag{29.3.21}
$$

We set these variables to be independent by definition; this suffices to fix all the remaining $n - 1 = g$ free parameters of the model. We also let times t_J comprise all the above *canonical variables*.

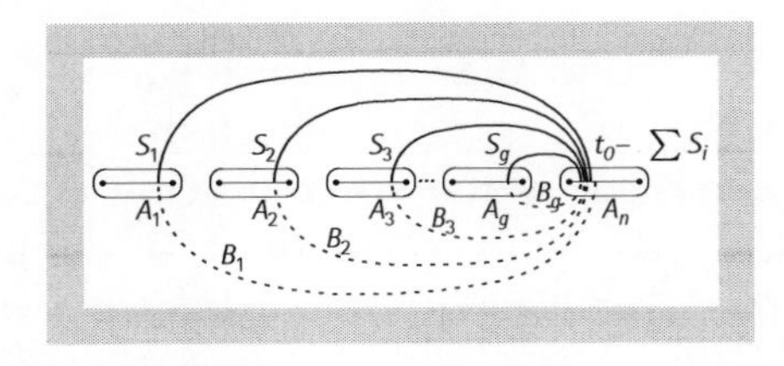

Fig. 29.3 Structure of cuts and contours for the hyperelliptic Riemann surface. Dotted are parts of contours on the second, nonphysical sheet.

We now consider the derivative of $W_1^{(0)}(x)$ w.r.t. t_J. Note that

$$\partial_{t_J}\sqrt{x-\mu_a}dx \sim \frac{dx}{\sqrt{x-\mu_a}}$$

at a dynamical branch point and

$$\partial_{t_J}\frac{f(x)dx}{\sqrt{x-a_\beta}} = \frac{f'(x)dx}{\sqrt{x-\mu_a}}$$

at a hard edge, so the derivative of the 1-form ydx w.r.t. any of the canonical variables is nonsingular at any of the branch points. The only singularities may arise at the point(s) of infinities, but there we have $\partial_{S_i}V'(x) = \partial_{t_0}V'(x) = 0$, $\partial_{t_k}V'(x)dx = x^{k-1}$, and using the asymptotic conditions (29.3.15), we obtain that $\partial_{S_i}W_1^{(0)}(x)$ is a holomorphic 1-differential ω_i, whose normalization is canonical, ($\oint_{A_i}\omega_j = \delta_{i,j}$), which follows from our choice of the canonical variables, so we have

$$\frac{\partial ydx}{\partial S_i} = \omega_i,\ i = 1,\dots,g;$$

$$\frac{\partial ydx}{\partial t_0} = \Omega_0; \tag{29.3.22}$$

$$\frac{\partial ydx}{\partial t_k} = \Omega_k,\ k = 1,2,\dots, \tag{29.3.23}$$

where Ω_k are the normalized ($\oint_{A_i}\Omega_k \equiv 0$) Whitham-Krichever 1-differentials with the only pole at $x = \infty_x$ at which

$$\Omega_k|_{x\to\infty_x} = x^{k-1}dx + O(x^{-2})dx,$$

while Ω_0 must have two simple poles: one at ∞_x and another at ∞_y.

Turning for an instant to the eigenvalue pattern, we can associate the variables S_i with 'occupation numbers' on the intervals of eigenvalue distributions and the variable t_0 with the (normalized) total number of eigenvalues.

Bergman bidifferential and two-point resolvent

We now find the second derivative $\dfrac{\partial^2\mathcal{F}_0}{\partial V(x_1)\partial V(x_2)} = \dfrac{\partial W_1^{(0)}(x_1)}{\partial V(x_2)}$. From (29.3.23) we obtain that it is given asymptotically by $\sum_{k=1}^{\infty} k\dfrac{x_1^{k-1}}{x_2^{k+1}}dx_1dx_2 \sim \dfrac{dx_1\,dx_2}{(x_1-x_2)^2}$ in the *local* coordinates near the infinite point. Because $\dfrac{\partial V'(x_1)}{\partial V(x_2)} = 1/(x_1-x_2)^2$ *globally*, we can conclude that these double poles cancel each other on the physical sheet (and add on another sheet), so we obtain another fundamental relation

$$\frac{\partial^2 \mathcal{F}_0}{\partial V(x_1)\partial V(x_2)} = B(x_1, \overline{x_2}), \tag{29.3.24}$$

where $B(x_1, \overline{x_2})$ is the *Bergman bi-differential* defined on the Riemann surface (whose points we denote by capital letters to distinguish them from the variable x), which is symmetric and its only singularity is the double pole at the coinciding arguments (see [Fay73]),

$$B(P, Q) = \left(\frac{1}{(\xi(P) - \xi(Q))^2} + O(1)\right) d\xi(P)d\xi(Q), \tag{29.3.25}$$

in some local coordinate $\xi(P)$. Here we have a freedom to add to (29.3.25) any bilinear combination of Abelian 1-differentials $d\omega_i$; we fix the normalization claiming vanishing the integrals over all A-cycles of $B(P, Q)$:

$$\oint_{A_i} B(P, Q) = 0, \text{ for } i = 1, \ldots, g. \tag{29.3.26}$$

Arguments x_1 and $\overline{x_2}$ of the Bergman bi-differential in formula (29.3.24) lie on different sheets of the Riemann surface, so the only singularities (double poles) of the two-point differential $W_2^{(0)}(x_1, x_2)$ are at the branch points (irrespective of their origin).

In the vicinity of a branch point μ_i, the bidifferential $B(P, Q)$ has the behaviour

$$B(P, Q)|_{Q\to\mu_i} = B(P, [\mu_i])\left(\frac{dq}{\sqrt{q-\mu_i}} + O(\sqrt{q-\mu_i})dq\right), \tag{29.3.27}$$

and we have the standard Rauch variational formulae [Rau59]:

$$\frac{\partial}{\partial \mu_i} B(P, Q) = \frac{1}{2} B(P, [\mu_i])\, B([\mu_i], Q), \quad i = 1, \ldots, 2n, \tag{29.3.28}$$

and

$$\oint_{B_i} B(P, Q) = 2\pi i\, \omega_i(P), \tag{29.3.29}$$

where μ_i is any simple branching point of the complex structure and ω_i are as above the canonically normalized holomorphic differentials.

We also use the primitive

$$dE_{Q,\overline{Q}}(P) = \int_Q^{\overline{Q}} B(P, \cdot) \tag{29.3.30}$$

of $B(P, Q)$ in the vicinity of a branch point μ_i (this primitive is in general a multivalued function of Q ramified over B-cycles). The points Q and $\overline{Q}$ are the two copies of the same point x on two y-variable sheets ramified over the corresponding branch point and the integration contour lies entirely in a small neighbourhood of μ_i.

$\mathcal{F}_0$ as prepotential of Seiberg-Witten-Whitham-Krichever system

Locality property We call the Seiberg-Witten-Whitham-Krichever (SWWK) system [Sei94] [Kri92] a moduli space of Riemann surfaces of genus g with a special point (∞), with the given meromorphic 1-differential dS, and with the *prepotential* F, which is a scalar function of the canonical variables $S_i = \oint_{A_i} dS$ and $t_k = \mathrm{res}_{\lambda=\infty}(dS\,\lambda^{-k})$ for $k > 0$. The prepotential and dS satisfy the conditions:

$$\frac{\partial F}{\partial S_i} = \oint_{B_i} dS; \qquad \frac{\partial dS}{\partial S_i} = \omega_i \tag{29.3.31}$$

and

$$\frac{\partial F}{\partial t_k} = \operatorname*{res}_{\lambda=\infty}(dS\,\lambda^k) \quad k > 0, \qquad \frac{\partial F}{\partial t_0} = \int_{\infty_x}^{\infty_y} dS. \tag{29.3.32}$$

We now show that the matrix model integral $\mathcal{F}_0$ is a SWWK system prepotential if we identify dS with the above 1-differential ydx. Indeed, the first formula of (29.3.32) just follows from the expansion of (29.3.13) into negative powers of x, the second formula in (29.3.31) was proved above, so it remains to prove only the formula for the derivatives w.r.t. S_i and t_0. In fact, this formula is a particular case of a more general statement.

Theorem 29.3.1 *For any object X (scalar function, differentials, etc.) that is 'local,' i.e. which depends only on the RS itself (on the branch points) and on the finite number of derivatives (of finite orders) of ydx at these branch points, we have*

$$\frac{\partial X}{\partial S_i} = \frac{1}{2\pi i}\oint_{B_i} \frac{\partial X}{\partial V(z)}, \qquad \frac{\partial X}{\partial t_0} = \int_{\infty_x}^{\infty_y} \frac{\partial X}{\partial V(z)} \tag{29.3.33}$$

All the scalar quantities $\mathcal{F}^{(h)}$ together with all the correlation functions (differentials) $W_n^{(h)}(x_1, \ldots, x_n)$ are *local objects for all matrix models* under consideration. For example, the Rauch identity (29.3.29) is formula (29.3.33) if we identify X with $y(x)dx$. Double application of this identity gives the celebrated expression relating $\mathcal{F}_0$ and the period matrix $\tau_{i,j}$ of the RS:

$$\frac{\partial^2 \mathcal{F}_0}{\partial S_i \partial S_j} = \oint_{B_i} d\omega_j(\lambda) \equiv \tau_{i,j}. \tag{29.3.34}$$

Residue formula and WDVV equations The WDVV equations [Wit90, Dij91b] is an (overdetermined) system of matrix equations [Mar96]

$$\mathcal{F}_I \mathcal{F}_J^{-1} \mathcal{F}_K = \mathcal{F}_K \mathcal{F}_J^{-1} \mathcal{F}_I, \quad \forall\ I, J, K \tag{29.3.35}$$

on the third derivatives

$$\|\mathcal{F}_I\|_{JK} = \frac{\partial^3 \mathcal{F}}{\partial t_I\, \partial t_J\, \partial t_K} \equiv \mathcal{F}_{IJK} \tag{29.3.36}$$

of some function $\mathcal{F}(\{t_I\})$. These equations can be often interpreted as associativity relations in some algebra for an underlying topological theory.

That $\mathcal{F}_0$ satisfies the WDVV equations as a function of the above *canonical* variables was shown in [Che03] for the one- and in [Ber03] for the two-matrix model. The proof is based on the *residue formula* for third derivatives of $\mathcal{F}_0$. The free energy then satisfies the WDVV equations if we keep the number of independent variables to be *equal* to the number of 'dynamical' branch points. That is, the presence of the hard-wall branch points reduces the total dimension of the WDVV system.

To obtain the formula for the third derivative $\partial^3\mathcal{F}_0/(\partial t_I \partial t_J \partial \mu_a) \equiv \mathcal{F}^{(0)}_{IJa}$ we first mention that the second derivatives of $\mathcal{F}_0$ depend on the RS *only*, not on the differential dS. So, their variation is governed completely by the variations of the branch points. We then use (29.3.28) to get that

$$W_3^{(0)}(x_1, x_2, x_3) = \frac{1}{2}\sum_i B(x_1, [\mu_i])\, B(x_2, [\mu_i]) \frac{\partial \mu_i}{\partial V(x_3)},$$

and to obtain the last missing derivative, $\partial\mu_i/\partial V(x_3)$, it suffices to invert the variational formula $\dfrac{\partial y(\xi) d\xi}{\partial V(x_3)} = B(\xi, x_3)$ in the vicinity of the branch point μ_i, which gives that

$$\frac{\partial \mu_i}{\partial V(x_3)} = \frac{B(x_3, [\mu_i])}{y([\mu_i])},$$

so, eventually, we obtain the generating function for derivatives in all the canonical variables:

$$\begin{aligned} W_3^{(0)}(x_1, x_2, x_3) &= \sum_{i=1}^{s} \frac{B(x_1, [\mu_i])\, B(x_2, [\mu_i])\, B(x_3, [\mu_i])}{2y([\mu_i])} \\ &= \sum_{i=1}^{s} \operatorname*{res}_{\xi=\mu_i} \frac{B(x_1, \xi)\, B(x_2, \xi) d\, E_{\xi,\bar{\xi}}(x_3)}{(y(\xi) - y(\bar{\xi}))d\xi}, \end{aligned} \tag{29.3.37}$$

where the summation ranges the dynamical branch points only and dE is the 1-differential (29.3.30). In fact, we can pass to summation over *all* branch points in the last line because the residues at $\lambda = a_\beta$ just vanish.

Given the residue formula (29.3.37), the WDVV equations (29.3.35) are satisfied if we choose exactly m canonical variables to construct the matrices $\mathcal{F}_I$. The associativity then follows automatically.

29.4 Higher expansion terms $\mathcal{F}_h$ and symplectic invariants

The recurrent procedure for finding corrections in $1/N$ in the one-matrix case was elaborated first in the one-cut case in [Amb92, Amb93]. A procedure for constructing *all* corrections in $1/N$ in a comprehensive way and in the most general, multicut case was developed in [Eyn04a], [Che06a] for the one-matrix model. It was then generalized to the two-matrix model [Che06b] where it was shown that the solutions for both one- and two-matrix models can be formulated in algebraic-geometrical terms for a relevant RS (the spectral curve) $\mathcal{C}(x, y) = 0$, which is hyperelliptic in the one-matrix model case and is a general algebraic curve in the two-matrix model case. Both these models are *local models*, i.e. each expansion term of these models depends on the special differential $y dx$ only through a finite number of derivatives at the branch points (the moments of the potential).

29.4.1 The iterative solution of the loop equation

The loop equations (29.3.11) and (29.3.18) can be solved using the same technique (see [Che06b] and Chapter 16 by N. Orantin in this volume); as a result, we have the recurrent relation on the resolvents,

$$W^{(h)}_{n+1}(x_0, J) = \sum_i \operatorname*{res}_{\xi \to \mu_i} \frac{d\, E_{\xi,\bar{\xi}}(x_0)}{2(y(\xi) - y(\bar{\xi}))d\xi} \Big[W^{(h-1)}_{n+2}(\xi, \bar{\xi}, J) + \sum_{s, I \subseteq J}{}' W^{(s)}_{|I|+1}(\xi, I)\, W^{(h-s)}_{n-|I|+1}(\bar{\xi}, J/I) \Big], \tag{29.4.1}$$

where the prime over the sum sign indicates that the sum ranges only stable correlation functions $W^{(a)}_b$ with $a \geq 0$, $b > 0$, and $2a + b - 2 > 0$.

The free energy $\mathcal{F}_h$ We now construct the operator that is in a sense 'inverse' of $\partial/\partial V(x)$. For a 1-differential ϕ let

$$H_x \cdot \phi := \operatorname*{res}_{\infty_x} V_1(x)\phi - \operatorname*{res}_{\infty_y}(V_2(y) - \kappa x y)\phi + t_0 \int_{\infty_x}^{\infty_y} \phi + \sum_{i=1}^{g} S_i \oint_{B_i} \phi. \tag{29.4.2}$$

Now, if ϕ was obtained by the action of $\partial/\partial V_1(x)$, then, recalling the locality condition and definition (29.3.7) of the loop insertion operator, we obtain

$$H_x \cdot \frac{\partial}{\partial V_1(x)} \cdot \theta = \left(\sum_k t^{(1)}_k \frac{\partial}{\partial t^{(1)}_k} + \sum_k t^{(2)}_k \frac{\partial}{\partial t^{(2)}_k} + \kappa \frac{\partial}{\partial \kappa} + \sum_i S_i \frac{\partial}{\partial S_i} + t_0 \frac{\partial}{\partial t_0} \right) \theta, \tag{29.4.3}$$

but the operator in the right-hand side is nothing but minus the operator of scaling by $\hbar$, that is, taking $\theta = \mathcal{F}_h$ and using (29.3.3) we obtain that $H_x \cdot \frac{\partial}{\partial V_1(x)} \cdot \mathcal{F}_h = -(2 - 2h)\mathcal{F}_h$ or

$$\mathcal{F}_h = \frac{1}{2h-2} H_x \cdot W_1^{(h)}(x), \quad h \neq 1, \tag{29.4.4}$$

whereas the case $h = 1$ requires special methods (see [Che04] for 1-matrix model and [Eyn04b] for 2-matrix model cases).

In the case of local models, all the integrations in (29.4.2) can be collapsed to the integrals in vicinities of the branching points, and we obtain $H_x\phi \to \sum_a \text{res}_{\mu_a} \Phi\phi$, where $\Phi(x) = \int_{\mu_a}^x y d\xi$.

29.4.2 Symplectic invariance

The first occurrence of symplectic invariance is that the above result for $\mathcal{F}_h$ does not depend on resolvents of which matrix, X or Y, we take as the basic one [Che06b]. We therefore have a symmetry for exchanging $x \leftrightarrow y$. This was generalized in [Eyn08] to formulating that *all the correction terms* $\mathcal{F}_h$ *are invariant under special changes of variables for* (y, x) *that preserve the symplectic form* $dx \wedge dy$. Note that this invariance holds only for the free-energy terms, all the correlation functions $W_n^{(h)}(x_1, \dots, x_n)$ are not symplectic invariants.

By definition, the original symplectic transformations are

- $\tilde{x} = x,\ \tilde{y} = y + R(x)$, where $R(x)$ is a rational function;
- $\tilde{x} = \frac{ax+b}{cx+d},\ \tilde{y} = \frac{(cx+d)^2}{ad-bc} y$;
- $\tilde{x} = y,\ \tilde{y} = -x$.

All these transformations preserve $\mathcal{F}_h$, which is therefore a symplectic invariant.

Construction of the recurrent solution of the loop equations together with the H-operator construction for producing the 'free energies' has proved to be successful not only in just matrix model applications. If we lose the algebraicity condition and consider x to be a variable in some open domain of a Riemann surface, we can develop a recurrent procedure for constructing symplectic invariants for generally non-algebraic curves such as those appearing in string model applications (see the contribution of M. Mariño to this volume Chapter 31.) for which, say, the characteristic equation (from mirror symmetry considerations) has the form $H(e^x, e^y) = 0$ with H being a polynomial function. We can then generalize the symplectic invariance property for any transformation of the form

- $\tilde{x} = f(x),\ \tilde{y} = \frac{1}{f'(x)} y,\ f(x)$ is analytical and injective in the image of x.

It was conjectured in [Bou09] that the open string amplitudes $W_k^{(h)}(z_1, \dots, z_k)$ and the closed string amplitudes $\mathcal{F}_h$ are merely given by the solutions (29.4.1) and (29.4.4) evaluated at a zero of dx on an open domain of the curve. The conjecture, albeit not yetproven , has passed numerous tests since then (see [Eyn09] for these and other applications of the symplectic invariance method; say, that the Lambert equation type curve $ye^{-y} = x$ generates the Hurwitz numbers [Bou07]).

- Decomposition formulae.

In the vicinity of every square-root branch point μ_i one can introduce a 'local' coordinate z_i together with the 'local' times $t_k^{(i)}$ such that $z_i(\xi) = -z_i(\bar{\xi})$, which results in only odd times $t_{2k+1}^{(i)}$, $k = 0, 1, \ldots$, contribute, and we therefore obtain a KMM partition function (29.2.15) $\mathcal{Z}_K(\{t_{2k+1}^{(i)}\})$ residing at every branch point. We must then intertwine these Kontsevich partition functions in order to obtain the 'global' partition function of 1MM or 2MM; the resulting formula reads:

$$\mathcal{Z}_{1MM/2MM} = e^{\sum_{i,j=1}^{2n}\sum_{k,l=0}^{\infty}\frac{\partial}{\partial t_k^{(i)}}A_{k,l}^{i,j}\frac{\partial}{\partial t_l^{(j)}}}\prod_{i=1}^{2n}\mathcal{Z}_K(\{t_{2k+1}^{(i)}\})$$

with some explicit expressions for coefficients $A_{k,l}^{i,j}$. This formula was first established in [Che97] for the one-cut solution of 1MM using the equivalence of the Virasoro conditions in both sides of this expression and it was generalized by Alexandrov, Mironov, and Morozov in [Ale07a] and [Ale07b] to the case of a multi-cut 1MM solution using the current algebras.

Acknowledgements

This work was supported by the RFBR grants 08-01-00501 and 09-01-92433-CE, by the Grant for Support for the Scientific Schools 195.2008.1, and by the Program Mathematical Methods for Nonlinear Dynamics.

References

[Ale07a] A. Alexandrov, A. Mironov, and A. Morozov, Theor. Math. Phys. **150** (2007) 179–192, hep-th/0605171.

[Ale07b] A. Alexandrov, A. Mironov, and A. Morozov, Physica **D235** (2007) 126–167, hep-th/0608228.

[Amb92] J. Ambørn, L. Chekhov, and Yu. Makeenko, Phys. Lett. **282B** (1992) 341–348.

[Amb93] J. Ambjørn, L. Chekhov, C. F. Kristjansen and Yu. Makeenko, Nucl. Phys. **B404** (1993) 127–172; Erratum ibid. **B449** (1995) 681.

[Ber03] M. Bertola, JHEP **0311**:062 (2003).

[Bou07] V. Bouchard and M. Mariño, Proceedings of Symposia in Pure Mathematics, AMS (2008) [arXiv:0709.1458 [math.AG]].

[Bou09] V. Bouchard, A. Klemm, M. Mariño, and S. Pasquetti, Commun. Math. Phys. **287** (2009) 117–178.

[Che97] L. Chekhov, Acta Appl. Mathematicae **48** (1997) 33–90; hep-th/9509001.

[Che03] L. Chekhov, A. Marshakov, A. Mironov, and D. Vasiliev, Phys. Lett. **562B** (2003) 323–338.

[Che04] L. Chekhov, Theor. Math. Phys. **141** (2004) 1640–1653.

[Che06a] L. Chekhov and B. Eynard, JHEP **0603**:014 (2006).
[Che06b] L. Chekhov, B. Eynard, and N. Orantin, JHEP **0612**:053 (2006).
[Dij91a] R. Dijkgraaf, E. Verlinde, and H. Verlinde, Nucl. Phys., **B348** (1991) 435.
[Dij91b] R.Dijkgraaf, E.Verlinde, and H.Verlinde, Nucl. Phys. **B352** (1991) 59–86.
[Eyn03] B. Eynard, JHEP 01 (2003) 051.
[Eyn04a] B. Eynard, *JHEP* **0411**:031 (2004); hep-th/0407261.
[Eyn04b] B. Eynard, A. Kokotov, and D. Korotkin, Nucl. Phys. **B694** (2004) 443–472.
[Eyn05] B. Eynard, J. Stat. Mech. 0510 (2005) P006.
[Eyn08] B. Eynard and N. Orantin, J. Phys. Math. Theor. **A41** (2008) 015203.
[Eyn09] B. Eynard and N. Orantin, J. Phys. **A42** (2009) 293001.
[Fay73] J. Fay, *Theta-Functions on Riemann Surfaces, Lect. Notes Math.* Vol. **352**, Springer, New York, 1973.
[Har86] J. Harer and D. Zagier, Invent. Math., **85** (1986) 457–485.
[Kon91] M. L. Kontsevich, Funkt. Anal. Pril., **25** (1991) 123–129.
[Kon92] M. L. Kontsevich, Commun. Math. Phys., **147** (1992) 1–23.
[Kri92] I. Krichever, Commun. Pure Appl. Math. **47**(4) (1992) 437–475.
[Mak91] Yu. Makeenko, Mod. Phys. Lett. (Brief Reviews) **A6** (1991), 1901–1913.
[Mar96] A. Marshakov, A. Mironov and A. Morozov, Phys. Lett. **B389** (1996) 43–52.
[Pen88] R. C. Penner, J. Diff. Geom. **27** (1988), 35–53.
[Rau59] H.E. Rauch, Commun. Pure Appl. Math. **12** (1959) 543–560.
[Sei94] N. Seiberg and E. Witten, Nucl. Phys. **B426** (1994) 19–52; [Erratum: ibid. **B430** (1994) 485].
[Sta93] M. Staudacher, Phys. Lett. **B305** (1993) 332–338.
[Str84] K. Strebel, *Quadratic Differentials*, Springer, Berlin 1984.
[Wit90] E. Witten, Nucl. Phys. **B340** (1990) 281–332.
[Wit91] E. Witten, Surv. Diff. Geom. **1** (1991) 243–310.

·30·

Two-dimensional quantum gravity

Ivan Kostov

Abstract

This chapter contains a short review of the correspondence between large N matrix models and critical phenomena on lattices with fluctuating geometry.

30.1 Introduction

One of the applications of the large N matrix models is that they can be used as machines for generating planar graphs. The concept of the $1/N$ expansion for matrix fields and its topological meaning was introduced by 't Hooft in [tH74]. The problem of counting planar graphs was related to the saddle-point solution of matrix integrals by Brézin, Parisi, Itzykson, and Zuber [BIPZ78].

The ensembles of planar graphs exhibit universal critical behaviour. By tuning the coupling constants one can achieve a regime which is dominated by planar graphs of diverging size. Seen from a distance, such planar graphs resemble surfaces with fluctuating metric. Therefore the sum over planar graphs can be considered as a discretization of the path integral of 2D quantum gravity [Pol81]. The ensembles of planar graphs were formulated as models of 2D lattice gravity in [Dav85b, Kaz85, ADF85] and further studied in [Dav85a, KKM85, BKKM86].

The 2D lattice gravity with matter fields was initially constructed as a Gaussian discretization of the path integral for the non-critical Polyakov string [Pol81] with D-dimensional target space. However such a discretization leads to an exactly solvable model only in the case of $D = 0$ and $D = -2$ [KKM85, BKKM86, KM87]. Another way to introduce a non-trivial target space is to consider solvable spin models on planar graphs. The Ising model on a dynamical lattice was solved, after being reformulated as a two-matrix model, by Kazakov and Boulatov [Kaz86, BK87]. This solution was used as a guideline by Knizhnik, Polyakov, and Zamolodchikov (KPZ), who obtained the exact spectrum of anomalous dimensions in the continuum theory of 2D gravity [KPZ88].

Soon after the the Ising model, all basic statistical models were solved on a dynamical lattice: the $O(n)$ model [Kos89a], the Q-state Potts model [Kaz88], the solid-on-solid (SOS) and ADE height models [Kos89b][Kos92b], as well as the six-vertex model [Kos00]. Particular cases of the Potts and $O(n)$ models are the tree/bond percolation [Kaz88] and self-avoiding polymer networks [DK88] [DK90]. In general, the theories of discrete gravity admit an infinite number of multi-critical regimes, which are first observed in the one-matrix model [Kaz89]. The multi-critical point in the ADS models and the $O(n)$ model were identified in [Kos92b, KS92].

Practically all solvable 2D statistical models whose critical behaviour is described by a $c \leq 1$ conformal field theory (CFT) can be transferred from the plane to a dynamical lattice and solved exactly. There can be several different discretizations of the same continuous theory. For example, all rational theories of 2D gravity can be obtained either by the ADE matrix models, or as the critical points of the two-matrix model [DKK93].

For statistical systems coupled to gravity it is possible to solve analytically problems whose exact solution is inaccessible on flat lattice. On the other hand, the solution of a spin model on a dynamical lattice allows us to reconstruct the phase diagram and the critical exponents of the same spin model on the flat lattice. Therefore the solvable models of 2D gravity represent a valuable tool to explore the critical phenomena in two dimensions. This is especially true in the presence of boundaries, where solution of the matrix model can give insight into the interplay between the bulk and boundary flows.

We review the solvable models of 2D lattice quantum gravity and the way they are related to matrix models. We put the accent on the critical behaviour and the solution in the scaling limit near the critical points. We start with a short description of the continuum world sheet theory, the Liouville gravity, and derive the KPZ scaling relation. Then we present in detail the simplest model of 2D gravity and the corresponding matrix model, followed by less or more sketchy presentation of the vertex/height integrable models on planar graphs and their mapping to matrix models. In this short text we do not discuss the numerous applications of matrix models to string theory.

30.2 Liouville gravity and Knizhnik-Polyakov-Zamolodchikov scaling relation

The continuum approach to critical phenomena in the presence of gravity was developed by Knizhnik, Polyakov, and Zamolodchikov [KPZ88]. The conformal weights from the Kac spectrum are converted by 'gravitational dressing' into new ones through the remarkable KPZ formula [KPZ88]. Subsequently, a more intuitive alternative derivation of the KPZ formula was proposed in [Dav88, DK89]. The conformal theory with quantum gravity has larger symmetry and hence simpler phenomenology: we are only allowed to ask questions invariant under diffeomorphisms. The anomalous dimensions in such a theory are perfectly meaningful but not the operator product expansion; only fields integrated over all points make sense. Therefore, instead of the correlation functions of the local operators we have to consider the corresponding susceptibilities. The way they depend on the temperature and the cosmological constant allows us to determine the scaling dimensions.

30.2.1 Path integral over metrics

Let $Z_{\text{matter}}[g_{ab}]$ be the partition function of the matter field on a two-dimensional worldsheet $\mathcal{M}$ with Riemann metric $ds^2 = g_{ab}\, d\sigma^a d\sigma^b$. For the moment assume that the world sheet is homeomorphic to a sphere. The partition function on the sphere, which we denote by $\mathcal{F}(\mu)$, is symbolically written as the functional integral

$$\mathcal{F}(\mu) = \int [\mathcal{D}g_{ab}]\, e^{-\mu A}\, Z^{(\text{sphere})}_{\text{matter}}[g_{ab}], \tag{30.2.1}$$

where the parameter μ, called the *cosmological constant*, is coupled to the area of the worldsheet,

$$A = \int_{\mathcal{M}} d^2\sigma \sqrt{\det g_{ab}}. \tag{30.2.2}$$

The functional measure $[\mathcal{D}g_{ab}]$ in the space of Riemann metrics is determined by the property that it is invariant with respect to the diffeomorphisms of the world sheet. Part of this symmetry is taken into account by the conformal gauge

$$g_{ab} = e^{2\phi}\, \hat{g}_{ab}, \tag{30.2.3}$$

where $\hat{g}_{ab}$ is some background metric, so that the fluctuations of the metric are described by the local scale factor ϕ. At the critical points, where the matter QFT is conformal invariant, the scale factor ϕ couples to the matter field through a universal classical action induced by the conformal anomaly c_{matter}. In addition, the conformal gauge necessarily introduces the Faddeev-Popov reparameterization ghosts with central charge $c_{\text{ghosts}} = -26$ [Pol81] . The effective action of the field ϕ is called *Liouville action* because the corresponding classical equation of motion is the equation for the metrics with constant curvature, the Liouville equation.

For a world sheet $\mathcal{M}$ with a boundary $\partial\mathcal{M}$ one can introduce a second parameter, the *boundary cosmological constant* μ_B coupled to the boundary length

$$\ell = \int_{\partial\mathcal{M}} \sqrt{g_{ab}\, d\sigma^a d\sigma^b}. \tag{30.2.4}$$

The partition function on the disk, which we denote by $\mathcal{U}(\mu, \mu_B)$, is defined as a functional integral over all Riemann metrics on the disk,

$$\mathcal{U}(\mu, \mu_B) = \int [\mathcal{D}g_{ab}]\, e^{-\mu A - \mu_B \ell}\, Z^{(\text{disk})}_{\text{matter}}[g_{ab}]. \tag{30.2.5}$$

Here we have assumed that the matter field satisfies certain boundary condition. We will also consider more complicated situations where the boundary is split into several segments with different boundary cosmological constants and different boundary conditions for the matter field.

30.2.2 Liouville gravity

Liouville gravity is a term for the two-dimensional quantum gravity whose action is induced by a conformal invariant matter field. The effective action of the Liouville, gravity consists of three pieces, associated with the matter, Liouville and ghost fields. For our discussion we will only need the explicit form of the Liouville piece $\mathcal{A}_{\mu,\mu_B}[\phi]$. This is the action of a Gaussian field with linear coupling to the curvature and exponential terms coupled to the bulk and boundary cosmological constants:

$$\mathcal{A}_{\mu,\mu_B}[\phi] = \frac{1}{4\pi} \int\limits_{\mathcal{M}} \left(g\,(\nabla\phi)^2 + Q\phi\hat{R}\right) + \frac{1}{2\pi} \int\limits_{\partial\mathcal{M}} d\sigma Q\phi\hat{K} + \mu A + \mu_B \ell. \tag{30.2.6}$$

The area and the length are defined by 30.2.2 and (30.2.4), $\hat{R}$ the background curvature in the bulk and by $\hat{K}$ the geodesic background curvature at the boundary. The coupling to the curvature is through the imaginary background charge $i\,Q$. We will assume that the coupling constant g of the Liouville field is larger than one: $g \geq 1$.

The Liouville coupling g and the background charge Q are determined by the invariance with respect to the choice of the background metric $\hat{g}_{ab}$. The background charge of the Liouville field must be tuned so that the total conformal anomaly vanishes:

$$c_{\text{tot}} \equiv c_\phi + c_{\text{matter}} + c_{\text{ghosts}} = (1 + 6Q^2/g) + c_{\text{matter}} - 26 = 0. \tag{30.2.7}$$

The balance of the central charge gives

$$c_{\text{matter}} = 1 - 6(g-1)^2/g. \tag{30.2.8}$$

The observables in Liouville gravity are integrated local densities

$$\mathcal{O} \sim \int_{\mathcal{M}} \Phi_h \, e^{2\alpha\phi}, \tag{30.2.9}$$

where Φ_h represents a scalar ($h = \bar{h}$) matter field[1] and V_α is a Liouville 'dressing' field, which completes the conformal weights of the matter field to $(1, 1)$. The balance of the conformal dimensions, $h + h_\alpha^{\text{Liouv}} = 1$, gives a quadratic relation

[1] The condition $\bar{h} = h$ follows from the fact that any local rotation can be 'unwound' by a coordinate transformation.

between the Liouville dressing charge and the conformal weight of the matter field:

$$h + \frac{\alpha(Q-\alpha)}{g} = 1. \qquad (30.2.10)$$

In particular, the Liouville interaction $e^{2\phi}$ dresses the identity operator. The balance of scaling dimensions (30.2.10) determines, for $\alpha = 1$ and $h = 0$, the value of the background charge Q:

$$Q = g + 1. \qquad (30.2.11)$$

Similarly, the boundary matter fields Φ_h^B with boundary dimensions h_B are dressed by boundary Liouville exponential fields,

$$\mathcal{O}^B \sim \int_{\partial\mathcal{M}} \Phi_h^B \, e^{\alpha^B \phi}, \qquad (30.2.12)$$

where the boundary Liouville exponent α^B is related to h^B as in (30.2.10).

The quadratic relation (30.2.10) has two solutions, α and $\tilde{\alpha}$, such that $\alpha \le Q/2$ and $\bar{\alpha} \ge Q/2$. The Liouville dressing fields corresponding to the two solutions are related by Liouville reflection $\alpha \to \bar{\alpha} = Q - \alpha$. Generically the scaling limit of the operators on the lattice is described by the smaller of the two solutions of the quadratic constraint (30.2.10). The corresponding field V_α has a good quasiclassical limit. Such fields are said to satisfy the Seiberg bound $\alpha \le Q/2$, see the reviews [DFGZJ95, GM93].[2] For a more recent review of Liouville gravity see [Na04].

The simplest correlation functions in Liouville gravity (up to 3-point functions of the bulk fields and 2-point functions of the boundary fields) factorize into matter and Liouville pieces. This factorization does not hold away from the critical points.

30.2.3 Correlation functions and KPZ scaling relation

The scaling of the correlation function with the bulk cosmological constant μ follows from the covariance of the Liouville action (30.2.6) with respect to translations of the Liouville field:

$$\mathcal{A}_{\mu,\mu_B}[\phi - \tfrac{1}{2}\ln\mu] = -\tfrac{1}{2}Q\chi \ln\mu \; + \mathcal{A}_{1,\,\mu_B/\sqrt{\mu}}[\phi] \qquad (30.2.13)$$

where $\chi = 2 - 2\#(\text{handles}) - \#(\text{boundaries})$ is the Euler characteristics of the world sheet. Applying this to the correlation function of the bulk operators $\mathcal{O}_j$ and the boundary operators $\mathcal{O}_k^B$, we obtain the scaling relation

[2] Originally it was believed that the second solution is unphysical, until it was realized that it describes a specially tuned measure over surfaces [Kle95].

$$\langle \prod_j \mathcal{O}_j \prod_k \mathcal{O}_k^B \rangle_{\mu,\mu_B} = \mu^{\frac{1}{2}\chi Q - \sum_j \alpha_j - \frac{1}{2}\sum_k \alpha_k^B} f(\mu_B/\sqrt{\mu}), \tag{30.2.14}$$

where f is some scaling function. We see that for any choice of the local fields the unnormalized correlation function dependence on the global curvature of the world sheet through a universal exponent $Q = 1 + g = 2 - \gamma_{\rm str}$, where $\gamma_{\rm str} = 1 - g$ is called the *string susceptibility exponent*. The scaling of the bulk and boundary local fields is determined by their *gravitational anomalous dimensions* Δ_i and Δ_j^B, related to the exponents α_i and α_j^B by

$$\Delta_i = 1 - \alpha_i, \qquad \Delta_i^B = 1 - \alpha_i^B. \tag{30.2.15}$$

Once the exponent $\gamma_{\rm str}$ and the gravitational dimensions Δ_i are known, the central charge $c_{\rm matter}$ and conformal weights h_i of the matter fields can be determined from (30.2.7) and (30.2.10):

$$h = \frac{\Delta(\Delta - \gamma_{\rm str})}{1 - \gamma_{\rm str}}, \qquad c_{\rm matter} = 1 - 6\frac{\gamma_{\rm str}^2}{1 - \gamma_{\rm str}} \qquad (\gamma_{\rm str} = 1 - g), \tag{30.2.16}$$

and a similar relation for the boundary fields. This relation is known as KPZ correspondence between flat and gravitational dimensions of the local fields [KPZ88, Dav88, DK89].

The 'physical' solution of the quadratic equation (30.2.10) can be parameterized by a pair of real numbers r and s, not necessarily integers,[3]

$$h_{rs}(g) = \frac{(rg - s)^2 - (g-1)^2}{4g}, \qquad \alpha_{rs}(g) = \frac{g + 1 - |rg - s|}{2}. \tag{30.2.17}$$

The corresponding gravitational anomalous dimensions $\Delta_{rs} = 1 - \alpha_{rs}$ are

$$\Delta_{rs}(g) = \frac{|rg - s| - (g - 1)}{2}. \tag{30.2.18}$$

Remember that our conventions are such that $g > 1$. The solution (30.2.17) applies also for the boundary exponents, after replacing $h \to h^B$, $\Delta \to \Delta^B$, $\alpha \to \beta$. It generalizes in an obvious way to the case when each boundary segment is characterized by a different boundary cosmological constant.

The string susceptibility exponent $\gamma_{\rm str}$ is a measurable quantity in discrete models of 2D gravity and is determined by the the scaling behaviour of the susceptibility on the sphere $u \equiv \partial_\mu^2 \mathcal{F} \sim \mu^{-\gamma_{\rm str}}$ at the critical points. At different critical points of the same model one can have different exponents $\gamma_{\rm str}$. Once $\gamma_{\rm str}$ is known, the matter central charge and the conformal weights of the local

[3] When r and s are integers, h_{rs} are the conformal weights of the the degenerate fields of the matter CFT.

fields can be determined by the KPZ relation (30.2.16), with the gravitational dimensions Δ_i and Δ_i^B defined through the scaling relation

$$\langle \prod_j \mathcal{O}_j \prod_k \mathcal{O}_k^B \rangle_{\mu,\mu_B} = \mu^{\frac{1}{2}\chi(2-\gamma_{\rm str})-\sum_j(1-\Delta_i)-\frac{1}{2}\sum_k(1-\Delta_k^B)} f(\mu_B/\sqrt{\mu}). \quad (30.2.19)$$

30.3 Discretization of the path integral over metrics

In order to built models of discrete 2D gravity it is sufficient to consider the ensemble of trivalent planar graphs which are dual to triangulations, as the one shown in Fig. 30.1. To avoid ambiguity the propagators are represented by double lines. The faces of such a 'fat' graph are bounded by the polygons formed by the single lines.

The local curvature of the triangulation is concentrated at the vertices r of the triangulation. The curvature R_r (or the boundary curvature K_r if the vertex is on the boundary) is expressed through the coordination number c_r, equal to the number of triangles meeting at the point r,

$$R_r = \pi \frac{6 - c_r}{3}, \qquad K_r = \pi \frac{3 - c_r}{3}. \quad (30.3.1)$$

The total curvature is equal to the Euler number $\chi(\Gamma)$:

$$\frac{1}{2\pi} \sum_{r \in \text{bulk}} R_r + \frac{1}{\pi} \sum_{r \in \text{boundary}} K_r = \chi(\Gamma). \quad (30.3.2)$$

For each fat graph Γ we define the area and the boundary length as

$$|\Gamma| \stackrel{\text{def}}{=} \#\,(\text{faces of } \Gamma)\ , \quad |\partial\Gamma| \stackrel{\text{def}}{=} \#\,(\text{external lines of } \Gamma)\,. \quad (30.3.3)$$

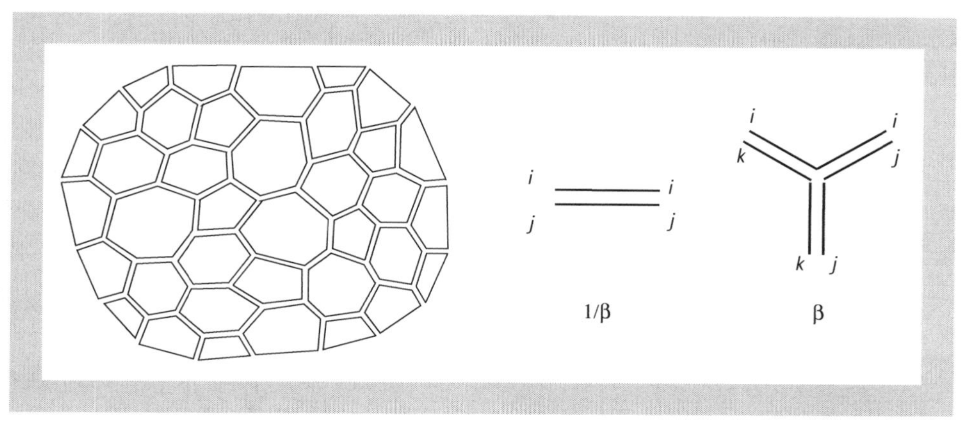

Fig. 30.1 Left: A trivalent planar graph with the topology of a disk and its dual triangulation. Right: Feynman rules for the matrix model for pure gravity (figures from [BHK09]).

Let us denote by {Sphere} and {Disk} the ensembles of fat graphs dual to triangulations respectively of the sphere and of the disk. Assume that the matter field can be defined on any triangulation Γ and denote its partition function by $Z_{\text{matter}}[\Gamma]$. Then the path integrals for the partition function of the sphere (30.2.1) and that of the disk (30.2.5) are discretized as follows:

$$\mathcal{F}(\check{\mu}) = \sum_{\Gamma \in \{\text{Sphere}\}} \frac{1}{k(\Gamma)} \check{\mu}^{-|\Gamma|} \; Z^{(\text{sphere})}_{\text{matter}}[\Gamma]. \tag{30.3.4}$$

$$\mathcal{U}(\check{\mu}, \check{\mu}_B) = \sum_{\Gamma \in \{\text{Disk}\}} \frac{1}{|\partial\Gamma|} \check{\mu}^{-|\Gamma|} \check{\mu}_B^{-|\partial\Gamma|} \; Z^{(\text{disk})}_{\text{matter}}[\Gamma]. \tag{30.3.5}$$

Here $\check{\mu}$ and $\check{\mu}_B$ are the the lattice bulk and boundary cosmological constants and $k(\Gamma)$ is the volume of the symmetry group of the planar graph Γ.

30.4 Pure lattice gravity and the one-matrix model

30.4.1 The matrix model as a generator of planar graphs

The solution of pure discrete gravity thus boils down to the problem of counting planar graphs, reformulated in terms of a matrix integral in [BIPZ78]. The matrix integral

$$\mathcal{Z}(N, \beta) = \int d\mathbf{M} \, e^{\beta\left(-\frac{1}{2}\text{Tr}\mathbf{M}^2 + \frac{1}{3}\text{Tr}\mathbf{M}^3\right)}, \tag{30.4.1}$$

where $d\mathbf{M}$ is the flat measure in linear space of $N \times N$ Hermitian matrices, can be considered as the partition function of a zero-dimensional QFT with Feynman rules given in Fig. 30.1 (right). The integral is divergent but makes sense as a formal perturbative expansion around the Gaussian measure. The logarithm of the partition function (30.4.1), the 'vacuum energy' of the matrix field, is equal to the sum of all connected diagrams.

The weight of a vacuum graph Γ is written, using the Euler relation, as

$$\text{Weight}(\Gamma) = \frac{1}{k(\Gamma)} \beta^{\#\text{vertices}-\#\text{lines}} N^{\#\text{faces}} = \frac{1}{k(\Gamma)} \beta^{\chi(\Gamma)} \, (N/\beta)^{|\Gamma|}. \tag{30.4.2}$$

Therefore we can identify the ratio $\check{\mu} = \beta/N$ with the lattice bulk cosmological constant. In the large N limit with the ratio β/N fixed,

$$N \to \infty, \qquad \beta/N = \check{\mu}, \tag{30.4.3}$$

only Feynman graphs dual to triangulations of the sphere ($\chi = 2$) contribute and their weight will be $\beta^2 \check{\mu}^{-|\Gamma|}$. As a result the sphere partition function (30.3.4) of discrete 2D pure gravity is equal to the vacuum energy of the matrix model in the the large N, or planar, limit

$$\mathcal{F}(\check{\mu}) = \lim_{N\to\infty} \frac{1}{\beta^2} \log \mathcal{Z}(N, \beta), \quad \check{\mu} = \beta/N. \tag{30.4.4}$$

It is convenient to consider the disk partition function (30.3.5) for any complex value of the boundary cosmological constant, which we denote by the letter z. The disk partition function for $\check{\mu}_B = z$ is evaluated, up to a trivial logarithmic term, by

$$\mathcal{U}(\check{\mu}, z) = -\frac{1}{\beta} \langle \operatorname{Tr} \log(z - \mathbf{M}) \rangle. \tag{30.4.5}$$

The derivative $W = -\partial_z \mathcal{U}$, which is the partition function on the disk with a marked point on the boundary, or the boundary one-point function on the disk, is the expectation value of the resolvent of the random matrix,

$$W(z) = \frac{1}{\beta} \langle \operatorname{tr} \mathbf{W}(z) \rangle, \qquad \mathbf{W}(z) \stackrel{\text{def}}{=} \frac{1}{z - \mathbf{M}}. \tag{30.4.6}$$

30.4.2 Solution at large N

The resolvent (30.4.6) is the only solution of a non-linear Ward identity, known also as a loop equation, which follows from the translational invariance of the matrix integration measure. For any matrix observable $\mathbf{F}$ we have the identity

$$\frac{1}{\beta} \langle \partial_{\mathbf{M}} \mathbf{F} \rangle = \langle \operatorname{tr}\big[V'(\mathbf{M})\, \mathbf{F} \big] \rangle, \quad \partial_{\mathbf{M}}\, \mathbf{F} \stackrel{\text{def}}{=} \partial F_{ij} / \partial X_{ij}. \tag{30.4.7}$$

Written for $\mathbf{F} = \mathbf{W}(z)$, the equation states

$$W^2(z) = \frac{1}{\beta} \big\langle \operatorname{tr}\big[V'(\mathbf{M}) \mathbf{W}(z) \big] \big\rangle. \tag{30.4.8}$$

Now, using the factorization property of the traces, $\langle \operatorname{tr} A \operatorname{tr} B \rangle = \langle \operatorname{tr} A \rangle \langle \operatorname{tr} B \rangle$, which is true up to $1/N^2$ corrections, we obtain that the resolvent satisfies a quadratic equation

$$W^2(z) - V'(z)\, W(z) = P(z), \tag{30.4.9}$$

where P is a polynomial. Therefore the function

$$J(z) \stackrel{\text{def}}{=} W(z) - \tfrac{1}{2} V'(z), \tag{30.4.10}$$

should be such that that $J^2(z)$ is a polynomial. The polynomial is determined by the asymptotics at infinity,

$$J(z) = -\tfrac{1}{2} V'(z) + (\check{\mu} z)^{-1} + O(z^{-2}). \tag{30.4.11}$$

The solution, which minimizes the effective potential is[4]

$$J(z) = (z - 1 + \tfrac{1}{2}(a+b))\sqrt{(z-a)(z-b)}, \tag{30.4.12}$$

where the positions of the branch points are determined as functions of $\check{\mu}$ by

$$(b-a)^2 = 8S(1-S),\ a+b = 2S,\ \tfrac{1}{2}S(1-S)(1-2S) = \frac{1}{\check{\mu}}. \tag{30.4.13}$$

It is also easy to determine the derivative $\partial_\mu J(z)$, which has the same analytic properties as $J(z)$, but simpler asymptotics at infinity,

$$\partial_{\check{\mu}} J(z) = \partial_{\check{\mu}} W(z) = -\check{\mu}^{-2} z^{-1} + \partial_{\check{\mu}} W_1 z^{-2} + O(z^{-3}), \tag{30.4.14}$$

where $W_1 = \langle \mathrm{tr}\mathbf{M} \rangle$. The solution is obviously

$$\check{\mu}^2 \partial_{\check{\mu}} J(z) = -\frac{1}{\sqrt{(z-a)(z-b)}}. \tag{30.4.15}$$

One can check that the compatibility of (30.4.15) and (30.4.12) gives (30.4.13). Comparing (30.4.15) and (30.4.14) we get

$$\partial_{\check{\mu}} \langle \mathrm{tr}\mathbf{M} \rangle = S. \tag{30.4.16}$$

30.4.3 The scaling limit

We are interested in the universal critical properties of the ensemble of triangulations in the limit when the size of the typical triangulation diverges. This limit is achieved by tuning the bulk and boundary cosmological near their critical values,

$$\check{\mu} = \check{\mu}_c + \epsilon^2 \mu, \qquad z = \check{\mu}_B^c + \epsilon x, \tag{30.4.17}$$

where ϵ is a small cutoff parameter with dimension of length and x is proportional to the renormalized boundary cosmological constant μ_B. The renormalized bulk and boundary cosmological constants μ and x are coupled respectively to the renormalized area A and length ℓ of the triangulated surface, defined as

$$A = \epsilon^2 |\Gamma|, \quad \ell = \epsilon |\partial\Gamma|. \tag{30.4.18}$$

In the continuum limit $\epsilon \to 0$ the universal information is contained in the singular parts of the observables, which scale as fractional powers of ϵ. This is

[4] The general solution of the quadratic equation (30.4.9) has four branch points and two cuts, and describes the situation where a fraction of the eigenvalues are in equilibrium near the local maximum of the cubic potential. The general solution has exponentially small weight compared with (30.4.12) and describes non-perturbative phenomena, which are beyond the scope of this review.

why the continuum limit is also called the scaling limit. In particular, for the partition functions (30.2.1) and (30.2.5) we have

$$\begin{aligned}
\mathcal{F}(\check{\mu}) &= \text{regular part} + \epsilon^{2(2-\gamma_{\rm str})}\, F(\mu), \\
\mathcal{U}(\check{\mu}, z) &= \text{regular part} + \epsilon^{2-\gamma_{\rm str}}\, U(\mu, x), \\
W(\check{\mu}, z) &= \text{regular part} + \epsilon^{1-\gamma_{\rm str}}\, w(\mu, x).
\end{aligned} \tag{30.4.19}$$

The powers of ϵ follow from the general scaling formula scaling (30.2.19) with $\chi = 2$ for the sphere and with $\chi = 1$ for the disk. The partition functions for fixed length and/or area are the inverse Laplace images with respect to x and/or μ. For example

$$w(\mu, x) = -\partial_x U(\mu, x) = \int_0^\infty d\ell\, e^{-\ell x}\, \tilde{w}(\mu, \ell). \tag{30.4.20}$$

Now let us take the continuum limit of the solution of the matrix model. First we have to determine the critical values of the lattice couplings. From the explicit form of the disk partition function, Eqs. (30.4.12)–(30.4.13), it is clear that the latter depends on μ only through the parameter $S = (a+b)/2$. The singularities in 2D gravity are typically of third order and one can show that S is proportional to the second derivative of the free energy of the sphere. This means that the derivative $\partial_{\check{\mu}} S$ diverges at $\check{\mu} = \check{\mu}_c$. Solving the equation $d\,S/d\check{\mu} = 0$ gives

$$\check{\mu}_c = 12\sqrt{3}. \tag{30.4.21}$$

Furthermore, according to (30.4.20), the large ℓ critical behaviour is encoded in the right-most singularity in the x plane. Therefore the critical boundary cosmological constant coincides with the *right* endpoint b of the eigenvalue distribution at $\check{\mu} = \check{\mu}_c$,

$$\check{\mu}_B^c = b(\check{\mu}^c) = (3+\sqrt{3})/6. \tag{30.4.22}$$

The function $J(z)$ is equal to $W(z)$ with the regular part subtracted. In the scaling limit (30.4.17)

$$J(\check{\mu}, z) \sim \epsilon^{3/2}\, w(\mu, x) \tag{30.4.23}$$

$$w(x) = (x+M)^{3/2} - \tfrac{3}{2} M\,(x+M)^{1/2}, \qquad M = \sqrt{2\mu}. \tag{30.4.24}$$

We have dropped the overall numerical factors which do not have universal meaning. With this normalization of w and μ, the derivative $\partial_\mu W$ is given by

$$\partial_\mu w(x) = -\frac{3}{4}\frac{1}{\sqrt{x+M}}. \tag{30.4.25}$$

The scaling exponent 3/2 in (30.4.23) corresponds, according to (30.2.16) and (30.2.19), to $\gamma_{\rm str} = 1 - 3/2 = -1/2$ and $c_{\rm matter} = 0$. Finally, it follows from (30.4.16) that $M \sim \partial_\mu^2 F$ and the free energy on the sphere scales as $F \sim \epsilon^5 \mu^{5/2}$.

This is not the only critical point for the one-matrix model. With a polynomial potential of highest degree $p+2$ one can tune p couplings to obtain a scaling behaviour $w(z) \sim x^{p+\frac{1}{2}}$. On the other hand, the derivative $\partial_\mu w \sim x^{-1/2}$ does not depend on p, which means that $M \sim \partial_\mu^2 F$ and x scale as $\mu^{1/(p+1)}$. These multicritical points were discovered by Kazakov [Kaz89]. The tricritical point was first correctly identified by Staudacher [Sta90] as the hard dimer problem on planar graphs, which is in the universality class of the Yang-Lee CFT with $c_{\rm matter} = -22/5$. The Liouville gravity with such matter CFT is studied in [Zam07]. In general, the p-critical point corresponds to a negative matter central charge given by (30.2.16) with $g = p + \frac{1}{2}$. In these non-unitary models of 2D gravity the constant μ is coupled not to the identity operator, but to the field with the most negative dimension.

30.5 The Ising model

This simplest statistical model is characterized by a fluctuating spin variable $\sigma_r = \pm 1$ associated with the vertices of the planar graph. The nearest neighbour spin-spin interaction is introduced by the energy

$$E[\{\sigma_r\}] = \sum_{<rr'>} K\, \sigma_r \sigma_{r'} + \sum_{r\in\Gamma} H\sigma_r \tag{30.5.1}$$

and the partition function on the graph Γ is given by

$$Z_{\rm Ising}[\Gamma] = \sum_{\{\sigma_r\}} e^{-E[\{\sigma_r\}]}. \tag{30.5.2}$$

The ferromagnetic vacuum is attained at $K \to +\infty$.

The Ising model coupled to 2D gravity discretized on planar graphs can be constructed by assigning a colour to all vertices which can take two values. The corresponding matrix model involves two random matrices, $\mathbf{M}_\sigma$, $\sigma = \pm$, and its partition function is [Kaz86]

$$\mathcal{Z}_{\rm Ising}(N, \beta; K, H) = \int d\mathbf{M}_+ d\mathbf{M}_-\; e^{-\beta {\rm Tr}\, V(\mathbf{M}_+, \mathbf{M}_-)}, \tag{30.5.3}$$

$$V(\mathbf{M}_+, \mathbf{M}_-) = \sum_{\sigma=\pm} \left(-\tfrac{1}{2}\mathbf{M}_\sigma^2 + \tfrac{1}{3} e^{\sigma H} \mathbf{M}_\sigma^3\right) + e^{-2K} \mathbf{M}_+ \mathbf{M}_-. \tag{30.5.4}$$

The model can be solved exactly in the large N limit and the solution exhibits the following critical behaviour. For generic K the sum over planar graphs diverges at some critical value of $\check{\mu} = \beta/M$. Again we can define the scaling limit as in (30.4.17). Along the critical curve $\check{\mu} = \check{\mu}_c(K, H)$ the partition function is dominated by infinite planar graphs and we have the 'pure gravity' critical behaviour $F \sim \mu^{5/2}$. For $H = 0$ and there is a critical value of the Ising coupling $K = K_c$ for which the critical behaviour of the free energy changes to $F \sim \mu^{7/3}$. At this point the correlations of the Ising spins on the infinite planar graph become long range and are described by a matter CFT with $c_{\text{matter}} = 1/2$ ($\gamma_{\text{str}} = -1/3$).

Boulatov and Kazakov [BK87] found that in the vicinity of the critical point $\check{\mu} = \check{\mu}_c$, $K = K_c$, $H = 0$, parameterized by

$$\check{\mu} - \check{\mu}_c \sim \epsilon^2 \mu, \quad K - K_0 \sim -\epsilon^{2/3} t, \quad H \sim \epsilon^{5/3} h, \tag{30.5.5}$$

the susceptibility $u = \partial_\mu^2 F$ satisfies the following simple algebraic equation:

$$\mu = u^3 + \tfrac{3}{2} t\, u^2 + \tfrac{3}{2} \frac{h^2}{(u+t)^2}. \tag{30.5.6}$$

The coefficients in the equation correspond to some normalization of the scaling couplings.

At the critical point $t = h = 0$ the susceptibility is $u = \mu^{1/3}$. The world sheet theory at the critical point is that of Liouville gravity with the Ising CFT (Majorana fermions) as a matter field. The primary fields of this CFT are the energy density ε with dimension $h_\varepsilon = 1/2$ and the spin field σ with dimension $h_\sigma = 1/16$. For the corresponding gravitational dimensions we have $1 - \Delta_\varepsilon = 1/3$ and $1 - \Delta_\sigma = 5/6$, which explains the scaling (30.5.5). The equation of state (30.5.6) describes a perturbation of the Liouville gravity with Ising matter CFT by the action

$$\delta\mathcal{A}(t, h) = t\, \varepsilon\, e^{2\phi/3} + h\, \sigma\, e^{5\phi/3}. \tag{30.5.7}$$

(The normalizations of the couplings are of course different.)

Apart from the Ising critical point there is a second non-trivial critical point at purely imaginary value of the magnetic field: the gravitational Yang-Lee edge singularity [Sta90, CGM90, ZZ06]. In order to localize the YL singularity let us analyse the equation of state (30.5.6). First we observe that assuming that $t > 0$ and rescaling

$$\hat{u} = \frac{u}{t}, \; \hat{h} = \frac{h}{t^{5/2}}, \; \hat{\mu} = \frac{\mu}{t^3} \tag{30.5.8}$$

we can bring (30.5.6) to the scaling form

$$\hat{\mu} = \hat{u}^3 + \tfrac{3}{2}\, \hat{u}^2 + \tfrac{3}{2} \frac{\hat{h}^2}{(\hat{u}+1)^2}. \tag{30.5.9}$$

There is a line of critical points where the susceptibility $\hat{u}$ develops a square root singularity, which is that of pure gravity ($\gamma_{\text{str}} = -1/2$). The equation of the critical line is obtained by differentiating the right-hand side of (30.5.9) in $\hat{u}$:

$$\hat{u}(\hat{u}+1)^4 = \hat{h}^2. \tag{30.5.10}$$

Expanding (30.5.9) near the solution $\hat{u} = \hat{u}_c(\hat{h})$,

$$\hat{\mu} = \hat{\mu}_c + \tfrac{3}{2}(1+5\hat{u}_c)(\hat{u}-\hat{u}_c)^2 + \dots, \tag{30.5.11}$$

we see that a tri-critical point occurs when $\hat{u}_c = -1/5$, which corresponds to purely imaginary magnetic fields $\hat{h}_{\text{LY}} = \pm i\, 4^2/5^{5/2}$ and $\hat{\mu}_{\text{LY}} = -7/50$. This is the Yang-Lee critical point for the gravitational Ising model. Near this point the equation of state has the form

$$\delta\hat{\mu} = \delta\hat{u}^3 \pm \delta\hat{h}\, \delta\hat{u} \tag{30.5.12}$$

where $\delta\hat{\mu} \sim \hat{\mu} - \mu_{\text{LY}}$, $\delta\hat{u} \sim \hat{u} - \hat{u}_{\text{LY}}$ and $\delta\hat{h} \sim i(\hat{h} - \hat{h}_{\text{LY}})$.

The free energy per unit volume on the critical line (30.5.10), where the size of the typical planar graph diverges, is equal to the critical value of the cosmological constant $\hat{\mu}_c(\hat{h})$. This is a meromorphic function of $\hat{h}$ whose Riemann surface consists of five sheets [Zam05]. The high temperature expansion goes in the even powers of $\hat{h} = h/t^{5/2}$ and is convergent for $|\hat{h}| < |\hat{h}_{\text{LY}}|$. On the 'high-temperature' sheet it has two simple branch points on the imaginary axis at $\hat{h} = \pm\hat{h}_{\text{LY}}$. The two YL cuts connect the high-temperature sheet to a pair of other sheets. These two sheets are in turn connected to another pair of sheets by two Langer cuts which go from $\hat{h} = 0$ to $\hat{h} = \pm\infty$ and describe the low temperature ($t < 0$) instability of the Ising model [FZ01, ZZ06].

30.6 The *O(n)* model ($-2 \leq n \leq 2$)

The local fluctuating variable in the $O(n)$ loop model [Nie82] is an n-component vector $\vec{S}(r)$ with unit norm, associated with the vertices $r \in \Gamma$. The partition function of the $O(n)$ model on the graph Γ can be represented by a sum over self-avoiding, mutually avoiding loops as the one shown in Fig. 30.2 (left), each counted with a factor of n:

$$Z_{O(n)}(T;\Gamma) = \sum_{\text{loops on } \Gamma} T^{-\text{length}}\, n^{\#\text{loops}}. \tag{30.6.1}$$

The temperature coupling T controls the length of the loops. The advantage of the loop gas representation (30.6.1) is that it makes sense also for non-integer n. The model has a continuous transition at $-2 \leq n \leq 2$. A standard parameterization of n in this interval is

$$n = 2\cos(\pi\theta), \qquad 0 < \theta < 1. \tag{30.6.2}$$

The $O(n)$ matrix model [Kos89a] generates planar graphs covered by loops in the same way as the one-matrix model generates empty planar graphs. The fluctuating variables are the Hermitian matrix $\mathbf{M}$ and an $O(n)$ vector $\vec{\mathbf{Y}}$, whose

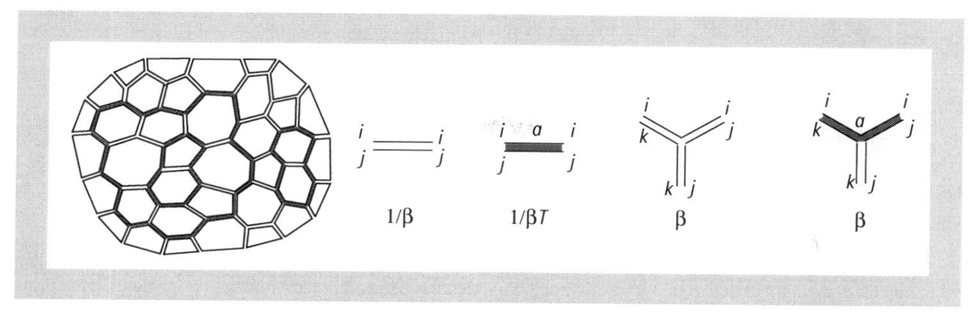

Fig. 30.2 Left: A loop configuration on a disk with ordinary boundary condition. Right: Feynman rules for the $O(n)$ matrix model (figures from [BHK09]).

components $\mathbf{Y}_1, \ldots, \mathbf{Y}_n$ are Hermitian matrices. All matrix variables are of size $N \times N$. The partition function is given by the integral

$$\mathcal{Z}_N^{O(n)}(T) \sim \int d\mathbf{M}\, d^n\mathbf{Y}\, e^{-\beta\, \mathrm{tr}\left(\frac{1}{2}\mathbf{M}^2 + \frac{T}{2}\vec{\mathbf{Y}}^2 - \frac{1}{3}\mathbf{M}^3 - \mathbf{M}\vec{\mathbf{Y}}^2\right)}. \tag{30.6.3}$$

This integral can be considered as the partition function of a zero-dimensional QFT with Feynman rules given in Fig. 30.2 (right).

The basic observable in the matrix model is the resolvent (30.4.6), evaluated in the ensemble (30.6.3), which is the derivative of the disk partition function with *ordinary boundary condition* for the $O(n)$ spins. In terms of loop gas, the ordinary boundary condition means that the loops in the bulk avoid the boundary as they avoid the other loops and themselves.

The matrix integral measure becomes singular at $\mathbf{M} = T/2$. We perform a linear change of the variables $\mathbf{M} = T\left(\frac{1}{2} + \mathbf{X}\right)$ which sends this singular point to $\mathbf{X} = 0$. After suitable rescaling of $\vec{\mathbf{Y}}$ and β, the matrix model partition function takes the canonical form

$$\mathcal{Z}_N^{O(n)}(T) \sim \int d\mathbf{X}\, d^n\mathbf{Y} e^{\beta \mathrm{tr}[-V(\mathbf{X}) + \mathbf{X}\vec{\mathbf{Y}}^2]}, \tag{30.6.4}$$

where $V(x)$ is a cubic potential $V(x) = -\frac{1}{3}T(x+\frac{1}{2})^3 + \frac{1}{2}(x+\frac{1}{2})^2$. Accordingly, we redefine the spectral parameter by

$$z = T\left(x + \tfrac{1}{2}\right). \tag{30.6.5}$$

Now the one-point function with ordinary boundary condition is

$$W(x) = \frac{1}{\beta}\left\langle \mathrm{tr}\mathbf{W}\right\rangle, \qquad \mathbf{W} \stackrel{\mathrm{def}}{=} \frac{1}{x - \mathbf{X}}. \tag{30.6.6}$$

After integration with respect to the $\mathbf{Y}$-matrices, the partition function takes the form of a Coulomb gas

$$\mathcal{Z}_N^{O(n)}(T) \sim \int \prod_{j=1}^{N} dx_i \, e^{-V(x_i)} \frac{\prod_{i<j}(x_i - x_j)^2}{\prod_{i,j}(x_i + x_j)^{n/2}}. \tag{30.6.7}$$

30.6.1 Loop equation

As in the case of the one-matrix model, the resolvent can be determined by Ward identities which follow from the translational invariance of the measure in (30.6.3). Introduce the singular part $J(x)$ of the resolvent by

$$W(x) = W_{\text{reg}}(x) + J(x), \quad W_{\text{reg}}(x) = \frac{2V'(x) - nV'(-x)}{4 - n^2}. \tag{30.6.8}$$

The function $J(x)$ satisfies the quadratic functional identity [Kos92b]

$$J^2(x) + J^2(-x) + n\,J(x)J(-x) = A + Bx^2 + Cx^4. \tag{30.6.9}$$

The coefficients A, B, C as functions of T, $\check{\mu}$ and $W_1 = \langle \mathrm{tr} X \rangle$ can be evaluated by substituting the large-x asymptotics

$$J(x) = -W_{\text{reg}}(x) + \frac{\check{\mu}^{-2}}{x} + \frac{\langle \mathrm{tr} X \rangle}{\beta x^2} + O(x^{-3}) \tag{30.6.10}$$

in (30.6.9). The solution of the loop equation (30.6.9) with the asymptotics (30.6.10) is given by a meromorphic function with a single cut $[a, b]$ on the first sheet, with $a < b < 0$. This equation can be solved by an elliptic parameterization and the solution is expressed in terms of Jacobi theta functions [EK95] [EK96] (see also [Kos06]).

The functional equation (30.6.9) becomes singular when $A = B = 0$. The condition $A = 0$ determines the critical value $\check{\mu}_c$ of the cosmological constant. The condition $B = 0$ determines the critical value T_c of the temperature coupling, for which the length of the loops diverges:

$$T_c = 1 + \sqrt{\tfrac{2-n}{6+n}}. \tag{30.6.11}$$

30.6.2 The scaling limit

We introduce a small cutoff ϵ and and define the renormalized coupling constants as

$$\check{\mu} - \check{\mu}_c \sim \epsilon^2 \mu, \quad T - T_c \sim \epsilon^{2\theta} t. \tag{30.6.12}$$

We also rescale the boundary cosmological constant and the resolvent as

$$x \to \epsilon x, \quad J(x) \to \epsilon^{1+\theta} w(x). \tag{30.6.13}$$

The scaling resolvent $w(x)$ has a cut $[-\infty, -M]$, where M is a function of μ and t. We can resolve the branch point at $x = -M$ by the hyperbolic map

$$x = M \cosh \tau. \tag{30.6.14}$$

Then (30.6.9) becomes a quadratic functional equation for the entire function $w(\tau) \equiv w[x(\tau)]$:

$$w^2(\tau + i\pi) + w^2(\tau) + n\, w(\tau + i\pi) w(\tau) = A_1 + B_1 t\, M^2 \cosh^2 \tau. \tag{30.6.15}$$

Here we have used the fact that near the critical temperature $B \sim T - T_c$ and the term Cx^4 vanishes. The unique solution, for some normalization of t, is

$$w(\tau) = M^{1+\theta} \cosh(1+\theta)\tau + t\, M^{1-\theta} \cosh(1-\theta)\tau. \tag{30.6.16}$$

The function $M = M(\mu, t)$ can be evaluated using the fact that the derivative $\partial_\mu w(x)$ depends on μ and t only through M. This is so if M satisfies the transcendental equation

$$\mu = (1+\theta) M^2 - t\, M^{2-2\theta}. \tag{30.6.17}$$

To summarize, the derivatives of the disk partition function in the scaling limit, $U(x, \mu, t)$, are given in the following parametric form:

$$\begin{aligned} -\partial_x U|_\mu &= M^{1+\theta} \cosh((1+\theta\tau) + t\, M^{1-\theta} \cosh(1-\theta)\tau, \\ -\partial_\mu U|_x &\sim M^\theta \cosh \theta\tau, \\ x &= M \cosh \tau, \end{aligned} \tag{30.6.18}$$

with the function $M(\mu, t)$ determined from the transcendental equation (30.6.17).

The function $M(t, \mu)$ plays an important role in the solution. Its physical meaning can be revealed by taking the limit $x \to \infty$ of the bulk one-point function $-\partial_\mu U(x)$. Since x is coupled to the length of the boundary, in the limit of large x the boundary shrinks and the result is the partition function of the $O(n)$ field on a sphere with two punctures, the susceptibility $u(\mu, t)$. Expanding at $x \to \infty$ we find

$$-\partial_\mu U \sim x^\theta - M^{2\theta} x^{-\theta} + \text{ lower powers of } x \tag{30.6.19}$$

(the numerical coefficients are omitted). We conclude that the string susceptibility is given, up to a normalization, by

$$u = M^{2\theta}. \tag{30.6.20}$$

The normalization of u can be absorbed in the definition of the string coupling constant $g_s \sim 1/\beta$. Thus the transcendental equation (30.6.17) for M gives the equation of state of the loop gas on the sphere,

$$(1+\theta)\, u^{\frac{1}{\theta}} - t\, u^{\frac{1-\theta}{\theta}} = \mu. \tag{30.6.21}$$

The equation of state (30.6.21) has three singular points at which the three-point function of the identity operator $\partial_\mu u$ diverges. The three points correspond to the critical phases of the loop gas on the sphere.

(1) At the critical point $t = 0$ the susceptibility scales as $u \sim \mu^\theta$. This is the dilute phase of the loop gas, in which the loops are critical, but occupy an insignificant part of the lattice volume. By the KPZ relation (30.2.16), the $O(n)$ field is described by a CFT with central charge (30.2.8) with $g = 1 + \theta$.

(2) The dense phase is reached when $t/x^\theta \to -\infty$. In the dense phase the loops remain critical but occupy almost all the lattice and the susceptibility has different scaling, $u \sim \mu^{\frac{1-\theta}{\theta}}$. The $O(n)$ field in the dense phase is described by a CFT with a lower central charge (30.2.8) with $g = 1 - \theta$. Considered on the interval $-\infty < t < 0$, the equation of state (30.6.21) describes the massless thermal flow relating the dilute and the dense phases [Kos06].

(3) At the third critical point $\partial_\mu M$ becomes singular but M itself remains finite. Around this critical point $\mu - \mu_c \sim (M - M_c)^2 + \cdots$, hence the scaling of the susceptibility is that of pure gravity, $u \sim (\mu - \mu_c)^{1/2}$.

30.6.3 Other boundary conditions

As well as the ordinary boundary condition, there is a continuum of integrable boundary conditions, which are generated by the anisotropic boundary term [DJS09]. The loop gas expansion on the disk with such a boundary condition can be formulated in terms of loops of two different colours with fugacities $n_{(1)}$ and $n_{(2)}$, such that $n_{(1)} + n_{(2)} = n$. The bulk is colour blind, while the boundary interacts differently with the loops of different colour. A loop of colour (α) acquires an additional weight factor $\lambda_{(\alpha)}$ each time it touches the boundary. The operator in the matrix model creating a boundary segment with anisotropic boundary condition is defined as follows. Decompose the vector $\vec{\mathbf{Y}}$ into a sum of an $n_{(1)}$-component vector $\vec{\mathbf{Y}}_{(1)}$ and an $n_{(2)}$-component vector $\vec{\mathbf{Y}}_{(2)}$. Then the disk one-point function with anisotropic boundary conditions is evaluated by

$$H(x, \lambda_{(1)}, \lambda_{(2)}) = \frac{1}{\beta}\langle \operatorname{tr}\mathbf{H}\rangle, \quad \mathbf{H} \stackrel{\text{def}}{=} \frac{1}{x - \mathbf{X} - \lambda_{(1)}\vec{\mathbf{Y}}^2_{(1)} - \lambda_{(2)}\vec{\mathbf{Y}}^2_{(2)}}. \tag{30.6.22}$$

Non-trivial boundary critical behaviour is achieved by tuning the two matter boundary couplings $\lambda_{(1)}$ and $\lambda_{(2)}$. The boundary phase diagram is qualitatively the same as the one on the flat lattice [BHK09]. The boundary two-point functions $D_0(x, y) = \frac{1}{\beta}\Big\langle \operatorname{tr}\big[\mathbf{W}(x)\mathbf{H}(y)\big]\Big\rangle$ can be found as solutions of functional equations.

The analysis of these equations away from the criticality allows us to draw the qualitative picture of bulk and boundary flows in the $O(n)$ model [BHK09].

30.7 The six-vertex model

The matrix model for Baxter's six-vertex model on planar graphs is a deformation of the $O(2)$ model. Its partition function is given by

$$\mathcal{Z}_N^{O(n)}(T) \sim \int d\mathbf{X}\, d\mathbf{Y} d\mathbf{Y}^\dagger\, e^{\beta \mathrm{tr}[-V(\mathbf{X})+e^{i\nu}\mathbf{X}\vec{\mathbf{Y}}\mathbf{Y}^\dagger+e^{-i\nu}\mathbf{X}\vec{\mathbf{Y}}^\dagger\mathbf{Y}]}. \tag{30.7.1}$$

The solution in the large N limit is found in [Kos00]. In the scaling limit the six-vertex matrix model describes a $c = 1$ string theory compactified at length ν.

30.8 The q-state Potts model $(0 < q < 4)$

The local fluctuating variable in the q-state Potts model takes q discrete values and the Hamiltonian has S_q symmetry. The partition function of the Potts model is a sum over all clusters with fugacity q and weight $1/T$ for each link occupied by the cluster. The corresponding matrix model is [Kaz88, Kos89c]

$$\mathcal{Z}_N(T) \sim \int d\mathbf{M}\, d^q\mathbf{Y}\, e^{-\beta\, \mathrm{tr}\left(\frac{1}{2}\mathbf{M}^2+\sum_{a=1}^{q}(\frac{T}{2}\mathbf{Y}_a^2-\frac{1}{3}(\mathbf{M}+\mathbf{Y}_a)^3\right)}. \tag{30.8.1}$$

The solution of the q-state Potts matrix model in the large N limit boils down to a problem of matrix mean field. After a shift $\mathbf{Y}_a \to \mathbf{Y}_a - \mathbf{M}$ the matrix integral can be written as

$$\mathcal{Z}_N^{q\ \mathrm{Potts}}(T) \sim \int d\mathbf{M}\, e^{-\beta\, \frac{T+q}{2T}\mathrm{tr}\mathbf{M}^2-\beta q\, W[\mathbf{M}])}, \tag{30.8.2}$$

$$e^{\beta W[\mathbf{M}]} = \int d\mathbf{Y}\, e^{\beta \mathrm{tr}(-\frac{1}{2T}\mathbf{Y}^2+\frac{1}{3}\mathbf{Y}^3+\frac{1}{T}\mathbf{M}\mathbf{Y})}. \tag{30.8.3}$$

The effective action W defined by the last integral depends only on the eigenvalues of the matrix $\mathbf{M}$. It was evaluated in the large N limit by methods used previously for the mean field problem in the large N gauge theory [BG80]. In the scaling limit one obtains, after a redefinition of the spectral parameter, the same functional equation as that for the $O(n)$ model with $n = q - 2$ [Kos89c]. Analysis of the solution is however more delicate than in the case of the $O(n)$ matrix model. The complete solution of the q-state Potts matrix model was found in [Dau94, ZJ01, EB99].

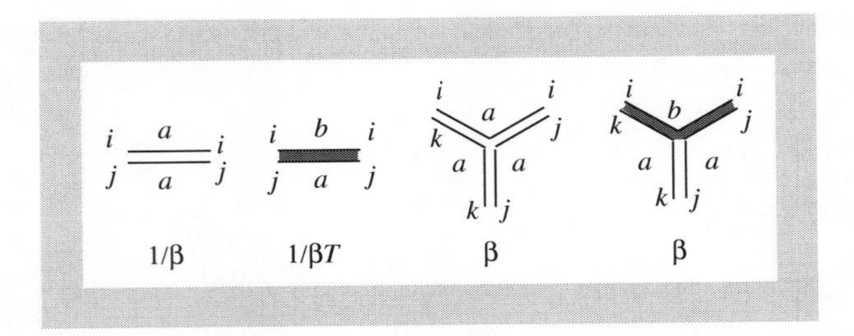

Fig. 30.3 Feynman rules for the matrix model with target space $\mathcal{G}$.

30.9 Solid-on-solid and ADE matrix models

The rational theories of 2D QG are discretized by ensembles of planar graphs embedded in ADE Dynkin diagrams [Kos90, Kos89b, Kos92b]. The corresponding matrix models are constructed in [Kos92a].

More generally, for a discrete target space representing a graph $\mathcal{G}$ one associates a Hermitian matrix $\mathbf{X}_a$ with each node $a \in \mathcal{G}$ and a complex matrix $\mathbf{Y}_{<ab>} = \mathbf{Y}^{\dagger}_{<ba>}$ with each link $<ab>$. The matrix model that generates planar graphs embedded in the graph $\mathcal{G}$ is defined by the partition function

$$\mathcal{Z}_N^{\mathcal{G}}(\beta, T) = \int \prod_{a \in \mathcal{G}} d\mathbf{X}_a \, e^{-\beta \mathrm{tr}\, V(\mathbf{X}_a)} \prod_{<ab>} e^{\mathrm{tr}(\mathbf{X}_a \mathbf{Y}_{<ab>} \mathbf{Y}_{<ba>} + \mathbf{Y}_{<ab>} \mathbf{X}_b \mathbf{Y}_{<ba>})}. \tag{30.9.1}$$

For trivalent graphs the potential V is the same as that of the $O(n)$ model. The Feynman rules are shown in Fig. 30.3. The propagators of the **Y**-matrices form a set of self-avoiding and mutually avoiding loops on the planar graph. Each such line is associated with a link $< ab >$ of the target space and separates two domains of different heights, a and b. The matrix model has an interesting continuum limit if the graph $\mathcal{G}$ is an ADE or $\hat{A}\hat{D}\hat{E}$ Dynkin diagram. The SOS model, which is a discretization of the Gaussian field with a charge at infinity [Nie82], corresponds to an infinite target space $\mathbb{Z}$ [Kos92b]. Matrix models with semi-infinite target spaces A_∞ and D_∞ were studied in [KP07].

References

[ADF85] J. Ambjorn, B. Durhuus, and J. Frohlich. Diseases of Triangulated Random Surface Models, and Possible Cures. *Nucl. Phys.*, B257:433, 1985.

[BG80] E. Brézin and D. J. Gross. The External Field Problem in the Large N Limit of QCD. *Phys. Lett.*, B97:120, 1980.

[BHK09] J.-E. Bourgine, K. Hosomichi, and I. Kostov. Boundary transitions of the $O(n)$ model on a dynamical lattice. 2009, hep-th/0910.1581.

[BIPZ78] E. Brézin, C. Itzykson, G. Parisi, and J. B. Zuber. Planar Diagrams. *Commun. Math. Phys.*, 59:35, 1978.

[BK87] D. Boulatov and V. Kazakov. The Ising Model on Random Planar Lattice: The Structure of Phase Transition and the Exact Critical Exponents. *Phys. Lett.*, 186B:379, 1987.

[BKKM86] D. Boulatov, V. Kazakov, I. Kostov, and A. Migdal. Analytical and Numerical Study of the Model of Dynamically Triangulated Random Surfaces. *Nucl. Phys.*, B275:641, 1986.

[CGM90] Č. Crnković, P. H. Ginsparg, and G. W. Moore. The Ising Model, the Yang-Lee Edge Singularity, and 2D Quantum Gravity. *Phys. Lett.*, B237:196, 1990.

[Dau94] J.-M. Daul. Q states Potts model on a random planar lattice. 1994, hep-th/9502014.

[Dav85a] F. David. A Model of Random Surfaces with Nontrivial Critical Behavior. *Nucl. Phys.*, B257:543, 1985.

[Dav85b] F. David. Planar Diagrams, Two-Dimensional Lattice Gravity and Surface Models. *Nucl. Phys.*, B257:45, 1985.

[Dav88] F. David. Conformal Field Theories Coupled to 2D Gravity in the Conformal Gauge. *Mod. Phys. Lett.*, A3:1651, 1988.

[DFGZJ95] P. Di Francesco, P. Ginsparg, and J. Zinn-Justin. 2D gravity and random matrices. *Physics Reports*, 254:1–133, 1995.

[DJS09] J. Dubail, J. Lykke Jacobsen, and H. Saleur. Exact solution of the anisotropic special transition in the o(n) model in two dimensions. *Physical Review Letters*, 103(14):145701, 2009.

[DK88] B. Duplantier and I. Kostov. Conformal spectra of polymers on a random surface. *Phys. Rev. Lett.*, 61:1433, 1988.

[DK89] J. Distler and H. Kawai. Conformal field theory and 2D quantum gravity. *Nuclear Physics B*, 321:509–527, 1989.

[DK90] B. Duplantier and Ivan K. Kostov. Geometrical critical phenomena on a random surface of arbitrary genus. *Nucl. Phys.*, B340:491–541, 1990.

[DKK93] J.M. Daul, V. Kazakov, and I. Kostov. Rational theories of 2D gravity from the two-matrix model. *Nucl.Phys.*, B409:311–338, 1993, hep-th/9303093.

[EB99] B. Eynard and G. Bonnet. The Potts-q random matrix model : loop equations, critical exponents, and rational case. *Phys. Lett.*, B463:273–279, 1999, hep-th/9906130.

[EK95] B. Eynard and C. Kristjansen. Exact Solution of the O(n) Model on a Random Lattice. *Nucl. Phys.*, B455:577–618, 1995, hep-th/9506193.

[EK96] B. Eynard and C. Kristjansen. More on the exact solution of the O(n) model on a random lattice and an investigation of the case $|n| > 2$. *Nucl. Phys.*, B466:463–487, 1996, hep-th/9512052.

[FZ01] P. Fonseca and A. Zamolodchikov. Ising field theory in a magnetic field: Analytic properties of the free energy. 2001, hep-th/0112167.

[GM93] P. H. 'Ginsparg and G. W. Moore. Lectures on 2-D gravity and 2-D string theory. 1993, hep-th/9304011.

[Kaz85] V. Kazakov. Bilocal Regularization of Models of Random Surfaces. *Phys. Lett.*, B150:282–284, 1985.

[Kaz86] V. Kazakov. Ising model on a dynamical planar random lattice: Exact solution. *Phys. Lett.*, A119:140–144, 1986.

[Kaz88] V. Kazakov. Exactly solvable Potts models, bond and tree percolation on dynamical (random) planar surface. *Nucl. Phys. B (Proc. Suppl.)*, 4:93–97, 1988.

[Kaz89] V. Kazakov. The Appearance of Matter Fields from Quantum Fluctuations of 2D Gravity. *Mod. Phys. Lett.*, A4:2125, 1989.

[KKM85] V. Kazakov, I. Kostov, and A. Migdal. Critical Properties of Randomly Triangulated Planar Random Surfaces. *Phys. Lett.*, B157:295–300, 1985.

[Kle95] I. R. Klebanov. Touching random surfaces and Liouville gravity. *Phys. Rev.*, D51:1836–1841, 1995, hep-th/9407167.

[KM87] I. Kostov and M. L. Mehta. Random surfaces of arbitrary genus: exact results for D=0 and −2 dimansions. *Phys. Lett.*, B189:118–124, 1987.

[Kos89a] I. Kostov. $O(n)$ vector model on a planar random surface: spectrum of anomalous dimensions. *Mod. Phys. Lett.*, A4:217, 1989.

[Kos89b] I. Kostov. The ADE face models on a fluctuating planar lattice. *Nucl. Phys.*, B326:583, 1989.

[Kos89c] I. K. Kostov. Random surfaces, solvable lattice models and discrete quantum gravity in two dimensions. *Nuclear Physics B (Proc. Suppl.)*, 10A:295–322, 1989. Lecture given at GIFT Int. Seminar on Nonperturbative Aspects of the Standard Model, Jaca, Spain, Jun 6–11, 1988.

[Kos90] I. Kostov. Strings embedded in Dynkin diagrams. 1990. Lecture given at Cargese Mtg. on Random Surfaces, Quantum Gravity and Strings, Cargese, France, May 27 - Jun 2, 1990.

[Kos92a] I. Kostov. Gauge invariant matrix model for the A-D-E closed strings. *Phys. Lett.*, B297:74–81, 1992, hep-th/9208053.

[Kos92b] I. Kostov. Strings with discrete target space. *Nucl. Phys.*, B376:539–598, 1992, hep-th/9112059.

[Kos00] I. Kostov. Exact solution of the six-vertex model on a random lattice. *Nucl. Phys.*, B575:513–534, 2000, hep-th/9911023.

[Kos06] I. Kostov. Thermal flow in the gravitational $O(n)$ model. 2006, hep-th/0602075.

[KP07] I. Kostov and V.B. Petkova. Non-rational 2D quantum gravity ii. target space cft. *Nucl.Phys.B*, 769:175–216, 2007, hep-th/0609020.

[KPZ88] V. G. Knizhnik, A. M. Polyakov, and A. B. Zamolodchikov. Fractal structure of 2D-quantum gravity. *Mod. Phys. Lett.*, A3:819, 1988.

[KS92] I. Kostov and M. Staudacher. Multicritical phases of the O(n) model on a random lattice. *Nucl. Phys.*, B384:459–483, 1992, hep-th/9203030.

[Na04] Y. Nakayama. Liouville field fheory: A decade after the revolution, *Int. J. Theor. Phys.*, A19, p. 2771–2930, 2004.

[Nie82] B. Nienhuis. Exact critical point and critical exponents of $o(n)$ models in two dimensions. *Phys. Rev. Lett.*, 49(15):1062–1065, Oct 1982.

[Pol81] A. M. Polyakov. Quantum geometry of bosonic strings. *Phys. Lett.*, B103:207–210, 1981.

[Sta90] M. Staudacher. The Yang-Lee edge singularity on a dynamical planar random surface. *Nucl. Phys.*, B336:349, 1990.

[tH74] G. t Hooft. A planar diagram theory for strong interactions. *Nuclear Physics B*, 72:461–473, 1974.

[Zam05] Al. Zamolodchikov. Thermodynamics of the gravitational ising model. A talk given at ENS, Paris (unpublished), 2005.

[Zam07] Al. Zamolodchikov. Gravitational Yang-Lee model: Four point function. *Theor. Math. Phys.*, 151:439–458, 2007, hep-th/0604158.

[ZJ01] P. Zinn-Justin. The dilute potts model on random surfaces. *Journal of Statistical Physics*, 98:245, 2001.

[ZZ06] A. Zamolodchikov and Al. Zamolodchikov. Decay of metastable vacuum in Liouville gravity. 2006, hep-th/0608196.

·31·

String theory

Marcos Mariño

Abstract

We give an overview of the relations between matrix models and string theory, focusing on topological string theory and the Dijkgraaf–Vafa correspondence. We discuss applications of this correspondence and its generalizations to supersymmetric gauge theory, enumerative geometry, and mirror symmetry. We also present a brief overview of matrix quantum mechanical models in superstring theory.

31.1 Introduction: strings and matrices

String theories are defined, in perturbation theory, by a conformal field theory (CFT) on a general Riemann surface Σ_g which is then coupled to two-dimensional gravity. If we write the CFT action as

$$S[\phi, h], \tag{31.1.1}$$

where $h_{\alpha\beta}$ is the two-dimensional metric on Σ_g and ϕ are the "matter" fields, then the basic object to compute in a string theory is the total free energy

$$F = \sum_{g=0}^{\infty} g_s^{2g-2} F_g, \qquad F_g = \int \mathcal{D}h\,\mathcal{D}\phi\, \mathrm{e}^{-S[\phi,h]} \tag{31.1.2}$$

where g_s is the string coupling constant, and the path integral in F_g is over field configurations on the Riemann surface Σ_g. We can also perturb the CFT with various operators $\{\mathcal{O}_n\}$ leading to a general action which we write

$$S[\phi, h, t] = S[\phi, h] + \sum_n t_n \mathcal{O}_n. \tag{31.1.3}$$

In this case, the free energies at genus g will depend as well on the couplings t_n, and we write $F_g(t)$ to indicate this explicit dependence.

Computation of the free energies in (31.1.2) is a phenomenal problem, and many approaches have been developed in order to solve it. In the continuum approach, one fixes diffeomorphism invariance and finds that there is a critical central charge $c = 26$ in the CFT for which the 2D metric *decouples* and one is left with an integration over a *finite* set of $3g - 3$ coordinates τ parameterizing the Deligne-Mumford moduli space $\mathcal{M}_g$ of Riemann surfaces:

$$F_g(t) \sim \int_{\mathcal{M}_g} \mathrm{d}\tau \int \mathcal{D}\phi\, \mathrm{e}^{-S[\phi,\tau,t]} \tag{31.1.4}$$

This is the so-called *critical string theory*. For $c \neq 26$ (noncritical string theories), the metric has, on top of the finite set of moduli τ, a dynamical degree of freedom – the Liouville field. Critical strings are manageable but still very hard to solve, since integrating over the moduli space remains a difficult problem. Noncritical strings for generic values of c are so far intractable.

Noncritical strings for $c < 1$ are however special, since one can not only solve for the Liouville dynamics, but also perform the resulting integrals over the moduli and obtain explicit results for $F_g(t)$ at all genera (see [DiF+95] for a review, as well as Chapter 30 of this handbook). This is achieved by using matrix modes to *discretize* the world sheet of the string. Non-trivial matter content with $c \leq 1$ can be implemented with a wise use of the matrix interactions. For example, the Hermitian one-matrix model

$$Z = \int \mathrm{d}M\, \mathrm{e}^{-\mathrm{tr}\ V(M)} \tag{31.1.5}$$

where $V(M)$ is a polynomial potential, makes it possible to implement the $(2, 2m-1)$ minimal models coupled to gravity by tuning the parameters of the potential. These are the famous Kazakov multicritical points. In order to recover string theory properly speaking, one has to take a *double-scaling limit*. The correlation functions and free energies of the matrix model become, in this limit, correlation functions and free energies of the string theory, and the deformations of the action by operators and couplings can be implemented by scaling appropriately the parameters of the potential. This was, historically, the first example of a closed relation between matrix models in the $1/N$ expansion and string theories, and continues to play an important role.

One of the results of the analysis of the $1/N$ expansion of matrix models is that the genus g free energies $F_g^{\mathrm{MM}}(t)$ can be calculated by using only data from the *spectral curve* of the matrix model. In [EO07], this result was formulated in a very elegant way by associating an infinite number of *symplectic invariants* $F_g^{\Sigma}(t)$ to any algebraic curve Σ. These invariants are defined recursively as residues of meromorphic forms on the curve Σ. When Σ is the spectral curve of a matrix model, the symplectic invariants F_g^{Σ} are the genus g free energies obtained in the large N expansion. The formalism also provides recursive relations for the computation of the $1/N$ expansion of correlation functions (see Chapters 16 and 29 of this handbook for a review of these results).

The formalism of symplectic invariants makes it possible to reformulate the relationship between $c < 1$ non-critical strings and matrix models, as follows. One can take the double-scaling limit of the matrix model at the level of the spectral curve, obtaining in this way a doubly scaled spectral curve. For example, for the $c = 0$ string, the relevant spectral curve is described by the algebraic equation

$$y^2 = (2x - \sqrt{t})^2(x + \sqrt{t}) \tag{31.1.6}$$

where t is the bulk cosmological constant. It can be shown that the formalism of [EO07] commutes with taking this limit, therefore the symplectic invariants of the double-scaled spectral curve give directly the string theory amplitudes $F_g(t) = F_g^{\Sigma}(t)$.

It turns our that the correspondence between matrix models and string theory which makes it possible to solve non-critical strings is also valid for an important type of string theory models, namely *topological strings*. Topological strings can be regarded as a generalization of non-critical strings, and they describe a topological sector of physical superstring theories. In terms of mathematical complexity, they provide an intermediate class of examples in between non-critical strings with $c < 1$ and full superstring theories. Topological strings also have two important applications: in physics, they make it possible to compute holomorphic quantities in a wide class of four-dimensional, supersymmetric gauge theories. They also provide a framework to formulate Gromov-Witten invariants and mirror symmetry, and therefore they have had an enormous impact in algebraic geometry.

The correspondence between topological string theory and matrix models was first unveiled by Dijgkraaf and Vafa in [DV02]. It is not valid for all types of topological strings, since it only applies to a special type of target spaces, namely local (i.e. non-compact) Calabi-Yau manifolds. Since topological strings on these non-compact spaces are related to a wide class of supersymmetric gauge theories, this correspondence makes it possible to use matrix model techniques to obtain exact solutions in four-dimensional quantum field theory. It also leads to a remarkable connection between matrix models, mirror symmetry, and enumerative geometry on non-compact Calabi-Yau manifolds. In this chapter we will review these developments.

The organization of this chapter is as follows. First, we will review some basic aspects of topological strings on Calabi-Yau manifolds. Then, we will state the Dijkgraaf–Vafa correspondence, we will sketch its connection to string dualities, and we will briefly explain how it can be used to compute superpotentials in certain supersymmetric gauge theories. We also explain how the correspondence extends to toric manifolds and leads to a matrix model approach to enumerative geometry, closely related to mirror symmetry. Finally, we mention applications of matrix quantum-mechanical models in superstring theory.

31.2 A short survey of topological strings

Topological strings are, as all string theories, CFTs coupled to 2D gravity. The underlying CFTs are *2D non-linear sigma models* with a *target space* X. This means that the basic fields are maps

$$x : \Sigma_g \to X. \tag{31.2.1}$$

It turns out that, in order to construct topological strings, the target must have a very precise structure – it must be a *Calabi-Yau manifold*. This will guarantee among other things the conformal invariance of the model. We will then start with a very brief review of these manifolds. See [Ho+03, CoK99] for more information.

31.2.1 Calabi-Yau manifolds

Mathematically, a Calabi-Yau (CY) manifold X is a complex manifold which admits a Kähler metric with vanishing Ricci curvature. The condition of vanishing curvature is called the *Calabi-Yau condition*. Let us spell out in detail the ingredients in this definition. Since X is complex, it has complex coordinates (in local patches) that we will denote by

$$x^I, \quad x^{\overline{I}}, \qquad I = 1, \cdots, d, \tag{31.2.2}$$

where d is the complex dimension of X. X is also endowed with a Riemannian metric which is *Hermitian*, i.e. it only mixes holomorphic with antiholomorphic coordinates, and in a local patch it has the index structure $G_{I\overline{J}}$. We also need X to be *Kähler*. This means that the *Kähler form*

$$\omega = \mathrm{i} G_{I\overline{J}} \mathrm{d}x^I \wedge \mathrm{d}x^{\overline{J}} \tag{31.2.3}$$

which is a real two-form, is *closed*:

$$\mathrm{d}\omega = 0 \tag{31.2.4}$$

It is easy to check that for a Kähler manifold the Christoffel symbols do not have mixed indices, i.e. their only non-vanishing components are

$$\Gamma^I_{JK}, \qquad \Gamma^{\overline{I}}_{\overline{J}\overline{K}}. \tag{31.2.5}$$

In addition, the Calabi-Yau condition requires the metric $G_{I\overline{J}}$ to be Ricci-flat:

$$R_{IJ} = 0. \tag{31.2.6}$$

It was conjectured by Calabi and then proved by Yau that, for compact Kähler manifolds, Ricci flatness is equivalent to a topological condition, namely that the first Chern class of the manifold vanishes

$$c_1(X) = 0. \tag{31.2.7}$$

The CY manifolds appearing in string theory usually have complex dimension $d = 3$, and they are called CY threefolds. Using Hodge theory, Poincaré duality, and the CY condition, one can see that the Hodge diamond of a simply connected CY threefold X, which encodes the Hodge numbers $h^{p,q}(X)$, $p, q = 1, 2, 3$, has the structure

$$\begin{array}{ccccccc} & & & 1 & & & \\ & & 0 & & 0 & & \\ & 0 & & h^{1,1}(X) & & 0 & \\ 1 & & h^{1,2}(X) & & h^{1,2}(X) & & 1 \\ & 0 & & h^{1,1}(X) & & 0 & \\ & & 0 & & 0 & & \\ & & & 1 & & & \end{array} \tag{31.2.8}$$

and therefore it only depends on two integers $h^{1,1}(X)$, $h^{1,2}(X)$ (we have assumed that X is simply connected). It also follows that the Euler characteristic of X is

$$\chi(X) = 2(h^{1,1}(X) - h^{1,2}(X)). \tag{31.2.9}$$

One of the most important properties of Calabi-Yau manifolds (which can actually be taken as their defining feature) is that they have a holomorphic, non-vanishing section Ω of the canonical bundle $K_X = \Omega^{d,0}(X)$. This form is unique up to multiplication by a non-zero number, and in local coordinates it can be written as

$$\Omega = \Omega_{I_1 \cdots I_d} \mathrm{d}x^{I_1} \wedge \cdots \wedge \mathrm{d}x^{I_d}. \tag{31.2.10}$$

Since the section is nowhere vanishing, the canonical line bundle is trivial, and this is equivalent to (31.2.7).

Calabi-Yau manifolds typically come in families, in the sense that once a Calabi-Yau manifold has been obtained, one can deform it by changing its parameters without violating the Calabi-Yau condition. There are two types of parameters in a Calabi-Yau family: the *Kähler parameters* and the *complex deformation parameters*. The first ones specify sizes (i.e. areas of embedded holomorphic curves) while the second ones specify the complex structure, i.e. the choice of holomorphic/anti-holomorphic splitting. It turns out that the number of Kähler parameters (which are real) is $h^{1,1}(X)$, while the number of complex deformation parameters (which are complex) is $h^{1,2}(X)$.

31.2.2 Topological sigma models

There are two different sigma models that can be used to construct topological strings, and they are referred to as type A and type B topological sigma models [LL92, W91]. Let us first explain what is common to both of them and how they couple to gravity.

First of all, both models are non-linear sigma models, hence they are based on a scalar, commuting field (31.2.1). Second, they are *topological field theories of the cohomological type*, i.e. they both possess a Grassmannian, scalar symmetry $\mathcal{Q}$ with the following properties:

(1) Nilpotency: $\mathcal{Q}^2 = 0$.
(2) The action is $\mathcal{Q}$-exact:

$$S = \{\mathcal{Q}, V\}. \tag{31.2.11}$$

(3) The energy-momentum tensor is also $\mathcal{Q}$-exact,

$$T_{\alpha\beta} = \frac{\delta S}{\delta h^{\alpha\beta}} = \{\mathcal{Q}, b_{\alpha\beta}\}. \tag{31.2.12}$$

The operator $\mathcal{Q}$ is formally identical to a BRST operator, and any operator of the form $\{\mathcal{Q}, \cdot\}$ (i.e. a *$\mathcal{Q}$-exact operator*) has a vanishing expectation value. Three consequences follow from this observation:

(1) The partition function does not depend on the background two-dimensional metric. Indeed,

$$\frac{\delta Z}{\delta h^{\alpha\beta}} = -\langle\{\mathcal{Q}, b_{\alpha\beta}\}\rangle = 0. \tag{31.2.13}$$

(2) The operators in the cohomology of $\mathcal{Q}$, i.e. the operators $\mathcal{O}$ which satisfy

$$\{\mathcal{Q}, \mathcal{O}\} = 0, \qquad \mathcal{O} \neq \{\mathcal{Q}, \Psi\} \tag{31.2.14}$$

lead to a topological sector in the theory, since their correlation functions are metric independent. These operators correspond to *marginal deformations* of the CFT, and in both the A and the B model, turning on a marginal deformation corresponds to a *geometric deformation* of the target CY manifold X.
(3) *The semiclassical approximation is exact*: if we explicitly introduce a coupling constant $\hbar$, we have

$$Z = \int \mathcal{D}\phi \, \mathrm{e}^{-\frac{1}{\hbar}S} \Rightarrow \frac{\delta Z}{\delta \hbar^{-1}} = -\langle\{\mathcal{Q}, V\}\rangle = 0. \tag{31.2.15}$$

In particular, we can evaluate the partition function (and the correlation functions of $\mathcal{Q}$-invariant operators) in the semiclassical limit $\hbar \to 0$.

In order to obtain a string theory we have to couple these topological sigma models to two-dimensional gravity. It turns out that topological sigma models are in this respect very similar to the critical bosonic string. Indeed, what characterizes critical strings is that the two-dimensional metric field essentially decouples from the theory, except for a finite number of moduli. In the BRST quantization of the critical bosonic string, this decoupling is the consequence of having a nilpotent BRST operator, $\mathcal{Q}_{\rm BRST}$, such that the energy-momentum tensor is $\mathcal{Q}_{\rm BRST}$-exact

$$T(z) = \{\mathcal{Q}_{\rm BRST}, b(z)\}. \tag{31.2.16}$$

As we have seen, topological sigma models possess a nilpotent Grasmannian operator which plays the role of $\mathcal{Q}_{\rm BRST}$. The property (31.2.12), formally identical to (31.2.16), guarantees that a similar decoupling will occur in topological sigma models.

Topological sigma models coupled to gravity are called *topological strings*. In analogy with bosonic strings, their free energies at genus g, $F_g(t)$, can be computed by a path integral like (31.1.4), where the t denote the parameters for marginal deformations. We will now describe the A and the B models in more detail, as well as the content of their free energies. For a full exposition, see for example [Ho+03, M05].

31.2.3 The type A topological string

The type A topological sigma model includes, on top of the field x, Grassmannian fields $\chi \in x^*(TX)$, which are scalars on Σ_g, and a Grassmannian one-form ρ_μ with values in $x^*(TX)$. This one-form satisfies a self-duality condition. The relevant saddle points or instantons in semiclassical calculations are *holomorphic maps* from Σ_g to X. These instantons are classified by topological sectors. If we consider a basis of $H_2(X, \mathbb{Z})$ denoted by Σ_i, $i = 1, \cdots, b_2(X)$, then

$$[x(\Sigma_g)] = \sum_{i=1}^{h^{1,1}(X)} n_i \Sigma_i \in H_2(X, \mathbb{Z}), \qquad n_i \in \mathbb{Z}_{\geq 0}. \tag{31.2.17}$$

In the A model, there are $b_2(X) = h^{1,1}(X)$ marginal deformations and they correspond to what are called *complexified Kähler parameters*. These can be understood as the volumes of the two-cycles Σ_i,

$$t_i^A = \int_{\Sigma_i} (\omega + \mathrm{i} B), \tag{31.2.18}$$

where $B \in H^2(X, \mathbb{Z})$ is the so-called B field and gives a natural complexification of these volumes.

The genus g free energies of the A model depend on the marginal deformation parameters (31.2.18). They have the structure

$$F_g^A(t^A) = \sum_{n_i} N_{g,n_i} \mathrm{e}^{-n_i t_i^A} \tag{31.2.19}$$

and they involve a sum over all saddle points (31.2.17) (there is also a simple contribution coming from trivial instantons with $n_i = 0$, but we have not included it in (31.2.19)). In this equation, N_{g,n_i} are the *Gromov-Witten invariants* at genus g and for the two-homology class (31.2.17). They are rational numbers that 'count' holomorphic maps at genus g in that class, and they can be computed by integrating an appropriate function over the space of collective coordinates of the instanton. They can be formulated in a rigorous mathematical way and they play a crucial role in modern algebraic geometry, see [CoK99, Ho+03] for an exposition.

31.2.4 The type B topological string

The type B topological sigma model includes, on top of the field x, Grassmannian fields $\eta^{\bar{I}}, \theta^{\bar{I}} \in x^*(T^{(0,1)}X)$, which are scalars on Σ_g, and a Grassmannian one-form on Σ_g, ρ^I_α, with values in $x^*(T^{(1,0)}X)$. In the B model, the only relevant saddle points are trivial ones, i.e. constant maps x. This makes the model much easier to solve, since integration over the collective coordinates of the saddle points becomes integration over the CY X. There are $h^{1,2}(X) = b_3(X)/2 - 1$ marginal deformations which can be interpreted as changes in the complex structure of the CY, and they are called *complex deformation parameters.* These parameters form a moduli space $\mathcal{M}$ of dimension $h^{1,2}(X)$. If the CY is presented through a set of algebraic equations, they roughly correspond to changing the parameters appearing in the equations.

The free energies F_g^B in the B model depend on the complex structures of the CY X, and mathematically they are closely related to the theory of variations of complex structures of X. A convenient parameterization of $\mathcal{M}$ is the following. Choose first a symplectic basis for $H_3(X, \mathbb{Z})$,

$$A_i, \quad B_j, \qquad i, j = 0, 1, \cdots, h^{1,2}(X). \tag{31.2.20}$$

The symplectic condition is that

$$A_i \cap B_j = \delta_{ij}. \tag{31.2.21}$$

We then define the *periods* of the Calabi-Yau manifold as

$$X_i = \int_{A_i} \Omega, \qquad \mathcal{F}_i = \int_{B_i} \Omega, \quad i = 0, \cdots, h^{1,2}(X) \tag{31.2.22}$$

where Ω is the form (31.2.10). A basic result of the theory of deformation of complex structures says that the X_i are (locally) complex projective coordinates for $\mathcal{M}$. They are called *special projective coordinates,* and since they parameterize $\mathcal{M}$ we deduce that the other set of periods must depend on them, *i.e.* $\mathcal{F}_i = \mathcal{F}_i(X)$. Moreover, one can show that there is a function $\mathcal{F}(X)$ which satisfies

$$\mathcal{F}_i(X) = \frac{\partial \mathcal{F}(X)}{\partial X_i} \tag{31.2.23}$$

and turns out to be a homogeneous function of degree two in the X_i.

Inhomogeneous coordinates can be introduced in a local patch where one of the projective coordinates, say X^0, is different from zero, and taking

$$t_i^B = \frac{X_i}{X_0}, \qquad i = 1, \cdots, h^{1,2}(X). \tag{31.2.24}$$

These coordinates on the moduli space of complex structures are called *flat coordinates*. The flat coordinates are non-trivial functions of the parameters z_i appearing in the algebraic equations defining X. One can rescale $\mathcal{F}(X)$ in order to obtain a function of the inhomogeneous coordinates t_i^B called the *prepotential*:

$$F_0^B(t^B) = \frac{1}{X_0^2} \mathcal{F}(X). \tag{31.2.25}$$

This is precisely the genus zero free energy of the type B topological strings on X, and it encodes all the relevant information about the B model on the sphere.

The higher genus amplitudes in the B model are also related to the variation of complex structures, but they are more difficult to describe. As first shown in [Be+94], they admit a non-holomorphic extension $F_g^B(t_B, t_B^*)$, and the original $F_g^B(t_B)$ are then recovered in the limit

$$F_g^B(t_B) = \lim_{t_B^* \to \infty} F_g^B(t_B, t_B^*). \tag{31.2.26}$$

The non-holomorphic dependence of these amplitudes is given by a recursive set of differential equations called the *holomorphic anomaly equations*. These equations do not determine the amplitudes uniquely, due to the lack of a complete set of boundary conditions. As we will see, in certain non-compact CY backgrounds, the higher genus topological string amplitudes of the type B topological string can be obtained from matrix models in a unique way.

31.2.5 Mirror symmetry

Mirror symmetry is an example of a string duality. It says that, given a CY manifold $\tilde{X}$, there is another CY manifold X such that the A model on $\tilde{X}$ is equivalent to the B model on X, and vice versa. This means that the free energies are simply equal:

$$F_g^A(t^A; \tilde{X}) = F_g^B(t^B; X), \tag{31.2.27}$$

after setting

$$t_i^B = t_i^A. \tag{31.2.28}$$

Notice that for mirror symmetry to make sense, one needs

$$h^{1,1}(\tilde{X}) = h^{1,2}(X), \tag{31.2.29}$$

so that the Hodge diamonds of X and $\tilde{X}$ are related by an appropriate reflection. Here, it is very important that the t^B in the B model are flat coordinates (31.2.24) obtained from the periods, and not the complex parameters z_i that appear in the algebraic equations defining X. For this reason, the non-trivial relation $t^A = t^B(z)$ between the complexified Kähler parameters in the A model and the complex parameters z_i is called the *mirror map*. Mirror symmetry makes it possible to compute the complicated instanton expansion of the A model in terms of period integrals in the B model, and it has been extensively studied in the last fifteen years, see [Ho+03, CoK99] for a detailed exposition.

31.3 The Dijkgraaf–Vafa correspondence

31.3.1 The correspondence

In its original formulation, the Dijkgraaf–Vafa correspondence is a statement about the genus g amplitudes of the type B topological string on some special, non-compact Calabi-Yau geometries. Such geometries are described by an equation of the form

$$u^2 + v^2 + z^2 - W'(x)^2 + R(x) = 0 \tag{31.3.1}$$

where $W(x)$, $R(x)$ are polynomials of degree $n+1$ and $n-1$, respectively. The simplest example occurs when $n = 1$, and

$$W(x) = \frac{1}{2}x^2, \qquad R(x) = \mu. \tag{31.3.2}$$

One then obtains the equation,

$$z^2 + u^2 + v^2 = \mu + x^2 \tag{31.3.3}$$

wich is the so-called *deformed conifold*. It is obtained as a deformation of the so-called conifold singularity

$$z^2 + u^2 + v^2 = x^2. \tag{31.3.4}$$

by turning on the parameter μ. The non-compact manifold (31.3.1) has a holomorphic three-form:

$$\Omega = \frac{1}{8\pi^2 \mathrm{i}} \frac{\mathrm{d}x\mathrm{d}z\mathrm{d}u}{v} \tag{31.3.5}$$

which is nowhere vanishing. Therefore, this space can be regarded as a non-compact Calabi-Yau threefold.

In the geometry (31.3.1) there are n compact three-cycles A_i, $i = 1, \cdots, n$, with the topology of a three-sphere. These cycles can be regarded as $\mathbb{S}^2$ fibrations over the cuts $\mathcal{C}_i = [x_{2i-1}, x_{2i}]$ of the curve

$$y^2(x) = W'(x)^2 - R(x) = \prod_{i=1}^{2n}(x - x_i), \tag{31.3.6}$$

where we have assumed that $R(x)$ is sufficiently generic so that the x_i are all distinct.

Consider the type B topological string on the non-compact CY manifold defined by (31.3.1), and let

$$t_i^B = \int_{A_i} \Omega \tag{31.3.7}$$

be the periods of the three-cycles A_i (in a non-compact CY threefold, one has $X_0 = 1$, so that $X_i = t_i^B$). We are interested in computing $F_g^B(t^B)$. To do this, let us consider a matrix model for a Hermitian $N \times N$ matrix defined by

$$Z = \frac{1}{\mathrm{vol}(U(N))} \int \mathrm{d}M \, \mathrm{e}^{-\frac{1}{g_s} \mathrm{tr}\, W(M)}, \tag{31.3.8}$$

where $W(x)$ is the potential appearing in (31.3.1). The most general solution to the loop equations of this matrix model is a *multi-cut solution*. This solution is characterized by a partition of N,

$$N = N_1 + \cdots + N_n, \tag{31.3.9}$$

where we put N_i eigenvalues near the i-th critical point of $W(x)$. At large N_i, these sets of eigenvalues fill up cuts $\mathbb{C}_i$, $i = 1, \cdots, n$. Different multi-cut solutions of the model are associated with different choices of a polynomial $R(x)$ of order $n - 1$, as in (31.3.1), and the *spectral curve* of the multi-cut solution to (31.3.8) has precisely the form (31.3.6). The *partial 't Hooft parameters*

$$S_i = g_s N_i = \frac{1}{4\pi \mathrm{i}} \oint_{\mathbb{C}_i} y(x) \mathrm{d}x, \tag{31.3.10}$$

can be computed in terms of periods of the spectral curve. The total free energy of the matrix model has a $1/N$ expansion of the form

$$F^{\mathrm{MM}}(S_i, g_s) = \log Z(N_i, g_s) = \sum_{g=0}^{\infty} F_g^{\mathrm{MM}}(S_i) g_s^{2g-2}. \tag{31.3.11}$$

According to the Dijkgraaf–Vafa correspondence, *the genus g amplitudes of the type B topological string on (31.3.1) are given by the genus g free energies of the multi-cut matrix model (31.3.8)*:

$$F_g^B(t_i^B) = F_g^{\mathrm{MM}}(S_i) \tag{31.3.12}$$

after identifying $t_i^B = S_i$. Notice that this identification implies in particular that the A-periods of Ω in the full geometry (31.3.1), as calculated in (31.3.7), must agree with the periods of the spectral curve (31.3.10). This is easily verified.

Notice that, in view of the results of [EO07], we can reformulate (31.3.12) by saying that

$$F_g^B(t_i^B) = F_g^\Sigma(S_i) \tag{31.3.13}$$

where F_g^Σ are the symplectic invariants associated to the spectral curve (31.3.6), i.e. to the 'non-trivial' part of the geometry (31.3.1).

The Dijkgraaf–Vafa correspondence (31.3.12) gives an answer in closed form for the perturbative amplitudes of the type B topological string, and when X is of the form (31.3.1). For $g = 0$ the correspondence follows from the equality of the special geometry solution for $F_0^B(t^B)$ and the saddle-point solution for the planar free energy of the matrix model. The equality (31.3.12) for $g = 1$ can be also verified by direct computation of both sides [KMT03, DST04]. For $g \geq 2$ there is no general proof of the equivalence, although it has been noticed that the matrix model free energies $F_g^{\mathrm{MM}}(S_i)$ can be modified in a natural way so as to satisfy the holomorphic anomaly equations characterizing the type B topological string [HK07, EMO07].

31.3.2 Relation to gauge/string dualities and geometric transitions

The Dijkgraaf–Vafa correspondence is an example of a *gauge/string duality*, relating a gauge theory in the $1/N$ expansion (in this case a matrix model, i.e. a zero-dimensional quantum gauge theory) to a string theory (in this case, a topological string theory). It can be also reformulated as an *open/closed string duality*, relating an open topological string theory to a closed topological string theory. We will sketch here the main argument, referring the reader to [DV02, M06] for details.

Let us consider the *singular* Calabi-Yau geometry given by

$$u^2 + v^2 + z^2 - W'(x)^2 = 0. \tag{31.3.14}$$

When $W'(x) = x$, this is the conifold singularity (31.3.4). Singularities in algebraic geometry can be avoided in two different ways. One way consists in *deforming* them, by adding lower-order terms to the equation. In this way one recovers the geometry (31.3.1). Another way consists in *resolving* the singularities by 'gluing' two-spheres to them. In this case, one obtains a geometry X_{res} with n two-cycles with the topology of $\mathbb{S}^2$. These two-spheres are in one-to-one correspondence with the critical points of $W(x)$. One can then consider the *open* type B topological string on this space. Open strings need boundary conditions, and in string theory the most general boundary conditions are provided by D-branes. Let us consider N D-branes on X_{res}. For each partition (31.3.9) we obtain a D-brane configuration in which N_i D-branes wrap the i-th two-sphere. An open string in X_{res} with this D-brane configuration will have in general $h_i \geq 0$ boundaries ending on the i-th two-sphere.

We are now interested in computing open string amplitudes for this system. Let F_{g,h_i} be the open string amplitude corresponding to a Riemann surface of genus g with h_i boundaries attached to the i-th sphere. The total free energy is then given by

$$F^{B,\,\text{open}}(N_i, g_s) = \sum_{g=0}^{\infty} \sum_{h_1,\cdots,h_n \geq 0} F_{g,h_1,\cdots,h_n} g_s^{2g-2+h} N_1^{h_1} \cdots N_n^{h_n}, \tag{31.3.15}$$

where g_s is the string coupling constant and $h = h_1 + \cdots + h_n$. Using the seminal results of Witten in [W95], one can show that the string field theory describing the dynamics of this system of open strings is precisely the matrix model (31.3.8). This means that

$$F^{B,\,\text{open}}(N_i, g_s) = \log Z(N_i, g_s) \tag{31.3.16}$$

where $Z(N_i, g_s))$ is the partition function (31.3.8) expanded around the multi-cut vacuum characterized by the partition (31.3.9).

The open/closed string duality states that the type B topological string theory on the *resolved* geometry X_{res} in the presence of N D-branes is equivalent to a *closed* string field theory on the *deformed* geometry (31.3.1). This means two things. First, the periods t_i^B parameterizing the moduli space of complex parameters of (31.3.1) are given by the partial 't Hooft parameters $g_s N_i$ in (31.3.10). Second, the closed string amplitudes $F_g^{\text{closed}}(t_i^B)$ are obtained by fixing the genus g of the open string theory and summing over all possible h_is:

$$F_g^{B,\,\text{closed}}(t_i^B) = \sum_{h_1,\cdots,h_n \geq 0} F_{g,h_1,\cdots,h_n}^{B,\,\text{open}} t_1^{h_1} \cdots t_n^{h_n}, \tag{31.3.17}$$

so that

$$F^{B,\,\text{closed}}(t_i^B, g_s) = F^{B,\,\text{open}}(N_i, g_s). \tag{31.3.18}$$

(31.3.12) follows from this.

We finally point out three important conceptual aspects of this open/closed string duality. First, it relates an open string theory on X_{res} (with N D-branes) to a closed string theory on the deformed counterpart (31.3.1), and therefore relates two different geometric backgrounds – namely, the two different desingularizations of (31.3.14). For this reason, it is sometimes called a *geometric transition*, and it is conceptually similar to other dualities of this type, like the celebrated AdS/CFT correspondence [Ah+00] or the Gopakumar-Vafa correspondence (see [M05] for a review). Second, the open string theory side is equivalent to a gauge theory (in this case, a matrix model), and to go from the open free energies to the closed free energies as in (31.3.17) one has to sum over all double-line diagrams in the gauge theory with fixed genus g, as in the large N expansion. For this reason, geometric transitions are also called *large N transitions*. Finally, in this duality the target space geometry (31.3.1) is essentially given by the spectral curve of the matrix model (31.3.6). In this sense, the Calabi-Yau geometry can be regarded as an *emergent* geometry made of a large number of gauge theory degrees of freedom, since the spectral curve encodes the distribution of eigenvalues of the matrix model in the large N limit.

We conclude this section with a dictionary of the relations obtained between matrix models and type B topological strings in the geometries (31.3.1).

Matrix model	Type B topological string
't Hooft parameters	flat coordinates
spectral curve	target CY geometry
planar free energy	prepotential
$1/N$ expansion	genus expansion

31.3.3 Applications to supersymmetric gauge theories

One of the most important applications of the Dijkgraaf–Vafa correspondence has been the calculation of exact superpotentials in $\mathcal{N}=1$ supersymmetric gauge theories. The canonical example, as discussed in [CIV01, DVb02], is $\mathcal{N}=1$, $U(M)$ Yang-Mills theory with a chiral superfield in the adjoint representation Φ. When there is no tree-level superpotential for Φ, this is in fact $U(M)$, $\mathcal{N}=2$ Yang-Mills theory. A tree-level superpotential for Φ of degree $n+1$, $W(\Phi)$, will break supersymmetry down to $\mathcal{N}=1$. The classical vacua of the theory are determined by the distribution of the eigenvalues of the matrix Φ over the n critical points of $W(\Phi)$. A partition

$$M = M_1 + \cdots + M_n, \tag{31.3.19}$$

where M_i eigenvalues are at the i-th critical point, corresponds to the classical gauge symmetry breaking pattern

$$U(M) \to U(M_1) \times \cdots \times U(M_n). \tag{31.3.20}$$

Quantum mechanically, these vacua are characterized by a gaugino condensate and confinement in each of the $SU(M_i)$ factors, and by dynamically generated scales Λ_i. Let

$$S_i = \frac{1}{32\pi^2} \mathrm{tr}_{\,SU(M_i)} W_\alpha^2 \tag{31.3.21}$$

be the gaugino superfield in the $SU(M_i)$ factor. The effective superpotential describing the quantum vacua associated to the classical pattern (31.3.20) can be computed as a function of the superfields S_i, $W_{\mathrm{eff}}(S_i)$. In [DV02, CIV01, DVb02] the following formula was proposed for $W_{\mathrm{eff}}(S_i)$:

$$W_{\text{eff}}(S_i) = \sum_{i=1}^{n} \left\{ M_i\, S_i \log(S_i/\Lambda_i^3) - 2\pi \mathrm{i} \tau S_i + M_i \frac{\partial F_0^{\text{MM}}(S_i)}{\partial S_i} \right\}. \tag{31.3.22}$$

In this formula,

$$\tau = \frac{\theta}{2\pi} + \frac{4\pi \mathrm{i}}{g^2} \tag{31.3.23}$$

is the classical, complexified gauge coupling. $F_0^{\text{MM}}(S_i)$ is the genus zero free energy of the Hermitian matrix model (31.3.8), corresponding to a multi-cut solution with partial 't Hooft parameters S_i (notice that the rank of the matrix model, N, is *a priori* unrelated to the rank M of the gauge group in the supersymmetric gauge theory).

This application of the Dijkgraaf–Vafa correspondence follows by relating the topological string to the *physical* type IIB superstring theory [Be+94, DVb02]. When the type IIB superstring is compactified on X_{res}, in the background of M D5-branes wrapping the two-spheres as in (31.3.19), one obtains an 'engineering' of the four-dimensional gauge theory we have just described. The effective superpotential, as a function of the gaugino superfields, can be computed from the genus zero open string amplitudes of the *topological* string on X_{res}. Since these amplitudes are computed by the matrix model, we obtain the relation (31.3.22). See [DVb02] for more details.

The relation (31.3.22) between matrix models and supersymmetric gauge theories can be also derived by using field-theory techniques [D+03, C+02, AFH04] and it has led to an extensive literature. In particular, it can be extended to many other $\mathcal{N}=1$ supersymmetric gauge theories, and it leads to powerful derivations of gauge theory dualities and exact solutions, like for example the Seiberg-Witten solution or the duality properties of softly broken $\mathcal{N}=4$ supersymmetric gauge theories.

31.4 Matrix models and mirror symmetry

The Dijkgraaf and Vafa correspondence, as it was originally formulated in [DV02], gives a matrix model description only for the type B topological string on geometries of the form (31.3.1). However, one can find a matrix model-like description for a different, important class of non-compact Calabi-Yau geometries. These geometries come in mirror pairs. In the type A topological string, they are *toric* Calabi-Yau manifolds. These are manifolds which contain an algebraic torus $\mathbb{T} \subset (\mathbb{C}^*)^r \subset X$ as an open set, and they admit an action of $\mathbb{T}$ which acts on this set by multiplication. Important examples of toric Calabi-Yau threefolds are the *resolved conifold*

$$\mathcal{O}(-1) \oplus \mathcal{O}(-1) \to \mathbb{P}^1 \tag{31.4.1}$$

and the spaces

$$K_S \to S, \tag{31.4.2}$$

where S is an algebraic surface and K_S is its canonical bundle. A simple example occurs when $S = \mathbb{P}^2$. The resulting Calabi-Yau

$$\mathcal{O}(-3) \to \mathbb{P}^2, \tag{31.4.3}$$

is also known as *local* $\mathbb{P}^2$. Toric Calabi-Yau geometries have a mirror description given by

$$u^2 + v^2 = P(x, y), \qquad x, y \in \mathbb{C}^* \tag{31.4.4}$$

where $P(x, y)$ is a polynomial in x, y. The locus

$$P(x, y) = 0 \tag{31.4.5}$$

defines an algebraic curve $\Sigma \subset \mathbb{C}^* \times \mathbb{C}^*$. Notice that this curve is the analogue of (31.3.6) in the geometries of the form (31.3.1). For example, the mirror curve to local $\mathbb{P}^2$ (31.4.3) is the cubic curve in $\mathbb{C}^* \times \mathbb{C}^*$ given by

$$y(y + x + 1) + zx^3 = 0 \tag{31.4.6}$$

where z is a complex deformation parameter. The flat coordinates parameterizing the moduli space of complex structures of the Calabi-Yau manifolds (31.4.4) reduce to periods on the curve (31.4.5), but since the variables now belong to $\mathbb{C}^* \times \mathbb{C}^*$ the meromorphic form to integrate is different, and one has

$$t_i^B = \oint_{\mathbb{C}_i} \log y \frac{\mathrm{d}x}{x}. \tag{31.4.7}$$

As we pointed out in the introduction, given any algebraic curve Σ one can construct a series of matrix-model-like quantities $F_g^{\Sigma}(t)$ associated to it, which are called the symplectic invariants of the curve. We can now regard the curve (31.4.5) as an algebraic curve Σ and calculate its symplectic invariants following the formalism of [EO07].[1] An extension of the Dijkgraaf–Vafa correspondence was conjectured in [M08, Bo+09] and states that the genus g free energies of the type B topological string theory on the CY (31.4.4) are given by the symplectic invariants of the curve Σ given by (31.4.5),

$$F_g^B(t_i^B) = F_g^{\Sigma}(t_i^B), \tag{31.4.8}$$

where t_i^B are the periods (31.4.7). By mirror symmetry, this computes the type A topological string amplitudes on toric manifolds, and in particular generating functionals of Gromov-Witten invariants of these manifolds. Extensive checks of this proposal were performed in [M08, Bo+09]. An important aspect of this

[1] The fact that the variables x, y now live in $\mathbb{C}^*$ leads to some modifications of this formalism, however. See [Bo+09] for details.

proposal is that one can also compute *open* string amplitudes on the geometries (31.4.4). These are given by the generating functionals associated to Σ which correspond to matrix model correlation functions. When applied to (31.4.4), and after using mirror symmetry, they can be seen to compute *open* Gromov-Witten invariants associated to an important class of D-branes in the mirror toric manifolds.

To summarize, the Dijkgraaf–Vafa correspondence can be extended to a large family of non-compact Calabi-Yau manifolds given by toric manifolds and their mirrors. The mirror geometry leads naturally to a spectral curve. The symplectic invariants and generating functionals associated to this curve by using matrix model techniques give topological string amplitudes and in particular they calculate Gromov-Witten invariants of toric Calabi-Yaus.

We conclude with some remarks on this correspondence between matrix models and topological strings on toric manifolds and their mirrors. First of all, this correspondence (in contrast to the Dikgraaf-Vafa correspondence) does not provide explicit matrix integral realizations of topological string theory. These are not needed to compute the genus g amplitudes, since they only depend on the spectral curve, which is provided by the mirror geometry (31.4.4). However, for certain toric Calabi-Yau manifolds (like the resolved conifold) there are explicit matrix models whose spectral curve coincides with the one featuring in their mirror geometry. These matrix models are called *Chern-Simons matrix models* and were introduced in [M04] to describe the partition function of Chern-Simons gauge theory on certain three-manifolds, see [A+04, M06, M05] for more details on these models. Second, the symplectic invariants of (31.4.5) give, through mirror symmetry, the genus g free energies of the A model on toric Calabi-Yau manifolds. These free energies can also be obtained as sums over partitions using the topological vertex formalism [A+05]. Indeed, it seems that the spectral curves (31.4.5) describe the saddle point of these sums in the limit of large partitions, and that the formalism of [EO07] can be applied as well to obtain an analogue of the topological $1/N$ expansion [E08]. This makes it possible to verify the correspondence of [M08, Bo+09], at least in some cases, by using the theory of the topological vertex. Finally, there are other, simpler enumerative problems which can be solved as well in terms of the matrix-model like formalism of [EO07], as for example the calculation of Hurwitz numbers of the sphere [BoM08].

31.5 String theory, matrix quantum mechanics, and related models

Matrix quantum mechanics, a close cousin of matrix models, also plays an important role in string theory. In matrix quantum mechanics the fundamental variables are time-dependent matrices, and the theory can be regarded as a one-dimensional quantum gauge theory. One of the most important applications of matrix quantum mechanics is the non-perturbative description of string theories in one dimension. For example, the $c = 1$ non-critical bosonic string can be defined by the double-scaling limit of a matrix quantum mechanics model with a single, Hermitian matrix in an appropriate potential, see [GM93] for a review and a list of references.

With the advent of D-branes, a more crucial role has been advocated for certain matrix quantum mechanical models, in the context of M-theory. The low-energy dynamics of a system of N D0-branes in type IIA theory is described by the dimensional reduction of $\mathcal{N} = 1$ supersymmetric Yang-Mills theory to $0 + 1$ dimensions. The resulting gauge-fixed Lagrangian can be written as

$$L = \frac{1}{2}\mathrm{tr}\left\{\dot{X}^a\dot{X}^a + \frac{1}{2}[X^a, X^b]^2 + \theta^T(\mathrm{i}\dot{\theta} - \Gamma_a[X^a, \theta])\right\}. \tag{31.5.1}$$

In this Lagrangian, the X^a are nine $N \times N$ matrices, θ are sixteen $N \times N$ Grasmann matrices forming a spinor of $SO(9)$, and Γ_a are the corresponding gamma matrices. According to a conjecture in [B+97], the large N limit of this Lagrangian should describe all of M-theory in light-front coordinates. There is evidence that certain aspects of eleven-dimensional supergravity are appropriately captured by this matrix quantum-mechanical model, see for example [T01] for a detailed review. In particular, linearized gravitational interactions, as well as non-linear corrections, can be obtained from this model. This is an important example of how gravitational space-time dynamics emerges as a large distance phenomenon from a well-defined quantum-mechanical system with matrix degrees of freedom.

Other important matrix quantum-mechanical models of D-brane dynamics include the so-called ADHM model, describing bound states of D4 and D0 branes. This model has important applications in instanton and BPS state counting in supersymmetric gauge theory and string theory, see for example [N05] for a review of the applications to instanton counting.

References

[A+04] M. Aganagic, A. Klemm, M. Mariño and C. Vafa, "Matrix model as a mirror of Chern-Simons theory," JHEP **0402**, 010 (2004) [arXiv:hep-th/0211098].

[A+05] M. Aganagic, A. Klemm, M. Mariño and C. Vafa, "The topological vertex," Commun. Math. Phys. **254**, 425 (2005) [arXiv:hep-th/0305132].

[Ah+00] O. Aharony, S. S. Gubser, J. M. Maldacena, H. Ooguri and Y. Oz, "Large N field theories, string theory and gravity," Phys. Rept. **323**, 183 (2000) [arXiv:hep-th/9905111].

[AFH04] R. Argurio, G. Ferretti and R. Heise, "An introduction to supersymmetric gauge theories and matrix models," Int. J. Mod. Phys. A **19**, 2015 (2004) [arXiv:hep-th/0311066].

[B+97] T. Banks, W. Fischler, S. H. Shenker and L. Susskind, "M theory as a matrix model: A conjecture," Phys. Rev. D **55**, 5112 (1997) [arXiv:hep-th/9610043].

[Be+94] M. Bershadsky, S. Cecotti, H. Ooguri and C. Vafa, "Kodaira-Spencer theory of gravity and exact results for quantum string amplitudes," Commun. Math. Phys. **165**, 311 (1994) [arXiv:hep-th/9309140].

[Bo+09] V. Bouchard, A. Klemm, M. Mariño and S. Pasquetti, "Remodeling the B-model," Commun. Math. Phys. **287**, 117 (2009) [arXiv:0709.1453 [hep-th]].

[BoM08] V. Bouchard and M. Mariño, "Hurwitz numbers, matrix models and enumerative geometry," in R. Donagi and K. Wendland (eds.), *From Hodge theory to integrability and TQFT*, p. 263, Proc. Sympos. Pure Math. **78**, Providence, 2008 [arXiv:0709.1458 [math.AG]].

[C+02] F. Cachazo, M. R. Douglas, N. Seiberg and E. Witten, "Chiral Rings and Anomalies in Supersymmetric Gauge Theory," JHEP **0212**, 071 (2002) [arXiv:hep-th/0211170].

[CIV01] F. Cachazo, K. A. Intriligator and C. Vafa, "A large N duality via a geometric transition," Nucl. Phys. B **603**, 3 (2001) [arXiv:hep-th/0103067].

[CoK99] D. Cox and S. Katz, *Mirror symmetry and algebraic geometry*, American Mathematical Society, Providence, 1999.

[DiF+95] P. Di Francesco, P. Ginsparg and J. Zinn-Justin, "2-D Gravity and random matrices," Phys. Rept. **254**, 1 (1995) [arXiv:hep-th/9306153].

[DV02] R. Dijkgraaf and C. Vafa, "Matrix models, topological strings, and supersymmetric gauge theories," Nucl. Phys. B **644**, 3 (2002) [arXiv:hep-th/0206255].

[DVb02] R. Dijkgraaf and C. Vafa, "A perturbative window into non-perturbative physics," arXiv:hep-th/0208048.

[D+03] R. Dijkgraaf, M. T. Grisaru, C. S. Lam, C. Vafa and D. Zanon, "Perturbative computation of glueball superpotentials," Phys. Lett. B **573**, 138 (2003) [arXiv:hep-th/0211017].

[DST04] R. Dijkgraaf, A. Sinkovics and M. Temurhan, "Matrix models and gravitational corrections," Adv. Theor. Math. Phys. **7**, 1155 (2004) [arXiv:hep-th/0211241].

[E08] B. Eynard, "All orders asymptotic expansion of large partitions," J. Stat. Mech. **0807**, P07023 (2008) [arXiv:0804.0381 [math-ph]].

[EMO07] B. Eynard, M. Mariño and N. Orantin, "Holomorphic anomaly and matrix models," JHEP **0706**, 058 (2007) [arXiv:hep-th/0702110].

[EO07] B. Eynard and N. Orantin, "Invariants of algebraic curves and topological expansion," Commun. Number Theory Phys. **1**, 347 (2007) [arXiv:math-ph/0702045].

[GM93] P. H. Ginsparg and G. W. Moore, "Lectures on 2-D gravity and 2-D string theory," arXiv:hep-th/9304011.

[Ho+03] K. Hori, S. Katz, A. Klemm, R. Pandharipande, R. Thomas, C. Vafa, R. Vakil, and E. Zaslow, *Mirror symmetry*, American Mathematical Society, Providence, 2003.

[HK07] M. x. Huang and A. Klemm, "Holomorphic anomaly in gauge theories and matrix models," JHEP **0709**, 054 (2007) [arXiv:hep-th/0605195].

[KMT03] A. Klemm, M. Mariño and S. Theisen, "Gravitational corrections in supersymmetric gauge theory and matrix models," JHEP **0303**, 051 (2003) [arXiv:hep-th/0211216].

[LL92] J. M. F. Labastida and P. M. Llatas, "Topological matter in two-dimensions," Nucl. Phys. B **379**, 220 (1992) [arXiv:hep-th/9112051].

[M04] M. Mariño, "Chern-Simons theory, matrix integrals, and perturbative three-manifold invariants," Commun. Math. Phys. **253**, 25 (2004) [arXiv:hep-th/0207096].

[M05] M. Mariño, *Chern-Simons theory, matrix models, and topological strings*, Oxford University Press, Oxford, 2005.

[M06] M. Mariño, "Les Houches lectures on matrix models and topological strings," in E. Brézin *et al.* (eds), *Applications of random matrices in physics*, p. 319, Springer-Verlag, New York, 2006 [arXiv:hep-th/0410165].

[M08] M. Mariño, "Open string amplitudes and large order behavior in topological string theory," JHEP **0803**, 060 (2008) [arXiv:hep-th/0612127].

[N05] N. A. Nekrasov, "Lectures on nonperturbative aspects of supersymmetric gauge theories," Class. Quant. Grav. **22**, S77 (2005).

[T01] W. Taylor, "M(atrix) theory: Matrix quantum mechanics as a fundamental theory," Rev. Mod. Phys. **73**, 419 (2001) [arXiv:hep-th/0101126].

[W91] E. Witten, "Mirror manifolds and topological field theory," in S.T. Yau (ed.), *Mirror symmetry I*, second edition, p. 121, AMS/International Press, Providence 1998 [arXiv:hep-th/9112056].

[W95] E. Witten, "Chern-Simons Gauge Theory As A String Theory," Prog. Math. **133**, 637 (1995) [arXiv:hep-th/9207094].

·32·

Quantum chromodynamics

J. J. M. Verbaarschot

Abstract

In this chapter we introduce chiral random matrix theories with the global symmetries of quantum chromodynamics (QCD). In the microscopic domain, these theories reproduce the mass and chemical potential dependence of QCD. Both spectra of the anti-Hermitian Dirac operator and spectra of the non-Hermitian Dirac operator at non-zero chemical potential are discussed.

32.1 Introduction

Applications of random matrix theory (RMT) to the physics of strong interactions have a long history (see Chapter 2). Random matrix theories were introduced to nuclear physics by Wigner to describe the level spacing distribution of nuclei [Wig55]. This paper inspired a large body of early work on RMT (see [Por65]). An important conceptual discovery was the large N approximation. It first appeared in the work of Wigner, and became an integral part of QCD through the seminal work of 't Hooft [tHo74], which showed that the limit of a large number of colours is dominated by planar diagrams with combinatorial factors that can be obtained from matrix integrals [Bre78]. This culminated in the RMT formulation of quantum gravity in 2D, which is a sum over random surfaces that can be triangulated by planar diagrams (see Chapter 30 and [DiF93] for a review).

The main focus of this chapter is to apply RMT to spectra of the Dirac operator both at zero chemical potential (first half of this chapter), when the Dirac operator is Hermitian, and at nonzero chemical potential (second half of this chapter), when the Dirac operator is non-Hermitian. Because the Euclidean Dirac operator can be interpreted as a Hamiltonian, this application is closer in spirit to the original ideas of Wigner than to the work of 't Hooft. However, we have benefited greatly from the mathematical techniques that were developed for large N QCD and 2D quantum gravity.

Applications of RMT to QCD have been reviewed extensively in the literature [Guh97b, Jan98b, Ver00, Ver05, Ake07]. These papers offer both additional details and different points of view. Phenomenological applications of RMT to QCD are not discussed in this chapter (see [Ver00, Ake07] for reviews).

32.1.1 Spontaneous symmetry breaking in RMT

One of the essential features of random matrix theory is spontaneous symmetry breaking. The real part of the resolvent

$$G(z) = \frac{1}{N}\left\langle \mathrm{Tr}\frac{1}{z+D}\right\rangle, \tag{32.1.1}$$

where the average is over the probability distribution of the ensemble of $N \times N$ anti-Hermitian random matrices, D, can be expressed as the replica limit

$$\mathrm{Re}\, G(z) = \lim_{n\to 0}\frac{1}{2nN}\frac{d}{dz}\log(Z_n(z)), \qquad \text{with} \quad z \in \mathcal{R}, \tag{32.1.2}$$

where the generating function is defined by

$$Z_n(z) = \langle \det{}^n(D+z)\det{}^n(z-D)\rangle. \tag{32.1.3}$$

For $z = 0$, it is invariant under $Gl(2n)$. This symmetry is broken spontaneously to $Gl(n) \times Gl(n)$ by a non-zero value of $\lim_{z\to 0}\lim_{N\to\infty}\mathrm{Re}\, G(z)$. The order of the limits is essential – the reverse order gives zero. This can be seen by expressing the resolvent in terms of eigenvalues of D.

When we integrate the resolvent over the contour C in the complex plane that is the boundary of $[-\frac{\epsilon}{2} < \mathrm{Re}\, z < \frac{\epsilon}{2}] \times [-\frac{1}{2}\Delta x < \mathrm{Im}\, z < \frac{1}{2}\Delta x]$ we obtain

$$N\oint_C G(z)dz = 2\pi i\, N_{\Delta x} = 2\pi i\rho(0)\Delta x, \tag{32.1.4}$$

where $N_{\Delta x}$ is the number of eigenvalues enclosed by the contour and $\rho(0)$ is the average spectral density around zero. In the limit $\epsilon \to 0$ the left-hand side is given by $2i\Delta x \mathrm{Re}\, G(\frac{\epsilon}{2})$. The discontinuity of the resolvent and $\rho(0)$ are thus related by

$$\lim_{\epsilon\to 0}\mathrm{Re}\,(G(\frac{\epsilon}{2})) = \frac{\pi\rho(0)}{N}. \tag{32.1.5}$$

This formula is known as the Banks-Casher formula [Ban80]. It relates the order parameter for spontaneous symmetry breaking to the spectral density.

When the spectrum of D is reflection symmetric, then $\det(z-D) = \det(z+D)$, and the generating function for the real part of the resolvent is given by

$$Z_n(z) = \langle \det{}^n(D+z)\rangle. \tag{32.1.6}$$

This is the case when the random matrix ensemble has an involutive symmetry, $ADA = -D$ with $A^2 = 1$, which is the case for the Dirac operator in QCD.

32.1.2 The QCD partition function

The Euclidean QCD partition function for N_f quarks with mass m_f is given by

$$Z_{\text{QCD}} = \prod_{f=1}^{N_f} \langle \det(D + m_f) \rangle. \tag{32.1.7}$$

The average weighted by the Yang-Mills action is over gauge fields in the Lie algebra of $SU(N_c)$, either in the fundamental or adjoint representation. The Dirac operator D is a function of the 'random' gauge fields. For strong interactions we have $N_c = 3$ with gauge fields in the fundamental representation.

Applications of RMT to QCD differ in several respects from other applications. First, the physical system itself is already a stochastic ensemble. Second, the QCD partition function is not a quenched average, but the fermion determinant describes the quark degrees of freedom. Third, the average is over Dirac operators with different rank. The reason is that, according to the Atiyah-Singer index theorem, the number of zero modes is equal to the topological charge of the gauge field configuration. Because statistical properties of Dirac spectra depend on the number of topological zero modes, we will treat each topological charge sector separately. Fourth, QCD is a quantum field theory that has to be regularized and renormalized. Note that the low-lying Dirac spectrum is both gauge invariant and renormalizable [Giu09].

32.2 Quantum chromodynamics and chiral random matrix theory

32.2.1 Symmetries of QCD

To analyse the global symmetries of QCD we consider the Dirac operator for a finite chiral basis. Then it is given by a matrix with the block structure

$$D = \begin{pmatrix} 0 & iA \\ -iA^\dagger & 0 \end{pmatrix}, \tag{32.2.1}$$

where A is a complex $N_+ \times N_-$ matrix. Therefore, all nonzero eigenvalues occur in pairs $\pm\lambda_k$. The number of zero eigenvalues is equal to $|N_+ - N_-|$ and is interpreted as the topological charge. Paired zeros may occur, but this is a set of measure zero and of no interest. The block structure of (32.2.1) is due to the axial $U(1)$ symmetry. The partition function also has a vector $U(1)$ symmetry related to the conservation of baryon charge. For $N_f > 1$ the Dirac operator is the direct sum of N_f one-flavour Dirac operators so that QCD has the axial flavour symmetry $U_A(N_f)$ and the vector flavour symmetry $U_V(N_f)$.

For QCD with $N_c \geq 3$ and gauge fields in the fundamental representation there are no other global symmetries. Because $SU(2)$ is pseudoreal, for $N_c = 2$ the Dirac operator has the anti-unitary symmetry [Ver94a]

$$[UK, D] = 0 \tag{32.2.2}$$

with K the complex conjugation operator and U a fixed unitary matrix with $U^2 = 1$. Then it is possible to find a basis for which A becomes real [Dys62], and for $m = 0$, we have that $\det D = \det^2 A$. For N_f massless flavours the quark determinant occurs in the partition function as $\det^{2N_f} A$. This enlarges the flavour symmetry group to $U(2N_f)$. The third case is when gauge fields are in the adjoint representation. Then the Dirac operator also has an anti-unitary symmetry but now $U^2 = -1$ [Ver94a] which allows to construct a basis for which the Dirac operator can be rearranged in self-dual quaternions [Hal95b]. The flavour symmetry is also enlarged to $U(2N_f)$.

Quantum chromodynamics in three Euclidean space-time dimensions does not have an involutive symmetry. In that case the flavour symmetry is $U(N_f)$ for gauge fields in the fundamental representation. For two colours in the fundamental representation and for any number of colours ≥ 2 in the adjoint representation the symmetry group is enlarged to $O(2N_f)$ [Mag99a] and $Sp(2N_f)$ [Mag99b], respectively.

32.2.2 Spontaneous symmetry breaking

The features that mostly determine the physics of QCD at low energy are spontaneous chiral symmetry breaking and confinement: confined quarks and gluons do not appear in the physical spectrum so that QCD at low energy is a theory of the weakly interacting Goldstone modes.

According to the Vafa–Witten theorem [Vaf83], global vector symmetries of vector-like gauge theories cannot be broken spontaneously. Therefore,

$$\Sigma \equiv |\langle \bar{\psi}^a \psi^a \rangle| = \left| \lim_{m_a \to 0} \lim_{V \to \infty} \frac{1}{V} \left\langle \mathrm{Tr} \frac{1}{D + m_a} \right\rangle \right| = \frac{\pi \rho(0)}{V}, \tag{32.2.3}$$

is flavour independent. Its absolute value follows from the Banks-Casher relation (with $\rho(0)$ the spectral density of D near zero and V the volume of space-time).

In Table 32.1 we give the symmetry-breaking patterns [Pes80, Vys85] for the theories mentioned above and for QCD in three dimensions. (See [Ver94b, Mag99a, Mag99b, Sza00] for additional discussions).

32.2.3 Chiral random matrix theory

Since the global symmetries of the Dirac operator are a direct consequence of its block structure and the reality properties of its matrix elements, it should be clear how to construct a RMT with the same global symmetries: just replace the nonzero matrix elements by an ensemble of random numbers. Such chiral random matrix theory (chRMT) is defined by [Shu92, Ver94a]

$$Z_{\mathrm{chRMT}}^{\beta,\nu}(\{m_k\}) = \int DA \prod_k \det \begin{pmatrix} m_k & A \\ A^\dagger & m_k \end{pmatrix} P(A), \tag{32.2.4}$$

where A is an $N_+ \times N_-$ matrix, and the integration is over the real (and imaginary) parts of A. The reality classes are denoted by the Dyson index β which is equal to the number of degrees of freedom per matrix element and is the same as in the corresponding QCD-like theory (see Table 32.1). The properties of the chRMT partition function do not depend on the details of the probability distribution. This is known as universality [Bre95, Jac96a, Ake96, Guh97a] and justifies to simply average over a Gaussian distribution,

$$P(A) = c e^{-N\Sigma^2 \mathrm{Tr}(A^\dagger A)} \qquad \text{with} \quad N = N_+ + N_-. \tag{32.2.5}$$

Both the Vafa–Witten theorem [Vaf83] and the Banks-Casher formula (32.1.5) apply to chRMT. The global symmetry breaking pattern and the Goldstone manifold are therefore the same as in QCD.

In chRMT, N is interpreted as the volume of space-time. This corresponds to units where $N/V = 1$ so that Σ can be written as $\Sigma V/N$ and Eq. (32.2.5) becomes dimensionally correct (the matrix elements of the Dirac operator and its eigenvalues have the dimension of mass). The normalization of (32.2.5) is such that Σ is the chiral condensate and satisfies the Banks-Casher relation.

Table 32.1 Classification of QCD-like theories in three and four dimensions. The Dyson index, β, is the number of degrees of freedom per matrix element ($\beta = 1$ and $\beta = 4$ are interchanged for staggered lattice fermions).

Theory	β	Symmetry G	Broken to H	chRMT
Fundamental $N_c \geq 3, d = 4$	2	$U(N_f) \times U(N_f)$	$U(N_f)$	chGUE
Fundamental $N_c = 2, d = 4$	1	$U(2N_f)$	$Sp(2N_f)$	chGOE
Adjoint $N_c \geq 2, d = 4$	4	$U(2N_f)$	$O(2N_f)$	chGSE
Fundamental $N_c \geq 3, d = 3$	2	$U(N_f)$	$U(N_f/2) \times U(N_f/2)$	GUE
Fundamental $N_c = 2, d = 3$	1	$Sp(2N_f)$	$Sp(N_f) \times Sp(N_f)$	GOE
Adjoint $N_c \geq 2, d = 3$	4	$O(2N_f)$	$O(N_f) \times O(N_f)$	GSE

32.2.4 Chiral Lagrangian

The low-energy limit of QCD is given by a chiral Lagrangian and necessarily has the same transformation properties as the QCD partition function. The Lorentz invariant chiral Lagrangian to $O(M)$ and $O(p^2)$ is given by

$$L = \frac{1}{4}F^2 \operatorname{Tr} \partial_\mu U \partial_\mu U^\dagger - \frac{1}{2}\Sigma \operatorname{Tr}[MU^\dagger + M^\dagger U], \tag{32.2.6}$$

where $U \in G/H$ with G the global symmetry group that is spontaneously broken to H (see Table 32.1), and F is the pion decay constant. In the domain

$$M \ll \frac{\pi^2 F^2}{\Sigma L^2} \ll \frac{\pi^2 F^4}{\Sigma} \tag{32.2.7}$$

the kinetic term factorizes from the chiral Lagrangian [Gas87]. In this domain, known as the microscopic domain [Ver94a], the mass dependence of the QCD partition function in the sector of topological charge ν is given by

$$Z^\nu(M) = \int_{U \in G/H} dU \det{}^\nu U e^{\frac{1}{2}\Sigma \operatorname{Tr}[MU^\dagger + M^\dagger U]}. \tag{32.2.8}$$

The first inequality in (32.2.7) can be rewritten as $1/M_\pi \gg L$, i.e. the pion Compton wavelength is much larger than the size of the box. The domain where QCD is described by the chiral Lagrangian (32.2.6) including the kinetic term, i.e. where $M_\pi \sim p \ll \Lambda_{\text{QCD}}$, will be called the chiral domain. The zero momentum partition function (32.2.8) and the chiral Lagrangian (32.2.6) are the first terms of the ϵ and p expansion [Gas87], respectively. Therefore these domains are also known as the ϵ domain or p domain corresponding to a counting scheme where $M \sim 1/V$ and $M \sim 1/\sqrt{V}$, in this order.

Chiral random matrix theory can be reformulated in terms of a non-linear σ-model even for finite size matrices [Shu92, Jac96a] (see also Chapter 7). In this formulation the partition function is an integral over two types of modes, would-be Goldstone modes with mass $\sim \sqrt{Nm}$ and massive modes with mass $\sim \sqrt{N}$ (the number of integration variables does not depend on N). To leading order in $1/N$ and m the Goldstone mode part factorizes from the partition function.

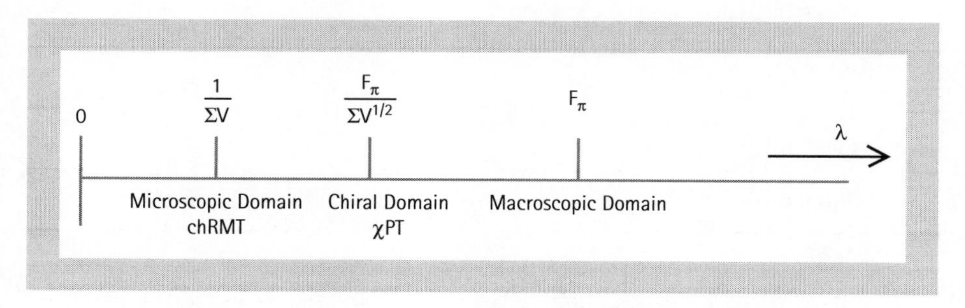

Fig. 32.1 Domains for the mass dependence of the QCD partition function.

Since the pattern of chiral symmetry breaking for QCD and chRMT is the same, their mass dependence is the same in the microscopic domain (32.2.7). The correction terms are of $O(Nm^2)$, so that chRMT results are universal for $m \ll 1/\sqrt{N}$. With the identification of N as V, this corresponds to the ϵ domain.

In lattice QCD, and QCD in general, the topological charge does not affect the block structure of the Dirac matrix but leads to non-trivial correlations between matrix elements. Then it is natural to work at fixed θ-angle (in fact at $\theta = 0$) with the partition function given by $Z(m, \theta)$. However, Dirac spectra depend on the topological charge, and in comparing lattice QCD and chRMT, Dirac spectra are sorted accordingly. In chRMT, topology is included by means of the block structure of the Dirac matrix right from the start, and it is natural to work at fixed topological charge. The two partitions functions are related by

$$Z(m, \theta) = \sum_{\nu=-\infty}^{\infty} e^{i\nu\theta} Z^{\nu}(m). \tag{32.2.9}$$

Since the spectrum of the Dirac operator depends on the topological charge it makes sense to introduce the topological domain [Leh09], as the domain where the average properties of the eigenvalues are sensitive to the topological charge. This domain is expected to coincide with the microscopic domain.

32.2.5 Generating function for the Dirac spectrum

The generating function for the Dirac spectrum is also given by (32.1.6) with the determinant of the physical quarks contained in the average. Therefore, z plays the role of a quark mass and the theory will have Goldstone bosons with squared mass $2z\Sigma/F^2$. Therefore, for physical quark masses, i.e. quark masses that remain fixed in the thermodynamic limit, we can choose [Ver95, Osb98b]

$$z \ll \frac{F^2}{\Sigma L^2} \equiv E_{\text{Th}}. \tag{32.2.10}$$

In this domain the z-dependence of the generating function at fixed topological charge ν is given by the zero momentum partition function (32.2.8) with quark masses equal to z. The energy scale in (32.2.10) is known as the Thouless energy. A similar conclusion was reached in [Jan98a]. The volume dependence of the Thouless energy has been confirmed by lattice simulations [Ber98b, Ber99]. The determinants in the generating function containing z have to be quenched which can be done by the replica trick (see Chapter 8) or the supersymmetric method (see Chapter 7). A supersymmetric version of the chiral Lagrangian is known and the zero momentum integral has been evaluated analytically [Osb98a, Dam98b].

The number of eigenvalues which is described by chRMT scales as $E_{\text{Th}}/\Delta\lambda = F^2 L^2/\pi$. Since $F \sim \sqrt{N_c}$, this number increases linearly with N_c [Nar04].

32.2.6 Chiral random matrix theory and the Dirac spectrum

In an influential paper that motivated the introduction of chRMT, Leutwyler and Smilga [Leu92] proposed to expand both the QCD partition function at fixed topology and its low-energy limit (i.e. at zero momentum) in powers of m and equate the coefficients. The resulting sum rules are saturated by eigenvalues in the microscopic domain which appear in the combination $\lambda_k V$. With eigenvalues that scale as $1/V$, the microscopic scaling limit of the spectral density is defined as [Shu92, Ver94a]

$$\rho_s(z) = \lim_{V\to\infty} \frac{1}{V\Sigma}\rho\left(\frac{z}{V\Sigma}\right) \tag{32.2.11}$$

with the microscopic scaling variable defined by $z = \lambda V\Sigma$. For $z \ll \sqrt{V}\Lambda^2$ the microscopic spectral density is given by chRMT. For all three values of the Dyson index, it can be expressed in terms of the Bessel kernel [For93]

$$K_a(x, y) = \sqrt{xy}\frac{xJ_{a+1}(x)J_a(y) - yJ_a(x)J_{a+1}(y)}{x^2 - y^2}. \tag{32.2.12}$$

The microscopic spectral density for $\beta = 2$ is given by [Ver93]

$$\rho_s^{\beta=2,a}(x) = \lim_{y\to x} K_a(x, y) = \frac{1}{2}x(J_a^2(x) - J_{a+1}(x)J_{a-1}(x)), \tag{32.2.13}$$

with $a = N_f + |\nu|$. The microscopic spectral density for $\beta = 1$ and $\beta = 4$ can be obtained by rewriting the partition function in terms of skew-orthogonal polynomials. For $\beta = 1$ we find [Ver93, For98, Ake08b]

$$\begin{aligned}
\rho_s^{\beta=1,a}(z) &= \frac{1}{4}J_a(z) \\
&\quad + \frac{z^a}{4}\int_0^\infty dw w^a \frac{(z-w)}{|z-w|}\left(\frac{1}{w}\frac{d}{dw} - \frac{1}{z}\frac{d}{dz}\right)(zw)^{-a+3/2}K_{a-2}(w, z) \\
&= \rho_s^{\beta=2,a}(z) + \frac{1}{2}J_a(|z|)\left(1 - \int_0^{|y|} dt J_a(t)\right)
\end{aligned} \tag{32.2.14}$$

with $a = 2N_f + |\nu|$, and for $\beta = 4$ the result is [Nag95, Ake08b]

$$\begin{aligned}
\rho_s^{\beta=4,a}(z) &= 2z^2\int_0^1 du u^2 \int_0^1 dv(1 - v^2)v^{-1/2}K_a(2uz, 2uvz) \\
&= \rho_s^{\beta=2,2a}(2z) - \frac{1}{2}J_{2a}(2z)\int_0^{2|z|} dt J_{2a}(t).
\end{aligned} \tag{32.2.15}$$

where $a = N_f + 2|\nu|$. Similar relations between the microscopic spectral density and the kernel for $\beta = 2$ exist for an arbitrary invariant probability potential [Sen98, Kle00], and can be exploited to show universality for $\beta = 1$ and $\beta = 4$ from the universality of the microscopic $\beta = 2$ kernel [Ake96] (see Chapter 6).

It is of interest to study the critical exponent of the average spectral density at a deformation of the Dirac operator for which it vanishes at $\lambda = 0$. This universality class differs from the microscopic scaling limit. The mean field value of this critical exponent is equal to 1/3 [Jac95], but other critical exponents can be obtained by fine tuning the probability distribution [Ake97, Ake02a, Jan02].

The microscopic spectral density (32.2.13) was first observed for lattice QCD Dirac spectra through the z dependence of the resolvent [Ver95]. The first direct comparison was made for staggered fermions with two fundamental colours [Ber97a] for which the Dyson index is $\beta = 4$ (Fig. 32.2, right). Results for QCD with $N_c = 3$ were obtained in [Dam98a, Goc98] (Fig. 32.2, middle). The left panel of Fig. 32.2 is for staggered fermions in the adjoint representation [Edw99].

The distribution of the smallest Dirac eigenvalue is given by $P^{\beta\nu}_{\min}(s) = -E'(s)$ with $E(s)$ defined as the probability that there are no eigenvalues in the interval $[0, s\rangle$. In Table 32.2 we summarize analytical results for the quenched case [Ede88, For93, Wil97, Dam00]. The result for $\nu = 0$ is particularly simple. Expressions for the kth smallest eigenvalue at arbitrary quark mass, topological charge, and Dyson index are known as well [Wil97, Nis98, Dam00]. A useful measure to compare lattice QCD results with chRMT predictions is the ratio of low-lying eigenvalues [Giu03, Fod09]. Agreement with the topology and mass dependence has become an important tool in lattice QCD to test the lattice implementation of chiral symmetry and topology.

The joint probability distribution of chRMT only depends on N_f and ν through the combination $2N_f + \beta\nu$. This property, known as flavour-topology duality [Ver97], has been observed in lattice QCD Dirac spectra [Fuk07].

Contrary to correlations of low-lying eigenvalues, bulk spectral correlations can be investigated by spectral averaging, and for large lattice volumes the Dirac spectrum of a single gauge field configuration is sufficient to obtain statistically

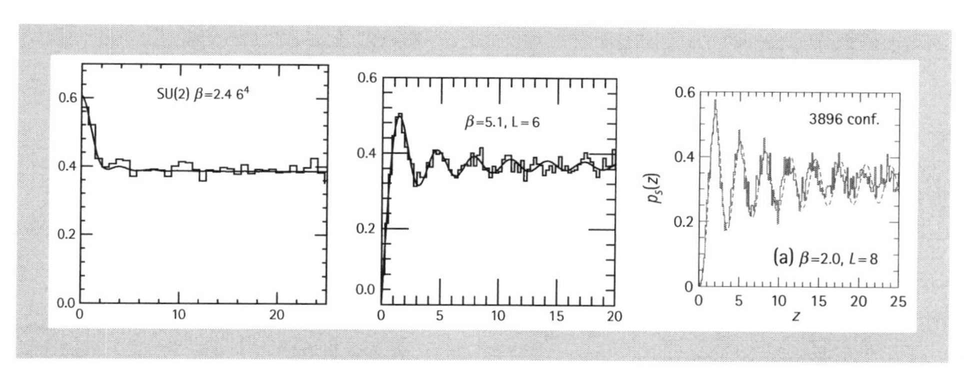

Fig. 32.2 Microscopic spectral density for $SU(2)$ gauge group in the adjoint representation (left, taken from [Edw99]) and in the fundamental representation (right, taken from [Ber98a]), and for $SU(3)$ gauge group in the fundamental representation (middle, taken from [Dam98a]).

Table 32.2 Results for the distribution of the smallest Dirac eigenvalue for the quenched case. In the last row a_j is an expression in terms of a sum over partitions of j. For explicit expressions we refer to [Ber98a].

$e^{\beta\zeta^2/8} P^{\beta\nu}_{\min}(\zeta)$	$\nu = 0$	$\nu = 1$	General ν ν odd for $\beta = 1$
$\beta = 1$	$\frac{1}{4}(2+\zeta)e^{-\zeta/2}$	$\frac{\zeta}{4}$	$\zeta^{(3-\nu)/2}\mathrm{Pf}[(i-j)\, I_{i+j+3}(\zeta)]_{i,j=-\nu/2+1,\dots,\nu/2-1}$
$\beta = 2$	ζ	$\frac{\zeta}{2} I_2(\zeta)$	$\frac{\zeta}{2}\det[I_{i-j+2}(\zeta)]_{i,j=1,\dots,\nu}$
$\beta = 4$	$\frac{1}{2}(e^{\zeta}(\zeta-1)+e^{-\zeta}(\zeta+1))$		$\zeta^{4\nu+3}(1+\sum_j a_j(\lvert\nu\rvert)\zeta^j)$

significant correlators. Excellent agreement with the Wigner-Dyson ensembles was obtained [Hal95a] without signature of a Thouless energy. It turns out [Guh98] that the Thouless energy scale is due to ensemble averaging. The conclusion is that there is no spectral ergodicity beyond the Thouless energy.

32.2.7 Integrability

For $\beta = 2$ the partition function of invariant random matrix theories can be interpreted as a partition function of non-interacting fermions which is an integrable system. This is the reason that the unitary matrix integral (32.2.8) obeys a large number of remarkable relations. The N_f flavour partition function for topological charge ν can be written as [Bro81a, Bro81b, Guh96, Jac96b]

$$Z^{\nu}_{N_f}(m_1, \cdots, m_{N_f}) = \frac{\det[x_k^{l-1} I_{\nu}^{(l-1)}(x_k)]_{k,l=1,\cdots,N_f}}{\Delta(\{x_k^2\})}, \qquad x_k = m_k V\Sigma. \quad (32.2.16)$$

In the limit $x_k \to x$ this partition function reduces to a Hankel determinant

$$Z^{\nu}_{N_f} = \det[(x\partial_x)^{k+l} I_{\nu}(x)]_{k,l=0,\cdots,N_f-1}. \quad (32.2.17)$$

Applying the Sylvester identity [For02] relating the determinant of a matrix to co-factors gives the Toda lattice equation [Kan02, Spl03a]

$$(x\partial_x)^2 \log Z^{\nu}_{N_f}(x) = 2N_f x^2 \frac{Z^{\nu}_{N_f+1}(x)\, Z^{\nu}_{N_f-1}(x)}{[Z^{\nu}_{N_f}(x)]^2}. \quad (32.2.18)$$

After taking the replica limit of this recursion relation we arrive at the following compact expression for the resolvent [Spl03a]

$$x\partial_x x G(x) = \lim_{N_f\to 0} \frac{1}{N_f}\partial_x \log Z^{\nu}_{N_f}(x) = 2x^2 Z^{\nu}_1(x) Z^{\nu}_{-1}(x). \quad (32.2.19)$$

This factorized form is a general property of the spectral density and correlation functions of RMTs with $\beta = 2$. The replica limit of a discrete recursion relation does not require analyticity in the replica variable and this way problems with the replica limit can be circumvented [Ver85, Kan02] (see Chapter 8).

As a consequence of integrability relations, the zero momentum partition function satisfies Virasoro constraints. They provide an efficient way to determine the coefficients of the small mass expansion [Dam99, Dal01].

In addition to the above relations we would like to mention: (i) The Toda lattice equation can be formulated for finite size random matrices using properties of orthogonal polynomials [Ake04]. (ii) The microscopic limit of partition functions in three and four dimensions satisfies $Z^{2N_f}_{\mathrm{QCD}_3}(\{x_k\}) = Z^{\nu=-1/2}_{N_f}(\{x_k\}) Z^{\nu=1/2}_{N_f}(\{x_k\})$ [Ake99, Ake00b, And04]. (iii) k-point spectral correlation functions can be expressed into partition functions with βk additional flavours [Ake98]. (iv) The correlation functions of invariant RMTs can be expressed in terms of the two-point kernel leading to relations between partition functions [Ake98]. (v) Interpreting the quark mass as an additional eigenvalue leads to relations between correlators of massive and massless partition functions [Ake00a].

32.3 Chiral random matrix theory at nonzero chemical potential

An important application of chRMT is to QCD at non-zero chemical potential μ. In that case the Dirac operator has no Hermiticity properties, and its eigenvalues are scattered in the complex plane. Because the determinant of the Dirac operator is complex, the QCD partition function at $\mu \neq 0$ is the average of a complex weight, and unless the chemical potential is small, it cannot be simulated by Monte-Carlo methods. For that reason chRMT has been particularly helpful in answering questions that could not be addressed otherwise. The following issues have been clarified: (i) The nature of the quenched approximation [Ste96b]. (ii) The relation between the chiral condensate and the spectrum of the Dirac operator for QCD with dynamical quarks [Osb05]. (iii) The expectation value of the phase of the fermion determinant [Spl06]. (iv) The low-energy limit of phase-quenched QCD and the spectrum of the Dirac operator [Tou00]. (v) The geometry of the support of the Dirac spectrum [Tou00, Osb08b].

The QCD partition function at nonzero chemical potential is given by

$$Z_{\mathrm{QCD}} = \langle \prod_{k=1}^{N_f} \det(D + m_k + \mu_k \gamma_0) \rangle. \tag{32.3.1}$$

where the average is over the Yang-Mills action. The chemical potential for different flavours is generally different. Two important special cases are: $\mu_k = \mu$, then μ is the baryon chemical potential, and the case for an even number of flavours $N_f = 2n$ with $\mu_k = \mu$ for $k = 1, \cdots, n$ and $\mu_k = -\mu$ for $k = n + 1, \cdots, 2n$. In the second case, the partition function is positive definite because

$$\det(D + m - \mu\gamma_0) = \det(D^\dagger + m + \mu\gamma_0) = \det{}^*(D + m + \mu\gamma_0), \quad (32.3.2)$$

and is known as the phase quenched partition function. For $n = 1$, μ can be interpreted as an isospin chemical potential [Son00].

32.3.1 Dirac spectrum

A particular useful tool for studying the spectrum of a non-Hermitian operator is the resolvent $G(z)$ (see (32.1.1)). This can be interpreted as the two-dimensional electric field at z of charges located at λ_k. The spectral density is given by

$$\rho(z, z^*) = \langle \sum_k \delta^2(z - \lambda_k) \rangle = \frac{1}{\pi} \frac{d}{dz^*} G(z). \quad (32.3.3)$$

When the eigenvalues are on the imaginary axis this picture illustrates that $G(z)$ has a discontinuity when z crosses the imaginary axis. When eigenvalues are not constrained by Hermiticity, because of level repulsion, they will scatter into the complex plane. Using the electrostatic analogy, the resolvent will be continuous. If z is outside the spectrum, $G(z)$ is analytic in z.

The resolvent cannot be expressed in terms of Eq. (32.1.6) but rather as a replica limit of the phase quenched partition function [Ste96b],

$$G(z) = \lim_{n \to 0} \frac{1}{n} \frac{d}{dz} \langle \det{}^n(D + \mu\gamma_0 + z) \det{}^n(D^\dagger + \mu\gamma_0 + z^*) \rangle. \quad (32.3.4)$$

Lattice QCD Dirac spectra were first calculated in [Bar86]. As remarkable features we note that the spectrum is approximately homogeneous, and that it has a sharp edge both of which are explained by chRMT.

32.3.2 Low-energy limit of QCD and phase quenched QCD

According to the definition of the grand canonical partition function, the free energy at low temperature does not depend on the chemical potential until it is equal to the lightest physical excitation (per unit charge) with charge conjugate to μ. For QCD this implies that the chiral condensate at zero temperature does not depend on μ until $\mu = m_N/3$ (with m_N the nucleon mass). The Dirac spectrum, is μ dependent, though, which seems to violate the Banks-Casher relation [Ban80]. This problem is known as the 'silver blaze problem' [Coh03].

At nonzero isospin chemical, μ_I, the critical chemical potential is equal to $\mu_I = m_\pi/2$. Beyond this point, pions will Bose condense. For light quarks, this phase transition can be studied by chiral perturbation theory, and for quark masses in the microscopic domain it is described by chRMT. At nonzero μ_I, the 'silver blaze problem' is that at zero temperature the chiral condensate remains constant until $\mu_I = m_\pi/2$, while the spectral density depends on μ_I. The solution is easy: according to the electrostatic analogy, the 'electric field', i.e. the chiral condensate, is constant outside a homogeneously charged strip. This implies that the width of the strip is determined by the relation $\mu_I = m_\pi/2$ [Gib86, Tou00], i.e. when the quark mass hits the boundary of the spectrum.

32.3.3 Chiral Lagrangian at nonzero chemical potential

Chiral symmetry remains broken at small nonzero chemical potential. Therefore, also in this case, the low-energy limit of QCD is given by a theory of weakly interacting Goldstone bosons corresponding to the spontaneous breaking of $U_L(N_f) \times U_R(N_f)$ to $U_V(N_f)$. The invariance properties of QCD should also hold for the Lagrangian that describes the low-energy limit of QCD. In particular, because the chemical potential is an external vector potential, it only enters in the combination of the covariant derivative [Kog99]

$$\nabla_\nu U = \partial_\nu U - [B_\nu, U], \qquad \text{with} \quad B_\nu = \text{diag}(\{\mu_k\})\delta_{\nu,0}. \tag{32.3.5}$$

Together with the mass term, the $O(p^2)$ chiral Lagrangian is thus given by

$$\mathcal{L} = \frac{F^2}{4}\nabla_\nu U \nabla_\nu U^\dagger - \frac{1}{2}\Sigma \text{Tr}(MU + MU^\dagger). \tag{32.3.6}$$

Eq. (32.3.6) shows that the chiral Lagrangian is determined by two constants which can be extracted from the density and the width of the Dirac spectrum.

At sufficient large chemical potential QCD may be in a colour-flavour locked phase with spontaneously broken colour-flavour symmetry. In this phase the chiral condensate vanishes, and terms quadratic in the quark mass, have to be taken into account in the chiral expansion. Universal results are obtained by scaling the Dirac eigenvalues with $\sqrt{V}$ and spectral sum rules have been derived both for QCD with three colours [Yam09] and QCD with two colours [Kan09].

32.3.4 Chiral random matrix theories at $\mu \neq 0$

In a suitably normalized chiral basis the Dirac operator at nonzero chemical potential has the block structure

$$D(\mu) = \begin{pmatrix} 0 & id + \mu \\ id^\dagger + \mu & 0 \end{pmatrix}. \tag{32.3.7}$$

A chiral random model at $\mu \neq 0$ is obtained [Ste96b] by replacing the matrix elements of d and $d^\dagger$ by an ensemble of random numbers exactly as in Section 32.2.3. Also in this case, because of expected universality [Ake02b], it is justified to simplify the model by choosing a Gaussian distribution.

As is the case for $\mu = 0$, we can distinguish three different non-Hermitian chRMTs [Hal97], with complex matrix elements ($\beta = 2$), with real matrix elements ($\beta = 1$), and with self-dual quaternion matrix elements ($\beta = 4$). They apply to the same cases as discussed in Table 32.1. The full classification is based on the Cartan classification of symmetric spaces [Zir96, Ber01, Mag07].

The random matrix model (32.3.7) is not unique. Adding μ in a different way results in the same chiral Lagrangian as long as the invariance properties of the matrix model remain the same. Contrary to the model (32.3.7), the model

$$D = \begin{pmatrix} 0 & id + \mu C \\ id^\dagger + \mu C^\dagger & 0 \end{pmatrix}, \tag{32.3.8}$$

introduced in [Osb04], has a representation in terms of eigenvalues so that methods that rely on the joint eigenvalue probability distribution can be used (C and d are complex random matrices with the same distribution).

A rerun of the arguments of [Gas87, Dam06] in the microscopic domain $m^2 V \ll 1$ and $\mu^4 V \ll 1$, shows that the partition function of the chiral Lagrangian (32.3.6) factorizes into a zero momentum part and a nonzero momentum part. The μ and mass dependence reside in the zero momentum part.

In the microscopic domain, global invariance properties for QCD at $\mu \neq 0$ also hold for chRMT, resulting in the same invariant terms. Therefore, in this domain, the QCD partition function is given by chRMT. Mean field studies only involve the zero momentum part of the chiral Lagrangian. Therefore, mean field results [Kog00] can be obtained from chRMT.

Applying the above arguments to the generating function for the Dirac spectrum we obtain the zero momentum partition function

$$\begin{aligned} Z_n^\nu(z, z^*; \mu) &= c \int_{U \in U(2n)} dU \det{}^\nu U e^{-\frac{VF^2\mu^2}{4}\mathrm{Tr}[U,B][U^\dagger,B] + \frac{1}{2}\Sigma V \mathrm{Tr}(MU + MU^\dagger)} \\ &\quad \text{with} \quad B = \Sigma_3 \quad \text{and} \quad M = \mathrm{diag}(z, z^*). \end{aligned} \tag{32.3.9}$$

At the mean field level the resolvent is independent of the replica index and the partition function can be analysed for $n = 1$. For the resolvent we find, $G(z) = \Sigma$, and $G(z) = \Sigma^2(z + z^*)/4\mu^2 F^2$ for $\mathrm{Re}\, z > 2\mu^2 F^2/\Sigma$ and $\mathrm{Re}\, z < 2\mu^2 F^2/\Sigma$, respectively. The eigenvalues are therefore distributed homogeneously inside a strip with width $4F^2\mu^2/\Sigma$. In agreement with lattice simulations [Bar86], the eigenvalue density has a sharp edge while the resolvent is continuous at this point. If z and z^* are interpreted as quark masses, the squared mass of the corresponding Goldstone bosons is equal to $m_G^2 = (z + z^*)\Sigma/F^2$. The condition $\mathrm{Re}\, z < 2\mu^2 F^2/\Sigma$ can then be written as $\mu = m_G/2$, in agreement with physical considerations.

32.3.5 Integrability of the partition function

Remarkably, as was the case at $\mu = 0$, the zero momentum partition function (32.3.9) can be rewritten in terms of a Hankel-like determinant [Spl03a, Spl03b]

$$Z_n^\nu(z, z^*, \mu) = D_n(zz^*)^{n(1-n)}\det_{k,l=0,1,\cdots,n-1}[(z\partial_z)^k(z^*\partial_{z^*})^l Z_1^\nu(z, z^*, \mu)]. \quad (32.3.10)$$

This form is responsible for the integrable structure of the partition function (32.3.9). Most notably, it satisfies the Toda lattice equation

$$z\partial_z z^*\partial_{z^*} \log Z_n^\nu(z, z^*, \mu) = \frac{\pi n}{2}(zz^*)^2 \frac{Z_{n+1}^\nu(z, z^*, \mu) Z_{n-1}^\nu(z, z^*, \mu)}{[Z_n^\nu(z, z^*, \mu)]^2}. \quad (32.3.11)$$

which is obtained by applying the Sylvester identity to the determinant in (32.3.10). This equation can be extended to imaginary chemical potential and the two-point correlation function [Spl03b, Dam05, Dam06]. For imaginary chemical potential there is no transition to a Bose condensed state and the n-dependent part of the free energy does not contribute. The non-trivial result in the free energy is $O(n^2)$ and gives the two-point correlation function.

The replica limit of the Toda lattice equation results in the spectral density

$$\rho(z, z^*, \mu) = \lim_{n\to 0} \frac{1}{n\pi}\frac{d}{dz}\frac{d}{dz^*} \log Z_n^\nu(z, z^*, \mu) = \frac{1}{2}zz^* Z_1^\nu(z, z^*, \mu) Z_{-1}^\nu(z, z^*, \mu), \quad (32.3.12)$$

which was derived in [Spl03b]. The fermionic partition function can be obtained by an explicit evaluation of the integral over $U(2)$. The result is given by

$$Z_1^\nu(z, z^*, \mu) = \frac{1}{\pi}e^{2VF^2\mu^2}\int_0^1 d\lambda\lambda e^{-2VF^2\mu^2\lambda^2} I_\nu(\lambda z\Sigma V) I_\nu(\lambda z^*\Sigma V). \quad (32.3.13)$$

Evaluation of the bosonic partition is more complicated. The inverse complex conjugated determinants can be written as a Gaussian integral after combining them into a Hermitian matrix (known as Hermitization [Jan97, Fei97]) and adding a mass $\sim \epsilon$. The $\log \epsilon$ divergence is due to a single eigenvalue close to z even if z is outside the Dirac spectrum [Spl03b, Spl08]. Because of the Vandermonde determinant, the probability of finding two eigenvalues close to z does not diverge. The partition function can be written as an integral over Goldstone bosons. Using the Ingham-Siegel integral [Fyo01], we obtain an integral over the noncompact manifold of positive definite Hermitian matrices,

$$Z_{-1}^\nu(z, z^*; \mu) = \lim_{\epsilon\to 0} C_\epsilon \int \frac{dQ}{\det^2 Q}\theta(Q) e^{\mathrm{Tr}[i\frac{V\Sigma}{2}\zeta^T(Q - IQ^{-1}I) - \frac{V}{4}F^2\mu^2[Q,\sigma_3][Q^{-1},\sigma_3]]},$$

$$\text{with} \quad I = \begin{pmatrix} 0 & 1 \\ -1 & 0 \end{pmatrix}, \qquad \zeta = \begin{pmatrix} \epsilon & z \\ z^* & \epsilon \end{pmatrix}. \quad (32.3.14)$$

The integral over Q can be performed analytically resulting in

$$Z_{-1}^\nu(z, z^*; \mu) = \frac{C_{-1}}{4\mu^2 F^2 V} e^{\frac{V\Sigma^2(y^2 - x^2)}{4\mu^2 F^2}} K_\nu(\frac{V\Sigma^2(x^2 + y^2)}{4\mu^2 F^2}). \quad (32.3.15)$$

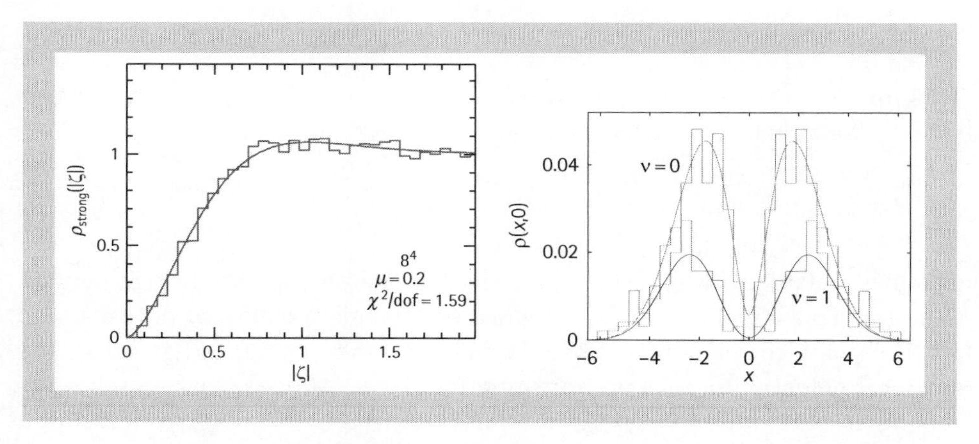

Fig. 32.3 Left: Radial microscopic spectral density for quenched QCD at $\mu \neq 0$ [Wet04]. Right: Spectral density of the overlap Dirac operator as a function of the distance, x, to the imaginary axis [Blo06].

In Fig. 32.3 we compare the expression (32.3.12) for the spectral density to quenched lattice data [Wet04, Blo06]. Results for nonzero topological charge (Fig. 32.3, right) were obtained using the Bloch-Wettig overlap Dirac operator.

Using superbosonization [Hac95, Bun07, Bas07], the fermionic and bosonic partition function can be combined into a supersymmetric partition function [Bas07] which can be used to derive the low-energy limit of the generating function of the QCD Dirac spectrum at $\mu \neq 0$.

32.3.6 Spectral density at $\mu \neq 0$ for QCD with dynamical quarks

Although the spectral density of the Dirac operator for QCD with dynamical quarks at $\mu \neq 0$ was first derived using complex orthogonal polynomials [Osb04], a simpler expression is obtained from the Toda lattice equation [Ake04],

$$\rho^{\nu}_{N_f}(z, z^*, \mu) \sim zz^* \prod_{f=1}^{N_f} (m_f^2 - z^2) \frac{Z^{n=-1}(z, z^*, \{m_f\}, \mu) Z^{n=1, N_f}(z, z^*, \{m_f\}, \mu)}{Z^{N_f}(\{m_f\})}. \tag{32.3.16}$$

In Fig. 32.4 we show a 3D plot of its real part. We can distinguish three phases in the Dirac spectrum. A phase with no eigenvalues, a phase with a constant eigenvalue density, and a phase with a strongly oscillating eigenvalue density that increases exponentially with the volume and has a period $\sim 1/V$. This region is absent in the (phase-)quenched case and is responsible for the discontinuity in the chiral condensate. These phases can be obtained [Osb08b] by means of a mean field study of a partition function with masses m, z, and z^*, similar to the analysis of three flavour QCD at $\mu \neq 0$ [Kog01] (see Fig. 32.4, right).

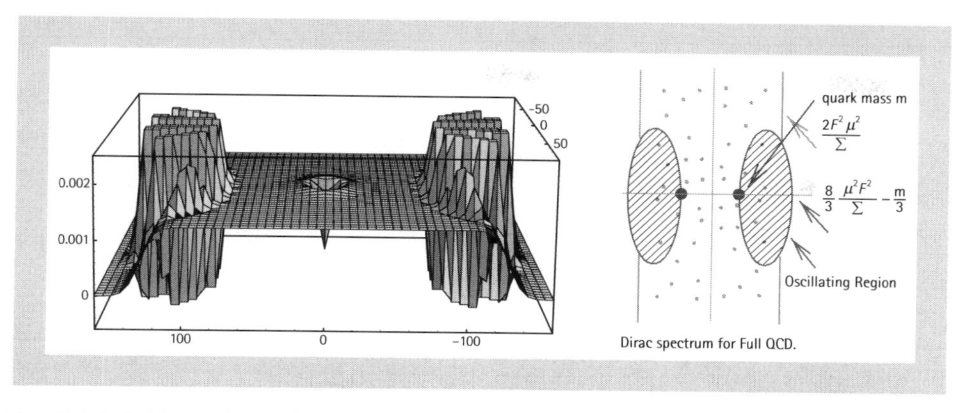

Fig. 32.4 Left: The real part of the spectral density of the QCD Dirac operator for one-flavour QCD at $\mu \neq 0$. For better illustration the z-axis has been clipped. Right: The phase diagram of the Dirac spectrum. This figure is reproduced in colour in the colour plates section.

In the microscopic domain, both the flat and oscillating regions give μ-dependent contributions the chiral condensate, which cancel in their sum [Osb05]. This solves the 'silver blaze problem' [Coh03]. It can be explained [Osb08a] in terms of orthogonality relations of the complex orthogonal polynomials.

32.3.7 The phase of the fermion determinant

Chiral random matrix theory can be used to study the complex phase of the fermion determinant. The average phase factor may be calculated with respect to the (phase) quenched or the two-flavour partition function. The phase quenched average

$$\langle e^{2i\theta}\rangle_{1+1^*} \equiv \left\langle \frac{\det(D + m + \mu\gamma_0)}{\det(D^\dagger + m + \mu\gamma_0)} \right\rangle_{1+1^*} = \frac{Z_{1+1}}{Z_{1+1^*}}. \qquad (32.3.17)$$

follows immediately from the expression from Z_{1+1^*} given in (32.3.9) (note that Z_{1+1} does not depend on μ).

The quenched average phase factor can be re-written in terms of a determinant of complex orthogonal polynomials. In the microscopic domain, the result is the sum of a polynomial in μ^2 and a part with an essential singularity at $\mu = 0$. Therefore it cannot be obtained from analytical continuation from an imaginary chemical potential which is polynomial in μ^2 [Dam05, Spl06, Blo08].

32.3.8 QCD at imaginary chemical potential

Random matrix models at imaginary chemical potential are obtained by replacing $\mu \to i\mu$ in the Dirac operator. Then the Dirac operator becomes anti-Hermitian with all eigenvalues on the imaginary axis. There are two ways of introducing an imaginary chemical potential, either as a multiple of the identity or as a multiple of a complex random matrix ensemble. Both models have the same symmetry properties and lead to the same universal partition function in the microscopic domain. Spectral correlation functions can be obtained by means of the Toda lattice equation [Dam06], or in the second model, by means of the method of bi-orthogonal polynomials [Ake08a].

Parametric correlations of Dirac spectra in the microscopic domain depend on two low-energy constants, F and Σ, which can be extracted from correlations of lattice QCD Dirac spectra [Dam06, Deg07, Ake08a].

32.4 Applications to gauge degrees of freedom

The Eguchi-Kawai model [Egu82] is the lattice Yang-Mills partition function with all links in the same direction identified. In the large N_c limit this model is an integral over $U(N_c)$-matrices. Although the original hope that Wilson loops of Yang-Mills theory are given by this reduced theory is incorrect, the model continues to attract a considerable amount of attention.

For $d = 4$ the Eguchi-Kawai model cannot be solved analytically, but for $d = 2$ it is known as the Brézin–Gross–Witten model [Bre80, Gro80]

$$Z = \int_{U \in U(N_c)} dU e^{\frac{1}{g^2}\mathrm{Tr}(U+U^\dagger)}, \tag{32.4.1}$$

and is identical to the zero momentum partition function (32.2.8) (see Chapter 17 for a discussion of such group integrals). In the large N_c limit this model undergoes a third-order phase transition at $g^2 N_c = 2$.

In the large N_c limit eigenvalues of Wilson loops can be analysed by means of RMT methods. It was shown [Dur80] that Wilson loops in two dimensions undergo a phase transition for $N_c \to \infty$ at a critical length of the loop. In one phase, its eigenvalues are distributed homogeneously over the unit circle, whereas in the other phase, they are localized at zero. This transition has been observed in lattice QCD [Nar06] and has been explained in terms of shock solutions of the Burgers equation [Bla08, Bla09, Neu08]. It can be studied by analysing spectra of products of unitary matrices [Gud03, Loh08].

32.5 Concluding remarks

Random matrix theory has changed our perspective of the QCD Dirac spectrum. Before the advent of chiral random matrix theory, the discrete structure of the Dirac spectrum was viewed as random noise that will go away in the continuum limit. Now we know that Dirac eigenvalues show intricate correlations that are determined by chiral random matrix theory with one or two low-energy constants as parameters. This implies that we can extract the chiral condensate and the pion decay constant from the distribution of individual eigenvalues.

Chiral random matrix theory primarily applies to the Dirac spectrum, and therefore we have to distinguish QCD at $\mu = 0$, when the Dirac operator is anti-Hermitian, and QCD at $\mu \neq 0$ with a non-Hermitian Dirac operator. In the first case the statistical properties of the low-lying Dirac eigenvalues are completely determined by the chiral condensate. In the second case they are determined by the chiral condensate and the pion decay constant. Since the non-Hermitian Dirac spectrum has the geometry of a strip, these constants are determined by the eigenvalue density and the width of the spectrum. The scale below which chRMT can be applied is set by the momentum-dependent terms in the chiral Lagrangian. Physically, it is the scale of the quark mass for which the Compton wave length of the Goldstone bosons is much larger than the size of the box. The scale of μ should also be well below the inverse box size.

For imaginary chemical potential, the Dirac operator is anti-Hermitian. Although spectral correlations are completely determined by the chiral condensate, this is not the case for parametric correlations for two different values of μ, which depend on both the chiral condensate and the pion decay constant.

There are two mechanisms to explain confinement in QCD: by condensation of monopoles, or by the disorder of gauge fields. The success of random matrix theory points to the second mechanism. It is our hope that the work discussed in this chapter will contribute to a solution of this problem.

Acknowledgments

This work was supported by U.S. DOE Grant No. DE-FG-88ER40388. Poul Damgaard and Kim Splittorff are thanked for a critical reading of the manuscript.

References

[Ake96] G. Akemann, P. Damgaard, U. Magnea and S. Nishigaki, Nucl. Phys. B **487**, 721 (1997).

[Ake97] G. Akemann, P. Damgaard, U. Magnea and S. M. Nishigaki, Nucl. Phys. B **519**, 682 (1998).

[Ake98] G. Akemann and P. Damgaard, Phys. Lett. B **432**, 390 (1998).

[Ake99] G. Akemann and P. Damgaard, Nucl. Phys. B **576**, 597 (2000).

[Ake00a] G. Akemann and E. Kanzieper, Phys. Rev. Lett. **85**, 1174 (2000).

[Ake00b] G. Akemann, D. Dalmazi, P. Damgaard and J. Verbaarschot, Nucl. Phys. B **601**, 77 (2001).

[Ake02a] G. Akemann and G. Vernizzi, Nucl. Phys. B **631**, 471 (2002).

[Ake02b] G. Akemann, Phys. Lett. B **547**, 100 (2002).

[Ake04] G. Akemann, J. C. Osborn, K. Splittorff and J. J. M. Verbaarschot, Nucl. Phys. B **712**, 287 (2005).

[Ake07] G. Akemann, Int. J. Mod. Phys. A **22**, 1077 (2007).

[Ake08a] G. Akemann and P. Damgaard, JHEP **0803**, 073 (2008).

[Ake08b] G. Akemann and P. Vivo, J. Stat. Mech. P09002 (2008).

[And04] T. Andersson, P. Damgaard and K. Splittorff, Nucl. Phys. B **707**, 509 (2005).

[Ban80] T. Banks and A. Casher, Nucl. Phys. **B 169**, 103 (1980).

[Bar86] I. Barbour *et al.*, Nucl. Phys. B **275**, 296 (1986).

[Bas07] F. Basile and G. Akemann, JHEP **0712**, 043 (2007).

[Ber97a] M. Berbenni-Bitsch, *et al.*, Phys. Rev. Lett. **80**, 1146 (1998).

[Ber98a] M. Berbenni-Bitsch, S. Meyer and T. Wettig, Phys. Rev. D **58**, 071502 (1998).

[Ber98b] M. Berbenni-Bitsch *et al.*, Phys. Lett. B **438**, 14 (1998).

[Ber99] M. Berbenni-Bitsch *et al.* Phys. Lett. B **466**, 293 (1999).

[Ber01] D. Bernard and A. LeClair, [arXiv:cone-mat/0110649].

[Bla08] J. Blaizot and M. Nowak, Phys. Rev. Lett. **101**, 102001 (2008).

[Bla09] J. Blaizot and M. Nowak, arXiv:0902.2223 [hep-th].

[Blo06] J. Bloch and T. Wettig, Phys. Rev. Lett. **97**, 012003 (2006).

[Blo08] J. Bloch and T. Wettig, JHEP **0903**, 100 (2009).

[Bre78] E. Brézin, C. Itzykson, G. Parisi and J. Zuber, Commun. Math. Phys. **59**, 35 (1978).

[Bre80] E. Brézin and D. Gross, Phys. Lett. B **97**, 120 (1980).

[Bre95] E. Brézin, S. Hikami and A. Zee, Nucl. Phys. B **464**, 411 (1996).

[Bro81a] R. Brower and M. Nauenberg, Nucl. Phys. B **180**, 221 (1981).

[Bro81b] R. Brower, P. Rossi and C. I. Tan, Nucl. Phys. B **190**, 699 (1981).

[Bun07] J. Bunder, K. Efetov, V. Kravtsov, O. Yevtushenko and M. Zirnbauer, J. Stat. Phys. **129**, 809 (2007).

[Coh03] T. Cohen, Phys. Rev. Lett. **91**, 222001 (2003).

[Dal01] D. Dalmazi and J. Verbaarschot, Phys. Rev. D **64**, 054002 (2001).

[Dam98a] P. Damgaard, U. Heller and A. Krasnitz, Phys. Lett. B **445**, 366 (1999).

[Dam98b] P. Damgaard, J. Osborn, D. Toublan and J. Verbaarschot, Nucl. Phys. **B 547**, 305 (1999).

[Dam99] P. Damgaard and K. Splittorff, Nucl. Phys. B **572**, 478 (2000).

[Dam00] P. Damgaard and S. Nishigaki, Phys. Rev. D **63**, 045012 (2001).

[Dam05] P. Damgaard, U. Heller, K. Splittorff and B. Svetitsky, Phys. Rev. D **72**, 091501 (2005).

[Dam06] P. Damgaard, U. Heller, K. Splittorff, B. Svetitsky and D. Toublan, Phys. Rev. D **73**, 105016 (2006).

[Deg07] T. DeGrand and S. Schaefer, Phys. Rev. D **76**, 094509 (2007).

[DiF93] P. Di Francesco, P. Ginsparg and J. Zinn-Justin, Phys. Rep. **254**, 1 (1995).

[Dur80] B. Durhuus and P. Olesen, Nucl. Phys. B **184**, 406 (1981).

[Dys62] F. Dyson, J.Math. Phys. **3**, 1199 (1962).

[Ede88] A. Edelman, SIAM J. Matrix. Anal. Appl. **9**, 543 (1988).

[Edw99] R. Edwards, U. Heller and R. Narayanan, Phys. Rev. D **60**, 077502 (1999) [arXiv:hep-lat/9902021].

[Egu82] T. Eguchi and H. Kawai, Phys. Rev. Lett. **48**, 1063 (1982).

[Fei97] J. Feinberg and A. Zee, Nucl. Phys. B **504**, 579 (1997).

[For02] P. Forrester and N. Witte, Comm. P. Appl. Math. **55**, 679 (2002).

[For93] P. Forrester, Nucl. Phys. **B402**, 709 (1993).

[For98] P.J. Forrester, T. Nagao and G. Honner, Nucl. Phys. **B533**, 601 (1999).

[Fod09] Z. Fodor, *et al.*, arXiv:0907.4562 [hep-lat].

[Fuk07] H. Fukaya *et al.* Phys. Rev. Lett. **98**, 172001 (2007).
[Fyo01] Y. Fyodorov, Nucl. Phys. B **621**, 643 (2002).
[Gas87] J. Gasser and H. Leutwyler, Phys. Lett. **188B**, 477 (1987).
[Gib86] P. Gibbs, Glasgow Preprint 86-0389 (1986).
[Giu03] L. Giusti, M. Luscher, P. Weisz and H. Wittig, JHEP **0311**, 023 (2003).
[Giu09] L. Giusti and M. Luscher, JHEP **0903**, 013 (2009).
[Goc98] M. Gockeler, *et al.*, Phys. Rev. D **59**, 094503 (1999).
[Gro80] D. Gross and E. Witten, Phys. Rev. D **21**, 446 (1980).
[Gud03] E. Gudowska-Nowak, R. Janik, J. Jurkiewicz and M. A. Nowak, Nucl. Phys. B **670**, 479 (2003).
[Guh96] T. Guhr and T. Wettig, J. Math. Phys. **37**, 6395 (1996).
[Guh97a] T. Guhr and T. Wettig, Nucl. Phys. B **506**, 589 (1997).
[Guh97b] T. Guhr, A. Müller-Groeling and H. Weidenmüller, Phys. Rept. **299**, 189 (1998).
[Guh98] T. Guhr, J. Ma, S. Meyer and T. Wilke, Phys. Rev. D **59**, 054501 (1999).
[Hac95] G. Hackenbroich and H. Weidenmüller, Phys. Rev. Lett. **74**, 4118 (1995).
[Hal95a] A. Halasz and J. Verbaarschot, Phys. Rev. Lett. **74**, 3920 (1995).
[Hal95b] A. Halasz and J. Verbaarschot, Phys. Rev. D **52**, 2563 (1995).
[Hal97] M. Halasz, J. Osborn and J. Verbaarschot, Phys. Rev. D **56**, 7059 (1997).
[Jac95] A. Jackson and J. Verbaarschot, Phys. Rev. D **53**, 7223 (1996).
[Jac96a] A. Jackson, M. Sener and J. Verbaarschot, Nucl. Phys. B **479**, 707 (1996).
[Jac96b] A. Jackson, M. Sener and J. Verbaarschot, Phys. Lett. B **387**, 355 (1996).
[Jan97] R. Janik, M. Nowak, G. Papp and I. Zahed, Acta Phys. Polon. B **28**, 2949 (1997).
[Jan98a] R. Janik, M. Nowak, G. Papp and I. Zahed, Phys. Rev. Lett. **81**, 264 (1998).
[Jan98b] R. Janik, M. Nowak, G. Papp and I. Zahed, Acta Phys. Polon. B **29**, 3957 (1998).
[Jan02] R. Janik, Nucl. Phys. B **635**, 492 (2002).
[Kan02] E. Kanzieper, Phys. Rev. Lett. **89**, 250201 (2002).
[Kan09] T. Kanazawa, T. Wettig and N. Yamamoto, JHEP **0908**, 003 (2009).
[Kle00] B. Klein and J. Verbaarschot, Nucl. Phys. B **588**, 483 (2000).
[Kog99] J. Kogut, M. Stephanov and D. Toublan, Phys. Lett. B **464**, 183 (1999).
[Kog00] J. Kogut, M. Stephanov, D. Toublan, J. Verbaarschot and A. Zhitnitsky, Nucl. Phys. B **582**, 477 (2000).
[Kog01] J. Kogut and D. Toublan, Phys. Rev. D **64**, 034007 (2001).
[Leh09] C. Lehner, M. Ohtani, J. Verbaarschot and T. Wettig, arXiv:0902.2640 [hep-th].
[Leu92] H. Leutwyler and A. Smilga, Phys. Rev. **D46**, 5607 (1992).
[Loh08] R. Lohmayer, H. Neuberger and T. Wettig, JHEP **0811**, 053 (2008).
[Mag99a] U. Magnea, Phys. Rev. D **61**, 056005 (2000).
[Mag99b] U. Magnea, Phys. Rev. D **62**, 016005 (2000).
[Mag07] U. Magnea, J. Phys. A: Math. Theor. **41**, 045203 (2008).
[Nag95] T. Nagao and P. Forrester, Nucl. Phys. B **435**, 401 (1995).
[Nar04] R. Narayanan and H. Neuberger, Nucl. Phys. B **696**, 107 (2004).
[Nar06] R. Narayanan and H. Neuberger, Phys. Lett. B **646**, 202 (2007).
[Neu08] H. Neuberger, Phys. Lett. B **670** (2008) 235.
[Nis98] S. Nishigaki, P. Damgaard and T. Wettig, Phys. Rev. D **58**, 087704 (1998).
[Osb98a] J.C. Osborn, D. Toublan and J. J. M. Verbaarschot, Nucl. Phys. **B 540**, 317 (1999).
[Osb98b] J. Osborn and J. Verbaarschot, Phys. Rev. Lett. **81**, 268 (1998).
[Osb04] J.C. Osborn, K. Splittorff, and J.J.M. Verbaarschot, Phys. Rev. Lett. **94** (2005) 202001.
J. Osborn, Phys. Rev. Lett. **93**, 222001 (2004).
[Osb05] J. Osborn, K. Splittorff and J. Verbaarschot, Phys. Rev. Lett. **94**, 202001 (2005).
[Osb08a] J. Osborn, K. Splittorff and J. Verbaarschot, Phys. Rev. D **78**, 065029 (2008).

[Osb08b] J. Osborn, K. Splittorff and J. Verbaarschot, Phys. Rev. D **78**, 105006 (2008).
[Pes80] M. Peskin, Nucl. Phys. **B175**, 197 (1980).
[Por65] C. Porter, *Statistical Theory of Spectra: Fluctuations*, Academic Press, New York, 1965.
[Sen98] M. Sener and J. Verbaarschot, Phys. Rev. Lett. **81**, 248 (1998).
[Shu92] E. Shuryak and J. Verbaarschot, Nucl. Phys. A **560**, 306 (1993).
[Son00] D. Son and M. Stephanov, Phys. Rev. Lett. **86**, 592 (2001).
[Spl03a] K. Splittorff and J. Verbaarschot, Phys. Rev. Lett. **90**, 041601 (2003).
[Spl03b] K. Splittorff and J. Verbaarschot, Nucl. Phys. B **683**, 467 (2004).
[Spl06] K. Splittorff and J. Verbaarschot, Phys. Rev. Lett. **98**, 031601 (2007).
[Spl08] K. Splittorff, J. Verbaarschot and M. Zirnbauer, Nucl. Phys. B **803**, 381 (2008).
[Ste96b] M. Stephanov, Phys. Rev. Lett. **76**, 4472 (1996).
[Sza00] R. Szabo, Nucl. Phys. B **598**, 309 (2001).
[tHo74] G. 't Hooft, Nucl. Phys. B **75**, 461 (1974).
[Tou00] D. Toublan and J. Verbaarschot, Nucl. Phys. B **603**, 343 (2001).
[Vaf83] C. Vafa and E. Witten, Nucl. Phys. B **234**, 173 (1984).
[Ver85] J. Verbaarschot and M. Zirnbauer, J. Phys. **A18**, 1093 (1985).
[Ver93] J. Verbaarschot and I. Zahed, Phys. Rev. Lett. **70**, 3852 (1993).
[Ver94a] J. Verbaarschot, Phys. Rev. Lett. **72**, 2531 (1994).
[Ver94b] J. Verbaarschot and I. Zahed, Phys. Rev. Lett. **73**, 2288 (1994).
[Ver95] J. Verbaarschot, Phys. Lett. B **368**, 137 (1996).
[Ver97] J. Verbaarschot, arXiv:hep-th/9710114.
[Ver00] J. Verbaarschot and T. Wettig, Ann. Rev. Nucl. Part. Sci. **50**, 343 (2000).
[Ver05] J. Verbaarschot, arXiv:hep-th/0502029.
[Vys85] M. Vysotskii, Y. Kogan and M. Shifman, Sov. J. Nucl. Phys. **42**, 318 (1985).
[Wet04] T. Wettig, private communication (2004).
[Wig55] E. Wigner, Ann. Math. **62**, 548 (1955).
[Wil97] T. Wilke, T. Guhr and T. Wettig, Phys. Rev. D **57**, 6486 (1998).
[Yam09] N. Yamamoto and T. Kanazawa, Phys. Rev. Lett. **103**, 032001 (2009).
[Zir96] M. Zirnbauer, J. Math. Phys. **37**, 4986 (1996).

·33·

Quantum chaos and quantum graphs

Sebastian Müller and Martin Sieber

Abstract

The spectral statistics of quantum chaotic systems have been conjectured to be universal and agree with predictions from random matrix theory. We discuss the origins of this universality from the viewpoint of periodic orbit theory. We also point out interesting analogies between periodic orbit theory and the sigma model, and review related work on quantum graphs.

33.1 Introduction

An important area of applications of random matrix theory is quantum chaos. In quantum chaos one considers systems that behave chaotically in the classical limit, implying that the classical motion depends sensitively on the initial conditions. Many examples for chaos can be found in atomic, molecular, and nuclear physics as well as in billiards. One then studies the quantum properties of such systems [Gut90, Sto99, Haa10]. Surprisingly, many of these properties turn out to be universal, and agree with the corresponding results from random matrix theory (RMT). For instance it was conjectured [Boh84] (see also [Ber77, Cas80]) that in the semiclassical limit the spectral statistics of fully chaotic quantum systems agrees with predictions from the corresponding ensembles of RMT. For systems without any symmetries one thus expects spectral correlation functions in line with those for the Gaussian unitary ensemble (GUE), and for (spinless) systems whose sole symmetry is time-reversal invariance one expects correlations as for the Gaussian orthogonal ensemble (GOE). In contrast integrable systems are expected to have Poissonian level statics, with energy levels behaving like uncorrelated random numbers [Ber77].

The random matrix conjecture is supported by broad experimental and numerical evidence. On the other hand, there are examples of chaotic systems (e.g. systems displaying arithmetic chaos) differing from the RMT predictions. This raises the question of the precise conditions for universal behaviour.

The focus of this chapter will be to review the present understanding of the origins of universal spectral statistics, for generic fully chaotic systems. Hence we will exclude many other interesting lines of research in quantum chaos, including work on transport phenomena as well as wavefunctions. In discussing spectral statistics our main goal is to make the ideas accessible for non-experts, rather than give a systematic overview of the literature.

The method of choice to understand the relation between classical chaos and energy spectra is semiclassics: In the semiclassical limit the level density of a chaotic system can be written as a sum over contributions from its classical orbits [Gut90]. Correlation functions of the level density then involve sums over multiplets of orbits. To evaluate these sums we need to understand universal mechanisms for correlations between classical periodic orbits [Arg93]. A systematic evaluation of these sums then leads to expressions familiar from RMT, and interesting analogies to the sigma model formalism (see Chapter 7).

We will start this chapter by reviewing the main classical features of chaotic systems. We will then introduce two semiclassical approximation techniques (Gutzwiller's periodic orbit theory and a refined approach incorporating the unitarity of the quantum evolution) and show how they can be used to understand universal spectral statistics; we will also comment on their relation to the sigma model. This is followed by a discussion of parallel developments for important model systems for quantum chaos, called quantum graphs.

33.2 Classical chaos

We start by summarizing some facts and definitions for classically chaotic systems which will be needed to understand their quantum behaviour. For simplicity we will consider systems with two degrees of freedom.

Ergodicity. A system is called *ergodic* if almost all infinitely long trajectories uniformly fill the energy shell, i.e. the surface of constant energy in phase space.

Hyperbolicity. A system is called *hyperbolic* if at almost every point on the energy shell we can specify *stable* and *unstable directions* transverse to the direction of the flow. The stable direction is chosen such that two trajectories separated initially by an infinitesimal deflection along the stable direction approach each other exponentially fast. In contrast initial separations along the unstable direction grow exponentially fast and shrink for large negative times. The rate of approach/divergence is called the *Lyapunov exponent* λ.

Throughout this chapter we will consider systems that are fully chaotic, i.e. both ergodic and hyperbolic.

Periodic orbits. Periodic orbits are exceptional as they cannot reach all points on the energy shell. However it is possible to make statistical statements about ensembles of long periodic orbits [Bow72]. For instance, *Hannay and Ozorio de Almeida's sum rule* [Han84] states that sums over orbits γ weighted according to their stability and their periods T_γ can be evaluated using

$$\sum_{\gamma} \frac{T_\gamma^2}{|\det(M_\gamma - 1)|} \delta_\varepsilon(T - T_\gamma) \sim T \quad \text{as} \quad T \to \infty. \tag{33.2.1}$$

Here δ_ε denotes a smoothed delta function and M_γ is the so-called stability matrix. This matrix maps infinitesimal initial deviations of neighbouring trajectories from the periodic orbit to deviations after one period. Equation (33.2.1) holds because the exponential decrease of the orbit weight for large periods is compensated by an exponential increase of the number of orbits.

Encounters. A long periodic orbit often comes close to itself in phase space. If such an *'encounter'* occurs, it is possible to change the connection of the close-by parts of the orbit as sketched in Figs. 33.1 and 33.2, and obtain a different periodic orbit. This means that the periodic orbits of classically chaotic systems are not independent. They come in groups (*'bunches'*) of orbits that are very similar to each other and differ noticeably only in the way they are connected inside the encounters [Sie01, Alt08].

We start from the example identified by Sieber and Richter in [Sie01] for time-reversal invariant systems. Figure 33.1 shows a schematic picture of an orbit (full line) that has one encounter region containing two 'stretches' (thick lines) that are almost mutually time reversed. They are very close in configuration space, but traversed with opposite directions. Outside the encounter the stretches are connected by so-called 'links' which are normally very long parts of the orbit (sketched by the left and right loops in the picture).

It is possible to change the connections inside the encounter such that afterwards the two links are joined differently as indicated by the dashed orbit. The avoided crossing of the thick stretches is thus replaced by a crossing. Outside the encounter the orbits look practically the same. This is a consequence of hyperbolicity: during the left link in Fig. 33.1 the orbit and its partner initially approach each other exponentially fast, since their deviation is almost in the stable direction. During the right link the same is true for the orbit and the time reversed of its partner. Also note that going through the same link with an opposite sense of motion is possible only for time-reversal invariant systems.

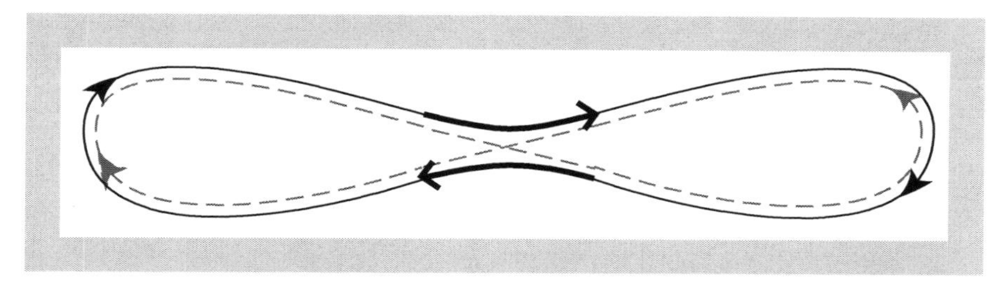

Fig. 33.1 Schematic sketch of an orbit pair (picture taken from [Mue05]).

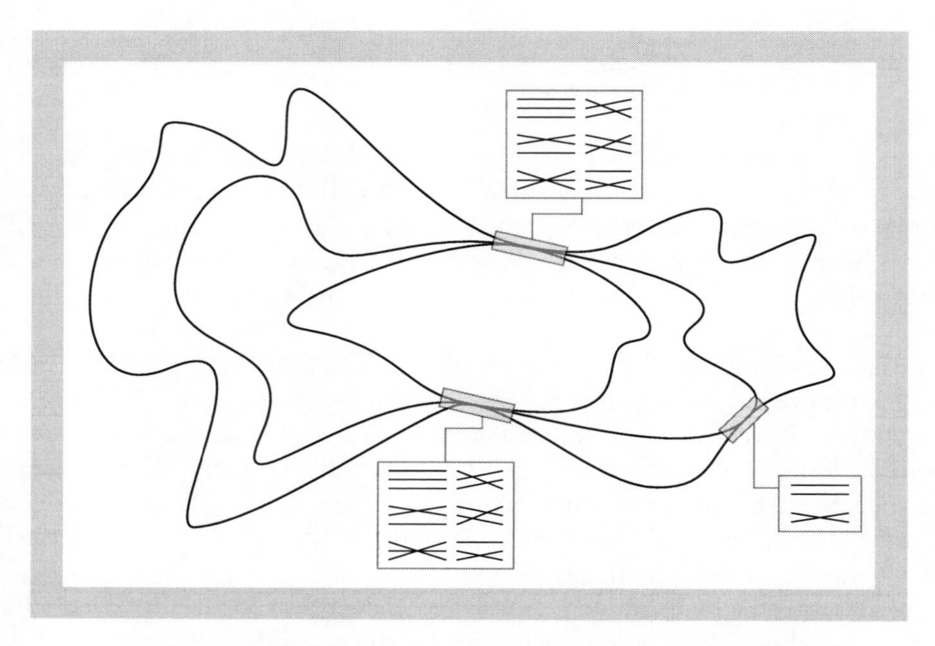

Fig. 33.2 More complicated example for a bunch of orbits (from [Alt08]).

There are many more complicated examples based on the same principle. Orbits can differ by their connections inside several encounters, each involving stretches that are either close in phase space $\rightrightarrows$ or almost mutually time-reversed $\rightleftarrows$. There can also be encounters with more than two close or almost time-reversed stretches, as in ⇶ or ⇶. The connections inside the encounter can then be changed as well, e.g. leading to ⋊. Figure 33.2 shows an example involving two encounters with three stretches and one encounter with two stretches. In each encounter, the connections can be chosen in any of the ways depicted in the inset. Some choices lead to periodic orbits. For others (e.g. choosing ⇶, ⇶ and $\rightrightarrows$) the picture decomposes into several disjoint periodic orbits. Such sets of disjoint orbits are referred to as pseudo-orbits. As for Fig. 33.1 all examples where the stretches in one of the encounters are almost mutually time reversed (as in $\rightleftarrows$ or ⇶) require time-reversal invariance.

33.3 Gutzwiller's trace formula and spectral statistics

The trace formula We now move on to discuss the quantum properties of chaotic systems. This is most conveniently done in the semiclassical limit, where $\hbar$ is much smaller than all classical actions in the system. The semiclassical limit is justified for the high-lying eigenstates of the Hamiltonian. In this limit, Gutzwiller [Gut90] used the Feynman path integral and stationary phase approximations to show that the level density $\rho(E)$ is given by a smooth term $\overline{\rho}(E)$, plus fluctuations given by a sum over classical periodic orbits γ,

$$\rho(E) = \sum_j \delta(E - E_j) \approx \overline{\rho}(E) + \frac{1}{\pi\hbar}\mathrm{Re}\sum_\gamma A_\gamma e^{iS_\gamma(E)/\hbar}. \tag{33.3.1}$$

Apart from $\hbar$, the right-hand side depends only on classical properties of the system: the smooth (Weyl) term can be written as $\overline{\rho}(E) = \frac{\Omega(E)}{(2\pi\hbar)^f}$ where f is the number of degrees of freedom of the system and $\Omega(E)$ is the volume of the energy shell with energy E. The fluctuations depend on the actions S_γ and stability amplitudes A_γ of the periodic orbits with energy E. The latter amplitudes are defined as $A_\gamma = \frac{T_\gamma e^{-i\mu_\gamma \frac{\pi}{2}}}{\sqrt{|\det(M_\gamma - 1)|}}$ which allows us to write the weight in (33.2.1) as $|A_\gamma|^2$. (For orbits that are repetitions of shorter orbits A_γ also has to be divided by the number of repetitions.) The amplitude contains a topogical phase factor involving the (integer) Maslov index μ_γ [Gut90, Haa10]. For convergence, E has to be taken with a positive imaginary part.

Universal spectral statistics We now want to use the Gutzwiller formula to show that chaotic systems have universal spectral statistics, referring to the original literature and [Haa10] for details. We will consider the correlation function

$$R(\epsilon) = \frac{1}{\overline{\rho}^2}\left\langle \rho\left(E + \frac{\epsilon}{2\pi\overline{\rho}}\right)\rho\left(E - \frac{\epsilon}{2\pi\overline{\rho}}\right)\right\rangle - 1 \tag{33.3.2}$$

where $\langle \ldots \rangle$ is an energy average. We want to explain why *in the limit* $\hbar \to 0$ *the spectral correlation functions* $R(\epsilon)$ *of generic fully chaotic systems are faithful to the predictions from the corresponding RMT ensembles.* Here *'fully chaotic'* means hyperbolic, ergodic, and (for technical steps skipped here [Mue05]) mixing. *'Generic'* means that there should be no systematic correlations between orbits apart from those discussed above (excluding e.g. arithmetic chaos).

The RMT predictions for $R(\epsilon)$ have been obtained in Chapter 6 and have the following form

$$R_{\mathrm{GUE}}(\epsilon) = -\left(\frac{\sin\epsilon}{\epsilon}\right)^2, \tag{33.3.3}$$

$$R_{\mathrm{GOE}}(\epsilon) = -\left(\frac{\sin\epsilon}{\epsilon}\right)^2 + \left(\int_0^\epsilon \frac{\sin y}{y}dy - \frac{\pi}{2}\mathrm{sgn}(\epsilon)\right)\left(\frac{\cos\epsilon}{\epsilon} - \frac{\sin\epsilon}{\epsilon^2}\right).$$

(Our ϵ corresponds to $\pi(x - y)$ in Chapter 6 and we subtract 1.) These predictions can be written as a sum of two series expansions in $1/\epsilon$, one of them with an additional oscillatory factor $e^{2i\epsilon}$,

$$R(\epsilon) \sim \mathrm{Re}\sum_{n=2}^{\infty}(c_n + d_n e^{2i\epsilon})\left(\frac{1}{\epsilon}\right)^n, \qquad (\epsilon > 0). \tag{33.3.4}$$

In the unitary case we have $c_2 = -\frac{1}{2}$ and $d_2 = \frac{1}{2}$ while all other coefficients vanish. In the orthogonal case we have $c_2 = -1$, $c_n = \frac{(n-3)!(n-1)}{2i^n}$ for $n \geq 3$, $d_2 = d_3 = 0$ and $d_n = \frac{(n-3)!(n-3)}{2i^n}$ for $n \geq 4$. Alternatively we can consider the Fourier transform of $R(\epsilon)$ w.r.t. ϵ, the spectral form factor $K(\tau)$. Then the non-oscillatory terms yield a series expansion in terms of τ. The oscillatory terms contribute only for $\tau > 1$ and cause the singularity of $K(\tau)$ at $\tau = 1$. We will first recover the non-oscillatory terms c_n. A treatment of the oscillatory contributions requires a refinement of the semiclassical approximation to be discussed in Section 33.4.

If we insert the Gutzwiller trace formula for the two level densities in Eq. (33.3.2) we obtain a double sum over periodic orbits. After some algebra this double sum can be brought to the form

$$R(\epsilon) = \frac{2}{T_H^2}\mathrm{Re}\left\langle \sum_{\gamma,\gamma'} A_\gamma A_{\gamma'}^* e^{i(S_\gamma(E)-S_{\gamma'}(E))/\hbar} e^{i(T_\gamma+T_{\gamma'})\epsilon/T_H} \right\rangle. \qquad (33.3.5)$$

Here we have used $\frac{dS_\gamma}{dE} = T_\gamma$ and thus $S_\gamma\left(E \pm \frac{\epsilon}{2\pi\bar{\rho}}\right)/\hbar \approx S_\gamma(E)/\hbar \pm \epsilon T_\gamma/T_H$, where T_H is the Heisenberg time $T_H = 2\pi\hbar\bar{\rho}$. Spectral statistics is thus determined by the *classical dynamics* (periodic orbits), combined with quantum mechanical *interference effects*. The interference between the contributions of two orbits is expressed by a phase factor involving the difference between the two classical actions $\Delta S = S_\gamma - S_{\gamma'}$, divided by $\hbar$. For small $\hbar$, the phase factor usually oscillates wildly, and therefore vanishes after averaging over the energy. These oscillations can only be avoided if ΔS is very small, at most of the order of $\hbar$. This means that we need to find pairs of orbits with small action differences. These orbit pairs need to be systematically correlated because a random distribution of the actions would also lead to a vanishing correlation function [Arg93].

Diagonal approximation The first step in this direction was taken in [Han84, Ber85]. In the diagonal approximation introduced in these papers one takes into account only γ' identical to γ and, for time reversal invariant systems, γ' time-reversed w.r.t. γ. The correlation function can thus be approximated by a single sum over $|A_\gamma|^2 e^{2iT_\gamma\epsilon/T_H}$. This sum is easily evaluated using (33.2.1). It yields

$$R_{\mathrm{diag}}(\epsilon) = \frac{2\kappa}{T_H^2}\mathrm{Re}\int_0^\infty dT\; T e^{2i\epsilon T/T_H} = -\frac{\kappa}{2\epsilon^2}. \qquad (33.3.6)$$

Here the degeneracy factor κ is defined as $\kappa = 1$ for systems without time-reversal invariance and $\kappa = 2$ for time-reversal invariant systems (due to the inclusion of mutually time-reversed orbits). Equation (33.3.6) gives the leading non-oscillatory contribution with the coefficient c_2 in agreement with RMT, both with and without time-reversal invariance.

Encounters To go beyond the leading order in $1/\epsilon$ we have to consider correlations between periodic orbits arising from encounters. For instance the next-to-leading contribution for time-reversal invariant systems arises from the pairs of orbits found by Sieber and Richter [Sie01, Sie02] differing in an encounter of two almost time-reversed stretches (or, in short, a 2-encounter).

The separations of the encounter stretches can be measured by a coordinate s in the stable direction and a coordinate u in the unstable direction [Spe03, Tur03].[1] Inside the encounter these coordinates decrease and increase exponentially, but their product is constant. We define the encounter as the region where the absolute values of both coordinates remain below a constant c (which can be chosen arbitrarily). The coordinates s and u determine the *action difference* between the partner orbits as the product $\Delta S = su$ [Tur03, Spe03]. Similarly the *duration* of the encounter (i.e. the time difference between its beginning and end) is obtained as $t_{\text{enc}}(s, u) = \frac{1}{\lambda} \ln \frac{c^2}{|su|}$. The *probablitity density* for encounters with given s and u occurring in orbits of period T reads (see [Mue04, Mue05], generalizing [Sie01, Sie02])

$$w_T(s, u) = \frac{T \int dt_1 dt_2 \delta(t_1 + t_2 + 2t_{\text{enc}} - T)}{2\Omega\, t_{\text{enc}}(s, u)}. \tag{33.3.7}$$

Here the factor $1/\Omega$ accounts for ergodicity. The integral goes over the possible durations t_1, t_2 of the two links. The delta function in (33.3.7) makes sure that the sum over the link duration and the overall duration $2t_{\text{enc}}$ of the two encounter stretches coincides with the period.

We can use these results to sum over all pairs of orbits γ, γ' differing in a 2-encounter. We use Hannay and Ozorio de Almeida's sum rule to replace the sum over γ by an integral over T. We then write the sum over γ' as an integral over the density of encounters in γ (with a factor 2 as each encounter gives rise to two mutually time-reversed choices for γ'). This yields

$$R_{2\text{enc}}(\epsilon) = \text{Re}\frac{4}{T_H^2}\left\langle \int dT\; T \int ds \int du\; w_T(s, u) e^{isu/\hbar} e^{2iT\epsilon/T_H}\right\rangle. \tag{33.3.8}$$

If we now insert (33.3.7) and generate the two factors T in (33.3.7) and (33.3.8) by taking two derivatives w.r.t. ϵ the integral nicely decomposes into three factors corresponding to the two links and to the encounter. We can express the contribution of 2-encounters to the correlation function as

$$R_{2\text{enc}}(\epsilon) = -\frac{1}{2}\text{Re}\frac{\partial^2}{\partial \epsilon^2}(-2i\epsilon)^{-2}\,(4i\epsilon) \tag{33.3.9}$$

[1] For definiteness we take the basis vectors along the unstable and stable directions to be mutually normalized such that their symplectic product is 1.

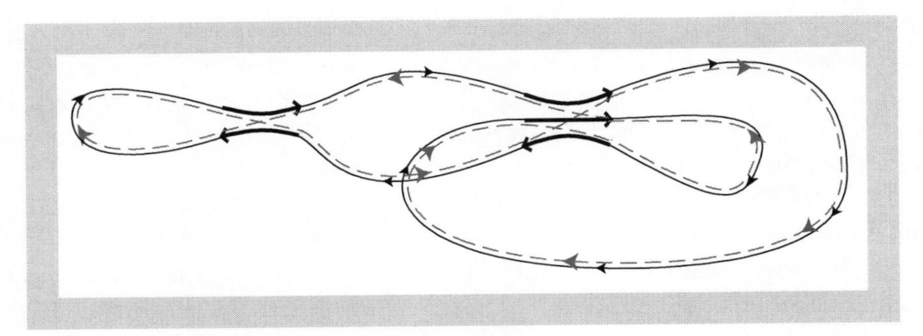

Fig. 33.3 Example for two orbits differing by their connections in an encounter of two stretches and an encounter of three stretches (from [Mue05]).

where the factors following the derivatives originate from the *links* $j = 1, 2$ [Mue07b]

$$\frac{1}{T_H}\int_0^{\infty} dt_j e^{2it_j\epsilon/T_H} = -\frac{1}{2i\epsilon} \tag{33.3.10}$$

and from the *encounter*

$$T_H^2 \int ds du \frac{1}{\Omega t_{\text{enc}}} e^{isu/\hbar} e^{4it_{\text{enc}}\epsilon/T_H} = 4i\epsilon. \tag{33.3.11}$$

The integral in (33.3.11) is easily obtained if we keep only the linear term in the Taylor expansion of the second exponential; it is shown in [Mue05, Mue07b] that the other terms vanish in the semiclassical limit. Inserting these results in (33.3.9) we obtain $R_{2\text{enc}}(\epsilon) = \text{Re}\,\frac{i}{\epsilon^3}$ in agreement with the GOE prediction $c_3 = i$. Note that even though this result is nonzero only for ϵ taken complex, after Fourier transformation it furnishes the term $-2\tau^2$ in the series expansion of $K(\tau)$. Since pairs as in Fig. 33.1 require time-reversal invariance, there are no corresponding contributions for systems without time-reversal invariance and we have $c_3 = 0$ as in the GUE prediction.

The rules derived above generalize to more complicated orbit pairs such as the one in Fig. 33.3 [Mue04, Mue05, Mue07b]. *In general, each link gives a factor* $-\frac{1}{2i\epsilon}$ *whereas each encounter of* l *stretches gives a factor* $2li\epsilon$. The contribution of each diagram of orbit pairs is then accessed by double derivatives as in Eq. (33.3.9). Before showing how these diagrams can be summed we want to extend the semiclassical approach to recover non-oscillatory contributions.

33.4 A unitarity-preserving semiclassical approximation

The spectral determinant The oscillatory contributions to $R(\epsilon)$ can be resolved if one refines the semiclassical approach by explicitly incorporating the unitarity of the quantum time evolution [Heu07, Kea07, Mue09] (see also [Kea93, Bog96] for earlier works). This refinement becomes easier if we characterize quantum spectra not through the level density $\rho(E)$ but through the spectral determinant $\Delta(E) = \det(E - H)$. A semiclassical approximation of $\Delta(E)$ can be obtained from the Gutzwiller trace formula [Vor88, Ber92]. It expresses $\Delta(E)$ as a sum over all pseudo-orbits (finite sets of orbits) Γ

$$\Delta(E) \propto \mathrm{e}^{-i\pi\overline{N}(E)} \sum_{\Gamma} F_{\Gamma}(-1)^{n_{\Gamma}} \mathrm{e}^{i S_{\Gamma}(E)/\hbar}. \tag{33.4.1}$$

In Eq. (33.4.1) $\overline{N}(E) = \int d E \overline{\rho}(E)$ is the smoothed number of energy levels below E. T_{Γ} and S_{Γ} are sums of the periods and actions of the orbits in the set Γ. Neglecting repetitions of orbits, F_{Γ} is the product of the stability factors A_{γ}/T_{γ} of the individual orbits, and n_{Γ} is the number of orbits in Γ. Note that the sum over Γ includes the empty set, contributing unity.

Equation (33.4.1) has an obvious flaw: since the energy levels are real, the spectral determinant $\Delta(E)$ should become real for real arguments. This is not at all clear from (33.4.1). However, one can improve the semiclassical approximation by explicitly incorporating the reality of the energy levels (i.e. the unitarity of the quantum mechanical time evolution). By demanding that $\Delta(E)$ should become real for real E, Berry and Keating (see [Ber92] and references therein) derived the following 'resummed' semiclassical approximation,

$$\Delta(E) \propto \mathrm{e}^{-i\pi\overline{N}(E)} \sum_{\Gamma\ (T_{\Gamma} < T_H/2)} F_{\Gamma}(-1)^{n_{\Gamma}} \mathrm{e}^{i S_{\Gamma}(E)/\hbar} + \text{c.c.} \tag{33.4.2}$$

Here the contribution of 'long' pseudo-orbits is replaced by the complex conjugate of the contribution of the 'short' pseudo-orbits. The border between 'short' and 'long' is taken at half the Heisenberg time. Because of its similarity with the Riemann-Siegel formula in number theory Eq. (33.4.2) is called the *Riemann-Siegel lookalike*. Having unitarity built in, the Riemann-Siegel lookalike allows us to go beyond the previous semiclassical approximation.

Generating function To be able to use the Riemann-Siegel lookalike formula in an approach to spectral statistics, we express the correlation function through a double derivative of a generating function involving four spectral determinants (compare to Chapter 7)

$$\begin{aligned} R(\epsilon) &= -2\,\mathrm{Re} \left.\frac{\partial^2 Z(\epsilon_A, \epsilon_B, \epsilon_C, \epsilon_D)}{\partial\epsilon_A \partial\epsilon_B}\right|_{\epsilon_{A,B,C,D}=\epsilon} - \frac{1}{2} \\ Z(\epsilon_A, \epsilon_B, \epsilon_C, \epsilon_D) &= \left\langle \frac{\Delta\left(E + \frac{\epsilon_C}{2\pi\overline{\rho}}\right)\Delta\left(E - \frac{\epsilon_D}{2\pi\overline{\rho}}\right)}{\Delta\left(E + \frac{\epsilon_A}{2\pi\overline{\rho}}\right)\Delta\left(E - \frac{\epsilon_B}{2\pi\overline{\rho}}\right)} \right\rangle. \end{aligned} \tag{33.4.3}$$

Here for reasons of convergence the two energy increments ϵ_A and ϵ_B in the denominator have to be taken with a positive imaginary part, and $\langle \ldots \rangle$ again denotes an energy average. We can now use the Riemann-Siegel lookalike formula (33.4.2) for the two determinants in the numerator. In the denominator the use of the Riemann-Siegel lookalike is prohibited by the imaginary parts of ϵ_A and ϵ_B, and we have to stick to the counterpart of Eq. (33.4.1) for inverse spectral determinants. After a little algebra, this leads to the semiclassical approximation of the generating function as a sum of two contributions [Kea07]

$$Z = Z^{(1)} + Z^{(2)}, \tag{33.4.4}$$

originating from different combinations of the summands in (33.4.2). The parts $Z^{(1)}$ and $Z^{(2)}$ inherit four sums over pseudo-orbits A, B, C, D from the four spectral determinants. After expanding S and $\overline{N}$ close to E we find [Heu07]

$$Z^{(1)} = \mathrm{e}^{i(\epsilon_A+\epsilon_B-\epsilon_C-\epsilon_D)/2} \times \Big\langle \sum_{A,B,C,D} F_A F_B^* F_C F_D^* (-1)^{n_C+n_D} \times \mathrm{e}^{i[(S_A(E)+S_C(E))-(S_B(E)+S_D(E))]/\hbar} \mathrm{e}^{i(T_A\epsilon_A+T_B\epsilon_B+T_C\epsilon_C+T_D\epsilon_D)/T_H} \Big\rangle. \tag{33.4.5}$$

$Z^{(2)}$ can be obtained from $Z^{(1)}$ by [Kea07]

$$Z^{(2)}(\epsilon_A, \epsilon_B, \epsilon_C, \epsilon_D) = Z^{(1)}(\epsilon_A, \epsilon_B, -\epsilon_D, -\epsilon_C). \tag{33.4.6}$$

The sum (33.4.4) is just what we need to resolve both non-oscillatory and oscillatory contributions to the correlation function. After identifying $\epsilon_{A,B,C,D}$ with ϵ as in Eq. (33.4.3), the factor $e^{i(\epsilon_A+\epsilon_B-\epsilon_C-\epsilon_D)/2}$ in $Z^{(1)}$ converges to unity; hence $Z^{(1)}$ resolves the non-oscillatory contributions. In $Z^{(2)}$ this factor is replaced by $e^{i(\epsilon_A+\epsilon_B+\epsilon_D+\epsilon_C)/2}$ and thus converges to $e^{2i\epsilon}$ if all energy increments are identified; hence we expect $Z^{(2)}$ to yield the oscillatory terms in $R(\epsilon)$.

Diagonal approximation The phase of $Z^{(1)}$ depends on the difference $\Delta S = (S_A + S_C) - (S_B + S_D)$ between the cumulative action of the pseudo-orbits A and C and the cumulative action of the pseudo-orbits B and D. In the limit $\hbar \to 0$ we expect systematic contributions only for ΔS at most of the order of $\hbar$. The easiest way to achieve small (even zero) ΔS is to just repeat the orbits of A, C inside B, D. If we restrict ourselves to this situation we obtain a suitable generalization of the diagonal approximation.

Evaluating the quadruple sum in (33.4.5) is still non-trivial. We can make our life easier if we remember that each of the four pseudo-orbit sums in (33.4.5) originates from an exponentiated sum over periodic orbits. Thus $Z^{(1)}$ could be written in terms of an exponentiated orbit sum as well, in the form $Z^{(1)} = \left\langle e^{\sum_\gamma f_\gamma} \right\rangle$ where the brackets remind us of the energy average involved. Now the diagonal approximation assumes that for systems without time-reversal invariance there are no correlations between periodic orbits. Restricting ourselves to the diagonal approximation we can thus use the central limit theorem to replace

$\left\langle e^{\sum_\gamma f_\gamma} \right\rangle = e^{\left\langle (\sum_\gamma f_\gamma)^2 \right\rangle/2}$. This brings back a double sum over orbits. Applying the sum rule then yields [Heu07]

$$Z^{(1)}_{\text{diag}} = e^{i(\epsilon_A+\epsilon_B-\epsilon_C-\epsilon_D)/2} \frac{(\epsilon_A + \epsilon_D)(\epsilon_C + \epsilon_B)}{(\epsilon_A + \epsilon_B)(\epsilon_C + \epsilon_D)}. \tag{33.4.7}$$

Equations (33.4.7), (33.4.6), and (33.4.3) now give the coefficients c_2 and d_2 in agreement with the GUE prediction. For time-reversal invariant systems the possibilty to revert orbits in time leads to a squaring of the ratio in (33.4.7); we then recover $c_2 = -1$ and $d_2 = 0$ in agreement with the GOE.

Encounters For the higher-order contributions we have to consider the effects of *encounters* on the pseudo-orbit sum in (33.4.5). The orbits in A and C may involve encounters where stretches of one or more orbits come close in phase space (modulo time reversal). We can then change the connections inside these encounters and include the reconnected orbits in either B or D (compare Fig. 33.4). The action difference between the original and the reconnected orbits will be small, leading to a systematic contribution to $Z^{(1)}$.

$Z^{(1)}$ can then be written as a product of $Z^{(1)}_{\text{diag}}$ and a factor accounting for all quadruplets of pseudo-orbits differing in encounters. The product form is needed to also incorporate quadruplets where some orbits of A, C are repeated in B, D while others are first reconnected in encounters. The contribution of each diagram as in Fig. 33.4 can be evaluated using diagrammatic rules similar to Section 33.3. The only difference is that the ϵ's in the link and encounter factors are replaced by linear combinations of ϵ_A, ϵ_B, ϵ_C and ϵ_D, depending to which pseudo-orbits the links and encounter stretches belong.

Summing over all diagrams of pseudo-orbit quadruplets one obtains full agreement with the predictions of RMT for the respective symmetry classes [Heu07, Mue09]. The sum can be performed using combinatorial techniques

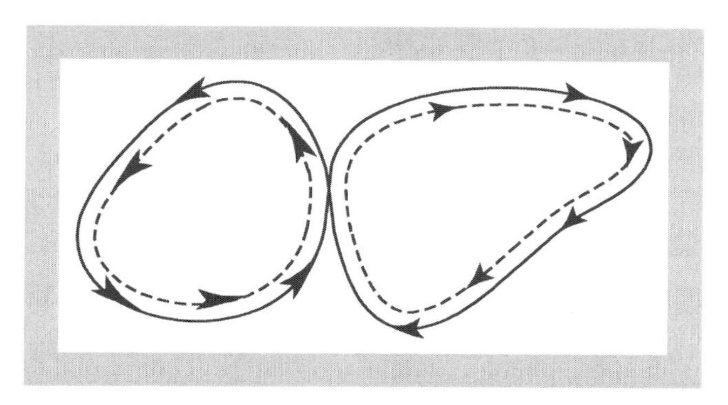

Fig. 33.4 Example for pseudo-orbits B, D obtained by changing connections inside A, C (from [Alt08]). We can take the orbit indicated by a solid line as the only element of A, and choose C to be the empty set. The two dashed orbits can be distributed in any way among B and D.

(at least for the unitary case). However in the following we want to sketch a different approach based on field-theoretic methods.

33.5 Analogy to the sigma model

To motivate this approach we note that schemes of orbits reconnected in encounters as in Fig. 33.1–33.4 have a great similarity to *Feynman diagrams*. The encounters can be interpreted as vertices. Just like vertices in field theory are connected by propagator lines, encounters are connected by links. Indeed the identification of encounters as the relevant mechanism was inspired by corresponding diagrams in the sigma model [Ale96, Whi99]. The sigma model gives a field-theoretical way of evaluating averages over potentials in disordered systems, or averages over the ensembles of RMT. In the supersymmetric version of the sigma model (Chapter 7) these averages are replaced by integrals over supermatrices of much smaller size.

We can exploit this analogy to show that the semiclassical expansion in terms of quadruplets of pseudo-orbits differing in encounters leads to results faithful to the predictions of RMT. We use that the integral over supermatrices σ given in Eq. (7.4.1) can be computed with the help of two stationary-phase approximations. In the first stationary-phase approximation the matrix size N is taken to infinity, leading to an integral over the saddle-point manifold given in Eq. (7.4.2). In addition we can perform a stationary-phase approximation where the energy increments ϵ (in particular their positive imaginary parts) are taken large. In this limit the integral breaks into contributions from two saddle points, the first (the standard saddle point) corresponding to our $Z^{(1)}$ and the second (the Andreev-Altshuler saddle point) corresponding to $Z^{(2)}$. The sigma-model expression for $Z^{(1)}$ can be written as [Mue09]

$$\begin{aligned} Z^{(1)} &= e^{i(\epsilon_A+\epsilon_B-\epsilon_C-\epsilon_D)/2} \left\langle\!\!\left\langle \sum_{V=0}^{\infty} \frac{1}{V!} \left[\sum_{l=2}^{\infty} \mathrm{Str}\left(i\hat{\epsilon}(\tilde{B}B)^l + i\hat{\epsilon}'(B\tilde{B})^l \right) \right]^V \right\rangle\!\!\right\rangle, \\ & \mathrm{Im}\,\epsilon_{A,B,C,D} \gg 1. \end{aligned} \tag{33.5.1}$$

Here $\hat{\epsilon}$, $\hat{\epsilon}'$ are defined as $\hat{\epsilon} \equiv \mathrm{diag}(\epsilon_A, \epsilon_C)$, $\hat{\epsilon}' \equiv \mathrm{diag}(\epsilon_B, \epsilon_D)$ and the double brackets indicate the Gaussian integral

$$\langle\!\langle \{\ldots\} \rangle\!\rangle = -\int d[B, \tilde{B}] \exp\left(\mathrm{Str}\left(i\hat{\epsilon}' B\tilde{B} + i\hat{\epsilon}\tilde{B}B\right)\right)\{\ldots\}. \tag{33.5.2}$$

The supermatrices B and $\tilde{B}$ parametrize σ in a way discussed in [Mue09]. The restriction $\mathrm{Im}\,\epsilon_{A,B,C,D} \gg 1$ implies that only terms surviving in the limit of large imaginary parts should be taken into account. The second contribution $Z^{(2)}$ is related to $Z^{(1)}$ as in semiclassics, see Eq. (33.4.6). We thus see that the breakup into two types of contributions has a clear counterpart in the sigma model, and we only need to show that the two expressions for $Z^{(1)}$ coincide.

This can be done if we write the supertrace in (33.5.1) in components and then use Wick's theorem. Wick's theorem allows us to express the Gaussian integral over a product as a sum over all ways to connect mutually complex conjugate factors (at present B_{kj} and $\tilde{B}_{jk} = \pm B^*_{kj}$) by *contraction lines*. For each summand the pairs of contracted elements can then be removed, inserting instead a factor depending on the corresponding elements of $\hat{\epsilon}$, $\hat{\epsilon}'$ in the exponent of (33.5.2). This factor coincides with the one obtained from every link in semiclassics. This suggests to identify links with contraction lines (forming the propagator lines of RMT). The remaining parts of the translation between semiclassics and the sigma model are technically more involved, and we refer to [Mue09]. In particular one can show that encounters should be identified with the supertraces in (33.5.1) (taking the role of vertices). All contributions can be shown to coincide one-to-one and hence the semiclassical results are bound to agree with RMT.

The ballistic sigma model One might ask whether there is a more direct way to derive a sigma model for the spectral statistics of individual chaotic systems. Indeed such a derivation (albeit still controversial) was proposed in [Muz95, And96]. The main idea in this approach, termed the ballistic sigma model, is to implement the average over energies in Eq. (33.4.3) in a way analogous to an average over random matrices or disorder potentials. The resulting expression for Z is similar to (7.4.1), but with matrices σ that depend on positions in configuration space, and an explicit dependence on the Hamiltonian of the system.

This way to approach the universality problem is more direct than establishing the equivalence of two perturbative expansions as above. In addition, by incorporating the Hamiltonian of the system it also includes corrections to universality outside the semiclassical limit. However the status of the ballistic sigma model remains controversial. While in the present chapter we are not able to do much justice to this ongoing debate we at least want to point out one interesting cross-fertilization between the ballistic sigma model and the semiclassical approach: in [Mue07a] it was shown that a perturbative expansion of the ballistic sigma model can be realized if one generalizes the notion of encounters to the ballistic sigma model. This helps to study how the ballistic sigma model reduces to the RMT one in the semiclassical limit.

33.6 Quantum graphs

It was discovered by Kottos and Smilansky that quantum graphs share many properties with quantum chaotic systems [Kot97, Kot99]. The density of states is semiclassically given by a trace formula in close analogy to the Gutzwiller trace formula and involves an exponentially proliferating number of periodic orbits. The spectral statistics of quantum graphs agree very closely with predictions of RMT, under certain conditions to be discussed in the following. But quantum graphs have advantages. The trace formula is exact rather than an approximation and quantum graphs are easier to treat by analytical methods. This has made them important model systems and many generic aspects of quantum chaotic systems have been investigated on the example of quantum graphs. For an extensive review see [Gnu06].

Graphs consist of a network of B one-dimensional lines, so-called bonds, that are connected at V vertices. The lengths of the bonds are denoted by L_b, and in the following it is assumed that they are rationally independent. On each bond one defines a coordinate x_b. The wavefunctions of quantum graphs satisfy a one-dimensional Schrödinger equation on the bonds, supplemented by matching conditions at the vertices. In most cases one considers the Schrödinger equation of a free particle, or vector potentials that are constant on each bond,

$$\left(\frac{1}{i}\frac{\mathrm{d}}{\mathrm{d}x_b} + A_b\right)^2 \psi_b(x_b) = k^2\psi_b(x_b). \tag{33.6.1}$$

The general solutions of these equations are

$$\psi_b(x_b) = \mathrm{e}^{i(k-A_b)x_b} a_{(b,+)} + \mathrm{e}^{i(k+A_b)(L_b-x_b)} a_{(b,-)}, \tag{33.6.2}$$

where $a_{(b,+)}$ and $a_{(b,-)}$ are constant. It is convenient to introduce for each bond b two directed bonds $\beta = (b,\omega)$ where $\omega = \pm 1$. The directed bond $(b,+)$ starts at $x_b = 0$ and ends at $x_b = L_b$, and vice versa for $(b,-)$. For each directed bond one can define an amplitude, the numbers $a_{(b,+)}$ and $a_{(b,-)}$ in (33.6.2). We further define length and vector potenial of β as $L_\beta = L_b$ and $A_\beta = \omega A_b$.

Let us denote by $s(\beta)$ and $e(\beta)$ the vertices at which a directed bond β starts and ends, respectively. The matching condition at each vertex i, $1 \le i \le V$, can be specified by a vertex scattering matrix $\sigma^{(i)}$ in the following way

$$a_{\beta'} = \sum_{\beta:\, e(\beta)=i} \sigma^{(i)}_{\beta',\beta}\, \mathrm{e}^{i(k+A_\beta)L_\beta}\, a_\beta, \qquad \forall \beta' \text{ such that } s(\beta') = i. \tag{33.6.3}$$

The vertex scattering matrices $\sigma^{(i)}$ are unitary and satisfy additional conditions so that the Laplace operator on the graph is self-adjoint [Kos99]. In the following we assume that the $\sigma(i)$ do not depend on k. Quantum graphs are invariant under time reversal if $\sigma^{(i)}_{\beta',\beta} = \sigma^{(i)}_{\hat\beta,\hat\beta'}$, where $\hat\beta$ runs in the opposite direction of β, and $A_\beta = 0$. The vertex scattering matrices can be collected in a $2B \times 2B$ graph scattering matrix

$$S_{\beta',\beta} = \begin{cases} \sigma^{(i)}_{\beta',\beta} & \text{if } s(\beta') = e(\beta) = i, \\ 0 & \text{else,} \end{cases} \tag{33.6.4}$$

and the conditions (33.6.3) can be combined into

$$\mathbf{a} = \mathcal{U}(k)\,\mathbf{a}, \qquad \mathcal{U}(k) = T(k)\; S, \tag{33.6.5}$$

where $\mathbf{a}$ is the vector of the amplitudes a_β, $\mathcal{U}$ is the unitary quantum evolution map, and T is a diagonal matrix, the bond propagation matrix, with elements $T_{\beta,\beta} = \mathrm{e}^{i(k+A_\beta)L_\beta}$. Eq. (33.6.5) has non-trivial solutions for $k > 0$ only if

$$\zeta(k) := \det(\mathbb{I} - \mathcal{U}(k)) = 0. \tag{33.6.6}$$

This is the quantization condition that determines the spectrum for $k > 0$.

Before we discuss spectral statistics let us consider the classical dynamics on graphs. The classical motion on a graph is non-deterministic. It is a Markov random walk on the set of the $2B$ directed bonds. The classical transition probabilities are chosen to be equal to the quantum transition probabilities, and they are given by the absolute squares of the $\mathcal{U}$ matrix elements

$$P_{\beta'\leftarrow\beta} = |\mathcal{U}_{\beta',\beta}(k)|^2 = |\,S_{\beta',\beta}|^2. \tag{33.6.7}$$

In the following we will only consider *dynamically connected graphs*. These are graphs whose directed bonds cannot be split into non-empty subsets such that the transition probabilities to go from one set to the other vanish.

The matrix of the transition probabilities, $\mathcal{M}_{\beta',\beta} = P_{\beta'\leftarrow\beta}$, defines a classical evolution operator (Frobenius-Perron operator) that describes the discrete evolution of probability densities on the directed bonds, $\rho(n+1) = \mathcal{M}\rho(n)$. Here $\rho(n)$ is the vector of the probabilities $\rho_\beta(n)$ to occupy the bond β at topological time n. The matrix $\mathcal{M}$ is bistochastic because of the unitarity of $\mathcal{U}$, i.e. columns and rows add up to one, and its eigenvalues satisfy $|\nu_l| \leq 1$. There is a unit eigenvalue $\nu_1 = 1$ with eigenvector $\rho^{\mathrm{inv}} = \frac{1}{2B}(1, \ldots, 1)^{\mathrm{T}}$. A graph is *ergodic* if the time-averaged occupation probability on a bond β is equal to $\rho_\beta^{\mathrm{inv}} = (2B)^{-1}$. This implies that ν_1 is the only unit eigenvalue, and is satisfied for all dynamically connected graphs. A graph is *mixing* if any initial probability distribution converges to the invariant distribution ρ^{inv} in the limit of long times. This property holds if ν_1 is the only eigenvalue with modulus one.

Although the motion is probabilistic one can define trajectories on graphs. These are sequences of directed bonds $(\beta_1, \ldots, \beta_n)$ such that $e(\beta_l) = s(\beta_{l+1})$ for all l. The number n of directed bonds is the topological length of the trajectory. The trajectory is closed if $e(\beta_n) = s(\beta_1)$, and a periodic orbit is an equivalence class of closed trajectories that are equal up to cyclic permutations of the bonds.

Let us come back to the quantum properties. A trace formula for the spectrum was derived by Roth [Rot85], and alternatively by Kottos and Smilansky who also pointed out the close similarity to the Gutzwiller trace formula [Kot97, Kot99]. The trace formula expresses the density of states as

$$\rho(k) = \sum_n \delta(k - k_n) = \bar{\rho}(k) + \frac{1}{\pi}\mathrm{Re}\sum_p \frac{\mathcal{L}_p \mathcal{A}_p}{r_p}\,\mathrm{e}^{i\mathcal{L}_p k}, \tag{33.6.8}$$

where p labels all periodic orbits and r_p is the repetition number of an orbit (if p is a multiple repetition of a shorter orbit). The length $\mathcal{L}_p$ of a periodic orbit is the sum of all its bond lengths, and its amplitude $\mathcal{A}_p$ is the product over the matrix elements $S_{\beta',\beta}\,\mathrm{e}^{i A_\beta L_\beta}$ for all visited vertices. The mean density of states is $\bar{\rho}(k) = \sum_b L_b/\pi$ where the sum runs over all B bonds.

The trace formula allows us to express the spectral correlation function $R(\epsilon)$ (see (33.3.2) with ρ as function of k and an average over all positive k) or its Fourier transform, the spectral form factor, in terms of a double sum over periodic orbits. The semiclassical limit for quantum graphs corresponds to letting the number of bonds B go to infinity, and it is expected that spectral statistics agree with RMT only in this limit.

For a semiclassical analysis in terms of periodic orbits it has been found convenient to use a different starting point. The unitary evolution operator $\mathcal{U}(k)$ has eigenvalues $\mathrm{e}^{i\phi_j}$ with eigenphases ϕ_j. In the limit of large graphs $B \to \infty$ and for moderate bond length fluctuations, the spectral statistics of the eigenvalues of k_j is equivalent to the statistics of the ϕ_j [Kot99, Gnu06]. The form factor of the eigenphases is given for discrete times by

$$K(\tau) = \frac{1}{B}\langle|\mathrm{Tr}(\mathcal{U}(k)^n)|^2\rangle, \tag{33.6.9}$$

where $\tau = n/B$ and the average is over all positive k. Note that $\mathcal{U}$ depends on k only via the B different phases e^{ikL_b} of the bond propagator. For rationally independent bond lengths L_b the flow $k \Rightarrow \{\mathrm{e}^{ikL_1}, \ldots, \mathrm{e}^{ikL_B}\}$ is ergodic on the B-dimensional torus defined by the B phases. This implies that the k-average can be replaced by an average over the B phases [Bar00]. This is one of the main properties of quantum graphs that facilitates an analytical approach.

The expansion of the trace of $\mathcal{U}^n$ in (33.6.9) can be interpreted as the sum over all periodic orbits of topological length n. The modulus square of this expression gives a double sum. After performing the average of k one arrives at a double sum that contains only pairs of orbits with identical lengths

$$K(\tau) = \frac{n^2}{B}\sum_{p,q}\frac{\mathcal{A}_p\mathcal{A}_q^*}{r_p r_q}\,\delta_{\mathcal{L}_p,\mathcal{L}_q}. \tag{33.6.10}$$

The quantities in this equation are defined after (33.6.8). For an orbit p the factor n/r_p is the number of cyclic permutations of bond sequences that correspond to p. In the limit $B \to \infty$ the contributions to the form factor come from very long orbits. Because of the rational independence of the bond lengths, orbits with identical length visit each bond b the same number of times. However, it is a complicated combinatorical problem to determine all possible orbit pairs with identical lengths. Instead, the semiclassical approach to spectral statistics of quantum graphs follows rather closely the diagrammatic approach for general chaotic systems that was discussed in preceding sections.

The leading order contributions comes from the diagonal approximation in which only pairs of identical orbits are considered, or those related by time inversion in the case of time-reversal invariant systems [Kot97, Kot99].

$$K^{\text{diag}}(\tau) \sim \kappa \frac{n^2}{B} \sum_p \frac{|\mathcal{A}_p|^2}{r_p^2} \sim \kappa \frac{n}{B} \text{Tr}(\mathcal{M}^n). \tag{33.6.11}$$

Here κ equals 2 and 1 for systems with and without time-reversal symmetry, respectively, and $\mathcal{M}_{\beta',\beta} = |S_{\beta',\beta}|^2$ was used. Orbits with repetition number $r_p > 1$ and self-retracing orbits are not treated correctly, but their contribution is negligible as $B \to \infty$. To obtain $K^{\text{diag}}(\tau) \sim \kappa\tau$ in agreement with the small τ expansion of the RMT results one needs $\text{Tr}(\mathcal{M}^n) \to 1$ as $B \to \infty$. This follows on the one hand from the fact that $\nu_1 = 1$ is the only eigenvalue of $\mathcal{M}$ with modulus one in mixing systems. On the other hand, because $n = \tau B$, one needs the additional condition that the gap between $\nu_1 = 1$ and the modulus of the next largest eigenvalue decreases slower than $1/B$. Tanner conjectured that this is the condition for universal RMT statistics in quantum graphs [Tan01]. Examples for systems where this condition is violated are star graphs; indeed the spectral statistics of star graphs can be shown to disagree with RMT [Ber01].

The evaluation of off-diagonal contributions to the form factor is closely related to that in general chaotic systems. The orbit correlations are described by the same type of diagrams. For graphs one specifies encounters by a vertex at which an orbit self-intersects. An l-encounter corresponds to a vertex that is traversed l times during a traversal of the periodic orbit. For each diagram one sums over all corresponding orbit pairs. The difficulty with this method is to ensure that each orbit pair is counted only once and not more. This is not straightforward, because the vertex at which an orbit self-intersects is not always unique. For example, for the orbit pair in Fig. 33.5 one can choose either vertex c or d as self-intersection. To solve this problem one can impose a restriction on the orbit sum such that the first and last vertex in one link, e.g. the link traversed in opposite directions, are different [Ber02]. In the example this selects d. Even with this restriction there is an overcounting if the other link is self-retracing, because in this case the orbit pairs are already included in the diagonal approximation. These exceptional contributions were subtracted in [Ber02], and the first off-diagonal correction to the form factor was obtained for time-reversal invariant systems in agreement with RMT [Ber02]. The third-order terms were calculated in [Ber03, Ber04]. A general method for all higher-order diagrams was outlined in [Ber06] for a special type of graphs, complete Fourier quantum

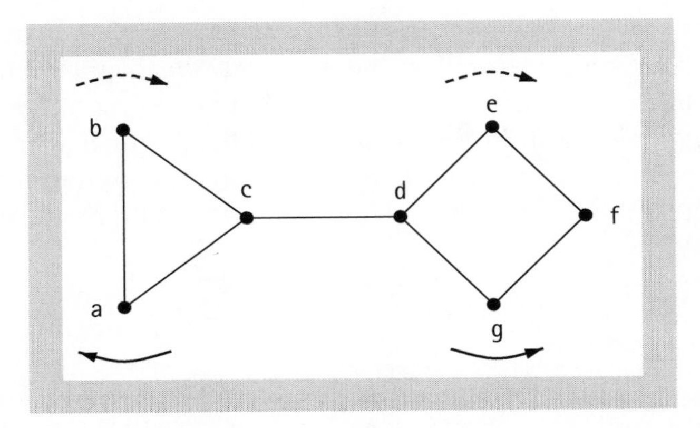

Fig. 33.5 A pair of orbits on a graph that differ in the sense of direction in one of the loops (full and dashed arrows).

graphs. The semiclassical methods can be extended to quantum graphs with Dirac operators [Bol03] as well as graphs in any of Zirnbauer's ten symmetry classes [Gnu04b].

As in quantum chaotic systems one can use an alternative approach to spectral statistics that is based on the supersymmetry method [Gnu04a, Gnu05]. In comparison to the ballistic sigma model it has a significantly firmer mathematical basis. The starting point is a representation of the spectral correlation function $R(\epsilon)$ as derivative of a generating function that is defined by quotients of spectral determinants as in (33.4.3). This generating function is expressed as an integral over commuting and anticommuting variables. Then the spectral average over the wavenumber k is performed by using the exact equivalence of the spectral average with an average of the B independent phases of the bond propagator, as discussed after (33.6.9). The phase-averaged correlation function can then be mapped onto a non-linear σ-model by using the so-called colour-flavour transformation. It results in an integral over supermatrices of appropriate symmetry. Although exact, it cannot be evaluated without further approximations. A first approximation, known as zero-mode approximation, is obtained by integrating over the saddle-point manifold of the integral. It results in the generating function that agrees with RMT. To check the validity of this approximation one considers correction terms that are obtained by expanding the action in the superintegral up to second order around the saddle points. This results in sufficient conditions for the validity of the random matrix result. They are stricter than Tanner's conjecture and require that the spectral gap of the matrix $\mathcal{M}$ decreases slower then $1/\sqrt{B}$ instead of $1/B$. As mentioned before the derivation also assumes rationally independent bond lengths, dynamically connected graphs and k-independent scattering matrices.

References

[Ale96] I. L. Aleiner and A. I. Larkin, Phys. Rev. B **54** (1996) 14423

[Alt08] A. Altland et al., in Path Integrals - New Trends and Perspectives, Proc. of 9th Int Conference, World Scientific (2008) 40

[And96] A. V. Andreev, O. Agam, B. D. Simons, and B. L. Altshuler, Phys. Rev. Lett. **76** (1996) 3947

[Arg93] N. Argaman et al., Phys. Rev. Lett. **71** (1993) 4326

[Bar00] F. Barra and P. Gaspard, J. Stat. Phys. **101** (2000) 283

[Ber77] M. V. Berry and M. Tabor, Proc. R. Soc. Lond. A **356** (1977) 375

[Ber85] M. V. Berry, Proc. R. Soc. Lond. A **400** (1985) 229

[Ber92] M.V Berry and J.P. Keating, Proc. R. Soc. Lond. A **437** (1992) 151

[Ber01] G. Berkolaiko, E. B. Bogomolny, and J. P. Keating, J. Phys. A **34** (2001) 335

[Ber02] G. Berkolaiko, H. Schanz, and R. S. Whitney, Phys. Rev. Lett. **88** (2002) 104101

[Ber03] G. Berkolaiko, H. Schanz, and R. S. Whitney, J. Phys. A **36** (2003) 8373

[Ber04] G. Berkolaiko, Waves Random Media **14** (2004) S7

[Ber06] G. Berkolaiko, in "Quantum Graphs and Their Applications", (G. Berkolaiko, R. Carlson, S. Fulling, and P. Kuchment, eds), Contemp. Maths. **415**, AMS, Providence, RI (2006) 35

[Bog96] E. B. Bogomolny and J. P. Keating, Phys. Rev. Lett. **77** (1996) 1472

[Boh84] O. Bohigas, M. J. Giannoni, and C. Schmit, Phys. Rev. Lett. **52** (1984) 1

[Bol03] J. Bolte and J. Harrison, J. Phys. A **36** (2003) 2747

[Bow72] R. Bowen, Amer. J. Math. **94** (1972) 1

[Cas80] G. Casati, F. Valz-Gris, and I. Guarneri, Lett. Nuovo Cim. **28** (1980) 279

[Gnu04a] S. Gnutzmann and A. Altland, Phys. Rev. Lett. **93** (2004) 194101

[Gnu04b] S. Gnutzmann and B. Seif, Phys. Rev. E **69** (2004) 056219 and 056220

[Gnu05] S. Gnutzmann and A. Altland, Phys. Rev. E **72** (2005) 056215

[Gnu06] S. Gnutzmann and U. Smilansky, Advances in Physics **55** (2006) 527

[Gut90] M. C. Gutzwiller, *Chaos in Classical and Quantum Mechanics*, Springer, New York (1990)

[Haa10] F. Haake, *Quantum Signatures of Chaos*, 3rd ed., Springer, Berlin (2010)

[Han84] J. H. Hannay and A. M. Ozorio de Almeida, J. Phys. A **17** (1984) 3429

[Heu07] S. Heusler, S. Müller, A. Altland, P. Braun, and F. Haake, Phys. Rev. Lett. **98** (2007) 044103

[Kea93] J. P. Keating, *The Riemann zeta-function and quantum chaology*, in Quantum Chaos, G. Casati, I. Guarneri, and V. Smilansky, eds., North-Holland, Amsterdam (1993) 145

[Kea07] J. P. Keating and S. Müller, Proc. R. Soc. A **463** (2007) 3241

[Kos99] V. Kostrykin and R. Schrader, J. Phys. A **32** (1999) 595

[Kot97] T. Kottos and U. Smilansky, Phys. Rev. Lett. **79** (1997) 4794

[Kot99] T. Kottos and U. Smilansky, Ann. Phys. **274** (1999) 76

[Mue04] S. Müller, S. Heusler, P. Braun, F. Haake, and A. Altland, Phys. Rev. Lett. **93** (2004) 014103

[Mue05] S. Müller, S. Heusler, P. Braun, F. Haake, and A. Altland, Phys. Rev. E **72** (2005) 046207

[Mue07a] J. Müller, T. Micklitz, and A. Altland, Phys. Rev. E **76** (2007) 056204

[Mue07b] S. Müller, S. Heusler, P. Braun, and F. Haake, New J. Phys. **9** (2007) 12 (see cond-mat/0610560v1 for an online appendix on spectral statistics)

[Mue09] S. Müller, S. Heusler, A. Altland, P. Braun, and F. Haake, New J. Phys. **11** (2009) 103025

[Muz95] B. A. Muzykantskii and D. E. Khmel'nitskii, JETP Lett. **62** (1995) 76

[Rot85] J.-P. Roth, in *Théorie du Potentiel*, G. Mokobodzki and D. Pinchon, eds., Springer, Berlin (1985) 521

[Sie01] M. Sieber and K. Richter, Physica Scripta **T90** (2001) 128

[Sie02] M. Sieber, J. Phys. A **35** (2002) L613

[Spe03] D. Spehner, J. Phys. A **36** (2003) 7269

[Sto99] H.-J. Stöckmann, *Quantum chaos: an introduction*, Cambridge University Press, Cambridge (1999)

[Tan01] G. Tanner, J. Phys. A **34** (2001) 8485

[Tur03] M. Turek and K. Richter, J. Phys. A **36** (2003) L455

[Vor88] A. Voros, J. Phys. A **21** (1988) 685

[Whi99] R.S. Whitney, I. V. Lerner, and R. A. Smith, Waves in Random Media **9** (1999) 179

·34·

Resonance scattering of waves in chaotic systems

Y. V. Fyodorov and D. V. Savin

Abstract

This is a brief overview of RMT applications to quantum or wave chaotic resonance scattering with an emphasis on non-perturbative methods and results.

34.1 Introduction

Describing bound states corresponding to discrete energy levels in *closed* quantum systems, one usually addresses properties of Hermitian random matrices (Hamiltonians) H. Experimentally, however, one often encounters the phenomenon of chaotic scattering of quantum waves (or their classical analogues) in *open* systems [Kuh05a]. The most salient feature of open systems is the set of resonances which are quasibound states embedded into the continuum. The resonances manifest themselves via fluctuating structures in scattering observables. A natural way to address them is via an energy-dependent scattering matrix, $S(E)$, which relates the amplitudes of incoming and outgoing waves. In such an approach, the resonances correspond to the poles of $S(E)$ located (as required by the causality condition) in the lower half-plane of the complex energy plane [Nus72].

The correspondence becomes explicit in the framework of the Hamiltonian approach [Ver85]. It expresses the resonance part of the S-matrix via the Wigner reaction matrix $K(E)$ (which is a $M\times M$ Hermitian matrix) as follows

$$S(E) = \frac{1 - i\,K(E)}{1 + i\,K(E)}, \qquad K(E) = \tfrac{1}{2} V^{\dagger}(E - H)^{-1} V, \tag{34.1.1}$$

where V is an $N\times M$ matrix of energy-independent coupling amplitudes between N levels and M scattering channels. By purely algebraic transformations, the above expression can be equivalently represented in terms of the $N\times N$ effective non-Hermitian Hamiltonian $\mathcal{H}_{\text{eff}}$ of the open system:

$$S(E) = 1 - i\,V^{\dagger}\frac{1}{E - \mathcal{H}_{\text{eff}}}V, \qquad \mathcal{H}_{\text{eff}} = H - \tfrac{i}{2} V V^{\dagger}. \tag{34.1.2}$$

The anti-Hermitian part of $\mathcal{H}_{\text{eff}}$ has a factorized structure which is a direct consequence of the unitarity condition (flux conservation): $S^{\dagger}(E)S(E) = 1$ for real E. It describes a coupling of the bound states via the common decay channels, thus converting the parential levels into N *complex* resonances $\mathcal{E}_n = E_n - \frac{i}{2}\Gamma_n$ characterized by energies E_n and widths $\Gamma_n > 0$. The eigenvectors of $\mathcal{H}_{\text{eff}}$ define the corresponding *bi-orthogonal* resonance states (quasimodes). They appear via the residues of the S-matrix at the pole positions.

Universal properties of both resonances and resonance states in the chaotic regime are analysed by replacing the actual non-Hermitian Hamiltonian with an RMT ensemble of appropriate symmetry.[1] In the limit $N \to \infty$ spectral fluctuations on the scale of the mean level spacing Δ turn out to be universal, i.e. independent of microscopical details such as the particular form of the distribution of H or the energy dependence of Δ. Similarly, the results are also independent of particular statistical assumptions on coupling amplitudes V_n^c as long as $M \ll N$ [Leh95a]. The amplitudes may therefore be equivalently chosen fixed [Ver85] or random [Sok89] variables and enter final expressions only via the so-called transmission coefficients

$$T_c \equiv 1 - |\overline{S}_{cc}|^2 = \frac{4\kappa_c}{(1+\kappa_c)^2}, \qquad \kappa_c = \frac{\pi \| V^c \|^2}{2N\Delta} \tag{34.1.3}$$

where $\overline{S}$ stands for the average (or optical) S matrix. The set of T_cs is assumed to be the only input parameters in the theory characterizing the degree of system openness. The main advantage of such an approach is that it treats on equal footing both the spectral and scattering characteristics of open chaotic systems and is flexible enough to incorporate other imperfections of the system, e.g. disorder due to random scatterers and/or irreversible losses [Fyo05].

In fact, the description of wave scattering outlined above can be looked at as an integral part of the general theory of linear dynamic open systems in terms of the input-output approach. These ideas and relations were developed in system theory and engineering mathematics many years ago, going back to the pioneering works by M. Livšic [Liv73]. A brief description of main constructions and interpretations of the linear open systems approach, in particular, a short derivation of Eq. (34.1.2), can be found in [Fyo00]. In what follows, we overview briefly selected topics on universal statistics of resonances and scattering observables, focusing mainly on theoretical results obtained via non-perturbative methods starting from the mid-nineties. For a more detailed discussion, including applications, experimental verifications as well as references, we refer to the special issue of J. Phys. A **38**, Number 49 (2005) on 'Trends in Quantum Chaotic Scattering' as well as to the recent reviews [Bee97, Guh98,

[1] Conventionally, H is taken from the GOE or GUE labelled by the Dyson's index $\beta = 1$ or $\beta = 2$, respectively, according to time-reversal symmetry being preserved or broken in the system (GSE, $\beta = 4$, is to be taken when spin becomes important).

Mel99, Alh00, Mit10]. Finally, it is worth mentioning that many methods and results discussed below can be extended beyond the universal RMT regime and the use of more ramified random matrix ensembles (e.g. of banded type) allows one to take into account effects of the Anderson localization. Discussion of those developments goes, however, beyond the remit of the present review, and an interested reader may find more information and further references in [Kot02, Fyo03b, Men05, Kot05, Oss05, Fyo05, Wei06, Mir06, Mon09].

34.2 Statistics at the fixed energy

Statistical properties of scattering observables considered at the given fixed value of the scattering energy E can be inferred from the corresponding distribution of $S = S(E)$. The latter is known to be uniquely parameterized by the average S-matrix and is distributed according to Poisson's kernel [Hua63, Mel85]:

$$P_{\overline{S}}(S) = \frac{1}{V_\beta}\left|\frac{\det[1 - \overline{S}^\dagger \overline{S}]}{\det[1 - \overline{S}^\dagger S]^2}\right|^{(\beta M+2-\beta)/2}, \tag{34.2.1}$$

where V_β is a normalization constant. Although this expression follows from Eq. (34.1.2) in the limit $N \to \infty$, as shown by Brouwer [Bro95], $P_{\overline{S}}(S)$ can be derived starting from statistical assumptions imposed on S without referencing to the Hamiltonian at all, as was initially done by Mello *et al.* [Mel85]. For a detailed account of the theory see the recent review [Mel99] and book [Mel04].

34.2.1 Maximum entropy approach

The starting point of that information-theoretical description is an information entropy $\mathcal{S} = -\int d\mu(S)\, P_{\overline{S}}(S) \ln P_{\overline{S}}(S)$ associated with the probability distribution $P_{\overline{S}}(S)$. The integration here is over the invariant (Haar) measure $d\mu(S)$ which satisfies the symmetry constraints imposed on S: unitary S should be chosen symmetric for $\beta = 1$ or self-dual for $\beta = 4$. Due to the causality condition, $S(E)$ is analytic in the upper half of the complex energy plane, implying $\overline{S^k} = \overline{S}^k$ for the positive moments, where the average is assumed to be performed over an energy interval. One can further assume that $S(E)$, for real E, is a stationary random-matrix function satisfying the condition of ergodicity, i.e. ensemble averages are equal to spectral averages, the latter being determined by the (given) optical scattering matrix $\overline{S}$. These two conditions together yield the so-called *analyticity-ergodicity* requirement $\int d\mu(S)\, S^k P_{\overline{S}}(S) = \overline{S}^k$, $k = 1, 2, \ldots$. The sought distribution is then assigned to an 'equal *a prior*' distribution of S, i.e. the one which maximizes the entropy $\mathcal{S}$, subject to the above constraints. The answer turns out to be exactly given by Eq. (34.2.1) [Mel85], with $V_\beta = \int d\mu(S)$ being the volume of the matrix space. Then for any analytic function

$f(S)$ its average is given by the formula $\int d\mu(S) f(S) P_{\overline{S}}(S) = f(\overline{S})$. Thus the system-specific details are irrelevant, except for the optical S-matrix.

The important case of $\overline{S} = 0$ (ideal, or perfect coupling) plays a special role, being physically realised when prompt or direct scattering processes described by $\overline{S}$ are absent. The case corresponds to Dyson's circular ensembles of random unitary matrices S distributed uniformly with respect to the invariant measure (note $P_0(S) = \text{constant}$). Thus statistical averaging amounts to the integration over unitary group, various methods being developed for this purpose [Mel85, Bro96]. The general situation of nonzero $\overline{S}$ is much harder for analytical work. In this case, however, one can use the following formal construction

$$S_0 = (t_1')^{-1}(S - \overline{S})(1 - \overline{S}^{\dagger} S)^{-1} t_1^{\dagger}, \tag{34.2.2}$$

where S_0 is drawn from the corresponding circular ensemble and the matrices t_1 and t_1' belong to a $2M \times 2M$ unitary matrix $S_1 = \left(\begin{smallmatrix} r_1 & t_1' \\ t_1 & r_1' \end{smallmatrix}\right)$ of the same symmetry. Evaluating the Jacobian of the transformation (34.2.2), one finds [Hua63, Fri85] that the distribution of S is exactly given by the Poisson's kernel (34.2.1) provided that we identify $r_1 = \overline{S}$. Physically, this can be understood by subdividing the total scattering process into the stage of prompt (direct) response S_1 occurring prior to the stage of equilibrated (slow) response S_0 [Bro96, Mel99], Eq. (34.2.2) being a result of the composition of the two processes. In this way one can map the problem with direct processes to that without and several useful applications have been considered in [Gop98, Sav01, Fyo05, Gop08].

In problems of quantum transport it is often convenient to group scattering channels into N_1 'left' and N_2 'right' channels which correspond to the propagating modes in the two leads attached to the conductor (with $M = N_1 + N_2$ channels in total). In this way the matrix S acquires the block structure

$$S = \begin{pmatrix} r & t' \\ t & r' \end{pmatrix} = \begin{pmatrix} u & 0 \\ 0 & v \end{pmatrix} \begin{pmatrix} -\sqrt{1-\tau^T\tau} & \tau^T \\ \tau & \sqrt{1-\tau\tau^T} \end{pmatrix} \begin{pmatrix} u' & 0 \\ 0 & v' \end{pmatrix}, \tag{34.2.3}$$

where r (r') is a $N_1 \times N_1$ ($N_2 \times N_2$) reflection matrix in the left (right) lead and t (t') is a rectangular matrix of transmission from left to right (right to left). Unitarity ensures that the four Hermitian matrices $tt^{\dagger}$, $t't'^{\dagger}$, $1 - rr^{\dagger}$ and $1 - r'r'^{\dagger}$ have the same set of $n \equiv \min(N_1, N_2)$ nonzero eigenvalues $\{T_k \in [0, 1]\}$. In terms of these so-called transmission eigenvalues the S matrix can be rewritten by employing the singular value decomposition for $t = v\tau u'$ and $t' = u\tau^T v'$ yielding the second equality in (34.2.3). Here τ is a $N_1 \times N_2$ matrix with all elements zero except $\tau_{kk} = \sqrt{T_k}$, $k = 1, \ldots, n$, u and u' (v and v') being $N_1 \times N_1$ ($N_2 \times N_2$) unitary matrices ($u^* u' = v^* v' = 1$ for $\beta = 1, 4$). For S drawn from one of Dyson's circular ensembles those unitary matrices are uniformly distributed (with respect to the Haar measure), whereas the joint probability density function (JPDF) of the transmission eigenvalues reads [Bar94, Jal94]

$$\mathcal{P}_\beta(\{\mathcal{T}_i\}) = \frac{1}{\mathcal{N}_\beta}|\Delta(\{\mathcal{T}_i\})|^\beta \prod_{i=1}^{n} \mathcal{T}_i^{\alpha-1}, \qquad \alpha \equiv \frac{\beta}{2}(|N_1 - N_2| + 1). \qquad (34.2.4)$$

Here and below $\Delta(\{x_i\}) = \prod_{i<j}(x_i - x_j)$ denotes the Vandermonde determinant, $\mathcal{N}_\beta$ being the normalization constant. Since the Landauer-Büttiker scattering formalism allows one to express various transport observables in terms of T_js [Bla00], Eq. (34.2.4) serves as the starting point for performing statistical analysis of quantum transport in chaotic systems, see [Bee97] for a review.

34.2.2 Quantum transport and the Selberg integral

The conductance $g = \mathrm{tr}(tt^\dagger) = \sum_{i=1}^{n} \mathcal{T}_i$ and the (zero-frequency) shot-noise power $p = \sum_{i=1}^{n} \mathcal{T}_i(1 - \mathcal{T}_i)$ are important examples of a linear statistic on transmission eigenvalues. Their mean values are provided by averaging over t or by making use of the mean eigenvalue density (derived recently in [Viv08]). However, the higher moments involve products of $\mathcal{T}_j$s and require us to integrate over the full JPDF. This task can be accomplished by recognizing a profound connection of (34.2.4) to the celebrated Selberg integral [Sav06a] which is a multidimensional generalization of Euler's beta-function.[2] The specific structure of its integrand yields a closed set of algebraic relations for the moments $\langle \mathcal{T}_1^{\lambda_1} \cdots \mathcal{T}_m^{\lambda_m} \rangle$ of given order $r = \sum \lambda_i$ with integer parts $\lambda_i \geq 0$ [Sav08].

As a particular example, we consider the relations for the second-order moments: $[\alpha + 2 + \beta(n-1)]\langle \mathcal{T}_1^2 \rangle = [\alpha + 1 + \beta(n-1)]\langle \mathcal{T}_1 \rangle - \frac{\beta}{2}(n-1)\langle \mathcal{T}_1 \mathcal{T}_2 \rangle$ and $\langle \mathcal{T}_1 \cdots \mathcal{T}_m \rangle = \prod_{j=1}^{m} \frac{\alpha + \beta(n-j)/2}{\alpha + 1 + \beta(2n-j-1)/2}$ at $m = 2$. These provide us straightforwardly with the well-known exact results for both the average and variance of the conductance (first obtained by different methods in [Bar94, Jal94, Bro96]) as well as with the average shot-noise power (hence, Fano factor $\frac{\langle p \rangle}{\langle g \rangle}$) [Sav06a]:

$$\langle g \rangle = \frac{N_1 N_2}{M + 1 - \frac{2}{\beta}}, \qquad \langle p \rangle = \langle g \rangle \frac{(N_1 - 1 + \frac{2}{\beta})(N_2 - 1 + \frac{2}{\beta})}{(M - 2 + \frac{2}{\beta})(M - 1 + \frac{4}{\beta})}. \qquad (34.2.5)$$

(We also note an interesting exact relation $\mathrm{var}(g) = \frac{2}{\beta}(N_1 N_2)^{-1}\langle g \rangle \langle p \rangle$.) The approach has been developed further to study full counting statistics of charge transfer [Nov07] as well as to obtain explicit expressions for the skewness and kurtosis of the charge and conductance distributions, and for the shot-noise variance [Sav08]. With symmetry index β entering the Selberg integral as a continuous parameter, the method allows one to treat all the three ensembles on

[2] $\int_0^1 dT_1 \cdots \int_0^1 dT_n |\Delta(T)|^{2c} \prod_{i=1}^{n} T_i^{a-1}(1 - T_i)^{b-1} = \prod_{j=0}^{n-1} \frac{\Gamma(1+c+jc)\Gamma(a+jc)\Gamma(b+jc)}{\Gamma(1+c)\Gamma(a+b+(n+j-1)c)}$, $\Gamma(x)$ being the gamma function, gives the definition of the Selberg integral valid at integer $n \geq 1$, complex a and b with positive real parts, and complex c with $\mathrm{Re}\, c > -\min[\frac{1}{n}, \mathrm{Re}\,\frac{a}{n-1}, \mathrm{Re}\,\frac{b}{n-1}]$. The constant $\mathcal{N}_\beta$ is given by this expression at $a = \alpha$, $b = 1$ and $c = \frac{\beta}{2}$. Chapter 17 of Mehta's book gives an introduction into the Selberg integral theory, see [For08] for its current status.

an equal footing and provides a powerful non-perturbative alternative to orthogonal polynomial [Ara98, Viv08] or group integration approaches [Bro96, Bul06].

It is further possible to combine the Selberg integral with the theory of symmetric functions and to develop a systematic approach for computing the moments of the conductance and shot-noise (including their joint moments) of arbitrary order and at any number of open channels [Kho09]. The method is based on expanding powers of g or p (or other linear statistics) in Schur functions s_λ. These functions are symmetric polynomials $s_\lambda(\mathcal{T}) = \det\{\mathcal{T}_i^{\lambda_j+n-j}\}_{i,j=1}^n / \det\{\mathcal{T}_i^{n-j}\}_{i,j=1}^n$ in n transmission eigenvalues $\mathcal{T}_1, \ldots, \mathcal{T}_n$ indexed by partitions λ.[3] In the group representation theory the Schur functions are the irreducible characters of the unitary group and hence are orthogonal. This orthogonality can be exploited to determine the 'Fourier coefficients' in the Schur function expansion by integration over the unitary group. The general form of this expansion reads:

$$\prod_{i=1}^{n}\left(\sum_{j=-\infty}^{+\infty} a_j \mathcal{T}_i^j\right) = \sum_{\lambda} c_\lambda(a) s_\lambda(\mathcal{T}), \qquad c_\lambda(a) \equiv \det\left\{a_{\lambda_k-k+l}\right\}_{k,l=1}^{n}. \tag{34.2.6}$$

The summation here is over all partitions λ of length n or less, including empty partition for which $s_\lambda = 1$. The Schur functions can then be averaged over the JPDF (34.2.4) with the help of integration formulae due to [Hua63]. One gets

$$\begin{aligned}\langle s_\lambda\rangle_{\beta=1} = c_\lambda \prod_{j=1}^{l(\lambda)} \frac{(\lambda_j+N_1-j)!}{(N_1-j)!}\frac{(\lambda_j+N_2-j)!}{(N_2-j)!}\frac{(M-l(\lambda)-j)!}{(\lambda_j+M-l(\lambda)-j)!}\\ \times \prod_{1\le i\le j\le l(\lambda)} \frac{M+1-i-j}{M+1+\lambda_i+\lambda_j-i-j},\end{aligned} \tag{34.2.7}$$

where the coefficient $c_\lambda = \frac{\prod_{1\le i<j\le l(\lambda)}(\lambda_i - i - \lambda_j + j)}{\prod_{j=1}^{l(\lambda)}(l(\lambda)+\lambda_j-j)!}$ has been introduced, and

$$\langle s_\lambda\rangle_{\beta=2} = c_\lambda \prod_{j=1}^{l(\lambda)} \frac{(\lambda_j+N_1-j)!}{(N_1-j)!}\frac{(\lambda_j+N_2-j)!}{(N_2-j)!}\frac{(M-j)!}{(\lambda_j+M-j)!}, \tag{34.2.8}$$

in the cases of orthogonal ($\beta = 1$) and unitary ($\beta = 2$) symmetry, respectively.

As an application of the method we consider the moments of $\epsilon g + p$. This quantity has a physical meaning of the total noise including both thermal and shot-noise contributions [Bla00]. From expansion (34.2.6) one finds [Kho09] the r-th moment of the total noise given by

[3] A partition is a finite sequence $\lambda = (\lambda_1, \lambda_2, \ldots, \lambda_m)$ of non-negative integers (called parts) in decreasing order $\lambda_1 \ge \lambda_2 \ge \ldots \ge \lambda_m \ge 0$. The weight of a partition, $|\lambda|$, is the sum of its parts, $|\lambda| = \sum_j \lambda_j$, and the length, $l(\lambda)$, is the number of its nonzero parts.

$$\langle(\epsilon g + p)^r\rangle = r! \sum_{m=0}^{r} (-1)^m (1+\epsilon)^{r-m} \sum_{|\lambda|=r+m} f_{\lambda,m} \langle s_\lambda \rangle, \quad (34.2.9)$$

where the second sum is over all partitions of $r+m$. Here the expansion coefficient $f_{\lambda,m} = \sum \det\{\frac{1}{k_i!(\lambda_i - i + j - 2k_i)!}\}$, where the summation indices k_i run over all integers from 0 to m subject to the constraint $k_1 + \ldots + k_{l(\lambda)} = m$. In the limit $\epsilon \to \infty$ Eq. (34.2.9) yields all the conductance moments $\langle g^r \rangle = r! \sum_{|\lambda|=r} c_\lambda \langle s_\lambda \rangle$ (note that $f_{\lambda,0} = c_\lambda$ introduced above) first obtained in [Nov08] at $\beta = 2$ (see [Osi08] for an alternative treatment). In the opposite case of $\epsilon = 0$ one gets all the moments of shot-noise, hence all cumulants by well-known recursion relations. Apart from producing lower-order cumulants analytically, the method can be used for computing higher-order cumulants symbolically by employing computer algebra packages. In particular, one can perform an asymptotic analysis of the exact expressions in the limit $N_{1,2} \gg 1$ to make predictions for the leading-order term in the $\frac{1}{N}$ expansion of higher-order cumulants [Kho09].

As concerns the corresponding distribution functions, explicit expressions for the conductance distribution $P_g^{(\beta)}(g)$ are available in particular cases of $n = 1, 2$ [Mel99]. At $N_2 = K \geq N_1 = 1$, Eq. (34.2.4) readily gives $P_{g,n=1}^{(\beta)}(g) = (\beta K/2) g^{\beta K/2-1}$, $0 < g < 1$. In the case of $N_1 = 2$ and arbitrary $N_2 = K \geq 2$, one can actually find an exact result at any positive integer β [Kho09] which reads: $P_{n=2}^{(\beta)}(g) = K g^{\beta K-1}[X_1 - (-1)^{(\beta K-1)/2} X_2\, \Theta(\text{g-1}) \sum_{j=0}^{\beta} \binom{\beta}{j} B_{1-g}(\frac{\beta}{2}(K-1)+j,\ 1-\beta K)]$ for $0 < g < 2$ and zero otherwise. Here $B_z(a,b)$ stands for the incomplete beta function, $X_1 = \frac{\Gamma[\beta(K+1)/2+1]\Gamma(\beta K/2)}{\Gamma(\beta/2)\Gamma(\beta K)}$ and $X_2 = \frac{\Gamma[\beta(K+1)/2+1]}{\Gamma(\beta)\Gamma[\beta(K-1)/2]}$. As both N_1 and N_2 grow, the distribution of the conductance or, generally, of any linear statistic rapidly approaches a Gaussian distribution [Bee93]. With exact expressions for higher-order cumulants in hand, one can obtain next-order corrections to the Gaussian law by making use of the Edgeworth expansion [Bli98]. Such approximations were shown to be fairly accurate in the bulk even for small channel numbers. Actually, one can derive an explicit representation for the distribution functions in terms of Pfaffians. This works in the whole density support including the spectral edges where the distributions have a power-law dependence. Such asymptotics can be investigated exactly for any β and $N_{1,2}$, including powers and corresponding pre-factors, see [Kho09] for details.

34.3 Correlation properties

The maximum-entropy approach proved to be a success for extracting many quantities important for studying electronic transport in mesoscopic systems, see Chapter 35. At the same time, correlation properties of the S-matrix at different values of energy E as well as other spectral characteristics of open systems related to the resonances turn out to be inaccessible in the framework of the maximum-entropy approach because of the single-energy nature of the latter. To address such quantities one has to employ the Hamiltonian approach based on (34.1.2). Being supplemented with the powerful supersymmetry technique of ensemble averaging, this way resulted in advances in calculating and studying two-point correlation functions of S-matrix entries and many other characteristics, as discussed below. However, exact analytical treatment of the higher-order correlation functions (needed, e.g. in the theory of Ericsson fluctuations, see Chapter 2, Section 2.8) remains yet an important outstanding problem.

34.3.1 S-matrix elements

The energy correlation function of S-matrix elements is defined as follows

$$C_S^{abcd}(\omega) \equiv \langle S_{\rm fl}^{ab*}(E_1) S_{\rm fl}^{cd}(E_2) \rangle = \int_0^\infty dt\, e^{2\pi i \omega t} \hat{C}_S^{abcd}(t), \tag{34.3.1}$$

where $S_{\rm fl} = S - \overline{S}$ stands for the fluctuating part of the S-matrix. Being interested in local fluctuations (on the energy scale of the mean level spacing), it is natural to measure the energy separation in units of Δ. In the RMT limit $N \to \infty$, function (34.3.1) turns out to depend only on the frequency $\omega = \frac{E_2 - E_1}{\Delta}$. The Fourier transform $\hat{C}_S^{abcd}(t)$ describes a gradual loss of correlations in time ($\hat{C}_S^{abcd}(t) = 0$ at $t < 0$ identically, as required by causality). Physically, it is related to the total current through the surface surrounding the scattering centre, thus, a decay law of the open system [Lyu78, Dit00].

The energy dependence becomes explicit if one considers the pole representation of the S-matrix which follows from (34.1.2): $S^{ab}(E) = \delta^{ab} - i \sum_n \frac{w_n^a \tilde{w}_n^b}{E - \mathcal{E}_n}$. Due to unitarity constraints imposed on S, the residues and poles develop nontrivial mutual correlations [Sok89]. For this reason the knowledge of only the JPDF of the resonances $\{\mathcal{E}_n\}$ (considered in the next section) is insufficient to calculate (34.3.1). The powerful supersymmetry method (see Chapter 7) turns out to be an appropriate tool to perform the statistical average in this case. In their seminal paper [Ver85], Verbaarschot, Weidenmüller, and Zirnbauer performed the exact calculation of (34.3.1) at arbitrary transmission coefficients in the case of orthogonal symmetry. The corresponding expression for unitary symmetry was later given in [Fyo05]. These results are summarized below.

The exact analytic expression for $C_S^{abcd}(\omega)$ can be represented as follows:

$$C_S^{abcd} = \delta^{ab}\delta^{cd} T_a T_c \sqrt{(1-T_a)(1-T_c)}\, J_{ac} + (\delta^{ac}\delta^{bd} + \delta_{1\beta}\delta^{ad}\delta^{bc}) T_a T_b P_{ab}. \tag{34.3.2}$$

(The same representation holds for $\hat{C}_S^{abcd}(t)$, so the argument can be omitted.) Here, the $\delta_{1\beta}$ term accounts trivially for the symmetry property $S^{ab} = S^{ba}$ in the presence of time-reversal symmetry. Considering expression (34.3.2) in the energy domain, the functions $J_{ac}(\omega)$ and $P_{ab}(\omega)$ can generally be written as certain expectation values in the field theory (non-linear zero-dimensional supersymmetric σ-model). In the $\beta = 1$ case of orthogonal symmetry, one has

$$J_{ac}(\omega) = \Big\langle\Big(\sum_{i=1}^{2}\frac{\mu_i}{1+T_a\mu_i}+\frac{2\mu_0}{1-T_a\mu_0}\Big)\Big(\sum_{i=1}^{2}\frac{\mu_i}{1+T_c\mu_i}+\frac{2\mu_0}{1-T_c\mu_0}\Big)\mathcal{F}_M\Big\rangle_\mu^{\text{goe}}$$

$$P_{ab}(\omega) = \Big\langle\Big(\sum_{i=1}^{2}\frac{\mu_i(1+\mu_i)}{(1+T_a\mu_i)(1+T_b\mu_i)}+\frac{2\mu_0(1-\mu_0)}{(1-T_a\mu_0)(1-T_b\mu_0)}\Big)\mathcal{F}_M\Big\rangle_\mu^{\text{goe}} \quad (34.3.3)$$

where $\mathcal{F}_M = \prod_c [\frac{(1-T_c\mu_0)^2}{(1+T_c\mu_1)(1+T_c\mu_2)}]^{1/2}$ is the so-called 'channel factor', which accounts for system openness, and $\langle(\cdots)\rangle_\mu^{\text{goe}}$ is to be understood explicitly as

$$\frac{1}{8}\int_0^\infty d\mu_1\int_0^\infty d\mu_2\int_0^1 \frac{d\mu_0\,(1-\mu_0)\mu_0|\mu_1-\mu_2|\,e^{i\pi\omega(\mu_1+\mu_2+2\mu_0)}}{[(1+\mu_1)\mu_1(1+\mu_2)\mu_2]^{1/2}(\mu_0+\mu_1)^2(\mu_0+\mu_2)^2}\,(\ldots).$$

In the $\beta = 2$ case of unitary symmetry, the corresponding expressions read:

$$J_{ac}(\omega) = \Big\langle\Big(\frac{\mu_1}{1+T_a\mu_1}+\frac{\mu_0}{1-T_a\mu_0}\Big)\Big(\frac{\mu_1}{1+T_c\mu_1}+\frac{\mu_0}{1-T_c\mu_0}\Big)\mathcal{F}_M\Big\rangle_\mu^{\text{gue}}$$

$$P_{ab}(\omega) = \Big\langle\Big(\frac{\mu_1(1+\mu_1)}{(1+T_a\mu_1)(1+T_b\mu_1)}+\frac{\mu_0(1-\mu_0)}{(1-T_a\mu_0)(1-T_b\mu_0)}\Big)\mathcal{F}_M\Big\rangle_\mu^{\text{gue}} \quad (34.3.4)$$

with $\mathcal{F}_M = \prod_c \frac{1-T_c\mu_0}{1+T_c\mu_1}$ and $\langle(\cdots)\rangle_\mu^{\text{gue}} = \int_0^\infty d\mu_1\int_0^1 d\mu_0 \frac{\exp\{i2\pi\omega(\mu_1+\mu_0)\}}{(\mu_1+\mu_0)^2}(\ldots)$.

There are several important cases when the above general expressions can be simplified further. We mention first elastic scattering when only a single channel is open (see [Dit92] and Eq. (19) in [Fyo05]). The profile of the correlation function is strongly non-Lorenzian in this case. This is related to the power-law behaviour of the form-factor at large times:[4] $\hat{C}_S^{abcd}(t) \sim t^{-M\beta/2-2}$ in the general case of M open channels. Such a power-law time decay is a typical characteristic of open chaotic systems [Lew91, Dit92]. Formally, it appears due to the dependence $\mathcal{F}_M \sim \prod_c(1+\frac{2}{\beta}T_c t)^{-\beta/2}$ of the channel factor at $t \gg 1$. Physically, such a behaviour can be related to resonance width fluctuations which become weaker as M grows, see Section 34.3.3 below. In the limiting case of the large number $M \gg 1$ of weakly open channels, $T_c \ll 1$, all the resonances acquire identical escape width $\sum_{c=1}^{M} T_c$, the so-called Weisskopf's width (in units of $\frac{\Delta}{2\pi}$), so that $\mathcal{F}_M$ becomes equal to $e^{-t\sum_{c=1}^{M} T_c}$. As a result, fluctuations in the S-matrix are essentially due to those in real energies of the closed counterpart of the scattering system [Ver86], with the final expression in this limit being

[4] We conventionally measure the time in units of Heisenberg time $t_H = 2\pi/\Delta$ ($\hbar = 1$). The dimensionless time is $t = \mu_0 + \frac{1}{2}(\mu_1+\mu_2)$ or $t = \mu_0 + \mu_1$ at $\beta = 1$ or $\beta = 2$, respectively.

$$C_S^{abcd}(\omega) = \frac{(\delta^{ac}\delta^{bd} + \delta_{1\beta}\delta^{ad}\delta^{bc})T_a T_b}{\sum T_c - 2\pi i\omega} + \delta^{ab}\delta^{cd}T_a T_c \int_0^\infty dt[1 - b_2(t)]e^{-(\sum T_c - 2\pi i\omega)t} \tag{34.3.5}$$

where $b_2(t)$ is the canonical two-level RMT form-factor.

As a straightforward application, we mention a prediction for the experimentally accessible elastic enhancement factor $\mathrm{var}(S^{aa})/\mathrm{var}(S^{ab})$, $a \neq b$, which follows from the S-matrix correlation function at $\omega = 0$. In particular, in the Ericsson's regime of many strongly overlapping resonances ($\sum T_c \gg 1$) the first (dominating) term in (34.3.5) yields the well-known Hauser-Feshbach relation. Very recently, the above results have been generalized to the whole crossover regime of gradually broken time-reversal symmetry (GOE–GUE crossover) and also tested in experiments with chaotic microwave billiards [Die09].

The above formulae can be also applied to *half-scattering* or *half-collision* processes (thus *decaying* quantum systems). Important examples of such events are represented by photo-dissociation or atomic autoionization where an absorption of photons excites the quantum system into an energy region (with chaotic dynamics) which allows the subsequent decay. The optical theorem can be used to relate the autocorrelation function of the photo-dissociation cross-sections to that of S-matrix elements (see [Fyo98] for an alternative approach). Another particularly interesting feature is the so-called Fano resonances resulting from intricate interference of long-time chaotic decays with short-time direct escapes. This can be again described in terms of the S-matrix correlations given above, see [Gor05] for explicit expressions, detailed discussion, and further references.

34.3.2 S-matrix poles and residues

Resonances Within the resonance approximation considered the only singularities of the scattering matrix are its poles $\mathcal{E}_n = E_n - \frac{i}{2}\Gamma_n$. Due to the unitarity constraint their complex conjugates $\mathcal{E}_n^*$ serve as S-matrix's zeros. These two conditions yield the following general representation:

$$\det S(E) = \prod_{k=1}^{N} \frac{(E - \mathcal{E}_k^*)}{(E - \mathcal{E}_k)}, \tag{34.3.6}$$

which can be also verified using (34.1.2) by virtue of $\det S(E) = \frac{\det(E - \mathcal{H}_{\mathrm{eff}}^\dagger)}{\det(E - \mathcal{H}_{\mathrm{eff}})}$. As a result, one can express the total scattering phaseshift $\phi(E) = \log\det S(E)$ (or other quantities, e.g. time-delays discussed below) in terms of resonances. Thus statistics of the $\mathcal{E}_n$s underlay fluctuations in resonance scattering.

To be able to systematically analyse statistical properties of N eigenvalues $\mathcal{E}_n$ of $\mathcal{H}_{\mathrm{eff}}$, it is natural to start with finding their JPDF $\mathcal{P}_M(\mathcal{E})$. Unfortunately, such a density is known in full generality only at $M = 1$. In this case the anti-Hermitian part of $\mathcal{H}_{\text{eff}}$ has only one nonzero eigenvalue which we denote $-i\kappa$. Assuming $\kappa > 0$ to be nonrandom, one can find in the GOE case [Fyo99]

$$\mathcal{P}^{\text{goe}}_{M=1}(\{\mathcal{E}_i\}) \propto \prod_{k=1}^{N} \frac{e^{-\frac{N}{4}|\mathcal{E}_k|^2}}{\sqrt{\operatorname{Im}\mathcal{E}_k}} \frac{|\Delta(\{\mathcal{E}_i\})|^2}{\prod_{m<n} |\mathcal{E}_m - \mathcal{E}_n^*|} \frac{e^{-\frac{N}{4}\kappa^2}}{\kappa^{N/2-1}} \delta\Big(\kappa + \sum_{i=1}^{N} \operatorname{Im}\mathcal{E}_i\Big), \tag{34.3.7}$$

whereas in the GUE case the JPDF is given by

$$\mathcal{P}^{\text{gue}}_{M=1}(\{\mathcal{E}_i\}) \propto \prod_{k=1}^{N} e^{-\frac{N}{2}\operatorname{Re}\mathcal{E}_k^2} |\Delta(\{\mathcal{E}_i\})|^2 \frac{e^{-\frac{N}{2}\kappa^2}}{\kappa^{N-1}} \delta\Big(\kappa + \sum_{i=1}^{N} \operatorname{Im}\mathcal{E}_i\Big). \tag{34.3.8}$$

Closely related formulae for randomly distributed κ were first derived in [Ull69] and, independently, in [Sok89] (see also Ref. [13] in [Fyo03]). Actually, exploiting a version of the Itzykson-Zuber-Harish-Chandra integral allows one to write the analogue of (34.3.8) at arbitrary M [Fyo99, Fyo03] but the final expression is rather cumbersome.

As is typical for RMT problems, the main challenge is to extract the n-point correlation functions corresponding to Eq. (34.3.7) or (34.3.8). The simplest, yet non-trivial statistical characteristics of the resonances is the one-point function, i.e. the mean density. Physically, it can be used to describe the distribution of the resonance widths Γ_n in a window around some energy E. An important energy scale in such a window is the mean separation Δ between neighbouring resonances along the real axis. One can show [Som99] that Eq. (34.3.7) implies the following probability density for the dimensionless widths $y_n = \pi\Gamma_n/\Delta$:

$$\rho^{\text{goe}}_{M=1}(y) = \frac{1}{4\pi}\frac{\partial^2}{\partial y^2}\int_{-1}^{1}(1-\lambda^2)\, e^{2\lambda y}\, F(\lambda, y)\, d\lambda \tag{34.3.9}$$

with $F(\lambda, y) = (g-\lambda)\int_g^\infty \frac{dp\, \exp(-py)}{(\lambda-p)^2\sqrt{(p^2-1)(p-g)}} \int_1^g \frac{dr\,(p-r)\exp(-ry)}{(\lambda-r)^2\sqrt{(r^2-1)(g-r)}}$ where the coupling constant $g = 2/T - 1 \geq 1$ is related to κ by Eq. (34.1.3). An important feature of this distribution is that it develops an algebraic tail $\propto y^{-2}$ in the case of perfect coupling $g = 1$. Such a behaviour was recently confirmed by experimental measurements of resonance widths in microwave cavities [Kuh08].

One can actually derive the distribution of the resonance widths for any finite number M of open channels. In the GOE case the corresponding expression is rather cumbersome (see Eq. (3) in [Som99]) but in the GUE case the M-channel analogue of (34.3.9) is relatively simple, being given by [Fyo97a]

$$\rho^{\text{gue}}_{M}(y) = \frac{(-1)^M}{(M-1)!}\, y^{M-1} \frac{d^M}{dy^M}\left(e^{-yg}\frac{\sinh y}{y}\right) \tag{34.3.10}$$

for equivalent channels (all $g_c = g$). This formula agrees with the results for chaotic wave scattering in graphs with broken time-reversal invariance [Kot00]. Note that in the limit of weak coupling $g \gg 1$ resonances are typically narrow and well isolated, $\Gamma_n/\Delta \sim T \ll 1$. In this regime the above expressions reduce to the standard Porter-Thomas distributions derived long ago by simple first-order perturbation theory (see discussion in Chapter 2 and [Por56]).

Actually, for the case of GUE symmetry and the fixed number $M \ll N \to \infty$ of open channels, one can find in full generality not only the distribution of the resonance widths but all the n-point correlation functions of resonances $\mathcal{E}_n$ in the complex plane. Assuming for simplicity a spectral window centred around $\mathrm{Re}\,\mathcal{E} = 0$, all correlation functions acquire the familiar determinantal form

$$\lim_{N\to\infty} \frac{1}{N^{2n}} R_n\left(z_1 = N\mathcal{E}_1, \ldots, z_n = N\mathcal{E}_n\right) = \det\left[K(z_i, z_j^*)\right]_{j,k=1}^{n}, \tag{34.3.11}$$

where the kernel is given by [Fyo99]

$$K(z_1, z_2^*) = F_M^{1/2}(z_1)\,F_M^{1/2}(z_2^*) \int_{-1}^{1} d\lambda\, e^{-i\lambda(z_1 - z_2^*)} \prod_{k=1}^{M} (g_k + \lambda), \tag{34.3.12}$$

with $F_M(z) = \sum_{c=1}^{M} \frac{e^{-2|\mathrm{Im}(z)|g_c}}{\prod_{s\neq c}(g_c - g_s)}$. In the particular case of equivalent channels the diagonal $K(z, z^*)$ reproduces the mean probability density (34.3.10).

Finally, we briefly describe the behaviour of S matrix poles in the semiclassical limit of many open channels $M \sim N \to \infty$. In such a case the poles form a dense cloud in the complex energy plane characterized by mean density $\rho(z)$ inside the cloud [Haa92, Leh95a]. The cloud is separated from the real axis by a finite gap. The gap's width sets another important energy scale, determining the correlation length in scattering fluctuations [Leh95a, Leh95b]. Considering the $M \to \infty$ limit of $K(z_1, z_2^*)$, one concludes that after appropriate rescaling statistics of resonances inside such a cloud are given by a Ginibre-like kernel

$$|K(z_1, z_2^*)| = \rho(z) \exp\{-\tfrac{1}{2}\pi\rho(z)|z_1 - z_2|^2\}, \quad z = \tfrac{1}{2}(z_1 + z_2). \tag{34.3.13}$$

This result is expected to be universally valid for non-Hermitian random matrices in the regime of strong non-Hermiticity [Fyo03], see also Chapter 18.

Nonorthogonal resonance states As was mentioned in the introduction, the resonance states (quasimodes) are identified in our approach with the eigenvectors of the non-Hermitian Hamiltonian $\mathcal{H}_{\mathrm{eff}}$. These eigenvectors form a *bi-orthogonal* rather than orthogonal set, the feature making them rather different from their Hermitian counterparts. More precisely, to any complex resonance energy $\mathcal{E}_k$ correspond to right $|R_k\rangle$ and left $\langle L_k|$ eigenvectors

$$\mathcal{H}_{\mathrm{eff}}|R_k\rangle = \mathcal{E}_k|R_k\rangle \quad \text{and} \quad \langle L_k|\mathcal{H}_{\mathrm{eff}} = \mathcal{E}_k\langle L_k| \tag{34.3.14}$$

satisfying the conditions of bi-orthogonality, $\langle L_k|R_m\rangle = \delta_{km}$, and completeness, $1 = \sum_k |R_k\rangle\langle L_k|$. The non-orthogonality manifests itself via the matrix $\mathcal{O}_{mn} = \langle L_m|L_n\rangle\langle R_n|R_m\rangle$. In the context of reaction theory such a matrix is known as the Bell-Steinberger non-orthogonality matrix which, e.g. influences branching ratios of nuclear cross-sections [Sok89] and also features in the particle escape from the scattering region [Sav97]. In the context of quantum optics $\mathcal{O}_{nn}$ yields the enhancement (Petermann's or excess-noise factor) of the line width of a lasing mode in open resonators whereas $\mathcal{O}_{n\neq m}$ describe cross-correlations between noise emitted in different eigenmodes (see Chapter 36).

This attracted an essential interest in statistical properties of $\mathcal{O}_{nm}$ which were first studied for Ginibre's complex Gaussian ensembles [Cha98]. For scattering ensembles, such a calculation at $\beta = 1$ can be performed successfully by resumming the perturbation theory in the regime of isolated resonances at $M = 1$ [Sch00] or in the opposite case of many open channels, $M \gg 1$, (strongly overlapping resonances) [Meh01]. The general non-perturbative result for the average density $\mathcal{O}(\mathcal{E}) = \langle \sum_n \mathcal{O}_n \delta(\mathcal{E}_n - \mathcal{E}) \rangle$ of the diagonal parts is available at any M only for $\beta = 2$ symmetry [Sch00]. For the off-diagonal parts, the correlator $\mathcal{O}(\mathcal{E}, \mathcal{E}') = \langle \sum_{n\neq m} \mathcal{O}_{nm} \delta(\mathcal{E}_n - \mathcal{E}) \delta(\mathcal{E}_m - \mathcal{E}') \rangle$ is known non-perturbatively only for the single-channel case, see [Fyo02, Fyo03].

34.3.3 Decay law and width fluctuations

The decay law of open wave chaotic systems is intimately related to fluctuations of the resonance widths. This important fact can be best quantified by considering the simplest yet non-trivial decay function, namely, the leakage of the norm inside the open system [Sav97] (we use units of $t_H = \frac{2\pi}{\Delta} = 1$ below):

$$P(t) \equiv \frac{1}{N} \left\langle \mathrm{Tr}\left(e^{i\mathcal{H}_{\mathrm{eff}}^{\dagger} t} e^{-i\mathcal{H}_{\mathrm{eff}} t} \right) \right\rangle = \frac{1}{N} \left\langle \sum_{n,m} \mathcal{O}_{mn} e^{i(\mathcal{E}_n^* - \mathcal{E}_m)t} \right\rangle. \tag{34.3.15}$$

In the case of the closed system $P(t) = 1$ identically at any time. Thus the entire time dependence of the 'norm-leakage' $P(t)$ is due to nonzero widths of the resonance states and their nonorthogonality.

The exact analytic expression for $P(t)$ obtained by supersymmetry calculations turns out [Sav97, Sav03] to be given by that for $P_{ab}(t)$, see Eqs. (34.3.3) and (34.3.4), where one has to put $T_a = T_b = 0$ appearing explicitly in the denominators. The typical behaviour $P(t) \sim \prod_c (1 + \frac{2}{\beta} T_c t)^{-\beta/2}$. In the so-called 'diagonal approximation', which neglects nonorthogonality of the resonance states and becomes asymptotically exact at large t, $P(t)$ gets simply related to the distribution of resonance widths by the Laplace transform: $P_{\mathrm{diag}}(t) = \frac{1}{N} \langle \sum_n e^{-\Gamma_n t} \rangle = \int_0^\infty d\Gamma \, e^{-\Gamma t} \rho(\Gamma)$. With Eq. (34.3.10) in hand, one can then analyse in detail the time evolution in decaying chaotic systems and the characteristic time scales [Sav97], see also [Cas97] for the relevant study. It is worth mentioning further applications of such a consideration to relaxation processes in open disordered conductors [Mir00] and non-linear random media [Kot04], to chaotic quantum decay in driven biased optical lattices [Glü02] and quasi-periodic structures

[Kot05], and also in the context of electromagnetic pulse propagation in disordered media [Cha03, Ski06].

34.4 Other characteristics and applications

34.4.1 Time delays

The time delay of an almost monochromatic wave packet is conventionally described via the Wigner-Smith matrix $Q(E) = -i\hbar S^{\dagger} dS/dE$, see e.g. [Fyo97a, dC02] for introduction and historical references. This matrix appears in many applications, e.g. describing charge response and charge fluctuations in chaotic cavities [Büt05]. In the considered resonance approximation of the energy-independent V the matrix $Q(E)$ can be equivalently represented as follows ($\hbar = 1$) [Sok97]:

$$Q(E) = V^{\dagger} \frac{1}{(E - \mathcal{H}_{\text{eff}})^{\dagger}} \frac{1}{E - \mathcal{H}_{\text{eff}}} V. \tag{34.4.1}$$

This gives the meaning to the matrix element Q_{ab} as the overlap of the internal parts $(E - \mathcal{H}_{\text{eff}})^{-1} V$ of the scattering wave function in the incident channels a and b. In particular, the diagonal entries Q_{cc} can be further interpreted as mean time delays in the channel c. Their sum $\tau_W = \frac{1}{M} \mathrm{Tr}\, Q$ can be related to the energy-derivative of the total scattering phase, $\tau_W = -\frac{i}{M} \frac{d}{dE} \log \det S$, and is known as the Wigner time delay. In view of (34.3.6), this quantity is solely determined by the complex resonances: $\tau_W(E) = \frac{1}{M} \sum_n \frac{\Gamma_n}{(E-E_n)^2 + \Gamma_n^2/4}$. Its average value is therefore given by the mean level density, $\langle \tau_W \rangle = 2\pi/M\Delta$. Fluctuations of τ_W around this value can be quantified in terms of the autocorrelation function of the fluctuating parts, $\tau_W^{\text{fl}} = \tau_W - \langle \tau_W \rangle$. The exact expression for $\beta = 1$ [Leh95b] and $\beta = 2$ [Fyo97a] cases are given by

$$\frac{\langle \tau_W^{\text{fl}}(E - \frac{\Delta}{2}\omega) \tau_W^{\text{fl}}(E + \frac{\Delta}{2}\omega) \rangle}{\langle \tau_W \rangle^2} = \begin{cases} 2\,\mathrm{Re} \left\langle (2\mu_0 + \mu_1 + \mu_2)^2 \right\rangle_{\mu}^{\text{goe}} \\ \mathrm{Re} \left\langle (\mu_0 + \mu_1)^2 \right\rangle_{\mu}^{\text{gue}} \end{cases}. \tag{34.4.2}$$

(The same shorthand as in Eqs. (34.3.3)–(34.3.4) has been used.) Actually, the corresponding result for the whole GOE-GUE crossover is also known [Fyo97b].

As is clear from definition (34.4.1), the time-delay matrix at a given energy E is in essence a two-point object. Therefore, its distribution cannot be obtained by methods of Section 34.2 and requires knowledge of the distribution functional of $S(E)$. The solution to this problem for the case of ideal coupling, $\bar{S} = 0$, was given by Brouwer *et al.* [Bro97], who found the time-delay matrix Q to be statistically independent of S. Considering the eigenvalues $q_1, \ldots, q_M$ of Q (known as the *proper* time delays), the JPDF of their inverses $\tilde{\gamma}_a = (\frac{2\pi}{\Delta}) q_a^{-1}$ was found to be given by the Laguerre ensemble as follows [Bro97]:

$$\mathcal{P}(\tilde{\gamma}_1, \ldots, \tilde{\gamma}_M) \propto \prod_{a<b} |\tilde{\gamma}_a - \tilde{\gamma}_b|^\beta \prod_{a=1}^{M} \tilde{\gamma}_a^{\beta M/2} e^{-\beta \tilde{\gamma}_a/2}. \tag{34.4.3}$$

The eigenvalue density can be computed by methods of orthogonal polynomials, explicit results being derived for $M \gg 1$ at any β and for $\beta = 2$ at any M. In the latter case, one can use (34.4.3) for further extracting the distribution of Q_{aa} and, finally, that of the Wigner time delay τ_W [Sav01] (this paper also provides those distributions at $\beta = 1, 4$ and $M = 1, 2$). Representation (34.4.1) together with the supersymmetry technique enables us to generalize the above results for the distribution of proper time delays to the general case of non-ideal coupling [Som01]. In this case, the distribution of the so-called partial time delays (related to Q_{cc}) is actually known in the whole GOE-GUE crossover [Fyo97b]. Finally, we note that in the special case $M = 1$ there exists a general relation between the statistics of the (unique in this case) time delay and that of eigenfunction intensities in the closed counterpart of the system [Oss05]. The RMT predicts that such intensities are distributed for $\beta = 1, 2, 4$ according to the Porter-Thomas distributions, more complicated formulae being available for the $\beta = 1$ to $\beta = 2$ crossover [Som94, Fal94]. In this way one can easily recover, e.g. the above distribution of the time delay in the crossover regime. The most powerful application of such a relation is however beyond the standard RMT for systems exhibiting Anderson localization transition, see [Oss05, Men05, Mir06].

34.4.2 Quantum maps and sub-unitary random matrices

The scattering approach can easily be adopted to treat open dynamical systems with discrete time, i.e. open counterparts of the so-called area-preserving chaotic maps. The latter are usually represented by unitary operators which act on Hilbert spaces of finite large dimension N, being often referred to as evolution, scattering or Floquet operators, depending on the given physical context. Their eigenvalues (eigenphases) consist of N points on the unit circle and conform statistically quite accurately the results obtained for Dyson's circular ensembles, see e.g. [Glü02, Oss03, Jac03, Sch09] for diverse models and physical applications. A general scattering approach framework for such systems was developed in [Fyo00] and we mention its gross features below.

For a closed linear system characterized by a wavefunction Ψ the 'stroboscopic' dynamics amounts to a linear unitary map such that $\Psi(n+1) = \hat{u}\Psi(n)$. The unitary evolution operator $\hat{u}$ describes the inner state domain decoupled both from input and output spaces. Then a coupling that makes the system open must convert the evolution operator $\hat{u}$ to a contractive operator $\hat{A}$ such that $1 - \hat{A}^\dagger \hat{A} \geq 0$. The equation $\Psi(n+1) = \hat{A}\Psi(n)$ now describes an irreversible decay of any initial state $\Psi(0) \neq 0$ when an input signal is absent. On the other hand, assuming a nonzero input and zero initial state $\Psi(0) = 0$, one can relate the (discrete) Fourier transforms of the input and output signals at a frequency ω to each other by a $M \times M$ unitary scattering matrix $\hat{S}(\omega)$ as follows:

$$\hat{S}(\omega) = \sqrt{1 - \hat{\tau}^\dagger\hat{\tau}} - \hat{\tau}^\dagger \frac{1}{e^{-i\omega} - \hat{A}} \hat{u}\hat{\tau}, \qquad \hat{A} = \hat{u}\sqrt{1 - \hat{\tau}\hat{\tau}^\dagger}, \tag{34.4.4}$$

where $\hat{\tau}$ is a rectangular $N \times M$ matrix with $M \le N$ nonzero entries $\tau_{ij} = \delta_{ij}\tau_j$, $0 \le \tau_i \le 1$. This formula is a complete discrete-time analogue of Eq. (34.1.2). In particular, one can straightforwardly verify unitarity and show that

$$\det \hat{S}(\omega) = e^{-i\omega N} \frac{\det\left(\hat{A}^\dagger - e^{i\omega}\right)}{\det(e^{-i\omega} - \hat{A})} = e^{-i\omega N} \prod_{k=1}^{N} \frac{\left(z_k^* - e^{i\omega}\right)}{(e^{-i\omega} - z_k)}, \tag{34.4.5}$$

where z_k stand for the complex eigenvalues of the matrix $\hat{A}$ (note $|z_k| < 1$). This relation is an obvious analogue of Eq. (34.3.6) and gives another indication of z_k playing the role of resonances for discrete time systems.

Generic features of quantized maps with chaotic inner dynamics are emulated by choosing $\hat{u}$ from one of the Dyson circular ensembles. By averaging Eq. (34.4.4) over $\hat{u}$, one easily finds $\hat{\tau}^\dagger\hat{\tau} = 1 - |\langle \hat{S} \rangle|^2$. Therefore, M eigenvalues $T_a = \tau_a^2 \le 1$ of $\hat{\tau}^\dagger\hat{\tau}$ play the familiar role of transmission coefficients. In the particular case of all $T_a = 1$ (ideal coupling), the non-vanishing eigenvalues of the matrix $\hat{A}$ coincide with those of a $(N-M)\times(N-M)$ sub-block of $\hat{u}$. Complex eigenvalues of such 'truncations' of random unitary matrices were first studied analytically in [Życ00]. The general ensemble of $N \times N$ random contractions $\hat{A} = \hat{u}\sqrt{1 - \hat{\tau}\hat{\tau}^\dagger}$ was studied in [Fyo03] starting from the following probability measure in the matrix space ($d\hat{A} = \prod d\mathrm{Re}\, A_{ij} d\mathrm{Im}\, A_{ij}$):

$$\mathcal{P}(\hat{A})d\hat{A} \propto \delta(\hat{A}^\dagger\hat{A} - \hat{G})d\hat{A}, \quad \hat{G} \equiv \mathbf{1} - \hat{\tau}\hat{\tau}^\dagger, \tag{34.4.6}$$

Equation (34.4.6) describes an ensemble of subunitary matrices $\hat{A}$ with given singular values. The question of characterizing the locus of complex eigenvalues for a general matrix with prescribed singular values is classical, and the recent paper [Wei08] provided a kind of statistical answer to that question.

34.4.3 Microwave cavities at finite absorption

The above consideration is restricted to the idealization neglecting absorption. The latter is almost invariably present, usually being seen as a dissipation of incident power evolving exponentially in time. In the approximation of *uniform* absorption all the resonances acquire, in addition to their escape widths Γ_n, one and the same absorption width $\Gamma_{\text{abs}} \equiv \gamma\frac{\Delta}{2\pi} > 0$. Operationally it is equivalent to a purely imaginary shift of the scattering energy $E \to E + \frac{i}{2}\Gamma_{\text{abs}} \equiv E_\gamma$. As a result, the S-matrix correlation function in the presence of the absorption is related to that without by replacing $\omega \to \omega + i\gamma/2\pi$ in (34.3.1). In the time domain, the corresponding form-factor then simply acquires the additional decay factor $e^{-\gamma t}$ (in this way the absorption strength γ can be extracted experimentally, e.g. for the microwave cavities [Sch03]).

Distribution functions undergo, however, more drastic changes at finite absorption, as the scattering matrix $S(E_\gamma)$ becomes subunitary. The mismatch between incoming and outgoing fluxes can be quantified by the reflection matrix

$$R = S(E_\gamma)^\dagger S(E_\gamma) = 1 - \Gamma_{\text{abs}} Q(E_\gamma). \tag{34.4.7}$$

This representation, with $Q(E_\gamma)$ from (34.4.1), is valid at arbitrary Γ_{abs} [Sav03]. The so-called reflection eigenvalues of R play an important role for the description of thermal emission from random media, as discovered by Beenakker, who also computed their distribution at perfect coupling ($T = 1$) (see Section 36.3). The above connection to the time-delay matrix at finite absorption allows one to generalize this result to the case of arbitrary coupling, $T \leq 1$ [Sav03].

In terms of quantum mechanics the quantity $1 - R_{aa}$ has the meaning of probability of no return to the incident channel. For a single channel scattering without absorption it should be identically zero, but finite absorption induces a 'unitary deficit' $1 - R_{aa} > 0$ which can be used as a sensitive probe for chaotic scattering. The quantity starts to play an even more important role in single-channel scattering from systems with intrinsic disorder. Indeed, for such systems the phenomenon of *spontaneous* breakdown of S-matrix unitarity was suggested as one of the most general manifestations of Anderson localization transition of waves in random media, see [Fyo03b] for further discussion, and [Kot02] for somewhat related ideas.

In the context of statistical electromagnetics, the matrix $Z = i\,K(E_\gamma)$ has the meaning of normalized cavity impedance [Hem05]. In view of (34.1.1), the joint distribution function of its real and imaginary parts fully determines the distribution of (non-unitary) S. It turns out that there exists a general relation (fluctuation-dissipation like) between this function and the energy autocorrelation function of resolvents of K at zero absorption [Sav05]. This relation enables us to further compute the exact distributions of complex impedances and of reflection coefficients $|S_{cc}|^2$ at any M and for the whole GOE-GUE crossover [Fyo05], and to explain the relevant experimental data [Kuh05b, Zhe06, Law08]. The corresponding study for the case of transmission is still an open problem.

Quite often, however, an approximation of uniform absorption may break down, and one should take into account *inhomogeneous* (or localized-in-space) losses which result in different broadening of different modes. The latter are easily incorporated in the model by treating them as though induced by additional scattering channels, see e.g. [Roz03, Sav06b, Pol09] for relevant applications.

References

[Alh00] Y. Alhassid, Rev. Mod. Phys. **72** (2000) 895.

[Ara98] J. E. F. Araújo and A. M. S. Macêdo, Phys. Rev. B **58** (1998) R13379.

[Bar94] H. U. Baranger and P. A. Mello, Phys. Rev. Lett. **73** (1994) 142.

[Bee93] C. W. J. Beenakker, Phys. Rev. Lett. **70** (1993) 1155.

[Bee97] C. W. J. Beenakker, Rev. Mod. Phys. **69** (1997) 731.

[Bla00] Ya. M. Blanter and M. Büttiker, Phys. Rep. **336** (2000) 1.

[Bli98] S. Blinnikov and R. Moessner, Astron. Astrophys. Suppl. Ser. **130** (1998) 193.

[Bro95] P. W. Brouwer, Phys. Rev. B **51** (1995) 16878.

[Bro96] P. W. Brouwer and C. W. J. Beenakker, J. Math. Phys. **37** (1996) 4904.

[Bro97] P. W. Brouwer, K. M. Frahm and C. W. J. Beenakker, Phys. Rev. Lett. **78** (1997) 4737; Waves Random Media **9** (1999) 91.

[Büt05] M. Büttiker and M. L. Polansky, J. Phys. A **38** (2005) 10559.

[Bul06] E. N. Bulgakov, V. A. Gopar, P. A. Mello and I. Rotter, Phys. Rev. B **73** (2006) 155302.

[Cas97] G. Casati, G. Maspero and D. Shepelyansky, Phys. Rev. E **56** (1997) R6233; K. M. Frahm, Phys. Rev. E **56** (1997) R6237.

[Cha03] A. A. Chabanov, Z. Zhang and A. Z. Genack, Phys. Rev. Lett. **90** (2003) 203903.

[Cha98] J. T. Chalker and B. Mehlig, Phys. Rev. Lett. **81** (1998) 3367;
B. Mehlig and J. T. Chalker, J. Math. Phys. **41** (2000) 3233.

[dC02] C. A. de Carvalho and H. M. Nussenzveig, Phys. Rep. **364** (2002) 83.

[Die09] B. Dietz *et al.*, Phys. Rev. Lett. **103** (2009) 064101.

[Dit92] F.-M. Dittes, H. L. Harney and A. Müller, Phys. Rev. A **45** (1992) 701.

[Dit00] F.-M. Dittes, Phys. Rep. **339** (2000) 215.

[Fal94] V. I. Fal'ko and K. B. Efetov, Phys. Rev. B **50** (1994) 11267.

[For08] P. J. Forrester and S. O. Warnaar, Bull. Am. Math. Soc. **45** (2008) 489.

[Fri85] W. A. Friedman and P. A. Mello, Ann. Phys. (N.Y.) **161** (1985) 276.

[Fyo97a] Y. V. Fyodorov and H.-J. Sommers, J. Math. Phys. **38** (1997) 1918.

[Fyo97b] Y. V. Fyodorov, D. V. Savin and H.-J. Sommers, Phys. Rev. E **55** (1997) R4857.

[Fyo98] Y. V. Fyodorov and Y. Alhassid, Phys. Rev. A **58** (1998) R3375.

[Fyo99] Y. V. Fyodorov and B. A. Khoruzhenko, Phys. Rev. Lett. **83** (1999) 65.

[Fyo00] Y. V. Fyodorov and H.-J. Sommers, JETP Lett. **72** (2000) 422.

[Fyo02] Y. V. Fyodorov and B. Mehlig, Phys. Rev. E **66** (2002) 045202(R).

[Fyo03] Y. V. Fyodorov and H.-J. Sommers, J. Phys. A **36** (2003) 3303.

[Fyo03b] Y. V. Fyodorov, JETP Letters **78** (2003) 250

[Fyo05] Y. V. Fyodorov, D. V. Savin and H.-J. Sommers, J. Phys. A **38** (2005) 10731.

[Glü02] M. Glück, A. R. Kolovsky and H. J. Korsch, Phys. Rep. **366** (2002) 103.

[Gop98] V. A. Gopar and P. A. Mello, Europhys. Lett. **42** (1998) 131.

[Gop08] V. A. Gopar, M. Martínez-Mares and R. A. Méndez-Sánchez, J. Phys. A **41** (2008) 015103.

[Gor05] T. Gorin, J. Phys. A **38** (2005) 10805.

[Guh98] T. Guhr, A. Müller-Groeling and H. A. Weidenmüller, Phys. Rep. **299** (1998) 189.

[Haa92] F. Haake *et al.*, Z. Phys. B **88** (1992) 359.

[Hem05] S. Hemmady *et al.*, Phys. Rev. Lett. **94** (2005) 014102.

[Hua63] L. K. Hua, *Harmonic Analysis of Functions of Several Complex Variables in the Classical Domains*, American Mathematical Society, Providence, RI, 1963.

[Jac03] Ph. Jacquod, H. Schomerus and C. W. J. Beenakker, Phys. Rev. Lett. **90** (2003) 207004.

[Jal94] R. A. Jalabert, J.-L. Pichard and C. W. J. Beenakker, Europhys. Lett. **27** (1994) 255.

[Kho09] B. A. Khoruzhenko, D. V. Savin and H.-J. Sommers, Phys. Rev. B **80** (2009) 125301.
[Kot00] T. Kottos and U. Smilansky, Phys. Rev. Lett. **85** (2000) 968.
[Kot02] T. Kottos and M. Weiss, Phys. Rev. Lett. **89** (2002) 056401.
[Kot04] T. Kottos and M. Weiss, Phys. Rev. Lett. **93** (2004) 190604.
[Kot05] T. Kottos, J. Phys. A **38** (2005) 10761.
[Kuh05a] U. Kuhl, H.-J. Stöckmann and R. Weaver, J. Phys. A **38** (2005) 10433.
[Kuh05b] U. Kuhl, M. Martínez-Mares, R. A. Méndez-Sánchez and H.-J. Stöckmann, Phys. Rev. Lett. **94** (2005) 144101.
[Kuh08] U. Kuhl, R. Höhmann, J. Main and H.-J. Stöckmann, Phys. Rev. Lett. **100** (2008) 254101.
[Law08] M. Lawniczak *et al.*, Phys. Rev. E **77** (2008) 056210.
[Leh95a] N. Lehmann, D. Saher, V. V. Sokolov and H.-J. Sommers, Nucl. Phys. A **582** (1995) 223.
[Leh95b] N. Lehmann, D. V. Savin, V. V. Sokolov and H.-J. Sommers, Physica D **86** (1995) 572.
[Lew91] C. H. Lewenkopf and H. A. Weidenmüller, Ann. Phys. (N.Y.) **212** (1991) 53.
[Liv73] M. S. Livšic, *Operators, Oscillations, Waves (Open Systems)*, Translations of Mathematical Monographs, Vol. 34, American Mathematical Society, Providence, RI, 1973.
[Lyu78] V. L. Lyuboshitz, Yad. Fiz. **27** (1978) 948, [Sov. J. Nucl. Phys. **27**, 502 (1978)].
[Meh01] B. Mehlig and M. Santer, Phys. Rev. E **63** (2001) 020105(R).
[Mel85] P. A. Mello, P. Pereyra and T. Seligman, Ann. Phys. **161** (1985) 254.
[Mel99] P. A. Mello and H. U. Baranger, Waves Random Media **9** (1999) 105.
[Men05] J. A. Mendez-Bermudez, T. Kottos Phys. Rev. B **72**, 064108 (2005)
[Mel04] P. A. Mello and N. Kumar, *Quantum Transport in Mesoscopic Systems*, Oxford University Press, 2004
[Mir06] A. D. Mirlin, Y. V. Fyodorov, F. Evers, and A. Mildenberger, Phys. Rev. Lett. **97** (2006) 046803.
[Mir00] A. D. Mirlin, Phys. Rep. **326** (2000) 259.
[Mit10] G. E. Mitchel, A. Richter, and H. A. Weidenmüller, Rev. Mod. Phys. **82** (2010) 2845.
[Mon09] C. Monthus and T. Garel, J. Phys. A **42** (2009) 475007.
[Nov07] M. Novaes, Phys. Rev. B **75** (2007) 073304.
[Nov08] M. Novaes, Phys. Rev. B **78** (2008) 035337.
[Nus72] H. M. Nussenzveig, *Causality and Dispersion Relations*, Academic Press, New York, 1972.
[Osi08] V. A. Osipov and E. Kanzieper, Phys. Rev. Lett. **101** (2008) 176804; J. Phys. A: Math. Theor. 42 (2009) 475101.
[Oss03] A. Ossipov, T. Kottos and T. Geisel, Europhys. Lett. **62** (2003) 719.
[Oss05] A. Ossipov and Y. V. Fyodorov, Phys. Rev. B **71** (2005) 125133.
[Pol09] C. Poli, D. V. Savin, O. Legrand and F. Mortessagne, Phys. Rev. E **80** (2009) 046203.
[Por56] C. E. Porter and R. G. Thomas, Phys. Rev. **104** (1956) 483.
[Roz03] I. Rozhkov, Y. V. Fyodorov and R. L. Weaver, Phys. Rev. E **68** (2003) 016204; *ibid.* **69** (2004) 036206.
[Sav97] D. V. Savin and V. V. Sokolov, Phys. Rev. E **56** (1997) R4911.
[Sav01] D. V. Savin, Y. V. Fyodorov and H.-J. Sommers, Phys. Rev. E **63** (2001) 035202(R).
[Sav03] D. V. Savin and H.-J. Sommers, Phys. Rev. E **68** (2003) 036211; *ibid.* **69** (2004) 035201(R).
[Sav05] D. V. Savin, H.-J. Sommers and Y. V. Fyodorov, JETP Lett. **82** (2005) 544.
[Sav06a] D. V. Savin and H.-J. Sommers, Phys. Rev. B **73** (2006) 081307(R).
[Sav06b] D. V. Savin, O. Legrand and F. Mortessagne, Europhys. Lett. **76** (2006) 774.
[Sav08] D. V. Savin, H.-J. Sommers and W. Wieczorek, Phys. Rev. B **77** (2008) 125332.

[Sch03] R. Schäfer, T. Gorin, T. H. Seligman and H.-J. Stöckmann, J. Phys. A **36** (2003) 3289.
[Sch00] H. Schomerus, K. M. Frahm, M. Patra and C. W. J. Beenakker, Physica A **278** (2000) 469.
[Sch09] H. Schomerus, J. Wiersig and J. Main, Phys. Rev. A **79** (2009) 053806.
[Ski06] S. E. Skipetrov and B. A. van Tiggelen, Phys. Rev. Lett. **96** (2006) 043902.
[Sok89] V. V. Sokolov and V. G. Zelevinsky, Nucl. Phys. A **504** (1989) 562.
[Sok97] V. V. Sokolov and V. G. Zelevinsky, Phys. Rev. C **56** (1997) 311.
[Som94] H.-J. Sommers and S. Iida, Phys. Rev. E **49** (1994) R2513.
[Som99] H.-J. Sommers, Y. V. Fyodorov and M. Titov, J. Phys. A **32** (1999) L77.
[Som01] H.-J. Sommers, D. V. Savin and V. V. Sokolov, Phys. Rev. Lett. **87** (2001) 094101.
[Ull69] N. Ullah, J. Math. Phys. **10** (1969) 2099.
[Ver85] J. J. M. Verbaarschot, H. A. Weidenmüller and M. R. Zirnbauer, Phys. Rep. **129** (1985) 367.
[Ver86] J. J. M. Verbaarschot, Ann. Phys. (N.Y.) **168** (1986) 368.
[Viv08] P. Vivo and E. Vivo, J. Phys. A **41** (2008) 122004.
[Wei08] Y. Wei and Y. V. Fyodorov, J. Phys. A **41** (2008) 502001.
[Wei06] M. Weiss, J. A. Mendez-Bermudez and T. Kottos, Phys. Rev. B **73** (2006) 045103.
[Zhe06] X. Zheng *et al.*, Phys. Rev. E **73** (2006) 046208.
[Życ00] K. Życzkowski and H.-J. Sommers, J. Phys. A **33** (2000) 2045.

·35·

Condensed matter physics

C. W. J. Beenakker

Abstract

This is a cursory overview of applications of concepts from random matrix theory (RMT) to condensed matter physics. The emphasis is on phenomena, predicted or explained by RMT, that have actually been observed in experiments on quantum wires and quantum dots. These observations include universal conductance fluctuations, weak localization, sub-Poissonian shot noise, and non-Gaussian thermopower distributions. The last section addresses the effects (not yet observed) of superconductivity on the statistics of the Hamiltonian and scattering matrix.

35.1 Introduction

Applications of random matrix theory (RMT) to condensed matter physics search for universal features in the electronic properties of metals, originating from the universality of eigenvalue repulsion. Eigenvalue repulsion is universal, because the Jacobian

$$J = \prod_{i<j} |E_j - E_i|^{\beta} \tag{35.1.1}$$

of the transformation from matrix space to eigenvalue space depends on the symmetry of the random matrix ensemble (expressed by the index $\beta \in \{1, 2, 4\}$) – but is independent of microscopic properties such as the mean eigenvalue separation [Meh67]. This universality is at the origin of the remarkable success of RMT in nuclear physics [Bro81, Wei09].

In condensed matter physics, the applications of RMT fall into two broad categories. In the first category, one studies thermodynamic properties of closed systems, such as metal grains or semiconductor quantum dots. The random matrix is the Hamiltonian H. In the second category, one studies transport properties of open systems, such as metal wires or quantum dots with point contacts. Now the random matrix is the scattering matrix S (or a submatrix, the transmission matrix t). Applications in both categories have flourished with the development of nanotechnology. Confinement of electrons on the nanoscale in wire geometries (quantum wires) and box geometries (quantum dots) preserves their phase coherence, which is needed for RMT to be applicable.

The range of electronic properties addressed by RMT is quite broad. The selection of topics presented in this chapter is guided by the desire to show those applications of RMT that have actually made an impact on experiments. For more complete coverage of topics and a more comprehensive list of references we suggest a few review articles [Bee97, Guh98, Alh00].

35.2 Quantum wires

35.2.1 Conductance fluctuations

In the 1960s, Wigner, Dyson, Mehta, and others discovered that fluctuations in the energy level density are governed by level repulsion and therefore take a universal form [Por65]. The universality of the level fluctuations is expressed by the Dyson–Mehta formula [Dys63] for the variance of a linear statistic[1] $A = \sum_n a(E_n)$ on the energy levels E_n. The Dyson–Mehta formula reads

$$\text{Var}\, A = \frac{1}{\beta}\frac{1}{\pi^2}\int_0^\infty dk\, |a(k)|^2 k, \tag{35.2.1}$$

where $a(k) = \int_{-\infty}^{\infty} dE\, e^{ikE} a(E)$ is the Fourier transform of $a(E)$. Equation (35.2.1) shows that: (1) The variance is independent of microscopic parameters; (2) The variance has a universal $1/\beta$-dependence on the symmetry index.

In a pair of seminal 1986 papers [Imr86, Alt86], Imry and Altshuler and Shklovskiĭ proposed to apply RMT to the phenomenon of universal conductance fluctuations (UCF) in metals, which was discovered using diagrammatic perturbation theory by Altshuler [Alt85] and Lee and Stone [Lee85]. Universal conductance Fluctuations is the occurrence of sample-to-sample fluctuations in the conductance which are of order e^2/h at zero temperature, *independent* of the size of the sample or the degree of disorder – as long as the conductor remains in the diffusive metallic regime (size L large compared to the mean free path l, but small compared to the localization length ξ). An example is shown in Fig. 35.1.

The similarity between the statistics of energy levels measured in nuclear reactions on the one hand, and the statistics of conductance fluctuations measured in transport experiments on the other, was used by Stone *et al.* [Mut87, Sto91] to construct a random matrix theory of quantum transport in metal wires. The random matrix is now not the Hamiltonian H, but the transmission matrix t, which determines the conductance through the Landauer formula

$$G = G_0 \text{Tr}\, tt^\dagger = G_0 \sum_n T_n. \tag{35.2.2}$$

[1] The quantity A is called a linear statistic because products of different E_ns do not appear, but the function $a(E)$ may well depend non-linearly on E.

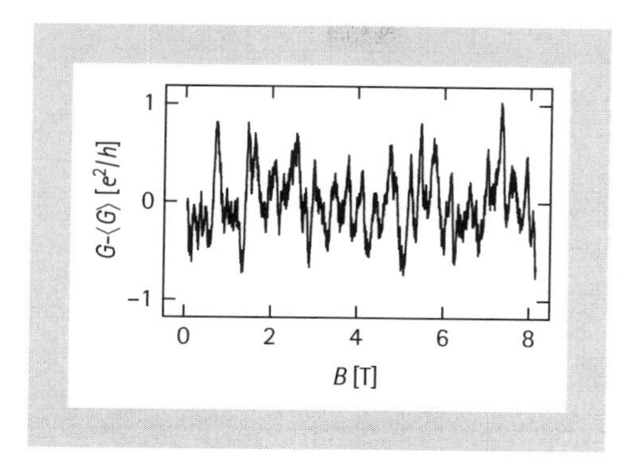

Fig. 35.1 Fluctuations as a function of perpendicular magnetic field of the conductance of a 310-nm-long and 25-nm-wide Au wire at 10 mK. The trace appears random, but is completely reproducible from one measurement to the next. The root-mean-square of the fluctuations is $0.3\, e^2/h$, which is not far from the theoretical result $\sqrt{1/15}\, e^2/h$ [Eq. (35.2.8) with $\beta = 2$ due to the magnetic field and a reduced conductance quantum $G_0 = e^2/h$ due to the strong spin-orbit scattering in Au]. Adapted from [Was86].

The conductance quantum is $G_0 = 2e^2/h$, with a factor of two to account for spin degeneracy. Instead of repulsion of energy levels, one now has repulsion of the transmission eigenvalues T_n, which are the eigenvalues of the transmission matrix product $tt^\dagger$. In a wire of cross-sectional area $\mathcal{A}$ and Fermi wave length λ_F, there are of order $N \simeq A/\lambda_F^2$ propagating modes, so t has dimension $N \times N$ and there are N transmission eigenvalues. The phenomenon of UCF applies to the regime $N \gg 1$, typical for metal wires.

Random matrix theory is based on the fundamental assumption that all correlations between the eigenvalues are due to the Jacobian $J = \prod_{i<j} |T_i - T_j|^\beta$ from matrix elements to eigenvalues. If all correlations are due to the Jacobian, then the probability distribution $P(T_1, T_2, \ldots T_N)$ of the T_ns should have the form $P \propto J \prod_i p(T_i)$, or equivalently,

$$P(\{T_n\}) \propto \exp\Big[-\beta\Big(\sum_{i<j} u(T_i, T_j) + \sum_i V(T_i)\Big)\Big], \qquad (35.2.3)$$

$$u(T_i, T_j) = -\ln |T_j - T_i|, \qquad (35.2.4)$$

with $V = -\beta^{-1} \ln p$. Equation (35.2.3) has the form of a Gibbs distribution at temperature β^{-1} for a fictitious system of classical particles on a line in an external potential V, with a logarithmically repulsive interaction u. All microscopic parameters are contained in the single function $V(T)$. The logarithmic repulsion is independent of microscopic parameters, because of its geometric origin.

Unlike the RMT of energy levels, the correlation function of the T_ns is not translationally invariant, due to the constraint $0 \le T_n \le 1$ imposed by unitarity of the scattering matrix. Because of this constraint, the Dyson–Mehta formula

(35.2.1) needs to be modified, as shown in [Bee93a]. In the large-N limit, the variance of a linear statistic $A = \sum_n f(T_n)$ on the transmission eigenvalues is given by

$$\mathrm{Var}\, A = \frac{1}{\beta} \frac{1}{\pi^2} \int_0^\infty dk \, |F(k)|^2 k \tanh(\pi k). \tag{35.2.5}$$

The function $F(k)$ is defined in terms of the function $f(T)$ by the transform

$$F(k) = \int_{-\infty}^{\infty} dx \, \mathrm{e}^{ikx} f\left(\frac{1}{1+\mathrm{e}^x}\right). \tag{35.2.6}$$

The formula (35.2.5) demonstrates that the universality which was the hallmark of UCF is generic for a whole class of transport properties, namely those which are linear statistics on the transmission eigenvalues. Examples, reviewed in [Bee97], are the critical-current fluctuations in Josephson junctions, conductance fluctuations at normal-superconductor interfaces, and fluctuations in the shot-noise power of metals.

35.2.2 Nonlogarithmic eigenvalue repulsion

The probability distribution (35.2.3) was justified by a maximum-entropy principle for an ensemble of quasi-1D conductors [Mut87, Sto91]. Quasi-1D refers to a wire geometry (length L much greater than width W). In such a geometry one can assume that the distribution of scattering matrices in an ensemble with different realizations of the disorder is only a function of the transmission eigenvalues (isotropy assumption). The distribution (35.2.3) then maximizes the information entropy subject to the constraint of a given density of eigenvalues. The function $V(T)$ is determined by this constraint and is not specified by RMT.

It was initially believed that Eq. (35.2.3) would provide an exact description in the quasi-1D limit $L \gg W$, if only $V(T)$ were chosen suitably [Sto91]. However, the generalized Dyson–Mehta formula (35.2.5) demonstrates that RMT is not exact in a quantum wire [Bee93a]. If one computes from Eq. (35.2.5) the variance of the conductance (35.2.2) (by substituting $f(T) = G_0 T$), one finds

$$\mathrm{Var}\, G/G_0 = \frac{1}{8}\beta^{-1}, \tag{35.2.7}$$

independent of the form of $V(T)$. The diagrammatic perturbation theory [Alt85, Lee85] of UCF gives instead

$$\mathrm{Var}\, G/G_0 = \frac{2}{15}\beta^{-1} \tag{35.2.8}$$

for a quasi-1D conductor. The difference between the coefficients $\frac{1}{8}$ and $\frac{2}{15}$ is tiny, but it has the fundamental implication that the interaction between the

Ts is not precisely logarithmic, or in other words, that there exist correlations between the transmission eigenvalues over and above those induced by the Jacobian [Bee93a].

The $\frac{1}{8} - \frac{2}{15}$ discrepancy raised the question what the true eigenvalue interaction would be in quasi-1D conductors. Is there perhaps a cutoff for large separation of the Ts? Or is the true interaction a many-body interaction, which cannot be reduced to the sum of pairwise interactions? This transport problem has a counterpart in a closed system. The RMT of the statistics of the eigenvalues of a random Hamiltonian yields a probability distribution of the form (35.2.3), with a logarithmic repulsion between the energy levels [Meh67]. It was shown by Efetov [Efe83] and by Altshuler and Shklovskiĭ [Alt86] that the logarithmic repulsion in a disordered metal grain holds for energy separations small compared to the inverse ergodic time $\hbar/\tau_{\text{erg}}$.[2] For larger separations the interaction potential decays algebraically [Jal93].

The way in which the RMT of quantum transport breaks down is quite different [Bee93b]. The probability distribution of the transmission eigenvalues does indeed take the form (35.2.3) of a Gibbs distribution with a parameter-independent two-body interaction $u(T_i, T_j)$, as predicted by RMT. However, the interaction differs from the logarithmic repulsion (35.2.4) of RMT. Instead, it is given by

$$u(T_i, T_j) = -\tfrac{1}{2}\ln|T_j - T_i| - \tfrac{1}{2}\ln|x_j - x_i|, \quad \text{with } T_n \equiv 1/\cosh^2 x_n. \tag{35.2.9}$$

The eigenvalue interaction (35.2.9) is different for weakly and for strongly transmitting scattering channels: $u \to -\ln|T_j - T_i|$ for $T_i, T_j \to 1$, but $u \to -\frac{1}{2}\ln|T_j - T_i|$ for $T_i, T_j \ll 1$. For weakly transmitting channels it is *twice as small* as predicted by considerations based solely on the Jacobian, which turn out to apply only to the strongly transmitting channels.

The nonlogarithmic interaction modifies the Dyson–Mehta formula for the variance of a linear statistic. Instead of Eq. (35.2.5) one now has [Bee93b, Cha93]

$$\text{Var}\, A = \frac{1}{\beta}\frac{1}{\pi^2}\int_0^\infty dk\, \frac{k|F(k)|^2}{1 + \text{cotanh}(\frac{1}{2}\pi k)}, \tag{35.2.10}$$

$$F(k) = \int_{-\infty}^{\infty} dx\, \mathrm{e}^{ikx} f\left(\frac{1}{\cosh^2 x}\right). \tag{35.2.11}$$

Substitution of $f(T) = T$ now yields $\frac{2}{15}$ instead of $\frac{1}{8}$ for the coefficient of the UCF, thus resolving the discrepancy between Eqs. (35.2.7) and (35.2.8).

[2] The ergodic time is the time needed for a particle to explore the available phase space in a closed system. In a disordered metal grain of size L and diffusion constant D, one has $\tau_{\text{erg}} \simeq L^2/D$. If the motion is ballistic (with velocity v_F) rather than diffusive, one has instead $\tau_{\text{erg}} \simeq L/v_F$.

The result (35.2.9) follows from the solution of a differential equation which determines how the probability distribution of the T_ns changes when the length L of the wire is incremented. This differential equation has the form of a multivariate drift-diffusion equation (with L playing the role of time) for N classical particles at coordinates $x_n = \operatorname{arcosh} T_n^{-1/2}$. The drift-diffusion equation,

$$l\frac{\partial}{\partial L}P(\{x_n\},L) = \frac{1}{2}(\beta N + 2 - \beta)^{-1}\sum_{n=1}^{N}\frac{\partial}{\partial x_n}\left(\frac{\partial P}{\partial x_n} + \beta P\frac{\partial \Omega}{\partial x_n}\right), \tag{35.2.12}$$

$$\Omega = -\sum_{i=1}^{N}\sum_{j=i+1}^{N}\ln|\sinh^2 x_j - \sinh^2 x_i| - \frac{1}{\beta}\sum_{i=1}^{N}\ln|\sinh 2x_i|, \tag{35.2.13}$$

is known as the DMPK equation, after the scientists who first studied its properties in the 1980s [Dor82, Mel88a]. (The equation itself appeared a decade earlier [Bur72].) The DMPK equation can be solved exactly [Bee93b, Cas95], providing the nonlogarithmic repulsion (35.2.9).

35.2.3 Sub-Poissonian shot noise

The average transmission probability $\bar{T} = l/L$ for diffusion through a wire is the ratio of mean free path l and wire length L. This average is not representative for a single transmission eigenvalue, because eigenvalue repulsion prevents the T_ns from having a narrow distribution around $\bar{T}$. The eigenvalue density $\rho(T) = \langle\sum_n \delta(T - T_n)\rangle$ can be calculated from the DMPK equation (35.2.12), with the result [Dor84, Mel89]

$$\rho(T) = \frac{Nl}{2L}\frac{1}{T\sqrt{1-T}}, \quad \text{for } T_{\min} \leq T < 1, \tag{35.2.14}$$

in the diffusive metallic regime $l \ll L \ll \xi$.[3] The lower limit $T_{\min}$ is determined by the normalization, $\int_0^1 dT\,\rho(T) = N$, giving $T_{\min} \approx 4e^{-L/2l}$ with exponential accuracy.

The transmission eigenvalue density is *bimodal*, with a peak at unit transmission (open channels) and a peak at exponentially small transmission (closed channels). This bimodal distribution cannot be observed in the conductance $G \propto \sum_n T_n$, which would be the same if all T_ns would cluster near the average $\bar{T}$. The shot noise power $\mathcal{S} \propto \sum_n T_n(1 - T_n)$ (the second moment of the time-dependent current fluctuations) provides more information.

The ratio of shot noise power and conductance, defined in dimensionless form by the Fano factor

$$F = \frac{\sum_n T_n(1-T_n)}{\sum_n T_n}, \tag{35.2.15}$$

[3] The localization length ξ also follows from the DMPK equation. It is given by $\xi = (\beta N + 2 - \beta)l$, so it is larger than l by a factor of order N.

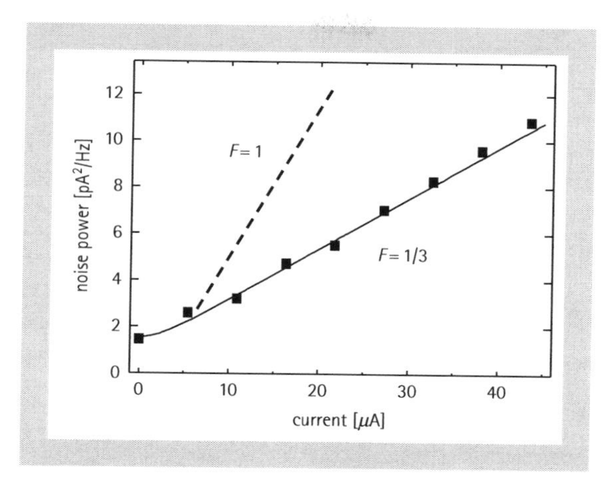

Fig. 35.2 Sub-Poissonian shot noise in a disordered gold wire (dimensions 940 nm × 100 nm). At low currents the noise saturates at the level set by the temperature of 0.3 K. At higher currents the noise increases linearly with current, with a slope that is three times smaller than the value expected for Poisson statistics. Adapted from [Hen99].

quantifies the deviation of the current fluctuations from a Poisson process (which would have $F = 1$). Since $\bar{T} \ll 1$, if all T_ns would be near $\bar{T}$ the current fluctuations would have Poisson statistics with $F = 1$. The bimodal distribution (35.2.14) instead gives sub-Poissonian shot noise [Bee92],

$$F \to 1 - \frac{\int dT\, \rho(T) T^2}{\int dT\, \rho(T) T} = 1 - \frac{2}{3} = \frac{1}{3}. \qquad (35.2.16)$$

(The replacement of the sum over n by an integration over T with weight $\rho(T)$ is justified in the large-N limit.) This one-third suppression of shot noise below the Poisson value has been confirmed experimentally [Ste96, Hen99], see Fig. 35.2.

35.3 Quantum dots

35.3.1 Level and wave function statistics

Early applications of random matrix theory to condensed matter physics were due to Gorkov and Eliashberg [Gor65] and to Denton, Mühlschlegel, and Scalapino [Den71]. They took the Gaussian orthogonal ensemble to model the energy level statistics of small metal grains and used it to calculate quantum size effects on their thermodynamic properties (see [Hal86] for a review). Theoretical justification came with the supersymmetric field theory of Efetov [Efe83], who derived the level correlation functions in an ensemble of disordered metal grains and showed that they agree with the RMT prediction up to an energy scale of the order of the inverse ergodic time $\hbar/\tau_{\rm erg}$.

Experimental evidence for RMT remained rare throughout the 1980s–basically because the energy resolution needed to probe spectral statistics on the scale of the level spacing was difficult to reach in metal grains. Two parallel advances in nanofabrication changed the situation in the 1990s.

One the one hand, it became possible to make electrical contact to individual metal particles of diameters as small as 10 nm [Del01]. Resonant tunnelling through a single particle could probe the energy level spectrum with sufficient accuracy to test the RMT predictions [Kue08] (see Fig. 35.3). On the other hand, semiconductor quantum dots became available. A quantum dot is a cavity of sub-micron dimensions, etched in a semiconducting two-dimensional electron gas. The electron wave length $\lambda_F \simeq 50\,\text{nm}$ at the Fermi energy in a quantum dot is two orders of magnitude greater than in a metal, and the correspondingly larger level spacing makes these systems ideal for the study of spectral statistics. The quantum dot may be disordered (mean free path l less than its linear dimension L) or it may be ballistic (l greater than L). Random matrix theory applies on energy scales $\hbar/\tau_{\text{erg}} \simeq (\hbar v_F/L)\min(1, l/L)$ irrespective of the ratio of l and L, provided that the classical dynamics is chaotic.

Resonant tunnelling through quantum dots has provided detailed information on both the level spacing distribution (through the spacing of the resonances) and on the wave function statistics (through the peak height of the resonances) [Alh00]. For resonant tunnelling through single-channel point contacts (tunnel probability Γ) the conductance peak height $G_{\max}$ is related to the wave function intensities I_1, I_2 at the two point contacts by [Bee91]

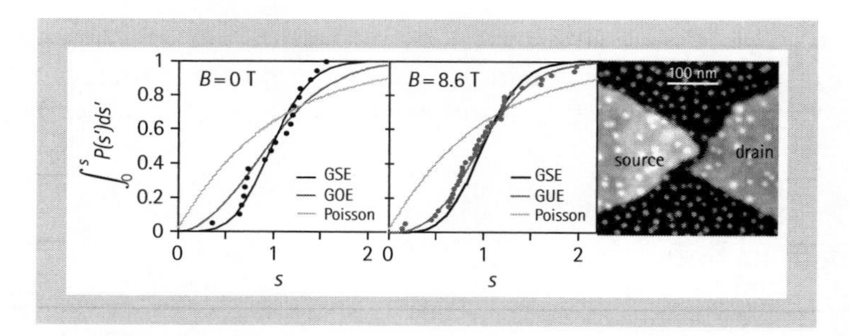

Fig. 35.3 Data points: integrated level spacing distribution of a single 10-nm-diameter gold particle (barely visible in the micrograph as a white dot touching source and drain electrodes), measured by resonant tunnelling in zero magnetic field and in a high magnetic field. The level spacings s are normalized by the mean level spacing δ (equal to 0.23 meV in zero field and reduced to 0.12 meV in high fields due to splitting of the spin degenerate levels by the Zeeman effect). The measured distributions are compared with Wigner's RMT prediction: $P(s) \propto s^\beta \exp(-c_\beta s^2)$ (with $c_1 = \pi/4$, $c_2 = 4/\pi$, $c_4 = 64/9\pi$), for the Gaussian orthogonal ensemble (GOE, $\beta = 1$), the Gaussian unitary ensemble (GUE $\beta = 2$), and the Gaussian symplectic ensemble (GSE, $\beta = 4$). The Poisson distribution $P(s) \propto e^{-s}$ of uncorrelated levels is also shown. A magnetic field causes a transition from the symplectic ensemble in zero field (preserved time reversal symmetry, broken spin rotation symmetry due to the strong spin-orbit coupling in gold), to the unitary ensemble in high fields (broken time reversal and spin rotation symmetries). Adapted from [Kue08].

$$G_{\max} = \frac{e^2}{h} \frac{\Gamma\delta}{4k_BT} \frac{I_1 I_2}{I_1 + I_2}. \tag{35.3.1}$$

(The intensities are normalized to unit average and δ is the mean energy level spacing. The thermal energy k_BT is assumed to be large compared to the width $\Gamma\delta$ of the resonances but small compared to δ.)

The Porter-Thomas distribution $P(I) \propto I^{\beta/2-1}e^{-\beta I/2}$ of (independently fluctuating) intensities I_1, I_2 in the GOE ($\beta = 1$) and GUE ($\beta = 2$) then gives the peak height distribution [Jal92, Pri93],

$$P(g) = \begin{cases} (\pi g)^{-1/2}e^{-g}, & \beta = 1, \\ g[K_0(g) + K_1(g)]e^{-g}, & \beta = 2, \end{cases} \tag{35.3.2}$$

with $g = (8k_BT/\Gamma\delta)(h/e^2)G_{\max}$ and Bessel functions K_0, K_1. A comparison of this RMT prediction with available experimental data has shown a consistent agreement, with some deviations remaining that can be explained by finite-temperature effects and effects of exchange interaction [Alh02].

35.3.2 Scattering matrix ensembles

In quantum dots, the most comprehensive test of RMT has been obtained by studying the statistics of the scattering matrix S rather than of the Hamiltonian H. The Hamiltonian H and scattering matrix S of a quantum dot are related by [Bla91, Got08]

$$\begin{aligned} S(E) &= \mathbf{1} - 2\pi i\, W^\dagger(E - H + i\pi W W^\dagger)^{-1} W \\ &= \frac{\mathbf{1} + i\pi W^\dagger(H - E)^{-1} W}{\mathbf{1} - i\pi W^\dagger(H - E)^{-1} W}. \end{aligned} \tag{35.3.3}$$

The $M \times (N_1 + N_2)$ coupling matrix W (assumed to be independent of the energy E) couples the M energy levels in the quantum dot to $N_1 + N_2$ scattering channels in a pair of point contacts that connect the quantum dot to electron reservoirs. The eigenvalue w_n of the coupling-matrix product $W^\dagger W$ is related to the transmission probability $\Gamma_n \in [0, 1]$ of mode n through the point contact by

$$\Gamma_n = \frac{4\pi^2 w_n M\delta}{(M\delta + \pi^2 w_n)^2}. \tag{35.3.4}$$

Equation (35.3.3) is called the Weidenmüller formula in the theory of chaotic scattering, because of pioneering work by Hans Weidenmüller and his group [Mah69].

A distribution function $P(H)$ for the Hamiltonian H implies a distribution *functional* $P[S(E)]$ for the scattering matrix $S(E)$. For electrical conduction at low voltages and low temperatures, the energy may be fixed at the Fermi energy E_F and knowledge of the distribution function $P(S_0)$ of $S_0 = S(E_F)$ is sufficient. For the Hamiltonian we take the Gaussian ensemble,

$$P(H) \propto \exp\left(-\beta(\pi/2\delta)^2 M^{-1} \operatorname{Tr} H^2\right), \tag{35.3.5}$$

and we take the limit $M \to \infty$ (at fixed δ, E_F, Γ_n), appropriate for a quantum dot of size $L \gg \lambda_F$. The number of channels N_1, N_2 in the two point contacts may be as small as 1, since the opening of the point contacts is typically of the same order as λ_F.

As derived by Brouwer [Bro95], Eqs. (35.3.3) and (35.3.5) together imply, in the large-M limit, for S_0 a distribution of the form

$$P(S_0) \propto |\mathrm{Det}(\mathbf{1} - \bar{S}^\dagger S_0)|^{-\beta N_1 - \beta N_2 - 2 + \beta}, \tag{35.3.6}$$

known as the *Poisson kernel* [Hua63, Lew91, Dor92]. The average scattering matrix[4] $\bar{S} = \int dS_0\, S_0\, P(S_0)$ in the Poisson kernel is given by

$$\bar{S} = \frac{M\delta - \pi^2 W^\dagger W}{M\delta + \pi^2 W^\dagger W}. \tag{35.3.7}$$

The case of ideal coupling (all Γ_ns equal to unity) is of particular interest, since it applies to the experimentally relevant case of ballistic point contacts (no tunnel barrier separating the quantum dot from the electron reservoirs). In view of Eq. (35.3.4) one then has $\bar{S} = 0$, hence

$$P(S_0) = \text{constant}. \tag{35.3.8}$$

This is the distribution of Dyson's circular ensemble [Dys62], first applied to quantum scattering by Blümel and Smilansky [Blu90].

The circular ensemble of scattering matrices implies for the $\min(N_1, N_2)$ nonzero transmission eigenvalues the distribution [Bee97]

$$P(\{T_n\}) \propto \prod_{n<m} |T_n - T_m|^\beta \prod_k T_k^{\frac{1}{2}\beta(|N_2 - N_1| + 1 - 2/\beta)}. \tag{35.3.9}$$

This distribution is of the form (35.2.3), with the logarithmic repulsion (35.2.4). There are no nonlogarithmic corrections in a quantum dot, unlike in a quantum wire.

35.3.3 Conductance distribution

The complete probability distribution of the conductance $G = G_0 \sum_{n=1}^{\min(N_1, N_2)} T_n$ follows directly from Eq. (35.3.9) in the case $N_1 = N_2 = 1$ of single-channel ballistic point contacts [Bar94, Jal94],

$$P(G) = \frac{\beta}{2G_0}(G/G_0)^{-1+\beta/2}, \quad 0 < G < G_0. \tag{35.3.10}$$

[4] The average $\bar{S}$ is defined by integration over the unitary group with Haar measure dS_0, unconstrained for $\beta = 2$ and subject to the constraints of time reversal symmetry for $\beta = 1$ (when S is symmetric) or symplectic symmetry for $\beta = 4$ (when S is self-dual). For more information on integration over the unitary group, see [Bee97, Guh98].

This strongly non-Gaussian distribution rapidly approaches a Gaussian with increasing $N_1 = N_2 \equiv N$. Experiments typically find a conductance distribution which is closer to a Gaussian even in the single-channel case [Hui98], due to thermal averaging and loss of phase coherence at finite temperatures.

In the limit $N \to \infty$ the variance of the Gaussian is given by the RMT result (35.2.7) for UCF – without any corrections since the eigenvalue repulsion in a quantum dot is strictly logarithmic. The experiment value in Fig. 35.4 is smaller than this zero-temperature result, but the factor-of-two reduction upon application of a magnetic field ($\beta = 1 \to \beta = 2$) is quite well preserved.

Without phase coherence the conductance would have average $G_0 N/2$, corresponding to two N-mode point contacts in series. Quantum interference corrects that average, $\langle G \rangle = G_0 N/2 + \delta G$. The correction δG in the limit $N \to \infty$, following from the circular ensemble, equals

$$\delta G = \frac{1}{4}\left(1 - \frac{2}{\beta}\right) G_0. \tag{35.3.11}$$

The quantum correction vanishes in the presence of a time-reversal-symmetry breaking magnetic field ($\beta = 2$), while in zero magnetic field the correction can be negative ($\beta = 1$) or positive ($\beta = 4$) depending on whether spin-rotation-symmetry is preserved or not. The negative quantum correction is called *weak localization* and the positive quantum correction is called *weak antilocalization*. An experimental demonstration [Cha94] of the suppression of weak localization

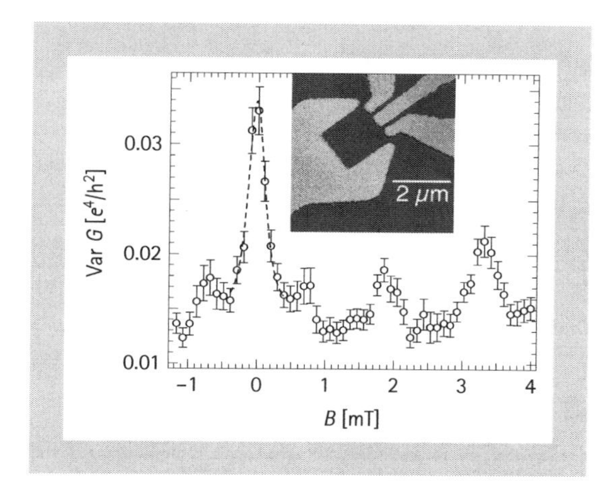

Fig. 35.4 Variance of the conductance of a quantum dot at 30 mK, as a function of magnetic field. The inset shows an electron micrograph of the device, fabricated in the two-dimensional electron gas of a GaAs/AlGaAs heterostructure. The black rectangle at the centre of the inset is the quantum dot, the grey regions are the gate electrodes on top of the heterostructure. Electrons can enter and exit the quantum dot through point contacts at the top and right corner of the rectangle. The side of the rectangle between these two corners is distorted to generate conductance fluctuations and obtain the variance. Adapted from [Cha95].

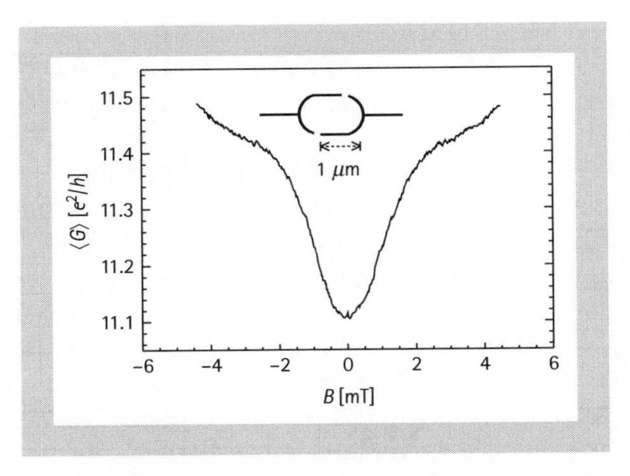

Fig. 35.5 Magnetoconductance at 50 mK, averaged over 48 quantum dots. The minimum around zero magnetic field is the weak localization effect. The inset shows the geometry of the quantum dots, which are fabricated in the two-dimensional electron gas of a GaAs/AlGaAs heterostructure. Adapted from [Cha94].

by a magnetic field is shown in Fig. 35.5. The measured magnitude δG of the peak around zero magnetic field is $0.2\, G_0$, somewhat smaller than the fully phase-coherent value of $\frac{1}{4}\, G_0$.

35.3.4 Sub-Poissonian shot noise

For $N_1 = N_2 \equiv N \gg 1$ the density of transmission eigenvalues for a quantum dot, following from Eq. (35.3.9), has the form

$$\rho(T) = \frac{N}{\pi} \frac{1}{\sqrt{T}\sqrt{1-T}}. \tag{35.3.12}$$

It is different from the result (35.2.14) for a wire, but it has the same bimodal structure. While the average transmission $\bar{T} = 1/2$, the eigenvalue density is peaked at zero and unit transmission.

This bimodal structure can be detected as sub-Poissonian shot noise. Instead of Eq. (35.2.16) one now has [Jal94]

$$F \to 1 - \frac{\int dT\, \rho(T)T^2}{\int dT\, \rho(T)T} = \frac{1}{4}. \tag{35.3.13}$$

An experimental demonstration is shown in Fig. 35.6.

35.3.5 Thermopower distribution

Knowledge of the distribution of the scattering matrix $S(E)$ at a single energy $E = E_F$ is sufficient to determine the conductance distribution, but other transport properties also require information on the energy dependence of S. The

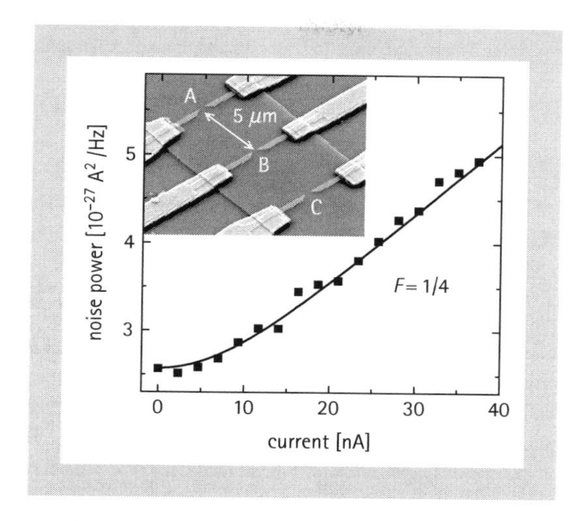

Fig. 35.6 Sub-Poissonian shot noise in a quantum dot at 270 mK. The slope at high currents corresponds to a one-quarter Fano factor, as predicted by RMT. The inset shows an electron micrograph of the device. The quantum dot is contained between point contacts A and B. (The gate labelled C is not operative in this experiment.) Adapted from [Obe01].

thermopower $\mathcal{P}$ (giving the voltage produced by a temperature difference at zero electrical current) is a notable example. Since $\mathcal{P} \propto d \ln G/dE$, we need to know the joint distribution of S and dS/dE at E_F to determine the distribution of $\mathcal{P}$.

This problem goes back to the early days of RMT [Wig55, Smi60], in connection with the question: What is the time delay experienced by a scattered wave packet? The delay times τ_n are the eigenvalues of the Hermitian matrix product $Q_{\text{WS}} = -i\hbar S^\dagger dS/dE$, known as the Wigner-Smith matrix in the context of RMT. (For applications in other contexts, see [Das69, Bla91, Got08].) The solution to the problem of the joint distribution of S and dS/dE (for S in the circular ensemble) was given in [Bro97]. The symmetrized matrix product

$$Q = -i\hbar S^{-1/2} \frac{dS}{dE} S^{-1/2} \tag{35.3.14}$$

has the same eigenvalues as Q_{WS}, but unlike Q_{WS} was found to be statistically independent of S. The eigenvalues of Q have distribution

$$\begin{aligned} P(\{\gamma_n\}) &\propto \prod_{i<j} |\gamma_i - \gamma_j|^\beta \prod_k \gamma_k^{\beta(N_1+N_2)/2} e^{-\beta\tau_H\gamma_k/2}, \\ \gamma_n &\equiv 1/\tau_n > 0. \end{aligned} \tag{35.3.15}$$

The Heisenberg time $\tau_H = 2\pi\hbar/\delta$ is inversely proportional to the mean level spacing δ in the quantum dot. Eq. (35.3.15) is known in RMT as the Laguerre ensemble.

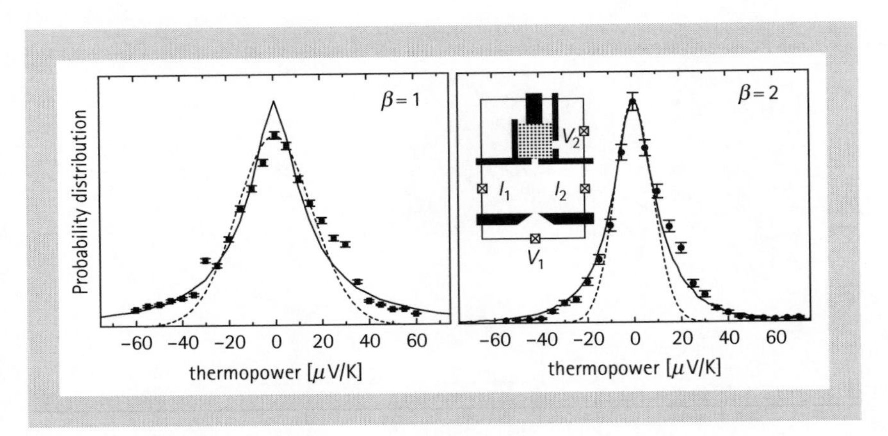

Fig. 35.7 Thermopower distribution for $\beta = 1$ ($|B| \leq 40\,\mathrm{mT}$) and $\beta = 2$ ($|B| \geq 50\,\mathrm{mT}$). Experimental results (dots), RMT results (solid line), and Gaussian fit (dashed line) are compared. The inset shows the experimental layout. The crosses denote Ohmic contacts to the two-dimensional electron gas and the shaded areas denote gate electrodes. The heating current is applied between I_1 and I_2, while the thermovoltage is measured between V_1 and V_2. The quantum dot is indicated by the dotted area. Adapted from [God99].

The thermopower distribution following from the Laguerre ensemble is strongly non-Gaussian for small $N_1 = N_2 \equiv N$. For $N = 1$ it has a cusp at $\mathcal{P} = 0$ when $\beta = 1$ and algebraically decaying tails $\propto |\mathcal{P}|^{-1-\beta} \ln |\mathcal{P}|$. Significant deviations from a Gaussian are seen in the experiment [God99] shown in Fig. 35.7, for $N = 2$.

35.3.6 Quantum-to-classical transition

Random matrix theory is a quantum mechanical theory which breaks down in the classical limit $h \to 0$. For electrical conduction through a quantum dot, the parameter that governs the quantum-to-classical transition is the ratio $\tau_E/\tau_{\mathrm{dwell}}$ of Ehrenfest time and dwell time [Aga00].

The dwell time τ_{dwell} is the average time an electron spends inside the quantum dot between entrance and exit through one of the two N-mode point contacts. It is given by

$$\tau_{\mathrm{dwell}} = \pi\hbar/N\delta. \tag{35.3.16}$$

The Ehrenfest time τ_E is the time up to which a wave packet follows classical equations of motion, in accord with Ehrenfest's theorem [Ber78, Chi71]. For chaotic dynamics with a Lyapunov exponent α,[5] it is given by [Sil03]

$$\tau_E = \alpha^{-1} \max\big[0, \ln(NW/\mathcal{A}^{1/2})\big]. \tag{35.3.17}$$

[5] The Lyapunov exponent α of chaotic motion quantifies the exponential divergence of two trajectories displaced by a distance $\Delta x(t)$ at time t, according to $\Delta x(t) = \Delta x(0)e^{\alpha t}$.

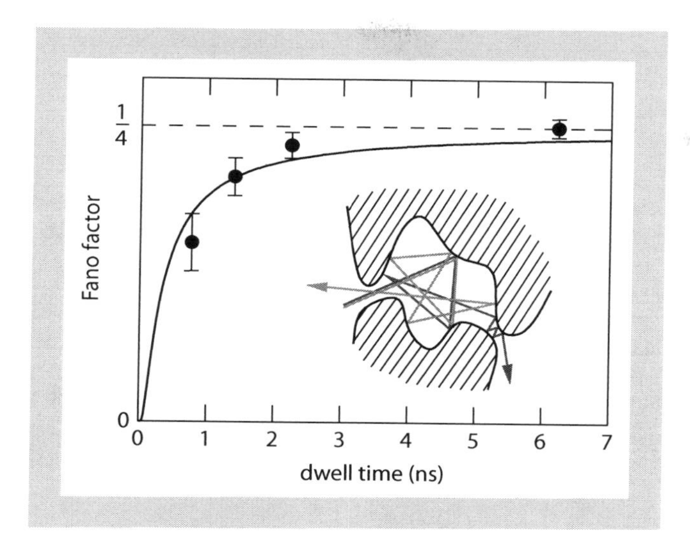

Fig. 35.8 Dependence of the Fano factor F of a ballistic chaotic quantum dot on the average time that an electron dwells inside. The data points with error bars are measured in a quantum dot, the solid curve is the theoretical prediction (35.3.18) for the quantum-to-classical transition (with Ehrenfest time $\tau_E = 0.27$ ns as a fit parameter). The dwell time (35.3.16) is varied experimentally by changing the number of modes N transmitted through each of the point contacts. The inset shows graphically the sensitivity to initial conditions of the chaotic dynamics. Adapted from [Aga00], with experimental data from [Obe02].

Here $\mathcal{A}$ is the area of the quantum dot and W the width of the N-mode point contacts.

The RMT result $F = 1/4$ holds if $\tau_E \ll \tau_{\text{dwell}}$. For longer τ_E, the Fano factor is suppressed exponentially [Aga00],

$$F = \tfrac{1}{4} e^{-\tau_E/\tau_{\text{dwell}}}. \tag{35.3.18}$$

This equation expresses the fact that the fraction $1 - e^{-\tau_E/\tau_{\text{dwell}}}$ of electrons that stay inside the quantum dot for times shorter than τ_E follow a deterministic classical motion that does not contribute to the shot noise. Random matrix theory applies effectively only to the fraction $e^{-\tau_E/\tau_{\text{dwell}}}$ of electrons that stay inside for times longer than τ_E. The shot noise suppression (35.3.18) is plotted in Fig. 35.8, together with supporting experimental data [Obe02].

35.4 Superconductors

35.4.1 Proximity effect

Figure 35.9 (lower right panel) shows a quantum dot with superconducting electrodes. Without the superconductor the energy spectrum of an ensemble of such quantum dots has GOE statistics. The proximity of a superconductor

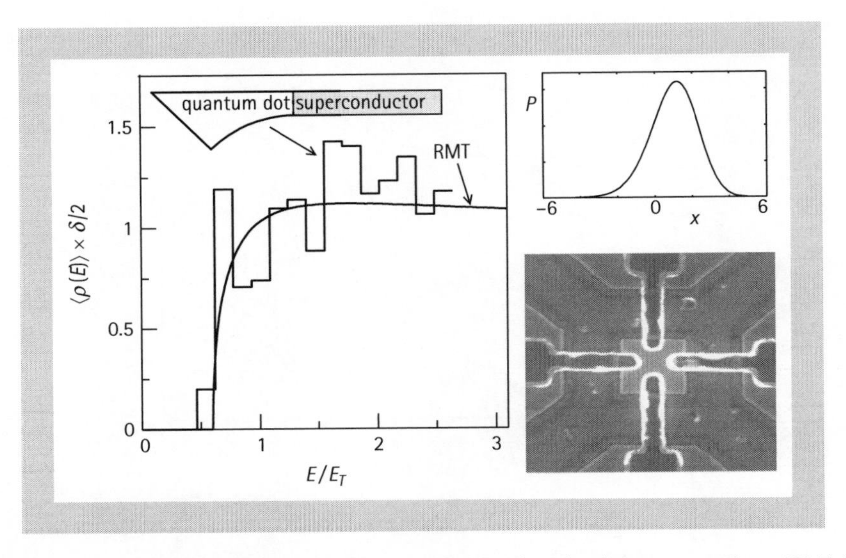

Fig. 35.9 *Left panel:* Average density of states (scaled by the Thouless energy $E_T = N\delta/4\pi$) of a quantum dot coupled by a ballistic N-mode point contact to a superconductor. The histogram is a numerical calculation for the geometry indicated in the inset (with $N = 20$), while the curve is the analytical prediction from RMT. Adapted from [Mel96]. *Upper right panel:* Probability distribution of the lowest excitation energy E_1, rescaled as $x = (E_1 - E_{\text{gap}})/\Delta_{\text{gap}}$. Adapted from [Vav01]. *Lower right panel:* Quantum dot (central square of dimensions 500 nm × 500 nm) fabricated in an InAs/AlSb heterostructure and contacted by four superconducting Nb electrodes. Device made by A.T. Filip, Groningen University. Figure from [Bee05].

has a drastic effect on the energy spectrum, by opening up a gap at the Fermi level. The RMT of this proximity effect was developed in [Mel96] (see [Bee05] for a review).

A quantum dot coupled to a superconductor has a discrete spectrum for energies below the gap Δ of the superconductor, given by the roots of the determinantal equation

$$\text{Det}\left[\mathbf{1} - \alpha(E)^2 S(E) S(-E)^*\right] = 0,$$
$$\alpha(E) = \frac{E}{\Delta} - i\sqrt{1 - \frac{E^2}{\Delta^2}}. \tag{35.4.1}$$

The scattering matrix S (at an energy E measured relative to the Fermi level) describes the coupling of the quantum dot to the superconductor via an N-mode point contact and is related to the Hamiltonian H of the isolated quantum dot by Eq. (35.3.3). At low energies $E \ll \Delta$ the energy levels can be obtained as the eigenvalues E_i of the effective Hamiltonian

$$H_{\text{eff}} = \begin{pmatrix} H & -\pi W W^T \\ -\pi W^* W^\dagger & -H^* \end{pmatrix}. \tag{35.4.2}$$

The Hermitian matrix $H_{\rm eff}$ is antisymmetric under the combined operation of charge conjugation ($\mathcal{C}$) and time inversion ($\mathcal{T}$) [Alt96]:

$$H_{\rm eff} = -\sigma_y H^*_{\rm eff}\sigma_y, \quad \sigma_y = \begin{pmatrix} 0 & -i \\ i & 0 \end{pmatrix}. \tag{35.4.3}$$

(An $M \times M$ unit matrix in each of the four blocks of σ_y is implicit.) The $\mathcal{CT}$-antisymmetry ensures that the eigenvalues lie symmetrically around $E = 0$. Only the positive eigenvalues are retained in the excitation spectrum, but the presence of the negative eigenvalues is felt as a level repulsion near $E = 0$.

As illustrated in Fig. 35.9 (left panel), the unique feature of the proximity effect is that this level repulsion can extend over energy scales much larger than the mean level spacing δ in the isolated quantum dot – at least if time reversal symmetry is not broken. A calculation of the density of states $\langle\rho(E)\rangle = \langle\sum_i \delta(E - E_i)\rangle$ of $H_{\rm eff}$, averaged over H in the GOE, produces a square root singularity in the large-N limit:

$$\langle\rho(E)\rangle \to \frac{1}{\pi}\sqrt{\frac{E - E_{\rm gap}}{\Delta^3_{\rm gap}}}, \quad E \to E_{\rm gap}, \quad N \to \infty, \tag{35.4.4}$$

If the point contact between quantum dot and superconductor is ballistic ($\Gamma_n = 1$ for $n = 1, 2, \ldots N$) the two energies $E_{\rm gap}$ and $\Delta_{\rm gap}$ are given by [Mel96]

$$E_{\rm gap} = \frac{\gamma^{5/2} N\delta}{2\pi} = 0.048\, N\delta, \quad \Delta_{\rm gap} = 0.068\, N^{1/3}\delta. \tag{35.4.5}$$

(Here $\gamma = \frac{1}{2}(\sqrt{5} - 1)$ is the golden number.) The gap $E_{\rm gap}$ in the spectrum of the quantum dot is larger than δ by a factor of order N.

35.4.2 Gap fluctuations

The value (35.4.5) of the excitation gap is representative for an ensemble of quantum dots, but each member of the ensemble will have a smallest excitation energy E_1 that will be slightly different from $E_{\rm gap}$. The distribution of the gap fluctuations is identical upon rescaling to the known distribution [Tra94] of the lowest eigenvalue in the GOE [Vav01, Ost01, Lam01]. Rescaling amounts to a change of variables from E_1 to $x = (E_1 - E_{\rm gap})/\Delta_{\rm gap}$, where $E_{\rm gap}$ and $\Delta_{\rm gap}$ parameterize the square-root dependence (35.4.4). The probability distribution $P(x)$ of the rescaled gap fluctuations is shown in Fig. 35.9 (upper right panel). The gap fluctuations are a mesoscopic, rather than a microscopic effect, because the typical magnitude $\Delta_{\rm gap} \simeq E^{1/3}_{\rm gap}\delta^{2/3}$ of the fluctuations is $\gg \delta$ for $E_{\rm gap} \gg \delta$. Still, the fluctuations are small on the scale of the gap itself.

35.4.3 From mesoscopic to microscopic gap

The mesoscopic excitation gap of order $N\delta$ induced by the proximity to a superconductor is strongly reduced if time reversal symmetry is broken by application of a magnetic field ($\beta = 2$). Because the repulsion of levels at $\pm E$ persists, as demanded by the $\mathcal{CT}$-antisymmetry (35.4.3), a microscopic gap around zero energy of order δ remains. An alternative way to reduce the gap from $N\delta$ to δ, without breaking time reversal symmetry ($\beta = 1$), is by contacting the quantum dot to a pair of superconductors with a phase difference of π in the order parameter. As shown by Altland and Zirnbauer [Alt96], the level statistics near the Fermi energy in these two cases is governed by the distribution

$$P(\{E_n\}) \propto \prod_{i<j} |E_i^2 - E_j^2|^\beta \prod_k |E_k|^\beta e^{-c^2 E_k^2}, \tag{35.4.6}$$

related to the Laguerre ensemble by a change of variables ($E_n^2 \to x_n$). (The coefficient c is fixed by the mean level spacing in the isolated quantum dot.) The density of states near zero energy vanishes as $|E|^\beta$. Two more cases are possible when spin-rotation symmetry is broken, so that in total the three Wigner-Dyson symmetry classes without superconductivity are expanded to four symmetry classes as a consequence of the $\mathcal{CT}$-antisymmetry.

35.4.4 Quantum-to-classical transition

The RMT of the proximity effect discussed so far breaks down when the dwell time (35.3.16) becomes shorter than the Ehrenfest time (35.3.17) [Lod98]. In order of magnitude,[6] the gap equals $E_{\text{gap}} \simeq \min(\hbar/\tau_E, \hbar/\tau_{\text{dwell}})$. In the classical limit $\tau_E \to \infty$, the density of states is given by [Sch99]

$$\langle \rho(E) \rangle = \frac{2}{\delta} \frac{(\pi E_T/E)^2 \cosh(\pi E_T/E)}{\sinh^2(\pi E_T/E)}, \tag{35.4.7}$$

with $E_T = N\delta/4\pi$ the Thouless energy. The density of states (35.4.7) (plotted in Fig. 35.10) is suppressed exponentially $\propto e^{-\pi E_T/E}$ at the Fermi level ($E \to 0$), but there is no gap.

To understand the absence of a true excitation gap in the limit $\tau_E \to \infty$, we note that in this limit a wave packet follows a classical trajectory in the quantum dot. The duration t of this trajectory, from one reflection at the superconductor to the next, is related to the energy E of the wave packet by $E \simeq \hbar/t$. Since t can become arbitrarily large (albeit with an exponentially small probability $e^{-t/\tau_{\text{dwell}}}$), the energy E can become arbitrarily small and there is no gap.

[6] More precisely, the gap crosses over between the RMT limit (35.4.5) for $\tau_E \ll \tau_{\text{dwell}}$ and the limit $E_{\text{gap}} = \pi\hbar/2\tau_E$ for $\tau_E \gg \tau_{\text{dwell}}$ [Vav03, Bee05, Kui09].

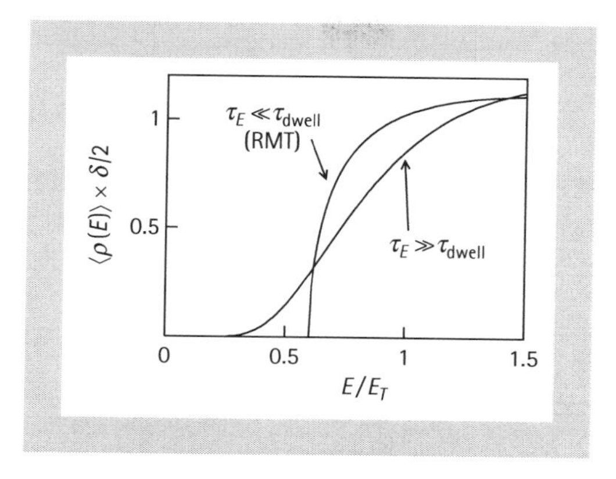

Fig. 35.10 Comparison of the density of states (35.4.7) with the RMT result (35.4.4). These are the two limiting results when the Ehrenfest time τ_E is, respectively, much larger or much smaller than the mean dwell time τ_{dwell}. From [Bee05].

References

[Aga00] O. Agam, I. Aleiner, and A. Larkin, Phys. Rev. Lett. **85** (2000) 3153.
[Alh00] Y. Alhassid, Rev. Mod. Phys. **72** (2000) 895.
[Alh02] Y. Alhassid and T. Rupp, Phys. Rev. Lett. **91** (2003) 056801.
[Alt85] B.L. Altshuler, JETP Lett. **41** (1985) 648.
[Alt86] B.L. Altshuler and B.I. Shklovskiĭ, Sov. Phys. JETP **64** (1986) 127.
[Alt96] A. Altland and M.R. Zirnbauer, Phys. Rev. Lett. **76** (1996) 3420; Phys. Rev. B **55** (1997) 1142.
[Bar94] H.U. Baranger and P.A. Mello, Phys. Rev. Lett. **73** (1994) 142.
[Bee91] C.W.J. Beenakker, Phys. Rev. B **44** (1991) 1646.
[Bee92] C.W.J. Beenakker and M. Büttiker, Phys. Rev. B **46** (1992) 1889.
[Bee93a] C.W.J. Beenakker, Phys. Rev. Lett. **70** (1993) 1155.
[Bee93b] C.W.J. Beenakker and B. Rejaei, Phys. Rev. Lett. **71** (1993) 3689.
[Bee97] C.W.J. Beenakker, Rev. Mod. Phys. **69** (1997) 731.
[Bee05] C.W.J. Beenakker, Lect. Notes Phys. **667** (2005) 131.
[Ber78] G.P. Berman and G.M. Zaslavsky, Physica A **91** (1978) 450.
[Bla91] J.M. Blatt and V.F. Weisskopf, *Theoretical Nuclear Physics*, Dover, New York 1991.
[Blu90] R. Blümel and U. Smilansky, Phys. Rev. Lett. **64** (1990) 241.
[Bro81] T.A. Brody, J. Flores, J.B. French, P.A. Mello, A. Pandey, and S.S.M. Wong, Rev. Mod. Phys. **53** (1981) 385.
[Bro95] P.W. Brouwer, Phys. Rev. B **51** (1995) 16878.
[Bro97] P.W. Brouwer, K.M. Frahm, and C.W.J. Beenakker, Phys. Rev. Lett. **78** (1997) 4737.
[Bur72] R. Burridge and G. Papanicolaou, Comm. Pure Appl. Math. **25** (1972) 715.
[Cas95] M. Caselle, Phys. Rev. Lett. **74** (1995) 2776.
[Cha93] J.T. Chalker and A.M.S. Macêdo, Phys. Rev. Lett. **71** (1993) 3693.
[Cha94] A.M. Chang, H.U. Baranger, L.N. Pfeiffer, and K.W. West, Phys. Rev. Lett. **73** (1994) 2111.
[Cha95] I.H. Chan, R.M. Clarke, C.M. Marcus, K. Campman, and A.C. Gossard, Phys. Rev. Lett. **74** (1995) 3876.
[Chi71] B.V. Chirikov, F.M. Izrailev, and D L. Shepelyansky, Physica D **33** (1988) 77.

[Das69] R. Dashen, S.-K. Ma, and H.J. Bernstein, Phys. Rev. **187** (1969) 345.
[Del01] J. von Delft and D.C. Ralph, Phys. Rep. **345** (2001) 61.
[Den71] R. Denton, B. Mühlschlegel, and D. J. Scalapino, Phys. Rev. Lett. **26** (1971) 707.
[Dor82] O.N. Dorokhov, JETP Lett. **36** (1982) 318.
[Dor84] O.N. Dorokhov, Solid State Comm. **51** (1984) 381.
[Dor92] E. Doron and U. Smilansky, Nucl. Phys. A **545** (1992) 455.
[Dys62] F.J. Dyson, J. Math. Phys. **3** (1962) 140.
[Dys63] F.J. Dyson and M.L. Mehta, J. Math. Phys. **4** (1963) 701.
[Efe83] K.B. Efetov, Adv. Phys. **32** (1983) 53.
[God99] S.F. Godijn, S. Möller, H. Buhmann, L.W. Molenkamp, and S.A. van Langen, Phys. Rev. Lett. **82** (1999) 2927.
[Gor65] L.P. Gorkov and G.M. Eliashberg, Sov. Phys. JETP **21** (1965) 940.
[Got08] K. Gottfried and T.-M. Yan, *Quantum Mechanics: Fundamentals*, Springer, New York 2008.
[Guh98] T. Guhr, A. Müller-Groeling, and H.A. Weidenmüller, Phys. Rep. **299** (1998) 189.
[Hal86] W.P. Halperin, Rev. Mod. Phys. **58** (1986) 533.
[Hen99] M. Henny, S. Oberholzer, C. Strunk, and C. Schönenberger, Phys. Rev. B **59** (1999) 2871.
[Hua63] L.K. Hua, *Harmonic Analysis of Functions of Several Complex Variables in the Classical Domains*, American Mathematical Society, Providence 1965.
[Hui98] A.G. Huibers, S.R. Patel, C.M. Marcus, P.W. Brouwer, C.I. Duruöz, and J.S. Harris, Jr., Phys. Rev. Lett. **81** (1998) 1917.
[Imr86] Y. Imry, Europhys. Lett. **1** (1986) 249.
[Jal92] R.A. Jalabert, A.D. Stone, and Y. Alhassid, Phys. Rev. Lett. **68** (1992) 3468.
[Jal93] R.A. Jalabert, J.-L. Pichard, and C.W.J. Beenakker, Europhys. Lett. **24** (1993) 1.
[Jal94] R.A. Jalabert, J.-L. Pichard and C.W.J. Beenakker, Europhys. Lett. **27** (1994) 255.
[Kue08] F. Kuemmeth, K.I. Bolotin, S.-F. Shi, and D.C. Ralph, Nano Lett. **8** (2008) 4506.
[Kui09] J. Kuipers, D. Waltner, C. Petitjean, G. Berkolaiko, and K. Richter, Phys. Rev. Lett. **104** (2010) 027001.
[Lam01] A. Lamacraft and B.D. Simons, Phys. Rev. B **64** (2001) 014514.
[Lee85] P.A. Lee and A.D. Stone, Phys. Rev. Lett. **55** (1985) 1622.
[Lew91] C.H. Lewenkopf and H.A. Weidenmüller, Ann. Phys. (N.Y.) **212** (1991) 53.
[Lod98] A. Lodder and Yu.V. Nazarov, Phys. Rev. B **58** (1998) 5783.
[Mah69] C. Mahaux and H.A. Weidenmüller, *Shell-Model Approach to Nuclear Reactions*, North-Holland, Amsterdam 1969.
[Meh67] M.L. Mehta, *Random Matrices*, Academic Press, New York 1991.
[Mel88a] P.A. Mello, P. Pereyra, and N. Kumar, Ann. Phys. (N.Y.) **181** (1988) 290.
[Mel89] P.A. Mello and J.-L. Pichard, Phys. Rev. B **40** (1989) 5276.
[Mel96] J.A. Melsen, P.W. Brouwer, K.M. Frahm, and C.W.J. Beenakker, Europhys. Lett. **35** (1996) 7; Physica Scripta **T69** (1997) 223.
[Mut87] K.A. Muttalib, J.-L. Pichard, and A.D. Stone, Phys. Rev. Lett. **59** (1987) 2475.
[Obe01] S. Oberholzer, E.V. Sukhorukov, C. Strunk, C. Schönenberger, T. Heinzel, K. Ensslin, and M. Holland, Phys. Rev. Lett. **86** (2001) 2114.
[Obe02] S. Oberholzer, E.V. Sukhorukov, and C. Schönenberger, Nature **415** (2002) 765.
[Ost01] P.M. Ostrovsky, M.A. Skvortsov, and M.V. Feigelman, Phys. Rev. Lett. **87** (2001) 027002.
[Por65] C.E. Porter, ed., *Statistical Theories of Spectra: Fluctuations*, Academic Press, New York 1965.
[Pri93] V.N. Prigodin, K.B. Efetov, and S. Iida, Phys. Rev. Lett. **71** (1993) 1230.

[Sch99] H. Schomerus and C.W.J. Beenakker, Phys. Rev. Lett. **82** (1999) 2951.
[Sil03] P.G. Silvestrov, M.C. Goorden, and C.W.J. Beenakker, Phys. Rev. B **67** (2003) 241301(R).
[Smi60] F.T. Smith, Phys. Rev. **118** (1960) 349.
[Ste96] A.H. Steinbach, J.M. Martinis, and M.H. Devoret, Phys. Rev. Lett. **76** (1996) 3806.
[Sto91] A.D. Stone, P.A. Mello, K.A. Muttalib, and J.-L. Pichard, in: *Mesoscopic Phenomena in Solids*, ed. by B.L. Altshuler, P.A. Lee, and R.A. Webb, North-Holland, Amsterdam 1991.
[Tra94] C.A. Tracy and H. Widom, Commun. Math. Phys. **159** (1994) 151; **177** (1996) 727.
[Vav01] M.G. Vavilov, P.W. Brouwer, V. Ambegaokar, and C.W.J. Beenakker, Phys. Rev. Lett. **86** (2001) 874.
[Vav03] M.G. Vavilov and A.I. Larkin, Phys. Rev. B **67** (2003) 115335.
[Was86] S. Washburn and R.A. Webb, Adv. Phys. **35** (1986) 375.
[Wei09] H.A. Weidenmüller and G.E. Mitchell, Rev. Mod. Phys. **81** (2009) 539.
[Wig55] E.P. Wigner, Phys. Rev. **98** (1955) 145.

·36·

Classical and quantum optics

C. W. J. Beenakker

Abstract

We review applications of random matrix theory to optical systems such as disordered wave guides and chaotic resonators. Applications to both classical optics and quantum optics are considered. These include speckle and coherent backscattering from a random medium, the effect of absorption on the reflection eigenvalue statistics in a multimode wave guide, long-range wave function correlations in an open resonator, direct detection of open transmission channels, the statistics of grey-body radiation, and lasing in a chaotic cavity.

36.1 Introduction

Optical applications of random matrix theory came later than electronic applications, perhaps because randomness is much more easily avoided in optics than it is in electronics. The variety of optical systems to which RMT can be applied increased substantially with the realization [Boh84, Ber85] that randomness is not needed at all for GOE statistics of the spectrum. Chaotic dynamics is sufficient, and this is a generic property of resonators formed by a combination of convex and concave surface elements. As an example, we show in Fig. 36.1 the Wigner level spacing distribution measured in a microwave cavity with a chaotic shape.

This is an example of an application of RMT to *classical* optics, because the spectral statistics of a cavity is determined by the Maxwell equations of a classical electromagnetic wave. (More applications of this type, also including sound waves, are reviewed in [Kuh05].) An altogether different type of application of RMT appears in *quantum* optics, when the photon and its quantum statistics play an essential role. Selected applications of RMT to both classical and quantum optics are presented in the following sections. The emphasis is on topics that do not have an immediate analogue in electronics, either because they cannot readily be measured in the solid state or because they involve aspects (such as absorption, amplification, or bosonic statistics) that do not apply to electrons.

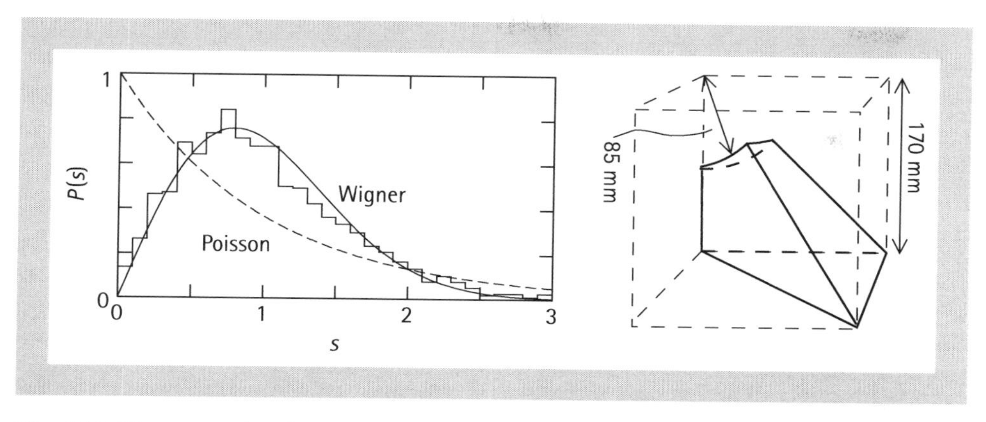

Fig. 36.1 Histogram: distribution of spacings s of eigenfrequencies measured in the chaotic microwave resonator shown at the right. (The resonator has superconducting walls, to minimize absorption.) The spacing distribution is close to the Wigner distribution $P(s) \propto s\exp(-\pi s^2/4\delta^2)$ [solid line] of the GOE, and far from the Poisson distribution $P(s) \propto e^{-s/\delta}$ [dashed line] of uncorrelated eigenfrequencies. The mean spacing has been set to $\delta = 1$, and non-chaotic 'bouncing-ball' resonances have been eliminated from the experimental histogram. Adapted from [Alt97].

Some of the concepts used in this chapter were introduced in the previous chapter on applications of RMT to condensed matter physics, in particular in Sections 35.2.2, 35.2.3, and 35.3.2.

36.2 Classical optics

36.2.1 Optical speckle and coherent backscattering

Optical speckle, shown in Fig. 36.2, is the random interference pattern that is observed when coherent radiation is transmitted or reflected by a random medium. It has been much studied since the discovery of the laser, because the speckle pattern carries information both on the coherence properties of the radiation and on microscopic details of the scattering object [Goo07]. The superposition of partial waves with randomly varying phase and amplitude produces a wide distribution $P(I)$ of intensities I around the average $\bar{I}$. For full coherence and complete randomization the distribution has the exponential form

$$P(I) = \bar{I}^{-1}\exp(-I/\bar{I}), \quad I > 0. \qquad (36.2.1)$$

For a description of speckle in the framework of RMT [Mel88a], it is convenient to enclose the scattering medium in a wave guide containing a large number N of propagating modes. The reflection matrix r is then an $N \times N$ matrix with random elements. Time-reversal symmetry (reciprocity) dictates that r is symmetric. Deviations of r from unitarity can be ignored if the mean free path l is much smaller than both the length L of the scattering medium and

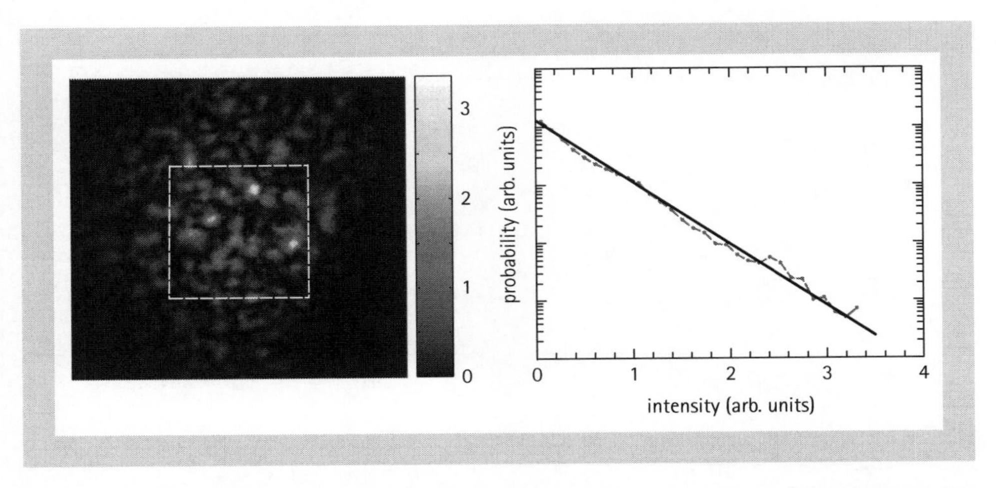

Fig. 36.2 *Left panel:* Speckle pattern produced by a laser beam behind a diffusor (full scale 45 mrad × 45 mrad). The vertical bar indicates the colour coding of the intensity, in arbitrary units. The average angular opening angle $\delta\alpha \approx 1.3$ mrad of a bright or dark spot (a 'speckle') is equal to $\lambda/\pi R$, with $\lambda = 830$ nm the wave length and $R = 200\,\mu$m the radius of the illuminated area on the diffusor. The envelope of the intensity pattern reflects the 18 mrad opening angle of the directional scattering from this type of diffusor. The intensity distribution $P(I)$ of the speckle pattern measured inside the white square is plotted in the right panel, and compared with the exponential distribution (36.2.1) (straight line in the semi-logarithmic plot). Figure courtesy of M.P. van Exter. This figure is reproduced in colour in the colour plates section.

the absorption length l_a. The RMT assumption is that r is distributed according to the circular orthogonal ensemble (COE), which means that $r = UU^T$ with U uniformly distributed in the group $\mathcal{U}(N)$ of $N \times N$ unitary matrices.

In this description, the reflected intensity in mode n for a wave incident in mode m is given by $I_{nm} = |r_{nm}|^2$. The intensity distribution can easily be calculated in the limit $N \to \infty$, when the complex matrix elements r_{nm} with $n \leq m$ have independent Gaussian distributions of zero mean and variance[1]

$$\langle |r_{nm}|^2 \rangle = \int_{\mathcal{U}(N)} dU \sum_{k,k'=1}^{N} U_{nk} U_{mk'} U^*_{nk'} U^*_{mk} = \frac{1+\delta_{nm}}{N+1}. \tag{36.2.2}$$

The resulting distribution of I_{nm} in the large-N limit has the exponential form (36.2.1), with an average intensity $\bar{I}_{nm} = (1+\delta_{nm})N^{-1}$ which is twice as large when $n = m$ than when $n \neq m$. This doubling of the average reflected intensity at the angle of incidence is the *coherent backscattering* effect [Akk07], illustrated in Fig. 36.3.

The RMT assumption of a COE distribution of the reflection matrix correctly reproduces the height of the coherent backscattering peak, but it cannot repro-

[1] For an introduction to such integrals over the unitary group, see [Bee97]. The factor $N+1$ in the denominator ensures that $\sum_{m=1}^{N} |r_{nm}|^2 = 1$, as required by unitarity, but the difference between N and $N+1$ can be neglected in the large-N limit.

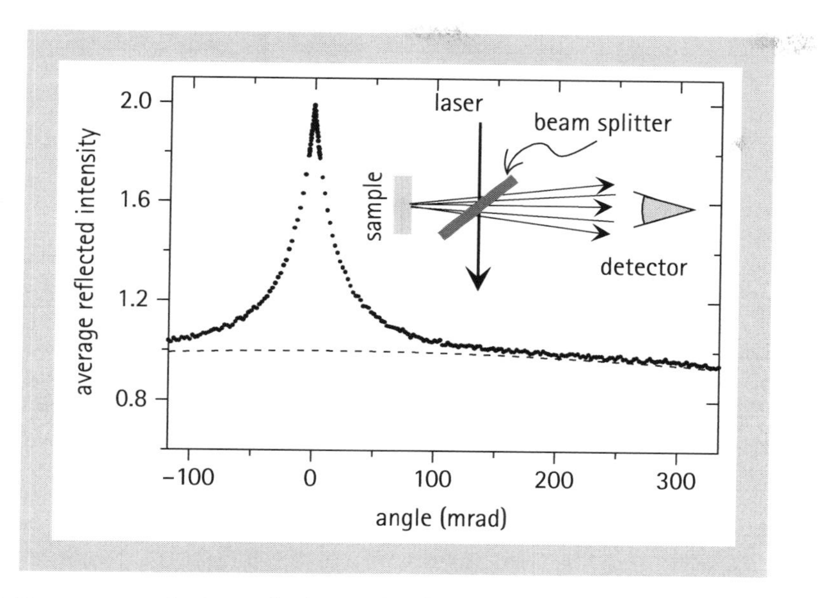

Fig. 36.3 Measurement of coherent backscattering from a ZnO powder. The sample is rotated to average the reflected intensity, which is plotted against the scattering angle. The measured peak due to coherent backscattering is superimposed on the diffuse scattering intensity (dashed curve, normalized to unity in the backscattering direction at zero angle). The relative height of the peak is a factor-of-two, in accordance with Eq. (36.2.2). The angular width is of order $1/kl \approx 40\,\text{mrad}$, for wave length $\lambda = 2\pi/k = 514\,\text{nm}$ and mean free path $l = 1.89\,\mu\text{m}$. The inset shows the optical setup. Use of a beam splitter permits detection in the backscattering direction, which would otherwise be blocked by the incident laser beam. Adapted from [Wie95]. This figure is reproduced in colour in the colour plates section.

duce its width [Akk88, Mel88b]. The Kronecker delta in Eq. (36.2.2) would imply an angular opening $\delta\alpha \simeq 1/kW$ of the peak (for light of wave number k in a wave guide of width W). This is only correct if the mean free path l is larger than W. In a typical experiment $l \ll W$ and the angular opening is $\delta\alpha \simeq 1/kl$ (as it is in Fig. 36.3).

36.2.2 Reflection from an absorbing random medium

An absorbing medium has a dielectric constant ε with a positive imaginary part. The intensity of radiation which has propagated without scattering over a distance L is then multiplied by a factor $e^{-\sigma L}$. The decay rate $\sigma > 0$ at wave number k is related to the dielectric constant by $\sigma = 2k\,\text{Im}\,\sqrt{\varepsilon}$.

The absence of a conservation law in an absorbing medium breaks the unitarity of the scattering matrix. The circular orthogonal ensemble, of uniformly distributed symmetric unitary matrices, should therefore be replaced by another ensemble. The appropriate ensemble was derived in [Bee96, Bru96], for the case of reflection from an infinitely long absorbing wave guide. The result is that the N eigenvalues $R_n \in [0, 1]$ of the reflection matrix product $rr^\dagger$ are distributed according to the Laguerre orthogonal ensemble, after a change of variables to $\lambda_n = R_n(1 - R_n)^{-1} \geq 0$:

$$P(\{\lambda_n\}) \propto \prod_{i<j} |\lambda_j - \lambda_i| \prod_k \exp[-\sigma l(N+1)\lambda_k]. \tag{36.2.3}$$

The distribution (36.2.3) is obtained by including an absorption term into the DMPK equation (35.2.12). This loss-drift-diffusion equation has the form [Bee96, Bru96]

$$l\frac{\partial P}{\partial L} = \frac{2}{\beta N + 2 - \beta} \sum_{n=1}^{N} \frac{\partial}{\partial \lambda_n} \lambda_n(1+\lambda_n)\Big[J\frac{\partial}{\partial \lambda_n}\frac{P}{J} + \sigma l(\beta N + 2 - \beta)\,P\Big],$$
$$\text{with } J = \prod_{i<j} |\lambda_j - \lambda_i|^{\beta}. \tag{36.2.4}$$

The drift-diffusion equation (35.2.12) considered in the electronic context is obtained by setting $\sigma = 0$ and transforming to the variables $x_n = \sinh^2 \lambda_n$.

In the limit $L \to \infty$ we may equate the left-hand side of Eq. (36.2.4) to zero, and we arrive at the solution (36.2.3) for $\beta = 1$ (unbroken time reversal symmetry). More generally, for any β, the distribution of the R_ns in the limit $L \to \infty$ can be written in the form of a Gibbs distribution at a fictitious temperature β^{-1},

$$P(\{R_n\}) \propto \exp\Big[-\beta\Big(\sum_{i<j} u(R_i, R_j) + \sum_i V(R_i)\Big)\Big], \tag{36.2.5}$$

$$u(R, R') = -\ln|R - R'|,$$
$$V(R) = \left(N - 1 + \frac{2}{\beta}\right)\left[\frac{\sigma l R}{1-R} + \ln(1-R)\right]. \tag{36.2.6}$$

The eigenvalue interaction potential $u(R, R')$ is *logarithmic*. This can be contrasted with the *nonlogarithmic* interaction potential in the absence of absorption, discussed in Section 35.2.2. Because $R_n = 1 - T_n$ without absorption, the interaction potential (35.2.9) of that section can be written as

$$u(R, R') = -\tfrac{1}{2}\ln|R - R'| - \tfrac{1}{2}\ln|x - x'|, \quad \text{with } R \equiv \tanh^2 x. \tag{36.2.7}$$

As calculated in [Mis96], the change in interaction potential has an observable consequence in the sample-to-sample fluctuations of the reflectance

$$\mathcal{R} = \mathrm{Tr}\, rr^{\dagger} = \sum_{n=1}^{N} R_n. \tag{36.2.8}$$

With increasing length L of the absorbing disordered waveguide, the variance of the reflectance drops from the value $\mathrm{Var}\,\mathcal{R} = 2/15\beta$ associated with the nonlogarithmic interaction (36.2.7) [compare Eq. (35.2.8)], to the value $\mathrm{Var}\,\mathcal{R} = 1/8\beta$ for a logarithmic interaction [compare Eq. (35.2.7)]. The crossover occurs when L becomes longer than the absorption length $l_a = \sqrt{l/\sigma_a}$, in the large-N regime $N \gg 1/\sqrt{\sigma l} \gg 1$.

36.2.3 Long-range wave function correlations

The statistics of wave function intensities $I = |\Psi(\boldsymbol{r})|^2$ in a chaotic cavity is described by the Porter-Thomas distribution [Por65],

$$P(I) = (2\pi\bar{I})^{-1/2} I^{-1/2} \exp(-I/2\bar{I}), \quad I > 0, \tag{36.2.9}$$

with $\bar{I}$ the average intensity. Equation (36.2.9) assumes time reversal symmetry, so Ψ is real (symmetry index $\beta = 1$). An experimental demonstration in a microwave resonator is shown in Fig. 36.4.

In the context of RMT, the distribution (36.2.9) follows from the GOE ensemble of the real symmetric $M \times M$ matrix H (the effective Hamiltonian), which determines the eigenstates of the cavity. The intensity I corresponds to the square of a matrix element O_{nm} of the orthogonal matrix which diagonalizes H, where the index n labels a point in discretized space and the index m labels a particular eigenstate. In the large-M limit the matrix elements of O have a Gaussian distribution, which implies Eq. (36.2.9) for the distribution of $I = O_{nm}^2$.

Different matrix elements O_{nm} and $O_{n'm}$ are independent, so the wave function has no spatial correlations in the RMT description. This is an approximation, but since the actual correlations decay on the scale of the wave length [Ber77], it is accurate to say that there are no *long-range* wave function correlations in a chaotic cavity.

The same absence of long-range correlations applies if time reversal symmetry is fully broken, by the introduction of a sufficiently strong magneto-optical element in the cavity [Sto99]. The intensity distribution changes from

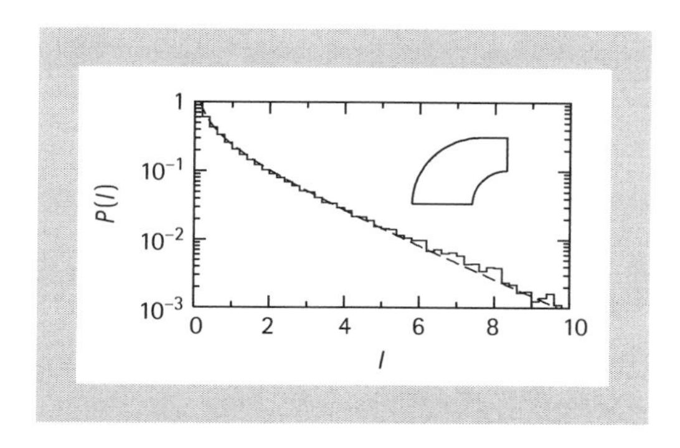

Fig. 36.4 Comparison of the Porter-Thomas distribution (36.2.9) (dashed line) of wave function intensities I in the GOE, with the intensity distribution measured on the two-dimensional microwave cavity shown in the inset. (The average intensity has been set to $\bar{I} = 1$.) Adapted from [Kud95].

the Porter-Thomas distribution (36.2.9) to the exponential distribution (36.2.1), but spatial correlations still decay on the scale of the wave length. *Partially* broken time reversal symmetry, however, has the striking effect of introducing wave function correlations that persist throughout the entire cavity. This was discovered theoretically by Fal'ko and Efetov [Fal94] for the crossover from GOE to GUE.

An altogether different way to partially break time reversal symmetry is to open up the cavity by attaching a pair of N-mode leads to it, and to excite a travelling wave from one lead to the other [Pni96]. Brouwer [Bro03] found that, if N is of order unity, the travelling wave produces relatively large long-range wave function correlations inside the cavity. As shown in Fig. 36.5, these correlations have been measured in a microwave resonator [Kim05].

Partially broken time reversal symmetry means that a wave function $\Psi(\boldsymbol{r})$ is neither real nor fully complex. Following [Lan97], the crossover from real to fully complex wave functions is quantified by the phase rigidity

$$\rho = \frac{\int d\boldsymbol{r}\, \Psi(\boldsymbol{r})^2}{\int d\boldsymbol{r}\, |\Psi(\boldsymbol{r})|^2}. \tag{36.2.10}$$

A real wave function has $\rho = 1$ while a fully complex wave function has $\rho = 0$.

As $|\rho|$ decreases from 1 to 0, the intensity distribution crosses over from the Porter-Thomas distribution (36.2.9) to the exponential distribution (36.2.1), according to [Pni96]

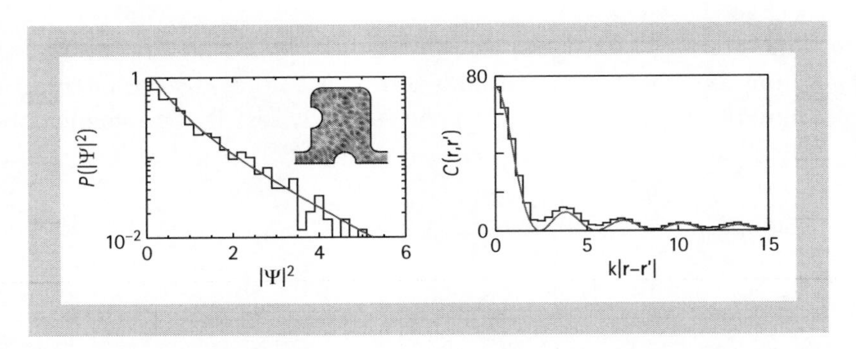

Fig. 36.5 *Left panel:* Distribution of the intensity $|\Psi(\boldsymbol{r})|^2$ of a travelling wave at a fixed frequency in the open two-dimensional chaotic microwave cavity shown in the inset (dimensions 21 cm × 18 cm). The wave function Ψ is the component of the electric field perpendicular to the cavity (normalized to unit average intensity), for a wave travelling from the right to the left lead. The measured values (histogram) are compared with the distribution (36.2.11) (solid curve), fitted to a phase rigidity $|\rho|^2 = 0.5202$. The grey scale plot in the inset shows the spatial intensity variations, with black corresponding to maximal intensity. *Right panel:* Correlator of squared intensity, for a single mode in both the right and left leads. The histogram shows the measured correlator, averaged over position in the cavity and frequency of the travelling wave. The solid curve is the theoretical prediction [Bro03], which tends to the nonzero limit 0.078 for $k|\boldsymbol{r} - \boldsymbol{r}'| \gg 1$. Adapted from [Kim05].

$$P(I|\rho) = \frac{1}{\bar{I}\sqrt{1-|\rho|^2}} \exp\left(-\frac{I/\bar{I}}{1-|\rho|^2}\right) I_0\left(\frac{|\rho|\, I/\bar{I}}{1-|\rho|^2}\right). \tag{36.2.11}$$

(The function I_0 is a Bessel function.) The notation $P(I|\rho)$ indicates that this is the intensity distribution for an eigenstate with a given value of ρ. The distribution $P(\rho)$ of ρ among different eigenstates, calculated in [Bro03], is broad for N of order unity.

For any given phase rigidity the joint distribution of the intensities $I \equiv I(\boldsymbol{r})$ and $I' \equiv I(\boldsymbol{r}')$ factorizes if $k|\boldsymbol{r} - \boldsymbol{r}'| \gg 1$. The long-range correlations appear upon averaging over the broad distribution of phase rigidities, since

$$P(I, I') = \int d\rho\; P(\rho)\, P(I|\rho)\, P(I'|\rho) \tag{36.2.12}$$

no longer factorizes.

36.2.4 Open transmission channels

The bimodal transmission distribution (35.2.14), first obtained by Dorokhov in 1984 [Dor84], tells us that a fraction l/L of the transmission eigenvalues through a random medium is of order unity, the remainder being exponentially small. A physical consequence of these open channels, discussed in Section 35.2.3, is the sub-Poissonian shot noise of electrical current [Bee92]. As expressed by Eq. (35.2.16), the shot noise power is reduced by a factor $1 - 2/3 = 1/3$, because the spectral average $\overline{T^2}$ of the transmission eigenvalues is 2/3 of the average transmission $\overline{T} = l/L$. If all transmission eigenvalues had been close to their average, one would have found $\overline{T^2}/\overline{T} \simeq l/L \ll 1$ and the shot noise would have been Poissonian.

The observation of sub-Poissonian shot noise is evidence for the existence of open transmission channels, but it is indirect evidence – because a theory is required to interpret the observed shot noise in terms of the transmission eigenvalues. In fact, one can alternatively interpret the sub-Poissonian shot noise in terms of a semiclassical theory that makes no reference at all to the transmission matrix [Nag92].

A direct measurement of the ratio $\overline{T^2}/\overline{T}$ would require the preparation of a specific scattering state, which is not feasible in electronics. In optics, however, this is a feasible experiment – as demonstrated very recently by Vellekoop and Mosk [Vel08]. By adjusting the relative amplitude and phase of a superposition of plane waves, they produced an incident wave with amplitude $E_n^{\rm in} = t_{m_0 n}^*$ in mode $n = 1, 2, \ldots N$ (for $N \simeq 10^4$). The index m_0 corresponds to an arbitrarily chosen 'target speckle' behind a diffusor, located at the centre of the square speckle pattern in Fig. 36.6. The transmitted wave has amplitude

$$E_m^{\rm out} = \sum_n t_{mn} E_n^{\rm in} = (tt^\dagger)_{mm_0}. \tag{36.2.13}$$

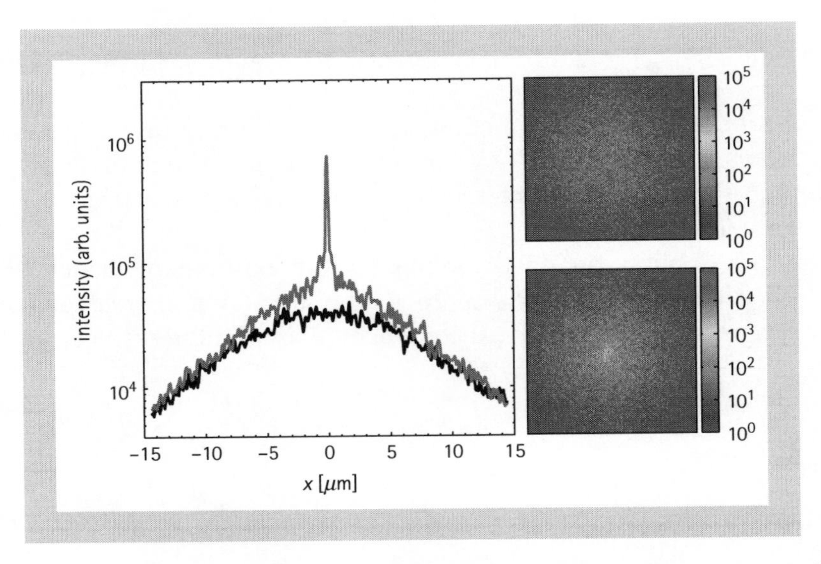

Fig. 36.6 *Right panels:* Speckle pattern (area $30\,\mu$m $\times$ $30\,\mu$m) behind a diffusor (a $11.3\,\mu$m layer of ZnO particles with mean free path $l = 0.85\,\mu$m), for a random incident wave front (top) and for a wave front optimized to couple to open transmission channels (bottom). The intensity of the bright speckle at the centre in the bottom panel is a factor of 750 greater than the background. *Left panel:* Intensity profile, integrated over the y-direction to average out the speckle pattern. The optimized wave front (red) has a peak, which the random wave front (black) lacks. Adapted from [Vel08]. This figure is reproduced in colour in the colour plates section.

As shown in [Vel08], this optimized incident wave front can be constructed 'by trial and error' without prior knowledge of the transmission matrix, because it maximizes the transmitted intensity at the target speckle (for a fixed incident intensity). The optimal increase in intensity is a factor of order $Nl/L \simeq 10^3$, as observed.

The total transmitted intensity is

$$I^{\text{out}} = \sum_m |E_m^{\text{out}}|^2 = (tt^\dagger tt^\dagger)_{m_0 m_0}. \tag{36.2.14}$$

The average transmitted intensity, averaged over the target speckle, gives the spectral average $\overline{T^2}$,

$$\overline{I^{\text{out}}} = \frac{1}{N}\sum_{m_0} I^{\text{out}} = \frac{1}{N}\text{Tr}\,(tt^\dagger)^2 = \overline{T^2}. \tag{36.2.15}$$

The average incident intensity is simply $\overline{I^{\text{in}}} = N^{-1}\text{Tr}\,tt^\dagger = \overline{T}$, so the ratio of transmitted and incident intensities gives the required ratio of spectral averages, $\overline{I^{\text{out}}}/\overline{I^{\text{in}}} = \overline{T^2}/\overline{T}$. The experimental results are consistent with the value $2/3$ for this ratio, in accordance with the bimodal transmission distribution (35.2.14).

36.3 Quantum optics

36.3.1 Grey-body radiation

The emission of photons by matter in thermal equilibrium is not a series of independent events. The textbook example is black-body radiation [Man95]. Consider a system in thermal equilibrium (temperature T) that fully absorbs any incident radiation in N propagating modes within a frequency interval $\delta\omega$ around ω. A photodetector counts the emission of n photons in this frequency interval during a long time $t \gg 1/\delta\omega$. The probability distribution $P(n)$ is given by the negative-binomial distribution with $\nu = Nt\delta\omega/2\pi$ degrees of freedom,

$$P(n) \propto \binom{n+\nu-1}{n} \exp(-n\hbar\omega/k_{\mathrm{B}}T). \tag{36.3.1}$$

The binomial coefficient counts the number of partitions of n bosons among ν states. The mean photocount $\bar{n} = \nu f$ is proportional to the Bose-Einstein function

$$f(\omega, T) = [\exp(\hbar\omega/k_{\mathrm{B}}T) - 1]^{-1}. \tag{36.3.2}$$

In the limit $\bar{n}/\nu \to 0$, Eq. (36.3.1) approaches $P(n) \propto \bar{n}^n/n!$, the Poisson distribution of independent photocounts. The Poisson distribution has variance $\mathrm{Var}\, n = \bar{n}$ equal to its mean. The negative-binomial distribution describes photocounts that occur in 'bunches', leading to an increase of the variance by a factor $1 + \bar{n}/\nu$.

By definition, a black body has scattering matrix $S = 0$, because all incident radiation is absorbed. If the absorption is not strong enough, some radiation will be transmitted or reflected and S will differ from zero. Such a 'grey body' can still be in thermal equilibrium, but the statistics of the photons which its emits will differ from the negative-binomial distribution (36.3.1). A general expression for the photon statistics of grey-body radiation in terms of the scattering matrix was derived in [Bee98]. The expression is simplest in terms of the generating function

$$F(\xi) = \ln \sum_{n=0}^{\infty} (1+\xi)^n P(n), \tag{36.3.3}$$

from which $P(n)$ can be reconstructed via

$$P(n) = \lim_{\xi \to -1} \frac{1}{n!} \frac{d^n}{d\xi^n} e^{F(\xi)}. \tag{36.3.4}$$

The relation between $F(\xi)$ and S is

$$F(\xi) = -\frac{t\delta\omega}{2\pi} \ln \mathrm{Det}\left[\mathbf{1} - (\mathbf{1} - SS^{\dagger})\xi f\right]. \tag{36.3.5}$$

If the grey body is a chaotic resonator, RMT can be used to determine the sample-to-sample statistics of S and thus of the photocount distribution. What is needed is the distribution of the so-called 'scattering strengths' $\sigma_1, \sigma_2, \ldots \sigma_N$, which are the eigenvalues of the matrix product $SS^\dagger$. All σ_ns are equal to zero for a black body and equal to unity in the absence of absorption. The distribution function $P(\{\sigma_n\})$ is known exactly for weak absorption (Laguerre orthogonal ensemble) and for a few small values of N [Bee01]. In the large-N limit, the eigenvalue density $\rho(\sigma) = \langle \sum_n \delta(\sigma - \sigma_n) \rangle$ is known in closed-form [Bee99], which makes it possible to compute the ensemble average of arbitrary moments of $P(n)$.

The first two moments are given by

$$\bar{n} = \nu f \frac{1}{N} \sum_{n=1}^{N} (1 - \sigma_n), \quad \mathrm{Var}\, n = \bar{n} + \nu f^2 \frac{1}{N} \sum_{n=1}^{N} (1 - \sigma_n)^2. \tag{36.3.6}$$

For comparison with black-body radiation we parameterize the variance in terms of the effective number ν_{eff} of degrees of freedom [Man95],

$$\mathrm{Var}\, n = \bar{n}(1 + \bar{n}/\nu_{\text{eff}}), \tag{36.3.7}$$

with $\nu_{\text{eff}} = \nu$ for a black body. Equation (36.3.6) implies a *reduced* number of degrees of freedom for grey-body radiation,

$$\frac{\nu_{\text{eff}}}{\nu} = \frac{\left[\sum_n (1 - \sigma_n)\right]^2}{N \sum_n (1 - \sigma_n)^2} \leq 1. \tag{36.3.8}$$

Note that the reduction occurs only for $N > 1$.

The ensemble average for $N \gg 1$ is

$$\nu_{\text{eff}}/\nu = (1 + \gamma)^2 (\gamma^2 + 2\gamma + 2)^{-1}, \tag{36.3.9}$$

with $\gamma = \sigma \tau_{\text{dwell}}$ the product of the absorption rate σ and the mean dwell time $\tau_{\text{dwell}} \equiv 2\pi / N\delta$ of a photon in the cavity in the absence of absorption. (The cavity has a mean spacing δ of eigenfrequencies.) As shown in Fig. 36.7 (red solid curve), weak absorption reduces ν_{eff} by up to a factor of two relative to the black-body value.

So far we have discussed thermal emission from absorbing systems. The general formula (36.3.5) can also be applied to amplified spontaneous emission, produced by a population inversion of the atomic levels in the cavity. The factor f now describes the degree of population inversion of a two-level system, with $f = -1$ for complete inversion (empty lower level, filled upper level). The scattering strengths σ_n for an amplifying system are > 1, and in fact one can show that $\sigma_n \mapsto 1/\sigma_n$ upon changing $\sigma \mapsto -\sigma$ (absorption rate $\mapsto$ amplification rate). As a consequence, Eq. (36.3.9) can also be applied to an amplifying cavity, if we change $\gamma \mapsto -\gamma$. The result (blue dashed curve in Fig. 36.7) is that the

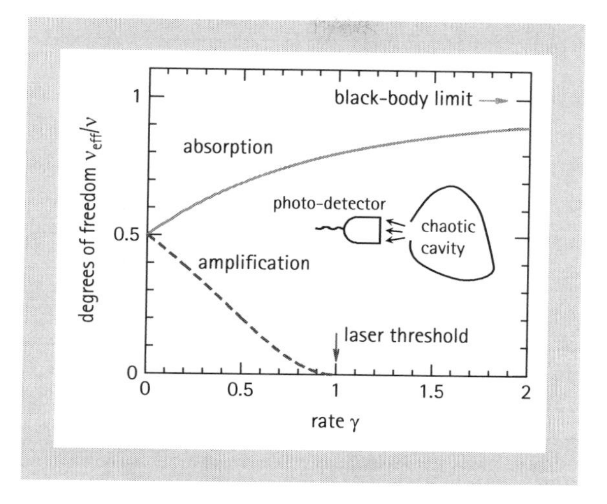

Fig. 36.7 Effective number of degrees of freedom as a function of normalized absorption or amplification rate in a chaotic cavity (inset). The black-body limit for absorbing systems (red, solid line) and the laser threshold for amplifying systems (blue, dashed line) are indicated by arrows. Adapted from [Bee98].

ratio $\nu_{\rm eff}/\nu$ decreases with increasing $\gamma = |\sigma|\tau_{\rm dwell}$ – vanishing at $\gamma = 1$. This is the laser threshold, which we discuss next.

36.3.2 RMT of a chaotic laser cavity

Causality requires that the scattering matrix $S(\omega)$ has all its poles $\Omega_m - i\Gamma_m/2$ in the lower half of the complex frequency plane. Amplification with rate $\sigma > 0$ adds a term $i\sigma/2$ to the poles, shifting them upwards towards the real axis. The laser threshold is reached when the decay rate Γ_0 of the pole closest to the real axis (the 'lasing mode') equals the amplification rate σ. For $\sigma > \Gamma_0$ the loss of radiation from the cavity is less than the gain due to stimulated emission, so the cavity will emit radiation in a narrow frequency band width around the lasing mode. If the cavity has chaotic dynamics, the ensemble averaged properties of the laser can be described by RMT.[2]

For this purpose, we include amplification in the Weidenmüller formula (35.3.3), which takes the form

$$S(\omega) = \mathbf{1} - 2\pi i\, W^{\dagger}(\omega - i\sigma/2 - \mathcal{H})^{-1} W. \tag{36.3.10}$$

The poles of the scattering matrix are the complex eigenvalues of the $M \times M$ matrix

[2] The statistical properties of a chaotic laser cavity are closely related to those of so-called random lasers (see [Cao05] for a review of experiments and [Tur08] for a recent theory). The confinement in a random laser is not produced by a cavity, but presumably by disorder and the resulting wave localization. (Alternative mechanisms are reviewed in [Zai09].)

$$\mathcal{H} = H - i\pi WW^{\dagger} = U \operatorname{diag}(\Omega_1 - i\Gamma_1, \ldots, \Omega_M - i\Gamma_M) U^{-1}, \tag{36.3.11}$$

constructed from the Hamiltonian H of the closed cavity and the $M \times N$ coupling matrix W to the outside. Because $\mathcal{H}$ is not Hermitian, the matrix U which diagonalizes $\mathcal{H}$ is not unitary. In the RMT description one takes a Gaussian ensemble for H and a non-random W, and seeks the distribution of eigenvalues and eigenvectors of $\mathcal{H}$. This is a difficult problem, but most of the results needed for the application to a laser are known [Fyo03].

The first question to ask, is at which frequencies does the laser radiate. There can be more than a single lasing mode, when more than a single pole has crossed the real axis. The statistics of the laser frequencies has been studied in [Mis98, Hac05, Zai06]. Only a subset N_{lasing} of the N_σ modes with $\Gamma_m < \sigma$ becomes a laser mode, because of mode competition. If two modes have an appreciable spatial overlap, the mode which starts lasing first will deplete the population inversion before the second mode has a chance to be amplified. For weak coupling of the modes to the outside, when the wave functions have the Porter-Thomas distribution, the average number of lasing modes scales as $\bar{N}_{\text{lasing}} \propto \bar{N}_\sigma^{2/3}$ [Mis98].

Once we know the frequency of a lasing mode, we would like to know its width. The radiation from a laser is characterized by a very narrow line width, limited by the vacuum fluctuations of the electromagnetic field. The quantum-limited linewidth, or Schawlow-Townes linewidth [Sch58],

$$\delta\omega = \tfrac{1}{2} K \Gamma_0^2 / I, \tag{36.3.12}$$

is proportional to the square of the decay rate Γ_0 of the lasing cavity mode and inversely proportional to the output power I (in units of photons/s). This is a lower bound for the linewidth when Γ_0 is much less than the linewidth of the atomic transition and when the lower level of the transition is unoccupied (complete population inversion). While Schawlow and Townes had $K = 1$, appropriate for a nearly closed cavity, it was later realized [Pet79, Sie89] that an open cavity has an enhancement factor $K \geq 1$ called the 'Petermann factor'.

The RMT of the Petermann factor was developed in [Pat00, Fra00]. The factor K is related to the nonunitary matrix U of right eigenvectors of $\mathcal{H}$, by

$$K = (U^{\dagger} U)_{00} (U^{-1} U^{-1\dagger})_{00}, \tag{36.3.13}$$

where the index 0 labels the lasing mode. (In the presence of time reversal symmetry, one may choose $U^{-1} = U^T$, hence $K = [(UU^{\dagger})_{00}]^2$.) If the cavity is weakly coupled to the outside, then the matrix U is unitary and $K = 1$, but more generally $K \geq 1$. The probability distribution $P(K|\Gamma_0)$ of the Petermann factor for a given value of the decay rate Γ_0 is very broad and asymmetric, with an algebraically decaying tail towards large K. For example, in the case $N = 1$ of a single-mode opening of the cavity, $P(K|\Gamma_0) \propto (K-1)^{-2-3\beta/2}$.

References

[Akk88] E. Akkermans, P.E. Wolf, R. Maynard, and G. Maret, J. Physique **49** (1988) 77.
[Akk07] E. Akkermans and G. Montambaux, *Mesoscopic Physics of Electrons and Photons*, Cambridge University Press, Cambridge 2007.
[Alt97] H. Alt, C. Dembowski, H.-D. Gräf, R. Hofferbert, H. Rehfeld, A. Richter, R. Schuhmann, and T. Weiland, Phys. Rev. Lett. **79** (1997) 1026.
[Bee92] C.W.J. Beenakker and M. Büttiker, Phys. Rev. B **46** (1992) 1889.
[Bee96] C.W.J. Beenakker, J.C.J. Paasschens, and P.W. Brouwer, Phys. Rev. Lett. **76** (1996) 1368.
[Bee97] C.W.J. Beenakker, Rev. Mod. Phys. **69** (1997) 731.
[Bee98] C.W.J. Beenakker, Phys. Rev. Lett. **81** (1998) 1829.
[Bee99] C.W.J. Beenakker, in *Diffuse Waves in Complex Media*, edited by J.-P. Fouque, NATO Science Series C531, Kluwer, Dordrecht 1999 [arXiv:quant-ph/9808066].
[Bee01] C.W.J. Beenakker and P.W. Brouwer, Physica E **9** (2001) 463.
[Ber77] M.V. Berry, J. Phys. A **10** (1977) 2083.
[Ber85] M. V. Berry, Proc. R. Soc. London A **400** (1985) 229.
[Boh84] O. Bohigas, M.-J. Giannoni, and C. Schmit, Phys. Rev. Lett. 52 (1984) 1.
[Bro03] P.W. Brouwer, Phys. Rev. E **68** (2003) 046205.
[Bru96] N.A. Bruce and J.T. Chalker, J. Phys. A **29** (1996) 3761.
[Cao05] H. Cao, J. Phys. A **38** (2005) 10497.
[Dor84] O.N. Dorokhov, Solid State Comm. **51** (1984) 381.
[Fal94] V.I. Falko and K.B. Efetov, Phys. Rev. Lett. **77** (1996) 912.
[Fra00] K. Frahm, H. Schomerus, M. Patra, and C.W.J. Beenakker, Europhys. Lett. **49** (2000) 48.
[Fyo03] Y. V. Fyodorov and H. J. Sommers, J. Phys. A **36** (2003) 3303.
[Goo07] J.W. Goodman, *Speckle Phenomena in Optics: Theory and Applications*, Roberts and Company, Englewood, Colorado 2007.
[Hac05] G. Hackenbroich, J. Phys. A **38** (2005) 10537.
[Kim05] Y.-H. Kim, U. Kuhl, H.-J. Stöckmann, and P.W. Brouwer, Phys. Rev. Lett. **94** (2005) 036804.
[Kud95] A. Kudrolli, V. Kidambi, and S. Sridhar, Phys. Rev. Lett. **75** (1995) 822.
[Kuh05] U. Kuhl, H.-J. Stöckmann, and R. Weaver, J. Phys. A **38** (2005) 10433.
[Lan97] S.A. van Langen, P.W. Brouwer, and C.W.J. Beenakker, Phys. Rev. E **55** (1997) 1.
[Man95] L. Mandel and E. Wolf, *Optical Coherence and Quantum Optics*, Cambridge University Press, Cambridge 1995.
[Mel88a] P.A. Mello, P. Pereyra, and N. Kumar, Ann. Phys. (N.Y.) **181** (1988) 290.
[Mel88b] P.A. Mello, E. Akkermans, and B. Shapiro, Phys. Rev. Lett. **61** (1988) 459.
[Mis96] T.Sh. Misirpashaev and C.W.J. Beenakker, JETP Lett. **64** (1996) 319.
[Mis98] T.Sh. Misirpashaev and C.W.J. Beenakker, Phys. Rev. A **57** (1998) 2041.
[Nag92] K.E. Nagaev, Phys. Lett. A **169** (1992) 103.
[Pat00] M. Patra, H. Schomerus, and C.W.J. Beenakker, Phys. Rev. A **61** (2000) 23810.
[Pet79] K. Petermann, IEEE J. Quantum Electron. **15** (1979) 566.
[Pni96] R. Pnini and B. Shapiro, Phys. Rev. E **54** (1996) R1032.
[Por65] C.E. Porter, ed., *Statistical Theories of Spectra: Fluctuations*, Academic Press, New York 1965.
[Sch58] A.L. Schawlow and C.H. Townes, Phys. Rev. **112** (1958) 1940.
[Sie89] A.E. Siegman, Phys. Rev. A **39** (1989) 1253.

[Sto99] H.J. Stöckmann, *Quantum Chaos: An Introduction*, Cambridge University Press, Cambridge 1999.

[Tur08] H. E. Türeci, L. Ge, S. Rotter, and A. D. Stone, Science **320** (2008) 643.

[Vel08] I.M. Vellekoop and A.P. Mosk, Phys. Rev. Lett. **101** (2008) 120601.

[Wie95] D.S. Wiersma, M.P. van Albada, and A. Lagendijk, Rev. Sci. Instrum. **66** (1995) 5473.

[Zai06] O. Zaitsev, Phys. Rev. A **74** (2006) 063803.

[Zai09] O. Zaitsev and L. Deych, J. Opt. **12** (2010) 024001.

·37·

Extreme eigenvalues of Wishart matrices: application to entangled bipartite system

Satya N. Majumdar

Abstract

This chapter discusses an application of the random matrix theory in the context of estimating the bipartite entanglement of a quantum system. We discuss how the Wishart ensemble (the earliest studied random matrix ensemble) appears in this quantum problem. The eigenvalues of the reduced density matrix of one of the subsystems have similar statistical properties to those of the Wishart matrices, except that their *trace is constrained to be unity*. We focus here on the smallest eigenvalue which serves as an important measure of entanglement between the two subsystems. In the hard edge case (when the two subsystems have equal sizes) one can fully characterize the probability distribution of the minimum eigenvalue for real, complex, and quaternion matrices of all sizes. In particular, we discuss the important finite size effect due to the *fixed trace constraint*.

37.1 Introduction

The different chapters of this book have already illustrated numerous applications of random matrices in a variety of problems ranging from physics to finance. In this chapter, I will demonstrate yet another beautiful application of random matrix theory in a bipartite quantum system that is *entangled*. Entanglement has of late become a rather fashionable subject due to its applications in quantum information theory and quantum computation. Entanglement serves as a simple measure of nonclassical correlation between different parts of a quantum system. The more the entanglement between two parts of a system, the better it is for the functioning of algorithms of quantum computation. This is so because, intuitively speaking, quantum states that are highly entangled contain more informations about different subparts of the composite system. In this chapter I will discuss the statistical properties of entanglement in a particularly simple model of the 'random pure state' of a bipartite system. We will see how random matrices come into play in such a system.

Indeed, historically the earliest studied ensemble of random matrices is the Wishart ensemble (introduced by Wishart [Wis28] in 1928 in the context of multivariate data analysis, much before Wigner introduced the standard Gaussian ensembles of random matrices in the physics literature). Wishart matrices have found wide applications in a variety of systems (see the discussion later and also Chapter 28 and Chapter 40 of this book). In this chapter, we will see that the Wishart ensemble (with a fixed trace constraint) also appears quite naturally as the reduced density matrix of a coupled entangled bipartite quantum system. The plan of this chapter, after a brief introduction to Wishart matrices, is to explore this connection more deeply with a particular focus on the statistics of the minimum eigenvalue which serves as a useful measure of entanglement.

Let us start with a brief recollection of the Wishart matrices. Consider a square $(N \times N)$ matrix W of the product form $W = XX^\dagger$ where X is a $(N \times M)$ rectangular matrix with real, complex, or quaternion entries and $X^\dagger$ its conjugate. The matrix W has a simple and natural interpretation. For example, let the entries X_{ij} of the X matrix represent some data, e.g. the price of the i-th commodity on, say, the j-th day of observation. So, there are N commodities and for each of them we have the prices for M consecutive days, represented by the $(N \times M)$ array X. Thus for each commodity, we have M different samples. The product $(N \times N)$ matrix $W = XX^\dagger$ then represents the (unnormalized) covariance matrix, i.e. the correlation matrix between the prices of N commodities. If the entries of X are independent Gassian random variables chosen from the joint distribution $P[\{X_{ij}\}] \propto \exp\left[-\frac{\beta}{2}\mathrm{Tr}(XX^\dagger)\right]$ (where the Dyson index $\beta = 1, 2$, or 4 corresponds respectively to real, complex, or quaternion entries), then the random covariance matrix W is called the Wishart matrix [Wis28]. This ensemble is also referred to as the Laguerre ensemble since its spectral properties involve Laguerre polynomials [Bro65, For93].

As mentioned earlier, since their introduction Wishart matrices have found an impressive list of applications. They play an important role in statistical data analysis [Wil62, Joh01], in particular in data compression techniques known as prinicipal component analysis (PCA) (see Chapter 28 and Chapter 40 of this book) with applications in image processing [Fuk90], biological microarrays [Hol00, Alt00], population genetics [Pat06, Nov08], finance [Bou01, Bur04], meteorology and oceanography [Pre88], amongst others. In physics, Wishart matrices have appeared in multiple areas: in nuclear physics [Fyo97], in low energy QCD and gauge theories [Ver94a] (see also Chapter 32 of this book), quantum gravity [Amb94, Ake97], and also in several problems in statistical physics. These include directed polymers in a disordered medium [Joh00], nonintersecting Brownian excursions [Kat03, Sch08], and fluctuating nonintersecting interfaces over a solid substrate [Nad09]. Several deformations of Wishart ensembles, with multiple applications, have also been studied in the literature [Ake08].

The Wishart matrix W has N non-negative random eigenvalues denoted by $\{w_1, w_2, \ldots, w_N\}$ ($w_i \geq 0$ for each i) whose spectral properties are well understood and some of them will be briefly reviewed in Section 37.2. These include the joint distribution of N eigenvalues, the average density of eigenvalues, and also the distribution of extreme eigenvalues (the largest and the smallest). In this chapter we will mostly be concerned with the distribution of the smallest eigenvalue $w_{\min} = \min(w_1, w_2, \ldots, w_N)$ in the particular case $M = N$ corresponding to the so-called *hard edge* (at the origin) case where the average $\langle w_{\min} \rangle \to 0$ as $N \to \infty$. In this case, the properties of the small eigenvalues (near $w = 0$) are governed by Bessel functions in the large N limit [Ede88, For93, Nag93, Ver94b, Nag95]. Such hard edge properties are absent in the traditional Wigner-Dyson Gaussian random matrix [Meh04] whose eigenvalues can be both positive and negative.

The reason we are interested in the smallest eigenvalue distribution of the Wishart matrix is because of its application in the seemingly unrelated quantum entanglement problem which is the main objective of this chapter. As we will see later, Wishart matrices will appear naturally as the reduced density matrix in a coupled bipartite quantum system that is in an *entangled random pure state*. There is a slight twist though: the Wishart matrix in this system satisfies a constraint, namely its *trace is fixed to unity*. This *fixed trace ensemble* is thus analogous to the *microcanonical* ensemble in statistical mechanics while the standard (unconstrained) Wishart ensemble is the analogue of the *canonical* ensemble in statistical mechanics (for other discussions on fixed trace ensembles see Chapter 14, Section 14.3.2 of this book). In particular, our emphasis will be on the distribution of the smallest eigenvalue $\lambda_{\min}$ in this fixed trace Wishart ensemble. This is because the smallest eigenvalue turns to be a very useful observable in this system which contains information about entanglement. For the special case $M = N$ (hard edge), we will see that the distribution of $\lambda_{\min}$ can be exactly computed *for all* N in this fixed trace ensemble in all three cases $\beta = 1$, $\beta = 2$ and $\beta = 4$. In particular, we will discuss how the fixed trace constraint modifies the distribution of $\lambda_{\min}$ from its counterpart in the canonical Wishart ensemble. We will see that the global fixed trace constraint gives rise to rather strong finite size effects. This is relevant in the quantum context where the subsystems can be just a few qubits. So, it is actually important to know the distribution of entanglement for finite size systems (the thermodynamic limit is not always relevant in this context). Hence, the fact that one can compute the distribution of the minimum eigenvalue exactly for all N (not necessarily large) in the presence of the fixed trace constraint becomes important and relevant.

The rest of this chapter is organized as follows. In Section 37.2, we briefly review some spectral properties of unconstrained Wishart matrices. In Section 37.3 we introduce the problem of the random pure state of an entangled quantum bipartite system. Its connection to Wishart matrices with a fixed trace constraint is established. Next we focus on the smallest eigenvalue and derive its probability distribution for the bipartite problem in Section 37.4. Finally we conclude in Section 37.5 with a summary and open problems.

37.2 Spectral properties of Wishart matrices: a brief summary

Let us first briefly recall some spectral properties of the $(N \times N)$ Wishart matrix $W = XX^\dagger$ with X being a rectangular $(N \times M)$ matrix with real ($\beta = 1$), complex ($\beta = 2$) or quaternion ($\beta = 4$) Gaussian entries drawn from the joint distribution $P[\{X_{ij}\}] \propto \exp\left[-\frac{\beta}{2}\mathrm{Tr}(XX^\dagger)\right]$. These results will be useful for the problem of the random pure state of the bipartite system to be discussed in the next section.

Joint distribution of eigenvalues: The N eigenvalues of W, denoted by $\{w_1, w_2, \ldots, w_N\}$, are non-negative and have the joint probability density function (pdf) [Jam64]

$$P[\{w_i\}] = K_{N,M}\, e^{-\frac{\beta}{2}\sum_{i=1}^N w_i} \prod_{i=1}^N w_i^{\alpha\beta/2} \prod_{j<k} |w_j - w_k|^\beta \tag{37.2.1}$$

where $\alpha = (1 + M - N) - 2/\beta$ and the normalization constant $K_{N,M}$ can be computed exactly [Jam64]. Without any loss of generality, we will assume $N \leq M$. This is because if $N > M$, one can show that $N - M$ eigenvalues are exactly 0 and the rest of the M eigenvalues are distributed exactly as in Eq. (37.2.1) with N and M exchanged. Note that while for Wishart matrices $M - N$ is a non-negative *integer* and $\beta = 1$, 2, or 4, the joint density in Eq. (37.2.1) is well defined for any $\beta > 0$ and $\alpha > -2/\beta$ (this last condition is necessary so that the joint pdf is normalizable). When these parameters take continuous values the joint pdf is called the Laguerre ensemble.

Coulomb gas interpretation, typical scaling, and average density of states: The joint pdf (37.2.1) can be written in the standard Boltzmann form, $P[\{w_i\}] \propto \exp\left[-\beta E(\{w_i\})\right]$ where

$$E[\{w_i\}] = \frac{1}{2}\sum_{i=1}^N (w_i - \alpha \log w_i) - \frac{1}{2}\sum_{j \neq k} \ln |w_j - w_k| \tag{37.2.2}$$

can be identified as the energy of a Coulomb gas of charges with positions $\{w_i\} \geq 0$. These charges repel each other via the 2D Coulomb (logarithmic) interaction (the second term in the energy), though they are restricted to live on the positive real line. In addition, these charges are subjected to an external potential which is linear+logarithmic (the first term in the energy). The external potential tends to push the charges towards the origin while the Coulomb repulsion tends to spread them apart. The first term typically scales as $w_{\mathrm{typ}} N$

where $w_{\rm typ}$ is the typical value of an eigenvalue, while the second term scales as N^2 for large N. Balancing these two terms one gets $w_{\rm typ} \sim N$ for large N. Indeed, this scaling shows up in the average density of states (average charge density) which can be computed from the joint pdf and has the following scaling for large N

$$\rho_N(w) = \frac{1}{N}\sum_{i=1}^{N}\langle\delta(w - w_i)\rangle \to \frac{1}{N} f_{\rm MP}\left(\frac{w}{N}\right) \qquad (37.2.3)$$

where the Marčenko-Pastur(MP) scaling function is given by [Mar67] (see also Chapter 28 Section 28.4.1 of this book)

$$f_{\rm MP}(x) = \frac{1}{2\pi x}\sqrt{(b-x)(x-a)}. \qquad (37.2.4)$$

Thus the charge density is confined over a finite support $[a, b]$ with the lower edge $a = (1 - c^{-1/2})^2$ and the upper edge $b = (1 + c^{-1/2})^2$ with $0 \leq c = N/M \leq 1$. For all $c < 1$, the average density vanishes at both edges of the MP sea. For the special case $c = 1$ (this happens in the large N limit when $M - N << O(N)$), the lower edge a gets pushed towards the hard wall at 0 (this is the so-called *hard edge limit*) and the upper edge $b \to 4$ and the average density simply becomes, $f_{\rm MP}(x) = \frac{1}{2\pi}\sqrt{(4-x)/x}$ for $0 \leq x \leq 4$. It diverges as $x^{-1/2}$ at the hard lower edge $x = 0$.

For later purposes, it is also useful to calculate the average value of the trace $Tr = \sum_{i=1}^{N} w_i$. Using the expression for the average density of states, it follows that for large N

$$\langle Tr\rangle = N\int_0^\infty w\,\rho_N(w)\,dw \to \frac{N^2}{c}. \qquad (37.2.5)$$

In particular, for $c = 1$ (i.e. $M - N << O(N)$), we have $\langle Tr\rangle = N^2$ in the large N limit.

Maximum eigenvalue: Let $w_{\rm max} = \max(w_1, w_2, \ldots, w_N)$ denote the maximum eigenvalue. On an average, it is located at the upper edge of the MP density of states. It then follows from Eq. (37.2.3) that $\langle w_{\rm max}\rangle = bN$ for large N. However, for large but finite N, the random variable $w_{\rm max}$ fluctuates, from one sample to another, around its average value bN. The typical fluctuations around its mean were shown to be $\sim O(N^{1/3})$ for large N [Joh00, Joh01] and the limiting distribution of these typical fluctuations are described by the well-known Tracy–Widom density [Tra94]. In other words, $w_{\rm max} = bN + c^{1/6}b^{2/3}N^{1/3}\chi$, where the random variable χ has an N-independent limiting pdf, $g_\beta(\chi)$ described by the Tracy–Widom function [Tra94]. In contrast, *atypically large*, e.g. $\sim O(N)$ fluctuations of $w_{\rm max}$ from its mean are *not described* by the TW density. Such large fluctuations play an important role in many practical applications such as in PCA [Viv07, Maj09a]. Far away from the mean bN, these atypically large fluctuations of $w_{\max}$ are instead described by large deviation functions associated with the pdf of $P(w_{\max}, N)$ and are of the form

$$P(w_{\max} = t, N) \sim \exp\left[-\beta N^2 \Phi_-\left(\frac{bN - t}{N}\right)\right] \quad \text{for} \quad t << bN \tag{37.2.6}$$

$$\sim \exp\left[-\beta N \Phi_+\left(\frac{t - bN}{N}\right)\right] \quad \text{for} \quad t >> bN. \tag{37.2.7}$$

The left rate function $\Phi_-(x)$ was computed explicitly for all c in [Viv07] extending a Coulomb gas approach developed originally in [Dea06] to compute the corresponding left rate function for Wigner-Dyson Gaussian matrices. On the other hand, the computation of the right rate function $\Phi_+(x)$ required a different approach and was recently obtained explicitly for all c [Maj09a]. The right rate function in the Wigner-Dyson Gaussian case was also obtained by a different, albeit rigorous, method in [Ben01]. One interesting point is that while the limiting TW density $g_\beta(\chi)$ depends on β, the rate functions $\Phi_\mp(x)$ are independent of β.

Minimum eigenvalue: Since in this chapter our main interest in the problem of bipartitite entanglement concerns the lowest eigenvalue of the reduced density matrix, we need to discuss, in some detail, the statistical properties of the minimum eigenvalue of the unconstrained Wishart ensemble. For the minimum eigenvalue, $w_{\min} = \min(w_1, w_2, \ldots, w_N)$, the situation is rather different for $c < 1$ and $c = 1$ cases. For $c < 1$, the lower edge of the MP sea is at $a = (1 - c^{-1/2})^2 > 0$, indicating that $\langle w_{\min}\rangle = aN$ in the large N limit and thus the typical value of $w_{\min} \sim O(N)$. The typical fluctuations of $w_{\min}$ around this mean value aN are again described by the TW density (appropriately rescaled). The large deviation functions describing atypical fluctuations, to our knowledge, have not been systematically studied as in the maximum eigenvalue case (though see [Che96] and references therein).

The situation, however, is quite different in the $c = 1$ case (when $M - N << O(N)$ where the lower edge of the MP sea $a \to 0$. This is the so-called *hard edge* case. We will see shortly that in this case the typical value of the minimum eigenvalue scales as $w_{\min} \sim 1/N$ for large N, to be contrasted with the behaviour $w_{\min} \sim aN$ for $c < 1$. There have been a lot of studies on the distribution of $w_{\min}$ in this hard edge $c = 1$ ($M - N << O(N)$) case, notably by Edelman [Ede88] and Forrester [For93, For94]. It has also found very nice applications in QCD (see e.g. Chapter 32, Section 32.2.6 of this book). Here, for simplicity, we will focus on the special case $M = N$ (such that $c = 1$ strictly for all N, and not just for large N). For other cases when $M - N \sim O(1)$ (so that $c = 1$ only in the large N limit), a summary can be found in Table 32.2 of Chapter 32 of this book (see also Section 4.2 of [Ake08] and references therein).

In this special case $M = N$, the cumulative distribution of the minimum, $Q_N(z) = \text{Prob}[w_{\min} \geq z, N]$, is known [Ede88] exactly *for all* N in all the three cases $\beta = 1$, $\beta = 2$, and $\beta = 4$. Note that, $Q_N(z) = \int_z^\infty \ldots \int_z^\infty P[\{w_i\}] \prod dw_i$ where $P[\{w_i\}]$ is the joint pdf given in Eq. (37.2.1). For $M = N$, this multiple integral $Q_N(z)$ can easily be performed for $\beta = 2$ by making a trivial shift $w_i \to w_i + z$ and one gets for all N

$$Q_N(z) = \exp[-Nz]; \quad \beta = 2 \tag{37.2.8}$$

For $\beta = 1$ and 4, the simple shift does not work. However, the integral $Q_N(z)$ can be calculated explicitly [Ede88]. For $\beta = 1$, one obtains for all N

$$Q_N(z) = \frac{\Gamma(N+1)}{2^{N-1/2}\Gamma(N/2)} \int_z^\infty y^{-1/2}\, e^{-Ny/2}\, U\left(\frac{N-1}{2}, -\frac{1}{2}, \frac{y}{2}\right) dy \tag{37.2.9}$$

where $U(p, q, z)$ is the confluent (Kummer) hypergeometric function [Abr72]. For $\beta = 4$, while Edelman does not provide an explicit expression for $Q_N(z)$, it is not difficult to obtain $Q_N(z)$ by using his Lemma 9.2 [Ede88] and one gets (see also [For94])

$$Q_N(z) = e^{-2Nz}\, {}_1F_1\left(-N; \frac{1}{2}; -z\right); \quad \beta = 4 \tag{37.2.10}$$

where

$${}_1F_1(p; q; z) = 1 + \frac{p}{q}\frac{z}{1!} + \frac{p(p+1)}{q(q+1)}\frac{z^2}{2!} + \ldots \tag{37.2.11}$$

is the degenerate hypergeometric function [Abr72].

The large N limit is interesting where in all three cases $\beta = 1$, 2, and 4, the cumulative distribution of the minimum $Q_N(z)$ approaches a scaling form: $Q_N(z) \to q_\beta(zN)$ where the scaling function $q_\beta(y)$ can be computed explicitly

$$\begin{aligned} q_1(y) &= \exp\left[-\sqrt{y} - y/2\right] && (37.2.12)\\ q_2(y) &= \exp[-y] && (37.2.13)\\ q_4(y) &= \frac{1}{2}\left[e^{-2y+2\sqrt{y}} + e^{-2y-2\sqrt{y}}\right]. && (37.2.14) \end{aligned}$$

Note in particular that the average value $\langle w_{\min}\rangle = \int_0^\infty Q_N(z)dz \to c_\beta/N$ for large N, where the prefactor

$$c_\beta = \int_0^\infty q_\beta(y)dy \tag{37.2.15}$$

can be computed explicitly in all three cases and one gets

$$c_1 = 2\left[1 - \sqrt{\frac{\pi e}{2}}\,\mathrm{erfc}(1/\sqrt{2})\right] = 0.68864.. \tag{37.2.16}$$

$$c_2 = 1 \tag{37.2.17}$$

$$c_4 = \frac{1}{2}\left[1 + \sqrt{\frac{\pi e}{2}}\,\mathrm{erf}(1/\sqrt{2})\right] = 1.20534.. \tag{37.2.18}$$

where $\mathrm{erf}(z) = \frac{2}{\sqrt{\pi}}\int_0^z e^{-u^2}\,du$ is the standard error function and $\mathrm{erfc}(z) = 1 - \mathrm{erf}(z)$. These results will be used in Section 37.4.

37.3 Entangled random pure state of a bipartite system

We now turn to the main problem of interest in this chapter, namely the properties of an entangled random state of a quantum bipartite system. We will see that Wishart matrices, albeit with a *fixed trace constraint*, play a central role in this problem.

As mentioned in the introduction, entanglement has been studied extensively in the recent past due to its central role in quantum information and possible involvement in quantum computation. In the context of quantum algorithms, it is often desirable to create states of large entanglement. A potential candidate for such a state with 'large entanglement' that is relatively simple to analyse turns out to be the 'random pure state' in a bipartite system [Hay06] which we will describe in detail shortly. Such a random pure state can also be used as a null model or reference point to which the entanglement of an arbitrary time-evolving state may be compared. Apart from the issue of bipartite entanglement, statistical properties of such random states are relevant for quantum chaotic or non-integrable systems. The applicability of random matrix theory and hence of random states to systems with well-defined chaotic classical limits was pointed out long ago [Boh84].

We start with a general discussion of entanglement in a bipartite setting without any reference to any specific statistical measure. The statistical properties will be discussed later when we introduce the 'random' state. For now, the discussion below holds for any quantum pure state. Let us consider a composite bipartite system $A \otimes B$ composed of two smaller subsystems A and B, whose respective Hilbert spaces $\mathcal{H}_A^{(N)}$ and $\mathcal{H}_B^{(M)}$ have dimensions N and M. The Hilbert space of the composite system $\mathcal{H}^{(NM)} = \mathcal{H}_A^{(N)} \otimes \mathcal{H}_B^{(M)}$ is thus NM-dimensional. Without loss of generality we will assume that $N \leq M$. Let $\{|i^A\rangle\}$ and $\{|\alpha^B\rangle\}$ represent two complete basis states for A and B respectively. Then, any arbitrary pure state $|\psi\rangle$ of the composite system can be most generally written as a linear combination

$$|\psi\rangle = \sum_{i=1}^{N} \sum_{\alpha=1}^{M} x_{i,\alpha} |i^A\rangle \otimes |\alpha^B\rangle \tag{37.3.1}$$

where the coefficients $x_{i,\alpha}$s form the entries of a rectangular $(N \times M)$ matrix $X = [x_{i,\alpha}]$. As an example of such a bipartite system, A may be considered a given subsystem (say a set of spins) and B may represent the environment (e.g. a heat bath).

Next we discuss the density matrix and the concept of entanglement. For a pure state, the density matrix of the composite system is simply defined as

$$\rho = |\psi\rangle \langle\psi| \tag{37.3.2}$$

with the constraint $\mathrm{Tr}[\rho] = 1$, or equivalently $\langle\psi|\psi\rangle = 1$. Note that had the composite system been in a statistically *mixed* state, its density matrix would have been of the form

$$\rho = \sum_k p_k |\psi_k\rangle \langle\psi_k|, \tag{37.3.3}$$

where $|\psi_k\rangle$s are the pure states of the composite system and $0 \leq p_k \leq 1$ denotes the probability that the composite system is in the k-th pure state, with $\sum_k p_k = 1$. A classical example of such a mixed state is when the system is in the canonical ensemble at given temperature T: in this case the density matrix is given by

$$\rho = \sum_E \frac{1}{Z} e^{-E/k_B T} |E\rangle \langle E| \tag{37.3.4}$$

where $Z = \sum_E e^{-E/k_B T}$ is the canonical partition function (k_B is the Boltzmann constant) and the pure state $|E\rangle$ denotes the energy eigenstate (with eigenvalue E) of the full system. In this chapter, we will not discuss the mixed state and will restrict ourselves only to the case of a *pure* state whose density matrix is given in Eq. (37.3.2).

The concept of entanglement is simple. A pure state $|\psi\rangle$ is called **entangled** if it is *not* expressible as a direct product of two states belonging to the two subsystems A and B. Only in the special case when the coefficients have the product form, $x_{i,\alpha} = a_i b_\alpha$ for all i and α, the state $|\psi\rangle = |\phi^A\rangle \otimes |\phi^B\rangle$ can be written as a direct product of two states $|\phi^A\rangle = \sum_{i=1}^{N} a_i |i^A\rangle$ and $|\phi^B\rangle = \sum_{\alpha=1}^{M} b_\alpha |\alpha^B\rangle$ belonging respectively to the two subsystems A and B. In this case, the composite state $|\psi\rangle$ is fully *separable* or *unentangled*. But otherwise, it is generically *entangled*.

Upon using the decomposition in Eq. (37.3.1), the density matrix of the pure state can be expressed as

$$\rho = \sum_{i,\alpha} \sum_{j,\beta} x_{i,\alpha}\, x^*_{j,\beta} |i^A\rangle\langle j^A| \otimes |\alpha^B\rangle\langle\beta^B| \tag{37.3.5}$$

where the Roman indices i and j run from 1 to N and the Greek indices α and β run from 1 to M. We also assume that the pure state $|\psi\rangle$ is normalized to unity so that $\mathrm{Tr}[\rho] = 1$. Hence the coefficients $x_{i,\alpha}$s must satisfy $\sum_{i,\alpha} |x_{i,\alpha}|^2 = 1$.

Given the density matrix of the pure composite state in Eq. (37.3.5), one can then compute the reduced density matrix of, say, the subsystem A by tracing over the states of the subsystem B

$$\rho_A = \mathrm{Tr}_B[\rho] = \sum_{\alpha=1}^{M} \langle \alpha^B | \rho | \alpha^B \rangle. \tag{37.3.6}$$

The reduced density matrix is important because if we measure any observable $\hat{O}$ of the subsystem A, its expected value is given by $\mathrm{Tr}[\hat{O}\,\rho_A]$. Thus, ρ_A is the basic physical object whose properties are directly related to measurements.

Using (37.3.5) one gets

$$\rho_A = \sum_{i,j=1}^{N} \sum_{\alpha=1}^{M} x_{i,\alpha}\, x_{j,\alpha}^* \, |i^A\rangle\langle j^A| = \sum_{i,j=1}^{N} W_{ij} |i^A\rangle\langle j^A| \tag{37.3.7}$$

where W_{ij}s are the entries of the $N \times N$ square matrix $W = XX^\dagger$. In a similar way, one can express the reduced density matrix $\rho_B = \mathrm{Tr}_A[\rho]$ of the subsystem B in terms of the square $M \times M$ dimensional matrix $\tilde{W} = X^\dagger X$. Hence we see how the Wishart covariance matrix $W = XX^\dagger$ appears in this quantum problem.

Let $\lambda_1, \lambda_2, \ldots, \lambda_N$ denote the N eigenvalues of $W = XX^\dagger$. Note that these eigenvalues are non-negative, $\lambda_i \geq 0$ for all $i = 1, 2, \ldots, N$. Now the matrix $\tilde{W} = X^\dagger X$ has $M \geq N$ eigenvalues. It is easy to prove that $M - N$ of them are identically 0 and N nonzero eigenvalues of $\tilde{W}$ are the same as those of W. Thus, in this diagonal representation, one can express ρ_A as

$$\rho_A = \sum_{i=1}^{N} \lambda_i \, |\lambda_i^A\rangle \, \langle \lambda_i^A| \tag{37.3.8}$$

where $|\lambda_i^A\rangle$s are the eigenvectors of $W = XX^\dagger$. A similar representation holds for ρ_B. It then follows that one can represent the original composite state $|\psi\rangle$ in this diagonal representation as

$$|\psi\rangle = \sum_{i=1}^{N} \sqrt{\lambda_i} \, |\lambda_i^A\rangle \otimes |\lambda_i^B\rangle \tag{37.3.9}$$

where $|\lambda_i^A\rangle$ and $|\lambda_i^B\rangle$ represent the normalized eigenvectors (corresponding to the same nonzero eigenvalue λ_i) of $W = XX^\dagger$ and $\tilde{W} = X^\dagger X$ respectively. This spectral decomposition in Eq. (37.3.9) is known as the Schmidt decomposition. The normalization condition $\langle\psi|\psi\rangle = 1$, or equivalently $\mathrm{Tr}[\rho] = 1$, thus imposes a constraint on the eigenvalues, $\sum_{i=1}^{N} \lambda_i = 1$.

Note that while each individual state $|\lambda_i^A\rangle \otimes |\lambda_i^B\rangle$ in the Schmidt decomposition in Eq. (37.3.9) is *separable*, their linear combination $|\psi\rangle$, in general, is *entangled*. This simply means that the composite state $|\psi\rangle$ can not, in general, be written as a direct product $|\psi\rangle = |\phi^A\rangle \otimes |\phi^B\rangle$ of two states of the respective subsystems. The spectral properties of the matrix W, i.e. the knowledge of the eigenvalues $\lambda_1, \lambda_2, \ldots, \lambda_N$, in association with the Schmidt decomposition in Eq. (37.3.9), provide useful information about how entangled a pure state is.

Measures of entanglement: It is useful to construct a measure of entanglement, i.e. a function of the eigenvalues λ_is whose value will tell us how entangled a pure state is. There are many ways of constructing such a measure. Its value should monotonically increase from the configuration of λ_is where the state is fully *separable* to the configuration where the state is *maximally* entangled. These two configurations, recalling that $\sum_i \lambda_i = 1$, are the following:

(i) *separable*: When one of the eigenvalues, say λ_1 is 1 and the rest are all identically zero. Then the state completely decouples as only one term, say the first term, is present in Eq. (37.3.9).
(ii) *maximally entangled*: When all eigenvalues are equal, i.e. $\lambda_i = 1/N$. In this case all N terms in Eq. (37.3.9) are present.

In Fig. (37.1), we present a cartoon for $N = 3$ for the purpose of illustration. In the three-dimensional space $(\lambda_1, \lambda_2, \lambda_3)$, any point on the triangular plane $\lambda_1 + \lambda_2 + \lambda_3 = 1$ with $\lambda_i \geq 0$ represents an allowed configuration. The three vertices, where the system gets completely factorized, represent the fully *separable* configurations (situation (i) above). On the other hand, the centroid $(1/3, 1/3, 1/3)$ represents the fully (maximally) entangled configuration (situation (ii) above).

If the system is in a given configuration $\{\lambda_i\}$ on the allowed plane $\sum_{i=1}^N \lambda_i = 1$ (and $\lambda_i \geq 0$), how much is it entangled? In other words, how do we measure the entanglement content of a given configuration of $\{\lambda_i\}$? This is usually done by defining the so-called entanglement entropy, a single scalar quantity associated with each configuration, i.e. each point on the plane $\sum_{i=1}^N \lambda_i = 1$ (and $\lambda_i \geq 0$). A standard and perhaps most studied measure of entanglement is the so-called von Neumann entropy [Ben96], $S_1 = -\sum_{i=1}^N \lambda_i \ln(\lambda_i) = -\mathrm{Tr}[\rho_A \ln(\rho_A)]$, which has its smallest value $S_1 = 0$ in configuration (i) and its maximum possible value $S_1 = \ln N$ in configuration (ii). Renyi entropy defined as $S_q = \ln\left(\sum_i \lambda_i^q\right)/(1-q)$ [Ren70] with the parameter $q > 0$ is a natural generalization that reduces to the von Neumann entropy when $q \to 1$. Again, for any $q > 0$, $S_q = 0$ at the 'separable' vertices (situation (i) above) and $S_q = \ln(N)$ at the 'maximally entangled' centroid (situation (ii) above). For $q = 2$, $\sum_{i=1}^N \lambda_i^2 = \exp[-S_2]$ is called the purity that has been widely studied (see [Fac06] and references therein). For other measures we refer the reader to the introduction in [Gir07]

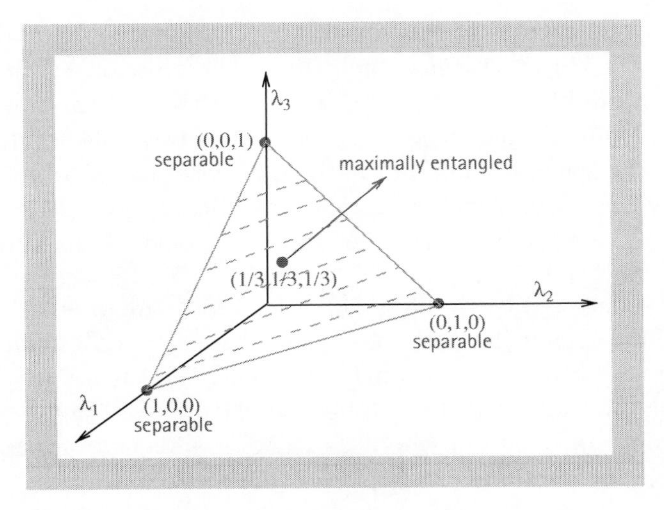

Fig. 37.1 A cartoon for $N = 3$. The system lives on the triangular plane $\lambda_1 + \lambda_2 + \lambda_3 = 1$. The vertices of the triangle represent the *separable* configurations and the centroid ($\lambda_1 = \lambda_2 = \lambda_3 = 1/3$) represents the *maximally entangled* configuration.

and also in [Per93]. Essentially, one can define any scalar quantity whose value increases monotonically as one moves from fully 'separable' to maximally 'entangled' configurations (e.g. as one moves from the vertices towards the centroid in Fig. (37.1)) (for more detailed prescriptions and requirements on the measure, see e.g. [Ved98]).

Important information regarding the degree of entanglement can also be obtained from the two extreme eigenvalues, the largest $\lambda_{\max} = \max(\lambda_1, \lambda_2, \ldots, \lambda_N)$ and the smallest $\lambda_{\min} = \min(\lambda_1, \lambda_2, \ldots, \lambda_N)$. Because of the constraint $\sum_{i=1}^{N} \lambda_i = 1$ and the fact that eigenvalues are all non-negative, it follows that $1/N \le \lambda_{\max} \le 1$ and $0 \le \lambda_{\min} \le 1/N$. Consider, for instance, the following limiting situations. Suppose that the largest eigenvalue $\lambda_{\max} = \max(\lambda_1, \lambda_2, \ldots, \lambda_N)$ takes its maximum allowed value 1. Then due to the constraint $\sum_{i=1}^{N} \lambda_i = 1$ and the fact that $\lambda_i \ge 0$ for all i, it follows that all the other $(N-1)$ eigenvalues must be identically 0. Thus it corresponds to configuration (i) above of the fully separable state. On the other hand, if $\lambda_{\max} = 1/N$ (i.e. it takes its lowest allowed value), it follows that all the eigenvalues must have the same value, $\lambda_i = 1/N$ for all i, again due to the constraint $\sum_{i=1}^{N} \lambda_i = 1$. This then corresponds to situation (ii) of a maximally entangled state. Thus, for instance, one can consider $-\ln(\lambda_{\max})$ as a measure of entanglement as it increases from its value 0 in the separable state to its maximal value $\ln(N)$ in the maximally entangled case. In fact, $-\ln(\lambda_{\max})$ is precisely the $q \to \infty$ limit of the Renyi entropy S_q.

In this chapter our particular interest is in the smallest eigenvalue $0 \le \lambda_{\min} \le 1/N$. When $\lambda_{\min}$ takes its maximal allowed value $1/N$, it follows again, from the constraint $\sum_{i=1}^{N} \lambda_i = 1$ and $\lambda_i \geq 0$ for all i, that all the eigenvalues must have the same value $\lambda_i = 1/N$. This will thus make the state $|\psi\rangle$ *maximally* entangled, i.e. situation (ii). In the opposite case, when $\lambda_{\min}$ takes its smallest allowed value 0, while it does not provide any information on the entanglement of the state $|\psi\rangle$, one sees from the Schmidt decomposition (37.3.9) that the dimension of the effective Hilbert space of the subsystem A reduces from N to $N-1$ (assuming that $\lambda_{\min}$ is non-degenerate). Indeed, if $\lambda_{\min}$ is very close to zero, one can effectively ignore the term containing $\lambda_{\min}$ in Eq. (37.3.9) and achieve a reduced Hilbert space, a process called 'dimensional reduction' which is often used in the compression of large data structures in computer vision [Wil62, Fuk90, Viv07]. Thus knowledge of $\lambda_{\min}$ and in particular its proximity to its upper and lower limits provides information on both the degree of entanglement as well as on the efficiency of the dimensional reduction process.

Random pure state: So far, our discussion has been valid for an arbitrary pure state in Eq. (37.3.1) with any fixed coefficient matrix $X = [x_{i,\alpha}]$. One can now introduce a statistical measure or distribution for the entries of the matrix X which, in turn, will induce a probability distribution for the eigenvalues λ_is of $W = XX^\dagger$ that appear in the Schmidt representation in Eq. (37.3.9). As a result, any measure of entanglement (e.g. the von Neumann entropy or the minimum eigenvalue $\lambda_{\min}$) will also have a statistical distribution associated with it. The main challenge then is to compute this probability distribution of the entanglement, given the measure on the entries of X.

So, what is an appropriate measure on X? Evidently, we can not choose any arbitrary measure on X. Indeed, in Eq. (37.3.1) we can just rotate (unitarily) the basis of the Hilbert space. Clearly physical properties of the system in a pure state should not depend on which basis we choose. Thus the joint probability distribution of the entries $x_{i,\alpha}$ in Eq. (37.3.1) should be invariant under a unitary (or orthogonal if we restrict X to be real) transformation $|\psi\rangle \to U|\psi\rangle$, where U represents a unitary operator. The only measure that remains invariant under a unitary rotation is the uniform measure over all pure states. This is the so called Haar measure where the coefficients $x_{i,\alpha}$s are uniformly distributed over all possible values satisfying the constraint $\sum_{i,\alpha} |x_{i,\alpha}|^2 = 1$ or equivalently, $P[\{X_{ij}\}] \propto \delta\left(\mathrm{Tr}(XX^\dagger - 1\right)$. The physical meaning of this Haar measure is clear: under unitary time evolution, and in the absence of any other conservation law (such as fixed energy as in the case of a standard microcanonical ensemble in statistical physics), the system visits all allowed normalized pure states that are equally likely, i.e. the Haar measure is a natural stationary measure under unitary evolution when ergodicity holds over all allowed pure states that are of the composite system. This is, in fact, the case in many physical situations when the system is described by a sufficiently complex 'time-dependent' Hamiltonian as in quantum chaotic systems [Ban02].

Given that the entries of X are distributed via the Haar measure $P[\{X_{ij}\}] \propto \delta\left(\mathrm{Tr}(XX^\dagger - 1\right)$, the next question is how are the eigenvalues of $W = XX^\dagger$ distributed? Noticing that the eigenvalues λ_is of $W = XX^\dagger$ are the same as Wishart eigenvalues, except with the additional constraint $\mathrm{Tr}(W) = \sum_{i=1}^{N} \lambda_i = 1$, it follows immediately from Eq. (37.2.1) that the joint pdf of λ_is is given by [Llo88, Zyc01]

$$P[\{\lambda_i\}] = B_{M,N}\delta\left(\sum_{i=1}^{N} \lambda_i - 1\right) \prod_{i=1}^{N} \lambda_i^{\frac{\beta}{2}(M-N+1)-1} \prod_{j<k} |\lambda_j - \lambda_k|^\beta \tag{37.3.10}$$

where the normalization constant $B_{M,N}$ is known explicitly [Zyc01]. Note that the exponential factor $e^{-\frac{\beta}{2}\sum_{i=1}^{N}\lambda_i}$ present in Eq. (37.2.1) becomes a constant due to the constraint $\sum_{i=1}^{N} \lambda_i = 1$ and hence is absorbed in the normalization constant $B_{M,N}$. The ensemble described in Eq. (37.3.10) can thus be seen as the microcanonical version of the canonical Wishart ensemble in (37.2.1).

When the coefficient matrix X is drawn from the Haar measure, we will refer to the state in Eq. (37.3.1) as a *random pure* state. Given that λ_is corresponding to the random pure state are distributed via the joint pdf (37.3.10), it follows that the associated observables such as the von Neumann entropy $S_1 = -\lambda_i \ln(\lambda_i)$, the maximum eigenvalue λ_{max}, the minimum eigenvalue λ_{min} etc. are also random variables. The main technical problem then is to evaluate the statistical properties (such as the mean, variance, or even the full probability distribution) of such observables.

There have been quite a few studies in this direction. For example, the average entropy $\langle S_1 \rangle$ (where the average is performed with the measure in Eq. (37.3.10)) was computed for $\beta = 2$ by Page [Pag95] and was found to be $\langle S \rangle \approx \ln(N) - \frac{N}{2M}$ for large $1 << N \leq M$. Noting that $\ln(N)$ is the maximal possible value of entropy of the subsystem A, the average entanglement entropy of a random pure state was concluded to be near maximal. Later, the same result was shown to hold for the $\beta = 1$ case [Ban02]. On the other hand, there have been only a few studies on the full probability distribution of the entanglement entropy. The distribution of the so-called G-concurrence [Cap06], a measure of entanglement, was computed exactly in the large N limit and was shown to have a point measure (delta function). For small N, the distribution of purity is known [Gir07]. On the other hand, for large N, the Laplace transform of the distribution of purity (for a positive Laplace variable) was computed in [Fac08, Pas09] which only gave partial information about the full purity distribution.

Recently, using a Coulomb gas approach, the full probability distributions of the Renyi and von Neumann entropies, as well as that of purity, were computed exactly in the large N limit by studying the associated Coulomb gas model via a saddle-point method [Nad10]. Interestingly, the pdf of the entropy exhibits two singular points which correspond to two interesting phase transitions in the Coulomb gas problem [Nad10, Fac08, Pas09]. Similar phase transitions in the Coulomb gas picture, leading to a nonsingular pdf of a physical observable, have also been noted recently in several other problems where random matrix theory is applicable: these include the pdf of the conductance and the shot noise power through a mesoscopic cavity [Viv08, Viv10, Osi08] (see also Chapters 35 and 36 of this book for applications of the RMT to quantum transport properties), the pdf of the number of positive eigenvalues (the so-called index) of Gaussian random matrices [Maj09b], nonintersecting Brownian interfaces near a hard wall [Nad09], and in information and communication systems [Kaz09], to name a few.

Here our focus is on the statistical properties of the minimum eigenvalue $\lambda_{\min}$ and for all values of $M = N$. For the special case $\beta = 2$ and $M = N$, the average value $\langle \lambda_{\min} \rangle$ was studied recently by Znidaric [Zni07]. He computed, by hand, $\langle \lambda_{\min} \rangle$ for small values of N and conjectured that $\langle \lambda_{\min} \rangle = 1/N^3$ for all N. Later, in [Maj08], the full probability distribution of $\lambda_{\min}$ was computed explicity for all $M = N$ and $\beta = 1$ and $\beta = 2$. Znidaric's conjecture for $\beta = 2$ then followed as a simple corollary [Maj08]. In the next section, I briefly outline this derivation and also provide a new result for the distribution of $\lambda_{\min}$ for $\beta = 4$.

37.4 Minimum Eigenvalue distribution for quadratic matrices

In this section, we compute the distribution of $\lambda_{\min}$ when the eigenvalues are distributed via Eq. (37.3.10). It is easier to compute the cumulative distribution

$$R_N(x) = \text{Prob}\left[\lambda_{\min} \geq x\right] = \text{Prob}\left[\lambda_1 \geq x, \lambda_2 \geq x, \ldots, \lambda_N \geq x\right]. \tag{37.4.1}$$

Using (37.3.10)

$$R_N(x) = B_{M,N} \int_x^\infty \cdots \int_x^\infty \delta\left(\sum_{i=1}^N \lambda_i - 1\right) \prod_{j<k} |\lambda_j - \lambda_k|^\beta \prod_{i=1}^N \lambda_i^{\beta/2(M-N+1)-1}\, d\lambda_i. \tag{37.4.2}$$

The challenge is to evaluate this multiple integral.

We proceed by introducing an auxiliary integral

$$I(x,t) = \int_x^\infty \cdots \int_x^\infty \delta\left(\sum_{i=1}^N \lambda_i - t\right) \prod_{j<k} |\lambda_j - \lambda_k|^\beta \prod_{i=1}^N \lambda_i^{\beta/2(M-N+1)-1}\, d\lambda_i. \tag{37.4.3}$$

If we can evaluate $I(x, t)$ for all t, then

$$R_N(x) = B_{M,N}\, I(x, 1). \tag{37.4.4}$$

To evaluate $I(x, t)$, it is natural to consider its Laplace transform

$$\int_0^\infty I(x, t)e^{-st}dt = \int_x^\infty \cdots \int_x^\infty e^{-s\sum_{i=1}^N \lambda_i} \prod_{j<k} |\lambda_j - \lambda_k|^\beta \prod_{i=1}^N \lambda_i^{\beta/2(M-N+1)-1}\, d\lambda_i. \tag{37.4.5}$$

Next, a change of variable $\lambda_i = \frac{\beta}{2s}\, w_i$ reduces it to

$$\int_0^\infty I(x, t)e^{-st}dt = \left(\frac{\beta}{2s}\right)^{-\frac{\beta MN}{2}} \int_{\frac{2sx}{\beta}}^\infty .. \int_{\frac{2sx}{\beta}}^\infty e^{-\frac{\beta}{2}\sum_{i=1}^N w_i} \prod_{j<k} |w_j - w_k|^\beta \prod_{i=1}^N w_i^{\alpha\beta/2} dw_i \tag{37.4.6}$$

where $\alpha = (1 + M - N) - 2/\beta$. Next we recognize the multiple integral, up to an overall constant, as the cumulative distribution $Q_N(2sx/\beta)$ of the minimum eigenvalue $w_{\min}$ in the unconstrained Wishart ensemble discussed previously. Thus, up to an overall constant A_1 independent of s, we have

$$\int_0^\infty I(x, t)e^{-st}dt = A_1 s^{-\beta MN/2}\, Q_N\left(\frac{2sx}{\beta}\right). \tag{37.4.7}$$

The program then is to invert this Laplace transform, compute $I(x, t)$ for all t and calculate $R_N(x)$ using Eq. (37.4.4). Henceforth, we will drop the overall constant which can be finally fixed from the normalization that $R_N(0) = 1$.

Thus, if we know the cumulative distribution of the Wishart minimum eigenvalue $Q_N(z)$, we can, at least in principle, determine the minimum eigenvalue distribution $R_N(x)$ for the random pure state problem. This is hardly surprising given the microcanonical to canonical correspondence between the two ensembles. In practice, however, it is non-trivial to invert the Laplace transform in Eq. (37.4.7). That indeed is the real challenge. We will see below that fortunately for $M = N$, where $Q_N(z)$ is given in Eqs. (37.2.9), (37.2.8), and (37.2.10) for $\beta = 1$, 2, and 4 respectively, this Laplace inversion can be carried out in closed form and one can compute $R_N(x)$ explicitly in all three cases $\beta = 1, 2$, and 4.

***The case* $\beta = 2$:** Let us start with the simplest case $\beta = 2$ with $M = N$. Here $Q_N(z) = e^{-Nz}$ from Eq. (37.2.8). Dropping the overall constant, Eq. (37.4.7) gives

$$\int_0^\infty I(x, t)e^{-st}dt = \frac{e^{-sNx}}{s^{N^2}}. \tag{37.4.8}$$

The Laplace inversion is trivial upon using the convolution theorem giving (up to an overall constant) $I(x, t) = (t - Nx)^{N^2-1}\Theta(t - Nx)$ where $\Theta(z)$ is the step function. Putting $t = 1$ and using (37.4.4) gives the exact distribution [Maj08]

$$R_N(x) = \text{Prob}\,[\lambda_{\min} \geq x] = (1 - Nx)^{N^2-1}\,\Theta\,(1 - Nx)\,. \tag{37.4.9}$$

Subsequently, the pdf is given by

$$P_N(x) = -\frac{d\,R_N(x)}{dx} = N(N^2 - 1)(1 - Nx)^{N^2-2}\,\Theta(1 - Nx). \tag{37.4.10}$$

A plot of this pdf can be found in Fig.37.2 for $N = 2$. Thus $P_N(x)$ in $x \in [0, 1/N]$ has the limiting behaviour

$$\begin{aligned} P_N(x) &\to N\,(N^2 - 1) \qquad \text{as} \quad x \to 0 \\ &\to N\,(N^2 - 1)\,(1 - Nx)^{N^2-2} \quad \text{as} \quad x \to 1/N. \end{aligned} \tag{37.4.11}$$

One can easily compute all the moments explicitly

$$\mu_k(N) = \langle \lambda_{\min}^k \rangle = \int_0^\infty x^k\,P_N(x)\,dx = \frac{\Gamma(k+1)\Gamma(N^2)}{N^k\,\Gamma(N^2 + k)}. \tag{37.4.12}$$

In particular, for $k = 1$, we obtain for all N

$$\mu_1(N) = \langle \lambda_{\min} \rangle = \frac{1}{N^3}, \tag{37.4.13}$$

proving the conjecture by Znidaric [Zni07]. Putting $k = 2$ in Eq. (37.4.12), we get the second moment $\mu_2 = \frac{2}{N^4(N^2+1)}$. Thus the variance is given by

$$\sigma^2 = \mu_2(N) - [\mu_1(N)]^2 = \frac{1}{N^6}\left(\frac{N^2 - 1}{N^2 + 1}\right). \tag{37.4.14}$$

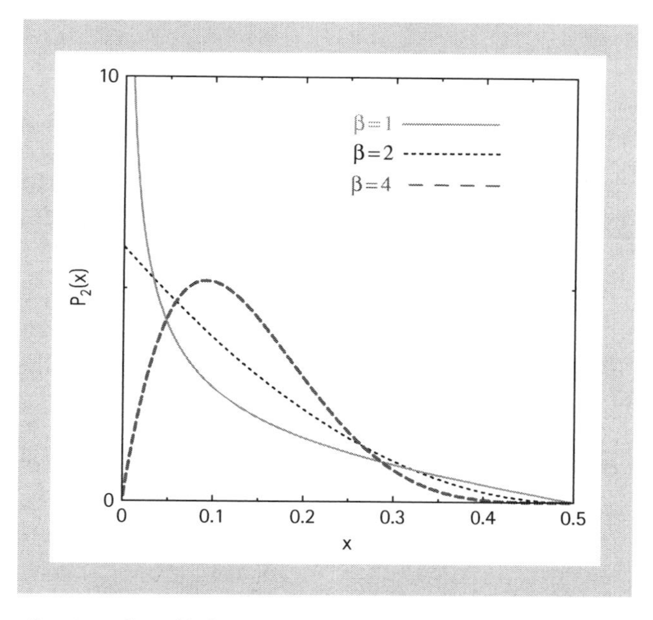

Fig. 37.2 The pdf of $\lambda_{\min}$ for $N = 2$ for $\beta = 1$, $\beta = 2$, and $\beta = 4$.

The case $\beta = 1$: The computation in this case proceeds as in the case $\beta = 2$, though the Laplace inversion is non-trivial. Omitting details [Maj08], we just quote here the main results. For $M = N$, the pdf of the minimum eigenvalue, $P_N(x) = -d\,R_N(x)/dx$ is nonzero in $x \in [0, 1/N]$ and is given by [Maj08]

$$P_N(x) = A_N\, x^{-N/2}\,(1 - Nx)^{(N^2+N-4)/2} \times {}_2F_1\left(\frac{N+2}{2}, \frac{N-1}{2}, \frac{N^2+N-2}{2}, -\frac{(1-Nx)}{x}\right) \tag{37.4.15}$$

where ${}_2F_1(a, bc, c, z)$ is the standard hypergeometric function [Abr72] and the constant A_N is given by

$$A_N = \frac{N\,\Gamma(N)\,\Gamma(N^2/2)}{2^{N-1}\,\Gamma(N/2)\,\Gamma((N^2+N-2)/2)}. \tag{37.4.16}$$

The limiting behaviour of $P_N(x)$ as $x \to 0$ and $x \to 1/N$ can be worked out

$$\begin{aligned} P_N(x) &\approx \left[\frac{\sqrt{\pi}\,\Gamma(N)\,\Gamma(N^2/2)}{2^{N-1}\,\Gamma^2(N/2)\,\Gamma((N-1)/2)}\right] x^{-1/2} \quad \text{as} \quad x \to 0 \\ &\approx A_N\, N^{-N/2}\,(1 - Nx)^{(N^2+N-4)/2} \quad \text{as} \quad x \to 1/N \end{aligned} \tag{37.4.17}$$

All moments can also be worked out explicitly [Maj08]. In particular, the average is given by

$$\mu_1(N) = \langle \lambda_{\min} \rangle = \frac{\sqrt{\pi}\,\Gamma(N)}{N\,\Gamma(N/2)\,\Gamma((N+5)/2)2^{N-1}}\, {}_2F_1\left(3, \frac{3}{2}, \frac{N+5}{2}, 1 - N\right). \tag{37.4.18}$$

Thus the expression for $\langle \lambda_{\min} \rangle$ for arbitrary N in the real ($\beta = 1$) case is considerably more complicated than its counterpart in Eq. (37.4.13) for the complex case. One finds, from Eq. (37.4.18), that $\mu_1(N)$ decreases with increasing N, e.g. $\mu_1(1) = 1$, $\mu_1(2) = (4 - \pi)/8$, $\mu_1(3) = (2 - \sqrt{3})/9$ etc. One can show [Maj08] that asymptotically for large N, $\mu_1(N)$ decays as

$$\mu_1(N) \approx \frac{c_1}{N^3} \tag{37.4.19}$$

where c_1 is precisely the constant in Eq. (37.2.16).

The case $\beta = 4$: For $\beta = 4$, we first substitute $Q_N(z)$ from (37.2.10) in (37.4.7), expand the hypergeometric function in power series as in (37.2.11) and then invert the Laplace transform term by term to get a series for $I(x, t)$. To transform each term, we make use of the convolution theorem. We then put $t = 1$ in the expression for $I(x, t)$ and compute $R_N(x)$ in (37.4.4). The overall constant is fixed by imposing $R_N(0) = 1$. This gives the explicit expression, valid for $0 \le x \le 1/N$,

$$R_N(x) = \Gamma(N+1)\Gamma(2N^2) \sum_{k=0}^{N} \frac{(2x)^k (1-Nx)^{2N^2-k-1}}{(N-k)!(2k)!\Gamma(2N^2-k)} \tag{37.4.20}$$

The pdf $P_N(x) = -d\,R_N(x)/dx$ of $\lambda_{\min}$ vanishes linearly at $x=0$ as $P_N(x) \to B_N\, x$ where $B_N = N(2N+1)(2N^2-1)(2N^2-2)/3$. This is in contrast to the $\beta = 1$ case (where $P_N(x)$ diverges as $x^{-1/2}$ as $x \to 0$) and also to the $\beta = 2$ case where $P_N(x)$ approaches a constant as $x \to 0$ (see Fig.37.2). At the upper edge, when $x \to 1/N$, the pdf vanishes as $P_N(x) \sim (1-Nx)^{2N^2-N-2}$.

All the moments can also be calculated explicitly for $\beta = 4$. For example, the average is given by

$$\mu_1(N) = \langle \lambda_{\min} \rangle = \frac{1}{2N^3} \sum_{k=0}^{N} \binom{N}{k} \frac{(k!)^2}{(2k)!} \left(\frac{2}{N}\right)^k. \tag{37.4.21}$$

One can extract the large N asymptotics of this sum. We first express $\binom{N}{k} = \Gamma(N+1)/\Gamma(N-k+1)$, then use the property of the Gamma function, $\lim_{z\to\infty} \Gamma(z+a)/\Gamma(z) \to z^a$, to obtain for large N

$$\mu_1(N) = \langle \lambda_{\min} \rangle = \frac{1}{2N^3} \sum_{k=0}^{\infty} \frac{k!}{(2k)!} 2^k \tag{37.4.22}$$

The sum can be exactly evaluated giving,

$$\mu_1(N) = \langle \lambda_{\min} \rangle \approx \frac{c_4}{N^3} \tag{37.4.23}$$

where c_4 is precisely the constant in Eq. (37.2.18).

Let us then summarize the behaviour of the pdf $P_N(\lambda_{\min})$ of $\lambda_{\min}$ in the three cases $\beta = 1$, $\beta = 2$, and $\beta = 4$ (see Fig. 37.2). At the lower edge $x \to 0$, $P_N(x)$ displays very different behaviour in the three cases. As $x \to 0$, $P_N(x)$ diverges as $x^{-1/2}$ for $\beta = 1$, approaches a constant for $\beta = 2$ and vanishes linearly for $\beta = 4$. On the other hand, at the upper edge $x \to 1/N$, $P_N(x)$ approaches 0 as a power law in all three cases, albeit with different powers, $P_N(x) \sim (1-Nx)^{\nu_\beta}$ where $\nu_1 = (N^2+N-4)/2$, $\nu_2 = N^2-2$ and $\nu_4 = 2N^2-N-2$.

In the large N limit and in the range $x << 1/N$ (far away from the upper edge) the cumulative distribution $R_N(x) = \int_x^{1/N} P_N(x')\,dx'$ approaches the scaling form $R_N(x) \to q_\beta(xN^3)$, where the scaling functions $q_\beta(y)$ are exactly the same as in the unconstrained Wishart case given respectively in Eqs. (37.2.12), (37.2.13), and (37.2.14). Thus, in this range and for large N, effectively the random variable $\lambda_{\min}$ in the bipartite problem behaves, in law, as the Wishart minimum eigenvalue scaled by a factor N^{-2}, i.e. $\lambda_{\min} \to w_{\min}/N^2$. This is also confirmed in Eq. (37.2.5), where we see that the average trace in the unconstrained Wishart ensemble scales as N^2. Thus, in the microcanonical ensemble, where the trace is constrained to be unity, it amounts to rescaling all the Wishart eigenvalues by a factor N^{-2} for large N. However, for finite N, the distributions are very different in the constrained and unconstrained ensembles, in particular near the upper edge $x = 1/N$. In other words, the distribution of $\lambda_{\min}$ exhibits strong finite size effects.

37.5 Summary and conclusion

In this chapter we have discussed an application of Wishart matrices in an entangled random pure state of a bipartite system consisting of two subsystems whose Hilbert spaces have dimensions M and N respectively with $N \leq M$. The N eigenvalues of the reduced density matrix of the smaller subsystem are distributed exactly as the eigenvalues of a Wishart matrix, the only difference being that the eigenvalues satisfy a global constraint: the trace is fixed to be unity.

We have studied the distribution of the minimum eigenvalue in this fixed-trace Wishart ensemble. For the hard edge case (when two subsystems have the same size $M = N$); we have shown that the minimum eigenvalue distribution can be computed exactly for all N in all three interesting physical cases $\beta = 1$, $\beta = 2$, and $\beta = 4$.

What does this exact distribution of $\lambda_{\min}$ tell us about the entanglement entropy of the bipartite system in a pure state? We have seen before that if $\lambda_{\min}$ is close to its maximally allowed value $1/N$ (set by the unit trace constraint), then that configuration is maximally entangled since all the eigenvalues contribute equally to the composition of the state. A measure of how close the random state is to this maximally entangled state can be estimated by computing the net measure of $\lambda_{\min}$ in a small range close to $1/N$, e.g. by the cumulative probability $R_N(1/N - \epsilon) = \int_{1/N-\epsilon}^{1/N} P_N(x)\, dx$ where $\epsilon << 1/N$. From our exact calculation, we see that the pdf of $\lambda_{\min}$, in all three cases, approaches zero as $\lambda_{\min}$ approaches its maximum possible value $1/N$, $P_N(\lambda_{\min} = x, N) \sim (1 - Nx)^{\nu_\beta}$ where $\nu_1 = (N^2 + N - 4)/2$, $\nu_2 = N^2 - 2$ and $\nu_4 = 2N^2 - N - 2$. This shows that the 'closeness to maximal entropy' measure $R_N(1/N - \epsilon) \sim (\epsilon\, N)^{\nu_\beta + 1}$. For $\epsilon << 1/N$, this measure is evidently very small. It was argued before [Pag95], on the basis of the computation of the only the first moment of the von Neumann entropy (not the full distribution), that a random state is almost maximally entangled. Our result shows that the *probability* that a random state is maximally entangled is actually very small. The same conclusion was also deduced recently on the basis of the large N computation of the full distribution of the Renyi entropy [Nad10]. Thus, the lesson is that conclusions based just on the first moment, may sometimes be a bit misleading.

Here we have discussed only the hard edge $M = N$ case where the two subsystems have equal sizes. Our results are of relevance for small systems such as when each subsystem consists of an identical number of qubits. It would be interesting to extend these calculations to the cases when $M \neq N$. In particular, in the context of thermodynamic systems where, for instance, one of the subsystems is a heat bath, one needs to study the opposite limit $N << M$. It would be interesting to estimate the distribution of the minimum eigenvalue and other measures of entanglement in that limit.

Finally, we have restricted ourselves here to 'random pure' states where all pure states are sampled equally likely. This is the Haar measure. So far, we have not discussed dynamics, i.e. the temporal unitary evolution of the system. Under any unitary evolution that is ergodic over the space of all pure states, the Haar measure is the unique stationary measure of the unitary evolution. This ergodicity holds provided one does not have any strict conservation law. For instance, the uniform measure *over all pure states* will not hold under the standard microcanonical scenario where the composite system has a fixed total energy E (eigenvalue of the Hamiltonian $\hat{H}$ of the composite system). Only those pure states with total energy E, *and not all pure states*, will be sampled by the system under unitary evolution. An appropriate 'stationary' measure is then the microcanonical measure which is uniform over all pure states belonging to the fixed E manifold. For such a measure, one can again define the reduced density matrix of the subsystem A and its eigenvalues. It would be interesting to study the statistics of the bipartite entanglement (von Neumann or the Renyi entropy or the minimum eigenvalue λ_{min}) in this microcanonical setting with a fixed total energy E. For such systems, some recent results of a very general nature (that do not require detailed knowledge of the Hamiltonian of the system) have been derived [Pop06, Gol06, Rei07] which say that any pure state, drawn from the uniform measure on the constrained manifold, will almost surely be 'maximally entangled' i.e. very close to the maximally entangled (centroid) configuration $\lambda_i = 1/N$ for all i, provided $M >> N$, i.e. the environment (subsystem B) is much bigger than the system (subsystem A). It would be interesting to compute explicitly the distribution of this 'typicality' i.e. the distance between the pure state (drawn from a uniform measure over the constrained manifold) and the maximally entangled state for some systems with specific Hamiltonians.

Acknowledgements

The discussion in this chapter is based on my joint work [Maj08, Nad10, Viv07] with O. Bohigas, A. Lakshminarayan, C. Nadal, M. Vergassola, and P. Vivo. It is my pleasure to thank them. I am particularly grateful to C. Nadal for carefully reading and correcting the manuscript. I also thank my other collaborators on related subjects in random matrix theory: A. Comtet, D.S. Dean, A. Scardicchio, and G. Schehr. After this article was accepted, I came across a recent preprint by

Y. Chen, D.-Z. Liu, and D.-S. Zhou (J. Phys. A: Math. Theor. **43** (2010) 315303.) where the distribution of $\lambda_{\min}$ was computed for $M > N$ and for all β. Also, the average density of eigenvalues for all finite N and M but for $\beta = 1$ was computed in a recent preprint of P. Vivo. (J. Phys. A: Math. Theor. **43** (2010) 405206.)

Note added in proof

I would like to thank P. Vivo for pointing out a relation between the results presented here for the minimum eigenvalue distribution in the fixed trace case to the distribution of the so called 'Demmel condition number' for unconstrained Wishart ensemble derived a while ago by A. Edelman (Math. Comput. **58**, 185 (1992)) for $\beta = 1$ and $\beta = 2$. Essentially the minimum eigenvalue in the fixed trace Wishart ensemble turns out to have the same distribution as the inverse square of the 'Demmel condition number' in the unconstrained Wishart case. Thus the new results presented here for $\beta = 4$ for the minimum eigenvalue will then also provide the exact 'Demmel condition number' distribution in the unconstrained case for $\beta = 4$.

References

[Abr72] M. Abramowitz and I.A. Stegun, *Handbook of Mathematical Functions* (Dover, New York, 1972)

[Ake97] G. Akemann, Nucl. Phys. **B507** (1997) 475

[Ake08] G. Akemann and P. Vivo, JSTAT **P09002** (2008)

[Alt00] O. Alter *et. al.* Proc. Nat. Acad. Sci. USA **97** (2000) 10101

[Amb94] J. Ambjorn, Yu. Makeenko, and C.F. Kristjansen, Phys. Rev. **D50** (1994) 5193

[Ban02] J. N. Bandyopadhyay and A. Lakshminarayan, Phys. Rev. Lett. **89** (2002) 060402

[Ben01] G. Ben-Arous, A. Dembo, and A. Guionnet, Probab. Theory Relat. Fields **120** (2001) 1

[Ben96] C.H. Bennet et. al. Phys. Rev. A bf 53 (1996) 2046

[Boh84] O. Bohigas, M.-J. Giannoni, and C. Schmit, Phys. Rev. Lett. **52** (1984) 1

[Bou01] J.-P. Bouchaud and M. Potters, *Theory of Financial Risks* (Cambridge University Press, Cambridge, 2001)

[Bro65] B.V. Bronk, J. Math. Phys. **6** (1965) 228

[Bur04] Z. Burda and J. Jurkiewicz, Physica **A344** (2004) 67

[Cap06] V. Cappellini, H.-J Sommers, and K. Zyczkowski, Phys. Rev. A **74** (2006) 062322

[Che96] Y. Chen and S.M. Manning, J. Phys. A.: Math. Gen. **29** (1996) 7561

[Dea06] D.S. Dean and S.N. Majumdar, Phys. Rev. Lett. **97** (2006) 160201; Phys. Rev. E **77** (2008) 041108

[Ede88] A. Edelman, J. Matrix Anal. and Appl. **9** (1988) 543; Linear Algebra Appl. **159** (1991) 55

[Fac06] P. Facchi, G. Florio, and S. Pascazio, Phys. Rev. A **74** (2006) 042331

[Fac08] P. Facchi *et. al.* Phys. Rev. Lett. **101** (2008) 050502

[For93] P.J. Forrrester, Nucl. Phys. **B402** (1993) 709

[For94] P.J. Forrester, J. Math. Phys. **35** (1994) 2539

[Fuk90] K. Fukunaga, *Introduction to Statistical Pattern Recognition* (Elsevier, New York, 1990)

[Fyo97] Y.V. Fyodorov and H.-J. Sommers, J. Math. Phys. **38** (1997) 1918

[Gir07] O. Giraud, J. Phys. A.: Math. Theor. **40** (2007) 1053

[Gol06] S. Goldstein, J.L. Lebowitz, R. Tumulka, and N. Zanghi, Phys. Rev. Lett. **96** (2006) 050403

[Hay06] P. Hayden, D. W. Leung and A. Winter, Comm. Math. Phys. **265** (2006) 95

[Hol00] N. Holter *et. al.* Proc. Nat. Acad. Sci. USA **97** (2000) 8409

[Jam64] A.T. James, Ann. Math. Stat. **35** (1964) 475

[Joh00] K. Johansson, Comm. Math. Phys. **209** (2000) 437

[Joh01] I.M. Johnstone, Ann. Statist. **29** (2001) 295

[Kat03] M. Katori, H. Tanemura, T. Nagao, and N. Komatsuda, Phys. Rev. **E68** (2003) 021112

[Kaz09] P. Kazakopoulos et al., IEEE Trans. Inform. Th **57** (2011) 1984.

[Llo88] S. Lloyd and H. Pagels, Ann. Phys. (NY) **188** (1988) 186

[Maj08] S.N. Majumdar, O. Bohigas, and A. Lakshminarayan, J. Stat. Phys. **131** (2008) 33

[Maj09a] S.N. Majumdar and M. Vergassola, Phys. Rev. Lett. **102** (2009) 060601

[Maj09b] S.N. Majumdar, C. Nadal, A. Scardicchio, and P. Vivo, Phys. Rev. Lett. **103** (2009) 220603

[Mar67] V.A. Marcenko and L.A. Pastur, Math. USSR-Sb **1** (1967) 457

[Meh04] M.L. Mehta, *Random Matrices* (Academic Press, 3rd Edition, London, 2004)

[Nad09] C. Nadal and S.N. Majumdar, Phys. Rev. **E79** (2009) 061117

[Nad10] C. Nadal, S.N. Majumdar, and M. Vergassola, Phys. Rev. Lett. **104** (2010) 110501, see also J. Stat. Phys. **142** (2011) 403.

[Nag93] T. Nagao and K. Slevin, J. Math. Phys. **34** (1993) 2075

[Nag95] T. Nagao and P.J. Forrester, Nucl. Phys. **B435** (1995) 401

[Nov08] J. Novembre and M. Stephens, Nature Genetics **40** (2008) 646

[Osi08] V. Al Osipov and E. Kanzieper, Phys. Rev. Lett. **101**, 176804 (2008); Note that the results for the tails of the conductance pdf stated in this paper were errorneous and the correct results can be found in [Viv08] and [Viv10].

[Pag95] D. N. Page, Phys. Rev. Lett. **71** (1995) 1291

[Pas09] A. De Pasquale, P. Facchi, G. Parisi, S. Pascazio, and A. Scardicchio, Phys. Rev. A **81** (2010) 052324.

[Pat06] N. Patterson, A.L. Preis, and D. Reich, PLos Gentics **2** (2006) 2074

[Per93] A. Peres, *Quantum Theory: Concepts and Methods,* (Kluwer Academic Publishers, Dordrecht, 1993)

[Pop06] S. Popescu, A.J. Short and A. Winter, Nature Physics, **2** (2006) 754

[Pre88] R.W. Preisendorfer, *Principal Component Analysis in Meteorology and Oceanography* (Elsevier, New York, 1988)

[Rei07] P. Reimann, Phys. Rev. Lett. **99** (2007) 160404

[Ren70] A. Renyi, *Probability Theory* (North Holland, Amsterdam, 1970)

[Sch08] G. Schehr, S.N. Majumdar, A. Comtet, and J. Randon-Furling, Phys. Rev. Lett. **101** (2008) 150601

[Tra94] C. Tracy and H. Widom, Commun. Math. Phys. **159** (1994) 151; **177** (1996) 727

[Ver94a] J.J.M. Verbaarschot, Phys. Rev. Lett. **72** (1994) 2531 and references therein

[Ver94b] J.J.M. Verbaarschot, Nucl. Phys. **B426** (1994) 559

[Ved98] V. Vedral and M.B. Plenio, Phys. Rev. A **57** (1998) 1619

[Viv07] P. Vivo, S.N. Majumdar, and O. Bohigas, J. Phys. A **40** (2007) 4317

[Viv08] P. Vivo, S.N. Majumdar, and O. Bohigas, Phys. Rev. Lett. **101** (2008) 216809

[Viv10] P. Vivo, S.N. Majumdar, and O. Bohigas, Phys. Rev. B. **81** (2010) 104202

[Wil62] S.S. Wilks, *Mathematical Statistics* (John Wiley & Sons, New York, 1962)

[Wis28] J. Wishart, Biometrica **20** (1928) 32

[Zyc01] K. Zyczkowski and H-J. Sommers, J. Phys. A: Math. Gen. **34** (2001) 7111

[Zni07] M. Znidaric, J. Phys. A: Math. Theor. **40** (2007) F105

·38·

Random growth models

P. L. Ferrari and H. Spohn

Abstract

The link between a particular class of growth processes and random matrices was established in the now famous 1999 article of Baik, Deift, and Johansson on the length of the longest increasing subsequence of a random permutation [BDJ99]. During the past ten years, this connection has been worked out in detail and has led to an improved understanding of the large scale properties of one-dimensional growth models. The reader will find a commented list of references at the end. Our objective is to provide an introduction highlighting random matrices. From the outset it should be emphasized that this connection is fragile. Only certain aspects, and only for specific models, can the growth process be re-expressed in terms of partition functions also appearing in random matrix theory.

38.1 Growth models

A growth model is a stochastic evolution for a height function $h(x, t)$, x space, t time. For the one-dimensional models considered here, either $x \in \mathbb{R}$ or $x \in \mathbb{Z}$. We first define the TASEP[1] (totally asymmetric simple exclusion process) with parallel updating, for which $h, x, t \in \mathbb{Z}$. An admissible height function, h, has to satisfy $h(x+1) - h(x) = \pm 1$. Given $h(x, t)$ and x_* a local minimum of $h(x, t)$, one defines

$$h(x_*, t+1) = \begin{cases} h(x_*, t) + 2 & \text{with probability } 1-q, \ \ 0 \le q \le 1, \\ h(x_*, t) & \text{with probability } q \end{cases} \tag{38.1.1}$$

independently for all local minima, and $h(x, t+1) = h(x, t)$ otherwise, see Figure 38.1 (left). Note that if $h(\cdot, t)$ is admissible, so is $h(\cdot, t+1)$.

There are two limiting cases of interest. At x_* the waiting time for an increase by 2 has the geometric distribution $(1-q)q^n$, $n = 0, 1, \ldots$. Taking the $q \to 1$ limit and setting the time unit to $1-q$ one obtains the exponential distribution of mean 1. This is the time-continuous TASEP for which, counting from the moment of the first appearance, the heights at local minima are increased

[1] Here we use the height function representation. The standard particle representation consists in placing a particle at x if $h(x+1) - h(x) = -1$ and leaving empty if $h(x+1) - h(x) = 1$.

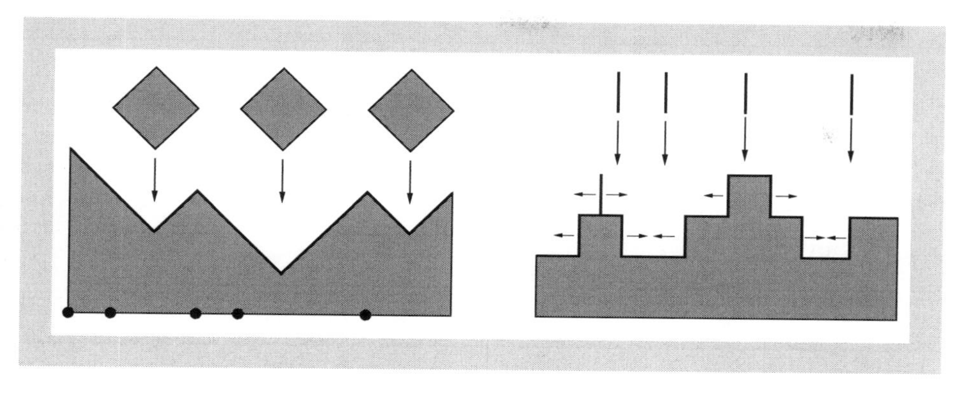

Fig. 38.1 Growth of TASEP interface (left) and continuous time PNG (right).

independently by 2 after an exponentially distributed waiting time. Thus h, $x \in \mathbb{Z}$ and $t \in \mathbb{R}$.

The second case is the limit of rare events (Poisson points), where the unit of space and time is $\sqrt{q}$ and one takes $q \to 0$. Then the lattice spacing and time become continuous. This limit, after a slightly different height representation (see Section 38.5 for more insights), results in the polynuclear growth model (PNG) for which $h \in \mathbb{Z}$, $x, t \in \mathbb{R}$. An admissible height function is piecewise constant with jump size ± 1, where an increase by 1 is called an up-step and a decrease by 1 a down-step. The dynamics is constructed from a space-time Poisson process of intensity 2 of nucleation events. $h(x, t)$ evolves deterministically through (1) up-steps move to the left with velocity -1, down-steps move to the right with velocity $+1$, (2) steps disappear upon coalescence, and (3) at points of the space-time Poisson process the height is increased by 1, thereby nucleating an adjacent pair of up-step and down-step. They then move symmetrically apart by the mechanism described under (1), see Figure 38.1 (right).

The TASEP and PNG have to be supplemented by initial conditions and, possibly by boundary conditions. For the former one roughly divides between macroscopically flat and curved. For the TASEP examples would be $h(x) = 0$ for x even and $h(x) = 1$ for x odd, which has slope zero, and $h(x + 1) - h(x)$ independent Bernoulli random variables with mean m, $|m| \leq 1$, which has slope m. An example for a curved initial condition is $h(0) = 0$, $h(x + 1) - h(x) = -1$ for $x = -1, -2, \ldots$, and $h(x + 1) - h(x) = 1$ for $x = 0, 1, \ldots$.

What are the quantities of interest? The most basic one is the macroscopic shape, which corresponds to a law of large numbers for

$$\frac{1}{t} h\big([yt], [st]\big) \tag{38.1.2}$$

in the limit $t \to \infty$ with $y \in \mathbb{R}$, $s \in \mathbb{R}$, and $[\cdot]$ denoting the integer part. From a statistical mechanics point of view the shape fluctuations are of prime concern.

For example, in the flat case the surface stays macroscopically flat and advances with constant velocity v. One would then like to understand the large scale limit of the fluctuations $\{h(x,t) - vt, (x,t) \in \mathbb{Z} \times \mathbb{Z}\}$. As will be discussed below, the excitement is triggered through non-classical scaling exponents and non-Gaussian limits.

In 1986 in a seminal paper Kardar, Parisi, and Zhang (KPZ) proposed the stochastic evolution equation [KPZ86]

$$\frac{\partial}{\partial t} h(x,t) = \lambda \Big(\frac{\partial}{\partial x} h(x,t) \Big)^2 + \nu_0 \frac{\partial^2}{\partial x^2} h(x,t) + \eta(x,t) \tag{38.1.3}$$

for which $h, x, t \in \mathbb{R}$. $\eta(x,t)$ is space-time white noise which models the deposition mechanism in a moving frame of reference. The nonlinearity reflects the slope-dependent growth velocity and the Laplacian with $\nu_0 > 0$ is a smoothing mechanism. To make the equation well defined one has to introduce either a suitable spatial discretization or a noise covariance $\langle \eta(x,t)\eta(x',t')\rangle = g(x-x')\delta(t-t')$ with $g(x) = g(-x)$ and supported close to 0. KPZ argue that, according to (38.1.3) with initial conditions $h(x,0) = 0$, the surface width increases as $t^{1/3}$, while lateral correlations increase as $t^{2/3}$. It is only through the connection to random matrix theory that universal probability distributions and scaling functions have become accessible.

Following TASEP, PNG, and KPZ as guiding examples it is easy to construct variations. For example, for the TASEP one could introduce evaporation through which heights at local maxima are decreased by 2. The deposition could be made to depend on neighbouring heights. Also generalizations to higher dimensions, $x \in \mathbb{R}^d$ or $x \in \mathbb{Z}^d$, are easily accomplished. For the KPZ equation the nonlinearity then reads $\lambda\big(\nabla_x h(x,t)\big)^2$ and the smoothening is $\nu_0 \Delta h(x,t)$. For the PNG model in $d = 2$, at a nucleation event one generates on the existing layer a new layer of height one consisting of a disk expanding at unit speed. None of these models seems to be directly connected to random matrices.

38.2 How do random matrices appear?

Let us consider the PNG model with the initial condition $h(x,0) = 0$ under the constraint that there are no nucleations outside the interval $[-t,t]$, $t \geq 0$, which is also referred to as a PNG droplet, since the typical shape for large times is a semicircle. We study the probability distribution $\mathbb{P}(\{h(0,t) \leq n\})$, which depends only on the nucleation events in the quadrant $\{(x,s) | \, |x| \leq s, s-t \leq x \leq t-s\,,\ 0 \leq s \leq t\}$. Let us denote by $\omega = (\omega^{(1)}, \ldots, \omega^{(n)})$ a set of nucleation events and by $h(0,t;\omega)$ the corresponding height. The order of the coordinates of the $\omega^{(j)}$s in the frame $\{x = \pm t\}$ naturally defines a permutation of n elements. It can be seen that $h(0,t;\omega)$ is simply the length of the longest increasing subsequence of that permutation. By the Poisson statistics of ω, the permutations are random and their length is Poisson distributed. By this reasoning, somewhat unexpectedly, one finds that $\mathbb{P}(\{h(0,t) \le n\})$ can be written as a matrix integral [BR01a]. Let $\mathcal{U}_n$ be the set of all unitaries on $\mathbb{C}^n$ and $\mathrm{d}U$ be the corresponding Haar measure. Then

$$\mathbb{P}(\{h(0,t) \le n\}) = e^{-t^2} \int_{\mathcal{U}_n} \mathrm{d}U \exp[t \operatorname{tr}(U + U^*)] . \tag{38.2.1}$$

(38.2.1) can also be expressed as a Fredholm determinant on $\ell_2 = \ell_2(\mathbb{Z})$. On ℓ_2 we define the linear operator B through

$$(Bf)(x) = -f(x+1) - f(x-1) + \frac{x}{t} f(x) \tag{38.2.2}$$

and denote by $P_{\le 0}$ the spectral projection onto $B \le 0$. Setting $\theta_n(x) = 1$ for $x > n$ and $\theta_n(x) = 0$ for $x \le n$, one has

$$\mathbb{P}(\{h(0,t) \le n\}) = \det(\mathbf{1} - \theta_n P_{\le 0} \theta_n) . \tag{38.2.3}$$

Such an expression is familiar from GUE random matrices. Let $\lambda_1 < \ldots < \lambda_N$ be the eigenvalues of an $N \times N$ GUE distributed random matrix. Then

$$\mathbb{P}(\{\lambda_N \le y\}) = \det(\mathbf{1} - \theta_y P_N \theta_y) . \tag{38.2.4}$$

Now the determinant is over $L^2(\mathbb{R}, \mathrm{d}y)$. If one sets $H = -\frac{1}{2}\frac{\mathrm{d}^2}{\mathrm{d}y^2} + \frac{1}{2N} y^2$, then P_N projects onto $H \le \sqrt{N}$.

For large N one has the asymptotics

$$\lambda_N \cong 2N + N^{1/3} \xi_2 \tag{38.2.5}$$

with ξ_2 a Tracy–Widom distributed random variable, that is,

$$\mathbb{P}(\{\xi_2 \le s\}) = F_2(s) := \det(\mathbf{1} - \chi_s K_{\mathrm{Ai}} \chi_s), \tag{38.2.6}$$

with det the Fredholm determinant on $L^2(\mathbb{R}, \mathrm{d}x)$, $\chi_s(x) = \mathbf{1}(x > s)$, and K_{Ai} is the Airy kernel (see (38.3.4) below). Hence it is not so surprising that for the height of the PNG model one obtains [PS00b]

$$h(0,t) = 2t + t^{1/3} \xi_2 \tag{38.2.7}$$

in the limit $t \to \infty$. In particular, the surface width increases as $t^{1/3}$ in accordance with the KPZ prediction. The law in (38.2.7) is expected to be universal and to hold whenever the macroscopic profile at the reference point, $x = 0$ in our example, is curved. Indeed for the PNG droplet $h(x,t) \cong 2t\sqrt{1 - (x/t)^2}$ for large t, $|x| \le t$.

One may wonder whether (38.2.7) should be regarded as an accident or whether there is a deeper reason. In the latter case, further height statistics might also be representable through matrix integrals.

38.3 Multi-matrix models and line ensembles

For curved initial data the link to random matrix theory can be understood from underlying line ensembles. They differ from case to case but have the property that the top line has the same statistics in the scaling limit. We first turn to random matrices by introducing matrix-valued diffusion processes.

Let $B(t)$ be GUE Brownian motion, to say $B(t)$ is an $N \times N$ Hermitian matrix such that, for every $f \in \mathbb{C}^N$, $t \mapsto \langle f, B(t) f\rangle$ is standard Brownian motion with variance $\langle f, f\rangle^2 \min(t, t')$ and such that for every unitary U it holds

$$U B(t) U^* = B(t) \tag{38.3.1}$$

in distribution. The $N \times N$ matrix-valued diffusion, $A(t)$, is defined through the stochastic differential equation

$$\mathrm{d}A(t) = -V'(A(t))\mathrm{d}t + \mathrm{d}B(t), \quad A(0) = A, \tag{38.3.2}$$

with potential $V : \mathbb{R} \to \mathbb{R}$. We assume $A = A^*$, then also $A(t) = A(t)^*$ for all $t \geq 0$. It can be shown that eigenvalues of $A(t)$ do not cross with probability 1 and we order them as $\lambda_1(t) < \ldots < \lambda_N(t)$. $t \mapsto \big(\lambda_1(t), \ldots, \lambda_N(t)\big)$ is the line ensemble associated to (38.3.2).

In our context the largest eigenvalue, $\lambda_N(t)$ is of most interest. For $N \to \infty$ its statistics is expected to be independent of the choice of V. We first define the limit process, the Airy$_2$ process $\mathcal{A}_2(t)$, through its finite-dimensional distributions. Let

$$H_{\mathrm{Ai}} = -\frac{\mathrm{d}^2}{\mathrm{d}y^2} + y \tag{38.3.3}$$

as a self-adjoint operator on $L^2(\mathbb{R}, \mathrm{d}y)$. The Airy operator has $\mathbb{R}$ as spectrum with the Airy function Ai as generalized eigenfunctions, $H_{\mathrm{Ai}}\mathrm{Ai}(y-\lambda) = \lambda\, \mathrm{Ai}(y-\lambda)$. In particular the projection onto $\{H_{\mathrm{Ai}} \leq 0\}$ is given by the Airy kernel

$$K_{\mathrm{Ai}}(y, y') = \int_0^\infty \mathrm{d}\lambda \mathrm{Ai}(y+\lambda)\mathrm{Ai}(y'+\lambda)\,. \tag{38.3.4}$$

The associated extended integral kernel is defined through

$$K_{\mathcal{A}_2}(y, \tau; y', \tau') = -(e^{-(\tau'-\tau)H_{\mathrm{Ai}}})(y, y')\mathbf{1}(\tau' > \tau) + (e^{\tau H_{\mathrm{Ai}}} K_{\mathrm{Ai}} e^{-\tau' H_{\mathrm{Ai}}})(y, y'). \tag{38.3.5}$$

Then, the m-th marginal for $\mathcal{A}_2(t)$ at times $t_1 < t_2 < \ldots < t_m$ is expressed as a determinant on $L^2(\mathbb{R} \times \{t_1, \ldots, t_m\})$ according to [PS02b]

$$\mathbb{P}\big(\mathcal{A}_2(t_1) \leq s_1, \ldots, \mathcal{A}_2(t_m) \leq s_m\big) = \det(\mathbf{1} - \chi_s K_{\mathcal{A}_2} \chi_s), \tag{38.3.6}$$

with $\chi_s(x, t_i) = \mathbf{1}(x > s_i)$. $t \mapsto \mathcal{A}_2(t)$ is a stationary process with continuous sample paths and covariance $g_2(t) = \mathrm{Cov}\big(\mathcal{A}_2(0), \mathcal{A}_2(t)\big) = \mathrm{Var}(\xi_2) - |t| + \mathcal{O}(t^2)$ for $t \to 0$ and $g_2(t) = t^{-2} + \mathcal{O}(t^{-4})$ for $|t| \to \infty$.

The scaling limit for $\mathcal{A}_2(t)$ can be most easily constructed by two slightly different procedures. The first one starts from the stationary Ornstein-Uhlenbeck process $A^{\text{OU}}(t)$ in (38.3.2), which has the potential $V(x) = x^2/2N$. Its distribution at a single time is GUE, to say $Z_N^{-1}\exp[-\frac{1}{2N}\text{tr}A^2]\mathrm{d}A$, where the factor $1/N$ results from the condition that the eigenvalue density in the bulk is of order 1. Let $\lambda_N^{\text{OU}}(t)$ be the largest eigenvalue of $A^{\text{OU}}(t)$. Then

$$\lim_{N\to\infty} N^{-1/3}\big(\lambda_N^{\text{OU}}(2N^{2/3}t) - 2N\big) = \mathcal{A}_2(t) \qquad (38.3.7)$$

in the sense of convergence of finite-dimensional distributions. The $N^{2/3}$ scaling means that locally $\lambda_N^{\text{OU}}(t)$ looks like Brownian motion. On the other side the global behaviour is confined.

The marginal of the stationary Ornstein-Uhlenbeck process for two time instants is the familiar 2-matrix model [EM98, NF98]. Setting $A_1 = A^{\text{OU}}(0)$, $A_2 = A^{\text{OU}}(t)$, $t > 0$, the joint distribution is given by

$$\frac{1}{Z_N^2}\exp\Big(-\frac{1}{2N(1-q^2)}\text{tr}[A_1^2 + A_2^2 - 2q\,A_1A_2]\Big)\mathrm{d}A_1\mathrm{d}A_2, \quad q = \exp(-t/2N). \qquad (38.3.8)$$

A somewhat different construction uses the Brownian bridge defined by (38.3.2) with $V = 0$ and $A^{\text{BB}}(-T) = A^{\text{BB}}(T) = 0$, that is

$$A^{\text{BB}}(t) = B(T+t) - \frac{T+t}{2T}B(2T), \quad |t| \le T. \qquad (38.3.9)$$

The eigenvalues $t \mapsto \big(\lambda_1^{\text{BB}}(t), \dots, \lambda_N^{\text{BB}}(t)\big)$ is the Brownian bridge line ensemble. Its largest eigenvalue, $\lambda_N^{\text{BB}}(t)$, has the scaling limit, for $T = 2N$,

$$\lim_{N\to\infty} N^{-1/3}\big(\lambda_N^{\text{BB}}(2N^{2/3}t) - 2N\big) + t^2 = \mathcal{A}_2(t) \qquad (38.3.10)$$

in the sense of finite-dimensional distributions. Note that $\lambda_N^{\text{BB}}(t)$ is curved on the macroscopic scale resulting in the displacement by $-t^2$. But with this subtraction the limit is stationary.

To prove the limits (38.3.7) and (38.3.10) one uses in a central way that the underlying line ensemble are determinantal. For the Brownian bridge this can be seen by the following construction. We consider N independent standard Brownian bridges over $[-T, T]$, $b_j^{\text{BB}}(t)$, $j = 1, \dots, N$, $b_j^{\text{BB}}(\pm T) = 0$ and condition them on non-crossing for $|t| < T$. The resulting line ensemble has the same statistics as $\lambda_j^{\text{BB}}(t)$, $|t| \le T$, $j = 1, \dots, N$, which hence is determinantal.

The TASEP, and its limits, also have an underlying line ensemble, which qualitatively resemble $\{\lambda_j^{\text{BB}}(t),\ j = 1, \dots, N\}$. The construction of the line ensemble is not difficult, but somewhat hidden. Because of lack of space we explain only the line ensemble for the PNG droplet. As before t is the growth time and x is space which takes the role of t from above. The top line is $\lambda_0(x, t) = h(x, t)$, h the PNG droplet of Section 38.1. Initially we add the extra lines $\lambda_j(x, 0) = j$, $j = -1, -2, \ldots$. The motion of these lines is completely determined by $h(x, t)$ through the following simple rules: (1) and (2) from above are in force for all lines $\lambda_j(x, t)$, $j = 0, -1, \ldots$. (3) holds only for $\lambda_0(x, t) = h(x, t)$. (4) The annihilation of a pair of an adjacent down-step and up-step at line j is carried out and copied instantaneously as a nucleation event to line $j - 1$.

We let the dynamics run up to time t. The line ensemble at time t is $\{\lambda_j(x, t), |x| \leq t, j = 0, -1, \ldots\}$. Note that $\lambda_j(\pm t, t) = j$. Also, for a given realization of $h(x, t)$, there is an index j_0 such that for $j < j_0$ it holds $\lambda_j(x, t) = j$ for all x, $|x| \leq t$. The crucial point of the construction is to have the statistics of the line ensemble determinantal, a property shared by the Brownian bridge over $[-t, t]$, $\lambda_j^{\mathrm{BB}}(x)$, $|x| \leq t$. The multi-line PNG droplet allows for a construction similar to the Brownian bridge line ensemble. We consider a family of independent, rate 1, time-continuous, symmetric nearest neighbour random walks on $\mathbb{Z}$, $r_j(x)$, $j = 0, -1, \ldots$. The j-th random walk is conditioned on $r_j(\pm t) = j$ and the whole family is conditioned on non-crossing. The resulting line ensemble has the same statistics as the PNG line ensemble $\{\lambda_j(x, t), j = 0, -1, \ldots\}$, which hence is determinantal.

In the scaling limit for $x = \mathcal{O}(t^{2/3})$ the top line $\lambda_0(x, t)$ is displaced by $2t$ and $t^{1/3}$ away from $\lambda_{-1}(x, t)$. Similarly $\lambda_N^{\mathrm{BB}}(x)$ for $x = \mathcal{O}(N^{2/3})$ is displaced by $2N$ and order $N^{1/3}$ apart from $\lambda_{N-1}^{\mathrm{BB}}(x)$. On this scale the difference between random walk and Brownian motion disappears but the non-crossing constraint persists. Thus it is no longer a surprise that

$$\lim_{t\to\infty} t^{-1/3}\big(h(t^{2/3}x, t) - 2t\big) + x^2 = \mathcal{A}_2(x)\,, \quad x \in \mathbb{R}\,, \tag{38.3.11}$$

in the sense of finite dimensional distributions [PS02b].

To summarize: for curved initial data the spatial statistics for large t is identical to the family of largest eigenvalues in a GUE multi-matrix model.

38.4 Flat initial conditions

Given the unexpected connection between the PNG model and GUE multi-matrices, a natural question is whether such a correspondence holds also for other symmetry classes of random matrices. The answer is affirmative, but with unexpected twists. Consider the flat PNG model with $h(x, 0) = 0$ and nucleation events in the whole upper half plane $\{(x, t), x \in \mathbb{R}, t \geq 0\}$. The removal of the spatial restriction of nucleation events leads to the problem of the longest increasing subsequence of a random permutation with involution [BR01b, PS00b]. The limit shape will be flat (straight) and in the limit $t \to \infty$ one obtains

$$h(0, t) = 2t + \xi_1 t^{1/3}\,, \tag{38.4.1}$$

where

$$\mathbb{P}(\xi_1 \leq s) = F_1(2^{-2/3}s). \tag{38.4.2}$$

The distribution function F_1 is the Tracy–Widom distribution for the largest GOE eigenvalue.

As before, we can construct the line ensemble and ask if the link to Brownian motion on GOE matrices persists. Firstly, let us compare the line ensembles at fixed position for flat PNG and at fixed time for GOE Brownian motions. In the large time (resp. matrix dimension) limit, the edges of these point processes converge to the same object: a Pfaffian point process with 2×2 kernel [Fer04]. It seems then plausible to conjecture that the two line ensembles also have the same scaling limit, i.e. the surface process for flat PNG and for the largest eigenvalue of GOE Brownian motions should coincide. Since the covariance for the flat PNG has been computed exactly in the scaling limit, one can compare with simulation results from GOE multi-matrices. The evidence strongly disfavours the conjecture [BFP08].

The process describing the largest eigenvalue of GOE multi-matrices is still unknown, while the limit process of the flat PNG interface is known [BFS08] and called the Airy$_1$ process, $\mathcal{A}_1$,

$$\lim_{t\to\infty} t^{-1/3}\big(h(t^{2/3}x, t) - 2t\big) = 2^{1/3}\mathcal{A}_1(2^{-2/3}x). \tag{38.4.3}$$

Its m-point distribution is given in terms of a Fredholm determinant of the following kernel. Let $B(y, y') = \mathrm{Ai}(y + y')$, $H_1 = -\frac{\mathrm{d}}{\mathrm{d}y^2}$. Then,

$$K_{\mathcal{A}_1}(y, \tau; y', \tau') = -(e^{-(\tau'-\tau)H_1})(y, y')\mathbf{1}(\tau' > \tau) + (e^{\tau H_1} B e^{-\tau' H_1})(y, y') \tag{38.4.4}$$

and, as for the Airy$_2$ process, the m-th marginal for $\mathcal{A}_1(t)$ at times $t_1 < t_2 < \ldots < t_m$ is expressed through a determinant on $L^2(\mathbb{R} \times \{t_1, \ldots, t_m\})$ according to

$$\mathbb{P}\big(\mathcal{A}_1(t_1) \leq s_1, \ldots, \mathcal{A}_1(t_m) \leq s_m\big) = \det(\mathbf{1} - \chi_s K_{\mathcal{A}_1}\chi_s), \tag{38.4.5}$$

with $\chi_s(x, t_i) = \mathbf{1}(x > s_i)$ [Sas05, BFPS07]. The Airy$_1$ process is a stationary process with covariance $g_1(t) = \mathrm{Cov}\big(\mathcal{A}_1(0), \mathcal{A}_1(t)\big) = \mathrm{Var}(\xi_1) - |t| + \mathcal{O}(t^2)$ for $t \to 0$ and $g_1(t) \to 0$ super-exponentially fast as $|t| \to \infty$.

The Airy$_1$ process is obtained using an approach different from the PNG line ensemble, but still with an algebraic structure encountered also in random matrices (in the the GUE-minor process [JN06, OR06]). We explain the mathematical structure using the continuous time TASEP as a model, since the formulae are the simplest. For a while we use the standard TASEP representation in terms of particles. One starts with a formula by Schütz [Sch97] for the transition probability of the TASEP particles from generic positions. Consider the system of N particles with positions $x_1(t) > x_2(t) > \ldots > x_N(t)$

and let $G(x_1, \ldots, x_N; t) = \mathbb{P}(x_1(t) = x_1, \ldots, x_N(t) = x_N | x_1(0) = y_1, \ldots, x_N(0) = y_N)$. Then

$$G(x_1, \ldots, x_N; t) = \det\left(F_{i-j}(x_{N+1-i} - y_{N+1-j}, t)\right)_{1 \le i,j \le N} \tag{38.4.6}$$

with

$$F_n(x, t) = \frac{1}{2\pi i} \oint_{|w|>1} dw \frac{e^{t(w-1)}}{w^{x-n+1}(w-1)^n}. \tag{38.4.7}$$

The function F_n satisfies the relation

$$F_n(x, t) = \sum_{y \ge x} F_{n-1}(y, t)\,. \tag{38.4.8}$$

The key observation is that (38.4.6) can be written as the marginal of a determinantal line ensemble, i.e. of a measure which is a product of determinants [Sas05]. The 'lines' are denoted by x_i^n with time index n, $1 \le n \le N$, and space index i, $1 \le i \le n$. We set $x_1^n = x_n$. Then

$$G(x_1, \ldots, x_N; t) = \sum_{x_i^n, 2 \le i \le n \le N} \left(\prod_{n=1}^{N-1} \det(\phi_n(x_i^n, x_j^{n+1}))_{i,j=1}^n\right) \det(\Psi_{N-i}^N(x_j^N))_{i,j=1}^N \tag{38.4.9}$$

with $\Psi_{N-i}^N(x) = F_{-i+1}(x - y_{N+1-i}, t)$, $\phi_n(x, y) = \mathbf{1}(x > y)$ and $\phi_n(x_{n+1}^n, y) = 1$ (here x_{n+1}^n plays the role of a virtual variable). The line ensembles for the PNG droplet and GUE-valued Brownian motion have the same determinantal structure. However in (38.4.9) the determinants are of increasing sizes which requires us to introduce the virtual variables x_{n+1}^n. However, from the algebraic perspective the two cases can be treated in a similar way. As a result, the measure in (38.4.9) is determinantal (for instance for any fixed initial conditions, but not only) and has a defining kernel dependent on $y_1, \ldots, y_N$. The distribution of the positions of TASEP particles are then given by a Fredholm determinant of the kernel. To have flat initial conditions, one sets $y_i = N - 2i$, takes first the $N \to \infty$ limit, and then analyses the system in the large time limit to get the Airy$_1$ process defined above.

A few remarks:

(1) For general initial conditions (for instance for flat initial conditions), the measure on $\{x_i^n\}$ is not positive, but some projections, like on the TASEP particles $\{x_1^n\}$, define a probability measure.

(2) The method can be applied also to step initial conditions ($x_k(0) = -k$, $k = 1, 2, \ldots$) and one obtains the Airy$_2$ process. In this case, the measure on $\{x_i^n\}$ is a probability measure.

(3) In random matrices a measure which is the product of determinants of increasing size occurs too, for instance in the GUE-minor process [JN06, OR06].

38.5 Growth models and last passage percolation

For the KPZ equation we carry out the Cole-Hopf transformation $Z(x, t) = \exp(-\lambda h(x, t)/\nu_0)$ with the result

$$\frac{\partial}{\partial t} Z(x, t) = -\left(-\nu_0 \frac{\partial^2}{\partial x^2} + \frac{\lambda}{\nu_0}\eta(x, t)\right) Z(x, t), \tag{38.5.1}$$

which is a diffusion equation with random potential. Using the Feynman-Kac formula, (38.5.1) corresponds to Brownian motion paths, x_t, with weight

$$\exp\left(-\frac{\lambda}{\nu_0}\int_0^t \mathrm{d}s\, \eta(x_s, s)\right). \tag{38.5.2}$$

In physics this problem is known as a directed polymer in a random potential, while in probability theory one uses directed first/last passage percolation. The spirit of the somewhat formal expression (38.5.2) persists for discrete growth processes. For example, for the PNG droplet we fix a realization, ω, of the nucleation events, which then determines the height $h(0, t; \omega)$ according to the PNG rules. We now draw a piecewise linear path with local slope m, $|m| < 1$, which starts at $(0, 0)$, ends at $(0, t)$, and changes direction only at the points of ω. Let $L(t; \omega)$ be the maximal number of Poisson points collected when varying over allowed paths. Then $h(0, t; \omega) = L(t; \omega)$. So to speak, the random potential from (38.5.2) is spiked at the Poisson points.

In this section we explain the connection between growth models and (directed) last passage percolation. For simplicity, we first consider the case leading to a discrete time PNG droplet, although directed percolation can be defined for general passage time distributions. Other geometries like flat growth are discussed later.

Let $\omega(i, j)$, $i, j \geq 1$, be independent random variables with geometric distribution of parameter q, i.e. $\mathbb{P}(\omega(i, j) = k) = (1 - q)q^k$, $k \geq 0$. An up-right path π from $(1, 1)$ to (n, m) is a sequence of points $(i_\ell, j_\ell)_{\ell=1}^{n+m-1}$ with $(i_{\ell+1}, j_{\ell+1}) - (i_\ell, j_\ell) \in \{(1, 0), (0, 1)\}$. The last passage time from $(1, 1)$ to (n, m) is defined by

$$G(n, m) = \max_{\pi:(1,1)\to(n,m)} \sum_{(i,j)\in\pi} \omega(i, j). \tag{38.5.3}$$

The connection between directed percolation and different growth models is obtained by appropriate projections of the three-dimensional graph of G. Let us see how this works.

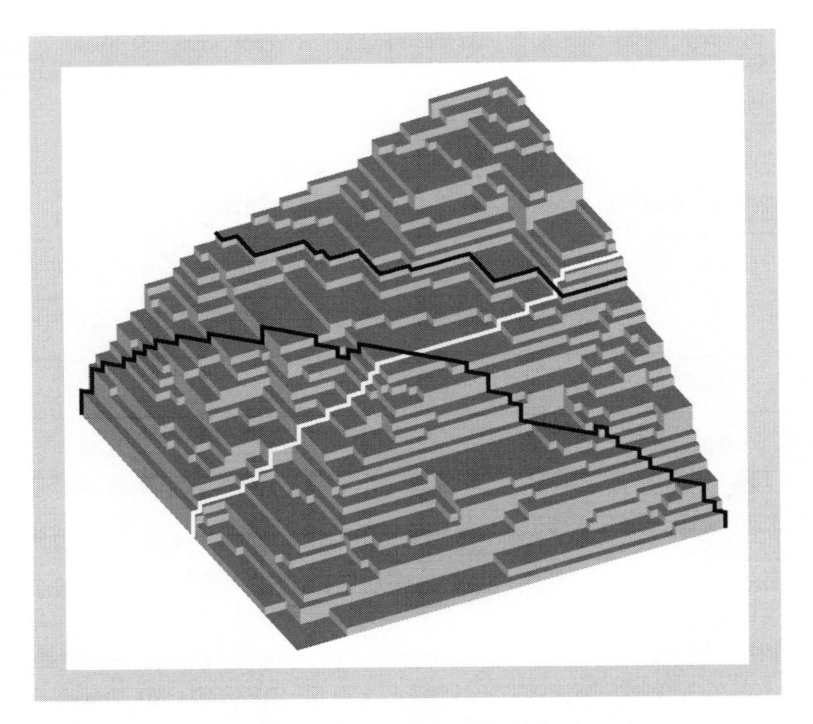

Fig. 38.2 Directed percolation and growth models.

Let time t be defined by $t = i + j - 1$ and position by $x = i - j$. Then, the connection between the height function of the discrete time PNG and the last passage time G is simply [Joh03]

$$h(x, t) = G(i, j). \tag{38.5.4}$$

Thus, the discrete PNG droplet is nothing other than the time-slicing along the $i + j = t$ directions, see Figure 38.2.

We can however also use a different slicing, at constant $\tau = G$, to obtain the TASEP at time τ with step initial conditions. For simplicity, consider $\omega(i, j)$ to be exponentially distributed with mean one. Then $\omega(n, m)$ is the waiting time for particle m to do its nth jump. Hence, $G(n, m)$ is the time when particle m arrives at $-m + n$, i.e.

$$\mathbb{P}(G(n, m) \leq \tau) = \mathbb{P}(x_m(\tau) \geq -m + n). \tag{38.5.5}$$

From the point of view of the TASEP, there is another interesting cut, namely at fixed $j = n$. This corresponds to looking at the evolution of the position of a given (tagged) particle, $x_n(\tau)$.

A few observations:

(1) Geometrically distributed random passage times correspond to discrete time TASEP with sequential update.

(2) The discrete time TASEP with parallel update is obtained by replacing $\omega(i, j)$ by $\omega(i, j) + 1$.

The link between last passage percolation and growth holds also for general initial conditions. In (38.5.3) the optimization problem is called point-to-point, since both $(1, 1)$ and (n, m) are fixed. We can generalize the model by allowing $\omega(i, j)$ to be defined on $(i, j) \in \mathbb{Z}^2$ and not only for $i, j \geq 1$. Consider the line $L = \{i + j = 2\}$ and the following point-to-line maximization problem:

$$G_L(n, m) = \max_{\pi: L \to (n,m)} \sum_{(i,j) \in \pi} \omega(i, j). \tag{38.5.6}$$

Then, the relation to the discrete time PNG droplet, namely $h(x, t) = G_L(i, j)$ still holds but this time h is the height obtained from flat initial conditions. For the TASEP, this means to have at time $\tau = 0$ the particles occupying $2\mathbb{Z}$. Also random initial conditions fit into the picture, this time one has to optimize over end-points which are located on a random line.

In the appropriate scaling limit, for large time/particle number one obtains the Airy$_2$ (resp. the Airy$_1$) process for all these cases. One might wonder why the process seems not to depend on the chosen cut. In fact, this is not completely true. Indeed, consider for instance the PNG droplet and ask the question of joint correlations of $h(x, t)$ in space-time. We have seen that for large t the correlation length scales as $t^{2/3}$. However, along the rays leaving from $(x, t) = (0, 0)$ the height function decorrelates on a much longer time scale, of order t. These slow decorrelation directions depend on the initial conditions. For instance, for flat PNG they are parallel to the time axis. More generally, they coincide with the characteristics of the macroscopic surface evolution. Consequently, except along the slow directions, on the $t^{2/3}$ scale one will always see the Airy processes.

38.6 Growth models and random tiling

In the previous section we explained how different growth models (TASEP and PNG) and observables (TASEP at fixed time or tagged particle motion) can be viewed as different projections of a single three-dimensional object. A similar unifying approach exists also for some growth models and random tiling problems: there is a $2 + 1$ dimensional surface whose projection to one less dimension in space (resp. time) leads to growth in 1+1 dimensions (resp. random tiling in two dimensions) [BF08a]. To explain the idea, we consider the dynamical model connected to the continuous time TASEP with step initial conditions, being one of the simplest to define.

In Section 38.4 we encountered a measure on set of variables $S_N = \{x_i^n, 1 \leq i \leq n \leq N\}$, see (38.4.9). The product of determinants of the ϕ_ns

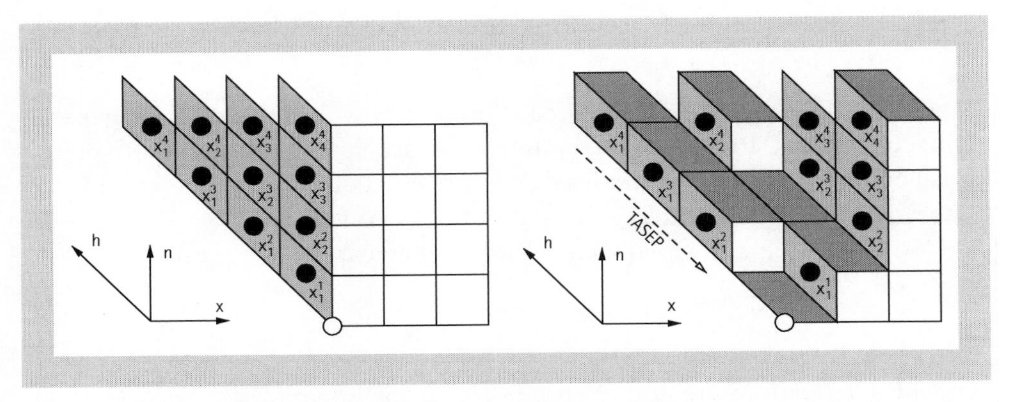

Fig. 38.3 Illustration of the particle system and its corresponding lozenge tiling. The initial condition is on the left.

implies that the measure is nonzero, if the variables x_i^n belong to the set S_N^{int} defined through an interlacing condition,

$$S_N^{\text{int}} = \{x_i^n \in S_N \mid x_i^{n+1} < x_i^n \leq x_{i+1}^{n+1}\}. \qquad (38.6.1)$$

Moreover, for TASEP with step initial conditions, the measure on S_N^{int} is a probability measure, so that we can interpret the variable x_i^n as the position of the particle indexed by (i, n). Also, the step initial condition, $x_n(t = 0) \equiv x_1^n(0) = -n$, implies that $x_i^n(0) = i - n - 1$, see Figure 38.3 (left).

Then, the dynamics of the TASEP induces a dynamics on the particles in S_N^{int} as follows. Particles independently try to jump on their right with rate one, but they might be blocked or pushed by others. When particle x_k^n attempts to jump:

(1) it jumps if $x_i^n < x_i^{n-1}$, otherwise it is blocked (the usual TASEP dynamics between particles with same lower index),
(2) if it jumps, it pushes by one all other particles with index $(i + \ell, n + \ell)$, $\ell \geq 1$, which are at the same position (so to remain in the set S_N^{int}).

For example, consider the particles of Figure 38.3 (right). Then, if in this state of the system particle (1, 3) tries to jump, it is blocked by the particle (1, 2), while if particle (2, 2) jumps, then also (3, 3) and (4, 4) will move by one unit at the same time.

It is clear that the projection of the particle system onto $\{x_1^n, 1 \leq n \leq N\}$ reduces to the TASEP dynamics in continuous time and step initial conditions, this projection being the sum in (38.4.9).

The system of particles can also be represented as a tiling using three different lozenges as indicated in Figure 38.3. The initial condition corresponds to a perfectly ordered tiling and the dynamics on particles reflects a corresponding dynamics of the random tiling. Thus, the projection of the model to fixed time reduces to a random tiling problem. In the same spirit, the shuffling

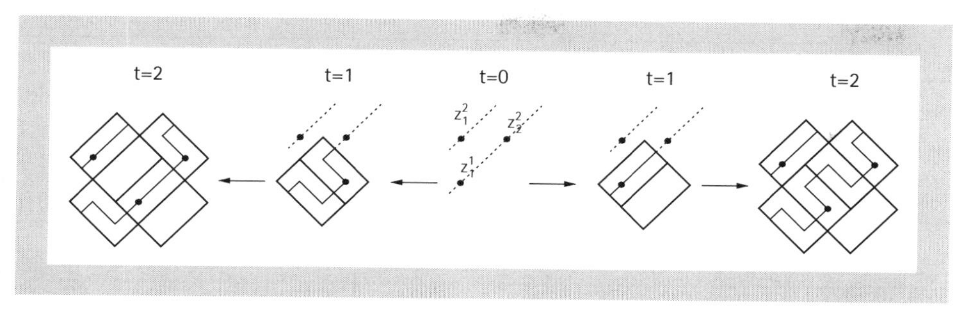

Fig. 38.4 Two examples of configurations at time $t = 2$ obtained by the particle dynamics and its associated Aztec diamonds. The line ensembles are also drawn.

algorithm of the Aztec diamond falls into place. This time the interlacing is $S_{\text{Aztec}} = \{z_i^n \mid z_i^{n+1} \leq z_i^n \leq z_{i+1}^{n+1}\}$ and the dynamics is on discrete time, with particles with index n not allowed to move before time $t = n$. As before, particle (i, n) can be blocked by $(i, n-1)$. The pushing occurs in the following way: particle (i, n) is forced to move if it stays at the same position of $(i-1, n-1)$, i.e. if it would generate a violation due to the possible jump of particle $(i-1, n-1)$. Then at time t all particles which are not blocked or forced to move, jump independently with probability q. As explained in detail in [Nor10] this particle dynamics is equivalent to the shuffling algorithm of the Aztec diamond (take $q = 1/2$ for uniform weight). In Figure 38.4 we illustrate the rules with a small-sized example of two steps. There we also draw a set of lines, which come from a non-intersecting line ensemble similar to the ones of the PNG model and the matrix-valued Brownian motions described in Section 38.3.

On the other hand, let $x_i^n := z_i^n - n$. Then, the dynamics of the shuffling algorithm projected onto particles x_1^n, $n \geq 1$, is nothing other than the discrete time TASEP with parallel update and step initial condition! Once again, we have a $1+1$ dimensional growth and a two-dimensional tiling model which are different projections of the same $2+1$-dimensional dynamical object.

38.7 A guide to the literature

There is a substantial body of literature and only a rather partial list is given here. Some of the techniques are discussed more completely in other chapters: determinantal point processes are covered in Chapter 11, two-matrix models in Chapters 15 and 16, and random permutations in Chapter 25. The guideline is ordered according to physical model under study.

PNG model A wide variety of growth processes, including PNG, are explained in [Mea98]. The direct link to the largest increasing subsequences of a random permutation and to the unitary matrix integral of [PS90] is noted in [PS00b,

PS00a]. The convergence to the Airy$_2$ process is worked out in [PS02b] and the stationary case in [BR00, PS04]. For flat initial conditions, the height at a single space-point is studied in [BR01a, BR01b] and for the ensemble of top lines in [Fer04]. The extension to many space-points is accomplished by [BFS08]. Determinantal space-time processes for the discrete time PNG model are discussed by [Joh03]. External source at the origin is studied in [IS04b] for the full line and in [IS04a] for the half-line.

Asymmetric simple exclusion (ASEP) As a physical model reversible lattice gases, in particular the simple exclusion process, were introduced by Kawasaki [Kaw72] and its asymmetric version by Spitzer [Spi70]. We refer to Liggett [Lig99] for a comprehensive coverage from the probabilistic side. The hydrodynamic limit is treated in [Spo91] and [KL99], for example. There is a very substantial body on large deviations with Derrida as one of the driving forces, see [Der07] for a recent review. Schütz [Sch97] discovered a determinantal-like formula for the transition probability for a finite number of particles on $\mathbb{Z}$. TASEP step initial conditions are studied in the seminal paper by Johansson [Joh00]. The random matrix representation of [Joh00] may also be obtained by the Schütz formula [NS04, RS05]. The convergence to the Airy$_2$ process in a discrete time setting [Joh05]. General step initial conditions are covered by [PS02a] and the extended process in [BFP10]. In [FS06b] the scaling limit of the stationary two-point function is proved. Periodic intial conditions were first studied by Sasamoto [Sas05] and widely extended in [BFPS07]. A further approach comes from the Bethe ansatz which is used in [GS92] to determine the spectral gap of the generator. In a spectacular analytic tour de force Tracy and Widom develop the Bethe ansatz for the transition probability and thereby extend the Johansson result to the PASEP [TW09, TW08].

2D tiling (statics) The connection between growth and tiling was first understood in the context of the Aztec diamond [JPS98], who prove the arctic circle theorem. Because of the specific boundary conditions imposed, for typical tilings there is an ordered zone spatially coexisting with a disordered zone. In the disordered zone the large scale statistics is expected to be governed by a massless Gaussian field, while the line separating the two zones has the Airy$_2$ process statistics.

a) Aztec diamond. The zone boundary is studied by [JPS98] and by [JN06]. Local dimer statistics are investigated in [CEP96]. Refined details are the large scale Gaussian statistics in the disordered zone [Ken00], the edge statistics [Joh05], and the statistics close to a point touching the boundary [JN06].
b) Ising corner. The Ising corner corresponds to a lozenge tiling under the constraint of a fixed volume below the thereby defined surface. The largest scale information is the macroscopic shape and large deviations [CK01].

The determinantal structure is noted in [OR03]. This can be used to study the edge statistics [FS03, FPS04]. More general boundary conditions (skew plane partitions) leads to a wide variety of macroscopic shapes [OR07].

c) Six-vertex model with domain wall boundary conditions, as introduced in [KZJ00]. The free energy including prefactors is available [BF06]. A numerical study can be found in [AR05]. The mapping to the Aztec diamond, on the free fermion line, is explained in [FS06a].

d) Kasteleyn domino tilings. Kasteleyn [Kas63] noted that Pfaffian methods work for a general class of lattices. Macroscopic shapes are obtained by [KOS06] with surprising connections to algebraic geometry. Gaussian fluctuations are proved in [Ken08].

2D tiling (dynamics) see Section 38.6. The shuffling algorithm for the Aztec diamond is studied in [EKLP92, Pro03, Nor10]. The pushASEP is introduced by [BF08b] and anisotropic growth models are investigated in [BF08a], see also [PS97] for the Gates-Westcott model. A similar intertwining structure appears for Dyson's Brownian motions [War07].

Directed last passage percolation 'Directed' refers to the constraint of not being allowed to turn back in time. In the physics literature one speaks of a directed polymer in a random medium. Shape theorems are proved, e.g. in [Kes86]. While the issue of fluctuations had been repeatedly raised, sharp results had to wait for [Joh00] and [PS02b]. Growths models naturally lead to either point-to-point, point-to-line, and point-to-random-line last passage percolation. For certain models one has to further impose boundary conditions and/or symmetry conditions for the allowed domain. In (38.5.3) one takes the max, thus zero temperature. There are interesting results for the finite temperature version, where the energy appears in the exponential, as in (38.5.2) [CY06].

KPZ equation The seminal paper is [KPZ86], which generated a large body of theoretical work. An introductory exposition is [BS95]. The KPZ equation is a stochastic field theory with broken time reversal invariance, hence a great theoretical challenge, see, e.g. [Läs98].

Review articles A beautiful review is [Joh06]. Growth models, of the type discussed here, are explained in [FP06, Fer08]. A fine introduction to random matrix techniques is [Sas07]. [KK10] provide an introductory exposition to the shape fluctuation proof of Johansson. The method of line ensembles is reviewed in [Spo06].

References

[AR05] D. Allison and N. Reshetikhin. Numerical study of the 6-vertex model with domain wall boundary conditions. *Ann. Inst. Fourier*, **55** (2005) 1847–1869.

[BDJ99] J. Baik, P.A. Deift, and K. Johansson. On the distribution of the length of the longest increasing subsequence of random permutations. J. Amer. Math. Soc., **12** (1999), 1119–1178.

[BFP10] J. Baik, P.L. Ferrari, and S. Péché. Limit process of stationary TASEP near the characteristic line. *Comm. Pure App. Math* **63** (2010), 1017–1070.

[BF06] P. Bleher and V. Fokin. Exact solution of the six-vertex model with domain wall boundary condition, disordered phase. *Comm. Math. Phys.*, **268** (2006) 223–284.

[BF08a] P.L. Ferrari and Alexei Borodin, *Anisotropic growth of random surfaces in 2 + 1 dimensions,* Comm. Math. Phys. **325** (2014) 603–684.

[BF08b] A. Borodin and P.L. Ferrari. Large time asymptotics of growth models on space-like paths I: PushASEP. *Electron. J. Probab.*, **13** (2008) 1380–1418.

[BFP08] F. Bornemann, P.L. Ferrari, and M. Prähofer. The Airy_1 process is not the limit of the largest eigenvalue in GOE matrix diffusion. *J. Stat. Phys.*, **133** (2008) 405–415.

[BFPS07] A. Borodin, P.L. Ferrari, M. Prähofer, and T. Sasamoto. Fluctuation properties of the TASEP with periodic initial configuration. *J. Stat. Phys.*, **129** (2007) 1055–1080.

[BFS08] A. Borodin, P.L. Ferrari, and T. Sasamoto. Large time asymptotics of growth models on space-like paths II: PNG and parallel TASEP. *Comm. Math. Phys.*, **283** (2008) 417–449.

[BR00] J. Baik and E.M. Rains. Limiting distributions for a polynuclear growth model with external sources. *J. Stat. Phys.*, **100** (2000) 523–542.

[BR01a] J. Baik and E.M. Rains. Algebraic aspects of increasing subsequences. *Duke Math. J.*, **109** (2001) 1–65.

[BR01b] J. Baik and E.M. Rains. The asymptotics of monotone subsequences of involutions. *Duke Math. J.*, **109** (2001) 205–281.

[BS95] A.L. Barabási and H.E. Stanley. *Fractal Concepts in Surface Growth.* Cambridge University Press, Cambridge, 1995.

[CEP96] H. Cohn, N. Elkies, and J. Propp. Local statistics for random domino tilings of the Aztec diamond. *Duke Math. J.*, **85** (1996) 117–166.

[CK01] R. Cerf and R. Kenyon. The low-temperature expansion of the Wulff crystal in the 3D-Ising model. *Comm. Math. Phys*, **222** (2001) 147–179.

[CY06] F. Comets and N. Yoshida. Directed polymers in random environment are diffusive at weak disorder. *Ann. Probab.*, **34** (2006) 1746–1770.

[Der07] B. Derrida. Non-equilibrium steady states: fluctuations and large deviations of the density and of the current. *J. Stat. Mech.*, **P07023** (2007).

[EKLP92] N. Elkies, G. Kuperbert, M. Larsen, and J. Propp. Alternating-Sign Matrices and Domino Tilings I and II. *J. Algebr. Comb.*, **1** (1992) 111–132 and 219–234.

[EM98] B. Eynard and M.L. Mehta. Matrices coupled in a chain. I. Eigenvalue correlations. *J. Phys. A*, **31** (1998) 4449–4456.

[Fer04] P.L. Ferrari. Polynuclear growth on a flat substrate and edge scaling of GOE eigenvalues. *Comm. Math. Phys.*, **252** (2004) 77–109.

[Fer08] P.L. Ferrari. The universal Airy_1 and Airy_2 processes in the Totally Asymmetric Simple Exclusion Process. In J. Baik et al. editors, *Integrable Systems and Random Matrices: In Honor of Percy Deift*, Contemporary Math., pages 321–332. Amer. Math. Soc., 2008.

[FP06] P.L. Ferrari and M. Prähofer. One-dimensional stochastic growth and Gaussian ensembles of random matrices. *Markov Processes Relat. Fields*, **12** (2006) 203–234.

[FPS04] P.L. Ferrari, M. Prähofer, and H. Spohn. Fluctuations of an atomic ledge bordering a crystalline facet. *Phys. Rev. E*, **69** (2004) 035102(R).

[FS03] P.L. Ferrari and H. Spohn. Step fluctations for a faceted crystal. *J. Stat. Phys.*, **113** (2003) 1–46.

[FS06a] P.L. Ferrari and H. Spohn. Domino tilings and the six-vertex model at its free fermion point. *J. Phys. A: Math. Gen.*, **39** (2006) 10297–10306.

[FS06b] P.L. Ferrari and H. Spohn. Scaling limit for the space-time covariance of the stationary totally asymmetric simple exclusion process. *Comm. Math. Phys.*, **265** (2006) 1–44.

[GS92] L-H. Gwa and H. Spohn. The six-vertex model, roughened surfaces and an asymmetric spin Hamiltonian. *Phys. Rev. Lett.*, **68** (1992) 725–728 and *Phys. Rev. A*, **46** (1992) 844–854.

[IS04a] T. Imamura and T. Sasamoto. Fluctuations of the one-dimensional polynuclear growth model in a half space. *J. Stat. Phys.*, **115** (2004) 749–803.

[IS04b] T. Imamura and T. Sasamoto. Fluctuations of the one-dimensional polynuclear growth model with external sources. *Nucl. Phys. B*, **699** (2004) 503–544.

[JN06] K. Johansson and E. Nordenstam. Eigenvalues of GUE minors. *Electron. J. Probab.*, **11** (2006) 1342–1371.

[Joh00] K. Johansson. Shape fluctuations and random matrices. *Comm. Math. Phys.*, **209** (2000) 437–476.

[Joh03] K. Johansson. Discrete polynuclear growth and determinantal processes. *Comm. Math. Phys.*, **242** (2003) 277–329.

[Joh05] K. Johansson. The arctic circle boundary and the Airy process. *Ann. Probab.*, **33** (2005) 1–30.

[Joh06] K. Johansson. Random matrices and determinantal processes. In A. Bovier et al. editors, *Mathematical Statistical Physics, Session LXXXIII: Lecture Notes of the Les Houches Summer School 2005*, pages 1–56. Elsevier Science, 2006.

[JPS98] W. Jockush, J. Propp, and P. Shor. Random domino tilings and the arctic circle theorem. *arXiv:math.CO/9801068* (1998).

[Kas63] P.W. Kasteleyn. Dimer statistics and phase transitions. *J. Math. Phys.*, **4** (1963) 287–293.

[Kaw72] K. Kawasaki. Kinetics of Ising models. In C. Domb and M. S. Green, editors, *Phase Transitions and Critical Phenomena*. Academic Press, 1972.

[Ken00] R. Kenyon. The planar dimer model with boundary: a survey. *Directions in mathematical quasicrystals, CRM Monogr. Ser.*, **13** (2000) 307–328.

[Ken08] R. Kenyon. Height fluctuations in the honeycomb dimer model. *Comm. Math. Phys.*, **281** (2008) 675–709.

[Kes86] H. Kesten. Aspects of first-passage percolation. *Lecture Notes in Math.*, **1180** (1986) 125–264.

[KK10] T. Kriecherbauer and J. Krug. Interacting particle systems out of equilibrium. *J. Phys. A: Math. Theor.* **43** (2010) 403001.

[KL99] C. Kipnis and C. Landim. *Scaling Limits of Interacting Particle Systems.* Springer Verlag, Berlin, 1999.

[KOS06] R. Kenyon, A. Okounkov, and S. Sheffield. Dimers and amoebae. *Ann. of Math.*, **163** (2006) 1019–1056.

[KPZ86] K. Kardar, G. Parisi, and Y.Z. Zhang. Dynamic scaling of growing interfaces. *Phys. Rev. Lett.*, **56** (1986) 889–892.

[KZJ00] V. Korepin and P. Zinn-Justin. Thermodynamic limit of the six-vertex model with domain wall boundary conditions. *J. Phys. A*, **33** (2000) 7053–7066.

[Läs98] M. Lässig. On growth, disorder, and field theory. *J. Phys. C*, **10** (1998) 9905–9950.

[Lig99] T.M. Liggett. *Stochastic Interacting Systems: Contact, Voter and Exclusion Processes.* Springer Verlag, Berlin, 1999.

[Mea98] P. Meakin. *Fractals, Scaling and Growth Far From Equilibrium.* Cambridge University Press, Cambridge, 1998.

[NF98] T. Nagao and P.J. Forrester. Multilevel dynamical correlation functions for Dysons Brownian motion model of random matrices. *Phys. Lett. A*, **247** (1998) 42–46.

[Nor10] E. Nordenstam. On the shuffling algorithm for domino tilings. *Electron. J. Probab.* **15** (2010) 75–95.

[NS04] T. Nagao and T. Sasamoto. Asymmetric simple exclusion process and modified random matrix ensembles. Nucl. Phys. B, **699** (2004) 487–502.

[OR03] A. Okounkov and N. Reshetikhin. Correlation function of Schur process with application to local geometry of a random 3-dimensional Young diagram. *J. Amer. Math. Soc.*, **16** (2003) 581–603.

[OR06] A. Okounkov and N. Reshetikhin. The birth of a random matrix. *Mosc. Math. J.*, **6** (2006) 553–566.

[OR07] A. Okounkov and N. Reshetikhin. Random skew plane partitions and the Pearcey process. *Comm. Math. Phys.*, **269** (2007) 571–609.

[Pro03] J. Propp. Generalized Domino-Shuffling. *Theoret. Comput. Sci.*, **303** (2003) 267–301.

[PS90] V. Perival and D. Shevitz. Unitary-matrix models as exactly solvable string theories. *Phys. Rev. Lett.*, **64** (1990) 1326–1329.

[PS97] M. Prähofer and H. Spohn. An Exactly Solved Model of Three Dimensional Surface Growth in the Anisotropic KPZ Regime. *J. Stat. Phys.*, **88** (1997) 999–1012.

[PS00a] M. Prähofer and H. Spohn. Statistical self-similarity of one-dimensional growth processes. *Physica A*, **279** (2000) 342–352.

[PS00b] M. Prähofer and H. Spohn. Universal distributions for growth processes in $1+1$ dimensions and random matrices. *Phys. Rev. Lett.*, **84** (2000) 4882–4885.

[PS02a] M. Prähofer and H. Spohn. Current fluctuations for the totally asymmetric simple exclusion process. In V. Sidoravicius, editor, *In and out of equilibrium*, Progress in Probability. Birkhäuser, 2002.

[PS02b] M. Prähofer and H. Spohn. Scale invariance of the PNG droplet and the Airy process. *J. Stat. Phys.*, **108** (2002) 1071–1106.

[PS04] M. Prähofer and H. Spohn. Exact scaling function for one-dimensional stationary KPZ growth. *J. Stat. Phys.*, **115** (2004) 255–279.

[RS05] A. Rákos and G. Schütz. Current Distribution and random matrix ensembles for an integrable asymmetric fragmentation process. *J. Stat. Phys.*, **118** (2005) 511–530.

[Sas05] T. Sasamoto. Spatial correlations of the 1D KPZ surface on a flat substrate. *J. Phys. A*, **38** (2005) L549–L556.

[Sas07] T. Sasamoto. Fluctuations of the one-dimensional asymmetric exclusion process using random matrix techniques. *J. Stat. Mech.*, **P07007** (2007).

[Sch97] G.M. Schütz. Exact solution of the master equation for the asymmetric exclusion process. *J. Stat. Phys.*, **88** (1997) 427–445.

[Spi70] F. Spitzer. Interaction of Markov processes. *Adv. Math.*, **5** (1970) 246–290.

[Spo91] H. Spohn. *Large Scale Dynamics of Interacting Particles.* Springer Verlag, Heidelberg, 1991.

[Spo06] H. Spohn. Exact solutions for KPZ-type growth processes, random matrices, and equilibrium shapes of crystals. *Physica A*, **369** (2006) 71–99.

[TW09] C.A. Tracy and H. Widom. Asymptotics in ASEP with step initial condition. *Comm. Math. Phys.*, **290** (2009) 129–154.

[TW08] C.A. Tracy and H. Widom. Integral Formulas for the Asymmetric Simple Exclusion Process. *Comm. Math. Phys.*, **279** (2008) 815–844.

[War07] Jon Warren. Dyson's Brownian motions, intertwining and interlacing. *Electron. J. Probab.*, **12** (2007) 573–590.

·39·

Random matrices and Laplacian growth

A. Zabrodin

Abstract

The theory of random matrices with eigenvalues distributed in the complex plane and more general 'β-ensembles' (logarithmic gases in 2D) is reviewed. The distribution and correlations of the eigenvalues are investigated in the large N limit. It is shown that in this limit the model is mathematically equivalent to a class of diffusion-controlled growth models for viscous flows in the Hele-Shaw cell and other growth processes of Laplacian type. The analytical methods used involve the technique of boundary value problems in two dimensions and elements of the potential theory.

39.1 Introduction

Applications of random matrices in physics (and mathematics) are known to range from energy level statistics in nuclei to number theory and from quantum chaos to string theory. Most extensively employed and best-understood are ensembles of Hermitian or unitary matrices (see e.g. [Meh91],[Mor94]). Their eigenvalues are confined either to the real axis or to the unit circle. In this paper we consider more general classes of random matrices, with no *a priori* restrictions to their eigenvalues. Such models are as yet less well understood but they are equally interesting and meaningful from both mathematical and physical points of view. (A list of the relevant physical problems and corresponding references can be found in, e.g., [Fyo97].)

The progenitor of ensembles of matrices with general complex eigenvalues is the statistical model of complex matrices with the Gaussian weight introduced by Ginibre [Gin65] in 1965. The partition function of this model is

$$Z_N = \int [D\Phi] \exp\left(-\frac{N}{t}\mathrm{tr}\,\Phi^\dagger\Phi\right). \qquad (39.1.1)$$

Here $[D\Phi] = \prod_{ij} d(\mathrm{Re}\,\Phi_{ij})d(\mathrm{Im}\,\Phi_{ij})$ is the standard volume element in the space of $N \times N$ matrices with complex entries Φ_{ij} and t is a (real positive) parameter. Along with the Ginibre ensemble and its generalizations we also consider ensembles of normal matrices [Cha92], i.e. such that Φ commutes with its Hermitian conjugate $\Phi^\dagger$.

Since one is primarily interested in statistics of eigenvalues, it is natural to express the probability density in terms of complex eigenvalues $z_j = x_j + iy_j$ of the matrix Φ. It appears that the volume element can be represented as

$$[D\Phi] \propto \prod_{i<j} |z_i - z_j|^2 \prod_i d^2 z_i. \tag{39.1.2}$$

If the statistical weight depends on the eigenvalues only, as it is usually assumed, the other parameters of the matrix (often referred to as 'angular variables') are irrelevant and can be integrated out giving an overall normalization factor. In this case the original matrix problem reduces to statistical mechanics of N particles with complex coordinates z_j in the plane. Specifically, the factor $\prod_{i<j} |z_i - z_j|^2$, being equal to the exponentiated Coulomb energy in two dimensions, means an effective 'repelling' of eigenvalues. This remark leads to the Dyson logarithmic gas interpretation [Dys62], which treats the matrix ensemble as a two-dimensional 'plasma' of eigenvalues in a background field and prompts us to introduce more general 'β-ensembles' with the statistical weight proportional to $\prod_{i<j} |z_i - z_j|^{2\beta}$.

It is also natural to consider matrix ensembles with statistical weights of a general form, $\exp\left(-\frac{N}{t}\mathrm{tr}\, W(\Phi^\dagger, \Phi)\right)$, with a background potential W. An important observation made in [Kos01] and developed in subsequent works [Wie03, Zab03, Teo05, Zab06] is that evolution of an averaged spectrum of such matrices as a function of t, as $N \to \infty$, serves as a simulation of Laplacian growth of water droplets in the Hele-Shaw cell (for different physical and mathematical aspects of the latter see [Ben86, Gus06, Gil]). To be more precise, in the limit $N \to \infty$, under some weak assumptions about the statistical weight, the eigenvalues are confined, with probability 1, to a compact domain in the complex plane. One can ask how its shape depends on the parameter t. The answer is: this dependence is exactly the Laplacian growth of the domain with zero surface tension. Namely, the edge of the support moves along the gradient of a scalar harmonic field in its exterior, with the velocity being proportional to the absolute value of the gradient. The general solution can be expressed in terms of the exterior Dirichlet boundary value problem.

This fact allows one to treat the model of normal or complex random matrices as a growth problem [Teo05]. The advantage of this viewpoint is two-fold. First, the hydrodynamic interpretation makes some of the large N matrix model results more illuminating and intuitively accessible. Second and most important, the matrix model perspective may help to suggest new approaches to the long-standing growth problems. In this respect, of special interest is the identification of finite time singularities in some exact solutions to the Hele-Shaw flows with critical points of the normal and complex matrix models.

39.2 Random matrices with complex eigenvalues

We consider square random matrices Φ of size N with complex entries Φ_{ij}. The probability density is assumed to be of the form $P(\Phi) \propto e^{\frac{1}{\hbar}\mathrm{tr}\, W(\Phi)}$, where the function $W(\Phi)$ (often called the potential of the matrix model) is a matrix-valued function of Φ and $\Phi^\dagger$ such that $(W(\Phi))^\dagger = W(\Phi)$ and $\hbar$ is a parameter introduced to stress a quasiclassical nature of the large N limit. The partition function is defined as an integral

$$Z_N = \int [D\Phi] e^{\frac{1}{\hbar}\mathrm{tr}\, W(\Phi)} \tag{39.2.1}$$

over the matrices with the integration measure $[D\Phi]$ to be specified below in this section.

We consider two ensembles of random matrices Φ with complex eigenvalues: ensemble $\mathcal{C}$ of *general complex matrices* (with no restrictions on the entries except for $\det\Phi \neq 0$) and ensemble $\mathcal{N}$ of *normal matrices* (such that $[\Phi, \Phi^\dagger] = 0$).

39.2.1 Integration measure

The integration measure has the most simple form for the ensemble of general complex matrices:

$$[D\Phi] = \prod_{i,j=1}^{N} d(\mathrm{Re}\, \Phi_{ij})\, d(\mathrm{Im}\, \Phi_{ij}).$$

This measure is additively invariant and multiplicatively covariant, i.e. for any fixed (non-degenerate) matrix $A \in \mathcal{C}$ we have the properties $[D(\Phi + A)] = [D\Phi]$ and $[D(\Phi A)] = [D(A\Phi)] = |\det A|^{2N}[D\Phi]$. It is also clear that the measure is invariant under transformations of the form $\Phi \to U^\dagger \Phi U$ with a unitary matrix U.

The measure for $\mathcal{N}$ is induced by the standard flat metric in $\mathcal{C}$, $||\delta\Phi||^2 = \mathrm{tr}\,(\delta\Phi\delta\Phi^\dagger) = \sum_{ij} |\delta\Phi_{ij}|^2$ via the embedding $\mathcal{N} \subset \mathcal{C}$. Here $\mathcal{N}$ is regarded as a hypersurface in $\mathcal{C}$ defined by the quadratic relations $\Phi\Phi^\dagger = \Phi^\dagger\Phi$. As usual in the theory of random matrices, one would like to integrate out the 'angular' variables and to express the integration measure through the eigenvalues only.

The measure for $\mathcal{N}$ through eigenvalues [Meh04, Cha92]. We derive the explicit representation of the measure in terms of eigenvalues in three steps:

1. Introduce coordinates in $\mathcal{N} \subset \mathcal{C}$.
2. Compute the inherited metric on $\mathcal{N}$ in these coordinates: $||\delta\Phi||^2 = g_{\alpha\beta} d\xi^\alpha d\xi^\beta$.
3. Compute the volume element $[D\Phi] = \sqrt{|\det g_{\alpha\beta}|} \prod_\alpha d\xi^\alpha$.

Step 1: Coordinates in $\mathcal{N}$. For any matrix Φ, the matrices $H_1 = \frac{1}{2}(\Phi + \Phi^\dagger)$, $H_2 = \frac{1}{2i}(\Phi - \Phi^\dagger)$ are Hermitian. The condition $[\Phi, \Phi^\dagger] = 0$ is equivalent to $[H_1, H_2] = 0$. Thus $H_{1,2}$ can be simultaneously diagonalized by a unitary matrix U:

$$H_1 = UXU^\dagger\,, \quad X = \text{diag}\,\{x_1, \ldots, x_N\}$$
$$H_2 = UYU^\dagger\,, \quad Y = \text{diag}\,\{y_1, \ldots, y_N\}\,.$$

Introduce the diagonal matrices $Z = X + iY$, $\bar{Z} = X - iY$ with diagonal elements $z_j = x_j + iy_j$ and $\bar{z}_j = x_j - iy_j$ respectively. Note that z_j are eigenvalues of Φ. Therefore, any $\Phi \in \mathcal{N}$ can be represented as $\Phi = UZU^\dagger$, where U is a unitary matrix and Z is the diagonal matrix with eigenvalues of Φ on the diagonal. The matrix U is defined up to multiplication by a diagonal unitary matrix from the right: $U \to U\,U_{\text{diag}}$. The dimension of $\mathcal{N}$ is thus

$$\dim(\mathcal{N}) = \dim(\mathcal{U}) - \dim(\mathcal{U}_{\text{diag}}) + \dim(\mathcal{C}_{\text{diag}}) = N^2 - N + 2N = N^2 + N$$

(here $\mathcal{U} \subset \mathcal{C}$ is the submanifold of unitary matrices).
Step 2: The induced metric. Since $\Phi = UZU^\dagger$, the variation is $\delta\Phi = U(\delta u \cdot Z + \delta Z + Z \cdot \delta u^\dagger)U^\dagger$, where $\delta u^\dagger = U\delta U^\dagger = -\delta u^\dagger$. Therefore,

$$||\delta\Phi||^2 = \text{tr}\,(\delta\Phi\delta\Phi^\dagger) = \text{tr}\,(\delta Z\delta\bar{Z}) + 2\,\text{tr}\,(\delta u Z\delta u\bar{Z} - (\delta u)^2 Z\bar{Z})$$

$$= \sum_{j=1}^{N} |\delta z_j|^2 + 2\sum_{j<k}^{N} |z_j - z_k|^2\,|\delta u_{jk}|^2.$$

(Note that δu_{jj} do not enter.)
Step 3. The volume element. We see that the metric $g_{\alpha\beta}$ is diagonal in the coordinates $\text{Re}(\delta z_j)$, $\text{Im}(\delta z_j)$, $\text{Re}(\delta u_{jk})$, $\text{Im}(\delta u_{jk})$ with $1 \le j < k \le N$, so the determinant of the diagonal matrix $g_{\alpha\beta}$ is easily calculated to be $|\text{det}g_{\alpha\beta}| = 2^{N^2-N}\prod_{j<k}^{N}|z_i - z_k|^4$. Therefore,

$$[D\Phi] \propto [DU]'\,|\Delta_N(z_1, \ldots, z_N)|^2 \prod_{j=1}^{N} d^2 z_j\,, \tag{39.2.2}$$

where $d^2z \equiv dxdy$ is the flat measure in the complex plane, $[DU]' = [DU]/[DU_{\text{diag}}]$ is the invariant measure on $\mathcal{U}/\mathcal{U}_{\text{diag}}$, and Δ_N is the Vandermonde determinant:

$$\Delta_N(z_1, \ldots, z_N) = \prod_{j>k}^{N}(z_j - z_k) = \det_{N\times N}(z_j^{k-1}). \tag{39.2.3}$$

The measure for $\mathcal{C}$ through eigenvalues. A complex matrix Φ with eigenvalues $z_1, \ldots, z_N$ can be decomposed as $\Phi = U(Z + R)U^\dagger$, where $Z = \text{diag}\,\{z_1, \ldots, z_N\}$ is diagonal, U is unitary, and R is strictly upper triangular, i.e., $R_{ij} = 0$ if $i \ge j$.

These matrices are defined up to a 'gauge transformation': $U \to U\, U_{\text{diag}}$, $R \to U^{\dagger}_{\text{diag}}\, R\, U_{\text{diag}}$. It is not so easy to see that the measure factorizes. This requires some work, of which the key step is a specific ordering of the independent variables. The final result is:

$$[D\Phi] \propto [DU]' \left(\prod_{k<l} d^2 R_{kl} \right) |\Delta_N(z_i)|^2 \prod_{j=1}^{N} d^2 z_j \,. \tag{39.2.4}$$

The details can be found in the Mehta book [Meh91].

39.2.2 Potentials

For the ensemble $\mathcal{N}$ the 'angular variables' (parameters of the unitary matrix U) always decouple after taking the trace $\operatorname{tr} W(\Phi) = \sum_j W(z_j)$, so the potential W can be a function of Φ, $\Phi^{\dagger}$ of a general form $W(\Phi) = \sum a_{nm} \Phi^n (\Phi^{\dagger})^m$. The partition function reads

$$Z_N = \int |\Delta_N(z_i)|^2 \prod_{j=1}^{N} e^{\frac{1}{\hbar} W(z_j)} d^2 z_j, \tag{39.2.5}$$

where we ignore a possible N-dependent normalization factor. From now on this formula is taken as the definition of the partition function.

The choice of the potential for the ensemble $\mathcal{C}$ is more restricted. For a general potential, the matrix U in $\Phi = U(Z + R)U^{\dagger}$ still decouples but R does not. An important class of potentials when R decouples nevertheless is $W(\Phi) = -\Phi\Phi^{\dagger} + V(\Phi) + \bar{V}(\Phi^{\dagger})$, where $V(z)$ is an analytic function of z in some domain containing the origin and $\bar{V}(z) = \overline{V(\bar{z})}$. In terms of the eigenvalues,

$$W(z) = -|z|^2 + V(z) + \overline{V(z)}. \tag{39.2.6}$$

In what follows, we call such a potential *quasiharmonic*. In this case, $\operatorname{tr}(\Phi\Phi^{\dagger}) = \operatorname{tr}(Z\bar{Z}) + \operatorname{tr}(RR^{\dagger})$, $\operatorname{tr}(\Phi^n) = \operatorname{tr}(Z + R)^n = \operatorname{tr} Z^n$, and so

$$\int_{\mathcal{C}} [D\Phi] e^{\frac{1}{\hbar} \operatorname{tr} W(\Phi)} = C_N \int |\Delta(z_i)|^2 \prod_k e^{\frac{1}{\hbar} W(z_k)} d^2 z_k, \tag{39.2.7}$$

where C_N is an N-dependent normalization factor proportional to the Gaussian integral $\int [DR] e^{-\frac{1}{\hbar} \operatorname{tr}(RR^{\dagger})}$.

As an example, let us consider the quadratic potential: $W(z) = -|z|^2 + 2\mathrm{Re}\,(t_1 z + t_2 z^2)$. The ensemble $\mathcal{C}$ with this potential is known as *the Ginibre-Girko ensemble* [Gin65, Gir85]. In this case the partition function (39.2.5) can be calculated exactly [DiF94]:

$$Z_N = Z_N^{(0)}(1-4|t_2|^2)^{-N^2/2} \exp\left(\frac{N}{\hbar}\frac{t_1^2\bar{t}_2 + \bar{t}_1^2 t_2 + |t_1|^2}{1-4|t_2|^2}\right), \tag{39.2.8}$$

where $Z_N^{(0)} = \hbar^{(N^2+N)/2}\pi^N \prod_{k=1}^{N} k!$ is the partition function of the model with $W = -|z|^2$.

Note that for quasiharmonic potentials the integral (39.2.5) diverges unless V is quadratic or logarithmic with suitable coefficients. The simplest way to give sense to the integral when it diverges at infinity is to introduce a cutoff, i.e. to integrate over a big but finite disk of radius R_0 centred at the origin. Just for technical simplicity we assume that a) V is a holomorphic function everywhere inside this disk, b) W has a maximum at the origin with $W(0) = 0$ and no other critical critical points inside the disk, c) At $|z| = R_0$ the potential W is bounded from above by a constant $B < 0$. The large N expansion is then well defined. For details and rigorous proofs see [Elb04, Hed04, Bal08].

39.2.3 The Dyson gas picture

The statistics of eigenvalues appears to be mathematically equivalent to some important models of classical statistical mechanics, with the eigenvalues being represented as charged particles in the plane interacting via 2D Coulomb (logarithmic) potential. This interpretation, first suggested by Dyson [Dys62] for the unitary, symplectic and orthogonal matrix ensembles, relies on rewriting $|\Delta_N(z_i)|^2$ as $\exp\left(\sum_{i\neq j}\log|z_i - z_j|\right)$. Clearly, the integral (39.2.5) looks then exactly as the partition function of the 2D Coulomb plasma (often called the Dyson gas) in the external field:

$$Z_N = \int e^{-\beta E(z_1,\dots,z_N)} \prod d^2 z_j\,, \tag{39.2.9}$$

where

$$E = -\sum_{i<j}\log|z_i - z_j|^2 - \frac{1}{\beta\hbar}\sum_j W(z_j). \tag{39.2.10}$$

Here β plays the role of inverse temperature (in (39.2.5) $\beta = 1$). The first sum is the Coulomb interaction energy, the second one is the energy due to the external field. For the Hermitian and unitary ensembles the charges are confined to lines of dimension 1 (the real line or the unit circle) but still interact as 2D Coulomb charges. So, the Dyson gas picture for ensembles of matrices with general complex eigenvalues distributed on the plane looks even more natural. It becomes especially helpful in the large N limit, where it allows one to apply thermodynamical arguments.

The Dyson gas picture prompts us to consider more general ensembles with arbitrary values of β ("β-ensembles"):

$$Z_N = \int |\Delta_N(z_i)|^{2\beta} \prod_{j=1}^{N} e^{\frac{1}{\hbar}W(z_j)} d^2 z_j \,. \tag{39.2.11}$$

In general they can not be defined through matrix integrals. As we shall see, the leading large N contribution has a simple regular dependence on β. However, the sub-leading corrections may depend on β in a rather non-trivial way.

39.3 Exact relations at finite *N*

Here we present some general exact relations for correlation functions valid for any values of N and β.

39.3.1 Correlation functions: general relations

The main objects of interest are correlation functions, i.e. mean values of functions of matrices. We shall consider functions that depend on eigenvalues only – for example, traces $\mathrm{tr}\, f(\Phi) = \sum_i f(z_i)$. Here, $f(\Phi) = f(\Phi, \Phi^\dagger)$ is any function of Φ, $\Phi^\dagger$ which is regarded as the function $f(z_i) = f(z_i, \bar{z}_i)$ of the complex argument z_i (and $\bar{z}_i$). (For the abuse of notation, in case of arbitrary β we write $\mathrm{tr}\, f := \sum_i f(z_i)$ although a matrix realization may be not available.) Typical correlation functions which we are going to study are mean values of products of traces: $\langle \mathrm{tr}\, f(\Phi) \rangle$, $\langle \mathrm{tr}\, f_1(\Phi)\, \mathrm{tr}\, f_2(\Phi) \rangle$ and so on. Clearly, they are represented as integrals over eigenvalues. For instance,

$$\langle \mathrm{tr}\, f(\Phi) \rangle = \frac{1}{Z_N} \int |\Delta_N(z_i)|^{2\beta} \left(\sum_{l=1}^{N} f(z_l) \right) \prod_{j=1}^{N} e^{\frac{1}{\hbar}W(z_j)} d^2 z_j \,.$$

A particularly important example is the density function defined as

$$\rho(z) = \hbar \sum_j \delta^{(2)}(z - z_j), \tag{39.3.1}$$

where $\delta^{(2)}(z)$ is the 2D δ-function. Note that in our units $\hbar$ has dimension of $[\text{length}]^2$ and $\rho(z)$ is dimensionless. As it immediately follows from the definition, any correlator of traces is expressed through correlators of ρ:

$$\langle \mathrm{tr}\, f_1 \ldots \mathrm{tr}\, f_n \rangle = \hbar^{-n} \int \langle \rho(z_1) \ldots \rho(z_n) \rangle f_1(z_1) \ldots f_n(z_n) \prod_{j=1}^{n} d^2 z_j \,. \tag{39.3.2}$$

Instead of correlations of density it is often convenient to consider correlations of the field

$$\varphi(z) = -\beta\hbar \sum_j \log |z - z_j|^2 \tag{39.3.3}$$

from which the correlations of density can be found by means of the relation

$$4\pi\beta\rho(z) = -\Delta\varphi(z), \tag{39.3.4}$$

where $\Delta = 4\partial\bar{\partial}$ is the Laplace operator. Clearly, φ is the 2D Coulomb potential created by the eigenvalues (charges).

Handling with multi-point correlation functions, it is customary to pass to their *connected parts*. For example, in the case of 2-point functions, the connected correlation function is defined as

$$\left\langle \rho(z_1)\rho(z_2)\right\rangle_c \equiv \left\langle \rho(z_1)\rho(z_2)\right\rangle - \left\langle \rho(z_1)\right\rangle\left\langle \rho(z_2)\right\rangle .$$

The connected multi-trace correlators are expressed through the connected density correlators by the same formula (39.3.2) with $\left\langle \rho(z_1)\dots\rho(z_n)\right\rangle_c$ in the right-hand side. The connected part of the $(n+1)$-point density correlation function is given by the linear response of the n-point one to a small variation of the potential. More precisely, the following variational formulas hold true:

$$\left\langle \rho(z)\right\rangle = \hbar^2 \frac{\delta \log Z_N}{\delta W(z)}, \qquad \left\langle \rho(z_1)\rho(z_2)\right\rangle_c = \hbar^2 \frac{\delta \left\langle \rho(z_1)\right\rangle}{\delta W(z_2)} = \hbar^4 \frac{\delta^2 \log Z_N}{\delta W(z_1)\delta W(z_2)}. \tag{39.3.5}$$

Connected multi-point correlators are higher variational derivatives of $\log Z_N$. These formulae follow from the fact that variation of the partition function over a general potential W inserts $\sum_i \delta^{(2)}(z - z_i)$ into the integral. Basically, they are linear response relations used in the Coulomb gas theory [For98].

39.3.2 Loop equations

The standard source of exact relations for correlation functions is the formal identity

$$\sum_i \int \frac{\partial}{\partial z_i}\left(\epsilon(\{z_j\})\, e^{-\beta E}\right) \prod_j d^2 z_j = 0, \tag{39.3.6}$$

where $\epsilon(\{z_j\})$ is any function of coordinates z_j bounded at infinity and E is given by (39.2.10). Introducing, if necessary, a cutoff at infinity one sees that the 2D integral over z_i can be transformed, by virtue of the Green's theorem, into a contour integral around infinity and so it does vanish. Being expressed in terms of correlation functions of local fields (such as ρ or φ), this identity yields, with a suitable choice of ϵ, certain exact relations between them. For historical reasons, they are referred to as *loop equations*.

Let us take

$$\epsilon(z_i) = \frac{X(\{z_l\})}{z - z_i}, \tag{39.3.7}$$

where $X(\{z_l\})$ is any symmetric function of $z_1, z_2, \ldots, z_N$, then the identity reads

$$\sum_i \int \left[\left(\frac{-\beta\, \partial_{z_i} E}{z - z_i} + \frac{1}{(z - z_i)^2} \right) X + \frac{\partial_{z_i} X}{z - z_i} \right] e^{-\beta E} \prod_j d^2 z_j = 0.$$

The singularity at the point z does not destroy the identity since its contribution is proportional to the vanishing integral $\oint d\bar{z}_i/(z_i - z)$ over a small contour encircling z. Plugging $\partial_{z_i} E$ in the first term and using the bracket notation for the mean value, we have:

$$\left\langle \left(\frac{1}{\hbar} \sum_i \frac{\partial W(z_i)}{z - z_i} + \beta \sum_{i \neq j} \frac{1}{(z - z_i)(z_i - z_j)} + \sum_i \frac{1}{(z - z_i)^2} \right) X + \sum_i \frac{\partial_{z_i} X}{z - z_i} \right\rangle = 0.$$

The second sum can be transformed by means of the simple algebraic identity

$$\sum_{i,j} \frac{1}{(z - z_i)(z - z_j)} = \sum_{i \neq j} \frac{2}{(z - z_i)(z_i - z_j)} + \sum_i \frac{1}{(z - z_i)^2}.$$

Finally, we arrive at the relation

$$\left\langle L(z) X + \sum_i \frac{\partial_{z_i} X}{z - z_i} \right\rangle = 0, \tag{39.3.8}$$

where we have introduced the special notation

$$L(z) = \frac{1}{\hbar} \sum_i \frac{\partial W(z_i)}{z - z_i} + \frac{\beta}{2} \left(\sum_i \frac{1}{z - z_i} \right)^2 + \left(1 - \frac{\beta}{2} \right) \sum_i \frac{1}{(z - z_i)^2}, \tag{39.3.9}$$

or, in terms of the fields ρ, φ (39.3.1), (39.3.3),

$$2\beta\hbar^2 L(z) = 2\beta \int \frac{\partial W(\xi)\rho(\xi) d^2\xi}{z - \xi} + (\partial\varphi(z))^2 + (2-\beta)\hbar\partial^2\varphi(z). \tag{39.3.10}$$

This quantity plays an important role. It is to be compared with the stress energy tensor in 2D CFT.

We call (39.3.8) the generating loop equation. It generates an infinite hierarchy of identities obeyed by correlation functions. The simplest one is obtained at $X \equiv 1$: $\langle L(z) \rangle = 0$. It reads:

$$\frac{1}{2\pi} \int \frac{\partial W(\zeta) \langle \Delta\varphi(\zeta) \rangle}{\zeta - z} d^2\zeta + \langle (\partial\varphi(z))^2 \rangle + (2-\beta)\hbar \langle \partial^2\varphi(z) \rangle = 0. \tag{39.3.11}$$

This identity gives an exact relation between one- and two-point correlation functions because the mean value $\langle(\partial\varphi(z))^2\rangle$ can be reproduced from the two-point correlation function $\langle\partial\varphi(z)\,\partial\varphi(z')\rangle$ by averaging over all possible directions of approaching $z' \to z$:

$$\langle(\partial\varphi(z))^2\rangle = \lim_{\varepsilon\to 0}\frac{1}{2\pi}\int_0^{2\pi}\langle\partial\varphi(z)\,\partial\varphi(z+\varepsilon e^{i\theta})\rangle\,d\theta\,. \tag{39.3.12}$$

Another interesting choice is $X = \varphi(\zeta)$, where $\zeta \neq z$, then $\partial_{z_i}X = \beta\hbar/(\zeta - z_i)$ and (39.3.8) yields

$$\left\langle L(z)\varphi(\zeta) + \frac{\partial\varphi(z) - \partial\varphi(\zeta)}{z-\zeta}\right\rangle = 0. \tag{39.3.13}$$

Acting by $\partial_{\bar\zeta}$, we get $\langle L(z)\bar\partial\varphi(\zeta)\rangle = \pi\beta\langle\rho(\zeta)\rangle/(\zeta - z)$. Further, acting by ∂_ζ to both sides, we obtain the relation

$$\langle L(z)\rho(\zeta)\rangle = \frac{\langle\rho(\zeta)\rangle}{(z-\zeta)^2} + \frac{\partial_\zeta\langle\rho(\zeta)\rangle}{z-\zeta}\,. \tag{39.3.14}$$

A similar but longer calculation gives a generalization of this identity for m-point functions:

$$\langle L(z)\rho(\zeta_1)\dots\rho(\zeta_m)\rangle = \sum_{i=1}^{m}\left(\frac{1}{(z-\zeta_i)^2} + \frac{1}{z-\zeta_i}\partial_{\zeta_i}\right)\langle\rho(\zeta_1)\dots\rho(\zeta_m)\rangle\,. \tag{39.3.15}$$

It has the form of the conformal Ward identity for primary field of conformal dimensions 1, with $L(z)$ playing the role of the (holomorphic component of) the stress energy tensor in CFT.

39.4 Large N limit

Starting from this section, we study the large N limit

$$N\to\infty\,,\quad \hbar\to 0\,,\quad \hbar N = t \quad \text{finite}. \tag{39.4.1}$$

We shall see that in this limit meaningful analytic and algebro-geometric structures emerge, as well as important applications in physics.

39.4.1 Solution to the loop equation in the leading order

It is instructive to think about the large N limit under consideration in terms of the Dyson gas picture. Then the limit we are interested in corresponds to a very low temperature of the gas, when fluctuations around equilibrium positions of the charges are negligible. The main contribution to the partition function then comes from a configuration, where the charges are 'frozen' at their equilibrium positions. It is also important that the temperature tends to zero simultaneously with increasing the number of charges, so the plasma can be regarded as a continuous fluid at static equilibrium. Mathematically, all this means that the integral is evaluated by the saddle-point method, with only the leading contribution being taken into account. As $\hbar \to 0$, correlation functions take their 'classical' values $\langle \varphi(z) \rangle = \varphi_0(z)$, $\langle \rho(z) \rangle = \rho_0(z)$, and multi-point functions factorize in the leading order: $\langle \partial\varphi(z)\, \partial\varphi(z') \rangle = \partial\varphi_0(z)\partial\varphi_0(z')$, etc. Then the loop equation (39.3.11) becomes a closed relation for φ_0:

$$\frac{1}{2\pi}\int \frac{\partial W(\zeta)\Delta\varphi_0(\zeta)}{\zeta - z}\, d^2\zeta + \left(\partial\varphi_0(z)\right)^2 = 0, \tag{39.4.2}$$

where we have ignored the last term because it has a higher order in $\hbar$. Applying $\bar\partial$ to the both terms, we get: $-\partial W(z)\Delta\varphi_0(z) + \partial\varphi_0(z)\Delta\varphi_0(z) = 0$. Since $\Delta\varphi_0(z) = -4\pi\beta\rho_0(z)$ (see (39.3.4)), we obtain

$$\rho_0(z)\left[\partial\varphi_0(z) - \partial W(z)\right] = 0. \tag{39.4.3}$$

This equation should be solved with the additional constraints $\int \rho_0(z) d^2z = t$ (normalization) and $\rho_0(z) \geq 0$. The equation tells us that either $\partial\varphi_0(z) = \partial W(z)$ or $\rho_0(z) = 0$. Applying $\bar\partial$ to the former, we get $\Delta\varphi_0(z) = \Delta W(z)$. This gives the solution for ρ_0:

$$\rho_0(z) = -\frac{\Delta W(z)}{4\pi\beta} \qquad \text{in the bulk.} \tag{39.4.4}$$

Here, 'in the bulk' means 'in the region where $\rho_0 > 0$'. The physical meaning of the equation $\partial\varphi_0(z) = \partial W(z)$ is clear. It is just the condition that the charges are in equilibrium (the saddle point for the integral). Indeed, the equation states that the total force experienced by a charge at any point z where $\rho_0 \neq 0$ is zero, i.e. the interaction with the other charges, $\partial\varphi_0(z)$, is compensated by the force $\partial W(z)$ due to the external field.

39.4.2 Support of eigenvalues

Let us assume that

$$\sigma(z) := -\frac{1}{4\pi}\Delta W(z) > 0. \tag{39.4.5}$$

For quasiharmonic potentials, $\sigma(z) = 1/\pi$. If, according to (39.4.4), $\rho_0 = \sigma/\beta$ everywhere, the normalization condition for ρ_0 in general can not be satisfied. So we conclude that $\rho_0 = \sigma/\beta$ in a compact bounded domain (or domains) only, and outside this domain one should switch to the other solution of (39.4.3), $\rho_0 = 0$. The domain D where $\rho_0 > 0$ is called *support of eigenvalues* or droplet of eigenvalues. In general, it may consist of several disconnected components. The complement to the support of eigenvalues, $\mathsf{D}^c = \mathbb{C} \setminus \mathsf{D}$, is an unbounded domain in the complex plane. For quasiharmonic potentials, the result is especially simple: ρ_0 is constant in D and 0 in D^c.

To find the shape of D is a much more serious problem. It appears to be equivalent to the inverse potential problem in 2D. The shape of D is determined by the condition $\partial\varphi_0(z) = \partial W(z)$ (valid at all points z inside D) and by the normalization condition. One can write them in the form

$$\begin{cases} \dfrac{1}{4\pi}\displaystyle\int_{\mathsf{D}} \frac{\Delta W(\zeta) d^2\zeta}{z-\zeta} = \partial W(z) \qquad \text{for all } z \in \mathsf{D} \\ \displaystyle\int_{\mathsf{D}} \sigma(\zeta) d^2\zeta = \beta t\,. \end{cases} \tag{39.4.6}$$

The integral over D in the first equation can be transformed to a contour integral by means of the Cauchy formula. As a result, the first equation reads:

$$\oint_{\partial\mathsf{D}} \frac{\partial W(\zeta) d\zeta}{z-\zeta} = 0 \qquad \text{for all } z \in \mathsf{D}. \tag{39.4.7}$$

This means that the domain D has the following property: the function $\partial W(z)$ on its boundary is the boundary value of an analytic function in its complement D^c.

The connection with the inverse potential problem is most straightforward in the quasiharmonic case, where $\partial W(z) = -\bar{z} + V'(z)$, $\Delta W(z) = 4\partial\bar{\partial} W(z) = -4$. The normalization then means that the area of D is equal to $\beta\pi t$. Assume that: i) $V(z) = \sum_k t_k z^k$ is regular in D (say a polynomial), ii) $0 \in \mathsf{D}$ (it is always the case when W has a maximum at 0), iii) D is connected. Then equation (39.4.7) acquires the form $\dfrac{1}{2\pi i}\displaystyle\oint_{\partial\mathsf{D}} \frac{\bar{\zeta} d\zeta}{\zeta - z} = V'(z)$ for $z \in \mathsf{D}$. Expanding it near $z = 0$, we get:

$$t_k = \frac{1}{2\pi i k}\oint_{\partial\mathsf{D}} \bar{\zeta}\zeta^{-k} d\zeta = -\frac{1}{\pi k}\int_{\mathsf{D}} \zeta^{-k} d^2\zeta\,. \tag{39.4.8}$$

We see that the 'coupling constants' t_k are *harmonic moments* of $\mathsf{D}^c = \mathbb{C} \setminus \mathsf{D}$ and the area of D is $\pi\beta t$. It is the subject of the inverse potential problem to reconstruct the domain from its area and harmonic moments. In general, the solution is not unique. But it is known that *locally*, i.e. for a small enough change $t \to t + \delta t$, $t_k \to t_k + \delta t_k$ there is only one solution.

As an explicitly solvable example, consider the Ginibre-Girko ensemble with the partition function (39.2.8). In this case the support of eigenvalues is an ellipse with the half-axes

$$a = \sqrt{t\,\frac{1+2|t_2|}{1-2|t_2|}}, \qquad b = \sqrt{t\,\frac{1-2|t_2|}{1+2|t_2|}},$$

centred at the point $z_0 = \dfrac{2t_1\bar{t}_2 + \bar{t}_1}{1 - 4|t_2|^2}$, and with the angle between the big axis and the real line being equal to $\frac{1}{2}\arg t_2$. For the model with the potential $W = -|z|^2 + 2\mathrm{Re}(t_3 z^3)$ the droplet of eigenvalues is bounded by a hypotrochoid [Teo05].

39.4.3 Small deformations of the support of eigenvalues

Coming back to the general case, let us examine how the shape of D changes under a small change of the potential W with $t = \hbar N$ fixed. It is convenient to describe small deformations $\mathsf{D} \to \tilde{\mathsf{D}}$, by the normal displacement $\delta n(\zeta)$ at a boundary point ζ (Fig. 39.1). Consider a small variation of the potential W in the condition (39.4.7). To take into account the deformation of the domain, we write, for any fixed function f,

$$\delta\left(\oint_{\partial\mathsf{D}} f(\zeta)d\zeta\right) = \oint_{\partial(\delta\mathsf{D})} f(\zeta)d\zeta = 2i\int_{\delta\mathsf{D}} \bar{\partial} f(\zeta)d^2\zeta \approx 2i\oint_{\partial\mathsf{D}} \bar{\partial} f(\zeta)\delta n(\zeta)|d\zeta|$$

(here $\delta\mathsf{D} = \tilde{\mathsf{D}} \setminus \mathsf{D}$) and thus obtain from (39.4.7):

$$\oint_{\partial\mathsf{D}} \frac{\partial\, \delta W(\zeta)d\zeta}{z-\zeta} + \frac{i}{2}\oint_{\partial\mathsf{D}} \frac{\Delta\, W(\zeta)\delta n(\zeta)}{z-\zeta}\,|d\zeta| = 0. \tag{39.4.9}$$

This integral equation for $\delta n(\zeta)$ can be solved in terms of the exterior Dirichlet boundary value problem. Given any smooth function $f(z)$, let $f^H(z)$ be its *harmonic continuation* from the boundary of D to its exterior, i.e. a unique function such that $\Delta f^H = 0$ in D^c and regular at ∞, and $f^H(z) = f(z)$ for all $z \in \partial\mathsf{D}$. Explicitly, a harmonic function can be reconstructed from its boundary value by means of the formula

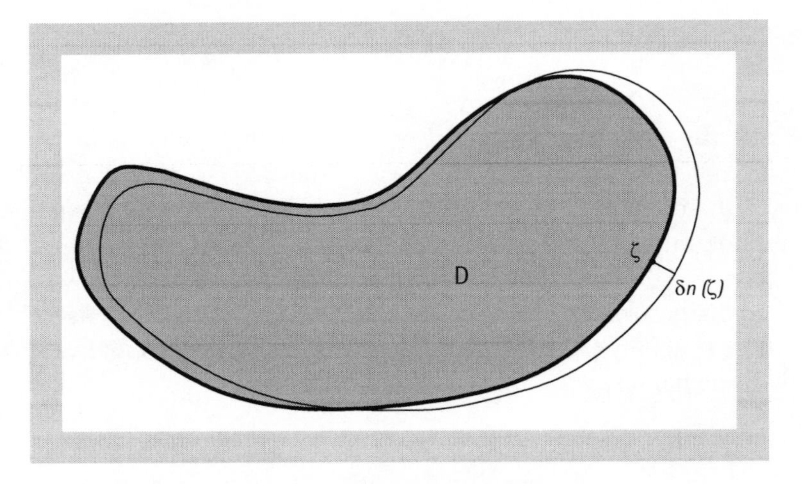

Fig. 39.1 The normal displacement of the boundary.

$$f^H(z) = -\frac{1}{2\pi}\oint_{\partial\mathsf{D}} f(\xi)\partial_n G(z,\xi)|d\xi|\,. \qquad (39.4.10)$$

(Here and below, ∂_n is the normal derivative at the boundary, with the outward pointing normal vector.) The main ingredient of this formula is the Green's function $G(z,\xi)$ of the domain D^c characterized by the properties $\Delta_z G(z,\zeta) = 2\pi\delta^{(2)}(z-\zeta)$ in D^c, $G(z,\zeta) = 0$ if z or $\zeta \in \partial\mathsf{D}$. As $\zeta \to z$, it has the logarithmic singularity $G(z,\zeta) \to \log|z-\zeta|$.

Consider the integral $\oint_{\partial\mathsf{D}} \frac{\partial(\delta W^H(\xi))d\xi}{z-\xi}$ which is obviously equal to 0 for all z inside D, subtract it from the first term in (39.4.9) and rewrite the latter as an integral over the line element $|d\xi|$. After this transformation (39.4.9) acquires the form

$$\frac{1}{2\pi i}\oint_{\partial\mathsf{D}} \frac{R(\xi)}{z-\xi}|d\xi| = 0 \qquad \text{for all } z \in \mathsf{D}, \qquad (39.4.11)$$

where

$$R(z) = \Delta W(z)\delta n(z) + \partial_n^-\left(\delta W(z) - (\delta W)^H(z)\right) \qquad (39.4.12)$$

is a real-valued function on the boundary contour $\partial\mathsf{D}$. The superscript indicates that the derivative is taken in the exterior of the boundary. By properties of Cauchy integrals, it follows from (39.4.11) that $R(\xi)/\tau(\xi)$, where $\tau(\xi) = d\xi/|d\xi|$ is the unit tangential vector to the boundary curve, is the boundary value of an analytic function $h(z)$ in D^c such that $h(\infty) = 0$. For $z \in \mathsf{D}^c$, this function is just given by the integral in the left-hand side. of (39.4.11). Variation of the normalization condition (the second equation in (39.4.6)) yields, in a similar manner:

$$\oint_{\partial\mathsf{D}} R(\xi)|d\xi| = 0 \qquad (39.4.13)$$

This relation implies that the zero at ∞ is at least of the second order.

The following simple argument shows that an analytic function with these properties must be identically zero. Let $w(z)$ be the conformal map from D^c onto the unit disk such that $w(\infty) = 0$ and the derivative at ∞ is real positive. By the well known property of conformal maps we have

$$\frac{dz}{|dz|}e^{i\arg w'(z)} = \frac{dw}{|dw|}$$

along the boundary curve. Therefore, $\tau(z) = i|w'(z)|\,w(z)/w'(z)$ and we thus see that

$$\frac{R(z)\,w'(z)}{i|w'(z)|w(z)}$$

is the boundary value of the holomorphic function $h(z)$. Since $w'(z) \neq 0$ in D^c, the function $g(z) = h(z)\, w(z)/w'(z)$ is holomorphic there with the *purely imaginary* boundary value $\frac{R(z)}{i|w'(z)|}$. Then the real part of this function is harmonic and bounded in D^c and is equal to 0 on the boundary. By uniqueness of a solution to the Dirichlet boundary value problem, $\mathrm{Re}\, g(z)$ must be equal to 0 identically. Therefore, $g(z)$ takes purely imaginary values everywhere in D^c and so is a constant. By virtute of condition (39.4.13) this constant must be 0 which means that $R(z) \equiv 0$.

Therefore, one obtains the following result for the normal displacement of the boundary caused by a small change of the potential $W \to W + \delta W$:

$$\delta n(z) = \frac{\partial_n^- (\delta W^H(z) - \delta W(z))}{\Delta W(z)}\,. \tag{39.4.14}$$

39.4.4 From the support of eigenvalues to an algebraic curve

There is an interesting algebraic geometry behind the large N limit of matrix models. For simplicity, here we consider models with quasiharmonic potentials.

The boundary of the support of eigenvalues is a closed curve in the plane without self-intersections. If $V'(z)$ is a rational function, then this curve is a real section of a complex algebraic curve of finite genus. In fact, this curve encodes the $1/N$ expansion of the model. In the context of Hermitian 2-matrix model such a curve was introduced and studied in [Sta93, Kaz03].

To explain how the curve comes into play, we start from the equation $\partial \varphi_0 = \partial W$, which can be written in the form $\bar{z} - V'(z) = G(z)$ for $z \in \mathsf{D}$, where $G(z) = \dfrac{1}{\pi} \displaystyle\int_{\mathsf{D}} \frac{d^2\zeta}{z - \zeta}$. Clearly, this function is analytic in D^c. At the same time, $V'(z)$ is analytic in D and all its singularities in D^c are poles. Set $S(z) = V'(z) + G(z)$. Then $S(z) = \bar{z}$ on the boundary of the support of eigenvalues. So, $S(z)$ is the analytic continuation of $\bar{z}$ away from the boundary. Assuming that poles of V' are not too close to $\partial \mathsf{D}$, $S(z)$ is well defined at least in a piece of D^c adjacent to the boundary. The complex conjugation yields $\overline{S(z)} = z$, so the function $\bar{S}(z) = \overline{S(\bar{z})}$ must be inverse to the $S(z)$: $\bar{S}(S(z)) = z$ ('unitarity condition'). The function $S(z)$ is called the *Schwarz function* [Dav74].

Under our assumptions, $S(z)$ is an algebraic function, i.e. it obeys a polynomial equation $P(z, S(z)) = 0$ of the form

$$P(z, S(z)) = \sum_{n,l=1}^{d+1} a_{nl} z^n (S(z))^l = 0,$$

where $\overline{a_{ln}} = a_{nl}$ and d is the number of poles of $V'(z)$ (counted with their multiplicities). Here is a sketch of proof. Consider the Riemann surface $\Sigma =$

$\mathsf{D}^c \cup \partial\mathsf{D} \cup (\mathsf{D}^c)^*$ (the *Schottky double* of D^c). Here, $(\mathsf{D}^c)^*$ is another copy of D^c, with the local coordinate $\bar{z}$, attached to it along the boundary. On Σ, there exists an anti-holomorphic involution that interchanges the two copies of D^c leaving the points of $\partial\mathsf{D}$ fixed. The functions z and $S(z)$ are analytically extendable to $(\mathsf{D}^c)^*$ as $\overline{S(z)}$ and $\bar{z}$ respectively. We have two meromorphic functions, each with $d+1$ poles, on a closed Riemann surface. Therefore, they are connected by a polynomial equation of degree $d+1$ in each variable. Hermiticity of the coefficients follows from the unitarity condition.

The polynomial equation $P(z, \tilde{z}) = 0$ defines a complex curve Γ with anti-holomorphic involution $(z, \tilde{z}) \mapsto (\bar{\tilde{z}}, \bar{z})$. The real section is the set of points such that $\tilde{z} = \bar{z}$. It is the boundary of the support of eigenvalues.

It is important to note that for models with non-Gaussian weights (in particular, with polynomial potentials of degree greater than two) the curve has a number of singular points, although the Riemann surface Σ (the Schottky double) is smooth. Generically, these are *double points*, i.e. the points where the curve crosses itself. In our case, a double point is a point $z^{(d)} \in \mathsf{D}^c$ such that $S(z^{(d)}) = \overline{z^{(d)}}$ but $z^{(d)}$ does not belong to the boundary of D. Indeed, this condition means that two different points of Σ, connected by the antiholomorphic involution, are stuck together on the curve Γ, which means self-intersection. The double points play the key role in deriving the nonperturbative (instanton) corrections to the large N matrix models results (see [Dav93] for details).

39.4.5 Free energy and correlation functions

Free energy. The leading contribution to the free energy in the large N limit is

$$F_0 = \lim_{\hbar \to 0} \left(\hbar^2 \log Z_N\right).$$

It is determined by the maximal value of the integrand in (39.2.9), i.e. by the extremum of the function E. In the continuous approximation, one can represent it as a functional of the density:

$$-\beta\hbar^2 E[\rho] = \beta \int\!\!\int \rho(z)\rho(\zeta) \log|z-\zeta| d^2z d^2\zeta + \int W(z)\rho(z) d^2z \qquad (39.4.15)$$

and find its minimum with the constraint $\int \rho\, d^2z = t$. Introducing a Lagrange multiplier λ, we get the equation

$$2\beta \int \log|z-\zeta|\rho(\zeta) d^2\zeta + W(z) + \lambda = 0$$

which is solved, as expected, by the function ρ_0 discussed in Section 39.4.1. Assuming that $W(0) = 0$, we find $\lambda = \varphi_0(0)$ and

$$F_0 = -\beta E[\rho_0] = -\frac{1}{\beta} \int_{\mathsf{D}}\!\int_{\mathsf{D}} \sigma(z) \log\left|\frac{1}{z} - \frac{1}{\zeta}\right| \sigma(\zeta) d^2z d^2\zeta. \qquad (39.4.16)$$

Some results for correlation functions. Here are some results for the correlation functions obtained by the variational technique. (For details of the derivation see [Wie03].) They are correct at distances much larger than the mean distance between the charges. The leading contribution to the one-trace function was already found in Section 39.4.1:

$$\langle \operatorname{tr} f(\Phi) \rangle = \frac{1}{\hbar\beta} \int_{\mathsf{D}} \sigma(z) f(z)\, d^2 z + O(1). \tag{39.4.17}$$

The connected two-trace function is:

$$\langle \operatorname{tr} f(\Phi) \operatorname{tr} g(\Phi) \rangle_c = \frac{1}{4\pi\beta} \int_{\mathsf{D}} \nabla f \nabla g\, d^2 z - \frac{1}{4\pi\beta} \oint_{\partial\mathsf{D}} f \partial_n g^H |dz| + O(\hbar) \tag{39.4.18}$$

(here $\nabla f \nabla g = \partial_x f \partial_x g + \partial_y f \partial_y g$). In particular, for the connected correlation functions of the fields $\varphi(z)$ this formula gives:

$$\langle \varphi(z)\varphi(z') \rangle_c = \begin{cases} -2\beta\hbar^2 \log \dfrac{|z-z'|}{r} + O(\hbar^3) & \text{inside } \mathsf{D} \\ 2\beta\hbar^2 \left(G(z,z') - G(z,\infty) - G(z',\infty) - \log \dfrac{|z-z'|}{r} \right) + O(\hbar^3) & \text{outside } \mathsf{D} \end{cases} \tag{39.4.19}$$

where G is the Green's function of the Dirichlet boundary value problem and

$$r = \exp\left[\lim_{\xi\to\infty} (\log|\xi| + G(\xi,\infty)) \right] \tag{39.4.20}$$

is the external conformal radius of the domain D (the Robin's constant). This result is valid if $|z - z'| \gg \sqrt{\hbar}$. The 2-trace functions are *universal*, i.e. they depend on the shape of the support of eigenvalues only and do not depend on the potential W explicitly. They resemble the two-point functions of the Hermitian 2-matrix model found in [Dau93]; they were also obtained in [Ala84] in the study of thermal fluctuations of a confined 2D Coulomb gas. The structure of the formulae indicates that there are local correlations in the bulk as well as strong long range correlations at the edge of the support of eigenvalues. (See [Jan82] for a similar result in the context of classical Coulomb systems).

39.5 The matrix model as a growth problem

39.5.1 Growth of the support of eigenvalues

When N increases at a fixed potential W, one may say that the support of eigenvalues grows. More precisely, we are going to find how the shape of the support of eigenvalues changes under $t \to t + \delta t$, where $t = N\hbar$, if W stays fixed.

The starting point is the same as for the variations of the potential, and the calculations are very similar as well. Variation of the conditions (39.4.6) yields

$$\oint_{\partial D} \frac{\Delta W(\zeta)\delta n(\zeta)}{z-\zeta} |d\zeta| = 0 \quad \text{for all } z \in \mathsf{D}, \qquad \oint_{\partial D} \Delta W(\zeta)\delta n(\zeta)\, |d\zeta| = -4\pi\beta\delta t\,.$$

The first equation means that $\Delta W(z)\delta n(z)\frac{|dz|}{dz}$ is the boundary value of an analytic function $h(z)$ such that $h(z) = -4\pi\beta\delta t/z + O(z^{-2})$ as $z \to \infty$. The solution for the $\delta n(z)$ is again expressed in terms of the Green's function in D^c:

$$\delta n(z) = -\frac{\beta\delta t}{2\pi\sigma(z)}\,\partial_n G(\infty, z). \tag{39.5.1}$$

For quasiharmonic potentials (with $\sigma = 1/\pi$), the formula simplifies:

$$\delta n(z) = -\frac{\beta}{2}\,\delta t\, \partial_n G(\infty, z). \tag{39.5.2}$$

Identifying t with time, one can say that the normal velocity of the boundary at any point z, $V_n(z) = \delta n(z)/\delta t$, is proportional to the gradient of the Green's function: $V_n(z) \propto -\partial_n G(\infty, z)$.

If the domain is connected, G can be expressed through the conformal map $w(z)$ from D^c onto the exterior of the unit circle:

$$G(z_1, z_2) = \log\left|\frac{w(z_1) - w(z_2)}{1 - w(z_1)\overline{w(z_2)}}\right|. \tag{39.5.3}$$

In particular, $G(\infty, z) = -\log|w(z)|$. As $|z| \to \infty$, $w(z) = z/r + O(1)$, where r is the Robin's constant which enters Eq. (39.4.19). It is easy to see that $\partial_n \log|w(z)| = |w'(z)|$ on $\partial\mathsf{D}$, so one can rewrite the growth law (39.5.2) as $\delta n(z) = \frac{\beta}{2}\,|w'(z)|\,\delta t$.

39.5.2 Laplacian growth

The growth law (39.5.2) is common to many important problems in physics. The class of growth processes, in which dynamics of a moving front (an interface) between two distinct phases is driven by a harmonic scalar field is known under the name *Laplacian growth*. The best known examples are viscous flows in the Hele-Shaw cell, filtration processes in porous media, electrodeposition and solidification of undercooled liquids. A comprehensive list of relevant papers published prior to 1998 can be found in [Gil].

Viscous flows in Hele-Shaw cell. To be specific, we shall speak about an interface between two incompressible fluids with very different viscosities on the plane (say, oil and water). In practice, the 2D geometry is realized in the Hele-Shaw cell – a narrow gap between two parallel glass plates. For a review, see [Ben86, Gus06]. The velocity field in a viscous fluid in the Hele-Shaw cell is proportional to the gradient of pressure p (Darcy's law):

$$\vec{V} = -K\nabla p \,, \qquad K = \frac{b^2}{12\mu} \,.$$

The constant K is called the filtration coefficient, μ is viscosity and b is the size of the gap between the two plates. Note that if $\mu \to 0$, then $\nabla p \to 0$, i.e. pressure in a fluid with negligibly small viscosity is uniform. Incompressibility of the fluids ($\nabla \vec{V} = 0$) implies that the pressure field is harmonic: $\Delta p = 0$. By continuity, the velocity of the interface between the two fluids is proportional to the normal derivative of the pressure field on the boundary: $V_n = -K\partial_n p$.

To be definite, we assume that the Hele-Shaw cell contains a bounded droplet of water surrounded by an infinite 'sea' of oil (another possible experimental setup is an air bubble surrounded by oil or water). Water is injected into the droplet while oil is withdrawn at infinity at a constant rate, as is shown schematically in Fig. 39.2. The latter means that the pressure field behaves as $p \propto -\log|z|$ at large distances. We also assume that the interface is a smooth closed curve Γ. As mentioned above, one may set $p = 0$ inside the water droplet. However, pressure usually has a jump across the interface, so p in general does not tend to zero if one approaches the boundary from outside. This effect is due to *surface tension*. It is hard to give realistic estimates of the surface tension effect from fundamental principles, so one often employs certain ad hoc assumptions.

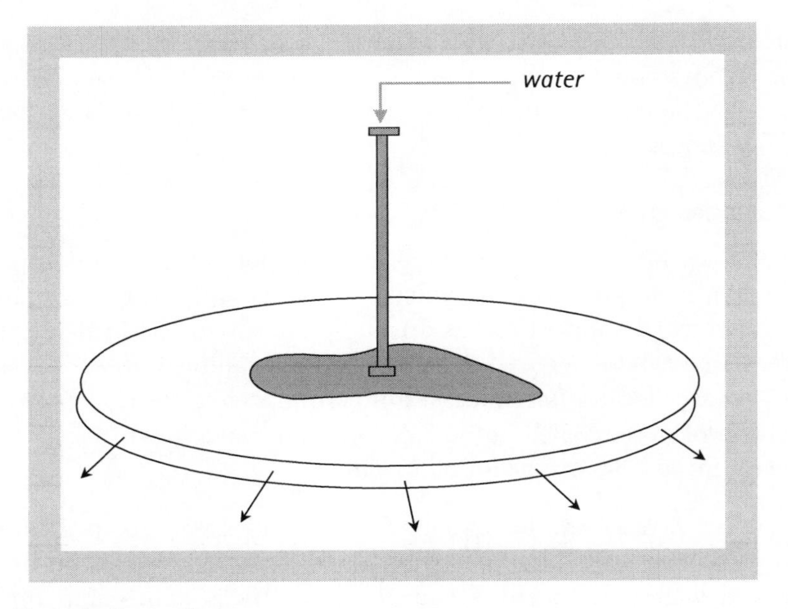

Fig. 39.2 The Hele-Shaw cell.
Source: A. Zabrodin, Matrix models and growth processes: from viscous flows to the quantum Hall effect help-th/0412219, in: Applications of Random Matrices in physics, pp. 261–318, Ed. E. Březin et al., Springer, 2006.

The most popular one is to say that the pressure jump is proportional to the local curvature, κ, of the interface.

To summarize, the mathematical setting of the Saffman-Taylor problem is as follows:

$$\begin{cases} V_n = -\partial_n p & \text{on } \Gamma \\ \Delta p = 0 & \text{in oil} \\ p \to -\log|z| & \text{in oil as } |z| \to \infty \\ p = 0 & \text{in water} \\ p^{(+)} - p^{(-)} = -\nu\kappa & \text{across } \Gamma \end{cases} \tag{39.5.4}$$

Here ν is the surface tension coefficient. (The filtration coefficient is set to be 1.) The experimental evidence suggests that when ν is small enough, the dynamics becomes unstable. Any initial domain develops an unstable fingering pattern. The fingers split into new ones, and after a long lapse of time the water droplet attains a fractal-like structure. This phenomenon is similar to the formation of fractal patterns in the diffusion-limited aggregation.

Comparing (39.5.2) and (39.5.4), we identify D and D^c with the domains occupied by water and oil respectively, and conclude that the growth laws are identical, with the pressure field being given by the Green function: $p(z) = G(\infty, z)$, and $p = 0$ on the interface. The latter means that supports of eigenvalues grow according to (39.5.4) with *zero surface tension*, i.e. with $\nu = 0$ in (39.5.4). Neglecting the surface tension effects, one obtains a good approximation unless the curvature of the interface becomes large. We see that the idealized Laplacian growth problem, i.e. the one with zero surface tension, is mathematically equivalent to the growth of the support of eigenvalues in ensembles of random matrices $\mathcal{N}$ or $\mathcal{C}$ and in general β-ensembles.

Finite-time singularities as critical points. As a matter of fact, the Laplacian growth problem with zero surface tension is ill-posed since an initially smooth interface often becomes singular in the process of evolution, and the solution blows up. The role of surface tension is to inhibit a limitless increase of the interface curvature. In the absence of such a cutoff, the tip of the most rapidly growing finger typically grows to a singularity (a cusp). In particular, a singularity necessarily occurs for any initial interface that is the image of the unit circle under a rational conformal map, with the only exception of an ellipse.

An important fact is that the cusp-like singularity occurs at a finite time $t = t_c$, i.e. at a finite area of the droplet. It can be shown that the conformal radius of the droplet r (as well as some other geometric parameters), as $t \to t_c$, exhibits a singular behaviour

$$r - r_c \propto (t_c - t)^{-\gamma}$$

characterized by a critical exponent γ. The generic singularity is the cusp (2, 3), which in suitable local coordinates looks like $y^2 = x^3$. In this case $\gamma = -\frac{1}{2}$.

A similar phenomenon has been known in the theory of random matrices for quite a long time, and in fact it was the key to their applications to 2D quantum gravity and string theory. In the large N limit, the random matrix models have *critical points* – the points where the free energy is not analytic as a function of a coupling constant. As we have seen, the Laplacian growth time t should be identified with a coupling constant of the normal or complex matrix model. In a vicinity of a critical point,

$$F_0 \sim F_0^{\mathrm{reg}} + a(t_c - t)^{2-\gamma},$$

where the critical index γ (often denoted by γ_{str} in applications to string theory) depends on the type of the critical point. Accordingly, the singularities show up in correlation functions. Using the equivalence established above, we can say that the finite-time blow-up (a cusp-like singularity) of the Laplacian growth with zero surface tension is a critical point of the normal and complex matrix models. Remarkably, the evolution can be continued to the post-critical regime as a dynamics of 'shock lines' [Lee09].

Acknowledgments

I am grateful to O. Agam, E. Bettelheim, I. Kostov, I. Krichever, A. Marshakov, M. Mineev-Weinstein, R. Teodorescu and P. Wiegmann for collaboration. The work was supported in part by RFBR grant 08-02-00287, by grant for support of scientific schools NSh-3035.2008.2 and by Federal Agency for Science and Innovations of Russian Federation (contract 02.740.11.5029).

References

[Ala84] A. Alastuey and B. Jancovici, J. Stat. Phys. **34** (1984) 557;
B. Jancovici, J. Stat. Phys. **80** (1995) 445

[Bal08] F. Balogh and J. Harnad, *Superharmonic perturbations of a Gaussian measure, equilibrium measures and orthogonal polynomials*, To appear in: Complex Analysis and Operator Theory (Special volume in honour of Bjorn Gustafsson), arXiv:0808.1770

[Ben86] D. Bensimon, L.P. Kadanoff, S. Liang, B.I. Shraiman, and C. Tang, Rev. Mod. Phys. **58** (1986) 977

[Cha92] L. Chau and Y. Yu, Phys. Lett. **A167** (1992) 452–458;
L. Chau and O. Zaboronsky, Commun. Math. Phys. **196** (1998) 203

[Dau93] J.-M. Daul, V. Kazakov and I. Kostov, Nucl. Phys. **B409** (1993) 311

[Dav74] P.J. Davis, *The Schwarz function and its applications*, The Carus Math. Monographs, No. 17, The Math. Assotiation of America, Buffalo, N.Y., 1974

[Dav93] F. David, Phys. Lett. **B302** (1993) 403–410, hep-th/9212106;
B. Eynard and J. Zinn-Justin, Phys. Lett. **B302** (1993) 396–402, hep-th/9301004;

V. Kazakov and I. Kostov, hep-th/0403152;
S. Alexandrov, JHEP 0405 (2004) 025, hep-th/0403116

[DiF94] P. Di Francesco, M. Gaudin, C. Itzykson and F. Lesage, Int. J. Mod. Phys. **A9** (1994) 4257–4351

[Dys62] F. Dyson, J. Math. Phys. **3** (1962) 140, 157, 166

[Elb04] P. Elbau and G. Felder, *Density of eigenvalues of random normal matrices*, math.QA/0406604

[For98] P. J. Forrester, Phys. Rep. 301, (1998) 235–270; *Log-Gases and Random Matrices*, London Mathematical Society Monographs no. 34, Princeton University Press, Princeton (2010)

[Fyo97] Y. Fyodorov, B. Khoruzhenko and H.-J. Sommers, Phys. Rev. Lett. **79** (1997) 557, cond-mat/9703152;
J. Feinberg and A. Zee, Nucl. Phys. **B504** (1997) 579–608, cond-mat/9703087;
G. Akemann, J. Phys. A **36** (2003) 3363, hep-th/0204246; Phys. Rev. Lett. **89** (2002) 072002, hep-th/0204068;
A.M. Garcia-Garcia, S.M. Nishigaki and J.J.M. Verbaarschot, Phys. Rev. **E66** (2002) 016132 cond-mat/0202151

[Gil] K.A. Gillow and S.D. Howison, *A bibliography of free and moving boundary problems for Hele-Shaw and Stokes flow*, http://www.maths.ox.ac.uk/ howison/Hele-Shaw/

[Gin65] J. Ginibre, J. Math. Phys. **6** (1965) 440

[Gir85] V. Girko, Theor. Prob. Appl. **29** (1985) 694

[Guh98] T. Guhr, A. Müller-Groeling and H. Weidenmüller, Phys. Rep. **299** (1998) 189–428, cond-mat/9707301

[Gus06] B. Gustafsson, A. Vasil'ev, *Conformal and Potential Analysis in Hele-Shaw Cells*, Birkhäuser Verlag, 2006.

[Hed04] H. Hedenmalm and N. Makarov, *Quantum Hele-Shaw flow*, math.PR/0411437

[Jan82] B. Jancovici, J. Stat. Phys. **28** (1982) 43

[Kaz03] V. Kazakov and A. Marshakov, J. Phys. A **36** (2003) 3107, hep-th/0211236

[Kos01] I. Kostov, I. Krichever, M. Mineev-Weinstein, P. Wiegmann and A. Zabrodin, *τ-function for analytic curves*, Random matrices and their applications, MSRI publications, eds. P. Bleher and A. Its, vol.40, p. 285–299, Cambridge Academic Press, 2001, hep-th/0005259

[Lee09] S.-Y. Lee, R. Teodorescu and P. Wiegmann, Physica **D238** (2009) 1113–1128

[Meh91] M.L. Mehta, *Random matrices and the statistical theory of energy levels*, 2-nd edition, Academic Press, NY, 1991

[Meh04] M.L. Mehta, *Random matrices* (Academic press, 3rd Edision, London, 2004)

[Mor94] A. Morozov, Phys. Usp. **37** (1994) 1–55, hep-th/9303139

[Sta93] M. Staudacher, Phys. Lett **B305** (1993) 332–338, hep-th/9301038

[Teo05] R. Teodorescu, E. Bettelheim, O. Agam, A. Zabrodin and P. Wiegmann, Nucl. Phys. **B704** (2005) 407–444, hep-th/0401165; Nucl. Phys. **B700** (2004) 521–532, hep-th/0407017

[Wie03] P. Wiegmann and A. Zabrodin, J. Phys. A **36** (2003) 3411, hep-th/0210159

[Zab03] A. Zabrodin, Ann. Henri Poincaré **4** Suppl. 2 (2003) S851–S861, cond-mat/0210331

[Zab06] A. Zabrodin, *Matrix models and growth processes: from viscous flows to the quantum Hall effect*, hep-th/0412219, in: "Applications of Random Matrices in Physics", pp. 261–318, Ed. E. Brézin et al., Springer, 2006

·40·

Financial applications of random matrix theory: a short review

Jean-Philippe Bouchaud and Marc Potters

Abstract

We discuss the applications of random matrix theory in the context of Financial markets and econometric models, a topic about which a considerable number of papers have been devoted to in the last decade. This mini-review is intended to guide the reader through various theoretical results (the Marčenko-Pastur spectrum and its various generalizations, random singular value decomposition, free matrices, largest eigenvalue statistics, etc.) as well as some concrete applications to protfolio optimization and out-of-sample risk estimation.

40.1 Introduction

40.1.1 Setting the stage

The Marčenko–Pastur 1967 paper [Mar67] on the spectrum of empirical correlation matrices is both remarkable and precocious. It turned out to be useful in many, very different contexts (neural networks, image processing, wireless communications, etc.) and was unknowingly rediscovered several times. Its primary aim, as a new statistical tool to analyse large dimensional data sets, only became relevant in the last two decades, when the storage and handling of humongous data sets became routine in almost all fields – physics, image analysis, genomics, epidemiology, engineering, economics, and finance, to quote only a few. It is indeed very natural to try to identify common causes (or factors) that explain the dynamics of N quantities. These quantities might be daily returns of the different stocks of the S&P 500, monthly inflation of different sectors of activity, motion of individual grains in a packed granular medium, or different biological indicators (blood pressure, cholesterol, etc.) within a population, etc., etc. (for reviews of other applications and techniques, see [Ver04, Ede05, Joh07]; see also Chapter 28 on multivariate statistics, Chapter 37 on extreme value statistics and entanglement, and Chapter 41 on information theory). We will denote by T the total number of observations of each of the N quantities. In the example of stock returns, T is the total number of trading days in the sampled data; but in the biological example, T is the size of the population. The realization of the ith quantity ($i = 1, \ldots, N$) at 'time' t ($t = 1, \ldots, T$) will be denoted r_i^t, which will be assumed in the following to be demeaned and standardized. The normalized $T \times N$ matrix of returns will be denoted as $\mathbf{X}$: $X_{ti} = r_i^t/\sqrt{T}$. The simplest way to characterize the correlations between these quantities is to compute the Pearson estimator of the correlation matrix:

$$E_{ij} = \frac{1}{T}\sum_{t=1}^{T} r_i^t r_j^t \equiv \left(\mathbf{X}^T\mathbf{X}\right)_{ij}, \tag{40.1.1}$$

where $\mathbf{E}$ will denote the empirical correlation matrix (i.e. on a given realization), that one must carefully distinguish from the 'true' correlation matrix $\mathbf{C}$ of the underlying statistical process (that might not even exist). In fact, the whole point of the Marčenko–Pastur result is to characterize the difference between $\mathbf{E}$ and $\mathbf{C}$. Of course, if N is small (say $N = 4$) and the number of observations is large (say $T = 10^6$), then we can intuitively expect that any observable computed using $\mathbf{E}$ will be very close to its 'true' value, computed using $\mathbf{C}$. For example, a consistent estimator of $\mathrm{Tr}\mathbf{C}^{-1}$ is given by $\mathrm{Tr}\mathbf{E}^{-1}$ when T is large enough for a fixed N. This is the usual limit considered in statistics. However, in many applications where T is large, the number of observables N is also large, such that the ratio $q = N/T$ is not very small compared to one. We will find below that when q is nonzero, and for large N, $\mathrm{Tr}\mathbf{E}^{-1} = \mathrm{Tr}\mathbf{C}^{-1}/(1-q)$. A typical number in the case of stocks is $N = 500$ and $T = 2500$, corresponding to 10 years of daily data, already quite a long strand compared to the lifetime of stocks or the expected structural evolution time of markets. For inflation indicators, 20 years' of monthly data produce a meagre $T = 240$, whereas the number of sectors of activity for which inflation is recorded is around $N = 30$. The relevant mathematical limit to focus on in these cases is $T \gg 1$, $N \gg 1$ but with $q = N/T = O(1)$. The aim of this paper is to review several random matrix theory (RMT) results that can be established in this special asymptotic limit, where the empirical density of eigenvalues (the spectrum) is strongly distorted when compared to the 'true' density (corresponding to $q \to 0$). When $T \to \infty$, $N \to \infty$, the spectrum has some degree of universality with respect to the distribution of the r_i^ts; this makes RMT results particularly appealing. Although the scope of these results is much broader (as alluded to above), we will gird our discussion to the applications of RMT to financial markets. A topic about which a considerable number of papers have been devoted in the last decade (see e.g. [Lal99, Lal00, Pot05, Bou07, Bir07b, Ple99a, Ple02, Bon04, Lil05, Lil08, Paf03, Bur04a, Bur04b, Bur05, Bur03, Mal04, Guh03, Ake10, Mar02, Fra05, Zum09, Bai07]). The following mini-review is intended to guide the reader through various results that we consider to be important, with no claim of being complete. We furthermore chose to state these results in a narrative style, rather than in a more rigorous lemma-theorem fashion. We provide references where more precise statements can be found.

40.1.2 Principal component analysis

The correlation matrix defined above is by construction an $N \times N$ symmetric matrix, that can be diagonalized. This is the basis of the well-known principal component analysis (PCA, see also Chapter 28 Section 28.1.1), aimed at decomposing the fluctuations of the quantity r_i^t into decorrelated contributions (the 'components') of decreasing variance. In terms of the eigenvalues λ_a and eigenvectors $\vec{V}_a$, the decomposition reads:

$$r_i^t = \sum_{a=1}^{N} \sqrt{\lambda_a} V_{a,i}\, \epsilon_a^t \tag{40.1.2}$$

where $V_{a,i}$ is the i-th component of $\vec{V}_a$, and ϵ_a^t are uncorrelated (for different as) random variables of unit variance. Note that the ϵ_a^t are not necessarily uncorrelated in the 'time' direction, and not necessarily Gaussian. This PCA decomposition is particularly useful when there is a strong separation between eigenvalues. For example if the largest eigenvalue λ_1 is much larger than all the others, a good approximation of the dynamics of the N variables r_i reads:

$$r_i^t \approx \sqrt{\lambda_1} V_{1,i}\, \epsilon_1^t, \tag{40.1.3}$$

in which case a single 'factor' is enough to capture the phenomenon. When N is fixed and $T \to \infty$, all the eigenvalues and their corresponding eigenvectors can be trusted to extract meaningful information. As we will review in detail below, this is not the case when $q = N/T = O(1)$, where only a subpart of the eigenvalue spectrum of the 'true' matrix $\mathbf{C}$ can be reliably estimated. In fact, since $\mathbf{E}$ is by construction a sum of T projectors, $\mathbf{E}$ has (generically) $(N-T)^+$ eigenvalues exactly equal to zero, corresponding to the $(N-T)^+$ dimensions not spanned by these T projectors. These zero eigenvalues are clearly spurious and do not correspond to anything real for $\mathbf{C}$.

It is useful to give early on a physical (or rather financial) interpretation of the eigenvectors $\vec{V}_a$. The list of numbers $V_{a,i}$ can be seen as the amount of wealth invested in the different stocks $i = 1, \ldots, N$, defining a portfolio Π_a, where some stocks are 'long' ($V_{a,i} > 0$) while other are 'short' ($V_{a,i} < 0$). The *realized* risk $\mathcal{R}_a^2$ of portfolio Π_a is often defined as the empirical variance of its returns, given by:[1]

$$\mathcal{R}_a^2 = \frac{1}{T}\sum_t \left(\sum_i V_{a,i}\, r_i^t\right)^2 = \sum_{ij} V_{a,i} V_{a,j} E_{ij} \equiv \lambda_a. \tag{40.1.4}$$

[1] Note, as will be discussed in detail below, the realized risk is not the same as the *expected* risk, computed with the 'true' covariance matrix $\mathbf{C}$.

The eigenvalue λ_a is therefore the risk of the investment in portfolio a. Large eigenvalues correspond to a risky mix of assets, whereas small eigenvalues correspond to a particularly quiet mix of assets. Typically, in stock markets, the largest eigenvalue corresponds to investing roughly equally in all stocks: $V_{1,i} = 1/\sqrt{N}$. This is called the 'market mode' and is strongly correlated with the market index. There is no diversification in this portfolio: the only bet is whether the market as a whole will go up or down, this is why the risk is large. Conversely, if two stocks move very tightly together (the canonical example would be Coca-Cola and Pepsi), then buying one and selling the other leads to a portfolio that hardly moves, being only sensitive to events that strongly differentiate the two companies. Correspondingly, there is a small eigenvalue of $\mathbf{E}$ with eigenvector close to $(0, 0, \ldots, \sqrt{2}/2, 0, \ldots, -\sqrt{2}/2, 0, \ldots, 0, 0)$, where the nonzero components are localized on the pair of stocks.

A further property of the portfolios Π_a is that their returns are uncorrelated, since:

$$\frac{1}{T}\sum_t \left(\sum_i V_{a,i}\, r_i^t\right)\left(\sum_j V_{\beta,j}\, r_j^t\right) = \sum_{ij} V_{a,i}\, V_{\beta,j}\, E_{ij} \equiv \lambda_a \delta_{a,\beta}. \qquad (40.1.5)$$

The PCA of the correlation matrix therefore provides a list of 'eigenportfolios', corresponding to uncorrelated investments with decreasing variance.

40.2 Return statistics and portfolio theory

40.2.1 Single asset returns: a short review

Quite far from the simple assumption of textbook mathematical finance, the returns (i.e. the relative price changes) of any kind of traded financial instrument (stocks, currencies, interest rates, commodities, etc.) are very far from Gaussian. The unconditional distribution of returns has fat tails, decaying as a power law for large arguments. In fact, the empirical probability distribution function of returns on shortish time scales (say between a few minutes and a few days) can be reasonably well fitted by a Student-t distribution (see e.g. [Bou03]):[2]

$$P(r) = \frac{1}{\sqrt{\pi}} \frac{\Gamma(\frac{1+\mu}{2})}{\Gamma(\frac{\mu}{2})} \frac{a^\mu}{(r^2 + a^2)^{\frac{1+\mu}{2}}} \qquad (40.2.1)$$

where a is a parameter related to the variance of the distribution through $\sigma^2 = a^2/(\mu - 2)$, and μ is in the range 3 to 5 [Ple99b]. We assume here and in the following that the returns have zero mean, which is appropriate for short

[2] On longer time scales, say weeks to months, the distribution approaches a Gaussian, albeit anomalously slowly (see [Bou03]).

enough time scales: any long term drift is generally negligible compared to σ for time scales up to a few weeks.

This unconditional distribution can however be misleading, since returns are in fact very far from IID random variables. In other words, the returns cannot be thought of as independently drawn Student random variables. For one thing, such a model predicts that upon time aggregation, the distribution of returns is the convolution of Student distributions, which converges far too quickly towards a Gaussian distribution of returns for longer time scales. In intuitive terms, the volatility of financial returns is itself a dynamical variable, that changes over time with a broad distribution of characteristic frequencies. In more formal terms, the return at time t can be represented by the product of a volatility component σ^t and a directional component ξ^t (see e.g. [Bou03]):

$$r^t = \sigma^t \xi^t, \tag{40.2.2}$$

where the ξ^t are IID random variables of unit variance, and σ^t a positive random variable with both fast and slow components. It is to a large extent a matter of taste to choose ξ^t to be Gaussian and keep a high frequency, unpredictable part to σ^t, or to choose ξ^t to be non-Gaussian (for example Student-t distributed) and only keep the low frequency, predictable part of σ^t. The slow part of σ^t is found to be a long memory process, such that its correlation function decays as a slow power-law of the time lag τ (see [Bou03, Muz00] and references therein):[3]

$$\overline{\sigma^t \sigma^{t+\tau}} - \overline{\sigma}^2 \propto \tau^{-\nu}, \qquad \nu \sim 0.1 \tag{40.2.3}$$

It is worth insisting that in Eq. (40.2.2), σ^t and ξ^t are in fact not independent. It is indeed well documented that on stock markets negative past returns tend to increase future volatilities, and vice versa [Bou03]. This is called the 'leverage' effect, and means in particular that the average of quantities such as $\xi^t \sigma^{t+\tau}$ is negative when $\tau > 0$.

40.2.2 Multivariate distribution of returns

Having now specified the monovariate statistics of returns, we want to extend this description to the joint distribution of the returns of N correlated assets. We will first focus on the joint distribution of *simultaneous* returns $\{r_1^t, r_2^t, \ldots, r_N^t\}$. Clearly, all marginals of this joint distribution must resemble the Student-t distribution (40.2.1) above; furthermore, it must be compatible with the (true) correlation matrix of the returns:

$$C_{ij} = \int \prod_k [dr_k]\, r_i r_j\, P(r_1, r_2, \ldots, r_N). \tag{40.2.4}$$

[3] The overline means an average over the volatility fluctuations, whereas the brackets means an average over both the volatility (σ^t) and the directional (ξ^t) components.

Needless to say, these two requirements are weak constraints that can be fulfilled by the joint distribution $P(r_1, r_2, \ldots, r_N)$ in an infinite number of ways. This is referred to as the 'copula specification problem' in quantitative finance. A copula is a joint distribution of N random variables u_i that all have a uniform marginal distribution in [0, 1]; this can be transformed into $P(r_1, r_2, \ldots, r_N)$ by transforming each u_i into $r_i = F_i^{-1}(u_i)$, where F_i is the (exact) cumulative marginal distribution of r_i. The fact that the copula problem is hugely under-constrained has led to a proliferation of possible candidates for the structure of financial asset correlations (for a review, see e.g. [Emb01, Mal06]). Unfortunately, the proposed copulas are often chosen because of mathematical convenience rather than based on a plausible underlying mechanism. From that point of view, many copulas appearing in the literature are in fact very unnatural [Chi10].

There is however a natural extension of the monovariate Student-t distribution that has a clear financial interpretation. If we generalize the above decomposition Eq. (40.2.2) as:

$$r_i^t = s_i\, \sigma^t \xi_i^t, \tag{40.2.5}$$

where the ξ_i^t are correlated Gaussian random variables with a correlation matrix $\widehat{C}_{ij}$ and the volatility σ^t is common to all assets and distributed as:

$$P(\sigma) = \frac{2}{\Gamma(\frac{\mu}{2})} \exp\left[-\frac{\sigma_0^2}{\sigma^2}\right] \frac{\sigma_0^\mu}{\sigma^{1+\mu}}, \tag{40.2.6}$$

where $\sigma_0^2 = 2\mu/(\mu - 2)$ in such a way that $\langle\sigma^2\rangle = 1$, such that s_i is the volatility of the stock i. The joint distribution of returns is then a multivariate Student P_S that reads explicitly:

$$P_S(r_1, r_2, \ldots, r_N) = \frac{\Gamma(\frac{N+\mu}{2})}{\Gamma(\frac{\mu}{2})\sqrt{(\mu\pi)^N \det \widehat{C}}} \frac{1}{\left(1 + \frac{1}{\mu}\sum_{ij} r_i (\widehat{C}^{-1})_{ij} r_j\right)^{\frac{N+\mu}{2}}}, \tag{40.2.7}$$

where we have normalized returns so that $s_i \equiv 1$. The matrix $\widehat{C}_{ij}$ can be estimated from empirical data using a maximum likelihood procedure. Given a time series of stock returns r_i^t, the most likely matrix $\widehat{C}_{ij}$ is given by the solution of the following equation:

$$\widehat{E}_{ij} = \frac{N+\mu}{T} \sum_{t=1}^{T} \frac{r_i^t r_j^t}{\mu + \sum_{mn} r_m^t (\widehat{E}^{-1})_{mn} r_n^t}. \tag{40.2.8}$$

Note that in the Gaussian limit $\mu \to \infty$ for a fixed N, the denominator of the above expression is simply given by μ, and the final expression reduces to the usual Pearson estimator.

This multivariate Student model is in fact too simple to describe financial data since it assumes that there is a unique volatility factor, common to all assets. One expects that in reality several volatility factors are needed. However, the precise implementation of this idea and the resulting form of the multivariate distribution (and the corresponding natural copula) has not been worked out in detail and is still very much a research topic [Chi10].

40.2.3 Risk and portfolio theory

Suppose one builds a portfolio of N assets with weight w_i on the ith asset, with (daily) volatility s_i. If one knew the 'true' correlation matrix C_{ij}, one would have access to the (daily) variance of the portfolio return, given by:

$$\mathcal{R}^2 = \sum_{ij} w_i s_i C_{ij} s_j w_j, \tag{40.2.9}$$

where C_{ij} is the correlation matrix. If one has predicted gains g_i, then the expected gain of the portfolio is $\mathcal{G} = \sum_i w_i g_i$.

In order to measure and optimize the risk of this portfolio, one therefore has to come up with a reliable estimate of the correlation matrix C_{ij}. This is difficult in general since one has to determine the order of $N^2/2$ coefficients out of N time series of length T, and in general T is not much larger than N. As noted in the introduction, typical values of $Q = T/N$ are in the range $1 \to 10$ in most applications. In the following we assume for simplicity that the volatilities s_i are perfectly known (an improved estimate of the future volatility over some time horizon can be obtained using the information distilled by option markets). By redefining w_i as $w_i s_i$ and g_i as g_i/s_i, one can set $s_i \equiv 1$, which is our convention from now on.

The risk of a portfolio with weights w_i constructed *independently* of the past realized returns r_i^t is faithfully measured by:

$$\mathcal{R}_E^2 = \sum_{ij} w_i E_{ij} w_j, \tag{40.2.10}$$

using the empirical correlation matrix $\mathbf{E}$. This estimate is unbiased and the relative mean-square-error one the risk is small ($\sim 1/T$). But when the w are chosen using the observed rs, as we show now, the result can be very different.

Problems indeed arise when one wants to estimate the risk of an optimized portfolio, resulting from a Markowitz optimization scheme, which gives the portfolio with maximum expected return for a given risk or equivalently, the minimum risk for a given return $\mathcal{G}$ (we will study the latter case below). Assuming $\mathbf{C}$ is known, simple calculations using Lagrange multipliers readily yield the optimal weights w_i^*, which read, in matrix notation:

$$\mathbf{w}_C^* = \mathcal{G} \frac{\mathbf{C}^{-1}\mathbf{g}}{\mathbf{g}^T\mathbf{C}^{-1}\mathbf{g}} \tag{40.2.11}$$

One sees that these optimal weights involve the inverse of the correlation matrix, which will be the source of problems, and will require a way to 'clean' the empirical correlation matrix. Let us explain why in detail.

The question is to estimate the risk of this optimized portfolio, and in particular to understand the biases of different possible estimates. We define the following three quantities [Pot05]:

- The 'in-sample' risk R_{in}, corresponding to the risk of the optimal portfolio over the period used to construct it, using $\mathbf{E}$ as the correlation matrix.

$$\mathcal{R}_{\text{in}}^2 = \mathbf{w}_E^{*T}\mathbf{E}\mathbf{w}_E^* = \frac{\mathcal{G}^2}{\mathbf{g}^T\mathbf{E}^{-1}\mathbf{g}} \tag{40.2.12}$$

- The 'true' minimal risk, which is the risk of the optimized portfolio in the ideal world where $\mathbf{C}$ is perfectly known:

$$\mathcal{R}_{\text{true}}^2 = \mathbf{w}_C^{*T}\mathbf{C}\mathbf{w}_C^* = \frac{\mathcal{G}^2}{\mathbf{g}^T\mathbf{C}^{-1}\mathbf{g}} \tag{40.2.13}$$

- The 'out-of-sample' risk which is the risk of the portfolio constructed using $\mathbf{E}$, but observed on the next (independent) period of time. The expected risk is then:

$$\mathcal{R}_{\text{out}}^2 = \mathbf{w}_E^{*T}\mathbf{C}\mathbf{w}_E^* = \frac{\mathcal{G}^2\mathbf{g}^T\mathbf{E}^{-1}\mathbf{C}\mathbf{E}^{-1}\mathbf{g}}{(\mathbf{g}^T\mathbf{E}^{-1}\mathbf{g})^2} \tag{40.2.14}$$

  This last quantity is obviously the most important one in practice.

If we assume that $\mathbf{E}$ is a noisy, but unbiased estimator of $\mathbf{C}$, such that $\overline{\mathbf{E}} = \mathbf{C}$, one can use a convexity argument for the inverse of positive definite matrices to show that in general:

$$\overline{\mathbf{g}^T\mathbf{E}^{-1}\mathbf{g}} \geq \mathbf{g}^T\mathbf{C}^{-1}\mathbf{g} \tag{40.2.15}$$

Hence for large matrices, for which the result is self-averaging, we have that $\mathcal{R}_{\text{in}}^2 \leq \mathcal{R}_{\text{true}}^2 \leq \mathcal{R}_{\text{out}}^2$, where the last inequality trivially follows from optimality. These results show that the out-of-sample risk of an optimized portfolio is larger (and in practice, much larger, see Section 40.5.2 below) than the in-sample risk, which itself is an underestimate of the true minimal risk. This is a general situation: using past returns to optimize a strategy always leads to over-optimistic results because the optimization adapts to the particular realization of the noise, and is unstable in time. Using the random matrix results of the next sections, one can show that for IID returns, with an *arbitrary* 'true' correlation matrix $\mathbf{C}$, the risk of large portfolios obeys: [Paf03], see also [Bur03]

$$\mathcal{R}_{\text{in}} = \mathcal{R}_{\text{true}}\sqrt{1-q} = \mathcal{R}_{\text{out}}(1-q). \tag{40.2.16}$$

where $q = N/T = 1/Q$. The out-of-sample risk is therefore $1/\sqrt{1-q}$ times larger than the true risk, while the in-sample risk is $\sqrt{1-q}$ smaller than the true risk. This is a typical data snooping effect. Only in the limit $q \to 0$ will these risks coincide, which is expected since in this case the measurement noise disappears, and $\mathbf{E} = \mathbf{C}$. In the limit $q \to 1$, on the other hand, the in-sample risk becomes zero since it becomes possible to find eigenvectors (portfolios) with exactly zero eigenvalues, i.e. zero in-sample risk. The underestimation of the risk turns out to be even stronger in the case of a multivariate Student model for returns [Bir07b]. In any case, the optimal determination of the correlation matrix based on empirical data should be such that the ratio $\mathcal{R}^2_{\text{true}}/\mathcal{R}^2_{\text{out}} \leq 1$ is as large as possible. The Markowitz optimal portfolio might not be optimal at all and may overallocate on small eigenvalues, which are dominated by measurement noise and hence unstable. There are several ways to clean the correlation matrix such as to tame these spurious small risk portfolios, in particular based on random matrix theory ideas. We will come back to this point in Section 40.5.2

40.2.4 Nonequal time correlations and more general rectangular correlation matrices

The equal time correlation matrix C_{ij} is clearly important for risk purposes. It is also useful to understand the structure of the market, or more generally the nature of the 'principle components' driving the process under consideration. A natural extension, very useful for prediction purposes, is to study a lagged correlation matrix between past and future returns. Let us define $C_{ij}(\tau)$ as:

$$C_{ij}(\tau) = \langle r_i^t r_j^{t+\tau} \rangle \tag{40.2.17}$$

such that $C_{ij}(\tau = 0) = C_{ij}$ is the standard correlation coefficient. Whereas C_{ij} is clearly a symmetric matrix, $C_{ij}(\tau > 0)$ is in general non-symmetric, and only obeys $C_{ij}(\tau) = C_{ji}(-\tau)$. How does one extend the idea of 'principle components', seemingly associated to the diagonalization of C_{ij}, to this assymetric case?

The most general case looks in fact even worse: one could very well measure the correlation between N 'input' variables X_i, $i = 1, \dots, N$ and M 'output' variables Y_a, $a = 1, \dots, M$. The Xs and Ys may be completely different from one another (for example, X could be production indicators and Y inflation indexes), or, as in the above example the same set of observables but observed at different times: $N = M$, $X_i^t = r_i^t$ and $Y_a^t = r_a^{t+\tau}$. The cross-correlations between Xs and Ys is characterized by a rectangular $N \times M$ matrix $\mathcal{C}$ defined as $\mathcal{C}_{ia} = \langle X_i Y_a \rangle$ (we assume that both Xs and Ys have zero mean and variance unity). If there is a total of T observations, where both X_i^t and Y_a^t, $t = 1, \dots, T$ are observed, the empirical estimate of $\mathcal{C}$ is, after standardizing X and Y:

$$\mathcal{E}_{ia} = \frac{1}{T} \sum_{t=1}^{T} X_i^t Y_a^t. \tag{40.2.18}$$

What can be said about these rectangular, non-symmetric correlation matrices? The singular value decomposition (SVD) answers the question in the following sense: what is the (normalized) linear combination of Xs on the one hand, and of Ys on the other hand, that have the strongest mutual correlation? In other words, what is the best pair of predictor and predicted variables, given the data? The largest singular value $c_{\max}$ and its corresponding left and right eigenvectors answer precisely this question: the eigenvectors tell us how to construct these optimal linear combinations, and the associated singular value gives us the strength of the cross-correlation: $0 \leq c_{\max} \leq 1$. One can now restrict both the input and output spaces to the $N-1$ and $M-1$ dimensional subspaces orthogonal to the two eigenvectors, and repeat the operation. The list of singular values c_a gives the prediction power, in decreasing order, of the corresponding linear combinations. This is called 'canonical component analysis' (CCA) in the literature [Joh08]; surprisingly in view of its wide range of applications, this method of investigation has been somewhat neglected since it was first introduced in 1936 [Hot36].

40.3 Random matrix theory: the bulk

40.3.1 Preliminaries

Random matrix theory (RMT) attempts to make statements about the statistics of the eigenvalues λ_a of large random matrices, in particular the density of eigenvalues $\rho(\lambda)$, defined as:

$$\rho_N(\lambda) = \frac{1}{N}\sum_{a=1}^{N} \delta(\lambda - \lambda_a), \tag{40.3.1}$$

where λ_a are the eigenvalues of the $N \times N$ symmetric matrix $\mathbf{H}$ that belongs to the statistical ensemble under scrutiny. It is customary to introduce the resolvent $G_H(z)$ of $\mathbf{H}$ (also called the Stieltjes transform), where z is a complex number:

$$G_H(z) = \frac{1}{N}\mathrm{Tr}\left[(z\mathbf{I} - \mathbf{H})^{-1}\right]. \tag{40.3.2}$$

In the limit where N tends to infinity, it often (but not always) happens that the density of eigenvalues ρ_N tends almost surely to a unique well-defined density $\rho_\infty(\lambda)$. This means that the density ρ_N becomes independent of the specific realization of the matrix $\mathbf{H}$, provided $\mathbf{H}$ is a 'typical' sample within its ensemble. This property, called 'ergodicity' or 'self-averaging', is extremely important for practical applications since the asymptotic result $\rho_\infty(\lambda)$ can be used to describe the eigenvalue density of a *single* instance. This is clearly one of the keys of the success of RMT.

Several other 'transforms', beyond the resolvent $G(z)$, turn out to be useful for our purposes. One is the so-called 'blue function' $B(z)$, which is the functional inverse of $G(z)$, i.e.: $B[G(z)] = G[B(z)] = z$. The R-transform is simply related to the blue function through [Ver04]:

$$R(z) = B(z) - z^{-1}. \tag{40.3.3}$$

The last object that we will need is more cumbersome. It is called the S-transform and is defined as follows [Ver04]:

$$S(z) = -\frac{1+z}{z}\eta^{-1}(1+z) \quad \text{where} \quad \eta(y) \equiv -\frac{1}{y}G\left(\frac{1}{y}\right). \tag{40.3.4}$$

In the following, we will review several RMT results on the bulk density of states $\rho_\infty(\lambda)$ that can be obtained using an amazingly efficient concept: matrix freeness [Voi92]. The various fancy transforms introduced above will then appear more natural.

40.3.2 Free matrices

Freeness is the generalization to matrices of the concept of independence for random variables. Loosely speaking, two matrices **A** and **B** are mutually free if their eigenbases are related to one another by a random rotation, or put differently, if the eigenvectors of **A** and **B** are almost surely orthogonal. A more precise and comprehensive definition can be found in, e.g. [Ver04], but our simplified definition, and the following examples, will be sufficient for our purposes. For an introdcution to free probability theory see Chapter 22 and references therein, as well as part of Chapter 13.

Let us give two important examples of mutually free matrices. The first one is nearly trivial. Take two fixed matrices **A** and **B**, and choose a certain rotation matrix **O** within the orthogonal group $O(N)$, uniformly over the Haar measure. Then **A** and $\mathbf{O}^T\mathbf{B}\mathbf{O}$ are mutually free. The second is more interesting, and still not very esoteric. Take two matrices $\mathbf{H}_1$ and $\mathbf{H}_2$ chosen independently within the Gaussian orthogonal ensemble (GOE), i.e. the ensemble symmetric matrices such that all entries are IID Gaussian variables. Since the measure of this ensemble of random matrices is invariant under orthogonal transformation, it means that the rotation matrix $\mathbf{O}_1$ diagonalizing $\mathbf{H}_1$ is a random rotation matrix over $O(N)$ (this is actually a convenient numerical method to generate random rotation matrices). The rotation $\mathbf{O}_1^T\mathbf{O}_2$ from the eigenbasis of $\mathbf{H}_1$ to that of $\mathbf{H}_2$ is therefore also random, and $\mathbf{H}_1$ and $\mathbf{H}_2$ are mutually free. More examples will be encountered below.

Now, matrix freeness allows one to compute the spectrum of the sum of matrices, knowing the spectrum of each of the matrices, provided they are mutually free. More precisely, if $R_A(z)$ and $R_B(z)$ are the R-transforms of two free matrices **A** and **B**, then:

$$R_{A+B}(z) = R_A(z) + R_B(z) \tag{40.3.5}$$

This result clearly generalizes the convolution rule for the sum of two independent random variables, for which the logarithm of the characteristic function is additive. Once $R_{A+B}(z)$ is known, one can in principle invert the R-transform to reach the eigenvalue density of $\mathbf{A} + \mathbf{B}$

There is an analogous result for the product of non-negative random matrices. In this case, the S-transform is multiplicative:

$$S_{AB}(z) = S_A(z)\, S_B(z) \tag{40.3.6}$$

In the rest of this section, we will show how these powerful rules allow one to establish very easily several well-known eigenvalue densities for large matrices, as well as some newer results.

40.3.3 Application: Marčenko and Pastur

Consider the definition of empirical correlation when the true correlation matrix is the identity: $\mathbf{C} = \mathbf{I}$. Then, $\mathbf{E}$ is by definition the sum of rank one matrices $\delta E^t_{ij} = (r^t_i r^t_j)/T$, where r^t_i are independent, unit variance random variables. Hence, $\delta \mathbf{E}^t$ has one eigenvalue equal to q (for large N) associated with direction $\mathbf{r}^t$, and $N - 1$ zero eigenvalues corresponding to the hyperplane perpendicular to $\mathbf{r}^t$. The different $\delta \mathbf{E}^t$ are therefore mutually free and one can use the R-transform trick. Since:

$$\delta G^t(z) = \frac{1}{N}\left(\frac{1}{z-q} + \frac{N-1}{z}\right) \tag{40.3.7}$$

Inverting $\delta G(z)$ to first order in $1/N$, the elementary Blue transform reads:

$$\delta B(z) = \frac{1}{z} + \frac{q}{N(1-qz)} \longrightarrow \delta R(z) = \frac{q}{N(1-qz)}. \tag{40.3.8}$$

Using the addition of R-transforms, one then deduces:

$$B_E(z) = \frac{1}{z} + \frac{1}{(1-qz)} \rightarrow G_E(z) = \frac{(z+q-1) - \sqrt{(z+q-1)^2 - 4zq}}{2zq} \tag{40.3.9}$$

which reproduces the well-known Marčenko and Pastur result for the density of eigenvalues (for $q < 1$) [Mar67]:

$$\rho_E(\lambda) = \frac{\sqrt{4\lambda q - (\lambda + q - 1)^2}}{2\pi\lambda q}, \qquad \lambda \in [(1-\sqrt{q})^2, (1+\sqrt{q})^2]. \tag{40.3.10}$$

The remarkable feature of this result is that there should be *no* eigenvalue outside the interval $[(1-\sqrt{q})^2, (1+\sqrt{q})^2]$ when $N \to \infty$. One can check that $\rho_E(\lambda)$ converges towards $\delta(\lambda - 1)$ when $q = 1/Q \to 0$, or $T \gg N$. Using $G_E(z)$, it is straightforward to show that $(1/N)\mathrm{Tr}\mathbf{E}^{-1} = -G_E(0)$ is given by $(1-q)^{-1}$ for $q < 1$. This was alluded to in Section 40.2.3 above. The Marčenko–Pastur is important because of its large degree of universality: as for the Wigner semi-circle (see Chapter 21, section 21.2.1), its holds whenever the random variables r_i^t are IID with a finite second moment (but see Section 40.3.4 below and Chapter 13 for other 'universality classes').[4]

40.3.4 More applications

The case of an arbitrary true correlation matrix

In general, the random variables under consideration are described by a 'true' correlation matrix $\mathbf{C}$ with some non-trivial structure, different from the identity matrix $\mathbf{1}$. Interestingly, the Marčenko–Pastur result for the spectrum of the empirical matrix $\mathbf{E}$ can be extended to a rather general $\mathbf{C}$, and opens the way to characterize the true spectrum ρ_C even with partial information $Q = T/N < \infty$. However, for a general $\mathbf{C}$, the different projectors $r_i^t r_j^t$ cannot be assumed to define uncorrelated directions for different t, even if the random variables r_i^t are uncorrelated in time and the above trick based on R-transforms cannot be used. However, assuming that the r_i^t are Gaussian, the empirical matrix $\mathbf{E}$ can always be written as $\mathbf{C}^{1/2}\hat{\mathbf{X}}[\mathbf{C}^{1/2}\hat{\mathbf{X}}]^T$, where $\hat{\mathbf{X}}$ is an $N \times T$ rectangular matrix of uncorrelated, unit variance Gaussian random variables. But since the eigenvalues of $\mathbf{C}^{1/2}\hat{\mathbf{X}}[\mathbf{C}^{1/2}\hat{\mathbf{X}}]^T$ are the same as those of $\mathbf{C}\hat{\mathbf{X}}\hat{\mathbf{X}}^T$, we can use the S-transform trick mentioned above, with $\mathbf{A} = \mathbf{C}$ and $\mathbf{B} = \hat{\mathbf{X}}\hat{\mathbf{X}}^T$ mutually free, and where the spectrum of $\mathbf{B}$ is by construction given by the Marčenko–Pastur law. This allows one to write down the following self-consistent for the resolvent of $\mathbf{E}$ [Sil95, Bur04a]:

$$G_E(z) = \int d\lambda\, \rho_C(\lambda) \frac{1}{z - \lambda(1 - q + qzG_E(z))}, \tag{40.3.11}$$

a result that in fact already appears in the original Marčenko–Pastur paper! One can check that if $\rho_C(\lambda) = \delta(\lambda - 1)$, one recovers the result given by Eq. (40.3.9). Equivalently, the above relation can be written as:

$$zG_E(z) = ZG_C(Z) \qquad \text{where} \qquad Z = \frac{z}{1 + q(zG_E(z) - 1)}, \tag{40.3.12}$$

which is convenient for numerical evaluation [Bur04a]. From these equations, one can evaluate $-G_E(0) = \mathrm{Tr}\mathbf{E}^{-1}$, which is found to be equal to $\mathrm{Tr}\mathbf{C}^{-1}/(1-q)$, as we mentioned in the introduction, and used to derive Eq. (40.2.16) above.

Note that while the mapping between the true spectrum ρ_C and the empirical spectrum ρ_E is numerically stable, the inverse mapping is unstable, a little bit like the inversion of a Laplace transform. In order to reconstruct the spectrum of $\mathbf{C}$ from that of $\mathbf{E}$ one should therefore use a parametric ansatz of ρ_C to fit the

[4] A stronger statement, to which we will return below, is that provided the r_i^t have a finite *fourth* moment, the largest eigenvalue of the empirical correlation matrix $\mathbf{E}$ tends to the upper edge of the Marčenko–Pastur distribution, $\lambda_+ = (1 + \sqrt{q})^2$.

observed ρ_E, and not try to directly invert the above mapping (for more on this, see [Bur04b, Kar08]).

Note also that the above result does *not* apply when **C** has isolated eigenvalues, and only describes continuous parts of the spectrum. For example, if one considers a matrix **C** with one large eigenvalue that is separated from the 'Wishart sea', the statistics of this isolated eigenvalue has recently been shown to be Gaussian [Bai05] (see also below), with a width $\sim T^{-1/2}$, much smaller than the uncertainty on the bulk eigenvalues ($\sim q^{1/2}$). A naive application of Eq. (40.3.12), on the other hand, would give birth to a 'mini-Wishart' distribution around the top eigenvalue. This would be the exact result only if the top eigenvalue of C had a degeneracy proportional to N.

The Student ensemble case

Suppose now that the r_i^t are chosen according to the Student multivariate distribution described in Section 40.2.2 above. Since in this case $r_i^t = \sigma_t \xi_i^t$, the empirical correlation matrix can be written as:

$$E_{ij} = \frac{1}{T} \sum_t \sigma_t^2 \xi_i^t \xi_j^t, \qquad \langle \xi_i \xi_j \rangle \equiv \hat{C}_{ij} \tag{40.3.13}$$

In the case where $\hat{\mathbf{C}} = \mathbf{1}$, this can again be seen as a sum of mutually free projectors, and one can use the R-transform trick. This allows one to recover the following equation for the resolvent of **E**, first obtained in the Marčenko–Pastur paper and exact in the large N, T limit:

$$\lambda = \frac{G_R}{G_R^2 + \pi^2 \rho_E^2} + \int ds\, P(s) \frac{\mu(s - q\mu G_R)}{(s - q\mu G_R)^2 + \pi^2 \rho_E^2} \tag{40.3.14}$$

$$0 = \rho \left(-\frac{1}{G_R^2 \pi^2 \rho_E^2} + \int ds\, P(s) \frac{q\mu^2}{(s - q\mu G_R)^2 + \pi^2 \rho_E^2} \right), \tag{40.3.15}$$

where G_R is the real part of the resolvent, and $P(s) = s^{\mu/2-1} e^{-s} / \Gamma(\mu/2)$ is the distribution of $s = \mu/\sigma^2$ in the case of a Student distribution; however, the above result holds for other distributions of σ as well, corresponding to the class of 'elliptic' multivariate distributions (see also Chapter 13, Section 13.4). The salient results are [Bir07b]: (i) there is no longer any upper edge of the spectrum: $\rho_E(\lambda) \sim \lambda^{-1-\mu/2}$ when $\lambda \to \infty$; (ii) but there is a lower edge to the spectrum for all μ. The case $\hat{\mathbf{C}} \neq \mathbf{1}$ can also be treated using S-transforms.

Instead of the usual (Pearson) estimate of the correlation matrix, one could use the maximum likelihood procedure, Eq. (40.2.8) above. Surprisingly at first sight, the corresponding spectrum $\rho_{ML}(\lambda)$ is then completely different [Bir07b], and is given by the standard Marčenko–Pastur result! The intuitive reason is that the maximum likelihood estimator Eq. (40.2.8) effectively renormalizes the returns by the daily volatility σ_t when σ_t is large. Therefore, all the anomalies brought about by 'heavy days' (i.e. $\sigma_t \gg \sigma_0$) disappear.

Finally, we should mention that another Student random-matrix ensemble has been considered in the recent literature, where instead of having a time-dependent volatility σ_t, it is the global volatility σ that is random, and distributed according to Eq. (40.2.6) [Boh04, Bur06, Ake10]. The density of states is then simply obtained by averaging over Marčenko–Pastur distributions of varying width. Note however that in this case the density of states is *not* self-averaging: each matrix realization in this ensemble will lead to a Marčenko–Pastur spectrum, albeit with a random width.

40.3.5 Random SVD

As we mentioned in Section 40.2.4, it is often interesting to consider non-symmetrical, or even rectangular correlation matrices, between N 'input' variables X and M 'output' variables Y. The empirical correlation matrix using T-long times series is defined by Eq. (40.2.18). What can be said about the singular value spectrum of $\mathcal{E}$ in the special limit $N,\ M,\ T \to \infty$, with $n = N/T$ and $m = M/T$ fixed? Whereas the natural null hypothesis for correlation matrices is $\mathbf{C} = \mathbf{1}$, which leads to the Marčenko–Pastur density, the null hypothesis for cross-correlations between *a priori* unrelated sets of input and output variables is $\mathcal{C} = \mathbf{0}$. However, in the general case, input and output variables can very well be correlated between themselves, for example if one chooses redundant input variables. In order to establish a universal result, one should therefore consider the *exact* normalized principal components for the sample variables Xs and Ys:

$$\hat{X}^t_a = \frac{1}{\sqrt{\lambda_a}} \sum_i V_{a,i} X^t_i; . \tag{40.3.16}$$

and similarly for the $\hat{Y}^t_a$. The λ_a and the $V_{a,i}$ are the eigenvalues and eigenvectors of the sample correlation matrix $\mathbf{E}_X$ (or, respectively $\mathbf{E}_Y$). We now define the normalized $M \times N$ cross-correlation matrix as $\hat{\mathcal{E}} = \hat{Y}\hat{X}^T$. One can then use the following tricks [Bou07]:

- The nonzero eigenvalues of $\hat{\mathcal{E}}^T\hat{\mathcal{E}}$ are the same as those of $\hat{X}^T\hat{X}\hat{Y}^T\hat{Y}$
- $\mathbf{A} = \hat{X}^T\hat{X}$ and $\mathbf{B} = \hat{Y}^T\hat{Y}$ are two mutually free $T \times T$ matrices, with N (M) eigenvalues exactly equal to 1 (due to the very construction of $\hat{X}$ and $\hat{Y}$), and $(T-N)^+$ ($(T-M)^+$) equal to 0.
- The S-transforms are multiplicative, allowing one to obtain the spectrum of $\mathbf{AB}$.

Due to the simplicity of the spectra of $\mathbf{A}$ and $\mathbf{B}$, the calculation of S-transforms is particularly easy [Bou07]. The final result for the density of nonzero singular values (i.e. the square-root of the eigenvalues of $\mathbf{AB}$) reads (see [Wac80] for an early derivation of this result, see also [Joh08]):

$$\rho(c) = \max(m + n - 1, 0)\delta(c - 1) + \mathrm{Re}\frac{\sqrt{(c^2 - \gamma_-)(\gamma_+ - c^2)}}{\pi c(1 - c^2)}, \tag{40.3.17}$$

where $n = N/T$, $m = M/T$ and $\gamma_\pm$ are given by:

$$\gamma_\pm = n + m - 2mn \pm 2\sqrt{mn(1 - n)(1 - m)}, \quad 0 \le \gamma_\pm \le 1 \tag{40.3.18}$$

The allowed cs are all between 0 and 1, as they should since these singular values can be interpreted as correlation coefficients. In the limit $T \to \infty$ at fixed N, M, all singular values collapse to zero, as they should since there is no true correlation between X and Y; the allowed band in the limit $n, m \to 0$ becomes:

$$c \in \left[\frac{|m - n|}{\sqrt{m} + \sqrt{n}}, \frac{m - n}{\sqrt{m} - \sqrt{n}}\right], \tag{40.3.19}$$

showing that for fixed N, M, the order of magnitude of allowed singular values decays as $T^{-1/2}$.

Note that one could have considered a different benchmark ensemble, where one considers two independent vector time series X and Y with true correlation matrices C_X and C_Y equal to $\mathbf{1}$. The direct SVD spectrum in that case can also be computed as the S-convolution of two Marčenko–Pastur distributions with parameters m and n [Bou07], This alternative benchmark is however not well suited in practice, since it mixes up the possibly non-trivial correlation structure of the input variables and of the output variables themselves with the *cross*-correlations between these variables.

As an example of applications to economic time series, we have studied in [Bou07] the cross correlations between 76 different macroeconomic indicators (industrial production, retail sales, new orders, and inventory indices of all economic activity sectors available, etc.) and 34 indicators of inflation, the composite price indices (CPIs), concerning different sectors of activity during the period June 1983–July 2005, corresponding to 265 observations of monthly data. The result is that only one, or perhaps two singular values emerge from the above 'noise band'. From an econometric point of view, this is somewhat disappointing: there seems to be very little exploitable signal in spite of the quantity of available observations.

40.4 Random matrix theory: the edges

40.4.1 The Tracy–Widom region

As we have alluded to several times, the practical usefulness of the above predictions for the eigenvalue spectra of random matrices is (i) their universality with respect to the distribution of the underlying random variables (see Chapters 6 and 21 on this universality in invariant and Wigner ensembles respectively) and (ii) the appearance of sharp edges in the spectrum, meaning that the existence of eigenvalues lying outside the allowed band is a strong indication against several null hypothesis benchmarks.

However, the above statements are only true in the asymptotic, $N, T \to \infty$ limit. For large but finite N one expects that the probability to find an eigenvalue is very small but finite. The width of the transition region, and the tail of the density of states was understood a while ago, culminating in the beautiful results by Tracy & Widom on the distribution of the *largest* eigenvalue of a random matrix. There is now a huge literature on this topic (see e.g. [Joh08, Joh00, Bai05, Pec03, Maj06]) that we will not attempt to cover here in detail. We will only extract a few interesting results for applications.

The Tracy–Widom result is that for a large class of $N \times N$ matrices (e.g. symmetric random matrices with IID elements with a finite fourth moment, or empirical correlation matrices of IID random variables with a finite fourth moment), the rescaled distribution of $\lambda_{\max} - \lambda_+$ (where λ_+ is the edge of the spectrum for $N \to \infty$) converges towards the Tracy–Widom distribution, usually noted F_1:

$$\text{Prob}\left(\lambda_{\max} \leq \lambda_+ + \gamma N^{-2/3} u\right) = F_1(u), \tag{40.4.1}$$

where γ is a constant that depends on the problem. For example, for the Wigner problem, $\lambda_+ = 2$ and $\gamma = 1$; whereas for the Marčenko–Pastur problem, $\lambda_+ = (1 + \sqrt{q})^2$ and $\gamma = \sqrt{q}\lambda_+^{2/3}$. For more details on the Tracy–Widom distribution see Chapter 21, Section 21.4.

Note that the distribution of the smallest eigenvalue $\lambda_{\min}$ around the lower edge λ_- is also Tracy–Widom, except in the particular case of Marčenko–Pastur matrices with $Q = 1$. In this case, $\lambda_- = 0$ which is a 'hard' edge since all eigenvalues of the empirical matrix must be non-negative. This special case is treated in, e.g. [Pec03].

Finally, the distance of the largest singular value from the edge of the random SVD spectrum, Eq. (40.3.17) above, is also governed by a Tracy–Widom distribution, with parameters discussed in details in [Joh10].

40.4.2 The case with large, isolated eigenvalues and condensation transition

The Wigner and Marčenko–Pastur ensembles are in some sense maximally random: no prior information on the structure of the matrices is assumed. For applications, however, this is not necessarily a good starting point. In the example of stock markets, it is intuitive that all stocks are sensitive to global news about the economy, for example. This means that there is at least one common factor to all stocks, or else that the correlation coefficient averaged over all pairs of stocks, is positive. A more reasonable null-hypothesis is that the true correlation matrix is: $C_{ii} = 1$, $C_{ij} = \overline{\rho}, \forall i \neq j$. This essentially amounts to adding to the empirical correlation matrix a rank one perturbation matrix with one large eigenvalue $N\overline{\rho}$, and $N - 1$ zero eigenvalues. When $N\rho \gg 1$, the empirical correlation matrix will obviously also have a large eigenvalue close to $N\rho$, very far above the Marčenko–Pastur upper edge λ_+. What happens when $N\overline{\rho}$ is not very large compared to unity?

This problem was solved in great detail by Baik, Ben Arous, and Péché [Bai05], who considered the more general case where the true correlation matrix has k special eigenvalues, called 'spikes'. A similar problem arises when one considers Wigner matrices, to which one adds a perturbation matrix of rank k. For example, if the random elements H_{ij} have a nonzero mean $\overline{h}$, the problem is clearly of that form: the perturbation has one nonzero eigenvalue $N\overline{h}$, and $N - 1$ zero eigenvalues. As we discuss now using free random matrix techniques, this problem has a sharp phase transition between a regime where this rank one perturbation is weak and is 'dissolved' in the Wigner semi-circle sea of eigenvalues, and a regime where this perturbation is strong enough to escape from the Wigner sea. This transition corresponds to a 'condensation' of the eigenvector corresponding to the largest eigenvalue onto the eigenvalue of the rank one perturbation.

Let us be more precise using R-transform techniques for the Wigner problem. Assume that the nonzero eigenvalue of the rank one perturbation is Λ, with a corresponding eigenvector $\vec{e}_1 = (1, 0, \ldots, 0)$. Using free matrix arguments or direct perturbation theory, one can show that the largest eigenvalue of the perturbed matrix reads [Fer07, Bir07a]:

$$\lambda_{\max} = \Lambda + \frac{1}{\Lambda} \quad (\Lambda > 1); \qquad \lambda_{\max} = 2 \quad (\Lambda \leq 1) \tag{40.4.2}$$

Therefore, the largest eigenvalue pops out of the Wigner sea precisely when $\Lambda = 1$. The statistics of the largest eigenvalue $\lambda_{\max}$ is still Tracy–Widom whenever $\Lambda < 1$, but becomes Gaussian, of width $N^{-1/2}$ (and not $N^{-2/3}$) when $\Lambda > 1$. The case $\Lambda = 1$ is special and is treated in [Bai05]. Using simple perturbation theory, one can also compute the overlap between the largest eigenvector $\vec{V}_{\max}$ and $\vec{e}_1$ [Bir07a]:

$$(\vec{V}_{\max} \cdot \vec{e}_1)^2 = 1 - \Lambda^{-2}, \quad (\Lambda > 1), \tag{40.4.3}$$

showing that indeed, the coherence between the largest eigenvector and the perturbation becomes progressively lost when $\Lambda \to 1^+$.

A similar phenomenon takes place for correlation matrices. For a rank one perturbation of the type described above, with an eigenvalue $\Lambda = N\rho$, the criterion for expelling an isolated eigenvalue from the Marčenko–Pastur sea now reads [Bai05]:

$$\lambda_{\max} = \Lambda + \frac{\Lambda q}{\Lambda - 1} \ (\Lambda > 1 + \sqrt{q}); \ \ \lambda_{\max} = (1+\sqrt{q})^2 \ (\Lambda \leq 1+\sqrt{q}). \quad (40.4.4)$$

Note that in the limit $\Lambda \to \infty$, $\lambda_{\max} \approx \Lambda + q + O(\Lambda^{-1})$. For rank k perturbation, all eigenvalues such that $\Lambda_r > 1 + \sqrt{q}$, $1 \leq r \leq k$ will end up isolated above the Marčenko–Pastur sea, all others disappear below λ_+. All these isolated eigenvalues have Gaussian fluctuations of order $T^{-1/2}$. For more details about these results, see [Bai05].

40.4.3 The largest eigenvalue problem for heavy-tailed matrices

The Tracy–Widom result for the largest eigenvalue was first shown for the GOE, but it was soon understood that the result is more general. In fact, if the matrix elements are IID with a finite fourth moment, the largest eigenvalue statistics is *asymptotically* governed by the Tracy–Widom mechanism. Let us give a few heuristic arguments for this [Bir07a]. Suppose the matrix elements are IID with a power-law distribution:

$$P(H) \sim_{|H| \to \infty} \frac{A^{\mu}}{|H|^{1+\mu}} \ \text{with} \ A \sim O(1/\sqrt{N}). \quad (40.4.5)$$

and $\mu > 2$, such that the asymptotic eigenvalue spectrum is the Wigner semicircle with $\lambda_\pm = \pm 2$. The largest element $H_{\max}$ of the matrix (out of $N^2/2$) is therefore of order $N^{2/\mu - 1/2}$ and distributed with a Fréchet law. From the results of the previous subsection, one can therefore expect that:

- If $\mu > 4$: $H_{\max} \ll 1$, and one recovers Tracy–Widom.
- If $2 < \mu < 4$: $H_{\max} \gg 1$, $\lambda_{\max} \approx H_{\max} \propto N^{\frac{2}{\mu} - \frac{1}{2}}$, with a Fréchet distribution. Note that although $\lambda_{\max} \to \infty$ when $N \to \infty$, the density itself goes to zero when $\lambda > 2$ in the same limit.
- If $\mu = 4$: $H_{\max} \sim O(1)$, $\lambda_{\max} = 2$ or $\lambda_{\max} = H_{\max} + 1/H_{\max}$, corresponding to a non-universal distribution for $\lambda_{\max}$ with a δ-peak at 2 and a transformed Fréchet distribution for $\lambda_{\max} > 2$.

Although the above results are expected to hold for $N \to \infty$ (a rigorous proof can be found in [Auf09]), one should note that there are very strong finite size corrections. In particular, although for $\mu > 4$ the asymptotic limit is Tracy–Widom, for any finite N the distribution of the largest eigenvalue has power-law tails that completely mask the Tracy–Widom distribution – see [Bir07a]. Similarly, the convergence towards the Fréchet distribution for $\mu < 4$ is also very slow. For an introduction to heavy-tailed ensembles, see Chapter 13.

40.5 Applications: cleaning correlation matrices

40.5.1 Empirical eigenvalue distributions

Having now all the necessary theoretical tools to hand, we turn to the analysis of empirical correlation matrices of stock returns. Many such studies, comparing the empirical spectrum with RMT predictions, have been published in the literature. Here, we perform this analysis once more, on an extended data set, with the objective of comparing precisely different cleaning schemes for risk control purposes (see next Section, 40.5.2).

We study the set of US stocks between July, 1993 and April, 2008 (3700 trading days). We consider 26 samples obtained by sequentially sliding a window of $T = 1000$ days by 100 days. For each period, we look at the empirical correlation matrix of the $N = 500$ most liquid stocks during that period. The quality factor is therefore $Q = T/N = 2$. The eigenvalue spectrum shown in Fig. 40.2 is an average over the 26 sample eigenvalue distributions, where we have removed the market mode and rescaled the eigenvalues such that $\int d\lambda \rho_E(\lambda) = 1$ for each sample. The largest eigenvalue contributes on average to 21% of the total trace.

We compare in Fig. 40.1 the empirical spectrum with the Marčenko–Pastur prediction for $Q = 1/q = 2$. It is clear that several eigenvalues leak out of the Marčenko–Pastur band, even after taking into account the Tracy–Widom tail, which have a width given by $\sqrt{q}\lambda_+^{2/3}/N^{2/3} \approx 0.02$, very small in the present case. The eigenvectors corresponding to these eigenvalues show significant structure, that correspond to identifiable economic sectors. Even after accounting for these large eigenvalues, the Marčenko–Pastur prediction is not very good, suggesting that the prior for the underlying correlation matrix $\mathbf{C}$ may carry more structure than just a handful of eigenvalue 'spikes' on top of the identity matrix [Bur04b, Mal04]. An alternative simple prior for the spectrum of $\mathbf{C}$ is a power-law distribution, corresponding to the coexistence of large sectors and smaller sectors of activity:

$$\rho_C(\lambda) = \frac{\mu A}{(\lambda - \lambda_0)^{1+\mu}} \Theta(\lambda - \lambda_{\min}), \tag{40.5.1}$$

with A and λ_0 related to $\lambda_{\min}$ by the normalization of ρ_C and by $\mathrm{Tr}\mathbf{C} = N$ (the latter requiring $\mu > 1$). Using Eq. (40.3.12) one can readily compute the dressed spectrum $\rho_E(\lambda)$. In Fig. 40.2, we show, on top of the empirical and Marčenko–Pastur spectrum, the 'bare' and the dressed power-law spectrum for $\mu = 2$. For later convenience, we parameterize the distribution using $a = \lambda_{\min} \in [0, 1]$, in which case $A = (1 - a)^2$ and $\lambda_0 = 2a - 1$ (note that $a = 1$ corresponds to the Marčenko–Pastur case since in this limit $\rho_C(\lambda) = \delta(\lambda - 1)$). The fit shown in Fig. 40.1 corresponds to $a = 0.35$, and is now very good, suggesting indeed that the correlation of stocks has a hierarchical structure with a power-law distribution for the size of sectors (on this point, see also [Mar02]). We should point out that

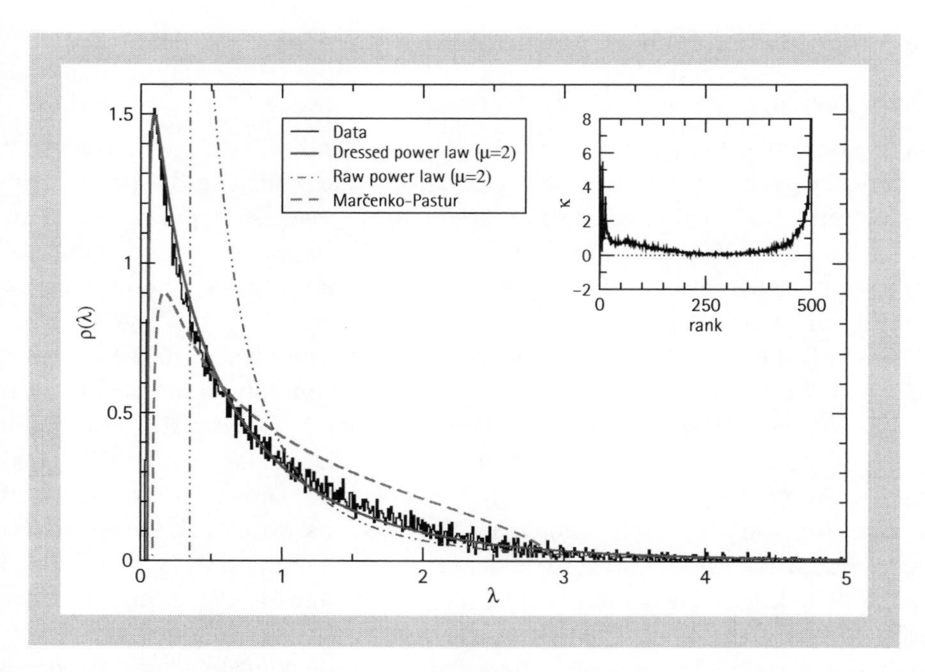

Fig. 40.1 Main figure: empirical average eigenvalues spectrum of the correlation matrix (plain black line), compared to (a) the Marčenko–Pastur prediction (dashed line) and the dressed power-law spectrum model (thick line). We also show the bare power law distribution with $\mu = 2$ and the optimal value of $\lambda_{\min}$ (dashed-dotted line). Inset: Kurtosis of the components of the eigenvectors as a function of the eigenvalue rank. One clearly sees some structure emerging at both ends of the spectrum, whereas the centre of the band is compatible with rotationally invariant eigenvectors.

a fit using a multivariate Student model also works very well for the Pearson estimator of the empirical correlation matrix. However, as noted in [Bir07b], such an agreement appears to be accidental. If the Student model was indeed appropriate, the spectrum of the most likely correlation matrix (see Eq. (40.2.8)) should be given by Marčenko–Pastur, whereas the data does not conform to this prediction [Bir07b]. This clearly shows that the Student copula is in fact not adequate to model multivariate correlations.

A complementary piece of information is provided by the statistics of the eigenvectors. Structure-less eigenvectors (i.e. a normalized random vector in N dimensions) have components that follow a Gaussian distribution of variance $1/N$. The kurtosis of the components for a given eigenvector gives some indication of its 'non-random' character (and is trivially related to the well-known inverse participation ratio or Herfindahl index). We show in the inset of Fig. 40.1 the excess kurtosis as a function of the rank of the eigenvectors (small ranks corresponding to large eigenvectors). We clearly see that both the largest and the smallest eigenvectors are not random, while the eigenvectors at the middle of the band have a very small excess kurtosis. As mentioned above,

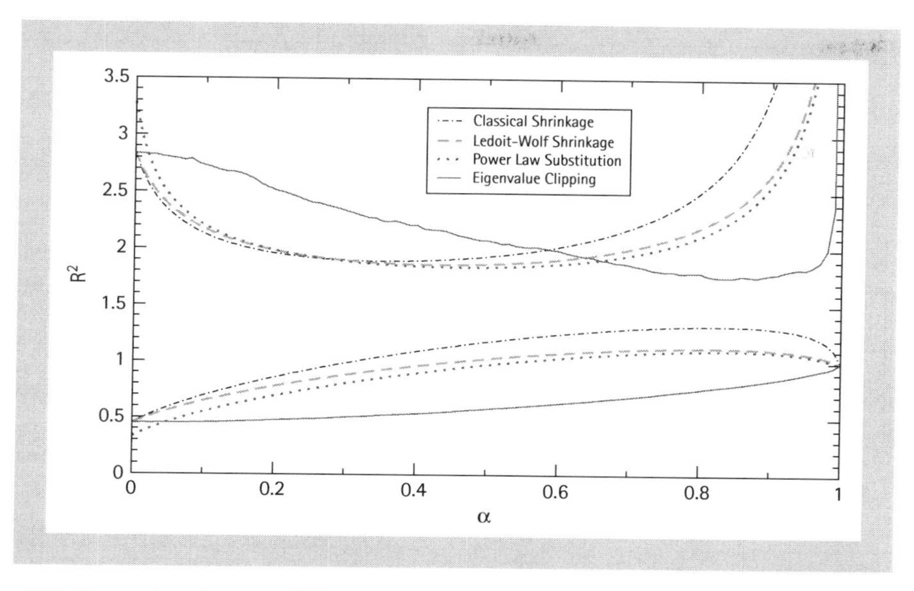

Fig. 40.2 Comparison between different correlation matrix cleaning schemes for Markowitz optimization. Top curves: out-of-sample squared risk $\mathcal{R}^2_{\text{out}}$ as a function of the cleaning parameter α (see Eq. 40.5.3). $\alpha = 0$ corresponds to the raw empirical correlation matrix, and $\alpha = 1$ to the identity matrix. The best cleaning corresponds to the smallest out-of-sample risk. The 'true' risk for this problem is $\mathcal{R}_{\text{true}} = 1$. Bottom curves: in-sample risk of the optimal portfolio as a function of α.

large eigenvalues correspond to economic sectors, while small eigenvalues correspond to long-short portfolios that invest on fluctuations with particularly low volatility, for example the difference between two very strongly correlated stocks within the same sector.

40.5.2 RMT inspired cleaning recipes

As emphasized in Section 40.2.3, it is a bad idea to at all use directly the empirical correlation matrix in a Markowitz optimization program. We have seen that the out-of-sample risk is at best underestimated by a factor $(1 - q)$, but the situation might be worsened by tail effects and/or by the non-stationarity of the true correlations. Since we know that measurement noise, induced by the finite size effects, significantly distort the spectrum of the correlation matrix, one should at the very least try to account for these noise effects before using the correlation matrix in any optimization program. With the above RMT results in mind, several 'cleaning schemes' can be devised. These cleaning schemes must of course be mathematically consistent (the cleaned matrix must correspond to a correlation matrix), and aim at optimizing some cleaning quality factor. Some will be given below.

The simplest recipe, first suggested and tested in [Lal00], is to keep unaltered all the eigenvalues (and the corresponding eigenvectors) that exceed the

Marčenko–Pastur edge $(1+\sqrt{q})^2$, while replacing all eigenvalues below the edge, deemed as meaningless noise, but a common value $\overline{\lambda}$ such that the trace of the cleaned matrix remains equal to N. We call this procedure eigenvalue clipping, and will consider a generalized version where the $(1-\alpha)N$ largest eigenvalues are kept while the $N\alpha$ smallest ones are replaced by a common value $\overline{\lambda}$.

A more sophisticated cleaning is inspired by the power-law distribution model described above. If the true distribution is given by Eq. (40.5.1), then we expect the kth eigenvalue λ_k to be around the value:[5]

$$\lambda_k \approx \lambda_0 + \left(A\frac{N}{k}\right)^{1/\mu} \stackrel{\mu=2}{\longrightarrow} 2\alpha - 1 + (1-\alpha)\sqrt{\frac{N}{k}} \tag{40.5.2}$$

The 'power-law' cleaning procedure is therefore to fix $\mu = 2$ and let α vary to generate a list of synthetic eigenvalues using the above equation Eq. (40.5.2) for $k > 1$, while leaving the corresponding kth eigenvector untouched. Since the market mode $k = 1$ is well determined and appears not to be accounted for by the power-law tail, we leave it as is.

We will compare these RMT procedures to two classical, so-called shrinkage algorithms that are discussed in the literature (for a review, see [Led04]; see also [Guh03, Dal08, Bai07] for alternative proposals and tests). One is to 'shrink' the empirical correlation matrix $\mathbf{E}$ towards the identity matrix:

$$\mathbf{E} \longrightarrow (1-\alpha)\mathbf{E} + \alpha\mathbf{1}, \qquad 0 \le \alpha \le 1 \tag{40.5.3}$$

An interpretation of this procedure in terms of a minimal diversification of the optimal portfolio is given in [Bou03]. A more elaborate one, due to Ledoit and Wolf, is to replace the identity matrix above by a matrix $\overline{\mathbf{C}}$ with 1s on the diagonal and $\overline{\rho}$ for all off-diagonal elements, where $\overline{\rho}$ is the average of the pairwise correlation coefficient over all pairs.

This gives us four cleaning procedures, two shrinkage and two RMT schemes. We now need to devise one or several tests to compare their relative merits. The most natural test that comes to mind is to see how one can improve the out-of-sample risk of an optimized portfolio, following the discussion given in Section 40.2.3. However, we need to define a set of predictors we use for the vector of expected gains $\mathbf{g}$. Since many strategies rely in some way or other on the realized returns, we implement the following investment strategy: each day, the empirical correlation matrix is constructed using the 1000 previous days, and the expected gains are taken to be proportional to the returns of the current day, i.e. $g_i = r_i^t/\sqrt{\sum r_j^{t2}}$. The optimal portfolio with a certain gain target is then constructed using Eq.(40.2.11) with a correlation matrix cleaned according to

[5] Note that actually $\lambda_{\min}$ is given by the very same equation with $k = N$, i.e. it is indeed the smallest eigenvalue for large N.

one of the above four recipes. The out-of-sample risk is measured as the realized variance of those portfolios over the next 99 days. More precisely, this reads:

$$\mathbf{w}^t = \frac{\mathbf{E}_\alpha^{-1}\mathbf{g}^t}{\mathbf{g}^{tT}\mathbf{E}_\alpha^{-1}\mathbf{g}^t}, \qquad (40.5.4)$$

where $\mathbf{E}_\alpha$ is the cleaned correlation matrix, which depends on a parameter α used in the cleaning algorithm (see for example Eq. (40.5.3) above). A nice property of this portfolio is that if the predictors are normalized by their dispersion on day t, the true risk is $\mathcal{R}^{t2}_{\text{true}} = 1$. The out-of-sample risk is measured as:

$$\mathcal{R}^{t2}_{\text{out}} = \frac{1}{99}\sum_{t'=t+1}^{t+99}\left[\sum_i \frac{w_i^t}{\sigma_i^t} r_i^{t'}\right]^2, \qquad (40.5.5)$$

where σ_i^t is the volatility of stock i measured over the last 1000 days (the same period used to measure $\mathbf{E}$). The out-of-sample risk is then averaged over time, and plotted in Fig. 40.2 as a function of α for the four different recipes. In all cases but Ledoit-Wolf, $\alpha = 1$ corresponds to the $\mathbf{E}_\alpha = \mathbf{1}$ (in the case of the power-law method, $\alpha = 1$ corresponds to $\rho_C(\lambda) = \delta(\lambda - 1)$). In this case, $\mathcal{R}^2_{\text{out}} \approx 25$ which is very bad, since one does not even account for the market mode. When $\alpha = 0$, $\mathbf{E}_0$ is the raw empirical matrix, except in the power-law method. We show in Fig 40.2 the in-sample risks as well. From the values found for $\alpha = 0$ (no cleaning), one finds that the ratio of out-of-sample to in-sample risk is ≈ 2.51, significantly worse that the expected result $1/(1-q) = 2$. This may be due either to heavy tail effects or to non-stationary effects. The result of Fig. 40.2 is that the best cleaning scheme (as measured from this particular test) is eigenvalue clipping, followed by the power-law method. Shrinkage appears to be less efficient than RMT-based cleaning; this conclusion is robust against changing the quality factor Q. However, other tests can be devised, that lead to slightly different conclusions. One simple variant of the above test is to take for the predictor $\mathbf{g}$ a random vector in N dimensions, uncorrelated with the realized returns. Another idea is to use to correlation matrix to define residues, i.e. how well the returns of a given stock are explained by the returns of all other stocks on the same day, excluding itself. The ratio of the out-of-sample to in-sample residual variance is another measure of the quality of the cleaning. These two alternative tests are in fact found to give very similar results. The best cleaning recipe now turns out to be the power-law method, while the eigenvalue clipping is the worst one. Intuitively, the difference with the previous test comes from the fact that the random predictor $\mathbf{g}$ is (generically) orthogonal to the top eigenvectors of $\mathbf{E}$, whereas a predictor based on the returns themselves has significant overlap with these top eigenvectors. Therefore, the correct description of the corresponding eigenvalues is more important in the latter case, whereas the correct treatment of strongly correlated pairs (corresponding to small eigenvectors) is important to keep the residual variance small.

In summary, we have found that RMT-based cleaning recipes are competitive and outperform, albeit only slightly, more traditional shrinkage algorithms when applied to portfolio optimization or residue determination. However, depending on the objective and/or on the structure of the predictors, the naive eigenvalue clipping method proposed in [Lal00] might not be appropriate. In view of both the quality of the fit of the eigenvalue distribution (Fig. 40.1) and the robustness of the results to a change of the testing method, our analysis appears overall to favour the power-law cleaning method. However, one should keep in mind that the simple-minded shrinking with $\alpha = 1/2$ is quite robust and in fact difficult to beat, at least by the above RMT methods that do not attempt to mop up the eigenvectors. This is a direction for future research (see [Dim09] for a step in that direction).

References

[Ake10] G. Akemann, J. Fischmann and P. Vivo, Physica A **389**, 2566–2579 (2010).

[Auf09] A. Auffinger, G. Ben Arous, and S. Péché, Ann. Inst. H. Poincaré Probab. Statist. **45** 589 (2009).

[Bai05] J. Baik, G. Ben Arous, and S. Péché. Ann. Probab., **33** (2005) 1643.

[Bai07] Z. Bai, H. Liu, and W. K. Wong, *Making Markowitz Portfolio Optimization Theory Practically Useful*, University of Hong-Kong working paper (2007).

[Bir07a] G. Biroli, J.-P. Bouchaud and M. Potters, EuroPhys. Lett. **78** 10001 (2007).

[Bir07b] G. Biroli, J.-P. Bouchaud, and M. Potters, Acta Phys. Pol. B **38**, 4009 (2007).

[Boh04] A. C. Bertuola, O. Bohigas, and M.P. Pato, Phys. Rev. E **70** 065102 (2004).

[Bon04] G. Bonanno, G. Caldarelli, F. Lillo, S. Micciche', N. Vandewalle, and R. N. Mantegna, Eur Phys J B **38**, 363 (2004).

[Bou03] J.-P. Bouchaud and M. Potters, *Theory of Financial Risk and Derivative Pricing* (Cambridge University Press, 2003).

[Bou07] J.-P. Bouchaud, L. Laloux, M. A. Miceli and M. Potters, Eur. Phys. J. B **55** (2007) 201.

[Bur03] Z. Burda, J. Jurkiewicz, M. A. Nowak, Acta Phys. Polon. B**34**, 87 (2003).

[Bur04a] Z. Burda, A. Görlich, A. Jarosz and J. Jurkiewicz, Physica A, **343**, 295–310 (2004).

[Bur04b] Z. Burda and J. Jurkiewicz, Physica A **344**, 67 (2004).

[Bur05] Z. Burda, J. Jurkiewicz and B. Waclaw, Phys. Rev. E **71** 026111 (2005).

[Bur06] Z. Burda, A. Goerlich, and B. Waclaw, Phys. Rev. E **74** 041129 (2006).

[Chi10] R. Chicheportiche, J.P. Bouchaud, *The joint distribution of stock returns is not elliptical*, arXiv 1009.1100.

[Dal08] J. Daly, M. Crane, and H. J. Ruskin, Physica A **387** 4248 (2008).

[Dim09] I. Dimov, P. Kolm, L. Maclin, and D. Shiber, *Hidden Noise Structure and Random Matrix Models of Stock Correlations*, arXiv:09091383.

[Ede05] A. Edelman and N. Raj Rao, *Random Matrix Theory*, Acta Numerica, 1 (2005).

[Emb01] P. Embrechts, F. Lindskog, and A. McNeil, *Modelling Dependence with Copulas and Applications to Risk Management*, in Handbook of Heavy Tailed Distributions in Finance. Edited by Rachev ST, Published by Elsevier/North-Holland, Amsterdam (2001).

[Fer07] D. Féral and S. Péché, *The largest eigenvalue of rank one deformation of large Wigner matrices*, Comm. Math. Phys. 272, 185 (2007); S. Péché, Probab. Theory Relat. Fields, **134** 127 (2006).

[Fra05] G. Frahm and U. Jaekel, *Random matrix theory and robust covariance matrix estimation for financial data* arXiv:physics/0503007.

[Guh03] T. Guhr and B. Kälber, J. Phys. A **36** 3009 (2003).

[Hot36] H. Hotelling, Biometrika, 28, 321–377 (1936).

[Joh00] K. Johansson. Comm. Math. Phys. **209** 437 (2000).

[Joh07] I. M. Johnstone, *High-dimensional statistical inference and random matrices*, in Proceedings of the International Congress of Mathematicians I 307–333. Eur. Math. Soc., Zurich (2007).

[Joh08] I. M. Johnstone, The Annals of Statistics **36**, 2638 (2008).

[Joh10] I. M. Johnstone, *Approximate Null Distribution of the Largest Root in Multivariate Analysis*, Submitted to the Annals of Applied Statistics, **36**, 2757 (2008).

[Kar08] N. El Karoui, *Spectrum estimation for large dimensional covariance matrices using random matrix theory*, Annals of Statistics, **36**, 2757 (2008).

[Lal99] L. Laloux, P. Cizeau, J.-P. Bouchaud and M. Potters, Phys. Rev. Lett. **83**, 1467 (1999).

[Lal00] L. Laloux, P. Cizeau, J.-P. Bouchaud and M. Potters, Int. J. Theor. Appl. Finance **3**, 391 (2000).

[Led04] O. Ledoit and M. Wolf, J. Multivariate Anal. **88** 365 (2004).

[Lil05] F. Lillo and R. N. Mantegna, Physical Review E **72** 016219 (2005).

[Lil08] M. Tumminello, F. Lillo, and R.N. Mantegna, *Correlation, hierarchies, and networks in financial markets* to appear in J. Econ. Behav. & Org.

[Maj06] S. N. Majumdar, *Random Matrices, the Ulam Problem, Directed Polymers*, in Les Houches lecture notes for the summer school on Complex Systems (Les Houches, 2006).

[Mal04] Y. Malevergne and D. Sornette, Physica A **331**, 660 (2004).

[Mal06] Y. Malevergne and D. Sornette, *Extreme Financial Risks: From Dependence to Risk Management*, Springer Verlag, 2006.

[Mar67] V. A. Marčenko and L. A. Pastur, Math. USSR-Sb, **1**, 457–483 (1967).

[Mar02] M. Marsili, Quant. Fin., **2**, 297 (2002).

[Muz00] J.-F. Muzy, J. Delour, and E. Bacry, Eur. Phys. J. B **17** 537 (2000).

[Paf03] Sz. Pafka and I. Kondor, Physica A, **319**, 487 (2003); Physica A, **343**, 623–634 (2004).

[Pec03] S. Péché, *Universality of local eigenvalue statistics for random sample covariance matrices*, Thèse EPFL, no 2716 (2003).

[Ple99a] V. Plerou, P. Gopikrishnan, B. Rosenow, L.A.N. Amaral, and H.E. Stanley, Phys. Rev. Lett. **83**, 1471 (1999).

[Ple99b] V. Plerou, P. Gopikrishnan, L.A. Amaral, M. Meyer, and H.E. Stanley, Phys. Rev. **E60** 6519 (1999); P. Gopikrishnan, V. Plerou, L. A. Amaral, M. Meyer, and H. E. Stanley, Phys. Rev. **E 60** 5305 (1999).

[Ple02] V. Plerou, P. Gopikrishnan, B. Rosenow, L.A.N. Amaral, T. Guhr, and H.E. Stanley, Phys. Rev. E **65**, 066126 (2002).

[Pot05] M. Potters, J.-P. Bouchaud, and L. Laloux, Acta Phys. Pol. B **36**, 2767 (2005).

[Sil95] J. W. Silverstein and Z. D. Bai, Journal of Multivariate Analysis **54** 175 (1995).

[Ver04] for a recent review, see: A. Tulino, S. Verdù, *Random Matrix Theory and Wireless Communications*, Foundations and Trends in Communication and Information Theory, **1**, 1–182 (2004).

[Voi92] D. V. Voiculescu, K. J. Dykema, and A. Nica, *Free Random Variables*, AMS, Providence, RI 1992.

[Wac80] K. W. Wachter, The Annals of Statistics **8** 937 (1980).

[Zum09] G. Zumbach, *The Empirical Properties of Large Covariance Matrices*, to appear in Quantitative Finance (2011).

·41·

Asymptotic singular value distributions in information theory

A. M. Tulino and S. Verdú

Abstract

The channel capacity of several important digital communication systems is governed by the distribution of the singular values of the random channel matrices. In this paper we give a survey of the major channel models arising in practical applications. While some of the results used in applications, such as the Marčenko-Pastur law, are classical results in random matrix theory, a number of new results on the asymptotic distribution of singular values have been obtained in the information theory literature.

41.1 The role of singular values in channel capacity

Random matrices play an important role in the modelling of wireless communications systems. In the last few years, information theory has made substantial use of existing results on the spectrum of random matrices; moreover, motivated by certain channel models, information theorists have obtained a number of new results in random matrix theory. Most of those results are related to the asymptotic distribution of the (square of) the singular values of certain random matrices that model data communication channels. Depending on the model, the dimension of the matrices may relate to number of transmitters, number of antennae, spreading factor in a spread-spectrum system, number of tones in an orthogonal frequency division multiplexing system, coding blocklength, etc. In this chapter, we give a brief up-to-date summary of some of the main applications of random matrices to the capacity of communication channels. A comprehensive survey of the literature on that topic until 2004 can be found in [Tul04].

The basic wireless channel model used in much of the information theoretic literature is a general linear system

$$\mathbf{y} = \mathbf{H}\mathbf{x} + \mathbf{n} \tag{41.1.1}$$

where

- $\mathbf{x}$ is the K-dimensional vector of the signal input,
- $\mathbf{y}$ is the N-dimensional vector of the signal output,
- the N-dimensional vector $\mathbf{n}$ is the additive Gaussian noise, whose components are independent complex Gaussian random variables with zero mean and independent real and imaginary parts with the same variance $\sigma^2/2$ (i.e. circularly distributed)
- $\mathbf{H}$, is the $N \times K$ random matrix describing the channel.

Model (41.1.1) encompasses a variety of channels of interest in wireless communications such as multiaccess channels, linear channels with frequency-selective and/or time-selective fading, multiantenna communication, multi-cellular systems with cooperative detection, crosstalk in digital subscriber lines, signal space diversity, etc. In each of these cases, N, K, and $\mathbf{H}$ take different meanings. For example, N and K may indicate the number of transmit and receive antennae, respectively, while the coefficients in $\mathbf{H}$ quantify the fading between each pair of transmit and receive antennae; or K may be the number of users and N the spreading gain in a code-division multiple-access (CDMA) system while $\mathbf{H}$ is the signature matrix which is often generated in a pseudo-random fashion. Naturally, the same linear model applies to problems that incorporate more than one of the above features. In those cases it is conceptually advantageous to deconstruct them into their essential ingredients. In each of those cases, different statistical models apply to $\mathbf{H}$ leading to different results and tools. The simplest $\mathbf{H}$-model that we will consider is the one embodied by a random matrix $\mathbf{H}$ whose entries are independent and identically distributed (iid) random variables. This $\mathbf{H}$-model is very relevant to several problems like randomly spread CDMA without fading, multi-antenna links with uncorrelated transmit and receive antennae, etc.

Claude Shannon in 1948 [Shan48] formulated the key quantity, which he called *channel capacity*, that describes the maximal rate of transmission of information that can be sustained with arbitrary reliability by a noisy communication channel. Under ergodicity assumptions, the channel capacity is given by the maximal mutual information between the channel inputs and outputs where the maximum is over all feasible input distributions.

Under the assumption that the realization of the matrix $\mathbf{H}$ is known at the receiver (this is the so-called 'coherent' communication model), and that the inputs are power-constrained independent and identically distributed (e.g. in a multiaccess channel) the capacity (in information units per received degree of freedom) of (41.1.1) is given by

$$\frac{1}{N} I(\mathbf{x}; \mathbf{y}|\mathbf{H}) = \frac{1}{N}\mathbb{E}\left[\log\det\left(\mathbf{I} + \gamma\mathbf{H}\mathbf{H}^{\dagger}\right)\right] \tag{41.1.2}$$

$$= \frac{1}{N}\sum_{i=1}^{N}\mathbb{E}\left[\log\left(1 + \gamma\lambda_i(\mathbf{H}\mathbf{H}^{\dagger})\right)\right] \tag{41.1.3}$$

with the transmitted signal-to-noise ratio (SNR)

$$\gamma = \frac{N\mathbb{E}[||\mathbf{x}||^2]}{K\mathbb{E}[||\mathbf{n}||^2]}, \tag{41.1.4}$$

and with $\lambda_i(\mathbf{HH}^\dagger)$ equal to the ith squared singular value of $\mathbf{H}$. This result reflects the fact that the ability of the channel to carry information is independent of the chosen basis, and only depends on the strength of the signal in the various dimensions.

In practice, the size of the matrix $\mathbf{H}$ is fixed and in fact it may be quite small. While (41.1.2) can be accurately approximated by Monte-Carlo simulation, such numerical computations, while useful, provide limited insight to the engineer. Much more useful are analytical results that show the effect of various design parameters on capacity. Fixed-sized analytical results are cumbersome and often unfeasible. Fortunately, it is often the case that as the dimensions of $\mathbf{H}$ grow without bound with a fixed aspect ratio, the limit of (41.1.2) admits closed-form expressions that depend on the aspect ratio, the signal-to-noise ratio, as well as the statistics of $\mathbf{H}$. The aspect ratio often has strong engineering significance. For example, in a CDMA system it is equal to the number of users per chip, while in the multiantenna setting it represents the ratio between the sizes of the transmit and the receive antenna arrays. The natural question is how accurate are those asymptotic results as predictors of the capacity of fixed-size systems. Fortunately, frequently even for rather small sizes (such as 4–8) the accuracy of the approximation is rather remarkable.

The empirical cumulative distribution function of the eigenvalues (also referred to as the *spectrum* or empirical distribution) of an $n \times n$ Hermitian matrix $\mathbf{A}$ is denoted by $\mathsf{F}^n_{\mathbf{A}}$ defined as[1]

$$\mathsf{F}^n_{\mathbf{A}}(x) = \frac{1}{n}\sum_{i=1}^{n} 1\{\lambda_i(\mathbf{A}) \le x\}, \tag{41.1.5}$$

where $\lambda_1(\mathbf{A}), \ldots, \lambda_n(\mathbf{A})$ are the eigenvalues of $\mathbf{A}$ and $1\{\cdot\}$ is the indicator function. Using this notation, we can express (41.1.2) as

$$\frac{1}{N} I(\mathbf{x};\mathbf{y}|\mathbf{H}) = \int_0^\infty \log(1+\gamma x)\, d\mathsf{F}^N_{\mathbf{HH}^\dagger}(x) \tag{41.1.6}$$

Therefore, if the asymptotic spectrum of $\mathbf{HH}^\dagger$ can be determined, the channel capacity can be obtained using (41.1.6). However, this is not the usual or preferred route to obtain such results. In fact, it is often the case that the limit of (41.1.6) admits a closed form expression even if an explicit expression for the asymptotic spectrum is unknown. Traditionally, asymptotic spectrum results

[1] If $\mathsf{F}^n_{\mathbf{A}}$ converges as $n \to \infty$, then the corresponding limit (asymptotic empirical distribution or asymptotic spectrum) is simply denoted by $\mathsf{F}_{\mathbf{A}}(x)$.

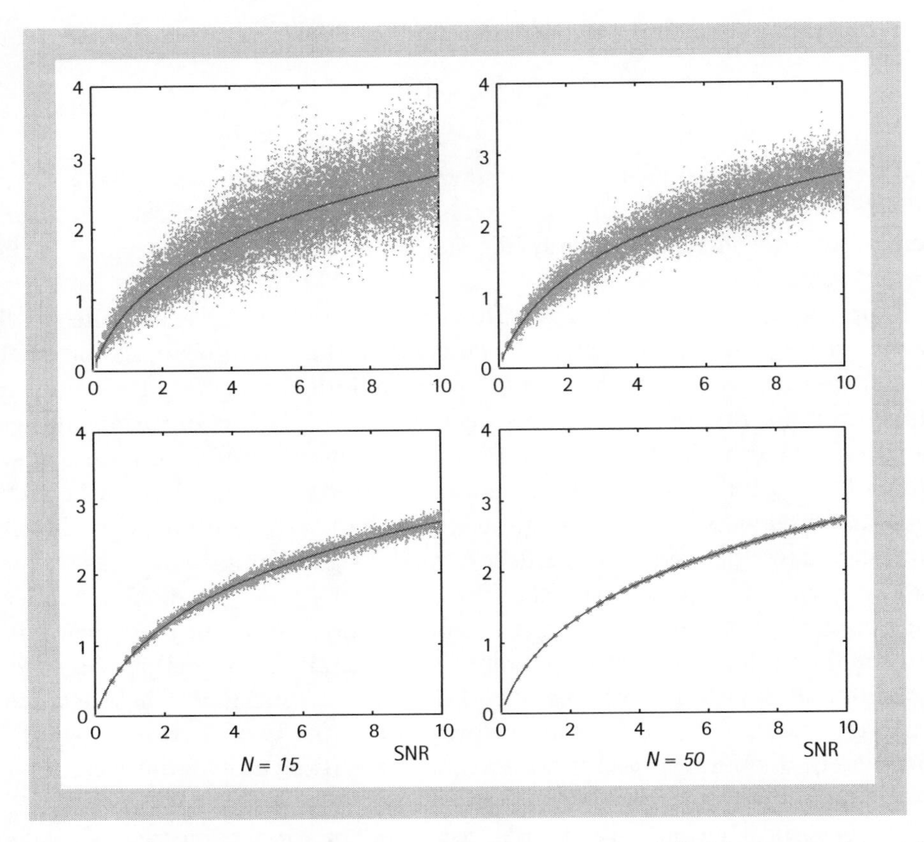

Fig. 41.1 Several realizations of (41.1.6) are compared to its asymptotic limit in the case of $\beta = 1$ for sizes: $N = 3, 5, 15, 50$

are given through the Stieltjes transform (or free-probability transforms such as the R-transform and the S-transform). In information theory, it has proven convenient to introduce two new transforms that we review in Section 41.2.

A more general setup than that leading to (41.1.2) is obtained by dropping the assumption that the input vector components are independent. In some practical communication systems, the transmitter may have partial knowledge of the realization or the statistics of $\mathbf{H}$ which can be exploited to increase mutual information by forcing dependence between the components of the input vector. In those cases, (41.1.2) generalizes to

$$\frac{1}{N} I(\mathbf{x};\mathbf{y}|\mathbf{H}) = \frac{1}{N} \max \mathbb{E}\left[\log\det\left(\mathbf{I} + \gamma \mathbf{H}\Sigma_X \mathbf{H}^\dagger\right)\right] \tag{41.1.7}$$

where the maximum is over a certain class of input covariance matrices Σ_X.

41.2 Transforms

In this section, we give the definitions of three transforms that are useful in expressing the asymptotic spectrum results. More information on these and other transforms arising in random matrices can be found in [Tul04, Section 2.2].

41.2.1 The Stieltjes transform

Let X be a real-valued random variable with distribution $F_X(\lambda)$. Its Stieltjes transform is defined for complex arguments as[2]

$$\mathcal{S}_X(z) = \mathbb{E}\left[\frac{1}{X-z}\right] = \int_{-\infty}^{\infty} \frac{1}{\lambda - z}\, d\,F_X(\lambda). \tag{41.2.1}$$

Although (41.2.1) is an analytic function on the complement of the support of F_X on the complex plane, it is customary to further restrict the domain of $\mathcal{S}_X(z)$ to arguments having positive imaginary parts. According to the definition, the signs of the imaginary parts of z and $\mathcal{S}_X(z)$ coincide.

$\mathcal{S}_X(z)$ can be regarded as a generating function for the moments of the random matrix whose averaged empirical eigenvalue distribution is F_X.

41.2.2 η-transform

In the applications of interest, it is advantageous to consider a transform which carries some engineering intuition, while at the same time is closely related to the Stieltjes transform.

Definition 41.2.1 *The η-transform of a non-negative random variable X is given by:*

$$\eta_X(\gamma) = \mathbb{E}\left[\frac{1}{1+\gamma X}\right] \tag{41.2.2}$$

where γ is a non-negative real number and thus $0 < \eta_X(\gamma) \leq 1$.

Either with analytic continuation or including the negative real line in the domain of definition of the Stieltjes transform, we obtain the simple relationship with the η-transform:

$$\gamma\eta_X(\gamma) = \mathcal{S}_X\left(-\frac{1}{\gamma}\right) \tag{41.2.3}$$

41.2.3 Shannon transform

Another transform which is motivated by applications is the following (see [Tul04] for more details):

[2] The Stieltjes transform is also known as the Cauchy transform and it is equal to $-\pi$ times the Hilbert transform when defined on the real line. As with the Fourier transform there is no universal agreement on its definition, as sometimes the Stieltjes transform is defined as $\mathcal{S}_X(-z)$ or $-\mathcal{S}_X(z)$.

Definition 41.2.2 *The Shannon transform of a non-negative random variable X is defined as*

$$\mathcal{V}_X(\gamma) = \mathbb{E}[\log(1 + \gamma X)]. \tag{41.2.4}$$

where γ is a non-negative real number.

The Shannon transform is intimately related to the Stieltjes and η transforms:

$$\frac{\gamma}{\log e}\frac{d}{d\gamma}\mathcal{V}_X(\gamma) = 1 - \frac{1}{\gamma}\mathcal{S}_X\left(-\frac{1}{\gamma}\right) \tag{41.2.5}$$

$$= 1 - \eta_X(\gamma) \tag{41.2.6}$$

Using $\mathcal{V}_X(0) = 0$, $\mathcal{V}_X(\gamma)$ can be obtained for all $\gamma > 0$ by integrating the derivative obtained in (41.2.5). The Shannon transform, provided it exists for all γ (a necessary condition is the finiteness of the entropy $H(\lfloor X \rfloor)$ [Wu10]) contains the same information as the distribution of X, either through the inversion of the Stieltjes transform or from the fact that all the moments of X are obtainable from $\mathcal{V}_X(\gamma)$.

41.3 Main results

In this section, we survey the main results on the limit of the empirical distributions of the eigenvalues of various random matrices of interest in information theory.

41.3.1 Independent identically distributed entries

Typical examples of channels where **H** has iid entries are:

- Unfaded randomly spread direct-sequence CDMA (DS-CDMA) channels with equal-power users [Ver98, (2.3.5)], where $\mathbf{H} = \mathbf{S}$ with $\mathbf{S}$ denoting the signature matrix whose entries are, in the simplest embodiment, chosen independently and equiprobably on $\{-1/\sqrt{N}, 1/\sqrt{N}\}$. One motivation for this is the use of 'long sequences' in commercial CDMA systems, where the period of the pseudo-random sequence spans many symbols.
- Multiantenna settings where both transmitter and receiver have arrays with sufficiently separated elements and the scattering environment is sufficiently rich. The pioneering analyses that ignited research on the topic [Tel99, Fos98] started with **H** having iid zero-mean complex Gaussian random entries (all antennae implicitly assumed identical and uncorrelated) which is referred to in the multi-antenna literature as the canonical model. Asymptotic (in the sense of the number of receiving and transmitting antennae) results for the capacity with full CSI can be found in [Chu02, Gra02, Mes02, Alf04, And00, Jay03]. Closed-form expressions have been obtained for the capacity of canonical multi-antenna channels, both

non-asymptotically [Shi03, Jan03] and asymptotically [Ver99, Rap00] in the number of antennae. This model also plays a role in the analysis of the total capacity of the Gaussian broadcast channel with multiple antennae at the transmitter [Hoch02, Vis03].

Theorem 41.3.1 *[Mar67, Wac78, Jon82, Bai99] Consider an $N \times K$ matrix $\mathbf{H}$ whose entries are independent zero-mean complex (or real) random variables with variance $\frac{1}{N}$ and fourth moments of order $O(\frac{1}{N^2})$. As $N, K \to \infty$ with $\frac{K}{N} \to \beta$, the empirical distribution of $\mathbf{H}^\dagger\mathbf{H}$ converges almost surely to a nonrandom limiting distribution with density*

$$f_\beta(x) = \left(1 - \frac{1}{\beta}\right)^+ \delta(x) + \frac{\sqrt{(x-a)^+(b-x)^+}}{2\pi\beta x} \tag{41.3.1}$$

where

$$a = (1-\sqrt{\beta})^2 \qquad\qquad b = (1+\sqrt{\beta})^2.$$

The distribution in (41.3.1) is called the Marčenko-Pastur law with ratio β. Its η-transform is

$$\eta(\gamma) = 1 - \frac{\mathcal{F}(\gamma,\beta)}{4\beta\gamma} \tag{41.3.2}$$

while its Shannon transform is:

$$\begin{aligned}\mathcal{V}(\gamma) &= \log\left(1+\gamma-\frac{1}{4}\mathcal{F}(\gamma,\beta)\right) + \frac{1}{\beta}\log\left(1+\gamma\beta-\frac{1}{4}\mathcal{F}(\gamma,\beta)\right)\\ &\quad -\frac{\log e}{4\beta\gamma}\mathcal{F}(\gamma,\beta)\end{aligned} \tag{41.3.3}$$

with

$$\mathcal{F}(x,z) = \left(\sqrt{x(1+\sqrt{z})^2+1} - \sqrt{x(1-\sqrt{z})^2+1}\right)^2. \tag{41.3.4}$$

Theorem 41.3.2 *[Tul05a] Let $\mathbf{H}$ be an $N \times K$ complex matrix whose entries are iid zero-mean random variables with variance $\frac{1}{N}$ such that $\mathbb{E}[|H_{i,j}|^4] = \frac{2}{N^2}$. As $K, N \to \infty$ with $\frac{K}{N} \to \beta$, the random variable*

$$\Delta_N = \log\det(\mathbf{I} + \gamma\mathbf{H}\mathbf{H}^\dagger) - N\mathcal{V}_{\mathbf{H}\mathbf{H}^\dagger}(\gamma) \tag{41.3.5}$$

is asymptotically zero-mean Gaussian with variance

$$\mathbb{E}\left[\Delta^2\right] = -\log\left(1 - \frac{(1-\eta_{\mathbf{H}\mathbf{H}^\dagger}(\gamma))^2}{\beta}\right) \tag{41.3.6}$$

where $\eta_{\mathbf{H}\mathbf{H}^\dagger}(\gamma)$ and $\mathcal{V}_{\mathbf{H}\mathbf{H}^\dagger}(\gamma)$ are given in (41.3.2) and (41.3.3).

41.3.2 Independent entries

In this section we review some of the results on the asymptotic empirical eigenvalue distribution of random matrices with independent and arbitrarily distributed entries. This model plays a fundamental role in describing a very broad range of multi-antenna channels. Specifically, a very general model for a multi-antenna channel is:

$$\mathbf{H} = \mathbf{U}_{\mathrm{R}}\tilde{\mathbf{H}}\mathbf{U}_{\mathrm{T}}^{\dagger} \tag{41.3.7}$$

where $\mathbf{U}_{\mathrm{R}}$ and $\mathbf{U}_{\mathrm{T}}$ are unitary, while the entries of $\tilde{\mathbf{H}}$ are independent zero-mean Gaussian. This model is advocated and experimentally supported in [Wei06], and its capacity is characterized asymptotically in [Tul05b]. This model encompasses most of the multi-antenna channel models considered in the literature. Examples include the virtual representation [Say02] where $\mathbf{U}_{\mathrm{R}}$ and $\mathbf{U}_{\mathrm{T}}$ are Fourier matrices and the so-called *separable* correlation model proposed by several authors [Shi00, Chi00, Ped00] where the channel matrix is given by

$$\mathbf{H} = \Theta_{\mathrm{R}}^{1/2}\mathbf{S}\Theta_{\mathrm{T}}^{1/2} \tag{41.3.8}$$

with $\mathbf{S}$ denoting a $n_{\mathrm{R}} \times n_{\mathrm{T}}$ iid Gaussian matrix with variance $\frac{1}{n_{\mathrm{T}}}$, and with Θ_{T} and Θ_{R}, denoting deterministic matrices, whose entries represent, respectively, the correlation between the transmit antennae and between the receiver antennae. In this case, $\mathbf{U}_{\mathrm{R}}$ and $\mathbf{U}_{\mathrm{T}}$ represent the eigenvector matrices of the correlation matrices Θ_{R} and Θ_{T}, respectively, and $\tilde{\mathbf{H}} = \mathbf{S} \circ \mathbf{P}$ with $\mathbf{P}$ the outer product of the vectors consisting of the eigenvalues of Θ_{R} and Θ_{T} [Tul05b, Tul06].

Diversity mechanisms such as polarization[3] and radiation pattern diversity[4] are becoming increasingly popular as they enable more compact arrays. The use of different polarizations and/or radiation patterns creates correlation structures that can be represented through (41.3.7). However, if antennae with mixed polarizations are used and there is no correlation, then (41.3.7) boils down to:

$$\mathbf{H} = \sqrt{\mathbf{P}} \circ \mathbf{S} \tag{41.3.9}$$

where $\circ$ indicates Hadamard (element-wise) multiplication, $\mathbf{S}$ is composed of zero-mean iid Gaussian entries with variance $\frac{1}{K}$ and $\sqrt{\mathbf{P}}$ is a deterministic matrix with non-negative entries.

[3] Polarization diversity: antennae with orthogonal polarizations are used to ensure low levels of correlation with minimum or no antenna spacing [Lee72, Som02] and to make the communication link robust to polarization rotations in the channel [Ber86].

[4] Pattern diversity: antennae with different radiation patterns or with rotated versions of the same pattern are used to discriminate different multipath components and reduce correlation.

Another very relevant framework, where (41.3.7) is relevant is the faded randomly-spread DS- or MC-CDMA setting. In the case of faded randomly-spread DS-CDMA the channel matrix $\mathbf{H}$ is [Ver98]:

$$\mathbf{H} = \mathbf{SA} \tag{41.3.10}$$

with $\mathbf{S}$ the $N \times K$ matrix containing the spreading sequences and with $\mathbf{A}$ a $K \times K$ diagonal matrix containing the complex fading coefficients.

For faded randomly-spread MC-CDMA the role of $\mathbf{H}$ is played by [Har97]:

$$\mathbf{H} = [\mathbf{C} \circ \mathbf{S}]\,\mathbf{A} \tag{41.3.11}$$

where $\mathbf{S}$ is the random signature matrix, as described in Section 41.3.1, $\mathbf{A}$ is a diagonal matrix of complex fading coefficients and $\mathbf{C}$ is an an $N \times K$ matrix whose (ℓ,k)-th entry denotes the fading for the ℓ-th sub-carrier of the k-th user.

The analysis of randomly-spread DS- or MC-CDMA in the asymptotic regime of number of users, K, and spreading gain, N, going to infinity with $\frac{K}{N} \to \beta$ provides valuable insight into the behaviour of multiuser receivers for large DS- and MC-CDMA systems employing pseudo-noise spreading sequences (e.g. [Tul04, Tse99, Gra98, Sha01, Kir00, Tse00, Vis01, Zha02, Li05, Han01, Li04, Eva00, Cha04, Deb03, Man03, Zai01, Som07, Pea04, Deb03]).

Before stating the results, we define:

Definition 41.3.1 *Consider an $N \times K$ random matrix $\mathbf{H}$ whose (i, j) entry has variance $\frac{\mathsf{P}_{i,j}}{N}$ with $\mathbf{P}$ an $N \times K$ deterministic matrix whose entries are uniformly bounded. For each N, let $v^N : [0, 1) \times [0, 1) \to \mathbb{R}$ be the* variance profile *function given by*

$$v^N(x, y) = \mathsf{P}_{i,j} \qquad \frac{i-1}{N} \le x < \frac{i}{N}, \quad \frac{j-1}{K} \le y < \frac{j}{K} \tag{41.3.12}$$

Whenever $v^N(x, y)$ converges uniformly to a limiting bounded measurable function, $v(x, y)$, we define this limit as the asymptotic variance profile *of* $\mathbf{H}$.

A very general result in random matrix theory states:

Theorem 41.3.3 *[Gir90, Gui00, Shl96] Let $\mathbf{H}$ be an $N \times K$ complex random matrix whose entries are independent zero-mean complex random variables (arbitrarily distributed) satisfying the Lindeberg condition (cf. [Tul04, Section 2.3.1 Eq (2.98)]) and with variances $\frac{\mathsf{P}_{i,j}}{N}$ where $\mathbf{P}$ is an $N \times K$ deterministic matrix whose entries are uniformly bounded and from which the asymptotic variance profile of $\mathbf{H}$, denoted $v(x, y)$, can be obtained as per Definition 41.3.1. The empirical eigenvalue distribution of $\mathbf{HH}^\dagger$ converges almost surely to a limiting distribution whose η-transform is [Tul04]:*

$$\eta_{\mathbf{HH}^\dagger}(\gamma) = \mathbb{E}\left[\, \Gamma_{\mathbf{HH}^\dagger}(\mathsf{X}, \gamma)\, \right] \tag{41.3.13}$$

where

$$\Gamma_{\mathbf{HH}^\dagger}(x,\gamma) = \frac{1}{1+\beta\gamma\mathbb{E}[v(x,\mathsf{Y})\Upsilon_{\mathbf{HH}^\dagger}(\mathsf{Y},\gamma)]} \tag{41.3.14}$$

$$\Upsilon_{\mathbf{HH}^\dagger}(y,\gamma) = \frac{1}{1+\gamma\,\mathbb{E}[v(\mathsf{X},y)\Gamma_{\mathbf{HH}^\dagger}(\mathsf{X},\gamma)]} \tag{41.3.15}$$

$\beta = \lim_{N\to\infty} \frac{K}{N}$ *and* X *and* Y *are independent random variables uniform in* $[0, 1]$.

Theorem 41.3.4 *[Tul05b] Let* $\mathbf{H}$ *be an* $N \times K$ *complex random matrix defined as in Theorem 41.3.3. The Shannon transform of the asymptotic spectrum of* $\mathbf{HH}^\dagger$

$$\begin{aligned}\mathcal{V}_{\mathbf{HH}^\dagger}(\gamma) &= \beta\mathbb{E}\left[\log(1+\gamma\,\mathbb{E}[v(\mathsf{X},\mathsf{Y})\Gamma_{\mathbf{HH}^\dagger}(\mathsf{X},\gamma)|\mathsf{Y}])\right] \\ &\quad +\mathbb{E}\left[\log(1+\gamma\beta\,\mathbb{E}[v(\mathsf{X},\mathsf{Y})\Upsilon_{\mathbf{HH}^\dagger}(\mathsf{Y},\gamma)|\mathsf{X}])\right] \\ &\quad -\gamma\beta\,\mathbb{E}\left[v(\mathsf{X},\mathsf{Y})\Gamma_{\mathbf{HH}^\dagger}(\mathsf{X},\gamma)\Upsilon_{\mathbf{HH}^\dagger}(\mathsf{Y},\gamma)\right]\log e\end{aligned} \tag{41.3.16}$$

with $\Gamma_{\mathbf{HH}^\dagger}(x,\gamma)$ *and* $\Upsilon_{\mathbf{HH}^\dagger}(y,\gamma)$ *given by (41.3.14) and (41.3.15).*

A central limit result analogous to Theorem 41.3.2 and 41.3.7, for the Shannon transform of a matrix $\mathbf{HH}^\dagger$ defined as in Theorem 41.3.3, has been proved in [Hac08]. Specifically, it is proven that when centred and properly rescaled, the random variable:

$$\Delta_N = \log\det(\mathbf{I} + \gamma\mathbf{HH}^\dagger) - N\mathcal{V}_{\mathbf{HH}^\dagger}(\gamma)$$

with $\mathbf{H}$ defined as in Theorem 41.3.3 and $\mathcal{V}_{\mathbf{HH}^\dagger}(\gamma)$ defined as in (41.3.16), satisfies a central limit theorem (CLT) and it has a Gaussian limit whose mean is zero and its variance can be expressed in terms of $v(x, y)$.

The η- and the Shannon-transform given in Theorem 41.3.3, as well as the asymptotic empirical eigenvalue distribution of $\mathbf{HH}^\dagger$, can be found in closed-form if some structure is imposed on the variance matrix $\mathbf{P}$ defined in Theorem 41.3.3. We give next a useful example of such a structure.

Definition 41.3.2 *[Tul04] An* $N \times K$ *matrix* $\mathbf{P}$ *is* asymptotically mean row-regular *if*

$$\lim_{K\to\infty} \frac{1}{K} \sum_{j=1}^{K} \mathsf{P}_{i,j} \le \alpha \tag{41.3.17}$$

is independent of i *for all* $\alpha \in \mathbb{R}$, *as the aspect ratio* $\frac{K}{N}$ *converges to a constant. A matrix whose transpose is asymptotically mean row-regular is called* asymptotically mean column-regular. *A matrix that is both asymptotically row-regular and asymptotically column-regular is called* asymptotically mean doubly-regular *and satisfies*

$$\lim_{N\to\infty} \frac{1}{N} \sum_{i=1}^{N} \mathsf{P}_{i,j} = \lim_{K\to\infty} \frac{1}{K} \sum_{j=1}^{K} \mathsf{P}_{i,j} \tag{41.3.18}$$

Example 41.3.1 *An $N \times K$ rectangular Toeplitz matrix $\mathsf{P}_{i,j} = \varphi(i-j)$ with $K \geq N$ is an asymptotically mean row-regular matrix. If either the function φ is periodic or $N = K$, then the Toeplitz matrix is asymptotically mean doubly-regular.*

A typical example of a random matrix $\mathbf{H}$ where the corresponding variance matrix $\mathbf{P}$ is asymptotically mean doubly-regular is the case where $\mathbf{H}$ describes a multi-antenna channel where the antennae at both transmitter and receiver are split on two orthogonal polarizations. Denoting by $\mathcal{X}$ the cross-polar discrimination (gain between cross-polar antennae relative to gain between co-polar antennae), $\mathbf{P}$ equals:

$$\mathbf{P} = \frac{2}{1+\mathcal{X}} \begin{pmatrix} 1 & \mathcal{X} & 1 & \mathcal{X} & \cdots \\ \mathcal{X} & 1 & \mathcal{X} & 1 & \cdots \\ 1 & \mathcal{X} & 1 & \mathcal{X} & \cdots \\ \mathcal{X} & 1 & \mathcal{X} & 1 & \cdots \\ \vdots & \vdots & \vdots & \vdots & \ddots \end{pmatrix}$$

Under the assumption that $\mathbf{H}$ is asymptotically mean doubly-regular, we have the following result:

Theorem 41.3.5 *[Tul05b] Define an $N \times K$ complex random matrix $\mathbf{H}$ whose entries are independent complex random variables satisfying the Lindeberg condition and with identical means and second moments with $\mathbf{P}$ an $N \times K$ deterministic asymptotically doubly-regular matrix whose entries are uniformly bounded. The asymptotic empirical eigenvalue distribution of $\mathbf{H}^\dagger\mathbf{H}$ converges almost surely to a distribution whose density is given by (41.3.1).*

41.3.3 Products of random matrices

The random-matrix models described in the next two theorems can be used to characterize the capacity of multi-antenna channels when the iid matrix $\mathbf{S}$ in the separable correlation model (41.3.8) is not Gaussian.

Theorem 41.3.6 *[Tul04] Let $\mathbf{S}$ be an $N \times K$ matrix whose entries are iid complex random variables with variance $\frac{1}{N}$. Let $\mathbf{T}$ be a $K \times K$ Hermitian non-negative random matrix, independent of $\mathbf{S}$, whose empirical eigenvalue distribution converges almost surely to a non-random limit. As $N, K \to \infty$ with $\frac{K}{N} \to \beta \in (0, \infty)$, the empirical eigenvalue distribution of $\mathbf{H}\mathbf{H}^\dagger = \mathbf{S}\mathbf{T}\mathbf{S}^\dagger$ converges almost surely to a distribution whose η-transform satisfies*

$$\beta = \frac{1-\eta}{1-\eta_{\mathbf{T}}(\gamma\eta)} \tag{41.3.19}$$

where for notational simplicity we have abbreviated $\eta_{\mathbf{STS}^\dagger}(\gamma) = \eta$. *The corresponding Shannon transform satisfies:*

$$\mathcal{V}_{\mathbf{STS}^\dagger}(\gamma) = \beta\mathcal{V}_{\mathbf{T}}(\eta\gamma) + \log\frac{1}{\eta} + (\eta - 1)\log e \tag{41.3.20}$$

Note that in the special case that $\mathbf{T}$ is a diagonal deterministic matrix then Theorem 41.3.6 becomes a special case of Theorem 41.3.3.

Theorem 41.3.7 *[Tul05a, Deb05] Let* $\mathbf{S}$ *and and* $\mathbf{T}$ *defined as in Theorem 41.3.6. As* $K, N \to \infty$ *with* $\frac{K}{N} \to \beta$,

$$\Delta_N = \log\det(\mathbf{I} + \gamma\mathbf{STS}^\dagger) - N\mathcal{V}_{\mathbf{STS}^\dagger}(\gamma) \tag{41.3.21}$$

with $\mathcal{V}_{\mathbf{STS}^\dagger}(\gamma)$ *given by (41.3.20), is asymptotically zero-mean Gaussian with variance*

$$\mathbb{E}[\Delta^2] = -\log\left(1 - \beta\mathbb{E}\left[\left(\frac{\mathsf{T}\gamma\eta_{\mathbf{HTH}^\dagger}(\gamma)}{1+\mathsf{T}\gamma\eta_{\mathbf{HTH}^\dagger}(\gamma)}\right)^2\right]\right)$$

where the expectation is over the non-negative random variable T *whose distribution is given by the asymptotic ESD of* $\mathbf{T}$.

In addition to the separable correlation model, the model described in the next result applies to the channel matrix $\mathbf{H}$ in the setting of downlink DS-CDMA with frequency-selective fading.

Theorem 41.3.8 *[Gir90, Bou96, Li04] Define* $\mathbf{H} = \mathbf{CSA}$ *where* $\mathbf{S}$ *is an* $N \times K$ *matrix whose entries are independent complex random variables satisfying the Lindeberg condition with identical means and variance* $\frac{1}{N}$. *Let* $\mathbf{C}$ *and* $\mathbf{A}$ *be, respectively,* $N \times N$ *and* $K \times K$ *random matrices such that the asymptotic spectra of* $\mathbf{D} = \mathbf{CC}^\dagger$ *and* $\mathbf{T} = \mathbf{AA}^\dagger$ *converge almost surely to compactly supported measures.*[5] *If* $\mathbf{C}$, $\mathbf{A}$ *and* $\mathbf{S}$ *are independent, the* η*-transform of* $\mathbf{HH}^\dagger$ *is [Tul04]:*

$$\eta_{\mathbf{HH}^\dagger}(\gamma) = \mathbb{E}\left[\,\Gamma_{\mathbf{HH}^\dagger}(\mathsf{D}, \gamma)\,\right] \tag{41.3.22}$$

where

$$\frac{1}{\Gamma_{\mathbf{HH}^\dagger}(d,\gamma)} = 1 + \gamma\beta d\,\mathbb{E}\left[\frac{\mathsf{T}}{1+\gamma\,\mathsf{T}\,\mathbb{E}\left[\mathsf{D}\,\Gamma_{\mathbf{HH}^\dagger}(\mathsf{D},\gamma)\right]}\right] \tag{41.3.23}$$

with D *and* T *independent random variables whose distributions are the asymptotic spectra of* $\mathbf{D}$ *and* $\mathbf{T}$ *respectively.*

[5] In the case that $\mathbf{C}$ and $\mathbf{A}$ are diagonal deterministic matrices, Theorem 41.3.8 is a special case of Theorem 41.3.3.

Theorem 41.3.9 *[Tul04, Tul05b, Tul05c] Let* $\mathbf{H}$ *be an* $N \times K$ *matrix as defined in Theorem 41.3.8. The Shannon transform of* $\mathbf{HH}^\dagger$ *is given by:*

$$\mathcal{V}_{\mathbf{HH}^\dagger}(\gamma) = \mathcal{V}_{\mathbf{D}}(\beta\gamma_d) + \beta\mathcal{V}_{\mathbf{T}}(\gamma_t) - \beta\frac{\gamma_d\gamma_t}{\gamma}\log e \tag{41.3.24}$$

where

$$\frac{\gamma_d\gamma_t}{\gamma} = 1 - \eta_{\mathbf{T}}(\gamma_t) \qquad \beta\frac{\gamma_d\gamma_t}{\gamma} = 1 - \eta_{\mathbf{D}}(\beta\gamma_d) \tag{41.3.25}$$

From (41.3.25), an alternative expression for $\eta_{\mathbf{HH}^\dagger}(\gamma)$ with $\mathbf{H}$ as in Theorem 41.3.8, can be obtained as

$$\eta_{\mathbf{HH}^\dagger}(\gamma) = \eta_{\mathbf{D}}(\gamma_d(\gamma)) \tag{41.3.26}$$

where $\gamma_d(\gamma)$ is the solution to (41.3.25).

Consider now a wireless link with $L-1$ clusters of scatterers each with n_ℓ scattering objects, $\ell = 1, \ldots, L-1$, such that the signal propagates from the transmit array to the first cluster, from there to the second cluster and so on, until it is received from the $(L-1)$th cluster by the receiver array. The matrix $\mathbf{H}$ describing such link can be written as the product of L random matrices $\mathbf{H}_\ell$ where the $n_\ell \times n_{\ell-1}$ matrix $\mathbf{H}_\ell$ describes the subchannel between the $(\ell-1)$th and ℓth clusters. The next theorem characterizes the asymptotic singular value distribution of such matrix.

Theorem 41.3.10 *[Mul02] Let* $\mathbf{H}_\ell$ *with* $1 \le \ell \le L-1$, *an* $n_\ell \times n_{\ell-1}$ *matrix and let consider the random matrix*

$$\mathbf{H} = \mathbf{H}_1 \cdots \mathbf{H}_L \tag{41.3.27}$$

If the matrices $\mathbf{H}_\ell$ *are independent with zero-mean iid entries with variance* $\frac{1}{n_\ell}$, *the* η*-transform of* $\mathbf{HH}^\dagger$ *is the solution to [Tul04]:*

$$\gamma\frac{\eta(\gamma)}{1-\eta(\gamma)} = \prod_{\ell=1}^{L}\frac{\beta_\ell}{\eta(\gamma)+\beta_{\ell-1}-1} \tag{41.3.28}$$

where $\beta_\ell = \frac{n_\ell}{n_L}$.

41.3.4 Sums of random matrices

Theorem 41.3.11 *[Cou09] Let* Q *be some fixed positive integer. Define* $\mathbf{H}_q = \mathbf{C}_q\mathbf{S}_q\mathbf{A}_q$ *with* $q = 1, \ldots, Q$ *where* $\mathbf{S}_q$ *is an* $N \times K_q$ *matrix whose entries are independent complex random variables (arbitrarily distributed) satisfying the Lindeberg condition with identical means and variance* $\frac{1}{N}$. *Let* $\mathbf{C}$ *and* $\mathbf{A}$ *be, respectively,* $N \times N$ *and* $K_q \times K_q$ *random matrices such that the asymptotic spectra of* $\mathbf{D}_q = \mathbf{C}_q\mathbf{C}_q^\dagger$ *and* $\mathbf{T}_q = \mathbf{A}_q\mathbf{A}_q^\dagger$ *converge almost surely to compactly supported measures.*

As $N, K_q \to \infty$ *with* $\frac{K_q}{N} = \beta_q$ *for all* $q = 1, \ldots, Q$, *then the empirical eigenvalue distribution of*

$$\mathbf{H}\mathbf{H}^\dagger = \sum_{q=1}^{Q} \mathbf{H}_q \mathbf{H}_q^\dagger \tag{41.3.29}$$

converges almost surely to a limiting distribution whose η*-transform is:*

$$\eta_{\mathbf{H}\mathbf{H}^\dagger}(\gamma) = \eta_{\mathbf{A}}(\gamma) \tag{41.3.30}$$

where

$$\mathbf{A} = \sum_{q=1}^{Q} \frac{1}{\beta_q} \left(1 - \eta_{\mathsf{T}_q}(\beta_q e_q(\gamma))\right) \mathbf{D}_q$$

and $e_q(\gamma)$ *for all* $q = 1, \ldots, Q$ *satisfies the set of equations*

$$e_q(\gamma) = \lim_{N \to \infty} \operatorname{tr}\left(\mathbf{D}_q (\mathbf{I} + \gamma \mathbf{A})^{-1}\right) \tag{41.3.31}$$

with T_q *a random variable whose distribution is the asymptotic spectrum of* $\mathbf{T}_q$ *respectively.*

41.3.5 Nonzero mean random matrices

In wireless line-of-sight applications, sometimes it is useful to allow nonzero mean matrices. If we assume that the rank of $\mathbb{E}[\mathbf{H}]$ is $o(N)$, then, using [Yin86], all results obtained in the previous sections still hold. The next result deals with the case where the deterministic component does have an effect.

Theorem 41.3.12 *[Doz07] Let* $\mathbf{S}$ *be an* $N \times K$ *matrix with iid entries with variance* $\frac{1}{N}$ *and* $\mathbf{H}_0$ *an* $N \times K$ *deterministic random matrix such that the spectrum of* $\mathbf{H}_0\mathbf{H}_0^\dagger$ *converges almost surely to a nonrandom limit. Let*

$$\mathbf{H} = \sigma \mathbf{S} + \sqrt{\tau} \mathbf{H}_0, \tag{41.3.32}$$

with $\tau > 0$ *a nonrandom positive value, then the ESD of* $\mathbf{H}\mathbf{H}^\dagger$ *converges, as* $K, N \to \infty$ *with* $\frac{K}{N} \to \beta$, *almost surely to a nonrandom limit whose* η*- transform satisfies*

$$\eta(\gamma) = \mathbb{E}\left[\left(\frac{\gamma \tau \mathsf{M}}{1 + \gamma \sigma^2 \eta(\gamma)} + (1 + \gamma \sigma^2 \eta(\gamma)) + \gamma \sigma^2 (\beta - 1)\right)^{-1}\right] \tag{41.3.33}$$

with M *a random variable whose distribution is the asymptotic spectrum of* $\mathbf{H}_0\mathbf{H}_0^\dagger$.

41.3.6 Random Vandermonde matrices

In this section we consider the following model which appears in some sensor networks and multiantenna multiuser communications [Rya09].

- Let $\mathbf{V}$ be an $(2n+1)^d \times m$, d-folded Vandermonde matrix, with

$$\mathbf{V}_{\nu(\ell),q} = \frac{1}{\sqrt{(2n+1)^d}} \exp\left(-2\pi \mathrm{i} \ell^{\mathsf{T}} \mathbf{x}_q\right) \tag{41.3.34}$$

where $d \geq 1$ is an integer and $\ell = [\ell_1, \ldots, \ell_d]^{\mathsf{T}}$ is a vector of integers, $\ell_k = -n, \ldots, n$, $k = 1, \ldots, d$; the vectors $\mathbf{x}_q$ are independent random variables uniformly distributed in [0,1) and the function

$$\nu(\ell) = \sum_{k=1}^{d} (2n+1)^{k-1} \ell_k, \tag{41.3.35}$$

maps the vector ℓ onto a scalar index.
- Let $\mathbf{C}$ and $\mathbf{A}$ be $(2n+1)^d \times (2n+1)^d$ and $m \times m$ diagonal matrices, respectively, mutually independent and independent of $\mathbf{V}$, such that the asymptotic spectra of $\mathbf{D} = \mathbf{C}\mathbf{C}^\dagger$ and $\mathbf{A} = \mathbf{A}\mathbf{A}^\dagger$ converge almost surely to compactly supported measures sufficiently well-behaved, such that all moments exist.

Typical wireless communications systems where the channel matrix is $\mathbf{H} = \mathbf{CVA}$ with $\mathbf{C}$, $\mathbf{V}$, and $\mathbf{A}$ defined as above include:

- A multiuser multiantenna system, where m transmitters send a signal over a flat fading channel while the receiver is equipped with a d-dimensional receive-antenna array with n^d antennae distributed on a regular d-dimensional grid.
- A network composed of m sensors, uniformly deployed in the region $[0, 1)^d$, where the sensing devices, possibly having different sensitivity, or different noise levels, send the detected values, along with some timing/localization information, to a central controller. In this case the vectors $\mathbf{x}_q$ are random variables and the central controller has to reconstruct the physical field from a set of irregularly spaced samples.
- A high resolution MIMO radar employed for detecting extended targets where the (i, j)-th element of $\mathbf{H} = \mathbf{CVA}$ describes the impulse response of the channel between the i-th receiving antennae and the j-th elementary scatterer at the target.

Theorem 41.3.13 *[Nor10] Let*

$$\mathbf{H} = \mathbf{CVA} \tag{41.3.36}$$

with $\mathbf{C}$, $\mathbf{V}$, *and* $\mathbf{A}$ *defined as above. As* $d \to \infty$, $n \to \infty$ *and* $m \to \infty$ *with constant ratio* $(2n+1)^d/m \to \beta$, *the* η*–transform and the Shannon-transform of the asymptotic averaged empirical spectrum of* $\mathbf{H}\mathbf{H}^\dagger = \mathbf{CVTV}^\dagger\mathbf{C}^\dagger$ *are given as in (41.3.22) and (41.3.24)*

41.3.7 Random Toeplitz matrices

The next results provide some data on the existence of the asymptotic eigenvalue distributions of random Toeplitz matrices as well as an explicit expression for the asymptotic eigenvalue distributions of a specific iid Toeplitz random matrix.

Theorem 41.3.14 *[Bry06] Consider an $N \times N$ symmetric Toeplitz matrix $\mathbf{T}_N$ defined by an iid vector X_i. When scaled by $\frac{1}{\sqrt{N}}$, the empirical eigenvalue distribution converges weakly to a non-random symmetric probability measure which does not depend on the distribution of X_1, and has unbounded support. The same result holds for symmetric Hankel matrices.*

In [Ham05] and [Bry06], the non-Gaussian behaviour of the asymptotic spectrum of $\frac{1}{\sqrt{N}}\mathbf{T}_N$ has been characterized. In particular, the ratio of the $2k$-th Toeplitz moment to the standard Gaussian moment tends to zero as $k \to \infty$.

Theorem 41.3.15 *[Mas07] If $\frac{1}{\sqrt{N}}\mathbf{T}_N$ is symmetric Toeplitz and the first row is a palindrome, then the asymptotic spectrum of $\frac{1}{\sqrt{N}}\mathbf{T}_N$ is the Gaussian law.*

41.3.8 Diagonal times circulant random matrices

A new and recent result in random matrix theory, which lies well outside the body of existing results, is obtained by solving a problem of considerable interest in fading communication channels.

Let $\mathbf{G} = \mathrm{diag}(\mathsf{G}_1, \ldots, \mathsf{G}_n)$ be a random diagonal matrix. Let $\mathbf{F}$ be the unitary Fourier matrix defined as

$$\mathbf{F}_{i,k} = \frac{1}{\sqrt{n}} e^{-\frac{j2\pi}{n}(i-1)(k-1)}. \tag{41.3.37}$$

and let Λ a real diagonal matrix with diagonal elements $[\Lambda]_{ii} = \lambda_i$. Then

$$\Psi = \mathbf{F}\Lambda\mathbf{F}^\dagger = \sum_{i=0}^{n-1} \lambda_i \mathbf{f}_i \mathbf{f}_i^\dagger \tag{41.3.38}$$

is a random circulant matrix. Since most mobile wireless systems are subject to both frequency-selective fading (e.g. due to multipath) and to time-selective fading (e.g. due to shadowing), we are interested in characterizing the asymptotic singular value distribution of the random matrix

$$\mathbf{H} = \mathbf{AFGF}^\dagger \tag{41.3.39}$$

where $\mathbf{A} = \mathrm{diag}(\mathsf{A}_1, \ldots, \mathsf{A}_n)$ is independent of $\mathbf{G}$.

Theorem 41.3.16 *[Tul10] Suppose that* $\{\mathsf{G}_i\}$ *and* $\{\mathsf{A}_i\}$*, are independent and iid with respective distributions* P_{G} *and* P_{A}*. For* $\gamma \geq 0$*, let*

$$0 \leq \alpha \leq \mathbb{E}\left[|\mathsf{A}|^2\right] \tag{41.3.40}$$

$$0 \leq \nu \leq \mathbb{E}\left[|\mathsf{G}|^2\right] \tag{41.3.41}$$

be defined by the solution to

$$\mathbb{E}\left[\frac{1}{1+\alpha\gamma|\mathsf{G}|^2}\right] = \frac{1}{1+\alpha\nu\gamma} = \mathbb{E}\left[\frac{1}{1+\nu\gamma|\mathsf{A}|^2}\right] \tag{41.3.42}$$

Then, the η *and Shannon transforms of* $\mathbf{HH}^\dagger = \mathbf{AFGG}^\dagger\mathbf{F}^\dagger\mathbf{A}^\dagger$ *are given, respectively, by*

$$\eta_{\mathbf{AFGG}^\dagger\mathbf{F}^\dagger\mathbf{A}^\dagger}(\gamma) = \frac{1}{1+\alpha\nu\gamma} \tag{41.3.43}$$

$$\mathcal{V}_{\mathbf{HH}^\dagger}(\gamma) = \mathbb{E}\left[\log\left(1+\alpha\gamma|\mathsf{G}|^2\right)\right] + \mathbb{E}\left[\log\left(1+\nu\gamma|\mathsf{A}|^2\right)\right] - \log(1+\alpha\nu\gamma) \tag{41.3.44}$$

Furthermore, the same result holds as long as either $\{\mathsf{G}_i\}$ *or* $\{\mathsf{A}_i\}$ *is allowed to be strongly mixing.*

In the special case where $|\mathsf{A}|^2$ takes on the values 0 or 1 with probability e and $1-\mathsf{e}$, we obtain

$$C_{\mathsf{e}}(\gamma) = \mathbb{E}\left[\log\left(1+(1-\hat{\mathsf{e}})\gamma|\mathsf{G}|^2\right)\right] + d(\mathsf{e}||\hat{\mathsf{e}}) \tag{41.3.45}$$

where the binary divergence is defined as

$$d(a||b) = a\log\frac{a}{b} + (1-a)\log\frac{1-a}{1-b} \tag{41.3.46}$$

and $\hat{\mathsf{e}} \geq \mathsf{e}$ is the (γ-dependent) solution to

$$\frac{\mathsf{e}}{\hat{\mathsf{e}}} = \mathbb{E}\left[\frac{1}{1+(1-\hat{\mathsf{e}})\gamma|\mathsf{G}|^2}\right] \tag{41.3.47}$$

The binary assumption on $|\mathsf{A}|^2$ models frequency-selective fading with on-off time-selective fading and erasure channels.

41.3.9 Products with Haar matrices

A model which is much easier to analyse and that leads to the same result as Theorem 41.3.16 is

$$\mathbf{H} = \mathbf{XUY} \tag{41.3.48}$$

where **U** is a Haar matrix (i.e. uniformly distributed on the set of unitary matrices) independent of **X** and **Y**, which are (possibly dependent) random matrices whose spectra converge almost surely to P_{A} and P_{G} respectively [Tul10].

Typical applications of (41.3.48) include:

- Downlink MC-CDMA systems with unequal power users and frequency selective fading. In this case, each of the K columns of **U** represents the code allocated to each user, while N coincides with the number of subcarriers.
- Precoded OFDM [Wan01] or spread OFDM [Deb03] in a single-user context.

References

[Alf04] G. Alfano, A. Lozano, A. M. Tulino, and S. Verdú, "Capacity of MIMO channels with one-sided correlation," in *Proc. IEEE Int. Symp. on Spread Spectrum Tech. and Applications (ISSSTA'04)*, pp. 515–519. Aug. 2004.

[And00] J. B. Andersen, "Array gain and capacity for known random channels with multiple element arrays at both ends," *IEEE J. on Selected Areas in Communications*, vol. 18, no. 11, pp. 2172–2178, Nov. 2000.

[Bai99] Z. D. Bai, "Methodologies in spectral analysis of large dimensional random matrices," *Statistica Sinica*, vol. 9, no. 3, pp. 611–661, 1999.

[Ber86] S. A. Bergmann and H. W. Arnold, "Polarization diversity in portable communications environment," *IEE Electronics Letters*, vol. 22, no. 11, pp. 609–610, May 1986.

[Bou96] A. Boutet de Monvel, A. Khorunzhy, and V. Vasilchuk, "Limiting eigenvalue distribution of random matrices with correlated entries," *Markov Processes and Related Fields*, vol. 2, no. 2, pp. 607–636, 1996.

[Bry06] W. Bryc, A. Dembo, and T. Jiang, "Spectral measure of large random Hankel, Markov and Toeplitz matrices," *Annals of Probability*, vol. 34, no. 1, pp. 745–778, May-June 2006.

[Chi00] D. Chizhik, F. R. Farrokhi, J. Ling, and A. Lozano, "Effect of antenna separation on the capacity of BLAST in correlated channels," *IEEE Communications Letters*, vol. 4, no. 11, pp. 337–339, Nov. 2000.

[Cha04] J. M. Chaufray, W. Hachem, and P. Loubaton, "Asymptotic analysis of optimum and sub-optimum CDMA MMSE receivers," *IEEE Trans. Information Theory* vol. 50, no. 11, pp. 2620–2638, 2004.

[Chu02] C. Chuah, D. Tse, J. Kahn, and R. Valenzuela, "Capacity scaling in dual-antenna-array wireless systems," *IEEE Trans. on Information Theory*, vol. 48, no. 3, pp. 637–650, Mar. 2002.

[Cou09] R. Couillet, M. Debbah, and J. W. Silverstein, "Asymptotic capacity of multi-user MIMO communications," *IEEE Information Theory Workshop (ITW 2009)*, pp. 16–20, Taormina, Italy, October 2009.

[Deb03] M. Debbah, W. Hachem, P. Loubaton, and M. de Courville, "MMSE analysis of certain large isometric random precoded systems," *IEEE Trans. on Information Theory*, vol. 49, no. 5, pp. 1293–1311, May 2003.

[Deb05] M. Debbah and R. Müller, "MIMO channel models and the principle of maximum entropy," *IEEE Trans. on Information Theory*, vol. 51, no. 5, pp. 1667–1690, May 2005.

[Doz07] B. Dozier and J. W. Silverstein, "Analysis of the limiting spectral distribution of large dimensional information-plus-noise type matrices," *Journal of Multivariate Analysis*, vol. 98, no. 6, pp. 1099–1122, 2007.

[Eva00] J. Evans and D. Tse, "Large system performance of linear multiuser receivers in multipath fading channels," *IEEE Trans. on Information Theory*, vol. 46, no. 6, pp. 2059–2078, Sep. 2000.

[Fos98] G. Foschini and M. Gans, "On limits of wireless communications in fading environment when using multiple antennas," *Wireless Personal Communications*, vol. 6, No. 6, pp. 315–335, Mar. 1998.

[Gir90] V. L. Girko, *Theory of Random Determinants*, Kluwer Academic Publishers, Dordrecht, 1990.

[Gui00] A. Guionnet and O. Zeitouni, "Concentration of the spectral measure for large matrices," *Electronic Communications in Probability*, vol. 5, pp. 119–136, 2000.

[Gra98] A. J. Grant and P. D. Alexander, "Random sequences multisets for synchronous code-division multiple-access channels," *IEEE Trans. on Information Theory*, vol. 44, no. 7, pp. 2832–2836, Nov. 1998.

[Gra02] A. Grant, "Rayleigh fading multi-antenna channels," *EURASIP J. on Applied Signal Processing*, vol. 3, pp. 316–329, 2002.

[Hac08] W. Hachem, P. Loubaton, and J. Najim, "A CLT for information-theoretic statistics of Gram random matrices with a given variance profile," *Ann. Appl. Probab.*, vol. 18, no. 6, pp. 2071–2130, 2008.

[Ham05] C. Hammond and S. J. Miller, "Distribution of eigenvalues for the ensemble of real symmetric Toeplitz matrices," *Journal of Theoretical Probability*, vol. 18, pp. 537–566, May 2005.

[Han01] S. Hanly and D. Tse, "Resource pooling and effective bandwidths in CDMA networks with multiuser receivers and spatial diversity," *IEEE Trans. on Information Theory*, vol. 47, no. 4, pp. 1328–1351, May 2001.

[Har97] S. Hara and R. Prasad, "Overview of multicarrier CDMA," *IEEE Communications Magazine*, vol. 35, no. 12, pp. 126–133, Dec. 1997.

[Hoch02] B. Hochwald and S. Vishwanath, "Space-time multiple access: Linear growth in the sum-rate," in *Proc. Allerton Conf. on Communication, Control and Computing*, Monticello, IL, Oct. 2002.

[Jan03] R. Janaswamy, "Analytical expressions for the ergodic capacities of certain MIMO systems by the Mellin transform," *Proc. IEEE Global Telecomm. Conf. (GLOBECOM'03)*, vol. 1, pp. 287–291, Dec. 2003.

[Jay03] S. K. Jayaweera and H. V. Poor, "Capacity of multiple-antenna systems with both receiver and transmitter channel state information," *IEEE Trans. on Information Theory*, vol. 49, no. 10, pp. 2697–2709, Oct. 2003.

[Jon82] D. Jonsson, "Some limit theorems for the eigenvalues of a sample covariance matrix," *J. Multivariate Analysis*, vol. 12, pp. 1–38, 1982.

[Kir00] Kiran and David N.C. Tse, "Effective interference and effective bandwidth of linear multiuser receivers in asynchronous systems," *IEEE Trans. on Information Theory*, vol. 46, no. 4, pp. 1426–1447, July 2000.

[Lee72] W. C. Y. Lee and Y. S. Yeh, "Polarization diversity system for mobile radio," *IEEE Trans. on Communications*, vol. 20, no. 5, pp. 912–923, Oct 1972.

[Li04] L. Li, A. M. Tulino, and S. Verdú, "Design of reduced-rank MMSE multiuser detectors using random matrix methods," *IEEE Trans. on Information Theory*, vol. 50, no. 6, pp. 986–1008, June 2004.

[Li05] L. Li, A. M. Tulino, and S. Verdú, "Spectral efficiency of multicarrier CDMA," *IEEE Trans. on Information Theory*, vol. 51, no. 2, pp. 479–505, Feb. 2005.

[Man03] A. Mantravadi, V. V. Veeravalli, and H. Viswanathan, "Spectral efficiency of MIMO multiaccess systems with single-user decoding," *IEEE J. on Selected Areas in Communications*, vol. 21, no. 3, pp. 382–394, Apr. 2003.

[Mar67] V. A. Marčenko and L. A. Pastur, "Distributions of eigenvalues for some sets of random matrices," *Math. USSR-Sbornik*, vol. 1, pp. 457–483, 1967.

[Mas07] A. Massey, S. J. Miller, and J. Sinsheimer, "Distribution of eigenvalues of real symmetric palindromic Toeplitz matrices and circulant matrices," *Journal of Theoretical Probability*, vol. 20, no. 3, pp. 16–20, September 2007.

[Mes02] X. Mestre, *Space processing and channel estimation: performance analysis and asymptotic results*, Ph.D. thesis, Dept. de Teoria del Senyal i Comunicacions, Universitat Politècnica de Catalunya, Barcelona, Catalonia, Spain, 2002.

[Mul02] R. Müller, "On the asymptotic eigenvalue distribution of concatenated vector-valued fading channels," *IEEE Trans. on Information Theory*, vol. 48, no. 7, pp. 2086–2091, July 2002.

[Nor10] A. Nordio, G. Alfano, C. Chiasserini, and A. M. Tulino, "Asymptotics of multi-fold Vandermonde matrices with random entries," *IEEE Transactions on Signal Processing*, in press.

[Pea04] M. Peacock, I. Collings, and M. L. Honig, "Asymptotic analysis of LMMSE multiuser receivers for multi-signature multicarrier CDMA in Rayleigh fading," *IEEE Transactions on Communications*, vol. 52, no. 6, pp. 964–972, 2004.

[Ped00] K. I. Pedersen, J. B. Andersen, J. P. Kermoal, and P. E. Mogensen, "A stochastic multiple-input multiple-output radio channel model for evaluations of space-time coding algorithms," *Proc. IEEE Vehicular Technology Conf. (VTC'2000 Fall)*, pp. 893–897, Sep. 2000.

[Rap00] P. Rapajic and D. Popescu, "Information capacity of a random signature multiple-input multiple-output channel," *IEEE Trans. on Communications*, vol. 48, no. 8, pp. 1245–1248, Aug. 2000.

[Rya09] O. Ryan and M. Debbah, "Asymptotic behavior of random Vandermonde matrices with entries on the unit circle," *IEEE Trans. on Information Theory*, vol. 55 , no. 7, pp. 3115–3147, July, 2009.

[Say02] A. Sayeed, "Deconstructing multi-antenna channels," *IEEE Trans. on Signal Processing*, vol. 50, no. 10, pp. 2563–2579, Oct. 2002.

[Sha01] S. Shamai and S. Verdú, "The effect of frequency-flat fading on the spectral efficiency of CDMA," *IEEE Trans. on Information Theory*, vol. 47, no. 4, pp. 1302–1327, May 2001.

[Shan48] C. E. Shannon, "A mathematical theory of communication," *Bell System Technical Journal*, vol. 27, no. 4, pp. 379–423, 623–656, May and October 1948.

[Shi03] H. Shin and J. H. Lee, "Capacity of multiple-antenna fading channels: Spatial fading correlation, double scattering and keyhole," *IEEE Trans. on Information Theory*, vol. 49, no. 10, pp. 2636–2647, Oct. 2003.

[Shi00] D. S. Shiu, G. J. Foschini, M. J. Gans, and J. M. Kahn, "Fading correlation and its effects on the capacity of multi-element antenna systems," *IEEE Trans. on Communications*, vol. 48, no. 3, pp. 502–511, Mar. 2000.

[Shl96] D. Shlyankhtenko, "Random Gaussian band matrices and freeness with amalgamation," *Int. Math. Res. Note*, vol. 20, pp. 1013–1025, 1996.

[Som02] P. Soma, D. S. Baum, V. Erceg, R. Krishnamoorthy, and A. Paulraj, "Analysis and modelling of multiple-input multiple-output (MIMO) radio channel based on outdoor measurements conducted at 2.5 GHz for fixed BWA applications," *Proc. IEEE Int. Conf. on Communications (ICC'02), New York City, NY*, pp. 272–276, 28 Apr.-2 May 2002.

[Som07] O. Somekh, B. M. Zaidel, and S. Shamai, "Spectral efficiency of joint multiple cell-site processors for randomly spread DS-CDMA systems," in *IEEE Transactions on Information Theory*, vol. 53, no. 7, pp. 2625–2637, 2007.

[Tel99] E. Telatar, "Capacity of multi-antenna Gaussian channels," *Euro. Trans. Telecommunications*, vol. 10, no. 6, pp. 585–595, Nov.-Dec. 1999.

[Tse99] D. Tse and S. Hanly, "Linear multiuser receivers: Effective interference, effective bandwidth and user capacity," *IEEE Trans. on Information Theory*, vol. 45, no. 2, pp. 641–657, Mar. 1999.

[Tse00] D. Tse and O. Zeitouni, "Linear multiuser receivers in random environments," *IEEE Trans. on Information Theory*, vol. 46, no. 1, pp. 171–188, Jan. 2000.

[Tul04] A. M. Tulino and S. Verdú, "Random matrix theory and wireless communications," *Foundations and Trends in Communications and Information Theory*, vol. 1, no. 1, 2004.

[Tul05a] A. M. Tulino and S. Verdú, "Asymptotic outage capacity of multiantenna channels," in *2005 IEEE Int. Conf. Acoustics, Speech and Signal Processing*, Philadelphia, PA, vol. 5, pp. 825–828, March 2005.

[Tul05b] A. M. Tulino, A. Lozano, and S. Verdú, "Impact of correlation on the capacity of multi-antenna channels," *IEEE Trans. Information Theory*, pp. 2491–2509, July 2005.

[Tul05c] A. M. Tulino, A. Lozano, and S. Verdú, "Impact of Antenna Correlation on the Capacity of Multiantenna Channels," *IEEE Trans. Information Theory*, vol. 51, no. 7, pp. 2491–2509, July 2005.

[Tul06] A. M. Tulino, A. Lozano, and S. Verdú, "Capacity-achieving input covariance for single-user multi-antenna channels," *IEEE Trans. Wireless Communications*, vol. 5, no. 3, pp. 662–671, March 2006.

[Tul10] A. M. Tulino, G. Caire, S. Shamai, and S. Verdú, "Capacity of channels with frequency-selective and time-selective fading," *IEEE Trans. on Information Theory*, vol. 56, no. 3, pp. 1187–1215. March 2010.

[Ver98] S. Verdú, *Multiuser Detection*, Cambridge University Press, Cambridge, UK, 1998.

[Ver99] S. Verdú and S. Shamai, "Spectral efficiency of CDMA with random spreading," *IEEE Trans. on Information Theory*, vol. 45, no. 2, pp. 622–640, Mar. 1999.

[Vis01] P. Viswanath, D. N. Tse, and V. Anantharam, "Asymptotically optimal water-filling in vector multiple-access channels," *IEEE Trans. on Information Theory*, vol. 47, no. 1, pp. 241–267, Jan. 2001.

[Vis03] H. Viswanathan and S. Venkatesan, "Asymptotics of sum rate for dirty paper coding and beamforming in multiple antenna broadcast channels," in *Proc. Allerton Conf. on Communication, Control and Computing*, Monticello, IL, Oct. 2003.

[Wac78] K. W. Wachter, "The strong limits of random matrix spectra for sample matrices of independent elements," *Annals of Probability*, vol. 6, no. 1, pp. 1–18, 1978.

[Wan01] Z. Wang and G. B. Giannakis, "Linearly precoded or coded OFDM against wireless channel fades?," *Proc. 3rd Workshop on Signal Processing Advances in Wireless Communications*, Mar. 2001, pp. 267–270.

[Wei06] W. Weichselberger, M. Herdin, H. Ozcelik, and E. Bonek, "A Stochastic MIMO channel model with joint correlation of both link ends," IEEE Trans. on Wireless Communications, vol. 5, no. 1, pp. 90–100, Jan. 2006.

[Yin86] Y. Q. Yin, "Limiting spectral distribution for a class of random matrices," *J. of Multivariate Analysis*, vol. 20, pp. 50–68, 1986.

[Wu10] Y. Wu, S. Verdú, "Rényi Information Dimension: Fundamental Limits of Almost Lossless Analog Compression," *IEEE Trans. Information Theory*, vol. 56, no. 8, pp. 3721–3747, August 2010.

[Zai01] B. M. Zaidel, S. Shamai, and S. Verdú, "Multicell uplink spectral efficiency of coded DS-CDMA with random signatures," *IEEE Journal on Selected Areas in Communications*, vol. 19, no. 8, pp. 1556–1569, Aug. 2001.

[Zha02] J. Zhang and X. Wang, "Large-system performance analysis of blind and group-blind multiuser receivers," *IEEE Trans. on Information Theory*, vol. 48, no. 9, pp. 2507–2523, Sep. 2002.

·42·

Random matrix theory and ribonucleic acid (RNA) folding

Graziano Vernizzi and Henri Orland

Abstract

In this article we review a series of recent applications of random matrix theory (RMT) to the problem of RNA folding. The intimate connection between these two fields of study lies in their common diagrammatic description: all secondary structures of an RNA molecule are represented by planar diagrams which are naturally interpreted as the Feynman diagrams of a suitable large-N matrix model. As a consequence, all subleading 1/N corrections of the matrix model correspond to diagrams that are not planar, and that can be mapped into folded configurations of the RNA molecule including pseudoknots.

Moreover, the standard topological expansion of the matrix model induces an elegant classification of all possible RNA pseudoknots, according to their topological genus. The RMT-based statistical description of the RNA-folding problem has been extended also to the folding of a generic homopolymer with saturating interaction. While reviewing several known findings, we extend previous results about the asymptotic distribution of pseudoknots of a phantom homopolymer chain in the limit of large chain length.

42.1 Introduction

The reader familiar with the rich versatility of random matrix theory, which is well documented in this book, will perhaps not be surprised by some intriguing applications in the realm of biology and biophysics. In this chapter we review in particular the application to the so-called *RNA folding problem*. We therefore start by providing a schematic introduction to the problem, but for more detailed reviews on RNA folding we refer the reader to [Tin99, Hig00].

One of the most exciting fields in modern computational molecular biology is the search for tools predicting the complex foldings of bio-polymers such as RNA [Sch04], ssDNA and proteins [Sno05], when homologous sequences are not available. In particular, the quest for an algorithm which can predict the spatial structure of an RNA molecule given its chemical sequence has received considerable attention from molecular biologists over the past three decades [Cou02]. One of the main reasons is that the three-dimensional structure of an RNA molecule is intimately connected to its specific biological function in the cell (e.g. for protein synthesis and transport, catalysis, chromosome replication, and regulation) [Tin99]. Some examples are tRNA molecules involved in translation, the RNA components of the ribosome, group I and group II introns, several ribozymes, and even the non-coding region of some messenger RNAs. When considering that all RNA molecules are ultimately coded inside the DNA genetic sequence, one easily realizes that the RNA three-dimensional structure is basically determined by the sequence of nucleotides along the sugar-phosphate backbone of the RNA. The chemical formula or sequence of covalently linked nucleotides along the molecule from the 5' to the 3' end is usually called the *primary structure*. The four basic types of nucleotides are adenine (A), cytosine (C), guanine (G), and uracil (U), but it is known that modified bases may appear [Lim94].

At high enough temperatures, or under high-denaturant conditions RNA molecules are free single-stranded swollen polymers without any particular structure, and their behaviour is entropy dominated. At room temperature, different nucleotides can pair by means of saturating hydrogen bonds. Namely, the standard Watson-Crick pairs are A-U and C-G with two and three hydrogen bonds respectively, while the G-U 'wobble' pair has two hydrogen bonds. Comparative methods show that 'non-canonical' pairings are also possible [Nag00], as well as higher-order interactions such as triplets, or quartets [Leo01]. In this chapter we consider only canonical base-pair interactions. Adjacent base pairs provide an additional binding 'stacking' energy, which is actually the origin of the formation of stable A-form helices, one of the main structural characteristics of folded RNAs. During this process, helices may embed unpaired sections of RNA, in the form of hairpins, loops, and bulges which give a destabilizing entropic contribution. It is precisely the cooperative formation of pairings, stacking of bases, and the hierarchical building of structural motifs [Bri97] which bring the RNA into its final folded three-dimensional configuration. The lack of a complete mathematical framework and the complexity of the physical interactions involved in this process, are the main reasons that the prediction of the actual spatial molecular structure of RNA (i.e. its shape) given its primary structure is still one of the main open problems of molecular biology [Tin99].

In order to rationalize the process, the concept of *secondary structure* of an RNA molecule has been introduced: it is essentially the list of paired bases of the RNA. It is very convenient to represent the nucleotides along the RNA backbone as points on an oriented circle (from the 5' end to the 3' end), and each base-pair is represented by an *arc* (or cord) joining the two interacting nucleotides, inside the circle (see Figure 42.1 on the left). Furthermore, the *tertiary structure* is defined as the actual three-dimensional arrangement of the entire molecule. The benefit of distinguishing between secondary and tertiary structures stems from the fact that the secondary structure of RNA molecules is stable under suitable ionic conditions of the solution, and it carries the main contribution to the free energy of a fully folded configuration, also including some of the steric

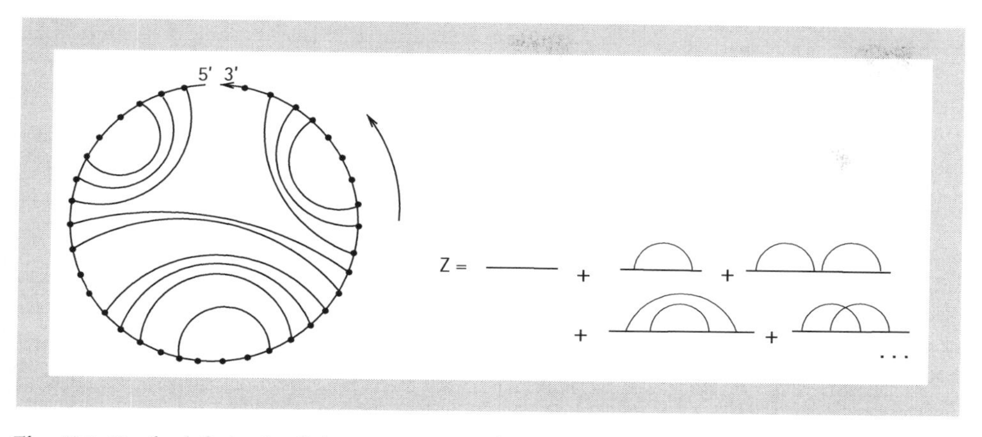

Fig. 42.1 On the left: in the disk representation the RNA backbone is represented by an oriented circle and the base pairs by arcs inside the circle. On the right: the virial expansion of the RNA partition function in Eq. (42.2.5) contains both planar and non-planar diagrams.

constraints. While that is not necessarily true for proteins, for RNA molecules such a distinction is truly meaningful since the folding process is hierarchical [Tin99, Bri97, Ban93, Lai94]. However one has to keep in mind that the list of binary contacts between different nucleotides contains only partial information on the total folding, since most information about distances in real three-dimensional space is lost. The importance of the secondary structure mostly relies in the fact that it may provide the scaffold for the final tertiary structure.

Several algorithms have been proposed for the prediction of RNA folding. They are based on: deterministic or stochastic minimization of a free energy function [Zuk84, Sch96], phylogenetic comparison [Gut94], kinetic folding [Fla00, Mir85, Isa00], maximal weighted matching method [Tab98], and several others (for a survey see [Zuk00]). Despite the large number of tools available for the prediction of RNA structures, one must recognize that no reliable algorithms exist for the prediction of the full tertiary structure of long RNA molecules. Most of the algorithms listed above deal with the prediction of just the RNA secondary structure, and the general case is far from being solved [Edd04].

To describe the full folding it is important to introduce the concept of RNA *pseudoknots* [Ple85]. One says that two base pairs form a pseudoknot when the parts of the RNA sequence spanned by those two base pairs are neither disjointed, nor have one contained in the other. They are conformations such that the associated disk diagram is not planar (i.e. there are intersecting arcs). Pseudoknots play an important role in natural RNAs [Wes92], for structural, regulatory and catalytic functions, such as ribosomal frameshifting in many viruses [She95], self-splicing introns [Ada04], catalytic domains in various ribozymes [Ras96], or also in ribonucleoprotein complexes such as telomerase [The05]. Their ubiquity manifests itself in the large variety of possible shapes and structures [Sta05], and their existence should not be neglected in structure prediction algorithms, as they account for 10%–30% on average of the total number of base pairs. Pseudoknots are often excluded in the definition of RNA secondary structure and many authors consider them as part of the tertiary structure. This restriction is due in part to the fact that RNA secondary structures without pseudoknots can in principle be predicted easily. One should also note that pseudoknots very often involve base-pairing from distant parts of the RNA, and are thus quite sensitive to the ionic strength of the solution. It has been shown that the number of pseudoknots depends on the concentration of Mg^{++} ions, and can be strongly suppressed by decreasing the ionic strength (thus enhancing electrostatic repulsion) [Mis98].

The most popular and successful technique for predicting secondary structures without pseudoknots is dynamic programming [Zuk84, Wat85], for which the memory and CPU requirements scale with the sequence length L as $O(L^2)$ and $O(L^3)$, respectively. Such energy-based methods have proven to be the most reliable (as, e.g. [Zuk03, Hof03]) even for long sequences (thousands of bases). They assign some energy to the base pairings and some entropy to the loops and bulges. In addition, they take into account stacking energies, and assign precise weights to specific patterns (tetraloops, multiloops, etc.) [Mat99]. The lowest free energy folds are obtained either by dynamic programming algorithms [Zuk81], or by computing the partition function of the RNA molecule [Mcc90]. The main drawback of these energy-based methods is that they deal solely with secondary structures and cannot take into account pseudoknots in a systematic way.

Recently, new deterministic algorithms that deal with pseudoknots have been formulated ([Riv99, Liu06, Isa00, Aku01, Lyn00, Orl02, Pil05a, Pil05b]; the list is not exhaustive). In this case the memory and CPU requirements generally scale as $O(L^4)$ and $O(L^6)$ respectively ($O(L^4)$ and $O(L^5)$ in [Lyn00], or $O(L^4)$ and $O(L^3)$ for a restricted model in [Aku01, Bel10]), which can be a very demanding computational effort even for short RNA sequences ($L \sim 100$). Moreover, the main limitation of these algorithms is the lack of precise experimental information about the contribution of pseudoknots to the RNA free energy, which is often excluded *a priori* in the data analysis (as also pointed out in [Isa00, Mir93, Gul99]). The increase of computational complexity does not come as a surprise. In fact the RNA-folding problem with pseudoknots has been proven to be NP-complete for some classes of pseudoknots [Aku01, Lyn00]. For that reason, stochastic algorithms might be a better choice to predict secondary structures with pseudoknots in a reasonable time and for long enough sequences.

In [Sch96, Abr90, Ver05a] stochastic Monte Carlo algorithms for the prediction of RNA pseudoknots have been proposed. In such stochastic approaches, the very irregular structure of the energy landscape (glassy-like) is the main obstacle: configurations with small differences in energy may be separated by high energy barriers, and the system may very easily get trapped in metastable states. Among the stochastic methods, the direct simulation of the RNA-folding dynamics (including pseudoknots) with kinetic folding algorithms [Isa00] is most successful. This technique allows one to describe the succession of secondary structures with pseudoknots during the folding process.

The approach of random matrix theory to the RNA folding problem puts a strong emphasis on the topological character of the RNA pseudoknots. It is based on a correspondence (first noticed by E. Rivas and S.R. Eddy in [Riv99]) between a graphical representation of RNA secondary structures with pseudoknots and Feynman diagrams. In [Riv99] the authors consider only a particular class of pseudoknots. Along the same direction, the authors of [Orl02] made the correspondence between RNA folding and Feynman diagrams more explicit by formulating a *matrix field theory* model whose Feynman diagrams give exactly all the RNA secondary structures with pseudoknots. What is remarkable in this new approach is that it provides an analytic tool for the prediction of pseudoknots, and all the diagrams appear to be naturally organized in a series, called *topological expansion*, where the first term corresponds to planar secondary structures without pseudoknots, and higher-order terms correspond to structures with pseudoknots. In the next section we review (and generalize in part) this approach.

42.2 A model for RNA-folding

A simplified framework (but rather standard) for describing the folding of an RNA molecule, is given by the statistic mechanical model of a polymer chain of L nucleotides in three dimensions with interacting monomers. One can characterize the configuration of a given RNA sequence $\{s\} = \{s_1, s_2, \cdots, s_L\}$, with $s_{\{i\}} \in \{A, U, G, C\}$, by the set $\{\mathbf{r}\} = \{\mathbf{r}_1, \mathbf{r}_2, \cdots, \mathbf{r}_L\}$, where $\mathbf{r}_k$ is the position of the k-th base, and by the $L \times L$ real symmetric contact matrix C:

$$C_{ij} = \begin{cases} 1 & \text{if } i, j \text{ are hydrogen-bonded} \\ 0 & \text{otherwise} \end{cases} . \tag{42.2.1}$$

The requirement that the hydrogen-bond interactions are saturating can be easily imposed by requiring that $\sum_i C_{ij} \leq 1, \forall i$, which means that any base can make a hydrogen-bond with at most one other base at a time. At such a coarse-grained level, the total energy $\mathcal{H}$ of a configuration is:

$$\mathcal{H} = \mathcal{H}_s(\{\mathbf{r}\}) + \mathcal{H}_\varepsilon(\{\mathbf{r}\}; C) + \mathcal{H}_r(\{\mathbf{r}\}), \tag{42.2.2}$$

where $\mathcal{H}_s$ takes into account the contributions from steric constraints, the stiffness of the chain, and bond potentials between consecutive monomers; $\mathcal{H}_\varepsilon$ takes into account hydrogen-bond pairings; $\mathcal{H}_r$ contains all remaining interactions, such as stacking energies, electrostatics etc. The first term in Eq. (42.2.2) is often of the form $\mathcal{H}_s = \sum_{i<j} b(r_{ij})$, where $b(r) = \delta(r - l)$ for a model in which the nucleotides are connected along the RNA molecule by rigid rods of length l, or $b(r) = e^{-(r-l)^2/6\sigma^2}$ for a model with elastic rods. Let $\varepsilon(a, b)$ denote the 4×4 real symmetric matrix of the attractive energies between nucleotides. Typical values are $\varepsilon(C, G) \simeq -3$ kcal/mol and $\varepsilon(A, U) \simeq -2$ kcal/mol, respectively, while for the weaker (G,U) pair $\varepsilon(G, U) \simeq -1$ kcal/mol (at room temperature 300 K $\simeq$ 0.6 kcal /mole $\simeq$ 1/40 eV). We define:

$$\mathcal{H}_\varepsilon(\{\mathbf{r}\}; C) = \sum_{i<j} \varepsilon(s_i, s_j) C_{ij} v(r_{ij}), \tag{42.2.3}$$

where $r_{ij} = |\mathbf{r}_i - \mathbf{r}_j|$ and $v(r_{ij})$ is the short-range space-dependent part of the hydrogen-bond interaction, which for simplicity can be taken as the Heaviside attractive potential $v(r) = -w\theta(R - r)$, w and R being the strength and range of the attraction, respectively. The RNA free energy $\mathcal{F} = -\log \mathcal{Z}/\beta$ at the absolute temperature $T = 1/K_B\beta$ (K_B being the Boltzmann constant) can be obtained from the partition function

$$\mathcal{Z} = \sum_{C_{ij}=1,0}' \int \prod_{k=1}^{L} d^3\mathbf{r}_k \, e^{-\beta\mathcal{H}}, \tag{42.2.4}$$

where the prime index over the sum indicates that the sum is over all real symmetric matrices with entries 0, 1, that satisfy the saturation constraint $\sum_i C_{ij} \le 1$. We perform a virial expansion of the partition function in terms of the Mayer functions $f_{ij} = \exp(-\beta\varepsilon(s_i, s_j)C_{ij}v(r_{ij})) - 1$:

$$\mathcal{Z} = \left[e^{-\beta\mathcal{H}_\varepsilon}\right] = \left[\prod_{i<j}(1 + f_{ij})\right] = [1] + \sum_{<ij>} [f_{ij}] + \sum_{<ijkl>} [f_{ij} f_{kl}] + \sum_{<ijkl>} [f_{il} f_{jk}] + \sum_{<ijkl>} [f_{ik} f_{jl}] + \cdots \tag{42.2.5}$$

where the square brackets $[g(\mathbf{r}, C)] \equiv \sum_C' \int d^3\mathbf{r} g(\mathbf{r}, C) e^{-\beta(\mathcal{H}_s + \mathcal{H}_r)}$ indicates the weighted sum and integration over all degrees of freedoms, $< ij >$ denotes all pairs with $j > i$, $< ijkl >$ all quadruplets with $l > k > j > i$, and so on. We note that the Mayer functions f_{ij} are vanishing if the base i and the base j are not interacting. This is needed to reproduce the free polymer chain model in the limit where all $\varepsilon(a, b)$ are zero. Moreover, all the terms corresponding to a cluster of three or more simultaneously interacting monomers are omitted (i.e. of type $f_{ij} f_{jk} \ldots$), since such terms are suppressed by the constrained sum $\sum_C'$. Similarly, all self-interacting terms f_{ii} are zero since a nucleotide does not attract itself. A graphical representation of the expansion is shown on the right of Fig. 42.1. The integration over the monomer coordinates in Eq. (42.2.5) accounts for the actual geometrical feasibility of a given folded configuration and also for the correct entropic factor associated with the formation of loops, bulges, multiloops. In [Orl02] part of these constraints were effectively introduced by the Heaviside function $\theta(|i-j|>4)$ in the energy term, meaning that an RNA molecule is not infinitely flexible and usually cannot pair nucleotides separated by less than four sites. We could enforce such a term here as well or, alternatively, one can choose $\mathcal{H}_r$ such that the persistence length of the RNA backbone is long enough to disallow configurations with $|i-j|\leq 4$. As shown in Fig. 42.1, each term in the sum Eq. (42.2.5) can be represented graphically by a suitable arc diagram. In such a representation the nucleotides are dots on an oriented horizontal line (which represents the RNA sugar backbone from the 5′ end to the 3′ end), and each base pairs is drawn as an arc – above that line – between the two interacting bases. Equation (42.2.5) contains contributions from both planar and non-planar diagrams. The configurations of an RNA molecule that correspond to non-planar diagrams are precisely the RNA pseudoknots.

Most works that deal with only RNA secondary structures, conceptually drop – by hand – all non-planar terms in Eq. (42.2.5). Similarly, most works dealing with RNA pseudoknots usually include *some* of the non-planar terms, but generally not all of them. In other words, very complex pseudoknotted structures are generally neglected by generic folding algorithms. It is here where random matrix theory enters the scene [Orl02]. Due to the unique ability of random matrix theory to classify and organize planar and non-planar Feynman diagrams (as done for instance in Chapter 27 of this book) and in large-N QCD, one can promote the sum in Eq. (42.2.5) as an integral over matrices. The central observation, originally made by 't Hooft [tHo74], is that there is a systematic relation between the topology of a Feynman diagram of a given matrix integral and the power of $1/N^2$ of its corresponding term in the asymptotic large-N expansion of the matrix integral. For instance, planar diagrams are of order $O(1)$, and non-planar diagrams are of higher order. The main idea introduced first in [Orl02] is to consider the following integral over matrices:

$$Z_{1,L}(N)=\frac{1}{A_{1,L}(N)}\int\prod_{i=1}^{L}\mathcal{D}\Phi_i e^{-\frac{N}{2}\sum_{ij}(V^{-1})_{ij}\,\mathrm{tr}(\Phi_i\Phi_j)}\frac{1}{N}\mathrm{tr}\prod_{l=1}^{L}(1+\Phi_l), \qquad (42.2.6)$$

where Φ_i, $i=1,\ldots,L$, are L independent $N\times N$ Hermitian matrices ($\Phi_i^{+}=\Phi_i$) and $\prod_{l=1}^{L}(1+\Phi_l)$ is the ordered matrix product $(1+\Phi_1)(1+\Phi_2)\cdots(1+\Phi_L)$. The normalization factor is:

$$A_{1,L}(N)=\int\prod_{i=1}^{L}\mathcal{D}\Phi_i e^{-\frac{N}{2}\sum_{ij}(V^{-1})_{ij}\,\mathrm{tr}(\Phi_i\Phi_j)}. \qquad (42.2.7)$$

The matrix V is a $L \times L$ symmetric matrix with elements V_{ij}, and it must be invertible in order for $Z_{1,L}(N)$ and $A_{1,L}(N)$ to exist. Expanding the integral in Eq. (42.2.6) by using the Wick theorem, we obtain:

$$Z_{1,L}(N) = 1 + \sum_{i<j} V_{ij} + \sum_{i<j<k<l} V_{ij}V_{kl} + \sum_{i<j<k<l} V_{il}V_{jk} + \ldots \\ + \frac{1}{N^2}\left(\sum_{i<j<k<l} V_{ik}V_{jl} + \ldots\right) + \frac{1}{N^4}\ldots. \tag{42.2.8}$$

which is a large-N asymptotic series. The relation with the expansion in Eq. (42.2.5) is evident. After identifying $V_{ij} = f_{ij}$, the two series coincide formally for $N = 1$, since $\mathcal{Z} = [Z_{1,L}(N)]$, whereas for $N > 1$ Eq. (42.2.8) contains additional information. All the planar structures are given by the $O(1)$ term of Eq. (42.2.8) while higher-order terms corresponds to RNA secondary structures with pseudoknots. A bonus of the matrix representation in Eq. (42.2.8) is a systematic and topological classification of all possible RNA pseudoknots. It is important to note that the random matrix theory approach to the study of RNA pseudoknots is analogous in flavour to the work presented in Chapter 27 of this book, where the authors introduce suitable matrix integrals for counting and enumerating the more complex class of *alternating knots* and *links*.

42.3 Physical interpretation of the RNA matrix model

An important remark made in [Orl02] is that the matrix theory defined by Eq. (42.2.6) has the same topological structure as 't Hooft's large-N quantum chromodynamics where there are L types of gluons, each gluon propagator is associated with a factor of $\frac{1}{N}$, and each colour loop is associated with a factor of N. It turns out that the overall diagram is associated with a power of $1/N^2$, which corresponds to a specific topological character (see also Chapter 26 and Chapter 32 of this book). In simpler words, if all m arcs in the arc-diagram representation are drawn by double lines, and c is the total number of closed loops obtained by following the single lines of the diagram (including the circular backbone), then the topological genus of the diagram is the integer $g = (1 + m - c)/2$. Such a result is a consequence of the Euler's theorem (for a derivation see [Ver04]). The genus g of a diagram is the minimum number of handles a sphere must have in order to be able to draw the diagram on it without any crossing arc. In the disk representation, the circle of the RNA-backbone becomes the boundary of a hole or *puncture* on the surface, and the arcs corresponding to the RNA base-pairs are drawn on that surface without crossing.

We note that a different way (but mathematically completely equivalent) to state the same combinatorial problem is to consider the RNA diagrams as representing the class of *permutation involutions* of L elements. More precisely, if the ith and jth base are connected, one can associate to it the transposition T_{ij}. The product of all transpositions in a given diagram generates a permutation involution $\pi = \prod T_{ij}$. For instance, the last two diagrams on the right of Figure 42.1 correspond to the permutations $\{2, 1, 4, 3\}$ and $\{3, 4, 1, 2\}$, respectively. The genus therefore can be obtained by counting the number of cycles c of the permutation $(\pi\sigma)$ where $\sigma = \{2, 3, 4, \ldots, L, 1\}$ is the cyclic shift-permutation. One can easily verify that the genus of the diagram is given again by $g = (1 + m - c)/2$ (see eq. 26.2.2 from Chapter 26 of this Handbook). From this point of view, the constrained contact matrix C introduced in the previous section, is actually the matrix representation of a permutation involution. This observation is beneficial, since one can obtain the total number $\gamma(L)$ of involution permutations [Ski90]:

$$\gamma(L) = L! \sum_{n=0}^{L/2} \frac{1}{2^n n!(L-2n)!} = (-2)^{L/2}\, U(-\frac{L}{2}, \frac{1}{2}, -\frac{1}{2}). \qquad (42.3.1)$$

Having identified the genus of each term in the series, one can define $\mu = 2 \log N/\beta$ and rewrite the topological asymptotic expansion as:

$$\begin{aligned} Z_{1,L}(N) &= 1 + \sum_{i<j} V_{ij} + \sum_{i<j<k<l} V_{ij} V_{kl} + \sum_{i<j<k<l} V_{il} V_{jk} + \ldots \\ &+ e^{-\beta\mu} \left(\sum_{i<j<k<l} V_{ik} V_{jl} + \ldots \right) + e^{-2\beta\mu} \times (\text{genus 2 pseudoknots}) \\ &+ \cdots = \sum_{C_{ij}}{}' e^{-\beta(\mathcal{H}(C) + \mu g(C))} \end{aligned} \qquad (42.3.2)$$

where $g(C)$ is the genus of the configuration C, and $\sum'_{C_{ij}}$ is the usual constrained sum over all possible contact matrices. One could speculate on the meaning of the topological chemical potential μ which is in analogy to the experimental known fact that the number of pseudoknots in RNA can be controlled by the amount of Mg^{++} ions in the solution (the lower the concentration, the lower the number of pseudoknots) [Mis98]. In Eq. (42.3.2) this effect is modelled by including possible topological fluctuations that are controlled by a topological chemical potential μ. In other words, we can define the new Hamiltonian [Ver05a, Zee05]: $\mathcal{H}_\mu = \mathcal{H} + \mu g$. From this definition one sees that the average genus of pseudoknots $< g(\mu) >= d\mathcal{F}/d\mu$ is a function of μ. Such a relation could be tested experimentally in principle. However, such an analogy is qualitative, and the exact dependence of the RNA free energy on the concentration of divalent ions is very likely not a simple linear function.

Nevertheless, the matrix representation shows how the topological chemical potential can be introduced to control the average number of pseudoknots in a given RNA sequence effectively. An explicit implementation can be found in [Bon10] and at http://eole2.lsce.ipsl.fr/ipht/tt2ne/tt2ne.php.

Furthermore, the topological asymptotic expansion induces a systematic classification of RNA structures, since it allows us to completely grasp the topological complexity of any pseudoknot with a single integer number, the genus. It can be viewed as a kind of 'quantum' number, which is reminiscent (although at a rather speculative level) of the superfold families CATH or SCOP [Ore97], that have proven so useful in protein structure classification. In the literature other possible classifications of RNA structures with pseudoknots have been proposed (see, e.g., [Con04]). However, the one proposed in [Ver05b] is the only one that is purely topological, i.e. independent of any three-dimensional embedding and which is based only on the classical topological expansion of closed bounded surfaces. This is also the reason why this expansion can be derived mathematically with standard tools of combinatorial topology.

42.4 Large-N expansion

The large-N expansion of the matrix integral is not as straightforward as it seems at first, since N appears implicitly in the size of the matrices Φ_i. In [Orl02, Pil05a] the successful strategy of writing exact recursion relations for the partition function containing only secondary structures [Wat85] has been generalized to higher-order terms, i.e. to structures with pseudoknots. In [Orl02] two elegant formulae have been obtained. First:

$$Z(1, L) = < M^{-1}(A)_{L+1,1} > , \tag{42.4.1}$$

where $M_{ij} = \delta_{ij} - \delta_{i,j+1} + i(V_{i-1,j})^{\frac{1}{2}} A_{i-1,j}$, and the average is with respect to the weight $e^{-NS(A)}/C$, with $S(A) = \frac{1}{2}\mathrm{tr}\, A^2 - \mathrm{tr} \log M(A)$ and $C = \int dA e^{-\frac{N}{2}\mathrm{tr}A^2}$. The second useful relation is:

$$Z(1, L+1) = Z(1, L) + \sum_{k=1}^{L} V_{L+1,k} < \frac{1}{N}\mathrm{tr}\Pi_{i=1}^{k-1}(1+\varphi_i)\frac{1}{N}\mathrm{tr}\Pi_{j=k+1}^{L}(1+\varphi_j) > \tag{42.4.2}$$

The first formula displays the N-dependence explicitly and it allows us to carry the $1/N$ expansion via a steepest-descent method to any order. The second formula can also be systematically expanded and provides recurrence equations for the partition function to any desired order in $1/N$, therefore also for structures including pseudoknots. In the next sections we will consider the first two terms in such an asymptotic expansion.

42.4.1 $O(1)$ term: RNA planar secondary structures

The first term of the expansion can be written as a well-known recurrence equation [Wat85, Mcc90, deG68, Wat78, Bun99, Mon01] which is often referred to as the "Hartree approximation":

$$Z(1, L+1) \simeq Z(1, L) + \sum_{j=1}^{L} Z(1, j)\, V_{j, L+1}\, Z(j+1, L+1) \tag{42.4.3}$$

It has the obvious interpretation that to lowest order the additive effect of including one extra nucleotide labelled by $L+1$ to the RNA heteropolymer can be described by pairing that nucleotide to the nucleotide labelled by j, which separates the heteropolymer into two segments, one from 1 to j and the other from $j+1$ to $L+1$. One then sums over all possible j. The geometric interpretation is simply that any arc in the disk diagram cuts the disk into two disjoint planar diagrams. Entropic contributions can be included easily by means of the "$[\cdots]$-integration" (see definition after Eq. (42.2.5)). Such a procedure has been exploited in [Mon01] for the particular case of a homogeneous homopolymer chain. Moreover, such a recurrence relation constitutes the kernel of most successful prediction algorithms for RNA secondary structures available today [Hof03], that allow a fast prediction even for thousand-base-long sequences.

42.4.2 $O(1/N^2)$ terms: RNA pseudoknots

The recursion relation for the partition function containing RNA pseudoknots of genus 1 is considerably more complicated than the case for pure secondary structures only. The explicit expression has been obtained in [Pil05a]:

$$\begin{aligned} Z_{1,L} = Z_{1,L-1} &+ \sum_{j=1}^{L-1} Z_{1,j-1} V_{j,L} Z_{j+1,L-1} + \frac{1}{N} \sum_{j=1,m,n}^{L-1} G_{1,n-1} \Gamma^{j}_{m,n} V_{j,L} G_{m+1,L-1} \\ &+ \sum_{j=1,m,n}^{L-1} G_{m+1,L-1} \Theta^{j}_{m,n} V_{j,L} G_{1,n-1} \end{aligned} \tag{42.4.4}$$

where $G_{a,b} = Z_{a,b}$ is the partition function of planar secondary structures (from monomer a to monomer b), and the explicit expression of the functions Γ and Θ can be found in the appendix of [Pil05a]. We just mention here that one can include entropic contributions by suitable averaging over the spatial degrees of freedom. This has been computed explicitly in [Isa00] for some simple pseudoknotted configurations of a flexible Gaussian chain. A systematic evaluation of entropic contributions for all pseudoknots is possible in principle by using the random matrix model approach.

In [Pil05a], while obtaining the recursion equations for RNA structures with pseudoknot of genus 1, the number of independent classes of pseudoknots with genus 1 was also determined. In particular it has been shown that the family of simplest RNA pseudoknots (i.e. with genus 1) contains only four types of pseudoknots: a) the H-pseudoknot (made of two helices A and B, and called also ABAB because of the order of helices appearing from the 5' to the 3' end), b) the Kissing-hairpin pseudoknot (with three helices, and also called ABACBC), c) the 'double pseudoknot' ABCABC (with three helices), and d) the pseudoknot ABCADBCD with four helices.

Other characterizations of pseudoknots have been proposed in the literature (e.g. [Ste04, Riv00, Luc03]) but the advantage of the approach in [Orl02] is that the classification is topological, i.e. independent from the way the diagram is drawn, and dependent only on the intrinsic complexity of the contact structure. Other interactions such as stacking energies of adjacent complementary base pairs and electrostatic interactions can be added without major changes. For instance, the stacking energies can be taken into account by defining a 16×16 interaction matrix $\varepsilon(s_i s_j; s_m s_n)$ between pairs of bases instead of the 4×4 matrix $\varepsilon(s_i, s_j)$ defined above. While such additions would improve a numerical algorithm based on this novel approach, in the following we actually focus on a simplified model that allows for the exact enumeration of all contact structures with fixed genus. Such a model is based on a simplified version of the model in [Orl02, Pil05a, Pil05b, Ver04] and is relevant for studying the behaviour of non-planar contributions from an enumerative point of view.

42.5 The pseudoknotted homopolymer chain

By neglecting momentarily all energy and entropic considerations, the matrix model approach allows us to attack also a much simpler fundamental problem, namely the exact *combinatorics and enumeration* of RNA secondary structures with any possible kind of pseudoknots. This has been done in general first in [deG68] and for pseudoknots in [Ver05b]. The simplifying assumptions are: (1) any possible pairing between nucleotides is allowed, independently from the type of nucleotides and from their distance along the chain; (2) all the pairings may occur with the same probability, i.e. the energies ε are constant for all pairs of nucleotides, meaning that the matrix V_{ij} has all constant entries equal to $v > 0$, i.e: $V_{ij} = v + \delta_{ij} a$. The real number a has been added in order for V^{-1} to exist, with the prescription to take the limit $a \to -v$ eventually, which effectively suppresses diagrams with self interactions, $V_{ii} = 0$. Under such simplifying assumptions, the multi-matrix integral in Eq. (42.2.6) can be recast in a single-matrix form. To that goal, it is useful first to rewrite Eq. (42.2.6) in terms of the matrix V instead of its inverse V^{-1}, by using the Hubbard-Stratonovich transformation, i.e. the identity:

$$\exp\left[-\frac{N}{2}\sum_{ij}(V^{-1})_{ij}\,\mathrm{tr}(\varphi_i\varphi_j)\right] = K\int\prod_{k=1}^{L}d\psi_k e^{-\frac{N}{2}\sum_{ij}V_{ij}\,\mathrm{tr}(\psi_i\psi_j)}e^{i\,N\sum_i\,\mathrm{tr}(\psi_i\varphi_i)}, \tag{42.5.1}$$

where ψ_i, $i = 1, \ldots, L$, are L $N\times N$ Hermitian matrices, and K is a normalization constant. Here and in the following we do not worry too much about overall constant factors, as they all simplify when performing the same transformations in the integrals in the numerator and denominator of Eq. (42.2.6) (i.e. also in $A_{1,L}(N)$). Secondly, we insert the explicit expression of V_{ij} and we obtain:

$$Z_{1,L}(N) = K_1\int\prod_{k=1}^{L}d\varphi_k\prod_{h=1}^{L}d\psi_h e^{-\frac{Na}{2}\mathrm{tr}\sum_i\psi_i^2}\times e^{-\frac{Nv}{2}\mathrm{tr}\left(\sum_i\psi_i\right)^2}e^{i\,N\sum_i\,\mathrm{tr}(\psi_i\varphi_i)} \times\frac{1}{N}\mathrm{tr}\prod_{l=1}^{L}(1+\varphi_l). \tag{42.5.2}$$

Thirdly, we use again an Hubbard-Stratonovich transformation for writing:

$$e^{-\frac{Nv}{2}\mathrm{tr}\left(\sum_i\psi_i\right)^2} = K_2\int d\sigma\, e^{-\frac{N}{2v}\,\mathrm{tr}\sigma^2}e^{i\,N\mathrm{tr}(\sigma\sum_i\psi_i)}, \tag{42.5.3}$$

where σ, is an $N\times N$ Hermitian matrix. After integrating over the matrices ψ_k and applying the shift $\varphi_h \to \varphi_h - \sigma$, eq. (42.5.2) becomes:

$$\int d\sigma\, e^{-\frac{N}{2v}\,\mathrm{tr}\sigma^2}\int\prod_{k=1}^{L}d\varphi_k\, e^{-\frac{N}{2a}\mathrm{tr}\sum_i\varphi_i^2}\frac{1}{N}\mathrm{tr}\prod_{l=1}^{L}(1+\varphi_l-\sigma). \tag{42.5.4}$$

The integral over the matrices φ_h can be easily done. In fact, since the product $\prod_{l=1}^{L}(1+\varphi_l-\sigma)$ does not contain any φ_l^2 term, the result is

$$Z_L(N) = \frac{1}{\tilde{A}(N)}\int d\sigma e^{-\frac{N}{2v}\,\mathrm{tr}\sigma^2}\frac{1}{N}\mathrm{tr}(1+\sigma)^L, \tag{42.5.5}$$

where the reflection $\sigma\to-\sigma$ has been applied, and the normalization factor $\tilde{A}(N) = \left(\frac{\pi v}{N}\right)^{\frac{N^2}{2}}2^{\frac{N}{2}}$. In [Ver05b] one can find the derivation of the explicit expression of the exponential generating function (EGF) of $Z_L(N)$ (reproduced also in [Niv09]):

$$G(t,N) \equiv \sum_{L=0}^{\infty}Z_L(N)\frac{t^L}{L!} = e^{\frac{vt^2}{2N}+t}\frac{1}{N}L_{N-1}^{(1)}\left(-\frac{vt^2}{N}\right), \tag{42.5.6}$$

where $L_N^{(1)}(z)$ is a generalized Laguerre polynomial. From this exact result we can extract all the coefficients $Z_L(N)$ (see [Ver05b] for a list up to $L = 8$). Their combinatorial meaning is straightforward: the power of v gives the number of arcs in the diagram, and the power of $1/N^2$ is the genus of the diagram. For instance for $L = 7$ one has $Z_7(N) = 1 + 21v + 70v^2 + 35v^3 +$

$(35v^2 + 70v^3)/N^2$ meaning that there are 21 planar diagrams with one arc, and 35 diagrams on the torus (i.e. genus one closed oriented surface) with two arcs. The total number of diagrams for each fixed genus can be obtained by putting $v = 1$, (for instance, the total number of diagrams on the torus for $L = 7$ is 105). Analogously, the total number of diagrams, irrespectively from the genus, can be obtained by putting also $N = 1$ (for instance, the number of diagrams for $L = 7$ with 3 arcs is 105). The general $1/N^2$ topological expansion of $Z_L(N)$ is:

$$Z_L(N) = \sum_{g=0}^{\infty} a_{L,g}(v) \frac{1}{N^{2g}}, \tag{42.5.7}$$

where the coefficients $a_{L,g}(v)$ give exactly the number of diagrams at fixed length L and fixed genus g. From the formulae in Eq. (42.5.6) and Eq. (42.5.7) one can obtain all the coefficients $a_{L,g}$ recursively. Moreover, by normalizing each $a_{L,g}$ by the total number of diagrams at fixed L, i.e. by $\mathcal{N} \equiv Z_L(1)$, one can obtain the distribution of the number of diagrams. A closed formula for the coefficients $a_{L,g}$ has been obtained in [Erb09a]:

$$a_{L,g} = \sum_{k=0}^{[L/2]} \sum_{j=k-2g}^{k} \frac{L! 2^{j-k} S_{j+1}^{(k+1-2g)}}{(L-2k)!(k-j)!j!(j+1)!} v^k, \tag{42.5.8}$$

where $S_j^{(m)}$ are the Stirling number of the first kind [Abr64]. As a consequence each term $Z_L(N)$ of the t-expansion of Eq. (42.5.6) is a polynomial in powers of $\frac{1}{N^2}$, the highest power being $\left[\frac{L}{4}\right]$ [Ver05b, Erb09a]. That means that for a given diagrams with L vertices, the genus cannot be greater than $\left[\frac{L}{4}\right]$, i.e. $a_{L,g>\left[\frac{L}{4}\right]}(v) = 0$. Moreover, the number of arcs is always greater or equal to twice the genus. In other words, in order for the homopolymer to have a pseudoknot with genus g, the length of the chain should be at least $L \geq 4g$ and it must have at least $m > 2g$ base-pairs. The large-N limit of Eq. (42.5.6) gives the EGF for planar diagrams (i.e. the EGF of secondary structures without pseudoknots). It reads [Abr64]:

$$\begin{aligned} \lim_{N\to\infty} G(t, N) &= \frac{e^t}{\sqrt{vt}} I_1\left(2\sqrt{vt}\right) = 1 + t + \frac{1}{2}(1+v)t^2 + \frac{1}{3!}(1+3v)t^3 \\ &\quad + \frac{1}{4!}(1+6v+2v^2)t^4 + O(t^5) \end{aligned} \tag{42.5.9}$$

where $I_1(x)$ is a modified Bessel function of the first kind. By putting $v = 1$ one counts the total number of diagrams (up to the factor $1/n!$), irrespective of the number of arcs. The first few terms of such an expansion are given by the sequence 1, 1, 2, 4, 9, 21, ... which is the sequence of Motzkin numbers: they precisely count the number of ways of drawing any nonintersecting chords among L points on a circle [Mot48]. We also note that Eq. (42.5.9) admits a series expansion in powers of v, which is valid in the full v-complex plane:

$$\lim_{N\to\infty} G(t, N) = e^t \sum_{k=0}^{\infty} \frac{(t^2 v)^k}{k!(k+1)!}. \tag{42.5.10}$$

It is also useful to introduce here the ordinary generating function (OGF) of the $Z_L(N)$:

$$F(t, N) \equiv \sum_{L=0}^{\infty} Z_L(N) t^L, \tag{42.5.11}$$

which can be obtained by computing the Laplace transform of the EGF. Strictly speaking, when $t \to \infty$ on the real axis, $G(t, N)$ in Eq. (42.5.6) is exponentially divergent. However it is convergent when $v < 0$ and therefore we proceed formally: we first compute the Laplace transform for $v \to -v$ and at the end we will rotate back the result to the positive axis $-v \to v$. From the definition of Laplace transform of a function $f(t)$:

$$\mathcal{L}[f(t)](x) \equiv \int_0^{+\infty} dt\, e^{-xt} f(t), \tag{42.5.12}$$

and from the property $\mathcal{L}[t^n](x) = \Gamma(n+1)/x^{n+1}$ we obtain

$$F(x, N) = \frac{1}{x}\mathcal{L}[G(t, N)]\left(\frac{1}{x}\right). \tag{42.5.13}$$

By applying, in order, the shift property of Laplace transform $\mathcal{L}[e^t f(t)](x) = \mathcal{L}[f(t)](x-1)$, then the identities [Pru98]: $\mathcal{L}[f(t^2)](x) = \int_0^\infty du \exp(-x^2/4u^2) \times \mathcal{L}[f(t)](u^2)/\sqrt{\pi}$, and $\mathcal{L}[L^{(1)}_{N-1}(at)](s) = [1-(1-a/s)^N]/a$, we obtain:

$$F(x, N) = -\frac{1}{2x\sqrt{2\pi v}} \int_0^{+\infty} \frac{d\tau}{\tau^{3/2}} e^{-\frac{(1-x)^2\tau}{2vx^2}} \left(1 - \left(\frac{1+\tau/N}{1-\tau/N}\right)^N\right) \tag{42.5.14}$$

where we have rotated back $v \to -v$ and we have introduced the integration variable $\tau = v/(2u^2)$. The integral expression in Eq. (42.5.14) is useful for large-N asymptotics expansions. In fact, from the expansion of the logarithm of the argument:

$$-2\tau + N \log\frac{1+\tau/N}{1-\tau/N} = 2\tau \sum_{k=1}^{\infty} \frac{1}{2k+1}\left(\frac{\tau}{N}\right)^{2k}, \tag{42.5.15}$$

one sees that the expansion is in even powers $1/N^2$. One therefore has:

$$e^{-2\tau}\left(\frac{1+\tau/N}{1-\tau/N}\right)^N = \sum_{g=1}^{\infty} \frac{1}{N^{2g}} A_g(\tau). \tag{42.5.16}$$

where the coefficients $A_g(\tau)$ can be obtained from the recurrence relation:

$$A'_g(\tau) = \tau^2 (A'_{g-1}(\tau) - 2A_{g-1}(\tau)). \tag{42.5.17}$$

This equation can be solved iteratively with the initial conditions $A_0(\tau) = 1$ and $A_{g\geq 1}(0) = 0$ that follow from (42.5.16). The solution of such a recurrence formula can be expressed in term of Stirling numbers [Erb09a]. We note that the coefficients $A_g(\tau)$ are related to the coefficients $a_{L,g}(v)$ in Eq. (42.5.8) by means of Eq. (42.5.14). By inserting this expansion into Eq. (42.5.14), all the integrals are elementary and provide all the coefficients F_g of the topological expansion $F(t, N) = \sum_{g=0}^{\infty} F_g(t)/N^{2g}$. For instance, at large-N the leading planar term is:

$$\begin{aligned} F_0(t) &= \frac{1 - t - \sqrt{(1-4v)(t-t_+)(t-t_-)}}{2t^2 v} = +t + (1+v)t^2 + \\ &+ (1+3v)t^3 + (1+6v+2v^2)t^4 + O(t^5) \end{aligned} \tag{42.5.18}$$

with $t_\pm = 1/(1 \pm 2\sqrt{v})$ [deG68]. Analogously, the term of order $1/N^{2g}$ is:

$$\frac{P_g(t)}{2r(t)^{3g-\frac{1}{2}}(vx^2)^{3g}}, \quad P_g(t) = \sum_{j=0}^{g-1} A_{g-1-j,j} \left(vx^2 r(t)\right)^j (6g-2j-3)!!, \tag{42.5.19}$$

where $r(t) = (1-4v)(t-t_+)(t-t_-)$. More explicitly, the first three terms are:

$$F_1(t) = \frac{t^4 v^2}{r(t)^{5/2}} = v^2 t^4 + 5v^2 t^5 + (15v^2 + 10v^3)t^6 + (35v^2 + 70v^3)t^7 + \ldots$$

$$F_2(t) = \frac{21t^8 v^4((1-t)^2 + t^2 v)}{r(t)^{11/2}} = 21v^4 t^8 + 189v^4 t^9 + (945v^4 + 483v^5)t^{10} + \ldots$$

$$F_3(t) = \frac{11t^12v^6\left(135(t-1)^4 + 558(t-1)^2 t^2 v + 158t^4 v^2\right)}{r(t)^{17/2}} = 1485v^6 t^{12} +$$

$$+ 19305v^6 t^{13} + \ldots.$$

The meaning of such expansions is straightforward: for instance, for a chain of length 7 there are 35 diagrams with two arcs that have genus 1; or for a chain of length 12 there are 1485 diagrams of genus three (and they must have necessarily 3 arcs), and so on. We note that all the terms in the asymptotic large-N expansion have a similar analytic structure in the t-complex plane. They are analytic at $t = 0$ with root-type singularities at $t = t_\pm$. The radius of convergence of the power series at the origin $t = 0$ is then t_+. It is around such a point where we can extract exact asymptotic expansions for chains at large-L (see Section 42.5.2).

42.5.1 Irreducible diagrams

In this section we introduce diagrams for RNA pseudoknots with special topological restrictions. A first important class is the set of *irreducible diagrams*. Even though the number of planar diagrams $a_{L,0}$ is exactly the number of secondary structures without pseudoknots, and even though the number of diagrams on a torus $a_{L,1}$ counts structures with one pseudoknot only, $a_{L,g}$ with $g > 2$ counts structures that contains both a single topologically complex pseudoknot, or a sequence of several simple pseudoknots at a time. In such cases the genus of a diagram is an additive quantity. For instance, if we consider a succession of two H-pseudoknots each one has genus 1, and the total genus of the structure is $g = 2$. In order to characterize the intrinsic complexity of a pseudoknot, it is thus desirable to define the notion of *irreducibility*. A diagram is said to be irreducible if it can not be broken into two disconnected pieces by cutting a single backbone-line. Any diagram can thus be recursively decomposed into irreducible parts uniquely. It is obvious that the genus of a non-irreducible diagram is the sum of the genii of its irreducible components (the *irreducible diagrams* are also known as "one-particle irreducible" diagrams for historical reasons [Orl02, Pil05a, Pil05b]). Let thus $w_{L,g}$ be the number of irreducible diagrams of a homopolymer of length L and genus g. We define the two OGF's:

$$W_L(N) = \sum_{g=0}^{\infty} w_{L,g} \frac{1}{N^{2g}}, \qquad \Sigma(t, N) = \sum_{L=0}^{\infty} W_L(N) t^L. \tag{42.5.20}$$

From the definition irreducibility one has the following Schwinger-Dyson equation:

$$Z_L(N) = Z_{L-1}(N) + \sum_{k\geq 2}^{L} W_k(N) Z_{L-k}(N), \quad L \geq 2, \tag{42.5.21}$$

and $W_0(N) = W_1(N) = 0$. Its diagrammatic meaning is depicted in Figure 42.2. By multiplying Eq. (42.5.21) by t^L and summing over L we obtain $F = 1 + tF + \Sigma F$, i.e.:

$$F(t, N) = (1 - t - \Sigma(t, N))^{-1}, \text{ or } \quad \Sigma(t, N) = 1 - t - 1/F(t, N). \tag{42.5.22}$$

From this functional relation and the expansion Eq. (42.5.20) we obtain all the $W_L(N)$:

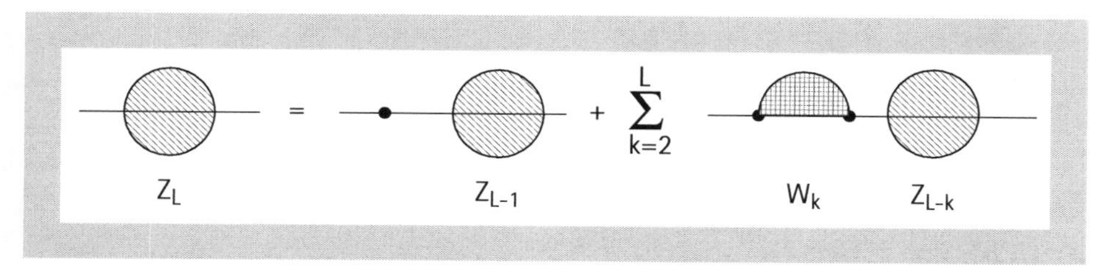

Fig. 42.2 Graphical representation of the Schwinger-Dyson equation, Eq. (42.5.21) that relates irreducible diagrams with generic diagrams. Any diagram Z_L either does not contain any base, or starts with a free leftmost base, or it is interacting with an other base which is part of an irreducible component of the diagram W_k.

$$W_2 = v\,, \quad W_3 = v\,, W_4 = v + v^2 + v^2/N^2\,, W_5 = v + 3v^2 + (3v^2)/N^2$$

$$W_6 = v + 6v^2 + 2v^3 + \frac{6v^2 + 8v^3}{N^2}\,, W_7 = v + 10v^2 + 10v^3 + \frac{10v^2 + 40v^3}{N^2}$$

If we want to focus only on the topological properties of the pseudoknotted configurations (i.e. topologies with $g \geq 1$), then it is useful to remove all the planar insertions from the diagrams. It means we want to enumerate all the diagrams by using a new 'dressed' backbone line τ which contains all the planar diagrams. All planar diagrams are given by $F_0(t)$ in Equation (42.5.18). We define the dressed planar propagator τ as

$$\tau(t) \equiv t\,F_0(t)\,, \text{ or } \quad t(\tau) = \tau/(1 + \tau + v\tau^2). \tag{42.5.23}$$

We have introduced a multiplicative factor t in order to have $\tau(0) = 0$, and it technically means that it is associated to the link between two bases along the backbone, and not to the bases themselves. The 'renormalized' OGF of all irreducible diagrams is defined as $\tilde{\Sigma}(\tau, N) \equiv \Sigma(t(\tau), N)$. Furthermore, by following [Orl02] we can also redefine the arc as a new 'renormalized' arc ν which contains all set of contiguous non-intersecting arcs (see the fourth diagram on the right of Figure 42.1). In fact, adding a line of pairing parallel to an existing one does not change the genus of the diagram. Therefore, all diagrams with parallel pairings are equivalent topologically. It is advantageous therefore to use a reduced representation of the diagrams, where each pairing line can be replaced by any number of parallel pairings. The equation defining ν is

$$\tau^3\nu = \tau^3 v + \tau^6 v^2 + \ldots = \frac{\tau^3 v}{1 - v\tau^2}. \tag{42.5.24}$$

We define thus the OGF of all renormalized irreducible diagrams (as in [Orl02]):

$$\Pi(\tau, N, \nu) = \tilde{\Sigma}(\tau, N)\big|_{v=\frac{\nu}{1+\tau^2\nu}}\,. \tag{42.5.25}$$

In principle, it is difficult to obtain a closed form for Π from the integral representation in Eq. (42.5.14). We show here how one can instead obtain an equivalent (and useful) formulation in terms of differential equations. First of all, we note that since $u(z) = L_N^{(1)}(z)$ satisfies the second-order differential equation $zu'' + (2 - z)u' + Nu = 0$, then also $G(t, N)$ in Eq. (42.5.6) satisfies:

$$t\frac{d^2G}{dt^2} + (3 - 2t)\frac{dG}{dt} + \left(t - 4tv - 3 - \frac{v^2t^3}{N^2}\right)G = 0, \tag{42.5.26}$$

with initial conditions $G(0, N) = G'(0, N) = 1$. By taking the Laplace transform of Equation (42.5.26) with respect to t, we obtain a differential equation for $F(t, N)$:

$$\frac{t^4 v^2}{N^2}\left(t^3\frac{d^3F}{dt^3}+9t^2\frac{d^2F}{dt^2}+18t\frac{dF}{dt}+6F\right)-\Big[(1-4v)$$

$$\left(t\frac{dF}{dt}+F\right)(t-t_-)(t-t_+)+(1-t)F-2\Big]=0, \qquad (42.5.27)$$

where $t_\pm \equiv 1/(1 \pm 2\sqrt{v})$, and with initial conditions $F(0, N) = F'(0, N) = 1$, $F''(0, N) = 2(1 + v)$. Finding the solution to the equation (42.5.27) at finite N is not a simple task. It can however be solved at large-N iteratively (as done in [Kos87, Mig83] for a similar problem). We insert the large-N expansion of $F(t, N)$:

$$F(t, N) = \sum_{g=0}^{\infty} F_g(t)\frac{1}{N^{2g}}, \qquad (42.5.28)$$

into the differential equation (42.5.27), and by comparing equal powers in $1/N$ we obtain a system of first-order differential equations that can easily be solved iteratively. It is easy to verify that the result is the same as the expansion obtained from the integral representation Eq. (42.5.14). The advantage here is that we can now write differential equations also for $\Sigma(t, N)$ $\tilde{\Sigma}(\tau, N)$ and $\Pi(\tau, N)$. Such equations are simply obtained first by substituting Eq. (42.5.22) into Eq. (42.5.27), and then by inserting the definitions of τ and ν. While the algebra is straightforward, we do not report here the lengthy results. For completeness, we mention that a second class of diagrams that is worth considering are the so-called 'non-nested diagrams' [Bon08]. To clarify the concept, let us consider for instance a pseudoknot that is composed of an H pseudoknot, embedded inside another H pseudoknot. A diagram is said to be embedded or *nested* in another, if it can be removed by cutting two lines while the rest of the diagram stays connected in a single component. Obviously the genus of a nested diagram is the sum of the genii of its nested components. As a result, to any non-nested diagram of genus g there corresponds a nested diagram of same genus, obtained by adding a pairing line between the first base and the last base of the diagram. A pseudoknot (with renormalized arcs and propagators) which is irreducible and non nested is said to be *primitive*. Clearly, all RNA structures can be constructed from primitive pseudoknots. The primitive diagram for planar secondary structures is obviously a single renormalized backbone line. We postpone the analytic study of primitive diagrams elsewhere.

42.5.2 Large-L asymptotics

The fundamental tool for deriving the large-L asymptotic expansions of the number of diagrams $a_{L,g}$, $w_{L,g}$, the normalization factor $\mathcal{N}$, and the average genus $< g >_L$ is described in [Fla09]. It is based on a general correspondence between the asymptotic expansion of a function near its dominant singularities and the asymptotic expansion of the function's coefficients. In particular, one can extract the asymptotic expansion of the coefficients $a_{L,g}$ at large-L by considering the corresponding generating function (with respect to L): $c_g(x) = \sum_{L=0}^{\infty} a_{L,g}\, x^L$. After $c_g(x)$ is computed and its singularities in the complex plane are determined, one then selects the dominant singularity (i.e. the one which is closer to the origin) and finds the asymptotic behaviour $\tilde{c}_g(x)$ of $c_g(x)$ around it. Finally, the power series expansion of $\tilde{c}_g(x)$ around $x = 0$ yields coefficients from which one can read the asymptotic behaviour of $a_{L,g}$ at large-L. From such a standard singularity analysis of the generating functions [Fla09], we obtain at large L:

$$a_{L,g} \sim v^{(6g-3)/4}(1+2\sqrt{v})^{L-3g+\frac{3}{2}}\frac{\kappa_g L^{3g-\frac{3}{2}}}{\Gamma(3g-\frac{1}{2})}, \tag{42.5.29}$$

where the constant κ_g is v-independent, and it is given by $\kappa_g = 1/(3^{4g-3/2}2^{2g+1}g!\sqrt{\pi})$. We also compute the distribution of the number of diagrams, by normalizing $a_{L,g}$ by the total number of diagrams at fixed L:

$$\mathcal{N} = \sum_g a_{L,g} = Z_L(1) = \frac{d}{dt^L}\bigg|_{t=0} G(t,1) = (-2)^{\frac{L}{2}}\, U\left(-\frac{L}{2},\frac{1}{2},-\frac{1}{2}\right) \tag{42.5.30}$$

where $U(a, b, Z)$ is the confluent hypergeometric function of the second kind. Such a result matches the observation made in Section 42.3, on the relation between RNA diagrams and the number of permutation involutions, Eq. (42.3.1). If one plots the distributions of diagrams as a function of L and g one will note that for any given $L >> 1$ most of the diagrams are not planar, and they actually mainly have a genus close to a characteristic value $\langle g\rangle_L$ [Ver05b]. The expectation value $\langle g(\mu)\rangle_L$ is defined by $\langle g(\mu)\rangle_L \equiv \sum_{g=0}^{\infty} a_{L,g}\, g e^{-\beta\mu g} / \sum_{g=0}^{\infty} a_{L,g} e^{-\beta\mu g}$. Such a value increases with L, and the exact behaviour can be evaluated in the following way. Consider first the sum (we set $v = 1$ for simplicity):

$$\sum_{g=0}^{\infty} a_{L,g}\, g e^{-\beta\mu g} \sim \sum_{g=0}^{L/4} g\, \frac{3^L L^{3g-\frac{3}{2}} e^{-\beta\mu g}}{3^{4g-\frac{3}{2}}2^{2g+1}g!\sqrt{\pi}} = \lambda \sum_{g=0}^{L/4} g\, \frac{(bL^3 e^{-\beta\mu})^g}{g!}, \tag{42.5.31}$$

where we introduced the asymptotic expansion Eq. (42.5.29), and $\lambda = 3^L L^{-3/2} 3^{3/2}/(2\sqrt{\pi})$, $b = 1/(3^4 2^2)$. We use now the identities:

$$\sum_{k=0}^{s} \frac{x^k}{k!} = e^x \frac{\Gamma(s+1,x)}{\Gamma(s+1)}, \quad \Gamma(a,x) = \int_x^{+\infty} dt\, e^{-t} t^{a-1}, \tag{42.5.32}$$

where $\Gamma(a, x)$ is the incomplete Gamma function, and we obtain:

$$I = \lambda \sum_{g=0}^{L/4} \frac{(bL^3e^{-\beta\mu})^g}{g!} = \lambda e^{bL^3e^{-\beta\mu}} \frac{\Gamma(\frac{L}{4}+1, bL^3e^{-\beta\mu})}{\Gamma(\frac{L}{4}+1)} = \frac{\lambda e^{bL^3e^{-\beta\mu}}}{\Gamma(\frac{L}{4}+1)} \int_{bL^3e^{-\beta\mu}}^{+\infty} dte^{-t}t^{\frac{L}{4}}. \tag{42.5.33}$$

The large-L limit of the integral can be computed by standard saddle-point analysis. The dominant contribution to the integral is given by the extremum $t = bL^3e^{-\beta\mu}$, and therefore $I \sim (bL^3e^{-\beta\mu})^{L/4}/\Gamma(\frac{L}{4}+1)$. Finally, since $\langle g(\mu)\rangle_L = \frac{b}{I}\frac{d}{db}I$, we obtain $\langle g(\mu)\rangle_L \sim \frac{L}{4}$, which is μ independent. In this derivation we kept μ fixed and finite while taking the large L limit. Such a result allow us also to extend all asymptotics for $F(t, N)$ and $a_{L,g}$ to $\Sigma(t, N)$ and $w_{L,g}$, respectively. We find that the asymptotic behaviour at large L is:

$$w_{L,g} \sim v^{(6g+1)/4}(1+2\sqrt{v})^{L-3g-\frac{1}{2}} \frac{\tilde{\kappa}_g L^{3g-\frac{3}{2}}}{\Gamma(3g-\frac{1}{2})}, \tag{42.5.34}$$

in a similar way. We could speculate on the possibility of such asymptotic result to be universal in the large-L limit and therefore have some relevance also for ssDNA folding, or even for more general heteropolymers. Although all important steric constraints are not taken into account, random matrix theory applied to the folding of an homogenous heteropolymer allows us to obtain exact analytical combinatorial results, in particular in the asymptotic regime of long chains. A different type of application has been obtained in the limit $N \to 0$ in [Erb09b], where the authors discuss also the relation of this model with the dense polymer phase of the $O(n)$ vector model in the limit $n \to 0$ (see also Chapter 30 of this book).

42.6 Numerical comparison

The findings from the simple combinatorial model reviewed above predict large average genus for long RNA molecules. However, real RNA molecules typically have a small genus: for instance, a simple H-type pseudoknot ($\sim$ 20 bases or more) or the classical kissing-hairpins pseudoknot ($\sim$ 30 bases or more) have genus 1, much less than the predicted theoretical value. Even tRNAs ($\sim$ 80 bases) that contain four helices, two of which are linked together by a kissing-hairpin pseudoknot, has still genus 1. Typical tmRNAs ($\sim$ 350 bases long) contain four H-type pseudoknots [Bur05], and its total genus is 4, far below the theoretical upper bound $L/4$. Of course, real RNA molecules are not infinitely flexible and stretchable homopolymers. Not only the bases i and j interact if they are sufficiently far apart along the chain (i.e. $|i-j| \geq 4$, [Zuk84]), but also the helices have a long persistence length ($\sim$ 200 base pairs [Keb95]) and this necessarily constrains the allowed configurations even more.

One can expect therefore that including all steric and geometrical constraints should considerably decrease the genus of allowed pseudoknots, compared to the purely combinatorial case [Ver05b] where the actual three-dimensional conformation was neglected. A numerical investigation of the effects of steric and geometric constraints on the genus distribution of pseudoknots topologies in homopolymers was done in [Ver06]. There, a model for self-avoiding random walks on a cubic lattice with long-range attractive saturating hydrogen bond interactions was considered. The numerical sampling, implemented by using the Monte Carlo growth method [Gar90], evidenced that at low temperature (in the globule phase) the average genus scales like $\langle g \rangle \sim (144 \pm 4)10^{-3} L$. The scaling is at about a 50% lower rate than the value $L/4$ computed in [Ver05b], but still too high when compared to the genus of real RNA molecules.

In order to establish quantitatively the relevance of the topological genus in real RNA molecules, a statistical survey was performed in [Bon08]. Thanks to the fact that nowadays several online databases containing RNA structures are available online, RNA pseudoknot structures can be obtained easily. In particular, all the RNA structures in two databases (namely PSEUDOBASE [Bat00] and the wwPDB, the Worldwide Protein Data Bank which contains some RNAs; the RNA structures in the latter are also listed in RNABase [Mur03]) were analyzed and classified according to their topology. The results show that among the four primitive pseudoknots of genus 1, two are quite common in the databases, namely the ABAB and ABACBC pseudoknot, one is very rare (the ABCABC), and the remaining one (ABCADBCD) has not been reported as of yet. Among all the 304 RNA pseudoknots in PSEUDOBASE one finds that (at the time of writing this chapter): there are 296 H pseudoknots (or nested H pseudoknots) of the ABAB type with genus 1; there are six kissing hairpin pseudoknots (or nested) of the ABACBC type with genus 1; there is one pseudoknot of the type ABCABC (number PKB71) with genus 1; there is one pseudoknot of the type ABCDCADB (number PKB75) with genus 2. Note that the pseudoknot PKB71, from the regulatory region of the E. coli alpha ribosomal protein operon is the only example of the ABCABC pseudoknot in the PSEUDOBASE database. It is not possible to say at the moment whether rare pseudoknots in databases are really rare, or whether they have been understudied or under-recognized.
From the statistical analysis of the pseudoknots in wwPDB, it has been found that large RNAs, such as RNA ribosomal 50s subunits (length larger than 2000), have total genii less than 18 [Bon08]. Such a value is much smaller than the value obtained from the combinatorial analysis $L/4 \approx 500$ [Ver05b], or from numerical simulations that include steric constraints $2000 \times 0.14 \simeq 280$ [Ver06]. Such an analysis also showed that most pseudoknots can be uniquely decomposed as a sequence of *primitive* pseudoknots concatenated sequentially and/or nested. Such primitive pseudoknots are generally of low genus of order 1 or 2 (the most complex primitive pseudoknots in the databases have genus 13).

The fundamental question that this analysis raises is: Why has evolution selected sequences that do not allow complex topologies? It is a challenge to relate information with topology. Clearly most of the information should be in the folds that are less stable, i.e. in sequences where the structure can fluctuate via minor environmental changes such as the specific addition of ions. The lack of reliable models to determine the energy of complex RNA conformations is the main obstacle for extending the random matrix theory analysis to more realistic RNA molecules in their folded state. However, the potential impacts of such field-theoretical method applied to the RNA-folding problem are surely promising, and perhaps can be extended to the problem of protein folding or other biologically relevant heteropolymers.

References

[Abr64] Abramowitz and Stegun, *Handbook of Mathematical Functions*, Dover, New York, 1964.

[Abr90] J.P. Abrahams, M. van den Berg, E. van Batenburg and C.W.A Pleij, Nucl. Acids Res. **18** (1990) 3035; A.P. Gultyaev, Nucleic Acids Res. **19** (1991) 2489; I.L. Hofacker, in *Monte Carlo Approach to Biopolymers and Protein Folding*, World Scientific, Singapore (1998), 171.

[Ada04] P.L. Adams, et al., Nature **430** (2004) 45.

[Aku01] T. Akutsu, Discr. Appl. Math. **104** (2001) 45.

[Ban93] A. Banerjee, J. Jaeger and D. Turner, Biochemistry **32** (1993) 153.

[Bat00] F.H.D. van Batenburg et al., Nucl. Acids Res., **28** (2000) 201.

[Bel10] S. Bellaousov and D.H. Mathews, RNA 16 (2010), 1870.

[Bon08] M. Bon, G. Vernizzi, H. Orland, A. Zee, J. Mol. Biol., 379 (2008) 900.

[Bon10] M. Bon and H. Orland, Physica A 389 (2010) 2987.

[Bun99] R. Bundschuh, T. Hwa, Phys. Rev. Lett. **83** (1999) 1479; D.K. Lubensky and D.R. Nelson, Phys. Rev. Lett. **85** (2000) 1572.

[Bur05] J. Burks et al., BMC Molecular Biology **6** (2005) 14.

[Bri97] P. Brion and E. Westhof, Ann. Rev. Biophys. Biomol. Struct. **26** (1997) 113.

[Con04] A. Condon et al., Theor. Comp. Science **320** (2004) 35; H.H. Gan, S. Pasquali and T. Schlick Nucl. Acids Res. **31** (2003) 2926; N. Kim, N. Shiffeldrim, H.H. Gan and T. Schlick J. Mol. Biol. **341** (2004) 1129; E.A. Rodland, J. Comp. Biol. **13** (2006) 1197.

[Cou02] J. Couzin, Science **298** (2002) 2296.

[deG68] P.G. de Gennes, Biopolymers **6** (1968) 715.

[Edd04] S.R. Eddy, Nat. Biotech. **22** (2004) 1457.

[Erb09a] M.G. dell'Erba and G.R. Zemba, Phys. Rev. E **80** (2009) 041926.

[Erb09b] M.G. dell'Erba and G.R. Zemba, Phys. Rev. E **79**(2009) 011913.

[Fla00] C. Flamm, W. Fontana, L. Hofacker and P. Schuster, RNA **6** (2000) 325; C. Flamm et al., RNA **7** (2001) 325.

[Fla09] P.Flajolet and A.M. Odlyzko, SIAM Journal on Algebraic and Discrete Methods, 3(2) (1990) 216–240; P. Flajolet and R. Sedgewick, *Analytic Combinatorics*, Cambridge University Press, 2009.

[Gar90] T. Garel and H. Orland, J. Phys. A: Math. Gen. **23** (1990) L621; P.G. Higgs and H. Orland, J. Chem. Phys. **95**, 4506 (1991).

[Gul99] A.P. Gultyaev, F.H.D. van Batenburg and C.W.A. Pleij, RNA **5** (1999) 609.

[Gut94] R.R. Gutell, N. Larsen and C.R. Woese, Microbiol. Rev. **58** (1994) 10; C.R. Woese, R.R. Gutell, R. Gupta, H.F. Noeller, Microbiol. Rev. **47** (1983) 621; P.Higgs, Phys. Rev. Lett **76** (1996) 704; R. Nussinov, G. Peickzenink, J. Griggs and D. Kleitman, SIAM J. Appl. Math. **35** (1978) 68.

[Hof03] I.L. Hofacker, Nucl. Acids Res. **31** (2003) 3429.

[Hig00] P.G. Higgs, Quart. Rev. Biophys. **33** (2000) 199.

[Isa00] H. Isambert and E.D. Siggia, Proc. Natl. Acad. Sci. USA **97** (2000), 6515; A. Xayaphoummine, T. Bucher, F. Thalmann and H. Isambert, Proc. Natl. Acad. Sci. USA **100** (2003) 15310; A. Xayaphoummine, T. Bucher and H. Isambert, Nucl. Acids Res. **33** (2005) W605.

[Keb95] P. Kebbekus, D.E. Draper and P. Hagerman, Biochemistry **34** (1995) 4354.

[Kos87] I.K. Kostov and M.L. Mehta, Phys. Lett. B 189 (1987) 118.

[Lai94] L. Laing and D. Draper, J. Mol. Biol. **237** (1994) 560.

[Leo01] N.B. Leontis and E. Westhof, RNA **7** (2001) 499.

[Lim94] P.A. Limbach, P.F. Crain and J.A. McCloskey, Nucl. Acids Res. **22** (1994) 2183.

[Liu06] H.J. Liu, D. Xu, , J.L. Shao and Y.F. Wang, Comp. Biol. Chem. **30** (2006) 72; H.W. Li and D.M. Zhu, Computing and combinatorics: proceedings, Lecture notes in computer science, **3595** (2005) 94; J.H. Ren, B. Rastegari, A. Condon and H.H. Hoos, RNA **11** (2005) 1494; J. Reeder and R. Giegerich, Bioinformatics **5** (2004) 104; J.H. Ruan, G.D. Stormo and W.X. Zhang, Nucl. Acids Res. **32** (2004) W146; R.M. Dirks and N.A. Pierce, J. Comp. Chem. **25** (2004) 1295; also J. Comp. Chem. **24** (2003) 1664; Y. Uemura, A. Hasegawa, S. Kobayashi and T. Yokomori, Theor. Comp. Science **210** (1999) 277; J.S. Deogun, R. Donis, O. Komina and F. Ma, Proceedings of APBC2004, Dunedin, New Zealand CRPIT, 29. Chen, Y.P.P., Ed. ACS. 239.

[Luc03] A. Lucas and K.A. Dill, J. Chem. Phys. **119** (2003) 2414.

[Lyn00] R.B. Lyngsø, C.N.S. Pedersen, J. Comp. Biol., **7** (2000) 409.

[Mat99] D.H. Mathews, J. Sabina, M. Zuker and D.H. Turner, J. Mol. Biol. **288** (1999) 911.

[Mcc90] J.S. McCaskill, Biopoly. **29** (1990) 1105.

[Mig83] A. A. Migdal, Phys. Rept. **102** (1983) 199.

[Mir85] A.A. Mironov, L.P. Dyakonova and A.E. Kister, J. Biomol. Struct. Dyn. **2** (1985) 953.

[Mir93] A.A. Mironov and V.F. Lebedev, BioSystems **30** (1993), 49.

[Mis98] V.K. Misra and D.E. Draper, Biopoly. **48** (1998) 113; J.P.D. Thirumalai, S.A. Woodson, PNAS USA **96** (1999) 96 6149; V.K. Misra, R. Shiman and D.E. Draper, Biopoly. **69** (2003) 118.

[Mon01] A. Montanari and M. Mézard, Phys. Rev. Lett. **86** (2001) 2178.

[Mot48] T.S. Motzkin, (1948), Bulletin of the American Mathematical Society **54** 352360.

[Mur03] V.L. Murthy and G.D. Rose Nucl. Acids Res. **31** (2003) 502.

[Nag00] U. Nagaswamy, N. Voss, Z. Zhang and G.E. Fox, Nucl. Acids Res. **28** (2000) 375.

[Niv09] I. Garg and N. Deo, Phys. Rev. E, **79** (2009) 061903; Pramana, **73** (2009) 533.

[Ore97] C.A. Orengo et al., Structure **5** (1997) 1093; A.G. Murzin, S.E. Brenner, T. Hubbard and C. Chothia, J. Mol. Biol. **247** (1995) 536.

[Orl02] H. Orland and A. Zee, Nucl. Phys. **B620** (2002) 456.

[Pil05a] M. Pillsbury, H. Orland and A. Zee, Phys. Rev. E **72** (2005) 011911.

[Pil05b] M. Pillsbury, J.A. Taylor, H. Orland and A. Zee, http://arXiv.org/cond-mat/0310505.

[Ple85] C.W. Pleij, K. Rietveld and L. Bosch, Nucl. Acids Res. **13** (1985) 1717.

[Pru98] A. P. Prudnikov, Yu. A. Brychkov, and O. I. Marichev, Integrals and Series (Gordon and Breach, New York, 1998.

[Ras96] T. Rastogi, T.L. Beattie, J.E. Olive and R.A. Collins, EMBO J **15** (1996) 2820–2825; A. Ke et al., Nature 429 (2004) 201.

[Riv99] E. Rivas and S.R. Eddy, J. Mol. Biol. **285** (1999) 2053.

[Riv00] E. Rivas and S.R. Eddy, Bioinformatics **16** (2000) 334.

[Sch96] M. Schmitz and G. Steger, J. Mol. Biol. **255** (1996) 254.

[Sch04] R. Schroeder, A. Barta and K. Semrad, Nat. Rev. Mol. Cell Biol. **5** (2004) 908.

[She95] L.X. Shen and I. Tinoco Jr, J. Mol. Biol. **247** (1995) 963.

[Ski90] Skiena, S. *Implementing Discrete Mathematics: Combinatorics and Graph Theory with Mathematica* Reading, MA: Addison-Wesley (1990).

[Sno05] C.D. Snow, E.J. Sorin, Y.M. Rhee and V.S. Pande, Annu. Rev. Biophys. Biomol. Struct. **34** (2005) 43.

[Sta05] D.W. Staple and S.E. Butcher, PLoS Biol. **3** (2005) e213.

[Ste04] A. Kabakçioğlu and A.L. Stella, Phys. Rev. E **70** (2004) 011802.

[Tab98] J.E. Tabaska, R.B. Cary, H.N. Gabow, G.D. Stormo, **14** (1998) 691.

[The05] C.A. Theimer, C.A. Blois and J. Feigon, Mol. Cell. **17** (2005) 671.

[tHo74] 't Hooft,G. *Nucl. Phys.* **B72** (1974) 461.

[Tin99] I. Tinoco Jr. and C. Bustamante, J. Mol. Biol. **293** (1999) 271.

[Ver04] G. Vernizzi, H. Orland, A. Zee, http://arxiv.org/abs/q-bio.BM/0405014.

[Ver05a] G. Vernizzi and H. Orland Acta Phys. Polon. B **36** (2005) 2821.

[Ver05b] G. Vernizzi, H. Orland and A. Zee, Phys. Rev. Lett. **94** (2005) 168103.

[Ver06] G. Vernizzi, P. Ribeca, H. Orland and A. Zee, Phys. Rev. E **73** (2006) 031902.

[Wat78] M. S. Waterman, Adv. Math. Suppl. Studies **1** (1978) 167; P. R. Stein and M. S. Waterman, J. Disr. Math. **26** (1978) 261; M. S. Waterman, Studies Appl. Math. **60** (1978) 91; M. S. Waterman and T. F. Smith, Math. Biosc. **42** (1978) 257; M.S. Waterman and T.F. Smith, Adv. Applied Maths. **7** (1986) 455.

[Wat85] M.S. Waterman and T.H. Byers, Math. Biosci. **77** (1985) 179; A.L. Williams and I. Tinoco Jr., Nucl. Acids Res. **14** (1986) 299; R. Nussinov and A.B. Jacobson, PNAS **77** (1980) 6309; M. Zuker, Science **244** (1989) 48; M. Zuker and P. Stiegler, Nucl. Acids Res. **9** (1981) 133; I.L. Hofacker et al., Monatshefte f. Chemie **125** (1994) 167; S. Wuchty, W. Fontana, I.L. Hofacker, P. Schuster, Biopoly. **49** (1999) 145.

[Wes92] E. Westhof and L. Jaeger, Curr. Opinion Struct. Biol. **2** (1992) 327.

[Zee05] A. Zee, Acta Phys. Polon. B **36** (2005) 2829.

[Zuk81] M. Zuker and P. Stiegler, Nucl. Acids Res. **9** (1981) 133.

[Zuk84] M. Zuker and D. Sankoff, Bull. Math. Biol. **46** (1984) 591.

[Zuk00] M. Zuker, Curr. Opin. Struct. Biol. **10** (2000) 303.

[Zuk03] M. Zuker, Nucl. Acids Res **31** (2003) 3406.

·43·

Complex networks

G. J. Rodgers and T. Nagao

Abstract

This chapter contains a brief introduction to complex networks, and in particular to small world and scale free networks. We show how to apply the replica method developed to analyse random matrices in statistical physics to calculate the spectral densities of the adjacency and Laplacian matrices of a scale free network. We use the effective medium approximation to treat networks with finite mean degree and discuss the local properties of random matrices associated with complex networks.

43.1 Introduction

Graphs are incorporated into many theories in physics, and as our society becomes ever more globalised and inter-connected, the study of graphs forms an integral part of the systematic study of a vast array of human, social, economic, technological, biological, and physical systems. The graph theory developed by Erdös and Rényi, [Erd60] considers vertices and edges linked with a fixed probability p. This yields networks with a Poisson degree distribution. Since then Watts observed that many real networks exhibit properties in which there is a small finite number of steps between any two vertices in the network, but some local order, so that if vertex A is connected to vertex B and B is connected to vertex C then there is a good chance that A is connected to C. Networks with these two properties are referred to as small world networks. Albert and Barabási [Alb02] observed that many real networks had a degree distribution that had a fat tail, which is that the degree distribution decays at a rate slower than exponential. Furthermore, Barabási and Albert [Bar99] introduced a model that built a network with a power-law degree distribution, and networks of this type have become known as scale free networks. These two classes of networks are the two best understood complex networks. Any network that is neither regular nor of Erdös and Rényi type is normally described as complex.

43.1.1 Small world networks

A small world is a loosely defined concept normally associated with systems in which most vertices are not neighbours of one another but where it is only a small finite number of steps between any two vertices, and where there is a degree of local order. The small world phenomenon, in which strangers are often linked by a mutual acquaintance, is captured in the phrase six handshakes from the president in which nearly everyone in the world is around six acquaintances from the president. Networks exhibiting the small world phenomenon include social networks, the world wide web (the network of webpages and html links) and gene expression networks. The local ordering property of small world networks is usually associated with regular networks such as a 2D square lattice. The finite number of steps between any two vertices is associated with completely connected graphs. In this sense small world networks can be thought of as being intermediate between completely connected graphs and regular lattices. Most scale free networks exhibit the small world phenomenon but few small world networks are scale free. In 1998 Watts and Strogatz [Wat98] characterised all graphs by the clustering coefficient C, defined by

$$C = 3\frac{\text{Number of triangles}}{\text{Number of connected triples}}, \tag{43.1.1}$$

and by the mean vertex-to-vertex distance. C is the average probability that two of one's friends are friends themselves ($C = 1$ on a fully connected graph, everyone knows everyone else). Watts and Strogatz observed that while Erdös and Rényi graphs had a small clustering coefficient and small mean vertex-to-vertex distance, many real graphs had a clustering coefficient larger than those seen in random graphs, but had a similar mean vertex-to-vertex distance. They then introduced a model that exhibited this property. Later, a related model was introduced by Newman and Watts [New99]. This is built as follows;

- Let N vertices be connected in a circle;
- Each of several neighbours is connected by a unit length edge;
- Then each of these edges is rewired with probability φ to a randomly chosen vertex.

This network has the following properties:

- $\varphi = 0$ is a regular lattice;
- $\varphi = 1$ is an Erdös and Rényi graph;
- Exhibits small world property for $0 < \varphi < 1$.
- Average shortest distance behaves as $\sim N$ for $\varphi = 0$ and $\sim \log N$ for $\varphi > 0$.

Obviously this model mechanism can be generalised to any regular 2D or 3D graph. Models of this type are difficult to formulate analytically, and only a few basic properties have been obtained analytically, in contrast to both Erdös and Rényi graph and scale free networks.

43.1.2 Scale free networks

A scale free network has a degree distribution that is asymptotically power-law. That is, the number $N(k)$ of vertices with degree k behaves like $k^{-\lambda}$ for large k with $2 < \lambda$. Many real world networks are scale free including the world wide web, the internet, the citation graph, the science collaboration graph, the actor collaboration graph, and the phone call graph. The latter is of course big business for telecommunications firms. Scale free networks are hugely heterogeneous structures, with, depending on the position of the cutoff in the network model, nodes with degrees ranging from 1 to $N^{\frac{1}{\lambda-1}}$. A list of cutoffs for different scale-free networks is given in [Dor08]. There are several different ways to build a model scale free graph. These include the Barabási-Albert model [Bar99], the static model introduced by Goh, Kahng and Kim [Goh01b], and the grown scale free graph with statistically defined modularity introduced by Tadić [Tad01]. In this chapter we consider uncorrelated networks, there is discussion of causality and homogeneity in complex networks in [Bia05, Bog06].

In this chapter we briefly introduce two models of scale free graphs, then use the replica method to examine the adjacency matrix and the Laplacian associated with a complex network built using the static model. We then show how the effective medium approximation can be used to treat a complex network with a finite mean degree. Finally we discuss extensions of this work to other local properties of random matrices and to other types of complex network.

43.2 Replica analysis of scale free networks

43.2.1 Degree distribution and spectral density

Let us consider a complex network with N nodes and examine the asymptotic behaviour in the limit $N \to \infty$. The connection pattern of the network is described by the adjacency matrix A, which is an $N \times N$ symmetric matrix with

$$A_{jl} = \begin{cases} 1, & \text{if } j\text{-th and } l\text{-th nodes are directly connected,} \\ 0, & \text{otherwise.} \end{cases} \tag{43.2.1}$$

In order to characterise the connection pattern, various statistical quantities are studied. One of the most common quantities is the degree distribution function. The number of edges attached to the j-th node

$$k_j = \sum_{l=1}^{N} A_{jl} \tag{43.2.2}$$

is called the degree. The degree distribution function $P(k)$ is defined as

$$P(k) = \left\langle \frac{1}{N} \sum_{j=1}^{N} \delta(k - k_j) \right\rangle, \tag{43.2.3}$$

where the brackets stand for the average over the probability distribution function of the adjacency matrices.

If the degree distribution function $P(k)$ obeys a power law, namely, if it is proportional to $k^{-\lambda}$ with a positive exponent λ in the limit $k \to \infty$, then the network is said to be scale free. The scale free property is one of the prominent universal features of social and biological networks [Bar99].

In spectral theory of networks, the Laplacian matrix is also of interest. The Laplacian matrix L is an $N \times N$ symmetric matrix defined as

$$L_{jl} = \begin{cases} k_j, & j = l, \\ -A_{jl}, & j \neq l. \end{cases} \tag{43.2.4}$$

The Laplacian matrix has non-negative eigenvalues and its smallest eigenvalue is always zero.

The spectral density of the adjacency (or Laplacian) matrix is an important quantity for characterizing the properties of the network. It is defined as

$$\rho(\mu) = \left\langle \frac{1}{N} \sum_{j=1}^{N} \delta(\mu - \mu_j) \right\rangle, \tag{43.2.5}$$

where $\mu_1, \mu_2, \cdots, \mu_N$ are the eigenvalues of the adjacency matrix A (or the Laplacian matrix L).

For scale free complex networks, it is expected that the spectral density of adjacency matrices obeys a power law $\rho(\mu) \propto \mu^{-\gamma}$ in the limit $\mu \to \infty$. Such a behaviour was first observed in numerical work on complex networks [Far01, Goh01a]. Then the relation between γ and λ was analytically specified as $\gamma = 2\lambda - 1$ in work on locally tree-like networks [Dor03, Dor04]. This relation was confirmed in a study on the static network model [Rod05].

In the following, after a brief description of a growth model, the static model of complex networks is introduced. Then the replica method in statistical physics is applied to the static model and the spectral densities of the adjacency and Laplacian matrices are evaluated.

43.2.2 Models of scale free networks

Barabási and Albert invented a growth model (BA model) of complex networks [Bar99], in which the n-th node with m edges is newly introduced at a time t_n, where $n = 1, 2, 3, \cdots$. Let us discuss the degree distribution of the BA model, neglecting the fluctuation of k_j (so that $\langle k_j \rangle = k_j$). Each of m edges is attached to the old nodes in the network, according to the rule of preferential attachment. That is, the attachment probability Π_j of the j-th old node is proportional to the degree:

$$\Pi_j = \frac{k_j}{\sum_{l=1}^{n-1} k_l}, \quad j = 1, 2, \cdots, n-1, \tag{43.2.6}$$

so that a difference equation

$$k_j(t_n) - k_j(t_{n-1}) = m\Pi_j = m\frac{k_j}{\sum_{l=1}^{n-1} k_l} \tag{43.2.7}$$

follows. As m edges are attached at each time when a node is introduced, an asymptotic estimate $\sum_{l=1}^{n-1} k_l \sim 2mn$ holds for large n. Let us suppose that the time difference $\Delta t = t_n - t_{n-1}$ is a constant. Then, in the continuous limit $\Delta t \to 0$ with a fixed $t = n\Delta t$, we obtain a differential equation

$$\frac{\partial k_j(t)}{\partial t} = \frac{k_j}{2t}. \tag{43.2.8}$$

One can solve this equation with the initial condition $k_n(t_n) = m$ as

$$k_j(t_n) = m\sqrt{\frac{n}{j}}. \tag{43.2.9}$$

That is, the degree of the j-th node is proportional to $j^{-1/2}$ at a fixed time.

In order to have a degree k_j smaller than k, we need to have $j > n(m/k)^2$. Therefore the number $\nu(k)$ of the nodes with the degrees smaller than k is $\nu(k) = \sum_{j>n(m/k)^2}^{n} 1 = n(1-(m/k)^2)$. Then the degree distribution function $P(k)$ is estimated as

$$P(k) = \frac{\partial}{\partial k}\frac{\nu(k)}{n} = 2\frac{m^2}{k^3}, \tag{43.2.10}$$

so that the BA model has the exponent $\lambda = 3$.

Goh, Kahng, and Kim introduced a static model (GKK model) simulating the growth model with a fixed number of nodes [Goh01b]. In the GKK model, the upper (lower) triangular elements of the adjacency matrix are assumed to be independently distributed. Suppose that there are N nodes and that the j-th node is assigned a probability

$$P_j = \frac{j^{-\alpha}}{\sum_{j=1}^{N} j^{-\alpha}} \sim (1-\alpha)N^{\alpha-1}j^{-\alpha}, \quad 0 < \alpha < 1. \tag{43.2.11}$$

In each step two nodes are chosen with the assigned probabilities and connected unless they are already connected. In order to have a mean degree p, as we shall see below, such a step is repeated $pN/2$ times. Then the j-th and l-th nodes are connected with a probability

$$f_{jl} = 1 - (1 - 2P_j P_l)^{pN/2} \sim 1 - e^{-pNP_j P_l}, \tag{43.2.12}$$

so that the adjacency matrix A of the network is distributed according to a probability distribution function

$$\mathcal{P}_{jl}(A_{jl}) = (1 - f_{jl})\delta(A_{jl}) + f_{jl}\delta(A_{jl} - 1), \quad j < l. \tag{43.2.13}$$

A useful asymptotic estimate [Kim05] in the limit $N \to \infty$ for averaging over $\mathcal{P}_{jl}(A_{jl})$ is

$$\ln\left\langle \exp\left(-i\sum_{j<l}^{N} A_{jl}t_{jl}\right)\right\rangle \sim pN\sum_{j<l}^{N} P_j P_l(e^{-it_{jl}} - 1) \tag{43.2.14}$$

where t_{jl} is a parameter independent of N. The remainder term is $O(1)$ for $0 < \alpha < 1/2$, $O((\ln N)^2)$ for $\alpha = 1/2$ and $O(N^{2-(1/\alpha)} \ln N)$ for $1/2 < \alpha < 1$. As a special case, we find an estimate

$$F_j(t) \equiv \ln\left\langle e^{-i\sum_{l=1}^{N} A_{jl}t}\right\rangle \sim pNP_j(e^{-it} - 1). \tag{43.2.15}$$

Now let us calculate the degree distribution. Using the definition (43.2.2) and asymptotic estimate (43.2.15), one obtains [Lee04]

$$\langle k_j \rangle = \sum_{l=1}^{N}\langle A_{jl}\rangle = i\frac{\partial}{\partial t}F_j(t)\bigg|_{t=0} \sim pNP_j, \tag{43.2.16}$$

so that the mean degree of the j-th node is proportional to $j^{-\alpha}$. Hence one can expect that the case $\alpha = 1/2$ approximates the BA model. As we mentioned before, the mean degree is $(1/N)\sum_{j=1}^{N}\langle k_j\rangle = p$. Moreover it follows from (43.2.3), (43.2.11) and (43.2.15) that the degree distribution function is

$$\begin{aligned} P(k) &= \frac{1}{2\pi N}\sum_{j=1}^{N}\int dt\, e^{ikt+F_j(t)} \\ &\sim \frac{1}{2\pi}\int dt \int_0^1 dx \exp\left\{ikt + p(1-\alpha)x^{-\alpha}(e^{-it} - 1)\right\}. \end{aligned} \tag{43.2.17}$$

Then, in the limit $k \to \infty$, we find

$$P(k) \sim \int_0^1 dx\, \delta\left\{k - p(1-\alpha)x^{-\alpha}\right\} = \frac{\{p(1-\alpha)\}^{1/\alpha}}{\alpha}\frac{1}{k^{1+(1/\alpha)}}. \tag{43.2.18}$$

Therefore the exponent λ of the GKK model is equal to $1 + (1/\alpha)$. When we put $\alpha = 1/2$, λ is identified with the corresponding exponent of the BA model, as expected.

43.2.3 Partition function

In order to calculate the spectral densities of the adjacency and Laplacian matrices of the GKK model, let us apply the replica method [Rod88, Bra88] in statistical physics (see Chapter 8). To begin with, we rewrite (43.2.5) in the form

$$\rho(\mu) = \frac{2}{N\pi}\mathrm{Im}\frac{\partial}{\partial\mu}\langle \ln Z(\mu + i\epsilon)\rangle, \quad \epsilon \downarrow 0 \tag{43.2.19}$$

in terms of the partition function

$$Z(\mu) = \int \prod_{j=1}^{N} d\phi_j \exp\left(\frac{i}{2}\mu\sum_{j=1}^{N}\phi_j^2 - \frac{i}{2}\sum_{jl}^{N} J_{jl}\phi_j\phi_l\right). \tag{43.2.20}$$

Here J is the adjacency matrix A or the Laplacian matrix L. Since there is a relation

$$\langle \ln Z\rangle = \lim_{n\to 0}\frac{\ln\langle Z^n\rangle}{n}, \tag{43.2.21}$$

we wish to evaluate

$$\langle Z^n\rangle = \int \prod_{j=1}^{N} d\vec{\phi}_j \exp\left(\frac{i}{2}\mu\sum_{j=1}^{N}\vec{\phi}_j^{\,2}\right)\left\langle \exp\left(-\frac{i}{2}\sum_{jl}^{N} J_{jl}\vec{\phi}_j\cdot\vec{\phi}_l\right)\right\rangle. \tag{43.2.22}$$

In term of the replica variables $\phi_j^{(k)}$, the vector $\vec{\phi}_j$ and the measure $d\vec{\phi}_j$ are defined as

$$\vec{\phi}_j = (\phi_j^{(1)}, \phi_j^{(2)}, \cdots, \phi_j^{(n)}), \quad d\vec{\phi}_j = d\phi_j^{(1)} d\phi_j^{(2)} \cdots d\phi_j^{(n)}. \tag{43.2.23}$$

Let us introduce

$$\tilde{c}_j(\vec{\phi}) = \delta(\vec{\phi} - \vec{\phi}_j) \tag{43.2.24}$$

and an auxiliary function $c_j(\vec{\phi})$ with a normalization

$$\int d\vec{\phi}\, c_j(\vec{\phi}) = 1. \tag{43.2.25}$$

Then the relation (43.2.14) yields an asymptotic estimate

$$\langle Z^n\rangle \sim \int \prod_{j=1}^{N} d\vec{\phi}_j \int \prod_{j=1}^{N} \mathcal{D}c_j(\vec{\phi}) \prod_{j=1}^{N}\prod_{\vec{\phi}} \delta(c_j(\vec{\phi}) - \tilde{c}_j(\vec{\phi}))e^{S_1+S_2} \tag{43.2.26}$$

in the limit $N \to \infty$. Here S_1 and S_2 are defined as

$$S_1 = \frac{i}{2}\mu\sum_{j=1}^{N}\int d\vec{\phi}\, c_j(\vec{\phi})\vec{\phi}^{\,2} \tag{43.2.27}$$

and

$$S_2 = \frac{pN}{2} \sum_{jl}^{N} P_j P_l \int d\vec{\phi} \int d\vec{\psi}\, c_j(\vec{\phi}) c_l(\vec{\psi}) (f(\vec{\phi}, \vec{\psi}) - 1) \tag{43.2.28}$$

with

$$f(\vec{\phi}, \vec{\psi}) = \begin{cases} e^{-i\vec{\phi}\cdot\vec{\psi}}, & \text{if } J \text{ is the adjacency matrix } A, \\ e^{-(i/2)(\vec{\phi}-\vec{\psi})^2}, & \text{if } J \text{ is the Laplacian matrix } L. \end{cases} \tag{43.2.29}$$

Now we consider the asymptotic estimate of

$$\int \prod_{j=1}^{N} d\vec{\phi}_j \prod_{j=1}^{N} \prod_{\vec{\phi}} \delta(c_j(\vec{\phi}) - \tilde{c}_j(\vec{\phi})) = \int \prod_{j=1}^{N} \mathcal{D}a_j(\vec{\phi}) \exp\left(\sum_{j=1}^{N} G_j \right)$$

with

$$G_j = 2\pi i \int d\vec{\phi}\, a_j(\vec{\phi}) c_j(\vec{\phi}) + \ln \int d\vec{\phi}\, e^{-2\pi i a_j(\vec{\phi})} \tag{43.2.30}$$

in the limit $N \to \infty$. The dominant contribution to the functional integral over $a_j(\vec{\phi})$ comes from the extremum satisfying $\delta G_j / \delta a_j = 0$. It follows that the asymptotic estimate of $\langle Z^n \rangle$ is rewritten as

$$\langle Z^n \rangle \sim \int \prod_{j=1}^{N} \mathcal{D}c_j(\vec{\phi}) e^{S_0 + S_1 + S_2} \tag{43.2.31}$$

with

$$S_0 = -\sum_{j=1}^{N} \int d\vec{\phi}\, c_j(\vec{\phi}) \ln c_j(\vec{\phi}). \tag{43.2.32}$$

43.2.4 Extremum condition

Let us next examine the functional integration over $c_j(\vec{\phi})$. It is dominated in the limit $N \to \infty$ by the extremum satisfying

$$\delta \left\{ S_0 + S_1 + S_2 + \sum_{j=1}^{N} \alpha_j \left(\int d\vec{\phi}\, c_j(\vec{\phi}) - 1 \right) \right\} = 0, \tag{43.2.33}$$

where α_j is the Lagrange multiplier ensuring the normalization of $c_j(\vec{\phi})$. This extremum condition can be rewritten in the form

$$c_j(\vec{\phi}) = \mathcal{A}_j \exp\left\{ \frac{i}{2}\mu \vec{\phi}^2 + pNP_j \sum_{l=1}^{N} P_l \int d\vec{\psi}\, c_l(\vec{\psi}) (f(\vec{\phi}, \vec{\psi}) - 1) \right\}, \tag{43.2.34}$$

where $\mathcal{A}_j$ is a normalization constant.

There have been several attempts to solve this extremum condition. In numerical work [Kue08] claimed that a superimposed Gaussian form

$$c_j(\vec{\phi}) = \int d\Pi_j(\omega) \frac{1}{(2\pi/\omega)^{n/2}} \exp\left(-\frac{\omega}{2}\vec{\phi}^2\right) \text{ with } \int d\Pi_j(\omega) = 1 \qquad (43.2.35)$$

gives good agreement with the results of numerical diagonalizations.

In the limit of large mean degree p, the extremum condition can be simplified, so that it can be treated analytically [Rod05, Kim07]. In order to see the simplification, one puts a (single) Gaussian ansatz

$$c_j(\vec{\phi}) = \frac{1}{(2\pi i \sigma_j)^{n/2}} \exp\left(-\frac{1}{2i\sigma_j}\vec{\phi}^2\right) \qquad (43.2.36)$$

into (43.2.34). It follows in the limit $n \to 0$ that

$$\exp\left(-\frac{1}{2i\sigma_j}\vec{\phi}^2\right) = \mathcal{A}_j \exp\left[\frac{i}{2}\mu\vec{\phi}^2 + pNP_j \sum_{l=1}^{N} P_l \left\{h_l(\vec{\phi}) - 1\right\}\right], \qquad (43.2.37)$$

where

$$h_l(\phi) = \begin{cases} \exp\left(-\frac{i\sigma_l}{2}\vec{\phi}^2\right), & \text{if } J \text{ is the adjacency matrix } A, \\ \exp\left(-\frac{i}{2(1-\sigma_l)}\vec{\phi}^2\right), & \text{if } J \text{ is the Laplacian matrix } L. \end{cases} \qquad (43.2.38)$$

Let us first consider the spectral density of the adjacency matrix. Introducing the scalings $\mu = O(p^{1/2})$, $\vec{\phi}^2 = O(p^{-1/2})$ and $\sigma_j = O(p^{-1/2})$, we take the limit of large p. Then we see from (43.2.37) that

$$\mu - \frac{1}{\sigma_j} - pNP_j \sum_{l=1}^{N} P_l \sigma_l = 0, \qquad (43.2.39)$$

which determines σ_j. It follows from (43.2.19), (43.2.21), and (43.2.31) that

$$\rho(\mu) \sim -\frac{1}{N\pi} \text{Im} \sum_{j=1}^{N} \sigma_j \sim -\frac{1}{\pi} \text{Im} \int_0^1 \sigma(x) dx \qquad (43.2.40)$$

gives the spectral density, where $\sigma(x) = \sigma_j$ with $x = j/N$.

Let us define the scaled variables $s(x) = \sqrt{p}\,\sigma(x)$ and $E = \mu/\sqrt{p}$. Then, using (43.2.11), we can rewrite (43.2.39) as

$$s(x)x^{-\alpha} = \frac{1}{Ex^{\alpha} - c} \text{ with } c = (1-\alpha)^2 \int_0^1 s(x)x^{-\alpha} dx, \qquad (43.2.41)$$

from which $s(x)$ is evaluated and the asymptotic expansions of the spectral density are derived. The results are [Rod05]

$$\rho(\mu) = \frac{1}{\pi\sqrt{p}}\left\{\frac{1}{1-a^2} - \frac{1+5a+18a^2+20a^3+16a^4}{8(1-a^2)^3(1+2a)(1+3a)}|E|^2\right\} + O(|E|^4) \tag{43.2.42}$$

for small $|E|$ and

$$\rho(\mu) \sim \frac{2}{\sqrt{p}}\frac{(1-a)^{1/a}}{a}\frac{1}{E^{1+(2/a)}} \tag{43.2.43}$$

for large E. Therefore the exponent γ is $1+(2/a)$, so that the relation with $\lambda = 1+(1/a)$ is $\gamma = 2\lambda - 1$, as anticipated from the analysis of locally tree-like networks [Dor03, Dor04].

We next analyse the spectral density of Laplacian matrices. Let us adopt the scalings $\mu = O(p)$, $\vec{\phi}^2 = O(p^{-1})$ and $\sigma_j = O(p^{-1})$. Then it follows from (43.2.37) that

$$\sigma_j = \frac{1}{\mu + i\epsilon - pNP_j}, \quad \epsilon \downarrow 0. \tag{43.2.44}$$

In terms of the scaled variable $\omega = \mu/(p(1-a))$, the spectral density $\rho(\mu)$ of the Laplacian matrix is derived from (43.2.40) and (43.2.44) as [Kim07]

$$\rho(\mu) = \begin{cases} \dfrac{\omega^{-1-(1/a)}}{p\,a(1-a)}, & \omega > 1, \\ 0, & \omega < 1. \end{cases} \tag{43.2.45}$$

The weighted versions of the adjacency and Laplacian matrices, such as the weighted Laplacian matrix

$$W_{jl} = \frac{L_{jl}}{(\langle k_j\rangle\langle k_l\rangle)^{\beta/2}} \tag{43.2.46}$$

with an exponent β, can be analysed [Kim07] in a similar way in the limit $p \to \infty$.

43.2.5 Effective medium approximation

In order to approximately treat a network with a finite mean degree p, we can employ a useful scheme called the effective medium approximation (EMA) [Sem02, Nag07, Nag08]. In the EMA, one substitutes the (single) Gaussian ansatz (43.2.36) into (43.2.27),(43.2.28), and (43.2.32) and considers the variational equation

$$\frac{\partial}{\partial\sigma_j}(S_0 + S_1 + S_2) = 0. \tag{43.2.47}$$

For the spectral density of adjacency matrices, the variational equation (43.2.47) turns into

$$\mu - \frac{1}{\sigma_j} - Np\,P_j \sum_{l=l}^{N} \frac{P_l\,\sigma_l}{1 - \sigma_j \sigma_l} = 0 \tag{43.2.48}$$

in the limit $n \to 0$. This equation can be solved by a numerical iteration method. Then we put the solution into (43.2.40) and obtain the EMA spectral density. The result is compared in Figure 43.1 with the spectral density of numerically generated adjacency matrices (averaged over 100 samples, $N = 1000$, $\alpha = 1/2$ and $p = 12$). The agreement is fairly good except around the origin.

In the limit $p \to \infty$ with a scaling $\sigma_j = O(p^{-1/2})$, the EMA Equation (43.2.48) becomes the extremum condition (43.2.39), as expected. Then one can formulate a perturbative method to analytically calculate the finite p correction to the solution of (43.2.39). The result for the spectral density takes the form

$$\rho(\mu) = \frac{1}{\sqrt{p}} \left\{ \rho_0(\mu) + \frac{1}{p}\rho_1(\mu) + O\left(\frac{1}{p^2}\right) \right\}. \tag{43.2.49}$$

For large E, we have already seen in equation (43.2.43) that the unperturbed term $p^{-1/2}\rho_0(\mu)$ is $O(E^{-1-(2/\alpha)})$, whereas the first order correction $p^{-3/2}\rho_1(\mu)$ turns out [Nag08] to be $O(E^{-3-(2/\alpha)})$. Hence the spectral density is dominated by the unperturbed term in the limit $E \to \infty$.

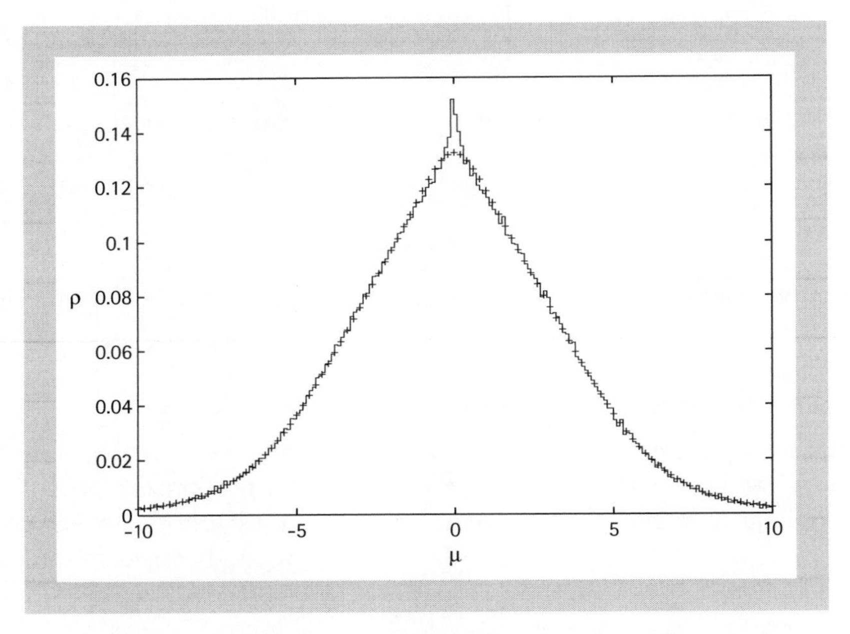

Fig. 43.1 The EMA spectral density (+) and the spectral density of numerically generated adjacency matrices (histogram) with $\alpha = 1/2$ and $p = 12$.

In the limiting case $\alpha = 0$, one obtains the adjacency matrices of classical random graphs [Erd60, Rod88, Bra88]. Then the dependence of σ_j on j can be omitted so that σ_j is set to be σ. The EMA equation becomes a simple cubic equation [Sem02]

$$\mu\sigma^3 + (p-1)\sigma^2 - \mu\sigma + 1 = 0 \tag{43.2.50}$$

and gives a closed analytic solution for the spectral density. In the limit $p \to \infty$ we obtain a semi-circular density

$$\rho(\mu) = \begin{cases} \dfrac{\sqrt{4p-\mu^2}}{2\pi p}, & -2\sqrt{p} < \mu < 2\sqrt{p}, \\ 0, & \mu < -2\sqrt{p} \text{ or } \mu > 2\sqrt{p}, \end{cases} \tag{43.2.51}$$

as expected from the theory of random matrices.

In the case of Laplacian matrices, the variational equation (43.2.47) takes the form

$$\mu - \frac{1}{\sigma_j} - Np\,P_j \sum_{l=1}^{N} \frac{P_l}{1 - \sigma_j - \sigma_l} = 0. \tag{43.2.52}$$

As before, the EMA spectral density can be evaluated from this equation by a numerical iteration method. It is compared with the spectral density of numerically generated Laplacian matrices (averaged over 100 samples, $N = 1000$, $\alpha = 1/2$ and $p = 12$) in Figure 43.2.

In the limiting case of classical random graphs ($\alpha = 0$), we find a quadratic equation

$$2\mu\sigma^2 + (p - \mu - 2)\sigma + 1 = 0, \tag{43.2.53}$$

which yields the EMA spectral density

$$\rho(\mu) = \begin{cases} \dfrac{\sqrt{8p - (\mu - p - 2)^2}}{4\pi\mu}, & p_- < \mu < p_+, \\ 0, & \mu < p_- \text{ or } \mu > p_+ \end{cases} \tag{43.2.54}$$

with $p_\pm = p \pm 2\sqrt{2p} + 2$.

43.3 Local properties

In the application of random matrices to quantum physics, it is known that the local spectral distributions are universal even if the global spectral density depends on the details of each system. It is interesting to see if such strong universality also holds for the matrices associated with complex networks.

Chung, Lu, and Vu rigorously analysed the largest eigenvalues of the adjacency matrices of scale free networks [Chu03a, Chu03b]. A variant of the static

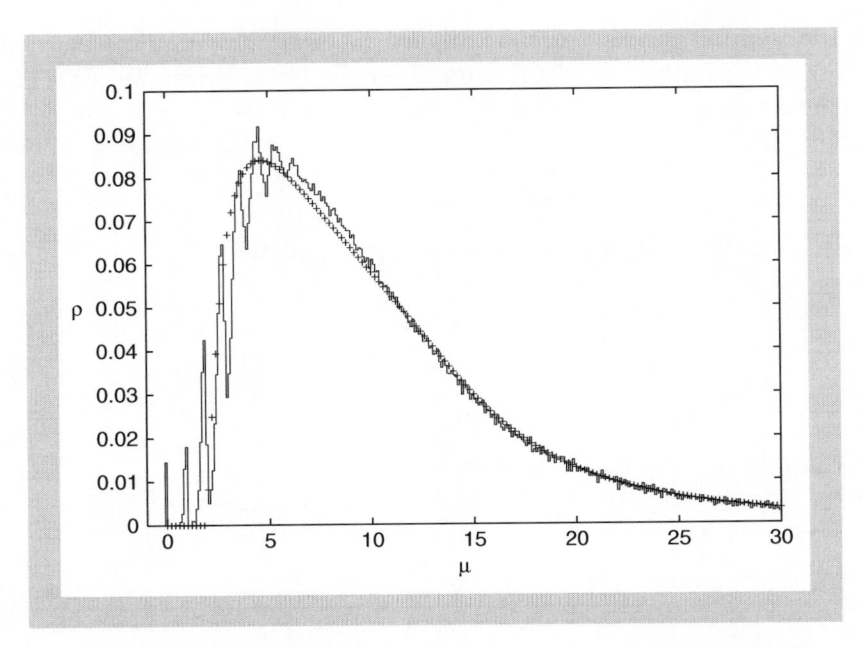

Fig. 43.2 The EMA spectral density (+) and the spectral density of numerically generated Laplacian matrices (histogram) with $a = 1/2$ and $p = 12$.

model was used in their work. If the exponent of the degree distribution is $\lambda > 5/2$, their result claims under certain conditions that the largest eigenvalues have power law distributions with the exponent $2\lambda - 1$.

Bandyopadhyay and Jalan, on the other hand, numerically analysed the eigenvalue correlation of the adjacency matrices of scale free as well as small world networks [Ban07, Jal07, Jal09]. The growth process of the BA model was used to generate scale free networks. Their results show that the eigenvalue spacing distributions and the spectral rigidity follow the predictions of Gaussian Orthogonal Ensemble (GOE) of random matrices (see Chapter 13). In the case of small world networks, they used the Watts-Strogatz model, in which a regular ring lattice is randomised [Wat98]. A transition toward the GOE behaviour is observed as a function of the fraction of randomised edges. Very recently [Car09], the deformation in the eigenvalue spacing distribution described in Chapter 13 has been found empirically in a small world network.

Finally, in [Mit09] Mitrović and Tadić studied numerically the spectral properties of the adjacency and Laplacian matrices in a wide class of scale free networks with mesoscopic subgraphs. They identify signals of cyclic mesoscopic structures in the spectra. For instance, the centres of the spectra are effected by minimally connected nodes and the number of distinct modules leads to additional peaks in the Laplacian spectra.

References

[Alb02] R. Albert and A.-L. Barabási, Rev. Mod. Phys. **74** (2002) 47
[Ban07] J.N. Bandyopadhyay and S. Jalan, Phys. Rev. **E76** (2007) 026109
[Bar99] A.-L. Barabási and R. Albert, Science **286** (1999) 509
[Bia05] P. Białas, Z. Burda and B. Wacław, AIP. Conf. Proc. **776** (2005) 14
[Bog06] L. Bogacz, Z. Burda and B. Wacław, Physica A **366** (2006) 587
[Bra88] A.J. Bray and G.J. Rodgers, Phys. Rev. **B38** (1988) 11461
[Car09] J. X. de Carvalho, S. Jalan and M. S. Hussein, Phys. Rev. **E79** (2009) 056222
[Chu03a] F. Chung, L. Lu and V. Vu, Proc. Natl. Acad. Sci. U.S.A. **100** (2003) 6313
[Chu03b] F. Chung, L. Lu and V. Vu, Ann. Comb. **7** (2003) 21
[Dor03] S.N. Dorogovtsev, A.V. Goltsev, J.F.F. Mendes and A.N. Samukhin, Phys. Rev. **E68** (2003) 046109
[Dor04] S.N. Dorogovtsev, A.V. Goltsev, J.F.F. Mendes and A.N. Samukhin, Physica **A338** (2004) 76
[Dor08] S.N. Dorogovtsev, A.V. Goltsev and J.F.F. Mendes, Rev. Mod. Phys. **80** (2008) 1275
[Erd60] P. Erdös and A. Rényi, Publ. Math. Inst. Hung. Acad. Sci. Ser. A **5** (1960) 17
[Far01] I.J. Farkas, I. Derényi, A.-L. Barabási and T. Vicsek, Phys. Rev. **E64** (2001) 026704
[Goh01a] K.-I. Goh, B. Kahng and D. Kim, Phys. Rev. **E64** (2001) 051903
[Goh01b] K.-I. Goh, B. Kahng and D. Kim, Phys. Rev. Lett. **87** (2001) 278701
[Jal07] S. Jalan and J.N. Bandyopadhyay, Phys. Rev. **E76** (2007) 046107
[Jal09] S. Jalan and J.N. Bandyopadhyay, Eur. Phys. Lett. **87** (2009) 48010
[Kim05] D.-H. Kim, G.J. Rodgers, B. Kahng and D. Kim, Phys. Rev. **E71** (2005) 056115
[Kim07] D. Kim and B. Kahng, Chaos, **17** (2007) 026115
[Kue08] R. Kühn, J. Phys. A: Math. Theor. **41** (2008) 295002
[Lee04] D.-S. Lee, K.-I. Goh, B. Kahng and D. Kim, Nucl. Phys. **B696** (2004) 351
[Mit09] M. Mitrović and B. Tadić, Phys. Rev. E **80** (2009) 026123
[Nag07] T. Nagao and T. Tanaka, J. Phys. A: Math. Theor. **40** (2007) 4973
[Nag08] T. Nagao and G.J. Rodgers, J. Phys. A: Math. Theor. **41** (2008) 265002
[New99] M. E. J. Newman and D. J. Watts, Phys. Rev.**E60** (1999) 7332
[Rod88] G.J. Rodgers and A.J. Bray, Phys. Rev. **B37** (1988) 3557
[Rod05] G.J. Rodgers, K. Austin, B. Kahng and D. Kim, J. Phys. A: Math. Gen. **38** (2005) 9431
[Sem02] G. Semerjian and L.F. Cugliandolo, J. Phys. A: Math. Gen. **35** (2002) 4837
[Tad01] B. Tadić, Physica A **293** (2001) 273
[Wat98] D.J. Watts and S.H. Strogatz, Nature **393** (1998) 440

Index

Printed and bound by CPI Group (UK) Ltd, Croydon, CR0 4YY